Biology and Conservation of Wild Felids

From top left, clockwise: Jaguar, African Lion, Cheetah, Andean cat, Scottish Wildcat, Tiger, and Puma. Drawn by Priscilla Barrett.

Biology and Conservation of Wild Felids

Edited by

DAVID W. MACDONALD AND ANDREW J. LOVERIDGE

OXFORD
UNIVERSITY PRESS

Great Clarendon Street, Oxford OX2 6DP

Oxford University Press is a department of the University of Oxford.
It furthers the University's objective of excellence in research, scholarship,
and education by publishing worldwide in

Oxford New York

Auckland Cape Town Dar es Salaam Hong Kong Karachi
Kuala Lumpur Madrid Melbourne Mexico City Nairobi
New Delhi Shanghai Taipei Toronto

With offices in

Argentina Austria Brazil Chile Czech Republic France Greece
Guatemala Hungary Italy Japan Poland Portugal Singapore
South Korea Switzerland Thailand Turkey Ukraine Vietnam

Oxford is a registered trade mark of Oxford University Press
in the UK and in certain other countries

Published in the United States
by Oxford University Press Inc., New York

First published 2010
Reprinted 2011, 2012 (twice)

British Library Cataloguing in Publication Data
Data available

Library of Congress Control Number: 2010924763

Typeset by SPI Publisher Services, Pondicherry, India
Printed in Great Britain
on acid-free paper by
CPI Group (UK) Ltd, Croydon, CR04YY

ISBN 978–0–19–923444–8 (Hbk.)
ISBN 978–0–19–923445–5 (Pbk.)

5 7 9 10 8 6

To Tom and Dafna Kaplan and their family,
in grateful recognition of their unique contribution
to felid conservation

Contents

Preface xi
List of contributors xv

Part I Reviews

1 *Dramatis personae*: an introduction to the wild felids 3
David W. Macdonald, Andrew J. Loveridge, and Kristin Nowell

2 Phylogeny and evolution of cats (Felidae) 59
Lars Werdelin, Nobuyuki Yamaguchi, Warren E. Johnson, and Stephen J. O'Brien

3 Felid form and function 83
Andrew C. Kitchener, Blaire Van Valkenburgh, and Nobuyuki Yamaguchi

4 Genetic applications in wild felids 107
Melanie Culver, Carlos Driscoll, Eduardo Eizirik, and Göran Spong

5 Felid society 125
David W. Macdonald, Anna Mosser, and John L. Gittleman

6 People and wild felids: conservation of cats and management of conflicts 161
Andrew J. Loveridge, Sonam W. Wang, Laurence G. Frank, and John Seidensticker

7 Many ways of skinning a cat: tools and techniques for studying wild felids 197
K. Ullas Karanth, Paul Funston, and Eric Sanderson

8 Felids *ex situ*: managed programmes, research, and species recovery 217
David Wildt, William Swanson, Janine Brown, Alexander Sliwa, and Astrid Vargas

9 Wild felid diseases: conservation implications and management strategies 237
Linda Munson, Karen A. Terio, Marie-Pierre Ryser-Degiorgis, Emily P. Lane, and Franck Courchamp

Part II Case Studies

10 Ecology of infectious diseases in Serengeti lions 263
Meggan E. Craft

11 African lions on the edge: reserve boundaries as 'attractive sinks' 283

Andrew J. Loveridge, Graham Hemson, Zeke Davidson, and David W. Macdonald

12 Tiger range collapse and recovery at the base of the Himalayas 305

John Seidensticker, Eric Dinerstein, Surendra P. Goyal, Bhim Gurung, Abishek Harihar, A.J.T. Johnsingh, Anil Manandhar, Charles W. McDougal, Bivash Pandav, Mahendra Shrestha, J.L. David Smith, Melvin Sunquist, and Eric Wikramanayake

13 Amur tiger: a case study of living on the edge 325

Dale G. Miquelle, John M. Goodrich, Evgeny N. Smirnov, Phillip A. Stephens, Olga Yu Zaumyslova, Guillaume Chapron, Linda Kerley, Andre A. Murzin, Maurice G. Hornocker, and Howard B. Quigley

14 An adaptive management approach to trophy hunting of leopards (*Panthera pardus*): a case study from KwaZulu-Natal, South Africa 341

Guy A. Balme, Luke T. B. Hunter, Pete Goodman, Hayden Ferguson, John Craigie, and Rob Slotow

15 Cheetahs and ranchers in Namibia: a case study 353

Laurie Marker, Amy J. Dickman, M. Gus L. Mills, and David W. Macdonald

16 Past, present, and future of cheetahs in Tanzania: their behavioural ecology and conservation 373

Sarah M. Durant, Amy J. Dickman, Tom Maddox, Margaret N. Waweru, Tim Caro, and Nathalie Pettorelli

17 Jaguars, livestock, and people in Brazil: realities and perceptions behind the conflict 383

Sandra M. C. Cavalcanti, Silvio Marchini, Alexandra Zimmermann, Eric M. Gese, and David W. Macdonald

18 The ecology of jaguars in the Cockscomb Basin Wildlife Sanctuary, Belize 403

Bart J. Harmsen, Rebecca J. Foster, Scott C. Silver, Linde E.T. Ostro, and C. Patrick Doncaster

19 Snow leopards: conflict and conservation 417

Rodney M. Jackson, Charudutt Mishra, Thomas M. McCarthy, and Som B. Ale

20 Pumas and people: lessons in the landscape of tolerance from a widely distributed felid 431

Tucker Murphy and David W. Macdonald

21 Long-term research on the Florida panther (*Puma concolor coryi*): historical findings and future obstacles to population persistence 453

Dave Onorato, Chris Belden, Mark Cunningham, Darrell Land, Roy McBride, and Melody Roelke

22 Reversing cryptic extinction: the history, present, and future of the Scottish wildcat 471

David W. Macdonald, Nobuyuki Yamaguchi, Andrew C. Kitchener, Mike Daniels, Kerry Kilshaw, and Carlos Driscoll

23 The changing impact of predation as a source of conflict between hunters and reintroduced lynx in Switzerland 493

Urs Breitenmoser, Andreas Ryser, Anja Molinari-Jobin, Fridolin Zimmermann, Heinrich Haller, Paolo Molinari, and Christine Breitenmoser-Würsten

24 Iberian lynx: the uncertain future of a critically endangered cat 507

Pablo Ferreras, Alejandro Rodríguez, Francisco Palomares, and Miguel Delibes

25 Cyclical dynamics and behaviour of Canada lynx in northern Canada 521

Mark O'Donoghue, Brian G. Slough, Kim G. Poole, Stan Boutin, Elizabeth J. Hofer, Garth Mowat, and Charles J. Krebs

26 Black-footed cats (*Felis nigripes*) and African wildcats (*Felis silvestris*): a comparison of two small felids from South African arid lands 537

Alexander Sliwa, Marna Herbst, and M. Gus L. Mills

27 Ocelot ecology and its effect on the small-felid guild in the lowland neotropics 559

Tadeu G. de Oliveira, Marcos A. Tortato, Leandro Silveira, Carlos Benhur Kasper, Fábio D. Mazim, Mauro Lucherini, Anah T. Jácomo, José Bonifácio G. Soares, Rosane V. Marques, and Melvin Sunquist

28 Highland cats: ecology and conservation of the rare and elusive Andean cat 581

Jorgelina Marino, Mauro Lucherini, M. Lilian Villalba, Magdalena Bennett, Daniel Cossíos, Agustín Iriarte, Pablo G. Perovic, and Claudio Sillero-Zubiri

Part III A Conservation Perspective

29 Felid futures: crossing disciplines, borders, and generations 599

David W. Macdonald, Andrew J. Loveridge, and Alan Rabinowitz

References 651

Index 739

Preface

To undertake a major edited book on the biology and conservation of an entire taxonomic family of wild carnivores might seem foolhardy. Having done so once (on the Canidae) then to do it again (for the Felidae) may be the hallmark of a slow-learner! Nonetheless, that is the task we have shouldered in producing the *Biology and Conservation of Wild Felids*, the sister volume of the *Biology and Conservation of Wild Canids* by David Macdonald and Claudio Sillero-Zubiri, which was published by Oxford University Press in 2004. We believe that the two, as a pair, are more than the sum of their parts.

This book had several beginnings, as did our own interests in felids. In 1978, as the culmination to a study of farm cats, one of us (David W. Macdonald) produced a BBC TV documentary on the intricacies of their sociality, which reported for the first time such then emerging phenomena as communal nursing and infanticide. There were two upshots of this documentary, *The Curious Cat* (World About Us). First, and this will not surprise those who have stood watch over the interminable sleepiness of felids, is that *Private Eye* magazine produced a lengthy satirical spoof on David W. Macdonald, who was portrayed radio-tracking a family of slugs in order to discover how long it would take them to circumnavigate the globe! Second, in the behaviour of those farm cats Lucky, his father Tom, mother Smudge, and aunts, Pickle and Domino were shadows of all the images that have made lion society rivetingly interesting to a generation of behavioural ecologists. Thus began a journey from domestic to wild felids, which initially dwelt—in the context of the preciously rare Scottish wildcat—on the technically and philosophically tricky question of what is a wildcat (Chapter 22, this volume). Then, in 1995, as the two of us sat by moonlight at a waterhole in Zimbabwe, chatting over a thermos flask of coffee while radio-tracking jackals, our great friend the late Lionel Reynolds confided in us his fears for the future of lions in Zimbabwe and beyond. Lionel, a pillar of the Zimbabwean Hunters and Guides Association, feared that excessive hunting quotas, pursuit of easy money, and a tangle of lions in conflict with communities, presaged sinister, but then unheeded, warnings for the lion's future. Thus began the WildCRU's (Oxford University's Wildlife Conservation Research Unit's) Hwange Lion Project, which celebrated its 10th anniversary in 2009 (see Chapter 11, this volume). Meanwhile, the WildCRU initiated projects on felids large and small, which now span Andean cats of Bolivia to the manul in Mongolia to the bay cat in Borneo, the jaguar to the cheetah to the tiger, and the leopards—spotted, clouded, and snow. Indeed, of the 128 world-leading felid experts contributing to this book, 18 are, or have been, from the WildCRU stable.

Thus, with 'the canid book' published in 2004, it seemed that the obvious next step was 'the felid book'. Learning from the former, the process began with an international conference—from 17 to 20 September 2007 some 300 delegates from around the world gathered in Oxford to discuss felids. At the same time, we organized two workshops, one an IUCN red-listing session workshop (with Mike Hoffmann and Jan Schipper), the second a think tank on conflict and its resolution. All these events were sponsored by Panthera which, at that time, was exploding on the scene as a major new force in felid conservation. Many of the speakers who attended the Felid Biology and Conservation Conference in Oxford are authors in this book. However, this book is not a conference proceedings, and its chapters are much more than the written version of conference speeches. Rather, the brainstorming excitement of the conference and associated workshops were part of a process of identifying the world leaders in felid biology, and the ideas that most excited them. These people have been assembled into carefully constructed teams to produce the two types of chapter that dominate this book. These, the Review chapters and the Case Studies, serve different functions, but together are intended to make the contents of the book both topical and enduring, both meticulously encyclopaedic and vibrantly empirical. Together, they offer a view that reveals both the woods and the trees.

Each Review chapter tackles an overarching theme in felid biology. We deliberately recruited at least three accomplished authors for each review, ideally from different institutions, perspectives, and even continents. Bringing together their heterogeneous experience, and sometimes conflicting styles and opinions, has been a major task for them, and for us as editors. However, the resulting syntheses are truly original, and both broad and deep. In contrast, the Case Studies are written by teams, sometimes assembled specially for this book, who have deep insight into a particular species or guild and a particular set of issues. We agonized over whether or not to include chapters specifically on domestic cats, and while no single chapter is dedicated to them, domestic cats feature heavily in many of the reviews and, in the case of the threat they now pose to wild cats, in the Case Study of Scottish wildcats (Chapter 22). In addition to the Review and Case Study Chapters, the book begins with a *dramatis personae* (Chapter 1) which introduces every living species of wild felid, and it ends, as did the sister volume on canids, with a substantial essay on conservation (Chapter 29). It is in this concluding chapter that the inescapable need for interdisciplinarity that is the hallmark of this book becomes most strident. Issues in felid conservation encompass not only the biological and social sciences, but lead us through dilemmas in governance and development to, ultimately, difficult judgements and thus politics.

Although this Preface comes first in the book, inevitably it was written last! The task of writing and editing the book leaves us with almost nothing left to say, except that when we started we thought we knew quite a lot about felids, but we surely know a lot more now—and we believe that for years to come this will be the experience of anybody who reads these pages. Above all, we are immensely grateful to everybody who has joined us on this journey—most importantly, the authors have worked tirelessly and fruitfully. They will tell you that as editors we have been relentless, demanding, intrusive, and nit-picking—they must often have been reminded of the words of the Bob Dylan song 'knowing you is such a drag!'—we apologize, and we hope that they, the authors, and you, the reader, will think with hindsight that our harassing has been worthwhile in the end!

In addition, we want to thank the core team of people who helped us bring both the conference and the book to fruition. In alphabetical order, the tireless endurance runners in this race have been Dawn Burnham, Susan Cheyne, Joelene Hughes, Kerry Kilshaw, Joanne Loveridge, and Rosalind Shaw—they are few in number which emphases the heavy burden that each has shouldered. Naming these few is not to diminish our gratitude for the huge effort of the many in the wider WildCRU team who were selflessly devoted to working behind, around, and in front of the scenes, in order to make the conference a success. The book is enhanced by its illustrations, and we are particularly grateful to Andy Rouse, who is extraordinarily generous in his support of the WildCRU in allowing us to use his stunning photos, and to Priscilla Barrett whose frontispiece to the book captures the beauty that drives us to care so deeply for the felids whose lives and travails are explained in these pages. At Oxford University Press, Ian Sherman has nurtured both canid and felid projects, and Helen Eaton has been tirelessly supportive. Closer to home, and mindful that when, all too rarely, at home we have been steadfastly and unsociably glued to editorial computer screens for well over a year, we thank our wives, Jenny and Joanne, and families for their tolerance.

Finally, our personal research in particular, and the WildCRU's felid initiatives in general, owe much to the support, vision, and friendship of Tom Kaplan, his family, and our friends Alan Rabinowitz, Luke Hunter, and others at Panthera and in the extended family of its Cat Advisory Council. As we conclude this book, the new Panthera Buildings at WildCRU are also being completed—creating a centre where future generations of felid conservationists will be trained for work in far distant, remote corners of the world. In acknowledgement of all that they have made possible, we are proud to dedicate this book to Tom and Dafna. Already, our diploma course has recruited so-called WildCRU Panthers from Albania, Bhutan, Bolivia, China, Iran, Kenya, Madagascar, South Africa, Tanzania, and Zimbabwe. Hopefully this book will be an inspiration and guide to these dedicated people and many others like them, and will create a foundation for the understanding and conservation of wild felids for many years to come.

David W. Macdonald & Andrew J. Loveridge
Wildlife Conservation Research Unit,
Department of Zoology
and Lady Margaret Hall
Oxford University

Andrew Loveridge and David Macdonald radio-collar an immobilised lion, Hwange National Park, Zimbabwe.

List of Contributors

Som B. Ale, University of Minnesota, Dept of Fisheries, Wildlife and Conservation Biology, College of Food, Agricultural and Natural Resource Sciences, St Paul, MN 55108, United States

Guy A. Balme, Panthera, 8 West 40th Street, 18th Floor, New York, NY 10018, United States

Chris Belden, United States Fish and Wildlife Service, South Florida Ecological Services Office, 1339 20th Street, Vero Beach, Florida 32960, United States

Magdalena Bennett, Departamento de Ecología, Pontificia Universidad Católica de Chile, Alameda 340, Santiago (6513677), Chile

Stan Boutin, Department of Biological Sciences, CW-405 Biological Sciences Building, University of Alberta, Edmonton, Alberta T6G 2E9, Canada

Urs Breitenmoser, Institute for Veterinary Virology, University of Berne, Länggassstrasse 122, CH-3012 Bern, Switzerland

Christine Breitenmoser-Würsten, KORA, Thunstrasse 31, 3074 Muri b. Bern, Switzerland

Janine Brown, Center for Species Survival, Smithsonian's National Zoological Park, Front Royal, VA 22630, United States

Tim Caro, Department of Wildlife, Fish and Conservation Biology, University of California, Davis, CA 95616, United States

Sandra M. C. Cavalcanti, Panthera, R. Guajuvira, 537 — Atibaia, SP, 12946–260, Brazil

Guillaume Chapron, Grimsö Wildlife Research Station, Swedish University of Agricultural Sciences, 73091 Riddarhyttan, Sweden

Daniel Cossíos, Département de Sciences Biologiques, Université de Montréal, Pavillon Marie Victorin., Montréal H3C 3J7, Canada

Franck Courchamp, Laboratoire Ecologie, Systématique et Evolution, University Paris-Sud, Orsay F–91405, France

Meggan E. Craft, Boyd Orr Centre for Population and Ecosystem Health, Division of Ecology and Evolutionary Biology, University of Glasgow, Glasgow, G12 8QQ, United Kingdom

John Craigie, Ezemvelo KwaZulu-Natal Wildlife, P.O. Box 13053, Cascades, 3202, South Africa

Melanie Culver, Arizona Cooperative Fish and Wildlife Research Unit, USGS, and School of Natural Resources, University of Arizona, Tucson, AZ, United States

Mark Cunningham, Florida Fish and Wildlife Conservation Commission, 1105 SW Williston Rd, Gainesville, FL 32601, United States

Mike Daniels, John Muir Trust, Tower House, Station Road, Pitlochry, PH16 5AN, United Kingdom

Zeke Davidson, Wildlife Conservation Research Unit, Department of Zoology, University of Oxford, The Recanati-Kaplan Centre, Tubney House, Abingdon Road, Tubney, OX13 5QL, United Kingdom

Tadeu G. de Oliveira, Depto. Biologia, Universidade Estadual do Maranhão & Instituto Pró-Carnívoros, Rua das Quaresmeiras, Qd-08 No. 14, São Luís, MA 65076–270, Brazil

Miguel Delibes, Department of Conservation Biology, Estación Biológica de Doñana, CSIC C/Américo Vespucio s/n 41092 Sevilla, Spain

Amy J. Dickman Wildlife Conservation Research Unit, Department of Zoology, University of Oxford, The Recanati-Kaplan Centre, Tubney House, Abingdon Road, Tubney, OX13 5QL, United Kingdom

Eric Dinerstein, World Wildlife Fund, United States, 1250 24th Street, NW, Washington D.C. 20037–1193, United States

C. Patrick Doncaster, School of Biological Sciences, University of Southampton, Southampton SO16 7PX, United Kingdom

Carlos Driscoll, Laboratory of Genomic Diversity, National Cancer Institute, Frederick, MD 21702, USA

Sarah M. Durant, Institute of Zoology, Zoological Society of London, Regents Park, London, NW1 4RY, United Kingdom

Eduardo Eizirik, Centro de Biologia Genômica e Molecular, Faculdade de Biociências, PUCRS, and Instituto Pró-Carnívoros, Porto Alegre, Brazil

Hayden Ferguson, Ezemvelo KwaZulu-Natal Wildlife, P.O. Box 13053, Cascades, 3202, South Africa

Pablo Ferreras, Instituto de Investigación en Recursos Cinegéticos, IREC (CSIC-UCLM-JCCM), Ronda de Toledo s/n, 13005, Ciudad Real, Spain

Rebecca J. Foster, Panthera, 8 West 40th Street, 18th Floor, New York, NY 10018, United States

Laurence G. Frank, Panthera, 8 West 40th Street, 18th Floor, New York, NY 10018, United States

Paul Funston, Department of Nature Conservation, Tshwane University of Technology, Private Bag X680, Pretoria, 0001, South Africa

Eric M. Gese, USDA/WS/National Wildlife Research Center, Department of Wildland Resources, Utah State University, Logan, UT 84322, United States

John L. Gittleman, Odum School of Ecology, University of Georgia, Athens, GA 30602, United States

Pete Goodman, Ezemvelo KwaZulu-Natal Wildlife, P.O. Box 13053, Cascades, 3202, South Africa

John M. Goodrich Wildlife Conservation Society, 2300 Southern Blvd, Bronx, NY, United States

Surendra P. Goyal, Wildlife Institute of India, P.O. Box 18, Chandrabani, Dehradun 248 001, Uttarakhand, India

Bhim Gurung, University of Minnesota, Department of Fisheries, Wildlife and Conservation Biology, 200 Hodson Hall, 1980 Folwell Avenue, St Paul, MN 55108, United States

Heinrich Haller, Swiss National Park, Neues Nationalpark Zentrum, 7530 Zernez, Switzerland

Abishek Harihar, Wildlife Institute of India, P.O. Box 18, Chandrabani, Dehradun 248 001, Uttarakhand, India

Bart J. Harmsen, Panthera, 8 West 40th Street, 18th Floor, New York, NY 10018, United States

Graham Hemson, Wildlife Conservation Research Unit, Department of Zoology, University of Oxford, The Recanati-Kaplan Centre, Tubney House, Abingdon Road, Tubney, OX13 5QL, United Kingdom

Marna Herbst, South African National Parks, Scientific Services, Kimberley, South Africa

Elizabeth J. Hofer, Box 5428, Haines Junction, Yukon Territory YOB 1LO, Canada

Maurice G. Hornocker, Selway Institute, Inc, P.O. Box 929, 40 Heronwood Lane, Bellevue, Idaho 83313, United States

Luke T. B. Hunter, Panthera, 8 West 40th Street, 18th Floor, New York, NY 10018, United States

Agustín Iriarte, Fundacion Biodiversity, Galvarino Gallardo 1552, Providencia, Santiago, Chile.

Rodney M. Jackson, Snow Leopard Conservancy, 18030 Comstock Ave, Sonoma, CA 95476, United States

Anah T. Jácomo, Jaguar Conservation Fund, C.P. 193 Mineiros, GO 75830–000, Brazil

A. J. T. Johnsingh, Nature Conservation Foundation, Mysore and WWF-India

Warren E. Johnson, Laboratory of Genomic Diversity, National Cancer Institute, Frederick, MD 21702–1201, United States

K. Ullas Karanth, Wildlife Conservation Society, New York & Centre for Wildlife Studies 26–2, Aga Abbas Ali Road (Apt: 403), Bangalore Karnataka-560 042, India

Carlos Benhur Kasper, PPG Biologia Animal-UFRGS, Rui Barbosa, Caixa Postal 121, Arroio do Meio, RS 95940–000, Brazil

Linda Kerley, Zoological Society of London, England, and Lazovski State Zapovednik, Lazo, Primorski Krai, Russia

Kerry Kilshaw, Wildlife Conservation Research Unit, Department of Zoology, University of Oxford, The Recanati-Kaplan Centre, Tubney House, Abingdon Road, Tubney, OX13 5QL, United Kingdom

Andrew C. Kitchener, Department of Natural Sciences, National Museums Scotland, Chambers Street, Edinburgh, EH1 1JF, United Kingdom

Charles J. Krebs, Department of Zoology, 6270 University Blvd, University of British Columbia, Vancouver, British Columbia V6T 1Z4, Canada

Darrell Land, FL Fish and Wildlife Conservation Commission, 298 Sabal Palm Road, Naples, FL 34114, United States

Emily P. Lane, Zoological Pathology and Research Program, National Zoological Gardens of South Africa, Pretoria, South Africa

Andrew J. Loveridge, Wildlife Conservation Research Unit, Department of Zoology, University of Oxford, The Recanati-Kaplan Centre, Tubney House, Abingdon Road, Tubney, OX13 5QL, United Kingdom

Mauro Lucherini, Grupo de Ecología Comportamental de Mamíferos, Departamento de Biología, Bioquímica y Farmacia, UNS-CONICET, San Juan 670, 8000 Bahía Blanca, Argentina

David W. Macdonald, Wildlife Conservation Research Unit, Department of Zoology, University of Oxford, The Recanati-Kaplan Centre, Tubney

House, Abingdon Road, Tubney, OX13 5QL, United Kingdom

Tom Maddox, Zoological Society of London, Regents Park, London, NW1 4RY, United Kingdom

Anil Manandhar, WWF-Nepal, P.O. Box 7660, Baluwatar, Kathmandu, Nepal

Silvio Marchini, Escola da Amazônia, Cristalino Ecological Foundation, Avenida Perimetral Oeste, 2001, Alta Floresta, MT 78580–00, Brazil

Jorgelina Marino, Wildlife Conservation Research Unit, Department of Zoology, University of Oxford, The Recanati-Kaplan Centre, Tubney House, Abingdon Road, Tubney, OX13 5QL, United Kingdom

Laurie Marker, Cheetah Conservation Fund, P.O. Box 1755, Otjiwarongo, Namibia

Rosane V. Marques, Unidade de Assessoramento Ambiental, Ministério Público do Rio Grande do Sul, Rua Andrade Neves 106, 10° andar, Porto Alegre, RS 90210–210, Brazil

Fábio D. Mazim, Instituto Pró-Pampa, Rua Padre Anchieta No 1277, sala 201, Pelotas, RS 96015–420, Brazil

Roy McBride, Ranchers Supply Inc., 26690 Pine Oak Road, Ochopee, Florida 34141, United States

Thomas M. McCarthy, Panthera and Snow Leopard Trust, 4649 Sunnyside Ave. N., Suite 325, Seattle, WA, 98103, United States

Charles W. McDougal, International Trust for Nature Conservation, C/o Tiger Tops Pvt Ltd, P.O. Box 242, Kathmandu, Nepal

M. Gus L. Mills, Kgalagadi Cheetah Project, P. Bag X5890, Upington, 8800, South Africa and Tony and Lisette Lewis Foundation, South Africa

Dale G. Miquelle, Wildlife Conservation Society Russia Program, 2300 Southern Blvd, Bronx, NY, United States

Charudutt Mishra, Snow Leopard Trust and Nature Conservation Foundation, 4649 Sunnyside Ave, Suite 325, Seattle, WA 98103, United States

Paolo Molinari, KORA, Thunstrasse 31, 3074 Muri b. Bern, Switzerland

Anja Molinari-Jobin, KORA, Thunstrasse 31, 3074 Muri b. Bern, Switzerland

Anna Mosser, Department of Ecology Evolution and Behaviour, University of Minnesota, 1987 Upper Buford Circle, St Paul, MN 55414 United States

Garth Mowat, Fish and Wildlife Branch, Ministry of Environment, #401 — 333 Victoria Street, Nelson, British Columbia V1L 4K3, Canada

Linda Munson, Department of Pathology, Microbiology, and Immunology, School of Veterinary Medicine, University of California, Davis, CA 95616, United States

Tucker Murphy, Wildlife Conservation Research Unit, Department of Zoology, University of Oxford, The Recanati-Kaplan Centre, Tubney House, Abingdon Road, Tubney, OX13 5QL, United Kingdom

Andre A. Murzin, Pacific Institute of Geography, Far Eastern Branch of Russian Academy of Sciences, Vladivostok, Primorski Krai, Russia

Kristin Nowell, CAT, P.O. Box 332, Cape Neddick, ME 03902, United States

Stephen J. O'Brien, Laboratory of Genomic Diversity, National Cancer Institute, Frederick, MD 21702–1201, United States

Mark O'Donoghue, Fish and Wildlife Branch, Yukon Department of Environment, Box 310, Mayo, Yukon Territory Y0B 1M0, Canada

Dave Onorato, FL Fish and Wildlife Conservation Commission, 298 Sabal Palm Road, Naples, FL 34114, United States

Linde E. T. Ostro, The New York Botanical Garden 200th str. Kazimiroff Boulevard, Bronx, NY 10458–5126, United States

Francisco Palomares, Department of Conservation Biology, Estación Biológica de Doñana, CSIC C/ Américo Vespucio s/n 41092 Sevilla, Spain

Bivash Pandav, WWF-International, C/o—WWF-Nepal, P.O. Box 7660, Baluwatar, Kathmandu, Nepal

Pablo G. Perovic, Instituto de Bio Y Geociencias, Museo de Ciencias Naturales, Universidad Nacional de Salta, Mendoza 2, Salta (4400), Argentina

Nathalie Pettorelli, Institute of Zoology, Zoological Society of London, Regents Park, London, NW1 4RY, United Kingdom

Kim G. Poole, Aurora Wildlife Research, 1918 Shannon Point Road, Nelson, British Columbia V1A 6K1, Canada

Howard B. Quigley, Director, Western Hemisphere Felid Programs, Panthera Foundation, P.O. Box 11363, Bozeman, MT 59719, United States

Alan Rabinowitz, Panthera, 8 West 40th Street, 18th Floor, New York, NY 10018, United States

Alejandro Rodríguez, Department of Conservation Biology, Estación Biológica de Doñana, CSIC, C/Américo Vespucio s/n, 41092 Sevilla, Spain

Melody Roelke, Laboratory for Genomic Diversity, National Cancer Institute, Frederick, Maryland, 21702, United States

Andreas Ryser, KORA, Thunstrasse 31, 3074 Muri b. Bern, Switzerland

Marie-Pierre Ryser-Degiorgis, Centre for Fish and Wildlife Health, Institute of Animal Pathology, Vetsuisse Faculty, University of Bern, Postfach 8466, CH–3001 Bern, Switzerland

Eric Sanderson, Living Landscapes Program, Wildlife Conservation Society, 2300 Southern Blvd, Bronx, NY 10460, United States

John Seidensticker, Conservation Ecology Center, Smithsonian's National Zoological Park, Washington DC 2008 United States

Mahendra Shrestha, Save the Tiger Fund, National Fish and Wildlife Foundation, 1133 Fifteenth St, NW, Suite 1100, Washington D.C. 20005, United States

Claudio Sillero-Zubiri, Wildlife Conservation Research Unit, Department of Zoology, University of Oxford, The Recanati-Kaplan Centre, Tubney House, Abingdon Road, Tubney, OX13 5QL, United Kingdom

Leandro Silveira, Jaguar Conservation Fund, C.P. 193 Mineiros, GO 75830–000, Brazil

Scott C. Silver, The Wildlife Conservation Society, 2300 Southern Blvd., Bronx, New York 10460, United States

Alexander Sliwa, Cologne Zoo, Riehler Str. 173, Cologne 50735, Germany

Rob Slotow, School of Life and Environmental Sciences, University of KwaZulu-Natal, Durban, 4041, South Africa

Brian G. Slough, 35 Cronkhite Road, Whitehorse, Yukon Territory Y1A 5S9, Canada

Evgeny G. Smirnov, Sikhote-Alin State Biosphere Zapovednik, Terney, Primorski Krai, Russia

J. L. David Smith, Department of Fisheries, Wildlife and Conservation Biology, University of Minnesota, St Paul, MN, United States

José Bonifácio G. Soares, Instituto Pró-Pampa, Av. Nossa Senhora das Graças 294, Arroio Grande, RS 96330–000, Brazil

Göran Spong, Department of Wildlife, Fish, and Environmental Studies, Swedish Agricultural University, Sweden

Phillip A. Stephens, Biological and Biomedical Sciences, University of Durham, South Road, Durham, DH1 3LE, United Kingdom

Melvin Sunquist, Department of Wildlife Ecology and Conservation, University of Florida, Gainesville, FL United States

William Swanson, Centre for Conservation and Research of Endangered Wildlife, Cincinnati Zoo and Botanical Garden, Cincinnati, OH 45220, United States

Karen A. Terio, Zoological Pathology Program, University of Illinois, Maywood, IL 60153, United States

Marcos A. Tortato, CAIPORA Cooperativa para Conservação da Natureza, Rua Desembargador Vitor Lima No 260, Ed. Madison Center, sala 513, Florianópolis, SC 88040–400 Brazil

Blaire Van Valkenburgh, Department of Ecology and Evolutionary Biology, 621 Young Drive, S., University of California, Los Angeles, Los Angeles, CA 90095–1606, United States

Astrid Vargas, Iberian Lynx Conservation Breeding Programme, El Acebuche Centre, Doñana National Park, Matalascañas, Huelva 21760, Spain

M. Lilian Villalba, Andean Cat Alliance, Achumani, calle 17 No 41, La Paz, Bolivia

Sonam W. Wang, Chief of Nature Conservation Division, GP Box no. 684, Thimpu, Bhutan

Margaret N. Waweru, Tanzania Wildlife Research Institute, Box 661, Arusha, Tanzania

Lars Werdelin, Department of Palaeozoology, Swedish Museum of Natural History, Box 50007, S-104 05 Stockholm, Sweden

Eric Wikramanayake, Conservation Science Program, World Wildlife Fund, United States, 1250 24th Street, NW, Washington D.C. 20037–1193, United States

David Wildt, Center for Species Survival, Smithsonian's National Zoological Park, Front Royal, VA 22630, United States

Nobuyuki Yamaguchi, Department of Biological and Environmental Sciences, University of Qatar, P.O. Box 2713 Doha, Qatar, Arab Emirates

Olga Yu Zaumyslova, Sikhote-Alin State Biosphere Zapovednik, Terney, Primorski Krai, Russia

Alexandra Zimmermann, North of England Zoological Society and the Wildlife Conservation Research Unit, Department of Zoology, University of Oxford, The Recanati-Kaplan Centre, Tubney House, Abingdon Road, Tubney, OX13 5QL, United Kingdom

Fridolin Zimmermann, KORA, Thunstrasse 31, 3074 Muri b. Bern, Switzerland

PART I

Reviews

CHAPTER 1

Dramatis personae: an introduction to the wild felids

David W. Macdonald, Andrew J. Loveridge, and Kristin Nowell

Pallas's cat © John Tobias

What is a felid?

For those in tune with the intricate sophistication of cat behaviour, it may seem dusty minimalism to reduce what is special about them to the shapes of the bones in their ears. Nonetheless, felids belong to the cat-branch of the order Carnivora, formally the suborder Feloidea (cat-like carnivores), and traditionally have been distinguished from the Canoidea (dog-like carnivores) by the structure of the auditory bulla. That said, there is a deeper point to be made about the uniformity of the 36 extant species of wild felid, despite the fact that they range in size over two orders of magnitude from the massive Siberian tiger, *Panthera tigris altaica* (200–325 kg) to the dainty black-footed cat, *Felis nigripes* (1–2 kg), of southern Africa and diminutive rusty-spotted cat, *Prionailurus rubiginosus* (~1 kg), of India and Sri Lanka. All belong to a single subfamily of extant forms, the Felinae, and although their behaviour may be writ large or small, it is remarkably similar across them—cats are very distinctly cats (Macdonald 1992). In palaeogeological terms, the Felinae radiated relatively recently and rapidly in the late Miocene ~13–14 million years ago, with extinct and extant genera of the family Felidae derived from a common ancestor ~27 million years ago (Werdelin *et al.*, Chapter 2, this volume).

Felids are at an extreme among carnivores precisely because of their unanimous adherence to eating flesh, generally of vertebrate prey. Perhaps because they all face the same tasks—capturing (generally by ambush), subduing, and consuming their prey, and because of their relatively recent evolution, the morphology of most felids is remarkably similar (Kitchener *et al.*, Chapter 3, this volume). Felids are expert stalkers and killers, with specialized claws for holding and handling struggling prey before delivering a killing bite. Limbs are relatively long, with five digits on

Figure 1.1 Protracted claw of an immobilized African lion showing protective sheath, sharp ventral surface, and pointed tip © Joanne Loveridge.

the fore and four digits on the hind feet. The highly curved, laterally compressed, protractile claws are protected in sheaths when at rest and extended when needed (Fig. 1.1). However, the cheetah's (*Acinonyx jubatus*) claws, though partially protractile, are exposed at rest and the points blunted through contact with the ground. In this species, the claws may act as 'running spikes' providing traction for rapid acceleration in pursuit of prey. Relative to other carnivores, felids have shortened faces and rounded heads with 28–30 teeth adapted for dispatching prey and cutting flesh. Cat skulls have a less keel-like sagittal crest than canids and hyaenids and have wide zygomatic arches to accommodate large jaw muscles (Smithers 1983).

Ultimately, as for other mammals (e.g. Crook 1970; Bradbury and Vehrencamp 1976; Kruuk 1978; Jarman and Jarman 1979; Macdonald 1983; Macdonald *et al.* 2004b), felid bodies, lifestyles, societies, and species assemblages are reflections of their ecology. This chapter will introduce felids in the context of their role as predators, often apex predators, and as constituents of carnivore guilds and assemblages within their respective regional environments. We briefly discuss their intra- and interspecific relationships with each other and other members of the carnivore guild and the anthropogenic and conservation threats they face. The *dramatis personae*, intended to introduce the species discussed throughout this book, takes the form of 36 species vignettes. Finally, we present an analysis of the past 60 years of felid research to identify trends and potential gaps in the field of conservation and biology of wild felids.

Biogeography of felids

The wild Felidae inhabit all continents apart from Australasia and Antarctica (see Table 1.1), and occur on numerous islands, large (Borneo) and small (Trinidad). They utilize habitats as diverse as boreal and tropical forests, savannahs, deserts, and steppe—but many, particularly the smaller tropical species, are forest specialists, and 32 species occur in closed forest and woodland habitats (Nowell and Jackson 1996). Twenty-one species, almost 60% of all living felids, occur on the Asian continent, 14 of which are endemic there. Tropical and temperate Asian regions have the greatest number of cat species (12, with 10 found only in this biome), and Europe and the cold continental regions of Asia have seven species (with four found only here). Hot–dry south-west Asia also has seven species, all shared with Africa, although in south-west Asia the lion *Panthera leo* and cheetah *Acinonyx jubatus* have only small relict populations, and the tiger *P. tigris*

Table 1.1 Biogeographic occurrences of felid species. Species shown in bold typeface are endemic to the region.

Old World	New World
Asia—Tropical–Temperate	***Neotropics (Central and South America)***
Sunda clouded leopard *Neofelis diardi*	**Pampas cat *Leopardus colocolo***
Clouded leopard *Neofelis nebulosa*	**Geoffroy's cat *Leopardus geoffroyi***
Tiger *Panthera tigris*	**Guiña *Leopardus guigna***
Borneo bay cat *Pardofelis badia*	**Andean cat *Leopardus jacobbta***
Marbled cat *Pardofelis marmorata*	**Oncilla *Leopardus tigrinus***
Asiatic golden cat *Pardofelis temminckii*	**Margay *Leopardus wiedii***
Leopard cat *Prionailurus bengalensis*	Ocelot *Leopardus pardalis*
Flat-headed cat *Prionailurus planiceps*	Jaguar *Panthera onca*
Rusty-spotted cat *Prionailurus rubiginosus*	Puma *Puma concolor*
Fishing cat *Prionailurus viverrinus*	Jaguarundi *Puma yagouaroundi*
Jungle cat *Felis chaus*	Bobcat *Lynx rufus*
Leopard *Panthera pardus*	
Asia—Eurasia	***Nearctic (North America)***
Pallas's cat *Otocolobus manul*	**Canada lynx *Lynx canadensis***
Lynx *Lynx lynx*	Bobcat *Lynx rufus*
Iberian lynx *Lynx pardinus*	Ocelot *Leopardus pardalis*
Snow leopard *Panthera uncia*	Puma *Puma concolor*
Leopard *Panthera pardus*	Jaguarundi *Puma yagouaroundi*
Jungle cat *Felis chaus*	Jaguar *Panthera onca*
Wildcat *Felis silvestris*	
Asia—south-west Asia	
Caracal *Caracal caracal*	
Cheetah *Acinonyx jubatus*	
Lion *Panthera leo*	
Leopard *Panthera pardus*	
Jungle cat *Felis chaus*	
Sand cat *Felis margarita*	
Wildcat *Felis silvestris*	
Africa	
Black-footed cat *Felis nigripes*	
African golden cat *Caracal aurata*	
Serval *Leptailurus serval*	
Caracal *Caracal caracal*	
Cheetah *Acinonyx jubatus*	
Lion *Panthera leo*	
Leopard *Panthera pardus*	
Jungle cat *Felis chaus*	
Sand cat *Felis margarita*	
Wildcat *Felis silvestris*	

became extinct there decades ago, while in Africa the jungle cat *Felis chaus* occurs only in the vicinity of Egypt's Nile River. Africa has just three endemic species out of a total of 10. The Old World has twice the number of species (24) as the New World (12), and unlike the Canidae, no felid species straddle the divide, although two genera (*Lynx* and *Panthera*), do occur in both realms, and the Old World cheetah is grouped with the New World *Puma* clade (Werdelin *et al.*, Chapter 2, this volume). In the New World, only one species, *L. canadensis*, is exclusive to the northern realm; most species occur in the neotropics, with four found only in South America. Three Neotropical species have marginal ranges north of Mexico, and the Nearctic bobcat *L. rufus* extends south into north-central Mexico.

A few species have very large ranges spanning several continents. Leopards (*Panthera pardus*) are found from the Russian Far East and parts of Eurasia through tropical Asia, the Middle East, and throughout sub-Saharan Africa. The wildcat (*Felis silvestris*) is widely distributed in Africa, Asia, and Europe (Macdonald *et al.*, Chapter 22, this volume). The puma (*Puma concolor*) ranges across both North and South America, although it was extirpated a century ago from most of eastern North America. In contrast, some felid species are highly specialized and confined to limited areas of habitat in just a few countries. The Andean cat (*Leopardus jacobita*) occurs only in association with rocky outcrops in the arid zones of the high Andes, typically above 4200 m, a specialist predator of chinchillids (*Lagidium* spp.; Marino *et al.*, Chapter 28, this volume). The Iberian lynx (*Lynx pardinus*) is similarly a specialized rabbit hunter and its distribution is limited by the distribution of its prey on the Iberian peninsula, where disease has greatly reduced rabbit populations and there has been extensive habitat loss (Ferreras *et al.*, Chapter 24, this volume). Their specializations, on prey and habitat, may expose felids to anthropogenic threat and environmental or climate change. The more generalist, widespread species may be more robust to these threats, but none is immune to them. Generally, less than 10% of cat ranges consist of protected areas (Nowell and Jackson 1996); it is clear that these emblematic and threatened predators often occur beyond the safety of reserves.

Felid ecology and diet

Much of the behavioural ecology of felids is reviewed by Macdonald *et al.* (Chapter 5, this volume) but, as a foundation, here we briefly summarize some fundamentals. The most direct interface between felids and their environment is their diet. Environmental factors such as rainfall, seasonality, and nutrient availability in the environment determine primary production and hence the biomass of prey species (e.g. Coe *et al.* 1976; East 1984). In turn, biomass and dispersion of prey species in the environment determines felid population size and density, population structure, and social behaviour (Carbone and Gittleman 2002; Karanth *et al.* 2004c; Macdonald *et al.*, Chapter 5, this volume; Miquelle *et al.*, Chapter 13, this volume). Prey biomass and dispersion also determine home range size and intraspecific home range overlap, with home range size and often overlap between conspecific ranges inversely correlated with prey biomass (Macdonald *et al.*, Chapter 5, this volume; Miquelle *et al.*, Chapter 13, this volume; O'Donoghue *et al.*, Chapter 25, this volume; Loveridge *et al.*, 2009b).

Carnivores forage optimally when they are able to predate upon the largest suitable prey species they can safely kill; thus for each felid species there is a modal mass (and spread of taxa) of prey eaten by each population. In addition, diet varies according to individual and species prey preferences, local prey species assemblages, temporal availability of prey, and presence of intra-guild competitors. Taking leopards as an example, their prey may vary from hares (*Lepus* spp.) to kudu (*Tragelaphus strepsiceros*). In Gabon, the leopards studied by Henschel *et al.* (2005) ate a broad spread of prey sizes, peaking at 5.1–50 kg and comprising mainly ungulates (Fig. 1.2). Hayward *et al.* (2006) summarize how the average leopard diet varies across five African and three Asian populations (Fig. 1.3).

Similarly, the diet of ocelots differs radically between nine sites (Fig. 1.4; Moreno *et al.* 2006). As is commonplace for carnivores, the profile of prey consumed varies not only between felid species, but may also vary within species between populations, localities, territories, individuals, years, and seasons. Where studies are sufficient, generalities emerge, as for leopard diet

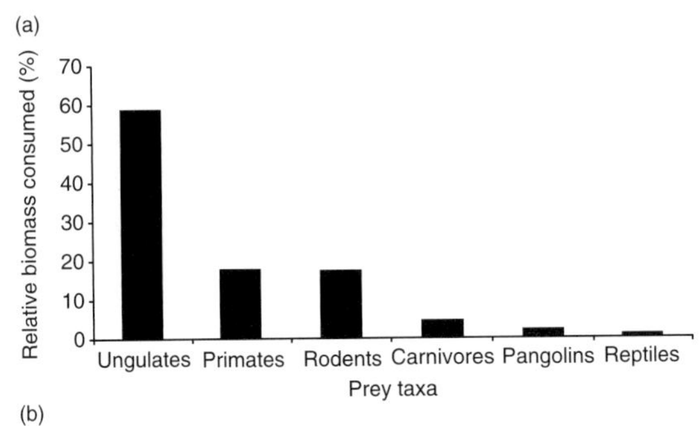

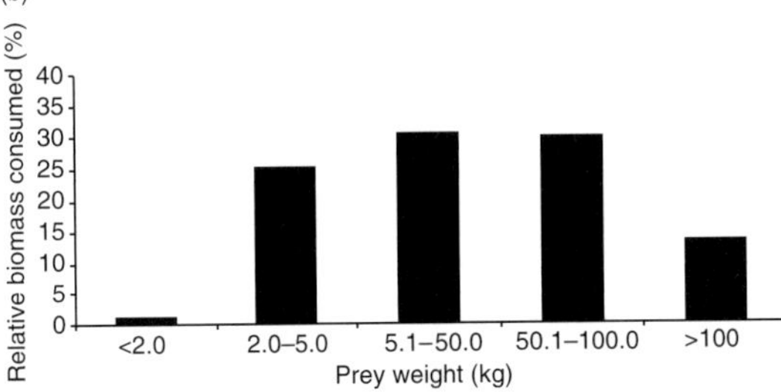

Figure 1.2 Representation of different (a) prey taxa; and (b) prey size classes in the diet of leopards. Calculations are based on a sample of 196 leopard scats, collected in the SEGC study area, Lopé National Park, Gabon, 1993–2001. (Taken from Henschel *et al.* 2005.)

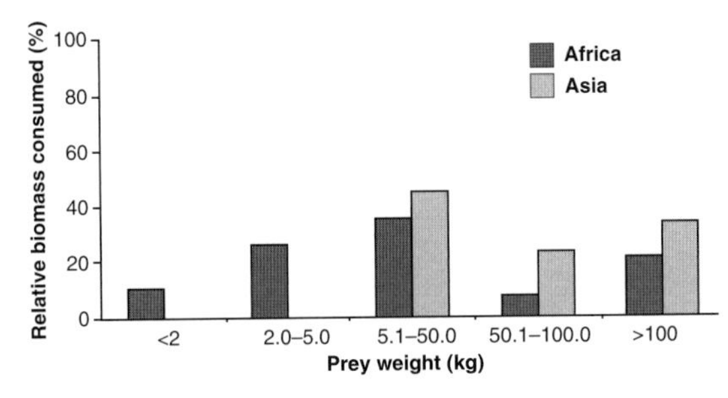

Figure 1.3 Total weight (kg) relative biomass consumed by African and Asian leopards. (Adapted from Hayward *et al.* 2006.)

from 41 localities (Fig. 1.5a), and for lions from 48 localities (Hayward and Kerley 2005) (Fig. 1.5b).

Diet may vary with season, over time, and with ecological conditions. In Patagonian Chile, Iriarte *et al.* (1991) analysed seasonal and yearly variation in puma diet between 1982 and 1988. Guanacos (*Lama guanicoe*) made up to 32% of prey items and accounted for 47% of the overall total biomass consumed by pumas. The proportion of guanaco remains in puma faeces increased from 9% to almost 30% of total prey items, paralleling an increase in the guanaco population from 670 to 1300 individuals in the study area.

Seasonality may affect populations differently, as illustrated by the patterns of prey availability experienced by the lions in the Serengeti plains and the nearby Ngorongoro Crater. Both have access to similar total annual prey biomass (12,000 kg/km^2; Hanby and Bygott 1987), but whereas this is resident in the Crater, on the plains it varies between 20,000 kg/km^2 in the wet season (November–May) and 1000 kg/km^2 in the dry season. Plains lions may have to switch to smaller, non-migratory prey in times of food shortage (Schaller 1972).

The proportion of prey species in the diet may differ from that in the environment due to various facets of their relative availability (such as size, habitat use, escape, and defensive behaviour), preference, or the impact of competitors (and even the size of the individual predator: juvenile puma eat smaller prey than do adults; Harveson 1997). Harveson *et al.* (2000) report that puma in southern Texas preyed on collared peccaries (*Pecari tajacu*) in proportion to their abundance, whereas by the same measure they selected for white-tailed deer (*Odocoileus virginianus*) and against feral hogs (*Sus scrofa*). Jaguars in Peru consumed collared peccary more frequently than expected, avoiding comparably abundant and similarly distributed white-lipped peccary and tapir (Emmons 1987; Weckel *et al.* 2006a, b).

In summary, the business of hunting is similar from the largest to the smallest felids. However, the relationship between the size of each felid species and that of its prey has reverberations throughout their behavioural ecology, and this is the topic of Macdonald *et al.* (Chapter 5, this volume).

Felid assemblages and communities

Intraspecific interactions

Most felids are solitary, with some notable exceptions (e.g. the lion, and to a lesser extent the cheetah, and wildcat). Conspecific interaction is generally limited to mating and the rearing of young (for females); when food is not scarce, a female cat is likely to be either pregnant or accompanied by dependent young (Sunquist and Sunquist 2002). However,

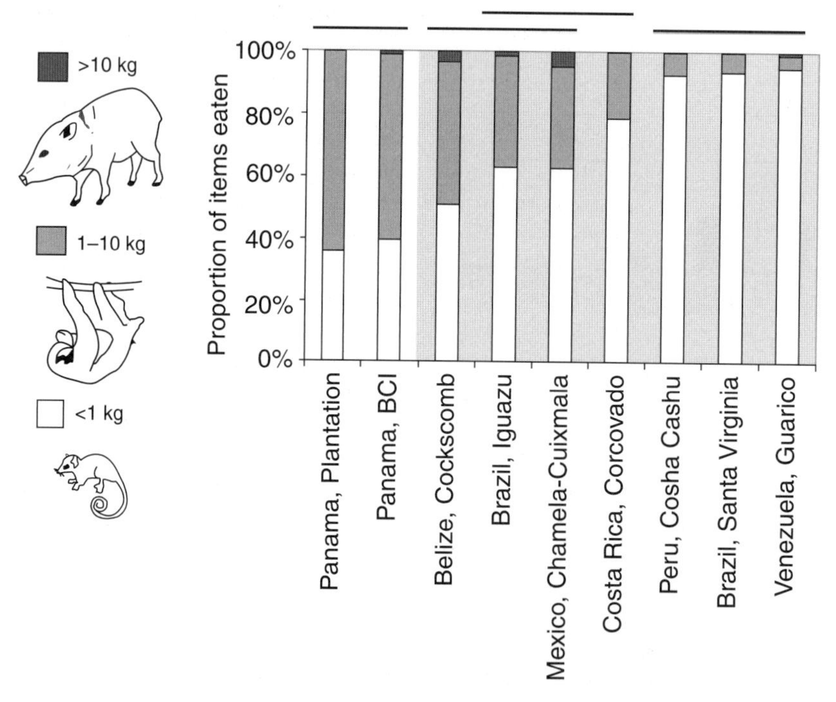

Figure 1.4 Weight of prey eaten by ocelots across nine sites. Top horizontal lines indicate the presence of robust sympatric jaguar populations. Horizontal lines above graph connect sites that are not significantly different from one another; distributions of prey size for all pairs of unconnected sites were significantly different (chi-square or Fisher's exact test on number of items detected in each size class, $P < 0.05$). *Sources*: Plantation and Barro Colorado Island (BCI), Panama (Moreno *et al.* 2006); Cockscomb, Belize (Konecny 1989); Iguazu, Brazil (Crawshaw 1995); Chamela-Cuixmala, Mexico (de Villa Meza *et al.* 2002); Corcovado, Costa Rica (Chinchilla 1997); Cosha Cashu, Peru (Emmons 1987); Santa Virginia, Brazil; and Guanico, Venezuela (Ludlow and Sunquist 1987). (Taken from Moreno *et al.* 2006.)

adults of some species collaborate in defence of territories, resources, mates, or offspring (Macdonald *et al.*, Chapter 5, this volume). Yet for many cats, conspecific encounters occur most frequently not between individuals, but between their signals, sent through territorial marking behaviour. Felids also compete for access to resources, ranges, mates, and/or reproductive opportunities, and deadly conspecific encounters have been recorded for some big cats. Among pumas, mature males may kill (but generally do not eat) other males (Anderson *et al.* 1992) or juveniles of both sexes (Harveson *et al.* 2000). Male African lions engage in combat for access to prides, and mortality among competing males is common (Schaller 1972). In Serengeti lions, adult female mortality rates are significantly correlated with the number of adult male neighbours, suggesting that females may be the target of attack by neighbouring males if they are not receptive to mating (Mosser 2008). However, for felids the most widely documented intraspecific mortality is infanticide, recorded for most pantherines (Davies and Boersma 1984; Bailey 1993; Smith 1993) and domestic cats (Macdonald *et al.* 1987; Macdonald *et al.*, Chapter 5, this volume). Typically, it is assumed to be by unrelated males, but Soares *et al.* (2006) speculated that the case of two jaguar cubs killed by their father was a pathology prompted by habitat fragmentation causing uncertainty over paternity.

Among lions, at a pride takeover, males famously either attack and kill, or otherwise cause the deaths of small cubs and evict larger ones they have not sired (Schaller 1972; Packer *et al.* 2001). Bertram (1975a) found a significant increase of mortality of lion cubs less than 24 months old in the first 4 months after a male takeover. (Of 11 takeovers there was only one in which small cubs survived.) Infanticide of tiger cubs is also associated with male territorial takeovers (Smith *et al.* 1987; Macdonald 1992). More than 50% of Serengeti lion cubs die

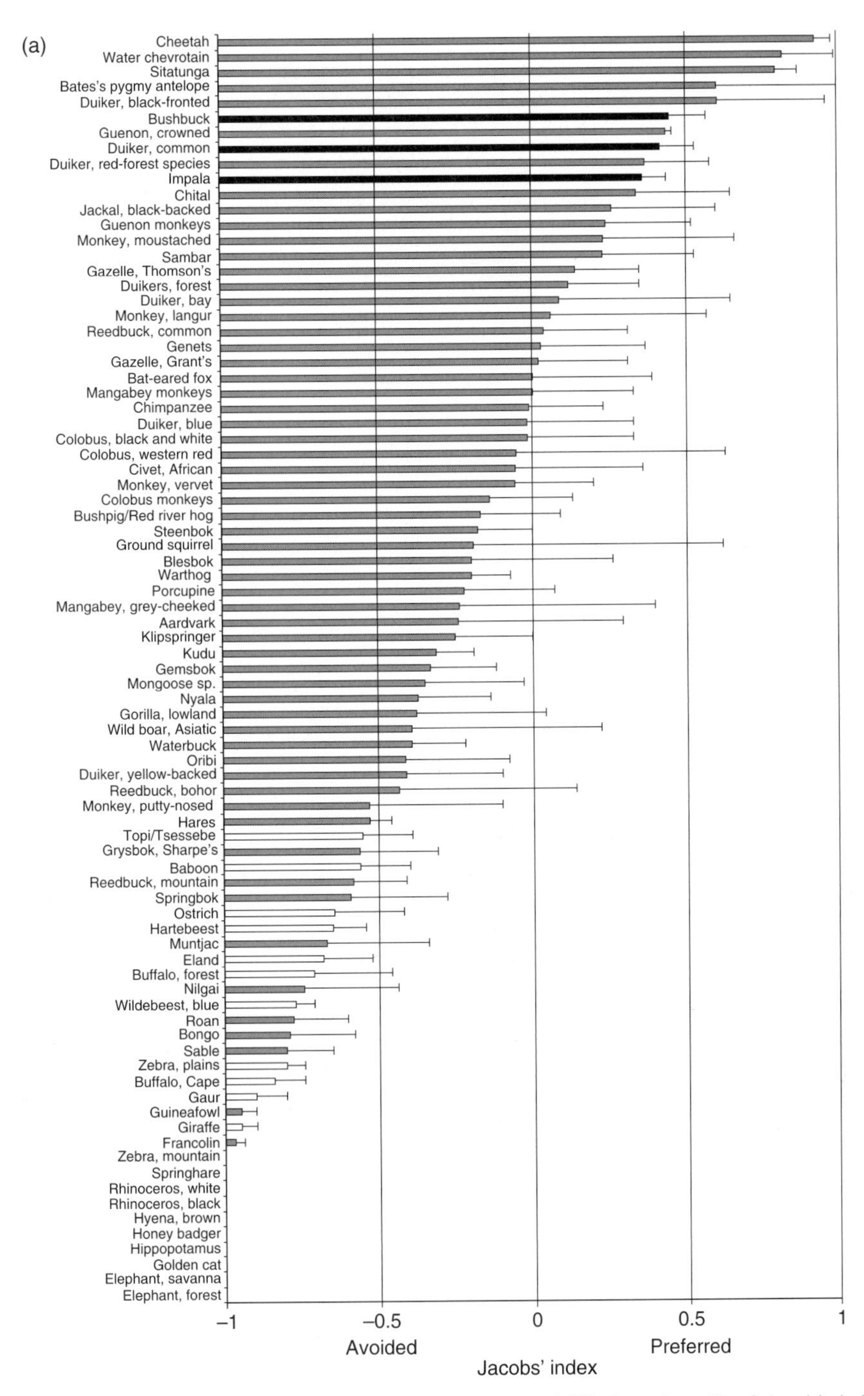

Figure 1.5 Dietary preferences determined with Jacob's index (mean ± 1 SE of species with >2 Jacob's index estimates) calculated for (a) leopards *Panthera pardus* from 41 populations; and (b) lions *Panthera leo* from 48 populations, all populations at differing prey densities for both species. Black bars represent species taken significantly more frequently than expected based on their abundance (preferred), grey bars indicate species taken in accordance with their relative abundance and unfilled bars show species taken significantly less frequently than expected based on their abundance (avoided). (Taken from Hayward *et al*. 2006a and Hayward and Kerley 2005.)

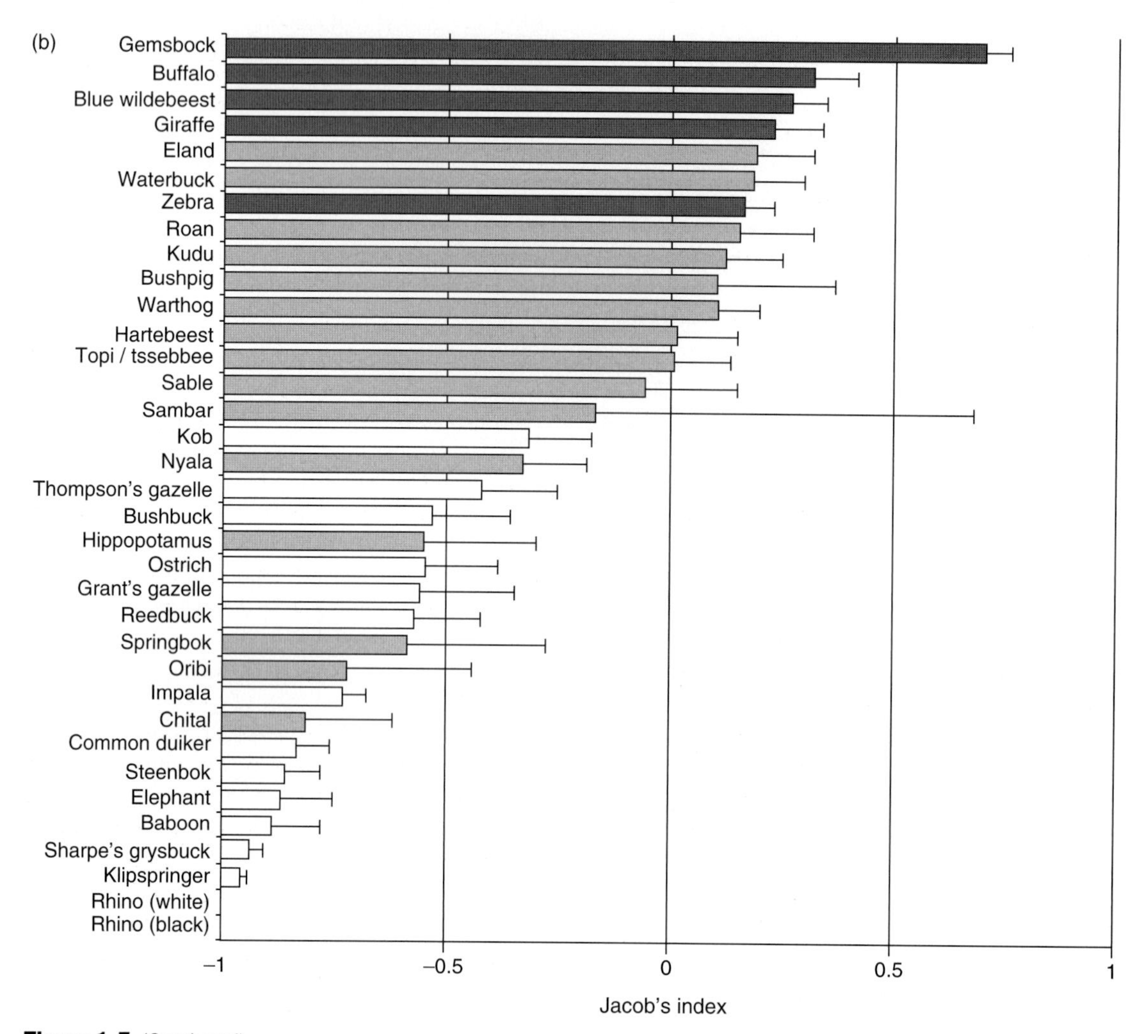

Figure 1.5 (Continued)

before their first birthday (Schaller 1972; Hanby and Bygott 1979) and females that lose their dependent offspring quickly resume mating activity, with the result that the synchronous loss of cubs at a takeover is followed by a synchronous resumption of mating activity. Consequently, the distribution of the duration of post-partum amenorrhoea across females reflects cub mortality for about the first 200 days after birth and reflects the approximate age of independence thereafter. Cubs have lower mortality when born synchronously (Bertram 1975) and mothers of single cubs are more likely to abandon them if they are born asynchronously than if they are born at the same time as other cubs in their pride (Rudnai 1974).

Community structure, interspecific relations, and intra-guild hostility

Interspecific competition is often a force that structures carnivore guilds and indeed widens communities of species (e.g. Rosenzweig 1966; Schoener 1974, 1984; Rautenbach and Nel 1978; Simberloff and Boecklen 1981). It is a phenomenon well described in the Canidae (Macdonald and Sillero-Zubiri 2004c), where interference competition by large canids has significant impacts on the distribution, density, and behaviour of smaller species. More subtle evidence of competition in the form of morphological character displacement has also been recorded in sympatric canid species (Van Valkenburg and Wayne 1994; Dayan *et al.* 1989). Some evidence for character displacement among the felids has also been found (Dayan *et al.* 1992). Kiltie (1988) examined jaw lengths in regional assemblages of felids and found that female jaw length was closely correlated with modal prey weight. Since felids use a killing bite to dispatch prey, species that feed upon larger prey should have concomitantly wider jaw gapes. Hence, jaw lengths (and probably other morphological parameters) appear to have evolved to maximize efficiency in handling and killing common prey species and to minimize overlap with adjacent-sized species within the assemblage. Evenness in distribution of jaw lengths in regional assemblages of felid species may possibly indicate partitioning of resources within the felid guild. Such even size ratios are particularly apparent among the large Neotropical felids (Kiltie 1984). While this may well be evidence of character displacement, jaw lengths could equally have evolved as a response to the size distributions of available prey. Intriguingly, for three pairs of sympatric species (jaguarundi, *Puma yagouaroundi* and margay, *Leopardus wiedii*; serval, *Leptailurus serval* and caracal, *Caracal caracal*; and Asiatic golden cat, *Pardofelis temminckii* and fishing cat, *Prionailurus viverrinus*) whose jaw lengths (and therefore presumably modal prey size) were indistinguishable; one of the pair was always dappled (spotted or striped) the other plain, which Kiltie (1988) suggested may be indicative of ecological differences (e.g. different requirements for crypsis, differing behaviours, or habitat use).

Being so similar in form and function, coexisting felids are destined to rivalry. This competition can take a number of forms, for instance exploitative competition where sympatric felids compete for resources, or more direct interference competition, where sympatric felids harass or kill non-conspecifics (Mills 1991). Competitive interactions can determine the presence or absence, abundance, behaviour, and distribution of carnivores within guilds of sympatric species.

Exploitative competition

Sympatric carnivores may have significant dietary niche overlap. Andheria *et al.* (2007) compared the diets of tiger (*Panthera tigris*), leopard (*P. pardus*), and dhole (*Cuon alpinus*) in Bandipur Tiger Reserve (India), each of which kill 11–15 species of vertebrate prey, but three abundant ungulate species provide 88–97% of the biomass consumed by each. Although there was some difference in emphasis (the largest ungulates, gaur *Bos gaurus* and sambar *Cervus unicolor*, provided 63% of biomass consumed by tigers, whereas medium-sized chital *Axis axis* and wild pig *Sus scrofa* formed, respectively, 65% of the biomass intake of leopards) (Fig. 1.6), dietary niche overlap among the three species was high (Pianka's index of 0.75–0.93) and one species, chital, comprised 33% of tiger diet and 39% of leopard diet. Similarly, high dietary overlap between tigers and leopards (Pianka's index of 0.97) was found by Wang and Macdonald (2009) in Jigme Singye Wangchuck National Park, Bhutan, though overlap was less marked between these felids and dholes (Pianka's index of 0.58 and 0.66 for tigers/dholes and leopards/dholes, respectively).

Similarly, the diets of lions, leopards, and cheetahs all overlapped by species and size in Mala Mala (RSA), but the distributions of size classes (relative to the predators' body sizes) were significantly different (Radloff and du Toit 2004). Depending on the perspective, one might be impressed by the niche separation or its overlap: considering the cline in weights from lions (male 188 kg, female 124 kg), leopards (male 61.3 kg, female 37.3 kg), and cheetahs (male 53.9 kg, female 43.0 kg), it is noteworthy that only the three heaviest cats kill buffalo or zebra. The mean weights of kudus killed are 193, 165, 100, 88.9, 110, and 57.8 kg, respectively; it is also noteworthy that they all eat kudu (albeit in neatly graded mean sizes)! Bertram (1979) found little evidence of

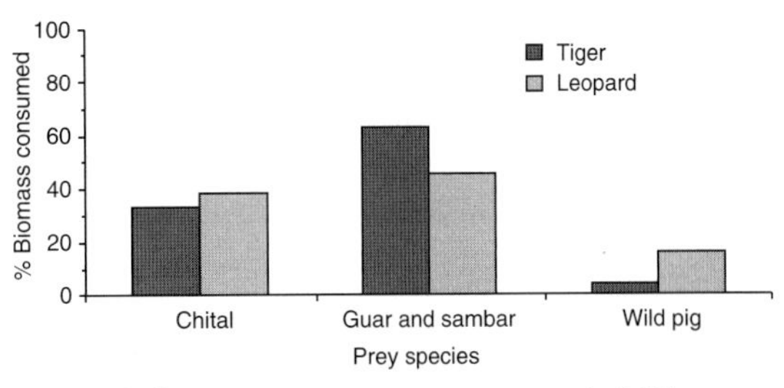

Figure 1.6 Percentage biomass consumed of different prey species by the tiger and leopard in the Bandipur Tiger Reserve, India. (Adapted from Andheria *et al.* 2007.)

exploitative competition among large felid species in the Serengeti, largely because they utilized prey of different sizes and employed different hunting techniques to capture prey. Pumas and jaguars are estimated to have a dietary overlap of up to 82% (Oliveira 2002). In the Paraguayan Chaco, jaguar and puma diet overlapped significantly in areas developed or over-exploited by people; however, in more pristine areas there was evidence of niche separation, with jaguars eating larger prey (Taber *et al.* 1997).

These wide niche overlaps may influence the choice of prey, behaviour, and distribution of the predators. For example, cheetahs unsurprisingly avoid lions, seeking out 'competition refuges' (Durant 2000a) and leopards allow tigers' first choice of habitats and prey (Eisenberg and Lockhart 1972; Seidensticker 1976a; but see Karanth and Sunquist 2000). Where there is a choice, leopards utilize smaller prey than do tigers (Karanth and Sunquist 1995). In Belize, female jaguars and male pumas overlap in size, suggesting the potential for competition. Camera-trapping data reveal that while activity patterns are similar, avoidance in space and time may minimize competitive interactions between the species (Harmsen *et al.*, Chapter 18, this volume). Schaller and Crawshaw (1980) found similar spatio-temporal avoidance between these species in the Brazilian Pantanal.

Absence or removal of larger competitors can result in competitive release for smaller species. In South America, margay, jaguarundi, oncilla (*Leopardus tigrinus*), and Geoffroy's cat (*L. geoffroyi*), occur at higher densities in the absence of larger ocelots, suggesting that ocelots may compete directly or indirectly with the smaller cats. However, ocelot numbers do not appear to be affected by the presence of much larger pumas or jaguars (Oliveira *et al.*, Chapter 27, this volume). In southern Africa, cheetah numbers are often higher on ranch lands where lions or spotted hyena have been extirpated than in protected areas, suggesting competitive release in the absence of larger competitors (Purchase and Vhurumuku 2005; Marker *et al.*, Chapter 15, this volume). Black-footed cats and African wild cats may benefit from the removal of larger competitors such as caracals and jackals (*Canis mesomelas*) (Sliwa *et al.*, Chapter 26, this volume).

Interference competition and intra-guild predation

The best-documented cause of interspecific mortality is intra-guild aggression (Wozencraft 1989; Palomares and Caro 1999). This peaks with 68% of cheetah cubs killed by lions, spotted hyenas, and leopards (Laurenson 1994, 1995b; Kelly and Durant 2000), and lions have negative impacts on the survival of females (Durant *et al.*, Chapter 16, this volume). Between 12% and 62% of bobcats (*Lynx rufus*) are killed by coyotes (*Canis latrans*) and pumas (Knick 1990; Koehler and Hornocker 1991), and 8% of lion cubs succumb to leopards and spotted hyenas (*Crocuta crocuta*) (Schaller 1972). In the Kalahari National Park, cheetahs lose competitive interactions with lions. However, lions had little effect on leopards and leopards did not affect cheetahs (Mills 1991). By contrast, in Hwange National Park a leopard was recorded killing and partially eating a cheetah (Davison 1967), G. Mills (personal communication) records similar behaviour in the Kalahari National Park (Fig. 1.7), and lions have been observed chasing leopards (A.J. Loveridge, personal observation). Perhaps the most extreme interspecific interaction is intra-guild predation. African leopards have been recorded as eating African golden cats, *Caracal aurata*, in central Africa (Henschel *et al.* 2005) and caracal in southern Africa (Hwange National Park; B. du Preez, personal communication). Caracals have been observed killing and partly eating African wildcats (G. Mills, personal comunication; Fig. 1.8). Leopards and pumas living in proximity to human settlements have a proclivity for predation on domestic cats (*Felis catus*; Martin and de Meulenaer 1988; Onorato *et al.*, Chapter 21, this volume).

Felids are part of wider carnivore guilds and interspecific competition is not limited only to sympatric felids. The interesting question is whether intra-family hostility is more punishing than inter-family competition. African wild dogs (*Lycaon pictus*) and dholes go out of their way to harass leopards (Davison 1967; Macdonald and Sillero-Zubiri 2004c;

Figure 1.7 A cheetah, killed and partly eaten by a leopard in the Kalahari. Circumstantial evidence indicated that the cheetah was injured prior to being attacked by the leopard. © M.G.L. Mills.

Figure 1.8 A caracal, having killed a wildcat, and carried it aloft a tree, in the Kalahari, prior to partly eating it. © M.G.L. Mills.

Venkataraman and Johnsingh 2004), perhaps because leopards not only compete for prey but also predate on adults and juveniles of these species. Spotted hyenas kleptoparasitize leopard kills and as a consequence leopards often cache their kills in trees out of reach of competitors. Lions and spotted hyenas compete fiercely for carcasses (Cooper 1991), and kill vulnerable adults and unprotected juveniles of the other species (A.J. Loveridge, personal observation).

Interactions with people: anthropogenic threats and conservation

The purpose of this book is twofold; first, to provide a compendium of knowledge on the biology of wild felids; second, to set that knowledge in the context of felid conservation, and to discuss how it can be used to greatest effect to safeguard their survival. These two purposes are each worthwhile in themselves, and in the concluding synthesis (Macdonald *et al.*, Chapter 29, this volume) we will explore the inextricable linkages between them. However, to set the scene we will briefly summarize here some of the threats posed by people to wild felids, and vice versa.

Anthropogenic mortality, human–wildlife conflict, prey depletion, and habitat loss

Mortality comes in diverse guises, but the contemporary reality for several felid species may be that humans are either inadvertently (e.g. road traffic accidents, Haines *et al.* 2006) or deliberately (e.g. hunting, Altrichter *et al.* 2006) behind much of it (Loveridge *et al.*, Chapter 6, this volume). Anthropogenic mortality (including vehicle collisions) predominated among adult Eurasian lynx, with starvation, intra- and interspecific killing, and disease having a minor role (Andrén *et al.* 2006). Six of a sample of 15 radio-collared adult bobcats died, leading Fuller *et al.* (1985) to suggest that in many areas, sources of mortality other than legal trapping or hunting (e.g. poaching, starvation, disease, and predation), may be substantial (53% of all deaths for studied populations) and should be incorporated into models of bobcat population change. The composite demographic data from radio-telemetry field studies indicate that if populations are stable when adult survival is about 60% (40% mortality), and about 50% of mortality is not harvest related, then on average, harvests averaging over about 20% of the population (40% mortality × 0.5) will be likely to result in declining populations. Of course, harvests of <20% of the population may also cause a decline if the rates of natural mortality are high, or if reproductive success is low. Annual losses to tiger populations in the Russian Far East that exceed 15% are predicted to lead to population declines (Chapron *et al.* 2008a).

A global craze for spotted cat fur coats in the 1960s was the genesis of the Convention on International Trade in Endangered Species (CITES), which regulates international trade (Nowell and Jackson 1996). Although some harvest and trade is well managed and legal, poaching and illegal trade for skins and body parts may present both a significant cause of mortality and a conservation threat for target species (Loveridge *et al.*, Chapter 6, this volume; Damania *et al.* 2008). Illegal trade in wildlife is not only increasing in magnitude, but is thought to be the second largest area of organized crime after the illegal drug trade (Angulo *et al.* 2009). Many of the felid species that feature in illegal trade, such as the tiger and other Asian cats are already rare and facing threats to their survival through habitat loss and isolation of existing populations (Rabinowitz 1999). The concern is that the demands of the illegal market may overwhelm the attempts of conservationists to protect these species in the wild unless international action is taken to curb trade.

In many cases, anthropogenic mortality is linked to conflict with people over space, depredation of livestock or game animals, and occasionally over man-eating incidents (Loveridge *et al.*, Chapter 6, this volume; Breitenmoser *et al.*, Chapter 23, this volume). Livestock owners, game farmers, and bereaved communities often respond to depredations (and other conflict situations) by poisoning, trapping, or shooting the predators responsible. Such control measures can be legal and well controlled or illegal and uncontrolled, the later potentially leading to population declines or extirpation.

Human activity can also indirectly impact felids by depletion of prey species populations. Carnivore population size is closely linked to abundance of prey species, therefore depletion of the prey base can have catastrophic impacts on felid populations (Karanth *et al.* 2004a). The bushmeat trade in some areas of central Africa has resulted in 'empty forest syndrome' where, although habitat has remained relatively intact, medium- to large-sized mammalian species have been decimated by over-hunting. Under these conditions, the prey base cannot support viable populations of species such as leopards (Henschel 2007).

However, one of the greatest threats to felids and biodiversity as a whole is loss of natural habitats: a global trend the rate of which is predicted to accelerate into the next millennium (Millennium Ecosystem Assessment 2005). Conversion of natural habitat to agricultural land and linked loss of natural prey is a major threat to all felid species and one that might be expected to cause the demise of many populations if present trends are not halted. Salvation may lie in continued protection of existing protected areas and creation of corridors that link these together (Weber and Rabinowitz 1996; Macdonald *et al.*, Chapter 29, this volume).

Conservation of felids

The IUCN Red List of Threatened Species (IUCN 2008) is considered to be the most authoritative index of species status, particularly for mammals, which was recently reassessed in a massive global exercise (Schipper *et al.* 2008). Threat categories are assigned based on quantitative criteria, with thresholds for population size, range size, rate of decline, or probability of extinction (e.g. the Iberian lynx qualifies as Critically Endangered by having a fragmented, declining population of fewer than 250 mature individuals). Over 44% of felids (16 of 36 species) are included in the top three categories of threat (Critically Endangered, Endangered, and Vulnerable) with an elevated extinction risk (Table 1.2). While larger mammals are significantly more threatened than small mammals, felids are not significantly more threatened than expected, although their threat level is relatively high compared to carnivores and mammals in general (25% of both groups being in

Table 1.2 Conservation status of cat species on the 2008 IUCN Red List of Threatened Species.

Critically Endangered	**Endangered**
Extremely high extinction risk	*Very high extinction risk*
Iberian lynx *Lynx pardinus*	Andean cat *Leopardus jacobita*
	Tiger *Panthera tigris*
	Snow leopard *Panthera uncia*
	Borneo bay cat *Pardofelis badia*
	Flat-headed cat *Prionailurus planiceps*
	Fishing cat *Prionailurus viverrinus*
Vulnerable	**Near Threatened**
High extinction risk	*Close to qualifying for a higher threat category*
Cheetah *Acinonyx jubatus*	African golden cat *Caracal aurata*
Black-footed cat *Felis nigripes*	Sand cat *Felis margarita*
Guiña *Leopardus guigna*	Pampas cat *Leopardus colocolo*
Oncilla *Leopardus tigrinus*	Geoffroy's cat *Leopardus geoffroyi*
Sunda clouded leopard *Neofelis diardi*	Margay *Leopardus wiedii*
Clouded leopard *Neofelis nebulosa*	Pallas's cat *Otocolobus manul*
Lion *Panthera leo*	Jaguar *Panthera onca*
Marbled cat *Pardofelis marmorata*	Leopard *Panthera pardus*
Rusty-spotted cat *Prionailurus rubiginosus*	Asiatic golden cat *Pardofelis temminckii*
Least Concern	
Relatively widespread and abundant	
Caracal *Caracal caracal*	
Jungle cat *Felis chaus*	
Wildcat *Felis silvestris*	
Ocelot *Leopardus pardalis*	
Serval *Leptailurus serval*	
Canada lynx *Lynx canadensis*	
Eurasian lynx *Lynx lynx*	
Bobcat *Lynx rufus*	
Leopard cat *Prionailurus bengalensis*	
Puma *Puma concolor*	
Jaguarundi *Puma yagouaroundi*	

the top three categories; Schipper *et al.* 2008). Dividing felids into eight clades (Werdelin *et al.*, Chapter 2, this volume) reveals that for three lineages the majority of species are threatened (*Panthera, Pardofelis*, and leopard cat lineages). Most of this group occurs in south and south-east Asia, as do threatened mammals in general (Schipper *et al.* 2008), although the only Critically Endangered felid, the Iberian lynx, occurs in Europe (Spain). The other species in its clade are classified as Least Concern, demonstrating that extinction risk is uneven across the remaining lineages. An additional 20 subspecies are included in the top three categories, although not all felid subspecies have yet been assessed. Leading threats to felids are similar to those found for mammals in general, being primarily habitat loss and degradation, followed by hunting pressure (Schipper *et al.* 2008).

Historically, wild felids have been hunted and persecuted (Tuck 2005), but also valued and revered (Callou *et al.* 2004). This dichotomy continues to be reflected in contemporary interactions between people and wild cats. On the one hand, tourists are enthralled by sightings of large free-ranging cats in national parks around the world while, beyond protected areas, rural people suffer depredations from the same species. Westerners may desire the conservation of the toothed and clawed creatures brought to the safety of their living rooms by television programmes, but living with large wild felids is anything but easy. The cultural and aesthetic values some people accord to charismatic species are often at odds with the conflict between them and the people who live with them. These issues are explored in Macdonald *et al.* (Chapter 29, this volume). Mechanisms for ameliorating this conflict include zonation and protection of wild habitats, compensation and protection of livelihoods, as well as integration of local communities into conservation activities and revenue-generating ecotourism.

Dramatis personae

Felid systematics have been well studied, but a consensus taxonomy, especially concerning the number of genera, has been slow to emerge. Speciation events have been relatively rapid and recent in the family, resulting in a sparse fossil record and few distinguishing morphological characteristics (Johnson *et al.* 2006b). Species vignettes are presented here following the recent well-supported revised classification of the Felidae based on molecular analysis of nuclear and mitochondrial DNA (Johnson *et al.* 2006b; O'Brien and Johnson 2007; Eizirik *et al.*, in prep; Werdelin *et al.*, Chapter 2, this volume). This taxonomic arrangement is in general agreement with other recent authorities (e.g. Wozencraft 2005; Macdonald 2006). Some notable differences include two species of clouded leopard (*Neofelis nebulosa* and *N. diardi*), one species of pampas cat (*Leopardus colocolo*), and the Chinese mountain cat as a subspecies of wildcat (*Felis silvestris bieti*).

Box 1.1 Legend for species distribution maps

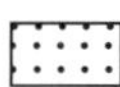
Current distribution

Former range

National boundaries

Subnational boundaries

Lakes, rivers, canals

Salt pans, intermittent rivers

Source: All maps from the IUCN Red List of Threatened Species. © IUCN 2008.

Clouded leopard *Neofelis nebulosa* (Griffith, 1821)

The clouded leopard, named for its elliptical pelage markings, held the unique position within the family Felidae of a 'small, big cat' (Sunquist and Sunquist 2002) until it was split into two species in 2006 (see *Neofelis diardi*). Both are now placed with the genus *Panthera* in the tribe Pantherini (Eizirik *et al.*, in prep). The skulls of clouded leopards are reminiscent of sabre-toothed machairodonts, their upper canine teeth being longer, relative to skull length, than those of any extant big cat (Christiansen 2006); measuring 3.8–4.5 cm (Guggisberg 1975). It is highly arboreal, its tail being *c.* 1 m long, although clouded leopards have been most frequently observed on the ground (Sunquist and Sunquist 2002). While they have been observed hunting primates in trees (Nowell and Jackson 1996), the only study to

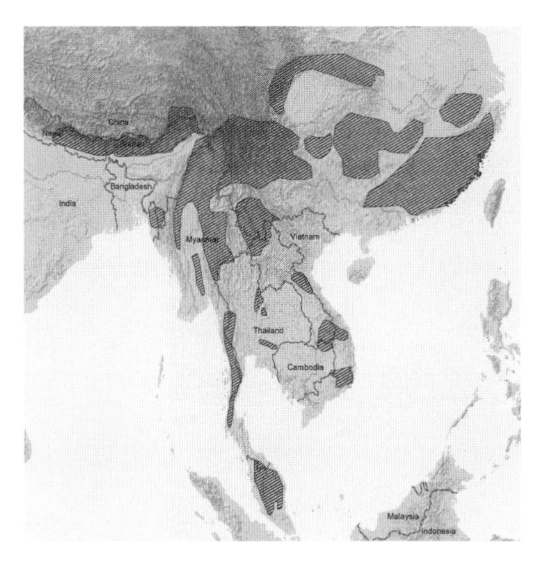

Map 1 Clouded leopard. © IUCN Red List 2008.

examine diet found hog deer (*Axis porcinus*), Asiatic brush-tailed porcupine (*Atherurus macrourus*), Malayan pangolin (*Manis javanica*), and Indochinese ground squirrel (*Menetes berdmorei*) (Grassman *et al.* 2005b). The clouded leopard, classified as Vulnerable (IUCN 2008), is found from the Himalayan foothills in Nepal through mainland south-east Asia into China (Map 1), where it has suffered heavy range loss (not shown on Map 1), and is extinct on Taiwan (Anonymous 1996). Although strongly associated with primary evergreen tropical rainforest, there are records from dry and deciduous forests, and secondary and logged forests. They have been recorded in the Himalayas up to 2500 m and possibly as high as 3000 m. Less frequently, they have been found in grassland and scrub, dry tropical forests, and mangrove swamps (Nowell and Jackson 1996). Radio-tracking in Thailand revealed a preference for forest (Austin *et al.* 2007), with both sexes occupying similar-sized home ranges of 30–40 km^2 (95% fixed kernel estimators) with intensively used core areas of 3–5 km^2 (Grassman *et al.* 2005b; Austin *et al.* 2007). Clouded leopards in Phu Khieu National Park travelled approximately twice the average daily distance (average 2 km: Grassman *et al.* 2005b) than those in Khao Yai National Park (Austin *et al.* 2007). While both studies found substantial home range overlap between males and females, Grassman *et al.* (2005b) also reported 39% overlap in the ranges of two males. Illegal trade is a serious threat, with large numbers of skins (largely of wild provenance) seen in markets, as well as bones for medicines, meat for exotic dishes, and live animals for the pet trade (Nowell 2007).

	Male		Female	
	Mean	Sample size	Mean	Sample size
Weight (kg)	16–18	*n* = 2	11.5–13.5	*n* = 2
Head/body length (mm)	980–1080	*n* = 2	820–940	*n* = 2

Refs: Austin and Tewes (1999); Grassman *et al.* (2005b)

Sunda clouded leopard *Neofelis diardi* (Cuvier, 1823)

The genetic differences between the Sunda clouded leopard and *N. nebulosa* are greater than those between well-accepted *Panthera* species (Buckley-Beason *et al.* 2006), and its pelage has smaller cloud-like markings (Kitchener *et al.* 2006). It is restricted to the islands of Sumatra and Borneo (Kitchener *et al.* 2007; Wilting *et al.* 2007a; Map 2). Analysis of mitochondrial DNA sequences suggests the Sumatran and Bornean populations deserve recognition at the subspecies level, and have been isolated from each other since the middle to late Pliocene (2.86 million years ago) (Wilting *et al.* 2007b). Like the mainland clouded leopard, it is highly arboreal. Holden (2001) found that on level or undulating terrain clouded leopards were seldom if ever caught on camera traps, suggesting considerable arboreality, although their spoor is recorded along logging roads and trails (Holden 2001; Gordon and Stewart 2007). Clouded leopards may be less arboreal on Borneo (Rabinowitz *et al.* 1987) than on Sumatra, where tigers and leopards are sympatric. They may also occur at higher densities: 6.4 adults (A. Hearn and J. Ross, personal communication in IUCN 2008) to 9.0 adults/100 km^2 (Wilting *et al.* 2006) on Borneo, compared to 1.29/100 km^2 on Sumatra (Hutajulu *et al.*

Plate A Sunda clouded leopard *Neofelis diardi*, caught on a camera trap. © Andreas Wilting.

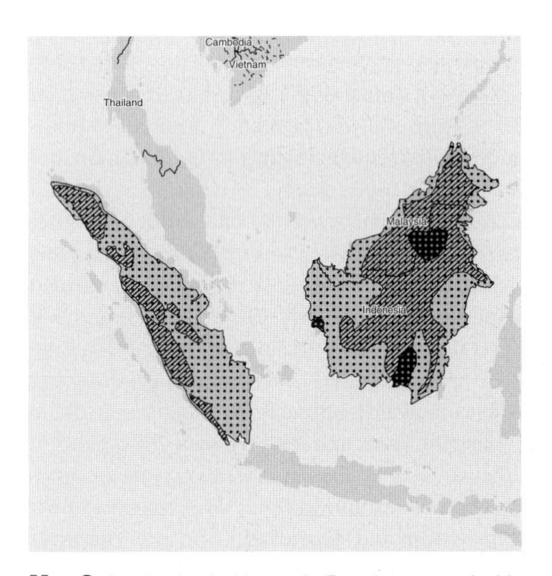

Map 2 Sunda clouded leopard. Former range coincides with areas where forest cover has been largely lost (GLC 2000, S. Cheyne, personal communication 2008). © IUCN Red List 2008.

2007). They are classified as Vulnerable by IUCN (2008). The Sumatran province of Riau lost 11% of its forest cover in 2005/6, and 65% over the past 25 years (Uryu *et al.* 2008), and if deforestation continues apace Borneo could lose its lowland forests by 2012–2018 (Rautner *et al.* 2005).

	All (range)
Weight (kg)	11–20
Head/body length (mm)	600–1000
Ref: Macdonald (2001a)	

Lion *Panthera leo* (Linnaeus, 1758)

Lions are uniquely social among wild felids, living in prides of up to 18 adult females and 1–9 adult males. Male lions are also unique for their dark manes, which are protective and signal fitness (West and Packer, in press). Lions occur at 1.2 adults/100 km^2 in southern African semi-desert to 40/100 km^2 in the Ngorongoro Crater, Tanzania (Hanby *et al.* 1995;

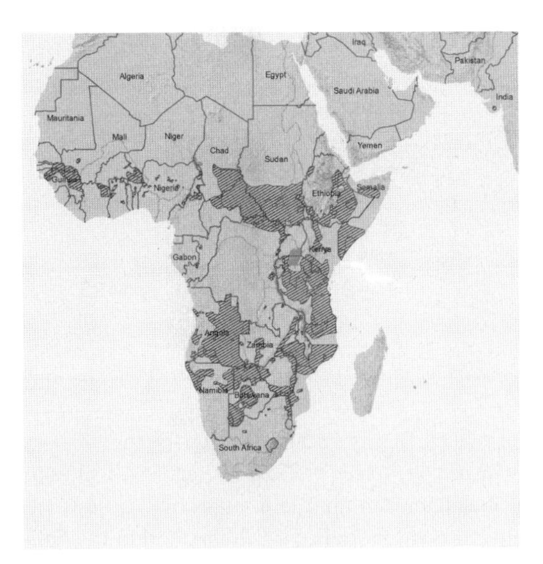

Map 3 Lion. © IUCN Red List 2008.

Castley *et al.* 2002). Pride ranges can vary widely even in the same region: for example, from 266 to 4532 km^2 in the Kgalagadi Transfrontier Park of South Africa (Funston 2001a), and 20–500 km^2 in the Serengeti (West and Packer, in press). Females do most of the hunting, and all pride members generally share the carcass, typically a large ungulate (Scheel and Packer 1995), although lions can and often do live on small prey (Sunquist and Sunquist 2002). Lion sociality is flexible, prides being smaller when persecuted by humans (Funston *et al.* 2007). In the more arid habitats of southern Africa, pride sizes are also smaller (West and Packer, in press), and in India's Gir Forest, males and females infrequently associate (Chellam 1993). The Gir Forest lions (*P. l. persica* ssp.: O'Brien *et al.* 1987b) are the only remaining population (estimated at ~175: IUCN 2008) in the wide historic Asian range of the lion. The Gir National Park and Wildlife Sanctuary is surrounded by cultivated areas and inhabited by pastoralists (Maldharis) and their livestock. Domestic cattle have historically been a major part of the Asiatic lion's diet, although the most common prey is the chital deer (Nowell and Jackson 1996). Outside this isolated Indian population, the lion is now found only south of the Sahara, primarily in savannah habitats, and primarily in eastern and southern Africa (77% of current lion range; Map 3). Lion status in 38% of its historic range is unknown, and known and probable range comprises just 22% of historic range (IUCN 2006a, b). The lion is considered regionally Endangered in west Africa (Bauer and Nowell 2004). The African lion population has been estimated at 23,000 (Bauer and van der

Plate B A male and a female lion (*Panthera leo*) in Hwange National Park, Zimbabwe, showing the marked sexual dimorphism characteristic of the species. © A. J. Loveridge.

	Male			Female		
	Mean	Range	Sample size	Mean	Range	Sample size
Weight (kg)						
East Africa[ab]	173.9	145.4–204.7	$n = 27$	117.7	90.0–167.8	$n = 15$
Southern Africa[b]	189.6	150.0–225.0	$n = 78$	103.8	83.0–165.0	$n = 118$
Head/body length (mm)						
East Africa[a]	1938	1840–2080	$n = 12$	1711	1600–1840	$n = 38$
Southern Africa[c]	1949	1835–2090	$n = 18$	1710	1425–1850	$n = 23$

Refs: [a] West and Packer (in press); [b] Smuts *et al.* (1980);
[c] A. Loveridge (unpublished data)

Merwe 2004) and 39,000 (Chardonnet 2002). An estimated 42% of major lion populations are declining (Bauer 2008). Genetic population models indicate that large populations (50–100 lion prides) are necessary to conserve genetic diversity and avoid inbreeding, which increases significantly when populations fall below 10 prides. Male dispersal is also important to conserving genetic variation (Björklund 2003). These conditions are rarely met, although there are at least 17 lion 'strongholds' >50,000 km^2 (Bauer 2008). The major threat facing lions is conflict with local people over life and livestock (IUCN 2006b). The lion is classified as Vulnerable, the Asiatic lion as Endangered (IUCN 2008).

Jaguar *Panthera onca* (Linnaeus, 1758)

The largest cat of the Americas, the jaguar is similar in appearance to the leopard *P. pardus*, but with a stockier build, more robust canines, and larger rosettes. Historically, the jaguar ranged from the south-western United States (where there are still some vagrants close to the Mexican border: Brown and Lopez-Gonzalez 2001) south through the Amazon basin to the Rio Negro in Argentina. It has been virtually eliminated from much of the drier northern parts of its range, as well as northern Brazil, the pampas scrub grasslands of Argentina, and throughout Uruguay, losing 54% of its historic range. However, it has high probability of survival in 70% of its current range (Map 4), comprised mainly of the Amazon basin rainforest, and adjoining areas of the Pantanal and Gran Chaco (Sanderson *et al.* 2002b). However, while ecological models suggest that the latter two are highly suitable for jaguars, the Amazon may be less so (Torres *et al.* 2007). Jaguar densities in the Brazilian Pantanal are estimated at 6.6–6.7 adults/100 km^2 (Soisalo *et al.* 2006), and in the Bolivian Gran Chaco 2.2–5/100 km^2 (Maffei *et al.* 2004a). In the Amazon basin in Colombia, jaguar density was estimated at 4.5/100 km^2 in Amacayacu National Park and 2.5/100 km^2 in unprotected areas (Payan 2008). In Madidi National Park in the Bolivian Amazon, density was estimated at 2.8/100 km^2 (Silver *et al.* 2004). Areas of tropical moist lowland forest in Central America also have high probability for long-term jaguar persistence (Sanderson *et al.* 2002b); density in Belize was estimated at 7.5–8.8/100 km^2 (Silver *et al.* 2004). Small home ranges were also reported there, with females averaging

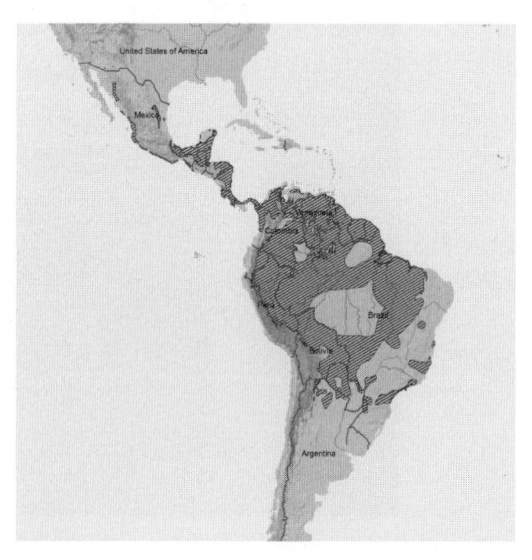

Map 4 Jaguar. © IUCN Red List 2008.

	Male			Female		
	Mean	Range	Sample size	Mean	Range	Sample size
Weight (kg)	104.5	68–121	$n = 26$	66.9	51–100	$n = 31$
Head/body length (mm)	1565	1260–1700	$n = 16$	1304	1160–1470	$n = 12$

Source: Hoogesteijn and Mondolfi (1996).

10 km^2 and males averaging 33 km^2, and extensive intra- and intersexual overlaps (Rabinowitz and Nottingham 1986). Larger home ranges for both males and females, up to ~150 km^2, have been reported elsewhere (Sunquist and Sunquist 2002). More than 85 prey species have been recorded, although large ungulates are preferred (Seymour 1989). People compete with jaguars for prey, and jaguars are frequently shot on sight, despite protective legislation (Nowell and Jackson 1996). An estimated 27% of jaguar range has a depleted wild prey base (WCS 2008). An ambitious programme is seeking to conserve a continuous north-to-south corridor through the species range (Rabinowitz 2007). The jaguar is classified as Near Threatened (IUCN 2008), but some of the most important Jaguar Conservation Units occur where their probability for long-term survival is low (Sanderson *et al.* 2002b). These include the Atlantic Forest subpopulation in Brazil, estimated at 200 ± 80 adults (Leite *et al.* 2002). Jaguar populations in the Chaco region of northern Argentina and Brazil, and the Brazilian Caatinga, are low density and highly threatened by livestock ranching and persecution (Altrichter *et al.* 2006; T. de Oliveira, personal communication 2008).

Leopard *Panthera pardus* (Linnaeus, 1758)

With a distribution that includes most of Africa and large parts of Asia (Map 5), ranging from desert to rainforest, the leopard is adaptable and widespread, but many subpopulations are threatened. In Africa, they are most successful in woodland, grassland savannah, and forest, but also occur widely in mountain habitats (up to 4600 m), coastal scrub, swampy areas, shrubland, semi-desert, and desert—although they have become very rare in the areas bordering the Sahara (Ray *et al.* 2005; Hunter *et al.*, in press). In south-west and central Asia, leopards are now confined chiefly to the more remote montane and rugged foothill areas (Breitenmoser *et al.* 2006a, 2007). Through India and south-east Asia, they occur in all forest types as well as dry scrub and grassland, and range up to 5200 m in the Himalaya (Nowell and Jackson 1996). Over 90 species have been recorded in leopard diet in sub-Saharan Africa, ranging from arthropods to adult male eland *Tragelaphus oryx* (Hunter *et al.*, in press). Preferred prey in Thailand were hog badger *Arctonyx collaris* (45.9% of prey items), muntjac *Muntiacus muntjak* (20.9%), and wild pig *Sus scrofa* (6.3%; Grassman 1998a). Densities vary with habitat, prey availability, and degree of threat, from fewer than 1/100 km^2 to over 30/100 km^2, with highest densities obtained in protected east and southern African mesic woodland savannahs (Hunter *et al.*, in press). Home ranges

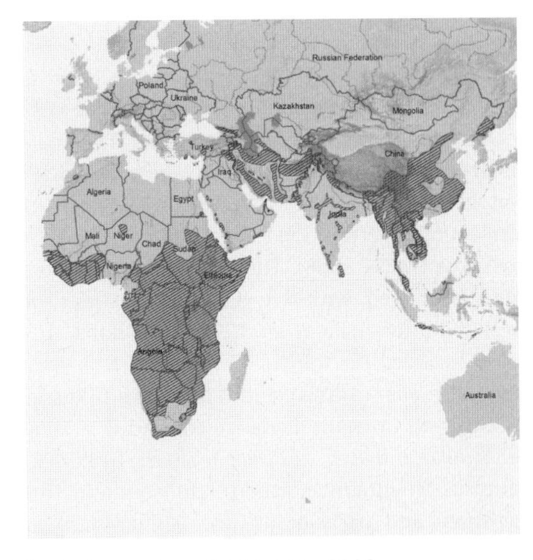

Map 5 Leopard. © IUCN Red List 2008.

Plate C A male African leopard (*Panthera pardus*), Hwange National Park, Zimbabwe. © A. J. Loveridge.

	Male			Female		
	Mean	Range	Sample size	Mean	Range	Sample size
Weight (kg)	53	34–69	*n* = 59	30.5	20.5–42	*n* = 58
Head/body length (mm)	1340	1160–1830	*n* = 59	1143	1050–1270	*n* = 58

Source: Hunter *et al*. (in press).

from two studies in Thailand were 8.8 km^2 (Grassman 1998a) and 11–17 km^2 (Rabinowitz 1989) for females, and 17.3–18 km^2 (Grassman 1998a) and 27–37 km^2 for males (Rabinowitz 1989). The major threats to leopards are habitat loss, prey base depletion, illegal trade in skins and other body parts, and persecution in retribution for real and perceived livestock loss (IUCN 2008). The leopard is classified as Near Threatened, with five genetically distinct subspecies (Miththapala *et al.* 1996; Uphyrkina *et al.* 2001) included on the IUCN Red List. *P. p. melas* (Java), *P. p. nimr* (Arabia), and *P. p. orientalis* (Russian Far East) are Critically Endangered, and *P. p. kotiya* (Sri Lanka) and *P. p. saxicolor* (eastern Turkey, the Caucasus Mountains, northern Iran, southern Turkmenistan, and parts of western Afghanistan) are Endangered (IUCN 2008).

Tiger *Panthera tigris* (Linnaeus, 1758)

The tiger is the largest felid (much larger in India and Russia than in south-east Asia; Kitchener 1999) and the only striped cat. Genetic analysis supports five classically described subspecies—*P. t. altaica* in the Russian Far East; *P. t. amoyensis* in South China (now possibly extinct in the wild; IUCN 2008); *P. t. corbetti* in south-east Asia; *P. t. sumatrae* on the Indonesian island of Sumatra; and *P. t. tigris* on the Indian subcontinent—as well as a new subspecies, *P. t. jacksonii*, restricted to Peninsular Malaysia (Luo *et al.* 2004). Tigers once ranged from Turkey in the west to the eastern coast of Russia (Nowell and Jackson 1996). Over the past 100 years tigers have disappeared from south-west and central Asia, from two Indonesian islands (Java and Bali) and from large areas of south-east and eastern Asia, and lost 93%

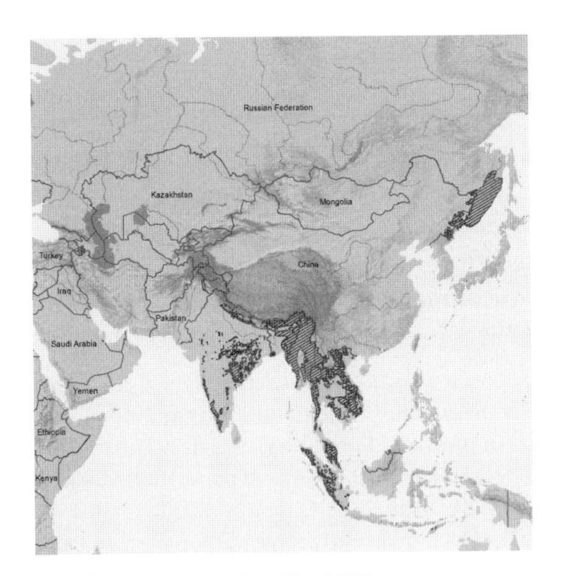

Map 6 Tiger. © IUCN Red List 2008.

of their historic range (Sanderson *et al.* 2006). Comparison of current range (~1.1 million km^2; Sanderson *et al.* 2006; Map 6) with an estimate a decade ago (Wikramanayake *et al.* 1998) suggests 41% shrinkage. While partly due to methodological distortions (Karanth *et al.* 2003; Sanderson *et al.* 2006), Dinerstein *et al.* (2007) consider tiger poaching and habitat loss to have caused major recent decline. The global population is only *c.* 3500–5000 tigers (IUCN 2008), compared to previous estimates of 5000–7000 (Seidensticker *et al.* 1999), although such direct comparison is unreliable due to improved precision of recent estimates (e.g. Miquelle *et al.* 2007; Jhala *et al.* 2008). Sanderson *et al.* (2006) still consider 77% of current range to consist of 'known and secured breeding populations of tigers in areas large enough for a substantive population'. Most tiger range (60%) is found in tropical and subtropical moist broadleaf forests, followed by temperate and broadleaf mixed forest (21%) and tropical and subtropical dry broadleaf forest (10%). Tigers also occur in coniferous forest, mangrove forest, and tropical grass and shrubland (Sanderson *et al.* 2006). Photos of tigers at elevations up to 4500 m have been obtained in Bhutan (Wang 2008). Tigers depend on a large ungulate prey base (Sunquist *et al.* 1999), and are capable of killing prey as large as adult Asian rhinos and elephants (Sunquist and Sunquist 2002). Tigers eat 18–27 kg of food in a single feeding session (Schaller 1967), and up to 35 kg (McDougal 1977). In Nepal's Chitwan National Park, a female spent an average of 3 days with a large kill (Sunquist 1981). Karanth *et al.* (2004a) estimate that tigers need to kill 50 large prey animals per year. Tigers are generally solitary, with adults maintaining exclusive

Plate D Tiger *Panthera tigris*. © Brian Courtenay.

	Male			Female		
	Mean	Range	Sample size	Mean	Range	Sample size
Weight (kg)	225	195–325	$n = 9$	121	96–160	$n = 5$
Head/body length (mm)	2300	1900–2900	$n = 5$	1663	1460–1770	$n = 5$

Source: Heptner and Sludskii (1992)

territories, or home ranges. Adult female home ranges seldom overlap, whereas male ranges typically overlap from one to three females (Sunquist and Sunquist 2002). Tiger home ranges are small where prey is abundant—for example, female home ranges in Chitwan averaged 20 km^2 (Smith *et al.* 1987) and 15–20 km^2 in India's Nagarhole National Park (Karanth 1993), while in the Russian Far East they are much larger, at 400 km^2 (Goodrich *et al.* 2007). Male tiger home ranges are 2–15 times larger than females (Sunquist and Sunquist 2002), and average 1379 $\pm$ 531 km^2 in the Russian Far East (Goodrich *et al.* 2007). Similarly, reported tiger densities range from 11.65 adult tigers/100 km^2 where prey is abundant (India's Nagarhole National Park) to as low as 0.13–0.45/100 km^2 where prey is more thinly distributed, as in Russia's Sikhote Alin Mountains (Nowell and Jackson 1996). With their substantial food requirements, tigers require a healthy large ungulate prey base, but these species are also under heavy human subsistence hunting pressure and competition from domestic livestock. Karanth and Stith (1999) consider prey base depletion to be the leading threat to tigers in areas of otherwise suitable habitat. Tiger attacks on livestock and people lead to local conflict. For example, 41 people were killed by tigers in the Sundarbans mangrove forest of Bangladesh during an 18-month period in 2001–03 (Khan 2004). Tigers are also commercially poached, for their skins as well as bones, used in traditional Asian medicine, particularly in China. While China's 1993 trade ban has greatly reduced illegal trade there, there are proposals to farm tigers to make tiger bone wine. Five thousand tigers are reputedly already in captivity, and tiger farming threatens to re-ignite consumer demand (Nowell and Xu 2007; Nowell, 2009). In 2007, the CITES enacted a decision stating that 'tigers should not be bred for trade in their parts and derivatives' (Nowell *et al.* 2007). The tiger is classified as Endangered (IUCN 2008).

Snow leopard *Panthera uncia* (Schreber, 1775)

The snow leopard's closest relative is the tiger (Johnson *et al.* 2006b; O'Brien and Johnson 2007; Eizirk *et al.*, in press), and it is also classified as Endangered (IUCN 2008), with an estimated population of 4080–6590 (McCarthy and Chapron 2003). Status in China, which makes up a large portion of snow leopard range (see Map 7), is little known. Snow leopards are restricted to the high arid mountain grasslands of central Asia, at 3000 to over 5000 m in the Himalaya and Tibetan plateau, but as low as 600 m in Russia or Mongolia. In the Sayan Mountains of Russia and parts of the Tien Shan range of China, they are found in open coniferous forest, but usually avoid dense forest (McCarthy and Chapron 2003). With their pale colouration, thick fur, and long luxuriant tail, they are adapted for cold, snowy, and rocky environments. They may occur in sympatry with tigers or leopards at high elevations in the Himalaya

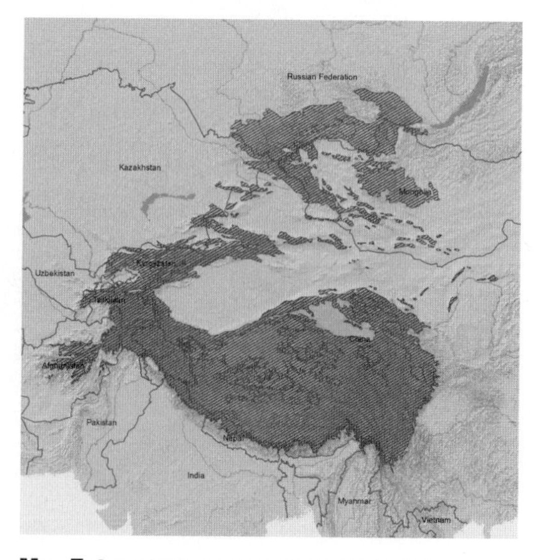

Map 7 Snow leopard. © IUCN Red List 2008.

Plate E Snow leopard *Panthera uncia* cubs. © Snow Leopard Conservancy.

	Male			Female		
	Mean	Range	Sample size	Mean	Range	Sample size
Weight (kg)	40.7	40–41.3	$n = 2$	34.3	30–38.6	$n = 2$
Head/body length (mm)	1210	1200–1220	$n = 2$	1173	1170–1175	$n = 2$

Source: McCarthy (2000)

(Wang 2008). Deep snow is not a barrier, although they prefer to use existing trails to avoid breaking through new snow (Sunquist and Sunquist 2002). The snow leopard's principal natural prey species are bharal or blue sheep (*Pseudois nayaur*) and ibex (*Capra sibirica*), which have a largely coincident distribution. They also take marmot (*Marmota* spp.), pika (*Ochotona* spp.), hares (*Lepus* spp.), small rodents, and game birds. Considerable predation is reported on domestic livestock. Annual prey requirements are estimated at 20–30 adult blue sheep, with radio-tracking data indicating such a kill every 10–15 days. Snow leopards may remain with their kills for up to a week. Snow leopard home ranges overlap widely between the sexes, and varied from 10 to 40 km^2 in relatively productive habitat in Nepal (Jackson and Ahlborn 1989). By comparison, home ranges are considerably larger (140 km^2 or greater) in Mongolia, where terrain is relatively open and ungulate prey densities lower (McCarthy *et al.* 2005). Densities range from 0.1 to 10 or more individuals/100 km^2 (Jackson *et al.*, Chapter 19, this volume). Major threats to the snow leopard include prey base depletion, illegal trade in skins and bones (used as a substitute for tiger bone medicine), conflict with local people, and lack of conservation capacity, policy, and awareness (Koshkarev and Vyrypaev 2000; McCarthy and Chapron 2003; Theile 2003).

Borneo bay cat *Pardofelis badia* (Gray, 1874)

The Borneo bay cat (Map 8) is not, as previously thought, a small island form of the Asiatic golden

	Male			Female		
	Mean	Range	Sample size	Mean	Range	Sample size
Weight (kg)				2.45	2.39–2.5	$n = 2$
Head/body length (mm)	650	620–670	$n = 2$[a]	525	452–591	$n = 3$

Ref: Kitchener *et al.* (2004)
[a] Sunquist and Sunquist (2002)

cat *P. temminckii*, the two having diverged *c.* 4 million years ago, well before the isolation of Borneo (Johnson *et al.* 1999; O'Brien and Johnson 2007). There are relatively few records for this species, but historically it probably occurred throughout Borneo (Azlan and Sanderson 2007; Meijaard 1997; IUCN 2008). The Borneo bay cat has been reported from hill, lowland, and swamp forest (Meijaard 1997; Azlan *et al.* 2003; Hearn and Bricknell 2003; Azlan and Sanderson 2007; Yasuda *et al.* 2007) and regenerating logged forest (Nowell and Jackson 1996; Hearn and Bricknell 2003; Kitchener *et al.* 2004; Meijaard *et al.* 2005a). Its diet is unknown, and it occurs in both a reddish and grey colour phase (Nowell and Jackson 1996; Sunquist and Sunquist 2002). Observations and camera trap photos have occurred at midday (Azlan *et al.* 2003; Yasuda *et al.* 2007), early morning (Hearn and Bricknell 2003), and at night (Dinets 2003; Meijaard *et al.* 2005a). Outside protected areas, habitat loss due to commercial logging and oil palm plantations is the major threat. Wildlife traders are aware of the species' rarity, and bay cats have been captured illegally from the wild for the skin and pet markets (Sunquist and Sunquist 2002; Kitchener *et al.* 2004; Azlan and Sanderson 2007). The Borneo bay cat is one of the few small cats classified as Endangered on the IUCN Red List of Threatened Species (IUCN 2008).

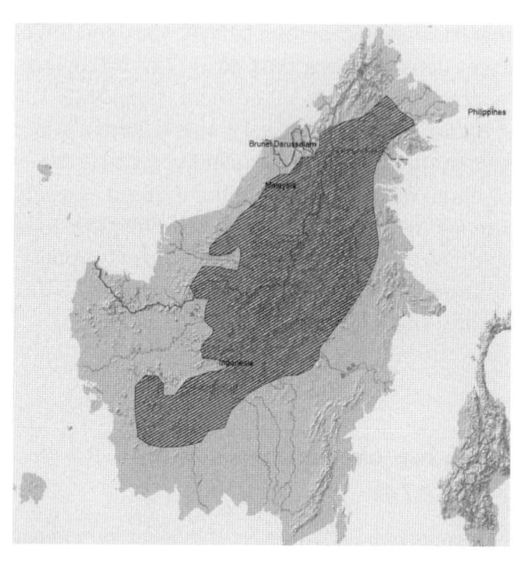

Map 8 Borneo bay cat. © IUCN Red List 2008.

Marbled cat *Pardofelis marmorata* (Martin, 1837)

The marbled cat resembles a miniature clouded leopard, with a tail equivalent to head–body length and markings so similar that Corbett and Hill (1992) grouped the two species in one genus, on the grounds that 'the unique and complex pattern of the pelage is unlikely to be independently derived or primitive'. Another similar trait is its relatively enlarged upper canines (Groves 1982). However, genetic analysis does not support a close relationship (Johnson *et al.* 2006b; Eizirik *et al.*, in prep). The marbled cat is found in tropical Indomalaya (Map 9), along the Himalayan foothills into southwest China, and on the islands of Sumatra and Borneo, and appears to be relatively rare (Nowell and Jackson 1996; Duckworth *et al.* 1999; Holden 2001; Sunquist and Sunquist 2002; Grassman *et al.* 2005a; Azlan *et al.* 2006; Lynam *et al.* 2006; Mishra *et al.* 2006; Yasuda *et al.* 2007), although a higher encounter rate was recorded in Cambodia (13 camera trap records, compared to 12 for the Asiatic golden cat and 4 for the clouded leopard; Duckworth *et al.* 2005). The marbled cat is primarily associated with moist and mixed deciduous–evergreen tropical forest (Nowell and Jackson 1996), and may prefer hill forest (Duckworth *et al.* 1999; Holden 2001; Grassman *et al.*

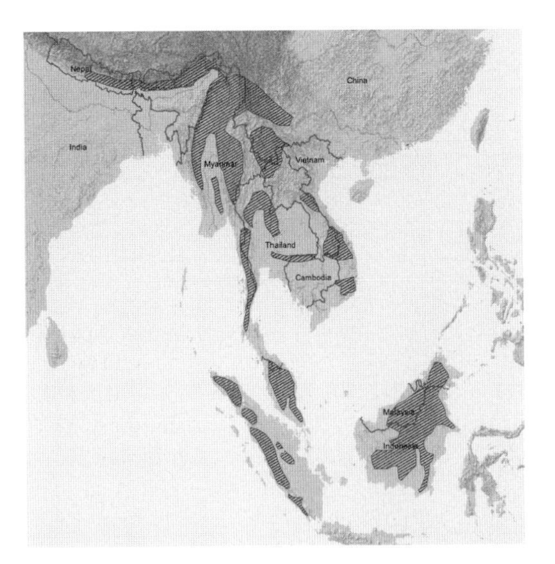

Map 9 Marbled cat. © IUCN Red List 2008.

2005a)—habitats undergoing the world's fastest deforestation rate (1.2–1.3% a year since 1990; FAO 2007), due to logging, oil palm, and other plantations, and human settlement and agriculture. Grassman and Tewes (2002) reported the observation of a pair of adult marbled cats in a salt lick in Thailand's Phu Khieu National Park, where Grassman *et al.* (2005a) estimated a home range of 5.3 km^2 for an adult female radio-tracked for 1 month. The marbled cat probably preys primarily on rodents, including squirrels (Nowell and Jackson 1996), and birds. Most camera trap records have been diurnal (Duckworth *et al.* 1999; Grassman and Tewes 2002). Although infrequently observed in the illegal Asian wildlife trade (Nowell and Jackson 1996), it is valued for its skin, meat, and bones; indiscriminate snaring is prevalent throughout much of its range (IUCN 2008). They have been reported as poultry pests (Nowell and Jackson 1996; Mishra *et al.* 2006). The marbled cat is classified as Vulnerable (IUCN 2008).

	Male		Female		
		Sample size	Mean	Range	Sample size
Weight (kg)			3.1	2.5–3.7	*n* = 2
Head/body length (mm)	525	*n* = 1	555	490–620	

Ref: Sunquist and Sunquist (2002)

Asiatic golden cat *Pardofelis temminckii* (Vigors and Horsfield, 1827)

The genus *Pardofelis* contains three cats that resemble other species. This species' doppelgänger is the African golden cat, which it resembles in size, appearance, and name. However, genetic analysis has determined that they are not closely related (Johnson *et al.* 2006b; O'Brien and Johnson 2007; Eizirik *et al.*, in prep). The Asiatic golden cat has a similar distribution to the congeneric marbled cat (Map 10). However, it has a larger range in China, like the clouded leopard, and it does not occur on the island of Borneo (Nowell and Jackson 1996; Sunquist and Sunquist 2002). The Asiatic golden cat is primarily found in forest, habitats, ranging from tropical and subtropical evergreen to mixed and dry deciduous forest (Nowell and Jackson 1996), and occasionally in shrub and grasslands (Choudhury 2007). Grassman *et al.* (2005a) found radio-collared golden cats used closed forest and more open habitats in proportion to their occurrence. In Sumatra's Kerinci Seblat National Park, all records for this species were from lowland forest, with none from montane forest, unlike the clouded leopard and marbled cat (Holden 2001). Mishra *et al.* (2006) also found clouded leopard and marbled cat, but no Asiatic golden cat, in the hill forests of India's western Arunachal Pradesh province. However, Wang (2007) obtained camera trap photos of the Asiatic golden cat at an elevation of 3738 m in Bhutan's Jigme Sigye Wangchuk

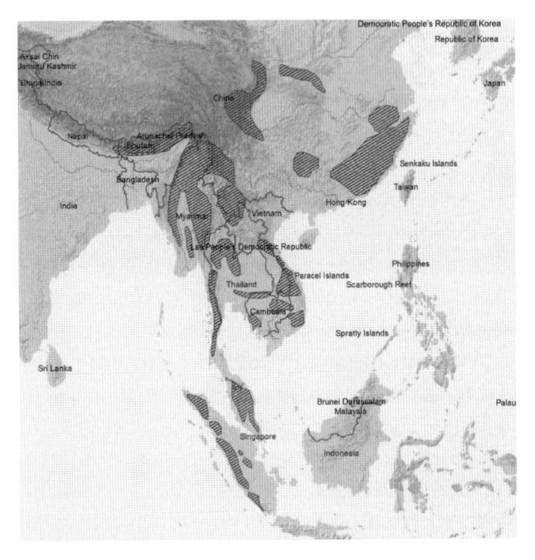

Map 10 Asiatic golden cat. © ICUN Red List 2008.

Plate F Asiatic golden cat *Pardofelis temminckii*. © Alex Silwa.

National Park in an area of dwarf rhododendron and grassland, an elevation record for the species. Activity readings from two radio-collared golden cats in Thailands's Phu Khieu National Park showed daytime and crepuscular activity peaks (Grassman *et al.* 2005a). Seven of 15 camera trap records in Sumatra's Kerinci Seblat National Park were diurnal (Holden 2001). An adult female Asiatic golden cat in Thailand's Phu Khieu National Park had a home range of 32.6 km^2, overlapped 78% by a male range of 47.7 km^2. Golden cat home ranges were 20% larger than those of clouded leopard, although they were similar in activity and mean daily distance moved (Grassman *et al.* 2005a). One scat contained the remains of Indochinese ground squirrel (Grassman *et al.* 2005a), others from Sumatra contained rat and muntjac remains, and one stomach from Thailand's Kaeng Krachan National Park contained a small snake (Grassman 1998b). It is capable of taking small ungulates (Sunquist and Sunquist 2002). While the reddish-gold pelage the cat is named for is the most common form, there are spotted (Wang 2007) and melanistic morphs (Holden 2001; Grassman *et al.* 2005a). The Asiatic golden cat is Near Threatened due to habitat loss, illegal hunting, and depletion of the wild ungulate prey base (IUCN 2008).

	Male	Sample size	Female	Sample size
Weight (kg)	13.5	$n = 1$	7.9	$n = 1$
Head/body length (mm)	910	$n = 1$	770	$n = 1$

Ref: Grassman *et al.* (2005a)

African golden cat *Caracal aurata* (Temminck, 1827)

While the Neotropical and Indomalayan regions have several sympatric forest-dependent felids, the golden cat is Africa's only one (Map 11). The African golden cat occurs mainly in primary moist equatorial forest, although it penetrates savannah areas along riverine forest. It also occurs in montane forest and alpine moorland in the east of its range (Nowell and Jackson 1996; Ray and Butynski, in press). Two studies of scats—from the Ituri forest of the Congo (Hart *et al.* 1996) and the Dzanga-Sangha forest of the Central African Republic (Ray and Sunquist 2001)—found that rodents made up the majority of prey items (frequency of occurrence: 70% and 62%, respectively), followed by small- and medium-sized duiker antelopes (25% and 33%) and primates (5%). Other reported prey items include birds, pangolin (*Manis gigantea*), and in southern Sudan a female with two kittens was observed hunting bats as they swooped over fallen mangoes (Ray and Butynski, in press). African golden cat remains were found in five of 196 leopard *Panthera pardus* scats from Gabon's Lopé National Park (Henschel *et al.* 2005); a single carcass killed by a leopard was found in the Ituri (Hart *et al.* 1996). The African golden cat is Near Threatened by deforestation, prey depletion caused by bushmeat offtake, and by-catch (IUCN 2008). Over 3 months at four sites in Lobeké, south-east Cameroon, 13 African golden cats were snared (T. Davenport, personal communication in Ray *et al.* 2005). Skins are seen in markets, for example in Yaoundé and Kampala, alongside medicinal herbs and fetishes (T. Davenport, personal communication in Ray and Butynski, in press).

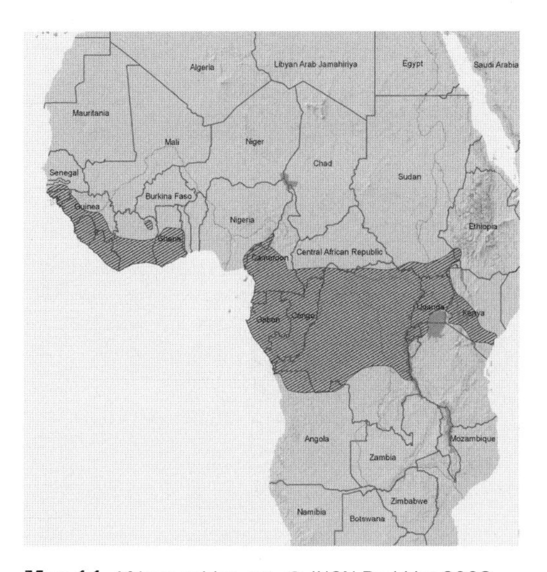

Map 11 African golden cat. © IUCN Red List 2008.

Caracal *Caracal caracal* (Schreber, 1776)

The caracal is a long-legged, medium-sized felid, tawny red with large black-backed ears tipped with prominent tufts of hair. It is widely distributed across the drier regions of Africa, central Asia, and south-west Asia into India (Map 12). Its historical range mirrors that of the cheetah, and both coincide with the distribution of several small desert gazelles (Sunquist and Sunquist 2002). Like cheetahs, caracals were captured and trained to hunt by Indian royalty for small game and birds (Divyabhanusinh 1995). Stuart (1982) recorded that between 1931 and 1952 an average of 2219 caracals per year were killed in control operations in the Karoo, South Africa. Similarly, Namibian farmers responding to a government questionnaire reported killing up to 2800 caracals in 1981 (Nowell and Jackson 1996; Stuart and Stuart, in press). Predation on small stock and introduced springbok was seasonal when alternative wild prey was scarce (Avenant and Nel 2002). No livestock were found in 200 caracal scats in the vicinity of South Africa's Mountain Zebra National Park where wild prey was abundant (Grobler 1981). The home

	Male			Female		
	Mean	Range	Sample size	Mean	Range	Sample size
Weight (kg)	11.0	8.0–14.0	$n = 6$	7.2	6.2–8.2	$n = 2$
Head/body length (mm)	766	616–935	$n = 18$	699	630–750	$n = 8$

Source: Ray and Butynski (in press)

Plate G Caracal *Caracal caracal*. © M. G. L. Mills.

	Male			Female		
	Mean	Range	Sample size	Mean	Range	Sample size
Weight (kg)	12.9	7.2–19	*n* = 77	10	7–15.9	*n* = 63
Head/body length (mm)	868	750–1080	*n* = 98	819	710–1029	*n* = 94

Ref: Stuart and Stuart (in press)

ranges of three males averaged 316.4 km^2 on Namibian ranch land (Marker and Dickman 2005). In Saudi Arabia, a radio-tracked male ranged over 270 km^2 to 1116 km^2 in different seasons (van Heezik and Seddon 1998), while in an Israeli study, home ranges of five males averaged 220.6 km^2 (Weisbein and Mendelssohn 1989). In the better-watered West Coast National Park of South Africa, two males averaged 26.9 km^2, completely overlapping smaller (7.39 km^2) female ranges (Avenant and Nel 1998). Based on spoor tracking in the Kalahari desert, Melville and Bothma (2007) found a hunting success rate of 10%. The caracal is listed as Least Concern, being widespread and relatively common, particularly in southern and eastern Africa, although there have been range losses in north and west Africa (Ray *et al.* 2005), and is of conservation concern in most of its Asian range (IUCN 2008).

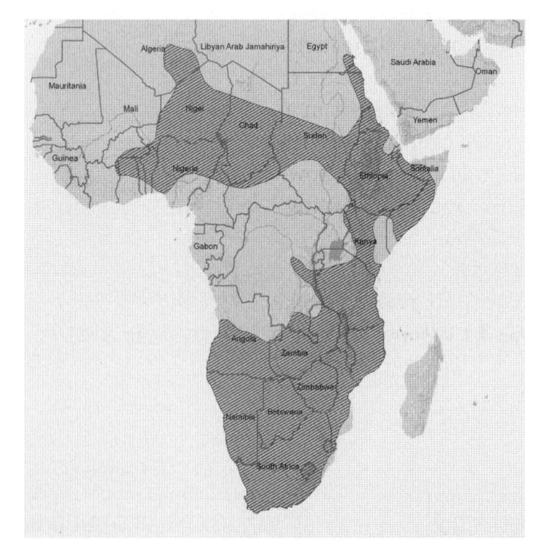

Map 12 Caracal. © IUCN Red List 2008.

Serval *Leptailurus serval* (Schreber, 1776)

Molecular phylogeny reveals that the serval is closely allied with the African golden cat and caracal (Johnson *et al.* 2006b, Eizerik *et al.* in press), diverging from a common ancestor ~5.4 million years ago (O'Brien and Johnson 2007). It is found in the well-watered savannah long-grass habitats of sub-Saharan Africa (Map 13), particularly associated with wetlands and other riparian vegetation, but also in dry forest and alpine grasslands up to 3800 m on Mount Kilimanjaro. In north Africa, they are very scarce in semi-desert and cork oak forest on the Mediterranean coasts of Morocco and possibly Algeria (Cuzin 2003; De Smet 1989, personal communication 2007). Extinct in Tunisia, a population has been re-introduced into Feijda National Park (K. De Smet in Hunter and Bowland, in press). Servals are able to tolerate agriculture provided cover is available (Hunter and Bowland, in press), and may also benefit from forest clearance and the resulting encroachment of savannah at the edges of the equatorial forest belt (Ray *et al.* 2005). With its long legs and large ears, the serval specializes on small mammals, particularly rodents, with 90% of the diet weighing <200 g (Sunquist and Sunquist 2002). Birds are the second most frequent prey item (Hunter and Bowland, in press). Frogs occurred in 77% of scats collected during a study in Tanzania's Ngorongoro crater, and one male ate at least 28 frogs in 3 h (Geertsema 1985). Servals have a characteristic high pounce when hunting, used even for the rare instances of larger prey, which can span 1–4 m and may be up to 2 m high (Nowell and Jackson 1996; Sunquist and Sunquist 2002; Hunter and Bowland, in press). Geertsema (1985) found a hunting success of 49% of almost 2000 pounces, with no significant difference between daylight and moonlit night. Servals killed 15–16 times/24 h, making 0.8 kills per hour during the day and 0.5 kills per hour at night. Juveniles killed more frequently, but hunted smaller prey with lower energy yields. Wetlands harbour comparatively high rodent densities compared with other habitat types, and form the core areas of serval home ranges (Geertsema 1985; Bowland 1990). The home ranges of two females and one male on farmland and an adjacent nature reserve in the KwaZulu-Natal Drakensberg were 19.8 km^2, 15.8 km^2, and 31.5 km^2, respectively (Bowland 1990). In Ngorongoro Crater, Tanzania, the home ranges of an adult female and adult male were 9.5 km^2 and 11.5 km^2, respectively (Geertsema 1985). The minimum density of servals in optimal habitat in Ngorongoro Crater was 42 animals/100 km^2 (Geertsema 1985). The serval is classified as Least Concern, being relatively abundant and widespread (and even expanding). However, degradation of wetlands is of concern, as is the level of skin trade in west Africa (Ray *et al.* 2005; Hunter and Bowland, in press). Servals are rare south of the Sahara in the Sahel region (Clement *et al.* 2007). Servals north of the Sahara are Critically Endangered, with only fewer than 250 mature individuals (IUCN 2008).

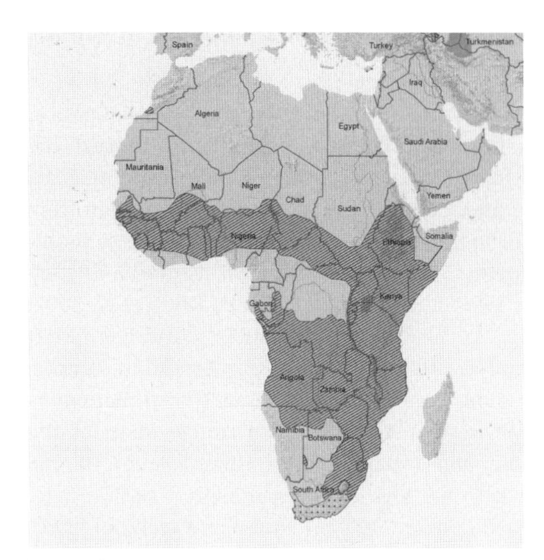

Map 13 Serval. © IUCN Red List 2008.

	Male			Female		
	Mean	Range	Sample size	Mean	Range	Sample size
Weight (kg)	11.2	9.8–12.4	$n = 6$	8.5	7–9.8	$n = 7$
Head/body length (mm)	853	750–920	$n = 5$	778	725–820	$n = 5$

Source: Bowland (1990)

Pampas cat *Leopardus colocolo* (Molina, 1782)

Named for the Argentine grasslands, the pampas cat is a cat of open habitats, ranging north through the dry forests and scrub grasslands of Bolivia, Paraguay, and Brazil, and up the Andes mountain chain from Chile to Ecuador and possibly marginally into south-western Colombia (Silveira 1995; Nowell and Jackson 1996; Ruiz-García *et al.* 2003; Dotta *et al.* 2007; Map 14). Yet Pereira *et al.* (2002) found few recent records for this species from the Argentine pampas, but more from a semi-arid climatic strip that enters north-western Argentina as a continuation of the Andes and expands south towards the Atlantic coast. In the high Andes, although it has been recorded at over 5000 m (Nowell and Jackson 1996), most records are from lower elevations than the Andean cat *L. jacobita*—in northern Argentina, the mean elevation for pampas cat records was 3567 ± 67 m, as compared to 4236 ± 140 m for the Andean cat (Perovic *et al.* 2003). While in the Andes the pampas cat is easily confused with the Andean cat (rusty-coloured oblong spots on the sides against a grey background), elsewhere the species looks different. In Brazil, it has barely discernible spotting, a shaggier, brownish coat, and black feet (García-Perea 1994; Silveira 1995). A melanistic form exists (Silveira *et al.* 2005). Silveira *et al.* (2005) suggest that the species' similarity to the domestic cat has caused under-recording, as camera traps reveal them to be common in Emas National Park (T. de Oliveira, personal communication 2008). On the basis of morphology, García-Perea (1994) proposed that the pampas cat consists of three species, but genetic analysis does not support this (Johnson *et al.* 1999; Eizirik *et al.*, in prep). Pampas cat prey includes small mammals and ground-dwelling birds (Nowell and Jackson 1996; Silveira *et al.* 2005) and, in the high Andes, mountain viscacha (*Lagidium* spp.) and small rodents (Walker *et al.* 2007; Napolitano *et al.* 2008). In Brazil's Emas National Park, home range size (90% MCP) was estimated at 19.47 ± 3.64 km^2 (Silveira *et al.* 2005). Primarily diurnal with some crepuscular and occasionally nocturnal activity, the pampas cat is listed as Near Threatened due to habitat loss and degradation, as well as retaliatory killing for poultry depredation hunting for traditional cultural purposes in the high Andes (IUCN 2008).

	Male		Female	
	Mean±SD	Sample size	Mean±SD	Sample size
Weight (kg)	3.9 ± 0.6	*n* = 10	4.0	*n* = 2
Head/body length (mm)	682 ± 8	*n* = 10	620 ± 3	*n* = 2

Ref: Silveira *et al.* (2005)

Map 14 Pampas cat. © IUCN Red List 2008.

Geoffroy's cat *Leopardus geoffroyi* (d'Orbigny and Gervais, 1844)

Like the pampas cat, the Geoffroy's cat is found in open habitats and the two are sympatric for much of their range, although Geoffroy's cat does not extend as far north (Map 15). The Geoffroy's cat is distributed throughout the pampas grasslands and dry Chaco shrub and woodlands, and around the alpine saline desert of north-western Argentina to 3300 m in the Andes (Nowell and Jackson 1996). Most of its range is arid or semi-arid (Pereira *et al.* 2006), but it also occurs in wetlands (Sunquist and Sunquist 2002). Lucherini *et al.* (2000) found that forest fragments tended to be used for faecal and scent-marking sites, while grasslands and marshes were used for hunting and resting. In three Argentinian studies cats used dense vegetation (Manfredi *et al.* 2007). Manfredi *et al.* (2004)

	Male			Female		
	Mean	Range	Sample size	Mean	Range	Sample size
Weight (kg)	6.2	4.3–7.8	$n = 4$	4.0	3.5–4.3	$n = 4$
Head/body length (mm)	683	630–740	$n = 4$	628	610–650	$n = 4$
Ref: Lucherini *et al.* (2000)						

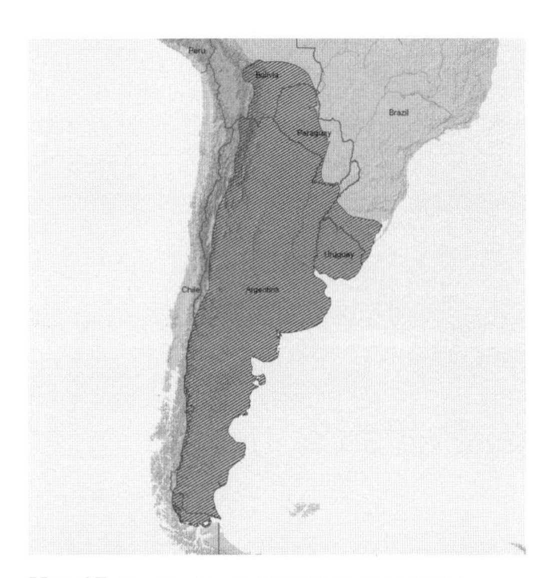

Map 15 Geoffroy's cat. © IUCN Red List 2008.

found local variation in diet in Argentina, consisting primarily of small rodents, supplemented with birds; small mammals made up 63% of food items in Lihue Calel National Park (Bisceglia *et al.* 2007). In Chile, rodents (including viscachas) and introduced hares were primary prey (Johnson and Franklin 1991; Branch 1995; Pereira *et al.* 2006). Fish and frog remains were found in the stomachs of Geoffroy's cats from Uruguay and Brazil (Sunquist and Sunquist 2002). In wet pampas grassland of Argentina, Manfredi *et al.* (2006) found mean home ranges of 2.5–3.4 km^2, with males' 25% larger than females'. In Chile's Torres del Paine National Park, in *Nothofagus* beech forest, home ranges were larger, at 2.3–6.5 km^2 for two females, and 10.9–12.4 km^2 for two males (Johnson and Franklin 1991). In Lihue Calel National Park, female home ranges averaged 2.5 km^2 during a drought period, in which six radio-collared cats died of starvation, and densities ranged from 2–36/100 km^2 but increased to 139.9 ± 35.5, 2 years later (Pereira *et al.* 2006; J. Pereira, personal communication 2008). In the Bolivian Chaco, densities ranged from 2–42/100 km^2 (Cuellar *et al.* 2006). An average of 116,000 pelts per year were exported from Argentina alone in the mid-1970s, and 55,000 per year in the early 1980s. However, little trade has taken place after 1988, and the species was upgraded to CITES Appendix I in 1992, prohibiting commercial trade (Nowell and Jackson 1996). Geoffroy's cats are still killed as livestock predators, and these pelts may be traded illegally. It is classified as Near Threatened due to habitat loss and fragmentation (IUCN 2008).

Guiña *Leopardus guigna* (Molina, 1782)

The guiña (kodkod) is the smallest felid in the Americas. It also has the smallest distribution, occurring only in central and southern Chile and marginally in adjoining areas of Argentina (Map 16). Its range is largely coincident with the Valdivian temperate moist forest, recognized as a Global 200 threatened

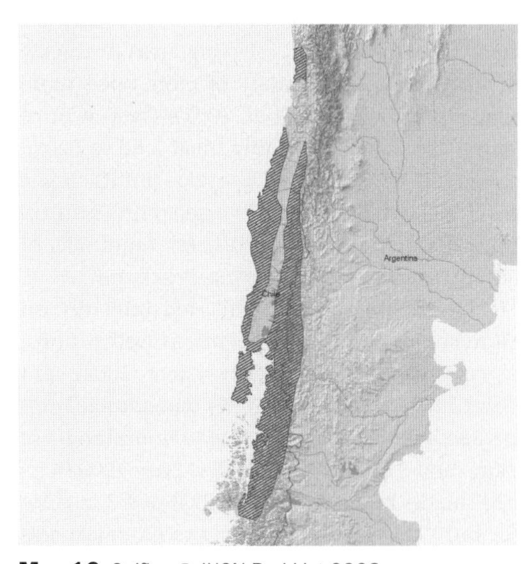

Map 16 Guiña. © IUCN Red List 2008.

Plate H Guiña *Leopardus guigna*. © Gerardo Acosta-Jammet.

eco-region (WWF 2006), and in the south it is also found in *Nothofagus* beech forest (Freer 2004), which has a high degree of endemism (Armesto *et al.* 1998). Native forest, which is preferred, is being lost to less suitable pine plantations (Acosta-Jamett *et al.* 2003; Acosta-Jamett and Simonetti 2007). Although the guiña is forest-dependent, selecting areas of thicket understory, they use a variety of more open scrub habitat types (Dunstone *et al.* 2002). Over most of its range, the guiña is the only small felid to occur, although it is sympatric in Argentina with its closest relative the Geoffroy's cat (O'Brien and Johnson 2007), which it resembles with its multitude of small black spots, in Argentina (Lucherini *et al.* 2001). Guiñas in southern Chile fed primarily on small mammals, especially rodents, as well as birds, and scavenge opportunistically (Freer 2004). On Chile's Chiloe Island, in a largely agricultural landscape, Sanderson *et al.* (2002c) found home ranges of 6.5 km^2 and 1.2 km^2 for females. Freer (2004) reported smaller home ranges (MCP95) of 1.3 km^2 for males and 1 km^2 for females from two national parks (Laguna San Rafael and Queulat) in southern Chile, where densities were 1 adult-subadult/km^2 (Dunstone *et al.* 2002). Much of their southern range is relatively free of human disturbance (Dunstone *et al.* 2002), but in central Chile there has been substantial habitat loss and there Acosta-Jamett *et al.* (2003) estimated *c.* 2000 individuals in 24 subpopulations. In southern Chile, 81.4% of 43 families considered it 'damaging or very damaging', although there was only a single recent report of a guiña killing 12 hens in a henhouse (Silva-Rodriguez *et al.* 2007), and on Chiloe Island, two out of five radio-collared cats were killed while raiding chicken coops (Sanderson *et al.* 2002c). It is classified as Vulnerable (IUCN 2008).

	Male		Female	
	Mean±SD	Sample size	Mean±SD	Sample size
Weight (kg)	1.8 ± 0.16	*n* = 7	1.4 ± 0.07	*n* = 6
Head/body length (mm)	412 ± 15	*n* = 7	394 ± 7	*n* = 6

Source: Dunstone et al. (2002)

Andean cat *Leopardus jacobita* (Cornalia, 1865)

The Andean cat is restricted to the arid, sparsely vegetated areas of the high Andes in Argentina, Bolivia, Chile, and Peru (Map 17), inhabiting rocky and steep terrain at elevations generally above 4000 m across most of its range (Perovic *et al.* 2003; Cossíos *et al.* 2007b; Napolitano *et al.* 2008; Villalba *et al.*, in press; Marino *et al.*, Chapter 28, this volume), but as low as 1800 m in the southern Andes (Sorli *et al.* 2006). Its distribution is similar to the historic range of the mountain chinchilla *Chinchilla brevicaudata* (Yensen and Seymour 2000), which was hunted to the brink of extinction for the fur trade a century ago (IUCN 2008), and the diet reveals a preference for another chinchillid, the mountain viscacha (Walker *et al.* 2007; Napolitano *et al.* 2008), which lives in patchily distributed small colonies and has also declined through hunting pressure. The Andean cat is rare compared to its close relative the pampas cat (Lucherini and Vidal 2003; Perovic *et al.* 2003; Cossíos *et al.* 2007b; Napolitano *et al.* 2008; Villalba *et al.*, in press), from which it is hard to distinguish in the high Andes (García-Perea 2002; Cossíos *et al.* 2007a; Palacios 2007; Villalba *et al.*, in press). Napolitano *et al.* (2008) found reduced genetic diversity in an Andean cat population in northern Chile, suggesting a 'smaller current or historic population size'. Based on genetic sampling in their study area, they estimated a density of one individual per 5 km^2. The Andean cat (as well as the pampas cat) is traditionally considered a sacred animal by indigenous Aymara and Quechua people. Throughout its range, dried and stuffed specimens are kept by local people for use in harvest festivals (Iriarte 1998; Sanderson 1999; Perovic *et al.* 2003; Villalba *et al.* 2004; Cossíos *et al.* 2007b; Villalba *et al.*, in press). Hunting for such cultural practices may represent a significant threat. Napolitano *et al.* (2008) found that the probability of finding sign of the Andean mountain cat decreased with proximity to human settlement. The Andean cat is one of the few small felids classified as Endangered (IUCN 2008). There are none known to be kept in captivity, and very few museum specimens.

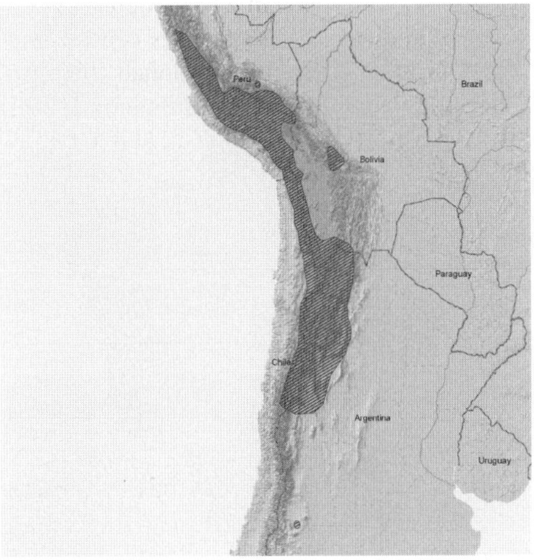

	Unknown sex	Female
Weight (kg)		4.5[a] *n* = 1
Head/body length (mm)	740–850[b] *n* = 11	

Sources: [a] Delgado *et al.* (2004); [b] García-Perea (2002), from museum skins of adults' unknown sex

Map 17 Andean cat. © IUCN Red List 2008.

Ocelot *Leopardus pardalis* (Linnaeus, 1758)

The ocelot is the only medium-sized cat found in the neotropics, and one of the most widespread and successful (Map 18). Historically, it ranged as far north as the American states of Arkansas and Arizona, but is now restricted to a small population of 80–120 in southern Texas (Sunquist and Sunquist 2002). In Mexico, it has disappeared from most of its historic range along the western coast, but still occurs on the eastern coast, and south through Central America into South America, where it is found in every country except Chile (IUCN 2008). The ocelot occupies a wide spectrum of habitats, including mangrove forests and coastal marshes, savannah grasslands and pastures, thorn scrub, and tropical forest of all types (primary, secondary, evergreen, seasonal, and montane, although it typically occurs at

	Male			Female		
	Mean	Range	Sample size	Mean	Range	Sample size
Weight (kg)	13.6	12–15.5	$n = 7$	9.8	9–11.3	$n = 7$
Head/body length (mm)	810	770–855	$n = 7$	770	740–795	$n = 7$

Source: Crawshaw (1995)

Map 18 Ocelot. © IUCN Red List 2008.

elevations below 1200 m; Nowell and Jackson 1996). Their prey is mostly terrestrial, nocturnal, and weighs less than 1 kg (Sunquist and Sunquist 2002), although larger prey is also taken. Diet varies with prey availability—in the seasonally flooded savannahs of Venezuela, ocelots fed intensively on land crabs when they became abundant during the wet season (Ludlow and Sunquist 1987). Home ranges vary from 1.8 to 30 km^2 for females and 5.4–38.8 km^2 for males (with much larger home ranges suggested by an ongoing study in the dry savannah of Brazil's Emas National Park). Variation is probably correlated with prey availability, and densities average 32 ± 22/100 km^2, typically much higher than sympatric small felids (Oliveira *et al.*, Chapter 27, this volume). As many as 200,000 animals were hunted yearly for the fur trade in the late 1960s and early 1970s. Concern over the impact of the Neotropical spotted cat skin trade was a major impetus for the establishment in 1975 of CITES, the Convention on International Trade in Endangered Species of Wild Flora and Fauna, which regulates wildlife trade between countries (Nowell and Jackson 1996). Although some illegal skin trade persists, habitat loss is the major threat to the ocelot, although it is classified as Least Concern due to its wide range and abundance in the Amazon (IUCN 2008).

Oncilla *Leopardus tigrinus* (Schreber, 1775)

Although it resembles a smaller version of the ocelot and margay, species formerly considered its closest relatives, genetic analysis groups the oncilla with the guiña and Geoffroy's cat (O'Brien and Johnson 2007). Its distribution is also broadly similar to the ocelot and margay (Map 19), but apparently patchier in the Amazon basin (Oliveira 2004), and with several

Map 19 Oncilla. © IUCN Red List 2008.

Plate I Oncilla *Leopardus tigrinus*. © Projecto Gatos do Mato Brasil.

large gaps—the Llanos grassland of Colombia and Venezuela, the Paraguayan Chaco, and southern Panama (IUCN 2008). Genetic divergence between populations in Costa Rica and southern Brazil is comparable to that between species in the *Leopardus* group, suggesting that the two populations have been isolated for ~3.7 million years. Both groups had relatively low levels of genetic diversity (Johnson *et al.* 1999). While in Central America and parts of northern South America it may be most common in montane cloud forest, it is mostly found in lowland areas in Brazil, being reported from rainforests to dry deciduous forest, savannahs, semi-arid thorny scrub, and degraded secondary vegetation in close proximity to human settlement (Oliveira *et al.* 2008). Although it has been collected as high as 4800 m (Cuervo *et al.* 1986), this is likely an outlier, as there are very few records at or slightly above 3000 m (T. de Oliveira, personal communication). Oncilla are nocturno-crepuscular, but with considerable daytime activity. Diet consists mostly of small mammals, birds, and lizards, with average prey size at <100 g, but does include larger sized prey (>1 kg). Home ranges are larger than predicted from body size (0.9–2.8 km^2 for females and 4.8–17 km^2 for males; Oliveira *et al.*, Chapter 27, this volume). Densities vary from 1/100 km^2 to 5/100 km^2, and in the Amazon may be as low as 0.01/100 km^2 (Oliveira *et al.* 2008). The oncilla is negatively impacted by ocelots, and may not be viable wherever ocelots are present ('ocelot effect'; Oliveira *et al.*, Chapter 27, this volume). With relatively higher densities outside protected areas, the oncilla is threatened by habitat loss and classified as Vulnerable on the IUCN Red List (IUCN 2008).

	Male			Female		
	Mean	Range	Sample size	Mean	Range	Sample size
Weight (kg)	2.6	1.8–3.5	*n* = 20	2.2	1.8–3.2	*n* = 15
Head/body length (mm)	506	430–591	*n* = 31	467	400–514	*n* = 27

Source: Sunquist and Sunquist (2002)

Margay *Leopardus wiedii* (Schinz, 1821)

The margay has broad feet, flexible ankles, and a long tail, all adaptations for arboreality (Nowell and Jackson 1996). In captivity, the margay is well known for its climbing and jumping acrobatics. One young margay 'would start running across a slack clothesline no more than half an inch in diameter, lose his balance, swing under it, then move paw over paw the rest of the way up a slight incline. Coming down, he would hook his paws around the line and slide down head first' (Wiley 1978, in Sunquist and Sunquist 2002). A wild radio-collared margay in Belize was observed feeding and travelling in trees, but also moved between hunting areas on the ground, up to 6 km/day (Konecny 1989). The margay is more forest-dependent than the ocelot and oncilla, reaching greatest abundance in lowland rainforest (Oliveira *et al.* 2008). It also occurs in dry deciduous forest (Nowell and Jackson 1996), but seems to be absent from the semi-arid caatinga scrub of Brazil, with the possible exception of some evergreen forest enclaves (Map 20). It appears to be intolerant of human settlement and altered habitat (IUCN 2008). One young adult male preferred primary to secondary forest, in a home range of 11 km^2 (Konecny 1989), and Crawshaw (1995) reported a male's range of 15.9 km^2 in subtropical forest of Brazil's Iguaçu National Park. Four males in El Cielo Biosphere Reserve, Tamaulipas, Mexico had an average range of 4.03 km^2, and a female 0.96 km^2 (Carvajal *et al.* 2007). Camera trapping suggests generally <5 individuals/100 km^2, with extremes of 20/100 km^2 (Oliveira *et al.* 2008). Predominantly nocturno-crepuscular (Oliveira 1998b), prey include birds, reptiles, fruit, and insects (Sunquist and Sunquist 2002), generally <200 g (Oliveira 1998b; but see Wang 2002). The margay is Near Threatened, primarily by deforestation (IUCN 2008).

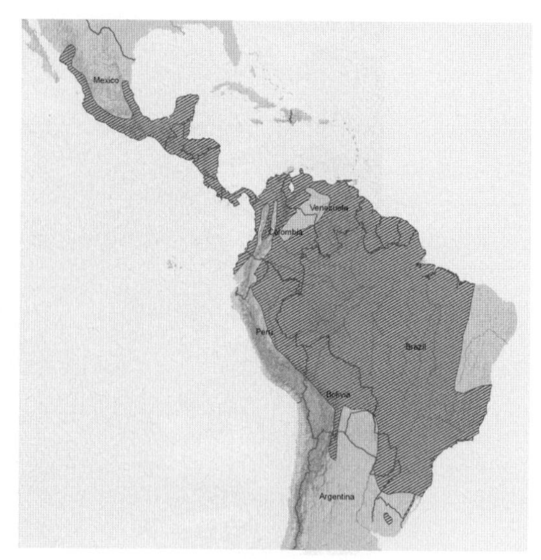

Map 20 Margay. © IUCN Red List 2008.

Canada lynx *Lynx canadensis* (Kerr, 1792)

Like the Iberian lynx, the Canada lynx is a lagomorph specialist, with its primary prey the snowshoe hare *Lepus americanus*. However, it is the only felid known to undergo prey-driven cyclic population declines. Documented from over a century of fur trade records, hare populations cycle through declines approximately every decade, and lynx populations follow with a lag of 1–2 years (Bulmer 1974; O'Donoghue *et al.*, Chapter 25, this volume). Lynx densities peak at 17–45/100 km^2, falling to 2/100 km^2(O'Donoghue *et al.* 2007). The cycle is driven by vegetation quality and both hare and lynx numbers, and its amplitude may be influenced by lynx-trapping pressure (Gamarra and Solé 2000). In most of Canada and Alaska (see Map 21 for current distribution), trapping of Canada lynx is managed for the fur trade. Trapping can reduce lynx populations,

	Male			Female		
	Mean	Range	Sample size	Mean	Range	Sample size
Weight (kg)	3.8	3.4–4	$n = 4$		2.6–3	$n = 2$
Head/body length (mm)	393	370–405	$n = 4$	400	390–410	$n = 3$

Source: Sunquist and Sunquist (2002)

Plate J Canada lynx *Lynx canadensis*. © Kim Poole.

	Male			Female		
	Mean	Range	Sample size	Mean	Range	Sample size
Weight (kg)	10.7	6.3–17.3	$n = 93$	8.6	5–11.8	$n = 91$
Head/body length (mm)	892	737–1067	$n = 96$	844	762–965	$n = 89$

Ref: Sunquist and Sunquist (2002)

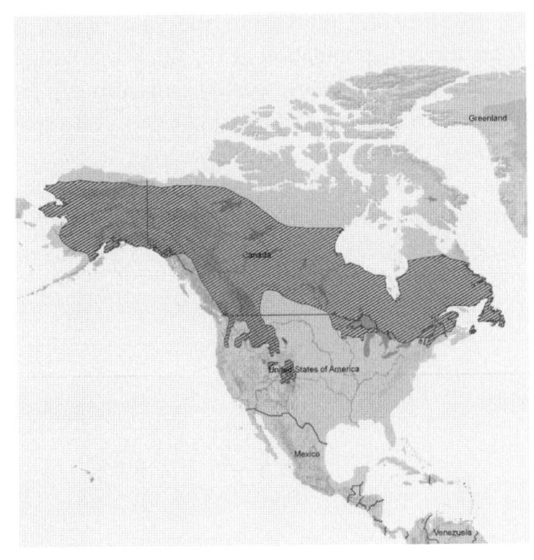

Map 21 Canada lynx. © IUCN Red List 2008.

especially when hare populations cyclically crash. During the cyclic low in the 1980s, most Canadian provinces and Alaska reduced harvests (Mowat *et al.* 2000). From 1980 to 1984, an average of 35,669 Canada lynx pelts were exported from the United States and Canada. That fell between 1986 and 1989 to an average annual export of 7360. Subsequently, annual exports from 2000 to 2006 averaged 15,387 (UNEP-WCMC 2008). Historical information suggests that lynx range and lynx–hare cycles have been largely stable in the north of their range (Mowat *et al.* 2000; Poole 2003). However, in the contiguous United States, lynx formerly occurred in 25 states (McKelvey *et al.* 2000a; Frey 2006), but now just 110,730 km^2 of critical lynx habitat has been proposed for designation in Maine, Minnesota, Washington, and the northern Rocky mountains (USFWS 2008b). A total of 204 lynx from Canada and Alaska have been released successfully since

1999 in the southern Rocky mountains of Colorado, where they are breeding (Nordstrom 2005) and have ranged up to 4310 m, with an average elevation of 3170 m (Wild *et al.* 2006). However, reintroduction of 83 lynx in the late 1980s in northern New York state failed (Sunquist and Sunquist 2002). In eastern Canada where lynx are rare and protected, the primary threat is considered to be interspecific competition from the eastern coyote, which has expanded its range into eastern North America in the past few decades (Parker 2001). In the contiguous United States, the primary threat is habitat fragmentation (Nordstrom 2005). Hybridization with bobcats *Lynx rufus* has been found by genetic analysis in Minnesota (Schwartz *et al.* 2003). The Canada lynx is classified as Least Concern (IUCN 2008).

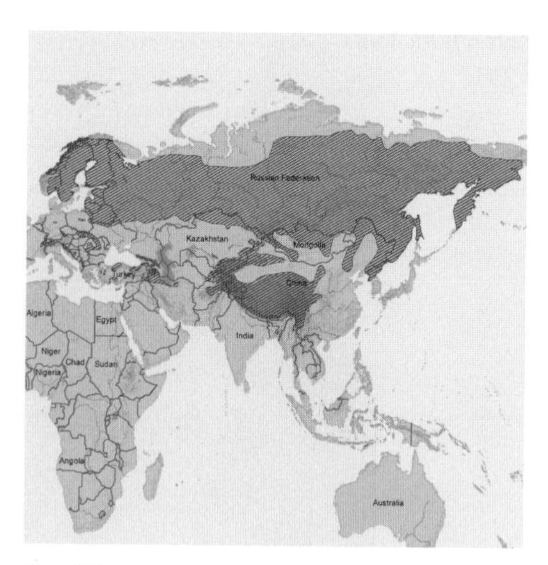

Map 22 Eurasian lynx. © IUCN Red List 2008.

Eurasian lynx *Lynx lynx* (Linnaeus, 1758)

The Eurasian lynx is the largest lynx, and the only one to take primarily ungulate prey, although they rely on smaller prey where ungulates are scarce (Nowell and Jackson 1996; Breitenmoser *et al.*, Chapter 23, this volume). Lynx kill ungulates ranging in size from the 15 kg musk deer to 220 kg adult male red deer, but show a preference for the smallest ungulate species in the community (Sunquist and Sunquist 2002). Home range averaged 248 km^2 for males ($n = 5$) and 133 km^2 for females ($n = 5$) in Poland's Bialowieza forest (Schmidt *et al.* 1997). Densities are typically one to three adults per 100 km^2, but up to 5/100 km^2 in eastern Europe (Sunquist and Sunquist 2002). The Eurasian lynx occurs from western Europe through the boreal forests of Russia, to central Asia and the Tibetan plateau (Map 22), with much range loss in Europe prior to reintroductions in Switzerland, Slovenia, Czech Republic, Austria, Germany, Italy, and France (Breitenmoser *et al.* 2000; IUCN 2008). The European lynx population (excluding Russia) is estimated at 8000. Populations in central and southern Europe are small and fragmented, although there are larger populations in Fennoscandia and the Baltic states (Breitenmoser *et al.* 2000), and a stronghold in southern Siberian woodland stretching through eastern Russia from the Ural mountains to the Pacific (IUCN 2008); the Russian population has been estimated at 30,000–35,000 (Matyushkin and Vaisfeld 2003). Its status in China is poorly known (IUCN 2008). In Mongolia, Matyushin and Vaisfeld (2003) estimate the lynx population at 10,000. While China and Russia exported nearly 20,000 skins annually in the late 1980s (Nowell and Jackson 1996), this trade has ended (UNEP-WCMC 2008). However, illegal skin trade remains the leading threat to the species, together with habitat loss and prey base depletion (Government of USA 2007a). It is classified as Least Concern (IUCN 2008).

	Male			Female		
	Mean	Range	Sample size	Mean	Range	Sample size
Weight (kg)	19.6	16.3–23.5	$n = 10$	17.3	14–21.5	$n = 12$
Head/body length (mm)	1000	760–1080	$n = 16$	900	850–1000	$n = 21$

Source: Heptner and Sludskii (1992)

Iberian lynx *Lynx pardinus* (Temminck, 1827)

The Iberian lynx is highly specialized on the European rabbit *Oryctolagus cuniculus*. It occurs only west of the Pyrenees on the Iberian Peninsula (Map 23), and its prey base has declined sharply due to habitat loss and disease (Ferreras *et al.*, Chapter 24, this volume). The Iberian lynx is the only cat species listed as Critically Endangered on the IUCN Red List (IUCN 2008). It occurs only in small isolated pockets in arid scrubland of southwestern Spain, but has not been detected in Portugal since the early 1990s, despite intensive surveys (Serra and Sarmento 2006). Surveys in Spain suggest between 84 and 143 adults surviving in two isolated breeding populations (in the Coto Doñana and near Andújar-Cardeña in the eastern Sierra Morena). The Doñana population numbers 24–33 adults and the Sierra Morena is the stronghold of the species, with an estimated 60–110 adults. None of the remaining potential populations in East Montes de Toledo, West Sistema Central, and some areas of central and western Sierra Morena is thought to include animals that breed regularly (IUCN 2008). The latest population estimates indicate a decline of more than 80% from the 880–1150 lynx estimated by surveys in 1987–1988. A similar decline of 80% was estimated for the period 1960–78 (Rodriguez and Delibes 1992). Current numbers are not viable (Delibes *et al.* 2000; IUCN 2008). In the Coto Doñana, total annual home ranges average 9.5 km^2 for adult females and 18.2 km^2 for adult males, with seasonal average adult density of 77/100 km^2 (Palomares *et al.* 2001). Previously considered conspecific with *L. lynx* by some authorities, the Iberian lynx is now accepted as a distinct species on the basis of both genetics (Johnson *et al.* 2006b; Eizirik *et al.*, in prep) and morphology (Werdelin 1981; Wozencraft 2005). The two species are estimated to have diverged ~1 million years ago (O'Brien and Johnson 2007), and were sympatric in central Europe during the late Pleistocene, but in modern times have not co-occurred (Nowell and Jackson 1996).

Map 23 Iberian lynx. © IUCN Red List 2008.

Bobcat *Lynx rufus* (Schreber, 1777)

Like its close relative *L. canadensis* the bobcat preys primarily on lagomorphs, but is less of a specialist. Rodents are commonly taken, and bobcats are capable of taking deer weighing up to 10 times their own body weight (Sunquist and Sunquist 2002). Bobcats occur mostly in the United States, but also into central Mexico and southern Canada (Map 24), where their range has been expanding northwards with forest clearance (Nowell and Jackson 1996; Sunquist and Sunquist 2002). While generally favouring low- and mid-elevations, in the western United States they have been trapped at elevations up to 2575 m (Nowell and Jackson

	Male			Female		
	Mean	Range	Sample size	Mean	Range	Sample size
Weight (kg)	12.8	11.1–15.9	$n = 6$	9.3	8.7–9.9	$n = 4$
Head/body length (mm)	787	747–820	$n = 6$	720	782–754	$n = 6$

Ref: Beltrán and Delibes (1993)

	Male			Female		
	Mean	Range	Sample size	Mean	Range	Sample size
Weight (kg)	9.3	5.6–11.4	$n = 21$	6.2	4.2–10.5	$n = 26$
Head/body length (mm)	735	603–927	$n = 21$	624	508–724	$n = 26$

Source: Young (1978)

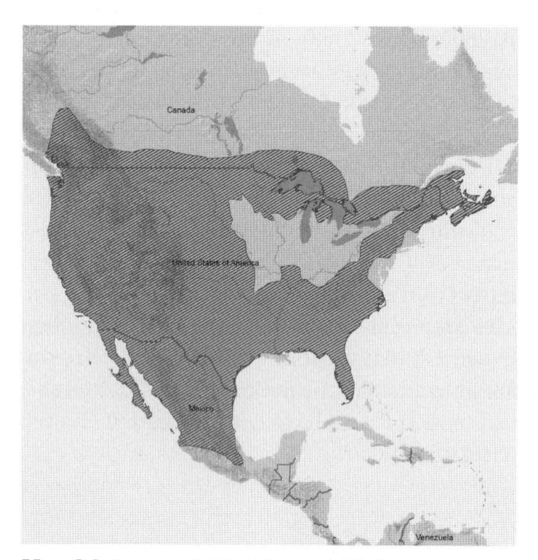

Map 24 Bobcat. © IUCN Red List 2008.

1996) and in western Mexico, radio-collared bobcats were located at 3500 m on the Colima Volcano (Burton *et al.* 2003). Bobcats are found in both forested and open habitats, which typically include areas with abundant rabbit and rodent populations, dense cover, and shelters that function as escape cover or den sites (Sunquist and Sunquist 2002). In Mexico, bobcats are found in dry scrub and grassland, as well as tropical dry forest including pine, oak, and fir (Monroy-Vilchis and Velazquez 2003; Arzate *et al.* 2007; C. Lopez-Gonzalez, personal communication 2007). Home ranges vary from 6 km^2 for females in southern California to 325 km^2 for male bobcats in upstate New York. Bobcats in northern and western portions of the United States are consistently larger than those in the south, possibly because the warmer climates provide a less variable prey base (Sunquist and Sunquist 2002). Density estimates include 48/100 km^2 in Texas (Heilbrun *et al.* 2006); 25/100 km^2 in Arizona (Lawhead 1984); <9/100 km^2 in Idaho (Knick 1990); and 11/100 km^2 in Virginia (minimal estimate; M. Kelly, personal communication 2007). Bobcat densities in the northern parts of their range are generally lower than in the south (Sunquist and Sunquist 2002). However, the first density estimate for bobcats in Mexico is low, at five individuals per 100 km^2 (Arzate *et al.* 2007). Habitat loss is viewed as the primary threat to bobcats in all three range countries (IUCN 2008). The bobcat is now the leading felid in the skin trade, with most exports coming from the United States. From 1990 to 1999, annual exports averaged 13,494; in 2000–06 the average climbed to 29,772, with a peak of 51,419 skins exported in 2006 (UNEP-WCMC 2008). The US government has repeatedly petitioned the CITES to delist the bobcat from Appendix II to Appendix III (Government of USA 2007b), but has been rejected by majority vote due to concerns about opening a loophole for illegal trade in other felids, all of which are listed on CITES Appendices I or II (Nowell *et al.* 2007). It is classified as Least Concern on the IUCN Red List (IUCN 2008).

Cheetah *Acinonyx jubatus* (Schreber, 1775)

Most famous for its speed (29 m/s; Sharp 1997), the cheetah has a number of adaptations, both morphological (Kitchener *et al.*, Chapter 3, this volume) and behavioural (Durant *et al.*, Chapter 16, this volume) that are unique among felids, and are specializations for catching fleet-footed antelope prey. In areas where large-scale ungulate migration patterns are still intact, such as the Serengeti plains of Tanzania (see map 25), wide-ranging solitary female cheetahs (average home range size: 800 km^2) pass through small temporary territories held by male coalitions (average

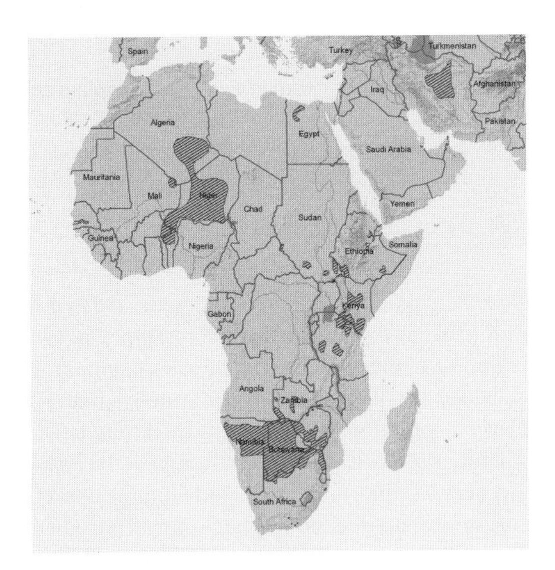

Map 25 Cheetah. © IUCN Red List 2008.

territory size: 50 km^2; Caro 1994). However, in areas where prey is non-migratory, male and females have overlapping ranges that are similar in size (Sunquist and Sunquist 2002). On Namibian farmlands, both cheetah sexes have very large home ranges (average 1642 km^2); however, intensively used core areas were just 14% of the total home range (Marker *et al.*, Chapter 15, this volume). Cheetahs are primarily active during the day, perhaps reducing competition (Caro 1994). Other large carnivores, especially lions and hyenas, steal cheetah kills and kill cheetahs. Cheetahs occur at lower densities than would be expected considering their energy needs (Anonymous 2007). On the Serengeti plains, cheetah densities range from 0.8/100 km^{2}to 1.0/100 km^2, but seasonally cheetahs can congregate at densities up to 40/100 km^2 (Caro 1994). Caro (1994) attributes lower cheetah densities to interspecific competition, but on Namibian farmlands, where lions and hyenas have been eradicated, cheetahs still occur at low densities (0.2/100 km^2), despite an abundant wild prey base (Marker *et al.*, Chapter 15, this volume). Namibia is believed to have the largest national cheetah population (2000; Purchase *et al.* 2007), despite decades of extensive trapping by farmers, with over 9500 cheetahs removed from 1978 to 1995 (Nowell 1996). Elsewhere in Africa, cheetahs have lost 76% of their historic range (Ray *et al.* 2005), and disappeared everywhere from their formerly extensive Asian range except for a small population in Iran (60–100 individuals, Hunter *et al.* 2007b; see Map 25 for current range). Cheetahs were captured and trained for hunting by Asian royalty, ranging from the Caucasus to India, for thousands of years (Divyabhanusinh 1995; Sunquist and Sunquist 2002). The species formerly had an even wider range, having originated in North America (with its closest living relatives, the puma and jaguarundi; Johnson *et al.* 2006b). The cheetah exhibits remarkably low levels of genetic diversity in comparison to other felids (O'Brien *et al.* 1985b) (but not compared to carnivores in general; Merola 1994; see Culver *et al.* Chapter 4, this volume). While causes of past low population sizes are unclear, the causes for the cheetah's current state of threat are well known: habitat loss and fragmentation, conflict with people, and depletion of their wild prey base (Marker 2002; Dickman *et al.* 2006a). Interspecific competition with larger predators leads to cheetahs achieving higher densities outside protected areas (Marker 2002). A recent analysis found that known resident cheetah populations in 'classic' cheetah habitat in eastern Africa only occur over approximately 350,000 km^2, with an estimated population of ~2500 (Anonymous 2007). The cheetah is classified as Vulnerable, and Critically Endangered in Iran and North Africa (IUCN 2008).

	Male		Female	
	Mean ± SD	Sample size	Mean ± SD	Sample size
Weight (kg)	41.4 ± 5.4	*n* = 23	35.9 ± 5.3	*n* = 19
Head/body length (mm)	1225 ± 7	*n* = 24	1245 ± 7.5	*n* = 16

Ref: Caro (1994)

Puma *Puma concolor* (Linnaeus, 1771)

Occurring from the mountains of British Columbia to Tierra del Fuego, the puma has the largest geographic range of the New World cats (Map 26); apparently larger than any terrestrial mammal in the western hemisphere (Sunquist and Sunquist 2002). However, it was extirpated over 100 years ago from the eastern half of its historic range in the United States and Canada. Numerous scattered sightings in

	Male			Female		
	Mean	Range	Sample size	Mean	Range	Sample size
Weight (kg)	58.9	56.2–64.4	*n* = 10	30.7	27.2–36.3	*n* = 11
Head/body length (mm)	1347	1260–1440	*n* = 10	1179	1110–1260	*n* = 11

Source: Sweanor (1990)

Map 26 Puma. © IUCN Red List 2008.

the Midwest suggest they may be recolonizing (Anonymous 2005b). DNA analysis of hair samples collected in Fundy National Park, New Brunswick, Canada identified two individual cougars (as they are known in eastern North America), one of North American and one of South American ancestries (Bertrand 2006). This and other evidence suggests that some of the hundreds of reported observations of cougars in eastern North America are escaped captive animals. The only area where panthers (as pumas are known in the south-eastern United States) are known to have survived historical extirpation is a single population in the Everglades swamp forest region of Florida, although it experienced a severe bottleneck (Culver *et al.* 2008), and currently numbers only about 100 (Onorato *et al.*, Chapter 21, this volume). Culver *et al.* (2000) suggest that the lack of genetic variation among North American pumas in general is the result of a late Pleistocene extinction event (as befell the North American cheetah), and later recolonization from the south. Pumas are found in a broad range of habitats, in all forest types as well as lowland and montane desert. Pumas are sympatric with jaguars in much of their Latin American range, and may favour more open habitats than their larger competitor (Vynne *et al.* 2007), although both can be found in dense forest (Sunquist and Sunquist 2002). Pumas are capable of taking large prey, but when available, small- to medium-sized prey are more important in their diet (in tropical portions of the range). This is true of wild prey as well as livestock (IUCN 2008). In North America, deer make up 60–80% of the puma's diet, and the mean weight of prey taken is 39–48 kg. In Florida, however, where deer numbers are low, pumas take smaller prey including feral pigs, raccoons, and armadillos, and deer account for only about one-third of the diet (Sunquist and Sunquist 2002). Bonacic *et al.* (2007) found hares to be the predominant prey (96%) in analysis of scats from the Mediterranean shrub ecoregion of Chile. Densities exceeding four adults per 100 km^2 do not appear to be common in North America (Sunquist and Sunquist 2002). In Chilean Patagonia, density was estimated at 6/100 km^2 (Franklin *et al.* 1999). Kelly *et al.* (2008) reported densities in three study sites as follows: Belize, 2–5/100 km^2; Argentina, 0.5–0.8/100 km^2; and Bolivia, 5–8/100 km^2. Although classified as Least Concern, pumas are threatened by habitat loss and fragmentation, poaching of their wild prey base, and persecution due to livestock depredation (IUCN 2008).

Jaguarundi *Puma yagouaroundi* (Lacépède, 1809)

With its elongated and low-slung body, the jaguarundi has been a taxonomic enigma (Sunquist and Sunquist 2002), but genetic analysis groups it with

	Male			Female		
	Mean	Range	Sample size	Mean	Range	Sample size
Weight (kg)	5.7	5–6.5	$n = 4$	4.1	3.7–4.4	$n = 3$
Head/body length (mm)	693	650–755	$n = 4$	596	550–635	$n = 3$

Ref: A. Caso (personal Communication) in Sunquist and Sunquist (2002)

Map 27 Jaguarundi. © IUCN Red List 2008.

the puma and the cheetah (Johnson *et al.* 2006b; O'Brien and Johnson 2007; Eizirik *et al.*, in prep). Probably extinct in the United States, it ranges from Mexico through Central America and the Amazon basin to central Argentina and Uruguay (IUCN 2008; Map 27). It occurs in a variety of habitats, from closed primary rainforest to open desert, scrub, and grassland, although in open areas it sticks to patches of dense cover (Nowell and Jackson 1996; Oliveira 1998a). Caso and Tewes (2007) found that radio-collared jaguarundis used mature forest 53% of the time and pasture-grassland 47% of the time. It is predominantly a lowland species ranging up to 2000 m, although it was reported up to 3200 m in Colombia (Cuervo *et al.* 1986). It is also predominantly diurnal (with activity peaks in late morning and late afternoon: Caso and Tewes 2007). Jaguarundis occur at relatively low densities of 1–5/100 km^2 in Brazil (Oliveira *et al.* 2008), but reach densities up to 20/100 km^2 in Tamaulipas, Mexico (A. Caso, personal communication 2007). Radio-collared jaguarundis in rainforest of Belize's Cockscomb Basin reserve had much larger home ranges than the sympatric jaguar. One female used a home range that varied between 13 and 20 km^2, while two males used home ranges of 100 and 88 km^2 (Konecny 1989). Home ranges in scrub land of Tamaulipas, Mexico, were smaller, averaging 9.6 km^2 for males and 8.9 km^2 for females. With 20 radio-collared animals in the study area, extensive inter- and intrasexual overlapping of home ranges was documented (Caso and Tewes 2007). Although with its wide range the species is classified as Least Concern, habitat loss is a threat, especially for the more open types, which are being converted to large-scale agriculture (IUCN 2008).

Jungle cat *Felis chaus* (Schreber, 1777)

The jungle cat, despite its name, is not strongly associated with the classic rainforest 'jungle' habitat, but rather with wetlands—habitats with water and dense vegetative cover, especially reed swamps, marsh, and littoral and riparian environments. Hence its other common and more applicable names, the swamp or reed cat. Its preferred microhabitat of water and dense ground cover can be found in a variety of habitat types, ranging from desert (where the cat occurs near oases or along riverbeds) to grassland, shrubby woodland, and dry deciduous forest, as well as cleared areas in moist forest (Nowell and Jackson 1996)—and thus the jungle cat has a broad but patchy distribution from Egypt's Nile River Valley through Asia to the Isthmus of Kra (Map 28). Small mammals, principally rodents, are the primary prey of the jungle cat. A study in India's Sariska reserve estimated that jungle cats catch and eat three to five rodents per day (Mukherjee *et al.* 2004). Birds rank second in importance, but in southern Russia waterfowl are the mainstay of

	Male			Female		
	Mean	Range	Sample size	Mean	Range	Sample size
Weight (kg)	8.1	5–12	$n = 11$	5.1	2.6–7.5	$n = 14$
Head/body length (mm)	763	650–940	$n = 13$	658	560–850	$n = 13$

Ref: Sunquist and Sunquist (2002)

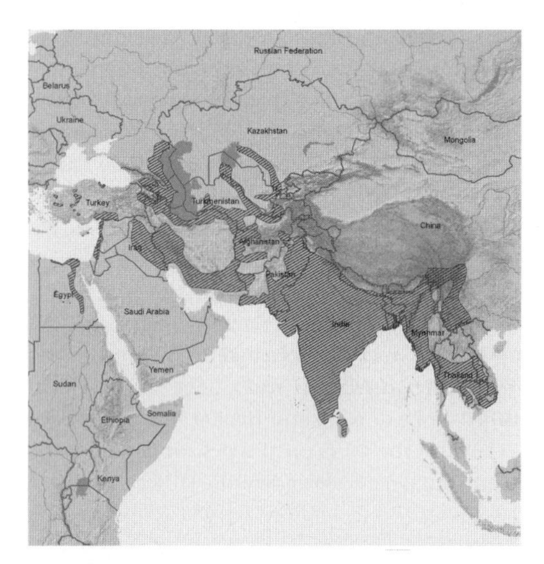

Map 28 Jungle cat. © IUCN Red List 2008.

jungle cat diet in the winter. In India, they have been seen to scavenge kills of large predators such as the Asiatic lion. In a study in southern Uzbekistan, the fruits of the Russian olive made up 17% of their diet in winter. While jungle cats specialize on small prey, they are large and powerful enough to kill young swine, subadult gazelles, and chital fawns (Sunquist and Sunquist 2002). Jungle cats can do well in cultivated landscapes (especially those that lead to increased numbers of rodents) and artificial wetlands. However, as population density in natural wetlands appears to be higher, ongoing destruction of natural wetlands, throughout its range but particularly in the more arid parts, still poses a threat to the species (Nowell and Jackson 1996). Unselective trapping, snaring, and poisoning around agricultural and settled areas have caused population declines in many areas throughout its range (Abu-Baker *et al.* 2003; Duckworth *et al.* 2005). The jungle cat is classified as Least Concern by IUCN (2008), as the species is widespread, and common in some parts of its range, particularly India (Mukherjee 1989). However, population declines and range contraction are of concern elsewhere, particularly Egypt (Glas, in press) and south-west Asia (Abu-Baker *et al.* 2003), the Caucasus (IUCN 2008), central Asia (Habibi 2004), and south-east Asia (Duckworth *et al.* 2005).

Sand cat *Felis margarita* (Loche, 1858)

The sand cat is large-eared, pale in colouration, and the only felid to occur exclusively in desert. The undersides of the feet are thickly furred, which may help it to move across shifting sands and protect the feet from high temperature sand (Nowell and Jackson 1996; Sunquist and Sunquist 2002; Sliwa, in press-a). The claws do not fully retract and are rather blunt—possibly due to the sand cat's digging behaviour (Sunquist and Sunquist 2002; Sliwa, in press-a). It is highly fossorial, known to Saharan nomads as 'the cat that digs holes' (Dragesco-Joffe 1993). During 9 months of radio-tracking four sand cats, on only a single occasion was a study animal observed in daytime outside its burrow, and then only 2 m away (Abbadi 1993). Sand cats use and enlarge burrows of other species as well as digging their own (Sliwa, in press-a). They cover their scats with sand, making diet study difficult. The only scats found by Abbadi (1993) during his study, 'despite painstaking searches on foot the day after our night watches', were inside the box traps that captured the cats. They contained the remains of Cairo spiny mouse (*Acomys cahirinus*) and gecko (*Stenodactylus* spp.). Sand-dwelling rodents made up the majority (65–88%) of stomach contents from carcasses collected in Turkmenistan and Uzbekistan in the 1960s (Schauenberg 1974). Sand cats have also been observed hunting birds and reptiles (Abbadi 1993; Dragesco-Joffe 1993), and will drink water

Plate K Sand cat *Felis margarita*. © Jameson Weston, Hogle Zoo.

readily but can survive on metabolic water (Sliwa, in press-a). They move long distances in a single night (5–10 km; Abbadi 1993). Annual ranges from Saudi Arabia are up to 40 km^2 (M. Strauss, personal communication 2008). The sand cat has a wide but apparently disjunct distribution through the deserts of northern Africa and south-west and central Asia (Hemmer *et al.* 1976; Nowell and Jackson 1996; see Map 29). Sightings have been reported in Libya and Egypt west of the Nile (Sliwa in press-a), but there are no historical records despite intensive collecting effort (Hemmer *et al.* 1976). It is classified as Near Threatened (IUCN 2008) because vulnerable arid ecosystems are being rapidly degraded by human settlement and activity, especially livestock grazing. The sand cat's small mammal prey base depends on adequate vegetation, and may

Map 29 Sand cat. © IUCN Red List 2008.

	Male			Female		
	Mean	Range	Sample size	Mean	Range	Sample size
Weight (kg)	2.78	2.0–3.2	*n* = 4	2.2	1.35–3.1	*n* = 5
Head/body length (mm)	499	460–570	*n* = 5	467	400–520	*n* = 6

Ref: Sunquist and Sunquist (2002)

experience large fluctuations due to drought (Sunquist and Sunquist 2002), or declines due to desertification and loss of natural vegetation.

Black-footed cat *Felis nigripes* (Burchell, 1825)

The black-footed cat, Africa's smallest felid, is endemic to the short grasslands of southern Africa (Map 30), where it is rare and classified as Vulnerable (IUCN 2008). Knowledge of its behaviour and ecology stems from a decade-long study on the Benfontein Game Farm in central South Africa, where more than 20 cats were radio-collared and habituated (Sliwa 2004; Sliwa 2007; Sliwa *et al.*, Chapter 26, this volume). Black-footed cats are solitary, except for females with dependent kittens, and during mating. Males have larger annual home ranges (20.7 km^2, $n = 5$) than females (10.0 km^2, $n = 7$). Male ranges overlap those of 1–4 females. Intra-sexual overlap varies from 12.9% for three males to 40.4% for five females (Sliwa 2004). In his 60 km^2 study area, Sliwa (2004) found the density of adult cats to be 0.17/km^2. Kittens are independent after 3–4

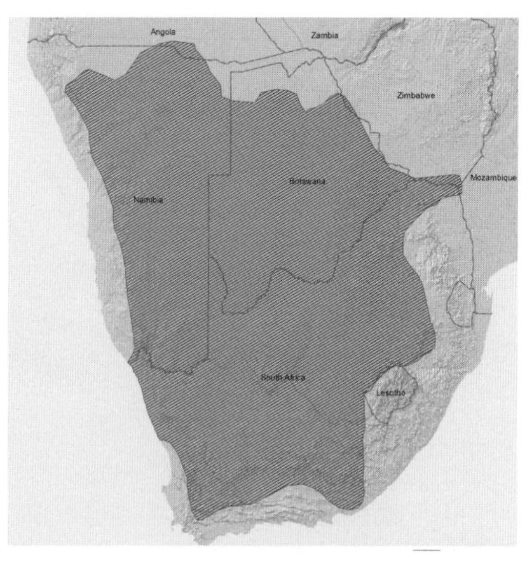

Map 30 Black-footed cat. © IUCN Red List 2008.

Plate L Black-footed cat *Felis nigripes*. © Alex Silwa.

	Male			Female		
	Mean	Range	Sample size	Mean	Range	Sample size
Weight (kg)	1.93	1.75–2.45	*n* = 8	1.3	1.1–1.65	*n* = 10
Head/body length (mm)	453	430–520	*n* = 8	397	370–420	*n* = 10
Ref: Sliwa (in press-b)						

months, but remain within the range of their mother for extended periods (Sliwa, in press-b). Sliwa (2006) observed 1725 prey items consumed by black-footed cats, with an average size of 24.1 g. Males fed on significantly larger prey (27.9 g) than did females (20.8 g). Mammals were the most common prey item (72%), followed by small birds weighing less than 40 g (26%), while invertebrates, amphibians, and reptiles were infrequently taken. Small rodents like the large-eared mouse *Malacothrix typica*, captured 595 times by both sexes, were particularly important during the reproductive season for females with kittens. Male black-footed cats showed less variation between prey-size classes consumed among climatic seasons. This sex-specific difference in prey size consumption may help to reduce intraspecific competition (Sliwa 2006). The black-footed cat is threatened primarily by habitat degradation by grazing and agriculture, as well as by poison and other indiscriminate methods of pest control (Nowell and Jackson 1996; Sliwa, in press-b).

Wildcat *Felis silvestris* (Schreber, 1777)

The wildcat has the widest distribution of any felid (Map 31), being found throughout the drier regions of Africa into Europe (including Scotland), south-west and central Asia, and Russia (Nowell and Jackson 1996). The house cat was domesticated from the wildcat. Archaeological evidence dates back to the origin of agriculture in the Eastern Crescent, and domestication is presumed to have started with a commensal relationship, with wildcats coming into villages to prey on rodents attracted to grain stores (Clutton-Brock 1999; Driscoll *et al.* 2007). Genetic analysis (Driscoll *et al.* 2007; Eizirik *et al.*, in prep) supports five wild subspecies plus a sixth, the domestic cat *F. catus*, which is mitochondrially monophyletic with *F.s. lybica* Forster, 1780 (North Africa and south-west Asia) (Driscoll *et al.* 2007). The taxanomic nomenclature of the domestic cat is complex and controversial and fully discussed in Macdonald *et al.* (Chapter 22, this volume). Other wild sub-species are *F.s. cafra* Desmarest, 1822 (sub-Saharan Africa);

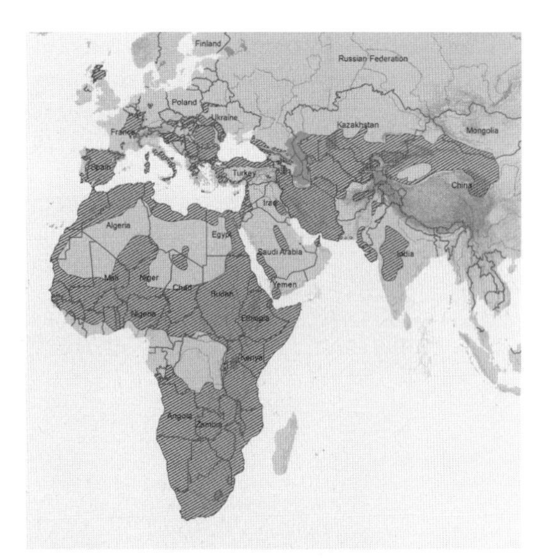

Map 31 Wildcat. © IUCN Red List 2008.

	Male			Female		
	Mean	Range	Sample size	Mean	Range	Sample size
Weight (kg)	4.9	4–6.2	*n* = 10	3.7	2.4–5.0	*n* = 10
Head/body length (mm)	601	545–665	*n* = 21	550	460–620	*n* = 15
Ref: Stuart *et al.* (in press)						

F.s. ornata Gray, 1830 (central Asia to India); *F.s. bieti* Milne-Edwards, 1872 (western China); and *F.s. silvestris* Schreber, 1775 (Europe). *F.s. bieti* has been previously considered a separate species *F. bieti* (Wozencraft 2005), and has an apparently restricted distribution on the eastern edge of the Tibetan plateau at elevations from 2500 to 5000 m (He *et al.* 2004). Female home ranges vary widely with habitat, from 52.7 km^2 in the United Arab Emirates (Phelan and Sliwa 2005) to 1–2 km^2 in France and Scotland (Stahl *et al.* 1988; Macdonald *et al.*, Chapter 22, this volume). The world's population of domestic cats was estimated at 400 million (Legay 1986), and interbreeding is the main threat to the wildcat (Macdonald *et al.* 2004b; IUCN 2008). Of the subspecies, only *F.s. bieti* shows no evidence of genetic introgression (Driscoll *et al.* 2007).

Pallas's cat *Otocolobus manul* (Pallas, 1776)

The cat first described by German explorer Peter Pallas was hardly a typical feline, with its short legs, shaggy fur, and its small rounded head and ears. In Mongolia, it is sometimes mistakenly killed by marmot hunters targeting one of the Pallas's cat's main prey species (IUCN 2008). While it is grouped in the tribe Felini with *Felis* and *Prionailurus*, the exact phylogenetic relationships are unclear and it is retained in the monospecific genus *Otocolobus* by Eizirk *et al.* (in press). Pallas's cat is also known by its Russian name, manul, in Mongolia, Russia, and the former Russian republics of central Asia, which make up the majority of its range (Map 32). It was recently recorded from north-west Iran (Aghili *et al.* 2008), although its current distribution in the trans-Caspian area is poorly known. It also occurs sparsely throughout the Tibetan Plateau, where an elevational record of 5050 m was reported recently (Fox and Dorji 2007). Populations in the south-west of its range (the Caspian Sea region, and Afghanistan and Pakistan) are diminishing, isolated, and sparse (Belousova 1993; Nowell and Jackson 1996; Habibi 2004). Typical habitat for the Pallas's cat is characterized by an extreme continental climate—little rainfall, low humidity, and a wide range of temperatures. They are rarely found in areas where the maximum mean 10-day snow cover depth exceeds 10 cm, and a continuous snow cover of 15–20 cm marks the ecological limit for this species (Sunquist and Sunquist 2002).

Plate M Pallas's cat *Otocolobus manul*. © East Azerbaijan Department of Environment of Iran Public Office.

	Male			Female		
	Mean	Range	Sample size	Mean	Range	Sample size
Weight (kg)	4.12	3.3–5.3	*n* = 25	4.02	3.05–5	*n* = 16
Head/body length (mm)		553 ± 17	*n* = 4	492 ± 20		*n* = 10

Ref: S. Ross (personal communication)

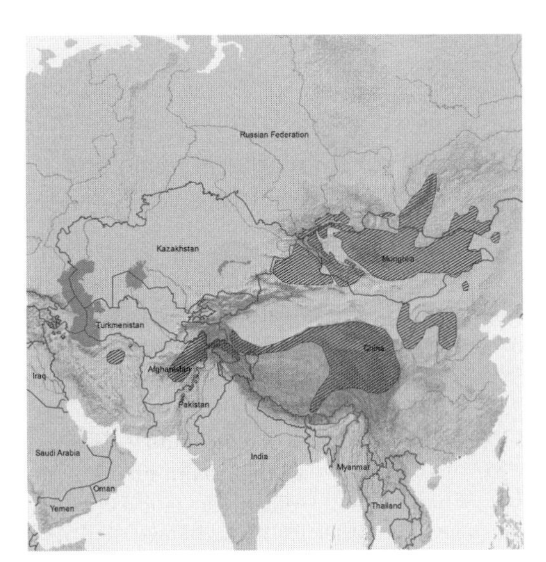

Map 32 Pallas's cat. © IUCN Red List 2008.

In the grass and shrub steppe of central Mongolia, annual home ranges were found to be strikingly large for a small felid, although it is not clear if such large ranges are typical for the species (Brown *et al.* 2003). Ongoing research there (S. Ross, personal communiction 2008) measured mean annual home ranges (95% MCP) at 27.1 ± 23.6 km^2 for adult females ($n = 10$) and 100.4 ± 101.2 km^2 for males ($n = 8$). Pallas's cats have a strong association with rocky, steep areas and were rarely found in open grasslands (where they may be more vulnerable to predation by sympatric carnivores; S. Ross, personal communication 2008). In China, Pallas's cats appear to be most numerous where pikas and voles are abundant and not living under deep snow cover (C. Wozencraft, personal communication). In Mongolia, preliminary analysis of scats indicated that gerbils (*Meriones* spp.) and jerboas (*Dipus sagitta* and *Allactaga* spp.) were the main prey, with lambs of the Argali sheep (*Ovis ammon*) taken during the spring (Murdoch *et al.* 2006). Body weight varies widely by season (lowest in winter) and phase in reproductive cycle (lowest for males when breeding and for females after raising kittens). Activity is predominantly crepuscular, although they can be active at any time (S. Ross, personal communication 2008). It is classified as Near Threatened due to prey-base depletion (poisoning and over-hunting), habitat degradation by livestock and agriculture, and illegal trade in skins and for traditional medicine (IUCN 2008).

Leopard cat *Prionailurus bengalensis* (Kerr, 1792)

The leopard cat is common and widespread throughout most of India west into Pakistan and Afghanistan, across most of China, and north to the Korean peninsula, and into the Russian Far East (Map 33). It also

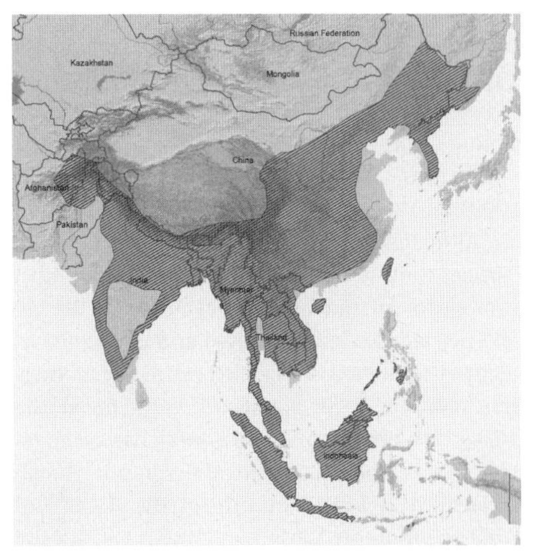

Map 33 Leopard cat. © IUCN Red List 2008.

occurs throughout south-east Asia, and on the islands of Sumatra, Java, Borneo, and Taiwan. The leopard cat is the only wild felid found in Japan, where it occurs on the small islands of Tsushima and Iriomote, and the Philippines, where it occurs on the islands of Palawan, Panay, Negros, and Cebu (IUCN 2008). An excellent swimmer, it is found on numerous small offshore islands of mainland Asia (Nowell and Jackson 1996; Sunquist and Sunquist 2002). Ranging up to 3000 m in the Himalayas, it occurs in habitats from tropical rainforest to temperate broadleaf and, marginally, coniferous forest, as well as shrub forest and successional grasslands. While the leopard cat is more tolerant of disturbed areas than other small Asian felids (Nowell and Jackson 1996; Sunquist and Sunquist 2002), it likely experiences higher mortality in such areas. Higher survival rates (92%) were recorded in a protected area with little human influence, compared with lower rates in areas with greater human activity (53–82%; Haines *et al.* 2004). Based on a large sample size of 20 radio-collared cats in Thailand's Phu Khieu Wildlife Sanctuary, mean home range size (95% MCP) was 12.7 km^2, larger than in other areas of Thailand (4.5 km^2; Grassman *et al.* 2005a), on Borneo (3.5 km^2; Rajaratnam *et al.* 2007), or on Japan's Iriomote island (Schmidt *et al.* 2003). There was no significant difference between male and female home range size. Open and closed forest habitats were used in proportion to their occurrence, and activity patterns showed crepuscular and nocturnal peaks. On Borneo, Rajaratnam *et al.* (2007) found that leopard cats hunted rodents in oil palm plantations, and used forest fragments for resting and breeding. Murids dominate the diet (85–90%; Grassman *et al.* 2005a; Rajaratnam *et al.* 2007). In China, commercial exploitation for the fur trade has been heavy, with annual harvests estimated at 400,000 in the mid-1980s (Nowell and Jackson 1996). Although commercial trade is much reduced, the species continues to be hunted throughout most of its range for fur and food, and captured for the pet trade. They are also widely viewed and persecuted as poultry pests. Leopard cats can hybridize with domestic cats, resulting in the popular domestic breed, the 'safari cat'. Hybridization in the wild has been reported, but is not considered a significant threat. It is classified as Least Concern on the IUCN Red List, but the Iriomote subspecies (Japan) is Critically Endangered, with a population of less than 100, and the West Visayan (the Philippines) leopard cat is Vulnerable due to habitat loss (IUCN 2008).

	Male		Female	
	Mean ± SD	Sample size	Mean ± SD	Sample size
Weight (kg)	2.9 ± 0.38	*n* = 17	2.3 ± 0.27	*n* = 8
Head/body length (mm)	572 ± 4.8	*n* = 16	533 ± 2.5	*n* = 8

Ref: Grassman *et al.* (2005a)

Flat-headed cat *Prionailurus planiceps* (Vigors and Horsfield, 1827)

The flat-headed cat takes its name from its unusually long, sloping snout and flattened skull roof, with small ears set well down the sides of its head. It has large, close-set eyes, and relatively longer and sharper teeth than its close relatives (Muul and Lim 1970; Groves 1982). Its claws do not retract fully into their shortened sheaths, and its toes are more completely webbed than the fishing cat's, with long, narrow-footed pads. Muul and Lim (1970), commenting on the cat's feet and other features, characterized it as the ecological counterpart of a semiaquatic mustelid. In captivity, they played for hours in basins of water, and Shigeki Yasuma observed a wild flat-headed cat playing in water (Nowell and Jackson 1996). The stomachs of two dead flat-headed cats contained mostly fish, and also shrimp shells. They may also take birds and small rodents, and have been reported to prey on domestic poultry (Nowell and Jackson 1996). The species is closely associated with wetlands, to a greater degree than the fishing cat, with a much smaller distribution, found only on the islands of Borneo and Sumatra and the Malayan peninsula (Map 34). Most collection records for the flat-headed cat are from swampy areas, oxbow lakes, and riverine forest (Nowell and Jackson 1996). They also occur in peat-swamp forest (Bezuijen 2000), and have been observed in recently logged forest (Bezuijen 2000, 2003; Meijaard *et al.* 2005b). All published observations of live animals have taken place at night, near water (Nowell and Jackson 1996; Bezuijen 2000, 2003; Meijaard *et al.* 2005b). Flat-headed cats are only found in lowland forest, and this habitat is disappearing with cultivation of oil palm,

	Male			Female		
	Mean	Range	Sample size	Mean	Range	Sample size
Weight (kg)	1.9	1.5–2.2	*n* = 5	1.7	1.5–1.9	*n* = 3
Head/body length (mm)	488	446–521	*n* = 6	470	455–490	*n* = 3

Ref: Sunquist and Sunquist (2002)

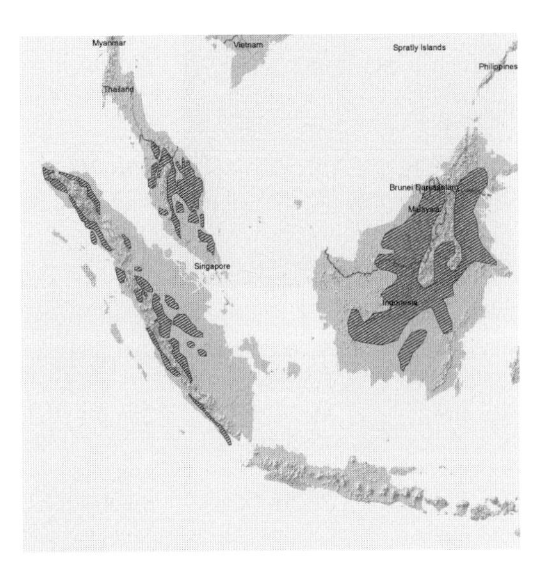

Map 34 Flat-headed cat. © IUCN Red List 2008.

logging, settlement, agriculture, and aquaculture. The flat-headed cat was upgraded from Vulnerable to Endangered on the IUCN Red List (IUCN 2008).

Rusty-spotted cat *Prionailurus rubiginosus* (I. Geoffroy Saint-Hilaire, 1831)

The world's smallest cat, its common name is taken from the elongated rust-brown spots that stripe the rufous grey fur of its back and flanks. Found only in India and Sri Lanka (Map 35), the rusty-spotted cat is poorly known. In India, it was long thought to be confined to the south, but recent records have established that it is found over much of the country (Sunquist and Sunquist 2002; Patel and Jackson 2005; Manakadan and Sivakumar 2006; Patel 2006; Vyas *et al.* 2007). Rusty-spotted cats occupy dry forest types as well as scrub and grassland, but are likely absent from evergreen forest in India (Nowell and Jackson 1996), although there are a few records from montane and lowland rainforest in Sri Lanka (Deraniyagala 1956; Nekaris 2003). While dense vegetation and rocky areas are preferred (Worah 1991; Kittle and Watson 2004; Patel 2006), rusty-spotted cats have been found in the midst of agricultural and settled areas (Nowell and Jackson 1996; Mukherjee 1998). They are highly arboreal (Sunquist and Sunquist 2002; Patel 2006), although most observations have been on the ground, at night (Mukherjee 1998; Kittle and Watson 2004; Patel 2006; Vyas *et al.* 2007). One cat was seen hunting frogs, but small rodents were the main prey reported from a series of observations by Patel (2006)—seeking out such prey is probably why the cats venture into cultivated areas, where they may interbreed with domestic cats. Outside Sri Lanka's Yala National Park, Kittle and Watson (2004) observed a rusty-spotted cat mating with a domestic cat and also saw a potential hybrid ('being slightly larger in size, with long legs and exhibiting unusual

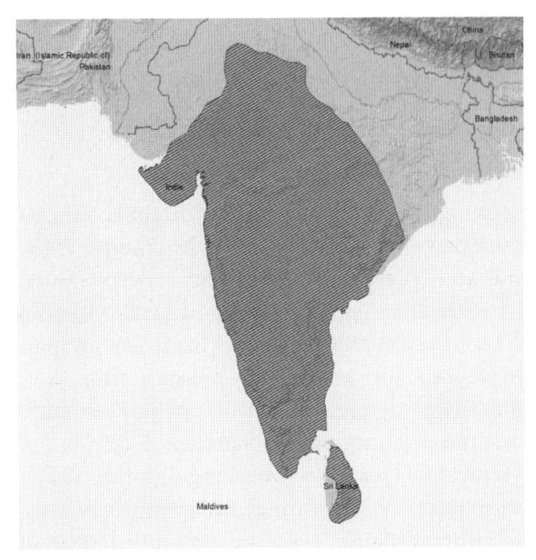

Map 35 Rusty-spotted cat. © IUCN Red List 2008.

Plate N Rusty spotted cat *Prionailurus rubiginosus* carrying a rat. © Vidya Athreya.

	Male			Female	
	Mean	Range	Sample size	Mean	Sample size
Weight (kg)	0.9	0.8–1.1	*n* = 3	0.9	*n* = 1
Head/body length (mm)	379 ± 37		*n* = 6	363 ± 34	*n* = 8

Refs: Head/body length, Deraniyagala (1956); Weight, Sunquist and Sunquist (2002)

markings on a paler background'). The rusty-spotted cat is classified as Vulnerable on the IUCN Red List of Threatened Species (IUCN 2008).

Fishing cat *Prionailurus viverrinus* (Bennett, 1833)

The fishing cat is well adapted for catching fish, its primary prey (Bhattacharyya 1989; Mukherjee 1989; Haque and Vijayan 1993). It has a deep-chested body, with short legs and tail, and small close-set ears. Like the flat-headed cat, its front feet are partially webbed, and its claw tips protrude from their sheaths even when retracted, thus giving a signature track imprint (Sunquist and Sunquist 2002). It is a strong swimmer and can cover long distances under water (Roberts 1977). Fishing cats are strongly associated with wetland. They are typically found in swamps and marshy areas, oxbow lakes, reed beds, tidal creeks, and mangrove areas. Along watercourses they have been recorded at elevations up to 1525 m, but most records are from lowland areas (Nowell and Jackson 1996). Although fishing cats are widely distributed through a variety of habitat types across Asia (Map 36), their occurrence tends to be highly localized and is still not well known. For example, in 2005, a fishing cat was run over by a vehicle in central India, well outside the known range of the species (Anonymous 2005a). In Sundaland, it has been confirmed to occur only on the island of Java, and is possibly replaced by *P. planiceps* on Borneo, Sumatra, and peninsular Malaysia (Melisch *et al.* 1996). On the island of Sumatra, previously reported camera trap records (Kawanishi and Sunquist 2003) were actually of leopard cats (J. Sanderson, personal communication 2008), and there are no museum specimens (Melisch *et al.* 1996). There is also no confirmed evidence of presence in peninsular Malaysia (Melisch *et al.* 1996; Kawanishi and Sunquist 2003). Fishing cats, unlike most other

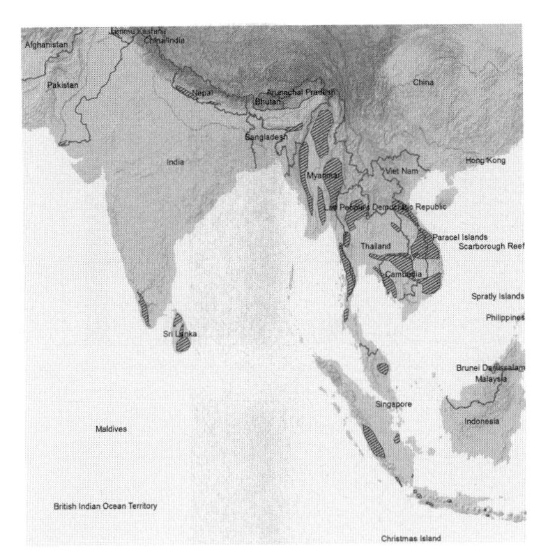

Map 36 Fishing cat. © IUCN Red List 2008.

small cats, may prey primarily on fish rather than small mammals. A 1-year study of scats in India's Keoladeo National Park found that fish comprised 76% of the diet, followed by birds (27%), insects (13%), and small rodents last (9%) (Haque and Vijayan 1993). Molluscs, reptiles, and amphibians are also taken (Mukherjee 1989; Haque and Vijayan 1993). However, they are capable of taking large mammal prey, including small chital fawns (Nowell and Jackson 1996; Sunquist and Sunquist 2002), and have been seen scavenging livestock carcasses and tiger kills (Nowell and Jackson 1996). The only radio-telemetry study took place in Nepal's Chitwan National Park in the early 1990s. Cats were active only at night and spent most of their time in dense tall and short grasslands, sometimes well away from water. Home ranges of three females were 4–6 km^2; that of a single male was larger at 16–22 km^2 (J. L. D. Smith, personal communication in Sunquist and Sunquist [2002]). Fishing cats have been observed in degraded habitats, such as near aquaculture ponds with little vegetation outside the Indian city of Calcutta (P. Sanyal, in Anonymous [2005a]). Locally common in some areas in eastern India and Bangladesh (Khan 2004), elsewhere fishing cats have become increasingly hard to find. The scarcity of recent fishing cat records suggests that over the past decade the species has undergone a serious and significant decline, throughout south-east Asia and in parts of India. This is largely attributed to wetland destruction and degradation, but indiscriminate trapping, snaring, and poisoning may also be to blame. Even in protected wetlands and former fishing cat study areas, researchers have been unable to document fishing cat presence (IUCN 2008). The fishing cat was upgraded from Vulnerable to Endangered on the 2008 IUCN Red List.

	Male Mean	Sample size	Female Mean	Sample size
Weight (kg)	16	$n = 1$	5.1–6.8	$n = 2$
Head/body length (mm)	660	$n = 1$	648–743	$n = 2$

Refs: Male, Haque and Vijayan (1993); Females, J. L. D. Smith (personal communication) in Sunquist and Sunquist (2002)

Study of felids

The value people place on wild animals will depend heavily on their knowledge of them and therefore science and the generation of information is itself a contribution to conservation (Macdonald *et al.*, Chapter 29, this volume). Furthermore, conservation action needs to be firmly underpinned by scientific data.

We searched, using all extant genus names as keywords, for felid papers in abstract databases (BIOSIS, CAB Abstracts, and Zoological records) for the years 1950–2008. The keywords searched came up with a total of 2110 published felid papers, 1811 of which mentioned a single felid species, 299 of which dealt with multiple felid species. Here, we use the number of published papers as a proxy for the degree of attention paid to the scientific studies of felids over time, for each species and within each topic within the field. This is surely a rough index of academic endeavour, and it does not include articles from the grey literature or papers to be found outside mainstream English language journals. As there are extensive bodies of knowledge published in other languages, such as Russian, Chinese, and Spanish, this may lead to some bias; nevertheless, some revealing trends emerge.

Overall, the rate of publication on wild felids has increased from a total of 41 in the 1950s to a total of 115 in the last decade, reflecting the growth of conservation biology and natural sciences in academia. The past 60 years has seen an exponential growth in publications on felid papers in ecology, behaviour, and conservation (Fig. 1.9a and b). More recent trends have seen growth in cutting-edge fields, such as genetics, as new tools and technologies have become available. Of these publications, 74% of papers dealing with human–felid conflict have been published in the past decade, making this the fastest-increasing area of research, perhaps due to a dawning realization that increasing human pressures on the natural world and the resultant conflict presents threats and dangers to wild predators.

While academic study may not necessarily translate to conservation action, it is comforting to note the massive increase in study of wild felids. Nevertheless, there are still lacunae in knowledge. Large species tend to receive more attention (Fig. 1.10a–d). This may be a reflection of their charisma, greater ease of observation, and the greater potential to attract funding to undertake studies on species that are well known and exciting to a wider public. Efforts to foster public knowledge and interest in the smaller, more cryptic, species may correct this imbalance. Equally, it should be acknowledged that there is a strong justification for the attention paid to the large species because of their potential as flagship and umbrella species for protecting natural ecosystems.

Felids in forested habitats tend to be less well studied. This may speak to the logistical difficulties of working and making observations of felid species in closed habitats. The small cats of Asia and South America are relatively poorly known and this is a major gap in knowledge, as close to 45% of extant felids are small (<10 kg) and occur in these two geographic regions. The apparent lack of attention paid to South American felids (Fig. 1.10c) may, in part, reflect a language bias in publications searched. Despite having smaller felid assemblages, greater numbers of field-based studies on felids were found for North America and Europe (Fig 1.10d), reflecting perhaps a longer history of formal publication in

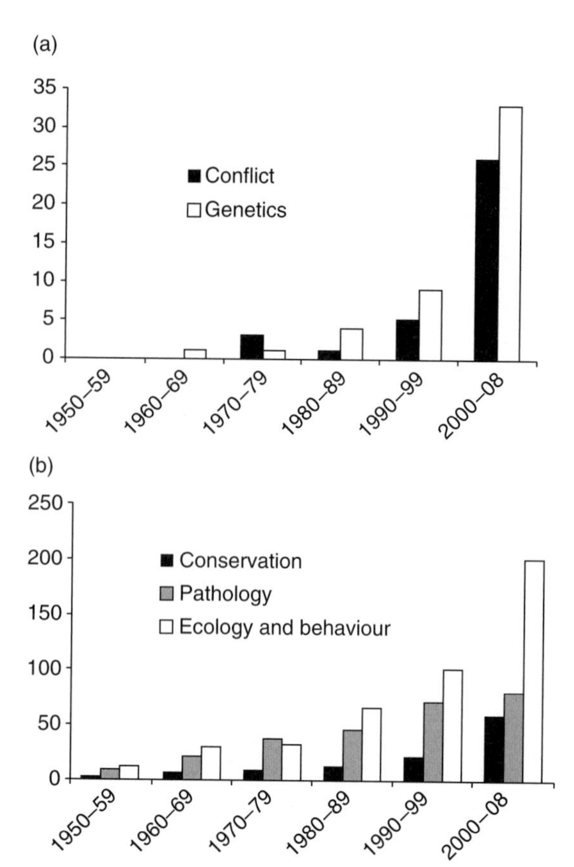

Figure 1.9a and b (a) graph showing increasing number of publications on felids in the fields of genetics and in human–wildlife conflict: both fields have received increasing attention in the last decade; (b) felid publications in conservation, pathology, and ecology/behaviour.

the natural sciences and greater resources and funding available for conservation and educational facilities and institutions.

Research effort varies not only geographically, but also in terms of species rarity. To analyse to what extent study effort has prioritized the most threatened species, we used the digital library maintained by the IUCN SSC Cat Specialist Group (www.catsglib.org), which in March 2008 contained 6123 documents pertaining to cats, including peer-reviewed journal articles, an extensive grey literature, as well as non-English materials. However, articles published in the group's biannual newsletter *Cat News* are not included in the database, and so research effort is still under-represented. We looked at

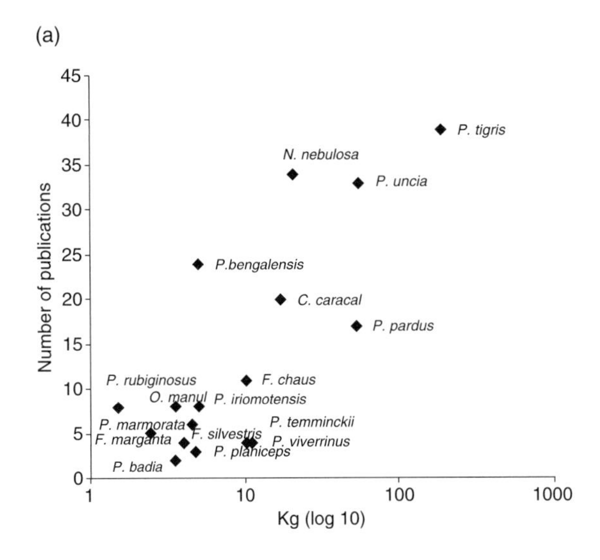

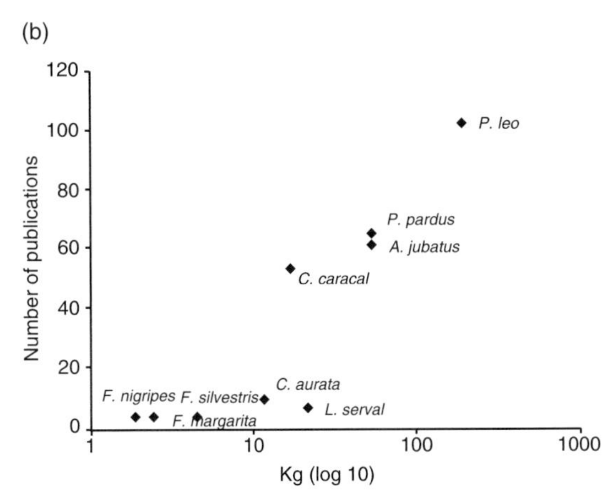

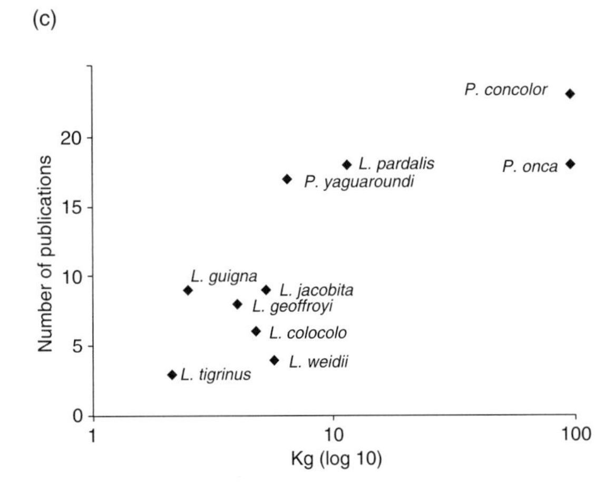

the number of references per species according to their classification on the IUCN Red List of Threatened Species (IUCN 2008).

As previously noted, big cats received predominant emphasis in research effort, with seven large species having twice the number of citations (4065) than the 29 smaller species (2051). However, big cats are relatively more threatened, with four (57%) listed in the top three threat categories, as compared to 38% of the smaller species. Overall, the top three categories of threat (Critically Endangered, Endangered, and Vulnerable) did have a slightly larger number of citations than the two lower threat categories (3192 vs. 2974), despite having fewer species (16 as compared to 20). However, this is largely due to the threatened status of the better-studied big cats. Threatened small cats have received very little research attention, particularly the small cats of south-east Asia. Figure 1.11 shows the uneven nature of study for the top two threat categories; the tiger has received the 'lion's share' of attention, reflecting not only its power of attraction for scientists, but also the extreme challenges its crowded Asian environment poses for conservation of this large predator. While research emphasis on big cats should aid their

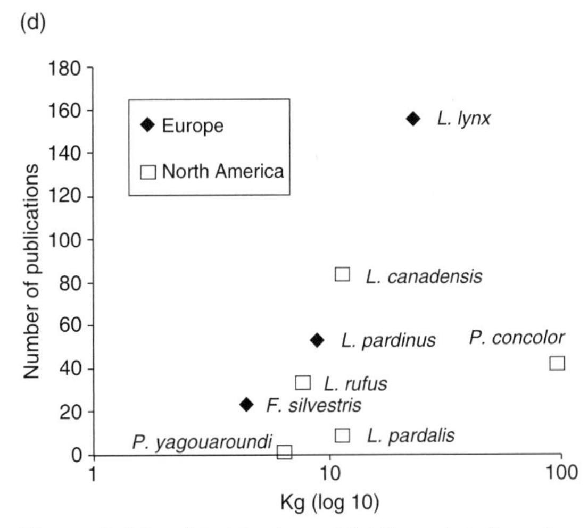

Figure 1.10a–d Publications of field-based studies of felids in (a) Asia; (b) Africa; (c) South America; and (d) North America and Europe, demonstrating that the larger species tend to recieve the most attention, while small-bodied species, particularly those favouring dense habitats tend to recieve less attention.

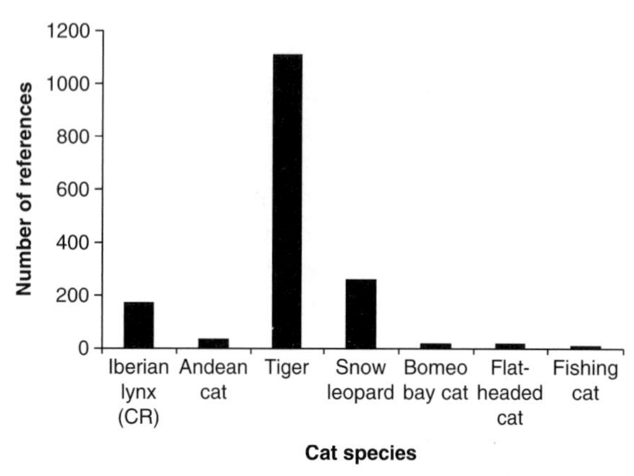

Figure 1.11 Number of references in the Cat Specialist Group library for Critically Endangered (CR) and Endangered cats. The tiger has received a disproportionate focus, although appropriate given its conservation challenges, while research effort for endangered small cats has lagged behind.

conservation, more work on small cats is needed, particularly the rarer species.

As Karanth *et al.* (Chapter 7, this volume) point out, the growth and increasing availability of new technologies and methodologies is likely to open up new and exciting areas of research, some of which will help to fill these gaps. Camera traps and technologies that allow remote recording of the presence and behaviour of cryptic species will illuminate the lives of elusive cats. Advances in genetics are likely to offer insights into the relatedness between species, populations, and individuals.

Finally, there is wide recognition that conservation of predatory species needs to be reconciled with human needs. Wild felids stand out as a family of species that engages our imaginations through their charisma, beauty, and wildness. However, these obligate carnivores also conflict strongly with human needs and activities, engendering equally passionate dislikes by those whose livelihoods they impact. It is crucial that these conflicts are addressed through education, that conservation is wisely planned, and that any utilization is well managed. This requires a broad grasp of relevant knowledge (the subject of the following nine reviews), and the deep insight of case studies (of which we present 18 exceptional examples). It will also require a radical approach to conservation, new heights in interdisciplinarity, and ingenious market mechanisms, and these are some of the strands presented in our concluding synthesis (Macdonald *et al.*, Chapter 29, this volume). Just as their precarious position aloft the food pyramid makes felids important umbrella species for conservation—and miners' canaries for biodiversity as a whole—their beauty, cultural significance, and charisma make them potent standard-bearers for nature on the world stage. It is therefore sobering to remember that while it takes whole communities, locally and internationally, to conserve a population of wild felids, it takes just one man with some poison to obliterate them. Therefore, the broad perspective of knowledge-based conservation, integrated with community development and an exhilarating hope for the future, will wither on the vine if local conflicts between people and felids are not resolved. The insights in this book provide the foundation for doing so, and make clear why it will be worth the effort.

Acknowledgements

D.W. Macdonald and A.J. Loveridge warmly acknowledge the enormous effort undertaken almost entirely by Kristin Nowell to compile the species vignettes included in this chapter. She in turn commends and depends upon the contributions of many experts associated with the IUCN SSC Cat Specialist Group and to the IUCN Red List of Threatened Species, which provided maps (coordinated by Jan Schipper and Mike Hoffmann) and inspiration for this chapter. We are also very grateful to Dr Ros Shaw for her meticulous work on the patterns of publication we report here, and also for her unwavering dedication to checking the proofs of this and all the other chapters in this book.

CHAPTER 2

Phylogeny and evolution of cats (Felidae)

Lars Werdelin, Nobuyuki Yamaguchi, Warren E. Johnson, and Stephen J. O'Brien

Artist's reconstruction of the sabre-toothed cat *Megantereon cultridens* stalking its prey. (Illustration courtesy of Mauricio Antón.)

Introduction

Cats, wild as well as domestic, fossil as well as living, are familiar to people around the world. The family Felidae has a worldwide distribution and has been associated with humans in various ways throughout history (Quammen 2004). Their functional morphology, ecology, and behaviour have been the subject of intense scrutiny by scientists for over 200 years. The fossil record of cats is extensive and some of its members are among the most recognizable of extinct animals. Despite all this, the phylogeny and evolution of the family Felidae, and even the content of the family, have remained poorly understood. In this review, we will first present the current state of knowledge with regard to the interrelationships of living Felidae and the timing of the radiation of modern cats. We will also present the fossil record of Felidae in broad outline, focusing first on describing the different groups of species and their characteristics, and then discussing the general patterns of cat evolution that we can deduce from current data. Provided with this overview, we will attempt to identify those areas most in need of further research in order to achieve the aim of a fuller understanding of felid evolution, especially that of the living felids and their ecological and functional relationship to the extinct sabre-toothed felids.

In this discussion, we will synthesize the available data, distinguishing as far as possible monophyletic groups of taxa, suggesting the most likely interrelationships of the fossil lineages, but also pointing out that there are many problem areas that need to be resolved. This section should be viewed as a challenge to investigators to use old data or discover new data to corroborate or refute the scenarios proposed herein. We end the paper with a small section demonstrating some evolutionary patterns among extant Felidae, suggesting that there is much to be gained from the deeper analysis of the current phylogenetic information.

Felid morphology is described and discussed elsewhere (Kitchener *et al.*, Chapter 3, this volume) and will not be reiterated here except as needed. Teeth of the upper jaw are referred to in upper-case letters (I, C, P, and M) and teeth of the lower jaw in lower-case letters (i, c, p, and m), followed by the appropriate number in the sequence. Character mapping on cladograms was carried out with Mesquite, version 1.12 (Maddison and Maddison 2004). Stratigraphic ages of taxa as given in the text and figures were obtained from either primary literature or (for North America) the Paleobiology Database (www.paleodb.org) and (for Eurasia) the NOW database (www.helsinki.fi/science/now/database.html).

Phylogeny

Many attempts have been made to investigate the interrelationships of Felidae. These have followed two broad approaches. Some, like Matthew (1910), Kretzoi (1929a, b) and Beaumont (1978) have incorporated both fossil and extant felids in their analyses, while others, such as Pocock (1917a), Herrington (1986), and Salles (1992) have focused exclusively on the living members of the family. A new era in felid phylogenetics was ushered in with the introduction of molecular evidence (Collier and O'Brien 1985; O'Brien *et al.* 1985a; Johnson *et al.* 1996), while the first study to use a total evidence approach was that of Mattern and McLennan (2000).

All of these approaches have had their problems. In the case of fossil studies, confounding factors have included the relatively poor fossil record, the problem of finding useful characters in fragmentary material and the convergence between Nimravidae and Felidae. Though previously included in the Felidae (Matthew 1910; Piveteau 1961), the former, Nimravidae, is now known to be diphyletic. Its Paleogene (65.5–23.0 million years ago [Ma]; Gradstein *et al.* 2004) members form a basal clade within either Feliformia or Carnivora as a whole (Neff 1983; Hunt 1987; Morlo *et al.* 2004), while its Neogene (23.0 Ma—recent) members are placed in a separate family, Barbourofelidae, with affinities to Felidae (see below). Morphological studies of extant felids have been hampered by the very uniform morphology of the members of the family, making it difficult to find and polarize characters for phylogenetic analysis. Molecular studies, on the other hand, have been particularly hampered by the apparently short timespan during which the clades of modern felids evolved. Thus, clades of closely related taxa have been identified but the interrelationships of these clades have been difficult to pinpoint.

Recently, two of us (Warren E. Johnson and Stephen J. O'Brien) published a phylogeny of Felidae based on a data set of 22,789 base pairs of DNA, including autosomal, Y-linked, X-linked, and mitochondrial gene segments (Johnson *et al.* 2006b). The results of this study, while not immutable, provide a firm basis for understanding the interrelationships and evolution of the extant Felidae. The results confirm some prior results, both molecular and morphological, while providing new insights and surprises.

The study distinguishes eight clades of extant felids (Fig. 2.1). The first of these to split off from the stem lineage is the *Panthera* lineage (genera *Neofelis* and *Panthera*) at *c.* 10.8 Ma (Fig. 2.1, node A). Most previous studies of felid phylogeny have placed *Panthera* as the crown group, but a few (Turner and Antón 1997; Mattern and McLennan 2000) also have the *Panthera* lineage as basal to other cats. Within this

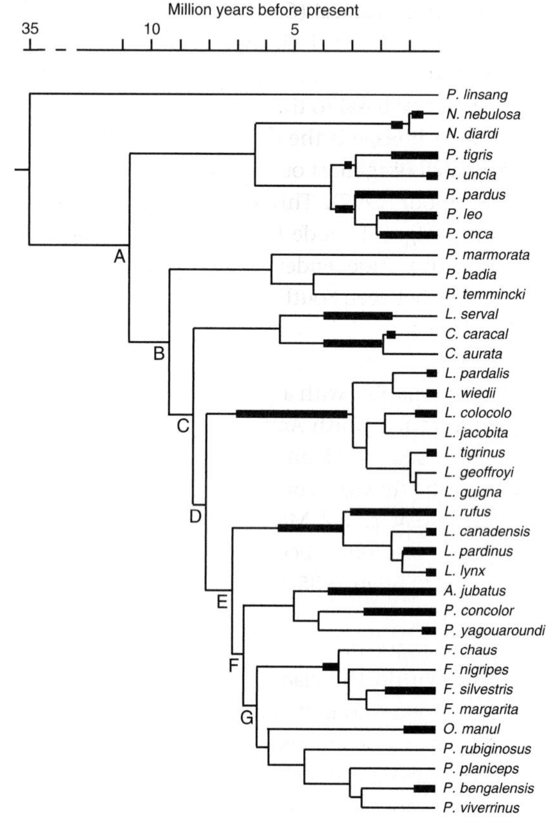

Figure 2.1 The phylogeny of the extant Felidae. Thick lines indicate the presence of a fossil record, thin lines indicate the absence of a fossil record. Node labels as in the main text. (Based on the work of Johnson *et al.* 2006b.)

lineage, the clouded leopard, *Neofelis*, with the two species *N. nebulosa* and *N. diardi* (Buckley-Beason *et al.* 2006; Kitchener *et al.* 2006) is placed basally, as would be expected from its distinctive morphology implying a long separate evolutionary lineage (Christiansen 2006), with the rest of the pantherines radiating within the last 4 million years.

The next clade to branch off, at *c.* 9.4 Ma (Fig. 2.1, node B), is the bay cat lineage (genus *Pardofelis*). This clade consists of the poorly known bay cat (*P. badia*), Asian golden cat (*P. temminckii*), and marbled cat (*P. marmorata*). The last mentioned species has been linked to the *Panthera* lineage (e.g. Herrington 1986) and this is reflected in its position here, as basal member of the clade branching off closest to the *Panthera* lineage.

The third lineage is the *Caracal* lineage, with two genera, *Caracal* and *Leptailurus*, incorporating three African species: caracal (*Caracal*), African golden cat (*C. aurata*), and serval (*S. leptail urus serval*). This lineage branches off at *c.* 8.5 Ma (Fig. 2.1, node C), with the serval basal to the other two species.

The next lineage is the ocelot lineage (genus *Leopardus*), including most of the South American small cats (Seymour 1999). This lineage branches off at *c.* 8.0 Ma (Fig. 2.1, node D). The beginning of this lineage is thus independent of the formation of the land bridge between South and North America about 3 Ma (Marshall *et al.* 1982). However, the radiation of the extant species within this lineage shows dates that are compatible with a single origin of the extant radiation from a North American ancestor, as previously proposed (Werdelin 1989).

The fifth lineage comprises the genus *Lynx*, splitting off at *c.* 7.2 Ma (Fig. 2.1, node E). This lineage has also often been linked to *Panthera* (e.g. Collier and O'Brien 1985; Salles 1992), but the recent more robust study by Johnson *et al.* (2006b) indicates that the relationship is more distant than previously thought. Within the clade, *L. rufus* is basal as has generally been thought, but *L. canadensis* and *L. lynx* are not reconstructed as sister taxa, unlike in previous analyses (Werdelin 1981).

The next lineage is the *Puma* lineage, including the genera *Puma* and *Acinonyx* which split off at *c.* 6.7 Ma (Fig. 2.1, node F). This lineage has previously been recognized in both morphological (Herrington 1986; Van Valkenburgh *et al.* 1990) and molecular (Johnson and O'Brien 1997) studies. It is worth noting that the puma and jaguarundi probably split before the Great American Biotic Interchange that followed the formation of the land bridge between South and North America (Marshall *et al.* 1982), and thus both are of North American origin.

The seventh and eighth lineages are the small cats of the Old World—the leopard cat and domestic cat lineages. They split from each other at *c.* 6.2 Ma (Fig. 2.1, node G). The former includes the genera *Otocolobus* and *Prionailurus* and the latter the genus *Felis*. The splits within the former are much deeper than within the latter, suggesting that the genus *Felis* may be oversplit. This is also the conclusion of Driscoll *et al.* (2007), who distinguish only four species in *Felis: F. chaus, F. nigripes, F. margarita*, and *F. silvestris*. The last mentioned species now also includes *F. ornata, F. bieti*, and *F. lybica*, making it one of the most widespread small cat species.

Most of the nodes in this phylogeny are robustly supported (Johnson *et al.* 2006b). A few, however, are still unstable, showing either low support or incongruence between different analyses and data sets. These as yet incompletely resolved nodes are: the relative positions of *Panthera leo, P. pardus*, and *P. onca*, as well as the relative positions of *P. tigris* and *P. uncia* within this clade; the position of *L. jacobita*; the position of *O. manul*; the position of *F. nigripes*; and the clade uniting *Felis* and *Prionailurus/Otocolobus* to the exclusion of *Puma/Acinonyx*.

The most notable fact about this phylogeny of extant cats lies in the short time intervals between the splits of the eight lineages. The radiation of lineages along the entire stem of the felid clade occurs within the Late Miocene (over a period of *c.* 6.3 Ma) and such a short space of time suggests the occurrence of some sort of functional or ecological release, but what that may be is at present unknown. We shall return to the fossil record of extant cats below.

The fossil record

According to available molecular data, the Felidae originated some time at or just after the end of the Eocene (Gaubert and Véron 2003). This accords well with the fossil record. The earliest forms placed in the felid lineage, *Proailurus* and possibly *Stenogale* and *Haplogale* (Hunt 1998; Peigné 1999), occur after the 'Grande Coupure' marking the Eocene/Oligocene boundary (*c.* 33.9 Ma; Gradstein *et al.* 2004). In the Mammals Paleogene (MP) level system of Paleogene terrestrial mammal stratigraphy in Europe, this boundary is placed between MP 20 and MP 21 (Schmidt-Kittler 1990). In the fissure fillings of the Quercy region, France, where most of our knowledge of early European carnivorans originates, feliforms are not known before MP 21 (Hunt 1998). Owing to the scarcity of their remains, modern excavations have yet to establish the first occurrence of the Felidae. What we know, however, suggests that some older known finds may be from the Early Oligocene, that is, before 28.4 Ma (Gradstein *et al.* 2004). Thus, the earliest felids appeared sometime between *c.* 35 Ma (age of the sister group) and 28.5 Ma (minimum age of the earliest fossils).

It is well established on morphological grounds, basicranial as well as dental, that *Proailurus*, known from the Quercy fissure fills, but also from excellent material from the Early Miocene site of Saint-Gérand-le-Puy, France, (MN 2 in the Neogene mammal zonation of Europe; 22.8–20 Ma) is a felid. Despite this, the morphological path leading to the felid condition is not well delineated. Hunt (1998) discusses changes to the auditory bulla seen in a variety of early feliforms, including *Haplogale* and *Stenogale*, and leading to the bulla of *Proailurus*. However, the placement of Asiatic linsangs (genus *Prionodon*) as the sister group to Felidae on molecular grounds by Gaubert and Véron (2003), instead of with the Viverridae, in which they have traditionally been placed, adds complexity to the story. Hunt (2001) placed *Prionodon* in a clade with 'true' viverrids, for example *Genetta*, on the basis of basicranial anatomy (but without consideration of other features). What this conflict between separate data sets consisting of non-overlapping characters means for our understanding of the fossil record of the precursors of Felidae and for the origins of the family has yet to be established.

Early felids

As noted, the earliest well-established felid is *Proailurus* (Figs. 2.2, letter A; 2.3, and 2.4). Peigné (1999) provides a discussion of the evolution of this species and its relationship to other early putative felids.

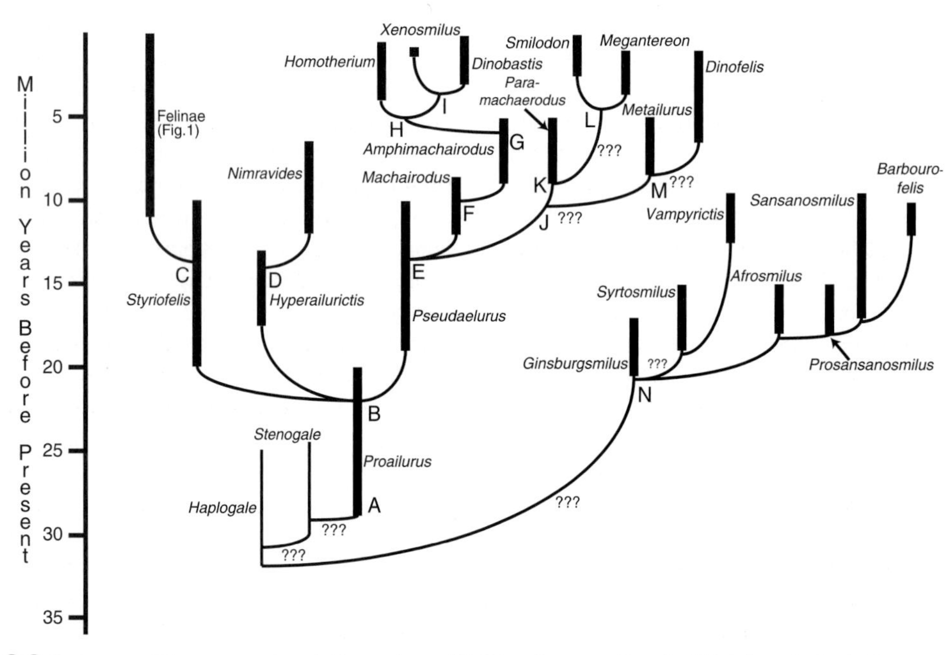

Figure 2.2 Summary of the proposed evolutionary tree of Felidae discussed herein. Thick lines indicate the presence of a fossil record, thin lines indicate the absence of a fossil record. Labels as in the main text and Table 2.1.

Table 2.1 The internal nodes of Fig. 2.2: content, place of origin, and age.

Letter	Content	Continent	Age, Ma (approximate)
A	Felidae *sensu stricto*	Europe	27
B	Pseudaelurine radiation and later felids	Eurasia	22
C	Felinae (radiation of extant felids)	Eurasia	14–13
D	*Nimravides*	North America	14
E	Machairodontinae	Europe	14–13
F	*Amphimachairodus* lineage	Eurasia	10
G	Homotheriini	Eurasia	6
H	Derived Homotheriini	Africa	5
I	North American Homotheriini	North America	4–3
J	*Paramachaerodus* lineage	Eurasia	11–10
K	*Paramachaerodus* and derivates	Eurasia	9
L	Smilodontini	Eurasia, Africa, and North America	5–4
M	Metailurini	Eurasia	9–8
N	Barbourofelidae	Eurasia and Africa	32

Proailurus (with three species, *P. lemanensis, P. bourbonnensis*, and *P. major*) is a medium-sized cat about the size of a bobcat, *L. rufus*. Dentally, it differs from living cats in the (variable) presence of p1, p2, m2, and P1, as well as the presence of a small metaconid and talonid on m1. Overall, the dentition is thus very similar to that of living felids, but includes some elements that have been fully reduced in the modern clade. Further, the auditory bulla of *Proailurus* has a ventral process of the petrosal promontorium (Hunt 1989, 1998). This process is lost in living felids. When it was lost in felid evolution has yet to be established, but it serves to distinguish at least the modern clade from the basally situated *Proailurus*. The geologically youngest *Proailurus* is from Laugnac, France, biostratigraphically placed in MN 2b (>20 Ma). In *Proailurus* we have (as far as it is known) an essentially modern felid except for a few minor details of the dentition, auditory bulla, and postcranium, which has shorter limbs than modern felids. Coupled with the molecular date for the divergence of *Prionodon* and Felidae, this suggests that there must have been a stem lineage of perhaps 5 Ma in the Early Oligocene leading up to the full felid morphology. *Haplogale* and *Stenogale* are likely to be members of that lineage (Hunt 1998; Peigné 1999), but the details of the process have not been worked out.

Proailurus is not known with certainty outside Europe. Hunt (1998) reports the presence of *Proailurus* sp. from the Hsanda Gol Formation, Mongolia. However, Peigné (1999) concludes, in our opinion correctly, that this specimen is better assigned to the Barbourofelidae. On the other hand, Hunt (1998) also describes the skull of a *Proailurus*-grade felid from the Ginn Quarry, Nebraska (Late Hemingfordian, *c*. 17–16.5 Ma). According to Hunt the basicranial structure of the Ginn Quarry felid is more plesiomorphic than that of European *Proailurus*. This suggests that phylogenetic diversification in Felidae had begun already in the Early Miocene and that North American '*Pseudaelurus*' (see below) may have evolved from a *Proailurus*-grade ancestor rather than from a migration of early *Pseudaelurus* into North America. If so, felids may have migrated into North America as early as the beginning of the Hemingfordian (*c*. 19 Ma), along with a number of other carnivoran taxa (Qiu 2003).

The next felids to evolve belong to the *Pseudaelurus* complex (Fig. 2.2, letter B; Fig. 2.5). This is a group of

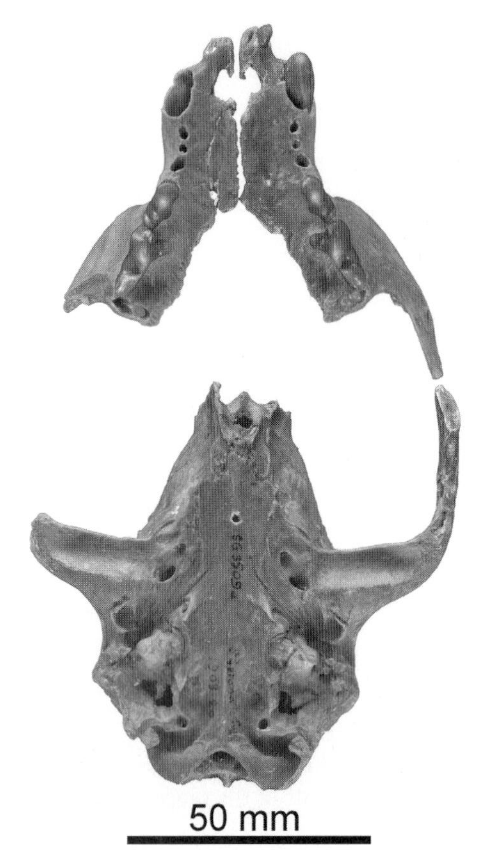

Figure 2.3 The skull of *Proailurus lemanensis*, MNHN SG 3509 (holotype) from Saint-Gérand-le-Puy, France, in ventral view. The anterior and posterior halves do not meet. (Photo courtesy of Stephane Peigné.)

species with representatives in Europe, Arabia, Asia, and North America. The interrelationships of the species included in *Pseudaelurus* and the relationship of this genus (or genera) to the radiations of the subfamilies Felinae (conical-toothed cats) and Machairodontinae (sabretooths) are a major challenge to felid palaeontology. *Pseudaelurus* is clearly a grade rather than a monophyletic clade, and this complex includes the ancestors of all subsequent felids. A number of generic names are available for parts of this complex, including *Styriofelis, Hyperailurictis, Miopanthera, Schizailurus*, and *Pseudaelurus* itself. We will consider the validity and applicability of these in the discussion below. A fuller knowledge of the interrelationships within this group would go a long way towards an understanding of the evolutionary patterns of the Felidae.

Pseudaelurus is first recorded from Wintershof-West in Germany (MN 3, 20–18 Ma; Dehm 1950). Hence, it does not overlap stratigraphically with *Proailurus* in Europe. Several reviews of *Pseudaelurus* have been published in the past decades (Heizmann 1973; Ginsburg 1983; Rothwell 2003) and we refer to them for a fuller discussion of evolutionary details.

Four species of *Pseudaelurus* are known from Europe. In the order of increasing size they are: *P. turnauensis (= P. transitorius), P. lorteti, P. romieviensis*, and *P. quadridentatus* (type species of the genus). They range in size from a modern wildcat to a lynx or small puma. Differences between them, apart from

Figure 2.4 Artist's reconstruction of *Proailurus lemanensis*, the first cat. (Illustration courtesy of Mauricio Antón.)

Figure 2.5 Artist's reconstruction of *Styriofelis lorteti*, a member of the stem lineage leading to the extant Felidae, together with the flying squirrel *Petaurista* sp. (Illustration courtesy of Mauricio Antón.)

size, are minute (Heizmann 1973). The first species to appear is the smallest, *P. turnauensis* (Dehm 1950). However, all three remaining species appear in MN 4 (18–17 Ma). This indicates a rapid radiation of the *Pseudaelurus* grade, suggesting a monophyletic origin of at least European *Pseudaelurus* from a single species of *Proailurus*. *P. lorteti* and *P. romieviensis* become extinct at the end of the Middle Miocene (*c.* 11.6 Ma), but *P. quadridentatus* and *P. turnauensis* survive into the Late Miocene (MN 9, *c.* 11.2–9.5 Ma). They thus overlap stratigraphically with the earliest documented Machairodontinae (*Miomachairodus pseudailuroides* from Turkey; Schmidt-Kittler 1976; Viranta and Werdelin 2003) (Fig. 2.2, letter E).

Pseudaelurus is poorly known from Asia, possibly due to a relative dearth of Middle Miocene localities on the continent. Two Chinese species are known. Cao *et al.* (1990) describe *P. guangheensis* from Gansu and Wang *et al.* (1998) describe *P. cuspidatus* from Xinjiang. In addition, Qiu and Gu (1996) describe material referred to *P. lorteti*. All this material is Middle Miocene in age. What the relationship is between the Chinese and European species has not been determined, nor has their relationship to the North American radiation of the grade.

The fossil record of *Pseudaelurus* in North America was recently reviewed by Rothwell (2003). There are five valid species: *P. validus* (stratigraphic range *c.* 17.5–16.5 Ma), *P. skinneri* (*c.* 17.5–17.1 Ma), *P. intrepidus* (*c.* 17.1–13.3 Ma), *P. stouti* (*c.* 15.2–12.7 Ma), and *P. marshi* (*c.* 16.4–12.7 Ma). Thus, *Pseudaelurus* appears later in North America and goes extinct sooner there than in Europe. This, and the cladistic analysis of Rothwell (2003), in which the three younger species (*P. intrepidus, P. stouti*, and *P. marshi*) form a clade with the two older species (*P. validus* and *P. skinneri*) as outgroups, are consistent with a single origin for North American *Pseudaelurus*.

Finally, a single record of *P. turnauensis* has been reported from Saudi Arabia (Thomas *et al.* 1982) in deposits now considered to be of MN 5 age (17.0–15.2 Ma). Material from Africa previously referred to *P. africanus* (Andrews 1914) is now referred to *Afrosmilus*, a barbourofelid (see Morales *et al.* [2001] and see below).

The endemic North American genus *Nimravides* undoubtedly originated from one of the above-mentioned North American species of *Pseudaelurus* (Baskin 1981; Beaumont 1990), probably *P. intrepidus* or *P. marshi*, which both have a prominent chin, also seen in *Nimravides* (Fig. 2.2, letter D). *Nimravides* differs from its putative ancestors only in relatively minor features: it has a more prominent chin, more elongated, serrated canines, a more reduced P4 protocone, and more developed P4 ectoparastyle. These are all features pointing towards a sabre-toothed morphology, not dissimilar to that seen in *M. pseudailuroides* and *Machairodus aphanistus* (see below), but evolved in parallel. Four species of *Nimravides* are known: *N. thinobates* (*c.* 11.0–9.6 Ma), *N. pedionomus* (*c.* 12.0–11.5 Ma), *N. hibbardi* (*c.* 7.0–6.4 Ma), and *N. galiani* (*c.* 11.6–10.7 Ma). Near the end of the Miocene, *Nimravides* became extinct, apparently without leaving descendant lineages. A North American felid of uncertain affinities that may possibly belong here is *Pratifelis martini* from the Late Miocene (*c.* 7–6 Ma) of Kansas (Hibbard 1934). This species has a distinctively enlarged m1 talonid and does not fit comfortably into any of the larger felid lineages.

Sabretooths

The further evolution of Felidae beyond the *Pseudaelurus* grade begins with *M. pseudailuroides* (Fig. 2.2, letter E). This taxon, which is at present known only from Turkey (Schmidt-Kittler 1976; Viranta and Werdelin 2003), has cheek teeth that are very similar to those of *P. quadridentatus*, but the upper canines are more flattened and have small crenulations on the mesial and distal faces that are not present in *Pseudaelurus* spp. (Schmidt-Kittler 1976, figs. 114a, 1c, 2, and 3, plate 5). In an important contribution, Schmidt-Kittler (1976) discusses the relationship between *M. pseudailuroides* and the *Pseudaelurus*-grade and how the morphological transition may have occurred. However, he does not pinpoint any specific relationships between taxa, nor does he extend his discussion to conical-toothed cats. *M. pseudailuroides* is at present known only from MN 7/8 and MN 9 (*c.* 12.5–9.5 Ma). The taxonomic status of the species and genus has been discussed several times. Beaumont (1978) made *Miomachairodus* a subgenus of *Machairodus*, and included *Machairodus robinsoni* from the early Late Miocene (*c.* MN 9) of Tunisia (Kurtén 1976) in the subgenus. On the other hand, Ginsburg *et al.* (1981) synonymized *M. pseudailuroides* with *M. aphanistus*, type species of the genus *Machairodus*. Morlo (1997) followed this, but suggested that *M. robinsoni* in that case be considered a separate genus. This discussion is far from settled, but at the very least shows that these forms grade into one another. Another early form about which there is taxonomic disagreement is *M. alberdiae* from MN 9 of Spain. Ginsburg (1999) considers this to be the most primitive *Machairodus*, but Morlo (1997) synonymizes it with *M. aphanistus*.

M. aphanistus was described by Kaup (1833) and was the first Miocene felid to be named. Its craniodental morphology was recently reviewed in detail (Antón *et al.* 2004). These authors found that the functional morphology of the killing bite in *M. aphanistus*, and characters related to this behaviour, were considerably more primitive than in later machairodonts from the Eurasian Late Miocene. They concluded that *Machairodus* should be restricted in content to Vallesian (*c.* 11.2–9.0 Ma) forms, while Turolian (*c.* 9.0–5.3 Ma) forms should be referred to *Amphimachairodus* (Fig. 2.2, letter F). Morlo and Semenov (2004) objected to this procedure, arguing that the evolution from *Machairodus* to *Amphimachairodus* was gradual and mosaic and that the two could not be generically distinct. However, making the distinction is taxonomically useful and in line with a trend in recent years of trying to restrict the usage of *Machairodus* to something other than a waste-basket taxon for any or all Miocene sabretooths (Beaumont 1978; Ginsburg *et al.* 1981; Ginsburg 1999).

Some time in the Vallesian, *Machairodus* probably migrated to North America, where it gave rise to *M. coloradensis* (*c.* 9.0–5.3 Ma). This is a fairly generalized species, similar to *M. aphanistus*. It is possible, if unlikely, that it evolved from the North American *Nimravides*. This would require extensive parallelism with *Machairodus*. The possibility has been noted before, however, and the generic name *Heterofelis* (Cook 1922) is available for this taxon.

The next stage in the evolution of the machairodont lineage is the genus *Amphimachairodus* (Fig. 2.2, letter G). This genus includes a number of closely related species that morphologically lead up to the Plio-Pleistocene tribe Homotheriini (Fig. 2.2, letter H), which includes the genera *Homotherium, Dinobastis*, and *Xenosmilus*. *Amphimachairodus* includes the species *A. giganteus* (Eurasia; *c.* 9–5.3 Ma), *A. kurteni* (Kazakhstan; *c.* 7.1–5.3 Ma), *A. kabir* (Chad and Libya; *c.* 7–5.5 Ma), and possibly *A. irtyschensis* (Russia; *c.* 7.1–5.3 Ma), though the latter may be a synonym of *A. giganteus*. Closely related is also *Lokotunjailurus emageritus* (Werdelin 2003b; *c.* 7.4–5.5 Ma), which lacks a number of the derived cranial features of *Amphimachairodus*, but is dentally the most derived of the group. *A. giganteus* is, as the name implies, characterized by very large size, extremely long upper canines and a derived mastoid region relative to that of *Machairodus*, implying modifications to the killing bite. The mastoid region is further evolved in *M. kurteni* and *M. kabir*, but has not yet reached the condition seen in *Homotherium*. Dentally, the upper incisor arcade is modified and the cheek dentition progressively simplified, with reduction of p3/P3, complete loss of the m1 talonid, and nearly complete loss of the P4 protocone. The dentition of *L. emageritus* is very close to that of primitive *Homotherium*, but the skull and skeleton of the former preclude it from the direct ancestry of that genus (Werdelin 2003b). *L. emageritus* has an extremely enlarged dew claw (absolutely and relative to the other claws) on the manus and this feature appears to be present also in *Homotherium* (Ballesio 1963).

The evolution of *Machairodus* and *Amphimachairodus* is paralleled in the sabretooth group by the evolution of the genus *Paramachaerodus* (Fig. 2.2, letters J and K). At least two and possibly as many as four species of this genus are known: *P. ogygius* (*c.* 9–7 Ma), *P. orientalis* (*c.* 8–6 Ma), *P. indicus* (age uncertain), and *P. maximiliani* (*c.* 7–5.3 Ma) (Salesa *et al.* 2003). The latter two may be synonymous, with each other and with *P. orientalis*. *Paramachaerodus* is much smaller than *Machairodus* and (especially) *Amphimachairodus* (*Paramachaerodus* is leopard, rather than lion-sized or larger in the case of *Amphimachairodus*). Clearly, this genus and its larger relatives were dividing up the prey-spectrum by size, though the details of this are not yet understood. New material from the early Late Miocene of Spain is doing much to clarify the taxonomic, functional, and ecological relationships between these Miocene sabretooths (Antón *et al.* 2004; Salesa *et al.* 2005).

A further lineage that is likely to at least in part belong among the sabretooths, despite lacking the typical craniodental attributes of this functional grade, is the tribe Metailurini (Fig. 2.6). This tribe as generally conceived includes the larger genus *Dinofelis* (Fig. 2.6a), with at least ten species (Werdelin and Lewis 2001), *Metailurus* (Fig. 2.6b), with at least four species, and *Stenailurus*, with one species (though the latter may be a synonym of *Metailurus*). *Dinofelis* is in many ways convergent on *Panthera*, but its evolution is not straightforward convergence. Instead, various species of *Dinofelis* are more or less pantherine-like, while the oldest and youngest species are the most sabretooth-like. The Metailurini is essentially a waste-basket for taxa that show some sabretooth features but can not be placed in either the *Machairodus* or the *Paramachaerodus* lineages. It is not clear that *Dinofelis* and *Metailurus* are closely related, nor what their respective antecedents are. Nor is it clear, although it seems likely, that *Metailurus* is a member of the subfamily

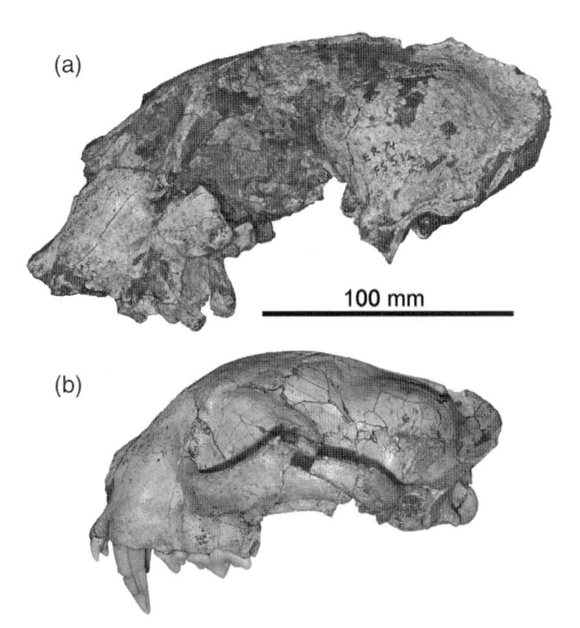

Figure 2.6 (a) Skull of *Dinofelis petteri*, KNM ER 2612 (holotype), Tulu Bor member, Koobi Fora Formation, Kenya; in left lateral view. (b) Skull of *Metailurus parvulus* PIU M3835, Locality 108, Baode Province, China; in left lateral view.

Machairodontinae (sabretooth cats). *Dinofelis*, however, shares several traits with derived sabretooths and can confidently be placed in this subfamily (Werdelin and Lewis 2001). Both of these genera originate in the Miocene and survive into the Plio-Pleistocene; *Metailurus* is mainly a Miocene genus, while *Dinofelis* has its main radiation in the Pliocene.

The Plio-Pleistocene sees the appearance of the two derived sabretooth tribes, Homotheriini and Smilodontini (Fig. 2.2, letters H and L; Fig. 2.7). The Homotheriini includes the genera *Dinobastis* (with at least one species, *D. serus*) and *Xenosmilus* (with one species, *X. hodsonae*) from North America and *Homotherium* (Fig. 2.7b) (with several species, including *H. crenatidens* and *H. problematicum*) from Eurasia and Africa. The relationships between these genera will be discussed below. The Smilodontini includes two genera: *Megantereon* (with at least five species: *M. cultridens, M. whitei, M. hesperus, M. falconeri*, and *M. ekidoit*) from Africa, Eurasia, and North America; and *Smilodon* (with three species: *S. gracilis, S. fatalis* [Fig. 2.7a], and *S. populator*) from North, Central, and South America.

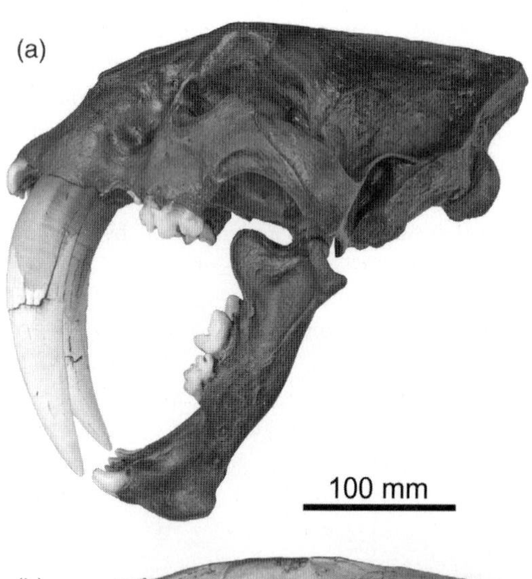

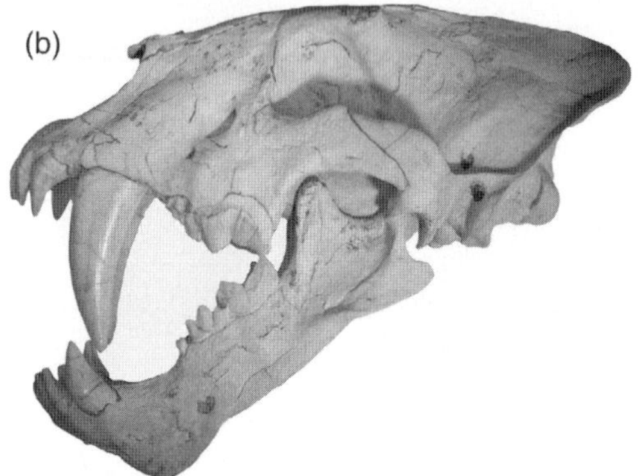

Figure 2.7 (a) Skull (cast) of *Smilodon fatalis* from Rancho La Brea, California, United States; in left lateral view. (b) Skull (cast) of *Homotherium* sp., unknown locality, China; in left lateral view.

Differences between Homotheriini and Smilodontini are substantial, both craniodentally and postcranially. The Homotheriini have relatively short, mediolaterally narrow upper canines with large crenulations on the anterior and posterior edges; their postcranial skeleton shows some adaptations to a cursorial lifestyle (except in *Xenosmilus*), with long, slender limbs and forequarters that are massive but not hyperdeveloped. The cheek dentition of Homotheriini is dominated by very large carnassials, which especially in *Homotherium* become larger in later forms, with the p4 also usurped into the cutting blade. The Smilodontini have very long, broad upper canines with minute serrations (lost in *Megantereon*). Their skeleton is very robust and the forequarters extremely massive. The cheek dentition is reduced, but the carnassials are not elongated to the extent seen in Homotheriini.

Conical-toothed cats

The conical-toothed cats, subfamily Felinae, comprise the common ancestor of all living cats and all of its descendants (Fig. 2.8). As the name implies, conical-toothed cats differ from sabretooths in having a more rounded canine cross-section. They are also united by a few other features, such as the relatively long lower canine. The interrelationships of the living members of this subfamily were discussed above. Their fossil history is much less well known than that of the sabre-toothed cats. This could be for three reasons: (1) they were predominantly adapted to environments in which fossilization is less likely than in the environments inhabited by sabre-toothed cats (i.e. the poor fossil record reflects a taphonomic bias; species that today occur in habitats in which fossilization potential can be considered fair [e.g. cheetahs and lynx], have a reasonably good fossil record, while species that today inhabit

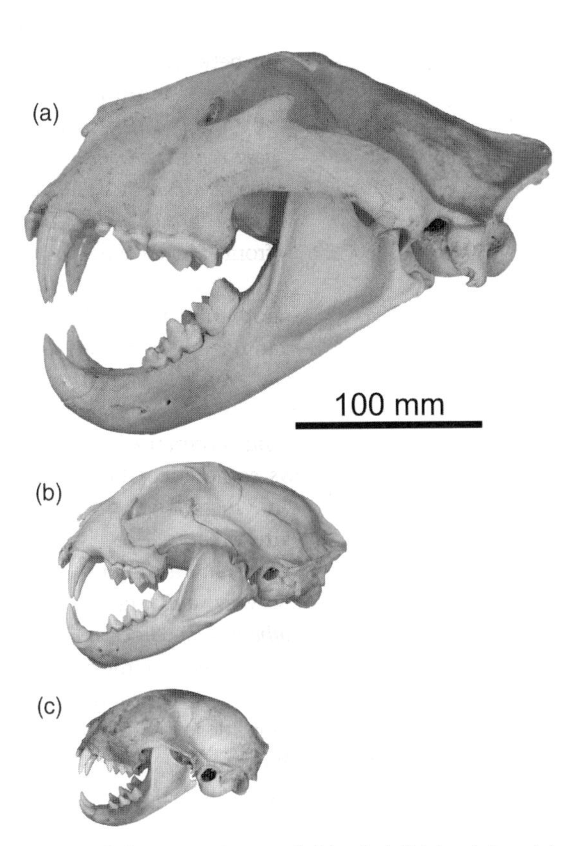

Figure 2.8 Skulls of extant Felidae in left lateral view: (a) Lion, *Panthera leo*; (b) Eurasian lynx, *Lynx lynx*; (c) Domestic cat, *Felis catus*.

tropical, wet forests [e.g. golden cats and clouded leopards] tend to have a very poor fossil record); (2) they were less common in the past than sabre-toothed cats (i.e. the poor fossil record reflects a true pattern that is an outcome of a consideration of intra-familial competition between sabre-toothed and conical-toothed cats); (3) they are more similar to each other in hard-tissue morphology than sabre-toothed cats (i.e. the poor fossil record reflects a bias in investigator perception; there are great similarities between all conical-toothed cats in, for example mandibular morphology, a region in which sabre-toothed cats exhibit a number of diagnostic differences). All three of these possibilities may be true to some extent. Finally, the poor fossil record of conical-toothed cats may also reflect the interests of researchers. Sabretooth cats are large, spectacular, and to some extent mysterious, at least as far as their feeding behaviour is concerned. Conical-toothed cats are often small, nondescript and closely similar to living forms that are comparatively well known ecologically and functionally. Hence, the former receive far more attention in the palaeontological literature than the latter.

Only one researcher, Helmut Hemmer, has focused almost exclusively on the fossil record of conical-toothed cats, and it is thus from his work (e.g. Hemmer 1974, 1976; Hemmer *et al.* 2001, 2004) that most of the information on the fossil record of this group is to be gleaned. In the following section, the fossil record of conical-toothed cats will be outlined, following the scheme of eight major lineages as found in the molecular phylogeny (Fig. 2.1). Focus will be on the earliest members of each lineage and/or species.

Some early conical-toothed cats cannot with confidence be included in any of the eight lineages. These include the first '*Felis*', '*F.*' *attica*, known from MN 11–MN 13 (*c.* 9.0–5.3 Ma) in western Eurasia. This species is a little larger than a wildcat. In morphology it is very similar to smaller species of *Pseudaelurus*, but it has a dentition that is reduced beyond the *Pseudaelurus* grade. It is noteworthy that the stratigraphic range of '*F.*' *attica* is younger than the estimated age of the base of the radiation of extant Felidae (Fig. 2.1), so that it may belong within that radiation rather than to the stem lineage. The same is true of '*F.*' *christoli*, another primitive cat, known from MN 13–MN 14 (*c.* 7.1–4.2 Ma) of Spain and France. In addition, there are significant collections of Late Miocene small cats from China that remain undescribed. This material may answer some questions regarding the early evolution of extant cats.

The clade with by far the best fossil record is the *Panthera* lineage. Despite this, it is also the clade with the longest ghost lineage (cladistically inferred lineage undocumented by fossils). According to molecular data (Johnson *et al.* 2006b) this lineage split off from the Felidae stem lineage about 10.8 Ma. However, the oldest fossils unequivocally assigned to the lineage are no older than 3.8 Ma (Barry 1987; Werdelin and Dehghani, in press), leaving a ghost lineage that is nearly twice as long as the documented lineage. The earliest fossil *Panthera* from Laetoli belong to two species: a lion-sized one and a leopard-sized one. They have been suggested to belong to the extant species (Turner 1990), but in fact differ from them morphologically (Werdelin and Dehghani, in press). The molecular dates suggest that they may belong to the stem lineage of these species and there is nothing in the fossils that would suggest otherwise.

The first definite lions are from Olduvai, Bed 1 (<2 Ma), which is also in line with the molecular data. The subsequent fossil history of lions is well known, with dispersal out of Africa across Eurasia and into North (and possibly South) America. These developments have been discussed by numerous people (Vereshchagin 1971; Hemmer 1974; Burger *et al.* 2004; Yamaguchi *et al.* 2004a). It is not until the middle Pleistocene that lions significantly extend their range outside Africa, and by about 500 ka they are found throughout Europe and parts of Asia north and east of the Black and Caspian Seas. By 300 ka their range had extended to encompass most of northern and eastern Asia except for the south-east and southern China, possibly due to competition with tigers (although this cannot be verified in the fossil record). At about this time, lions probably crossed the Bering Strait into North America, where they are known from Illinoian (<310 ka) and later deposits. In the Sangamonian, after the retreat of the Illinoian glaciers, lions could spread further into North America and, arguably, also northern South America. Lions became extinct in the Americas and large parts of Asia at the end of the latest glaciation. Further range contraction occurred in historic times.

Lion taxonomy has long been controversial. Some authorities place all fossil lions in the modern species, *P. leo*, while others recognize a number of extinct species, for example *P. spelaea*, the cave lion, and *P. atrox*, the North American lion. Burger *et al.* (2004) analysed mtDNA cytochrome *b* sequences of some cave lions and found them to form a monophyletic clade distinct from living lions. Until more data from a broader range of fossil lions have been studied, the question of whether lions conform better to a one-species or a multiple-species model must remain open.

The other members of the *Panthera* lineage are less well known in the fossil record and some aspects of their evolution are at present controversial. The characters linking fossils with extant species are often of uncertain value and the material commonly limited. The snow leopard, *P. uncia*, and clouded leopard, *N. nebulosa*, are, for example, only known from isolated fossil teeth, and it is doubtful whether this is sufficient for specific attribution.

The jaguar, *P. onca*, has been traced back to the 'European jaguar', *P. gombaszoegensis*, which is considered by some to be a subspecies of the extant species (Hemmer *et al.* 2001). If *P. toscana* can be included in this species, as suggested by Hemmer, it is first known from the latest Pliocene and survived into the Middle Pleistocene. During this time it was mainly distributed across western Eurasia.

The earliest leopards, *P. pardus*, are known from Africa. As noted, Laetoli (*c.* 3.8–3.4 Ma) includes a leopard-sized pantherine. Hemmer *et al.* (2004) have suggested that these remains should be referred to the *Puma* lineage, but the fossils provide no support for this hypothesis (Werdelin and Dehghani, in press). The oldest unequivocal leopards in Africa are from about 2 Ma, and the first leopards appear in Eurasia about 1 million years later.

Tiger remains are known from the Lower Pleistocene of South-east Asia (Kurtén, 1962). However, the oldest member of the tiger lineage is generally considered to be *P. palaeosinensis* (Fig. 2.9) from (probably) Upper Pliocene sediments in northern China (Zdansky 1924). However, renewed study (Christiansen 2008) indicates that its specific relationship to tigers is tenuous at best. Metrically, the specimen is not particularly close to any extant *Panthera*. It is generally agreed that fossil tigers have not been

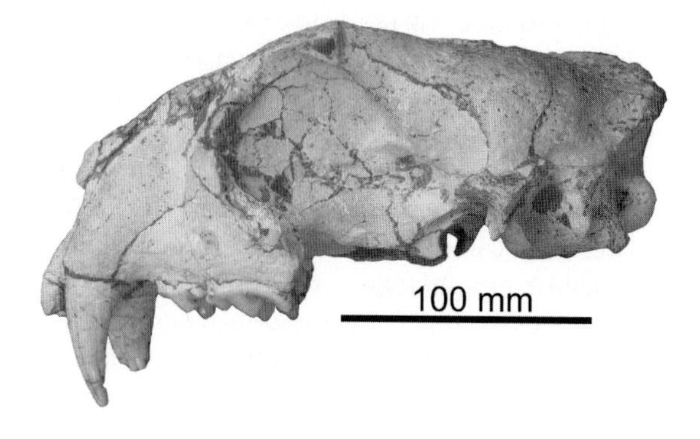

Figure 2.9 Skull of *Panthera palaeosinensis*, PIU M3654 (holotype), one of the earliest fossil *Panthera*, in left lateral view.

recorded outside Asia (but see Herrington 1987; Groiss 1996).

The bay cat lineage is not known with certainty in the fossil record. The *Caracal* lineage is represented in the fossil record by specimens dating back *c.* 4 Ma. These specimens group into two distinct size classes, large and small. Given the molecular ages of these lineages, these may represent members of the caracal/golden cat stem lineage and serval stem lineage, respectively. Whether any or all of the fossils, which are known from a number of sites in eastern and southern Africa, are conspecific with the extant forms is not determinable on the basis of the available material, which consists mainly of isolated teeth and fragmentary jaws. An intriguing recent suggestion is that *'Felis' issiodorensis*, a species generally referred to the genus *Lynx* (Werdelin 1981) should instead be referred to *Caracal* (Morales *et al.* 2003b). This conclusion is based on the observation that the metric analyses of Werdelin (1981) showed that specimens identified as belonging to *L. issiodorensis* were more similar to specimens of *Caracal* than to specimens of *Lynx*. This possibility deserves further study, but it is well to remember that it is just as likely that the similarities between *Caracal* and *L. issiodorensis* are shared ancestral characters.

The fossil record of the ocelot lineage is relatively poor. This record has recently been reviewed by Seymour (1999) with updates by Prevosti (2006). The South American record of the group is limited, and with the exception of some remains of *Leopardus colocolo* from Argentina in sediments dating as far back as *c.* 0.5–1 Ma, and the enigmatic *'Felis'vorohuensis* of about the same age, all records are latest Pleistocene in age. North American fossils unequivocally referable to this lineage are also from the Late Pleistocene (Werdelin 1985). The inferred age of the radiation of the extant taxa at *c.* 2.9 Ma (Fig. 2.1) is younger than previous estimates and compatible with a radiation from a single immigration event into South America (Werdelin 1989). However, this leaves a long ghost lineage back to the reconstructed age of the node leading to this group at *c.* 8.0 Ma. A number of North American taxa have been proposed at one time or another as members of this ghost lineage, including *'F.'lacustris*, *'F.'rexroadensis*, *'F.'longignathus*, and *'F.'proterolyncis* (e.g. Werdelin 1985; Seymour 1999). The first of these is likely to belong to the *Puma* lineage, but the relationships of the others are unclear. They may belong to the *Lynx* or ocelot lineages, or be on the backbone of the phylogeny between them. The earliest members of several of these taxa are Late Miocene (*c.* 7–6 Ma) in age.

The short phylogenetic distance between the ocelot and *Lynx* lineages may explain why several taxa mentioned above could be assigned to either. The genus *Lynx* is well represented in the fossil record, both in Eurasia and North America (Werdelin 1981). In light of the above, it is likely that the earliest fossil members of the lineage are Late Miocene in age. The earliest record of unequivocal *Lynx* in the fossil record has been considered to be *L. issiodorensis* from the Pliocene and Pleistocene of western Europe (but see the opinion of Morales *et al.* [2003a], as discussed above). This species is not, however, found on the African continent as previously suggested (Hendey 1974; Werdelin 1981). The only record of the genus on that continent is the Pleistocene *L. thomasi* from Morocco (Geraads 1980).

The *Puma* lineage has a long, if uneven, fossil record. The oldest fossils unequivocally belonging to this lineage are specimens referred to *Acinonyx* sp. from Laetoli (*c.* 3.8–3.4 Ma) (Barry 1987; Werdelin and Dehghani, in press). These specimens are about the size of the modern species but differ slightly in morphology. The cheetah subsequently has a continuous though sparse fossil record in Africa. The genus *Acinonyx* has a long history in Eurasia. The 'giant' species *A. pardinensis* appeared in western Europe a little over 3 Ma. This form is also found in China (as *A. pleistocaenicus*) and India (as *A. brachygnathus*). It was about the size of a small lion, though considerably lighter. In most other respects it displayed typical characters of *Acinonyx*, though the skull does not show the extreme vaulting seen in *A. jubatus*. During the later Pliocene there is a marked size reduction in Eurasian cheetahs, leading Thenius (1953) to describe the younger form as a separate species, *A. intermedius*. However, some Pleistocene specimens are as large as the Pliocene ones and we agree with Viret (1954) and Kurtén (1968) that the difference probably does not warrant specific separation. The Eurasian cheetah became extinct in the early Middle Pleistocene. The North American 'cheetah', *Miracinonyx*, with two species, *M. inexpectatus* and *M. studeri* (Adams 1979; Van Valkenburgh *et al.* 1990), is not the sister taxon to *Acinonyx* (Barnett *et al.* 2005).

Instead, it apparently evolved its cheetah-like features independently, from puma-like ancestors. The oldest members of this lineage are *c.* 2.5 Ma. However, the oldest '*F.'lacustris* is somewhat older than this. An interesting specimen of about the same age is the *Felis* sp. of Gustafson (1978) from the Blancan of Oregon, which may also belong to this lineage. The presence of *Puma* in Europe has also been suggested, in the form of *P. pardoides* (Hemmer *et al.* 2004). The oldest of this material is of Pliocene age and may be the oldest material of *Puma* on record. The suggestion that *Puma* is present at Laetoli is hardly tenable, however (Werdelin and Dehghani, in press). The oldest fossil jaguarundi is less than 0.5 Ma.

The leopard cat lineage is very poorly known in the fossil record. A few fossils probably pertaining to this lineage and possibly to *Prionailurus bengalensis* have been found in Middle Pleistocene sites in South-east Asia (Hemmer 1976). In addition, fossils tentatively referred to *O. manul* have been recorded from Kamyk, Poland (Kurtén 1968). These may be more than 1 Ma.

The fossil record of the domestic cat lineage is not poor, but much of it is hidden beneath the general designation of *Felis* sp., since the species are all but indistinguishable on the basis of incomplete remains. The oldest '*Felis* sp.' that definitely belongs to this lineage is from Kanapoi, Kenya, dated to >4 Ma (Werdelin 2003a). If the molecular dates are correct, this material belongs to a member of the stem lineage of *Felis*. Further specimens belonging to this lineage occur intermittently in the African fossil record. A species of some interest that may be the oldest member of the *F. silvestris* group is *F. lunensis* from Europe. This species goes back at least to the Early Pleistocene and possibly to the Late Pliocene. Specimens referable to *F. chaus* have been found in Holocene strata of Java (outside the modern range of the species; Hemmer 1976). No specimens definitely referable to *F. nigripes* or *F. margarita* have been found in the fossil record.

Barbourofelidae

Finally, we must touch upon the family (or subfamily) Barbourofelidae (Fig. 2.2, letter N), which consists of a number of derived sabre-toothed forms (though not all may be sabre-toothed—see below). Traditionally, they have been seen as Neogene members of the Nimravidae, a group that itself has been the subject of much phylogenetic discussion. The nimravids were once known as 'paleo-felids' because of their felid-like craniodental morphology. They are known from the Late Eocene to Late Oligocene of North America and Europe and include genera such as *Nimravus, Hoplophoneus*, and *Eusmilus*. Studies of basicranial morphology have, however, clearly shown that nimravids are not felids (Neff 1983; Hunt 1987). They are therefore placed in the family Nimravidae. In its original conception, Nimravidae also included the barbourofelids, Miocene sabretooths with representatives both in North America and Europe (Schultz *et al.* 1970). These, however, have a basicranial morphology, including an ossified bulla, that differs from those in both Nimravidae and Felidae. Therefore, Morales *et al.* (2001) proposed removing them from the Nimravidae and placing them as the subfamily Barbourofelinae within the Felidae. This proposal was amended by Morlo *et al.* (2004), who proposed raising Barbourofelinae to full family status as the Barbourofelidae, which is the path followed here. The Nimravidae are likely to be basal Carnivora, while the Barbourofelidae are either the sister-group to Felidae or the sister-group to other Aeluroidea (Fig. 2.2, letter N). Because of their phylogenetic and ecomorphological closeness to Felidae, their fossil record is outlined here.

In Africa, the likely centre of origin of Barbourofelidae, the family is known from a number of genera (Morales *et al.* 2001; Morlo *et al.* 2004). *Afrosmilus* has two east African species, *A. africanus* (Fig. 2.10) and *A. turkanae*, both *c.* 18–17 Ma. *Ginsburgsmilus*, the most primitive member of the family, has a single

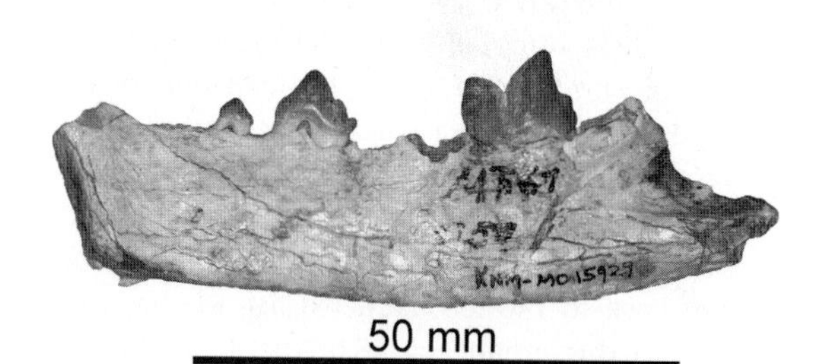

Figure 2.10 Left horizontal mandibular ramus of *Afrosmilus turkanae*, KNM MO 15929, Moruorot, Kenya, a barbourofelid. Note the well-developed metaconid at the posterior end of the tooth—a diagnostic difference between Barbourofelidae and Felidae (Morlo *et al.* 2004).

east African species, *G. napakensis* (*c.* 20.5–17 Ma). *Syrtosmilus* with one species, *S. syrtensis* (*c.* 19–15 Ma), and *Vampyrictis* with one, *V. vipera* (*c.* 12.5–9.5 Ma), are North African representatives of the family.

In Europe, Barbourofelidae is known from several genera, *Prosansanosmilus*, with two species, *P. peregrinus* (MN 4, *c.* 18–17 Ma) and *P. eggeri* (MN 5, *c.* 17–15.2 Ma), *Sansanosmilus* with two species, *S. palmidens* (Fig. 2.11) (MN 5–MN 7/8, *c.* 17–11.2 Ma) and *S. jourdani* (MN 6–MN 9, *c.* 15.2–9.5 Ma), and *Afrosmilus* with one European species, *A. hispanicus* (MN 5, *c.* 17–15.2 Ma). These show a temporal progression towards larger and more sabretooth forms, though they are generally less extreme in their adaptations than the North American *Barbourofelis* spp. *Sansanosmilus* is also known from the Middle Miocene of China, though it is less common there than in Europe.

In North America, the Barbourofelidae consists of the single genus *Barbourofelis*, with five species, *B. fricki* (*c.* 10 Ma), *B. loveorum* (*c.* 11–9.8 Ma), B. *morrisi* (*c.* 11.5 Ma), *B. osborni* (*c.* 11.5 Ma), and *B. whitfordi* (*c.* 12–11.5 Ma). They are all extreme sabretooth ecomorphs, with long sabres, large mental flanges and short, stout limbs (where known).

Finally, two species from southern Africa must be mentioned, *Diamantofelis ferox* (the size of a small puma) and *Namafelis minor* (lynx-sized) (Morales *et al.* 1998, 2003a). Both are from the late Early–earliest Middle Miocene of Arrisdrift, Namibia (*c.* 17–15.2 Ma). These species are not, as far as is known, sabre-toothed in morphology, as neither has a squared-off symphyseal region, but they do share other mandibular and dental features with species of *Afrosmilus*. *D. ferox* has a short and deep mandible, while that of *N. minor* is longer and more slender. The oldest true felid from Africa is a small specimen from Songhor, Kenya (*c.* 18–17 Ma), probably referable to *Pseudaelurus sensu lato*. Thus, since Felidae is rare or non-existent in Africa at this time, whereas Barbourofelidae is known from a number of sites and regions, and given the morphological similarities between them, it should at least be considered whether the Arrisdrift species might be 'conical-toothed' barbourofelids.

Figure 2.11 Artist's reconstruction of the head of *Sansanosmilus palmidens*, a barbourofelid. (Illustration courtesy of Mauricio Antón.)

The Barbourofelidae was a relatively short-lived group (*c.* 20.5–9.5 Ma), within which the vast majority of species were specialized sabretooths. Their extinction in the early Late Miocene may be tied to the spread of sabre-toothed Felidae at this time.

Discussion

In this section, we will attempt to draw some conclusions from the review above. We advocate the use of generic names for fossils that maximizes the number of monophyletic taxa by splitting up at least those genus-level groups that are obviously para- or polyphyletic. In so doing we hope to create a more consistent framework for future studies. Unfortunately, many of the assertions made in the following discussion are at present untested, though we hope that it will be possible to test them in the future. We aim to erect a series of hypotheses to establish the basic level of understanding of felid evolution, that of interrelationships. When the interrelationships of fossil felids have been better established, a foundation for the understanding of the ecological, biogeographical, and functional patterns of felid evolution will have been laid, in much the same way as the current phylogeny of living felids provides such a foundation for study of their radiation.

We will also point out areas where we know too little, which is especially true of the fossil record of the living felids, which at present has not much to contribute to an understanding of the modern radiation. Johnson *et al.* (2006b) estimate that fossil representation in the modern cat radiation is about 24%, leaving large areas unknown (cf. Fig. 2.1). Fig. 2.2 provides a graphical summary of the discussion in the following section.

Early cats

The origins of the family Felidae are relatively uncontroversial, though that may merely be because the gap between the earliest unequivocal felids and their ancestors among Carnivoramorpha is relatively substantial. Thus, *Proailurus* is unquestionably a felid and *Stenogale* and *Haplogale* are likely to belong to this family as well. All of these genera are undoubtedly closer to crown group Felidae than is the extant sister taxon, *Prionodon*. Aspects of this early evolution are covered by Hunt (1998) and Peigné (1999) and require no further elaboration here.

The subsequent radiation of Felidae in the Early–Middle Miocene is far more complex, however. The genus *Pseudaelurus* comprises 11 named species, 4 from Europe and Arabia, 2 from China, and 5 from North America. Unanswered questions surrounding this radiation include: Does *Pseudaelurus* have a single origin? What are the interrelationships of the European species to each other? What is the relationship between Chinese and European species? From which species did North American *Pseudaelurus* originate? Which species of *Pseudaelurus* belong to which lineages of later, more derived felids?

Answers to some of these questions have been proposed in the past, whereas some have rarely been discussed, if at all. An example of the latter is the first question raised above: Does *Pseudaelurus* have a single origin? This has tacitly been assumed in discussions of felid evolution in the past. However, the data in favour of this hypothesis are largely circumstantial. The presumed ancestor, *Proailurus*, has a limited geographic distribution and *Pseudaelurus* from Europe is older than *Pseudaelurus* on other continents, arguing for a single origin and subsequent dispersal. There is, on the other hand, no phylogenetic framework in which this has been demonstrated to be the most parsimonious hypothesis. It is certainly also possible that different species of *Proailurus* or related genera gave rise to different species of *Pseudaelurus*, rendering the latter polyphyletic. The Ginn Quarry felid discussed above (Hunt 1998) makes such a scenario more plausible. At present, there does not seem to be any way to resolve this issue definitively and the monophyletic origin of *Pseudaelurus* is assumed here as a working hypothesis.

On the other hand, *Pseudaelurus* is undoubtedly paraphyletic, with different species groups giving rise to different descendant taxa. The paraphyletic nature of *Pseudaelurus* has been recognized for a long time, though perhaps the first to do so explicitly was Kretzoi (1929b), and this was also implicitly acknowledged by Viret (1951) before being elaborated on by Beaumont (1964, 1978). Beaumont (1964), like Kretzoi before him, split *Pseudaelurus* into a number of genera at the bases of several subsequent radiations. Though Beaumont (1978) reduced these to subgenera, his Figure 2 remains the fullest envisioning of felid evolution to this day. If we ignore the subgenera, he split *Pseudaelurus* into three genera: *Pseudaelurus* (Gervais 1850, type species *P. quadridentatus*), *Schizailurus* (Viret 1951, type species *P. lorteti*), and *Hyperailurictis* (Kretzoi 1929b, type species *P. intrepidus*). The first-mentioned includes only the type species, while the second includes *P. turnauensis* in addition to *P. lorteti*. The third includes all North American species of *Pseudaelurus* listed above. However, it should be noted that *Schizailurus* is an objective junior synonym of *Miopanthera* Kretzoi (1938) (based on the same type species), and this, in turn is a subjective junior synonym of *Styriofelis* Kretzoi (1929a; type species *F. turnauensis*). Thus, the latter name is the senior valid synonym and is used here.

Beaumont (1978) places *Pseudaelurus* at the base of the radiation of sabre-toothed cats, *Styriofelis* at the base of the radiation of conical-toothed cats, and *Hyperailurictis* at the base of the North American radiation, as well as the radiation of 'intermediate' forms such as *Metailurus*, *Stenailurus*, and *Dinofelis*. The radiation of these three genera takes place at letter B in Fig. 2.2. The evidence for this scenario is not particularly strong, as it is based mainly on the somewhat more sabretooth-like characteristics of *P. quadridentatus*, as opposed to the clearly conical-toothed features of *Styriofelis lorteti* and *Styriofelis turnauensis*. Although the generic separation between these taxa has generally not been considered in reviews of *Pseudaelurus*, (Heizmann 1973; Ginsburg 1983), the separation has been implicitly acknowledged by several other workers (e.g. Morlo 1997). The status of the North American pseudaelurines (*Hyperailurictis*) as a distinct, generic-level clade is somewhat more secure, as these species are relatively derived, very similar to each other, and also very similar to their presumed descendant *Nimravides*.

This scenario provides possible answers to several of the questions posed above, apart from the question of the relationship of *Pseudaelurus* to later, more derived felids. Among European '*Pseudaelurus*', *S. lorteti* and *S. turnauensis* are closely related and more distant from *P. quadridentatus* (the status of *P. romieviensis* is unclear). The North American *Hyperailurictis* did not evolve from any of them, though it may be related to one or both of the Chinese species. It is more likely, however, that *Hyperailurictis* descended from a felid similar to the Ginn Quarry felid described by Hunt (1998). The wholly unanswered question is the relationship between European and Chinese pseudaelurines.

A reasonable consensus, however, is that *Styriofelis* gave rise to the radiation of modern cats (Fig. 2.2, letter C). Aside from *S. lorteti* and *S. turnauensis*, no species definitely belonging to the stem lineage are known (but see *Felis attica* and see below).

In North America there is little doubt that *Hyperailurictis* gave rise to *Nimravides* (Fig. 2.2, letter D). Whether the former is mono- or paraphyletic is not known at this time. If it is paraphyletic, further nomenclatural complications may arise, but these need not concern us here. *Nimravides* seems to have gone extinct without leaving descendants, though it is just possible that *M. coloradensis* evolved from this genus rather than being an immigrant from Eurasia. The close similarity between *M. coloradensis* and the Eurasian early Late Miocene *M. aphanistus* argues against this, however.

The relationship between *Hyperailurictis* and *Dinofelis*, *Metailurus* and *Stenailurus* is far less well established and is not followed here. These genera are usually grouped together as the Metailurini, though the monophyly of this tribe has not been satisfactorily demonstrated. This is clearly an Old World group, with the evolution of *Metailurus* centred in Eurasia and that of *Dinofelis* in Africa. This presents some biogeographic problems for an origin from *Hyperailurictis* in North America as suggested by Beaumont (1964, 1978). It is very tempting instead to associate this group with the Chinese pseudaelurines, though these are so poorly known that this remains pure speculation at present. Here we will consider the Metailurini to belong to the sabretooth cats (but see below), and thus a part of the radiation at letter E of Fig. 2.2.

Upper Miocene to Pleistocene cats

Pseudaelurus, sensu stricto, gave rise to the radiation of sabretooth cats that first appeared in the late Middle Miocene of Eurasia (and possibly Africa) and spread across the world in the Late Miocene (Fig. 2.2, letter E). There has been much controversy surrounding sabretooths (subfamily Machairodontinae) and considerable confusion regarding taxonomy and the allocation of specimens ever since Cuvier (1824) placed the first sabretooth specimens in the genus *Ursus*. Numerous genera and species have been named over the years and the course of evolution of the group has been poorly understood. Part of the problem has been the focus on *Smilodon*, a late and highly derived sabretooth, as the exemplar species in discussions of the functional morphology and evolution of the group (e.g. Bohlin 1940; Simpson 1941; Miller 1969; Akersten 1985). However, considerable progress in understanding these issues has come in recent years with the study of the excellently preserved material from the carnivore trap site of Batallones-1 in the Cerro de Batallones, Spain (e.g. Antón *et al.* 2004; Salesa *et al.* 2005). These studies show that the functional morphology of sabretooths was not uniform across taxa, evolved over time, and is compatible with a gradual origin from *Pseudaelurus*-grade forms.

Nevertheless, numerous questions regarding the systematics and evolution of sabretooth cats remain. Some of these are: What is the relationship of *Dinofelis* and *Metailurus* to Machairodontinae? What are the evolutionary patterns within the paraphyletic *Amphimachairodus* group? What is the relationship between *Homotherium* and *Dinobastis*? How did the Smilodontini evolve and which taxa are their ancestors? What did sabretooths feed on and how? How and why did sabretooths become extinct?

The relationship of *Dinofelis* and *Metailurus* to Machairodontinae (and to each other) has always been controversial. Some authors, (e.g. Beaumont 1978; Werdelin and Lewis 2001), have considered them to be members of the Machairodontinae with slight to moderate sabretooth adaptations, while others (Kretzoi 1929b; Hendey 1974) have considered them to be conical-toothed cats with a tendency to develop sabretooth adaptations. The main feature they share with Machairodontinae is a reduced lower canine relative to the upper canine. *Dinofelis* further shares with Machairodontinae a deep groove or pit supero-medial to the trochlear notch of the ulna (Werdelin and Lewis 2001). This feature seems not to be present in *Metailurus* (Roussiakis *et al.* 2006). Thus, it appears likely that *Dinofelis* belongs in the Machairodontinae, but the position of *Metailurus* is equivocal. This also, of course, makes the relationship between the two genera uncertain. Thus, the position of this group at letter J of Fig. 2.2 is problematic, as is the placement, even existence, of the node at letter M.

Amphimachairodus is clearly paraphyletic, as *Homotherium* evolved from within this species group. This is reflected in the intermediate position of letter G (Fig. 2.2), between letter F where *Amphimachairodus* splits off from a similarly paraphyletic *Machairodus*, and letter H, at the base of the monophyletic Homotheriini. What is not clear is exactly which species gave rise to Homotheriini (Fig. 2.2, letter H). *L. emageritus* from Kenya has a more derived dentition than any species currently assigned to *Amphimachairodus*, but is too primitive in other respects and too derived in a few to be the ancestral taxon. Of the species of *Amphimachairodus*, *A. kurteni* seems the most derived dentally, but *A. kabir* (if the material from Sahabi belongs there; cf. Sardella and Werdelin 2007) has the most derived mastoid region. Whichever of these (or some as yet unknown taxon) is ancestral to *Homotherium*, *Amphimachairodus* as presently conceived becomes paraphyletic. To resolve this issue, the detailed relationships of *Amphimachairodus* spp. need to be better understood.

The relationship between *Homotherium* and *Dinobastis* (and *Xenosmilus*) is particularly interesting (Fig. 2.2, letter I). Traditionally, they are synonymized in the genus *Homotherium* (Turner and Antón 1997). However, early North American homotheriines such as that from the Delmont Local Fauna, South Dakota (Martin and Harksen 1974) (*c.* 2.9–2.6 Ma) differ considerably from contemporary forms in Eurasia (see, e.g., Ficcarelli 1979), suggesting a long, separate evolution. In addition, *Homotherium* and *Dinobastis* differ in a number of aspects of their morphology. As an example, the upper canine of *Dinobastis* is smaller than that of *Homotherium* in specimens of approximately equal skull size (Werdelin and Sardella 2006, plate 1, fig. 1). This is an area that deserves further in-depth study.

Regardless of which species in the *Amphimachairodus* group is closest to *Homotherium*, it is nearly universally acknowledged that there is, broadly conceived, an ancestor–descendant relationship between the two genera. However, the origins of the other major Plio-Pleistocene sabretooth lineage, the Smilodontini (Fig. 2.2, letter L), is much less clear. This group consists of the genera *Megantereon* and *Smilodon*, which share features such as reduced or absent serrations on the teeth and extremely long and relatively mediolaterally broad upper canines compared to Homotheriini (the latter probably a plesiomorphic feature). It is tempting to associate them with the other Miocene sabretooth lineage, *Paramachaerodus* (Turner and Antón 1997) (Fig. 2.2, letter K), but the morphological distance between that genus and Plio-Pleistocene Smilodontini is considerable and the hypothesized relationship is not based on any clear synapomorphies. Another question germane to this issue is the difference between Smilodontini and Homotheriini: why is it there and what does it mean for the functional morphology and ecology of the respective groups? One answer would be that the former were closed-habitat taxa and the latter open-habitat taxa, but can such a simplistic view be maintained? Martin (1980) and Martin *et al.* (2000) discuss some of these questions, but more research needs to be done on the functional differences between Homotheriini and Smilodontini, and in particular on the latest Miocene species of *Paramachaerodus* (*P. orientalis* and *P. maximiliani*), to understand their ecology and feeding behaviour, and whether these can be directly related to those of Smilodontini.

The extinction of Homotheriini and Smilodontini occurs at different times on different continents. In

Africa, both *Homotherium* and *Megantereon* became extinct some time before 1.4 Ma (with *Dinofelis* lingering on another 500 ka). In Europe, *Homotherium* became extinct at *c.* 0.5 Ma (the recent record of a Late Pleistocene *Homotherium* from North Sea sediments [Reumer *et al.* 2003] needs to be corroborated by further material before its implications can be fully assessed) and *Megantereon* at *c.* 1 Ma. In North America, on the other hand, both tribes survive into the latest Pleistocene, with the last occurrence of *Dinobastis* from Friesenhahn Cave (Texas) at *c.* 11,000 BP and the last occurrence of *Smilodon* from Rancho La Brea (California) at *c.* 13,000 BP. We don't fully understand why these dates differ so much between continents. The differences may reflect the different first appearance datums on each continent of advanced hominid competitors in sufficient numbers to affect the populations of sabretooths, through direct or indirect competition for resources. Or they could be the result of major faunal changes on each continent brought about by human interference, climatic change, or a combination of the two. In building and testing these scenarios, it is also important to consider the conical-toothed cats and their impact on their sabretooth competitors, for example, the relatively rapid range expansion of lions from Africa through Eurasia during the Middle–Late Pleistocene (Yamaguchi *et al.* 2004a), understanding of which has been hampered by the poor fossil record of the conical-toothed cats.

Possibly no subject in mammal palaeontology has been more debated than that of sabretooth feeding adaptations. How did they use their canines? What did they feed on? What was their killing behaviour like? Questions like these have been posed and answered numerous times since sabretooths were first discovered (see Kitchener *et al.*, Chapter 3, this volume). To answer these questions, it is important to realize that this ecomorphology is not restricted to felids and their carnivoran relatives among nimravids and barbourofelids. The package (with variations) is also present in some creodonts, an extinct order of mammals that lived from the Paleocene to the Miocene (genera *Apataelurus* and *Machaeroides*, Early–Middle Eocene of North America), in marsupials (genus *Thylacosmilus*; Miocene–Early Pleistocene of South America) and in various groups of synapsid 'reptiles' of the Late Palaeozoic, for example, Gorgonopsia (Kemp 2004). Despite this, it is among felids, nimravids, and barbourofelids that the adaptation appears to have been most successful. Recent work on early felid sabretooths (Salesa *et al.* 2003; Antón *et al.* 2004; Salesa *et al.* 2005) has begun to close the functional gap between sabre-toothed and conical-toothed cats. This and other lines of evidence, such as the meandering evolutionary history of *Dinofelis* from more sabretooth to less sabretooth and back (Werdelin and Lewis 2001), suggest that the ecomorphology of the feeding apparatus in felids is more of a continuum than a dichotomy. The implications of this for understanding the ecology of sabretooths and competition between sabretooths and conical-toothed cats are in need of detailed investigation.

One possible implication of the feeding apparatus of sabre-toothed and conical-toothed cats being on a continuum is that there may have been more direct competition between the two groups than previously thought. Previous models tend to emphasize the difference, with sabretooths specializing in larger prey than similar-sized conical-toothed cats. However, more recent analyses suggest that perhaps the two groups focused on very similar prey. In Africa, sabretooths are fairly common fossils and conical-toothed cats rare until around the time when the number of fossils of sabretooths decreases (Werdelin and Lewis 2005). This can be explained if sabretooths were dominant in the most commonly sampled habitats and competitively excluded conical-toothed cats. Support for such an idea can be found at Laetoli. This site (or at least the Laetolil Member, Upper Beds) is unique among eastern African sites in not being near a large body of standing water. It is also unique among sites in having a large number of fossils of conical-toothed cats and very few fossils of sabretooths. Further research on competition between sabretooths and conical-toothed cats is needed, as is research on the competitive structure of the carnivore guild as a whole.

The single most important issue impeding an increased understanding of the evolution of conical-toothed cats is the extensive ghost lineage between the oldest fossil members of the *Panthera* lineage and the common ancestor of all Felinae. Two explanations for this gap in the fossil record immediately spring to mind: a poor fossil record in the earliest Pliocene and the possibility that the

Panthera lineage (and the Felinae as a whole) evolved in an environment that is not conducive to the process of fossilization. Both of these factors are undoubtedly in play, but it is hard to escape the impression that pantherines are present in the fossil record prior to 4 Ma, but that they are misidentified for as yet unknown reasons. To either identify these fossils or explain why they have not been found is the most pressing issue in felid palaeontology and evolution and without progress here it will not be possible to move towards a fuller reconciliation of the fossil record with the molecular evidence for felid evolution as presented by Johnson *et al.* (2006b).

The position of Barbourofelidae is, of course, very uncertain, since there is no consensus at present on how closely related it is to the Felidae. Here, we have opted for the view that it split off from the stem lineage leading to Felidae after *Prionodon* but before the evolution of *Proailurus* (Fig. 2.2, letter N). This leaves an extensive barbourofelid stem lineage that is at present entirely unknown.

Evolutionary patterns

The availability of a phylogeny of extant Felidae makes it possible to consider evolutionary patterns within the family in the absence of fossils. Such studies have been attempted in the past (Ortolani and Caro 1996; Werdelin and Olsson 1997; Ortolani 1999; Mattern and McLennan 2000), but given that the current phylogeny (Fig. 2.1) is fully resolved and, we believe, better corroborated than older hypotheses, this work is worth reconsidering. Further, since the current hypothesis is based on molecular data, it is possible to study morphological character evolution without the need to discuss possible circularity in the results. Some such uses of the phylogeny were presented by Johnson *et al.* (2006b) and O'Brien and Johnson (2007) (intercontinental migrations, ghost lineage analysis), and we will only briefly present two further examples of the sort of work that can and should be done on felid evolution using the phylogeny as a baseline. For other examples based on previously proposed phylogenetic hypotheses, see in particular Mattern and McLennan (2000).

Werdelin and Olsson (1997) presented a phylogenetic study of coat patterns in Felidae using a selection of then-current phylogenetic hypotheses as the baseline. Their conclusion was that 'most transformations of coat pattern originate from the flecked pattern, which we consider to be primitive for the Felidae as a whole' (Werdelin and Olsson 1997, p. 399). The current phylogeny has some substantial differences from the phylogenies used in that study, so the question arises whether the conclusions hold up. Fig. 2.12 shows coat pattern mapped on the current phylogeny. The data are identical to those in Werdelin and Olsson (1997) except for *P. tigris*, which has been recoded from vertical stripes to rosettes, as we believe that what appear to be vertical stripes in the tiger's coat in reality are enormously vertically elongated rosettes. This is indicated through examination of various coat pattern anomalies in tigers and can be more simply seen by holding up an image of a tiger pelt nearly parallel to one's line of sight. One difference from the previous results is immediately obvious: under the current phylogeny, the primitive coat pattern for Felidae as a whole is large blotches. This coat pattern is present in only two genera: *Neofelis*, clouded leopards and *Pardofelis*, marbled cat. Both are basal within their clades, and these clades are basal within the family and hence the primitive condition is reconstructed as large blotches. Above the node leading to *Pardofelis*, however, flecks are primitive as they were in the previous study. If we consider the number and direction of the state changes in the cladogram (Fig. 2.13), we can also see that changes to and from flecks are still the dominant transformations, though not quite as dominant as previously thought. Thus, the new phylogeny corroborates the main thrust of the results of Werdelin and Olsson (1997), but also leads to some modifications of specific parts of their conclusions.

In a second demonstration of possible phylogenetic reconstructions, we mapped habitat (as open or closed), activity pattern (diurnal or nocturnal), and pupil shape (slit-like or rounded in the contracted state) in all felids. The data are partly from Mattern and McLennan (2000) and partly original. Many species occur in both open and closed habitats and the mapping reflects this, not showing any clear phylogenetic associations of open- or closed-habitat specialists (Fig. 2.14), although the *Panthera* and domestic cat lineages are dominated by open-habitat taxa and have this habitat reconstructed as primitive for the

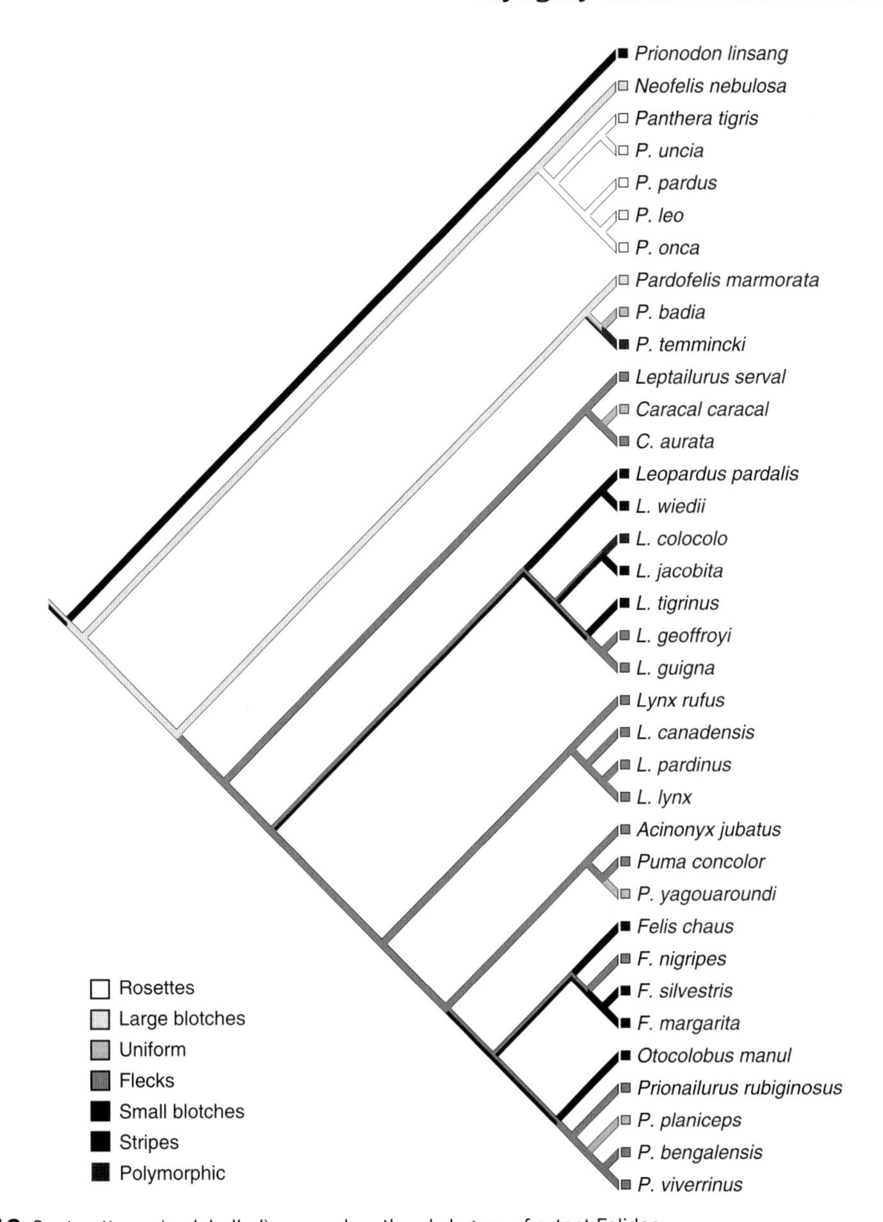

Figure 2.12 Coat patterns (as labelled) mapped on the phylogeny of extant Felidae.

respective clades. Likewise, there are no clear phylogenetic patterns underlying activity patterns in modern felids (mapping not shown). Rounded pupils, on the other hand, only occur in three clades, the *Panthera*-lineage, where all species except the two *Neofelis* (A. Kitchener, personal communication) have rounded pupils, the *Puma*-lineage, where all three species have rounded pupils, and the leopard cat lineage, where the single species *O. manul* has rounded pupils.

The question of the occurrence and causes behind slit-like or rounded pupils has been intermittently discussed in the literature without a consensus being reached (see Kitchener *et al.*, Chapter 3, this volume, for a discussion of some recent research). One suggestion that has been considered is that slit-like pupils allow the pupil to be more completely closed than rounded pupils (Walls 1942). This would suggest that the former would be more useful in the

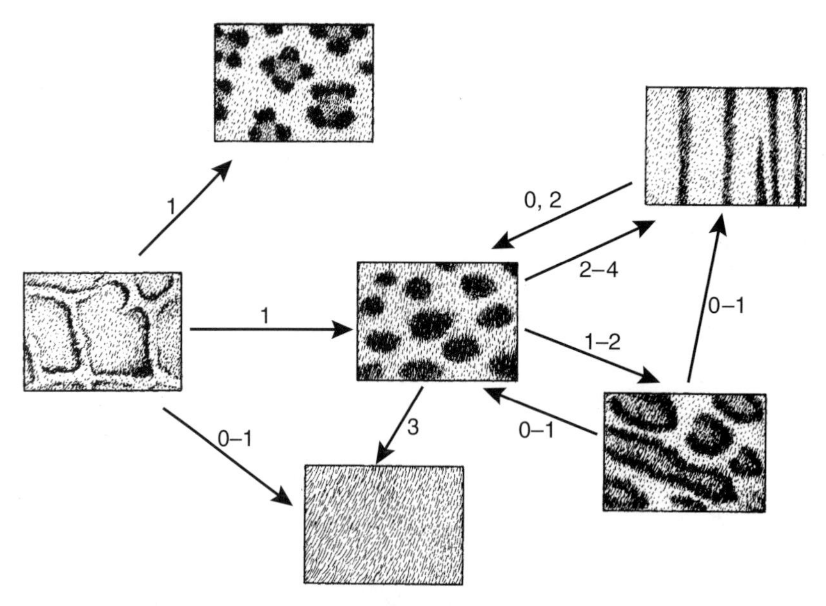

Figure 2.13 The coat pattern transformations implied by the mapping in Fig. 2.12. The majority of transformations involve flecks (centre pattern). Thus, clockwise from bottom, there are three transformations from flecks to uniform, one transformation from large blotches to flecks, two to four transformations from flecks to stripes, zero or two (no reconstruction allows for one) transformations from striped to flecks, one to two transformations from flecks to small blotches, and zero to one transformations from small blotches to flecks. Remaining reconstructed transformations are zero to one transformation from large blotches to uniform, one transformation from large blotches to rosettes, and zero or one transformations from small blotches to stripes. No other transformations are allowed by the phylogeny of extant Felidae.

brighter light of day: that is, slit-like pupils should be preferentially present in diurnal species. However, a comparison between the patterns does not corroborate this idea (not shown). There seems to be no correlation at all between pupil shape and activity pattern. However, if we compare habitat and pupil shape (Fig. 2.14), we find that with the exception of *Puma yagouaroundi*, which, if the fossil record of this clade is taken into account, must be considered secondarily adapted to closed habitats, rounded pupils never occur in closed-habitat specialist species. All the taxa with rounded pupils are either open-habitat species or occur in a variety of habitats. Further, all three nodes where there is a change from slit-like to rounded pupils are also nodes where there is a shift from closed-to open-habitat preference. What this means in functional terms is beyond the scope of this chapter, but the results point to a fruitful avenue of research. These very tentative results must be corroborated by more in-depth study and statistical testing. More generally, phylogenetically based studies such as the ones discussed above can direct future research and provide tests of functional hypotheses that could otherwise not be investigated due to a lack of independent data. The existence of a well-corroborated phylogeny such as that in Fig. 2.1 is a powerful tool for future research on felid evolution.

Final words

This chapter presents one possible scenario for the evolution and interrelationships of cats. Some of this work, such as that which has led to the phylogeny of Johnson *et al.* (2006b), is strongly corroborated by and based on considerable amounts of data. The fossil record of Felidae is uneven. Some groups, such as parts of the Machairodontinae, have a fairly extensive fossil record, while others, such as the lineage leading to the extant radiation, are much more

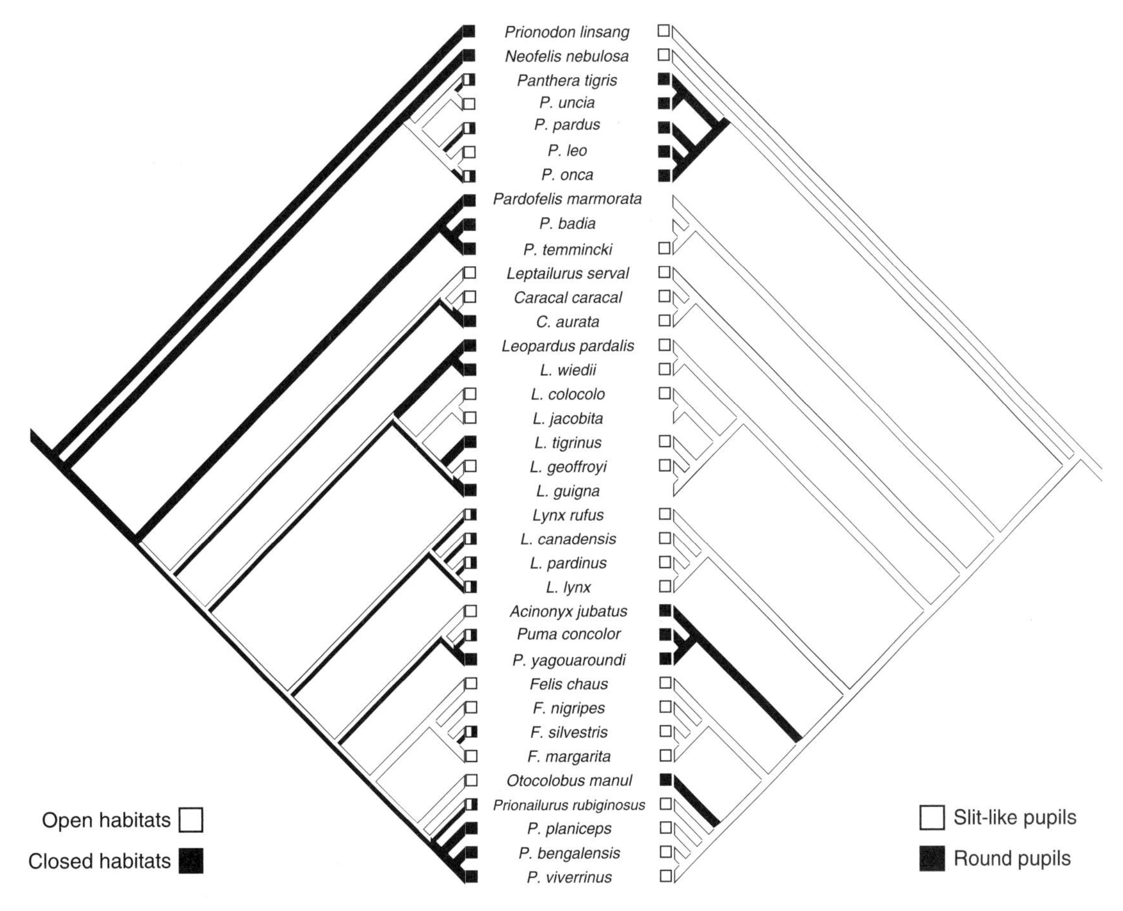

Figure 2.14 Habitat type preference and pupil shape mapped on the phylogeny of extant Felidae.

poorly documented. In no case, however, can the fossil record be said to be adequate, either in quantity or quality. Nor can the fossil record of Felidae be said to have been adequately studied. Some areas, such as the functional morphology of sabretooths, have been investigated over and over, while others, such as the stem lineage of modern cats, have been relatively neglected. Overall, the phylogeny and evolution of fossil Felidae have been neglected in favour of studies of their functional morphology and ecology. Given the limited resources available for this work, this is understandable, as the latter topics have proven more tractable and have yielded interesting and significant results. But if our understanding of the group is to progress, we must try to address such pressing issues as the fossil record of living cats, the origins of Smilodontini, and the relationship of Barbourofelidae to Felidae. This will require extending the work of Johnson *et al.* (2006b) into the realm of fossils, by comparing the fossil record with the results obtained from the phylogeny of extant cats on aspects such as continental migration (O'Brien and Johnson 2007), to see if the timing of intercontinental migrations of fossil cat groups can be matched up with those postulated for the extant cats based on phylogeny and geology.

It must be understood that developing a phylogeny, or even the simpler task of testing some aspect of the scenario developed herein, requires more than a superficial glance at the record and doing a phylogenetic analysis of the first few characters that come to mind. It will require developing new characters and looking at the fossil record in new ways. If the fossil record and phylogeny of extant Felidae can be better integrated, we can expect to develop a significantly better understanding of the evolution of this fascinating group and its conditions for existing, thereby not only enhancing current knowledge, but also building a better platform for the conservation of the many endangered species of Felidae today.

Acknowledgements

We would like to thank David Macdonald for inviting us to contribute to this volume, as well as for his encouragement and a careful reading of the manuscript. We thank Mauricio Antón for letting us use his reconstructions of extinct felids, Stéphane Peigné for the photographs of *Proailurus*, Mikael Axelsson for photographs of felid skulls, Mats Wedin for help with MacClade, and Margaret Lewis and Susan Cheyne for reading and commenting on the manuscript. Lars Werdelin acknowledges support from the Swedish Research Council. We thank Andrew Loveridge, Blaire van Valkenburgh, Andrew Kitchener, and an anonymous reviewer for their thorough readings that helped us clarify and correct parts of the text, making the whole far more accessible to a broad audience than it otherwise would have been.

CHAPTER 3

Felid form and function

Andrew C. Kitchener, Blaire Van Valkenburgh, and Nobuyuki Yamaguchi

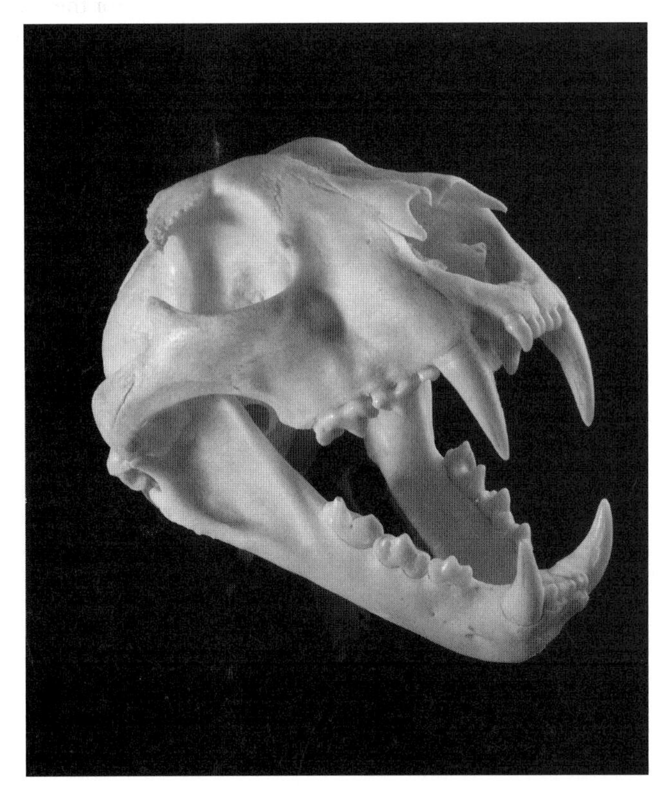

Skull of a snow leopard *Panthera uncia*. © National Museums Scotland.

Introduction

The felid body plan has remained remarkably unchanged over the past 20 million years or so, despite a 100-fold range in body weights from the tiny rusty-spotted, *Prionailurus rubiginosus*, and flat-headed cats, *Prionailurus planiceps*, weighing in at about 1 kg, to the mighty lion, *Panthera leo*, and tiger, *Panthera tigris*, the biggest males of which may reach 300 kg (Kitchener 1991; Turner and Antón 1997). The constraints imposed by being an ambush predator, whatever the prey size, have limited the adaptive radiation of the modern cats into other niches to a handful of examples, including a cursorial hunter (cheetah, *Acinonyx jubatus*), two piscivores (fishing cat, *Prionailurus viverrinus*, and flat-headed cat) and three arboreal specialists (margay, *Leopardus wiedii*; marbled cat, *Pardofelis marmorata*; and clouded leopards, *Neofelis* spp.), although most cats have some scansorial ability, including the ancestral cat, *Pseudaelurus* and the earliest well-known cat, *Proailurus lemanensis* (Turner and Antón 1997).

Sabretooth cats, however, represent a significant and distinct adaptive type no longer seen today. How the elongated and mediolaterally compressed canines of these extinct cats were used in predation remains enigmatic (Martin 1980; Werdelin *et al.*, Chapter 2, this volume), although it is erroneously suggested that clouded leopards may provide clues (Christiansen 2006; but see Emerson and Radinsky 1980). However, this adaptation evolved independently in at least four mammalian lineages, including creodonts (e.g. *Apataelurus*), borhyaenid marsupials *Thylacosmilus* spp., nimravids (e.g. *Hoplophoneus*), and felids (e.g. *Smilodon*), and is clearly a successful adaptation under many circumstances (Emerson and Radinsky 1980; Martin 1980; Kitchener 1991; see Text Box 3.3).

In this chapter we explore the form and function of felids to show how the felid body plan is adapted to typical felid hunting, including waiting in ambush (pelage colouration and markings), and detecting (senses), stalking (skeleton), catching (paws and jaws), killing (teeth and claws), processing, and eating (teeth and tongue) prey (Table 3.1). We will also look in detail at the few specialist cats to see how the basic cat body plan has evolved to broaden the functional and ecological niche of the Felidae.

Our review is based on Ewer (1973) and Kitchener (1991), but we update these previous works with the latest anatomical, physiological, molecular, and functional analyses of current and past felids. In particular, the refinement of molecular and morphological phylogenies, coupled with the development of ancient DNA (aDNA) techniques, provide new opportunities to determine which character states are ancestral or derived, and whether apparently similar morphologies are the result of adaptive radiation or convergence between more distantly related lineages (e.g. Bininda-Emonds *et al.* 1999; Johnson *et al.* 2006b; Werdelin *et al.*, Chapter 2, this volume).

Finally, we look at morphological adaptations for communication in the Felidae by reviewing variation in pelage characters used in visual signalling, laryngeal anatomy in relation to vocalizations, and the vomeronasal organ and the variety of scent glands used in olfactory communication compared with those of other carnivorans.

Hunting and feeding adaptations

Pelage

Pelage colouration and markings are important in preventing a hunting cat from being seen while waiting in ambush or during the final stalk. Camouflage is also important, particularly for smaller felids, in concealment from larger predators. In general, cats' pelages help them to hide in two main ways. Firstly, crypsis, where pelage colouration matches the general habitat colouration, is typical of cats from open habitats, for example the sand cat, *Felis margarita*, from deserts of North Africa, the Arabian Peninsula and Central Asia, and the African wildcat, *Felis silvestris lybica*, from arid bush and semi-deserts throughout Africa and parts of the Arabian Peninsula. Secondly, disruptive colouration occurs, where a bold pattern of dark stripes, spots,

Table 3.1 Morphological adaptations for hunting are used for different purposes during the progress of a typical hunting and feeding sequence.

Hunting behaviour/ adaptation	Pelage	Senses	Skeleton	Paws and claws	Teeth and jaws	Tongue
Ambush	x					
Detecting prey	x	x				
Stalking and running	x	x	x	x		
Catching		x	x	x	x	
Killing			x	x	x	
Processing				x	x	
Eating				x	x	x

and/or blotches against a paler ground colour, helps forest cats disappear in dappled light produced by sunlight falling through vegetation. However, most cats show both, depending on their habitat. Therefore, cryptic sand cats and African wildcats also have indistinct stripes (and young pumas, *Puma concolor*, lions, and jungle cats, *Felis chaus*, are spotted, which may also have a cryptic function), and the ground colour of forest cats may match their environment closely. A common mutation in felids—melanism—may enhance crypsis. Black leopards, *Panthera pardus*, are apparently more prevalent in humid tropical forests of Southeast Asia, where closed tree canopies may allow little light to fall to the forest floor, making a uniform dark colouration more cryptic than spots (Pocock 1930, 1932; Guggisberg 1975). However, reports of truly black leopards are virtually non-existent in Africa (Pocock 1930, 1932; Guggisberg 1975), suggesting that it is important geographically where the mutation occurs, if it is to influence the proportion of melanism in a leopard population. Recently, Eizirik *et al.* (2003) investigated melanism in felids and found at least four independent genetic origins in extant cats, involving six different loci, depending on the species. Melanism is dominant in some species, for example jaguar, *Panthera onca*, and jaguarundi, *Herpailurus yagouaroundi*, whereas it is recessive in others, for example leopard, *Panthera pardus*, and domestic cat, *Felis catus*.

Ortolani and Caro (1996) and Ortolani (1999) examined the correlation between coat colour patterns in carnivorans, including cats, with specific environmental variables. Using only four character states and analysing only adult coat patterns, they found that a spotted coat pattern is ancestral for felids. Dark spots, which are common in felids, were significantly associated with closed habitats, arboreal locomotion, and feeding on ungulates, which would be expected for felids (Ortolani 1999). Vertical stripes and bands were significantly associated with grasslands and terrestrial locomotion, which is consistent with the tiger's and pampas cats,' *Lynchailurus* spp., pelage patterns.

Several authors have explored the evolution of the varying pelage patterns of cats. Using developmental data and comparisons between closely related species, Weigel (1961) hypothesized that all coat patterns derived from large spots, which broke down to varying degrees by developing a lighter centre, and then splitting into smaller spots as rosettes and, finally, into individual spots. Stripes could have been derived from any intermediate stage. However, Kingdon (1977) felt that small spots fused to form rosettes and bigger spots, as did Werdelin and Olsson (1997), who also pointed out that it is not clear how Weigel's hypothetical spotted ancestor was deduced (see also Werdelin *et al.*, Chapter 2, this volume).

Murray (1988) suggested a developmental model that requires a pre-pattern to be formed by diffusing morphogens (e.g. an activator and an inhibitor) and this is then reflected in the resulting coat pattern (Werdelin and Olsson 1997). The actual pattern that develops is predicted by the size and shape of the animal when the pre-pattern is formed, so that this simple model can apparently explain all mammalian coat patterns (but see Bard 1981). However, there is no phylogenetic aspect to this model and it is virtually unfalsifiable.

Using six basic coat patterns (flecks [small spots], rosettes, vertical stripes, small blotches, blotches, and uniform [i.e. no pattern]) and a reanalysed phylogenetic tree based on Salles (1992), Werdelin and Olsson (1997) showed that flecks are the primitive felid character state, from which the other five main coat patterns evolved (Werdelin *et al.*, Chapter 2, this volume). Even when using slightly different phylogenies (e.g. Collier and O'Brien 1985; Herrington 1986), the same evolutionary pattern persisted. Werdelin *et al.* (Chapter 2, this volume) provide a reanalysis of the data using the most recent phylogeny (Johnson *et al.* 2006b).

Pelage also insulates cats from their ambient climate, which is particularly important when waiting in ambush in the cold. Some species, for example the tiger, which have a wide geographical distribution, have variable fur length depending on their geographical origin (Mazák 1996; Table 3.2). Fur length may vary seasonally through the annual temperature cycle (Table 3.2). Fur density (hairs/cm^2) is also important in trapping an insulating layer of air close to the skin, but unfortunately data are limited for most felids. For example, the European wildcat, *Felis silvestris silvestris*, may have 5500–24,000 hairs/cm^2 in summer, which increases to 10,000–30,000 hairs/cm^2 in winter (Piechocki 1990). However, Haltenorth (1953) gives much lower values of 1800–4700 hairs/cm^2 in summer and 6700–8800

Table 3.2 Geographical and seasonal variation in fur length in tigers (Piechocki 1990).

Geographical origin	Summer fur length (mm)	Winter fur length (mm)
Amur	15–17	40–60
Chinese	8	30
Indian	8–15	17–25
Indochinese	10–15	15–20
Sumatran[a]	10–15	
Javan[a]	8–12	
Balinese[a]	7–10	

[a] No seasonal variation.

hairs/cm^2 in winter. Heptner and Sludskii (1992) give further comparative data for fur length, density, and ratio of underfur to guard hairs for various Russian felids (Table 3.3), but the data are sometimes uncertain, owing to problems in translation, lack of information about season, and the effect of preservation on fur density. Generally, fur density increases with decreasing body size (e.g. 8000 cm^{-2} in the Caucasian wildcat, *Felis silvestris caucasica*; cf. 2500 cm^{-2} in the tiger), perhaps as a function of greater surface area to body volume ratios, and species that live in the coldest climates have denser and longer fur than those in warmer climates (e.g. 9000 cm^{-2} in the lynx cf. 2500 cm^{-2} in the caracal, *Caracal caracal*).

Table 3.3 Comparative data on fur density, ratio of underfur to guard hair numbers, and hair lengths in felids (Heptner and Sludskii 1992).

Species	Density/cm^2	Underfur: guard hair ratio	Guard hair length (mm)	Top hair 1 length (mm)	Top hair 2 length (mm)	Top hair 3 length (mm)	Top hair 4 length (mm)	Underfur length (mm)
Felis silvestris caucasica	8000	10	44	41	39	37	36	34
Felis silvestris caudata	3500	8	51	43	39	37	35	35
Felis margarita thinobia (winter)	4532	a	55.3	49	47.7	43.9	38.4	30
Felis chaus	4000	12	61	56	47	43	39	30
Otocolobus manul manul	9000	10	69	47	44	43	42	40
Prionailurus bengalensis euptilurus	7000	30	47	49	49	39	35	34
Caracal caracal michaelis	2500	4–5	39	34	36	32	a	28
Lynx lynx	9000	12–13	51	38	39	38	35	31
Acinonyx jubatus	2000	6	35	33	32	32	31	25
Panthera uncia	4000	8	54	48	47	51	49	43
Panthera pardus (Central Asia)	3000	4	30–40	26–30	25–31	24–29	22–27	20–24
Panthera tigris	2500	1.4	27–48	31–40	29–38	25–33	25–31	23–24

[a] Missing data.

Senses

Most cats hunt using sight and/or sound to detect their prey, but some cats locate prey using their sense of smell (Kitchener 1991). Cats are also well endowed with whiskers, which they may use to navigate in the dark and to help them capture and kill prey.

Vision

Many cats are nocturnal or crepuscular hunters, but may be active during the day. Therefore, eyes must be sensitive enough to function in darkness, but also in broad daylight. For example, domestic cats can see light one-sixth as intense as we can, but are also able to see in daylight (Ewer 1973). Cats' eyes show a number of adaptations to allow this dual functionality. By increasing pupil size, cats' eyes can let in more light at night than ours, but they need a bigger, more convex lens and more convex cornea to refract the light sufficiently to focus on the retina. Therefore, compared with human eyes, the anterior chamber of the eye is bigger, and this may be accentuated in nocturnal species, as evidenced by the lynx (*Lynx* spp.) relative to the more diurnal puma (Walls 1942).

The retinae of domestic cats' eyes consist predominantly of rods, sensory cells adapted to detecting low light intensities, but they have a cone-rich patch at the centre of the retina for detecting higher light levels (Ewer 1973; Hughes 1977, 1985). Domestic cats' eyes are dichromatic, for detecting green and blue light, but it seems unlikely that they sense colour as we do; instead, they probably retain cones to be able to see in daylight, when their rods would be bleached by sunlight. However, some researchers claim to have found a small proportion of red cones in some individuals, suggesting individual variation in the level of trichromatic vision (see Jacobs [1993] for a discussion). Intriguingly, Spanish wildcats' eyes have very much higher densities of cones than those of domestic cats (>100% more in the *area centralis*; up to 100,000 mm^{-2} cf. 35–40,000 mm^{-2} in domestic cats), suggesting well-developed dichromatic colour vision, whereas rod densities in the *area centralis* of the domestic cat were 400% greater than those of the wildcat's (200,000 mm^{-2} cf. <50,000 mm^{-2} in wildcats; Williams *et al.* 1993). Therefore, we cannot assume that all cats have similar capabilities for daylight and colour vision, and why wildcats and domestic cats appear to have such different colour vision is unknown.

Another adaptation for increasing the light sensitivity of cats' eyes is a reflective layer of cells behind the rods and cones of the retina. This *tapetum lucidum* acts as a mirror that reflects incoming light back on to the sensory cells, thereby maximizing the stimulation of the retina and resulting in a brighter image with reduced resolution owing to blurring (Ewer 1973; Kitchener 1991). However, some cats may overcome this problem by having multifocus lenses, comprising concentric zones with different focal lengths (Malmström and Kröger 2006), each of which focuses on different relevant spectral ranges of the retina. Malmström and Kröger (2006) found that the domestic cat and, to a lesser extent, the Eurasian lynx, *Lynx lynx*, have multifocus lenses, but the Amur tiger does not.

Human eyes have a slight depression in the centre of the retina, the fovea or yellow spot, where cone density is highest, so that we have a high degree of visual acuity in the centre of our field of view. Moving away from the fovea, cone densities fall and rod densities increase before the density of receptor cells falls away at the periphery. Domestic cats do not have a concentrated fovea, but instead display a horizontally dispersed visual streak, which allows them to more easily detect prey movements against the horizon, an adaptation that reaches an extreme in the cheetah (Hughes 1977, 1985).

Given the predominance of rods over cones, cats have sacrificed visual acuity for greater sensitivity under low light levels. Moreover, the rods in cats' eyes show a greater degree of superposition, that is, combining inputs from several receptor cells (Ewer 1973), resulting in a visual acuity some ten times less than ours. Another consequence of combining maximum nocturnal sensitivity with good daylight vision is the need to vary considerably the quantity of light entering the eyes. The pupils of cats' eyes are able to expand considerably to allow maximum light to enter them, but a normal round pupil would be difficult to contract sufficiently in bright sunlight. Therefore, most cats have elliptical pupils with interlacing ciliary muscles that pull across each other, leaving just pin holes at each end of the closed-up pupil (Walls 1942; Ewer 1973). However, all cats of the *Panthera* and *Puma* lineages, and the Pallas's cat, *Otocolobus manul*, have round pupils, although these species tend to be diurnal hunters or from open habitats (Table 3.4). Malmström and Kröger (2006) found that slit-like pupils were associated with multifocus lenses. If a cat were to have a round pupil and multifocus lens, the outer concentric zones could not be used as the pupil contracted in brighter light. A slit-like pupil ensures that a cross section of the multifocus lens is available for vision under all light conditions. Malmström and Kröger (2006) found that the Eurasian lynx was intermediate in its lens structure and pupil shape between the slit-like pupil of the domestic cat and the round pupil (and single-focus lens) of the Amur tiger, but offered no explanation for the difference.

Finally, to pounce successfully on prey with paws and jaws, and for arboreal species to leap in trees, cats need good stereoscopic vision for judging distances. Indeed, they have the greatest degree of stereoscopy (120°) among mammals after that of arboreal anthropoid primates and humans (>138°; Heesy 2004). The overall visual field of domestic cats exceeds 180° (Hughes 1976), giving them an extensive field of peripheral vision for detecting potential prey and predators.

Table 3.4 The distribution of elliptical and round pupils in the genera of the Felidae.

Elliptical pupil	Round pupil
Felis	*Panthera*
Prionailurus	*Puma*
Leptailurus	*Herpailurus*
Caracal	*Acinonyx*
Profelis	*Otocolobus*
Catopuma	
Pardofelis	
Lynx	
Neofelis	
Leopardus	
Oncifelis	
Lynchailurus	

Hearing

Many cats rely on hearing to detect and capture prey when it cannot be seen easily. For example, the serval, *Leptailurus serval*, has very large mobile pinnae (external ears), which help detect the direction of movements of rodents in tall grasses or even burrowing underground (Ewer 1976; Geertsema 1976, 1985). Domestic cats hear over a range from 200 Hz up to 100 kHz (Ewer 1973), whereas our upper limit is around 20 kHz. However, to hear at 100 kHz the sound intensity would have to be considerable, so that effectively domestic cats can hear up to about 78 kHz (at 60dB SPL) and the jaguarundi up to about 65 kHz (Ewer 1973; Heffner and Heffner 1985). Notably, the high-frequency hearing of cats may allow them to detect the ultrasonic communication of rodents, which occurs typically in the range of 20–50 kHz (Sales and Pye 1974; Peters and Wozencraft 1989; Kitchener 1991).

Huang *et al.* (2000) investigated 11 species of felid, ranging from the sand cat (3 kg) to the tiger (180 kg), to test whether acoustic performance was broadly comparable for a structurally similar middle ear, and to see how this varied with increasing skull length. They found that middle ear frequency response patterns shifted to lower frequencies with increasing skull size. In most cats there is a distinct notch or minimum in the response pattern at 2–5 kHz, which probably results from the felid middle ear being characteristically divided by a bilaminar bony septum, linked by a small foramen. The frequency at which this notch occurs and the reflectance-minimum frequency (best sound energy transmission between ear canal and middle ear) decreased twofold (*c.* 1 octave) between the smallest and largest species. Therefore, as a result of scaling, bigger cats are able to hear fainter lower-frequency sounds better than smaller species. But what is the behavioural significance of these differences? Lions use low-frequency roaring (<200 Hz) to defend their very large group home ranges and to communicate with widely dispersed group members (Larom *et al.* 1997). Being more sensitive to low-frequency sounds, lions communicate more effectively over greater distances. Smaller species with smaller home ranges do not communicate by roaring and the cost of having a relatively larger head (and ears) would be too high and perhaps maladaptive. Larger cats prey on larger animals, which should produce lower-frequency sounds, so that scaling in sensitivity of low-frequency hearing is not unexpected.

Speculating as to why felids have a bony septum in their middle ears, Huang *et al.* (2000) predicted that without a bilaminar septum, the notch in the middle ear frequency response pattern would occur at about 10 kHz, rather than 4 kHz in the domestic cat. However, the shape of the external ear and ear canal are important for locating sounds in space, particularly in the 8–15 kHz region of the external ear transfer function, so that a higher frequency notch in middle ear response could affect cats' ability to locate sounds important in detecting prey and auditory communication with other cats.

The sand cat did not obey the scaling observed for other felids (Huang *et al.* 2000). It has large widely spaced pinnae (although apparently not exceptionally large for its size) for detecting faint sounds of rodents, but its auditory bullae are also vastly expanded, similar to those of many desert mammals, including foxes and rodents (Fig. 3.1). Desert and savannah/steppe cats have much larger auditory bulla volumes (ABVs) (1.22–1.91 cm^3 for Pallas' and sand cats, respectively) than those of forest cats (0.63–0.95 cm^3, including tigrina, *Leopardus tigrinus*, kodkod, *Oncifelis guigna*, and leopard cat, *Prionailurus bengalensis*; wildcat, *Felis silvestris*, ABV was also measured, but it is unclear which subspecies; Huang *et al.* 2002). A larger bulla volume might reduce air resistance to vibrating auditory ossicles so that they resonate at specific resonant frequencies corresponding to sounds made by movements of prey or predators (Webster 1962; Savage 1977). Although the areas of the pinnae of the sand cat (mean = 14.4 cm^3) and domestic cat (mean = 12.7 cm^3) are similar, the pinnae are more widely spaced in sand cats, probably allowing for a greater ability to pinpoint sounds (Huang *et al.* 2002). Moreover, the sand cat's ear canal length (mean = 28.3 mm, cf. 14.2 mm), bony ear canal cross-sectional area (mean = 64.3 mm^2, cf. 29.6 mm^2), and middle ear air-space volume (mean = 1.91 cm^3, cf. 0.90 cm^3) were about twice those of the domestic cat, resulting in a fivefold increase in acoustic-input admittance compared with that of the domestic cat in the low-frequency range (<0.5 kHz). There was no notch in the middle ear frequency response pattern, probably owing to the larger foramen in the bilaminar septum, but the adaptive advantage of this, if any, is unknown (Huang *et al.* 2002). Overall, the sand cat has increased auditory sensitivity of about 8dB for frequencies <2 kHz. In deserts, high-frequency sounds are attenuated rapidly by the low-humidity atmosphere, so the sand cat's enhanced hearing sensitivity means that it can hear low-frequency sounds (<500 Hz) of prey, predators, and conspecifics 400 m further away than a domestic cat under similar conditions. Pallas's cats and black-footed cats also inhabit desert or steppe habitats and have substantially larger auditory bullae and bony ear canal areas (both >33% more than expected from allometric relationship in the Felidae) compared with those of other felids, suggesting a similar hearing adaptation to that of the sand cat. Caracal and lion also show greater than expected (>15%) ABVs and bony ear canal areas (Huang *et al.* 2002).

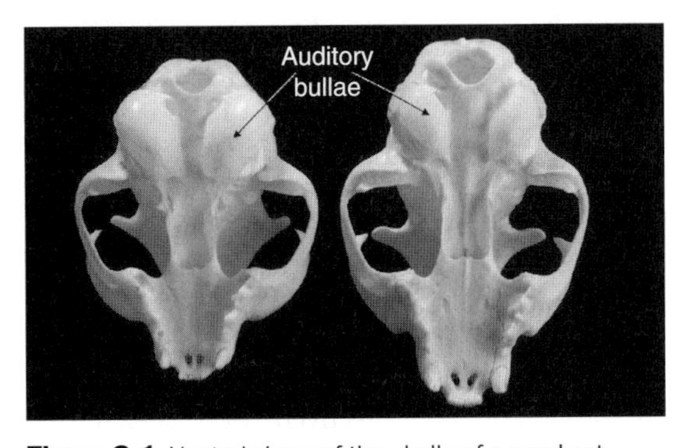

Figure 3.1 Ventral views of the skulls of a sand cat, *Felis margarita* (*left*) and a domestic cat, *Felis catus* (*right*). Note the much greater size of the auditory bullae of the sand cat's skull. © National Museums Scotland.

Smell

The sense of smell is used far less than hearing in hunting prey, but is important in intraspecific communication. However, ocelots, *Leopardus pardalis*, use their sense of smell to track down prey (Emmons 1988), and scent trails to hidden food are commonly used in zoos as a form of environmental enrichment for this species.

Compared with those of other carnivorans (e.g. dogs: mean 5% of cranial volume; viverrids: mean

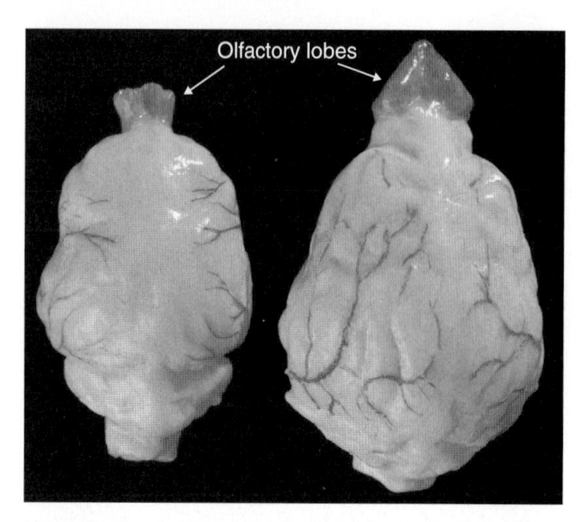

Figure 3.2 Dorsal views of cranial endocasts from the skulls of a clouded leopard, *Neofelis nebulosa* (*left*) and a maned wolf, *Chrysocyon brachyurus* (*right*). Note the much smaller olfactory lobes in the clouded leopard's endocast. © National Museums Scotland.

4.7%), cats have relatively small (mean 2.9%) olfactory lobes (Radinsky 1975; Fig. 3.2). However, although the surface area of the ethmoturbinals (which bear the olfactory epithelium) of domestic dogs is much larger (150 cm^2 in a German shepherd dog) compared with that of domestic cats (13.9 cm^2) and even taking into account that dogs are normally bigger than cats, the density of olfactory neurones in cats is more than six times greater than that of dogs (Dodd and Squirrel 1980). Therefore, dogs only have about twice the number of olfactory neurones as cats. Compared with those of other felids, the fishing cat has relatively small olfactory lobes (1.9% cranial volume; Radinsky 1975), which are more similar in relative size to those of aquatic otters, suggesting that smell is less important for communication or hunting in this semiaquatic felid (Gittleman 1991).

Touch

Whiskers or vibrissae are specialized touch-sensitive hairs, which are thicker than body hairs and embedded more deeply in the skin (Sunquist and Sunquist 2002). The whiskers sit in fluid-filled sacs, which have a rich nerve supply, so that if anything touches the whiskers it is detected instantly. Carnivorans typically have four main groups of whiskers; on the cheeks (genal), above the eyes (superciliary), on the muzzle (mystacial), and below the chin (inter-ramal), but cats lack the inter-ramal group (Pocock 1917b). The superciliary and genal whiskers are important for protecting the eyes, but the mystacial whiskers are probably most important for capturing prey, by ensuring that killing bites are correctly oriented. When pouncing on prey and just prior to capture, cats appear to extend their mystacial whiskers like a net to feel for the neck of the prey and their pupils expand at the same time (Leyhausen 1979). In domestic cats, the superior colliculus of the midbrain receives visual, acoustic, proprioreceptive, vestibular, and tactile inputs, which are organized spatially over its surface. Stein *et al.* (1976) found that superficial layers of the superior colliculus are exclusively visual, the intermediate layers are visual, somatic, and acoustic, and the deeper layers are non-visual. The tactile field was found to be in register with the visual field, such that the magnified representation of the *area centralis* overlapped with the magnified tactile representation of the face, so that cats are able to interpret their surroundings and interact with potential prey using a combined visuo-tactile sense (Stein *et al.* 1976).

Skeleton

Typically, cats either come across or flush prey when patrolling their home ranges or they lie in ambush by trails or burrow entrances waiting for prey to appear (Kitchener 1991). Once the prey has been detected (usually visually), most cats stalk by crouching low to the ground and approaching slowly, making use of any intervening cover. When within striking distance, the cat rushes forward or pounces on the prey. Some cats may also hunt arboreally, or hunt out of trees for terrestrial prey, or may hunt in water.

Although short, muscular limbs may provide powerful leaps and rapid acceleration over short distances, longer limbs allow for greater velocity over longer distances. However, long limbs are much less powerful and extremely long limbs cannot hold prey effectively. Therefore, limb length in cats is a compromise between long length for maximum speed over distance and short length for a powerful grasp, particularly in larger felids preying on large prey (Meachen-Samuel and Van Valkenburgh 2009a). In comparison to the humerus and femur, the lower leg is longer, the stance digitigrade, the metapodials are bound together by tendons and ligaments, and the carpals and tarsals are usually tightly bound together by ligaments, allowing for little movement between them (Ewer 1973). Long limbs do not necessarily indicate cursoriality; the serval has relatively long limbs compared with those of similar-sized cats, but this seems to be an adaptation for hunting in tall grass for rodents (Geertsema 1985), cf. the long-legged maned wolf, *Chrysocyon brachyurus*.

Arboreal cats have a flexible ankle joint, which extends and rotates through 180°, so that they can hold on with their hind feet alone when descending a tree head first, or even hang on to a branch with a single hind paw (Ewer 1973; Taylor 1989). This adaptation has evolved several times in carnivorans, including procyonids and the Madagascan euplerid, the fossa, *Cryptoprocta ferox* (Taylor 1989), and two or three times in felids, that is, the unrelated New World margay, and Old World clouded leopards and marbled cats (Andrew C. Kitchener [personal observation] both genera; Werdelin *et al.* Chapter 2, this volume; Fig. 3.3). However, it is unclear whether this adaptation is plesiomorphic for felids, which are generally believed to have evolved in forested habitats, since it was not found in *Pseudaelurus* and other basal felids (Turner and Antón 1997). Unlike canids, cats have a flexible elbow joint, which allows inward rotation of their paws to grab and manipulate prey (Andersson 2004).

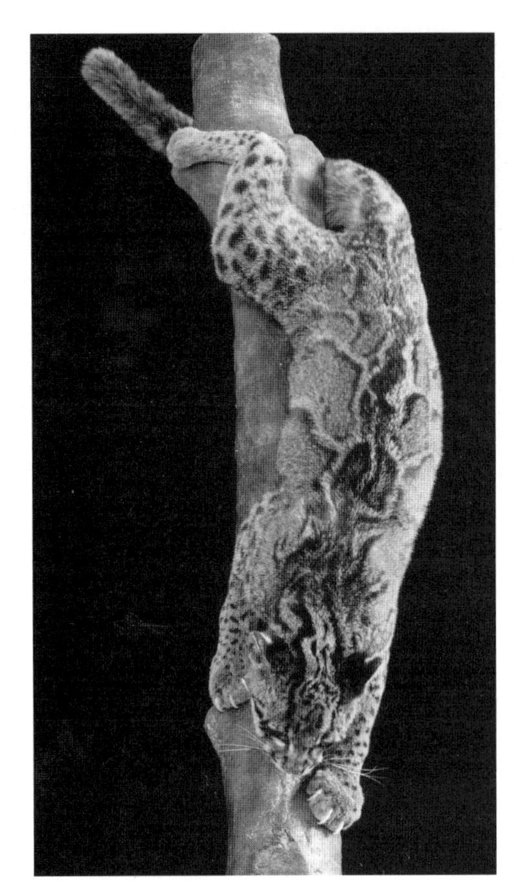

Figure 3.3 The marbled cat, *Pardofelis marmorata*, is able to descend trees head first, because its ankle joints can extend through 180° so that its claws can hold on to tree trunks. © National Museums Scotland.

Excluding the cheetah, cats' limbs are not specialized for cursoriality, unlike those of many ungulates, but cats may achieve greater running speeds than expected for their limb lengths, owing to other skeletal adaptations, allowing an extended stride length for greater speed. Stride length may be increased by extending and flexing the supple vertebral column. The intervertebral articulations are flexible about the long axis of the vertebral column, allowing cats to twist suddenly in three dimensions when in pursuit of prey taking evasive action, or when falling from low heights on to their feet. The clavicle is vestigial, freeing the shoulder joint to swing further back and forth, thereby permitting greater extension of the forelimbs. Scapular shape also reflects the lifestyle of the cat; cheetahs have narrower, rectangular-shaped scapulae for running, but the leopard has a broader, more fan-shaped scapula that enhances climbing (Fig. 3.4). The two muscles originating on the inner scapula of the cheetah, the *serratus magnus* and the *levator anguli scapula* muscles, occupy a deep, narrow strip for pulling the scapula back and forth while running, while in the leopard these muscles occupy a broader strip on the border of the scapula, allowing better forelimb adduction when climbing. The lion's scapular muscles are intermediate in their origination, as befits its intermediate climbing and running abilities (Hopwood 1947).

In a study of seven large felids, Gonyea (1976) found that forest dwellers, for example clouded

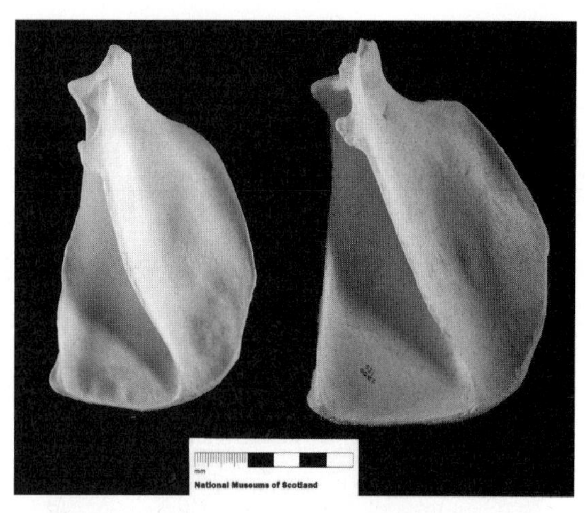

Figure 3.4 A comparison of the scapulae of a cheetah, *Acinonyx jubatus* (*left*) and a leopard, *Panthera pardus* (*right*). Note the much broader scapula of the leopard © National Museums Scotland.

leopard and jaguar, have relatively shorter forelimbs than those of open-habitat species, for example cheetah and lion (forelimb length <60% of back length). Moreover, these same two species were distinguished by having shorter radii relative to their humeri (radius length $<0.87 \times$ humerus length) than those of the more open-habitat lion and cheetah (radius length > 98% of humerus length). The cheetah and its extinct North American relatives in the genus *Miracinonyx* are the only big cats in which the radius is longer than the humerus (Van Valkenburgh *et al.* 1990). The metacarpals are also relatively shorter in forest felids; <9.0% of presacral vertebral column length (PVC) in jaguar and clouded leopard compared with >10.0% in lion, puma, and cheetah. A comparison of hind limb lengths reveals that the jaguar and clouded leopard are similar in having short hind limbs relative to those of other species (<67.5% of PVC, cf. >74% in cheetah), but differ in the proportions of the tibia and femur. The jaguar has an unusually short tibia (<90% of femur length), whereas in the clouded leopard the two elements are near equal in length. Intriguingly, the cheetah and snow leopard are the only species in which tibia length exceeds femur length. They also share a relatively long lumbar region, occupying about 35% of PVC. Although these two species are clearly not closely related (Johnson *et al.* 2006b), and while clearly the elongate tibia and lumbars represent adaptations for speed in the cheetah, they are more difficult to explain in the snow leopard. Unfortunately, Gonyea (1976) did not assess how these various proportions vary with body mass or phylogeny, making interpretation difficult. Day and Jayne (2007) tested whether within the felids, a more erect limb posture has developed with increasing body weight by measuring the body and limb proportions and analysing film of locomotion of nine felids species, ranging from the domestic cat (<4 kg) to the lion and tiger (192 kg). Despite the 50-fold range in body weights, there was no correlation with limb posture, and limb proportions did not vary between species, so that Day and Jayne (2007) concluded that the moderately erect limb posture of all felids is evolutionarily conservative and derives from a relatively large ancestral felid. However, it should be noted that the velocity of locomotion was broadly constant across all species, so that given their differences in body size, larger cats were moving relatively more slowly. This study also failed to find differences in limb and body proportions *contra* Gonyea (1976), Van Valkenburgh *et al.* (1990), and Andrew C. Kitchener (personal observation).

Relative to head and body length, tail length varies considerably among felids. Long tails appear to assist with balance either while climbing, leaping among branches, or turning sharply at high speed. For example, arboreal cats (e.g. the clouded leopards, marbled cats, and margay), have very long tails (62.3–106.7% of head and body length [HB]; Kitchener 1991). The snow leopard also has a long tail (69.2–84.0% HB; Kitchener 1991), which probably functions similarly when leaping between rocks and ascending or descending steep slopes and may also act as a muffler to insulate face and paws from the cold at high altitude when resting. Finally, the cheetah's long tail (53.7–54.5% HB; Kitchener 1991) is an important counterbalance when pursuing fast-running prey, which is apt to twist and turn sharply (Text Box 3.1). Tail length is shortest in lynx (15.9–29.5% HB; Kitchener 1991), which may reflect their distribution in northern latitudes, where cold winters may cause frostbite to longer extremities, but caracals and some sabretooth cats have short tails, so that there may be some other functional cause of short tail length. Cats that hunt aquatic prey, for example flat-headed cat, fishing cat (Text Box 3.2), and serval, tend to have short tails (26.1–40.3% HB; Kitchener 1991), as these may be a hindrance in water or swampy conditions.

When stalking prey, the tips of cats' tails may twitch uncontrollably, for example domestic cat and leopard. This might seem to be maladaptive behaviour, but it has been suggested that the twitching tail may distract or even attract prey, helping the cat hunt more successfully. However, this theory remains to be properly tested.

Paws and claws

Soft paw pads ensure a silent, firm grip during the stalking and final approach to prey, and when climbing with or without claws. Alexander *et al.* (1986) investigated the mechanics of the soft foot pads of domestic cats' paws and found that they placed their paws on the ground in such a way as to avoid a chattering vibration caused by the viscoelastic properties of the foot pads. Cheetahs have hard ridges on their foot pads that are thought to aid traction (see Box 3.2).

For some cats, there is a risk that their environment will not support their weight. Snow and sand are soft substrates, which may deter predators because they sink into them. For example, soft snow more than 0.5 m deep prevents Amur tigers from hunting successfully, leading to starvation in extreme cases (Heptner and Sludskii 1992). To spread its weight over the snow, the Canada lynx, *Lynx canadensis*, has maximized foot surface area by having elongated phalanges that are more than half the length of their metacarpals (Van Valkenburgh 1987), as well as having foot pads covered with long fur for insulation, to create a sort of snowshoe (Fig. 3.5). The sand cat displays a similar adaptation, but the fur covering the pads prevents sinking in soft sand and insulates the foot pads from extremes of both hot and cold (Fig. 3.5). Interestingly, elongated phalanges relative to metacarpals are also found in arboreal cats (margay, clouded leopard) to maximize paw size for grasping when climbing (Van Valkenburgh 1987).

Cats are renowned for their protractile claws. Both the terminal and penultimate phalanges of felids (and some other carnivorans, e.g., viverrids and related families) are modified to allow passive retraction and active protraction. The penultimate phalanges have shallow lateral depressions, in which the terminal phalanges sit in the retracted state, which Bryant *et al.* (1996) call hyper-retraction. The claws remain hidden within fleshy sheaths, being held in place by ligaments and tendons, so that they are not blunted by contact with the substrate (Gonyea and Ashworth 1975). The shape of the terminal phalanx produces a curving claw that would eventually grow back into the toe pad, but its method of growth allows the claw to split along its ventral side to retain a sharp point (Fig. 3.6). Cats use scratching posts, usually trees with soft bark, to maintain sharp claws. The ever-sharp claws are quickly protracted

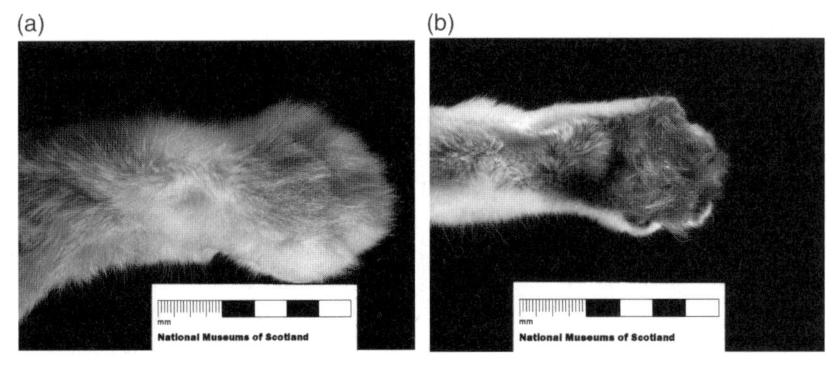

Figure 3.5 The paw pads of the (a) Canada Lynx, *Lynx canadensis*, and (b) the sand cat, *Felis margarita* are covered with fur. This fur protects the lynx's paws from extreme cold and helps the foot to act like a snowshoe, whereas it helps the sand cat to move over hot shifting sands. © National Museums Scotland.

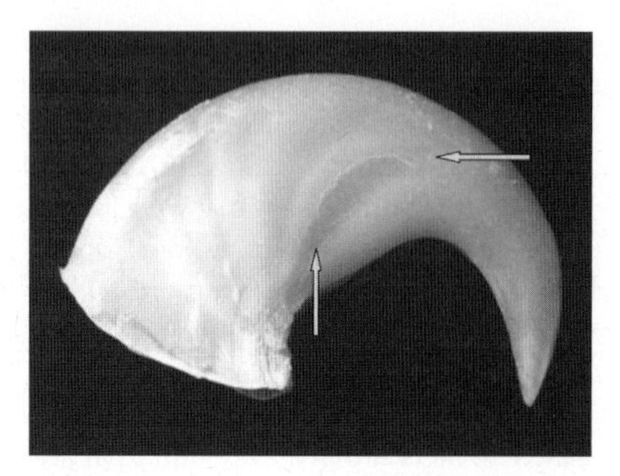

Figure 3.6 Lateral view of a claw of a leopard. Note the curved lines running across its surface, indicating planes of weakness where the claw would normally break to keep it naturally sharp. © National Museums Scotland.

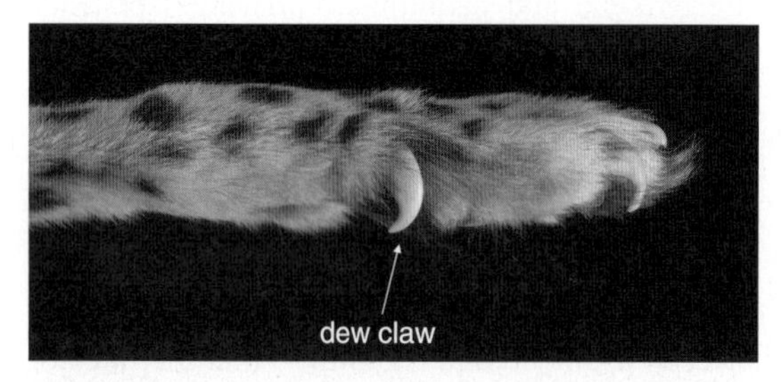

Figure 3.7 Lateral view of the left paw of a cheetah, *Acinonyx jubatus*, showing the greatly expanded dew claw, which is used to bring down prey. © National Museums Scotland.

like switchblades for grasping prey, fighting conspecifics, or climbing trees. The dew claws of the forelimbs do not reach the ground or protract, but are used in grasping prey (Bryant *et al.* 1996). Cheetahs and pumas have relatively large dew claws compared with those of other big cats (Meachen-Samuels and Van Valkenburgh 2009b), the cheetah having relatively the biggest of all (mean dew claw dorsoventral ungual height 140% of second digit's, cf. 130% in puma; 110% in tiger, lion, and leopard; Londei 2000; Fig. 3.7). The hind claws are generally less recurved and are more blade-like (Bryant *et al.* 1996), but may be used in lacerating or even disembowelling captured prey when raked by hind feet.

Skulls, jaws, and teeth

Relative to other carnivorans, felids have foreshortened skulls with roundish profiles. Their orbits are large and the zygomatic arches flare away from the cranium, providing space for strong jaw-closing muscles (the *temporalis* complex) that originate on the sides of the braincase (Radinsky 1981). These features scale differently with body size; small cats have relatively larger brains and eyes, but smaller jaw muscles than large cats. The negative allometry of brain and eye size is common among mammals (Jerison 1973) and the positive allometry of the feeding apparatus scales as expected. The braincase of smaller cats provides sufficient surface area for the attachment of the *temporalis* and *masseter* muscles, but as skull dimensions increase, the braincase surface area increases more slowly than jaw muscle volume for equivalent scaling. Therefore, larger cats evolved prominent sagittal and occipital crests for additional surface area for jaw muscle attachments (Fig. 3.8). However, the skulls of small cats do not simply scale up to the sizes of large cats (Werdelin 1983), which may reflect developmental modularity, but this needs to be tested.

The short snouts of felids relative to those of canids and most other carnivorans confer two advantages in feeding. Firstly, when one upper canine is loaded more heavily than its opposite during a strong bite, it will tend to twist the skull upwards on that side. This effect is more pronounced in long-snouted species, for example canids, and is resisted by thickening the skull in areas of maximum torsional stress. However, bone is energetically expensive for an organism to produce and Covey and Greaves (1994) determined that cranial thickening is not necessary if the ratio of length to width of the jaws does not exceed $\pi/2$ (Fig. 3.9). Although the longer-jawed canids are at this limit, the shorter-jawed felids are well within it, suggesting that other factors are important in constraining felid jaw length. For example, a short jaw length may allow a greater distance between the canines for a more effective killing bite for larger prey or more effective use of the incisors for stripping muscle from bone. The second advantage of a short snout is increased mechanical advantage of the jaw muscles for a bite with the canines and incisors. By bringing the canine teeth closer to the jaw joint or fulcrum, felids decrease the moment arm of resistance and enhance bite force. Cats kill in one of two ways. A nape bite is typical for smaller prey and involves dislocation of the cervical vertebrae by the canines, which are equipped with large nerve fibres

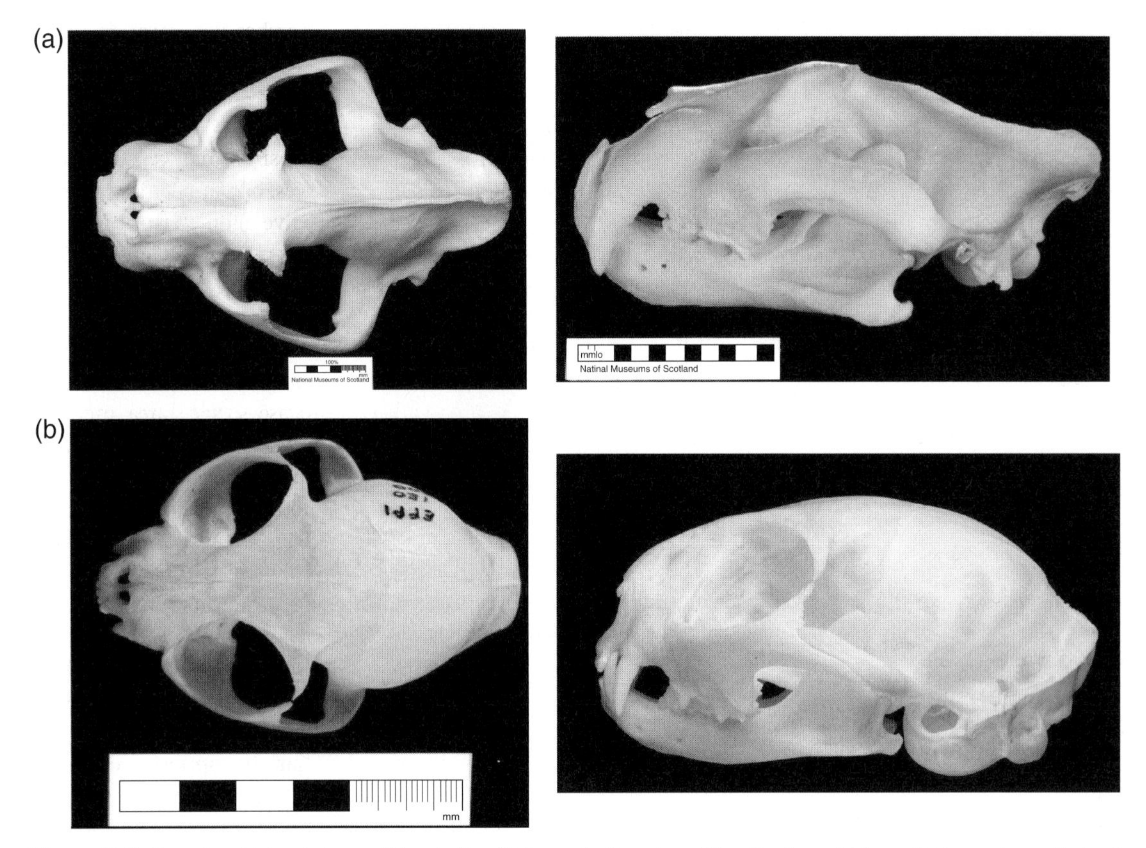

Figure 3.8 Dorsal and lateral views of the skulls of (a) a male Sumatran tiger, *Panthera tigris*; and (b) a rusty-spotted cat, *Prionailurus rubiginosus*. The larger tiger skull has a relatively smaller cranium for the attachment of the jaw muscles, so that sagittal and occipital crests are commonly seen in larger cats to provide for a greater surface area for jaw muscle attachment. © National Museums Scotland.

that are thought to allow them to feel their way during the bite (Leyhausen 1979), although this has never been tested experimentally. For larger prey, a throat-or snout-covering bite is used, both of which typically kill by suffocation and may not even break the skin of the prey. In both small and large cats the ability to kill quickly is enhanced by maximizing bite strength. For big cats using the throat bite, a maximal bite strength is required for this to be effective.

The strong jaw muscles and forceful biting of felids (see Wroe *et al.* [2005] for estimates of bite forces) affect the geometry of their mandibles as well as their skulls (Meachen-Samuels and Van Valkenburgh 2009b). Killing bites load the jaws anteriorly, bending the dentary bone downwards in the sagittal plane. Uneven loading of the lower canines can create torsional stresses, which can be resisted by deepening the dentary and thickening the cortical bone along its length. Modelling carnivoran lower jaws as beams, Biknevicius and Ruff (1992) showed that the jaws of large felids are more strongly buttressed against mediolateral and dorsolateral bending than those of canids, and more strongly buttressed against mediolateral loads than those of hyaenids.

Felids have fewer teeth (30) compared with canids and ursids (42), as a result of shorter jaws and bigger canines (see Fig 10a and b). In particular the number and diversity of cheek teeth is reduced by loss of post-carnassial molars and the most anterior premolars, leaving a dentition that is highly specialized for carnivory. Their carnassials are simplified into two opposing blades for slicing through skin and other tissues, but a minimal crushing function is produced by occlusion between the protocone of the upper carnassial

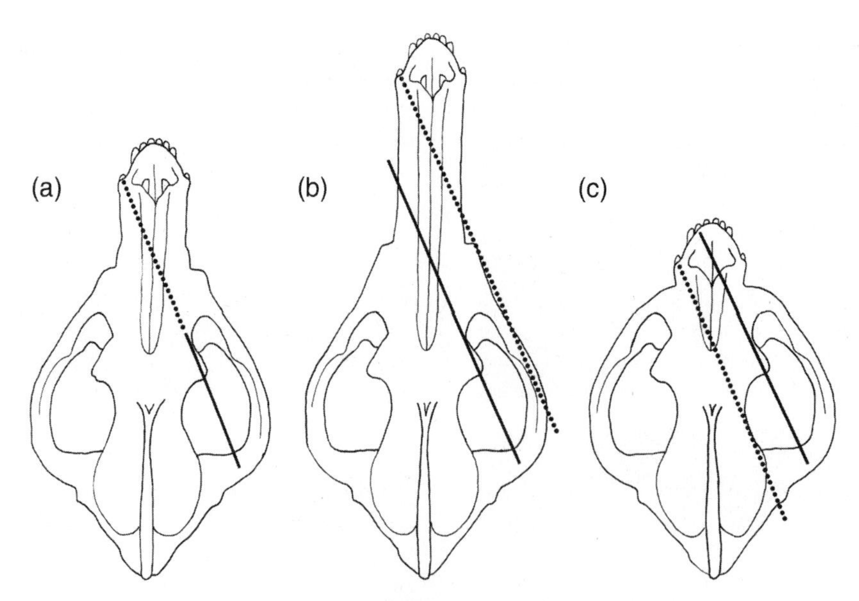

Figure 3.9 For these three skulls the length:width ratio is equal to (a) $\pi/2$; (b) $>\pi/2$; and (c) $<\pi/2$. The torsional helices are represented by lines from the left canine and the opposing jaw joint, assuming single-canine loading. (After Covey and Greaves 1994.)

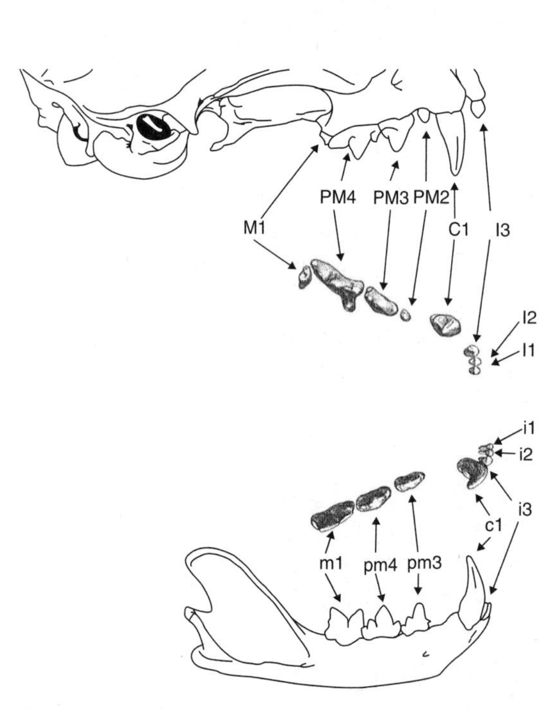

Figure 3.10a The dentition of the wildcat, *Felis silvestris*, in lateral and crown view, showing the typical reduced dentition of felids compared with that of other carnivorans. (After Miller 1912a.)

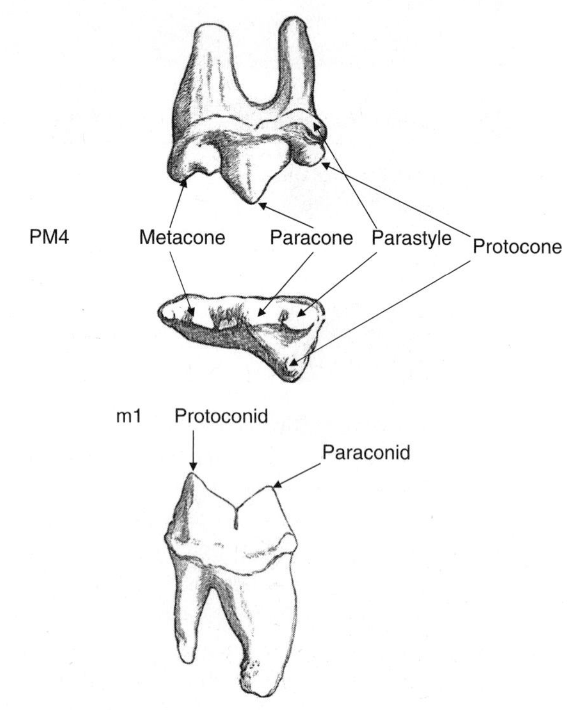

Figure 3.10b In particular, the carnassial teeth of the tiger, *Panthera tigris*, shows how a parastyle has developed on the fourth upper premolar (PM4) and the protoconid and paraconid have been expanded in the first lower molar (m1) to increase the sectorial or blade-like functioning of these teeth for cutting more efficiently through soft body tissues. (After Flower and Lydekker 1891.)

and lower fourth premolar in all species but the cheetah, in which the carnassial protocone is absent. The premolars are narrow laterally with a larger central cusp flanked in front and behind by smaller cusps. The flat-headed cat is unusual in having very tall, pointed premolars that are believed to be specialized for grasping slippery fish (Fig. 3.11). The largest teeth are the upper canines, which tend to be roundish in cross section in all but most sabretooths (see Text Box 3.3) and to some extent the clouded leopards, and unlike the knife-like canines of canids (Van Valkenburgh and Ruff 1987). The round cross-sectional shape of cat canines reduces overall sharpness (although the posterior canine edge may be blade-like and sharp to enhance penetration), but increases resistance to bending and fracture from loads in any direction. Because cats kill with a single strong bite, their canine teeth may encounter bone or be unpredictably loaded by struggling prey. Indeed, despite having robust canines, 20% of skulls of larger felids in a survey had broken at least one tooth and over half of these were canine teeth (Van Valkenburgh 1988).

Cats' teeth have fewer different functions than those of other carnivorans, perhaps reflecting the cats' obligate carnivory. Canines are clearly important for killing and incisors play an important role in plucking fur and feathers from carcasses prior to feeding, and pulling meat from carcasses during feeding. The carnassials have traditionally been thought of as the principal teeth for slicing through muscle, but studies of feeding in several large African carnivorans, including lion and cheetah, revealed that they were used primarily for slicing through tough skin and pieces of muscle too large to swallow, and that incisors are more important for removing meat from bone (Van Valkenburgh 1996). Mixtures of muscle and bone, for example ribcage or pelvis, were processed using both carnassials and adjacent premolars. To allow the carnassials to engage, the jaws must shift sideways, because the lower tooth rows are narrower and fit between the upper tooth rows. This lateral movement is facilitated by the cylindrical mandibular condyle (Fig. 3.12). When canines are being used, the carnassials do not engage,

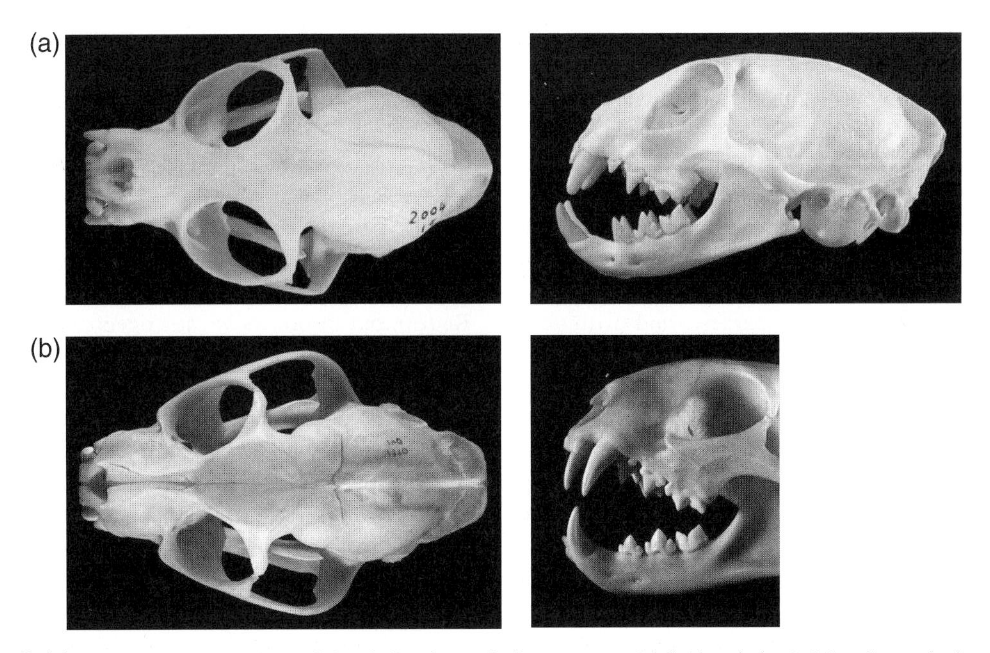

Figure 3.11 Dorsal and lateral views of the skulls of two piscivorous cats: (a) flat-headed cat, *Prionailurus planiceps*; and (b) fishing cat, *Prionailurus viverrinus*, showing narrow skull with extended toothrow with well-developed premolars for grasping fish. © National Museums Scotland.

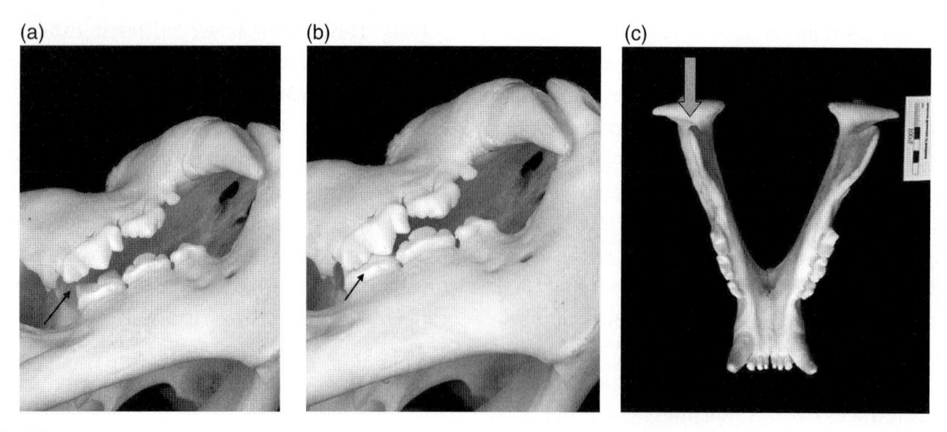

Figure 3.12 Ventral view of the skull of a Sumatran tiger showing how cylindrical mandibular condyles allow mandible to slide from side to side to (a) avoid carnassials accidentally hitting each other during killing bite; and (b) to allow carnassials to engage when cutting through skin and other tissues. Dorsal view of mandible to (c) shows cylindrical shape of mandibular condyle. © National Museums Scotland.

thereby preventing damage to these important teeth during killing.

Tongue

A cat's tongue is covered by horny papillae, which are used as a comb when it grooms. It is vital that cats retain a well-maintained pelage to insulate them from the cold or heat. Old cats in zoos that are unable to groom properly soon develop severely matted fur. Moreover, a cat's tongue may be useful in feeding because the horny papillae may function as a rasp to pull meat off bones.

The domestic cat has similar taste buds with taste receptor cells (TRCs) to those of humans, but is unable to taste sweet tastes and has a low sensitivity to salt. In mammals TRCs respond to salty, sweet, sour, bitter, and umami (savoury) flavours (Chandrashekar *et al.* 2006), but the domestic cat's sense of taste reflects its obligate carnivory (Bradshaw *et al.* 1996). The taste buds of the facial nerve respond to amino acids (i.e. umami TRCs) that people describe as 'sweet' (N.B. not the sweetness of sugars, which cats cannot detect; see below), including L-proline, L-cysteine, and L-alanine, but are strongly inhibited by those that people described as 'bitter', including L-tryptophan, L-isoleucine, and L-phenylalanine, suggesting that cats and people are experiencing similar tastes in response to these amino acids. Monophosphate nucleotides accumulate in prey body tissues after death and also inhibit the amino-acid-responding units of the facial nerve, so that cats can detect the freshness of a scavenged prey item. Cats are able to detect salt (sodium and potassium chlorides), but only at very high concentrations (>0.05 M), and it has been suggested that the high salt content of their prey means that unlike herbivores they do not need to be aware of the sodium content of their food (Bradshaw *et al.* 1996). The mammalian sweet-taste receptor is based on a dimer of two proteins (T1R2 and T1R3) that are the products of two genes (*Tas1r2* and *Tas1r3*), but in domestic cats, tigers, and cheetahs the *Tas1r2* gene has a 247-base-pair microdeletion that prevents this gene expressing itself normally (Li *et al.* 2005). As a result, unlike some other carnivorans, cats cannot taste sweet foods such as sugars, which reinforces their adaptation as obligate carnivores. However, Konecny's finding (1989) that fruit occurred in 14% of 27 wild margay scats is of interest. It is possible that other felids may have the ability to taste sugars, unless they are eaten as a source of water, a particular nutrient and/or roughage; lions, leopards, and cheetahs in the Kalahari Desert eat tsama melons, *Citrullus lanatus*, and other moisture-rich plants as a source of water (Bothma 1998).

Gut

Cats have very short intestines, compared with those of other mammals. Meat is comparatively quick and easy to digest, so that only a short small intestine is required. Interestingly, European wildcats have shorter intestines than those of domestic cats; for example, male Scottish wildcats, *Felis silvestris grampia*, have a mean intestinal length of 1.583 m compared with a mean of 2.094 m in male domestic cats (Kitchener 1995). This difference is usually explained as being the result of domestic cats becoming increasingly adapted to a less carnivorous diet (Kitchener 1995). However, the European wildcat is not an ancestor to the domestic cat and there are too few data from African or Asian wildcats to determine whether this difference is due to phenotypic plasticity or phylogeny. Feral domestic cats and wildcats in Scotland have an equally carnivorous diet, comprising mainly rabbits, *Oryctolagus cuniculus*, but their intestinal lengths remain significantly different.

Box 3.1 Cheetah

The cheetah has probably the most highly derived body plan of all felids, but is a member of the puma lineage, including also puma and jaguarundi (Johnson *et al.* 2006b). The cheetah is adapted to cursorial diurnal hunting, although it is able also to hunt at night by stalking smaller prey (Hamilton 1986). Its prey are typically gazelles—*Gazella* spp., *Nanger* spp., and hares, *Lepus* spp.—which are able to run fast, twisting and turning sharply. Hayward *et al.* (2006b) examined the prey preferences of cheetahs by analysing data from 21 studies in six countries. They concluded that cheetahs were adapted morphologically to capture prey weighing 23–56 kg (modal prey weight = 36 kg), which they can kill with minimal risk of injury and eat before kleptoparasites arrive.

Cheetahs usually live in open savannas and deserts (but may also inhabit woodland savannas), where they must be able to see prey clearly against the horizon. The visual streak on their retinae is very long and narrow to reflect this greater sensitivity to movements of prey, competitors, and predators against the horizon (Hughes 1977).

Cheetahs have been recorded running up to 64 mph (*c.*103 km/h; Sharp 1997), but may be able to achieve greater speeds. Cheetahs have relatively the longest limbs of any felid in order to achieve a longer stride length and hence faster speed (forelimb 66.8% and hind limb 74.1% of PVC; Gonyea 1976), but their limbs are relatively shorter than those of cursorial ungulates. Hence they retain a limb short enough to be powerful enough to knock over prey and long enough for sprinting fast. Cheetahs manage to capture their ungulate prey because they can accelerate rapidly and can achieve a longer stride length by flexing and extending their vertebral columns. Hildebrand (1959, 1961) estimated that the vertebral column was responsible for 11% of the stride length of 6.9 m at 56 km/h. The loosely attached forelimbs allow the scapulae to contribute towards relatively longer legs and hence stride length, adding 12 cm or 2% at 56 km/h (Hildebrand 1959, 1961). Hildebrand and Hurley (1985) modelled the legs of mammals as three-segment oscillating systems in order to estimate their kinetic and potential energies when running. For cheetahs, the action of the legs and their greater distal mass compared with those of another speedy mammal, the pronghorn, *Antilocapra americana*, resulted in higher energy costs to move them, which may explain why cheetahs cannot run at speed for more than a few hundred metres. In contrast, Taylor *et al.* (1974), found no difference in the energy costs of running between a cheetah, a gazelle, and a goat, *Capra hircus*.

The cheetah's paws and claws are also adapted for cursoriality. There are strong ridges along the paw pads, which are said to act like the tread on car tyres and the claws become blunt, being used as running spikes for better grip (Ewer 1973). Cheetah claws are less protractile than other cats' and they are not hidden within fleshy sheaths when retracted (Pocock 1916; Gonyea and Ashworth 1975; Russell and Bryant 2001). However, the cheetah has very large, recurved and sharp dew claws,

(continued)

Box 3.1 (Continued)

which are used to snag and pull prey off balance (Fig. 3.7); cheetahs typically knock their prey over when capturing them (Londei 2000). Interestingly, the cheetah has a reduced supinatory (grasping) ability compared to that of other cats (Andersson 2004), which probably reflects its adaptation to a cursorial method of hunting.

Being diurnal and dependent on vision to detect prey, a tactile sense is not so important, so that cheetahs have poorly developed whiskers. Cheetahs have very wide nasal passages, which are important in allowing them to recover their breath and prevent their brains overheating after hot pursuit. Taylor and Rowntree (1973) found that cheetahs were unable to dissipate all the heat that they generated while running after prey and that after only about 500 m their body and brain temperatures were near their lethal limits. Having wide nasal passages to cool down quickly is very important for cheetahs, but coupled with short jaws means that cheetahs have only small canines with short roots. However, the cheetah's carnassials are the most sectorial found in all living felids for slicing quickly through skin and muscle before competing carnivorans can steal their prey. The jaws are so short that there is no diastema to allow canines to penetrate fully. Therefore, they usually kill prey with a throat bite that may either occlude the trachea or possibly pinch the carotid artery. A recent multivariate statistical analysis of skull measurements of big cats has confirmed the morphological differences that distinguish the cheetah from other big cats, including narrow teeth, small canines, and a wide brain case for its size (O'Regan 2002). The broader skull seems to be a characteristic that has been retained from a smaller ancestor.

Palaeontological evidence showed that the cheetah and the extinct American cheetahs, *Miracinonyx* spp., are closely related and share a similar body plan for cursorial hunting (e.g. Adams 1979). However, a recent study of aDNA (Barnett *et al.* 2005) showed that the two genera had evolved independently from less cheetah-like ancestors within the puma lineage, so that their morphological similarity is due to convergence, rather than common ancestry. Presumably, *Miracinonyx* was adapted to hunting pronghorns (Family Antilocapridae), which are morphologically convergent with highly cursorial Old World gazelles.

Box 3.2 Piscivores

The fishing cat and flat-headed cat are two closely related felids that have evolved to specialize in catching fish and other aquatic vertebrates (Lekagul and McNeely 1988; Breeden 1989; Sunquist and Sunquist 2002). Although they retain the basic cat body plan, they show some adaptations for a semiaquatic hunting niche. Unlike other cats, these two species have a relatively narrow and long skull and jaws (Muul and Lim 1970; Lekagul and McNeely 1988), which is an adaptation common to piscivorous vertebrates, for example, river dolphins and gharial. Presumably this provides a longer tooth row to grab hold of slippery prey and less resistance when the head is plunged into water to grip prey, which may or may not have been grasped by one or both forepaws. The result of a longer tooth row is that the upper first and second premolars are well developed, particularly in the flat-headed cat, as are the corresponding lower premolars (Fig. 3.11; Muul and Lim 1970). The orbits on the flat-headed cat's skull are set forward in the same plane to give a high degree of stereoscopic vision to judge distances when pouncing on aquatic prey.

The fishing cat and flat-headed cat both have short tails compared with most felids' (flat-headed cat 26.8–31.7% HB; fishing cat 33.3–40% HB; cf. congenors: leopard cat 41.4–53.9% HB; rusty-spotted cat 54.0–65.3% HB; Sunquist and Sunquist 2002); long tails would presumably be an encumbrance in the water, unless they could be used for locomotion. Much is made of their feet being webbed (Lekagul and McNeely 1988; Sunquist and Sunquist 2002), but Pocock (1917b) illustrated the forefeet of a bobcat, *Lynx rufus*, and a fishing cat, and those of the latter appear only marginally more webbed. Although webbing would be an advantage when swimming, it would impede the feet when the claws are spread to strike prey. The claws of both the fishing cat and flat-headed cat are normally retracted, but the fleshy sheaths are not well developed, so that the tips of the claws protrude when at rest. However, it is unclear why these cats differ from other felids in this respect. Overall the flat-headed cat shows greater adaptation to piscivory than the fishing cat, including its behaviour. When a captive juvenile flat-headed cat was given a large bowl of water, it jumped in immediately and began to play (Muul and Lim 1970). It submerged its head completely to depths of about 12 cm to seize pieces of fish, and sometimes played in the water for hours.

There is an interesting convergence in the reddish pelage between the flat-headed cat and the aquatic genet, *Genetta piscivora*, of Central Africa. Both are semiaquatic piscivores of similar size. Red light attenuates quickly in water with increasing depth (Braun and Smirnov 1993), so perhaps this pelage colouration helps both these unusual piscivores remain hidden from their fish and amphibian prey, which may have well-developed trichromatic colour vision. While this is speculation on our part, it should be noted that the bay cat, *Catopuma badia*, and jaguarundi also have reddish pelages, but they are polymorphic for coat colour, so that if there is any selection for red pelage colouration in these two species, it is not related to a semiaquatic niche.

Box 3.3 Sabretooth cats

Although they seem bizarre by comparison with modern felids, sabretooths were the common form of large cat-like predator for most of the past 40 million years (Turner and Antón 1997; Werdelin *et al.*, Chapter 2, this volume). Found on every continent except Australia and Antarctica, mammalian sabretooths evolved repeatedly and independently, first as creodonts, then nimravids, and finally among felids and even marsupials (Emerson and Radinsky 1980; Van Valkenburgh 2007). While never speciose, there were usually two or three present in any ecosystem. Their widespread occurrence and persistence indicates their success as predators and dispels any notion of them as evolutionary failures owing to their 'extreme' evolutionary design.

Among felids two distinct forms of sabretooth evolved within the subfamily Machairodontinae (Fig. 3.13). Dirk-toothed cats, for example *Smilodon*, have very elongate, knife-like canines, with tiny or no serrations, whereas scimitar-toothed cats, for example *Homotherium*, have shorter, coarsely serrated canines that are broader anteroposteriorly (Martin 1980). In addition, dirk-toothed felids were ambush predators with relatively robust, stout limbs, compared with the more lion-like, gracile limbs typical of scimitar-toothed felids (Anyonge 1996). Dirk-toothed and scimitar-toothed species often coexisted in the same geographical areas, and the dissimilarities in dental and skeletal morphology imply differences in hunting behaviour and/or habitat choice that probably minimized competition between them. Interestingly, at least one scimitar-toothed lineage evolved somewhat stout limbs late in its history (*Xenosmilus hodsonae*; Martin *et al.* 2000).

(continued)

Box 3.3 (Continued)

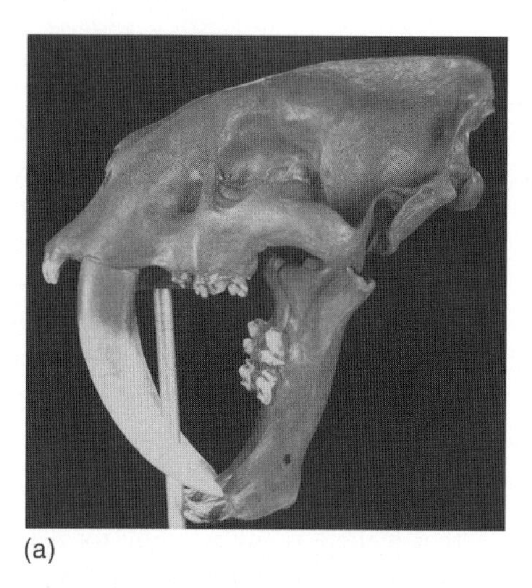

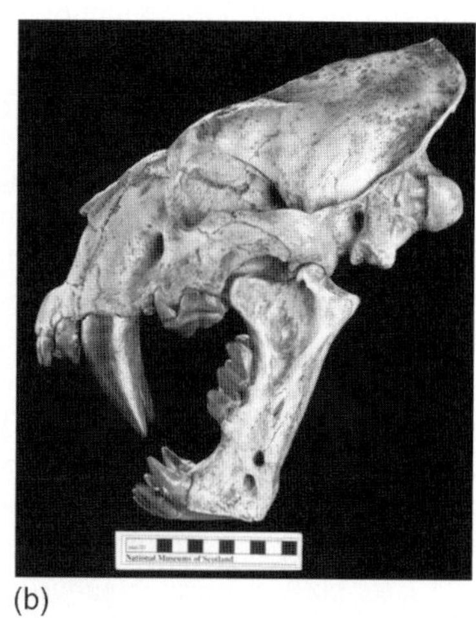

(a) (b)

Figure 3.13 Scimitar-toothed cats such as (a) *Smilodon fatalis*, have short, broad upper canines with a fine serrated following edge, whereas dirk-toothed cats, such as (b) *Machairodus giganteus*, have short, narrow canines with a coarse, serrated edge. © National Museums Scotland.

All sabretooth cats and non-cats converge on a similar suite of craniodental modifications, suggesting that there are limited ways to evolve a successful sabretooth (Emerson and Radinsky 1980; Van Valkenburgh 2007; Slater and Van Valkenburgh 2008). Initially, there is selection for longer upper canine teeth, which demands an increase in jaw gape so that upper and lower canine tips can separate sufficiently for biting (Fig. 3.14). A wider gape results from the jaw joint descending and the coronoid process on the dentary is reduced, with the latter resulting in decreased mechanical advantage for the *temporalis* musculature. To compensate for loss of leverage, there is a vertical reorientation of the *temporalis* fibres and enlargement of neck muscles that pull the skull downwards, driving the canines into the prey. Incisor teeth became large and procumbent and were likely to have been important in both killing and feeding (Biknevicius *et al.* 1996). The cheek teeth were reduced in number and extremely blade-like, with no adaptations for bone crushing or cracking. In addition to these skull and dental changes, machairodonts had relatively long, muscular necks that probably functioned to position and stabilize the head for a precise killing bite (Antón and Galobart 1999).

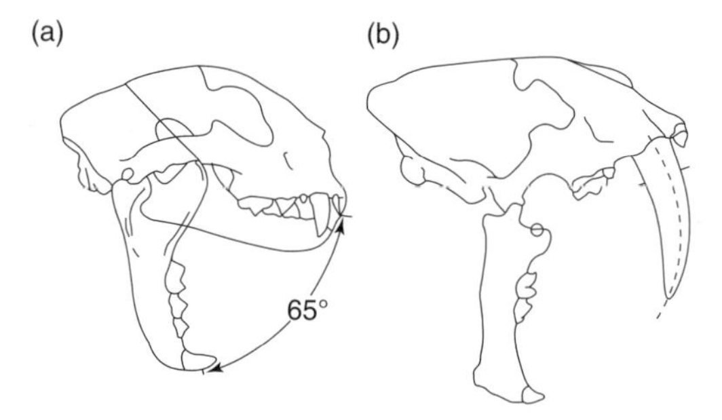

Figure 3.14 A schematic diagram to compare the jaw gapes of (a) modern felids (*Puma concolor*); and (b) sabretooth felids (*Smilodon fatalis*). Note the much wider gape of the sabretooth (>90°), compared with that of the modern felid (*c.* 65°). (After Kitchener 1991a.)

The advantages of having sabre-like canines are not absolutely clear, but most agree that they allowed individuals to kill larger prey rapidly. Predators are at increased risk of injury or death when attacking very large prey and the ability to end the struggle quickly would be favoured. The orientation of the killing bite was probably determined largely by the need to immobilize dangerous prey promptly without breaking the slender upper canines. To avoid canine fracture, sabretooths would have bitten areas of the body where contact with bone was unlikely and death would be swift. The best choice would have been a bite to the ventral throat that could sever critical vessels or the trachea, avoiding contact with cervical vertebrae. The bite force of a sabretooth was probably greater than that of living big cats. The cross-sectional geometry of the mandible of *Smilodon* reveals similar strength and rigidity to that of living spotted hyaenas, *Crocuta crocuta*, which are known for their powerful bone-cracking jaws (Biknevicius and Van Valkenburgh 1996, but see McHenry *et al.* 2007).

Cat communication

Making signals

Aspects of the colouration and markings of felid pelages may have an important role in visual communication. For example, the long, dark ear tufts of lynx and caracals, *Caracal caracal*, accentuate the movement of ears when signalling visually. Some felids have much smaller ear tufts, for example jungle cat and Chinese steppe cat, *Felis bieti*. The caracal, in particular, uses its ears in semaphore as a means of intraspecific communication (Kingdon 1977). Interestingly, felids with distinct or incipient ear tufts tend to have short tails, suggesting that selection for ear tufts or against longer tails may affect the primary means of visual signalling in these felids. In species lacking ear tufts, but which have long ringed tails with prominent black tips, for example European wildcat, the tail is held vertically when two cats greet each other, emphasizing its contrasting colouration. Ortolani (1999) used a phylogenetic approach to investigate how spots, stripes, contrasting tail tips, and dark eye patterns correlate with various behavioural and ecological factors in carnivorans. Dark tail tips were associated strongly with diurnal activity, grasslands, terrestrial locomotion, small body size, and preying on small mammals, and ungulates. White tail tips were also associated with grasslands, but also preying on birds and small mammals, and having raptors as predators. It has also been suggested that white or dark tail tips may be important for kittens/cubs to follow their mothers in tall grasses (e.g. cheetah) or closed habitats (e.g. Temminck's golden cat, *Catopuma temminckii*; Leyhausen 1979). Highly contrasting ringed tails were associated with nocturnal activity, closed habitats (mostly forests) and arboreal locomotion.

Making sounds

Cats communicate with each other using a wide variety of vocalizations from roars and mewing to spitting and purring (Peters and Wozencraft 1989). One of the key ways in which big and small cats have been defined is by their vocalizations and the morphological differences thought to be responsible for them. Therefore, big cats, *Panthera* spp., roar, but cannot purr, but small cats can purr continuously, but cannot roar (Peters and Hast 1994). Pocock (1916; 1917b) suggested that differences in the structure of the hyoid accounted for these differences in vocalizations. In most cats the hyoid is suspended from the skull by a series of bones from the ceratohyal through the epihyal and stylohyal to the tympanohyal, which connects to the tympanic bulla (Hast 1989). In the lion, leopard, jaguar, tiger, and snow leopard, the epihyal is replaced by an elastic ligament that allows the larynx to move away from the pharynx, and hence permits roaring. This anatomical difference is frequently quoted in the literature to explain roaring in big cats and is used as a taxonomic distinction for the genus *Panthera*.

However, detailed anatomical investigations of a wide variety of felids showed that hyoid structure was not correlated with ability to roar (Hast 1989; Peters and Hast 1994). Indeed, Peters and Hast (1994) discussed how poorly defined the roar is and concluded from sonograms that it was a structured call series shown only by the lion and in part by the jaguar and leopard. Tigers and snow leopards cannot roar, although tigers show the main call and grunt element of true roars. Peters and Hast (1994) found that the *Panthera* cats (excluding the snow leopard) have large vocal folds with large fibro-elastic pads (Fig. 3.15) and hypothesized that the folds vibrate to produce low-frequency sounds that are amplified by the

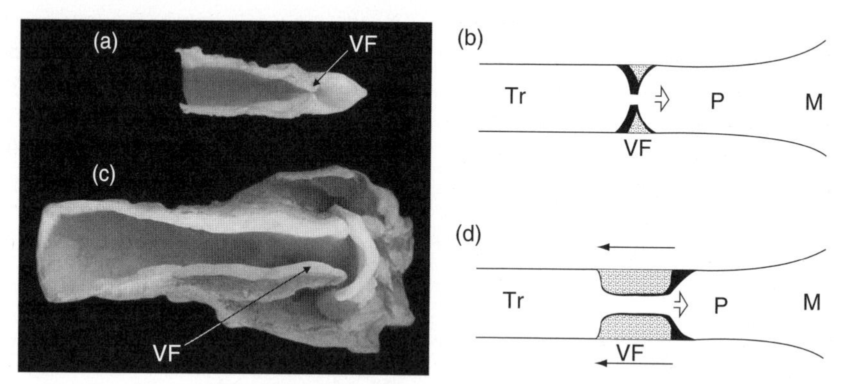

Figure 3.15 Transverse sections through the larynges of (a) a cheetah, *Acinonyx jubatus*; and (c) a lion, *Panthera leo*, to show differences in the shape and thicknesses of the vocal folds. It is believed that the thicker vocal folds of big cats resonate to produce roaring, whereas the thinner vocal folds of smaller cats resonate to allow continuous purring. Schematic diagrams of the vocal tracts of (b) a small cat with sharp-edged vocal folds; and (d) a big cat with vocal folds with large fibro-elastic pads show clearly the differences between the cheetah and the lion. Key: ← elastic ligament in place of epihyoid; M, mouth; P, pharynx; Tr, trachea; VF, vocal folds. Photo © National Museums Scotland. (After Peters and Hast 1994.)

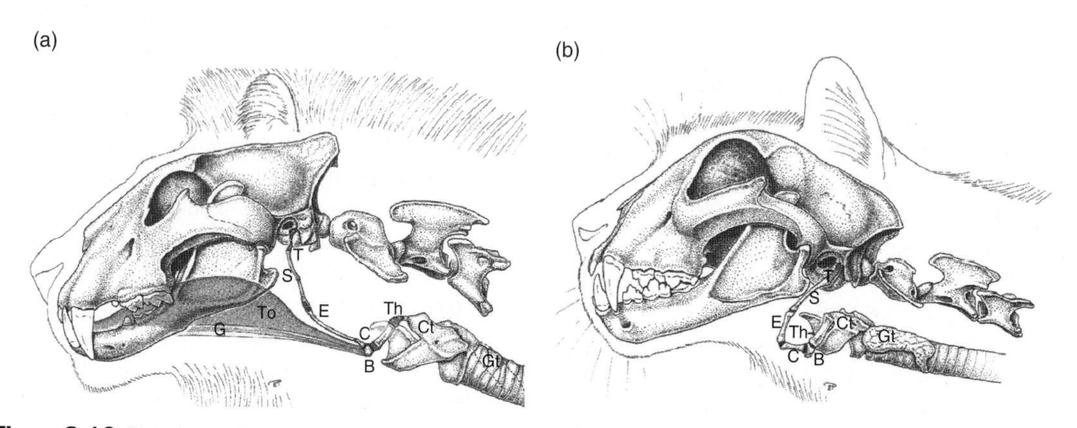

Figure 3.16 Diagram to show hyoid apparatus, larynx, and cranial part of trachea of (a) lion; and (b) cheetah. Key: G, *M. geniohyoideus*; To, tongue; T, *Tympanohyoideum*; S, *Stylohyoideum*; E, *Epihyoideum*; C, *Ceratohyoideum*; B, *Basihyoideum*; Th, *Thyrohyoideum*; Ct, *Cartilago thyroidea*; Gt, *Glandula thyroidea*. (After Weissengruber *et al.* 2002.)

long larynx, and bell-shaped pharynx and mouth, resulting in very loud low-frequency vocalizations, including roars. After examining the detailed anatomy of the hyoid apparatus and pharynx of lion, tiger, jaguar, cheetah, and domestic cat, Weissengruber *et al.* (2002) found that in the big cats, Pantherinae, the larynx lies more caudally, descending to this position during ontogeny (Fig. 3.16). They concluded that there were two components of vocal anatomy that define roaring acoustically. Firstly, long or heavy vocal folds lead to a low-pitch roar (low fundamental frequency) and secondly an elongated vocal tract results in the baritone timbre of roaring (lowered formant frequencies). The elastic epihyal allows the larynx to lower the formant frequencies and does not affect a cat's ability to purr, as thought previously.

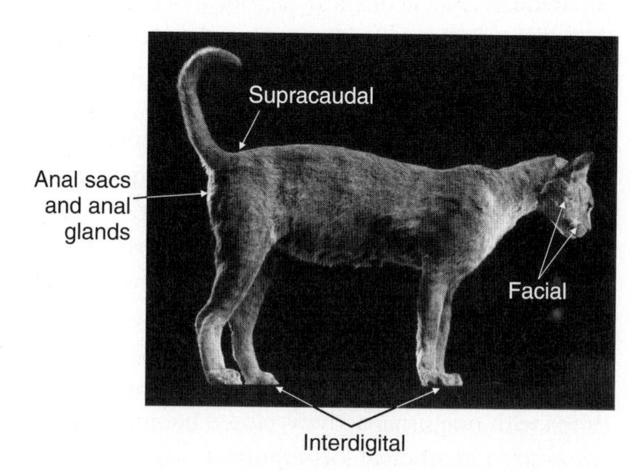

Figure 3.17 A female African golden cat, *Caracal aurata*, showing the positions of the principal scent glands of felids. © National Museums Scotland.

Smaller cats, including the snow leopard, have small, pointed vocal folds that cannot vibrate at such low frequencies and hence they cannot roar. In contrast, purring occurs through vibration of the smaller pointed vocal folds of smaller cats; domestic cats may purr continuously for up to 2 h with a fundamental frequency of about 25 Hz (Sissom *et al.* 1991). Peters and Hast (1994) could not refute the supposed correlation between a fully ossified hyoid and purring in smaller cats, because they could not examine all species. However, Weissengruber *et al.* (2002) concluded that purring in small cats is triggered by very rapid twitching of the *vocalis* muscle, which runs with the vocal folds, but the thick vocal pads of big cats damp this twitching, making it difficult or impossible for them to purr.

Detecting and making smells

Being predominantly solitary hunters, most felids have a dispersed social system (with variably overlapping home ranges) that is highly dependent on olfactory communication (Kitchener 2000). Felids most often use two free materials for scent marking that are produced routinely as waste products; urine and faeces. Urine is typically sprayed backwards against prominent objects within the home range, and males have short backwardly retractile penises to allow them to direct their urine in this way. Cats are able to distinguish individuals, different sexes, reproductive status, etc., from scent marks, which are probably the principal means by which cats communicate. Typically, females advertise their readiness for mating by spraying urine against tree trunks, rocks and other prominent features within their home range during pro-oestrous. This allows time for the male to find the female and detect her stage in the oestrous cycle. Because the sense of smell is relatively poorly developed compared with that of other carnivorans, felid scent marks tend to be a combination of both a visual (e.g. scratch marks) and an olfactory mark.

The general olfactory sense organs of felids have already been described above. However, the vomeronasal organ, situated in a small pit sagittally just behind the upper incisors plays an important role in communication between felids (Døving and Trotier 1998). Males display the flehmen response when sampling scent marks; they grimace and draw in the scent through the mouth into the vomeronasal organ, so that the reproductive state of the female can be assessed. The use of the vomeronasal organ and flehmen is not restricted to males and to detecting females in oestrous, but it is a clear example of their use.

In addition to urine and faeces, cats mark their environments using a variety of scent glands (Macdonald 1985; Fig. 3.17):

Facial—sebaceous glands on the chin, cheeks, lips, and around the mystacial and genal whiskers, which are rubbed against objects and other cats.

Supracaudal—a large sebaceous gland (0.85 × 0.5 mm in a female domestic cat, 1.84 × 1 mm in a reproductive male) above the root of the tail, coupled with tubular apocrine glands, which is rubbed against objects and conspecifics. Recently, this gland has been reclassified as a hepatoid gland in domestic cats; hepatoid glands characteristically produce secretions rich in proteins and hydrophobic lipids (Shabadash and Zelikina 1997).

Anal sacs—two sac-like glands that open laterally via ducts into the end of the rectum, the apocrine (and possibly sebaceous) secretions of which may be mixed with faeces or deposited on objects on their own. Recently, Bininda-Emonds *et al.* (2001) analysed the proportions of glycolipids, neutral lipids, and phospholipids from felid anal sac secretions and found that their composition varies between big (including puma and cheetah) and small cats.

Anal glands—sebaceous glands in the skin surrounding the anus that produce lipid-rich secretions lacking trimethylamine (cf. canids).

Interdigital glands—it is claimed that these glands are between the toes, but histological investigation is required to confirm this. Their secretion may be left when walking or used when cats are using scratching posts to leave a combined visual/olfactory signal.

Conclusion

As we have discussed above, the morphology of cats has evolved to produce one of the most efficient killing machines and sophisticated communicators in the natural world. As evolutionary forces continue to prevail, so cats will undoubtedly continue to evolve, and yet the very successful basic felid design may not change dramatically in the future, just as it has not done so since modern felids first emerged more than 20 million years ago.

Acknowledgements

We thank David Macdonald for inviting us to contribute to this volume and also thank him and Andrew Loveridge for their editorial work. Also, we thank Lars Werdelin, David Macdonald, and Andrew Loveridge for useful comments on an earlier version of this manuscript. We are grateful to Neil McLean for his excellent photographs and to the Trustees of the National Museums Scotland for permission to reproduce them. We are also grateful to the following publishers for permission to reproduce figures: NRC Research Press (Fig. 3.9), Elsevier (Fig. 3.15) and Wiley-Blackwell (Fig. 3.16) and to the National Museums Scotland for permission to reproduce photographs.

CHAPTER 4

Genetic applications in wild felids

Melanie Culver, Carlos Driscoll, Eduardo Eizirik, and Göran Spong

Molecular genetics shows that the Caspian tiger, long thought extinct, lives on in the Amur tigers of the Russian Far East.

Introduction

Genetic applications are becoming increasingly important in research and management of wild populations for a number of reasons. First, a rapid development of genetic techniques continually increases the usefulness of sources containing little or degraded DNA (e.g. hair, urine, and faeces), greatly facilitating the study of elusive species like felids, which often occur at low population densities in remote areas. Second, new types of genetic markers and equipment have dramatically improved the level of resolution and the quantity of genetic data that can be efficiently extracted (e.g. automated microarrays and rapid whole genome sequencing). Third, the interpretation of genetic data is constantly improved by the development of new statistical methods and models. Together, these improvements create an increasingly powerful toolbox for addressing a wide range of questions in natural populations.

Measures of genetic variance form the basis of all genetic analyses. Ultimately, all genetic variation stems from random mutations in the genome. But at the population level, genetic diversity is determined by several interacting processes: (1) the rate of novel mutations; (2) gene migration, the rate of gene flow from other populations; (3) natural selection, the differential reproduction of genotypes; and (4) genetic drift (stochastic changes in allele frequencies). Typically, genetic data quantify nucleotide variation in mitochondrial or nuclear DNA (genetic variance diversity). This variation can be single nucleotide variation (SNPs), allelic variation (variants of the same gene region, e.g., microsatellites), or whole sequences (e.g. mitochondrial DNA [mtDNA]; see Box 4.1). Because they differ in a number of characteristics (e.g. variance,

Box 4.1 Types and sources of DNA defined for felid genetics

DNA types

Mitochondrial DNA (mtDNA)—the genome of the circular mitochondrial molecule, maternally inherited

Genes—mammals have 37 genes coded for in the mitochondrial molecule

Control region—the single non-gene region of the mitochondria, evolves more rapidly than the genes

Nuclear DNA—the genome contained in the nucleus of a eukaryotic organism; maternal and paternal inheritance

Genes—there are approximately 30,000 genes in the mammalian nuclear DNA

Repetitive DNA—regions of the nuclear (or organelle) DNA that have tandemly repeated oligonucleotide sequences

Minisatellite DNA—section of DNA with tandemly repeated oligonucleotide sequences of 10–100 base pairs in length

Microsatellite DNA—section of DNA with tandemly repeated oligonucleotide sequences of 2–10 base pairs in length

Viral DNA—the genome of a viral organism

Tissue types

Invasive collection methods—any method to obtain samples for DNA extraction that causes disturbance to the organism; invasive DNA samples include:

Blood

Tissue snip (ear, tail, and other)

Saliva (buccal) swab

Biopsy dart (skin)

Non-invasive collection methods—any method to obtain DNA samples for DNA extraction that do not cause disturbance to the organism; non-invasive DNA samples include:

Museum skin, bone, tooth

Road kill tissue

Faeces, Urine

Shed hair

patterns of inheritance, and mutational rate), they are used for slightly different purposes (primarily determined by the spatial and temporal scale of the investigation). However, broadly speaking the use of genetic data falls into three main categories: (1) diagnostic, to identify genes, individuals, sex, populations, species (e.g. Nagata *et al.* 2005; McKelvey *et al.* 2006); (2) correlative, where the genetic data are compared to some other demographic, geographic, ecological, or behavioural variable (e.g. factors affecting genetic variation and structure, dispersal, territorial behaviour, and reproduction; e.g. Packer *et al.* 1991a; Spong *et al.* 2002); or (3) predictive, to infer processes or predict phenotype from genotype or separate genetic from environmental effects on phenotypic expressions (e.g. selection, reproductive success, individual dispersal, and genetic or demographic population dynamics; e.g. Spong and Creel 2001; Geffen *et al.* 2007).

In this chapter, we review the contribution of genetic studies to our knowledge of wild felid behavioural and evolutionary biology, as well as its importance for conservation and management. We have structured the chapter according to the level of study, beginning at the species level (taxonomy and speciation, phylogeography, and hybridization), to the population level (population genetics, bottlenecks, inbreeding, population structure, and dispersal), to the individual level (dispersal, kinship, and parentage), to the genomic level (mutation, selection, and Major Histocompatibility Complex [MHC]; see Box 4.2). Each section begins with a brief conceptual background, followed by representative case studies and discussion.

Taxonomy and speciation

One of the greatest contributions of genetics to wildlife conservation is through resolving taxonomic uncertainties. Cryptic species are those with few morphological differences, but identified as separate taxonomic units based on other types of data (e.g. genetic, behavioural, and vocalizations). Without genetic studies cryptic species could remain unknown or go extinct

Box 4.2 Definitions of methods for investigating felid genetics

Allele—different forms of a gene at the same locus

Allograft—tissue transplant from one individual to another of the same species

Allozyme—various forms of an enzyme; different alleles at a single enzyme locus

Diploid—a double set of homologous chromosomes

DNA fingerprinting—see VNTR below

DNA sequencing—sequence of nucleotide pairs along the DNA molecule

Electrophoresis—separating molecules based on different rates of travel in an electric field

Evolutionarily Significant Unit (ESU)—a population of organisms that is distinct for the purpose of conservation

Feline Immunodeficiency Virus (FIV)—a lentivirus that affects domestic cats worldwide

Genomics—sequence analysis of an entire genome for a species, and subsequent comparison of genomic sequence across individuals of that species or to genomes of other similar species

Haplotype—unique combination of linked mutations on a section of DNA

Locus—the position of a specific gene on a chromosome

Major Histocompatibility Complex (MHC)—group of genes encoding the class I and II MHC molecules, which form the basis of immune systems

Microsatellite genotype—all fragment lengths generated from PCR of a microsatellite locus and subsequent electrophoresis

Polymerase Chain Reaction (PCR)—technique to make many copies of a specific region of DNA using oligonucleotide primers on each end of sequence to amplify

Restriction Fragment Length Polymorphism (RFLP)—variation in length of DNA fragments generated by endonuclease digestion

mtDNA RFLP—restriction fragment length polymorphism found within the mitochondrial genome

Nuclear RFLP—restriction fragment length polymorphism found within the nuclear genome

Single Nucleotide Polymorphism (SNP) 'snips'—DNA sequence variant where one nucleotide differs from one individual to the next

Variable Number of Tandem Repeat (VNTR)—region of DNA whose alleles contain different numbers of tandemly repeated minisatellite oligonucleotide sequences

without ever being recognized. Conversely, genetic data may prompt the pooling of phenotypic variants to the same species or population, reducing the number of recognized species. Genetic analyses can also be helpful in shedding light on the evolutionary history and taxonomic position of single species. For example, Johnson *et al.* (1998) used mtDNA sequences to resolve the phylogenetic positioning of the Andean mountain cat, *Leopardus jacobita*, relative to other felids.

One recent example of a species addition is the discovery of a distinct clouded leopard species occurring on the islands of Borneo and Sumatra (*Neofelis diardi*). The previously described species, *N. nebulosa*, is now considered restricted to the Asian mainland. Two studies using mtDNA (see Box 4.1), microsatellite markers, nuclear DNA sequences, and cytogenetics revealed a deep historical partition between these populations, warranting species-level status (Buckley-Beason *et al.* 2006; Wilting *et al.* 2007a). This finding was further supported by a morphometric investigation of pelage features (Kitchener *et al.* 2006). In this case, molecular markers thus led to the description of a new species.

But genetic analyses may also lead to the opposite, that is, that populations or phenotypes previously considered to be different species, are found to be so genetically similar that they are more accurately classified as one species. For example, in the *Felis silvestris* complex (including *F. silvestris, F. catus, F. libyca*, and *F. bieti*), the genetic differentiation is very low, suggesting that they are more appropriately classified as a single species (Driscoll *et al.* 2007).

Because conservation efforts often focus at the species or subspecies level (e.g. IUCN red list), a population's protection status may be directly related to its taxonomic status. But since evolution is a slow, gradual process, species assignments are not always unambiguous. During a speciation event, there will be a period where the genetic difference is deemed insufficient to warrant subspecies or species status. But, such variation may still comprise local adaptations of significant ecological and evolutionary importance, perhaps leading to speciation (the *Felis silvestris* complex may be an example of this). Where to draw the line for the level of genetic differentiation worth protecting is often debatable, and as the resolution and population coverage of genetic analyses increase, such dilemmas will become increasingly common.

Hybridization

The process of hybridization, that is, gene flow between genetically differentiated species or populations, has been recognized as an important factor shaping genetic diversity and evolutionary history of groups of organisms (Allendorf *et al.* 2001). It is the opposite of the speciation processes discussed above and even though most cat species seem to constitute well-defined evolutionary units, a few examples of hybridization have emerged from genetic analyses.

The most well-known felid example involves the domestic cat (*Felis catus*), which hybridizes with free-ranging populations of the native wildcat (*F. silvestris*) throughout Eurasia and Africa (Driscoll *et al.* 2007; Macdonald *et al.*, Chapter 22, this volume). Although the occurrence of hybridization can be high, recent genetic studies have been able to differentiate between wild and domestic demographic units (e.g. Pierpaoli *et al.* 2003; Driscoll *et al.* 2007).

Another example of felid interspecies hybridization involves three South American cats of the genus *Leopardus*: Geoffroy's cat (*Leopardus geoffroyi*), the tigrina or oncilla (*L. tigrinus*), and the pampas cat (*L. colocolo*). Analyses based on mtDNA and Y-chromosome DNA sequences identified natural hybrids between *L. tigrinus* and *L. colocolo* sampled in central Brazil (Johnson *et al.* 1999). Recent studies found another hybrid zone between *L. tigrinus* and *L. geoffroyi* in southern Brazil (Eizirik *et al.* 2006; Trigo *et al.* 2008). Multiple hybrid individuals were identified, supporting the inference that hybridization is quite extensive in some areas where these species are sympatric. Moreover, in the case of *L. tigrinus* vs. *L. geoffroyi*, most hybrid individuals exhibit a genetic composition incompatible with F1 status, implying that hybrids can breed successfully. No evidence of hybridization was found between the largely sympatric *L. colocolo* and *L. geoffroyi*.

Yet another example of contemporary hybridization is found in the border regions of Canada and the United States, where anthopogenic disturbances (depressing population densities) may be the main cause of hybridization between Canada lynx (*Lynx canadensis*) and bobcat (*L. rufus*) (Schwartz *et al.* 2004; Homyack *et al.* 2008). Although the level of hybridization found was low, all hybrid individuals came from areas where *L. canadensis* was rare, suggesting that individuals in recovering, low-density populations may have difficulties finding conspecific mates. This process may thus impede the recovery of the less common species, with important implications for the management of both species (Schwartz *et al.* 2004). This finding may be of general importance for restoration efforts of closely related species complexes elsewhere.

As seen in the last example, population densities and distributions affect the prevalence of hybridization. Species with limited interactions in the past, may encounter each other more often due to changing ecological conditions (habitat destruction) or demographic changes (e.g. shrinking population density causing a lack of mates). Understanding the processes affecting species abundance and distribution (phylogeography), and how these affect the prevalence of hybridization, are therefore of direct relevance for conservation.

Phylogeography

In phylogeography, distribution data (past or present) and genetic data are combined to infer the history of a genetic lineage within a single species or cluster of related species (e.g. Johnson *et al.* 1999; Culver *et al.* 2000; Eizirik *et al.* 1998 and 2001). But phylogeographic studies also allow the detection of subspecies or evolutionarily significant units (ESUs; see Box 4.2) and can reveal the processes shaping geographic and temporal patterns of variation in natural populations (Avise *et al.*

1987; Avise 2000). While early efforts only used mtDNA data, a wider range of genetic data are now used, and analytical methods include population genetic models (e.g. Luikart *et al.* 2003; Knowles 2004). From a conservation standpoint, phylogenetic analyses may aid population assessments and provide critical information of how best to direct conservation efforts.

By assessing levels of diversity in different areas, early studies investigated genetic structure of felid populations and evaluated the magnitude of differentiation among geographically defined populations of lions (*Panthera leo*), tigers (*P. tigris*), and cheetahs (*Acinonyx jubatus*) (Newman *et al.* 1985; O'Brien *et al.* 1987a, b). Based on these data, two genetically distinct cheetah subspecies were defined (Newman *et al.* 1985). All these studies employed allozyme electrophoresis and mtDNA restriction fragment length polymorphism (RFLP; see Box 4.2), which were the most informative genetic approaches available at the time.

Miththapala *et al.* (1996) employed mtDNA-RFLPs, allozymes, variable number of tandem repeats (VNTRs) (see Box 4.2), and morphology to investigate the evolutionary history and population structure of the leopard, *Panthera pardus*. Genetic analyses of 96 individuals, representing the 27 previously described subspecies, indicated the existence of only six phylogeographic groups, and recommended retention of eight operational subspecies (Africa, central Asia, Java, India, Sri Lanka, north China, south China, and eastern Russia). However, by adding data from microsatellite markers (see Box 4.1) Uphyrkina *et al.* (2001) found additional diversity in one region. The central Asian group from Miththapala *et al.* (1996) was hence further split into two clusters.

But before nuclear DNA became widely used, DNA sequencing superseded RFLP as the preferred method for surveying mtDNA diversity. Such sequence data were used in a comparative phylogeographic study of two Neotropical felid species, *Leopardus pardalis* ($n = 39$) and *L. wiedii* ($n = 24$), where high levels of genetic diversity and pronounced population structure were found (Eizirik *et al.* 1998). The Amazon river was inferred to be an important historical barrier to gene flow; 10 subspecies were reduced to three to four phylogeographic groups in each species. More recent studies combine mtDNA sequencing with nuclear microsatellite markers to improve the resolution. This approach is currently the mainstream standard for phylogeographic analyses (see below).

A reduction in the number of ESUs was also the outcome of a study of four other Neotropical species (*L. tigrinus*, $n = 32$; *L. geoffroyi*, $n = 38$; *L. guigna*, $n = 6$; and *L. colocolo*, $n = 22$) which used mtDNA sequences for all four species, and added nuclear microsatellite genotypes for two species—*L. geoffroyi* and *L. guigna* (Johnson *et al.* 1999). All four species exhibited some population subdivision, with deeper structure (compatible with subspecies-level designation) found only for *L. tigrinus* and *L. colocolo*. Previously, all four species had subspecies-level designations (2–8 subspecies each). The Amazon basin is likely a major barrier for *L. tigrinus*, leading to a separation between Central American and Brazilian populations. Microsatellite markers were utilized for the two species showing the least genetic differentiation among populations (*L. geoffroyi* and *L. guigna*), corroborating the view that they are largely panmictic and contain no major phylogeographic partitions.

Similar patterns have been found in pumas (*Puma concolor*) and jaguars (*Panthera onca*), previously separated into a whole suite of subspecies based on morphological characters. But genetic data (mitochondrial gene sequences and nuclear microsatellite genotypes) from pumas throughout their range in North and South America ($n = 294$; Culver *et al.* 2000) indicate the existence of only six phylogeographic groups (North America, Central America, and four groups in South America), rather than the previous 32 subspecies described. These populations are mainly separated by major watercourses (including the Amazon river), and to a lesser degree the Andes Mountains. Similarly, a study of jaguars ($n = 44$) from throughout their range in the Americas found no major phylogeographic differences (Eizirik *et al.* 2001), in contrast to the previously recognized eight subspecies. Although less so than for pumas, the Amazon river still forms an important barrier for jaguars (Eizirik *et al.* 2001). Given the matrilineal inheritance of mtDNA, an interpretation of the mtDNA (which shows a difference across the Amazon) vs. microsatellite data (where a much smaller difference across the Amazon is observed) is that the Amazon river restricts female rather than male dispersal. Using molecular dating approaches, it was estimated that extant puma and jaguar populations have rather recent demographic history of about 0.3 million years ago (Ma). Both species display more genetic diversity in South America than in the more northerly portions of their ranges, likely derived from an origin in the south followed by subsequent demographic expansions to the north. In the case of the puma, Culver *et al.* (2000) inferred from the geographic location of the ancestral puma mtDNA haplotypes that extant lineages originated in the Brazilian highlands and from there dispersed throughout South America and into North America.

Some tiger subspecies (*Panthera tigris*) have an even more recent evolutionary origin based on mtDNA sequences and microsatellites, complemented by nuclear gene sequences from the MHC (see Box 4.2; Luo *et al.* 2004). Extant tiger lineages arose only 0.07 to 0.1 Ma. Tigers may thus have the shallowest phylogeographic history of all cats. Luo *et al.*'s study (2004) included 134 individuals from the five extant subspecies, and found low variation for all markers but with significant genetic differentiation among the subspecies. The study also found the Indochinese tiger to be comprised of two distinct lineages, validating a split into one northern Indochina and one Malaysian population.

Using 32 museum samples of modern lions and six samples of holarctic lions, Barnett *et al.* (2006b) found that lions from sub-Saharan Africa were basal and contained the highest genetic diversity, with three geographic groups representing the evolutionary history of lions (Asia and northern Africa, central Africa, and southern Africa). Despite the limited number of samples, the authors suggest that five phylogeographic groups should be recognized; central, southern, western, and south-eastern Africa, with a fifth group comprising Asia and northern Africa. As more samples are added from other areas, these groups may need to be split further or merged to accurately reflect the phylogeographic history of lions.

In American lynx (*Lynx canadensis*), mtDNA and microsatellite data from 183 individuals suggested that two types of barriers have affected this species in the past, one being physical (the Rocky Mountains), and the other more gradual, suggested to be the ecological transition from the Continental to Atlantic ecoregions (Rueness *et al.* 2003a). Stenseth *et al.* (2004a) suggested that these genetic partitions agreed with the demographic synchrony observed within each of these regions, and proposed a model in which spatially structured 'climatic forcing' would drive the ecology and evolution of this species. Climate forcing refers to a model developed to link the climate pattern with the ecological and evolutionary patterns, under the assumption that these patterns are more synchronized within rather than between different climate zones.

Most of the phylogeographic case studies described earlier have led to a reduction in ESUs, be it at the species or population level, relative to previous descriptions. This ongoing reshaping of felid phylogeny is thus directly affecting conservation strategies, by identifying relevant operational units for management (often coinciding with ESUs), realistically reflecting the evolutionary history of each species.

Feline immunodeficiency virus (FIV) sequences as tool for high-resolution phylogeography

The Feline Immunodeficiency Virus (FIV; see Box 4.2), a virus related to the AIDS-causing human immunodeficieny virus (HIV), has been the subject of many genetic and epidemiological studies in wild felids (Munsen *et al.*, Chapter 9 this volume), of which some have used FIV DNA sequences to resolve features of gene flow among host populations (by examining the relatedness of the virus in nearby populations).

An initial evolutionary study (Olmsted *et al.* 1992) presented phylogenetic analyses of FIV DNA sequences in multiple felid species to reveal deep divergence between FIV strains from domestic cats (*F. silvestris catus*) and those found in wild felids such as lions, cheetahs, and pumas. A later study (Brown *et al.* 1994) showed that lions in Africa had three distinct and highly divergent strains of FIV sequences; however, each strain was fairly widespread throughout Africa—probably indicating that all had been present for a long time, leading to geographic homogenization. In this case the population-level signal may have been lost.

In 1996, Carpenter *et al.* examined FIV lineages in pumas throughout North and South America using viral gene sequences. FIV is present in all parts of the species range, but South America has low rates of infection, with only Brazil represented. A phylogenetic analysis of FIV sequences indicated two separate groups; one group includes the extreme east and west coasts of North America (Florida and California), whereas the other group represents the rest of western North America and all of South America. Carpenter *et al.* (1996) suggest the unlikely relationship between the disjunct Florida and California sequences could indicate that these viral sequences are ancestral and a second wave of FIV infection spread widely throughout North and South America but did not reach those two states. Carpenter *et al.* (1996) further examined western North American pumas and found a striking amount of diversity in the Wyoming samples (all from Yellowstone National Park, centrally located within western North America). Intriguingly, pumas were eliminated from Yellowstone in the early 1900s and the high diversity of sequences from that area could indicate that pumas recolonizing Yellowstone have multiple origins. This study exemplifies how the natural history of puma FIV viral sequences can reflect the evolutionary history of pumas themselves.

A subsequent study using puma FIV sequences (Biek *et al.* 2006b) found evidence of population structure in northern Rocky Mountain pumas. A significant division between the northern areas (British Columbia, Alberta, and northern Montana) and southern areas (southern Montana and Wyoming) was observed in phylogenetic analyses of puma viral sequences. Also, spatial structure of viral sequences was quite pronounced, with some sequences restricted to small geographic regions. Based on estimated mutation rates for these viral sequences, this structure likely represents movements of the past 20–80 years, corresponding to population density lows due to human persecution. The existence of a widespread lineage throughout the northern Rocky Mountains likely comes from the first lineage to recover. Because pumas often disperse widely, this pattern of viral sequences is expected to be visible for a brief window of time, after which the viral lineages become mixed and uninformative. Mixing of genotypes from these regions of North America was evident using microsatellites, markers, with slightly lower resolution relative to viral DNA sequences (Culver *et al.* 2000; Biek *et al.* 2006a). Timing of the peak resolution will depend on the evolutionary rate of the pathogen and the species' dispersal behaviour.

Population genetics

The need to understand how genes are distributed within and between populations led to the development of population genetics. Conceptually, it is the extension of Mendelian genetics (transmission of genes within pedigrees) to the population scale. Classical F-statistics (Wright 1931) have proved useful for describing how genetic variation is structured in a population (within and between individuals, groups, and populations). In theory, it can also be used to infer rates of the processes causing the genetic structuring (e.g. gene flow/dispersal). However, most population genetic models assume that genetic processes are at equilibrium (rarely the case) and the results from such models are therefore more a reflection of the evolutionary past than of the present. Using population genetics to infer contemporary rates (typically the relevant timescale for managers and conservation programmes) should therefore be done with great caution (see examples below).

Basic population genetic models rest on the concept of an ideal population, assuming (1) non-overlapping generations; (2) equal individual mating success; (3) random mating; (4) even sex ratio; (5) constant population size; (6) no mutation; and (7) no selection. Natural populations do not meet these assumptions. However, with information on how the population under study deviates from these assumptions (Wright 1943; Crow and Kimura 1970; Lande and Barrowclough 1987), it is possible to calculate the effective population size (*Ne*), a metric of pivotal importance for population genetic processes. It is defined as the size of an ideal population that would lose genetic variance at the same rate as the population under study (Wright 1931). It therefore reflects the number of animals in a population whose genes are actually transmitted to the next generation, affecting the degree of genetic drift. Genetic drift (really *allelic* drift) is the stochastic fluctuation of allele frequencies in a population over time. Genetic drift eliminates, or fixes, alleles proportional to *Ne*. Thus, a population with a large *Ne* has a lower risk of loosing alleles than a population with a small *Ne*, even if the number of individuals is identical between the two populations. As allelic variation is lost, adaptive potential may also be eroded.

A straightforward application of population genetics is to describe the genetic structure of a population at a single point in time. Here classical F-statistics (Wright 1931) can quantify how variation is distributed among subpopulations in order to determine genetic management units. For example, extensive gene flow between jaguar populations in Colombia, Guatemala, Paraguay, Peru, Bolivia, Venezuela, and Brazil suggest that it could be considered a large continuous population (Eizirik *et al.* 2001; Ruiz-Garcia *et al.* 2006). Conversely, a study of the Scandinavian lynx found significant structuring and the authors suggested that the two subpopulations should be separately managed (Hellborg *et al.* 2002). In a study of California pumas, the effect of natural barriers to dispersal could be seen clearly in the genetic structure, and the importance of preserving dispersal corridors was highlighted (Ernest *et al.* 2003). Improved habitat connectivity is perhaps most urgently needed for the fragmented remnants of the Iberian lynx (Johnson *et al.* 2004, Ferreras *et al.*, Chapter 24, this volume), where small subpopulations are becoming more vulnerable as dispersal opportunities are decreasing due to habitat loss.

Population genetic models may also be helpful for the management of small populations, for example, to predict genetic processes (i.e. rate of genetic drift and level of inbreeding). The Iriomote cat (*Prionailurus bengalensis iriomotensis*) is a subspecies of the Asian leopard cat endemic to the 116 square-mile Japanese island of Iriomote (Johnson *et al.* 1999). The current population is estimated at *c.* 100 individuals and has likely been small over geological time. An analysis of four individuals using mtDNA found only a single base difference in a single cat, a divergence of 0.1%, and zero variation over 18 microsatellite loci when five Iriomote cats were surveyed. The Iriomote cat is probably the least variable wild felid known. Johnson *et al.* (1999) estimate that it diverged from its closest phylogenetic relative, *P. b. euptilura*, as recently as 0.18 Ma, about the same time a land bridge connected Iriomote with the mainland.

Population genetics can also be used to infer population metrics. Since *Ne* directly affects the cumulative genetic variation (e.g. number of alleles, allele range, allele frequencies, and other metrics), it can be estimated from genetic data alone. Such estimates are based on assumptions about the mutational rate and have, for example, been used to estimate the *Ne* of pumas in the Wyoming basin (Anderson *et al.* 2004) and leopards (*Panthera leo*) (Spong *et al.* 2000b). But caution is advisable because the results are strongly dependent on several factors that are difficult to quantify: firstly, the assumed mutational rate, which is species-specific; second, any measure of current genetic variation is strongly dependent on the population history, and the size of this bias is typically unknown. Lastly, gene flow from neighbouring populations regions may seriously affect the estimate. A genetically based estimate thus assumes that the population size has been more or less constant over time, and that the current population status is representative of the past. More often than not, this is unlikely to be the case, especially for species of conservation concern.

An illustrative example of how the results from population genetic models may depend on the underlying assumptions are the efforts to reconstruct the recent population history of the Eurasian lynx, which increased rapidly in numbers after having been hunted to the brink of extinction. Despite a recent bottleneck and rapid expansion, described by Hellborg *et al.* (2002), genetic evidence from mtDNA sequences and microsatellite loci for 276 individuals suggested that lynx recolonized Scandinavia by separate events in the northern and southern regions. Rueness *et al.* (2003b) pursued this further and found a pronounced spatial pattern of isolation in the extant population. The latter authors considered it unlikely that this structure would have had time to develop after the expansion, meaning that the current population must have originated from multiple sources. The southern population may have persisted in isolation, whereas, in populations north of the Arctic circle there is evidence of recent immigration from Finland and the Baltic region. This conclusion was challenged by Pamilo (2004), who suggested that a single-source population could also have resulted in the observed genetic structure for Scandinavian lynx populations, under the assumption of drift-migration equilibrium (the balance between alleles lost by genetic drift and gained by migration or dispersal of individuals into the population). Finally in 2006, Jorde *et al.* showed that the hypothesis of multiple recolonization events and Pamilo's hypothesis of a single recolonization, are equally likely, bringing us back to square one.

Population genetics works well for quantifying the genetic structure and may prove useful for resolving population histories. However, population genetic models based on the concept of an ideal population and the assumption of genetic equilibrium are clearly unsuitable for many questions of conservation interest. Recent theoretical developments (e.g. Peery *et al.* 2008) using genetically derived pedigree information are better at quantifying contemporary population processes. As the resolution of genetic analyses increase, inferences across multiple generations and between distantly related individuals will become more accurate, making these individually based models even more powerful. Such models can already be used to track contemporary changes in behavioural and demographic patterns (e.g. dispersal and reproductive patterns, or rapid population declines) with great precision.

Bottlenecks

A bottleneck is a significant reduction in population size followed by a significant increase. They are profound population events on many levels (e.g. disrupting social structures, behaviours, and ecological interactions); we restrict our discussion to the genetic implications of bottlenecks. It is often possible to infer a bottleneck from the signature left in the genetic profile of a post-bottleneck population (primarily caused by an increased influence of stochastic events). Moreover, the signature varies with the duration and size of the bottleneck, the timing, and the rate of recovery, making it possible to reconstruct historic population events with some detail.

The cheetah was the first felid species found to have low levels of genetic variation. In 1983, O'Brien *et al.* found monomorphism at 47 allozyme loci in 55 South African cheetahs (*A. j. jubatus*), and unusually low heterozygosity (0.013) in 155 fibroblast proteins, suggestive of a historic bottleneck event. Newman *et al.* (1985) surveyed allozyme variation at 25 loci in eight cat species (leopard, tiger, lion, serval *Leptailurus serval*, caracal *Caracal caracal*, clouded leopard, ocelot *Leopardus pardalis*, and margay *L. wiedii*). The number of polymorphic loci ranged from three in clouded leopard to 10 in ocelot, suggesting that the low variation found in cheetahs was indeed unusual.

O'Brien *et al.* (1985b) showed that 14 reciprocal skin grafts between unrelated cheetahs were accepted. Skin grafting is a powerful assessment of MHC variation because a single mismatch at any of the three class I or three class II MHC loci in the domestic cat is sufficient to cause tissue rejection (see MHC discussion below). In 1987b, O'Brien *et al.* extended their allozyme survey to include 30 east African cheetahs (*A. j. raineyi*) over 49 proteins, again finding low heterozygosity (0.0004–0.014) similar to that found in southern Africa. These results also revealed that the cheetah comprised an eastern and a southern subspecies. Since both showed dramatically reduced variation, the authors concluded that the bottleneck event preceded the split, likely during the Pleistocene. A later effort dated the bottleneck to between 6000 and 20,000 years ago, consistent with a Late-Pleistocene event (Menotti-Raymond and O'Brien 1993). In 2002, Driscoll *et al.* used reconstitution of microsatellite allelic variation to date the bottleneck to ~12,000 years ago and the split between *A. j. raineyi* and *A. j. jubatus* to ~4500 years ago. A separate computation of the subspecies split based on $(\partial\mu)2 = 2\mu G$ (μ = mutation rate and G = generations) results in an estimate of 4253 years ago, in close agreement with the previous estimate.

However, the best-documented bottleneck is based on observational data. In the Ngorongoro Crater, Tanzania, a disease outbreak reduced the resident lion population to a mere nine females and one male in 1962 (Gilbert *et al.* 1991). In 1964–65, seven males from the neighbouring Serengeti immigrated into the crater but no lions have immigrated since. The current study population of 75–125 lions descended from 15 founders. In this population, where the bottleneck lasted one generation, including one immigration event from a large population, O'Brien *et al.* (1987b) found that, as compared to Serengeti lions, Crater lions showed diminished heterozygosity at seven allozyme loci. Additionally, Yuhki and O'Brien (1990a) found that Crater lions have only one-third of the MHC variation of Serengeti lions.

Another isolated lion population is found in the Gir Forest, India, where the last 300 individuals of a once wide-ranging Asiatic lion population reside. In surveys of 46 to 50 loci in Gir lions, O'Brien *et al.* (1987a, b) found no allozyme variation in 28 individuals. In contrast, African lions of the Ngorongoro Crater and Serengeti had moderate variation ($P = 0.07$ and 0.11, respectively). Yuhki and O'Brien (1990a) found no RFLP variation in class I MHC and Gilbert *et al.* (1991) report an extreme lack of heterozygosity for multilocus DNA fingerprints. However, based on simulation estimates, Halley and Hoelzel (1996) concluded that even a bottleneck of ~20 individuals for the observed period was insufficient to account for this extremely low diversity. In 2002, Driscoll *et al.* confirmed the low genetic variation and also found a high variance in allele size distribution in 90 microsatellite loci, and suggested an initial bottleneck about 2680 years ago and a secondary bottleneck about 100 years ago.

The Florida puma population, once part of a contiguous North American population, was isolated in southern Florida and reduced to *c.* 30 individuals by the 1920s. O'Brien *et al.* (1990) identified an indigenous Florida mtDNA matriline in the Big Cypress Swamp (BCS), present in all BCS pumas with no variation. Similar to the Ngorongoro Crater lions, the pre-bottleneck diversity of Florida pumas can be estimated from the neighbouring population. Roelke *et al.* (1993b) surveyed 41 allozyme loci, showing that BCS pumas have severely reduced variation as measured by percentages of polymorphic loci (*P* 4.9%) and heterozygous loci (*H* 1.8%) (when compared to outbred pumas from six western North American subspecies (*P* 27% and *H* 1.8–6.7%) though BCS pumas do retain more variation than cheetahs (*P* 2.0–4.0% and *H* 0.04–1.4%). Comparisons of minisatellite diversity followed the same trend, with BCS pumas having a mean difference of 9.5%, which is 85% less than in western pumas. A recent study estimated the exact size of the Florida puma bottleneck by comparing genetic variation in pre-bottleneck (early 1890s) museum samples to post-bottleneck (late 1980s) samples (Culver *et al.* 2000). Pre-bottleneck samples had three times higher heterozygosity levels and six times higher mtDNA diversity than post-bottleneck samples. These data estimated the bottleneck *Ne* to be approximately two individuals. Given the ratio of *Ne* to census population size in pumas, this translates into a census size of 6.2 during bottleneck generations and 41 during non-bottleneck generations for Florida pumas.

The Siberian tiger population has been affected by anthropogenic effects of hunting and habitat loss, leading to a census size of *c.* 30 individuals at the turn of the twentieth century. Newman *et al.* (1985) noted that captive Siberian tigers surveyed were virtually monomorphic for three of five allozyme loci, all of which were polymorphic in Bengal tigers. The authors interpreted this as being either a result of founder effects and inbreeding related to captivity, or as divergence from parent stock and subsequent period of reproductive isolation. Luo *et al.* (2004, 2006) found no mtDNA variation in captive and wild Siberian tigers. Russello *et al.* (2004) argue that the predominant cause of the current genetic impoverishment in the modern Siberian tiger population was caused by a near-extinction event. However, Driscoll *et al.* (2009) suggest that the Siberian tiger is naturally monomorphic due to female philopatry, causing a filtering of matrilines as tiger populations expanded through the relatively narrow Gansu Corridor.

From an immediate conservation viewpoint, a rapid reduction or low population size is primarily of demographic concern (Lande 1988). However, it is worth noting that loss of genetic variation at the species level is irreversible (over ecological timescales). Moreover, if the low population size persists, inbreeding depression may eventually become a conservation concern. As we will see in the next section, several of the bottleneck events mentioned above have indeed led to inbreeding depression.

Inbreeding

Inbreeding occurs when related individuals mate, producing individuals with genetic material with a common descent (measured by the inbreeding coefficient, *F*, which is the probability that both alleles in a locus share the same descent). The occasional mating between relatives is expected in natural systems, but may reach unusually high levels in highly fragmented or small populations (e.g. during a bottleneck). But although inbreeding amplifies the effect of drift and selection, it does not, *per se*, affect the genetic diversity of a population; it only brings together allelic variants that are identical by descent.

Inbreeding depression occurs when inbred individuals show a reduction in fitness (e.g. reduced viability, fertility, longevity, and growth rate). Inbreeding depression may be due to the expression of deleterious recessive alleles at homozygote loci (identical alleles) affecting fitness (dominance hypothesis), or if variance in fitness is determined at loci where heterozygotes (different alleles at the same locus) are more fit than homozygotes (overdominance hypothesis). Inbreeding was previously considered a problem mainly in small and isolated populations, with high levels of inbreeding. However, recent studies have shown major costs of mild inbreeding in large populations, perhaps suggesting that the effect of overdominance is more important than previously recognized. Whatever the mechanisms, the deleterious effects of inbreeding are well documented and a few of these examples come from wild felids.

Cheetahs show physiological and morphological signs of inbreeding effects (Wildt *et al.* 1983). Compared to 29.1% abnormal sperm and 77% motility in domestic cats, cheetah sperm was 71% abnormal and only 54% motile, as well as being 10 times less concentrated. Wildt *et al.* (1987a) found similar patterns in Gir lions, and concluded that as genetic variation decreased, testosterone levels and ejaculate quality declined, with sperm structure as the most vulnerable component. Munson *et al.* (1996) suggested that, based on testicular histomorphology, Ngorongoro Crater lions also have reduced spermiogenesis (as compared to Serengeti lions). Although simulations suggest that the Crater lions may have been deficient in variation prior to the 1962 crash, most of the reduction in diversity occurred during the last bottleneck. Crater lions with lower *H* showed impaired reproduction relative to the Serengeti. Levels of *H* correspond closely with annual reproductive rates within the Crater, but differences in ecological factors may explain some of this (Packer *et al.* 1991a, b).

Comparative reproductive analysis by Wildt *et al.* (1987a), Howard *et al.* (1990), and Baron *et al.* (1994) reported BCS pumas to have the poorest seminal quality traits ever recorded in any felid, with a 94.3% average frequency of malformed sperm, and fewer motile sperm than males from other populations (0.54×10^6 vs. 2.19×10^6). Moreover, 56% of males examined between 1978 and 1992 were reported as cryptorchid (i.e. failure of one or both testes to descend to the scrotum). Cryptorchidism led to lower circulating testosterone, while unaffected males produced normal levels (Barone *et al.*, 1994). Cryptorchidism has not been seen in any other non-domestic felid species. Finally, Roelke *et al.* (1993b) also observed a previously unreported cardiac condition, atrial septal defect (ASD), in three of 49 examined Florida pumas between 1978 and 1992, and reported that 80% of Florida pumas have detectable heart murmurs compared to <4% in other North American pumas.

However, the fitness costs of inbreeding will vary among species. Some species may have had deleterious mutations purged by selection or drift (O'Brien 1994). For example, Gir lions have lower *H* (2.9) than BCS Florida pumas (average 14.2), yet show fewer signs of inbreeding (Onorato *et al.*, Chapter 21, this volume). Likewise, when comparing allozyme variation, cheetahs have lower levels than BCS pumas (4.9 and 1.8) but, again, appear healthy.

But inbreeding depression can be difficult to detect in wild populations. Data from other groups (e.g. birds [Kruuk *et al.* 2002] and ungulates [Slate *et al.* 2000]) illustrate that single inbreeding events in large populations may lead to pronounced inbreeding depression, but that these effects may be difficult to detect without long-term data sets with good pedigree data. Individual reproductive success is typically affected by several ecological and individual factors, making it difficult to discern the signature of inbreeding depression in small data sets. Moreover, in small or bottlenecked populations all individuals may be more or less equally inbred, causing inbreeding depression, but without discernible variance in individual reproductive success.

Dispersal

Dispersal is a phenomenon of fundamental importance. At the population level it affects species distribution and abundance, social structure, and gene flow; at the individual level it affects survival and reproductive success; and at the genetic level it affects gene flow and genetic structure. Yet, despite this prevailing influence of dispersal, our understanding is lacking because dispersal is difficult to estimate, mainly due to the logistical difficulties associated with tracking dispersers.

A few studies have published extensive observational data on individual dispersal (e.g. Packer and Pusey 1993; Palomares *et al.* 2000; Zimmermann *et al.* 2005b). However, even in the comprehensive lion study by Packer and Pusey (1993) the origin of most immigrants and the final dispersal location of emigrants were unknown, preventing an unbiased estimate of dispersal distance. But, genetic data can offer a solution to this dilemma. Large-scale genetic sampling can be deployed to identify individual movements (see section on 'genetic monitoring') or alternatively the genetic effects of dispersal can be quantified. Genetic isolation by distance patterns is found in nearly all populations, and the steeper the slope, the shorter the average dispersal distance. And the sampling can be done at a single point in time, avoiding the spatial and temporal biases associated with observational data.

To date, the potential of genetic applications to elucidate felid dispersal patterns has not been tapped. A few studies have estimated gene flow using traditional F-statistics (e.g. Spong *et al.* 2000b; Eizirik *et al.* 2001; Ruiz-Garcia *et al.* 2006). While this type of approach can be useful in estimating the (historic) average levels of gene flow between populations, it is not sensitive to the movement rate. Moreover, it cannot provide direct estimates of dispersal distances or frequency (i.e. the proportion of individuals that disperse), often the parameters of most interest for conservation. However, individually based models allow the estimation of actual dispersal distance. In lions, genetically based estimates (Spong and Creel 2001) showed that dispersal distances averaged only 1–2 pride ranges, agreeing well with observational data (Packer and Pusey 1993; Funston *et al.* 2003). The short dispersal distance of this species may be the result of a trade-off to avoid inbreeding, while minimizing the risks of dangerous intraspecific encounters. Finally, genetic data have also been used to investigate sex patterns in dispersal. For example, Biek *et al.* (2006a) used genetic data to infer a male-biased dispersal in pumas, while no such sex bias in dispersal was found in genetic data from lynx (Campbell and Strobeck 2006).

Better data on dispersal in felids is long overdue and we now have the tools to accomplish this. Of particular interest would be data on individual dispersal, factors affecting it, and its consequences on fitness. Such estimates would be highly useful for understanding basic felid behavioural ecology and population dynamics and a great aid for many conservation efforts.

Kinship

Early efforts to look at genetic relationships between individuals were limited by low genetic resolution and the need to calibrate the genetic output against known pedigrees. For most felids, the latter is exceedingly difficult to obtain. The African lion forms a notable exception by being amenable to direct observations, and was therefore the subject of the first effort. With genetic material from known mother–offspring and sibling relationships, Gilbert *et al.* (1991) used DNA fingerprinting to reveal the kinship patterns within lion prides. Genetic methods that quantify variation at single genetic loci, inherited according to simple Mendelian rules (e.g. microsatellite loci), have now replaced behaviourally/genetically inferred pedigrees. Since the advent of markers that show strict Mendelian inheritance (e.g. microsatellites and SNPs), the kinship between two individuals can be inferred from their degree of pair-wise band sharing compared to the population average. Such estimates may be biased or show high variance in small data sets or when mating strongly departs from random, but they typically work well in larger data sets where the social system is accounted for in the analyses (Queller and Goodnight 1989; Ritland 2005).

Despite these improvements, few studies have used genetic methods to address questions at the individual level in felids. Studies of African lions form the bulk of individually based studies. In lions, genetic data have been combined with observational and experimental data to investigate questions relating to cooperation, territoriality, and reproductive success (e.g. Gilbert *et al.* 1991; Packer 1991a; Packer and Pusey 1993; Grinnell 1995; Spong *et al.* 2002; Spong and Creel 2004). Additionally, local kinship structure has also been detected in female pumas (Biek *et al.* 2006a), and in female bobcats, where it correlated with behavioural interactions and ranging patterns (Janecka *et al.* 2006). It seems likely that spatial grouping of kin will eventually be found in many other solitary species too, with important implications for ecological and evolutionary processes, such as territoriality, dispersal, and reproduction. An increased effort to collect (non-invasive) genetic samples should prove an effective way to dramatically improve our understanding of local genetic structure and relatedness patterns of less-studied felids.

Parentage

Reproductive success is the very definition of individual fitness and its quantification is thus of central importance for our understanding of a wide range of evolutionary and ecological processes. Variation in reproductive success also affects population dynamics and demography and is thus of direct conservation interest. But while maternal reproductive success can often be derived from association patterns between mother and offspring (e.g. Packer *et al.* 1991a; Kelly 2001), accurate estimates of male reproductive success almost invariably require genetic data.

Despite the importance of estimates of reproductive success, there have been very few published data on parentage in wild felids. One study of African lions (Gilbert *et al.* 1991) included estimates of the degree of reproductive sharing among coalitions of lion males. In a case of intraspecific killing of two juveniles by an adult male jaguar, paternity testing suggested that the father was responsible (Soares *et al.* 2006). Natoli *et al.* (2007), in a study of feral cats, cite a high frequency of multiple paternity, suggesting that this may also be common in native wild cats under some circumstances.

The lack of information on parentage in felids may be partially explained by the need for high quality genetic data and comprehensive population coverage before reproductive success can be estimated with precision. The reliability of parent–offspring assignments is dependent on the level of genotyping errors and the proportion of the putative parents sampled. The void of information on the distribution of reproductive success in felids is unfortunate. However, the improvement of both quantity and quality promised by automated microarray applications should resolve this problem. For example, estimates of the distribution of paternity would be highly useful for demographically based estimates of *Ne* (likely to be more accurate than those based on genetic data alone).

Mutation, selection, and MHC

Mutation and selection

Mutation, allele selection, genetic drift, and gene migration are the four key factors in the evolution of genomes. Drifts and migrations are more easily studied in natural populations through examination of population subdivision, gene flow, and *Ne* (see 'Population genetics', p. 113, and 'Dispersal', p. 117), whereas mutation and selection are more typically studied in a laboratory setting. Occasionally, however, important lessons are learned about the process of genetic mutation and natural selection from wild felid populations. Additionally, the development of new genetic techniques (see 'Genetic monitoring' p. 122) is making the prospect of studying evolutionary processes directly at the genetic level in wild felids more feasible.

Melanism and coat-colour variants

In a study by Eizirik *et al.* (2003), the genes *ASIP* and *MC1R* were found responsible for melanism in three different felid species. Jaguars with the melanistic phenotype were found to have a 15-base pair deletion in the *MC1R* gene, whereas melanistic jaguarundis had a different deletion of 24 base pairs. Domestic cat individuals contained a two-base pair deletion in the *ASIP* gene. However, melanism in five other felids (leopard, *Panthera pardus*; Geoffroy's cat, *Leopardus geoffroyi*; oncilla *L. tigrinus*; pampas cat, *L. colocolo*; and Asian golden cat, *Pardofelis temmincki*) did not involve these mutations in the genes *ASIP* and *MC1R*, implying that at least one other variant is involved. At least four different evolutionary episodes of melanism are thought to have occurred in the Felidae (Eizirik *et al.* 2003). The parallel occurrences of mutations for melanism, and subsequent increase in frequency, suggest that the melanistic phenotype is adaptive, at least in some of these cases.

Molecular clocks

The ability to use DNA to estimate timing of historical events for a species, such as number of years since a founder event or bottleneck, is based on the notion of molecular clocks. In non-coding regions of the DNA, mutations accumulate at a metronomic rate over time (since they are not subjected to selection), and the number of mutations accumulated can hence be used to infer the evolutionary timing of historical events.

In 1994, Lopez *et al.* observed a recent transposition (insertion) of part of the mtDNA molecule (also known as cytoplasmic mtDNA or 'cymt') into the nuclear genome of the domestic cat. The transposition involved a 7.9 kb piece of mtDNA (from a total of 17 kb). Since the nuclear genome undergoes much slower evolutionary change, once the cymt DNA was inserted into the nuclear genome, its rate of accumulated mutations slowed to the nuclear rate. By counting the number of cymt DNA mutations, relative to the homologous region of nuclear mtDNA or 'numt', it was possible to not only estimate the timing of this genome reorganizing event (at 1.8–2.0 Ma) but also to provide a means to directly estimate the evolutionary rate for the mtDNA (cymt) molecule. Based on the timing of this event, together with accumulated cymt mutations since this event (relative to the rare numt mutations), Lopez *et al.* (1997) were able to calculate the evolutionary rate for each cymt gene contained in the numt insertion (15 genes and the control region). The mtDNA mutation rates derived by Lopez *et al.* (1997) provide an alternate to the fossil record dates, which can be utilized by those studying wild felids. Since mtDNA is commonly used for a variety of genetic studies, including molecular dating, where accurate estimates of evolutionary rates are key, this advantage is not trivial. In addition to the domestic cat, the nuclear genome of several other members of the domestic cat lineage contained this same numt event (black-footed cat, *F. nigripes*; sand cat, *F. margarita*; wildcat, *F. silvestris*; and African wildcat, *F. silvestris lybica*).

Repetitive region DNA mutations

Repetitive regions can contain a base unit as small as two base pairs (as in microsatellites; see Box 4.1), or 80 base pairs (as in minisatellites; see Box 4.1) and as large as several thousand base pairs (as in the numt insertion). Menotti-Raymond and O'Brien (1995) suggested that hypervariable regions are useful to estimate the extent of genetic diversity in natural populations. They also suggested that hypervariable regions (both minisatellites and microsatellites) could be used to calibrate the timing of historical population bottlenecks by using estimated mutational rates.

Traditionally, differences in microsatellite alleles are determined by the difference in the number of repeat units. Culver *et al.* (2001), instead, examined DNA sequence differences for microsatellite repeats in pumas throughout North and South America. It is assumed that alleles with the same number of repeats usually contain the same repeat structure; however, they found a small proportion of microsatellites containing the same number of repeats had repeat structures that differed. For example, in a compound repeat with AT repeated 10 times adjoining CG repeated 10 times (AT_{10}, CG_{10}), a comparable allele could be found with 20 repeats where the structure of the repeat differs (AT_9, CG_{11}). This phenomenon is known as homoplasy and occurred at a rate of 2.7% in the puma data set. Culver *et al.* (2001) suggested that these microsatellite markers are still useful for phylogenetic analyses of adequately sampled species because homoplasy is expected to occur at some level (given the mutational mechanism of microsatellites). However, homoplasy was not detected at the vast majority of same-size microsatellite alleles, and these microsatellites (including those with and without homoplasy) were an effective set of markers to resolve population structure for puma populations in a prior study (Culver *et al.* 2000).

MHC and selection

MHC is composed of many genes, including those that code for class I and class II MHC molecules, both of which bind to antigens. Class I molecules are membrane proteins that present antigens to T-cells which subsequently recognize the antigen as 'self' or 'non-self' (involved in acceptance or rejection of tissue transplantation). Class II molecules are found on cells that present foreign antigens to T-cells, to elicit an immune response against the foreign antigen (e.g. virus or other infective agent). Selection is thought to be an active force maintaining high genetic diversity at MCH genetic loci to aid recognition of a wide variety of antigens.

Class I MHC

Human and mouse MHC genetic regions have been well studied for decades. In 1988 and 1989, Yuhki and O'Brien used RFLP and DNA sequence analysis (respectively) to study feline class I MHC. They found many similarities between feline and human class I; both have a single heterozygous functional class I region and both conserved (non-polymorphic) and polymorphic regions are similar. Polymorphic regions are 82% similar between cats and humans and 79% similar between cats and mice.

Yuhki and O'Brien (1990b) examined RFLP polymorphism at class I MHC for bottlenecked and non-bottlenecked populations of wild felids. Using two cheetah populations (South and East Africa) and three lion populations (Serengueti, Ngorongoro Crater, and Asiatic) they found that populations lacking genetic variation at class I MHC are those populations lacking genetic variation at allozyme loci. Therefore, the level of genetic variation at class I MHC genes appears to reflect population history. The authors also suggested that reduced MHC diversity may increase vulnerability to future disease outbreaks. For Asiatic lions, sequence analysis performed by Sachdev *et al.* (2005) showed higher levels of class I MCH diversity in wild compared to captive-bred Asiatic lions.

Class II MHC

Yuhki and O'Brien (1997) examined class II MHC alleles, using sequence analysis, in four wild felids (tiger cat, *Leopardus tigrinus*; Iriomote cat, *Prionailurus bengalensis*; Geoffroy's cat, *L. geoffroyi*), and domestic cat, *F. silvestris catus*. They found five distinct class II lineages (a set of 'linked' alleles that are inherited together as a group) in the domestic cat, with each wild felid species closely associated to one of the five lineages. This suggests that these five domestic cat class II lineages predated the divergence of the domestic cat and the four wild felids. DNA sequences in this study are also consistent with selection acting to maintain variation at nucleotide sites involved in recognition of new antigens, and against variation at sites where nucleotide conservation is desirable. Finally, Drake *et al.* (2004) examined DNA sequence variation of class II MHC in cheetahs (a region typically holding high levels of variation). They analysed data for 25 individuals and found only four different genotypes. Of the 25 individuals, only one (4%) had a genotype not previously seen in cheetahs and 15 (60%) all shared the same genotype.

Genomics

Genomic level analyses (where most or all of a species' genome is sequenced) are now within the financial reach of typical felid projects (see Box 4.2). New ultra-high throughput sequencing, enabled by massive PCR parallelization and microarray fibre optic detection rigs (e.g. pyrosequencing), has drastically reduced the cost of genome scale sequencing. Such information can be used to extract tens of thousands of SNP markers to select from when designing a microarray setup (Morin *et al.* 2004). The sheer number of markers on these arrays greatly improves the resolution and precision of genotyping efforts, while micro-scale reaction volumes simultaneously reduce costs. Moreover, in combination with information (e.g. candidate genes) from the published genome sequence of the domestic cat (*F. silvestris catus*), markers can also be chosen for diagnostic purposes of gene variants or gene expression patterns. More complex questions, such as examining the fitness consequences of specific genetic variants or genotype—phenotype correlations (e.g. morphological or behavioural) may soon be possible in wild populations (Luikart *et al.* 2003). One important area will be the improved potential to track evolutionary patterns, for example, as a result of ecological change (Stockwell *et al.* 2003).

An alternative, if sequencing of the study species is unachievable, is to base marker designs on the domestic cat sequence (possible since cats share a relatively recent evolutionary origin). This means that all felid species are more or less genome-enabled and can benefit from the new genetic tools developed to identify features in this genome (Kohn *et al.* 2006).

Another application of the technique is to use it as a diagnostic tool for species detection (see below). The 454-sequencing chemistry amplifies any DNA found in the PCR reaction, producing several hundreds of thousands of short DNA sequences (~100–300 base pairs). In a single species application, these segments are stitched together to complete the genome. But when screening samples contain multiple species, the sequences are instead used to identify species by comparing them to published sequence data. Since 454 sequencing is a qualitative approach, a single 100 base pair amplification can be sufficient to identify a species. This makes the method well suited for non-invasively collected samples, which often have small amounts or poor quality DNA. These methods thus hold enormous potential for the type of monitoring projects outlined in the next section.

Genetic monitoring

An individual's genotype can be used as a unique tag for repeated sampling (similar to capture—mark—recapture). Such genetic monitoring, where individuals or populations are sampled repeatedly over time, can be used to quantify many basic individual and population metrics (e.g. population size and dynamics, reproductive success, and dispersal). It is typically more effective, more accurate, and cheaper than traditional methods (Schwartz *et al.* 2007). Examples of genetic monitoring studies include population size estimates (Palomares *et al.* 2002a; Pires and Fernandes 2003; Bhagavatula and Singh 2006; Swanson and Rusz 2006), sex determination (Spong *et al.* 2000a; Kurose *et al.* 2005; Perez *et al.* 2006), and species identification and abundance (Singh *et al.* 2004; Kurose *et al.* 2005; Nagata *et al.* 2005). Furthermore, genetic monitoring programmes are often amenable to non-invasively collected samples (e.g. hair, faeces, and urine), which is logistically appealing for monitoring elusive, solitary felids. For example, hairs used by birds to reinforce or insulate their nests can be used to efficiently census rare and elusive mammals (Toth 2008). Genetic methods can also be used to analyse diets and disease, typically from faecal analysis. While morphological analysis of scat contents often identifies to genus level, molecular identification can often identify individual species (Dee *et al.*, in review).

Ernest *et al.* (2000) used puma scats to perform genetic mark—recapture and estimate minimum population size for pumas in Yosemite Valley, California. Microsatellite markers were used to perform both species and individual identification and resulted in one individual 'captured' three times in different locations and a minimum population size estimate of nine individuals from the scat analysis only (seven other individuals were known from live captures for a total of 16 individuals). Farrell *et al.* (2000) used scat analysis and mtDNA sequence to detect the presence of three sympatric felids, and one canid, in Venezuela (jaguars, ocelot, puma, and crab-eating fox, *Cerdocyon thous*). Palomares *et al.* (2000a) used scat analyses to monitor one of the rarest and most endangered felids, the Iberian lynx. They were able to identify Iberian lynx's presence from three different areas and had a 92% success at PCR amplification from scat samples, then using mtDNA sequencing for species identification. Kurose *et al.* (2005) used mtDNA analyses of scats to determine the presence and abundance of two felid species (Tsushima leopard cat, *Prionailurus bengalensis*, and domestic cat) and three other carnivore species (marten, *Martes melampus*; Siberian weasel, *Mustela sibirica*; and domestic dog, *Canis familiaris*) on the Tsushima Islands of Japan. Additionally, they determined the sex of each sample using DNA sequencing of the SRY gene (located on the Y-chromosome). DNA analysis of hair for species identification was performed by Mills *et al.* (2000) using PCR amplification of mtDNA, followed by RFLP analysis, to determine the distribution of lynx and bobcat in the United States. The restriction fragments yielded a unique banding pattern for each felid species (including bobcat, *Lynx rufus*; lynx, *L. lynx*; puma, and domestic cat) allowing an efficient and more cost-effective method to identify species genetically.

Strictly speaking, most of the studies mentioned above exemplify genetic assessments, rather than genetic monitoring, since they lack the temporal component required for monitoring. However, many long-term studies have the data required to perform temporal analyses and many wildlife projects are now explicitly collecting samples for the purpose of monitoring population dynamics, reproductive success, and individual dispersal patterns. Analyses of these data sets promise to produce many interesting results about felid behavioural biology in the near future.

Concluding remarks

As shown in this chapter, genetic data can be used to address a wide range of questions about evolution, ecology, behaviour, and conservation. Indeed, it may often be the only, or most efficient, choice. Analysis costs continually decrease and whole genome sequencing and high-resolution microarray applications are now within the budget range of many field projects. Moreover, analyses can be outsourced to commercial laboratories, nearly invariably producing data faster, of better quality, and cheaper than other options (not least from saving equipment and infrastructure costs). Although the possibilities of genetic methods will probably always exceed their level of application in wild populations, recent developments promise to narrow this gap considerably. So although studies quantifying genetic variation between and within species, populations, and individuals have already dramatically improved our understanding of past and current felid biology, the recent leap in genetic resolution means that much remains to be discovered. Everything makes more sense in the light of genetic data.

CHAPTER 5

Felid society

David W. Macdonald, Anna Mosser, and John L. Gittleman

The father, Mpofu (left), and one of his three sons in a coalition in Hwange N.P. © David W. Macdonald.

Felids in context

In some respects, the 37 species of Felidae encompass great diversity, for example, in size, colouration, vocalizations, and in the habitats they occupy. Yet, all are unmistakably feline, and their behavioural ecologies are largely variations on a theme. In particular, most generally operate alone, seemingly more resonant with felid individuality than felid sociality. True, lions (*Panthera leo*) and, under some conditions, cheetahs (*Acinonyx jubatus*) and domestic cat (*Felis silvestris catus*), can be conspicuously gregarious and seemingly cooperative, but individuals of most felid species do not generally spend their time in company. That, however, is not to say that their lives are not lived within a societal framework. The purpose of this chapter is to describe their socialities, and to explain them.

Prey size to body size

What societies exist among the Felidae, and what selective pressures have fashioned them? We might expect the answers to differ within a family including 37 species, ranging in size between 1.6 kg black-footed cats (*F. nigripes*) and 219 kg tigers (*P. tigris*) (male weights) (Macdonald *et al.*, Chapter 1, this volume), even if they all adhere to an instantly recognizable uniformity of body form and thus, perhaps, ecological function (Macdonald 1992).

Among the Carnivora there is evidence that societies reflect their ecological circumstances (Kruuk 1976; Macdonald 1983; Gittleman 1989). Carbone *et al.* (1999) discovered that members of this order fall into two broad dietary groups: smaller carnivores that feed on very small prey (invertebrates and small vertebrates) and larger carnivores that hunt large vertebrates—a dichotomy which they explained in terms of alternative options for maximizing net energy gain from hunting. Exploring this discontinuity, Macdonald *et al.* (2004a) found that, among the Canidae, different types of society were associated with particular segments of the prey-to-predator size spectrum. For example, pack hunters generally lay above a 21.5 kg discontinuity (Carbone *et al.* 1999), whereas species forming spatial groups whose members generally forage alone tended to hunt the smallest prey (<2 kg), with an intermediate category lying between. An exploration of the sizes of felids and their prey may, therefore, provide the first dimension of a framework for understanding their societies, and perhaps also for considering whether felids are different to canids and, if so, why.

We compared cats and dogs using the method of Carbone *et al.* (1999), repeating the analysis identifying the discontinuity between consumers of small and large prey for two reasons. First, we were able to use data from an additional two cat taxa (Asian leopard, *P. pardus* and Sumatran tiger, *P. t. sumatrae*). Second, the cut-off point in prey size was based on all Carnivora families and we are focusing here only on a comparison between cats and dogs. A segmented regression suggested that a prey size of *c.* 6 kg maximized the fit of the prey size/carnivore mass model, and the data plotted on Fig. 5.1a also suggest a discontinuity at around the same prey size (only one prey mass category between 2.65 and 16.5 kg is observed, that for the African golden cat [6.23 kg], which therefore lies just above the boundary delimiting 'large prey' species).

Of the 35 species of cat represented in Fig. 5.1b, 12 fall comfortably above the discontinuity (prey sizes > 6 kg; predator sizes > 20 kg). The broad relationship, that larger carnivores eat larger prey, is obvious. The scaling parameter for the data in Fig. 5.1a is, overall, 1.60 (CI 1.32–1.88) and there is no evidence that it differs between the two families ($F_{1,60} = 0.77$, $P = 0.38$).

When we look separately at the small- and large-prey eaters, however, a different pattern emerges. If we consider just canids and felids eating small prey (48 taxa in total), we find that the prey size–predator size relationship differs significantly between cats and dogs ($F_{1,43} = 4.7$, $P = 0.035$ for the interaction term between family and predator size). The felid slope for the small prey group is 1.03 (SE 0.24, $P < 0.001$, $R^2 = 46.0.0\%$, $n = 22$) and the canid slope is 0.16 (SE 0.62, $P = 0.62$; $R^2 = 1.2\%$, $n = 23$). For large prey, the felid and canid slopes cannot be differentiated (data are few and cats bigger, $F_{1,13} = 2.73$, $P = 0.13$), but the pattern of felids fitting the power law more closely than do canids is the same (the felid slope is 1.01 SE 0.13, $P = 0.0001$, $R^2 = 85.9\%$, $n = 11$), the canid slope is 0.15 (0.70, $P = 0.84$, $R^2 = 1.5\%$, $n = 4$), but the number of species in this range is rather few. Carnivores are distinct but not homogenous. The closer link between their body size and that of their principal prey may arise because felids are more dedicatedly carnivorous than canids. Note that some of these patterns then are associated with the evolutionary history of canids and felids. Nevertheless, associations of body size, prey size, and sociality cut across families (though more careful analyses should be undertaken when information is fully available on population variability within these taxonomic families). The other obvious difference is that the biggest canids are individually much smaller than the biggest felids, and tackle very large prey not by evolving larger body size but by evolving collective packing power. For canids, designed for a generalist life and endurance running, the constraints of overheating may rule out larger body size, and the fact that their packs can fission into smaller components (an option not open in the same way to a large felid) enables even the largest of them to adhere less strictly to the prey size/predator size exponent (both by shifting group size and thereby shifting their collective body mass, and by shifting between carnivory and omnivory).

Carbone *et al.* (1999) estimated that the relationship scales for all carnivores as prey mass = 1.19 × predator mass, and Carbone *et al.* (2007) conjectured that this was a consequence of both energy expenditure and intake scaling to three-quarters with body mass. Larger carnivores achieve a higher net gain rate by concentrating on large prey, but because more energy is required to pursue, subdue, and possibly defend, large prey, this leads to a twofold step

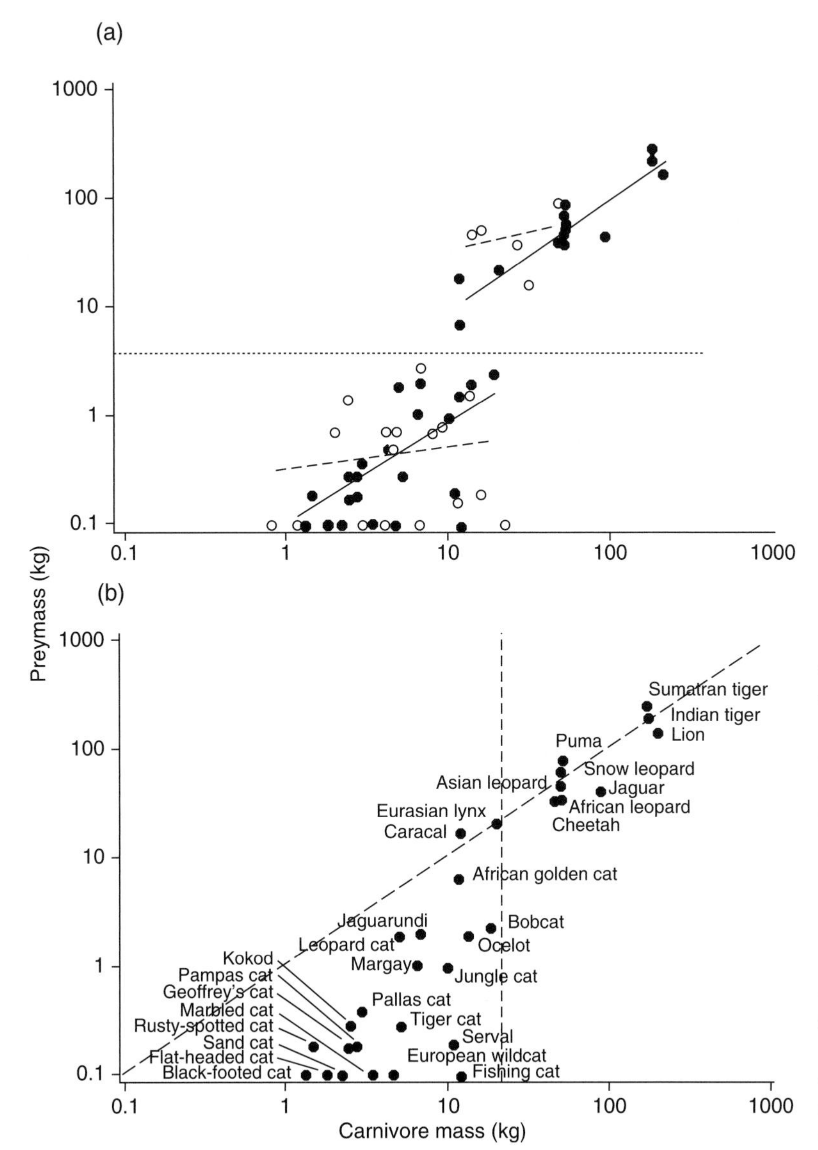

Figure 5.1 (a) Mass of most common prey versus felid and canid mass. Least squares regression lines for the two carnivore groups shown (felids solid line, closed points; canids dotted line, open points), fitted separately for consumers of small and large prey. (For further information on the canids, see Macdonald *et al.* 2004a.) Horizontal line denotes point separating prey masses into two groups while minimizing sum of mean squares. (After Carbone *et al.* 1999.) (b) Mass of most common prey compared to mass of felid species.

increase in energy expenditure, as well as increased intake. Despite the important categorical distinction between large and small carnivores, plots such as Fig. 5.1 are clearly a blunt instrument with which to probe behavioural ecology, partly because analyses like these average both sexual dimorphism and intraspecific variations in body size (Hoogesteijn and Mondolfi 1996), and deal with predominant prey size rather than the spectrum of prey eaten, but largely because intraspecific variation in ecology is ignored. Taking a canid example, the masses of the predominant prey and home ranges of the red fox (*Vulpes vulpes*) vary over two orders of magnitude (the masses of the foxes themselves vary threefold across their geographic range); it would seem prudent to expect intraspecific variation among felids too. Intraspecific variation will be a principal thread of this chapter, but Fig. 5.1b nonetheless provides a useful introductory ready reckoner to felids and their prey.

Body size to home range size

Not only will felid body size affect the prey that they can, and do, tackle, but it will also determine the quantity of prey needed to sustain them. Since, in turn, the home ranges (and thus biomass) of the prey will be determined by their own body sizes, larger predators are likely generally to require larger home ranges—a fundamental ecological truth embodied in the power law that reveals mammalian, and specifically carnivore, home range sizes to scale approximately with body size to the power of 1.25, which is significantly greater than that predicted from metabolic needs alone (Reiss 1988). Explanations for this have included environmental productivity scaling negatively with size and that larger species may tend to share their home range with more conspecifics (Reiss 1988; Jetz *et al.* 2004). Interspecific analyses have demonstrated positive correlations between the metabolic needs (individual mass and group size) and home range sizes for several different carnivore species (Gittleman and Harvey 1982; Mace *et al.* 1982; Swihart *et al.* 1988; Kelt and van Vuren 2001). Large carnivore species which often consume large prey have the largest home ranges and the lowest densities relative to their metabolic needs (Gittleman and Harvey 1982; Swihart *et al.* 1988; Kelt and van Vuren 2001).

Macdonald *et al.* (2004a) argued that the extent to which range area was not predicted by group metabolic needs (GMN, the residuals for Fig. 5.2) illuminated the relationship among predators, their social systems and the dispersion of their prey. Overall, the scaling parameter for Fig. 5.2 is 0.91 (CI 0.63–1.19). There is strong evidence, however, that, adjusting for GMN, range area differs among families ($F_{7,49} = 4.57$ and $P = 0.005$). There is no evidence that slopes differ among families ($F_{7,42} = 0.99$ and $P = 0.99$). Only the hyenas have larger home ranges than felids, adjusting for GMN. We argue that this is due to the cost of eating larger prey. The common slope for all families lies (coincidentally) exactly on the three-quarters expected from metabolic allometry, but with little statistical precision (0.75 CI 0.41–1.08). While the evidence that this differs among families is weak, it is highest for felids, 1.15 (CI 0.53–1.77). However, the two subspecies of wildcat (*F. silvestris* sp.), and the serval (*Leptailurus serval*), eat very small

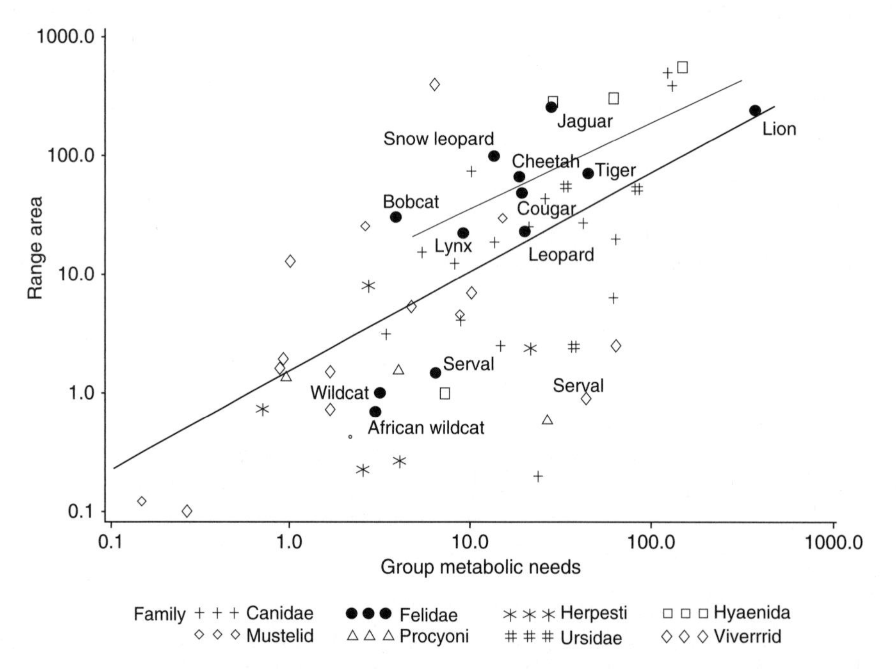

Figure 5.2 Log home range scales allometrically with log GMN. Solid line shows overall least squares regression line. Dotted line shows line for large cats (omitting serval and wildcats).

prey, at high density and probably dispersed evenly at the spatial scale that the cats feed—this is especially true for the serval, which is larger but feeds on grassland *Arvicanthus* spp.

The relative uniformity and high density of rodent distribution may explain why these three cats have smaller ranges than predicted from GMN. Without their inclusion, the slope for cats (0.63 CI 0.27–0.99) is close to that for carnivores in general. The predicted ranges for cats are, however, high compared with those for carnivores in general, which we speculate to be a result of the larger cats preying on herding species, which are by definition spatially aggregated.

Socio-spatial organization

The link between the sizes of prey, predator, and home range may be part of a syndrome of interacting factors that shape felid societies. Not only do the different energetic corollaries of eating small and large prey lead to the aforementioned dichotomy in predatory economics, but the size of prey is also one aspect of its dispersion (into larger or smaller packets) and this, among other measures of resource dispersion, has been suggested as a determinant of social organization (Macdonald 1983; Johnson *et al.* 2001a). Although it is commonplace to hear mention of the 'typical felid social system', it is far from easy to identify what system is indeed typical, not least because adequate descriptions of felid societies are rare (Sunquist and Sunquist 2002).

The crucial concept resonates in the title of Paul Leyhausen's paper (1965) 'The communal organisation of solitary mammals'. Considering the evident intricacy of olfactory communication, behavioural ecologists have been oddly disinclined to remember that a largely solitary *modus operandi* does not preclude an individual from a complex social life. Sandell (1989) and Macdonald (1992) are among those who have sought to categorize variations in felid spatial organization, which may be thought of in terms of the extent to which they form spatial groups. The term 'spatial group', coined by Carr and Macdonald (1986), describes the situation where the overlap of individual's home ranges is sufficiently non-random to suggest social congruency (or aggregation at a clumped resource). Clearly, if the ranges of two cats are highly congruent there is a suspicion that they constitute a sociologically linked spatial group (like a tower of plates being spun by a juggler) and the less the overlap, progressively the less it may be appropriate to consider them as being held together spatially by social ties. However, while the spectrum from congruency to tessellation clearly necessitates a different characterization in spatial terms, it does not indicate that the ranges are any less woven into a social fabric. The concept of spatial groups is highly relevant to felid socioecology because the spatial arrangements of so many felids involve various degrees of overlap, raising the questions of what level of overlap is diagnostic of social structure (they might merely be random), what determines these variations and what are their sociological consequences.

Felid spatial organization can be characterized by the extent of overlap within and between the home ranges of females and males. Simply, at the first level, the options are fourfold because, for each sex, individuals' home ranges may be either exclusive or overlapping, and for each sex that overlap may be so great as to approximate congruency and thus spatial groups (see Table 5.1). Furthermore, in each quadrant there may be differences between the sexes in scale (in practice, it is generally the home ranges of males that are various multiples of the size of those of females), and in the congruency between male and females ranges, and the movements of individuals may be variously dependent (in practice, either they may avoid each other, or travel together).

Of course, range overlap does not by itself form a reliable indicator of tendency to sociality. Individuals may overlap in home ranges without showing a tendency towards sociality, and there are numerous ways in which territorial overlap may be facilitated by environmental factors. As noted above, we expect

Table 5.1 Four possible social group arrangements based on the level of intersexual and intrasexual home range overlap.

		Females	
		Exclusive	Overlapping
Males	Exclusive	EE	EO
	Overlapping	OE	OO

greater overlap in larger species because they are more widely spaced and it becomes increasingly difficult to defend territory boundaries (Jetz *et al.* 2004). Home range size and density scale differently with mass because of this. The former deviates from that expected based on metabolic scaling (Jetz *et al.* 2004) as range exclusivity becomes less frequent with increasing body size. While there are large-scale patterns in overlap independent of social processes, we can nevertheless deduce a great deal of the social intricacies of species from fine-scale observation on their space use and interactions among individuals.

Thus, for example, where both males and females maintain exclusive ranges (defined, following Sandell 1989 as <10% overlap) the spatial organization is EE, for example Iberian lynx, *Lynx pardinus* (Ferreras *et al.* 2004); or, where males' ranges overlap but those of females do not, OE, for example pumas, *Puma concolor* (Logan and Sweanor 2001). Where the ranges of males are exclusive but those of females are not, the spatial organization is described as EO, for example cheetahs (Caro 1994); and where the ranges of both sexes overlap, OO, for example ocelots, *Leopardus pardalis* (Dillon 2005 but see Crawshaw and Quigley 1989) even to a level of congruency that they are plainly spatial groups, for example, lions (Schaller 1972). Although inadequate in the face of its many nuances, this categorization provides a starting point for our review of felid society.

Scottish wildcats (*F. s. grampia*) studied by Daniels *et al.* 2001 are OE, male ranges (median 4.59 km^2), overlapped each other by about 20–30%. Adult females (median home range 1.77 km^2) were largely exclusive and overlapped with each other only slightly (10%) but shared approximately 70% of their home range with one adult male and 36% with juvenile females. Similarly, juvenile males (median home range 1.59 km^2) and juvenile females (median home range 0.63 km^2) had little overlap with one another but shared about 70% of their home range with adult males or females. Nonetheless, radio-collared cats were seldom found together, overlap occurring in space but not in time (see also Biro *et al.* 2004). The whole system was in constant movement and readjustment (see drifting ranges as described by Doncaster and Macdonald 1991), adult females being the most faithful to their ranges over time and juvenile males the least faithful (67% and 38%, respectively, overlap in consecutive monthly or seasonal home ranges).

Bobcats (*L. rufus*) in Alaska (McCord and Cardoza 1982) and Idaho (Bailey 1972) have similarly OE arrangements. The home ranges of females are largely exclusive (McCord and Cardoza 1982), whereas those of males overlap widely (Lariviere and Walton 1997), and typically each male's range overlaps with those of three females than which it is generally two to five times larger (Kitchings and Story 1984). In areas of overlap, residents avoid each other (Lariviere and Walton 1997), and transients avoid residents and often use suboptimal habitat (McCord and Cardoza 1982). At a larger scale, leopards too are OE (although intraspecific variation is such that in Mizutani and Jewell's 1998 study of three males, two overlapped very closely and were largely exclusive of the third). Bailey (1993) working in Kruger National Park, Republic of South Africa, characterized the leopard's spatial organization as three-layered: the fundamental layer of resident female ranges, on which is superimposed the layer of larger ranges occupied by resident males, on which are further superimposed ranges of young or aged transients.

Similar blurring of the categories is illustrated by both female and male ocelots, for both of which the depictions in R. Kays and R. Moreno (personal communication) include cases of both rather neat tesselation and considerable overlap, so the general picture is OO. Overlap may, or may not, be evidence of amicability—encounters between male ocelots can be fatal (Moreno 2005). The anonymous polygons disguise the subtleties of sociology, for example, the overlapping ranges may be kin, as Smith *et al.* (1987) demonstrate for tigers, which seem to lean towards an OO system, suggesting clusters of overlapping mothers and daughters.

These shifts raise the caution that many depictions of home range may blur the sociological reality because spatial organizations are dynamic (Doncaster and Macdonald 1991). For example, Iberian lynx, an EE species, are close to spatial monogamy, where individuals were alone on 95.9% of occasions, but from season to season the territories of both sexes drifted in location by 10–50% (Ferreras *et al.* 2004). Furthermore, spatial categorization may change seasonally, as in Zezulak's finding (1981) that the winter ranges of male and female bobcats were up to 41% and 70%, respectively, smaller than their summer ranges. One community of dominant male feral domestic cats studied by Liberg *et al.* (2000), swapped from exclusive to overlapping ranges as their movements converged during the breeding season on the same colonies of females. Because female house cats essentially have a point resource (associated with their owner's house), the impact of their dispersion on the configuration of males' ranges is especially clear (Liberg 1980). Members of a female matriline overlap (but are exclusive of other matrilines), whereas the ranges of dominant males vary between EO and OO. This system closely parallels that of African lions, which are famously social (Schaller 1972) to the extent that they are often, although conspicuously incorrectly, referred to as the only truly social felids. Prides of up to 18 adults (generally related females and their dependent offspring [Gilbert *et al.* 1991; Spong and Creel 2001]), defend a shared permanent home range that can persist for many generations (McComb *et al.* 1994) and are defended by scent marking and roaring. Coalitions of (exceptionally up to seven) males, whose ranges are congruent, overlap perfectly one to four such prides (Schaller 1972). In this OO system of congruent spatial groups, there may be extensive overlap at the peripheries, but not the cores, of neighbouring prides where home ranges are large, but less so where they are small (van Orsdol 1981). In the Serengeti, the mean number of adults recorded in each pride was 1.5 males and 4.8 lionesses (Schaller 1972). Larger prides undergo fission–fusion dynamics, temporarily splitting into several sub-prides that may be scattered throughout the pride range.

For most felids, accounts of their social systems are at best fragmentary, a small number of tracked cats giving an incomplete picture, and only a partial and unfocused image of their movements. This is set to be rectified by the global postioning system (GPS) revolution, as illustrated by Loveridge *et al.*'s study of lions (2009b) in Hwange (Loveridge *et al.*, Chapter 11, this volume) and Cavalcanti and Gese's study of jaguars (2009) (*P. onca*) in the Pantanal (Cavalcanti *et al.*, Chapter 17, this volume). For the jaguars, while the annual print-out of locations for adult females initially suggests considerable overlap of home ranges, detailed scrutiny reveals exclusive (0% overlap) wet season (63.9 km^2) ranges and widely overlapping (26–38% overlap) dry season (62.0 km^2) ones (Fig. 5.3a). In contrast, the ranges (averaging 140.0 km^2 and 165.8 km^2 in the wet and dry seasons, respectively) of at least five males overlapped very widely, several to the extent that they appear to comprise a congruent spatial group (each male's range overlapping with all or most of the ranges of two females; Fig. 5.3b). The larger dry-season ranges, and altered spatial arrangements, might be due to seasonally altered resource dispersion. For example, the smaller home ranges of jaguars during the wet season might reflect their prey concentrating on islands of dry land in the seasonal Pantanal landscape (see Crawshaw and Quigley 1991). However, although cattle were indeed confined to dry islands, the caiman and peccaries on which jaguars largely preyed were, if anything, more widespread during the wet season (whereas their clumping around water in the dry season might cause jaguar ranges to overlap more then). So, both social networks and the details of resource dispersion remain unknown, but the marvellous data generated by GPS tracking at least allows the right questions to be formulated.

None of these sociological mysteries can be resolved until molecular studies reveal the familial ties underlying, in Leyhausen's prescient words, the communal organization of these solitary mammals (e.g. Spong 2002). Furthermore, this needs to be done while exploring the limits to, and drivers of, the intraspecific variation characteristic of each species (e.g. Sunquist 1981; Smith *et al.* 1987; Smith 1993). For example, in the Serengeti, female cheetahs occupy large, overlapping ranges, whereas territorial males maintain smaller areas (Caro 1994; Durant *et al.*, Chapter 16, this volume; Table 5.2).

In contrast, in Namibia, there were no statistically significant differences between sexes or social groupings in either wet or dry season in the sizes of vast home ranges averaging 1651.1 km^2 (Marker *et al.*, 2008a). Long-distance movements by cheetah elsewhere are due to migratory prey (Broomhall 2001) or pressure to escape intra-guild harassment (Durant 2000b), but neither applies in Namibia. In the case of the Serengeti cheetah, and effectively for the first time for a wild felid other than lions, Gottelli *et al.* (2007) have used molecular techniques to demonstrate that female cheetahs are promiscuous, with

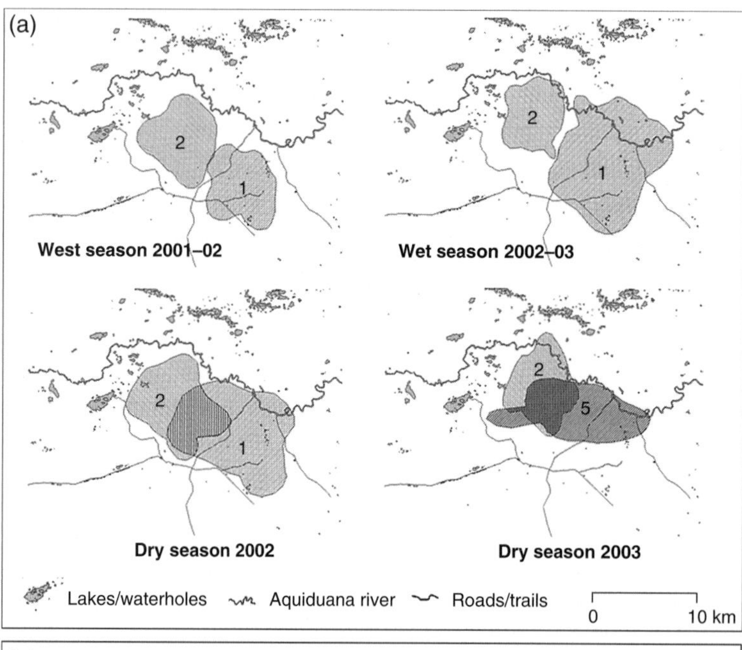

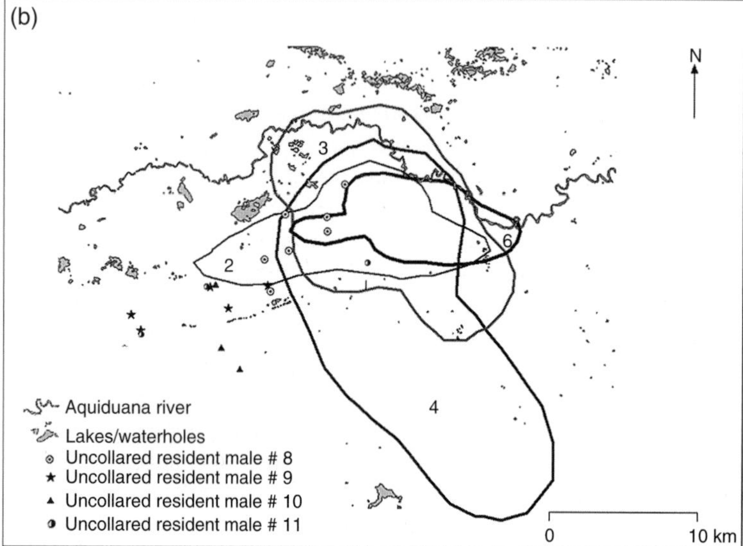

Figure 5.3 (a) Seasonal home ranges (90% adaptive kernel) of female jaguars #1, #2, and #5, October 2001–April 2004, southern Pantanal, Brazil. (b) Home ranges of radio-collared male jaguars #2, #3, #4, and #6 during the dry season 2003, southern Pantanal, Brazil. Other symbols show the locations of uncollared resident males also in the same area and photographed by camera traps. (From Cavalcanti and Gese 2009.)

high levels of multiple paternity and no evidence of mate fidelity between reproductive seasons. Females also appear to mate with unrelated males within the same reproductive cycle. The low success rate in paternity assignments indicates that males living outside the study area, and potentially outside the park, contribute substantially to cheetah reproduction. This finding reinforces the crucial role that high mobility plays in cheetah ecology and their conservation. The understanding of the mating system of this cheetah population will aid in the development of future management plans aimed at maintaining genetic diversity of cheetahs within and outside protected areas.

This, however, remains to be done for the remaining 32/37 (86.5%) of felid species. For them, too,

Table 5.2 Average home range sizes of different felids. (Adapted from Mizutani and Jewell 1998 and Schmidt *et al*. 2003.)

Species	Home range males (km^2)	Home range females (km^2)	Overlap female (%)	Overlap male %	Overlap male: female	Reference
Leopard cat [a]	7.5	6.6	Exclusive	—	1–3	Rabinowitz (1990); Grassman (2000)
	4.5	2.5				
Tiger	60	17	Exclusive	Exclusive	High	Schaller (1967); Sunquist (1981)
	67	21	Low	High	High	Karanth and Sunquist (2000)
Ocelot	10.6	3.4	Exclusive	Less exclusive	—	Ludlow and Sunquist (1987); Nowell and Jackson (1996)
Bobcat	26	15	5	3	—	Wassmer *et al*. (1988)
European wildcat	60.4	28.5	10	Exclusive	Yes	Biro *et al*. (2004); Stahl *et al*. (1988)
Scottish wildcat	41.9	19.7	10	20–30	Yes	Daniels *et al*. (2001); Macdonald *et al*. (2004b)
Iberian lynx	25	20	15	20–30	1:1	Ferreras *et al*. (1997, 2004)
Eurasian lynx	248	133	30	6	62	Schmidt *et al*. (1997);
			31	28		Jedrzejewski *et al*. (2002)
Black-footed cat	20.7	10	40.4	12.9	60.2–66.7	Sliwa (2004)
Leopard	451	118	46	43	—	Stander *et al*. (1997c);
	14	32.8	Low	Low	Some	Mizutani and Jewell (1998)
Puma	83	29	—	<45	—	Hemker *et al*. (1984); Scognamillo *et al*. (2003)
Jaguar	79	50	30	90 (not temporally)	Yes	Crawshaw and Quigley (1991)
Iriomote cat	2.8–5.8	1.4–4.0	High or little	High or little	—	Schmidt *et al*. (2003)
African lion[b]	0.4–82.7	0.4–82.7	High	High	—	Spong and Creel (2004)
Snow leopard	1590	13–141	High	High	—	McCarthy *et al*. (2005b)
Cheetah		883			—	Caro (1994)
Gordon's wildcat	19.24	21.36	—	—	—	Phelan and Sliwa (2005)

[a] Grassman (2000) proposed that male leopard cats overlap with several females but very little with other males.

[b] For a pride.

where there is any evidence at all, it suggests that social arrangements change with ecological circumstances. For example, some bobcats are classically OE (Bailey 1972; McCord and Cardoza 1982), but Zezulak (1981) reports up to 36% overlap between females in an OO population in California, and Nielson and Woolf's study (2001) in southern Illinois found an unusual pattern of spatial organization in bobcats whereby there were relatively high levels of intrasexual home range overlap for males and females, with the exception of core areas, which remained nearly exclusive. What factors, behavioural and ecological, drive felid societies and, in particular, are associated with them forming groups? This is the topic of the next part of this essay.

Spatial ecology

Movement patterns are so intricate that averages and categories risk obscuring biologically important detail and variation. Here, we summarize intraspecific variation in home range sizes, and dispersal patterns.

Intraspecific variation in home range size

Measures of intraspecific variation in felid home range sizes are limited by the paucity of studies, rather than the flexibility of the animals. Bobcat home ranges vary from 0.6–201.0 km^2, whereas those of lynx from 10–243 km^2 (bobcat density from 0.1–2.74 cats/100 km^2, lynx density from 1.7–5.6 cats/100 km^2; bobcat sex ratio from 40 males/100 females to 210 males/100 females and lynx sex ratio from 93 males/100 females to 137 males/100 females [McCord and Cardoza 1982]). Marker and Dickman (2005) reviewed 24 studies of leopards, with average home range sizes varying between 9 and 2,182 km^2 for males and 8 and 489 km^2 for females, respectively, in the Kalahari and Kenya (Hamilton 1981; Bothma *et al.* 1997). For North American pumas, home ranges vary between 194.4 and 575.4 km^2 (males) and 103.7–268.0 km^2 (females) (Dixon 1982). In the San Andreas Mountains, New Mexico, the home ranges of male pumas (averages from 136.4–236.4 km^2) were fourfold the area of females' (averages 52.6–69.0 km^2), but the range sizes of neither sex varied significantly between years of high or low density and this was true in a treatment area where pumas were experimentally reduced, and where they were unmanipulated (Table 5.3; Logan and Sweanor 2001).

Domestic cats provide among the more comprehensive panoramas of intraspecific variation, their population density and home range sizes both varying over three orders of magnitude (population density < 1 to >2000 km^2, home range is 0.001–2.0 km^2 in females and up to 10 km^2 in males; Kerby and Macdonald 1988; Liberg *et al.* 2000). Among the

Table 5.3 Annual home range size (mean ± SD) for adult pumas during years of relatively low and high deer density, San Andreas Mountains, New Mexico. Treatment area held all radio-collared mule deer, and bighorn sheep and pumas in this area were experimentally removed to study population dynamics. The reference area is an area where no manipulations of the puma population occurred. (Taken from Logan and Sweanor 2001.)

Low deer density			High deer density		
Sex	*n*	HR size (km^2)	Sex	n	HR size (km^2)
Treatment area					
M	10	140.8 ± 51.3	M	5	136.4 ± 42.4
F	10	56.0 ± 29.0	F	8	69.0 ± 29.0
Reference area					
M	15	236.4 ± 86.6	M	10	197.4 ± 87.6
F	13	52.6 ± 27.3	F	13	68.6 ± 36.1

Table 5.4 Dispersal age and dispersal distances (km) covered by different felid species.

Species	Dispersal age (months)		Dispersal distance (km)		Sex ratio (M:F)	Source
	Male	Female	Male	Female		
Bobcat	18	18–24	14.4–78.4	14.4–78.4	1:1	Kamler *et al.* (2000)
Canadian lynx	10	10	47	83	1:1	Brand and Keith (1979); Parker *et al.* (1983)
Amur tiger	18.8	18.8	≥5	≥5	1:1	Kerley *et al.* (2003)
Puma	15.2	15.2	30	100	1:1.13	Sweanor *et al.* (2000), Lambert *et al.* (2006)
Wildcat	5	12	6	3	1:1	Tryjanowski *et al.* (2002)
Lions	24	NA	>200	0	1:1	Smuts *et al.* (1980); Funston *et al.* (2003)
Cheetah	18	18–24	20	NA	1:2	Eaton (1974); Caro (1994); Beekman *et al.* (1999)
Iberian lynx	12–24	12–24	23	30	1:1	Ferreras *et al.* (2004)
Scandinavian lynx	12–24	12–24	450	450	1:1	Rueness *et al.* (2003a)
Tiger	19–28	19–28	33	10	1:1	Smith (1993)
Domestic cat	12–24	12–24	5	≤1.5	1:1	Liberg (1980)

Note: NA, not available.

dimensions of variation, adult females can be solitary or gregarious; when gregarious, their groups may vary in size and social structure and, in terms of social dynamics, the tenor of individual relationships may vary within and between age–sex classes. The ranges of males overlap extensively, and are generally at least threefold larger than the relatively less-overlapping ranges of female groups.

Dispersal and recruitment

Most felids live solitarily and hence most shift from their natal range before the next generation of their siblings is born. In so far as age at maturity, and interbirth intervals, are greater in larger mammals, the modal age at dispersal is older in larger species. Felids under 10 kg disperse, and breed, as yearlings, those of 10–35 kg generally disperse in time to breed as 2-year-olds, whereas the largest felids disperse at 3–4 years (Table 5.4). For example, many bobcats (Bailey 1981) and lynx disperse at 10 months (Brand and Keith 1979, Parker *et al.* 1983), Iberian lynx between 12 and 24 months (Ferreras *et al.* 2004), pumas at 15.2 ± 1.6 SD months (Sweanor *et al.* 2000), Amur tigers on average at 18.8 months (Kerley *et al.* 2003) and Serengeti lionesses at 3–4 years (Packer and Pusey 1983a). However, this artificially tidy summary ignores the reality that among many species of felids dispersal encompasses anything from a shuffling next door to a long-distance march into the unknown—these two having sociologically very different consequences (the recorded maximum beeline distance for Scandinavian lynx being 450 km; Rueness *et al.* 2003a). Mares *et al.* (2008) present an interesting case study of an ocelot gradually distancing itself socially from its parents. Furthermore, the decision on when, or even whether, to disperse will vary according to an individual assessment of the marginal costs and benefits of doing so (Macdonald and Carr 1989), so we expect inter-population differences in dispersal. For example, most studies of bobcats report that they disperse before the litters of the following year are born, with dispersal movements lasting from 21–76 days and straight-line dispersal distances ranging from 14.4 to 78.4 km (Crowe 1975; Griffith *et al.* 1980; Kitchings and Story 1984; Anderson 1987). However, in a high-density population (0.4–0.5 bobcats/km^2) at Fort Riley, Colorado, Kamler *et al.* (2000) found that males dispersed when they were nearly 1.5 years old, and females did so when they were 1.5–2 years old (see also Bailey 1981; Anderson 1987) and attributed this postponement to the abundance of prey (eastern cottontails, *Sylvilagus floridanus*) and absence of trapping. Similarly, young pumas disperse at 11–13 months in Texas (Harveson 1997), 17–24 months in New Mexico (Logan and Sweanor 1999), and 10–33 months in Alberta, Canada (Hemker *et al.* 1984; Ross and Jalkotzy 1992; Beier and Barrat 1993; Machr 1997; Sweanor *et al.* 2000). Logan *et al.* (1996) report that the average female puma gives birth to her first litter when aged 29 months. In general, transients, predominantly young or sexually immature individuals, form a floating population that coexists with resident territory holders and, in the cases of lynx and bobcats may use suboptimal habitat and lack an established home range (Bailey 1972; McCord and Cardoza 1982).

There is debate as to whether there is a general sex bias in felid dispersal (Mowat *et al.* 2000). For example, among lynx (*L. canadensis*), Campbell and Strobeck (2006) used molecular scatology to conclude that there was no significant difference in dispersal of males and females in Alberta, Canada. In contrast, Janečka *et al.* (2007) found that relatedness among females ($r = 0.050$; ± 0.042, 95% CI) was significantly higher than among males ($r = -0.075$, ± 0.031) among bobcats in southern Texas, suggesting male-biased dispersal (among 9 radio-collared females and 12 radio-collared males, 2 females and 6 males dispersed from the study site). Male pumas in the San Andreas Mountains of New Mexico dispersed over distances between 47 and 214.9 km, whereas for females the distances were 5.6–78.5 km (excluding those breeding philopatrically; Sweanor *et al.* 2000). Among the tigers in Chitwan, Nepal, which disperse at 19–28 months old, the beeline distances of males averaged 33 km and those of females 10 km (Smith 1993). This is an area where their home ranges averaged 6 km^2 and thus the mean dispersal distances are 1.5–6 home range diameters—a more sociologically meaningful metric of dispersal than beeline distance.

In their Serengeti study area, 72% of the female cheetahs, but only 54% of the males, had been known since birth, from which Durant *et al.* (2004) deduced that female immigration was lower. In so far as earlier studies had concluded that permanent dispersal of adult cheetahs out of the study area was unusual (Laurenson 1992; Caro 1994) the frequency of dispersal may be changing in this population (Durant *et al.*, Chapter 16, this volume).

Among lions, all males disperse (generally during a takeover by incoming males), whereas most females do not (and the third that do merely shuffle next door before their fourth birthday; Packer and Pusey 1983a; Pusey and Packer 1987), triggered either by an incoming male coalition or by the adult females of their pride giving birth to new cubs. Male lions rarely show interest in mating prior to dispersal, rarely reside in prides where their daughters have matured, and rarely return to breed in their natal pride. However, having dispersed and taken over a new territory, some males may subsequently voluntarily move on, or at least readjust their ranges, for example to annex an additional pride, or to abandon one of several simultaneously held prides to spend more time with those in which their cubs are younger and hence more vulnerable to infanticide. Coalition males may also abandon one pride for another. In such cases, they will abandon small cubs only in exchange for a much larger numbers of lionesses, but they abandon more females for fewer when their daughters by the former are reaching maturity.

Among felids there appears to be an incipient tendency to female philopatry, and where this is expressed approximations to spatial groups (see above), or clusters of kin may develop, for example among pumas (Logan and Sweanor 2001) and tigers (Smith *et al.* 1987). This may involve females reconfiguring their ranges to accommodate maturing daughters, for example, cheetahs among which Caro (1994) reported that once daughters become solitary, their ranges overlap that of their mother by 61.8% (there are clear parallels of neighbourhood settlement and enduring family ties in some 'solitary' canids, e.g., Macdonald and Courtenay 1996). For lionesses, the distinction between dispersal and pride fission is hazy and, in the Selous (where home ranges average 52.4 ± 26.3 km^2), pride splits have no consistent effect on the average levels of genetic relatedness between prides, and following fission the memberships of the resultant prides seldom fall outside the size range that confers maximum reproductive success per female (Spong and Creel 2004).

Intraspecific variation in society

Caro (1994) tabulated studies of range size and overlap for different felids, revealing that the extent of intraspecific variation in range size was a reflection of the number of studies of a given species. He also characterized each study in terms of whether female ranges overlapped or were exclusive. For both bobcats and tigers four studies were split three to one for exclusive to overlapping, whereas four studies of pumas were split three to one for overlapping to exclusive. Both jaguars and ocelots had two studies each, and in one each the females' ranges were overlapping and in the other they were exclusive.

Intraspecific variation in social behaviour is most dramatic in domestic cats and lions. Some rural domestic cats are arranged as solitary females occupying largely exclusive ranges overlain by the larger and substantially overlapping ranges of solitary males (Corbett 1979), others, around clumped food resources, may live in colonies of 70 or more, largely arranged around female matrilines (Liberg 1980; Dards 1983; Macdonald *et al.* 1987), and within large multi-male/multi-female groups there is widespread polygynandry (Natoli *et al.* 2000), although Say *et al.* (1999) report lower variability in male reproductive success in urban (high-density) than in rural (low-density) environments.

The pride sizes of lionesses average between four and six (Schaller 1972; Hanby and Bygott 1979; Stander 1991) but can be as large as 20 (van Orsdol *et al.* 1985), and social dynamics vary between populations. For example, males of both Indian lions (*P. l. persica*) and African lions in the Kalahari desert spend little time with lionesses and rarely hunt or feed with them; in both cases the principal prey of the females is small and not readily shared (Rao and Eswar 1980; Owens and Owens 1984; Kumara and Singh 2004).

The ethology of felid societies

Social odours

Sources of social odours in felids are anal and facial glands, urine and faeces, and odours associated with ground-scraping and bark-scratching (reviewed by Macdonald 1985). Mechanisms for deploying odour include spraying urine (Sanders 1963; Schaller 1967; Bailey 1974; Smith *et al.* 1989; Kitchener 1991), defecating (Rabinowitz and Nottingham 1986; Smith *et al.* 1989), ground scrapes (Seidensticker *et al.* 1973; Rabinowitz and Nottingham 1986; Smith *et al.* 1989), scratches on tree bark (Smith *et al.* 1989), and object rubbing (Smith *et al.* 1989). These mechanisms of olfactory communication are not only important for social function in felid societies, they are also critical for understanding the history of the Felidae and reveal consistent systematic characters for sorting out phylogenetic relationships (Bininda-Emonds *et al.* 2001). The literature is rife with unwarranted assumptions as to the generally unknown communicative function of most of these odours. Not surprisingly, marking generally occurs within an individual's home range, where, among many possible functions, it may enable individuals to avoid each other (Lariviere and Walton 1997) or contribute to territoriality (Sliwa 2004), for example by scent matching (Gosling 1982). Both sexes of all felids deploy urine by spraying or squatting, and at least in domestic cats these odours have an individual signature (Passanisi and Macdonald 1990). Adult domestic cats may spray more when travelling (Corbett 1979) or, among females, when hunting (Panaman 1981), and it seems likely that scent enables individuals to timeshare their use of space. Facial rubbing on conspecifics is widespread, perhaps universal, in felid societies, and at least in the case of domestic cats the flux of this behaviour is the main barometer of social structure, with rubbing flowing predominantly from subordinate to dominant individuals (Macdonald *et al.* 1987). Considering the diversity of roles demonstrated for social odours in mammals in general (Brown and Macdonald 1985), and other carnivores in particular (Macdonald 1985), and considering the obvious attention all species of felids pay to odours, scent doubtless plays diverse functions in many aspects of their societies.

Building on Wemmer and Scow (1977), Mellen (1993) quantified rates of scent marking in various captive small cats. Both males and females cheek rubbed and neck rubbed, generally at higher frequency when reproductively active (see Mellen [1989]). Head rubbing was associated with novel odours (Wemmer and Scow 1977). Flehmen (curling of the upper lip) has been seen in both sexes of many felid species, especially among males during a female's oestrus (e.g. Wright and Walters 1980; Hart and Hart 1985). Typically, a urine mark is approached, sniffed, whereupon the cat raises its head, drops its lower jaw, wrinkles its muzzle and appears almost immobilized while breathing slowly. Both sexes of most species (but not female servals; Mellen 1989) have been seen to spray urine, but males generally do so at a higher rate than females. Urination, and especially spraying, is often associated with scraping with the hind feet, and quivering of the tail. Scratching (perhaps sharpening) with claws occurred at a slightly higher rate among males than females and was associated with urine marking. Despite Mellen's compendious quantification of their occurrence (1989), little is understood of the function of most scent-marking activity by felids in the wild.

Domestic cats can either bury their faeces or leave them in conspicuous latrines especially along trails in the vicinity of good or dense hunting areas. In the mangrove swamps of Borneo, Malay houses are built atop stilts above the water and houses are connected by walkways which offer good locations for latrines: neat piles of scat are seen along these walkways (Macdonald, personal observation). Ocelot latrines have been found at road intersections (Ludlow and Sunquist 1987; Chinchilla 1997) and one, in tall Vietnam grass (*Saccharum spontaneum*), was visited by four individuals (Moreno and Giacalone 2006). Territory marking is but one of a myriad number of functions that scent could serve and a broad range of functions may operate simultaneously; because marking is seen in aggressive contexts, it does not imply that the odour expresses aggression (Macdonald 1980). A selection of observations on the deployment of scent is given on Table 5.5.

Felids select carefully where to place their marks. Marker *et al.* (Chapter 15, this volume), describe 'play trees' where cheetahs scent mark. One case study of the mysterious detail of scratch marking comes from Nath and Macdonald's observations of tigers (unpublished) in Bandhavgarh National Park. There, of 2320 trees sampled in random plots, only 108 had identifiable scratch marks by tigers and a total of 41 tree species were identified from these plots. Scratch marks were infrequent on the common *Shorea robusta* (1.85%, $n = 2/108$) but frequent on the much less abundant *Boswelia serrata* (37.04%, $n = 40/108$; Fig. 5.4a). This indicates that the tigers are preferentially selecting for *B. serrata*. Of the two *S. robusta* trees marked, both were marked by cubs. Diameter at breast height (DBH) of trees with scratch markings varied from 52 to 172 cm (mean 125.38, SD 3.39: Fig. 5.4b). Of marked trees, 75% were over 90 cm DBH. Tigers left their travel path to reach trees for marking: 38.8% of marking trees were 4–41 m from the path (Fig. 5.4c; Nath and Macdonald, unpublished data).

The interaction between felids and plants is famously illustrated by catnip, *Nepeta cataria*, which is highly attractive to cats—a response inherited as a dominant autosomal gene (Todd 1962). The cheek, chin, and body rubbing and rolling shown to catnip is similar to the behaviour shown by oestrus female domestic cats in the presence of males, and appears to take their mind off prey and on to sex (Palen and Goddard 1966).

Vocal communication

Felid vocal repertoires have been studied for 36 species, and involve 3–10 distinct vocalizations (Peters 2002), but the only species renowned for long-range communication is the lion, where roaring is a conspicuous behaviour of pride males (Funston *et al.* 1998). Yowling of domestic tomcats is also long-range (*c.* 200–400 m; Dards 1983). Among male lions, roaring is the prerogative of those holding prides, and they roar only within their territories and when prepared to fight (Grinnel and McComb 2001): apparently disclosing their whereabouts is prohibitively costly for nomads. Members of larger coalitions are quicker to enter battle, and males are more hesitant to advance on the taped voices of

Table 5.5 Overview of scent marking by felids.

Species	Location	Type of scent marking	Female use	Male use	Both sexes at same place	Time of greatest activity	Location of scent marking	Specific locations	Source
Ocelot	Barro Colorado Island	Latrine	Y	Y	Y	Mating period	Throughout territory	Guard posts and trails	Moreno and Giacalone (2006)
Ocelot	Panama	Latrine	Unreported	Y	Unreported	Mating period	Throughout territory	Guard posts and trails	Moreno and Giacalone (2006)
Tiger	Nepal	Urine spray	Y	Y	Unreported	Mating period	Throughout territory	Forest roads and trails	Sunquist (1981)
Tiger	Bangladesh	Urine spray	Y	Y	Unreported	Mating period	Throughout territory	Trails	Khan (2004)
Domestic cat	England	Urine spray	Y	Y	Y	Mating period	Throughout territory	High spots and trails	Bradshaw and Cameron-Beaumont (2000)
Black-footed cat	South Africa	Urine spray	Y	Unreported	Unreported	Mating period	Throughout territory	Borders and areas of intense use	Molteno *et al.* (1998)
Domestic cat	Scotland	Defecating	Y	Y	Y	Unreported	Throughout territory	Along trails near good hunting areas	Corbett (1979)
Puma	Mexico	Ground scrapes	N	Y	N	Unreported	Throughout territory	Ridgelines, channel bottoms and under trees and at kill sites	Logan and Sweanor (2001)
Puma	Idaho, United States	Scat mounds	Y	N	N	Unreported	Throughout territory	Near kill-cache sites	Logan and Sweanor (2001)

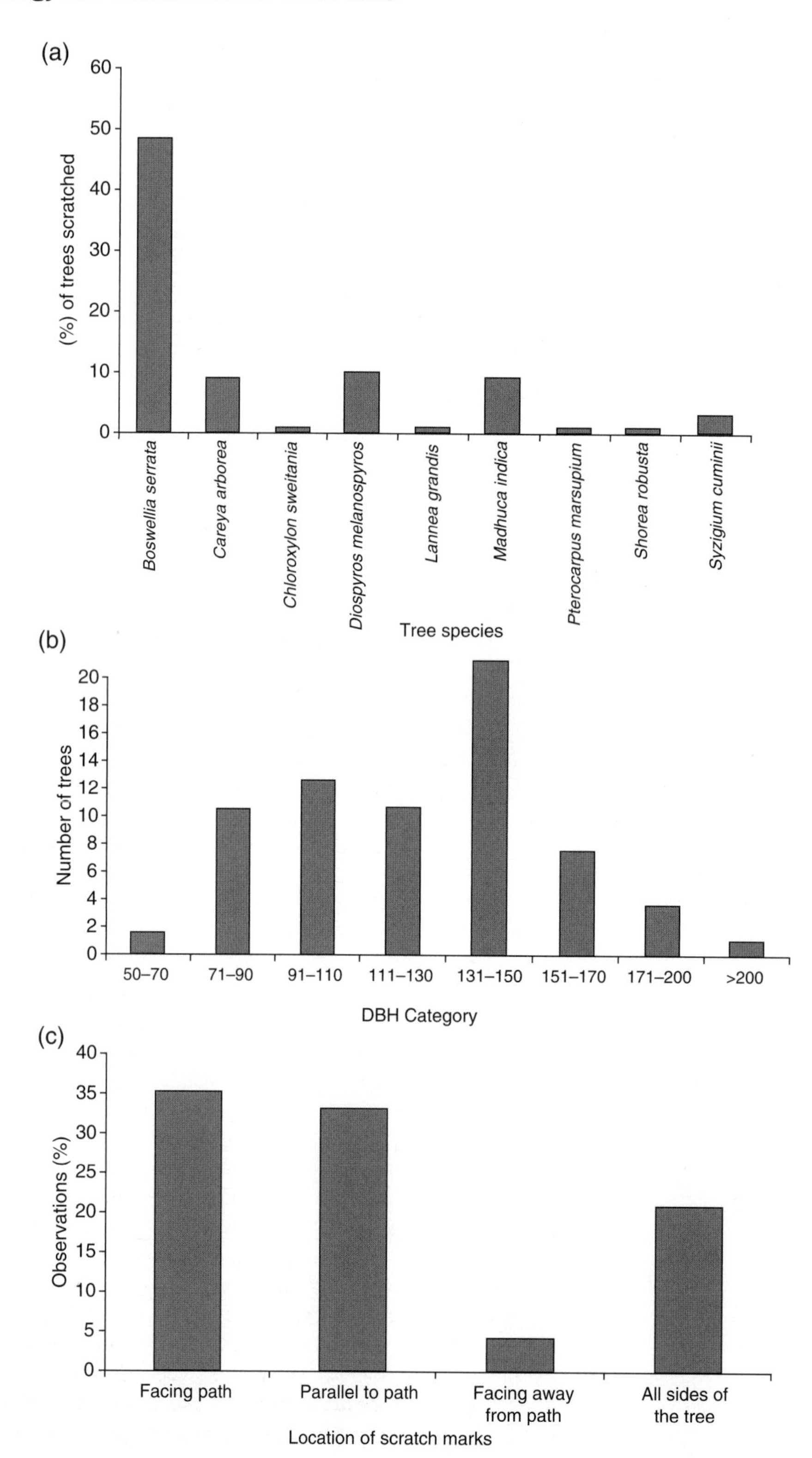

Figure 5.4 The trees at which tigers scratch are selected with respect to (a) species; (b) DBH (cm); and (c) orientation with respect to paths.

larger coalitions. McComb *et al.*'s playback experiments (1994) revealed roaring behaviour as a proxy for resource-holding potential among females (Heinsohn 1997). Defending lionesses were more likely to approach the voices of a lone intruding female than those of three or more intruders. Cogniscent of their own numbers, outnumbered lionesses roar to summon support from their pride.

Peters (2002) notes that cat calls occur in diverse contexts (females nursing/young suckling, juvenile, and/or adult individuals huddled up to one another, mutual grooming, friendly approach, and courtship), of which the 'common denominator' is an amicable, relaxed mood. This has resulted in the general view that purring indicates contentment (Kiley-Worthington 1984). However, domestic cats may purr in labour or when hurt (Leyhausen 1979; Beaver 1992), leading to the suggestion that it is an appeasement signal or an auto-communicatory signal, comparable to humming in humans. Leyhausen (1979) argues that purring (which all felids except the six pantherines do [Kitchener *et al.*, Chapter 3, this volume]) originally signalled that the situation is pleasurable to the sender, but can also indicate that the sender is harmless and 'helpless'. Bradshaw and Cameron-Beaumont (2000) conclude 'Purring may therefore function as a "manipulative" contact- and care-soliciting signal, possibly derived from its (presumed) function in the neonate.' The fundamental frequency of the highest amplitude component of purring is at the lower limit of felid hearing (Heffner and Heffner 1985), so it may be a tactile as well as auditory signal.

Social dynamics and hierarchy

Two felids are gregariously sociable, the domestic cat (Macdonald *et al.* 1987) and the lion (Schaller 1972), with tigers (McDougal 1977; Sunquist 1981; Bragin 1986), and cheetahs (Caro 1994) forming smaller, less conspicuous associations. Of these, the social dynamics have been most intensively studied in farm cats living semi-independently of people, among which the flow of social interactions between individuals reveals that they are socially, and to a large extent spatially, structured according to matrilineal genealogies (Kerby and Macdonald 1988; Macdonald *et al.* 2000c). Among farm cats there is structure among the age–sex classes in who sits with whom (e.g. juvenile males tend to sit together). The dynamics of sociality within each lineage were also structured, such that while grooming was largely reciprocal (although at different frequencies between different age–sex categories and different dyads of cats), facial rubbing was sometimes asymmetrical, generally flowing towards matriarchs and older females, and towards breeding males—and in this sense was socially centripetal (Macdonald *et al.* 1987; Fig. 5.5a, b). Relatedness affected social ties, with adult females preferring their own offspring to their nephews and nieces as nearest neighbours, although they preferred the latter to non-relatives (Fig. 5.5c). Lineages were also structured, with some dominating central resources, others living more peripherally, the latter having significantly lower reproductive success. These structures all suggest a centripetally organized society around clusters of closely related females, with spatially central matrilines within which there are socially central females (which form closer social ties with close relatives than with other adult females [Macdonald *et al.* 2000c]). Males too can be described as central and peripheral, but with the opposite implications to females regarding their social standing. The mature, presumably reproductive, males lived in peripheral association with matrilines, but were tolerated around central resources where they dominated central males that were young and still largely associated with their natal matrilines (Macdonald *et al.* 2000c).

Within these generalities, which prevail in colonies of widely different sizes, demography and social dynamics may nonetheless vary. Kerby and Macdonald (1988) contrasted large (25–30 adults), medium (5–15 adults), and small (4–9 adults) colonies. In the large and medium colonies, adult sex ratios were, respectively, 1:1.58 and 1:0.78, and adult–juvenile ratios among males, respectively, 1:0.44 and 1:0.59. It seemed that in the colony with fewer adult females and more juvenile males (potential competitors), adult males increased aggression towards juvenile males, with the effect that their subordination was prolonged. Kerby and Macdonald (1988) described the integration of kittens and juveniles into their lineages, with both male and female kittens preferring the company of their same-sex littermates over

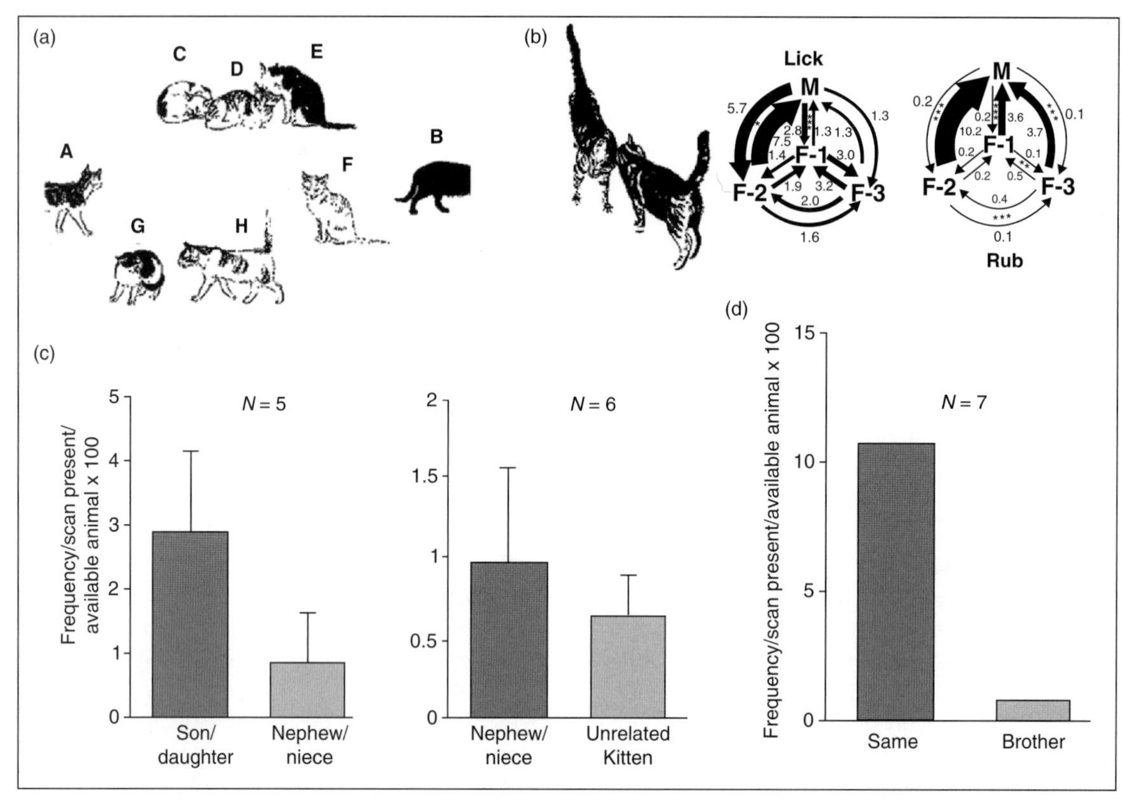

Figure 5.5 A methodology for quantifying felid social dynamics. (From Macdonald *et al.* 1987; Macdonald *et al.* 2000b.) (a) The presence or absence of individuals was logged every 5 min (e.g. cat A arrives and B leaves), as was each cat's proximity to all others (e.g. cats C, D, and E are all within 1 m of each other but not of cat F) and their behaviours (cat F is sitting alone, C sleeps with D and E, D sits with C and E, and E licks D, and H makes an aggressive approach to G). (b) Sociograms of the relationships between adult cats at a small colony when the members consisted of an adult male (M), an adult female (F1), and her two daughters (F2 and F3). (Adapted from Macdonald *et al.* 1987.) (c) Adult females significantly preferred their own sons/daughters to nephews/nieces as nearest neighbours (Wilcoxon Signed Rank Test: $Z = 2.023$, $P = 0.0431$), and did not discriminate their nephews/nieces from unrelated kittens (Wilcoxon Signed Rank Test: $Z = 0.734$, $P = 0.4631$). (d) Male juveniles significantly preferred their male littermates to their brothers of the different ages as nearest neighbours (Wilcoxon Signed Rank Test: $Z = 2.023$, $P = 0.0431$). They also significantly preferred related male juveniles to related male kittens as nearest neighbours (Wilcoxon Signed Rank Test: $Z = 2.366$, $P = 0.0180$). (Adapted from Macdonald *et al.* 2000b, c.)

the company of same-sex older (maternal) siblings (Macdonald *et al.* 2000c).

Remarkably little quantitative ethology has been published on lions or other gregarious groupings of felids. Societies of many cats have been roughly classified as either solitary or sociable, polygynous or promiscuous (Macdonald 1992), but there has been little study of relationships between individuals in these systems. Social ties have been explored quantitatively only for the domestic cat. Macdonald (1992) emphasized the striking similarities between the societies of farm cats and lions, such as the matrilineal female associations, nursing coalitions, infanticide by males and male dispersal (Macdonald *et al.* 2000c), which makes the differences between them particularly interesting. No nomadic male coalitions have been reported in domestic cats, which Macdonald *et al.* (2000c) attribute to differences in the size, dispersion, and defendability of females and prey. Within a domestic cat lineage, social ties between mothers and their own offspring appear to be very strong and kittens have strong ties with their littermates, with these ties being maintained after they become juveniles. Furthermore, Macdonald (1991) observed one large colony of farm cats in which cliques of male littermates remained conspicuously affiliated as young adults, for example dyads walking in bodily contact while facial rubbing with tails entwined (Fig. 5.5d). Social dynamics of this type, in ancestral felids, under selective pressure, may provide the basis for the evolution of coalitions between male littermates.

Behavioural selective pressures for group living

Cooperative hunting

Ethology of predation

With a few exceptions, perhaps the fishing cat (*Prionailurus vivverinus*) when fishing, the ethology of predation is remarkably consistent throughout the Felidae (Ewer 1968). First the cat crouches, and then hurries towards the prey with the body flat to the ground (Leyhausen's 'slink-run' [1965]). Depending on cover, the cat pauses, assumes the ambush position, crouched low with the sole of the foot flat to the ground, forepaws directly under the shoulders, whiskers spread and ears turned forward, and head turning to follow the prey's every movement. Depending on cover and distance, a second slink-run and ambush may follow, or the cat may stalk, edging forward, preparing for the final ambush, with heels now raised from the ground, hind legs shift back, often moving alternately, tail tip twitching. Then a short sprint, flat to the ground, and final 'spring', also flat to the ground, the fore quarters thrust forward to seize the prey, the hind feet remaining firmly planted, providing stability. Access to cover is as important as is prey density in determining food intake (Elliot *et al.* 1977; van Orsdol 1984; Stander and Albon 1993; Funston *et al.* 2001b; Hopcraft *et al.* 2005). Hopcraft *et al.* (2005) found that Serengeti lion kills were associated with landscape features that provided cover (vegetation and erosion terraces), rather than with estimated prey density, suggesting that terrain that facilitates stalk-and-ambush hunting is an important resource (Fig. 5.6).

Rewards of cooperation

In comparison to single males, members of coalitions of male cheetahs in the Serengeti plains preferred larger prey (>20 kg), but they displayed scant coordination when hunting, and if anything appeared to spend longer than singletons in pursuit of larger prey (Caro 1994). Coalitions of three had higher rates of food intake, but this appeared to be a by-product of selection for larger prey, rather than a result of successful group cooperation. In Namibia, Marker *et al.* (2008b) found that coalitions of male cheetahs selected for dense, prey-rich habitat, but

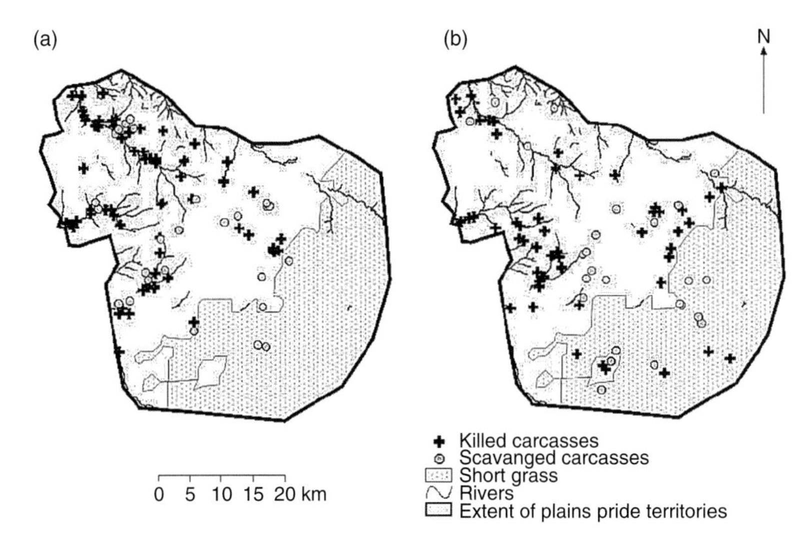

Figure 5.6 The short grass plains in relation to the distribution of kills and carcasses obtained by scavenging by plains lions during (a) the dry season; and (b) the wet season. When migratory herds are on short grass plains during the wet season, the majority of kills are still located in areas with more cover (long grass plains, woodlands, and near rivers). (Taken from Hopcraft *et al.* 2005.)

this preference was not apparent for other social groupings. In summary, it is not obvious that benefits of hunting together are among the factors encouraging male cheetah to form coalitions.

Although it is a beguiling intuition that lions hunting together benefit from their apparent cooperation, it has not been straightforward to test. The seemingly simple answer revealed by Schaller's initial calculation (1972) that a group of lionesses needed only half as many hunting attempts per kill as the six averaged by a single lioness, has descended into progressively more intricacy following Caraco and Wolf (1975) and Caraco *et al.*'s explorations of the per capita rewards of hunting in company (1980). Schaller (1972) had concluded that the optimum group size for maximum hunting success and intake of food was two. Reanalysing Schaller's data (1972), Caraco and Wolf (1975) concluded that group size was constrained by a minimum physiological requirement but that observed group sizes did not necessarily maximize food intake. When only small prey were available (Thomson's gazelle, *Eudorcas thomsonii*) group size was constrained to two, but when large prey were abundant, a range of group sizes met physiological requirements and lions usually fed in groups larger than two. They therefore concluded that, when not constrained by basic energy requirements, group size was strongly influenced by other social benefits. Later analyses also suggested that larger foraging groups might minimize variance in food intake, and thus the risk of starvation (Caraco *et al.* 1980; Rubenstein 1982; Packer 1986).

As evidence accumulated, in 1984 van Orsdol considered communal foraging a primary benefit of sociality in lions, but by 1990 Packer *et al.* concluded that cooperative hunting is not a selective pressure driving group living in lions. To resolve whether hunting together is truly cooperative in so far as benefiting all (or even some) of the participants, requires a full economic analysis in some common currency (ultimately energy) of all components of the hunt, including effort expended, risks run, and rewards accrued. Increasingly intricate, parallel studies of these questions for lions and African wild dogs, *Lycaon pictus*, culminating in net gain models, led to different conclusions. For wild dogs, individual net benefits from hunting together increase up to pack sizes of at least 10 individuals (Rasmussen *et al.* 2008). A crucial factor in the economic analysis is that the energy which lionesses spend to stalk their prey does not vary appreciably with group size (Scheel and Packer 1991), whereas with wild dogs it does (Rasmussen *et al.* 2008).

The effectiveness of group hunting among lions probably differs with circumstances, between and within populations, and between individuals. Thus, to compare populations, Stander (1992b) found that cooperative hunting was advantageous for lionesses in Etosha, seemingly because the absence of cover worsens the opportunities for stalking. In contrast, cover is more abundant in the Serengeti ecosystem, where prides may comprise up to 12 females and their cubs. There, Packer (1986) found that hunting success increased, and distances between kills decreased, with increasing hunting group size. Scheel and Packer (1991) described variation in the roles taken by individuals in 64 communal hunts of four different prey species. Individuals sometimes refrained from participation, sometimes conformed (joining hunting parties in which all participants behaved similarly), and sometimes pursued (joined parties in which individual roles varied). Males refrained more, and pursued less, than did females, and refraining was more common when the target was easier to capture (individuals were more likely to refrain when hunting warthog, *Phacochoerus aethiopicus*, than zebra, *Equus* sp., or buffalo, *Syncerus caffer*). The average probability of cheating (refraining from joining a group hunt), however, was still high even for large prey (e.g. 22% when hunting zebra).

Analysis of the relationship between adult female subgroup size and food intake from both hunting and scavenging (Packer *et al.* 1990) showed that during times of prey abundance group size had no relation to food intake (Packer and Ruttan 1988). In times of prey scarcity solitaries and subgroups of five to six do equally well, while subgroups of two to four suffer in terms of food intake. Despite this disadvantage, lion prides of two to four females remained in the largest group size possible. Social grouping in lions was not tightly associated with increased foraging success, especially for smaller prides (Fig. 5.7).

A different balance of selective pressure weighs on males depending on whether they are territorial or dispersing. The necessity of defending their prides restricts the capacity of territorial males to hunt beyond their territories, and there is a great incentive

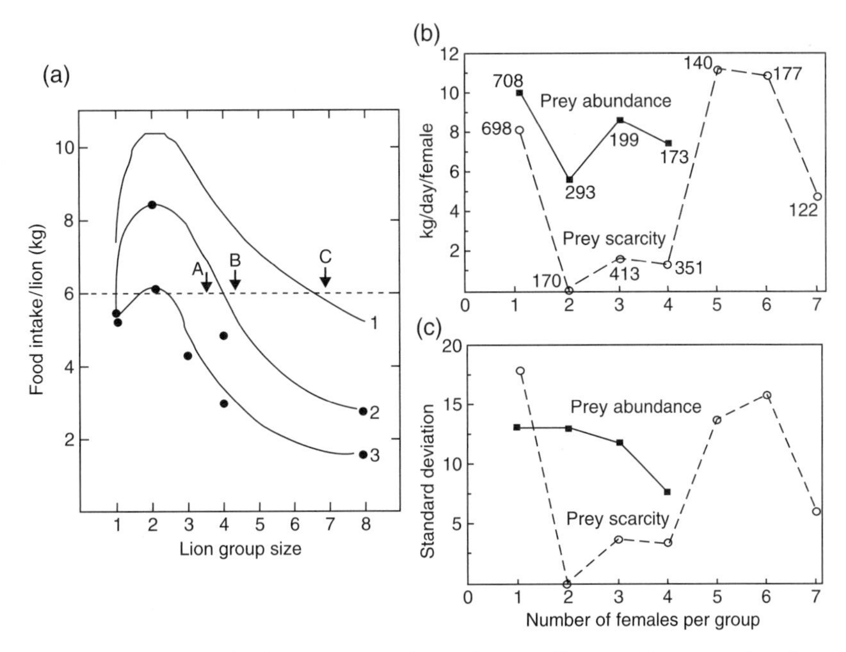

Figure 5.7 (a) Line 1 shows the hypothesized wet-season intake (mean edible prey biomass [kg] per lion per chase per 24 h as a function of lion group sizes preying on wildebeest) in woodland-plains border region, where vegetation cover increases lion capture efficiencies. Line 2 gives the same calculations for eastern plains and westerns plains. Line 3 represents the mean edible prey biomass (kg) per lion per three chases (predicted daily index) as a function of lions preying on Thompson's gazelle. The observed mean lion group sizes (Schaller 1972) are given for eastern plains (A), western plains (B) and border region (C). Dashed line at intake of 6 kg represents daily physiological minimum requirement for an individual lion, males and females averaged. (Adapted from Figs. 2 and 4 in Caraco and Wolf 1975.) (b) Daily capita per food intake for female lionesses as a function of the number of females per group when prey is abundant and scarce. (Taken from Packer *et al.* 1990.) (c) Variance in daily capita per food intake for female lionesses as a function of the number of females per group when prey is abundant and scare. (Taken from Packer *et al.* 1990.)

for dispersers to bulk-up to maximize their chances as contenders in pride takeovers. These pressures may underlie Funston *et al.*'s findings (1998) that non-territorial males in the Kruger National Park scavenged less from pride females, and those in larger coalitions killed more buffalo (a dangerous but rewarding prey), than did territorial males (Fig. 5.8). The overall rate of scavenging from a kill made by another carnivore (with no help from the scavenger), however, is considerably lower in Kruger National Park (15% of carcasses) as compared to the Serengeti National Park (40% of carcasses), potentially reflecting an underlying difference in prey availability or carcass detectability or carnivore competition.

While the debate continues as to the role of cooperative hunting in the sociality of lions, or indeed of other social carnivores (Packer *et al.* 1990; Caro 1994), the fact remains that the evidence for this selective pressure being strident in the evolution of any felid's sociality is not great.

Cooperative defence of prey

A factor related to, but distinct from, cooperative hunting is the retention of kills. Although few felids cache food with the fervour commonplace in most canids (Macdonald 1976), leopards famously stash kills in trees and/or caves (Brain 1981; Mills 1990; Skinner and Smithers 1990; Estes 1991). This may explain bone accumulations found in caves in Kenya (Simons 1966), Tanzania (Sutcliffe 1973), and Namibia (Brain 1981). Scottish wildcats may sweep leaves

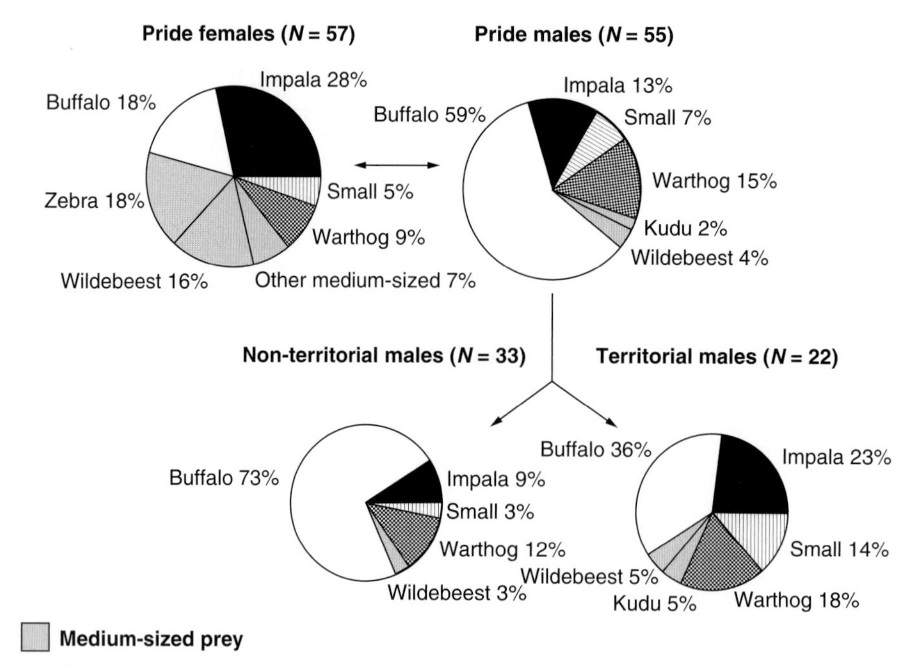

Figure 5.8 Proportional representation of the prey killed by the lion group types during continuous observations. Species such as impala, buffalo, and warthog were frequently caught by all lion group types and thus each was treated individually in the analysis. Medium-sized (100–230 kg) ungulates are herd living, and included wildebeest (*Connochaetes* sp.), zebra, kudu (*Tragus strepsiceros*), waterbuck (*Kobus ellipsiprymnus*), and giraffe (*Giraffa camelopardalis*) (although they are large they were stalked in the same way). Small prey do not live in herds and included steenbok (*Raphicerus campestris*), duiker (*Sylvicapra grimmia*), scrub hare (*Lepus saxatilis*), and porcupine (*Hystrix africaeaustralis*). (Taken from Funston *et al*. 1998.)

over kills, as may domestic cats and tigers, but most felids probably rely on discretion, or eating fast, rather than strength of numbers, to thwart kleptoparasites (Macdonald 1976).

In contrast, lions rely heavily on cooperative defence of prey. Although spotted hyenas (*Crocuta crocuta*) are unable to steal kills from lions when territorial males are present (Mills 1990; Cooper 1991), lionesses alone cannot repel hyenas if outnumbered by more than four to one, with the consequence that, where spotted hyenas occur at much higher population densities than lions, lionesses lose a lot of meat to them (Cooper 1991). Packer (1986) and Kissui and Packer (2004) report, however, that in the Ngorogoro Crater the loss of meat to hyenas may be negligible: females rarely surrendered carcasses to hyenas before most of the meat was consumed, and only lost 2 out of 98 fresh kills to hyenas.

Cooperative defence of mates

The males of two felids, lions and cheetahs, form conspicuously cooperative coalitions. The formation of coalitions in cheetahs does not appear to be directly related to defence of females (see below), but this is clearly the case for male lions. Among lions, males form coalitions of up to seven that act as a unit in competition with other coalitions (Packer 1986). Once subadult males leave their natal pride, the size of their age-set influences their reproductive success and bonds between age-mates are often permanent. Large cohorts of males are able to take over a new pride sooner, enjoy a longer reproductive life, and have higher reproductive success than males in smaller age-cohorts (Bygott *et al.* 1979; Packer and Pusey 1982, 1983a; Packer *et al.* 1988). Larger coalitions oust smaller ones from prides and chase nomadic coalitions from their prides' ranges.

Although such encounters typically involve only chasing, severe injuries may be inflicted. This pugilistic imperative among males means that their reproductive life is confined to a narrower range of ages than is that of females (Schaller 1972; Bertram 1975a; Bygott *et al.* 1979). Successful males become resident in their first pride when they are about 4 years old (Pusey and Packer 1987), and their first cubs are born about 6 months later (Packer and Pusey 1983a). Male coalitions remain in individual prides for an average of only 26 months, but some large coalitions gain residence in a succession of prides (Pusey and Packer 1987). The success of larger coalitions is sufficiently high that the individual fitness of males increases strikingly with increasing coalition size (Fig. 5.9a and b). Coalitions cooperate in fighting to take over, and thereafter to defend, prides of lionesses from rival nomadic coalitions (the latter seek to usurp the resident males, and kill their cubs). Their cooperation enables cohort (s) of their cubs to be reared (Schaller 1972; Packer 1986). Coalitions are often, but not exclusively, kin. Packer and Pusey (1982) report that 42% of the coalitions with known origins contained unrelated males. Thus the competitive advantages of cooperation overcome the potential costs of sharing reproductive benefits with non-relatives, but only for smaller coalitions. Reproductive skew within a coalition increases with coalition size, and unrelated males are unknown in coalitions of four or more (Packer *et al.* 1991a; Fig. 5.9c). Kin selection therefore underlies cooperation in coalitions of four or more.

Female cheetahs may adopt cubs if their own litter is of a similar age (Caro 1994). This can result in an increased survival of male adolescents through the presence of a sister (Durant *et al.* 2004). Observational data has shown that adolescent males rely strongly on their sisters to catch prey for the group (Caro 1994). Temporary associations of adolescent sibling cheetahs have a higher survival rate compared to singletons (X^2 coefficient = 1.09), and males especially benefit from the presence of a sister (X^2 coefficient = 2.09; Durant *et al.* 2004). These adolescent groups are less vulnerable to attacks by male cheetahs and hyenas and, because they catch larger prey, they do not suffer from reduced food intake (Caro 1994).

Cooperative care of young

Within canid social systems reproductive suppression commonly leads to non-breeding adults being available to act as helpers (among which females, may even lactate). This system is unknown in felids, where reproductive suppression has not been recorded. Further, the opportunity for communal denning, and thus cooperative care and communal suckling, also widespread in canids, exists most readily within the spatial organization of only two species of felid, lions and domestic cats, and both display this behaviour. Macdonald *et al.* (1987) and Macdonald (1991) first documented communal denning in domestic cats (indeed, filming one female chewing through the umbilical cord as her adult daughter's kittens were born; see Crawford 1979), and document apparently indiscriminate communal suckling within matrilines (Table 5.6). They also document infanticide by neighbouring males (on one occasion a roving male killed all save three kittens from three litters, which unusually, had no females in sight, before two mothers returned and together drove him away). This recalls Herodotus's study of temple cats in *c.* 450 BC, where he observed that to obtain the 'companionship' of nursing females, males practised 'a curious artifice'. 'They seize the kittens, carry them off, and kill them, but do not eat them afterwards. Upon this the females, being deprived of their young, and longing to supply their place, seek the males once more.'

Macdonald *et al.* (1987) report that communal denning was common: over 3 years, 32 of 40 possible nursing coalitions were seen, though the females differed greatly in the extent of their participation. Costs of communal denning were highlighted by the deaths of kittens belonging to two females due to contagious disease (feline distemper). Pusey and Packer (1994) record very similar behaviour among lionesses, demonstrating that communal nursing is most common among close kin, and that females with small litters tend to share milk more, but that there is scant evidence for reciprocity in the pattern of communal suckling (Fig. 5.10). They calculate that communal nursing confers no net nutritional benefit on the cubs and therefore argue that the behaviour has evolved secondarily to other factors—primarily defence against infanticide—making

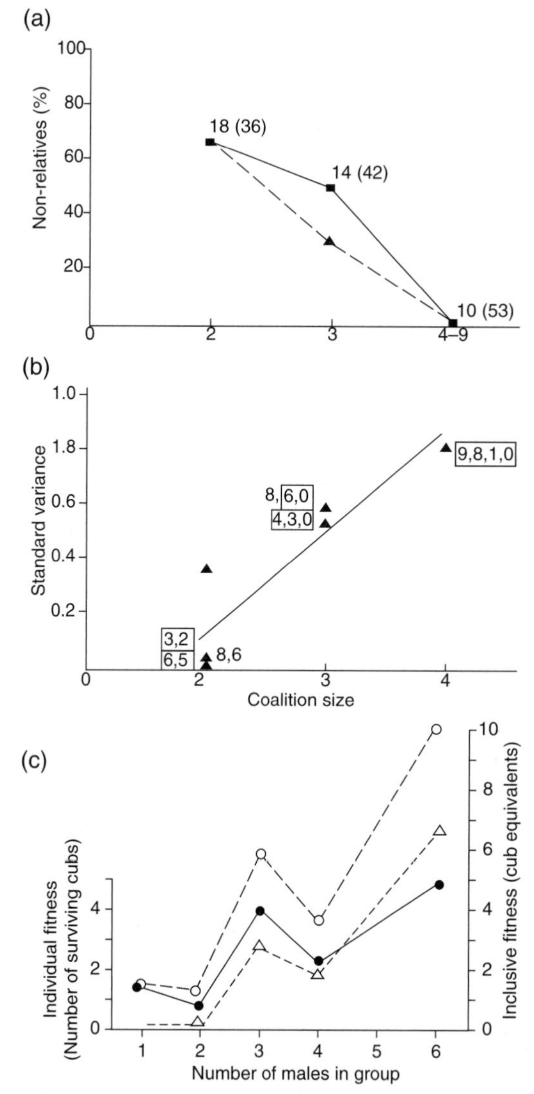

Figure 5.9 (a) Percentage of coalitions showing non-relatives (■) and of males lacking related partners (▲). Data include all coalitions for which demographic or band-sharing information was obtained. Coalitions containing unrelated males are significantly smaller than those composed exclusively of relatives ($Z = 3.122$, $n_1 = 19$, $n_2 = 23$, $P < 0.002$, two-tailed Mann–Whitney U-test; if the analysis is restricted to coalitions for which we have demographic data, $Z = 3.300$, $n_1 = 14$, $n_2 = 18$, $P < 0.001$). Numbers by each point give the number of coalitions of that size; numbers in parentheses give the number of males. The percentage of males lacking related partners declines even more sharply with increasing coalition size because four of the seven trios containing non-relatives were composed of two relatives and an unrelated third. In these cases, only one of the three males lacked a related partner. (b) Standardized variance in individual reproductive success within each coalition. Numbers by each point indicate the number of cubs fathered by each male in that coalition and a box around these points indicates that the males are close relatives. For example, all four males in the coalition of four were relatives: the most successful male in this coalition fathered nine cubs, and the remaining males fathered eight, one, and zero cubs, respectively. For the regression of standardized variance against coalition size, $r^2 = 0.8370$, $P < 0.001$. Probability for the regression is found by simulation. Simulations were performed by random generation of reproductive success for the 18 males according to the observed distribution of coalition sizes and the observed number of cubs fathered by each of the seven coalitions. The observed regression of standardized variance against coalition size was exceeded in only 2 of 10,000 runs. (Taken from Packer *et al.* 1991a.) (c) Individual fitness and inclusive fitness of individual male lions in groups of different sizes. The line for mean individual fitness (●) is derived from Fig. 1c (Packer et *al.* 2001) assuming equally shared paternity of the cubs born during the males' tenure, and multiplying by the different probabilities of achieving tenure from all. (See Table 1, Packer et *al.* 2001.) The line for a male's mean inclusive fitness (○) is the sum of first, the number of cubs he has fathered (i.e. his mean individual fitness) and second, the number of cubs fathered by his companions multiplied by 0.22 (i.e. devalued by their coefficient of relatedness to him). The lowest line (△) shows the inclusive fitness of a male who fathers no cubs of his own.

Table 5.6 The means (and SD) of durations of suckling bouts on SM and DO by each kitten, recorded between 59 and 183 days after SM gave birth and 40–164 days after DO gave birth. (Taken from Macdonald *et al.* 1987.)

		Female		
		SM	DO	Total
Kitten	SA(TA)	294 ± 291 n = 27	264 ± 297 n = 46	275 ± 293 n = 73
	OR(SM)	173 ± 170 n = 13	260 ± 279 n = 51	242 ± 262 n = 64
	CA(DO)	185 ± 176 n = 17	216 ± 259 n = 55	202 ± 239 n = 75
	Total	243 ± 240 n = 57	245 ± 277 n = 152	

it advantageous to keep the young together in a crèche. Indeed, the likelihood of females forming a crèche, and the number of participants, was related to patterns of food availability. There may, however, be energetic and evolutionary consequences of communality: age at sexual maturity occurs later and litter weight (litter size times birth weight) is heavier in communal carnivores (including African lions) compared to solitary species (Gittleman 1985).

Incoming male lions kill their predecessors' cubs. Infanticide accounts for 27% of cub mortality in the Serengeti (see below), and prides of two or more females have a better chance of fighting it off. Therefore, the higher reproductive success of larger prides with 3–10 members is partly due to their lowered risk of being taken over. Infanticide by neighbouring females is poorly understood. Suggestions have been made that neighbouring female lions may accidentally kill dependant cubs in group encounters and consume the infants for protein but 'deliberate' infanticide of cubs by non-pride females is difficult to demonstrate (Packer and Pusey 1983a).

Cooperative defence of territory

In cheetahs, related males form coalitions of up to three. Durant *et al.* (2004) show that the success of these coalitions is density-dependent; coalition males had higher rates of survival only when there were more coalitions in the population. Coalitions are more successful in defending territory (Caro 1994), and presumably the benefits of cooperative territorial defence are only reaped when there is stronger competition among the males in the population (although the level of female infidelity revealed by Gotelli *et al.* [2007] leaves a question mark over the reproductive pay-off).

As to the consequences of cooperation among lionesses, in the Serengeti and Ngorongoro Crater, Packer (1986) reported that a lioness's reproductive success depends on the number of her female companions. These lionesses lived in prides of up to 18 members, all of which breed at a similar rate (Packer *et al.* 2001). Those in prides of 3–10 adults have higher individual fitness than those in smaller prides. Several lines of evidence suggest that the success of large prides, and indeed the basis of their sociality, may be due to long-term cooperative defence of shared territory. Encounters between prides occur weekly, over half escalate to a chase, and the larger group almost always wins (Packer *et al.* 1990). In the Serengeti, larger prides not only reside in higher-quality habitat, but they out-compete smaller prides for access to that habitat (Mosser 2009; Table 5.7. and Fig. 5.11). Larger prides are significantly more likely to shift their ranges into better-quality habitat and are also more likely to maintain control of overlapping territory.

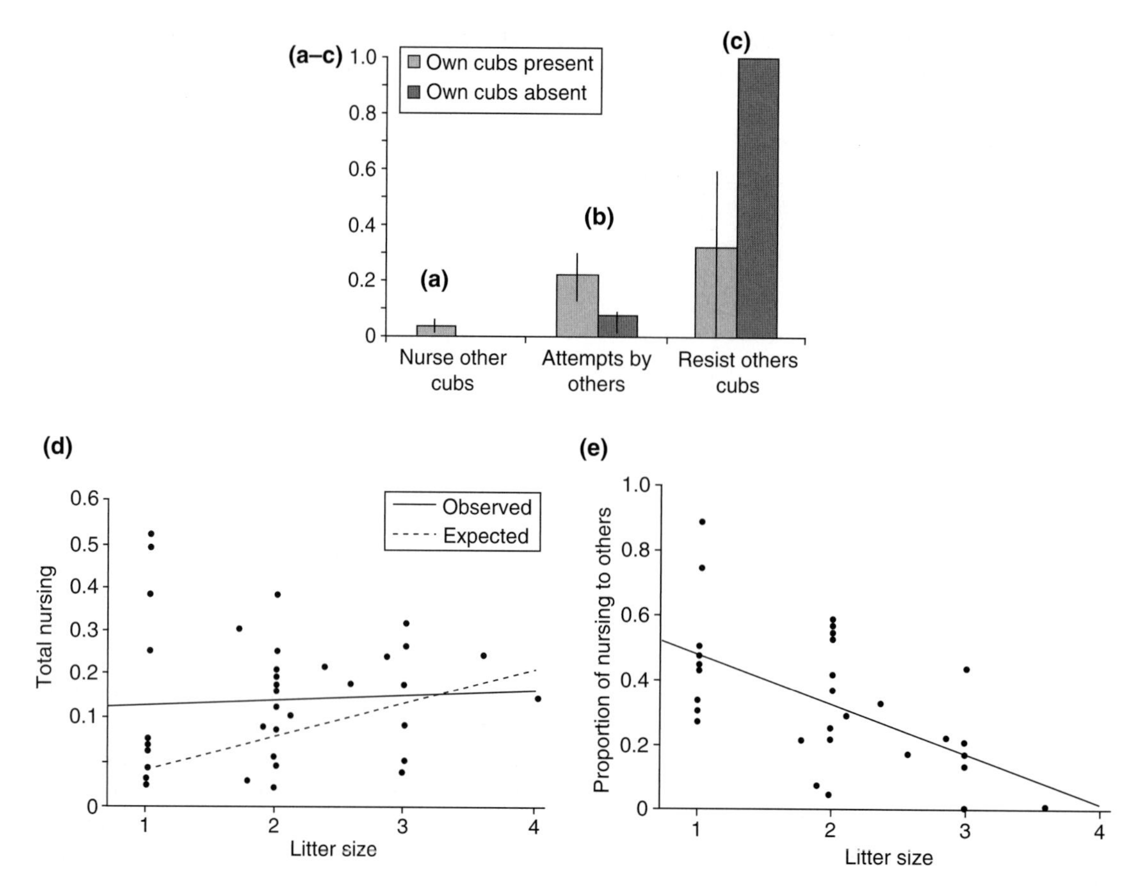

Figure 5.10 (a) Proportion of time female lions nursed each cub of another female when their offspring were present or absent ($T = 0$, $N = 6$ females, $P < 0.05$); (b) number of independent nursing attempts received per hour by female lions from each other cub when their offspring were present or absent ($T = 0$, $N = 6$ females, $P < 0.05$); and (c) proportion of nursing attempts from other cubs the females resisted when their offspring were present or absent ($T = 0$, $N = 2$ females, ns). (Adapted from Fig. 2 in Pusey and Packer 1994.) (d) Rate of total nursing (nurse all cubs) by females with different litter sizes. Narrow line represents regression line of total nursing against own litter size ($R^2 = 0.3$, $N = 35$ females, ns). Dotted line shows the expected rate of nursing if females only nursed their own cubs. (e) The proportion of a female's total nursing that goes to other cubs plotted against her own litter size ($R^2 = 0.34$, $N = 35$ females, $P = 0.002$). (Taken from Pusey and Packer 1994.)

The impact of neighbours on a pride of lionesses could be through any or all of the various behavioural selective pressures favouring group living—an abundance of neighbours might result in disrupted hunts, stolen kills, harassed or killed cubs, or territorial disputes. Indeed, Mosser (2009), found a strong negative relationship in the Serengeti between a pride's reproductive success and the numbers of neighbouring prides by which it was surrounded. The benefits of lion sociality appear to break down in prides of more than 10 adult females (Packer *et al.* 1988). The reason for this decrease in reproductive success is not known, although smaller belly sizes in larger prides suggests that it is due to intra-group food competition (Mosser 2008), but the benefits of cooperative defence of offspring and territory may not outweigh the costs of reduced food intake in very large prides.

Similarly, dominant, larger, central lineages of female farm cats were more often present at locations that were central with respect to resources and had higher reproductive success than did peripheral lineages (Kerby and Macdonald 1988; Fig. 5.12).

Table 5.7 Change in territory quality, from one 2-year time-step to the next, versus pride variables, in multivariate mixed model (SAS 9.2) with repeated measures on the same pride accounted for within the model. Prides that recruited or lost an adult female improved or declined in territory quality, respectively, and prides with more males than their immediate neighbours improved in territory quality. (Data from A. Mosser 2009.)

Variable	Effect	SE	Z	*P*-value
Change in pride size: adult females	0.1765	0.086	2.06	0.039
Log relative pride size: males	2.7781	0.980	2.83	0.005

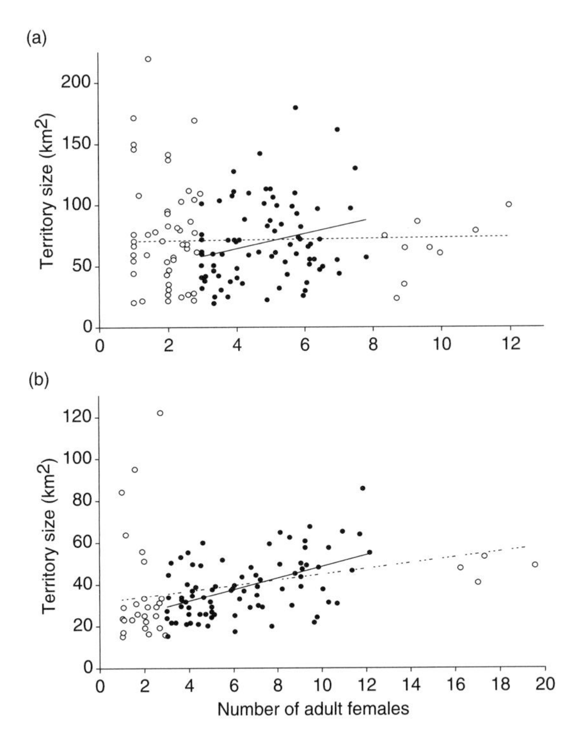

Figure 5.11 Territory size versus pride size for (a) plains and (b) woodlands habitat. Territory size was estimated using 75% kernel density contours. Territories are significantly larger in plains habitat (plains average = 70 km^2, woodlands average = 38 km^2, ANOVA: *F* 142, 117 (as subscript to F) = 76.97 $P < 0.0001$) and only vary with pride size in woodlands habitat when all pride sizes are included (dashed lines). Small prides are sometimes forced to abandon their territories and so occasionally have very large home ranges, while large prides typically gain access to the highest quality areas and need not defend very large territories. If small and large prides (open circles) are excluded from the regression model, territory size is significantly correlated with pride size in both habitats (solid lines). (Data from A. Mosser 2009.)

Ecological factors affecting felid behaviour and the power of intraspecific variation

The foregoing section makes clear that there can be advantages to cooperation among some felids, so why do not more of them opt for it. The answer may lie in their ecology. The burden of this essay is that felid societies are reflections of their ecology and, as for canids (Macdonald *et al.* 2004a) and indeed for other mammals (e.g. Crook 1970; Bradbury and Vehrencamp 1976; Kruuk 1976; Jarman and Jarman 1979; Macdonald 1983), this proposition is powerfully supported for felids, as in general, by the algebra of intraspecific comparisons (e.g. Lott 1991). Ideally, for at least several contrasting species, one would explore the extents of intraspecific variation in the light of all of a suite of relevant sociological and ecological parameters. However, the state of knowledge of felid behavioural ecology is inadequate to support such a comprehensive intraspecific comparison, even for one species. However, since the extent of variation with respect to some parameters is partly known for some species, and partly known for other parameters for other species, these fragments are reviewed by Macdonald *et al.* (Chapter 1, this volume) as part of an incomplete jigsaw puzzle. Details therein include diet, community structure, and intra-guild hostility, and intra- and interspecific mortality. Mortality patterns, also a selective force affecting sociality, are reviewed with respect to disease and conflict in Munsen *et al.* (Chapter 9, this volume) and Loveridge *et al.* (Chapter 6, this volume).

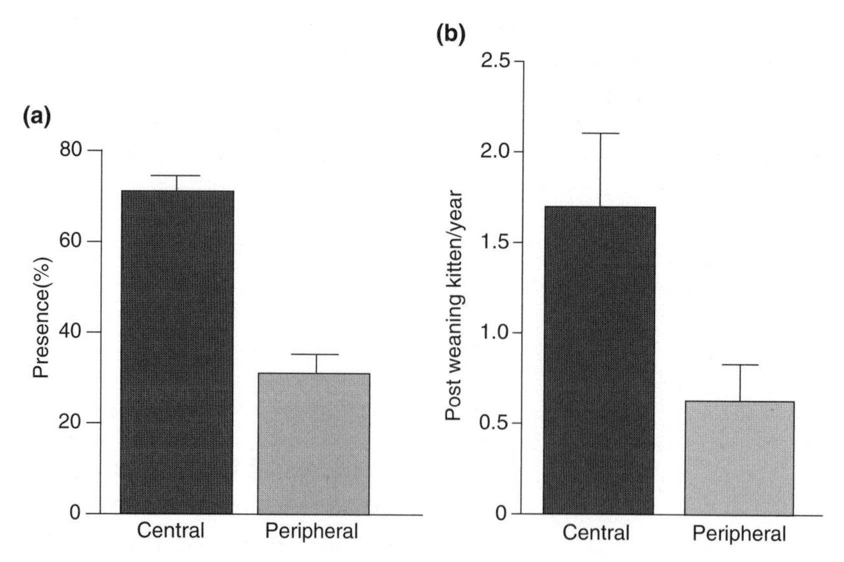

Figure 5.12 Dominant, larger, central lineages of female farm cats were more often present at locations central with respect to resources, rather than towards the periphery of a farmyard. Female cats with nest sites in the central area had a higher yearly reproductive success with more kittens surviving than peripheral lineages (central: $n = 11$, peripheral: $n = 10$, Mann–Whitney U-test $P < 0.05$). (Adapted from Kerby and Macdonald 1988.)

Individual behaviour as the engine of intraspecific variation

Individuals adapt to changing circumstances. Among bobcats, Lovallo and Anderson (1996) observed seasonal shifts in home range size and use, particularly among reproductive females (whose ranges nonetheless always remained exclusive). We might expect seasonal shifts in food dispersion to affect societies, and Corbett (1979) described a case of crofter's cats, provisioned only in winter. In summer two females lived solitarily, the dominant forcing the subordinate to hunt in suboptimal habitats, but in winter, when food was clumped, they fed together. Such seasonal changes are altitudinally dramatic for snow leopards (*P. uncia*) as they follow the altitudinal shifts of their natural prey, blue sheep (*Pseudois nayaur*), between steep (>20°), rugged terrain at lower elevations in winter and higher ones in summer (Oli 1997; Mishra *et al.* 2003a; McCarthy *et al.* 2005b). Although pumas preying on non-migratory animals have generally stable home ranges (Hopkins 1989; Sweanor 1990), those in populations feeding on migratory species, for example, mule deer (*Odocoileus hemionus*) and elk (*Cervus elaphus*), move seasonally in the wake of these prey (Rasmussen 1941; Seidensticker *et al.* 1973; Anderson *et al.* 1992a). This was the case in California, where Pierce *et al.* (1999) distinguished two strategies. Some pumas migrate together, slowly following mule deer between their winter and summer ranges on the eastern scarp of the Sierra Nevada, and at both ends of the journey the same community of individuals occupy home ranges that overlap widely. Other pumas from the eastern scarp decamped rapidly to the western side of the Sierra Nevada, where they shared distinct summer ranges with a different community of resident pumas. The herd-followers occupied large, overlapping, annual home ranges (95% adaptive kernel; mean = 817 km^2), whereas the decampers occupied smaller distinct summer (mean = 425 km^2) and winter (mean = 476 km^2) ranges (some showing strong site fidelity in successive years).

Comparable decisions faced the lions studied by Hemson (2003) in Makgadikgadi National Park, Botswana. There, as the dry season advanced, herds of zebra and wildebeest from the east migrated to the Boteti River in the west. Most lions from the east followed them, insinuating themselves among the lions resident in the west, but a minority stuck it out in the east—preying on the seasonally constant availability of resident ungulates (gemsbok, *Oryx gazella*; ostrich, *Struthio camelus*; kudu, hartebeest, *Alcelaphus buselaphus*; springbok, *Antidorcas marsupialis*; and impala, *Aepyceros melampus*) and domestic livestock (Loveridge *et al.*, Chapter 11, this volume). The home ranges of commuting prides averaged 480.1 km^2 overall (536 km^2 in the lean season, and 206 km^2 in the good season), while those of non-migratory prides averaged 112.4 km^2. That the lifestyle of some lions (the seasonal commuters) was transformed by following the herds, whereas that of others (the residents) was not (although they did kill fewer livestock when migratory species were present) raises the question of what mechanisms determined the lifestyle category into which each individual fell. Crudely, the answer doubtless hinges on the profitability of hunting migratory versus resident prey (of which the mean body mass was approximately half that of migratory species), but this sort of tautologous platitude raises questions about the cost factors that explain why both the killing and subsequent consumption of domestic stock was disproportionately low (but nonetheless grievous to their owners) and, more interestingly, how these decisions are made, sociologically, at the level of individuals.

The case of lions

Hemson *et al.*'s commuting lions (in press) illustrate both the flexibility of lion behaviour, and Macdonald's suggestion (1983, 1992) that it might be interpreted through the Resource Dispersion Hypothesis (RDH) in terms of the impact of water resources on prey. This is developed in terms of inter- and intra-population comparisons, respectively by Loveridge *et al.* (2007a) and Loveridge *et al.* (in 2009b). Indeed, lions, having been studied in more locations than any other felid (see Funston *et al.* 1998), provide the greatest opportunity to solve the ecological algebra of intraspecific variation and thus, perhaps, to shed light on the ecological factors shaping felid societies in general.

In Hwange National Park, Zimbabwe, lion pride home range size (90% kernal) is 388 km^2 $\pm$ SE 35 km^2 (Loveridge *et al.*, in 2009b) which is approximately one order of magnitude larger than comparable estimates from Ngorongoro Crater, Tanzania (45 km^2; Hanby *et al.* 1995), Selous Game Reserve, Tanzania (52.4 $\pm$ 26.3 km^2; Spong 2002) and Maasai Mara National Reserve, Kenya (71 $\pm$ 20 km^2; Ogutu and Dublin 2002), and also larger than estimates from Serengeti Plains, Tanzania (200 km^2; Hanby *et al.* 1995), or from Kruger National Park, South Africa (107 km^2; Funston and Mills 2006), but is similar to estimates from Etosha National Park, Namibia (404 $\pm$ 183 km^2; Stander 1991). Van Orsdol *et al.* (1985) showed how, between populations, lion home range size was generally inversely related to lean season prey abundance, and the importance of access to lean season water was emphasized by Orford *et al.*'s conclusion (1988) that the location of pride territories in Etosha is determined by the availability of permanent water. To explore this inter-population variation, Loveridge *et al.* (2007a) reviewed studies of lions in the context of broadly concurrent estimates of prey abundance to reveal a positive exponential relationship with slope 0.89 across three orders of magnitude in both prey biomass and lion population density (Fig. 5.13a). This relationship, which concerned only sites not perturbed by anthropogenic mortality, was very much tighter than a similar relationship including perturbed sites. Lions do not exist at prey biomasses much less that 100 kg/km^2 at which less than 1 lion/161.3 km^2 exists. Although points with very similar prey biomass have several-fold differences in lion density, this is a remarkably tight plot considering the colossal variety of habitats encompassed.

If all else were equal across the spectrum of environments distinguished by this vast range of prey biomasses, then lion home ranges would also simply follow inversely the same relationship, and for inter-population averages this is the case (Fig. 5.13b).

The slope of the relationship between lion density and herbivore biomass is significantly lower than 1, indicating that at prey-rich locations there are slightly more resources per individual. However, both relationships are close to what would be expected in both lion range and density scale to achieve the same resource availability per animal. The scatter about the lines in Fig. 5.13a and b suggests that along the continuum of prey biomasses all is not equal in terms of lion socioecology. In terms of prey dispersion, the obvious candidates to explain this are variation in either or both of the dispersion of prey or their community structure (i.e. distribution of species compositions and thus their social systems, for example, herd sizes, escape behaviour, habitat use, and body sizes).

Ultimately, the quantity and quality, and potentially dispersion, of resources in an African savannah ecosystem, are influenced by the interactions of soil characteristics and rainfall on primary production (Bell 1982), which are known to influence herbivore biomass (Coe *et al.* 1976; East 1984; Fritz and Duncan 1994) and herbivore community structure (Fritz *et al.* 2002), and therefore indirectly carnivore numbers (East 1984; Grange and Duncan 2006). Savannahs are classified according to the richness of their nutrients, from eutrophic to dystrophic (illustrated, in terms of lion studies, respectively by the Serengeti-Mara and Hwange ecosystems).

Logically, a change in prey biomass per unit area can be brought about only by either changing their body sizes or their packing, or both. For many of the lion's prey species, packing can mean either the number and/or size of their herds. Indeed, the body sizes of ungulates characteristic of different biomes vary (Fritz and Loison 2006), and in dystrophic savannahs the poor nutrient profile leads to low-density communities dominated by large prey such as African buffalo, giraffe, and African elephant, *Loxodonta africana* (Fritz *et al.* 2002). Developing a line of thought initiated by Hemson (2003), Loveridge *et al.* (2007a) summarize evidence that both the number of herds per unit area and their mean size vary exponentially with prey biomass density, respectively with exponents of 0.64 and 0.49 (Fig. 5.13c and d).

The dispersion of herds is not the same thing as the dispersion of individuals (e.g. if, as prey density rose, only herd size increased, their dispersion would be unchanged, whereas that of individuals would increase). Perhaps the dispersion characteristics of prey hold the explanation for the slight tendency for lion density to be lower than null expectation at high prey density. All three dimensions of prey (body size, herd size, and number) clearly have major implications for lion foraging, and thus ranging behaviour, and all are likely to be affected by a fourth covariate—namely the dispersion of surface water. In arid and semi-arid ecosystems, such as Hwange, herbivore habitat selection is largely influenced by the distance to water (Valeix *et al.*, 2009a). For example, in the dry season, surface water may occur in only a few waterholes, and these may be widely spaced, having a great impact on the dispersion of herds of prey (e.g. Bergström and Skarpe 1999; Redfern *et al.* 2003). Furthermore, some large species, such as buffalo, are more water-dependent than smaller ones, such as kudu, and consequently may tend to congregate more around waterholes (Redfern *et al.* 2003) and hence be more predictably dispersed under dry conditions.

The packet size of individual prey typical of dystrophic savannahs is large, each buffalo, giraffe, or young elephant feeding several lions (a version of the 'rich patch case' of RDH; Carr and Macdonald 1986). Furthermore, dystrophic areas are characterized by fewer and smaller herds, and hence the dispersion of prey herds is likely to be widespread. A simple prediction is thus that in dystrophic savannahs, lions will live at low population densities with large home ranges in large prides. In contrast, as prey biomass density increases, lions can occupy relatively smaller home ranges, hunting prey characterized by a greater preponderance of medium-sized ungulates packed into the environment in larger, more numerous herds. There comes a point for the lions at which further range contraction is unfeasible, whatever the biomass density of prey. This point may be reached when, for example, the smallest number of hunting sites needed by the minimum social unit cannot be shrunk further. This is illustrated by the dispersion of waterholes (and equally applicable in either dystrophic or eutrophic environments). For example, it might be necessary, in an environment with many waterholes, for lions to monopolize a sufficient number of them so that wind direction for hunting at one was always favourable, or to rotate their attentions between several to avoid prey predicting their whereabouts (this being analogous to Kruuk and Macdonald's suggestion [1985] that even a single spotted hyena needs a home range whose radius exceeded the average bee-line length of a successful chase). The minimum social unit of lions, in this context, may be more than one; due to the requirements of defending

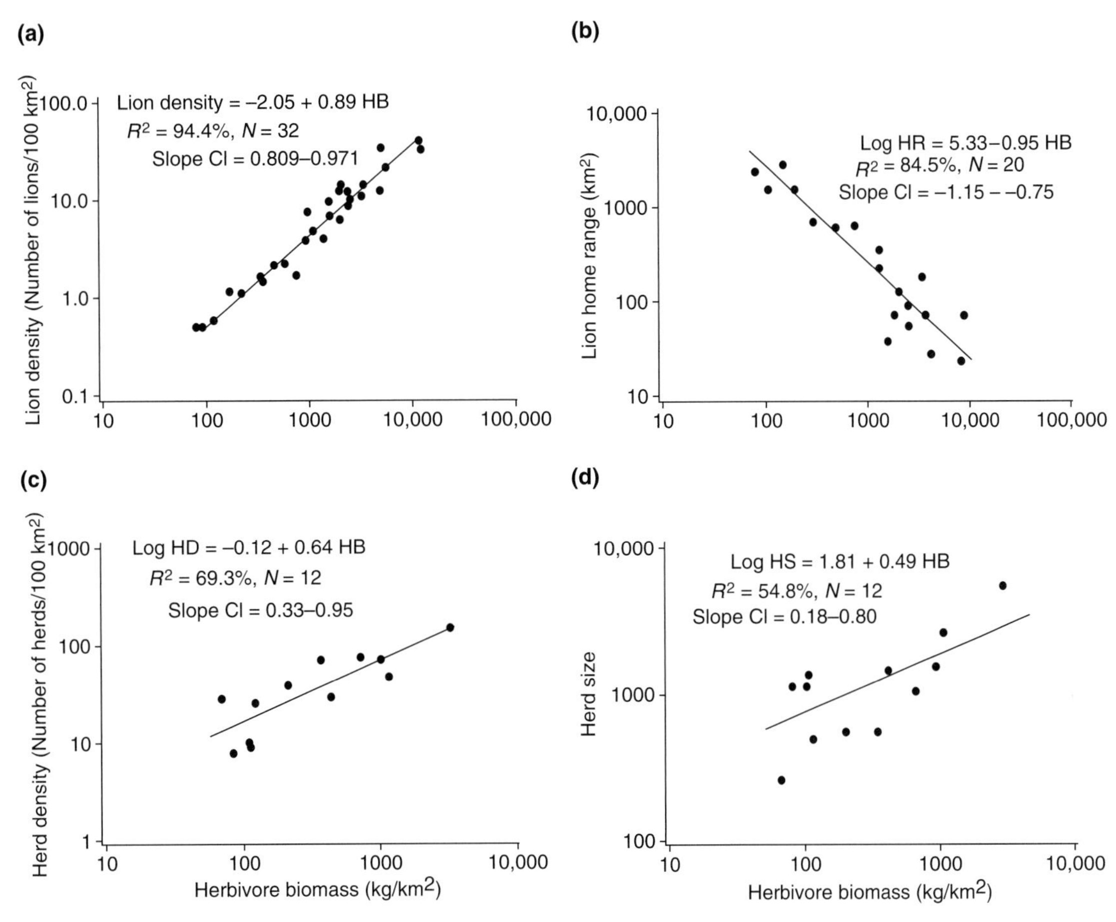

Figure 5.13 (a) Prey biomass and lion density—data from sites not perturbed by anthropogenic activities. Both axes are on a log scale. (b) Prey biomass and lion home range size. Both axes are on a log scale. (c) Herd density versus lean season prey biomass. Both axes on a log scale. (d) Mean herd size versus lean season prey biomass. Both axes are on a log scale. (Taken from Loveridge *et al.* 2007a.)

kills and cubs, or of hunting efficiently, it might be four or five adult lionesses. In that case, we would expect home range size to diminish with increasing prey biomass density until that point at which four or five adult lionesses had satisfactory food security. At some point, it is relevant to consider the notion of expansionism (*sensu*; Kruuk and Macdonald 1985) in so far as lions might face a choice of whether or not to expand their home ranges to accommodate larger groups (because the sociological rewards are overwhelming). Among species whose societies are based on spatial groups, there are different relationships between group size and home range size (Macdonald 1983), including cases where larger social groups occupy larger home ranges—this is expansionism, which contrasts with a contractionist strategy in which animals maintain the smallest economically defensible home range, living in larger groups only if resources therein permit (Kruuk and Macdonald 1985). Macdonald (1983) drew attention to early evidence of expansionism in the relationship between the sizes of Serengeti lion groups and the home ranges they occupied. In a similar vein, at increasing prey biomass densities, the minimum defendable home range needed by the minimum social unit may be so well supplied with prey as to

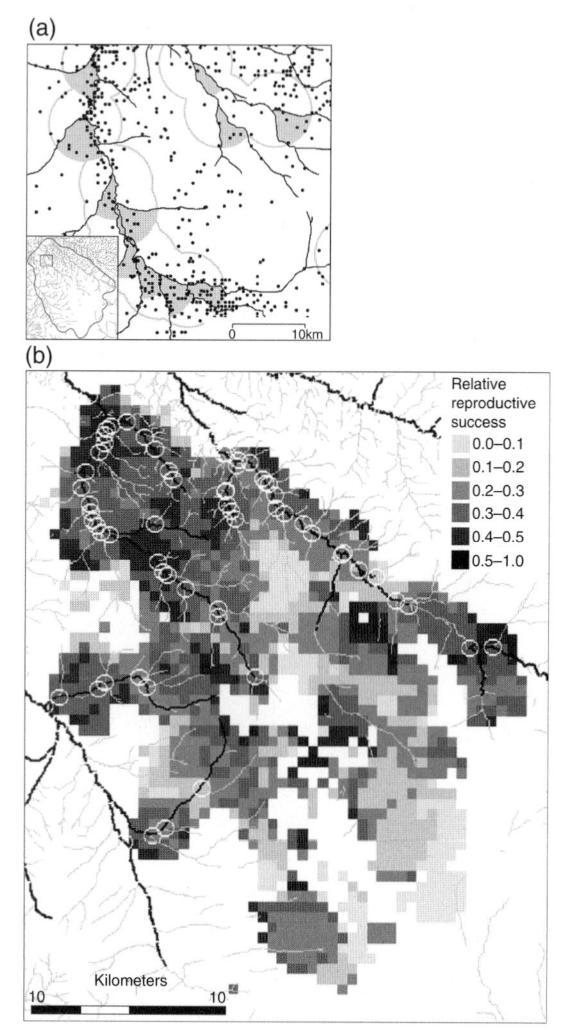

Figure 5.14 (a) Position of lion kills (black circles) with respect to river confluences. Of kills recorded <1 km of a confluence (grey circles), more prey were captured between the converging tributaries (grey shaded areas) than expected by chance ($\chi^2 = 96.8$, $N = 1141$, $P < 0.001$). The map shown here is just a portion of the full study area (inset). (b) Long-term (38-year) average relative reproductive success. Major rivers are shown in black, tributaries are light grey, and confluences are white open circles. White areas indicate grid squares with insufficient ranging data. Note that the legend scale is not evenly distributed and is extended for the lower values to better illustrate the observed patterns. Reproductive success was significantly correlated with proximity to river confluences. (Taken from Mosser 2009.)

accommodate additional lions at little or no further cost. To explore these issues, the interaction between lion hunting behaviour and patch richness (herd size and composition), numbers and aggregation/predictability (e.g. around waterholes) will have to be untangled, as will the factors causing males to overlap more than one pride (Loveridge *et al.* 2007c).

Sociological generalizations are the emergent properties of the responses of individual lions to the resource landscape, and two studies provide insight from different ends of the eutrophic–dystrophic continuum (Mosser 2008, Loveridge *et al.*, 2009b).

The Serengeti is at the eutrophic end of the spectrum, and in the dry season, surface water is scarce there, largely confined to river confluences (Knighton and Nanson 2000; Wolanski and Gereta 2001; Hopcraft 2002), and these sites not only attract prey and provide the cover from which to hunt them, but also provide shelter and den sites for new cubs. Confluences are valuable hunting locations (Fig. 5.14a; Hopcraft *et al.* 2005; see also Spong 2002), and Mosser (2009) demonstrated that of kills made within 1 km of a confluence, significantly more were made between the arms of two tributaries (Fig. 5.14a). Mosser also found that reproductive success was significantly correlated with proximity to river confluences (Fig. 5.14b).

Confluences therefore constitute rich patches which have the attributes, according to the RDH, to create conditions under which groups may develop where resources are dispersed such that the smallest economically defensible territory for a minimum social unit can also sustain additional animals (Macdonald 1983). As Mosser (2008) concludes, RDH predicts that such rich patches will lay the foundation for the formation of spatial groups, and the benefits of collective defence of these areas would have provided strong selection for a social felid in the heterogeneous savannah landscape. Indeed, in the Serengeti, territory quality was determined by landscape characteristics (distance from confluences territory quality is the average distance to confluence for grid squares within a territory), and larger prides (and those with fewer neighbours) were also more likely to gain territory quality, providing an incentive for expansionism (see above).

Loveridge *et al.* (2009b) studied the impact of variation in prey biomass and pride size at the *within-population* scale on lion home range size. They monitored 12 female and 8 male lions between 2002 to 2005 in the northern half of the *c.* 15,000 km^2 of semi-arid dystrophic savannah comprising Hwange National Park, where the annual rainfall average of 606 mm varies widely (CV $\approx$ 30%). Drought was worst in 2003 (the second of 2 drought years), bad in 2005 and mild in 2004. They asked how the biomasses of prides (females and young) and male coalitions related to seasonal core (50% kernel density estimator) and total (90%) home range sizes, in the contexts of the encounter rate per kilometre (in terms of biomass) for each prey species (monitored with road counts in December [wet season], May–June [early dry season] and September–October [late dry season]).

Prey abundance in the northern part of the Park was constant in the wet season. Counter-intuitively, prey abundance in the study area increased in the late dry season as annual rainfall decreased (Fig. 5.15). Prey abundance was consistently higher in the late dry season for a given year, particularly for a dry year (Fig. 5.15). The counter-intuitive increase in herbivores in drier weather arises because under these conditions ungulates migrate from the south of the Park, where no surface water persists, to the north, where there are pumped waterholes.

Mean seasonal home range size was 388 km^2 for females (SE = 35 km^2; min = 35 km^2; max = 981 km^2) and 478 km^2 for males (SE = 50 km^2; min = 71 km^2; max = 1002 km^2). Mean seasonal core size was 153 km^2 for females (SE = 15 km^2; min = 13 km^2; max = 384 km^2) and 182 km^2 for males (SE = 24 km^2; min = 20 km^2; max = 426 km^2). Annual rainfall significantly affected the home range sizes of females, with female home range size significantly larger in 2004 than in 2003 (Fig. 5.16a), but not males' (Fig. 5.16b). Home range size and home range core size did not change significantly between seasons of the same year for either sex (Fig. 5.16).

Loveridge *et al.*'s results (2009b) strongly suggest that lionesses in Hwange National Park may be expansionists (Fig. 5.17).

Once controlled for pride biomass, lioness home range sizes were inversely related to the prey kilometric biomass they contained. The significant increase in home range size with decreasing prey kilometric biomass was more pronounced in years of good rainfall. These effects on female home range sizes were driven by the kilometric biomass of buffaloes (R^2 = 0.47, P = 0.020) and elephants of less than 4 years old (P = 0.041, R^2 = 0.16) in the late dry season, and by kudus (R^2 = 0.53, P = 0.012) in the early dry season.

Additionally, female seasonal home range size was negatively related to the density of waterholes in the home range (R^2 = 0.67, P = 0.0007), with the result that lionesses' ranges were larger in the centre of the park (where the pumps were least numerous) than in the north. The prides that were dependent on the lowest density of waterholes had the largest ranges.

In terms of the RDH framework set out above, the interpretation is that the wetter the year the lower the prey abundance and, disproportionately, the more homogeneously they are dispersed; it being wetter, the less they are compelled to congregate at a small number of waterholes. The drier the year, the higher the abundance of prey and, disproportionately, the more clumped they are because, it being drier, they have to crowd around the very small number of waterholes (ultimately 1–2/lion range, around which the lions have themselves, for this very reason, configured their home ranges). It remains to be seen whether more widely spread waterholes would lead to larger home ranges that would either be more overlapping and/or support larger prides.

Turning to males, average home range size was not influenced by the number of males in the coalition but the relationship approached statistical significance ($F_{1,6}$ = 4.05, P = 0.09). There was a negative relationship between male average home range size and the average prey kilometric biomass (Fig. 5.18a). There was no significant relationship between average male home range size and either the average kilometric biomass of any particular prey species (all $P \geq 0.05$), or the density of waterholes in the average home range. However, the average male home range size decreased as the density of prides encompassed within the male home range increased (Fig. 5.18b).

In summary, the results showed that males differed from females in their ranging behaviour. Male home range size was larger than that of females and appeared affected by both prey abundance and pride density. It is likely that home range sizes of males

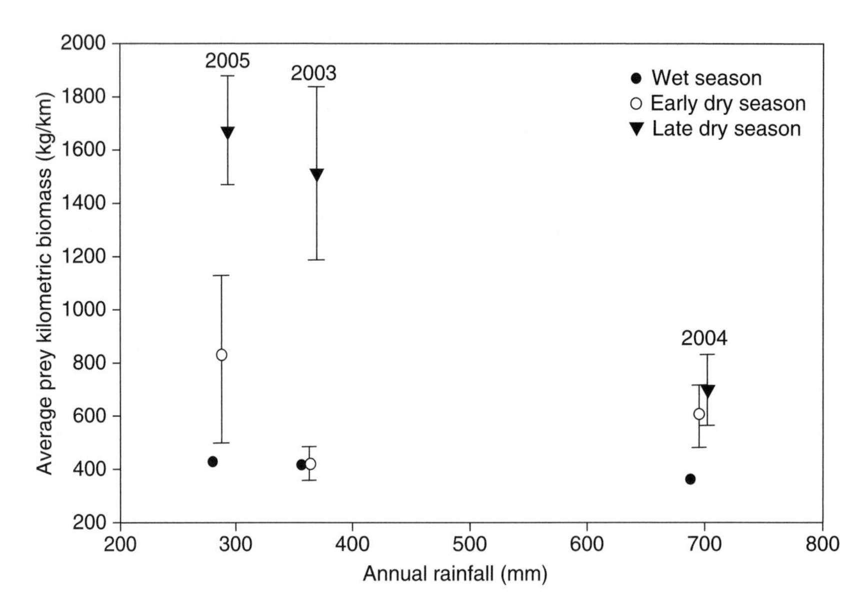

Figure 5.15 Relationship between annual rainfall and average prey kilometric biomass in the northern sector of Hwange National Park. (Taken from Loveridge *et al*., 2009b.)

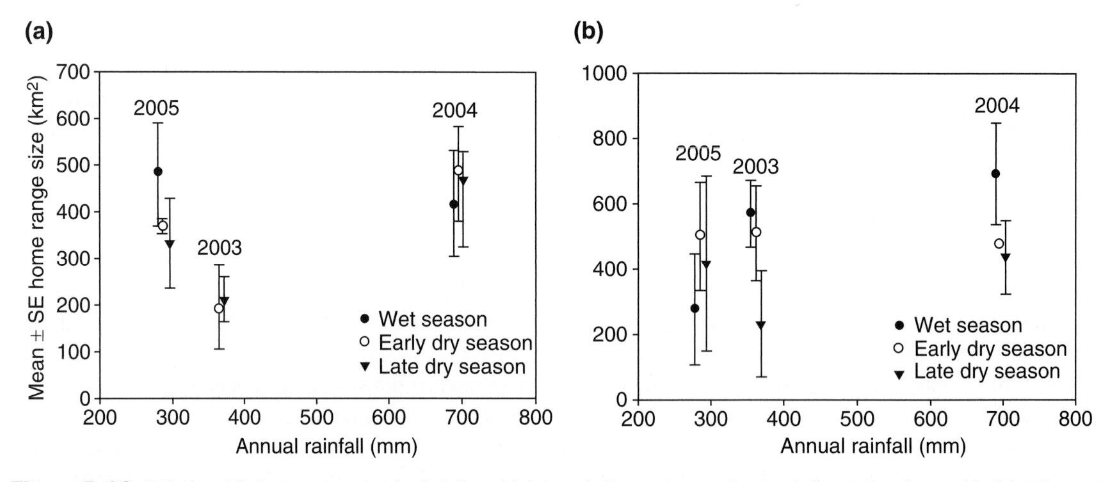

Figure 5.16 Relationship between annual rainfall and (a) female home range size (no information is provided for the wet season 2003 because positional data from the GPS Simplex radio-collars for female began in February 2003); (b) male home range size in the northern sector of Hwange National Park. (Taken from Loveridge *et al*. 2009b.)

were determined indirectly by prey abundance, through association with female prides, and larger ranges are a reproductive strategy for males because they configure their home ranges not only to secure food but also to acquire and defend access to females (Schaller 1972). This is corroborated by the fact that lions have larger home ranges when the density of female prides is lower, suggesting that lions extend their home range in order to cover a minimum number of prides.

Synthesis

At the outset, we asked whether location on the predator-to-prey size graph predicted social

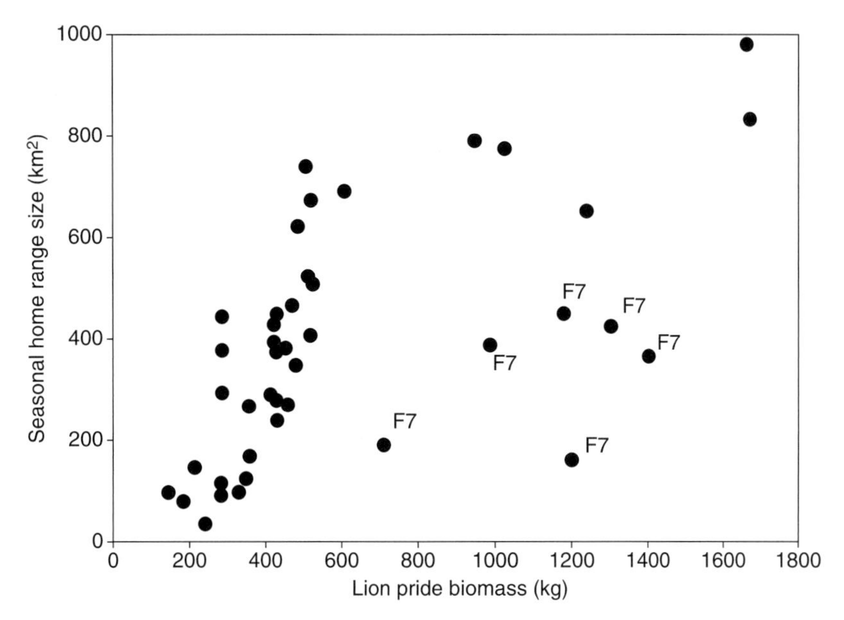

Figure 5.17 Relationship between lion pride biomass and female home range size. F7 is indicated as there were a priori grounds for considering her as an outlier. (Taken from Loveridge *et al.* 2009b.)

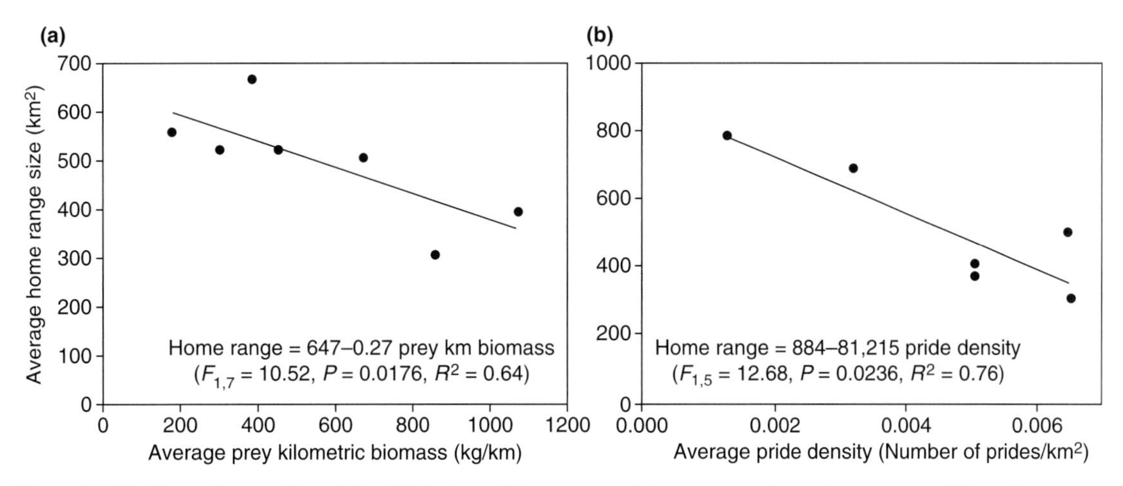

Figure 5.18 Shows the negative relationship between (a) average male home range size and average prey kilometric biomass (kg/km); (b) average male home range size and density of prides within his home range. (Taken from Loveridge *et al.* 2009b.)

organization for felids as it generally does for canids. The answer is not really: the two most gregarious are, respectively, huge lions and small farm cats, and there is rather little evidence of the social intricacies that lie between. It may be relevant that Macdonald *et al.* (2000c), taking account of several lifestyle variables, calculated that among felids only for plains-dwelling lions are the predominant individual prey generally sufficiently big that an adult daughter could routinely survive solely by parasitizing her mother's food—a condition also true, *mutatis mutandis*, for domestic cats.

For felids as a whole, we speculate that the scatter in Fig. 5.2 around the relationship between log prey size and the residuals of the prey size to home range size relationship will turn out to be at least partly explicable in terms of prey dispersion. Such graphs are bedevilled with intraspecific variation, but we suspect that understanding the social systems of felids will be in terms of the patchiness of prey, or of some other valuable resource. Hitherto, rather little effort has been made to devise biologically meaningful indices of prey dispersion, or to incorporate the spatial dispersion of prey into taxonomic comparisons. Hence, we do not know the relative importance of the spatial clumping of resources as a shaper of ecological patterns in cats. By comparison with prey density, biomass, and productivity (all of which can be relatively easily inferred from prey body mass and density), resource dispersion is both highly variable and difficult to quantify.

The foregoing discussion of lions suggests a framework for thinking about the societies of other felids in terms of the cascading consequences of resource dispersion. For example, Marker and Dickman's plots of leopard density (2005) and home range sizes suggest that at low prey densities leopard home ranges are both larger and overlap more, and Seidensticker *et al.* (1973) showed that female pumas in Idaho had almost entirely overlapping ranges when prey were concentrated in a single area but redistributed themselves into more exclusive home ranges when prey were more homogenously dispersed. The starting point is the extent to which prey are dispersed in a way that is economically defendable or in some sense sharable. For the biggest cats, very large prey may be shared at no cost (especially in the presence of kleptoparasites). This clearly applies to lions and their biggest prey, and may sometimes apply to tigers. However, the next size-category down (jaguars, leopards, and pumas) kill prey that can sustain them for several days. Macdonald *et al.* (2000c) argue that there is no advantage in these species sharing prey, but the fact that they could share prey prompts the question of why they do not. For example, why do not leopards form prides for just the same reasons lions do in the Serengeti? An obvious answer is that they could not defend a kill from lions. The same might be said for pumas living under the competitive thumb of jaguars, but why do not jaguars form groups? Perhaps because they live in jungle where water is not limiting and prey are thus more homogeneously distributed, prey size more variable, and conditions favouring spatial groups less frequent. Thus, a blend of resource dispersion and intra-guild hostility dictates the opportunity for group living. Why, then, do not tigers behave like lions? Because they live in less arid, more homogeneous landscapes, where prey are less aggregated (at lean season waterholes) and hence competitors more spread out, and cover more dense (so that protecting the kill for long enough to eat it is less problematic). In short, it was a combination of (1) aridity (hence the importance of waterholes); and (2) grasslands (hence the lack of cover), that started the ancestral lion from a tiger-like lifestyle on a trajectory to modern lion society (Yamaguchi *et al.* 2004a). All smaller cats are prevented by larger competitors from evolving towards a diet of larger, shareable prey, and thus constrained to more solitary, territorial existence.

Despite this variation, Macdonald *et al.* (2000c) conclude that the social matriline is the ubiquitous common denominator of all domestic cat societies, and Macdonald (1992) generalizes this when arguing that each Carnivore family carries its own sociological phylogenetic inertia into whatever ecological arena it must adapt. In the case of felids, that phylogenetic predisposition is for societies based on the female matriline. For almost all 'solitary' species of felid, the influence of genetic ties on spatial relationships remains unknown.

Acknowledgements

We thank Susan Cheyne and Kerry Kilshaw for great assistance in synthesizing this material, Paul Johnson for significant input throughout, Chris Carbone for helpful comments, and Andy Loveridge, Marion Valeix, and Paul Johnson for endless discussions of lion socioecology.

CHAPTER 6

People and wild felids: conservation of cats and management of conflicts

Andrew J. Loveridge, Sonam W. Wang, Laurence G. Frank, and John Seidensticker

A lioness killed in an illegal wire snare by poachers. ©. A.J. Loveridge.

Introduction

Wild felids and people have a complex and often paradoxical relationship. On the one hand, humankind admires and reveres felids; cats appear as cultural icons and symbols across the ages. In addition, there is a growing awareness of the value of wild felids as key components of ecosystems, tourist attractions generating income, umbrella species for conserving ecosystems, and flagships for engendering public support for conservation. These positive values sometimes contrast strongly with the relationship between wild felids and people in areas where they coexist. Human conflicts with wild cats, overexploitation of felid and prey populations, and habitat loss and fragmentation have extirpated felid populations and still threaten many more. The ways in which people value and interact with organisms and their habitats is at the heart of conservation. This chapter explores some of the interrelationships between people and wild felids, where human actions threaten felid populations.

Why conserve wild felids?

We preserve carnivores for aesthetic, symbolic, spiritual, ethical, utilitarian, and ecological reasons. Felids are culturally valued and are important as cultural icons and symbols. Felids are widely depicted in art, from Stone Age petroglyphs and cave paintings to more modern depictions of cats as art, reminding us that humans have interacted with felids for as long as we have been humans. Human culture is enriched by the symbolic use of felid images. Felids adorn many currencies, are heraldic symbols on coats of arms and appear widely on the badges of national sports teams. They were revered by many cultures, for instance, the lion-headed goddess Sekhmet represented war and disease in ancient Egypt. The jaguar (*Panthera onca*) features significantly in Central American cultures. Similarly, the tiger (*P. tigris*) is culturally important in South and East Asian cultures (Weber and Rabinowitz 1996).

Perhaps because of their place in our consciousness and the ease with which they are recognized, felids are used as umbrella and flagship species to conserve habitat, benefiting both felids and biodiversity generally. Large carnivores are often highly mobile and viable populations require large tracts of suitable habitat and adequate prey populations. Thus, strategies for the protection of large felids also offer protection for large 'functioning ecosystems' (Soulé and Simberloff 1986; Noss *et al.* 1996; Seidensticker *et al.* 1999). It is much easier to motivate the public and governments to protect a charismatic carnivore than a species that is seldom seen and outwardly unremarkable. Examples of this include the protection, through an initiative called *Paseo Panthera* (the Path of the Panther), of biological corridors on the Panamanian Isthmus, used by jaguars, pumas (*Puma concolor*), and other wildlife. 'Tiger Reserves' protect significant areas of biodiversity in India, as well as halting the decline of tigers, and tigers are used as a flagship to promote conservation efforts along border regions in the Indian subcontinent and Indochina (Weber and Rabinowitz 1996; Rabinowitz 1999; Seidensticker *et al.*, Chapter 12, this volume; Sunquist and Sunquist 2002).

Moral philosophers and animal welfarists are united in their arguments that every species has an intrinsic conservation value that implies it has a right to survive. Intrinsic conservation values have served as an impetus for conservation. However, intrinsic value in itself has been insufficient to secure successful conservation because humans are motivated more by economic self-interest than by ideas (Kellert *et al.* 1996). Economic benefits can provide powerful incentives to local communities to protect biodiversity. Felids are economic assets and, when used sustainably through tourism, trophy hunting, or commercial exploitation can contribute substantially to both their own conservation and that of their habitats. For example, African lions (*Panthera leo*) earned Amboseli National Park, Kenya, US $27,000 per lion per year in tourism revenues (Western and Henry 1979). Trapping of furbearers or unregulated trophy hunting can lead to controversial declines in populations. However, where it is carried out on a well-managed and sustainable basis, it can provide incentives for long-term protection, a good example being sustainable harvests of Canada lynx (*Lynx canadensis*) furs. Finally, felids play an important and often regulatory role within the ecosystems they inhabit (Mills and Biggs 1993; Karanth and Stith 1999; Grange and Duncan 2006). This ecological role needs to be factored into any valuation we make of felids in the wild. It must also be included in any assessment we make in valuing our relationship with nature and the pristine habitats that should form the baseline for all conservation efforts.

How people impact felids: anthropogenic threats to felid populations

Habitat loss is a global phenomenon, affecting all species. The earth's human population has increased from 3 billion to 6 billion since 1960. The global economy has increased sixfold and food production increased by 2.5 times. Nearly 25% of the earth's terrestrial surface is now under cultivation. The past 40–50 years have seen a sharp increase in the amount of land converted to agriculture, and projections suggest that further conversion is to be expected in the future (Millennium Ecosystem Assessment 2005). Conversion of natural habitat to agricultural land, urban development, and destruction or fragmentation of habitats through logging, building infrastructure (e.g. dams, roads, power lines, oil, and mineral extraction), and other human activity has serious impacts on wild felid populations. Equally important is loss of prey populations through over-hunting, retaliatory persecution, habitat loss, or fragmentation. Predator population densities are closely correlated with prey-population biomass, thus loss of prey species (either through overexploitation, such as the bushmeat trade, or habitat destruction) causes linked declines in felid populations (Karanth *et al.* 2004c; Henschel 2007).

Against a backdrop of global habitat loss, felids face a number of proximate anthropogenic threats. One of the most important ways in which people impact felids is through increasing rates of non-natural mortality (Fig. 6.1). For felid populations below their carrying capacity, anthropogenic mortality is thought to be largely additive rather than compensatory to natural levels of mortality (Lindzey *et al.* 1992). By contrast, in highly protected populations anthropogenic mortality is rare, for instance only one (7%) of 14 leopard (*Panthera pardus*) deaths recorded in Kruger National Park, South Africa, by Bailey (1993) was due to poaching; the rest were due to intraspecific fights, predation, and starvation. In general, anthropogenic mortality in many studied felid populations is high. While this may in part be due to the fact that conservation biologists choose to study populations that are under threat, it is also an indicator of the impact that humans have on felid populations. Common sources of anthropogenic mortality are legal hunting and trapping, poaching, problem animal control (both legal and illegal), and vehicle accidents. Furthermore, felid populations can also be exposed to diseases carried by domestic carnivores, often as a result of human encroachment into wild habitats. In some populations, anthropogenic mortality can be extremely high and lead to population declines. For instance in Laikipia, Kenya, 17 of 18 tagged lions which died were killed in retribution for livestock raiding, with the population declining by about 4% per annum (Woodroffe and Frank 2005). Similarly, lion density was significantly reduced by conflict with local people on group ranches surrounding Masai Mara Reserve, Kenya (Ogutu *et al.* 2005). In a study of Amur Tigers, all seven recorded deaths of individually recognized adult females were due to poaching and 57% of cub mortality in this population was anthropogenic (Kerley *et al.* 2002, 2003).

Utilization of felid populations varies substantially in intensity, with mortalities from this source ranging from around 7% in Eurasian lynx (*Lynx lynx*) in Switzerland (Schmidt-Posthaus *et al.* 2002a) to 27 (62.7%) of 43 deaths of radio-collared lions around Hwange National Park, Zimbabwe (Loveridge *et al.*, Chapter 11, this volume). Trophy hunting made up between 11.7% and 50% of recorded mortality in pumas (Logan *et al.* 1984; Lindzey *et al.* 1988; Cunningham *et al.* 2001), 37.5% in radio-tagged leopards in Natal, South Africa (Balme and Hunter 2004), and trapping

Figure 6.1 Lioness killed by a steel wire snare. Wire snares set for both predators and prey species may increase levels of mortality within predator populations. (Photograph courtesy of P. Lindsey and Sango Ranch, Save Valley Conservancy.)

made up around 45–66% of mortality of marked bobcats (*L. rufus*) in Mississippi and Maine, United States (Litvaitis *et al.* 1987; Chamberlain *et al.* 1999). Trapping mortalities made up 19 (70%) of 27 recorded deaths of Canada lynx (*L. canadensis*) in the Northwest Territories (Poole 1994). However, for many exploited populations few demographic data are available and this is particularly true of some populations hunted and trapped for the fur trade (e.g. spotted cats in South America in the 1960s and 1970s).

Conflict with people can lead to high levels of mortality, particularly where predator eradication occurs. For instance, an average of 29 (range 10–42) lions are destroyed each year in farmland around Etosha National Park, Nambia, where they pose a threat to livestock, with 563 lions killed over a 20-year period (Stander 2005b). In puma populations monitored in Arizona and Utah, United States, 47% and 24%, respectively of marked animals that died were killed in defence of livestock (Lindzey *et al.* 1988; Cunningham *et al.* 2001). Around 3500 pumas per year are killed by people (through sport-hunting, protection of livestock, and vehicle accidents) in the western United States (Papouchis 2004).

Poaching can contribute substantially to mortality within a population. Ferreras *et al.* (1992) found that 42% of deaths of Iberian lynx (*Lynx pardinus*) in Donana National Park, Spain, were due to illegal poaching. Poaching accounts for a significant proportion of tiger mortality in India and Russia (Kumar and Wright 1999; Miquelle *et al.* 2005a) and Kenney *et al.* (1994, 1995) suggest that even moderate levels of poaching over relatively short time periods (~6 years) can lead to massive population declines of up to 95% in this species. In the Russian Far East, Chapron *et al.* (2008b) suggest that tiger populations cannot recover if annual mortality rates exceed 15%. Finally, road accidents can also be significant sources of mortality in some felid populations. Forty-five per cent of ocelot (*Leopardus pardalis*) mortality in Texas (Haines *et al.* 2005) and 48% of recorded puma deaths in Florida were due to road accidents (Taylor *et al.* 2002).

In this chapter we identify two key areas where felid populations face anthropogenic threats. Firstly, we explore the way felids impact people's lives and livelihoods and in turn the way in which people retaliate. We then consider the impacts of trade and overexploitation on the conservation status of felid populations. In each section, we offer a synthesis of the potential conservation and management solutions.

Conflicts between felids and people

Key areas of conflict between people and felids are through depredation on domesticated animals or game species and, less frequently, when large felids kill or injure people. Livestock raiding or human deaths often lead to retaliatory killing of the felids responsible. We will discuss, in the sections below, the issues of livestock depredation and human killing, the consequences for felid populations, and potential management solutions in situations where such depredations occur.

Consequences of conflict for felid populations

Loss of human life or livelihood provides the impetus for people to attempt localized or sometimes wide-scale eradication of predators. Historically, eradication of carnivores has been a state-supported priority, incentivized by rewards and bounties. From 1860 to 1875, 4708 tigers and leopards were reported killed in India (Boomgaard 2001). In the United States, over 1160 pumas were culled from 1987 to 1990 in an attempt to limit livestock losses in ranching areas (Johnson *et al.* 2001b). Wide-scale lethal control of predators still occurs in some frontier areas. For example, Michalski *et al.* (2006a) found that 110–150 jaguars and pumas were killed over a 12-month period by professional predator hunters in southern Amazonia, Brazil.

Contemporary government policies regarding predator management are generally more enlightened. However, in areas where depredations are perceived to impact livelihoods, both legal and illegal retaliatory killing of felids can extirpate populations. Frank *et al.* (2006) describe the decimation of the lion population over large areas of Kenyan Masailand, where lion populations had previously been secure. The increasingly negative attitudes to wild predators in Masailand are at least nominally due to livestock depredation. However, limited involvement by local people in wildlife tourism and lack of access to the revenue generated, coupled with ready access to agricultural poisons, may also be motivations behind increasing levels of illegal predator control. In Kenya, today, the use of agricultural insecticide Furadan (carbofuran) to kill predators is probably the single greatest source of lion mortality; in Laikipia and Kajiado districts alone, a minimum of 70 lions are known to have been poisoned since 2001 (L.G. Frank, S. Maclennan, and A. Cotterill, unpublished data) and populations of vultures and other scavenging birds are plummeting throughout the country (S. Thomsett, personal communication).

Predation by felids on ungulate game species

Large felids can come into conflict with people over competition for wild prey species. Predators were even persecuted as vermin in some national parks until as late as the 1950s, because carnivores were thought to suppress 'game' numbers (Davison 1967; Smuts 1976). A worldwide meta-analysis of human–predator conflict found that predators are reported to kill from 0.02% to 2.6% of livestock, but 9% of game species annually (Graham *et al.* 2005). Where wild ungulates are utilized by people for either commercial gain (commercial hunting and game farming) or for leisure (sport hunting), competition and conflict may occur between users and large felids. For example, cheetahs in Namibia kill economically valuable wild ungulates on commercial game farms, leading to persecution of the cats by farm owners (Marker *et al.*, Chapter 15, this volume). Similarly, hunters in the Russian Far East come into conflict with tigers over real or perceived competition for ungulates (Miquelle *et al.* 1999a). European lynx are persecuted when they compete with sport hunters for prized ungulate trophy species (Breitenmoser *et al.*, Chapter 23, this volume).

Livestock depredation by felids

Felids are obligate carnivores. Humans and felids come into conflict in ecosystems, where a high proportion of ungulate biomass is made up of domesticated species. For the most part, the smaller felid species do not cause economically significant losses to domestic livestock, although species such as serval (*Leptailurus serval*), wildcat (*Felis silvestris*), and guiña (*Leopardus guigna*) may occasionally prey on small stock or poultry (Sanderson *et al.* 2002c; Sunquist and Sunquist 2002). Generally, species ranging in size from the caracal (*Caracal caracal*; ~12 kg) to the size of the tiger (~235 kg) have been found to be the most problematic stock raiders. With the exception of the caracal and Eurasian lynx (~23 kg) all of the most import livestock-killing felids are 50 kg or heavier (Inskip and Zimmermann 2009). Table 6.1 provides some examples of livestock depredation by felids. Comparison of trends between studies is complicated by geographic and ecological differences, varying spatial and temporal scales of the studies, and reporting of widely different parameters and data. This makes rigorous evaluation of trends difficult; however, there are a number of broad patterns that emerge and these are discussed in more detail in the following text.

Patterns of livestock depredation

Different size classes of livestock are vulnerable to different suites of felid predators. Smaller felids, such as lynx and caracals, prey on the smaller size classes of livestock such as sheep and goats, while the largest felids prey on the full size range of livestock. Intermediate-sized felids, such as pumas and leopards prey on smaller livestock and juveniles of the larger species.

The spatial distribution of livestock depredation is often uneven. Stahl *et al.* (2001a, b) found that in the French Jura Mountains, there were consistent 'hot spots' of lynx predation on sheep. The areas were often also areas with high roe deer (*Capreolus capreolus*) densities, suggesting that lynx populations may have been sustained by reservoirs of natural prey. Similarly, in southern and central Brazil, livestock were most vulnerable to puma and jaguar predation within or in close proximity to forest fragments or riverine forest, habitats that provide felids with cover and residual levels of natural prey (Mazzoli *et al.* 2002; Palmeira *et al.* 2008). In southern Africa, livestock predation by lions and leopards is often higher close to the boundaries of protected areas (Fig. 6.2). This was found to be the case adjacent to Khutse Game Reserve, Botswana (Schiess-Meier *et al.* 2007) and Chirisa Safari Area, Zimbabwe (Butler 2000). In

Table 6.1 Examples of studies of depredation by large felids on livestock, illustrating the nature, extent, and economic impact of the problem across four continents. (See notes at end of table for livestock species codes.)

Location	Dates	Predator species	Livestock species	Livestock damage	% total livestock depredated	Other causes of livestock loss	Economic impact	Details
Botswana, Kweneng district[a] (38, 123 km^2)	2000–02	Lion, leopard, and cheetah	C, G	857 leopard attacks, 588 lion attacks, and 211 cheetah attacks.	0.34% of all livestock in area (1.0–11% depending on farm size)	2.8–12.6% lost to disease, starvation, and accident		Losses mostly of free-ranging livestock. Depredation 10 times higher close to reserve. Lions and leopards caused 64% damage
Kenya, ranches adjacent to Tsavo East NP[b] (690 km^2)	1996–99	Lion and cheetah	G, S, C and Do, Ca	433 head in 312 attacks, lion responsible for 85.9%, and cheetah 2.9%	2.4% of stock holdings		Value of total loss (2.6% of herd's economic value): US $8749/year; Cost per lion: US$290 p.a.	Seasonal peak in depredation corresponded to low-prey density during wet season. Problem lions shot, of which 66% were subadult males
Kenya, Laikipia[c] (1443 km^2)	1999–2000	Lion, leopard, and cheetah	C, G, S	Lions killed 143 cattle, 396 shoats; leopards killed 38 cattle, 58 shoats; cheetah killed 94 shoats	0.7–0.8% cattle and 1.4–2.1% sheep	2.5% cattle, 8.2% sheep/ goats to disease and stock theft	US$360 to maintain a lion, US$211 for a leopard. US$40/ household lost on group ranches (11% of annual income)	Lethal control by commercial ranchers of stock raiders was focused on culprits. Subsistence pastoralists less tolerant. Lion population declining by 4% p.a.
Kenya, Group Ranches, adjacent to Masai Mara GR[d] (1500 km^2)	2003–04	Lion, leopard	C, G, S	130 attacks, 147 stock animals killed Lions responsible for 15% of attacks, leopards 32%	Overall loss of 0.2% cattle and 0.6% shoats. Of cattle losses, 57% due to lions		Loss of US$1866/year to lion, US$984 to leopard	Attacks more frequent in wet season, and correlated with lower prey density. Dogs effective at deterring attacks

Kenya, Loldaiga Ranch, Laikipia[e] (2000 km²)	1989–95	Leopard, Lion, and cheetah	C, S	Each year lions kill on average 15.8 cattle, 6.3 sheep; leopards 4.3 cattle, 10.5 sheep; cheetah kill 9.8 sheep	Total loss to felids: 0.45% cattle and 0.67% sheep	Losses to theft larger than predation losses	Lions cost US $14,400/year (US $0.72/ha/year); leopards US$5000/ year (US$200/leopard, US$0.25/ha/year)	Losses and costs of tolerating predators were low. High wildlife density may buffer against livestock killing
Kenya, Kitengela Conservation Unit[f]	1974–75	Lion	C, G, S	11 cattle and 14 shoats killed 1974–75	Total loss to felids: 0.05% cattle and 0.07% shoats			1970–75, lions killed 92 head. 19 lions shot, 2 speared in retaliation for stock killing
Cameroon, villages adjacent to Waza NP[g] (1700 km²)	2002	Lion	C, G	118 (±0.75) head/year		Loss to disease similar to depredation	US$37–1115 (mean US$589)/owner/year	Predation was seasonal. Higher close to NP boundary, where predation occurred all year round
North-east Namibia, Tsumkwe District[h] (4869 km²)	1992–95	Lion and leopard	C, H, D, P	Lions killed 20 cattle and 5 horses; leopards killed 43 cattle (calves), 32 dogs, and 11 chickens	4.3% cattle, 12.2% horses lost to lions; 9.2% cattle, 23% dogs, and 2.4% poultry lost to leopards		N$83/village to lions; N $55/village to leopards	95% attacks at night when stock left in bush. 14 lions translocated, 2 shot; 6 leopards translocated
Zimbabwe, Gokwe communal land[i] (33 km²)	1993–96	Lion and leopard	G, C, Do, S, Pi	Lions killed 81 head; leopards killed 30 head (mostly goats and cattle)	Predation by all predators reached 5% of livestock	5.8% lost to disease and starvation	US$13/household/ year (12% annual income); lions cause US$2640, leopards US $580 worth of damage/ year	Attacks mostly in dry season. Lions responsible for 34% predation, 57.5% of economic loss, and killed 3.7 animals per attack (most from Kraals)

(Continued)

Table 6.1 (Continued)

Location	Dates	Predator species	Livestock species	Livestock damage	% total livestock depredated	Other causes of livestock loss	Economic impact	Details
Bhutan, Jigme Singye Wangchuck National Park[j]	2000	Tiger and leopard	C, Y, H, S, Pi	76 domestic animals (of which 63 cattle) lost to predators	2.3% of livestock lost (to all predators including dhole and black bear)		Overall: US$44.72/ person/year (17% per capita income in area). For households affected: 84.5% of cash income	Tiger and leopard cause 82% of economic loss. Predation appeared higher close to forests and when stock poorly protected
India, Gir Wildlife Sanctuary and surrounds[k]	1986–91	Lion	C, B	Average of 1265 head/ year inside GWS, 1691 head/year outside	0.4% of stock in 'tehsils' adjoining Gir	High levels of starvation during drought		Very high densities of livestock kept in areas around Gir. People are relatively tolerant of losses
India, Indian Trans-Himalaya[l] (30 km^2 and 12,000 km^2)	1995–96 and 2002–03	Snow leopard	Y, C, G, S	1.6 head/ household 1995–6; 0.6–1.1 head/ household 2002/3	12% of livestock holdings/year in region (1995–6)		US$128/family (50% average per capita income)	High losses in areas with low natural prey. Livestock accounts for 40–58% of snow leopard diet
India, Bhadra Tiger Reserve[m] (495 km^2)	1997–99	Tiger and leopard	Mostly C	219 head killed in 131 incidents	12.9% of livestock in household	3.1% lost to starvation and disease	16% of annual income/ household	Losses lower in cropping season when people out in fields
Nepal, Annapurna Conservation Area[n]	1988–89	Snow leopard	G, S, C, Y	0.6–0.7 animals/ household/ year (mostly goats)	2.6% of total stock		25% (US$42) of annual per capita income	People highly intolerant. >7 snow leopards known to have been killed in 10 years

France, Jura Mountains[o] (6000 km^2)	1984–98	European lynx	S	1782 livestock killed (mostly sheep); 92–194 sheep/year (1990–98)	0.14–0.59% of sheep at regional level			High predation in 0.3–4.5% total area. Attacks increased as lynx population increased (3 attacks in 1984, 188 in 1989). 10 problem lynx removed
Southern Brazil, Santa Caterina[p] (138 km^2)	1993–95	Puma	C, S, G, Pi	Pumas killed 9 cattle, 172 sheep and goats and 5 pigs	0.1% of cattle holdings, 23.8% of shoats, and 3% pigs	2.2% cattle, 2.6% shoats, 3.8% swine lost to disease, theft, and accidents	Cattle US$1890 (0.27% total value), sheep US$5900 (32% value), goats US$4332 (38% value)	Managed flocks had lower losses than free-ranging flocks. >27 pumas killed 1988–95, often illegally
Brazil, Alto Florista, Mato Grosso[q] (34,200 km^2)	2001–04	Jaguar and puma	C	72% of predation by jaguars, rest by pumas	0.26–1.24% of total cattle		Loss US$145–885/ranch/year, depending on property and herd size	Peaks in predation in dry season during calving period. 2002–4, 185–240 pumas and jaguars killed. Predation correlated to presence of forest fragments and riparian corridors
Central-western Brazil, cattle ranch[r] (200 km^2)	1998–2002	Puma and jaguar	C	309 cattle killed by predators	0.4% of cattle lost to predators	1.7% lost to disease, malnutrition, accident, and mismanagement	0.3% of total livestock value	Most predation by pumas, proportional to calves available. Some retributive killing of predators occurs

Notes: B, domestic buffalo; C, cattle; Ca, camel; D, dog; Do, donkey; G, goats; H, horse; P, poultry; Pi, pig; S, sheep; Y, yak.
[a] Schiess-meier *et al*. (2007); [b] Patterson *et al*. (2004); [c] Frank (2005), Ogada *et al*. (2003); [d] Kolowski and Holecamp (2006); [e] Mizutani (1999); [f] Rudnai (1979); [g] Bauer (2003), Van Bommel (2003); [h] Stander (1997a, b); [i] Butler (2000); [j] Wang and Macdonald (2006); [k] Saberwal (1990), Srivastav (1997); [l] Bagchi and Mishra (2006), Mishra (1997); [m] Madhusudan (2003); [n] Oli *et al*. (1994); [o] Stahl (2001a, b); [p] Mazzoli *et al*. (2002); [q] Michalski *et al*. (2006); [r] Palmiera (2008).

Figure 6.2 A village headman displays the remains of a cow killed by lions, Tshlotsho Communal Land, adjacent to Hwange National Park. © A.J. Loveridge.

northern Cameroon, levels of lion depredation were inversely correlated with distance to Waza National Park (Van Bommel *et al.* 2007).

Temporal patterns of livestock depredation are often predictable and in many cases vary with levels of available natural prey. In many areas of Africa, livestock depredation peaks during the season when natural prey is most difficult for predators to acquire. This is usually the wet or rainy season, when prey are in better condition and often more widely dispersed (Patterson *et al.* 2004; Bauer and Iongh 2005; Kolowski and Holecamp 2006). Hemson (2003) found that in the Makagadigadi ecosystem, Botswana, livestock-raiding lions killed fewer livestock during periods when migratory wildebeest (*Connochaetes taurinus*) and zebra (*Equus burchelli*) were present in their ranges (Loveridge *et al.*, Chapter 11, this volume). In Laikipia, Kenya, livestock losses to lions and leopards were lower during drought years when wild prey were in poor condition and therefore easier for predators to capture (Ogada *et al.* 2003).

Peaks in livestock predation may also be due to seasonal availability or vulnerability of livestock. In areas with marked seasonal variation in climate, livestock may be in poor condition during the lean season, which may make them easier prey for felids during this period (Hoogesteijn *et al.* 1993; Butler 2000). Neonates of domesticated species are often more susceptible to predation than adults. This vulnerability can result in increased levels of depredation during the lambing or calving period. Peak losses to jaguars and pumas on cattle ranches in southern and central Brazil and central Venezuela often coincide with the calving season (Polisar *et al.* 2003; Michalski *et al.* 2006a; Palmeira *et al.* 2008).

Seasonal changes in husbandry or herding practices may also result in periods of vulnerability for livestock. In France, peak losses (May–November) of sheep to lynx coincided with periods when sheep were kept in fenced fields for the entire 24-hour period, presumably increasing their availability to lynx and therefore the probability of depredation (Stahl *et al.* 2001b). Villages around Bhadra Tiger Reserve, India, experienced lower levels of livestock loss during the harvest season, when it is thought that increased human activity in surrounding fields may also inadvertently provide protection for stock (Madhusudan 2003).

Economic impact of livestock depredation

Impacts of felid depredation on livestock vary, depending on the scale of livestock ownership, husbandry techniques, livestock type, stocking density, and density of predators. Small-scale, subsistence livestock owners are often disproportionately impacted by losses, in part because they may lack resources to provide effective protection for their stock. Loss of even an individual animal to small-scale livestock owners has a proportionally higher impact on herds or flocks than the same loss to owners with more stock. For this reason, Ikeda (2004) found that small-scale yak (*Bos grunniens*) herders in Kanchenjunga Conservation Area, Nepal, were more heavily impacted by snow leopard (*Panthera uncia*) predation on their herds than wealthier, medium- to large-scale owners.

Snow leopard predation in the Spiti Region, India, amounted to losses of between 25% and 52% of annual per capita income for small-scale farmers (Oli *et al.* 1994; Mishra 1997). In Africa, subsistence farmers in Kenya and Zimbabwe lost between 11% and 12% of annual income to lion and leopard depredation (Butler 2000; Ogada *et al.* 2003). Villages on Koyake Group Ranch adjacent to the Maasai Mara Reserve, Kenya, lost approximately US$1890 to lions and US$984 to leopard depredation over 14 months (Kolowski and Holecamp 2006). Tiger depredation on livestock around Bhadra Tiger Reserve, cost villagers 16% of annual income (Madhusudan 2003). Around Jigme Singye Wangchuck National Park, Bhutan, tigers and leopards caused losses of 17% of annual per capita income, however for households actually affected by losses this represented up to 84% of annual cash income (Wang and Macdonald 2006; Wang 2008). These losses are significant impacts on the livelihoods of people, who often have few alternative means of earning a living.

In contrast, large ranches lose relatively less to depredation. Large ranches in Alta Floresta, Brazil, lost relatively fewer cattle to depredation than intermediate-sized ranches, with no ranch losing more than 1.24% of its herd (Michalski *et al.* 2006a). On large-scale ranches in Kenya, it is estimated that the cost of maintaining a lion was US$290 in southern Kenya and US $360 in Laikipia (Patterson *et al.* 2004; Frank *et al.* 2005), while the cost of maintaining a leopard cost ranchers US$200–211/year in Laikipia (Mizutani 1997; Frank *et al.* 2005). The cost of tolerating occasional depredation was considered to be low by these relatively wealthy ranchers (Mizutani 1999). Losses due to depredation by carnivores are often much smaller than losses due to disease, poor husbandry, or theft. On a Laikipia ranch, average loss of stock to predators over a 23-year period was 34 cattle and 66 sheep a year. However, annual losses to disease were 2.5 times higher for cattle and nearly 4 times higher for sheep. Losses attributed to theft were similar to losses to carnivores (Mizutani 1997). In Kweneng district, Botswana, 0.34% of livestock was lost to carnivores each year (although losses to individual owners were higher at 1.0–11% of holdings). However losses to disease, starvation, and accident accounted for 2.8–12.6% of stock (Schiess-Meier *et al.* 2007). On central Brazilian ranches, 0.4% of stock was lost to predators, while 1.7% was lost to disease, malnutrition, and mismanagement (Palmeira *et al.* 2008). In contrast, around Bhadra Tiger Reserve losses of stock to predators were around 4 times higher than losses to disease or starvation (Madhusudan 2003). In Gokwe, Zimbabwe, levels of loss of livestock to predators (~5%) were similar to losses to starvation and disease (Butler 2000).

Attitudes to livestock depredation

People's attitudes to and tolerance of depredations by wild felids vary widely. Antipathy towards carnivores may be a result of historical or cultural attitudes, as well as based on past experiences and personal values. Although generalizations are difficult across widely variable socio-cultural circumstances and it is often difficult to distinguish between the underlying reasons for negative perceptions people hold towards predators, there is some evidence that levels of tolerance livestock owners have for predators are related to magnitude and impact of losses. In the Pantanal, Brazil, tolerance of jaguars by ranchers was at least partially explained by levels of livestock loss (Zimmermann *et al.* 2005a). However, other studies indicate that personal and societal beliefs and perceptions may be equally important in shaping responses to conflict with predators (see Cavalcanti *et al.*, Chapter 17, this volume; Murphy and Macdonald, Chapter 20, this volume).

Economic circumstance may also dictate levels of tolerance. Subsistence farmers in Annapurna Conservation Area, Nepal, who suffered significant losses to depredation held highly negative attitudes towards snow leopards and considered total extermination of these felids the only solution to the problem (Oli *et al.* 1994). However, in the Indian Trans-Himalaya, where people had alternative incomes and lower dependence on livestock, they were more tolerant of snow leopards (Bagchi and Mishra 2006). Subsistence pastoralists on group ranches in Laikipia, Kenya, were more likely to attempt to eradicate predators than neighbouring commercial ranchers (Woodroffe and Frank 2005). In this case, commercial ranchers tolerate losses to large felids both because of a cultural appreciation of wildlife and because wild felids have value to ecotourism operations (Mizutani 1997, 1999). In other areas, people are willing to tolerate moderate losses. In India, villagers often graze livestock on government forest lands and are willing to tolerate moderate loss as the price of access to grazing resources (Srivastav 1997; Karanth and Gopal 2005).

Human injury and fatality caused by felids

Perhaps the most striking relationship people have with felids is that large felids occasionally prey upon, kill, or injure people. Evidence that hominids have always suffered depredation by felids comes from the Swartkrans Caves in South Africa, where the excavated skull of a hominid child bears the tooth marks of a leopard (Brain 1969; de Ruiter and Berger 2000). Carnivores from the families Felidae, Canidae, and Ursidae are all on record as killing people (Quigley and Herrero 2005). Due to the need to overpower and kill their victims, human-killing felids are invariably the largest species, with lions and tigers being the most dangerous and widely publicized (Table 6.2). Famous historical cases of man-eating big cats include the man-eating lions of Tsavo, which, in 1898, are reputed to have killed around 128 construction workers on a railway line being built from Mombasa to Nairobi, Kenya (Patterson *et al.* 2003). Other historical examples include man-eating tigers and leopards in northern India, one of which (a tigress) reputedly killed 436 people before being hunted and shot (Corbett 1946). Boomgaard (2001) reports that in India, in 1875 alone, tigers and leopards killed 1039 people. However, human deaths due to large cats are not a thing of the past. Human deaths and injuries caused by lions are relatively common in central and southern Tanzania, with around 563 Tanzanians killed between 1990 and 2004 (Packer *et al.* 2005b). Similar numbers of people are killed and injured in the Sundarbans of India and Bangladesh, with 294 people killed between 1984 and 2001 in the Indian Sundarbans (Siddiqi and Choudhury 1987; Karanth and Gopal 2005) and 79 people killed between 2002 and 2007 in the Bangladesh Sundarbans (Khan 2007).

Human attitudes to man-eating and their consequences

At a global scale human killing by felids is relatively rare, much rarer than human deaths due to traffic accidents for instance. However, because of the terrible emotional consequences attached to the loss of human life, human killing puts conservationists in a difficult moral dilemma. It makes conservation and protection of big cats difficult for conservationists to justify to local people who suffer losses. Baldus (2004, 2006) gives an example of a man-eating incident in the Mkongo Division, Rufiji district, adjacent to Selous Game Reserve in Tanzania, that illustrates the traumatic disruption to people's lives. In this case, at least 34 people were killed and 10 injured by lions over a period of 20 months, with most incidents occurring in crop-lands south of the Rufiji River, which divides the district. Eventually, almost the entire human population had fled to the northern banks of the river to escape the lions, leaving homes and crops unguarded. Leaving aside the emotional and psychological consequences of living alongside man-eating carnivores, the perceptions and attitudes of people towards wildlife and conservation of large carnivores are seriously impacted. In poor countries, compensation for loss of human life is often grossly inadequate. In Tanzania, compensation for loss of human life to wild animals is around US$30–50 (Baldus 2004). Apparent undervaluation of human life may further compound ambivalent attitudes of local communities towards conservationists and conservation managers, reduce public acceptance of big cat conservation, and undermine conservation efforts. Indiscriminate retaliatory killing of cats often results from incidences of human killing. In the case of the Rufiji maneater, wire snares were set throughout the area to catch the culprit, killing at least eight lions (which may or may not have been maneaters) and probably many other non-target species, before the actual maneater was eventually shot (Baldus 2006). Attempts to eliminate maneaters may impact the viability of local populations of carnivores.

The magnitude of the problem

Table 6.2 provides some examples of accidental killing, injury, or predation on people by felids. Felids from the genus *Panthera* are the most consistent maneaters, although this behaviour is rarely recorded in *P. onca* (Rabinowitz 2005). Pumas (*Puma concolor*) are known to attack and kill people, but incidences are extremely rare, with 10 deaths and 48 injuries throughout the

Table 6.2 Summary of studies on depredation on humans by large felids, showing number and rate of attacks in various areas. The type of attack is classified as either being accidental (A) or as the result of depredation (D).

Location (area km^2)	Dates	Species	Human density	Human injuries	Human deaths	Attacks/year/ 1000 km^2 *	Type attack	Details
Tanzania, Njombe District[a] (2000 km^2)	1944–52	African lion			1000	55.5	A/D	At least 30 lions killed (17 confirmed maneaters). Lions completely extirpated by 1950s
Tanzania, Tenduru District[a] (1560 km^2)	1986	African lion			42	26.9	D	43 lions shot
Tanzania, Lindi District[a] (800 km^2)	1999–2000	African lion			24	15.0	D	Man-eating stopped after seven subadult lions were shot
Tanzania, Mkongo Division, Rufiji District[b] (350 km^2)	August 2002–April 2004	African lion	37.8/ km^2	10	35	50	D	1991–2004 at least 58 people killed. Most people killed were guarding crops. 3.5-year-old male maneater killed
Tanzania[c]	1990–2004	African lion		308	563	—	A/D	Attacks most common in districts with low density of natural prey
Mozambique, Niassa Game Reserve[d] (25,000 km^2)	2000–06	African lion	1.0/ km^2	17	11	0.19	A/D	~50% attacks occurred in villages, 32% in fields. One lion killed More than 34 people killed and 37 injured by lions 1974–2007
Zambia, Lupande GMA[e] (4800 km^2)	1991	African lion			3	0.62	D	Three lions shot. Lions had old wounds from wire snares, but otherwise healthy. Pride displaced into marginal habitat outside the park by a larger pride
Uganda[f]	1923–94	African lion		69	206	—	A/D	Men attacked more often than women, due to engagement in hunting activity. ~75% of lion attacks fatal. 376 ‘problem’ lions killed
Uganda, Queen Elizabeth National Park[g] (955 km^2)	1990–2000	African lion			32	3.4	A/D	People killed when herding or moving through park at night, and when drunk
Kenya, Tsavo East region[h]	1898	African lion	Low		128	10.0	D	Railway workers targeted by at least two male lions. Work was temporarily halted until lions shot. One lion had cranial deformities and broken teeth
India, Gir Wildlife Sanctuary[i] (1400 km^2)	1978–91	Asiatic lion		165	28	~10.6	D	82% people killed adjacent to reserve, largely by dispersing subadults. Man-eating may have been as a result of drought, low prey density. Similar trend recorded in 1901–04 drought
Uganda[f]	1923–94	Leopard		75	37	—	A/D	One-hundred-and-fourteen ‘problem’ leopards killed over period. Only 32.5% leopard attacks fatal

(Continued)

Table 6.2 (Continued)

Location (area km^2)	Dates	Species	Human density	Human injuries	Human deaths	Attacks/year/ 1000 km^2 *	Type attack	Details
India, Gharwhal region[j]	1996–97	Leopard			36	—	D	Women and children (and one drunk) killed by leopard
United States[k]	1890–90	Puma		43	10	—	A/D	30% of cases in vicinity of Vancouver Island, increasing incidence of attacks due to human encroachment into habitat
Russia, Sikhote-Alin Zapovednik[l] (4000 km^2)	1970–2001	Amur tiger	2.8/ km^2	37	14	0.01	A	People often killed/injured when encounter tigers accidentally while hunting. 81.7% of tiger mortality due to people (1992–2001)
India, Kanha Tiger Reserve, Madhya Pradesh[m]	1985–2001	Tiger		25	22	—	A	Attacks thought to be largely accidental
India, Sundarban Tiger Reserve[m] (~4300 km^2)	1985–2001	Tiger		57	294	5.1	D	Tigers appeared to target humans as prey
Bangladesh, Sundarbans[n] (5770 km^2)	1953–83	Tiger			554	3.4	D	Many deaths not reported due to illegal use of the forest
Bangladesh, Sundarbans[o] (5770 km^2)	2002–06	Tiger			79	2.7	D	At least 13 tigers killed in retaliation. Use of dogs to protect/alert forest users being tested
Nepal, Chitwan National Park[p] (750 km^2)	1980–2005	Tiger	52.4/ km^2	Not given	88	0.47	A/D	Tiger attacks have increased over study period, perhaps due to increasing tiger and human population. 50% people killed collecting forest products. Thirty-seven known maneaters (of which 15 killed, 6 captured for zoos, and 4 translocated)
Sumatra[q]	1978–97	Tiger		30	147	—	A/D	Attacks declined in mid-1980s–90s due to declining tiger populations. Twenty-eight man-eating tigers shot/poisoned and 20 translocated. 73% of man-eating tigers were male

Notes: [a] Baldus (2004); [b] Baldus (2006); [c] Packer *et al*. (2005); [d] Begg *et al*. (2007); [e] Yamazaki and Bwalya (1999); [f] Treves and Naughton-Treves (1999); [g] West African working group report (2001); [h] Patterson (2004); [i] Saberwal (1994); [j] Kruuk (2002); [k] Beier (1991); [l] Miquelle (2005); [m] Karanth and Gopal (2005); [n] Siddiqi and Choudhury (1987); [o] Khan (2007); [p] Gurung *et al*. (2006); [q] Nyhus and Tilson (2004b).
* Not calculated for entire countries or regions due to varying/sporadic nature of human killing.

United States and Canada from 1890 to 1990 (Beier 1991). Lions and tigers, the largest living cats, are responsible for most incidents, with episodes of leopards as maneaters being occasionally reported, particularly in northern India (Corbett 1946; Athreya 2006; Athreya *et al.* 2008). Levels of human killing by felids vary widely. Human fatalities to tigers were 0.01 people/1000 km^2/year in the Russian Far East, where human population density was low and interactions with tigers rare and usually accidental (Miquelle *et al.* 2005a). This contrasts strongly with the 50 people/1000 km^2/year killed by lions specifically targeting people as prey items in Rufiji district (Baldus 2004, 2006). By their nature, incidents of human killing by felids tend to be clumped in space and time, so compilations of incidents across large spatial scales or long time periods tend to under-represent the local impact. Likewise, reports that include incidents over a short time-span and localized scale may overstate the global importance of man-eating as a behavioural trait of felids, while at the same time capturing the traumatic impacts such incidents have on human communities living alongside populations of large felids.

It is not always possible to unambiguously distinguish between those incidents that were the result of accidental encounters between a person and large felid and those where the person was viewed as a prey item by the felid in question. Karanth and Gopal (2005) compare the examples of Kahna Tiger Reserve, India, where human deaths are relatively low (but not infrequent), with the Sundarban Tiger Reserve, India, where human killing by tigers is frequent. In Kahna, tigers occasionally come into contact with people in the densely populated areas surrounding the park, leading to occasional human deaths when tigers are disturbed or accidentally encounter people harvesting forest products. However in the Sundarban Tiger Reserve it is clear, both from the high incidence of human deaths and the circumstances of the attacks, that people are the intended prey of some tigers (Table 6.2).

Characterization of human–felid conflict

Situations when livestock depredation or man-eating are more likely to occur fall into a number of broadly defined circumstances. These are scenarios when encounters between people or livestock and felids are more frequent because either felids disperse from existing or emerging populations or when people enter felid habitat. Conflict may also be more likely to occur when levels of natural prey have been depleted. These three situations are discussed in more detail below.

Dispersal of felids into marginal, human-dominated habitat

Ironically, the consequence of successful protection of populations of large felids is that surplus individuals may need to disperse out of habitats or refugia already occupied by conspecifics. If the surrounding areas provide no connectivity to other habitat patches and/or are marginally suitable habitat, already used by people, conflict with human populations in these areas can result.

Examples of this include tigers dispersing into buffer areas surrounding Chitwan National Park, Nepal. This has led to 88 human deaths between 1980 and 2005, with many of the dispersing tigers being old individuals displaced from their ranges (McDougal 1993; Gurung *et al.* 2006a). A similar example of competition for space leading to displacement into areas of human settlement occurred in Lupande Game Management Area, Zambia, where a pride of lions, displaced from South Luangwa National Park by a stronger pride, killed three people (Yamazaki and Bwalya 1999). Conflicts between people and large felids over livestock loss are common on the boundary of many protected areas in Africa and Asia. Large felids disperse into marginal boundary areas and kill livestock. For example, dispersing lions kill livestock around Etosha and Hwange National Parks (Stander 2005b; A.J. Loveridge, personal observation). Similar conflicts between lions, people, and livestock occur in the boundary areas of Gir Forest, India (Saberwal *et al.* 1994). Problems of dispersing carnivores from emerging populations have been experienced in western Europe, where reintroduced European lynx have come into conflict with sheep farmers (Kaczensky 1999).

Human encroachment on felid habitat

People and their livestock enter (legally and illegally) large felid habitat for a number of reasons, the most common being for hunting or extraction of forest products, use of grazing resources, and occasionally for other reasons such as leisure activities, tourism, or as refugees (Johnson *et al.* 2001b; Kruuk 2002). These activities bring people into contact with felids and may result in either accidental killing of people (and/or felids) or provide opportunities for felids to prey upon people or livestock. In the Sundarbans of Bangladesh, there are no permanent settlements in the forest but people use the forest for extraction of forest products both legally and illegally. Here some tigers apparently regularly prey upon people, with fluctuations in numbers of fatalities correlated with periods of high use of the forest (Khan 2007). Similarly, injury or death caused by tigers in the Russian Far East appears to occur when people accidentally encounter tigers while hunting in the forest (Kerley *et al.* 2002; Miquelle *et al.* 2005a). Livestock losses are high in areas where livestock are grazed in or in close proximity to forested habitats with resident felid populations (Wang and Macdonald 2006; Palmeira *et al.* 2008). In Norway, sheep are grazed unprotected in forested habitats, leading to high losses to carnivores (Swenson and Andrén 2005).

Another reason people encroach on felid habitat is to settle in these areas. Nyhus and Tilson (2004b) record the increasing incidence of livestock loss and man-eating by tigers in Sumatra during the 1970s–80s when settlement of the island by transmigrants was encouraged and people encroached on tiger habitat during the massive transformation of the lowland rainforests into oil palm plantations. However, incidents declined during the mid-1980s as tiger habitat was destroyed and tiger populations on the island declined. This pattern of conflict followed by population decline has been closely documented for tigers on the Indonesian islands of Java and Sumatra and in India between 1600 and 1950 by Boomgaard (2001). Similar patterns of depredation by tigers may have occurred across south-east Asia as human populations increased and colonized forest habitats throughout the nineteenth century (McDougal 1993). Rapid fragmentation and conversion of forested habitats to agricultural land in Amazonia bring people and felids into contact and result in high levels of conflict (Michalski *et al.* 2006a). Similarly, the increasing encroachment of human residential development into puma habitat in North America may increase the chances of people encountering pumas and account for the increases in puma attacks on people in the past decade-and-a-half (Cougar Management Guidelines Working Group 2005). People also become vulnerable when circumstance forces them to intrude into big cat habitat, as illustrated by the example of Mozambican refugees, fleeing into South Africa, via the Kruger National Park, being killed and eaten by lions (Kruuk 2002).

Prey depletion causes felids to switch to alternative prey

Evidence from some studies suggests that felids have a preference for natural prey. This may be because availability of domesticated species is not reflected by their numerical abundance, as livestock is often protected and livestock-raiding animals risk higher levels of mortality in human-dominated environments. Eurasian lynx in Norway did not select habitat patches with high livestock densities, but preferred areas with high roe deer density, and preyed predominantly on roe deer, despite the fact that sheep densities were eight times higher (Moa *et al.* 2006; Odden *et al.* 2006). Rabinowitz and Nottingham (1986) found that jaguars prefer natural prey to domestic stock. Hemson (2003) found that lions appeared to 'prefer' wild prey, at least during periods of high wild ungulate abundance. In this case, they killed livestock less than expected based on availability in the good season and in proportion to availability in lean season. Accordingly, it was suggested that maintaining suitable levels of natural prey in protected-area buffer zones may serve to limit depredation on livestock.

A common suggestion in reports of depredation on both livestock and people is that levels of natural prey have been depleted and this may provide an explanation for these behaviours. While this has rarely been quantified, a number of examples appear to support the suggestion and may explain a proportion of the cases. For example, Asiatic lions attacked people around Gir Wildlife Sanctuary, when natural prey densities declined after a prolonged drought (Saberwal *et al.* 1990; Saberwal *et al.* 1994). In the case of man-eating lions in central and southern Tanzania, their natural prey was depleted, forcing the lions to switch to prey (bush pigs, *Potamochoerus porcus*) found in proximity to human settlement.

This may increase the likelihood of encounters with people guarding their crops, potentially leading to man-eating events (Packer *et al.* 2005b). Hoogesteijn (2002) and Crawshaw (2004) found that loss of natural habitat and prey predisposes jaguars to kill livestock. Similarly, losses of domestic stock to snow leopards were high, making up around 58% of snow leopard diet, in areas with low abundance of natural prey (Mishra 1997; Bagchi and Mishra 2006)

Mitigation of conflict

Characteristics of problem animals

Defining a large cat as a problem requiring a solution depends on both where it is and how it is behaving. Linnell *et al.* (1999) distinguish between two kinds of 'problem animal'. The first are animals that are in the 'wrong place' (generally in a wide-scale matrix of habitats, where not all individuals have access to livestock within their home ranges). Individuals dispersing out of protected areas or individuals coming into contact with livestock or human intruders into their habitat might fall into this category. The second 'type' of problem animal is one that kills more livestock (or people) per encounter than do conspecifics. Animals with old injuries or habitual livestock raiders (Stander 1990) could be included in this definition.

There are a number of factors that appear to predispose individual felids to becoming problem animals. Male felids appear to be more likely to become stock raiders (Stander 1990; Cunningham *et al.* 2001; Funston 2001a; Odden *et al.* 2002). Odden *et al.* (2002) found that male Eurasian lynx were not only more likely to kill livestock than females, they were also more likely to make multiple kills during a raiding incident. Specific male behaviours may also be factors that influence whether or not livestock is taken (Linnell *et al.* 1999). Bunnefeld *et al.* (2006) found male lynx were more likely to be found in proximity to human habitation than females with kittens, which may explain both higher male mortality and predisposition to livestock killing by males of this species. Similarly, because felids are markedly sexually dimorphic, with males being larger than females, one might expect that male felids would have a greater size range of livestock available to them, leading to higher rates of predation. However, there are few data to directly support this suggestion.

Higher rates of livestock killing by male felids could also be due to larger male home ranges and wider dispersal distances and therefore an increased likelihood of encountering livestock (Linnell *et al.* 2001). This is certainly the case in lions, a species where subadult males often disperse widely. Patterson *et al.* (2004) found that 66% of stock-raiding lions in southern Kenya were subadult males, while Funston (2001a) records that of 100 livestock-raiding lions in ranch land adjacent to Kgalagadi Transfrontier National Park, 59 were subadult males. In this case, a further 31% of raiders were females with cubs, suggesting the possibility that increased nutritional requirements of lactating females or females feeding dependent young may predispose them to stock raiding.

In a study on the boundary of Etosha National Park, Namibia, Stander (1990) was able to distinguish between habitual and occasional stock-raiding lions. Around 70% of habitual raiders were adult males, while 54% of occasional raiders were subadult individuals. He found that 11 of 12 'occasional' raiders translocated did not reoffend; however, 'habitual' raiders quickly returned to their home ranges and livestock raiding unless they were translocated >100 km.

Infirmity and disability have often been cited as a possible reason for man-eating and livestock depredation and there are a compelling number of examples of maneaters and livestock raiders with old injuries or damaged teeth. This may or may not cause the animal difficulty in capturing or subduing wild prey. For example, over 50% of 17 man-eating tigers from Chitwan had old injuries caused by gunshot wounds, intraspecific fights, or had worn or broken teeth (Gurung *et al.* 2006a). One of the two 'man-eating lions of Tsavo' had deformities of skull and jaw (Peterhans and Gnoske 2001) and the maneater of Rufiji was a 3.5-year-old male with severe abcessation in its lower jaw (Baldus 2006). A number of the notorious man-eating tigers and leopards shot by Corbett were old and had been suffering from disability (Corbett 1946). In post-mortem examinations of Belize jaguars, Rabinowitz (1986) found that of 13 livestock raiders, 10 had old injuries (many caused by old gunshot wounds), while in a sample of 17 non-livestock killers all were healthy. Hoogesteijn *et al.* (1993) report similar trends in Venezuelan jaguars, where 10 of 19 cattle killers suffered from old gunshot injuries. The notion that old age or disability could be the cause of aberrant predation on humans or livestock is appealing. While this appears to be the case in some incidences, it is equally likely that old or infirm individuals lose competitive intraspecific interactions in prime habitats and are thus displaced from core areas and into marginal habitat, where they come into contact with people and livestock. Habitual livestock raiders may have been more likely to have been shot at in the past. Indeed, Patterson *et al.* (2003) found that the skulls of lions killed as problem animals around Tsavo National Park, Kenya, were no more likely to show signs of disability, or tooth breakage than those in museums, which were presumed to have been collected at random.

Solving and mitigating human–felid conflicts

Designing conservation policy and building management capacity

In developed countries where there is a strong commitment to carnivore conservation, compensation schemes have been established to mitigate losses, coupled with selective removal of individual problem predators. In many developing countries, lack of attention by governments or low and delayed compensation discourages livestock owners from participating in similar schemes where these exist. The only perceived solution is to kill the predator, and often many non-target animals are killed in the process. People who fear for their safety or perceive that they are at economic risk will not support conservation efforts.

Conservation practitioners need to design frameworks for implementing a comprehensive response to problem animals. Such frameworks often include creation of professional problem animal response teams and crafting of national policies and protocols for response to conflicts between humans and large felids. Response teams should be professional, well trained and adequately equipped, and have the experience and confidence to handle the range of situations that are encountered and fully understand the consequences of their actions. Many countries have no written policies for dealing with problem carnivores; situations are handled on a case-by-case basis with no guidelines and without the proper staff training, equipment, and resources. A national policy provides a set of protocols that clearly delineate appropriate actions for the various situations likely to be encountered and guide the process of determining the appropriate response options. While every situation is unique, a general protocol that empowers local wildlife managers to make decisions, makes them accountable, and provides response teams with guidelines to follow for choosing the appropriate course of action is a critical first step in dealing quickly and efficiently with problem animals.

Because local support for, or at least tolerance of, large felids is one of the key factors determining the fate of all wild populations, elimination of real or perceived threats is pivotal. The ability to retain a problem animal within the wild population will depend on the abilities of the response personnel, effective liaison with local communities, and the severity of the problem, as well as whether official interventions are adequately directed, resourced, and supported.

Lethal control

Predator reduction or elimination, either through state-sponsored predator control or unregulated killing, has historically been the method of choice for protecting livestock. In this way, large felid populations have been extirpated from large areas of North and South America, Africa, Europe, and Asia. While complete eradication of predators is no longer practised in many areas, reduction of predator numbers or limitation of population growth through lethal control methods may be compatible with management plans that include zonation of predator management.

Where complete eradication of predators is not desirable, lethal control of specific problem animals can be used to deal with livestock raiding or incidents of human death or injury. However, it is sometimes difficult to identify the individual animal responsible and indiscriminate lethal control may kill many non-target individuals or species, particularly if wire snares or poison are used. Control methods that specifically target problem animals, such as toxic collars on vulnerable livestock (Burns *et al.* 1996), use of dogs or skilled trackers to follow problem animals, and shooting or trapping of culprits when they return to recently made kills may provide more focused control of problem individuals (Odden *et al.* 2002; Woodroffe and Frank 2005).

Trophy hunting is sometimes advocated as a means to reduce predator densities or to target specific problem animals. Trophy hunting has been shown to temporarily reduce population densities in some felid species (e.g. puma, Lindzey *et al.* 1992; Eurasian lynx, Herfindal *et al.* 2005a; Stoner *et al.* 2006) depending on the sustained level of the hunting pressure.

Felid populations often recover quickly from population reductions, if there are nearby source populations (e.g. lions, Smuts 1978; Eurasian lynx, Stahl *et al.* [2001a]). Therefore, unless hunting is used on a consistent basis to limit population growth, the benefits to livestock owners are likely to be temporary. Furthermore, vacant home ranges are often filled by dispersing subadults, a demographic group that appear more likely to become problem animals. There is evidence that ranges made vacant may be divided between multiple subadult dispersers resulting (at least temporarily) in increased population density (Laing and Lindzey 1993). This may escalate rather than resolve conflicts. This being the case, sustainable sport hunting which seeks to maintain viable predator populations (and therefore hunting opportunities) may be incompatible with predator reduction (or eradication) in livestock raising areas, unless livestock owners are prepared to tolerate some losses.

In an alternative approach to reducing population density, Anderson (1981) describes proactive culling of potential livestock-killing lions in Hluhluwe-Umfolozi National Park, South Africa. It was found that animals dispersing from the Park and causing livestock losses on surrounding farms were predominantly subadult males (18–42 months old, 59 of 79 problem lions destroyed). Proactive culling of subadult males in the Park resulted in a greatly reduced incidence of livestock loss on surrounding farmland. However, highly targeted interventions such as this may only be practical in intensively managed situations, such as small-fenced reserves where animals are often regularly monitored and individually recognized.

Reducing felid population densities through hunting can provide temporary benefits to livestock owners through reduced depredation. However, use of hunting as a tool to selectively target specific problem animals appears to be a limited and temporary solution (Stahl *et al.* 2001a; Herfindal *et al.* 2005a). Furthermore, identification of the problem animal after the fact is not always straightforward and in situations where tourist hunting opportunities are sold commercially, bogus claims of problem animals need to be guarded against. Lethal control of specific problem animals may be better undertaken by professional problem animal control personnel, rather than the sport-hunting public; the former is likely to be a more targeted and timely response. This is particularly the case when dealing with man-eating predators (Cougar Management Guidelines Working Group 2005).

Where problem animals are part of a rare or endangered population, lethal control may not be the option favoured by conservationists. Nevertheless, particularly in cases where people's lives and livelihoods are at risk, removal of a problem animal by lethal control may be the most expedient and efficient option available. Unsuccessful or ill-conceived interventions or unwillingness on the part of managers or conservationists to deal effectively with problem animals may lead to a perception that the welfare of animals is valued over the lives and livelihoods of local people. In this case, support for conservation efforts may be undermined (Tilson and Nyhus 1998).

Translocation

Other options for removal of problem animals include translocation back into protected areas, to zoos or other protected sites. This intervention has been used in a number of areas, particularly with lions and leopards in Africa, but success is equivocal due to high post-release mortality, extensive movements, and homing behaviour of translocated animals. In Tsumkwe district, eastern Nambia, Stander *et al.* (1997a) translocated six livestock-raiding leopards a total of 12 times to sites 10–135 km from their ranges. All individuals returned within 2 days and killed livestock again in an average of 8.2 (range 1–20) months. Man-eating leopards translocated from Pune district, India, to other forests and reserves continued to reoffend at the new sites (Athreya 2006) and similar behaviour by leopards is described in east Africa (Hamilton 1976). Of 14 stock-raiding male lions (two territorial adults, two non-territorial adults, and 10 subadults) translocated distances of 50–85 km from their home ranges in the Kgalagadi Transfrontier Park, South Africa, all except three subadults returned and killed livestock again within 170 ± 37 days. Most were repeatedly translocated (between two and six times) but either returned or were destroyed while killing livestock (Funston 2001a). Similarly, 40% of the 54 lions translocated from around Etosha National Park returned to their home ranges (Stander 2005b) and for the same reason capture and repatriation of stock-raiding Asiatic lions failed to control livestock raiding around Gir Wildlife Sanctuary, India (Saberwal *et al.* 1990). Similar post-release behaviour has been observed in other large felids. Rabinowitz (1986) reports that two livestock-killing jaguars translocated to Cockscomb Basin National Park, Belize, soon left the park. One continued to kill livestock and was shot after 5 weeks; the other disappeared. An experimental translocation of 14 cougars in New Mexico resulted in 25% mortality within 3 months and, overall, nine animals died during the study period. Two mature males moved back ~477 km to their original ranges in 166 and 469 days. Eight cougars moved >80 km back towards their original range before establishing new home ranges (Ruth *et al.* 1998).

Translocations are most likely to be successful if problem animals are moved long distances across significant landscape barriers to areas with reasonable prey densities and few livestock or people. However, such areas are infrequently available and tend to have already established populations of the species being translocated. Translocation of non-territorial, subadult individuals, which would disperse anyway in a natural setting, appears to be the most likely to be successful. In pumas, translocation was most effective for animals aged 12–24 months and least effective for older individuals (Ruth *et al.* 1998). Based on the results of lion translocation operations in Namibia, Stander (1990) recommended that habitual problem animals be destroyed while occasional raiders, particularly subadults, could be rehabilitated by translocation. While in this case translocation activities reduced the number of lions that had to be destroyed, it required extensive record-keeping and expertise to capture, mark, and re-identify stock raiders. This kind of intensive management is rarely possible in developing countries. Although translocation of problem felids is in general unsuccessful as an intervention, livestock-killing cheetahs (*Acinonyx jubatus*) and problem Amur tigers have been successfully translocated. In Zimbabwe, in the mid-1990s, 21 cheetahs were moved from the south of the country ~450 km to Matusadona National Park in the north, and successfully established a population there (Purchase 1998; Purchase and Vhurumuku 2005). Of four Amur tigers captured and translocated after killing domestic livestock or attacking people in the Russian Far East (Fig. 6.3), two were successful in that they caused no further conflict with people, killed wild prey, and survived their first winter (Goodrich and Miquelle 2005).

As well as translocating problem animals to alternative wild habitats, some problem animals are occasionally moved into captive settings. Five of 37 man-eating tigers from Chitwan were captured and housed in zoos (Gurung *et al.* 2006a), but facilities to house large numbers of such animals are limited and quickly become saturated (e.g. for tigers in Nepal, India, Malaysia, and Indonesia; leopards in India and Sri Lanka; and pumas in North America). Furthermore, wild-caught felids do not generally settle well in captivity (Karanth and Gopal 2005). However, movement into captive settings might be justified for highly endangered species, especially if release of offspring into the wild could enhance reintroduction programmes at a later date.

In addition to limited success, translocation operations are costly and require high levels of technical expertise and logistical support. Funston (2002) estimates that repatriation of lions that broke out of Kgalagadi Transfrontier Park cost in excess of US$218 per lion. Karanth and Gopal (2005) note that tigers are often injured or disabled during translocation, often through damaging their teeth in steel cages that are not adequately designed (Fig. 6.4). On balance, translocation does not appear either particularly successful or a viable tool for management, particularly in countries with limited resources.

Protection of livestock and improved livestock husbandry

Historically, traditional livestock husbandry practices have sought to limit the availability of livestock to predators. However, in areas where predators have been extirpated this knowledge has often been lost.

Figure 6.3 An Amur tiger is released back into the wild by WCS and Inspection Tiger after being rescued from a poacher's snare. © J. Goodrich, Wildlife Conservation Society.

This is the case over much of western Europe, where traditional herding methods such as herd guarding by shepherds and guard dogs and use of secure enclosures have been largely abandoned. This has contributed to conflicts between farmers and re-emerging predator populations such as Eurasian lynx (Kaczensky 1999; Stahl *et al.* 2001b). Where livestock are grazed extensively without supervision or protection, losses to predators are liable to be high. For example, in Norway sheep graze unsupervised in forested habitats, where they suffer high levels of predation to lynx and other carnivores. In contrast, sheep are usually kept in fenced enclosures in Sweden and loss to predators is limited (Odden *et al.* 2002; Swenson and Andrén 2005). Similarly, free-roaming cattle in central and Amazonian Brazil and central Venezuela were subject to relatively high losses to jaguars and pumas (Polisar *et al.* 2003; Michalski *et al.* 2006a; Palmeira *et al.* 2008). By contrast, in southern Brazil, managed flocks and herds suffered lower levels of depredation than those that were free-roaming (Mazzoli *et al.* 2002) and around Jigme Singye Wangchuck National Park, Bhutan, livestock losses were high in villages that did not have stables or corrals, or where livestock was grazed in forest habitats (Wang and Macdonald 2006).

Protective bomas (corrals; Fig. 6.5) and herd guarding (often by young men and boys) are widely used across Africa to reduce livestock losses (Frank *et al.* 2005). Schiess-Meier *et al.* (2007) found that over a 3-year period only three of 2272 livestock-killing incidents (largely by lions and leopards) took place inside bomas in central Botswana. Similarly, Ogada *et al.* (2003) found that livestock losses were reduced with attentive herding, strongly built bomas, and the presence of guard dogs and people. Enclosures constructed of poles or wicker were superior to those built of thorny *Acacia* branches or wire mesh, because they reduced the chances of stock panicking at the sight of a predator and breaking out of the protective enclosure. Similar reductions in depredation resulted from

Figure 6.4 An African leopard trapped in a steel cage trap. Care needs to be taken in design of such traps to avoid damage to the felid's teeth and claws. In general, trapping and translocation of large felids is not a successful or viable management intervention. © A.J. Loveridge.

stone and wire mesh corrals built to protect sheep against snow leopards in Ladakh, India (Jackson *et al.* 2002). However, Kolowski and Holecamp (2006) found that on ranches adjacent to the Masai Mara Reserve, Kenya, leopards preferentially targeted stock in enclosures, particularly those built of poles, but that in general fences, guard dogs, and human activity deterred predators. Guard dogs can be used to protect small livestock (Linnell *et al.* 1996). In Nambia, farmers using livestock-guarding dogs reported a 73% reduction in livestock loss to cheetahs (Marker *et al.* 2005a). Similarly, llamas (*Lama glama*) and domestic buffalo (*Bubalus bubalis*) have been used to protect flocks and herds from pumas and jaguars (Crawshaw 2004; Hoogesteijn and Hoogesteijn 2008). Visual and acoustic repellents, such as bright lights (Sechele and Nzehengwa 2002) and noises (e.g. shotgun blasts, playbacks of bio-acoustic sounds such as sounds of barking dogs; Koehler *et al.* 1990), may be useful in deterring carnivores. However, carnivores habituate quickly to unusual stimuli and it is not clear whether such methods are effective in the long term. Deterrents such as electronic shock collars and taste aversion work reasonably well under experimental conditions, but have not been widely applied to manage livestock raiding (Frank and Woodroffe 2002).

Improved animal husbandry may also reduce livestock losses to predators. Free-roaming cattle in areas such as the Pantanal, Brazil, and Llanos of Venezuela are often in poor condition, increasing their vulnerability to jaguar predation (Quigley and Crawshaw 1992; Rabinowitz 2005). Vulnerable stock such as calves and sheep often require more intensive management. Peak losses to livestock owners often occur during the calving or lambing seasons (Polisar *et al.* 2003; Michalski *et al.* 2006a; Palmeira *et al.* 2008). In areas of South America where puma and jaguar predation is a problem, use of pastures away from forest fragments and close to human habitation is recommended for calving enclosures (Quigley and Crawshaw 1992; Palmeira *et al.* 2008). In addition, management of breeding in cattle herds aimed at limiting birthing to a short period may allow for intensive protection for part of the year and also act to swamp predation (Crawshaw 2004; Palmeira *et al.* 2008). In areas of Norway where sheep are vulnerable to lynx predation, Linnell *et al.* (1996) recommend shifting to cattle farming, as cattle are not preyed upon by lynx.

Losses in productivity often far exceed losses to predators. Cattle raised under semi-wild conditions have extremely low productivity, with pregnancy rates reaching only 40–50% and have very high

Figure 6.5 Livestock kept at night in a secure pole 'bomas' or corrals suffer lower levels of depredation than those left out to graze. © J.E. Hunt.

neonatal mortality and mortality to disease (Hoogesteijn 2002). Improved husbandry not only limits opportunities for predation by felids, but also reduces losses to disease, accident, and theft and therefore also the potential for losses due to poor management to be blamed on felid depredation (Hoogesteijn 2002).

Compensation

Damage compensation schemes provide one approach to mitigate damage caused by carnivores. Payment of compensation for damages has the effect of spreading the economic burden and financial risk between those who must live alongside carnivores and those who wish to see wildlife protected. Compensation for actual carnivore depredation (sometimes called post-damage compensation) is widely used in Norway, Sweden (Swenson and Andrén 2005), and India (Karanth and Gopal 2005). However, while it is clear that compensation may alleviate some of the costs and potentially promote tolerance of damage caused by predators, this approach suffers from some major disadvantages. Verification of damage (to eliminate fraudulent claims and over-estimates of loss) can be time-consuming and expensive and lead to animosity between conservation managers and livestock owners. Compensation without the requirement to demonstrate adoption of approved preventative measures (as is required in Sweden; Swenson and Andrén 2005) can lead to a perverse incentive to neglect livestock. There may also be incentives to wrongly attribute losses due to poor husbandry, accident, or disease to carnivores in order to claim compensation. Furthermore, compensation may act as an agricultural subsidy, providing incentives to expand farming activities with the potential for conversion of additional natural habitat (Nyhus *et al.* 2005). Long time delays between occurrence of damage and payment of compensation and bureaucratic difficulties in making claims may also be problematic. Saberwal (1990) found that 81% of villagers suffering losses to lions around Gir Wildlife Sanctuary did not bother to make claims, due to procedural bureaucracy. Similarly, failure to fully compensate for losses may hamper the effectiveness of compensation schemes. For instance, around Bhadra Tiger Reserve, where significant losses of livestock to tigers occurred, low levels of compensation were received (3% of estimated losses, with only 20% of claims successful; Madhusudan 2003). Corrupt or poorly managed compensation schemes may deepen antipathies and mistrust of managers and conservation efforts. Furthermore, to provide a sustainable long-term solution significant financial resources are required (Miquelle *et al.* 2005a). For instance, in Norway, compensation costs 5 million Euros/year (Linnell and Brøseth 2003). Schemes are rarely feasible in countries with limited conservation resources unless supported by international sponsorship or non-governmental organization (NGO) funding. Unfortunately, few evaluations of post-damage compensation schemes have been undertaken (Nyhus *et al.* 2003; Nyhus *et al.* 2005). It is not known if they are cost-effective and there is little evidence that compensation payments are effective in improving people's tolerance or reducing attempts to eradicate predators. In one of the few analyses available, Hazzah (2006) found that compensation did not improve attitudes to lions or prevent retaliatory killing on Mbirikani Group Ranch, Kenya.

Private insurance schemes to cover costs of carnivore damage have also been attempted; however, in many cases rural farmers are unwilling to cover the relatively expensive premiums required (Nyhus *et al.* 2005). This unwillingness may stem in part from concepts of ownership of wildlife and the perception that compensation is the responsibility of the government or wildlife management institution. However, success of insurance-based schemes has been claimed in improving tolerance of snow leopards in Pakistan, although this scheme was subsidized through ecotourism revenues (Hussain 2003b). Viability of insurance compensation schemes may rely on innovative financial mechanisms such as this.

An alternative to post-damage compensation is conservation performance payments or compensation in advance (Schwerdtner and Gruber 2007; Zabel and Holm-Müller 2008). Such payments are made based on the expectation that damage is likely to occur and is usually linked to successful conservation outcomes (e.g. reduced levels of carnivore mortality and protection of carnivore habitat). In the Sami reindeer (*Rangifer tarandus*) husbandry area of Sweden, conservation performance payments are made on the basis of the number of certified reproductions of lynx and wolverines (*Gulo gulo*). The level of payment is calculated from the expected losses due to each carnivore during its lifetime and amount to around US$29,000 per individual offspring (Zabel and Holm-Müller 2008). The advantage is that payments are made regardless of damage; therefore, livestock protection efforts are not distorted in that there is no disincentive to protect livestock from predators. Similarly, because payment is made on the basis of expected damage, there is no time-lag in payments, which eliminates uncertainty and promotes transparency and trust. Nevertheless, transaction costs remain high, as carnivore populations still need to be independently assessed and agreement needs to be reached over levels of payment. In common with post-damage compensation payments, viability of schemes requires significant funds over extended periods. Care needs to be taken that payments do not encourage new immigration into the area or subsidize increases or expansion in livestock ownership.

Alternative livelihoods, benefit sharing, and stakeholder participation

Antipathy towards carnivores may be a result of historical and cultural attitudes as well as based on experiences of loss or damage. These attitudes may be difficult to change or ameliorate. However, there are examples where stakeholder attitudes to both conservation and presence of carnivores have been improved by access to alternative revenue or livelihood choices, benefit sharing, and local participation.

Co-management and stakeholder participation appear to improve attitudes towards conservation efforts. In reindeer husbandry areas of Sweden, carnivore surveys jointly undertaken by reindeer owners and conservation managers reduced conflicts over estimates and encouraged transparency and co-operation (Swenson and Andrén 2005). Participation of local communities in prioritization, conservation planning, and implementation enhanced local protection of snow leopards as well as improving the feasibility and sense of ownership of livestock protection initiatives in Hemis National Park, India (Jackson and Wangchuk 2004). Likewise, inclusion of local landowners in conservation meetings in the Pantanal, Brazil, improved perceptions of jaguar conservation efforts and engendered ownership of processes and decision-making (Rabinowitz 2005).

Oli *et al.* (1994) found that villagers who benefited from ecotourism were more willing to tolerate the presence of snow leopards. There is a higher tolerance of this species in areas where people are less dependent on livestock and have access to alternative livelihoods (Mishra *et al.* 2003; Bagchi and

Mishra 2006). In Namibia, ranchers who raise livestock using ecologically sound management techniques gained access to niche markets and premium prices for 'predator friendly beef'. Benefits from initiatives such as this may increase tolerance for cheetahs and other predators on ranch land (Marker and Dickman 2005; Marker *et al.*, Chapter 15, this volume). On commercial ranches in Laikipia that benefit from ecotourism or have wealthy foreign investors, landowners are more tolerant of depredation on livestock than are ranches that gain nothing from tourism. Similarly, traditional pastoralists do not tolerate predators at all: lions dispersing into these areas from commercial ranches are promptly poisoned (L. Frank, unpublished data). However, residents of communal areas that hope to profit from future wildlife tourism profess greater tolerance than those who are not planning conservation-based enterprises (Romañach *et al.* 2007). Stander *et al.* (1997a) describe an innovative ecotourism venture that combined use of traditional skills, revenue generation, and ecotourism. In the Kaudom area of Namibia, local people experienced losses of livestock to leopards. A pilot ecotourism venture took advantage of the local Ju/Hoansi people's exceptional tracking and bush-craft skills. They were able to track and find leopards for paying tourists in 92.9% of attempts. The two villages engaged in leopard-tracking tourism earned N$10,005 (~US$2140) in a year. This revenue amounted to 12 times the value of livestock lost to leopard over the entire district.

However, in many cases claims that revenue from ecotourism or tourist hunting alleviate conflict and improve attitudes towards conservation are almost entirely untested (Walpole and Thouless 2005). Incentive schemes are often heavily subsidized by external bodies (Mishra *et al.* 2003; Miquelle *et al.* 2005a) and internalizing costs to ensure sustainability is often challenging with uncertain outcomes. Wildlife revenues appear to work well in incentivizing protection of habitat when landowners are reasonably wealthy and where owners have land tenure and control over resource access and use (Bond *et al.* 2004). However, there is less incentive where resource ownership is centrally or communally controlled, land tenure is less secure, or where livelihood alternatives are absent. Ecotourism revenues can be mismanaged, misappropriated, or subverted by elites within communities, which may undermine the effectiveness of initiatives (Walpole and Thouless 2005). Furthermore, revenues earned through ecotourism may be enthusiastically accepted in the short term, but in the long term may be invested in development or expansion of agriculture (Murombedzi 1999). The key here is that benefits from ecotourism schemes need to be clearly linked to the need for sustainable conservation of ecosystems and natural resources.

Zonation of land use

Not all landowners or communities will tolerate predators, and certain livestock management practices and human-dominated or urban landscapes are incompatible with the presence of predatory felids. Thus geographic differentiation of land-use areas, ranging from complete protection, through areas where predators are tolerated and population management and/or utilization occurs to areas where predators are not tolerated, seems a sensible way of focusing conservation efforts and resources, while at the same time recognizing the importance of people's livelihoods.

The goals of zoning conservation of carnivores are to conserve viable populations of predators and to minimize, or at least mitigate, conflicts with people (Linnell *et al.* 2005). Limiting the interface between people and large carnivores can serve to reduce the areas where conflict occurs. This allows prioritization of mitigation efforts and efficient use of conservation resources in these areas. Enforced zoning schemes can be used to prohibit certain human activities, though forms of resource use or extraction that do not result in habitat modification such as fishing, regulated logging, regulated harvest of ungulates, leisure activities, and tourism may be compatible with conservation of felids (particularly those that do not pose a threat to human life). In some cases, the presence of charismatic felids may increase the desirability of an area as a leisure or tourist destination. The conservation of smaller species may be compatible with land uses that exclude the presence of larger predators. For instance, small felids pose no threat to livestock and coexistence may be possible if no habitat modification occurs.

Many southern and east African protected area networks have been designed to incorporate areas of strict protection (national parks or reserves), and areas where wildlife can be sustainably utilized (often through trophy hunting). In addition, community-managed areas (known as game management or wildlife management areas) have been established in some countries (e.g. Botswana, Zambia, Zimbabwe, and Namibia), where wildlife is tolerated and utilized for the benefit of communities (Lewis and Alpert 1997; Hulme and Murphree 2001). Utilization areas are often set up as buffer zones between national parks and agricultural areas. Zonation in other regions is often less clear-cut. For instance, many areas of Europe are already inhabited by people. In these areas, initiatives to prioritize conservation of predators have not always met with success and have in some instances been abandoned (Linnell *et al.* 2005).

Establishment of reserves without adequate separation (e.g. with fences or natural barriers such as large rivers) of people and wildlife has led to intense conflict on the peripheries. Clearly, management of conflict in areas where predators and people are expected to coexist makes zonation of predator conservation a challenge. In areas where people and wildlife share the same landscape, it would appear that coexistence is most likely to be successful under circumstances where disadvantages due to conflicts are outweighed by benefits of tolerating wildlife.

Some conservation practitioners have advocated separation of people and predators as the most viable solution, particularly in areas where human densities are high. Nyhus and Tilson (2004b) reported that Way-Kambas Reserve in Sumatra successfully protected tigers and eliminated conflicts with the surrounding human communities. This was largely because the reserve excluded people and reserve borders are largely formed by rivers that created boundaries seldom transgressed by either tigers or people. Karanth and Madhusudan (1997) and Karanth and Gopal (2005) support the voluntary relocation of people from the vicinity of tiger reserves to reduce conflicts and protect tiger populations. Such voluntary relocations to make space for expanding tiger populations have met with success in the Terai Arc Landscape, Nepal (Seidensticker *et al.*, Chapter 12, this volume).

In the western United States, zonation provides the framework for management of puma populations (Laundre and Clark 2003; Stoner *et al.* 2006). Here, subpopulations within the meta-population to be managed are designated as 'source' or 'sink' populations. Sport-hunting opportunities are provided in the sink areas, while source areas are protected from utilization. Management using a landscape of sources and sinks to provide both adequate protection for the species and opportunity for utilization may prove to be an effective strategy for conserving wide-ranging predators. However, understanding of a species' behavioural ecology is crucial. Predators, particularly large felids, are wide ranging and minimum protected-area sizes need to take this into account if edge effects are not to have negative impacts on population viability (Woodroffe and Ginsburg 1998; Loveridge *et al.*, Chapter 11, this volume).

Exploitation of felid populations through trade and utilization

Human cultures have always prized products derived from large carnivores. Traditional cultures and fashionable society value felid furs to advertize personal status and trade in furs has had significant impacts on felid populations. Felid products are used to 'cure' illness, to ward off misadventure, and to bring good luck. For instance, tiger bone as an ingredient of Traditional Asian Medicines (TAM) is thought to cure rheumatism, weakness, and paralysis (Mills and Jackson 1994). In south-western Nigeria, serval (*Leptailurus serval*) flesh and tongue are believed to cure leprosy and rheumatism, while leopard skin is used to treat snake bites (Sodeinde and Soewa 1999). There are recent reports about a growing illegal market for tiger meat as an exotic cuisine (Damania *et al.* 2008). Utilization of felids through trophy or sport hunting can provide motivations for conserving habitats and wildlife populations. We discuss the exploitation of felid populations through trophy hunting, fur trade, and for traditional medicines, and the potential threats and impacts these have on conservation of these populations.

Trophy hunting

Hunting large felids for sport has occurred for thousands of years, with records of lion hunts going back to the time of Pharaoh Amenhoteb III in 1400 BC and medieval tapestries depicting European nobles hunting lions and other felids (Guggisberg 1962). Indian and Nepalese royalty hunted extensively and records show the astounding number of tigers they killed for sport. The Maharajah of Surgujah is reputed to have shot 1150 tigers in his lifetime, followed closely by the Mararaj of Gwalior and his guests, who accounted for 900. British civil servants and military officers on leave hunted frequently, with some individuals credited with killing hundreds of tigers. The British Royal Family did their bit when visiting India and Nepal, with King George V killing 39 tigers in 11 days in 1911 and the Prince of Wales shooting seven over a 4-day period in 1922 (Sankhala 1978; Mountford 1981). During the 1960s, demand for tiger-hunting trophies accelerated as tiger populations dwindled and as it became apparent that hunting tigers would become officially prohibited (Sunquist and Sunquist 2002).

In colonial Africa, carnivores were viewed as vermin and killed wherever possible (Fig. 6.6); this hunting was, at least partially, motivated by 'the thrill of the chase' particularly among the colonial elite hunting in the more remote areas of east and southern Africa. Early hunting expeditions to the Serengeti typically killed large numbers (sometimes hundreds) of lions (Turner 1987), often in the mistaken belief that removal of carnivores would 'protect' ungulate populations, thereby providing improved opportunities for hunting. Along with loss of habitat, decline in prey populations and predator eradication initiatives, trophy hunting may have contributed to historical declines in some large felid populations. Indeed, I.R. Pocock (1939: p. 219) wrote: 'In all parts of the world occupied by Europeans where lions occur, the disappearance of lions is merely a question of time.'

Historically, hunting of carnivores was often characterized by lack of any controls or regulation (at least for the elite) and in some senses was indistinguishable, in motivation and effect, from predator eradication activities that were prevalent at the time. Contemporary trophy hunting (also known as sport, tourist, or recreational hunting) tends to be more restrained and harvests are often strictly regulated where competent management authorities are in place. In addition, trophy hunters are often concerned about and promote conservation values (Lindsey *et al.* 2006). Felids are popular quarry species where trophy hunting is legal, and large charismatic species attract large trophy fees (usually collected by government or management authorities). Trophy hunters pay premium prices to hunting guides or operators to hunt large felids. For example, puma hunts in New Mexico cost US$2000–3000 per hunt

Figure 6.6 Early settlers in east and southern Africa eliminated many of the larger predators from the land they colonized. (Photograph courtesy of National Archives of Zimbabwe.)

(Logan *et al.* 2004) and before hunting was officially banned, hunters paid US$11,200 to hunt a snow leopard in Mongolia (Anonymous 1989). In Botswana, in 2006, tourist hunters paid US$140,000 to hunt a male lion, with US$100,000 of that going to the community in the district it was killed (J. Rann, personal communication), although lion hunting has since been banned in Botswana (Fig. 6.7).

Revenues generated from trophy hunting can benefit local communities (if explicitly channelled) and provide employment for specialists that provide hunting-related services (e.g. hunting guides and taxidermists). This can provide justification and support for conservation of populations of felids and other species at local and national scales. Revenues can also be used to manage populations, protect habitats, or to compensate livestock owners for losses, thereby improving levels of tolerance and acceptance of predators (Loveridge *et al.* 2007b). Furthermore, habitats are often set aside and protected for trophy-hunting activities. For instance, in addition to national parks and protected areas, 1.4 million km^2 are used and, to varying degrees, protected for trophy hunting in sub-Saharan Africa (Lindsey *et al.* 2007). Crawshaw (2004) argues that allowing trophy hunting of problem jaguars may provide incentives for landowners to manage habitats to encourage self-sustaining populations of this species.

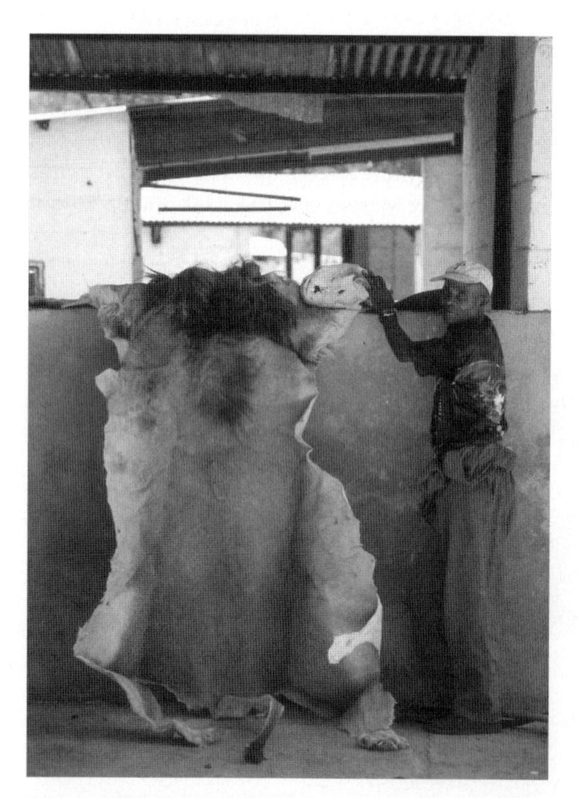

Figure 6.7 The skin and skull of a trophy-hunted lion is displayed in a hunter's skinning shed. © J.E. Hunt.

However, trophy hunting can have demographic consequences for felid populations. Adult males are often targeted and high male turnover can result in high levels of infanticide, which may in turn reduce reproductive success within the population. If severe, this can lead to population decline (Greene *et al.* 1998; Whitman *et al.* 2004). Gross (2008) found that heavy hunting of male pumas in Washington, DC, United States, led to severe perturbation including few resident adult males, influx of immigrants from surrounding home ranges (leading to widely varying puma densities), increased infanticide, and more puma–human conflicts. However, felid populations are relatively resilient to moderate levels of harvest (e.g. 10% of adult male lions; Greene *et al.* 1998), recovering quickly if immigration from surrounding habitat or nearby populations is possible. Smuts (1978) found that populations of African lions in Kruger National Park, South Africa, recovered within 17 months (largely through immigration) in areas that had been experimentally depopulated by culling of 129 lions from three experimental areas. However, the social disruption caused by removals persisted beyond the 17-month experimental period (Smuts 1978). Relatively rapid recoveries occurred in lion population in Hwange National Park, after trophy hunting had been suspended in surrounding hunting concessions (Loveridge *et al.*, Chapter 11, this volume). However, Lindzey *et al.* (1992) found that recovery of an experimentally manipulated puma population did not occur within 2 years when 28% of harvestable age animals were removed from the population.

Trophy hunters and hunting operators (outfitters) do not always behave ethically or responsibly. There are often strong financial incentives to overexploit hunted resources, and hunting activities often occur in remote areas with little or no official oversight. For example, Spong *et al.* (2000a) found that in a sample of trophy-hunted leopards (n = 77) shot in Selous Game Reserve, Tanzania, 28.6% were female despite hunting quotas for leopards being officially restricted to males. In this case, poor supervision of hunting activities allowed hunters to shoot females and potentially overexploit this resource. Effective regulation of hunting activities to ensure compliance with hunting quotas and regulations is often required to ensure that trophy hunting of felids and other species remains sustainable.

Despite examples of overexploitation (Spong *et al.* 2000a; Yamazaki 1996; Loveridge *et al.* 2007c), trophy hunting appears to have relatively low impacts on felid populations if there are large, protected source populations and if management practices and quota setting are informed and regulated by carefully monitoring the hunted and source populations. Protection of habitat and the potential for benefit streams for local communities and economies might outweigh behavioural and demographic impacts on felid populations, although clearly these should not be ignored if the overall goal is to sustain the target populations over the long term.

Exploitation of felids for their fur

Felid furs are often beautifully patterned and humans have, for millennia, used furs for manufacturing clothing, adornments, and household decoration. The wearing of felid furs often confers status and is a symbol of the wealth and power of the owner. Masai *morans* (young warriors) who have killed a lion wear lion mane headdresses to signify their courage. The fashion of wearing fur coats made of spotted cat pelts gained immense popularity among the wealthy of western Europe and the United States in the 1960s and 1970s. The resulting overexploitation of some wild populations of spotted cats prompted the enactment of the Convention on International Trade in Endangered Species of Wild Flora and Fauna (CITES) in 1975. Recently, tiger and leopard skins and clothing made from skins have become popular among the newly wealthy in China and Tibet (EIA 2004; EIA-WPSI 2006), driving an unprecedented onslaught of poaching in source countries, especially India and Nepal.

Demand for felid skins has driven widespread exploitation and sometimes extirpation of felid populations. One famous example is the extraction of over 80,000 ocelot and 15,000 jaguar skins from the Brazilian Amazon in the early to mid-1960s for the fashion fur trade (Smith 1976). This massive export of skins prompted the Brazilian government to ban export of wildcat skins in 1967. Myers (1973) estimates that the fur trade was worth around US$30 million in the late 1960s and early 1970s. The demand for spotted cat skins was so high that, in Neotropical South America, hunting for cat skins became a way of life for many local people in remote areas (Payan and Trujillo 2006). Hunting provided a lucrative livelihood and extraction of forest products was well organized through intermediaries and middlemen. During this period, in Columbia, a *Tigrillada* (cat hunter) was typically paid around US$130–350 for a jaguar skin, which became worth US$520 to a middleman once it reached Bogotá, before being eventually sold to a New York furrier for US$2500 and made into a coat worth US$20,000 in a New York Boutique (Nowell and Jackson 1996; Payan and Trujillo 2006).

The quantities of spotted cat skins extracted from South American range states from the 1960s to 1980s are staggering. In 1970, 140,000 ocelot and margay (*Leopardus weidii*) skins were traded in US markets, while 84,493 oncilla (*L. tigrinus*) skins were traded in 1983 alone (McMahan 1986). Between 1976 and 1979, 341,588 Geoffroy's cat (*Oncifelis geoffroyi*) and 78,239 Pampas cat (*O. colocolo*) skins were exported from Buenos Aires (Mares and Ojeda 1984). Old World spotted cats were also popular with furriers, including the leopard, snow leopard, clouded leopard, tiger, and cheetah (Nowell and Jackson 1996). Nine thousand, one hundred and sixty-two leopard skins were exported from Uganda between 1924 and 1960, leading to special protection of the species in that country in 1960 (Treves and Naughton-Treves 1999). Seventeen thousand, four hundred and ninety leopard skins were exported to the United States in 1968 and 1969 and demand for leopard skins was estimated at around 50,000 skins per year in the early 1970s. Furthermore, because leopards were often hunted on an unmanaged and often illegal basis by tribesmen or subsistence hunters (Fig. 6.8), many skins were rejected by middleman traders as being damaged or poorly cured, implying that losses to wild populations were even higher than figures suggest (Myers 1976).

Figure 6.8 Demand for leopard skins led to widespread hunting of the species, with at least some of the illegal trade in skins supplied through a network of illegal trappers and hunters. (Photograph courtesy of National Archives of Zimbabwe.)

Concern over the magnitude of trade in spotted cats and its potential impact on wild populations (Koford 1973; Myers 1973, 1976; Theile 2003) led to moratoria on trade in some species (Nowell and Jackson 1996). Protective legislation was introduced in many range states and CITES legislation restricting trade in many of the spotted cats, particularly the larger species, was enacted in 1975. The European Union (EU) banned all imports of Latin American cats in 1986 and many of the smaller Latin American cats were placed on CITES Appendix I between 1989 and 1992. Although trade in spotted cats still exists (1000 spotted cat skins were seized in Argentina in 1990; Anonymous 1992), trade restrictions have led to a steady decline in trade (Figs. 6.9 and 6.10). For instance, 30,000 margay skins were exported in 1977 (most from Paraguay to western Europe) but trade in this species was reduced to 138 skins by 1985 (McMahan 1986). Similarly, in 1969, at the peak of the trade in spotted cats, it is estimated that 61,000 leopards were killed for their skins, with no more than 6000 killed by 1988 (Martin and de Meulenaer 1988).

However, the fur trade shifted to other species as substitutes, such as the bobcat (*Lynx rufus*), and leopard cat (*Prionailurus bengalensis*), with both demand and prices increasing dramatically in the late 1970s (Fuller *et al.* 1985; McMahan 1986). As a result of high demand, prices for bobcat pelts rose from US$15 a pelt in 1973 to a high of US$164 a pelt in 1978 (Fuller *et al.* 1985), with resultant harvests reducing populations of this species to dangerously low numbers (Hornocker 2008). The modern fashion fur trade relies on pelts from the Canadian and Eurasian lynx (*L. canadensis* and *L. lynx*), bobcat, and the leopard cat (Nowell and Jackson 1996). Canadian lynx furs have been exploited since the 1700s (Elton and Nicholson 1942). Although trapping is currently well regulated, populations may be unsustainably trapped during cyclic population declines (Poole 1994; Slough and Mowat 1996). Bobcat populations currently appear stable, suggesting that fur harvests are biologically sustainable; however, very high volumes of leopard cat skins exported by China (over 200,000 skins in 1987) have been cause for concern (Nowell and Jackson 1996; Sunquist and Sunquist 2002).

While markets in Europe and the United States have declined due to strict legislation and negative sentiments about use of furs in fashion items, other markets have opened up, particularly in a more prosperous eastern Europe and an increasingly wealthy China. The illegal trade in tiger and Asiatic leopard (*P. pardus fusca*) skins appears to have been increasing throughout Indochina since the 1990s (Rabinowitz 1999; EIA 2004; Nowell and Xu 2007). Poaching of wildlife for the fur trade appears to be well organized and lucrative. Increasing demand for skins is thought to be driven by wealthy Han Chinese who value felid

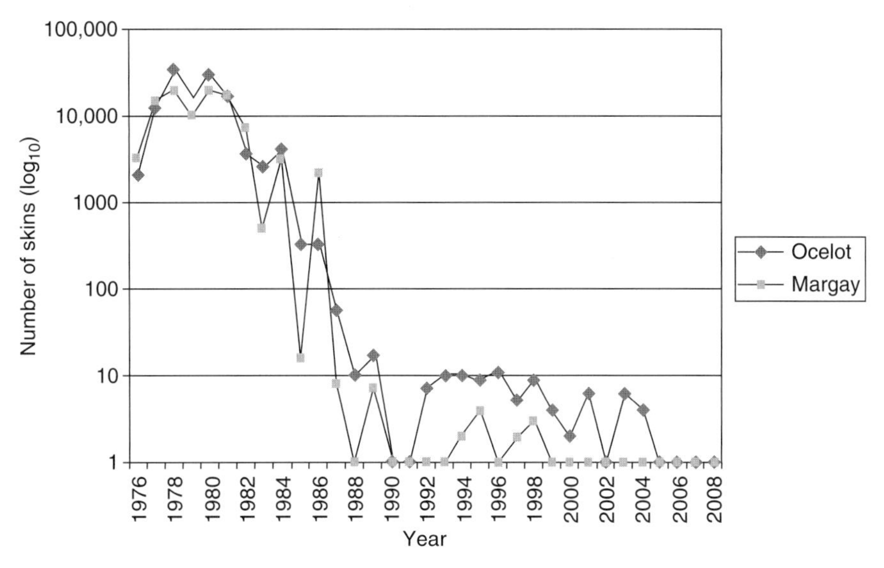

Figure 6.9 Trends in international trade in skins of two species of Neotropical spotted cat (ocelot and margay). Reduction in demand and imposition of trade bans by the EU and CITES in 1986 and 1989, respectively has limited the trade. These data are likely to be incomplete, as only official exports are recorded, but, unless seized, illegal ones are not. (Source of data: CITES–UNEP.)

Figure 6.10 Jaguar skin from the Venezuelan Chaco. © A. Taber.

(particularly tiger) skins for the prestige of owning an expensive and exotic item and because they are thought to bring good luck. There also appears to be a market in curios purchased by Western tourists (EIA 2004). Use of tiger and leopard skins in traditional Tibetan '*chubas*' (ceremonial gowns) became increasingly fashionable through the 1990s as a result of growing wealth among urban Tibetans. Skins are also used at traditional ceremonies such as weddings, and a ceremonial tent made of 108 whole tiger skins was seen in 2006 at the Litang Horse Festival, in the Chinese Province of Sichuan (EIA-WPSI 2006).

However, concerns raised by environmental groups have seen this use decline markedly (EIA-WPSI 2006;

Banks and Wright 2007). Symbolic burning of tiger and leopard skin robes occurred in Tibet in 2006 following condemnation of use of wildlife products by the Dalai Lama. This may have resulted in both greatly reduced demand and prices for these items (Huggler 2006), although Chinese officials, who discourage the Dalai Lama's influence, ordered Tibetans to continue to wear their traditional apparel, including clothing adorned with tiger and leopard skin. Trade in tiger and leopard skins remains highly lucrative. A trader in India is able to sell a tiger skin for US$1500 (having paid a local poacher in the region of US$15), while the same skin sells in China for US$16,000. A consignment of 31 tiger, 581 leopard, and 788 otter skins, seized in Tibet in 2003 while being smuggled from India to China, was thought to be worth an astonishing US$1.2 million (EIA 2004; EIA-WPSI 2006).

Though legislation is in place to protect tigers and most other Asian cats, wildlife agencies frequently lack the resources for effective enforcement of the laws. Poaching penalties are harsh, but the likelihood of apprehension remains low and that of conviction even lower (Damania *et al.* 2008). Hunting of tigers for trade is widespread and trade is thought to be more prevalent now than in previous decades. This may exacerbate declines due to logging, habitat loss, and expansion of human populations (Rabinowitz 1999). The 783 tiger and 2766 leopard skins seized in India between1994 and 2006 are thought to represent a fraction of skins on the illegal market, suggesting that the extent of illegal poaching of felids for their skins may be extensive. EIA-WPSI (2006) consider the illegal trade in tiger skins between India, Nepal, and China to be a substantial threat that may have driven recent declines in Indian tiger populations. Having depleted many tiger populations, commercial poachers have turned to other Asian big cats: Asian lions, leopards, snow leopards, and clouded leopards. Having poached tiger populations in Cambodia, Myanmar, and Thailand, commercial poachers have intensified and focused their efforts on Malaysia and now tigers and other wildlife there are under heavy pressure (Damania *et al.* 2008).

Poaching is usually undertaken by skilled local hunters using reusable steel traps, cable snares, or poisoned bait. The cost of poaching a wild tiger is small, unlikely to exceed US$ 100–200, even taking into account opportunity costs of time and expected penalties. A large number of poachers operate under near open-access conditions. The carcass is sold to traders who capture the bulk of the profit by smuggling tiger parts into urban centres of East Asia. All parts of the tiger can be traded with a total retail value in the region of US $10,000–70,000 (Damania *et al.* 2008).

In India, wildlife poaching and related crime is well organized, with extensive networks of suppliers and buyers (Kumar and Wright 1999). The highly profitable nature of the trade and lack of serious policing and judicial disincentives in many range states do little to curtail the trade in Asian cat skins. Although countries like China appear to be enforcing wildlife trade laws more vigorously (Nowell and Xu 2007), fighting wildlife crime is not a major priority in many South-east Asian range states. Agencies charged with environmental protection are often under-resourced and clear policies and political will are often non-existent. Punishment for trading in wildlife products provides little deterrent, and fines for possession of illegal skins are paltry in relation to the value of a single skin on the illegal market. Traders can adjust their margins to accommodate change in judicial pressure, competition, or demand at the retail end of the market and thus frustrate initiatives to diminish incentives to poach (Bulte and Damania 2005). Furthermore, prosecution of wildlife criminals is frequently delayed and rates of conviction are low. For instance, in India, which has a well developed institutional structure for conservation, only 14 convictions have been achieved in 748 cases of confiscation of wildcat skins (EIA 2004). The trade also spans a number of countries (Fig. 6.11), with sources in India, Indonesia, Laos, Thailand, Myanmar, and Vietnam and markets on the Chinese

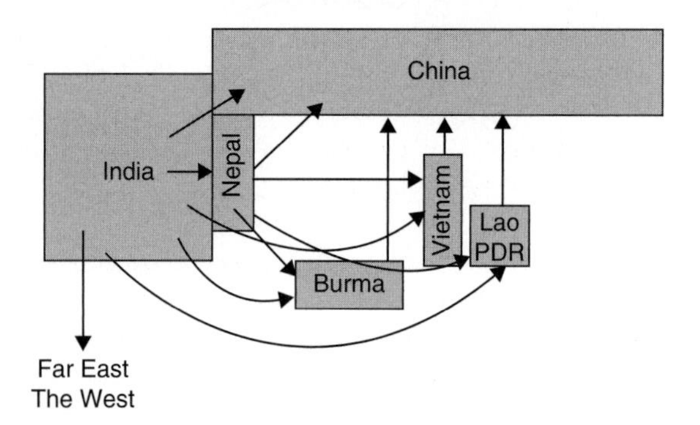

Figure 6.11 Illegal trade routes for tiger and leopard skins in Asia. (From Baker *et al.* 2006.)

Mainland, Hong Kong, Taiwan, South Korea, and Japan (Li and Wang 1999; Rabinowitz 1999; Baker *et al.* 2006; EIA-WPSI 2006). Control of this illegal trade requires regional and international commitment and transnational cooperation to tackle transboundary trade.

Use of felids in TAM

Conservationists have been increasingly concerned by trade in wildlife products used in Traditional Asian Medicines (TAM) and remedies. Trade in tiger parts and derivatives has been banned around the world for more than a decade and law-abiding practitioners of Traditional Chinese Medicine (TCM) now use alternatives. Yet, the illegal trade continues. The World Federation of Chinese Medicine Societies (WCMS) has declared that tiger parts are not necessary for human health care and that alternatives are plentiful, affordable, and effective. Poachers continue to kill tigers to satisfy a stubborn demand for tiger bones to make health tonics. While loss of habitat may be the major long-term cause of declines of species such as the tiger, across East and South Asia, poaching of tigers and other felids is the important short-term threat to survival of populations (Mills and Jackson 1994; Hemley and Mills 1999; Rabinowitz 1999; Wingard and Zahler 2006; Damania *et al.* 2008). Bones of other felids (clouded leopards, leopards, snow leopards, and lions) are also used as substitutes for tiger bone and the trade may therefore also impact populations of these species (Wingard and Zahler 2006). The major Chinese investment and engagement in Africa may present a new threat to lion populations, as it is nearly impossible to distinguish between lion and tiger bones.

Trade in tiger parts, particularly bone, reached epic proportions in the 1980s and 1990s, decimating wild populations. For example, South Korean customs records from before South Korea acceded to CITES show that 8951 kg of tiger bone were imported from 1970 to 1993, with around half the bone derived from Indonesia. There has since been a concerted effort to control trade in tiger bone. Efforts have focused on international legislation, raising public awareness, dialogue with TAM specialists and users, and a commitment among TAM specialists to search for alternative medicinal products to replace the use of tiger bone (Hemley and Mills 1999). Domestic and international trade in tiger parts is banned in most countries and some countries like South Korea have virtually eliminated the trade through vigorous prosecution. However, the residual trade has proved much harder to control and eliminate. In regions where enforcement and prosecution have been indifferent, tiger products are still obtainable. In a survey of Yunnan Province, China, Li and Wang (1999) found felid parts to be widely available and similar surveys in Sumatra, Indonesia, found little evidence of declining trade in tiger products, despite extensive efforts to raise awareness (Shepherd and Magnus 2004; Ng and Nemora 2007). However, a recent survey of 518 TAM outlets in China found that only 2.5% carried tiger bone products, suggesting that use may be declining in Mainland China (Nowell and Xu 2007). However, because few data are available from before the 1993 ban, quantification of declining trade is difficult.

There are a number of captive bred populations of tigers, leading to the suggestion that products from 'farmed' tigers could be sold commercially to replace products from the wild. However, the consensus among conservationists is that trade in farmed tigers is likely to stimulate demand and provide the means to launder tiger parts derived from poaching, thereby reversing recent successes in reducing the magnitude of the trade (Hemley and Mills 1999; Gratwicke *et al.* 2007; Nowell and Xu 2007). A recent survey measured attitude towards consumption of tigers in six Chinese urban areas. Of 1880 respondents, 43% had consumed some product alleged to contain tiger products. The results indicated that while urban Chinese people are generally supportive of tiger conservation, there is a huge residual demand for tiger products which could potentially be filled by supplying parts from both wild and farmed tigers if the ban on trade in tiger parts is lifted in China (Gratwicke *et al.* 2008b).

Management of trade and exploitation

The Convention on Biological Diversity (CBD) explicitly recognizes sustainable utilization as a component of conservation of biodiversity and ecosystems (Convention on Biological Diversity 2003). The key to sustainable utilization is that use is responsibly managed and that populations are well monitored to ensure that overexploitation does not occur. However, sustainable utilization is often challenging in situations where there is an inequity in distribution of wealth, weak environmental controls, apathetic national and international legislation, and endemic corruption. Smith *et al.* (2003) found that poor governance, corruption, institutional failure, and social and economic upheaval reduce the likelihood of successful conservation of biodiversity. Overexploitation of wildlife resources often goes hand in hand with uncontrolled habitat conversion (Rabinowitz 1999; Nyhus and Tilson 2004a) and poaching of tigers is often linked to existing human–tiger conflicts, with illegal trade being a by-product of conflict (Kumar and Wright 1999; Johnson *et al.* 2006a). Household economic and food insecurity has been shown to predispose people to illegal use of wildlife. In Zambia, improvement in food security reduced poaching and improved commitment to conservation initiatives (Lewis and Jackson 2005).

Sustainable use of felid populations is clearly possible, as evidenced by a sustainable fur trade in North America (Nowell and Jackson 1996). Key to this has been clear policies, effectively applied legislation, and population monitoring and research. These conditions do not exist in the developing world. Trophy hunting has the potential to be highly sustainable given its capacity for high revenue generation and protection of habitats and populations, provided institutional mechanisms are in place to assure its sustainability. It is also governed by international trade agreements, such as CITES, and through government conservation institutions. Also, sport hunters as a demographic grouping are often active in promoting the conservation of the ecosystems they utilize.

Trade in cat furs and the medicinal trade are less easy to control through international trade restrictions. However, systematic commercial hunting for spotted cat skins has seen a steady decline since the imposition of trade restrictions by the United States, EU, and CITES, indicating that international action can have a positive impact. In addition, consumer awareness and sensibilities surrounding use of cat skins greatly reduced demand and aided the demise of spotted cat trade in the Southern Hemisphere. This example may provide a template for controlling overexploitation of other felid populations, such as trade in Asian cat skins. If the downturn in demand seen for spotted cat skins is to be experienced for the Asian cats, a combination of international and national legislations, consumer awareness, and education and engagement with the TAM industry is a first step in seeking to reduce trade in tiger and other felid bones (Hemley and Mills 1999). The importance of both strictly applied legislation and public education is borne out by the decline in use of felid parts in China, where a concerted effort has been made to eliminate trade and raise awareness (Nowell and Xu 2007).

Control of unsustainable trade in felids appears to depend on strong international controls, the willingness and capacity of governments to implement and police conservation policy and provide disincentives to poaching and illegal trade, and the action of local and international conservation lobbyists and pressure groups. Consumer awareness and education and engagement of stakeholders also appear crucial in reducing or at least curtailing demand for felid products.

Conclusions

The world's ever-growing human population has dramatically degraded most natural ecosystems, causing an extinction event unprecedented in the last 65 million years. Large carnivores are the first species to disappear under the onslaught of human populations (Woodroffe 2001), having been eradicated first from Europe, then decimated by Europeans in North America and more recently in Asia and Africa. Today they persist only in large protected areas and in diminishing numbers in the least human-dominated of natural ecosystems. Barring a sudden and unlikely reversal in human population trend, the march towards expanding economies, or attitudes towards the natural world, persistence of the larger felid species into the next century is a challenge we must all face. The outlook is dire for those who cherish wild cats. Smaller felids that do not conflict with man and are not commercially overexploited are likely to persist only where their habitats are not deforested, overgrazed, or converted to agriculture.

The threats posed by humans to large felids are soluble if people sharing their landscapes have either economic or ethical incentives to preserve them. Like the ban on ivory trading that saved African elephant populations to date, the successful ban on trade in tropical spotted cat skins shows that conservationists can have a significant influence on international measures to preserve economically important species in the developing world. Moreover, demand was reduced when wearing spotted cat furs became socially unacceptable, suggesting that moral pressures on society may be equally effective. However, the continuing decline of two of the most iconic felid species, the tiger (Chapron *et al.* 2008a; Damania *et al.* 2008) and lion (IUCN 2006b), provide a lens through which to view the challenges facing conservationists when underlying conservation threats are not readily subject to legal or social measures.

Although conflict with people, habitat loss, and reduced availability of wild prey are a significant problem for some tiger populations, poaching to supply markets for traditional medicine and skins in Asia poses the greatest threat to the species (Dinerstein *et al.* 2007; Chapron *et al.* 2008a). It is unlikely that the trend will reverse in the absence of effective legal measures to control trade and the sort of moral pressure that made wearing of spotted cat furs unpopular in the West. Intransigent superstition and the dictates of fashion, increasing prosperity, and cultural inertia in Asia, ensure an increasing and unprecedented demand by consumers for skins and bones. Meanwhile, pervasive corruption at all levels defeats legal sanctions. Demand for tiger products can only be reduced through education about alternatives and development of a sense of personal responsibility for nature and wildlife and improved practice in wildlife conservation. Unless consumers can be convinced that wildlife is more valuable roaming in forests than dismantled in apothecaries and upscale homes, tigers and many other species seem doomed. However, it seems unlikely that such a shift in attitude can be imposed by Western conservationists. It will only accompany change in deeply held cultural values, a slow process compared to the short time left for some wild tiger populations.

In contrast, African lions are declining largely due to continuing poverty and cultural modernization in Africa. At a time when increasing numbers of impoverished people and their livestock are devastating semiarid ecosystems, pastoralist people are also abandoning traditional and effective methods of livestock protection in favour of agricultural pesticides that allow them to eliminate entire populations of predators at very low cost. Again, a major factor is ineffective law enforcement due to lack of resources, lack of interest, and pervasive corruption. However, while tigers are poached because their parts are so valuable, lions are speared, trapped, and poisoned largely because they lack any economic value to the humans that share their habitat. While both tourism and sport hunting are highly lucrative and that money could potentially be used to offset the costs of livestock depredation, very little profit from tourism and hunting reaches the people whose precious livestock are killed by lions. Thus lions and other wild animals are nothing but an expensive nuisance to rural people, with the inevitable result that they are being eradicated. Unless pastoralists and other rural Africans can earn significant income from wildlife, there is no reason for them to conserve it. International fascination with these charismatic animals gives them potentially great economic value. However, this value is largely inaccessible to the rural people who bear the brunt of living alongside large carnivores. Big cats will survive in the wild only if they become more valuable alive than dead to the people who share their landscapes.

Conflict between people and large carnivores has been a consistent theme throughout human history. As human populations burgeon, increasing pressure on remaining populations of carnivores will occur. Our understanding of the processes and patterns of use and conflict may determine our ability to mitigate overexploitation and conflicts and ensure survival of predatory carnivores in a world increasingly dominated by the needs and aspirations of the human species.

CHAPTER 7

Many ways of skinning a cat: tools and techniques for studying wild felids

K. Ullas Karanth, Paul Funston, and Eric Sanderson

Researchers radio-collaring a sedated tigress after it was free-darted using the 'Beat method' in Nagarahole, India. © Fiona Sunquist.

Introduction

The 36 species of extant wild felids (Macdonald *et al.*, Chapter 1, this volume) are the subject of great scientific interest because, in general, little is known about many of them. Wild cats also preoccupy conservationists because most are threatened by anthropogenic pressures such as hunting, prey depletion, and habitat destruction (Nowell and Jackson 1996; Sunquist and Sunquist 2002). In short, there are pressing motives to understand wild cat ecology and behaviour, and to apply that knowledge to conservation. However, this knowledge gap remains precisely because wild cats are difficult subjects to study.

Wild cats are elusive. Unlike primates, for example, it is difficult to habituate individual cats in order to observe their behaviour. Even when such habituation has been achieved, observation becomes difficult at night or when the cat eludes the observer. Most wild cats are nocturnal and prefer to live in cover, making observations difficult even where they are reasonably abundant. Because humans are diurnal primates, we tend to observe cats during the day, mostly when they are at rest.

Wild cat habitats are often remote and difficult to access because of mountains, rivers, swamps, deserts, or dense jungles. In addition, a few big cat species are dangerous and can inflict lethal injuries if not handled carefully.

Although early human cultures often viewed cats as mysterious or magical beasts (Nowell and Jackson 1996), sharing habitats with felids has given traditional cultures deep insights into felid natural history (Bothma and le Riche 1986; Stander *et al.* 1997b). Hunters from later civilized societies viewed cats from a more anthropocentric perspective as they observed and described them. Even within the framework of modern scientific enquiry, well into the twentieth century, naturalists could only employ simple tools such as binoculars, traps, guns, and spoor-tracking to study cats.

The past 50 years, however, witnessed a rapid expansion of the biologists' arsenal: new technologies like chemical immobilization, radio-telemetry, photography, and genetic analyses, in combination with satellites and computers, have revolutionized cat studies. It is less well recognized that advances in the allied fields of biostatistics and information technology have fuelled contemporaneous progress in study design and data analyses. As a result, there are now more than the proverbial 'nine ways of skinning a cat': old-style natural history has morphed into modern wildlife science. To benefit fully from this progress, biologists must now balance deeper field observations with more rigorous analyses.

In this chapter, we provide a concise overview of the diversity of field and analytical methods that are now available to facilitate studies of wild cats. The field tools we describe complement the more laboratory-oriented methods covered elsewhere in this book (Culver *et al.*, Chapter 4, this volume; Wildt *et al.*, Chapter 8, this volume). We present these tools and techniques in the context of addressing four common objectives of studying cats: as individuals (behaviour and intraspecific relations); as members of entire biological communities (interspecific and localized habitat relationships); as functioning 'populations' of their species (population and meta-population dynamics); and cats as parts of wider habitats (their landscape ecology). At each level, we first describe the field methods employed to collect the data and thereafter cover study design and data analysis issues.

Although we have tried to cover several techniques, for reasons of brevity, deeper treatment is provided only for some representative methods. These examples are somewhat biased towards methods that we are more familiar with. Moreover, some cat species, such as puma, tiger, lion, and the cheetah, bobcat, and lynx have been studied more extensively than others. Furthermore, studies of felids have been more numerous in North America, Europe, and Africa than elsewhere. To the extent possible, we have tried to provide a balanced coverage of this diversity.

Fundamentally, most tools employed to study wild cats are a subset of those used to study mammals in general. These are well described in standard reference books (e.g. Schemnitz 1980; Bookhout 1994; Wilson *et al.* 1996), which provide the context for this review. Similarly, appropriate analytical methods are also covered by many standard references (Thompson *et al.* 1998; Boitani and Fuller 2000; Williams *et al.* 2002a). References to methods for studying rare and elusive animal species (Thompson 2004) are particularly relevant to studies of felids.

Study of cats as individuals

Behavioural studies typically emphasize intraspecific relationships among individuals in spatiotemporal terms, elucidating how they interact with each other (see reviews by Ewer 1973; Leyhausen 1979; Macdonald 1983; Macdonald *et al.* 2000b).

Almost all cat species tend to hold and defend spatially well-defined territories, at least during the reproductive stage of their life histories (Ewer 1973; Sunquist and Sunquist 2002). Thus measurements of size, shape, and spatial arrangements of home ranges are of primary concern for behavioural studies of cats. Studies of daily or seasonal movements, associations between individuals, philopatry, dispersal, and habitat preferences also require measurement of use of space (Sunquist and Sunquist 2002).

Other behavioural questions examine how wild cats budget their time for hunting, finding mates, resting, or travelling through the day. These require measurement of specific activities in relation to time. Studies of felid social organization, mating systems, rearing of cubs, and land tenure involve both spatial and temporal measures. Behavioural interactions among cats in terms of communication, aggressive interactions, dominance ranking, etc. are of interest, particularly in relation to the more 'social' species (Macdonald 1983). To meet all these objectives, biologists must be able to individually recognize cats to follow their fates across time and space (McGregor and Peake 1998; Sunquist and Sunquist 2002). Thus, recognizing individuals either from natural marks or artificially using visual tags or radio-collars becomes a critical need.

Field methods for studying cats as individuals

Techniques of capturing and handling of wild cats

Before cats can be individually 'marked' for study, they must be captured and restrained safely to facilitate tagging, morphometric measurements, or collection of biological samples. The capture systems should be efficient, safe, selective, and minimize stress to the animal. Given the aggressiveness of wild cats, they must be chemically immobilized.

Ideally, the drugs should rapidly induce reliable and reversible immobilization. In earlier studies of lions, drugs such as phencyclidine were regularly used (Melton *et al.* 1987). Due to their unsatisfactory side-effects there was a general progression towards ketamine–zylazine combinations (Herbst *et al.* 1985; Stander and Morkel 1991), which are still used widely (e.g. Beltran and Tewes 1995). For lions tiletamine–zolazepam (Zoletil) has become a favoured drug (Kreeger *et al.* 2002). A more recent medetomidine–tiletamine–zolazapam combination offers even smoother induction, shorter recovery periods, and satisfactory analgesia and muscle relaxation (Fahlman *et al.* 2005; Jacquier *et al.* 2006). These new drugs have wider safety margins and require relatively low doses for relaxed individuals. However, doses that are 2–3 times higher may be required for cats that are agitated after being trapped or chased. Thorough reviews of immobilization of felids can be found in McKenzie and Burroughs (1993), Lewis (1994), Fowler (1995), and Kreeger *et al.* (2002).

Cages or box traps (Schemnitz 1980; Proulx 1999) are often used effectively (Haines *et al.* 2004, 2006a) to catch many cats. These traps typically have one or two doors that are activated by a foot treadle or a tripwire. Steel grill traps often result in cats breaking their teeth or hurting themselves. Traps made from alternative materials like wood, fibreglass, and aluminium avoided injuries during leopard captures (Karanth, unpublished data). Covering the trap to avoid exposure of the cat to surrounding stimuli, and minimizing the time spent in it can dramatically reduce injuries. A general disadvantage of traps is that they result in low capture rates for wary cats. In virtually all trap designs, some form of bait or lure is required to attract the cat into the trap. Pre-baiting is a very good way to increase capture rates and minimize the capture of non-target species. Jackson *et al.* (1990) describe how snow leopards were caught by placing several pre-baited traps along regular trails.

Foot snares made of wire or cables (Fig. 7.1) have also been used to capture a wide range of cat species (snow leopards, Jackson *et al.* 1990; lynx, Breitenmoser and Haller 1993; puma, Logan *et al.* 1999; tigers, Goodrich *et al.* 2001; and lions, Frank *et al.* 2003). Snares are particularly efficient when catching big cat species that either avoid conventional traps or hurt themselves in them (Frank *et al.* 2003). Leg-hold traps, which are essentially modified gin-traps with padded jaws to minimize risk of injury, are useful for catching smaller cat species. Such snares or traps are difficult to use in habitats, where the captured cat is at risk from attacks by other animals.

Other unique site and species-specific capture methods have been developed: Stander *et al.* (1996) employed local tribesmen, who used bows and arrows to immobilize lions and leopards that were too wary to approach in vehicles. In the Americas, jaguars and pumas are located, chased, and treed by

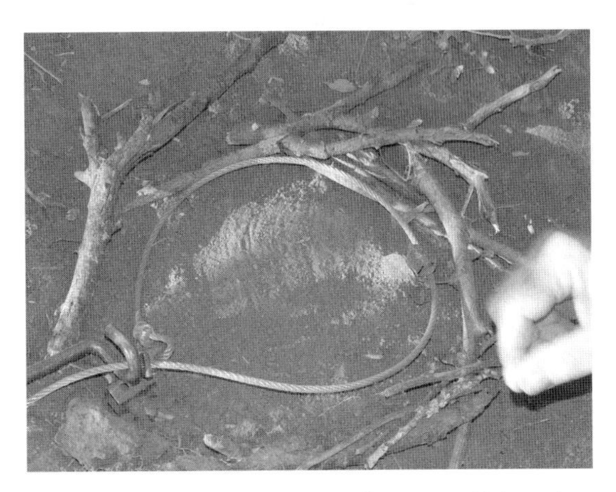

Figure 7.1 Close-up of a humane foot snare which can be used to capture wild felids, including tigers, lions, snow leopards, lynx, and pumas. © P.J. Funston.

hunting dogs before being darted (Crawshaw and Quigley 1991). This method appears to be highly stressful to the cats that are caught.

Although typically only one or two cats are captured at a time, Smuts *et al.* (1977) developed a mass-capture technique for lions in Kruger National Park, South Africa, in which entire prides of lions were immobilized at night while feeding on large carcasses. More than 1000 lions were thus immobilized, including an instance of simultaneous capture of 21 individuals out of a pride of 23.

A very site-specific capture method is the free darting of tigers from trees, practised in Nepal (Sunquist 1981; Smith *et al.* 1983) and India (Karanth and Sunquist 2000). In these studies, tigers feeding on buffalo bait were encircled within a funnel-shaped 'corral' of white cloth (which provides a psychological barrier the tiger will not cross) without them becoming aware. They were then pushed off their kills by trained elephants and into the funnel and towards the shooter positioned on a tree. When it works this 'Beat' method is almost stress-free, as the tiger is not even aware that it has been darted. However, application of such unique capture techniques requires highly skilled personnel and appropriate field conditions. In any free darting, risks associated with high ambient temperature, proximity to large water bodies, or the presence of other dangerous animals in the vicinity, need to be considered.

Cats are commonly immobilized using a compressed air or carbon dioxide rifle that effectively delivers a syringe dart at distances up to 30 m. Such silent, low-impact dart rifles have now largely superseded earlier noisy powder-charged guns that caused minor trauma on impact. For immobilizing cats inside traps, blowpipes or pole syringes can also be used. As far as possible, direct approach on foot towards any large cat caught in a snare or trap should be avoided in favour of darting from a vehicle or hide to ensure personnel safety and reduce stress on the cat.

Recently, Ryser *et al.* (2005) developed a tele-injection system to dart the trap-shy European lynx, which comprised a blowgun remotely controlled and guided by means of two built-in cameras and a swivelling two-way pan-tilt head. This system can be operated from a distance of up to 400 m. Here, as with free darting, having a carcass as bait is often needed to prevent the cat from moving away after darting.

Radio-telemetry

Because of difficulties in observing individual cats or in tracking their spoor, radio-telemetry has emerged as perhaps the most important tool for the study of wild-cat behaviour (Amlaner and Macdonald 1980; Schemnitz 1980; Kenward 1987; Bookhout 1994; Wilson *et al.* 1996). This method involves attachment of a radio-transmitter to the cat, which broadcasts signals that can be received by stationary or mobile receivers at remote locations (Fig. 7.2). Transmitters on different cats broadcast on different radio frequencies, permitting the researcher to locate and track several animals at a time. The most common information obtained from radio-telemetry is the location of the cat, but

Figure 7.2 Researchers in Hwange National Park, Zimbabwe, use radio-telemetry to locate radio-collared lions © A.J. Loveridge.

other data such as levels of activity, physiological state, or mortality can also be gathered using appropriate built-in sensors. Telemetry data can answer a wide range of behavioural questions related to use of space, time, and intraspecific social relationships that are virtually impossible to answer otherwise. Some telemetry-based studies have been able to determine track paths of pumas at night through frequent triangulation and later inferring behaviours by following spoor (Beier *et al.* 1995). However, in difficult terrain, researchers often can only collect radiolocation data without ever seeing the animal, as with Iriomote cats (Schmidt *et al.* 2003).

Cats are well suited for radio-collaring because they have relatively large heads and narrower necks (Fig. 7.3). Most radio-collars weigh less than about 3–5% of the cat's body mass. The standard technology employs Very High Frequency (VHF) transmitters embedded in a waterproof casing attached to a durable collar belt or webbing-buckle apparatus. The signals are detected by an antenna and heard through a receiver carried by the biologist or mounted at a fixed station. The cats are 'located' on a map by triangulation of compass bearings obtained to these signal directions (White and Garrott 1990) or directly via a Global Positioning System (GPS). In some cases, homing in on the cat directly permits visual observations and more accurate locations (Amlaner and Macdonald 1980).

Another telemetry option is to surgically implant transponders. However, this approach is not especially suited to field study of felids due to potential surgical complications and reduced signal strength resulting in lower detection rates. For example, Funston (unpublished data) found that a lioness with an implant could only be detected from ground tracking at <500 m, as opposed to a 2–5 km detection range for a VHF radio-collar.

A related issue concerns the platform used to mount the receiver. In the case of VHF telemetry, the researchers typically track cats on foot, ride on domestic animals (including elephants in India), drive cars, off-road vehicles, boats, or aircraft. Less often triangulation from fixed stations is also employed. It is critical to choose a tracking platform that best provides data in the context of the study questions and environmental features at the site.

Technological advances in recent years have yielded many sophisticated derivations of conventional VHF radio-telemetry (see Rodgers 2001), including Ultra High Frequency (UHF) and GPS transmitters (Hulbert 2001), and Argos satellite transmission (Schwartz and Arthur 1999) (Fig. 7.4). GPS animal-tracking options include downloading of location data from collars via UHF transmissions, GPS logger collars, where the data are retrieved only after the collar has been released from the animal using a drop-off mechanism, and GPS/GSM collars, where the location data are transmitted using the local cellular telephone network and received as SMS messages or via the internet by the researcher.

Figure 7.3 Researchers in Mashatu Game Reserve, Botswana, fit an immobilized lion with a radio-collar. © P.J. Funston.

Figure 7.4 Argos satellite GPS collar fitted to an adult male lion in Hwange National Park. Data are transmitted from the collar to the satellite at pre-programmed intervals and can be downloaded via the internet. © A.J. Loveridge.

GPS collars are not appropriate if study animals occur in small areas and need to be regularly tracked for behavioural observations (e.g. Mills and Shenk 1992; Funston *et al.* 1998). However, only GPS collars can provide a sufficient number of locations for studying big cat movements across extensive landscapes (Anderson and Lindzey 2003; McCarthy *et al.* 2005). Some manufacturers offer the option of fitting an automatic or remotely triggered collar release mechanism for situations when recapture of the animal is difficult or risky (Hemson 2002; Cain *et al.* 2005; Lewis *et al.* 2007). Despite their ability to generate large amounts of location data in comparison with conventional telemetry, the performance of GPS tracking systems still poses some challenges such as larger collar size, higher costs, the inability to directly observe the study animal, and relatively poor performance in densely forested landscapes. Therefore, sometimes researchers combine GPS and conventional VHF tracking systems in a single collar to overcome these problems. We have briefly summarized the advantages and disadvantages of different options available for tracking felids in Table 7.1.

Direct visual observation of behaviour

Although many cat species can be individually recognized from natural markings, in the field this is possible only in very open habitats, such as using coat patterns on cheetahs (Caro 1994) or whisker spot patterns on the muzzles of lions (Pennycuick and Rudnai 1970). Researchers on the Serengeti Lion Project could identify over 400 lions from natural markings or scars using identification cards that could be checked in the field during sightings (Packer *et al.* 2005a). Such direct observational methods, however, cannot be used for cats that live in dense cover.

More often, direct observations are made after cats are located using telemetry. The Serengeti Lion Project, Tanzania, has been able to maintain intensive daily monitoring of prides for over 40 years (Packer *et al.* 2005a), but has only spent a small fraction of this time observing nocturnal hunting behaviour (Scheel and Packer 1991). Extensive nocturnal (>500 nights) direct observations of lions in Kruger National Park, South Africa, resulted in important new insights into lion-feeding ecology (Funston *et al.* 1998), predator–prey relationships (Mills and Shenk 1992), and social behaviour (Funston *et al.* 2003). Similarly, Stander (1992b) observed nocturnal hunting behaviour of lions on the open plains of Etosha National Park, Namibia, and found that individual lions have specific roles while stalking prey.

Good binoculars with a wide field of vision, for example, 8 × 32, are ideal for daytime observation of most felids. Night-time observations require either

Table 7.1 Comparison of advantages and disadvantages of different telemetry options for felids.

Collar options	Mode of data retrieval	Estimated cost/ collar (US$)	Advantages	Disadvantages
VHF	VHF receiver	300–500	1. Cost-effective 2. Ideal for local direct observation studies	1. Relatively few data points 2. Expensive fieldwork
GPS	Logger retrieved from collar	2000–3000	1. Data-rich	1. Expensive 2. Depends on collared animal being located and recaptured 3. Not suitable for local direct observation studies
	UHF download	4000–6000	1. Data-rich 2. Adds a degree of fieldwork to the research effort, allowing other data to be collected 3. Can be used in conjunction with local direct observation studies	1. Expensive 2. Depends on researchers getting close enough to successfully and safely download data
	GSM land-based cellphone network	5000–7000	1. Data-rich 2. Allows data collection in areas of inhospitable terrain and in circumstances where study animals move out of study areas for extended periods of time	1. Expensive 2. Depends on the study animal being close enough to cellphone towers 3. Little opportunity to collect additional field data
	Iridium satellite phone	7000–8000	1. Data-rich 2. Allows almost uninterrupted data collection while based at a remote location	1. Expensive 2. Depends on the Iridium service provider providing continuous and reliable coverage 3. Very little opportunity to collect additional field data
Argos	Satellite transmission	6000–10,000	1. Data-rich 2. Allows long distance movement data to be collected	1. Very expensive 2. Very little opportunity to collect additional field data

binoculars with good light-gathering ability, (for example, 10 × 50, 9 × 63) night-vision goggles or binoculars (preferably military quality), video cameras with light-intensifying lenses, or low-intensity spotlamps, that can be softened with a red filter (Mills and Shenk 1992; Funston *et al.* 1998). Caution should always be taken not to dazzle the cat or the prey with a spotlamp, even when a red filter is used. Scanning ahead for prey may disrupt potential hunts. Once accustomed to the vehicle, foraging cats can often be followed at relatively close distances of 20–100 m. The black-footed cat, one of the smallest living species of cats (average mass 1.6 kg), has been successfully studied at night in semi-arid Karoo scrub habitats in South Africa through nocturnal observations (Sliwa 2004; Sliwa *et al.*, Chapter 26, this volume).

Observing felid tracks

Another alternative to direct observations of cats is to follow tracks of individual cats to reconstruct their behaviour. Such tracking can be done only in areas where suitable substrates like fine soil, sand (Bothma and le Riche 1986), or snow (Smallwood and Fitzhugh 1995) facilitate it. By following tracks of foraging cats, a wide range of data on prey-encounter frequencies, hunting success, prey species selection, home range use, and social interactions can be gathered (Bothma and le Riche 1986; Stander *et al.* 1997b). However, these techniques require special field skills, such as those possessed by indigenous 'San' trackers of southern Africa.

Callback techniques

The technique of playing back pre-recorded calls of cats, prey species, or of competing predators has been used in studies of cat behaviour (McComb *et al.* 1994; Durant 2000b). Callback techniques work best in open habitats, and enable researchers to perform interesting behavioural experiments on how cats interact with their own kind and with other predators. However, excessive use of playback calls can disrupt normal social and spacing behaviours or lead to habituation, factors that need to be considered when designing studies. Because callback techniques are more relevant to population surveys of cats, they are described later in that context.

Analytical methods for studying cats as individuals

If direct observations can be made of individually identified cats (either tagged or naturally marked), classical techniques for sampling and quantifying behaviour using methods such as time-activity budgets and ethograms (Curio 1976; Leyhausen 1979) can be employed in the analysis of behavioural data. Social and reproductive behaviour has been quantitatively analysed based, at least partially, on such visual observations of lions (Scheel and Packer 1991; Funston *et al.* 1998; Packer *et al.* 2001) and cheetahs (Caro *et al.* 1989; Laurenson 1995a). Macdonald *et al.* (2000a) provide a good overview of behavioural study methods and data analysis.

Telemetry data from wild cats are usually in the form of spatial locations, activity–inactivity metrics, and associated time and environmental information. Such spatiotemporal data can be modelled in a number of ways to understand wild cat behaviour. Telemetry data on individual animals can be used to analyse long-distance migration, as well as localized movement rates at shorter time intervals. For subadult cats, age, timing, and distances of dispersal can be measured (Smith 1993; Samuel and Fuller 1994). White and Garrot (1990) discuss several statistical methods of analysing such movement data.

Location data from telemetry are often used to estimate the size of cat home ranges and territories. Starting with the simple minimum convex polygon approach of earlier years, more sophisticated approaches, which assume that the observed locations come from different underlying statistical distributions, have been developed (Kenward 1990; White and Garrot 1990; Samuel and Fuller 1994) to assess animal home range size and the varying intensities of range use. Specific software for statistical modelling of spatial locations and estimating home range are available (e.g. RANGES, Kenward 1990; CALHOME, Kie *et al.* 1996; KERNELHR, Seaman *et al.* 1998), although the older convex polygon approach is still used.

Quantitative analyses of telemetry data have documented activity, habitat selectivity, and social organization in a number elusive cat species. Such VHF radio-telemetry was pioneered by studies of mountain lions in North America (Seidensticker *et al.* 1973; Lindzey *et al.* 1988), followed by radio-tracking studies on a number of cat species such as: tigers (Sunquist 1981; Smith 1993; Karanth and Sunquist 2000), leopards (Mizutani and Jewell 1998; Balme *et al.* 2007), snow leopards (McCarthy *et al.* 2005), jaguars (Soisalo and Cavalcanti 2006; Azevedo and Murray 2007a), bobcat (Benson *et al.* 2006), lynx (Ferreras *et al.* 2004), leopard cats (Grassman *et al.* 2005a; Rajaratnam *et al.* 2007), caracal (van Heezik and Seddon 1998), Irimote cat (Schmidt *et al.* 2003), ocelots (Harveson *et al.* 2004), Geoffrey's cat (Pereira *et al.* 2006), kodkod (Dunstone *et al.* 2002), and black-footed cats (Sliwa 2004).

Telemetry studies have addressed topics such as home range size and spatial organization (Harveson *et al.* 2004; Grassman *et al.* 2005a; Rajaratnam *et al.* 2007), philopatry, reproduction and dispersal (Smith 1993; Kerley *et al.* 2003; Ferreras *et al.* 2004; Zimmermann *et al.* 2005b), reproduction (Kerley *et al.* 2003), and other aspects of felid behaviour. Telemetry data also permit investigation and analysis of intraspecific communication mechanisms such as scent marking (Smith *et al.* 1989) and vocalizations (McComb *et al.* 1994; Durant 2000b).

Study of cats in biological communities

In community studies of cats, questions about prey selection, feeding ecology, and interspecific competition and niche separation are usually addressed. Study objectives may range from understanding basic biology (e.g. Schaller 1972; Karanth and Sunquist 1995) to ecological questions (Balme *et al.* 2007; Cooper *et al.* 2007) or evolutionary investigations (Durant 1998; Bissett and Bernard 2007). The availability and vulnerability of prey is the most important factor that drives all other aspects of felid community ecology. Although methods to assess prey availability and its relationship to cat community dynamics are thus important, details of these are beyond the scope of this chapter.

The location and activity data generated from radio-telemetry often come with ancillary data on the cat's prey species, presence of other predators, or about the physical environment. For example, Balme *et al.* (2007) used radio-telemetry to examine how relative importance of prey abundance versus vulnerability dictated habitat choice in leopards. Sunquist (1981) observed that tigers opted not to use short grass habitats after summer fires, because they found it harder to capture axis deer. Seidensticker (1976) and Karanth and Sunquist (2000) observed temporal avoidance of tiger habitats by leopards using radio-telemetry. Because many of the methods used to study cats in their communities are similar to those covered earlier (see above), only specific additional methods are described here.

Additional field methods for studying cats in biological communities

Assessing cat diets and prey composition from studies of kills and scats

Data from kills made by cats obtained either opportunistically or from radio-tracking, provide basic foundations for studying feeding ecology in larger cats. Although kill data are biased by higher detection rates for larger kills, when combined with ancillary data on prey vulnerability such as age, sex, physical condition, or health status (Schaller 1972; Mills 1996; Karanth and Sunquist 2000) they are extremely valuable. For example, researchers in Kruger National Park and adjoining reserves advanced the understanding of predator–prey relationships significantly by using kill records collected by park rangers (Radloff and du Toit 2004). They used >4000 kill records to show that predator's size determines its preferred prey and thus facilitates coexistence within large predator guilds. Mills *et al.* (1995) found from >7000 kill records that lions switched prey species under changing environmental conditions, thus impacting buffalo populations during droughts, while preferring wildebeest and zebra in wetter periods. Owen-Smith and Mills (2006) determined the causal processes governing the population changes shown in 11 ungulate species over a 20-year period in Kruger by combining >20,000 kill records with environmental data. If unbiased kill records can be readily obtained, even determination of prey preference is possible (Hayward and Kerley 2005).

Predation patterns of cheetahs have been examined in the light of optimal foraging and economic decision theory (Cooper *et al.* 2007). Influence of kleptoparasitism (Hunter *et al.* 2007a) has been examined based on direct observation of cheetahs at kills. Mills (1992, 1996) argues that following predators in a vehicle and observing them hunt gives the most accurate information on the food habits of carnivores. However, such a direct observation is rarely practical for most cat species.

How frequently cats need to kill can be estimated from modelling studies (Tambling and du Toit 2005), carnivore metabolic needs (Schaller 1972; Peel and Montagu 1999), or daily monitoring of stomach size indices (Bertram 1975b). Other approaches to quantifying food habits include reconstruction of hunts from tracks (Bothma and le Riche 1986; Stander *et al.* 1997b).

Scat analysis is the most commonly used method to understand the feeding ecology of cats. Because several scats are produced from a single meal, and cats defecate conspicuously to signal their presence, they present the biologist with a rich (if somewhat fragrant) lode of dietary information. The species that deposited the scat can be identified from traditional bile acid assays (Major *et al.* 1980) or the more recent genetic methods (Fernández *et al.* 2006). A major advantage of scats over kills for accurately assessing cat diets lies in the fact that smaller prey are also detectable. However, biases resulting from differential digestion rates and variable rates of ingestion of hair relative to meat are sometimes problematic issues. Useful overviews of methods of extracting information from examination of scats have been provided by Putman (1984), Reynolds and Aebischer (1991), and in other general texts (Schemnitz 1980; Wilson *et al.* 1996).

Traditional scat analyses simply report frequencies of occurrence of different prey types in the diet of a cat, which is not sufficient to understand the relative importance of different prey. If the prey types killed vary greatly in size, such frequency data overemphasize the importance of smaller prey items, because of their relatively larger surface-to-volume ratio. Ackerman *et al.* (1984) conducted a study that measured scat production per live animal prey of different sizes consumed by captive pumas, to generate regression equations that relate frequency of occurrence to number of scats produced per unit weight of live prey eaten. These equations permit estimation of relative biomass and relative numbers of different prey types consumed from frequency of occurrence data. In combination with estimates of prey availability, such analyses permit inferences about prey selectivity patterns (Karanth and Sunquist 1995) using the free software programme SCATMAN (Hines and Link 1994). Numerous scat analysis studies have documented prey profiles and selectivity in big cats (lions, Schaller 1972; tigers, Karanth and Sunquist 1995) and small cats (leopard cats, Rajaratnam *et al.* 2007; Grassman *et al.* 2005a).

Assessing prey availability

Although the availability of their prey is the most important driver of felid ecology, description of methods used to assess the abundances of the wide taxonomic range of prey animals taken by felids are well beyond the scope of this chapter. We direct the reader to texts that cover methods used for estimation of animal abundance in general (Burnham *et al.* 1980; Bookhout 1994; Wilson *et al.* 1996; Thompson *et al.* 1998; Williams *et al.* 2002a).

Additional analytical methods for studying cats in biological communities

Spatial and activity data on wild cats obtained from direct observations or telemetry can be quantitatively analysed to examine different aspects of predation. For example, hunting behaviour in leopards (Balme *et al.* 2007) and cheetahs (Bissett and Bernard 2007), lions (Hopcraft *et al.* 2005), location of kills in pumas (Anderson and Lindzey 2003), and prey availability for bobcats (Benson *et al.* 2006) have been studied in this way. These kinds of analyses also enable examination of interspecific spatiotemporal separation among cat species (Karanth and Sunquist 2000; Scognamillo *et al.* 2003). The relationship between wild cats and their habitats can be analysed using resource selection functions based on logistic regressions (Hopcraft *et al.* 2005), habitat suitability models (Scott *et al.* 2002), or using other landscape ecological tools (see below). Similarly, analysis of species co-occurrence patterns derived from occupancy modelling methods (MacKenzie *et al.* 2006) provide yet another analytical tool for looking at how wild cats interact with their surroundings.

Study of cats in populations and meta-populations

The number of wild cats in a discrete population is potentially set by four key factors: prey availability, intraspecific spacing mechanisms, interspecific competition with other predators or scavengers, and direct hunting pressure. Cat population density in turn influences other factors such as prey densities, inter- and intraspecific competition, and offtakes of cats by human hunters. Consequently, population size is a parameter of paramount interest to scientists and conservationists.

Changes in population density are driven by changes in vital rates such as mortality, survival, emigration, and recruitment (which may include *in situ* reproduction as well as immigration). Other phenomena such as transience and temporary emigration are also important drivers of population dynamics in cats (see Williams *et al.* [2002a] for a comprehensive technical treatment of animal population analysis). When cat meta-population dynamics are considered, the rate of movement of individuals between populations becomes an additional parameter of interest. Because most cats are not easily seen or counted but can be 'individually caught' in some manner, capture–recapture sampling provides a useful approach to understand their population dynamics.

Field methods for studying cat populations

Data from marking, radio-tagging, and hunting records

Physical capture and ear-tagging of individuals for identification was employed to study lion population size and demographics in the Serengeti (Schaller 1972; Packer *et al.* 2005a). Radio-telemetry has been used to estimate numbers, survival rates, and other demographic parameters in studies of tigers (Smith 1993) and pumas (Lindzey *et al.* 1988). However, because of logistical and practical problems, physical capture and tagging is typically difficult to use as a tool for demographic studies, particularly with larger species.

Sometimes current records of legal or illegal hunting levels generate data for population studies (Ferreras *et al.* 1992; Loveridge *et al.* 2007c). Historical hunting and trapping records of lynx for the substantial fur trade in North America or historical statistics of bounty and trophy hunting of big cats in India (Rangarajan 2001) have provided insights into past population status of wild felids.

More often than not, however, neither physical capture nor past hunting records are practical data sources for demographic studies. Therefore, some form of non-invasive method of 'catching' individual cats must be employed. In this context, non-invasive captures, such as those from photographic or DNA-based identifications, have assumed importance.

Photographic captures from camera trapping

In recent years, automated camera traps activated by the animals themselves have revolutionized wild cat studies, as telemetry did in the 1960s. 'Photographic captures' of individual cats, together with information on the date, time, and capture location provide the basic data for population analyses (Karanth *et al.* 2004b). Although camera traps were first employed in the early 1900s, they became practical tools for wildlife scientists only in the 1980s. Relatively inexpensive (<US$100), camera traps can now be deployed in large numbers necessary for population studies. As a result, non-invasive camera-trapping studies of wild cats have burgeoned.

Already these studies have documented the present occurrence or absence of many species of wild cat from across their range after several data-deficient decades. In some of these studies, a metric known as 'trapping rate'—the number of times a cat species is photographed for a standardized effort (say 1000 trap-nights)—has been used to compare their relative densities across space or time. However, because capture probabilities themselves vary across space and time (Thompson *et al.* 1998; Williams *et al.* 2002a), trapping rate is not usually a reliable measure of relative abundance.

For many cat species that are individually recognizable from natural marks, capture–recapture methods have become a practical alternative to study population dynamics. Demographic studies of tigers in India (Karanth *et al.* 2004c, 2006) have shown the exciting potential of this approach for understanding wild cat population dynamics across both space and time. Following

(a)

(b)

Figure 7.5 (a) Researchers test the set-up of camera traps; and (b) image of a tiger captured by a camera trap in Nagarahole, India. © (a): Eleanor Briggs; (b): K.U. Karanth/WCS.

camera-trap studies of tigers (Fig. 7.5), work on other naturally marked species such as jaguars (Maffei *et al.* 2004a; Soisalo and Cavalcanti 2006; Weckel *et al.* 2006b), leopards (Henschel and Ray 2003), and ocelots (Di Bitetti *et al.* 2006), exemplifies this major growth area in wild cat population studies. Even species-specific protocols for camera-trap surveys have been developed for tigers (Karanth and Nichols 2002), leopards (Henschel and Ray 2003), and snow leopards (Jackson *et al.* 1990). A comprehensive account of camera-trap methods is provided by O'Connell *et al.* (in press).

Other non-invasive 'individual capture' methods

'Captures' of individual cats, in the form of DNA extracted from their scats or hair samples (Waits 2004) can also be used to estimate demographic parameters.

However, laboratory techniques are only now being developed for extracting DNA and reliably identifying individuals in cat species.

Based on the fact that dogs have superior sensory abilities that permit them to distinguish unique scent specific to individual animals in other mammal species, they have been trained to individually identify tigers from scat samples (Kerley and Salkina 2007). This may develop into yet another practical cat 'capture' method in future.

Attempts have also been made to identify individual big cats from the shape and configuration of their paw prints or track-sets (Smallwood and Fitzhugh 1995; Riordan 1998; Sharma *et al.* 2005) under controlled conditions. However, applying these methods in the field for population assessment is more complicated (Grigione *et al.* 1999; Lewison *et al.* 2001; Karanth *et al.* 2003). The 'tiger pugmark census' of India is a well-documented case study of an unreliable track-based individual identification method used for counting of cats. This technique, developed by an Indian forester, was applied by all forestry staff, across thousands of square kilometers of tiger habitats in India in attempt to 'census' all of India's tigers. Bereft of biological or statistical basis, the method generated 'tiger numbers' that bore no logical relationship to reality, and seriously undermined the vigorous auditing of tiger conservation for over three decades (Karanth *et al.* 2003).

Counts of tracks and scats produced by cats

The larger cat species generate a lot of detectable spoor such as tracks and scats. Where substrate is conducive, biologists have tried to derive indices of relative density for cats from track encounter rates. Track counts have been used to produce population abundance estimates for pumas (Smallwood and Fitzhugh 1995; Beier and Cunningham 1996), and lion and leopard counts have been formulated on sandy Kalahari soils (Stander 1998; Funston *et al.* 2001a). A rigorous example of quantitative analysis of snow track count data on tigers in Russia has been provided by Stephens *et al.* (2006).

A prerequisite for using track counts to reliably estimate population size is calibration. In the Kgalagadi Transfrontier Park, researchers first determined actual abundance of lions by physical capture and tagging of over 95% of all lions. A strong correlation between actual and track density was then found (Funston *et al.* 2001a), enabling them to predict lion densities from track surveys in other areas. Similarly, Stander (1998) first identified all the leopards in a study area in the Kaudum Game Reserve in Namibia, based on the assumption that experienced San trackers could identify these animals individually from characteristics of their tracks (Stander *et al.* 1997b). The key element of both these studies is that the counts were calibrated against known cat population size. Such counts require skilled trackers (Stander *et al.* 1997b) as well as appropriate environmental conditions for track detection. In a few cases, where track counts have been calibrated against abundances derived using other means, they appear to have produced reasonable results (Stander 1998; Funston *et al.* 2001a). However, such a relationship could not be established for an index based on encounter rates with tiger scats and tiger densities derived from camera trapping (Karanth, unpublished data). Track survey trails could be biased and non-representative (Smallwood and Schonewald 1998) and strong assumptions about unvarying track detection probabilities are always inherent to such counts (Karanth *et al.* 2003). Overall, the reliability of track counts depends on many key assumptions being met (Smallwood and Fitzhugh 1995; Stander 1998; Stephens *et al.* 2006). The most important assumption of there being a positive, monotonic, unvarying relationship between the index counts obtained in the field and true abundances of cats is usually not tested.

Playback and call-in surveys

Pre-recorded calls of cats can be played at night to attract and count them using spotlights. This unique technique is largely applicable to big cats in African savannahs. Reliable playback or call-in field surveys have been developed for lions (Smuts 1976; Mills *et al.* 2001). During a survey in Kruger National Park 444 lions were called in; however, 72 leopards also arrived, showing its potential for attracting other big cats. Call-in surveys have also been used to generate assumed total counts (Smuts 1976) or indices of abundance (Ogutu and Dublin 1998). The density estimates determined from the call-in survey in Maasai Mara (29.4 ± 0.9 [5% CI] lions/100 km^2) was close to census records (30.6 ± 1.4 [95% CI] lions/100 km^2), suggesting the method was reliable in this context. In Kruger, lions responded well in winter but not in summer (Smuts 1982). Generally, such call-in methods have limited application in population level studies of cats for several reasons: they work only in open habitats for a few big cat species, may attract non-target species, and the resulting data are difficult to handle using standard population estimation models because of untested assumptions about the proportion of animals that responded.

Analytical methods for studying cat populations

Most often, a count of cats from any field technique provides only a sample from the population of interest, rather than a total count. Therefore, estimating detection probabilities, that is, the proportion of animals likely to be counted during the field survey, becomes a critical parameter. Unfortunately, robust methods such as distance sampling that require visual detection of animals with concurrent recording of sighting distances to estimate this parameter (Wilson *et al.* 1996; Thompson *et al.* 1998) are not useful for most felids because of their elusiveness.

Typically, field surveys of cats therefore use physical capture and tagging, radio-tagging, photographic captures, or faecal DNA-based captures of individual cats. Thereafter, capture probabilities must be estimated from the frequencies in which identified individuals are caught in repeated sample counts in field surveys. Consequently, the capture–recapture sampling approach (Thompson *et al.* 1998; Williams *et al.* 2002a) forms the bedrock of rigorous demographic studies of cats (Karanth *et al.* 2004c).

Broadly, two kinds of capture–recapture analytic models, respectively known as closed and open models, are used. Closed models assume that there were no births, deaths, immigration, and emigration during the survey, which can usually be achieved by keeping the survey period short (say a few weeks). Closed capture–recapture models offer great flexibility by realistically dealing with biological phenomena such as individual cats having different capture probabilities, becoming trap-shy after first capture, or capture probabilities varying during the survey due to environmental factors (Karanth and Nichols 2002; Karanth *et al.* 2004c). The major advantage of these methods is that population size (abundance) can be estimated without the need to capture all individuals in the population.

Once the population size is estimated, a buffer strip is added around the trap array to estimate the sampled area and thereby population density. The buffer width is estimated using one of several ad hoc approaches, all of which have some problems (Soisalo and Cavalcanti 2006). However, a new analytic approach called 'spatially explicit capture–recapture sampling' developed by Royle *et al.* (2009) provides a superior option for analysing such capture data by incorporating additional spatial information.

Camera trapping can be combined with powerful capture–recapture model approaches (Thompson *et al.* 1998; Williams *et al.* 2002a) to estimate abundance and other population parameters. Karanth *et al.* (2004c) validated their model predicting tiger density as a function of prey density in a macro-ecological spatial study that spanned 11 sites, covering 3024 km^2 of varied forest habitats across India. They sampled 11 tiger populations using 8677 days of camera-trapping effort, photo-capturing 167 distinct individuals. Using closed capture–recapture analyses (Williams *et al.* 2002a) they estimated that capture probabilities at these sites varied from 0.04 to 0.22 per sample, allowing them to estimate the proportion of tigers that were caught. The estimated tiger densities ranged from 3.3 to 16.8 tigers/100 km^2 across these sites and matched well with densities predicted based on independently derived prey abundance estimates.

When long-term population dynamics are to be explored, cats pose some additional biological challenges to capture–recapture surveys. They may not be in the surveyed area for part of the period (temporary emigration) or new cats may be just passing through on their way elsewhere (transience). Cats may be born, die, move into, and permanently move out of the population over the long term. However, these probabilities can be quantified by open capture–recapture models. Robust design analyses (Williams *et al.* 2002a) combine information from short-term closed model surveys and long-term open model ones.

Karanth *et al.* (2006) examined the long-term dynamics of a tiger population in Nagarahole, India using a robust design survey. This study, spanning 10 years, resulted in the photo-capture of 74 individual tigers based on 5725 trap-days of field effort. Using likelihood-based open model capture–recapture analysis (Williams *et al.* 2002a), their study estimated an overall annual survival rate of 0.77 (SE 0.051), or a net loss of 23% tigers/year from dispersal and permanent emigration. Despite this loss rate, the tiger population showed an impressive stability over a long time, with an estimated annual increase of about 3%.

Dynamic models (Williams *et al.* 2002a) of cat populations can provide heuristic understanding of population processes useful for making predictions. Such models of tiger populations have provided insights into conservation and management issues (Karanth and Stith 1999; Linkie *et al.* 2006). Refinements to dynamic population models include spatially explicit versions that link demographics to movements of individuals (Ahearn *et al.* 2001; Carroll and Miquelle 2006). The effect of trophy hunting on lion populations was modelled using a similar approach (Whitman *et al.* 2007). However, the predictive power of dynamic models depends substantially on the values set for key parameters such as density, survival, and recruitment and characterizing how these parameters vary in different parts of the landscape. Consequently, estimation of such parameters using capture–recapture models described earlier remains central to understanding cat population dynamics.

Study of cats across landscapes and regions

Wild cats can also be studied at wider spatial scales, going beyond the level of individuals, communities, or populations that we examined earlier. Such studies may cover the entire species range or extensive landscapes within it, addressing questions about patterns of distribution of wild cat species at regional or continental scales.

From an evolutionary perspective, how and where a cat species evolved and spread over its geographic range is of interest (e.g. Kitchener and Dugmore 2000). Across historical timescales, the focus shifts to range contractions due to human impacts. From a conservation perspective, how ecological and anthropogenic factors affect population connectivity may be of concern (Sanderson *et al.* 2006). All the above questions involve measuring spatial distributions of cats and linking these to other relevant factors such as geological, climatic, and historical data, collected via remote sensing or through ground surveys. The key spatial data of interest are locations of cats either in the past or in the present.

Field methods and techniques for studying cats in landscapes

Field- or information-based surveys of cats

Landscape-scale ground surveys of cats involve major investments and are difficult to execute. The tiger distribution surveys in the Russian Far East (Hayward *et al.* 2002; Carroll and Miquelle 2006) and India (reviewed by Karanth *et al.* 2003) are examples of such efforts. More often, information on the regional distribution of cats is based on records of hunter offtakes or the skin and fur trade. Reliable accounts from the past can provide historical distributional and range data on wild cats.

Questionnaire surveys or structured search for expert opinion is a rapid and efficient tool for gathering spatial data on cats (Fig. 7.6). Such data can be summarized within a common data structure. Examples of such atlases are in map-database combinations developed for tigers (Sanderson *et al.* 2006) and jaguars (Sanderson *et al.* 2002b) through range-wide expert opinion and data summaries. At the simplest level, the basic data include identity of the person or team conducting the survey, when and where the survey was conducted, the search effort and field methods involved, and the evidence collected. Frequently, regional experts are also

Figure 7.6 Based on their expert knowledge, felid biologists map the spatial distribution of lions as part of the African Lion Range-Wide Priority setting exercise, Johannesburg, South Africa, January 2006. © A.J. Loveridge.

asked to define the current range of a species and identify core populations targeted by management actions.

These spatial databases can be structured to allow many different sources of information (e.g. track/sign, telemetry, reports, historical records, etc.) to be incorporated. In some cases, measures of data quality or reliability are also assigned. However, such distributional surveys are more difficult to conduct for cat species which are small, cryptic, and not readily recognized.

Global Positioning Systems (GPS)

Data on wild cats gathered through any kind of survey need to be recorded as spatial locations. Although traditional land survey tools are used to fix locations, GPS based on satellite triangulation have rapidly replaced them. Relatively low-cost GPS receivers can now be purchased from a wide range of vendors. Vegetation canopies and topographic features do restrict the receipt of signals and can compromise the precision of the measured location. Standing outside the canopy when first powering up a GPS receiver can facilitate subsequent tracking of satellite signals beneath forest canopies.

Remotely sensed data

Remote sensing commonly refers to observations of electromagnetic radiation reflected back from the earth to a sensor mounted on a satellite or aircraft. Most remote sensing is accomplished in an imaging mode, where a sensor makes numerous observations over a fixed grid, resulting in an image of the land surface. Depending on the type of electromagnetic radiation and its interaction with the surface, useful information about land cover type, its structural characteristics, and its landscape patterns can be obtained. Multi-temporal remote sensing can be used to determine changes in cat habitats due to land cover or land use changes. 'Mosaicking', or combining of multiple images, can provide broad spatial coverage across the geographic range of a cat species.

Primary characteristics of a remote sensor include its spatial grain (i.e. how large are the picture elements or pixels), its spatial extent (i.e. how large an area the sensor images in a single pass), the portions of electromagnetic radiation spectrum measured (commonly referred to as 'bands'), and its repeat interval. Sensors can be passive (i.e. measuring solar radiation reflected from the surface, as in optical and thermal sensors) or active (i.e. radiation of fixed frequency and polarity is emitted by the sensor and the return is measured, as in radar sensors).

Other considerations include radiometric sensitivity, atmospheric corrections, platform movements (for aircraft), image availability, and cost.

Because plants reflect light in the near-infrared bands, comparison of visible and near-infrared bands are of particular use in separating vegetated areas from bare soil, water, or rock and are the basis for the ubiquitous normalized difference vegetation index (NDVI). Optical sensors are sensitive to cloud cover and may require corrections for atmospheric conditions. Panchromatic optical bands on high spatial resolution satellites can have pixel sizes as small as 66 cm across, but at the cost of limited spatial extents (typically 10 km across) and high financial cost. Traditional land cover sensors like the Landsat series and SPOT have spatial resolutions of 10–30 m and extents from 60 to 185 km. Sensors like MODIS and AVHRR have coarser spatial resolutions (250–1000 m), but broader extents suitable for regional studies (1000–1500 km). The costs of images vary from totally free to over US$5000 per image, depending on sensor, period, and other considerations.

Radar sensors emit active radiation in microwave frequencies with known polarities. As a result, they can image the land by piercing through clouds. Microwave frequencies also interact with moisture in vegetation and soil and are used for structural measurement in the form of digital elevation models (DEM). Interpretation of satellite imagery requires application of techniques like supervised or unsupervised classification, calculation of vegetation indices, or other more advanced techniques. Increasingly, standard interpreted products are generated 'on-the-fly' for important sensors, for example, the biweekly global NDVI, land-cover change, and fire observation products from MODIS.

Remotely sensed land cover and topography data are very useful to understand how cats use their environment and to measure human impacts on cat habitats. For example, the human footprint analysis includes remotely sensed measures of land cover, access, and power infrastructure to estimate human influence globally at 1 km^2 resolution (Sanderson *et al.* 2002a). Range-wide analyses of cat distributions and conservation priorities (e.g. Sanderson *et al.* 2006) typically include a combination of remotely sensed and other geographic data, combined in a geographic information systems (GIS) platform.

Analytic methods for studying cats at range-wide level

Data integration and analysis: GIS

GIS are computer software tools that combine map analysis, database capabilities, and visualization of geographic data into a single package. GIS enables users to integrate a broad array of data based on a common reference system of spatial locations. Data can be collected at all scales, whether locations of an individual cat to the locations of populations across the range. Spatial data are represented as layers with two basic components: the map itself, showing configuration of geographic features, and other ecological data gathered from surveys, which are linked to geographic features. Analyses can then be performed on map information or the survey data or in combination. GIS are powerful analytical tools because of their versatility in combining any data with a spatial reference and due to widespread availability of remotely sensed information. They also provide sophisticated, although easily misunderstood, visualization tools for geographic information. A number of landscape ecological studies of tigers (Sanderson *et al.* 2006), lynx (Fernández *et al.* 2006; Niedzialkowska *et al.* 2006), and lions (Ogutu and Dublin 2004) have made imaginative use of such data.

Two kinds of geographic data are usually represented in a GIS: vector data or raster data. The vector data model is based on points, lines, and polygons, akin to a standard road map. In the study of cats, common vector data are locations recorded from radio-tracking, sign surveys, etc., with lines and polygons representing cat movement or ranges. Raster data, based on a grid system although less spatially precise, are better suited for representing large-scale spatial gradients, for example, as measured from remotely sensed data. Species distributions are typically based on the raster data models.

GIS requires that all data be comparable in a common, well-defined coordinate system. The spatial references that GIS uses to link all the data together are in overlays, like layers of a cake. If the different layers are out of alignment because they are expressed in different coordinate systems, they cannot be meaningfully compared. Common coordinate systems used in felid studies are latitude–longitude and Universal

Transverse Mercator (UTM) coordinates. The latter is a family of related map projections defined by 36 different zones around the world and north and south of the Equator. The UTM system is suitable for studies with extents less than 1000 km across. The most commonly used datum in use today is WGS84, which places its origin at the centre of the earth.

Powerful GIS software can build datasets, analyse them singly or in comparison to each other, and also visually display the results. Standard map analyses include the rapid measurement of distance, direction, and area; the spatial combination of two or more map layers; interpolation and extrapolation of spatial patterns; and query, selection, and summary. In combination with analysis techniques that work on data tables (e.g. sort, query, recalculation, etc.), these techniques are the basis of application of GIS to studies of felid biogeography and spatial ecology.

Field surveys provide essential information about locations of individual cats and their populations. It is often of interest to extrapolate such information to unsurveyed areas using species distribution models. Because such models are abstractions of reality, they help us to understand ecological relationships and can lead to new hypotheses about cat ecology. Such models include inverse distance modelling and kriging (Millington *et al.* 2001). Known or hypothesized relationships between ecological variables such as a cat's habitat selectivity and prey can be integrated in a GIS model to predict its potential spatial distribution. Harveson *et al.* (2004) synthesized 8 years of radio-telemetry data on ocelots with soil and vegetation maps using GIS to identify potential restoration habitats for ocelots in Texas. Various models of habitat suitability (Ray and Burgman 2006) have been built and assessed for jaguars (Hatten *et al.* 2005) and lions (Ogutu and Dublin 2004).

Bukit Barisan Selatan National Park (BBSNP) in south-west Sumatra marks the current southern range limit of tigers. Sumatra generally faces rampant deforestation which is constricting the available habitat for cats, including the tiger; moreover, BBSNP is particularly under pressure because of its long, narrow shape that creates a boundary of nearly 700 km for a 3568 km^2 park (Fig. 7.7). To study the effects of deforestation on tigers and other large mammals, a team of field scientists, landscape ecologists, and GIS analysts recently published an integrated analysis of camera-trapping data, remote sensing analysis, and predictive GIS modelling to measure the extreme peril faced by tigers (Kinnaird *et al.* 2003).

The team began with a baseline study of the rate of deforestation between 1985 and 1999, through interpretation of satellite images acquired over the park at six time points. They show that forest loss within the park averaged ~2%/year and that the forest loss within a 10 km buffer outside the park, though slower, effectively isolated the park from other forested areas during the study period. Through GIS analysis, Kinnaird *et al.* (2003) show that lowland forest was lost faster than montane forest (by a factor of six) and that forest on gentle slopes was lost faster

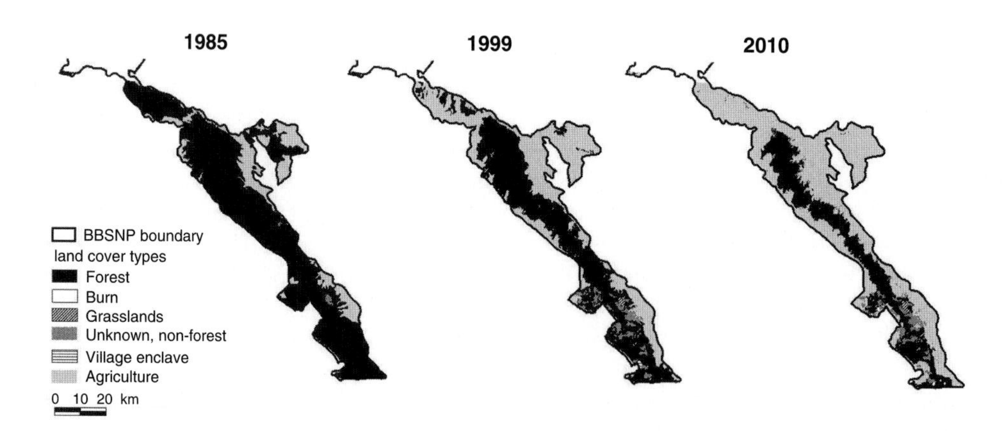

Figure 7.7 GIS analysis enabled (Kinnaird *et al.* 2003) to both measure past deforestation patterns (1985, 1999) and project future deforestation (2010) for Bukit Barisan Selatan National Park, Sumatra, Indonesia.

than forest on steep slopes (by a factor of 16). Further, they built a predictive model of forest loss using predictors of elevation, slope, and distance to road to estimate that, if trends continued, 70% of the area of BBSNP would have been deforested by 2010. These estimates compare well with a higher resolution image analysis of the same area conducted subsequently for 1972 to 2002 (Gaveau *et al.* 2007).

To connect these dramatic figures on deforestation to their ecological consequences for tigers in the BBSNP landscape, the authors took advantage of a spatially distributed camera-trap survey conducted in 1999. For each of the camera-trap locations, they used GIS to measure the distance from each camera to the edge of forest. The data were then analysed to show that tigers were avoiding the forest edge by at a distance of 2 km. Using a series of 2 km internal buffers, they defined core habitat for tigers in the centre of BBSNP, and then showed that the total amount of core habitat was declining even faster than the overall rate of deforestation, because of the long, narrow shape of the park. The consequence is a reduced number of female territories that can be contained within the core habitat, which implies increased intraspecific competition for habitat, decline in total amount of preferred habitat for the population, and increased use of suboptimal areas where conflicts with people are more likely.

When field data on locations of cats are available, a powerful array of statistical models can be used to investigate habitat relationships under the umbrella of resource selection functions (Boyce and McDonald 1999; Hopcraft *et al.* 2005) or habitat suitability models (Fernández *et al.* 2006). These techniques are based on estimating the value of various potential predictor variables at the observed locations using GIS analysis, then investigating the statistical relationship between the observed locations and the predictor variables based on multivariate logistical regression equations, bioclimatic or envelope models, and other advanced algorithms.

Spatially explicit models that integrate data from cat radio-locations, prior ecological knowledge, and remotely sensed information using GIS and viability analyses have resulted in some interesting studies on tigers (Ahearn *et al.* 2001; Carroll and Miquelle 2006). GIS-based models have also been used to extrapolate the distribution of cat populations based on inputs of survey location, land cover type and human influence (Sanderson *et al.* 2006).

Occupancy modelling and estimation

Spatial distribution of animals was traditionally measured by dividing the landscape or region of interest into some sort of a grid of habitat patches, and thereafter using field survey data on presence versus absence of the species of interest within each cell. However, such measurement of animal distribution systematically underestimates true habitat occupancy, because the cat's presence may sometimes go undetected, even when it is present in surveyed patches. This has been a fundamental drawback of studies of animal distribution using presence versus absence surveys (Karanth and Nichols 2002; Williams *et al.* 2002a; MacKenzie *et al.* 2006).

Recently, ecologists have developed a more robust approach known as occupancy modelling that essentially applies the capture–recapture sampling approach to overcome the problem of animal presence going undetected (Williams *et al.* 2002a; MacKenzie *et al.* 2006). These new methods that actually estimate probability of site occupancy, from replicated field surveys are just beginning to be used in the analysis of spatial data on wild cats. Because habitat occupancy surveys can use cat signs such as tracks, scats, visits to scent stations, or even historical data on cat distribution (Karanth *et al.* 2009), they have great potential for rigorously measuring processes like range contraction, expansion, patch colonization, and extinction.

Future challenges in studying wild felids

As the foregoing review shows, methods that felid ecologists use to study and understand wild cats have progressed enormously during the past five decades. However, during the same period, despite substantial conservation efforts, most wild cat species have become increasingly threatened. Wild cats have been hunted out, their prey base depleted, and habitats degraded, pushing many species into endangerment. Both human needs and greed have driven this decline of wild felids. There is thus a new sense of urgency to the felid conservation scientist's mission. How well we accomplish the mission will depend on how well we will apply the tools of our trade. In this context, there are some key challenges to be faced and opportunities to be grasped.

Studying some key aspects of wild cat ecology (e.g. phenomena such as predatory behaviour, social organization, land-tenure, or mortality) demand data that can only come from invasive studies that involve physical capture, immobilization, and tagging of individual wild cats. There is no other way such data can be gathered. Yet, in many countries, animal welfare concerns or simple ignorance now make it difficult to conduct such studies. The fact that properly conducted invasive studies carry only minor risks to individual cats, whereas they contribute hugely to the survival of the entire species, is sometimes lost sight of. After all, much of successful bird conservation (be it the study of migration flyways across the globe or the management of duck hunting in North America) has been possible only because of data derived from invasive methods such as capture and banding, that apparently enjoy wide public support. There is need for felid biologists to educate the public more widely than they have done so far.

Technology has progressed rapidly in recent times, vastly improving the computers, digital cameras, televisions, and a bizarre new array of communication devices we now use. Some of these tools are virtually unrecognizable from those of a decade ago. However, amidst all these leapfrogging technologies, a few key tools that are really critical to studies of felids have remained relatively unchanged: the cumbersome VHF telemetry equipment and low-tech camera traps that biologists are still compelled to use are some examples. Although fieldwork that cat biologists conduct undoubtedly excites the public, it has not yet attracted the attention of techno-entrepreneurs who can improve our key tools.

Most of the field data from cats are not generated within the framework of standard experimental statistics, which involves randomization, replication, treatments, and effects. Yet, felid biologists often try to shoehorn their largely observational field data into standard statistical analyses designed for such experimental data. Quantitative approaches to the design and conduct of field studies of cats and the subsequent analyses and interpretation of data have not kept pace with the synergistic developments in the fields of biostatistics, quantitative ecology, information theory, and the development of low-cost computing power during the past quarter-century. Because analytical approaches more appropriate for field data are now available (Williams *et al.* 2002a), cat researchers must try to explore and embrace these. On the other side of the coin, felid biologists now have access to an array of sophisticated laboratory tools ranging from molecular genetics to remote-sensing and GIS, which can comfortably keep them occupied between the four walls of the laboratory. Sometimes, we worry that easy recourse to these laboratory tools may even be dimming the cat biologist's ardour for traditional fieldcraft, which is still critical to getting sound data. Loss of field skills, the increasing inability to read stories that cats write on jungle pathways, is palpable not just among wildlife biologists, but also among people in local communities living next to wild cats. This erosion of skills deserves serious introspection among felid biologists: we should once again draw inspiration from classical field studies, such as those of Hornocker (1970) and Schaller (1972).

We are hopeful that felid biologists will overcome these and other challenges, and as a result these wonderful animals so worthy of study and interest will survive and thrive into the future centuries.

Acknowledgements

We would like to thank Ms Divya Vausdev, who helped us in the preparation of this manuscript. We are grateful to David Macdonald, Andrew Loveridge, and three anonymous reviewers for their very useful comments and suggestions. K.U. Karanth and E.W. Sanderson are grateful to the Wildlife Conservation Society, New York, for supporting their work on wild cats over the years.

CHAPTER 8

Felids *ex situ*: managed programmes, research, and species recovery

David Wildt, William Swanson, Janine Brown, Alexander Sliwa, and Astrid Vargas

Iberian lynx dam with captive-born cubs. © Luis D. Klink.

Introduction

The first priority in species conservation is to ensure that wild individuals exist and contribute to viable populations in nature. However, there is also value in managing *ex situ* populations. Wild cats are popular attractions in zoos, inspiring the public to learn about this taxon and the importance of protecting native habitats. *Ex situ* populations contribute to important scientific studies in establishing biomedical norms and discovering biological phenomena in veterinary medicine, behaviour, and physiology. For instance, many of the drug regimens for field anaesthesia and what constitutes normative (or abnormal) haematological traits were developed initially in captive felids (Lamberski and West 2007). Certainly, most of what is known about felid reproduction (Wildt *et al.* 1998) and phylogeny (Johnson *et al.* 2006b) has been derived from studying zoo animals. *Ex situ* approaches allow hypothesis-based studies under relatively controlled conditions that can produce non-confounded insights into biological mechanisms. There are many examples, from the fascinating finding of ovarian suppression in paired cheetahs (Wielebnowski *et al.* 2002b; see below) to wide species variations in reproductive seasonality, including some extremes in sensitivity to photoperiod (Brown *et al.* 2002; see below). Such phenomena would probably never have been discovered without studying captive animals.

Finally, *ex situ* populations can assist in retaining existing genetic diversity as insurance to support wild populations, especially those in danger of extinction. There are two programme types, the first being the zoo-oriented Species Survival Plan© (SSP; Association of Zoos and Aquariums 2008) and European Endangered Species Programme (EEP; European Association of Zoos and Aquaria 2008). These efforts are organized by the regional Felid Taxon Advisory Group (e.g. Felid Taxon Advisory Group 2008) comprised of curators, veterinarians, and research scientists, who identify numbers of 'cat spaces' available in accredited zoos and then create a regional collection plan. This roadmap is comprised of a subset of the 36 wild felid species believed to deserve captive management attention, while also being chosen because of founder availability, exhibit potential, and educational and research values. Currently, North America and Europe collectively manage breeding programmes for 12 of 24 cat species included on the International Union for Conservation of Nature (IUCN) Red List (IUCN 2007; Swanson *et al.* 2007) as well as three species not listed at risk (i.e. of 'least concern'; Table 8.1). Within regions, animals are shared to avoid inbreeding while striving to retain gene diversity for the next 50–100 years. Most of these programmes have no linkage to the field or to a specific recovery plan.

The second programme type is similar to the first, except that captive management is instigated in response to a recovery programme for wild conspecifics that are facing threats in nature. Animals are maintained purposely *ex situ* to (1) serve as a hedge to avoid species extinction; (2) collect data applicable to conservation; and/or (3) sustain the species by propagation, including producing offspring for release or reintroduction. *Ex situ* breeding programmes that are tightly linked to *in situ* recovery are rare for felids, with contemporary examples being the Iberian lynx (Vargas 2006; see below; Ferreras *et al.*, Chapter 24, this volume) and the Amur leopard (Hötte 2007). Iberian lynx conservation breeding efforts began in 2004 with the goal of returning captive-born individuals to areas of historical occupancy by 2010 (Vargas *et al.* 2008). Captive-born Amur leopards, managed by EEP/SSP programmes, are being considered for reintroduction in the Russian Far East (Hötte 2007). Revitalization of the endangered Florida panther population has also benefited from study of captive pumas, albeit in the absence of a breeding/reintroduction programme (USFWS 2008a; Onorato *et al.*, Chapter 21, this volume). There is also some linkage between *ex situ* and *in situ* populations of cheetahs, Pallas's cats, and black-footed cats through the use of frozen semen (Crosier *et al.* 2006; Swanson *et al.* 2007; see below).

This chapter has three main purposes, the first being to examine the process of developing an *ex situ* breeding programme that can contribute to a self-sustaining and genetically viable felid population. This includes a brief appraisal on the potential of reintroducing felids to bolster wild populations. Because reproduction is the underpinning discipline for *ex situ* breeding, our second intention is to review advances in the reproductive sciences that have contributed significantly to new information and understanding, many of which have practical applications. Finally, we provide an important contemporary example that illustrates the challenges to developing a real-world, high-priority *ex situ* programme in support of *in situ* conservation.

When is an *ex situ* breeding programme required?

Managed *ex situ* collections are unnecessary when wild populations are thriving in healthy, multiple habitats or when captive space or founders are limited and/or when no high-priority research has been identified. In contrast, in the face of a rapidly declining wild population, it is always prudent to at least consider the value of captive breeding, especially if there is an uncontrollable catastrophe (e.g. a devastating disease epidemic) or when *in situ* conservation is lacking or ineffective. There is a generic *ex situ* policy produced by the IUCN that is largely based on animal numbers in the wild and geographic area of occupancy (IUCN 2002). But these guidelines are of limited use for felids because of the lack of quality information on current status and distribution for many wild populations. Furthermore, in

Table 8.1 Cat species managed in an SSP and/or EEP in accredited North American and European zoos.

Cat species			
Common name	Latin name	Management programme	IUCN status[a]
Asian golden cat	*Catopuma temminckii*	EEP	Vu
Black-footed cat	*Felis nigripes*	SSP/EEP	Vu
Cheetah	*Acinonyx jubatus*	SSP/EEP	Vu
Clouded leopard	*Neofelis nebulosa*	SSP/EEP	Vu
Fishing cat	*Prionailurus viverrinus*	SSP/EEP	Vu
Geoffroy's cat	*Oncifelis geoffroyi*	EEP	NT
Jaguar	*Panthera onca*	SSP	NT
Lion	*Panthera leo*	SSP/EEP	Vu
Leopard	*Panthera pardus*	SSP/EEP	LC
Margay	*Leopardus wiedii*	EEP	LC
Ocelot	*Leopardus pardalis*	SSP	LC
Pallas's cat	*Otocolobus manul*	EEP	NT
Sand cat	*Felis margarita*	SSP/EEP	NT
Snow leopard	*Uncia uncia*	SSP/EEP	En
Tiger	*Panthera tigris*	SSP/EEP	En

[a] IUCN classification categories (En, endangered; Vu, vulnerable; NT, near threatened; and LC, least concern) for felids based on species level (IUCN 2007).

contrast to large-sized felids (e.g. tiger, lion, puma, and cheetah), long-term monitoring data on most cats less than 20 kg are deficient or non-existent. In many regions, *in situ* priorities for felid conservation often are not addressed or even recognized. Furthermore, natural delays associated with protracted habitat protection, connection, and restoration would likely contribute to inevitable losses in populations or species.

Therefore, there are legitimate reasons to consider seriously the option of *ex situ* programmes to minimize loss of felid species in the wild. One way of assessing the need for captive breeding is through a Population and Habitat Viability Assessment (PHVA; Beissinger and Westphal 1998), a participatory workshop attended by diverse stakeholders with a common goal of ensuring species persistence and/or recovery. Such meetings are often convened by a government agency, with invitees including local, regional, or federal administrators and managers, scientists (from the field, academia, breeding centres, and zoos), public awareness specialists, and local people (as recovery programmes generally affect human communities). The process assesses the likelihood of a species surviving in nature (or, alternatively, declining to extinction). Understanding the specific risks allows the development of mitigation strategies. Captive breeding is sometimes recommended for species recovery, as was the case for the Iberian lynx (Conservation Breeding Specialist Group 1998; see below). *Ex situ* propagation is, however, never recommended as a sole corrective strategy, but rather one of multiple potential interventions. Sometimes, PHVA analyses reveal that *ex situ* management is unnecessary. This was the case with the Florida panther, a subspecies already at critically low numbers (<50 specimens in nature) and with extremely low genetic diversity (USFWS 2008a). Biomedical assessments of free-ranging panthers revealed cardiac defects, high proportions of abnormally shaped sperm, and cryptorchidism (a genetic defect where one or both testes are retained in the abdominal cavity with bilaterally cryptorchid males being sterile; also see below). In this situation, there was no point in beginning a captive breeding programme. The subspecies had experienced a population bottleneck so severe that the result was serious anatomical, health-related defects (Roelke *et al.* 1993b). This lack of viable founders helped guide subsequent conservation initiatives to focus solely on the *in situ* population and hybridization with another puma subspecies (see below).

What is the potential for reintroducing captive-born felids?

The feasibility of reintroduction always arises early in the context of recovery-related *ex situ* management. Although there are formal guidelines offered by the IUCN (1995), reintroduction of captive-born wildlife remains largely experimental and has not been systematically studied in felids. The World Association of Zoos and Aquariums' conservation strategy emphasizes that reintroductions can benefit natural biological systems (Tilson and Christie 1999; World Association of Zoos and Aquariums 2008). However, success is low and the entire process extraordinarily complicated and expensive. One analysis concluded that only 11% of reintroduction efforts with captive-born wildlife (16 of 145 programmes) resulted in established, self-sustaining populations (Beck *et al.* 1994). The reason is due to species-specific requirements and a wide variety of biological complexities, ranging from developing efficient feeding and social skills to recognizing appropriate shelter, to avoiding predators and humans (Miller *et al.* 1994; Griffin *et al.* 2000; Vargas *et al.* 2000). Some species require pre-release conditioning to increase survival (Biggins *et al.* 1998), whereas others rely on post-release support and acclimatization (Stanley Price 1989). Concerns extend to political, social, and economic impacts of animal releases.

Christie and Seidensticker (1999) have summarized logistical challenges of reintroducing captive-born felids. First is assurance that there is (1) control of the cause of the original species decline; (2) adequate habitat and food sources; (3) long-term programme monitoring; and (4) in the case of large and dangerous felids, protection of, or compensation to, local people. Although captive-born animals may be less able to survive in nature compared to wild counterparts (Griffith *et al.* 1989; Beck *et al.* 1994; McPhee 2004), hunting behaviour is instinctive in felids, and skills can be learned with pre-release training (Christie and Seidensticker 1999). The experimental release of 'western' pumas in Florida (related to Florida panther recovery) demonstrated that all animals had the capacity to kill prey within a week of release whether wild-caught or captive-born. As summarized by Christie and Seidensticker (1999), the captive-born pumas were actually more successful than wild-caught counterparts at settling into a home range, adapting to a social system, reproducing, and surviving. However, pumas born *ex situ* experienced more animal/human conflict, which predictably would be one of the most significant challenges to successful felid reintroduction programmes.

Some examples of carnivores born *ex situ* and then introduced into nature include the Mexican (Arizona Game and Fish Department 2006) and red (Phillips 1995) wolves, black-footed ferret (Miller *et al.* 1994; Vargas *et al.* 2000), Eurasian lynx (Breitenmoser *et al.* 2006b), and European wildcat (Hartmann 2006). Such releases require facilities and protocols that reduce exposures of offspring to humans with aversive reinforcement (taste, odour, and electric shock; Christie and Seidensticker 1999). Rearing also should include expansive enclosures comprised of natural features, with opportunities to kill live prey followed by acclimatization to the reintroduction area before release (Vargas *et al.* 2000). Although expensive and challenging, reintroduction of felids could be feasible for certain species under the right circumstances (Christie and Seidensticker 1999), especially if socio-political implications are carefully considered (IUCN 1995).

What are the major factors associated with starting and implementing an *ex situ* management programme?

Cost and facility plans

Ex situ management programmes incur substantial costs for buildings, enclosures, food, animal and veterinary care, and research. Availability of financial resources should not compromise parallel *in situ* recovery efforts. There is no better way to foster mistrust and destructive competition than to require field and captive teams to battle for the same source of financial support. Actual design of the *ex situ* facility can benefit from close interaction with accredited zoos and breeding centres that have decades of experience building state-of-the-art facilities that promote breeding in some of the rarest species on earth. There is published or anecdotal evidence for successful reproduction in captivity for 32 of the 36 wild felid species. Important information is available through various sources (Association of Zoos & Aquariums 2008; Cat Specialist Group 2008; Conservation Breeding Specialist Group 2008; Consortium of Aquariums, Universities and Zoos 2008; European Association of Zoos and Aquaria 2008; World Association of Zoos and Aquariums 2008).

Long- versus short-term programmes

Conventional zoo-based captive breeding programmes for felids are considered long term, being designed and managed to maintain at least 90% of extant gene diversity for one to two centuries (Soulé *et al.* 1986). The minimum number of recommended founders is greater than (or equal to) 20 individuals, each expected to contribute 7–12 offspring (Lacy 1994). Ultimately, the total number of animals managed depends upon genetic variation in the founder stock and species-specific considerations, including generation interval, fecundity, and effective population size (Ballou *et al.* 1995; Frankham *et al.* 2009). To illustrate species variation and a sense of scale, there are 251, 147, 59, and 28 individuals in the cheetah, Amur tiger, fishing cat, and black-footed cat North American SSPs, respectively.

By contrast, *ex situ* breeding that is linked to a species recovery plan is best considered short term, largely related to its purpose of being insurance against extinction, generating new knowledge, and producing offspring for re-establishing or supporting existing wild populations. While having a more immediate and pragmatic impact than long-term zoo management programmes, conservation-linked captive breeding may in reality be long-standing because recovery in nature can be a protracted, seemingly endless challenge. This is especially relevant for felids because of serious habitat availability issues and ongoing human pressures. A pertinent example in another taxon is the black-footed ferret, where *ex situ* breeding has been obligatory for more than 20 years to produce enough kits for reintroduction to meet wild population targets (Conservation Breeding Specialist Group 2003b).

Facility location and organizational management

It is prudent to establish the breeding programme in the range country and near the habitat of the species of concern. This improves the chances that *ex situ* and *in situ* activities will be mutually supportive and interactive. It facilitates moving animals from the wild to captivity with minimal stress and expense while avoiding the legal complications of exporting and importing endangered species. Such efforts also stimulate local enthusiasm and pride in the targeted species. Perhaps most important is the opportunity to train local people to become the professional experts who ultimately are responsible for conserving the species. This is compatible with recommendations of the Convention on Biological Diversity (Article 9) that states programmes should 'adopt measures for the *ex situ* conservation of components of biological diversity, preferably in the country of origin'. Even so, it is worth noting that the IUCN (2002) has recommended locating captive programmes outside the original range country if the population is threatened by uncontrollable natural or unnatural disruptions. Thus, under these conditions or in the absence of any community capacity, it may be prudent to consider alternative sites.

The leader of the *ex situ* management team must have excellent communication skills, be able to implement partnerships with diverse stakeholders, and be empowered to make decisions and take action. The overall *ex situ* management programme, including its goals and activities, should be coordinated under the umbrella of a broader recovery programme for the target species. Meanwhile, routine operations, including recommended matings (to sustain gene diversity) and research, are best organized in a zoo-based SSP- or EEP-type style. The *ex situ* team formally solicits disciplinary advisers representing the areas of genetics, population biology, behaviour, husbandry, nutrition, veterinary medicine, and reproductive sciences. These specialists provide guidance or, preferably, directly participate with the team in conducting priority research and recommending management changes. (The types of progress that can be made by this coordinated, multidisciplinary approach between managers and scientists are illustrated below.) The advisers meet as a group at least annually to review progress and plan the next high-priority actions. It is also sensible to have external programme evaluations every 3–5 years.

Genetics, founders, and population biology

Felids are highly sensitive to inbreeding depression. For example, sib-to-sib matings in the domestic cat increase the number of malformed ejaculated sperm by more than 40% in first generation inbred offspring (Pukazhenthi *et al.* 2006). Lost heterozygosity in the Florida panther resulted in a cascade of anomalies (Roelke *et al.* 1993b; Onorato *et al.*, Chapter 21, this volume). Approximately 94% of ejaculated sperm were pleiomorphic, with the most frequent defect being an acrosomal malformation (a sperm structure critical for fertilization; Barone *et al.* 1994). There was a high incidence of cryptorchidism (affecting 85% of cubs) and cardiac abnormalities, especially an atrial septal defect that caused heart murmurs and mortalities. Florida panthers were also seroprevalent for various infectious pathogens (especially feline panleukopenia virus, feline calicivirus, and puma lentivirus [feline AIDS]), all consistent with homogeneity of genes involved in immune defence (Roelke *et al.* 1993b). Because some of these abnormalities were later reduced or eliminated by hybridization with another puma subspecies, the overall lesson is that 'genes matter'. Therefore, any breeding plan must implement strict genetic management guidelines to avoid inbreeding.

Number of founders required for a breeding programme is related to expected duration of need. Longer-term expectations require more founders. It is important to ensure that extraction of founders does not compromise reproductive fitness of the wild population, and such information is attainable from computer modelling (Palomares *et al.* 2002b). If a population is in free-fall and extinction a possibility, delaying capture of founders can have profound consequences. Such was the case with the black-footed ferret that only existed in nature at a perilously low 18 individuals, captured from the wild. Some of these were siblings, resulting in seven true founders, and only five of these ferrets are currently represented in captivity, resulting in overall levels of genetic diversity. And, while the *ex situ* population has grown remarkably (to >6200 ferrets over >20 generations), the lack of heterozygosity may be responsible for recent observations of occasional partial uterine and renal aplasia, systolic heart murmurs, hydrocephalus, hydronephrosis, amyloidosis, cryptorchidism, and consistently poor sperm quality (Bronson *et al.* 2007; Santymire *et al.* 2007). In retrospect, it would have been sensible to implement the captive programme when the *in situ* population was larger, thereby allowing the rescue of more heterogeneity. In general, most *ex situ* management programmes need 20–30 unrelated founders to ensure retaining adequate genetic diversity (Frankham *et al.* 2009).

Management tactics to ensure natural behaviours, well-being, and health

Animals in any *ex situ* programme must be intensively managed to promote optimal health, reproduction, and genetic diversity. A key husbandry challenge for felid recovery is to strike a balance between fostering natural behaviours (hunting, territoriality, and social interactions) and creating an environment to encourage mating. Natural behaviours may erode faster in captivity than gene diversity, which may be lost due to poor breeding management (McPhee 2004). A more 'domesticated' captive population might increase reproductive success, but be less adapted for reintroduction. For instance, long-term survival of reintroduced black-footed ferrets was 10 times higher (20% versus 2%) for captive-born animals raised in naturalistic pens compared to conspecifics reared in indoor cages (Biggins *et al.* 1998). Clark and Galef (1980) demonstrated that Mongolian gerbils raised in standard cages become tame, have accelerated growth, and experience early sexual maturity. In contrast, gerbils reared in tunnel systems revert in one generation to a wild-type phenotype (aggressive behaviour, difficult to restrain, slower development, and increased adrenal weight relative to body weight). It is well established that *ex situ* facilities can be designed to encourage natural behaviours (Mellen and Shepherdson 1997; Shepherdson *et al.* 1998). For felids, providing enclosures with abundant hiding places and opportunities to mark territories, climb, gnaw on large bones, and exercise on lure courses are important considerations (Mellen and Shepherdson 1997; Shape of Enrichment 2008).

There are detailed husbandry manuals for many of the large cat species managed within the North American Felid Taxon Advisory Group and appropriate SSPs (e.g. tiger, cheetah, clouded leopard, and jaguar); others are being developed by the European Association of Zoos and Aquaria's Felid Taxon Advisory Group (e.g. all leopard subspecies, Eurasian lynx, and Iberian lynx, among others). These documents contain information on natural history, taxonomy, health, nutrition, reproduction, behaviour, enrichment, and enclosure design, all supported by relevant bibliographies. Although there are obvious species specificities, husbandry recommendations for one species are often applicable to others until specific needs are identified for the target species. Some husbandry manuals are available online through regional zoo associations (European Association of Zoos and Aquaria 2008; World Association of Zoos and Aquariums 2008) or can be acquired by contacting the Felid Taxon Advisory Group (2008) or SPP or EEP coordinators.

A crucial component of management is a preventative disease programme that includes on-site, state-of-the-art veterinary personnel and facilities. Because felids are vulnerable to a wide range of insidious pathogens, appropriate quarantine facilities and protocols are essential, including barriers that prevent captive individuals from interacting with wild conspecifics or other predators living in adjacent native habitat. No less important is the capacity to conduct safe anaesthesia, effective curative medical protocols, vaccination (and/or vaccine development), disease research, and inevitable post-mortem assessments. There is a vast array of resources for helping understand the implications of disease in *ex situ* programmes, as well as screening methods and risk assessments (Conservation Breeding Specialist Group 2003a). Detailed procedures are available in felid husbandry protocols (SSP and/or EEP) with highly specific recommendations for the cheetah, Amur leopard, tiger, and black-footed cat. Information on felid veterinary specialists, manuals, and recommendations are available (American Association of Zoo Veterinarians 2008; Veterinary Specialist Group 2008).

Certain medical procedures (and many research activities) require anaesthesia. Appropriate protocols are paramount to avoid animal mortality, and even small cats can inflict serious injury on humans. As a taxon, felids respond well to injectable anaesthetics alone, especially healthy individuals (Swanson *et al.* 2007). For minor procedures (e.g. morphological measurements, venepuncture, and skin biopsy), a surgical plane of anaesthesia is usually unnecessary, and the animal can be handled safely after injecting dissociative anaesthetic (e.g. ketamine or tiletamine) alone or combined with a sedative (e.g. diazepam, midazolam, or zolazepam) or a reversible alpha-2 agonist (e.g. xylazine or medetomidine; Kreeger *et al.* 2002; Grassman *et al.* 2004). For electroejaculation or insertion of intra-peritoneal transponders, deep anaesthesia is needed, requiring higher dosages of injectable combination agents or similar induction followed by inhalant gas anaesthesia (e.g. isoflurane, sevoflurane, or halothane; Swanson *et al.* 2007). Analgesics (e.g. butorphanol, buprenorphin, or morphine) are recommended for all procedures potentially producing pain. Drug dosages vary, and there is no standard 'cookbook' anaesthetic protocol that is applicable across species. For example, tiletamine–zolazepam provides effective anaesthesia for many felids, but may be unsuitable in the tiger, puma, and jaguarundi due to inability to achieve proper anaesthetic level or post-recovery neurological complications. Other factors also influence drug effectiveness (age, health status, activity and stress levels, and captive versus wild environments). Because of these complexities, the presence of an experienced veterinarian is always recommended, especially for animal assessment prior to drug delivery and for physiological monitoring until recovery.

An especially important role for nutrition

Most wild felids managed *ex situ* in zoo-based SSP and EEP programmes are fed commercially prepared carnivore rations using detailed nutritional protocols (Felid Taxon Advisory Group 2008; Howard and Allen 2008). Food consists of horse meat or beef, balanced with vitamin and mineral premixes that are available as frozen or canned products. Outside these highly organized breeding programmes, however, it is common that felids are fed substandard diets, often with muscle meat as the sole food source (Howard and Allen 2008). Diets comprised of high fat, low protein, and imbalanced vitamins and minerals are well known to cause obesity, bone instability, joint laxity, eye lesions, cardiac defects, decreased fertility, poor lactation, and lowered cub survival in felids (Howard and Allen 2008). These problems are commonly encountered in zoos in developing countries that ironically often maintain incredibly valuable specimens, most being wild-born (i.e. founders). These felids are often fed poor foods because of inadequate information limited access to high-quality ingredients, or lack of financial resources to purchase commercial diets.

The profound impact of diet on felid health and reproduction has been demonstrated. One survey of 44 zoos in Latin America involved 185 males of 8 endemic cat species (Swanson *et al.* 2003). Only ~20% of these individuals (virtually all wild-born) had ever reproduced in captivity, and less than 30% were producing viable sperm numbers and quality (Swanson *et al.* 2003). The predominant food source was nutrient-deficient chicken heads and necks, beef, or horse meat (without bones and any vitamin and mineral supplementation). A subset of the ocelot, margay, and tigrina males (in Brazilian zoos) was fed the chicken neck diet supplemented with a vitamin–mineral premix and exhibited an increased sperm concentration by ~300% and a reduced incidence of sperm malformations by ~40% (Morais *et al.* 2002). A similar study was conducted in Thailand on 31 wild, zoo-maintained felids of 4 endemic species, all fed non-supplemented chicken diets. The simple, inexpensive provision of a human multivitamin preparation improved sperm concentration by ~250% across species (Howard and Allen 2008). Although as yet untested, the benefits are likely to have been derived from the added fat-soluble vitamins A and E (Howard and Allen 2008).

There are two interesting sets of observations indicative of a nutritional impact on female reproduction. In a study of clouded leopards in Thailand, improving diets by adding whole prey, chicken meat, and vitamin and mineral supplements increased ovarian cycling in six of eight females within 6 months (based on faecal hormone patterns; see below), with one of the females conceiving and carrying the pregnancy to term (Pelican *et al.* 2007). Swanson *et al.* (2002), while conducting oocyte recoveries for *in vitro* fertilization (IVF; see below), discovered that more eggs were successfully fertilized when recovered from ocelot and tigrina donors fed a vitamin- and mineral-enriched (69%, 96 of 140 oocytes) than non-supplemented diet (30%, 13 of 43 oocytes).

Advances in reproductive sciences to benefit biological understanding and management

Of the disciplines contributing to *ex situ* felid programmes, the most extensive new knowledge has been gained in the reproductive sciences. This work has largely focused on addressing the similarities, and especially the differences, in reproductive mechanisms across the Felidae. Much of this information has also been of practical value in exploring new ways to improve management and genetic health of *ex situ* collections.

Reproductive studies began on wild felids in the late 1970s and early 1980s, at the same time zoos began genetically managing species to avoid inbreeding. While some cats bred at will (e.g. lion and tiger), others failed to reproduce consistently (e.g., cheetahs), and still others displayed bizarre behaviours (e.g. the male clouded leopard often killing or maiming a designated mate, even one in oestrus). There was early interest in the potential of using assisted reproduction techniques, such as artificial insemination (AI), IVF, and embryo transfer, to meet the new ideals for maintaining gene diversity. These first attempts were based on exploiting techniques being rapidly developed to enhance livestock production and to overcome human infertility. But early efforts to boost reproduction in the cheetah (relying almost exclusively on cattle technologies) failed and ultimately triggered the conclusion that a 'cheetah is not a cow' and that the priority should be to understand species specificities (Wildt *et al.* 2001). There was a refocus on species uniqueness (rather than the quick fix), and the result has been a fascinating new array of information. For example, conventional wisdom is that cats are seasonal breeders and induced ovulators (ova being released in response to multiple copulations). However, the ocelot and margay display distinctive ovarian cycles throughout the year, in contrast to the tightly regulated Pallas's cat that generally cycles for only 2 months annually (Moreira *et al.* 2001; Brown *et al.* 2002; see below). Furthermore, occasional-to-frequent spontaneous ovulation (in the absence of copulation) has been documented in the clouded leopard, lion, leopard, fishing cat, Pallas's cat, and margay, as well as certain genetic lines of domestic cat (Pelican *et al.* 2005; Brown 2006). Data on the physiology of reproduction are now available on 29 of the 37 wild felid species (Wildt *et al.* 1998; Brown 2006).

Assessing fertility potential and reproductive condition

Advances have been made using four tools: laparoscopy, ultrasonography, electroejaculation, and hormonal assessment. Laparoscopy involves minor surgery to insert a narrow fibre-optic telescope into the abdominal cavity to make direct observations (Harrison and Wildt 1980). This allows characterization and photography of felid reproductive tract morphology and guiding of sperm deposition for intrauterine AI and oocyte harvest for IVF (see below). Ultrasonography, which is non-invasive and indirect, allows similar evaluations and is also useful for assessing testicular integrity (Hildebrandt *et al.* 2003). Both techniques have been used in felids to identify ovarian activity, pregnancy, foetal development, ovarian or paraovarian cysts, uterine cystic hyperplasia, and pyometra. Both also can be used to examine the liver, kidney, and spleen and to guide needles when obtaining biopsies. These procedures require anaesthesia, although pregnancy diagnosis via ultrasound has been possible in behaviourally conditioned felids through cage bars (Suedmeyer 2002).

Electroejaculation has been used extensively in anaesthetized felids to collect viable spermatozoa. Expected values for ejaculate volume and sperm concentration, motility, and structural integrity are available for 27 wild felid species (Howard 1993; Swanson *et al.* 2003). Electroejaculation has been used to assess reproductive status of free-ranging lions (Wildt *et al.* 1987a), cheetahs (Wildt *et al.* 1987b), Florida panthers (Barone *et al.* 1994), jaguars (Morato *et al.* 2001), black-footed cats (Swanson *et al.* 2007), and Pallas's cats (Swanson *et al.* 2007; Fig. 8.1). Used appropriately, this technique is safe, with no adverse consequences on natural breeding performance (Wildt *et al.* 1993). While often recognized for its role in assisted breeding (see below), this procedure has been most valuable for generating new information on testicular function. Especially interesting has been the phenomenon of teratospermia, a condition in which 15 felid species and subspecies consistently ejaculate more than 60% abnormally shaped sperm (Pukazhenthi *et al.* 2006). Defects range from a simple bending of the spermatozoon's midpiece or flagellum to extensive deconstruction of the mitochondrial sheath and acrosome. Malformed sperm are more prevalent in species or populations with diminished genetic diversity. As an example, outbred lions of the Serengeti ecosystem with abundant genetic variation produce more structurally normal sperm (~75%) than inbred counterparts, either isolated in the Ngoronogoro Crater (an extinct volcanic caldera; ~50%) or derived from a remnant population of wild lions in western India (~43%; Wildt *et al.* 1987a; for more detail on linkage between gene diversity and sperm pleiomorphisms, see Pukazhenthi *et al.* 2006). Teratospermic males have normal pituitary function, but often low circulating testosterone. Sperm (including morphologically normal-appearing cells) from teratospermic donors have less ability to undergo the acrosome reaction (needed for fertilization), bind, penetrate, and decondense in the oocyte's cytoplasm. In short, these sperm do not participate in cellular fertilization. Additionally, hyperactive spermatogenesis has recently been discovered in one domestic cat genotype with teratospermia (Pukazhenthi *et al.* 2006). These males produce many more sperm than normospermic controls, suggesting that teratospermic males have evolved a compensatory capacity to increase sperm output to overcome a genetically induced reduction in structurally normal sperm numbers. Such a mechanism makes sense given that some threshold number of sperm is required in felids to achieve fertilization. Most felids exhibit a high frequency of copulatory activity. This strategy could provide the male not only with the opportunity to induce multiple ovulations in the female, but also to deposit an adequate number of morphologically normal sperm in the vagina, thereby increasing reproductive success.

The most popular tool in felid reproductive science has been non-invasive hormonal monitoring because it generates substantial information without animal anaesthesia or disturbance (Brown 2006). The only requirements are the capacity to recover relatively fresh (up to 24 h old) voided faeces and access to a laboratory that has the resources and skills to measure gonadal (ovarian or testicular) hormonal metabolites. There are significant data now for 16 felid species on normal versus abnormal oestrous cyclicity, pregnancy versus pseudopregnancy (ovulation and progesterone production in the absence of pregnancy), effect of photoperiod on seasonality, and the impact of suboptimal husbandry on reproductive success. Findings have confirmed that some species (e.g. ocelot and cheetah) show waves of oestrogen activity throughout the year in the absence of progesterone increases (Fig. 8.2a) because ovulation only occurs after mating. Others exhibit inexplicable interruptions in cyclicity (e.g. cheetah) or bouts of spontaneous ovulation (in the absence of a male), as confirmed by sustained rises in progesterone excretion after an oestrogen surge (e.g. margay; Fig. 8.2b). Male gonadal hormones are also measurable by analysing faecal androgen metabolites (Brown *et al.* 1996), including documenting seasonality in the Pallas's cat (Brown *et al.* 2002; Newell-Fugate *et al.* 2007).

Endocrine findings also have management implications. For example, faecal oestrogen analyses have provided the first evidence of mutual reproductive suppression in female cheetahs maintained as pairs. Re-initiation of normal cycles occurs upon separating the females, even when animals remain in close visual and olfactory proximity (Fig. 8.3; Wielebnowski *et al.* 2002b). This finding is logical considering that cheetahs are solitary in nature. Housing cheetah females as singletons is now a routine breeding management strategy. Another interesting observation has been made in Pallas's cats exposed to a 'Festival of Lights' at a zoo for about a month before the normal expected breeding season (Brown *et al.* 2002). This species has evolved a tight and short breeding season that clearly is disrupted by excessive artificial lighting (Fig. 8.4). In this case, the problem was eliminated simply by moving the animals to an isolated area the next year where normal cycles occurred, and young were born (Brown *et al.* 2002).

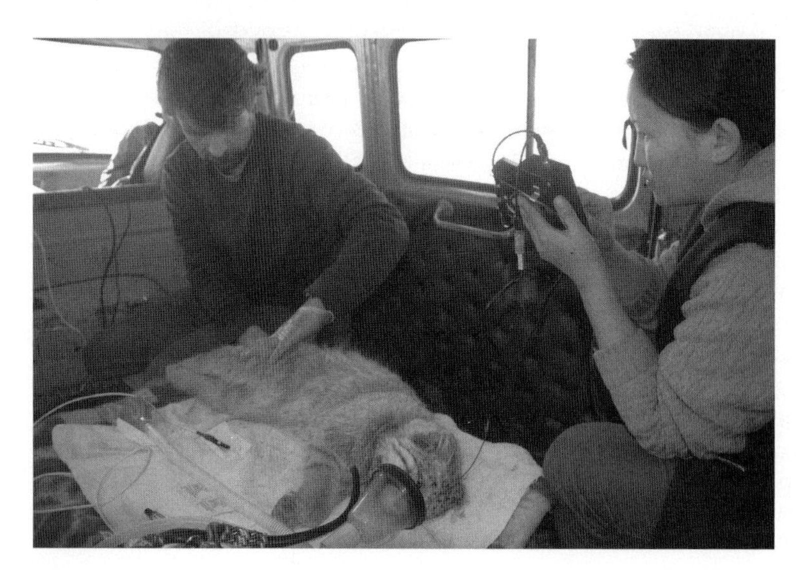

Figure 8.1 Semen collection of wild Pallas's cats in Mongolia to assess reproductive status and cryopreserve spermatozoa. © William Swanson.

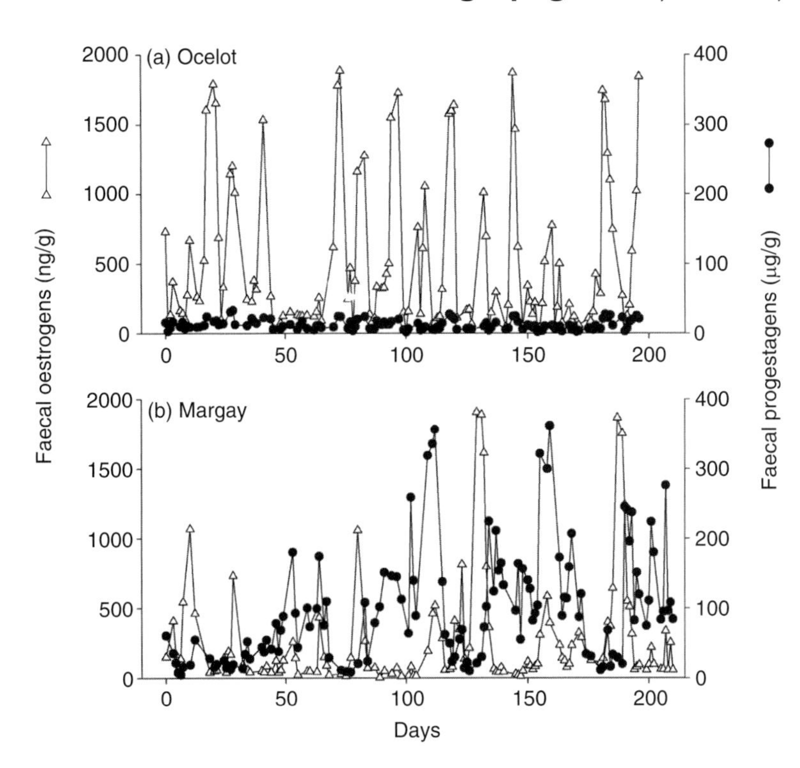

Figure 8.2 Longitudinal profiles of faecal oestrogens (open triangles) and progestagens (closed circles) in (a) an ocelot, representing an induced ovulator; and (b) a margay, representing a spontaneous ovulator. (Adapted from Moreira *et al.* 2001.)

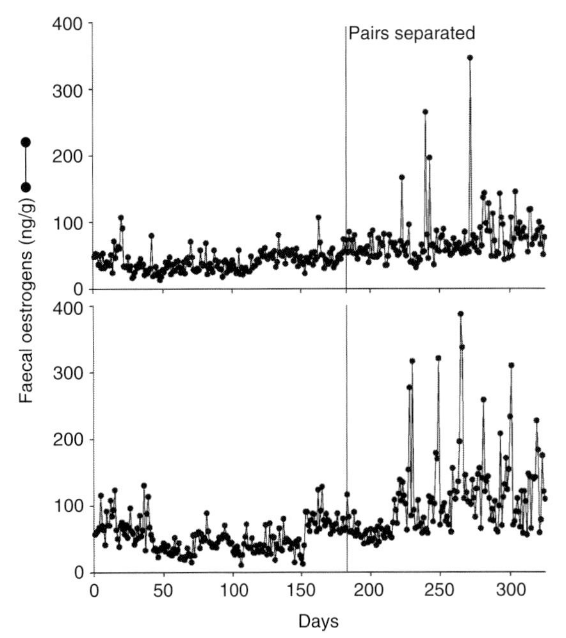

Figure 8.3 Longitudinal profiles of faecal oestrogens in two cheetah females housed together for 6 months and then housed separately. Females showed little ovarian activity while paired, but both began cycling after being moved to individual enclosures. (Adapted from Wielebnowski *et al.* 2002b.)

Endocrine monitoring has also become routine for evaluating ovarian responses to gonadotrophin (pituitary-type hormone) therapies used to induce ovulation for AI. Equine chorionic gonadotrophin (eCG) and human chorionic gonadotrophin (hCG) are typically used to stimulate follicular development and induce ovulation, respectively (Howard 1999). In cases of incorrect dosing, felids can be hyperstimulated by eCG and hCG, as is evident from excessive oestrogen excretion (Fig. 8.5; Graham *et al.* 2006). While laparoscopic observations reveal what appear to be normal ovarian follicles, faecal hormone monitoring provides more realistic assessments of animal physiology, such as steroidogenically abnormal function that contributes to poor fertilization and embryo development (Graham *et al.* 2000, 2006). Such findings have led to the design and implementation of safer and more effective gonadotrophin treatments as well as new methods to control ovarian activity in spontaneously ovulating felids (e.g. clouded leopard, fishing cat, and margay). Unexpected natural ovulations can derail effective ovulation induction for assisted breeding due to the disruptive influences of the endogenous progesterone. New approaches are

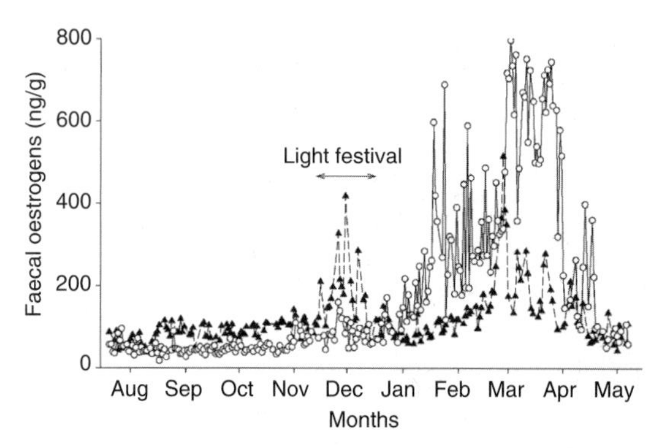

Figure 8.4 Longitudinal profiles of faecal oestrogens in two female Pallas's cats housed at different institutions, showing the effects of photoperiod on ovarian activity. One female was exposed to normal fluctuations in photoperiod (open circles), while the other female was exposed to an additional 5 h of artificial light per day during an annual holiday event (Festival of Lights) that disrupted normal reproductive activities (closed triangles). (Adapted from Brown *et al.* 2002.)

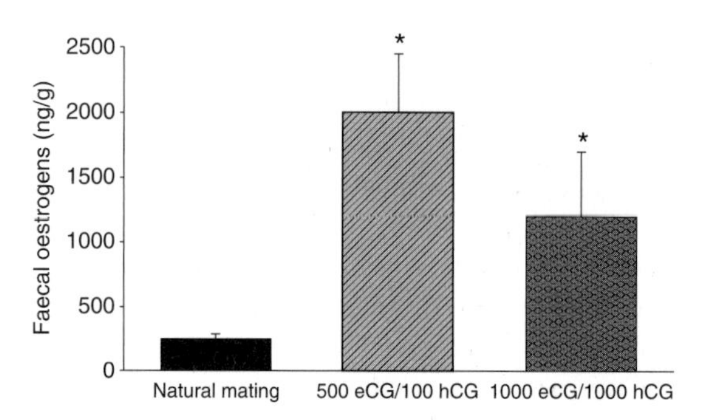

Figure 8.5 Mean (± SEM) concentrations of faecal oestrogen concentrations in female tigers after natural mating or treatment with two dose regimens of eCG and hCG to induce ovulation (* $P < 0.05$). (Adapted from Graham *et al.* 2006.)

being developed to suppress ovarian activity by giving progestin or a gonadotrophin-releasing hormone analog (Pelican *et al.* 2005, 2006b) with the theory being that a quiescent ovary will respond most consistently to gonadotrophin treatment. The pre-emptive progesterone suppression approach has shown promise in the domestic cat, clouded leopard, and fishing cat.

Another emerging endocrine approach involves assessing adrenal function, or stress, in felids. For example, excreted faecal corticoids transiently rise and fall in female cheetahs in response to anaesthesia, translocation, and introduction of a male (Terio *et al.* 1999). The clouded leopard, long believed to be a secretive felid, excretes more than twice as much corticoid when on public display compared to being housed 'off-exhibit' in isolation with nest boxes (Fig. 8.6a; Wielebnowski *et al.* 2002a). Likewise, housing this species in close proximity to tigers or leopards (i.e. potential predators) increases stress hormone excretion (Fig. 8.6b). The clouded leopard is also recognized as an arboreal species, which likely

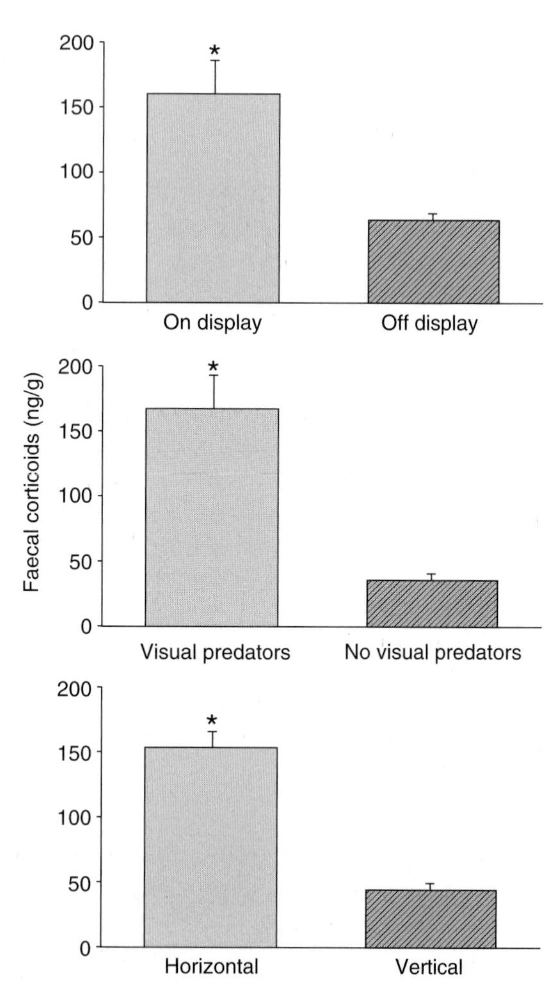

Figure 8.6 Mean (± SEM) faecal corticoid concentrations in clouded leopards housed: (a) on or off public display; (b) with or without visual contact with potential predator species; and (c) in enclosures with horizontal versus (preferred) vertical heights (* $P < 0.05$). (Adapted from Wielebnowski *et al.* 2002a.)

explains why providing enclosures with more vertical height reduces corticoid production compared to conspecifics limited to mostly horizontal space (Fig. 8.6c). The converse occurs when margays or tigrinas are moved from large, enriched enclosures to small, barren cages, with faecal corticoid concentrations and the incidence of agitated behaviours increasing significantly (Moreira *et al.* 2007). Thus, this tool offers an objective means of evaluating the quality of *ex situ* management methods while measuring the effects of altering facilities, husbandry, or enrichment opportunities.

Assisted reproduction

All of the above tools can contribute to assisted breeding, which is the application of an assortment of fertility techniques to enhance propagation and genetic management. The most recognized of these are AI, IVF, and embryo transfer. With few exceptions, the practical value of assisted reproduction for managing felid populations remains theoretical (Pukazhenthi and Wildt 2004; Swanson 2006; Howard and Wildt 2009). It is always preferable (and generally easier) to promote natural, rather than artificial, breeding. However, cats are well known for 'mate preferences', with some individuals resisting copulation in captivity. There are also situations where it is desirable to breed certain individuals to maintain genetic diversity and/or avoid the physical transfer of animals between geographic locations. However, these technologies require specialized skills, sophisticated equipment, and substantial experience (Pukazhenthi and Wildt 2004; Swanson 2006; Howard and Wildt 2009). For example, norms in the reproductive cycle, type of ovulation (induced versus spontaneous), hormonal therapies for provoking ovulation, semen-processing methods, and the ability to deposit sperm deep within the uterus are all necessary prerequisites for AI, which is the most basic assisted-breeding technique. These factors can, and often do, vary by species, although some generic protocols are available for felids (Howard 1999; Howard and Wildt 2009). In general, the process involves an intramuscular injection of the female with eCG, followed several days later by an hCG injection to cause ova release into the oviduct (where fertilization occurs). Felids differ markedly in sensitivity to these gonadotrophins, with the effect unrelated to body mass. For example, consistently successful ovulation in the cheetah (averaging 40 kg body mass) occurs with 200 IU eCG compared to the need for 500 IU in the smaller ocelot (9 kg) and only 75 IU for the mid-sized clouded leopard (13 kg). Sperm are recovered by electroejaculation, evaluated under a microscope, diluted and washed in culture medium, and then, by laparoscopic viewing, deposited through the abdominal wall into the cranial aspect of the uterine horn. Previous studies have demonstrated the ineffectuality of inseminating felid sperm into the vagina, cervix, or caudal uterine horn. This is probably due to inherently low concentrations of high-quality spermatozoa, and because anaesthesia reduces reproductive tract contractility, which normally helps propel sperm. Anaesthesia also can block ovulation in cats, so AI must commence only after egg release has occurred. AI has been used to produce offspring in the leopard cat, ocelot, tigrina, clouded leopard, snow leopard, puma, tiger, and cheetah (Howard 1999; Howard and Wildt 2009). Most successes have been associated with technique development rather than genetic management, although more than 10 AI litters have been produced in the cheetah.

Embryo transfer requires more complex expertise than AI, ranging from creating the embryo in the test tube (*in vitro*) to preparing the surrogate recipient to accept a 'foreign' embryo. No wildlife species is being genetically managed using embryo transfer (with or without IVF). The primary challenges are related to our abysmal knowledge about felid embryology and the appropriate surrogate for those embryos (Pukazhenthi and Wildt 2004). A serious question must be addressed. Must a conspecific surrogate population be managed or can the embryo recipient be another felid species? Embryo transfer is prevalent in agri-industry because a genetically valuable cow that normally produces one offspring per year can be hormonally stimulated to produce many ovulations (and thus embryos) at one time. These embryos can be manually transferred to gestate in the uterus of genetically 'common' cows, thereby freeing the original donor to undergo the process repeatedly (to produce an extraordinary number of calves in her lifetime). This specific embryo transfer process itself is inconsistent with most *ex situ* breeding programmes for wildlife since all individuals in the managed population should be (at least theoretically) genetically valuable. Embryo transfer is also impractical because there would be a mandate to maintain large numbers of surrogate felids to receive harvested embryos. What would be the source and how would such a recipient population be financially supported? Early enthusiasm for the idea of embryos from one species gestating in another has dissipated because it is now clear that interspecies embryo transfer (e.g. tiger embryos placed into lions) is unlikely to be practical (Pukazhenthi and Wildt 2004). It has become apparent that the biological mechanisms that ensure survival of transplanted embryos into a foreign species uterus are more sophisticated and fragile than once appreciated. There could also be legitimate concerns about behavioural and social competence of produced offspring (i.e. would a tiger born to and raised by a lion act like the former or the latter?). While there are these philosophical and practical limitations to embryo transfer in felids, these studies are providing a wealth of new information on basic embryogenesis. There is also exciting potential for the movement of whole animal genomes between countries to promote genetic management (see below). To date, offspring have been produced in the tiger, ocelot, caracal, fishing cat, serval, and African wildcat by IVF/embryo transfer (Pope *et al.* 2006; Swanson 2006).

Gamete rescue, sex sorting, and cloning

All animals produce surplus germ plasm that is lost at death. Although intra-ovarian oocytes and intra-testicular sperm are immature, it is possible to recover, artificially grow these cells in the laboratory, and then produce embryos by IVF. Such oocytes from post-mortem specimens of leopard, lion, puma, and cheetah have been matured *in vitro* to produce apparently healthy embryos despite an excision-to-culture interval as long as 36 h (Johnston *et al.* 1991). Embryos produced in this fashion have also resulted in offspring in the domestic cat (Gomez and Pope 2005). Viable sperm have been recovered from the felid epididymis (including Iberian lynx and Florida panthers killed by cars) for as long as 24 h post-mortem if the testes are maintained on ice.

Studies are also in progress to control sex of offspring by sorting sperm by gender using a flow cytometer that separates cells on the basis of DNA content for X- versus Y-bearing gametes. In theory, an inseminate could be created comprised mostly of X-bearing sperm to increase the chances of female embryo and then cub production. This technology has shown promise for species with inherently sperm-dense ejaculates (i.e. cattle and marine mammals; Pukazhenthi and Wildt 2004). Current methods have a low probability of success in felids because of the taxon's tendency to produce ejaculates with low sperm concentrations (and comparatively high proportions of cellular malformations; see above).

Nuclear transfer, (i.e., cloning) which has received extraordinary media attention, is another highly sophisticated technique that requires the nucleus (DNA) to be removed from a donor cell and then placed in an enucleated recipient cell. If an embryo forms and implants after transfer to a recipient, then the offspring has the same genetic complement as the original donor except for the mitochondrial DNA that is derived from the recipient. The domestic cat has been cloned (Shin *et al.* 2002) as well as the African wildcat, a feat that was also significant because these wildcat kittens were born to domestic cat surrogates by post-embryo transfer (Gomez *et al.* 2006). Current cloning approaches have negligible conservation value because of extreme inefficiency and failure to contribute to the most sought after goal—maintaining population level gene diversity. Also, significant foetal resorption, abortion, abdominal organ exterioriation, respiratory failure, and septicaemia have been associated with cloning studies in felids (Gomez *et al.* 2006). Nonetheless, this research is important for beginning to understand nuclear reprogramming, cytoplasmic inheritance, and recipient oocyte activation and function. Such studies could eventually lay the foundation for re-deriving genomes and diversity of long-dead felids from cryopreserved somatic tissues.

Sperm and embryo cryopreservation

The advantages of creating organized genome resource banks (GRBs), or frozen repositories, of wildlife biomaterials (sperm, oocytes, embryos, tissue, blood products, and DNA) have been reviewed (Wildt *et al.* 1997; Holt *et al.* 2003). Accessibility to this type of resource has been critical in sorting felid phylogeny, monitoring gene variation, determining lineages, screening for pathogens and parasites, and for dealing with crises, such as the canine distemper epizootic that devastated East African lions (Roelke-Parker *et al.* 1996; Johnson *et al.* 2006b). In theory, sperm could also be frozen from wild males and used via AI to support captive breeding without removing more animals from nature (Swanson *et al.* 2007; see below) and to reduce the total number of males needed in captivity (Wildt *et al.* 1997). Such ideas have remained mostly speculative because felid germ plasm is hypersensitive to cryopreservation and does not survive well using standard livestock methodologies. Recent studies have revealed that cat sperm are highly sensitive to cooling, even before the actual freezing process (Pukazhenthi and Wildt 2004). Massive damage is incurred at the sperm's acrosome, which erodes or prevents subsequent fertilizing ability. Thus, modern studies emphasize classical cryobiological investigations of sperm biophysical properties (e.g. membrane composition and permeability to water and cryoprotective agents), basic studies that inevitably will lead to more effective practical protocols. Nonetheless, offspring have been produced in the leopard cat, ocelot, and cheetah by AI with thawed sperm (Howard 1999; Howard and Wildt 2009). Noteworthy has been the production of pregnancies from sperm collected and frozen from free-living cheetahs in Africa that were imported and used for AI in the North American Cheetah SSP (Wildt *et al.* 1997). Three of six females produced young after insemination with 6–16 million motile sperm per AI.

Embryo cryopreservation has the added benefit of allowing storage of the full genetic complement of the sire and dam. Extensive studies of the domestic cat have helped generate a rudimentary understanding of embryo cryopreservation that, in turn, has resulted in living births in the African wildcat (Pope 2000), caracal (Pope *et al.* 2006), and ocelot (Fig. 8.7; Swanson 2001; see below). These milestones have led to interesting field experiments on the potential of sperm and embryo cryopreservation for contributing to metapopulation genetic management. Because cryopreserved cheetah sperm have been proven biologically competent, the Smithsonian's National Zoo has developed a partnership with the Cheetah Conservation Fund (CCF, Namibia) to develop a cheetah GRB. One of CCF's goals is to rescue wild cheetahs that are considered pests and are trapped by farmers. This has presented the opportunity to collect and cryopreserve sperm from wild individuals, with more than 200 viable sperm samples from ~100 males in the CCF-based GRB. Besides beginning to ensure countrywide cheetah gene diversity, other advantages have included building laboratory and equipment infrastructure and local training in biomedical techniques and basic research (Crosier *et al.* 2006). A mobile laboratory has been developed to permit handling and preserving semen (and animals) effectively under rigorous field conditions where temperatures can reach 45°C. Meanwhile, strategies are being developed to share the sperm bank with organized *ex situ* cheetah programmes to allow importing new genes (in lieu of living animals) but with *in situ* conservation.

There is a similar strategy for the Pallas's cat, a priority species for the Felid Taxon Advisory Group (2008) because of a lack of research and conservation attention, as well as more than 50% mortality in captive-born kittens due to toxoplasmosis. This interest by zoos has generated collaboration and funding for Pallas's cat studies in Mongolia, where the species is rarely exposed to *toxoplasma*, and there are opportunities to learn from studying wild specimens. Zoo-supported efforts have resulted in recent assessments of Pallas's cat population density, predator–prey relationships, and nutritional needs, including the advanced training of a local graduate student. Because population modelling revealed that the Pallas's Cat SSP population will lose significant genetic diversity over the next 50 years, the field studies were also an important opportunity for collecting and banking sperm (Swanson *et al.* 2007). The reproductive biology of the species has been characterized and sperm samples already banked from 10 wild males. Besides value for future AI studies, these frozen specimens could be used to produce embryos by IVF. Introducing two new founders every 5 years by this approach would allow the retention of 90% of genetic diversity without ever requiring the extraction of more animals from the wild to support zoos (Swanson *et al.* 2007).

Embryo technology is also a component of a similar effort to link geographic interest in ocelot genetic management. The subspecies *Leopardus pardalis mitis* is endemic to southern Brazil, with wild populations adversely affected by expanding human populations and concurrent habitat loss and fragmentation in São Paulo State. Brazilian zoos maintain more than 200 ocelots (mostly wild-born, confiscated specimens), but originally with no effective management programme for breeding or retaining genetic diversity. There is an Ocelot SSP in North America comprised of ~100 largely generic (i.e. intermixed subspecies) individuals. As a means of building capacity in the range country and to secure new genes for the SSP, the Brazilian Ocelot Consortium was formed that included a Brazilian non-governmental organization, the Ocelot SSP, and 10 North American zoos. Collaborations focused on improving the *ex situ* environment and starting captive breeding in Brazil. Diets were corrected, facilities improved, enrichment provided, and training conducted in population management and reproductive technologies (Moreira *et al.* 2007). The result has been the first ever established population management plan for any felid species in Latin America. Ocelots are being produced under scientific guidelines with a subset of eight animals exported to the United States to become founders of a re-invigorated SSP programme. A complementary strategy now in development is the transfer of embryos, rather than living animals, between countries (Swanson *et al.* 2007). This approach appears feasible based upon a trial in Brazil where 24 IVF ocelot embryos (Fig. 8.7a) cryopreserved in 1999 and 2000 were thawed in 2007 and then transferred laparoscopically to the oviducts of eight synchronized recipient ocelots. To date, 5 of 10 embryo transfers have resulted in pregnancies and the births of 4 viable kittens (Fig. 8.7b) and 1 stillborn (due to dystocia) kitten.

(a)

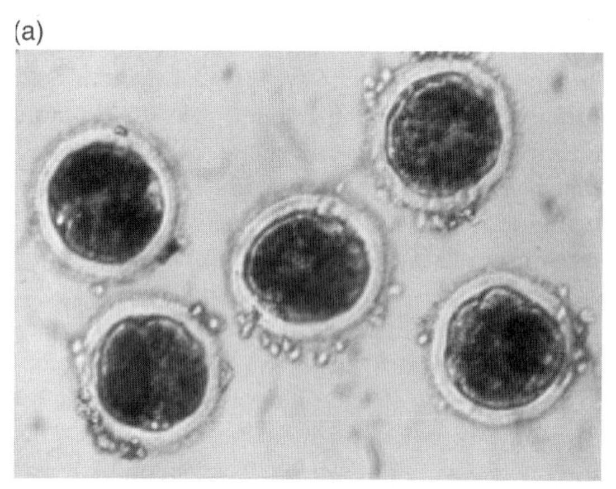

(b)

Figure 8.7 Assisted reproduction in felids: (a) ocelot embryos generated via IVF of *in vivo* matured oocytes using fresh spermatozoa; and (b) ocelot kitten born following transfer of frozen-thawed embryos. © William Swanson.

Contemporary example of a critically needed *ex situ* breeding programme: Iberian lynx

Throughout this chapter, we have advocated for safeguarding existing genetic diversity while generating new biological information via managed *ex situ* populations of selected felid species. We conclude by providing a modern example of a programme designed and implemented to help conserve the world's most endangered felid, the Iberian lynx. This species, in drastic decline due to habitat loss, scarce prey, and an alarming increase in disease-related problems (Delibes *et al.* 2000), is a model of why the principles, tools, and recommendations described in this chapter are necessary to increase the chances of successful species recovery.

The reduction in wild Iberian lynx populations has been recognized since the 1950s (Rodríguez and Delibes 1992). A 1998 PHVA workshop generated numerous species survival recommendations, including the need for captive breeding (Conservation Breeding Specialist Group 1998). An Iberian Lynx Captive Breeding Action Plan emerged from a subsequent meeting, but implementation was delayed for 5 years because of controversy and disagreements (Delibes and Calzada 2005). Eventually the programme was launched in 2003 in partnership between two responsible administrations with financial support provided and an *ex situ* management team hired—none too early as species status was dire (a situation that continues to the modern day). For example, Iberian lynx in the best known, protected habitat (Doñana National Park) experienced an epizootic of feline leukaemia in 2006 and 2007 that infected 12 free-ranging specimens, 6 of which died due to virus-induced immunosuppression (López *et al.* 2007). Lynx in this park also have 30% less genetic diversity than counterparts in the Sierra Morena mountains, and both remnant populations have shown recent signs of genetic drift (Godoy 2006). Most recently, 16 of 23 (wild and captive) lynx have experienced indications of kidney disease (Jiménez *et al.* 2008).

With such challenges, the captive breeding programme provides a modicum of comfort and control. Overall organization is under the umbrella of a Spanish and Portuguese governmental collaboration with participation by relevant regional (provincial) governments of Spain. The *ex situ* programme goal is to maintain a genetically and demographically healthy reservoir population that supports *in situ* conservation via education, research, and future animal releases into native habitat. Reintroduction is recognized as urgent because extreme delays will contribute to continuing habitat loss. Presently, there are approximately 200 Iberian lynx believed to be living in nature. As a result, the source population for creating a captive breeding population was limited. Population modelling using the programme PM2000 (Pollak *et al.* 2004) revealed that it would be impossible to maintain 90% of existing gene diversity for 100 years because the wild population could not withstand extraction of the required 12 founders per year for 5 years (Lacy and Vargas 2004). At the time of the analysis (2004), there were six wild-caught lynx already in the El Acebuche captive facility (constructed in 1991 in Doñana National Park, Southeastern Spain). Therefore, modelling suggested the potential of being able to maintain 85% of existing genetic diversity for the next 30 years. The outcome eventually would be the *ex situ* management of 60 individuals (30 males and 30 females) as breeding stock (Lacy and Vargas 2004). This goal could be attained by adding four founders (mostly cubs or juveniles) per year for 5 years, as well as one extra founder every 2 years (from the lynx that normally enter rescue centres) for the entire duration of the programme (Table 8.2). This level of extraction rate presumably would have no impact on the viability of the wild populations according to the model designed by Palomares *et al.* (2002b). Following this scheme, the programme could achieve its population target of 60 individuals (Table 8.2) by 2009. Reintroductions could begin in 2010 provided that adequate habitat (from biological, social, and logistical perspectives) was prepared to host releases of captive-born lynx. Modelling predicts the availability of 12–13 captive-born lynx annually from 2011 through 2019 (Table 8.2). Furthermore, habitat requirements for establishing a self-sustaining Iberian lynx population have been estimated at 10,000 ha per release area. Presently, there are two such areas in Andalusia that comply with IUCN criteria for reintroduction, which (because of a growing *ex situ* lynx population) is now on the horizon.

Consistent with the need for more basic species information, the lynx *ex situ* programme emphasizes basic and applied research in reproductive biology, behaviour, genetics, and veterinary medicine. For example, our husbandry and breeding efforts have established or reconfirmed normative data on sexual maturity (males, 3 years; females, 2–3 years), oestrous length (3–7 days), copulations per pairing (mean, 28), gestation interval (63–66 days), litter size (mean, 2.5 cubs; range, 1–4), lactation interval (3.5 months), and point at which cubs successfully kill rabbit prey (80–140 days). Faecal hormone monitoring has demonstrated that females experience ovarian cycles from January through to May, whereas males maintain testosterone year-round (Jewgenow *et al.* 2006b; Pelican *et al.* 2006a). Species peculiarities have also been revealed, for example, oestrogen metabolite concentrations that are 10-fold greater than in other felid species (Brown 2006; Pelican *et al.*

Table 8.2 Growth projections for the Iberian Lynx Breeding Programme (updated from Lacy and Vargas 2004), including potential number of lynxes available for future reintroductions (Vargas *et al.* 2007).

Year	*n* (in captivity)	Capture of founders	Availability of potential reintroduction candidates	Cumulative releases
2004	12	4	0	0
2005	19	4 + 1	0	0
2006	28	4	0	0
2007	39	4 + 1	0	0
2008	50	4	0	0
2009	62	1	0	0
2010	73		8	8
2011	72	1	13	20
2012	73		12	32
2013	72	1	13	44
2014	73		12	56
2015	72	1	13	68
2016	73		13	80
2017	72	1	12	92
2018	73		13	104
2019	72	1	12	116

2006a). Additionally, progestin excretion profiles are unusually lengthy, largely because ovarian corpora lutea (sites of earlier ovulations) remain active much longer than in most other felids (F. Goeritz, Goritz *et al.* 2009). Prolonged non-pregnant luteal activity has also been described in the closely related European lynx (*Lynx lynx*; Jewgenow *et al.* 2006a), suggesting an idiosyncratically conserved mechanism for lynx in general.

The *ex situ* population is also being used to explore a novel means of diagnosing pregnancy. Because all ovulating lynx produce rising progesterone (regardless of conception), conventional hormone monitoring is not useful for identifying a gestating female. However, increasing concentrations of the hormone relaxin (in blood or urine) are indicative of pregnancy in felids (Van Dorsser *et al.* 2007). A unique non-invasive means of collecting blood has been developed using blood-sucking Triatomine bugs (*Rhodnius prolixus* or *Dipetalogaster maxima*) placed in specially drilled hiding holes in the lynx's cork nest box (Voigt *et al.* 2007). As much as 3 ml of blood can be extracted per bug, more than adequate for the assay. Urine is captured using special collection devices distributed throughout the animal's enclosure. Pregnant females express a positive relaxin signal from 32 to 50 days post-copulation (Braun *et al.* 2007; Jewgenov *et al.* 2009) of a 64 day gestation, thereby allowing managers to prepare for an impending parturition.

Certainly, the most intriguing experience in the lynx breeding programme was discovering a sensitive period when cubs become highly competitive to the point of siblicide (Vargas *et al.* 2005; Antonevich *et al.* 2009). Spontaneous aggression erupted at 44 days of age in the first Iberian lynx litter born in captivity. The largest cub (a female) in a litter of three was killed by a brother who delivered lethal bites to the larynx and skull. Agonistic behaviours have been observed in five of six subsequent Iberian lynx litters of two or more cubs (from days 37 through 74 of age), with the most intensive fighting during the 6th and 7th post-natal weeks (Antonevich *et al.* 2009). The same phenomenon has been observed in the Eurasian lynx by Russian scientists at the Tcherngolovka facility, who recorded aggressive behaviours in 16 of 31 litters, with deaths occurring in 4 cases (Naidenko and Antonevich 2006;

Naidenko and Antonevich 2009). The highest prevalence of agonistic behaviours in Eurasian lynx cubs occurred at 36–64 days of age, with the greatest frequency during the 7th post-natal week. These fights are occurring at a time when cubs are transitioning from the dam's milk to solid food. The 'winning' Eurasian lynx cub grows faster and becomes the dominant individual in the litter (Naidenko and Antonevich 2006). Although growth differences have not been detected in Iberian lynx young, the winning cub always is the first to kill live prey and to eat at daily food presentation (while littermates wait until the winner has finished). Thus, these bouts of early aggression serve to establish a hierarchy within the litter. This phenomenon has also influenced programme management, as staff must be vigilant and prepared to break up aggression at feeding during the sensitive period. Although siblicide has not been directly observed in nature, a 1-month-old Iberian lynx cub was found in the wild in 2003 with severe injuries compatible with bites from another cub. It may also be relevant that free-ranging Iberian lynx generally give birth to three cubs, but 70% of the females are usually observed with only two young by 3 months post-parturition (Palomares *et al.* 2005).

By August 2009, the breeding programme for the Iberian lynx included 39 males and 37 females ($n = 76$) and was 2 years ahead of Action Plan forecasts (Table 8.2; Lacy and Vargas 2004; Vargas *et al.* 2007, 2008). Twenty-seven litters have been produced (five prematurely) with 40 surviving young (Vargas *et al.* 2009). Additional wild-caught founders have been added (on the basis of scientific estimates) that will not compromise wild population status. Two GRBs (known as 'biological resource banks' in Spain) are currently functioning, both maintaining lynx biomaterials (blood products, tissue, and germ plasm) in duplicate as a safety measure (León *et al.* 2006; Roldán *et al.* 2006). The programme management team works closely with *in situ* conservation authorities, managers, and researchers. For example, skills and resources are being shared to address increases in disease concerns for lynx and prey, and projects are coordinated to monitor, for example, genetic status of the wild and captive populations. Most importantly, our multidisciplinary research is generating new insights into the (often unique) biology of this species, while our intensive management is on course to meet the overall recovery goal of attempting lynx reintroduction by 2010. Finally, the programme carries out a capacity-building plan aimed at preparing professionals for working on endangered species conservation as well as sensitizing the general public and decision makers about the importance of conserving habitat for the recovery of this charismatic felid. In this fashion, the guidelines and experiences offered in this chapter illustrate how an *ex situ* breeding programme can be integrated to complement the challenge of sustaining wild felids in nature, which is always the highest priority.

Acknowledgements

The authors thank their research teams and numerous institutional partners and colleagues for their dedication in collecting and interpreting both the experiences and data shared in this chapter, which have improved understanding and managing of wild felid species.

CHAPTER 9

Wild felid diseases: conservation implications and management strategies

Linda Munson, Karen A. Terio, Marie-Pierre Ryser-Degiorgis, Emily P. Lane, and Franck Courchamp

Diseased lion in the Serengeti. © L. Munson.

Introduction

Disease is an inevitable part of life and has the potential to determine the condition and future of individuals, population dynamics, community structures, and even species evolution. Potential felid pathogens, whether arthropods, protozoa, viruses, bacteria, or fungi, are present in many environments (endemic) or circulate periodically among felids and other mammalian hosts (epidemic). While these organisms have the potential to cause disease, infection does not necessarily lead to disease unless critical organ functions of the host are impaired. Only rare accounts of clinical disease in wild felids were reported until 1994 when the canine distemper epidemic swept through the Serengeti ecosystem, killing hundreds of lions (*Panthera leo*) and many other felids (Roelke-Parker *et al.* 1996). This epidemic alerted the global conservation community of the impact that infectious disease could have on free-ranging felids and raised the questions: Could an epidemic cause the extinction of wild felid populations and could less catastrophic diseases threaten felid conservation efforts? This chapter explores those factors that may influence the ecology of felid diseases in contemporary environments and presents case studies of diseases that have affected wild felid populations.

Epidemiology of felid diseases

Because of a long history of co-evolution, felids are generally well adapted to the presence of endemic infectious agents in stable ecosystems and develop immunity, either individually or as a population. So disease usually does not occur. Even non-endemic pathogens that periodically circulate through felid habitats cause minimal harm if a portion of the population has immunity from previous exposure or if the pathogen is of low virulence (the pathogenicity or ability of a parasite to cause disease). Epidemics occur when traditional host/parasite relationships are disrupted or when highly virulent pathogens are introduced. Because many ecosystems inhabited by wild felids have deteriorated and historic geographic barriers have been bridged, most felid populations are at risk of exposure to new pathogens that may imperil population health. Fragmented habitats also impede recruitment and over time may reduce genetic variation, so that disease resistance and fecundity may be impaired and population recovery from disease may not occur. Therefore, felid conservation will increasingly need to include assessments of disease risks and strategies for disease management in order to be successful.

Risk of infection and disease

The outcome of contact between a felid and an infectious agent depends on both host and agent factors. Whether disease results from infection in an individual animal is determined by the quantity (dose) and virulence of the microbe acquired by the host, as well as the ability of the host to control the infection. This ability depends on the immunocompetency of the host (ability to mount a protective immune response) and its immune status (whether it has protective antibodies from previous exposure to this pathogen). In animals with poor immune function, even small quantities of low virulence microbes may result in disease. Poor immune function can result from stress or recent infection with immunosuppressive viruses, such as canine distemper virus (CDV) or feline parvovirus (FPV: also known as feline panleukopenia virus). Juvenile felids are more likely to develop disease when infected, because the passive immunity acquired from their mother through milk usually wanes between 6 and 12 weeks before their own immunity to specific pathogens has developed (Suzan and Ceballos 2005). Some pathogens have limited virulence by themselves, but if a host is simultaneously infected with one or more pathogens, then morbidity (clinical disease) and mortality could occur (Munson *et al.* 2008).

Role of felid behaviour in pathogen acquisition

The ecology of infectious diseases within any taxonomic group is invariably linked to behaviour. Social behaviour increases opportunities for transmission, as evidenced by the epidemics of rabies and canine distemper in the highly social African wild dog (*Lycaon pictus*; Alexander and Appel 1994; Alexander *et al.* 1996; Kat *et al.* 1996). Among wild felids, the lion is the most notably social, in contrast to relatively asocial felid species, such as the leopard (*Panthera pardus*) or cheetah (*Acinonyx jubatus*), which are less likely to acquire an infectious agent through direct contact. Interactions within lion prides would explain the rapid transmission of CDV among Serengeti lions during the 1994 epidemic (Roelke-Parker *et al.* 1996), whereas leopards and cheetahs that typically live in low-density populations or as solitary individuals were only rarely affected during this epidemic (Roelke-Parker *et al.* 1996). The high prevalence of feline immunodeficiency virus (FIV) in many lion populations is also likely to be attributable in part to their social behaviour (Troyer *et al.* 2005). It is interesting to note that although viruses such as CDV, feline herpes virus (FHV), and feline corona virus (FCoV; also called feline infectious peritonitis [FIP] virus) are thought to require contact for infection and are not known to persist in the environment, there is evidence of widespread exposure to these viruses in free-ranging felid species (Table 9.1), suggesting transmission occurs through means other than contact or that greater contact occurs among species which are historically considered to be asocial. For generalist or vector-borne pathogens, felid asociality does not matter because the felid may contact the agent in prey, from the environment, or through encounters with other species.

Other behaviours such as territory marking, use of play trees by cheetahs, or latrines by Iberian lynx (*Lynx pardinus*) may play a role in transmission of infectious agents that are transmitted in faeces, such as FPV and FCoV, by concentrating these agents at sites visited by felids. Climatic events, such as

Table 9.1 Prevalence and geographic distribution of free-ranging felids with antibodies against selected viruses. Percentages are the proportion of the population tested (*N*) that had antibodies against CDV, FCoV, FPV, FHV, FeLV, and FIV.

Species	Location	Years	*N*	CDV (%)	FCoV (%)	FCV (%)	FPV (%)	FHV (%)	FeLV (%)	FIV (%)	Reference
African lion	Tanzania	1984–91	311		57	70	68	99	0	91	Hofmann-Lehmann *et al.* (1996)
African lions	Kenya	1995–96	55	55							Kock *et al.* (1998)
African lions	Uganda	1998–99	14	79	0	79	36	64	0	71	Driciru *et al.* (2006)
African lions	Botswana	2003–05	21	5	0	9.5	0	100	0	71.4	Ramsauer *et al.* (2007)
African lions	South Africa					0	84				Spencer (1991)
African lions	Southern Africa	1997	127	15	23	5	6	97	0		van Vuuren *et al.* (1997)
Leopards			9	33	11	0	11	11	0		
African lions	South Africa									83	Spencer *et al.* (1992)
	Namibia									0	
African lions	Namibia						0				Spencer and Morkel (1993)
Cheetahs	Namibia	1992–98	72	24	29	65	48	12	0	0	Munson *et al.* (2004a)
Cheetah	Southern Africa Namibia	1999–2001	146		84						Kennedy *et al.* (2003)
Cheetah	Africa		25		32						Heeney *et al.* (1990)
Wild cats	Saudi Arabia	1998–2000	62		10	25	5	5	3	6	Ostrowski *et al.* (2003)
Sand cats					0	0	0	0	8	0	
Iriomote cat	Japan	1983–88	21		82	58	0	0	5	0	Mochizuki *et al.* (1990)
Leopard cats	Taiwan Vietnam	1995–97	11			27	82	9	0	0	Ikeda *et al.* (1999)
Wildcats	Europe	1996–97	51		4	16	2	4	74	0	Leutenegger *et al.* (1999)

(*Continued*)

Table 9.1 (Continued)

Species	Location	Years	*N*	CDV (%)	FCoV (%)	FCV (%)	FPV (%)	FHV (%)	FeLV (%)	FIV (%)	Reference
Wildcats	Scotland	1992–97	50		6	26		16	10	0	Daniels *et al.* (1999)
Wildcats	Scotland	1987–89	33		0				6	0	McOrist *et al.* (1991)
Wildcats	France	1996–98	38						23.7	7.9	Fromont *et al.* (2000)
Wildcats	France	NS	3–8			60	40	25	38	0	Artois and Remond (1994)
Iberian lynx	Spain	1989–2000	37–48	0	0	4.5	2.6	11.5	0	0	Roelke *et al.* (2008)
Iberian lynx	Spain	2003–06	46	19.1	26.6	26.7	20	33.3	0	0	Meli *et al.* (2006)
Eurasian lynx	Sweden	1993–99	102–106		0	0	1	0	0	0	Ryser-Degiorgis *et al.* (2005)
Canada lynx	Western North America	1993–2001	215	3	6	1	8	1			Biek *et al.* (2002)
Puma	Canada	1984–2001	15	0	0			47	0		Philippa *et al.* (2004)
Canada lynx			5	20	0			60	0		
Florida panther	United States	1978–91	38		19	56	78	0	0	37	Roelke *et al.* (1993a)
Puma	California	1987–90	58		28	17	93	19	0	0	Paul-Murphy *et al.* (1994)
Bobcats	United States	1992–95	26		4	46					Riley *et al.* (2004)
Puma	Brazil	1998–2004	21		5	29	48	29	10	5	Filoni *et al.* (2006)
Ocelot											
Little spotted cat											
Geoffroy's cat	Bolivia	2001–05	9	0		44.4	44.4	11.1			Fiorello *et al.* (2007)
Ocelot			10	0		100	0	0			
Jaguarundi			1	0		0	0	0			

NS = Not stated.

droughts, can affect behaviour by extending home ranges and increasing contact among animals at water sources, thereby enhancing opportunities for novel pathogen exposure. Pathogens infecting the central nervous system, such as rabies virus or CDV, can make a normally asocial felid lose avoidance behaviours and attract aggression, thereby increasing pathogen transmission (Roelke-Parker *et al.* 1996).

Possible anthropogenic influences

Over the past 50 years, the global human footprint has forever changed the environments in which felids have evolved. Domestic dogs in some regions play a major role in amplifying and transmitting important infectious agents, such as rabies virus and CDV (Cleaveland *et al.* 2000; Suzan and Ceballos 2005). The contribution of domestic cats (*Felis catus*) to virus and parasite transmission to wild felids is less well documented, but also likely to be a concern. Domestic cats and wild felids are susceptible to the same infectious agents, and either feral cats or outdoor pets can increase the number of infected hosts in an ecosystem (Millán *et al.* 2009). The risk of interactions between wild felids and domestic cats is higher for populations suffering habitat loss and fragmentation in areas densely populated by humans. For example, in California (United States), interactions among pumas (*Puma concolor*), bobcats (*Lynx rufus*), and domestic cats are common (Riley *et al.* 2007). Domestic cats (with their microbial flora) are also being moved to new geographic locations, and the infectious agents they harbour may be more virulent in naive wild felids that have not evolved in their presence, exemplified by the feline leukemia virus (FeLV) epidemic in Florida panthers (*Puma concolor*) described below and recent cases of Notoedric mange in pumas and bobcats (Pence *et al.* 1982; Maehr *et al.* 1995; Uzal *et al.* 2007).

Another anthropogenic influence on disease ecology is the alteration of habitat. Expansion of highways into felid habitat can provide avenues for pathogen introduction from domestic dogs and cats, as well as result in significantly greater numbers of mortalities from vehicular trauma. Habitat restriction and degradation (through fencing or human encroachment) limits dispersal and genetic flow, and can increase population densities, felid encounter rates, lethal aggression, concentrations of pathogens, and stress (which in turn can affect immune function). Building artificial waterholes in drought-prone regions can have a similar effect by attracting abnormally dense animal populations, thereby promoting pathogen transmission and concentration in the environment (Jager *et al.* 1990; Lindeque and Turnbull 1994). Encroaching human development also increases the risk of exposure to environmental pollution from industrial waste (e.g. mercury toxicity in Florida panthers [Roelke 1990]) and accidental rodenticide poisoning (Riley *et al.* 2007; Uzal *et al.* 2007).

Disease risks during translocations and reintroductions

When felid conservation includes translocation or reintroduction, the risk of exposure to novel pathogens and their vectors in the new environment needs to be considered (Munson and Karesh 2002). Many endemic infectious agents have minimal pathogenicity in species that have evolved in the same environment, but may be highly lethal to naive hosts from a non-endemic area introduced into the region. For example, *Cytauxzoon felis* infections are usually innocuous in free-ranging lynx, bobcats, and pumas from endemic regions, but can be highly fatal in felids that evolved in non-endemic sites (Jakob and Wesemeier 1996; Peixoto *et al.* 2007). Exposure in captivity to *Toxoplasma gondii* has proven fatal to naive Pallas's cats moved from Mongolia, where their exposure to this parasite was minimal, an evolutionary history that may explain their exceptional susceptibility (Kenny *et al.* 2002; Ketz-Riley *et al.* 2003a; Brown *et al.* 2005). The stress of translocation and introduction may also affect immune function and render a felid more susceptible to disease from both endogenous and exogenous agents.

The potential of contaminating a new environment with the microbial ecosystem present in the translocated felids (endogenous viruses, haemoparasites, bacteria, or vectors) also should be considered. Animals stressed by the translocation process are more likely to shed microbes, increasing the risk of pathogen introduction. Shedding of FCoV by cheetahs recently moved to a zoological park is thought to have caused a FIP epidemic in the resident cheetahs that had no previous exposure to this virus (Evermann *et al.* 1983). Because of these risks, animals translocated to a new wild environment or captive reserve should be tested for pathogens of concern before and after moving. If pathogens are detected, the benefits of translocation should be weighed against the risks, and preventative measures (e.g. treatment for mange or other parasites) undertaken when possible.

Disease risks in captivity

If current trends in habitat loss and human conflicts continue, many felid conservation programmes may require captive populations or temporary holding of individuals to assure genetically diverse, sustainable populations (Wildt *et al.*, Chapter 8, this volume). The Iberian lynx exemplifies a species at high risk of extinction with the need for a captive breeding programme. However, the captive setting is not without disease risks, and in fact, may pose additional challenges to maintaining felid health. In captivity, populations are dense, expression of normal behaviours is limited, and confinement or human contact may result in chronic anxiety, which in turn can affect fitness. Captivity stress is suspected to contribute to the poor health of some felids in captivity, despite their receiving excellent medical care.

A species that appears particularly susceptible to the adverse health effects of captivity stress is the cheetah. Self-sustaining populations of captive cheetahs have been difficult to maintain worldwide because chronic degenerative and inflammatory diseases (gastritis, systemic amyloidosis, hepatic veno-occlusive disease, glomerulosclerosis, and neurodegenerative diseases) cause premature death and considerable morbidity in breeding age animals (Munson 1993; Papendick *et al.* 1997; Bolton and Munson 1999; Munson *et al.* 1999a, b, 2005; Munson and Citino 2005; Robert 2008). These diseases only rarely occur in free-ranging cheetahs and are usually only mild (Munson *et al.* 2005). Captive cheetahs also develop severe chronic disease from common feline infectious agents, such as FCoV (Evermann *et al.* 1983, 1988), FHV (Junge *et al.* 1991; Munson *et al.* 2004b), FPV, and *Helicobacter* (Munson 1993; Munson *et al.* 1999b; Steinel *et al.* 2000), in contrast to free-ranging cheetahs that have also been infected with these agents without evidence of disease (Munson *et al.* 2004a; Terio *et al.* 2005). Although poor health in cheetahs was initially attributed to their lack of genetic diversity as a species (O'Brien *et al.* 1985b), the low prevalence of disease in the free-ranging Namibian cheetahs with similar genetic background (Munson *et al.* 2005) suggests that the captive environment contributes significantly to their poor health.

How stress affects health

Stressors elicit behavioural and physiological responses that have evolved to benefit survival by activating multiple interacting pathways which modulate metabolic, cardiovascular, neuromuscular, and immunological functions, with the goal of assisting the animal to escape the threat quickly ('fight or flight' response) and then return the animal to homeostasis (Breazile 1987). A stress response is usually not detrimental if acute and short-lived, but a prolonged response to persistent stressors, whether perceived or real, may be harmful. One important stress response involves the hypothalamic–pituitary–adrenal axis, where the release of hormones from the brain and pituitary gland stimulate the production and release of corticosteroids from the adrenal gland. With chronic stress, this axis is continually activated, and corticosteroids become persistently elevated, resulting in profound effects on the immune system and metabolic homeostasis (Elenkov and Chrousos 1999, 2006). Continual high corticosteroid levels cause marked immunosuppression through destruction of lymphocytes and modulation of gene expression for cytokines, inflammatory mediators, and lymphocyte receptors that determine how the host's immune system responds to infection. An ineffective immune response to pathogens and metabolic derangements, such as amyloidosis, are the result (Cowan and Johnson 1970; Germann *et al.* 1990; Elenkov and Chrousos 1999).

These immunomodulating and metabolic effects of chronic stress likely explain the high prevalence of some diseases in captive cheetahs, such as severe gastritis, herpesviral dermatitis, and amyloidosis (Papendick *et al.* 1997; Munson *et al.* 1999b, 2003). The health of other captive felid species, such as the black-footed cat (*Felis nigripes*; Terio *et al.* 2008), may also be impacted by chronic stress. In order to optimize the health of captive animals, felid stress responses need to be objectively assessed under different management conditions. Measuring chronic stress in wildlife is challenging, because any handling or manipulation to obtain blood (where stress hormones are circulating) will incite an acute stress response. In felids, most circulating steroids are excreted in the faeces (Brown *et al.* 1996; Terio *et al.* 1999, 2003), so faecal corticosteroid assays provide a convenient non-invasive method for measuring one important aspect of the stress response in wild animals. Another advantage of measuring corticoid metabolites in faeces instead of serum cortisol is that diurnal changes in cortisol levels do not complicate the interpretation of results. Using this non-invasive method, the stress response in captive cheetahs was found to be significantly higher than in free-ranging cheetahs and was persistently elevated after translocation (Terio *et al.* 2004; Wells *et al.* 2004). High corticosteroids have also been documented in captive clouded leopards (*Neofelis nebulosa*), another species with health problems in captivity, and dramatic decreases in corticoids occurred with exhibit modifications, suggesting an environmental cause for corticoid elevations (Wielebnowski *et al.* 2002a). As captive breeding programmes become essential for felid conservation, these methods provide a means to measure the impact of captive management on each species. It is as yet unknown whether human encroachment, tourism, translocation, or habitat restriction that prevents dispersal and increases population densities are stressors that may affect the health of free-ranging felids, but we should be mindful of lessons learned through animals in captivity.

Other determinants of immune function

Factors other than chronic stress can influence the ability to mount an effective immune response and survive infection when exposed to pathogens. Viruses infecting lymphocytes, such as FPV, CDV, or FeLV, can result in a marked transient or persistent immunosuppression (Pedersen 1988; Barker and Parrish 2001; Williams 2001). Animals infected with or recovering from these viral infections may succumb to co-infections with endemic infectious agents that are usually tolerated (Munson *et al.* 2008). In contrast, FIV does not appear to have an immunosuppressive effect in free-ranging felids (Pedersen and Barlough 1991).

Loss of major histocompatibility (MHC) gene diversity has the potential to affect immunocompetence, but a clear association between loss of MHC diversity and susceptibility to disease has not been established. The MHC Class I and II genes encode receptors that determine the response of the innate and adaptive immune systems to pathogens. MHC polymorphism theoretically broadens the ability of an individual to respond to different pathogens (Yuhki and O'Brien 1990b), so MHC diversity at the population level would provide more assurance that some individuals would survive an epidemic. What is not clear is how much diversity is necessary. Species such as the cheetah, which have little genetic diversity (O'Brien *et al.* 1985b; O'Brien and Evermann 1988), have survived infection by numerous potential pathogens in the wild (Munson *et al.* 2004a) without evidence of disease (Munson *et al.* 2005). Until more is known about the association of MHC diversity and disease susceptibility in felids, conserving maximum genetic diversity would be wise, not only for health, but also for general fitness.

Potential felid pathogens

Felid-specific viruses and haemoparasites

Infectious diseases pose the greatest threat of catastrophic losses to wildlife populations and are therefore the focus of most disease surveillance. Infectious agents that have the potential to cause disease in felids are either felid-specific or more generalist pathogens that affect a wide variety of hosts. Felid viruses include FCV, FCoV, FPV, FHV, FeLV, and FIV. Felid haemoparasites that have been identified in free-ranging felids include *Babesia felis, Babesia leo,* and other as yet unclassified *Babesia, Hepatozoon felis, Cytauxzoon felis, Bartonella henselae, Mycoplasma haemofelis* (formerly *Haemobartonella felis*), *Candidatus*

M. haemominutum, *Candidatus* M. turicensis, and *Ehrlichia* spp. (Rotstein *et al.* 2000; Molia *et al.* 2004; Filoni *et al.* 2006; Willi *et al.* 2007; Munson *et al.* 2008).

Tables 9.1 and 9.2 provide a summary to date of serological, morphological, or molecular evidence of free-ranging felid exposure to infectious agents (by geographic region and species). These data indicate that infection has occurred (or is present), but not that disease was the result of this infection. What is evident from these data is that exposure to feline viruses and haemoparasites is common in wild felid populations, although the worldwide distribution of two viruses, FeLV and FIV, are noteworthy. With the exception of the European wildcat, in which a high percentage of seropositive cats have been noted (Table 9.1), FeLV does not appear to be endemic in most free-ranging populations. Despite the high prevalence in the wildcat, FeLV-associated disease has not been reported. Therefore, it is likely that most free-ranging felid species have not developed resistance to this virus through co-evolution and may develop fatal disease due to the immunosuppressive and carcinogenic effects of the virus when exposed to FeLV by domestic cats (Marker *et al.* 2003d; Cunningham *et al.* 2008). In contrast to FeLV, FIV infection is widespread in at least 19 species of free-ranging felids, predominantly in Africa and South and Central America, without evidence of disease (Troyer *et al.* 2005). Free-ranging species determined to have species-specific FIV include lions in South and East Africa, but not in Namibia or Asia; leopards in Africa, but not in Asia; pumas; cheetahs in East Africa, but not Namibia; jaguaroundis (*Herpailurus yagouaroundi*); ocelots (*Leopardus pardalis*); and Pallas's cats (*Otocolobus manul*). Other free-ranging felid species in which FIV antibodies have been detected and/or infection confirmed by polymerase chain reaction (PCR) are tigers (*Panthera tigris*), jaguars (*Panthera onca*), margays (*Leopardus wiedii*), and Geoffroy's cats (*Oncifelis geoffroyi;* Troyer *et al.* 2005). There is evidence of cross-species transmission between pumas and bobcats (Franklin *et al.* 2007), free-ranging felids and domestic cats (Nishimura *et al.* 1999), captive felids and domestic cats, and among different species of captive felids (Troyer *et al.* 2005; Franklin *et al.* 2007). Interestingly, FIV infection does not occur in European free-ranging felids, except for a few cases reported in wildcats from France (Fromont *et al.* 2000). Wildcats from several countries (McOrist *et al.* 1991; Artois and Remond 1994; Daniels *et al.* 1999; Leutenegger *et al.* 1999; Troyer *et al.* 2005), Eurasian lynx (Ryser-Degiorgis *et al.* 2005), and Iberian lynx (Troyer *et al.* 2005; Meli *et al.* 2006; Roelke *et al.* 2008) have no evidence of FIV infection or exposure.

Generalist pathogens affecting felids

Some of the more important felid pathogens, such as rabies virus, CDV, and *Mycobacterium* spp., are cosmopolitan in their distribution and have reservoirs outside felids, including their prey species. Important felid pathogens transmitted by prey include *Mycobacterium* spp., *Bacillus anthracis, Yersinia pestis, Notoedres cati*, and *Sarcoptes scabiei*. Many of these generalist pathogens also affect humans, so great care should be taken when handling sick or dead felids.

Tuberculosis, the disease caused by *Mycobacterium* infection, is principally acquired by ingesting infected prey, but felid-to-felid transmission also occurs from aspiration of bacteria aerosolized by coughing (De Vos *et al.* 2001; de Lisle *et al.* 2002). Tuberculosis is usually a slowly progressive disease providing ample opportunities for widespread dissemination within a population. For that reason, it may pose a more serious threat to populations than epidemic viral diseases. Anthrax, the disease caused by ingestion or inhalation of *Bacillus anthracis*, can be acquired from infected prey or from the environment where the bacteria persist for long periods. Anthrax is suspected to be the reason why viable cheetah populations have been difficult to maintain in areas of Etosha National Park, Namibia, where anthrax is a persistent problem (Lindeque and Turnbull 1994; Lindeque *et al.* 1996). Plague, the disease caused by *Yersina pestis*, is acquired through rodents or their fleas, and serious population threats occur when rodent populations flourish or other prey are sparse due to climatic or anthropogenic influences (Gasper and Watson 2001). Mange, the disease caused by *Notoedres cati* or *Sarcoptes scabiei* mites, is another disease acquired through contact with prey or conspecific felids (Jeu and Xiang 1982; Mörner

Table 9.2 Prevalence and geographic distribution of free-ranging felids with evidence of infection or exposure to selected haemoparasites and haemobacteria.

Species	Location	Years	*N*	*Babesia* spp. (%)	*Cytauxzoon* spp. (%)	*Bartonella* spp. (%)	*Ehrlichia* spp. (%)	*Mycoplasma* spp. (%)	Reference
Cheetahs	Namibia, South Africa		40	8					Bosman *et al*. (2007)
Lion			56	50					
Caracal			2	0					
Black-footed Cat			8	63					
Leopard			1	100					
Serval			3	33					
Lions	Tanzania		123	100					Averbeck *et al*. (1990)
Cheetahs									
Lions	South Africa		47	100					Lopez-Rebollar *et al*. (1999)
Lions	South and East Africa	1982–2002	113			17			Molia *et al*. (2004)
Cheetahs	East Africa		22			46			
African lions	Botswana	2003–05	21				0		Ramsauer *et al*. (2007)
Wildcat	Switzerland	2003–06	32		31.3				Meli *et al*. (2006)
Eurasian lynx			35		25.7				
Iberian lynx	Spain		46	0	34.8	22%			
Iberian lynx	Spain	2004–06	20		15				Millán *et al*. (2007)
Pallas's cat	Mongolia	2000	4		100				Ketz-Riley *et al*. (2003b)

(Continued)

Table 9.2 (Continued)

Species	Location	Years	*N*	*Babesia* spp. (%)	*Cytauxzoon* spp. (%)	*Bartonella* spp. (%)	*Ehrlichia* spp. (%)	*Mycoplasma* spp. (%)	Reference
Ocelot	Brazil		1					100	Willi *et al*. (2007)
Iberian lynx	Spain		35					37	
Eurasian lynx	Switzerland		36					44	
European wildcat	France		31					39	
Lions	Tanzania		45					98	
Puma	California	1985–96	74			35			Yamamoto *et al*. (1998)
Bobcat			62			53			
Puma	North and South America	1984–99	479			19.4			Chomel *et al*. (2004)
Bobcats			91			23.1			
Puma	Florida	1997–98	35			20			Rotstein *et al*. (2000)
Bobcats	United States		4		100				Glenn *et al*. (1983)
Bobcats	United States	1992–95	26			74			Riley *et al*. 2004)
Puma	Brazil	1998–2004	21			90	5		Filoni *et al*. (2006)
Ocelot									
Little spotted cat		2004							
Puma	California	1987–90	58			90			Paul-Murphy *et al*. (1994)
Puma	Florida	1983–97	91		39				Rotstein *et al*. (1999)

1992; Bailey 1993; Mwanzia *et al.* 1995; Linnell *et al.* 1998; Ryser-Degiorgis *et al.* 2002; Uzal *et al.* 2007). Predisposing factors, such as poor general health, lack of immunity due to no prior exposure (Pence and Ueckermann 2002), lack of innate genetic resistance (Pence and Windberg 1994; Nimmervoll 2007), nutritional, social, or environmental stressors (Zumpt and Ledger 1973; Bailey 1993; Pence and Windberg 1994; Mwanzia *et al.* 1995; Bornstein *et al.* 2001; Pence and Ueckermann 2002), and chronic toxicity (Riley *et al.* 2007) are suspected to influence disease severity and outcome.

Free-ranging felids may be accidental hosts of several disease agents, such as those causing avian influenza (AI), cowpox, pseudorabies, African horse sickness (AHS), and bovine spongiform encephalopathy (BSE; Baxby *et al.* 1982; Alexander *et al.* 1995; Kirkwood *et al.* 1995; Lezmi *et al.* 2003; Keawcharoen *et al.* 2004; Thanawongnuwech *et al.* 2005). Cowpox, BSE, and AI infections were acquired in captivity from being fed infected food sources (rodents, beef, and chickens, respectively). In free-ranging felids, AHS was presumably acquired from prey. Felid-to-felid transmission has not been reported for any of these viruses, apart from AI.

Distinguishing infection from disease

When an infectious agent is endemic and the felid has evolved in that environment, disease is the exception, not the rule. Morbidity and mortality from felid-specific pathogens have only rarely been documented in free-ranging felids, and typically affect only individual animals, rather than populations (Table 9.3). In contrast, reports of clinical or fatal disease in captive felids from infectious agents are more numerous. Whether this reflects a propensity for captive animals to develop disease, increased exposure due to closer contact among animals and their infectious agents in captivity, exposure in captivity to new pathogens, or the closer surveillance of individual animal health in captivity is not known. Regardless, the greater potential of developing disease in the captive setting should be taken into account when considering captive breeding or zoo sanctuaries as a conservation tool for endangered felids.

The most commonly employed infectious disease surveillance tool in free-ranging felid populations is the serosurvey, a method with limitations that need to be recognized. Considerable misunderstanding has arisen from using serosurveys for disease surveillance because the distinction between infection (past or current) and disease is usually not recognized, leading to a misconception that infectious diseases are widespread in free-ranging felid populations. For serosurveys, serum or plasma is tested for antibodies that react against infectious agents (or closely related agents) indicating that an animal has been infected by viruses or other pathogens within their lifetime, mounted a detectable immune (antibody) response, and survived infection (Fig. 9.1). Thus, antibody surveys are the antithesis of disease surveys in that in most cases they measure the prevalence of exposed animals that either never developed disease or developed non-fatal disease. Only in rare instances when serum is collected during active infection is the fate of the animal that mounted an antibody response yet to be determined. Serosurveys do not provide information on the pathogenicity (virulence) of an infectious agent, because animals that died from infection are not detected. Antibodies can persist for months to years after clearing of infection (Fig. 9.1), so an antibody titre does not indicate when an animal was infected. Antibodies can also be passively transferred between mother and offspring, so antibody titres in animals under 5 months old may not represent true exposure (Fig. 9.2).

In closely monitored populations, serology can be used to investigate potential associations between mortalities and infectious agents by overlying temporal patterns of seroconversion in juvenile and yearling felids with mortalities. If mortalities increased at the time young animals had evidence of viral exposure (seroconverted), then mortalities as a result of that virus should be suspected in that population. Using this approach, CDV, FPV, FCoV, FCV, and FHV exposures were investigated as potential causes of mortality in the Serengeti lion population over a 10-year period, and only CDV was found associated with mortality events (Packer *et al.* 1999).

Surveillance for infectious agents (entire organisms, their antigens, or nucleic acids) requires examining blood or tissues microscopically, by ELISA, or by molecular methods (PCR or *in situ* hybridization).

Table 9.3 Clinical diseases or deaths from infectious agents reported in captive (C) or free-ranging (FR) felids. Whether the disease affected only a few individual animals (cases) or was an epidemic is indicated, as well as the impact on wild populations.

Name of clinical disease	Causative agent	Impact on wild population	Individual cases or epidemic	Species	Reference
Feline panleukopenia	FPV	No	Cases	Puma[FR] Eurasian lynx[FR] Bobcats[FR] Snow leopard[C] Lion[C] Cheetah[C]	Wassner *et al*. (1988); Fix *et al*. (1989); Valicek *et al*. (1993); Mochizuki *et al*. (1996); Steinel *et al*. (2000); Barker and Parrish (2001); Schmidt-Posthaus *et al*. (2002b)
Herpes viral conjunctivitis, keratitis, rhinitis, pneumonia, or dermatitis	FHV	No	Cases	Cheetah[C]	Junge *et al*. (1991); van Vuuren *et al*. (1999); Munson *et al*. (2004a)
Calici viral conjunctivitis, keratitis, or rhinitis	FCV	No	Cases	Lion[C] Cheetah[C]	Kadoi *et al*. (1997); Munson and Citino (2005)
Feline infectious peritonitis	FCoV	No	Epidemic	Cheetah[C]	Evermann *et al*. (1983); Van Rensburg and Silkstone (1984); Evermann *et al*. (1988); Munson (1993); Evermann and Benfield (2001)
Leukemia	FeLV	Yes	Cases and epidemics	Florida panthers[FR] Iberian lynx[FR] Cheetah[C]	Marker *et al*. (2003d); Cunningham *et al*. (2008)
Leukemia virus-induced immunosuppression	FeLV	Yes	Cases	Iberian lynx[FR] Puma[FR] Bobcat[C]	A. Vargas, personal communication; Jessup *et al*. (1993); Cunningham *et al*. (2008)

Canine distemper	CDV	Yes	Cases and epidemics	Lions[FR] Leopards[FR] Cheetahs[FR] Bobcat[FR] Canada lynx[FR] Lions[C] Tiger[C] Leopards[C] Jaguars[C]	Blythe *et al*. (1983); Gould and Fenner (1983); Fix *et al*. (1989); Appel *et al*. (1994); Roelke-Parker *et al*. (1996); Kock *et al*. (1998); McBurney *et al*. (1998); Munson (2001); Munson *et al*. (2008)
Toxoplasmosis	*Toxoplasma gondii*	No	Cases	Pallas's cat[C] Cheetahs[C]	Van Rensburg and Silkstone (1984); Kenny *et al*. (2002); Brown *et al*. (2005)
Plague	*Yersina pestis*	Yes	Cases	Canada lynx[FR]	Wild *et al*. (2006)
Anthrax	*Bacillus anthracis*	Yes	Cases	Cheetahs[FR] Cheetahs[C] Lions[FR]	Jager *et al*. (1990); Lindeque and Turnbull (1994); Tubbesing (1997); De Vos and Bryden (1997)
Tuberculosis	*Mycobacterium* spp.	Yes	Cases and epidemics	Bobcat[FR] Lion, leopard, and cheetah[FR] Iberian lynx[FR] Lion, snow leopard, and tiger[C]	Helman *et al*. (1998); Briones *et al*. (2000); Bruning Fann *et al*. (2001); Montali *et al*. (2001); Perez *et al*. (2001); De Vos *et al*. (2001); Lantos *et al*. (2003); Aranaz *et al*. (2004); Cleaveland *et al*. (2005); Cho *et al*. (2006); Michel *et al*. (2006); Monies *et al*. (2006)
Cryptococcosis	*Crytococcus neoformans*	No	Cases	Cheetah[C, FR]	Berry *et al*. (1997); Bolton *et al*. 1999; Munson *et al*. (1999b); Millward and Williams (2005)
Blastomycosis	*Blastomyces dermatitidis*	No	Cases	Lion, tiger, cheetah, and snow leopard[C]	Storms *et al*. (2003)
Notoedric mange	*Notoedres cati*	Yes	Cases and epidemics	Eurasian lynx[FR] Bobcat[FR] Leopard[FR] Cheetah[FR] Ocelot[FR] Pumas[FR] Snow leopard[C]	Young *et al*. (1972a); Zumpt and Ledger (1973); Fletcher (1978); Pence *et al*. (1982); Wassner *et al*. (1988); Bailey (1993); Maehr *et al*. (1995); Pence *et al*. (1995); Pence and Ueckermann (2002); Ryser-Degiorgis *et al*. (2002); Riley *et al*. (2007); Uzal *et al*. (2007)

(Continued)

Table 9.3 (Continued)

Name of clinical disease	Causative agent	Impact on wild population	Individual cases or epidemic	Species	Reference
Sarcoptic mange	*Sarcopes scabei*	Yes	Cases and epidemics	Eurasian lynx[FR] Cheetah[FR] Leopards[FR] African lion[FR] Serval[FR]	Young *et al*. (1972a,b); Zumpt and Ledger (1973); Young (1975); Caro *et al*. (1987); Holt and Berg (1990); Mörner (1992); Mwanzia *et al*. (1995); Pence and Ueckermann (2002); Ryser-Degiorgis *et al*. (2002)
Babesiosis	*Babesia* spp.	Yes	Epidemic	Lion[FR]	Jakob and Wesemeier (1996); Nietfeld and Pollock (2002); Penzhorn (2006)
Cytauxzoonosis	*Cytauxzoon* sp.	No	Cases	Bobcat[FR] Lions[C] Tiger[C]	Peixoto *et al*. (2007); Munson *et al*. (2008)

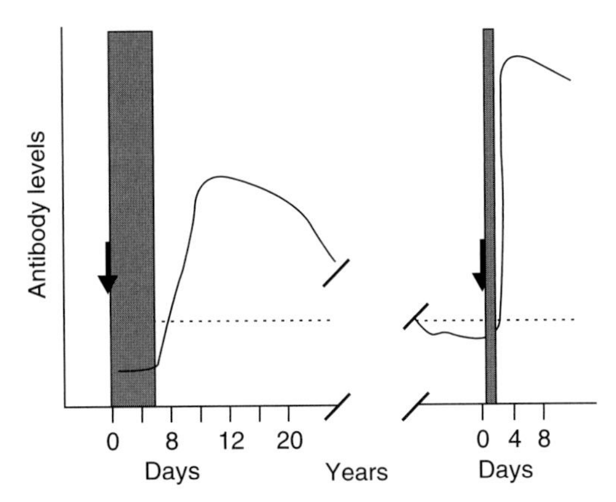

Figure 9.1 Production of antibodies after natural exposure to a pathogen. The arrows indicate first and subsequent exposure to a pathogen. Note that there is a delay (lag phase indicated by grey) between first exposure and the increase in antibody levels, but upon subsequent exposure to the same pathogen, this lag phase is shorter. The dotted line indicates a hypothetical minimum level of antibody detection by an assay. (Adapted from Janeway and Travers 1997.)

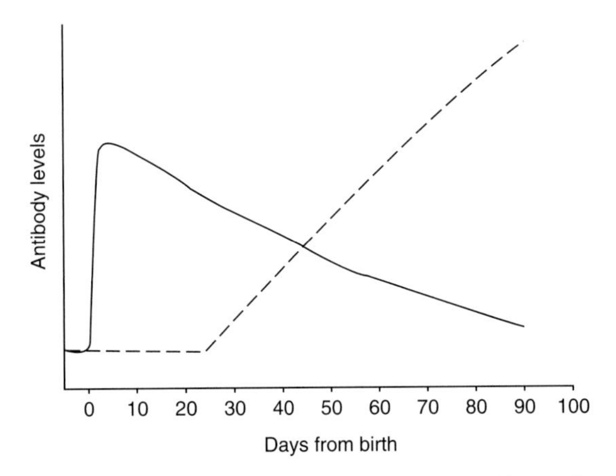

Figure 9.2 Antibody levels in neonatal kittens. Maternal antibodies (solid line) are transferred to kittens through colostrum (milk) immediately after birth and decline over time. Antibody levels are the lowest between 20 and 30 days, after which antibody levels increase as a result of *de novo* synthesis (dashed line) by the kitten. (Adapted from Yamada *et al.* 1991.)

Similar to serological assays, these methods only provide evidence of infection and not disease. Organisms detected in tissues are not necessarily pathogenic, because many species tolerate low levels of endemic parasites. Correlated clinical or necropsy data are necessary to establish virulence. It is important to remember that molecular techniques identify DNA, and not viable organisms; therefore positive results only indicate the presence of an organism, not its viability or pathogenicity. As molecular tools are increasingly used to detect infectious agents in free-ranging felids, many 'new' pathogens may be discovered and misinterpreted as 'emerging' threats. These molecular findings should be interpreted with caution, because many agents may not be new pathogens but rather newly detected infectious agents that have been present in felid populations for years. Examples of such agents are *Ehlichia*, *Bartonella*, and *Hemoplasma* (Yamamoto *et al.* 1998; Molia *et al.* 2004; Filoni *et al.* 2006).

The most sensitive and specific tools for determining if clinical disease or death was due to infectious agents are physical examination and necropsy, procedures that are rarely performed on free-ranging felids. For populations at risk, accurate data on causes of mortality and subclinical diseases that may have contributed to mortalities can be obtained by conducting comprehensive pathology investigations on all animals that die and by establishing a database of findings. These investigations include a full necropsy with histopathology and ancillary testing for micro-organisms and toxins. The logistics and costs of such surveys at the population level have limited their application. Among felids, only the cheetah, Florida panther, and Eurasian lynx have a comprehensive pathology surveillance and database established; a pathology survey has recently been established for the Iberian lynx.

Manifestations of disease in wild felids

Table 9.3 provides a summary of infectious diseases reported to date in captive and free-ranging felids. When disease does occur in wild felids from infection with felid-specific viruses or parasites, the clinical signs and pathology are usually similar to those reported in domestic cats. For detailed descriptions of felid diseases, Greene's *Infectious Diseases of the Dog and Cat* (2006) is an excellent resource. One exception to this generalization is FIV. The name 'feline immunodeficiency virus' implies a virulent pathogen in felids, but the designation arose from its homology with human immunodeficiency virus and not its effect on wild feline hosts. The high prevalence of FIV in healthy free-ranging felid populations, such as African lions, suggests an ancient introduction of a virus that has become host-adapted (Carpenter and O'Brien 1995; Troyer *et al.* 2004, 2005). Robust populations, such as the Serengeti lions, have evidence of persistent infections since at least 1984 without associated increased mortality in this population (Olmsted *et al.* 1992; Packer *et al.* 1999). Also, FIV-positive lions in this population did not have a lower mean age than FIV-negative lions (Packer *et al.* 1999). The few studies endeavouring to link FIV infection with disease (Bull *et al.* 2003; Roelke *et al.* 2006) had other factors confounding interpretation of immune function, such as recent CDV infection or captivity stress. Ongoing research may disclose subtle influences of FIV on immune function, but the weight of evidence suggests that healthy free-ranging felid populations can coexist with infection.

The lesions of anthrax in felids warrant attention because they differ from the 'classical' lesions of anthrax in ungulates. Most felids are resistant to the septicemic form of anthrax and develop instead localized infections of the oral cavity and throat that are only fatal if the swelling from infection causes asphyxiation. In contrast to most felids, cheetahs appear to be particularly susceptible to acute fatal septicemic anthrax (U. Tubbesing, personal communication), similar to ungulates, and may not develop oral cavity lesions. Although death from anthrax is rare in most other felid species, 22 lions died during a 1993 anthrax epidemic in Kruger National Park (KNP; 7% of total animal deaths), an event that led wildlife veterinarians to consider lions more vulnerable than previously thought due to their feeding behaviour and dense populations (De Vos and Bryden 1997).

Diseases caused by non-infectious agents

Surveillance for infectious agents fails to detect non-infectious causes of morbidity and mortality that can have important impacts on felid populations. For example, genetic diseases are now being recognized in felid populations with low genetic diversity (Munson *et al.* 1996; Cunningham *et al.* 1999; Marker and Dickman 2004a; Jiménez *et al.* 2008). Environmental contaminants are also emerging threats due to increasing overlap of free-ranging felid and human habitats. Accidental rodenticide toxicities (e.g. brodifacoum; Riley *et al.* 2007; Uzal *et al.* 2007) and heavy metal exposures, such as mercury and lead (Roelke 1990), are increasing concerns for felids whose range contacts human habitations. Bioaccumulation of environmental toxins in top predators such as felids makes exposure to toxic levels more probable. As environmental contamination becomes more widespread, health assessment of felid populations should include assays for toxins.

Degenerative diseases such as chronic renal disease (glomerulosclerosis) and degenerative spinal disease (ankylosing spondylosis) are common in older captive felids (Munson 1993; Munson *et al.* 1999b; Kolmstetter *et al.* 2000), but not in free-ranging animals (Munson *et al.* 2005). Although this difference may in part be due to the potentially greater longevity of captive animals, a study comparing age-matched cohorts from captive and free-ranging cheetah populations refutes this hypothesis (Munson *et al.* 2005). One unusual finding in a free-ranging population is the high prevalence of glomerulonephritis in Iberian lynx (Jiménez *et al.* 2008). This disease can be an outcome of chronic inflammation or infection; however, these predisposing conditions have not been present in all affected animals, so the cause is thought perhaps to have a genetic basis.

Case examples of felid diseases with population consequences

Most populations can risk losing individual animals to disease without long-range detrimental outcomes to their conservation. However, some diseases have affected significant proportions of a population and impaired conservation efforts. The following case examples illustrate diseases with a broader impact on felid populations or diseases of concern due to their potential effects on endangered felids.

Tuberculosis epidemics

Mycobacterium spp. are cosmopolitan generalist pathogens that have emerged as a significant threat to some free-ranging felid populations due to the widespread infection of their prey. Why the prevalence and severity of infection in some prey species has increased is not known, but is presumed by some to be an outcome of increased contact between domestic livestock and wildlife and/or of inadequate management of infected wild ungulate species (Vicente *et al.* 2007). Additionally, fragmented habitat and proximity to human settlements increases the risk of felids contacting livestock in areas where livestock are an important source of infection. Diminishing habitat may result in greater exposure of free-ranging felids to *Mycobacterium tuberculosis*-infected animals in the future.

The epidemic of tuberculosis in the KNP, South Africa, illustrates the potential impact on felids (De Vos *et al.* 2001; Michel *et al.* 2006). *Mycobacterium bovis* was likely introduced to South Africa in imported European cattle breeds in the eighteenth and nineteenth centuries and over time spread from infected cattle to a variety of free-ranging ruminants. Infection of African buffalo (*Syncerus caffer*), one of the main prey species for lions, was first confirmed in 1990 and occurred from close contact between buffalo and domestic cattle across the border of the southeastern KNP. Since then the disease has spread steadily northwards within the KNP. Buffalo are considered fully susceptible maintenance hosts in which infection is now endemic. Infection of free-ranging felids is considered sporadic and accidental, with felids serving as spillover or amplifying hosts, not as significant reservoirs for mycobacterial infection (de Lisle *et al.* 1995; Ragg *et al.* 1995; Briones *et al.* 2000; Buddle *et al.* 2000; Coleman and Cooke 2001; de Lisle *et al.* 2002; Dean *et al.* 2006; Monies *et al.* 2006). Whereas buffalo are the main source of infection for lions, the source for solitary predators such as cheetahs and leopards has not been identified. Ingestion of infected smaller antelope or scavenging from buffalo carcasses is possible (De Vos *et al.* 2001; Michel *et al.* 2006).

Iberian lynx is another species affected by tuberculosis. Fallow deer (*Dama dama*) are believed to be the main source of *Mycobacterium bovis* infection for free-ranging lynx, although wild boar (*Sus scrofa*) and red deer (*Cervus elaphus*) may also serve as reservoirs (Hunter 1996; Aranaz *et al.* 2004; Parra *et al.* 2006; Vicente *et al.* 2006). The decline of its main prey species (rabbit) due to habitat degradation enhances the risk of consuming alternative potentially infected prey such as fallow deer.

Current data indicate that tuberculosis infection has had significant adverse effects on lion populations (Keet *et al.* 1996, 1997; Martin-Atance *et al.* 2006). In individual lions, infection causes characteristic nodular lesions (granulomas) in the lymph nodes, lungs, bones, and rarely other tissues, which result in weight loss, lameness, and respiratory disease (Keet *et al.* 1996, 1997; de Lisle *et al.* 2002; Dean *et al.* 2006; Kirberger *et al.* 2006). At the population level, the impact is seen in changing prey population dynamics and hunting success rates, altered population demographics, including social structure within prides, faster territorial male coalition turnover, higher infanticide, altered sex ratios and birth rates, and reduced survival and breeding success, as well as eviction of prides from their territories (D.Keet, personal communication). The long-term conservation and population effects of this epidemic are under intense study. The projected impact of tuberculosis on the sustainability of free-ranging lion, leopard, and cheetah populations in KNP has yet to be determined. It is not known whether lion will become established as reservoir hosts, which propagate and maintain the infection in the ecosystem, or stay simply as spillover hosts who acquire infection from another source.

The *Mycobacterium bovis* epidemic in African lion in KNP highlights the difficulties of controlling tuberculosis in a free-ranging situation despite the demonstrated adverse population effects. Management strategies are hampered by the complexity of the epidemiological matrix. Since *Mycobacterium* infections do not become self-sustaining in felid populations (with the possible exception of lions), control of felid spillover infections may ultimately be best achieved by controlling the disease in reservoir or maintenance hosts, such as the African buffalo. However, decisions must take into consideration the entire ecosystem and its interface with domestic livestock and humans on the borders of conservation areas (Corner 2006). A measure of tuberculosis control in the KNP may be achieved in the future by selected buffalo culling, and the effect of buffalo culling on the prevalence of tuberculosis in lions will be closely monitored. However, early data suggest that a 'capture-test and slaughter' policy for buffalo in the similarly affected Umfolozi-Hluhluwe Park may not reduce infection rates, as disease transmission may be enhanced in the captured stressed buffalo and tests of low sensitivity may fail to detect infected animals (Michel *et al.* 2006).

Reliable, effective, economical, and practical diagnostic, treatment, and prevention measures for tuberculosis are not yet available (Michel *et al.* 2006). Whether or not to treat infected felids with antibiotics or therapeutic vaccines and/or isolate infected animals is controversial, because the risk of intraspecies transmission among felids is generally low (Briones *et al.* 2000; Buddle *et al.* 2000). In the case of critically endangered species such as the Iberian lynx, treating individuals may be justified by the value of each animal to species survival. Control through vaccination in endemic areas may be a future option, but effective vaccines are currently not available. Furthermore, vaccination would confound antibody screening tests in living animals, tests that are needed to prevent translocation of infected animals. At this time, preventing exposure is the only strategy for controlling tuberculosis in wild felids.

Mange in free-ranging felids

Mange is a cosmopolitan ectoparasitic skin disease that can cause dramatic but usually short-term effects on free-ranging populations (Pence and Ueckermann 2002). While mortality rates in felids have generally been low in comparison with observed morbidity, death has been the most common outcome of infected animals in some cheetah, bobcat, and Eurasian lynx populations (Young *et al.* 1972a; Bailey 1993; Ryser-Degiorgis *et al.* 2002, 2005; Riley *et al.* 2007).

Mite infection causes thickening and crusting of the skin, as well as hair loss. Skin lesions caused by *Notoedres cati* are typically located on the head, neck, and shoulders (Young *et al.* 1972a, b; Fletcher 1978; Pence *et al.* 1982; Maehr *et al.* 1995; Pence *et al.* 1995; Pence and Ueckermann 2002; Riley *et al.* 2007), but may extend over the entire body (Peters and Zwart 1973; Ryser-Degiorgis *et al.* 2002; Riley *et al.* 2007). In contrast, lesions caused by *Sarcoptes scabiei* are usually located on the tail, flanks, and withers before affecting the head (Peters and Zwart 1973; Bornstein *et al.* 2001). Severely affected animals may become listless, emaciated, cold-intolerant, and/or develop abnormal behaviour, such as loss of fear of humans and/or increased diurnal activity (Overskaug 1994; Ryser-Degiorgis *et al.* 2002; Riley *et al.* 2007), and eventually die from the infection.

How mange can suddenly spread through a large geographic area, affecting many different species including wild felids, is illustrated by the Scandinavian epidemic. The first case of sarcoptic mange in Swedish wildlife was diagnosed in 1972 in red foxes. In the following years, mange became the most common disease of foxes in Sweden and soon reached Norway. A number of other, mainly carnivorous, species were also affected, indicating that the mite involved in this outbreak was not specific to red foxes (Mörner 1992; Mörner *et al.* 2005). After 1980, mange was observed in Eurasian lynx in both countries (Holt and Berg 1990; Mörner 1992). During the next 20 years, approximately 70 more Swedish lynx, including entire family groups, were diagnosed with mange. Most animals succumbed to the disease, whereas less extensive skin lesions were found in lynx that died of other causes (M.-P. Ryser-Degiorgis and T. Mörner, unpublished data). During the first decade of the mange epidemic in lynx, the Swedish lynx population decreased, and mange was considered a probable important contributing factor (Mörner 1992). However, the population recovered from a low of 600 lynx in 1980 to 1500 lynx in 1999, a period when mange had become the most common infectious disease in this population, but legal hunting was restricted (Ryser-Degiorgis *et al.* 2005). So the impact of mange at the population level appeared minor, and population recovery was thought to be due to hunting protection.

In 1999, mange was observed for the first time in the reintroduced Swiss Eurasian lynx population (Ryser-Degiorgis *et al.* 2002), increasing the overall percentage of infectious causes of mortality (Schmidt-Posthaus *et al.* 2002b). It was of concern that mange could have an immediate impact on this small, geographically limited and inbred population.

However, only sporadic cases have been recorded since that time (M.-P. Ryser-Degiorgis, unpublished data), and no indication of population decline has been noticed (von Arx *et al.* 2004). Overall, mange does not appear as a serious threat to stable, free-ranging felid populations, since epidemics seem to have no long-term effect on population abundance. However, mange may have devastating short-term effects on small remnant, genetically compromised, or fragmented populations (Pence and Ueckermann 2002), so monitoring is recommended as part of felid conservation projects.

As for other diseases, control of mange is difficult in free-ranging populations. Capture and treatment of individual cases may be warranted in threatened populations (Pence and Ueckermann 2002), and has been successful (Young *et al.* 1972a; Pence *et al.* 1982; Maehr *et al.* 1995). However, eradication of mange is only feasible if the primary source of infection is known and can be eliminated (Bornstein *et al.* 2001). For animal translocations, ectoparasitic treatment is highly recommended if the source population is affected by mange, even if the translocated individuals do not show any signs of disease, because the time between infection and visible disease (incubation period) can be several weeks and some infected animals do not show clinical signs (healthy carriers; Young *et al.* 1972b; Bornstein *et al.* 1997; Pence and Ueckermann 2002). Furthermore, the stress of translocation may impair resistance and lead to the development of clinical mange (Peters and Zwart 1973; Zumpt and Ledger 1973). Considering all the above-mentioned factors, general conservation strategies (habitat restoration and population reinforcement) would be the best measures for preventing a population effect due to mange in free-ranging felids.

Canine distemper virus: factors influencing mortalities during epidemics

Historically, CDV was not considered a threatening pathogen to felids. Before 1991, CDV was suspected to have caused death in a few individual zoo cats (Blythe *et al.* 1983; Gould and Fenner 1983; Fix *et al.* 1989), but felids in general were thought to be fairly resistant. Then in 1991 and 1992, three epidemics of canine distemper occurred in numerous lions, tigers, leopards, and jaguars in United States zoological parks (Appel *et al.* 1994). Two years later, CDV caused the death or disappearance of approximately one-third of the lion population and numerous other carnivores in the Serengeti ecosystem (Roelke-Parker *et al.* 1996; Kock *et al.* 1998). The same year, canine distemper was reported in Canadian bobcats (Munson 2001) and Canada lynx (McBurney *et al.* 1998). Overall since 1991, CDV infection has been associated with clinical disease or death in at least six species of free-ranging or captive felids at numerous locations, suggesting evolution of the virus worldwide as a felid pathogen (Munson 2001). Genetic characterization of CDV from the 1994 lion epidemic detected differences from viruses isolated from previous fatal cases (Carpenter *et al.* 1998), but whether these differences contributed to the increased pathogenicity for felids could not be determined with certainty, because of the naturally high mutation rates of morbilliviruses.

The magnitude of the CDV epidemic in the Serengeti lion population demonstrated how quickly disease could impact a healthy felid population and the difficulty of detecting an epidemic in the wild. Despite close monitoring by an experienced team of ecologists and wildlife veterinarians, hundreds of lions disappeared and less than 1% of the population was seen with clinical signs or retrieved as carcasses. Another high mortality epidemic occurred in the nearby Ngorongoro Crater lion population in 2001, reducing their numbers by 40%, yet only two carcases were found (Munson *et al.* 2008; Craft *et al.*, Chapter 10, this volume).

The extent of mortalities during the 1994 and 2001 Serengeti epidemics was unusual for CDV, so other contributing factors were suspected. Retrospective serosurveys of Serengeti lion and Namibian cheetah populations disclosed numerous 'silent CDV epidemics' (infection without mortalities) over the past 25 years (Munson *et al.* 2004a, 2008), indicating that CDV infection was not always fatal in felids, even after the virus had evolved to be more pathogenic in the early 1990s. Although the lion population was highly susceptible to CDV infection in 1994 (no anti-CDV antibodies were detected in the lions in the years immediately before the epidemic [Packer *et al.* 1999] and the population was very dense at the time of the epidemic, providing greater opportunities for pathogen transmission within prides), these risk factors were also present during the 'silent' epidemics. What was unique about the 1994 and 2001 Serengeti CDV epidemics was that many lions had unusually high levels of *Babesia* infections, sufficient to cause death in some lions (Roelke-Parker *et al.* 1996; Munson *et al.* 2008). Low levels of *Babesia* haemoparasitism are common in healthy lions (Penzhorn *et al.* 2001; Penzhorn 2006), but the super-infections seen in sick and dead lions during the 1994 and 2001 CDV epidemics were unusual. The same species of *Babesia* were present during high mortality epidemics as were present during periods between epidemics, indicating that mortalities were not due to introduction of a novel species (Munson *et al.* 2008).

Finding fatal secondary pathogens is not uncommon in CDV-infected animals (Williams 2001), but the magnitude and epidemic proportion of babesiosis was highly unusual. Although CDV-induced immunosuppression may explain in part why *Babesia* superinfection occurred, CDV-induced immunosuppression would also have occurred during the silent CDV epidemics when fatalities were not noted. What differed between the 1994 and 2001 epidemics and other CDV epidemics were the extreme drought conditions immediately preceding both fatal CDV epidemics that resulted in debility and massive die-offs of buffalo, a major prey species for lions (Munson *et al.* 2008). When the rains resumed, dormant ticks (the intermediate hosts for *Babesia*) emerged en masse, and the buffalo and lions became heavily infested (Fyumagwa *et al.* 2007). So the convergence of CDV-induced immunosuppression with climatic events favouring high burdens of *Babesia*-infected ticks would account for the extent of mortalities during the 1994 and 2001 CDV epidemics.

The catastrophic epidemics in lions may portend future threats to felids if exposure to CDV occurs more frequently through contact with domestic animals and the biological or mechanical vectors of parasites increase due to global climate change. Altered ecosystems may favour proliferation of other pathogens with which felids have historically coexisted, such as *Toxoplasma gondii, Cytauxzoon* spp., and possibly *Neospora caninum* (Glenn *et al.* 1983; Cheadle *et al.* 1999; Ferroglio *et al.* 2003; Spencer *et al.* 2003; Millán *et al.* 2007). The lion case study also cautions us to be more vigilant of complex causes of epidemics, rather than seeking a single pathogen as the cause.

Low genetic diversity, disease, and anthropogenic factors

Several felid species have low genetic diversity, and this loss of diversity has been associated with decreased fitness, poor reproduction, and increased mortality. The lion population from the Ngorongoro Crater, Tanzania, exemplifies a population in which low genetic diversity and restricted habitat with close proximity to human populations has enhanced the impact of disease and impeded population survival. The Crater has historically been able to support a population of 124 lions; however, three epidemics of CDV (complicated by babesiosis) since 1994 reduced the population to fewer than 40 lions, and population recovery has not occurred (Munson *et al.* 2008). In contrast, the outbred Serengeti lion population quickly increased in number after the 1994–95 CDV-*Babesia* epidemic (Kissui and Packer 2004). Reduced testicular function and poor sperm quality of the Crater lions likely contributes to this slow population recovery (Wildt *et al.* 1987a; Munson *et al.* 1996). Whether low MHC diversity (Packer *et al.* 1991b) accounts for their greater CDV susceptibility and higher *Babesia* burdens than Serengeti lions is not known (Munson *et al.* 2008). Failure of this population to recover may portend the fate of other felids as genetic diversity is lost.

The Florida panther provides another example of the impact of genetic loss on health (Onorato *et al.*, Chapter 21, this volume). The Florida panther population has developed significant health problems over the past two decades, which have impeded their recovery despite considerable conservation efforts. These diseases probably would not have been detected without the close health surveillance that is a component of this programme. Two health problems, cryptorchidism (intra-abdominal testis) and atrial septal defects (a heart defect), likely have a genetic basis and manifested as a result of the loss of genetic diversity in this population (Roelke *et al.* 1993b; Cunningham *et al.* 1999). Cryptorchidism was found in 58% of 52 panthers examined and has had a profound impact on population recovery, because it prevents the development of sperm (Roelke *et al.* 1993b). Atrial septal defects were first noted in 1990 and have been identified in 21% of necropsied panthers (Roelke *et al.* 1993b; Cunningham *et al.* 1999). This heart defect has contributed to mortalities and will persist in the population if inherited in a recessive manner. Cowlicks and kinked tails are other common non-lethal defects in this population (Roelke *et al.* 1993b), reflecting additional adverse genes expressed through loss of diversity. Expression of these traits has decreased considerably since Texas pumas were introduced for breeding in the population (D. Onorato, personal communication).

As the Florida panther population slowly recovers and its habitat continues to diminish, conflicts with the expanding human populations increase. Significant losses have occurred from vehicular trauma (Taylor *et al.* 2002), and evidence of nutritional stress has emerged in the population (Duckler and Van Valkenburgh 1998). An additional manifestation of human encroachment has been the appearance of FeLV in the population (Cunningham *et al.* 2008). From 2000 to 2007, FeLV was detected in 5 of 131 pumas (5 of 11 in one area). All five viremic pumas died within 24 weeks of diagnosis. Three died from (presumed) FeLV-related signs (anemia or septicemia) and two from intraspecific aggression (Cunningham *et al.* 2008). A similar FeLV epidemic causing significant mortality may now be emerging in the Iberian lynx population (A. Vargas, personal communication; Meli *et al.* 2006; Roelke *et al.* 2008). The first case was identified in 2006 (Meli *et al.* 2006), and subsequently all infected lynx have died (A. Vargas, personal communication). There was no evidence of FeLV exposure in this population prior to 2001 (Roelke *et al.* 2008), indicating that the population had no resistance to infection. The relatively rapid clinical course in Florida panthers and Iberian lynx (in comparison to domestic cats) is reminiscent of the fatal FeLV-associated leukemia that developed in a wild-caught Namibian cheetah within a year of exposure to an infected cat (Marker *et al.* 2003d). The virulent behaviour of FeLV in Florida panthers, Iberian lynx, and cheetahs suggests that these species have not evolved with this virus. Loss of MHC diversity resulting in increased susceptibility has also been hypothesized. The long-term impact of this introduced virus on these highly endangered species is of great concern.

Managing felid diseases

Identifying disease threats

Before any disease management action is taken, it is essential to know the true threats to a population. For an infectious agent to be considered a disease threat, it should be clearly associated with characteristic clinical signs or determined by pathology to be associated with lesions caused by that agent in more than one animal in the population. Reliable diagnostic tests (microbial cultures, PCR amplification, or blood smears) should be conducted to confirm the identity of an infectious agent suspected to cause these signs and lesions. When morbidity or mortality occurs, the exposure status of the rest of the population to this pathogen should be determined by conducting serological tests on current and archived blood samples, so that the overall threat to the population can be assessed (Munson and Karesh 2002). The presence of pathogen-specific antibodies indicates that the population has been previously exposed and may have some protection from developing disease when reinfected. For organisms such as *Mycobacteria, Toxoplasma*, or FCoV that persist in the presence of specific antibodies, a positive serological test also may indicate that the animal is currently infected. The use of serologic test results should be weighed carefully in final management decisions, because most tests were designed to measure domestic cat antibodies and have not been validated for wild felids, so may yield false negative or false positive results.

Subclinical tuberculosis in wild felids is an example of a disease, the *ante mortem* diagnosis and control of which has been hampered by lack of reliable tests. Serological tests are generally insensitive (yield false negative results) and have not been validated for felids. Lymphocyte stimulation tests and gamma interferon tests used for other species also have not yet been validated for felids, so their results are not reliable (Mangold *et al.* 1999; de Lisle *et al.* 2002; Grobler *et al.* 2002). Skin tests using crude purified protein derivatives often give false positive results, do not indicate if the species of *Mycobacterium* responsible for a positive test is pathogenic, and require keeping an animal captive for 48 h. Therefore, reliable tests are not currently available for assuring that a felid is free of tuberculosis before translocating it from an endemic to non-endemic area. A definitive diagnosis of *Mycobacterium* infection in free-ranging felids still requires a necropsy, histopathology, and bacterial culture.

Considering disease dynamics in management strategies

Once a disease is considered a threat to the population, disease control strategies can be proposed. These strategies are based on risk assessments that include information not just on species susceptibility to infection, but also on the nature of infection in individuals (the route of infection, distribution of lesions, excretion routes, and necessary infective dose), the dynamics of infection in the population (proportions of infected and susceptible animals, infectiousness of animals, conditions for transmission, and viability of excreted pathogens; Cleaveland *et al.* 2007), the geographical distribution of the species under study, and interaction with potentially infected domestic animals and other wildlife (Corner 2006). Disease risk modelling continues to be the best method for predicting threats in free-ranging populations. However, the strength of these models is constrained by lack of knowledge regarding the prevalence and behaviour of infectious agents in free-ranging populations, susceptible species and reservoirs, and pathogen persistence in the environment (Cleaveland *et al.* 2007). Most published information on felid pathogen dynamics was derived from research conducted in a controlled laboratory or clinical setting, so further research is needed in natural environments to test and strengthen these models. For example, most models assessing the risk of canine distemper to African carnivore populations are based on the premise that the principal reservoir host is the domestic dog. However, hyaenas, jackals, and other free-ranging carnivores could be reservoir hosts and play a larger role in the ecology of CDV in the wild (Harrison *et al.* 2004). Additional studies on the ecology of felid infectious diseases are therefore needed to improve risk assessment models. It is likely that many feline viruses and haemoparasites have self-sustaining sylvatic cycles and indigenous wildlife reservoirs and are not dependent on domestic cats (or dogs) for persistence.

Protecting populations by isolation or vaccination

Disease risk modelling should take into account that some level of disease is normal in all populations and contributes to population stability. Even epidemics can be tolerated when populations are large. Most felid populations rebound from catastrophic declines due to their fecundity. However, stochastic modelling may indicate that a disease is a true threat to population viability, and preventing disease will benefit an endangered population in which each animal becomes important for population recovery. Disease threats to endangered populations can sometimes be managed by preventing exposure (physical separation) or preventing infection (vaccination). Preventing exposure in free-ranging populations requires physically impeding contact between reservoir or transmission hosts and the endangered felid, or preventing infection in the reservoir or transmission hosts (e.g. domestic cats or dogs) through a rigorous barrier vaccination programme. Both methods are costly and often unrealistic. Cleaveland *et al.* (2002) review the advantages and disadvantages of various management strategies used for controlling diseases for which dogs are the reservoir host; these strategies would also apply to diseases for which domestic cats are reservoirs.

Preventing infection through vaccination has several limitations and should be undertaken with caution. The decision to vaccinate should proceed with the understanding that vaccination usually only provides temporary protection to an individual or a population (if a large portion is vaccinated); however, achieving 100% protection of a population is unlikely. If vaccination programmes are halted, the population may be more susceptible to infection, because vaccine protection will be lost across the entire population at approximately the same time. Vaccination may also interfere with natural selection for disease resistance to similar, less virulent endemic pathogens. So preventing infection may temporarily benefit a population, but leave it more vulnerable in the future as the population evolves without exposure.

Prior to the initiation of vaccination programmes, vaccines should be proven safe in a small subset of the population, and an antibody response should be demonstrated (as a surrogate for efficacy, because pathogen challenge studies cannot be conducted in wild felids). It should be remembered that once a vaccination programme has begun, serology can no longer be used as a monitoring tool, because most serologic tests cannot distinguish between wild-type and vaccine viral responses. Modified live feline vaccines developed for domestic cats incite a stronger and longer-lived immune response than killed virus vaccines, but have been suspected to cause disease in some captive felids (Munson and Citino 2005). Bioengineered subunit vaccines have proven safe and effective in other carnivores, but have limited availability at this time, are costly, and require direct injection. Because of these limitations and cautions, vaccination is not currently practical in free-ranging felid populations without strong evidence that a feline virus is truly threatening the population. If infectious diseases become a significant threat to felids in the future, then development of oral vaccines delivered in baits would be warranted.

Captive sanctuaries and captive breeding programmes are an important conservation strategy for some highly endangered species, such as the Iberian lynx, but carry a greater risk of an epidemic due to unnaturally dense populations and (if those sanctuaries are at zoological parks) congregation of species originating from different ecological niches throughout the globe. Because of that risk, vaccination is warranted in the captive setting. However, it should be remembered that stress in recently caught, captive-held wild animals may compromise immune function and interfere with an effective immune response to vaccination. Because of these risks, at least two separate populations at epidemiologically distant sites should be established as a safety net to prevent extinction from disease.

Conclusions: integrating disease threats into felid conservation

Wild felid conservation may increasingly require understanding and managing disease threats. Although diseases are natural phenomena in healthy populations and do not need to be controlled under normal conditions, the historically stable relationship between felids and potential pathogens continues to be disrupted by increasing anthropogenic influences on ecosystems and climate. Habitat loss and fragmentation have isolated some populations genetically and increased interactions with domestic species. The consequences of genetic loss may be manifested as increased disease susceptibility in populations, while invasion of wild felid habitats by domestic dogs and cats increases the number of susceptible hosts for pathogen amplification. Additionally, the dynamics of disease vectors and contacts among felids will become increasingly unpredictable due to global climate change. Together these factors increase the potential for epidemics to affect wild felid populations globally.

Predicting and preventing epidemics to conserve all 37 felid species will require a better understanding of the ecology and pathogenesis of infectious diseases. Research is urgently needed in areas as diverse as infection detection, pathogen identification, practical treatment and prevention, interspecific infection pathways, and disease dynamics. Only with accurate diagnostic tools and strong predictive models will we be able to prevent disease from emerging as a significant factor in wild felid survival.

Linda Munson sadly died in May 2010

PART II

Case studies

CHAPTER 10

Ecology of infectious diseases in Serengeti lions

Meggan E. Craft

Male lion (*Panthera leo*) in the Serengeti National Park. © Meggan Craft.

Introduction

Diseases that affect lions (*Panthera leo*) are often connected with much larger ecosystem processes. Pathogens often infect more than one host species; these multi-host pathogens (e.g. rabies and canine distemper virus [CDV]) link lions to domestic animals and also to populations of endangered wildlife (e.g. African wild dogs (*Lycaon pictus*); Cleaveland *et al.* 2002). Second, each host species is commonly infected by more than one pathogen, which can change expected disease transmission rates and virulence (Graham *et al.* 2007). Such multi-host/multi-pathogen systems are difficult to study in the wild because information on the full range of hosts is lacking, and because pathogens may interact differently with each other and each host species. Third, environmental perturbations can change the interplay between the host and pathogen and trigger disease outbreaks in ways that can only be understood through long-term monitoring. Fortunately, in East Africa's Serengeti ecosystem, data on lions have been collected for over 40 years and the study lies within a framework of long-term studies of other predatory species, herbivores, human populations, and climatic conditions (Sinclair *et al.* 2008). Hence data collected from the Serengeti Lion Project provides the rare opportunity to tackle complex issues of disease dynamics in wild animal populations, with the ultimate aim of conserving lions in their natural habitat.

This chapter will begin with an exploration of a wide variety of diseases in Serengeti lions. It will highlight differences between endemic and epidemic pathogens, show that pathogenicity is often difficult to discern and could vary by ecosystem (e.g. bovine tuberculosis [bTB]), illustrate that some pathogens still fit the one-host, one-pathogen traditional disease model (e.g. feline immunodeficiency virus [FIV]), and highlight that co-infections are not always harmful to the host (e.g. trypanosomes). The second part of the chapter will provide an in-depth case study on the dynamics of CDV in the Serengeti lion population, and will conclude with new research synthesizing biology and epidemiology through the use of detailed mathematical models.

Study system

The Serengeti Lion Project is an important study system for insights into infectious disease ecology, though it is best known for seminal research on lion social behaviour, ecology, and genetics (Schaller 1972; Bertram 1975a, 1976; Bygott *et al.* 1979; Pusey and Packer 1987, 1994; Packer *et al.* 1988, 1990, 1991b, 1998, 2001, 2005a). Continuous demographic data have been collected on lions residing in a 2000 km^2 area of the Serengeti National Park (SNP; Tanzania, East Africa) since March 1966 and in the 250 km^2 floor of the nearby Ngorongoro Crater since 1963 (Fig. 10.1; Packer *et al.* 1998). In each Serengeti study pride, one female is fitted with a radio-collar, whereas the Crater lions are located by opportunistic sightings. The 25,000 km^2 Serengeti ecosystem is dominated by migratory ungulates that move with the seasonal rains (Sinclair 1995). The Serengeti lion population is estimated at around 3500 individuals and has considerable genetic diversity (Gilbert *et al.* 1991). In contrast, the Ngorongoro Crater provides a small oasis of persistent food and water (Hanby *et al.* 1995; Packer *et al.* 1999). The Crater population currently comprises about 60 individuals, and enjoys consistently high food availability (Hanby *et al.* 1995; Kissui and Packer 2004). However, the Crater lions have passed through a population bottleneck of <15 adults with no immigration since 1969, and show considerable signs of inbreeding (e.g. sperm abnormalities and lack of genetic diversity; O'Brien *et al.* 1987b; Wildt *et al.* 1987a; Packer *et al.* 1991b).

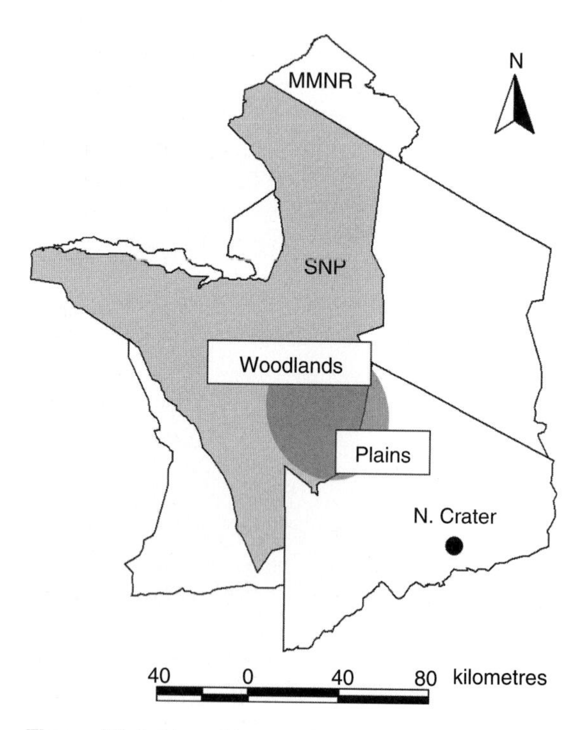

Figure 10.1 Map of Serengeti National Park (SNP) and the surrounding protected areas in East Africa, including the Masai Mara National Reserve, Kenya (MMNR) and Ngorongoro Crater (N. Crater). Lions have been studied continuously inside the N. Crater (black circle) and in the SNP (gray oval) since the 1960s. The Serengeti study area is divided into two habitat types: woodland in the north and plains to the south.

While it has been feasible to study the entire lion population of the Crater floor, the Serengeti lion study area is restricted to the southeastern quarter of the region (Fig. 10.1). The Serengeti study area includes two contrasting habitats: woodlands that are dominated by *Acacia* and *Commiphora* trees and by the open grass plains (Packer *et al.* 2005a). Because lion prides are highly territorial, resident lions are not able to access the migratory prey year round. Hence, prides of lions residing in the woodlands have year-round access to resident prey, while the plains prides face 'feast or famine', depending on the location of the migratory herds (Scheel and Packer 1995). As a result, lions live at higher densities in the woodlands and lower densities on the plains (Scheel and Packer 1995; Packer *et al.* 2005a).

Serengeti lions live in gregarious groups (prides) composed of 1–21 related females, their dependent offspring, and a resident coalition of 1–9 males. Prides are territorial and infrequently contact their neighbours (Packer *et al.* 1992); inter-pride encounters can be deadly (Schaller 1972; McComb *et al.* 1993; Grinnell *et al.* 1995). When prides grow too large, cohorts of young females split off and form a neighbouring pride (Pusey and Packer 1987) and are more tolerant of their non-pride relatives (VanderWaal *et al.* 2009). Coalitions of males can be resident in more than one pride (Bygott *et al.* 1979), and distribute their time between their various prides (Schaller 1972). In contrast, nomads are adult male and female lions that do not maintain a territory and move great distances though the ecosystem (Schaller 1972). Residents and nomads occasionally interact during mating, territorial defence, and at kills.

The disease facet of the Serengeti Lion Project benefits from a valuable archive of biological samples dating back to 1984, collected from individually identified, known-age lions. Lions are identified through natural markings (Pennycuick and Rudnai 1970), and each sample comes from a lion with a detailed life history. Information exists on individual ranging patterns, relatedness, and birth dates (normally accurate to 1 month), and contact patterns with conspecifics and other species can also be inferred. Blood is collected by tranquillizing an animal and then drawing blood, while faecal samples are collected during opportunistic sightings. The stored blood and/or faecal samples can then be used for retrospective surveys testing for a pathogen or exposure to a certain pathogen through the detection of antibodies (serology). Serological results can be interpreted as follows: by estimating the proportion of lions that were seropositive at a certain age in a particular year, age-seroprevalence curves can inform whether diseases persist or are transient in the lion population (endemic vs epidemic; Packer *et al.* 1999). In addition, the timing of epidemic disease outbreaks can be estimated by plotting each year's annual seroprevalence rate (Packer *et al.* 1999). Because snapshot studies of exposure can be misleading, these data must be collected through the highs and lows in prevalence over prolonged periods (Cleaveland *et al.* 2007).

Endemic and epidemic diseases: viruses and parasites

Diseases can be classified as endemic or epidemic, based on their persistence times in a population. Endemic diseases are consistently prevalent in a population and often demonstrate low virulence and long infectious periods (Anderson and May 1979). Macroparasites are a classic example of endemic diseases. These are parasites in which the life cycle usually occurs via transmission of free-living infective stages that pass from one host to the next. Macroparasites normally cause chronic morbidity in the host, instead of mortality (Hudson *et al.* 2002). The Serengeti and Ngorongoro lion populations are infected with gastrointestinal endoparasites, which can be identified in faecal matter. In one study of 112 Serengeti and Ngorongoro Crater lions, 15 parasite taxa were identified (although 3 taxa were likely acquired from prey; Muller-Graf 1995). Parasite species were aggregated in individual hosts, where a few lions were heavily infected and most lions were lightly infected (Muller-Graf 1995). This aggregation is consistent with other macroparasite studies in following the '20–80 rule' (20% of the population harbours 80% of the parasites; Anderson and May 1991; Shaw *et al.* 1998; Hudson *et al.* 2002). In a more detailed study of the cestode *Spirometra* spp. (the most common lion intestinal parasite), lions living in the Crater were more heavily infected than lions living in the Serengeti (Muller-Graf *et al.* 1999). It was difficult to assess whether these differences could be attributed to ecological differences (e.g. swampy vs dry habitat and abundant vs sparse prey) or from genetic differences (inbred vs outbred). Cubs less than 9 months in both locations were already heavily infected with *Spirometra* when sampled, and there were no significant correlations between individual parasite load and rainfall season, age, or sex of the lion, reproductive status, or pride size. In another cross-sectional study of 33 lions, over 19 species of parasites were identified in the lion population (although some were likely acquired from prey; Bjork *et al.* 2000). Again, the number of gastrointestinal parasite species per lion did not change significantly with respect to sex, habitat, or age of the lion.

The ecology of endemic diseases in natural populations is difficult to study without controlled experiments and interventions, as it seems lions are constantly infected and the lion population and their gastrointestinal parasites likely remain near equilibrium. In order to detect the effects of genetic (inbred vs outbred) or ecological factors (e.g. high vs low prey densities) on levels of parasite infections, there would have to be a much wider array of study populations, covering a wider range of genetic or ecological factors (Muller-Graf *et al.* 1999). To assess whether macroparasites in any way 'regulate' the lion population, either lions or their parasites would have to be experimentally perturbed to detect any effects on morbidity and mortality (Tompkins *et al.* 2002). Long-term eradication/control of the lions' gastrointestinal parasites would not be feasible and experimental infections would be impossible to justify.

In studies of disease in wildlife, it is often difficult to detect sick animals, locate carcasses for post-mortem exams, and to isolate the viral pathogens. In contrast, serological studies can be performed by screening large numbers of living individuals. However, serology merely determines whether an individual possesses antibodies to a certain pathogen and was therefore exposed at some time in the past; a positive serological result is unlikely to indicate current infection. As Serengeti lions have been known to live up to 20 years, a serological cross-sectional study could potentially be misleading. For example, the proportion of lions with antibodies for CDV in 1985 could have been attributed to a constantly circulating (endemic) disease with increased

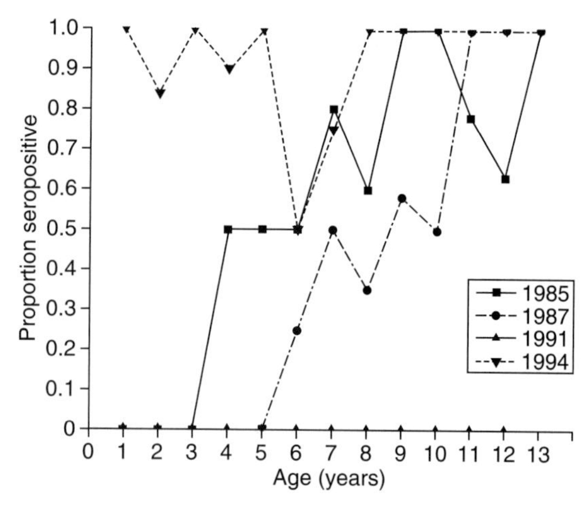

Figure 10.2 Canine distemper virus (CDV) age-seroprevalence patterns in the Serengeti lion study population from cross-sectional samples in 1985, 1987, 1991, and 1994. (From Cleaveland *et al.* 2007.)

cumulative exposure in adults, when, in fact, CDV had been absent from the population for several years, and the older age classes retained antibodies from an earlier epidemic (Fig. 10.2).

Feline herpesvirus (FHV) is a clear example of an endemic viral disease in the Serengeti lion population. FHV is a highly contagious respiratory disease spread by direct contact with an infected cat (Gaskell *et al.* 2006). Across a 10-year span, 372 out of 374 lions were positive for herpesvirus, including 8 small cubs under 1 year old (Fig. 10.3; Packer *et al.* 1999). Because herpesvirus is consistently circulating and cubs are immediately infected, this is a pattern of chronic infection. Although FHV has been implicated in the death of a captive lion in Germany (Wack 2003), no signs of clinical disease have been attributed to FHV in the Serengeti or in other wild felid populations (Spencer and Morkel 1993; Packer *et al.* 1999; Driciru *et al.* 2006; Ramsauer *et al.* 2007). However, since 100% of the Serengeti population is infected, it is difficult to compare infected and uninfected hosts to assess potential impacts of infection status on fecundity or survival (Packer *et al.* 1999).

In contrast to endemic diseases, epidemic diseases cause distinct rises and falls in patterns of seroprevalence, often briefly infect a population, and have the potential to inflict high mortality (May and Anderson 1979). Epidemic viruses frequently sweep

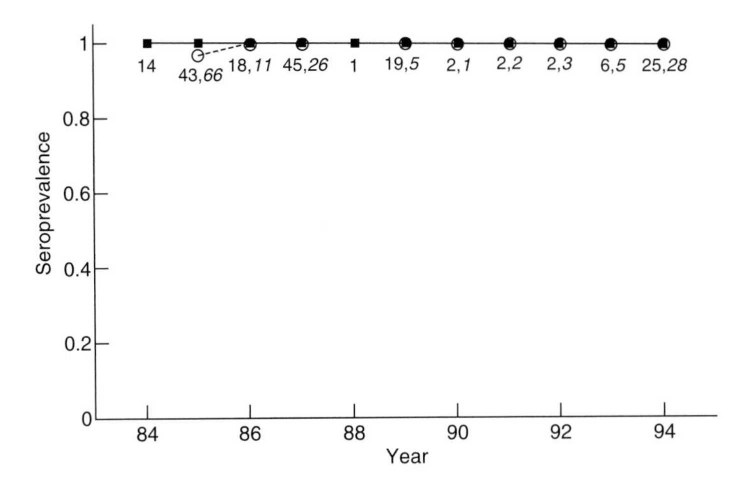

Figure 10.3 Annual seroprevalence rates in the Serengeti for feline herpesvirus (FHV) for young lions (dotted line and circles) and for adults (solid lines and squares), with respective sample sizes (immatures in italics). (From Packer *et al.* 1999.)

through a population, burn out because of lack of susceptibles, and then invade again once the susceptible population increases to a critical density threshold. Coronavirus, parvovirus, and calicivirus are all epidemic viruses in the Serengeti lion population, showing periods of high exposure, followed by 4–9 years of declining seroprevalence before another period of high exposure (Fig. 10.4; Packer *et al.* 1999). Whereas endemic diseases continuously infect the youngest age classes (Figs 10.3 and 10.6), epidemics can best be identified from temporal gaps in infection in exposure in the youngest age classes (Figs 10.2 and 10.4). For example, feline coronavirus (FCoV) is spread by indirect faecal–oral routes (and possibly aerosolized routes), can infect domestic dogs, and can turn into the more pathogenic feline infectious peritonitis in domestic cats (Addie and Jarrett 2006). FCoV was found in extremely low levels in South African and Namibian lions (Spencer 1991; Spencer and Morkel 1993), but was found in 57% of Serengeti lions (Hofmann-Lehmann *et al.* 1996). In Serengeti lions, coronavirus serostatus varied significantly across years in juveniles, but not in adults (Fig. 10.4a). Young lions in Ngorongoro Crater were positive for coronavirus, but the sample size was too small to detect variation among years. Using the unique combination of age and serology data, the timing of a coronavirus epidemic in Serengeti could be inferred by comparing the seroprevalence data with the age of each study animal. For example, of all lions sampled between 1984 and 1988, only those animals born in July 1984 or before were seropositive for coronavirus (Fig. 10.5b). Packer *et al.* (1999) interpret this to mean that 1984 was the end-date for an epidemic; other coronavirus end-dates were estimated in 1988 and 1993 (Figure 10.4).

Similar to patterns of coronavirus exposure, parvovirus (feline panleukopenia virus [FPV]/canine parovirus [CPV]) seroprevalence in adults did not differ significantly across years, but varied significantly in younger animals (Fig. 10.4b). Parvovirus is a multi-host virus most commonly spread by indirect contact through environmental contamination, where parvovirus can remain infectious up to a year at room temperature (Greene and Addie 2006). Parvovirus is a suspected cause of wolf pup (*Canis lupus*) mortality (Mech and Goyal 1993), but does not seem to cause morbidity or mortality in lions. An age–seroprevalence curve for lions tested between 1984 and 1991 revealed that the last period of exposure in Serengeti lions during this time frame was in 1985 (Fig. 10.5a), with other likely outbreaks ending in 1976 and 1992 (Hofmann-Lehmann *et al.* 1996; Packer *et al.* 1999).

Finally, feline calicivirus (FCV) is an upper respiratory infection similar to FHV, is spread by direct contact, and some strains can cause high mortality in domestic kittens (Gaskell *et al.* 2006). When calicivirus seroprevalence was plotted by year of birth for lions sampled between 1991 and 1994, no lion born after March 1990 was positive, indicating that 1990 was a likely end-date for a calicivirus epidemic

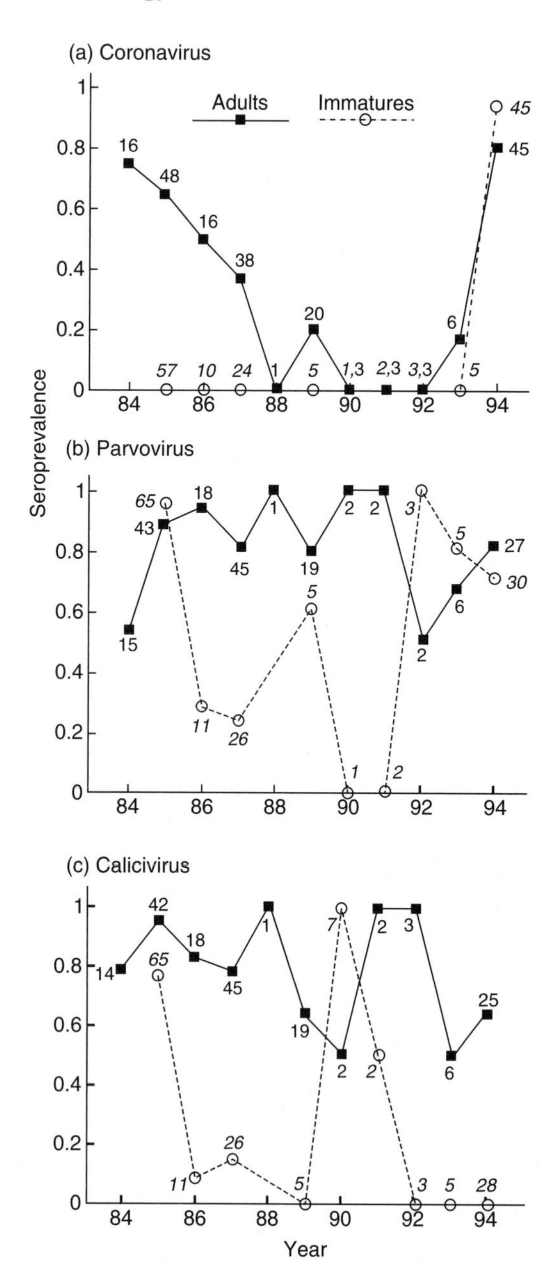

Figure 10.4 Annual seroprevalence rates in the Serengeti for (a) coronavirus; (b) parvovirus; and (c) calicivirus for young lions (dotted line and circles) and for adults (solid lines and squares), with sample sizes represented as numbers in the graph. (From Packer *et al.* 1999.)

(Fig. 10.5b, c), with other likely end-dates of 1980 and 1985 (Fig. 10.5b; Packer et al. 1999). Like other serological studies of lions from other ecosystems, there were no consistent signs of clinical disease, excess mortality, or decreases in lion fecundity due to infections from coronavirus, parvovirus, or calicivirus (Spencer 1991; Spencer and Morkel 1993; Hofmann-Lehmann *et al.* 1996; Packer *et al.* 1999; Driciru *et al.* 2006).

A low-impact pathogen or an insidious threat: bovine tuberculosis

Mycobacterium bovis, the causative agent of bTB, is a bacterium of growing concern in African wildlife (Michel *et al.* 2006). In Kruger National Park, South Africa, an epidemic of bTB infection in African buffalo (*Syncerus caffer*) has been moving northwards across the park since 1990 (Michel *et al.* 2006). Monitored lions from the same area have become infected (Cleaveland *et al.* 2005; Michel *et al.* 2006), and although the Kruger lion population currently seems stable (S. M. Ferreira and P. J. Funston, personal communication), both lions and buffalos show mortality and morbidity. In contrast, between 1985 and 2000 none of the 19 lions sampled in the Ngorongoro Crater were seropositive for bTB and only 8 of 184 (4%) Serengeti lions were seropositive for bTB (Cleaveland *et al.* 2005). The Tanzanian samples were collected over a period of 15 years and, although one of the positive lions was sampled in 1984, there were no significant differences between average prevalences from 1984 to 1996 and 1997 to 2000. This indicates that bTB has been rare in the population for a long time, and is not spreading quickly. While clinical signs were seen in four out of eight seropositive animals, and seropositive animals survived for a shorter (but non-significant) amount of time than did non-infected individuals, with such a low sample size of positives, it is difficult to quantify the pathogenicity of bTB in Serengeti lions (Cleaveland *et al.* 2005). Cleaveland *et al.* suspect that exposure in lions was due to eating infected prey.

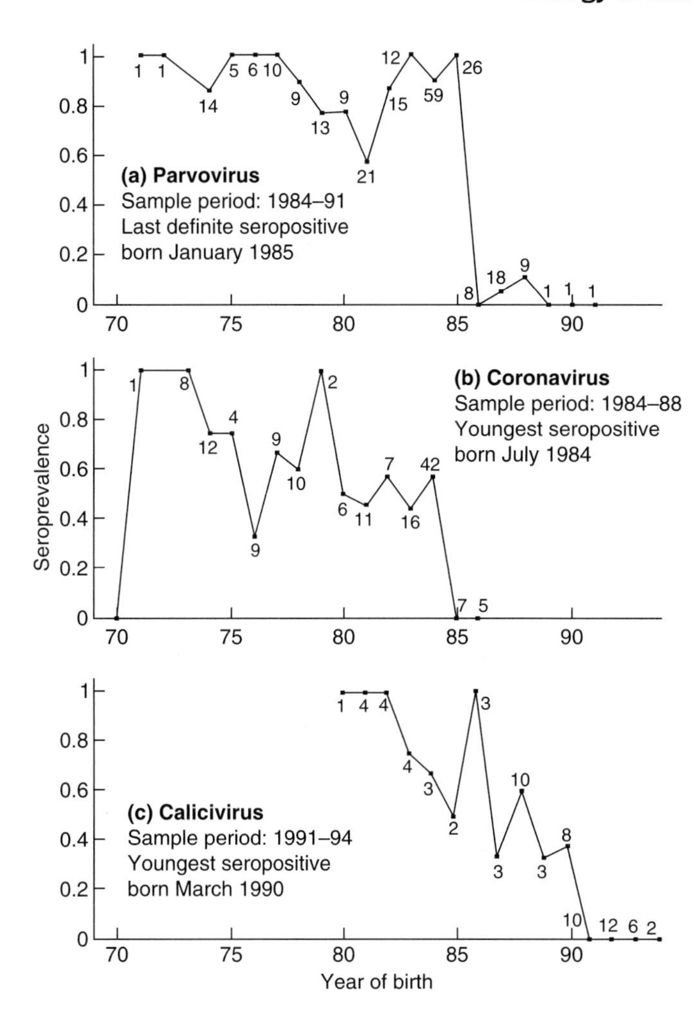

Figure 10.5 Representative seroprevalence curves with sample sizes indicated inside the graph. (a) Seroprevalence data for parvovirus is plotted according to year of birth for lions sampled between 1984 and 1991. Prevalence ±4 years of 1985 varied significantly. (b) For lions sampled between 1984 and 1988, coronavirus prevalence varies significantly for animals born ±4 years of 1984. (c) Seroprevalence for calicivirus for lions sampled between 1991 and 1994. Prevalence varies significantly for animals born within ±4 years of 1990. (From Packer *et al.* 1999.)

One-host, one-pathogen: feline immunodeficiency virus

Serengeti lions are infected with a lentivirus, FIV, which is genetically homologous and functionally analogous to human immunodeficiency virus (HIV; Brown *et al.* 1994). Like HIV, once infected, FIV permanently infects the host. FIV has species-specific strains, and the strain infecting lions is named FIV-Ple (Olmsted *et al.* 1992). In contrast to recent work showing frequent FIV transmission from bobcats to pumas in the United States (Franklin *et al.* 2007), phylogenetic analysis suggest that cross-species transmission is unlikely between lions and other large African carnivores (Troyer *et al.* 2005).

Although lions have been likely hosts to FIV since the late Pleistocene (Brown *et al.* 1994; Pecon-Slattery *et al.* 2008), not all populations of lions are currently infected. African lions in Namibia and Asiatic lions in India test negative for FIV. On the other hand, FIV is extremely prevalent in East African and South African lions, and the incidence in these populations is higher than in any other wild or domestic felid population (Olmsted *et al.* 1992; Driciru *et al.* 2006). Serengeti lions not only have exceptionally high incidence of FIV (84–93%), but because these high levels are consistently maintained over many years, FIV is endemic in the Serengeti (Olmsted *et al.* 1992; Brown *et al.* 1994; Hofmann-Lehmann *et al.* 1996; Packer *et al.* 1999; Troyer *et al.* 2005). FIV infection rates do not change significantly by sex or across years and this is true for both adult and juvenile lions (Fig. 10.6; Packer *et al.* 1999). Although not all juvenile lions test positive, the vast majority of lions in Serengeti and Ngorongoro are infected by 4 years of age. Because the entire adult Serengeti and Ngorongoro population is FIV-positive, there is no control group in which to assess effects of infection status on fecundity (Packer *et al.* 1999).

There are differences in rates of infection with respect to habitat type; lions inhabiting the Serengeti plains show lower rates of infection than lions inhabiting the woodlands or Ngorongoro Crater (Fig. 10.6;

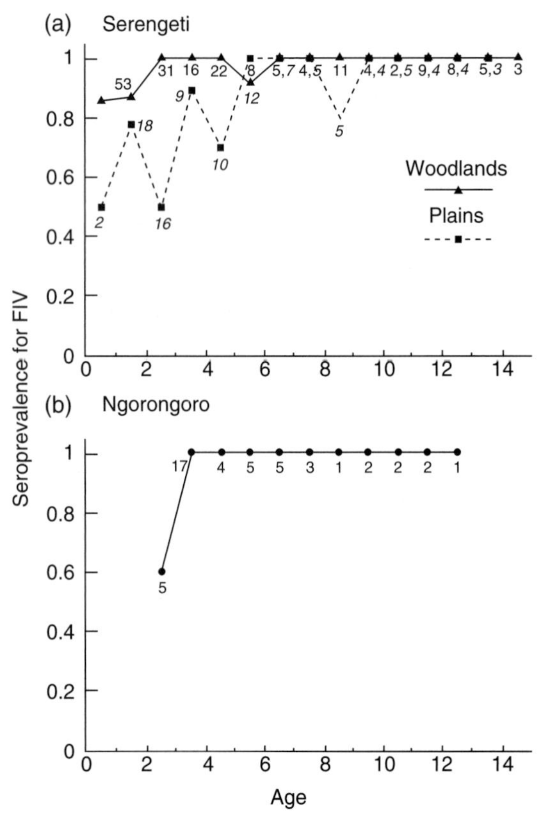

Figure 10.6 Age-prevalence curves for FIV averaged across 1984–94 in (a) Serengeti woodlands and plains; and (b) Ngorongoro Crater. Yearly sample size is indicated beside each point in the figure. Lions in the Serengeti plains show a lower rate of infection than the other two habitats ($T = 3.54$, $P < 0.001$), even when controlling for age in a multivariate analysis (effect of age: $T= 5.94$, $P< 0.001$; effect of Serengeti plains: $T = 3.98$, $P < 0.001$). (From Packer *et al.* 1999.)

Packer *et al.* 1999). Prides in the three habitats live in varying densities and have different within- and between-pride contact patterns. While all of these factors can influence infection rates, the differing rates of infection in various habitat types are unexplained.

Although FIV-Fca causes immunosuppression and mortality in domestic cats (*Felis catus*; Yamamoto *et al.* 1988; Ackley *et al.* 1990), there are no obvious signs of immunodeficiency or disease in Serengeti lions, nor in other wild feline species (Olmsted *et al.* 1992; Roelke-Parker *et al.* 1996; Packer *et al.* 1999; Ramsauer *et al.* 2007). FIV infection had no age or sex-specific effects on determining host longevity (Hofmann-Lehmann *et al.* 1996; Packer *et al.* 1999). Even FIV co-infection with other viruses (i.e. CDV, FHV, FCV, feline parvovirus, and FCoV) did not reduce host longevity (Packer *et al.* 1999).

Lions can be co-infected with different strains of FIV. Six FIV-Ple strains or subtypes occur throughout Africa (Antunes *et al.*, 2008). These subtypes (based on *pol* gene sequence divergences) come from a common FIV-Ple lion ancestor, but are distinct from each other (Brown *et al.* 1994). There is high genetic diversity within and between the subtypes (Brown *et al.* 1994). Three different clades are present in the Serengeti population (Brown *et al.* 1994), and multiple subtypes are found within the same pride, and within the same individual (Troyer *et al.* 2004). In a recent study, 43% of FIV-positive individuals in the Serengeti were infected with multiple strains of FIV (Troyer *et al.* 2004). Thus, co-infection with one subtype of FIV does not necessarily confer immunity against secondary infection from another subtype (Troyer *et al.* 2004).

It is unknown whether FIV is spread primarily through horizontal transmission (neighbour-to-neighbour) or from vertical transmission (parent–offspring). Both in domestic cats and in lions, horizontal transmission seems to be the major route of infection and likely occurs during biting (Yamamoto *et al.* 1988; Brown *et al.* 1994). As evidence of horizontal transmission, two male Serengeti lions born in 1982 and in 1986 both tested negative for FIV in 1987 but tested positive in 1989 (Brown *et al.* 1994). In addition, FIV-positive cubs were born from FIV-negative mothers, and FIV-negative cubs were born from FIV-positive mothers (Brown *et al.* 1994). There was evidence of both between-pride and within-pride transmission though phylogenetic analysis of sequences (Troyer *et al.* 2004). The strains were well mixed across all prides: 6 out of 13 prides were infected with all three strains. In contrast, one pride showed evidence of monophyletic clustering. Although closely related lions often had closely related viral sequences (e.g. three sibling pairs and a mother–daughter pair), indicating a common viral ancestor, it is hard to determine if these individuals were infected by each other (vertical transmission in case of the parent–offspring) or by another lion (Troyer *et al.* 2004). The identification of transmission routes can be difficult because closely related lions are often found in close association with each other.

According to phylogenetic analyses, lions have been infected with FIV-Ple virus for long periods of time, and the three FIV subtypes diverged a long time ago—maybe as far back as the radiation of the genus *Panthera* (Olmsted *et al.* 1992; Brown *et al.* 1994). Because there is no evidence for immune pathology or mortality from this 'ancient' virus, the interactions between lions and FIV-Ple could be an example of 'modern symbiosis' or commensalism between a host and a pathogen (Olmsted *et al.* 1992; Brown *et al.* 1994). This attenuation makes the FIV-Ple host–virus relationship a great contrast to the high pathogenicity of the recent HIV epidemic (Carpenter and O'Brien 1995).

Co-infections are not always harmful: trypanosomes

Trypanosomes are protozoan parasites that cause substantial economic and public health problems in sub-Saharan Africa due to the morbidity and mortality of 'sleeping sickness' in humans and livestock (Schmunis 2004; Shaw 2004). Because trypanosomes have extremely complex antigenic surface proteins that change composition every 3–5 days, they can evade attack by the host's immune system (Morrison *et al.* 2005). Trypanosomes have a long history of infection in the Serengeti; in fact, the presence of trypanosomiasis influenced the initial gazetting of the Serengeti National Park. The tsetse fly (*Glossina* spp.), the vector that transmits the disease, kept the Serengeti un-inhabitable for humans and their livestock, yet available for wildlife (Matzke 1979). Sleeping sickness still circulates in the Serengeti, as seen in 2001 when 20 cases were reported and 4 people died from trypanosome infection (Mlengeya *et al.* 2002). Trypanosomiasis is thought to be maintained in livestock in other parts of sub-Saharan Africa (Welburn *et al.* 2001) and in wildlife populations in the Serengeti (Kaare *et al.* 2007).

Lions are likely infected with trypanosomes in one of two ways: through the bite of an infected tsetse fly, or via oral inoculation when eating infected prey. Tsetse flies need shade and woody vegetation to survive and thus do not inhabit the plains or the Crater. In a study of 123 Serengeti and Ngorongoro lions, trypanosomes were found by microscopy in 32 individuals (Averbeck *et al.* 1990). While trypanosome prevalence did not vary with respect to the lion's age or sex, prevalence was highest in the Serengeti woodlands, lower in the Serengeti plains, and absent in the Crater (Table 10.1). Prevalence of trypanosome infection correlated with increasing levels of tsetse flies (Table 10.1), suggesting infection directly from tsetse bites. Two out of 29 plains lions, however, did become infected, suggesting that lions might also be

Table 10.1 Prevalence of trypanosome infection in lions as detected by microscopy in four habitat types of Serengeti and Ngorongoro in order of decreasing occurrence of tsetse flies (see Fig. 10.1 for geographic locations).

Lion habitat type	Occurrence of tsetse flies	Prevalence (%) of *Trypanosoma* spp.
Serengeti woodlands	Common	50 (26/52)
Serengeti woodlands/plains border	Rare	11 (3/28)
Serengeti plains	Absent	7 (2/29)
Ngorongoro Crater	Absent	0 (0/10)

Source: Adapted from Averbeck *et al.* (1990).

inoculated by consuming infected prey that migrate to the plains from the woodlands. Prey such as wildebeest (*Connochaetes taurinus*), Grants gazelle (*Gazella granti*), Thompson gazelle (*G. thomsonii*), and warthogs (*Phacochoerus africanus*) are known to be infected with trypanosomes (Baker 1968; Kaare *et al.* 2007). However, plains lions sometimes make short forays to tsetse fly habitat during droughts (Averbeck *et al.* 1990; Packer *et al.* 1990), making it difficult to infer the mode of trypanosome transmission.

The advent of sensitive and specific molecular techniques has allowed the identification of multiple *Trypanosoma* spp. that co-infect the Serengeti lions: *Trypanosoma congolense*; *T. brucei rhodesiense*, the causative agent of human sleeping sickness; and the non-pathogenic *T. brucei brucei* (Welburn *et al.*, 2008). Welburn *et al.* 2008 identify different age-prevalence patterns of exposure to *T. brucei* and *T. congolense* (Fig. 10.7). Lions are rapidly exposed at a young age to *T. brucei*, and then prevalence decreases, while prevalence of *T. congolense* increases steadily with age. Because *T. congolense* is more common, and more genetically diverse than *T. brucei*, Welburn *et al.* conclude that *T. congolense* infection confers protective immunity against infection with *T. brucei*. In addition, cross-immunity likely explains why lions are not infected with the human-pathogenic *T. brucei rhodesiense* after the age of six. Lions do not show increased mortality due to infection. Welburn *et al.* are the first to suggest acquired immunity

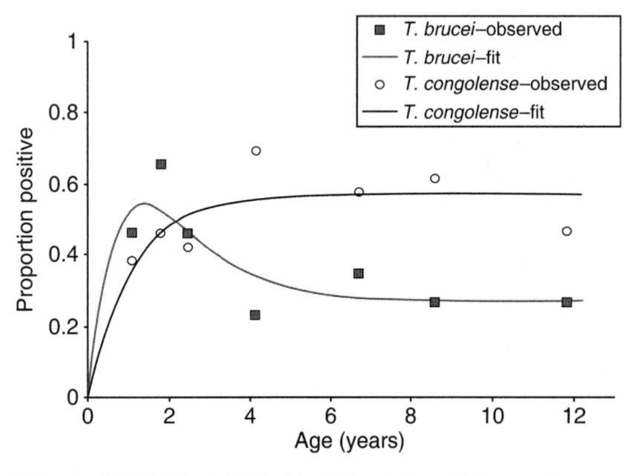

Figure 10.7 Model fitted to *T. brucei* and *T. congolense* age-prevalence curves in Serengeti lions. (For detailed description of model, see Welburn *et al.* 2008.)

to natural infection for trypanosomes, and more broadly, this study is a useful way to rethink the assumption that all co-infections necessarily harm the host.

Case study: canine distemper virus

Co-infection increases virulence in a multi-host pathogen

Infectious disease was not a major research focus when the Serengeti Lion Project was founded in 1966. However, things changed in 1994 with the observation of six lions experiencing violent symptoms such as *grand-mal* seizures and three lions with myoclonus (recurrent twitching; Roelke-Parker *et al.* 1996). Over a period of 8 months, one-third of the study lions died; a huge deviation from normal mortality rates, and a sign of a previously unappreciated threat from infectious disease (Roelke-Parker *et al.* 1996). When CDV was identified as the causative agent, it was the first time that CDV had been detected in wild lions (Appel and Summers 1995). At the end of the outbreak, CDV had spread extensively across the Serengeti ecosystem, infecting 85% of survivors (Roelke-Parker *et al.* 1996).

Domestic dogs (*C. familiaris*) were the likely source of infection into the lion population. In 1992 and 1993, CDV was circulating in the high-density domestic dog population to the northwest of the park and was not present elsewhere in the ecosystem (Fig. 10.8a; Roelke-Parker *et al.* 1996; Cleaveland *et al.* 2000). While it made intuitive sense that domestic dogs were the source of CDV, the exact mechanism of transmission between dogs and lions remained a mystery. As CDV is transmitted by aerosol or droplet exposure (or possibly by eating an infected carcass; Appel 1987; Greene and Appel 2006), and domestic dogs and lions do not occupy the same habitat, it seemed unlikely that a dog could transmit CDV directly to a lion, suggesting an intermediate link, such as spotted hyenas, which are known to 'commute' long distances and to enter agricultural areas outside the national park (Hofer and East 1993). Another question remained unanswered: Was this the first time that lions had been exposed to CDV?

A retrospective serological study showed discrete periods of CDV exposure in the study population (as evidenced by declining CDV seroprevalence levels in the 1980s, reflecting earlier exposure possibly from a

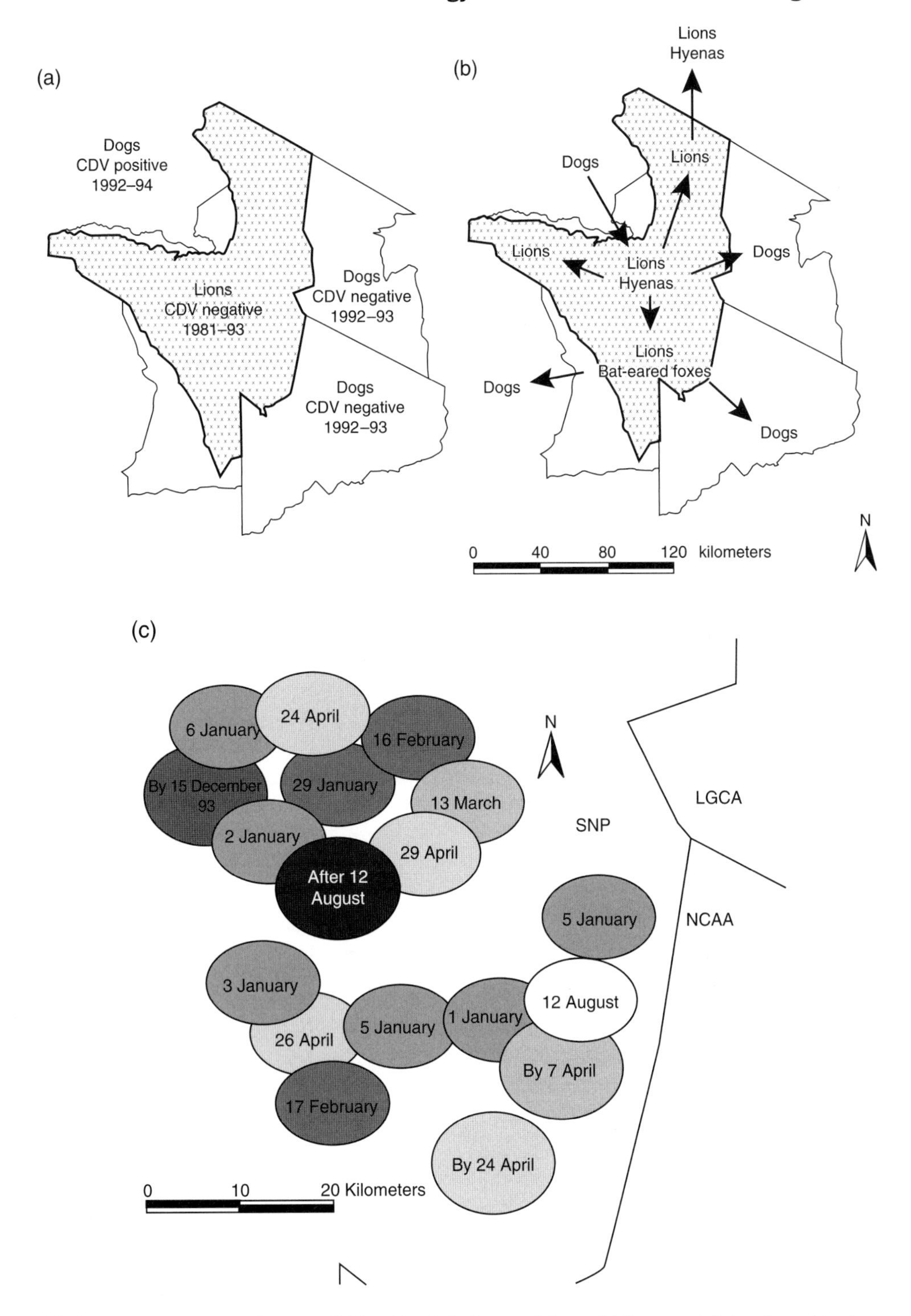

Figure 10.8 (a) Status of CDV before the 1994 epidemic in Serengeti lions. (b) Its spread during the epidemic as reconstructed using serological or viral evidence. (From Cleaveland *et al.* 2008). (c) Its spread among the lion study population. Each oval represents a lion pride. The time course was determined by either the date of first observed death in a pride or date of sampling for the first seropositive individual in the pride. Prides infected early in the epidemic are shaded dark, those infected in the epidemic grade through to white. One pride remained uninfected (black). (From Craft *et al.* 2008.)

1981 outbreak), although no symptoms or excess mortality were observed during these earlier periods (Roelke-Parker *et al.* 1996; Packer *et al.* 1999). Why then was the 1994 outbreak so harmful to the lion population? Was this simply a new, more virulent strain of CDV (Packer *et al.* 1999)?

Then 40% of Crater lions died in 10 weeks in 2001, and 10 out of 10 sampled lions tested positive for CDV antibodies (Kissui and Packer 2004; Munson *et al.* 2008). Retrospective serological results of stored lion samples showed that Ngorongoro lions were exposed to distemper at least once in the past (before 1984, most likely 1980), but had not died or shown symptoms in the earlier period (Packer *et al.* 1999). In total, out of at least seven CDV outbreaks in the Serengeti and Crater lion populations since 1975, lions only experienced symptoms and high mortality in the Serengeti in 1994 and in the Crater in 2001.

New results indicate that the two periods of mass mortalities were due to a convergence of biotic and abiotic conditions to create a 'perfect storm' where CDV exacerbated the impacts of a tick-borne pathogen (Munson *et al.* 2008). Lions are consistently infected with low levels of *Babesia*, a tick-borne parasite that can be transferred from herbivores. Severe droughts led to large-scale starvation and mass mortalities in African buffalo (*Syncerus caffer*) in the Serengeti in 1993 and the Crater in 2000; the weakened buffalo reached unprecedented heights in the lions' diet and exposed the lions to high levels of *Babesia* infection. CDV is immunosuppressive, and the outbreaks in early 1994 and early 2001 allowed already high levels of *Babesia* to overwhelm the co-infected lions. Serengeti prides in 1994 showed no increase in mortality if they were only exposed to CDV or only to high levels of *Babesia* (Munson *et al.* 2008).

Levels of *Babesia* were consistently higher in the Crater than the Serengeti. Fyumagwa *et al.* (2007) trace the build-up in Ngorongoro Crater to the 1970s (Fig. 10.9), when management authorities embarked upon a policy of fire suppression and evicted the pastoralist Masai from the Crater floor, removing the effects of fire, allowing the grass to grow taller, and increasing tick survival. Meanwhile, the buffalo population grew in size, and hence the numbers of tick-infested buffalos also increased, especially during the El Niño wet years (1997–98). This was followed by the drought of 1999–2000, which caused the death of buffalos, wildebeest, and rhinos, and the consequent die-off in the Ngorongoro lion population (due to disease rather than drought; Fig. 10.9; Fyumagwa *et al.* 2007). Although the Serengeti lion population was large enough to recover from the CDV/*Babesia* die-off and return to its original population size by the middle of 1997 (Packer *et al.* 1999, 2005a), frequent outbreaks of disease seem to have kept the Crater population below carrying capacity for the past 14 years (Kissui and Packer 2004).

Integrating biology and epidemiology into models

In order to manage disease threats effectively (e.g. to prevent CDV/*Babesia* from causing mass mortalities again), it is important to understand which populations maintain multi-host pathogens in the greater Serengeti ecosystem (Cleaveland *et al.* 2007, 2008). The maintenance population is the species, or set of species, in which the infection can independently persist (Haydon *et al.* 2002). In light of this goal, a mass vaccination programme was initiated in 2003, vaccinating >35,000 domestic dogs per year for CDV, parvovirus, and rabies with an aim of reducing disease transmission to Serengeti wildlife (Cleaveland *et al.* 2007). By 2008, the programme appeared to have successfully eliminated canine rabies from wildlife in the Serengeti ecosystem (Lembo *et al.* 2008) and it is not yet clear whether there has been any impact on parvovirus exposure in the lions; however, CDV struck the Serengeti lions in 2006 (Munson *et al.* 2008).

Because CDV still seems to be a threat to lions (despite the dog vaccinations), identification of a maintenance population is crucial. Some researchers claim that CDV can be maintained solely within the lion population (without transmission from other species such as hyenas or jackals), fuelled by occasional spillover from the domestic dog population (Guiserix *et al.* 2007). If so, lion-to-lion transmission alone should account for the observed dynamics of the 1994 CDV outbreak inside the Serengeti National Park. To test this hypothesis, an empirically parameterized network model was constructed to represent the demographic, spatial, and contact structure of the Serengeti lion population before the 1993–94 outbreak (Craft *et al.* 2009). In contrast to Guiserix's model, where all lions in the ecosystem have an equal chance of contacting other lions, the network model explicitly defines the different lion social groups and assigns contacts between groups according to network adjacencies.

Lion network and contact structure

The observed population structure and contact patterns of the Serengeti lions were estimated using empirical data from the Lion Project (Table 10.2 and 10.3). As described in Craft *et al.* 2009, the network model placed $N_P = 180$ prides and $N_N = 180$ coalitions of nomads at random locations in an $A = 10{,}000$ km^2 region of the Serengeti. Prides were assigned to be adjacent according to the estimated adjacency model (M_{adj}). A fraction of adjacent pairs of prides (Ψ) were randomly assigned to have recently split. Each pride was given a group size (X_P) drawn from an empirical distribution. Contacts between prides occurred at an average of $C_P = 4.55$ contacts per 2-week period per pride, as estimated from a study in which 16 lionesses were observed continuously for a total of 2213 h (Packer *et al.* 1990; Scheel and Packer 1991). Contacts between pairs of prides occurred stochastically at rates weighted by a logistic function of the network distance between the centroids of their territories ($M_{contact}$).

Coalitions of resident males and nomads were treated separately from prides of females and cubs. Male coalitions were represented as single units that increase connectivity between prides. Each territorial coalition belonged to either one or two prides; an estimated fraction η of all prides shared their territorial coalition with one of their adjacent prides, and every other pride had a territorial coalition to itself. If a territorial coalition was associated with more than one pride, it would switch between prides according to ς, the territorial male migration rate. Nomadic lions were assigned group sizes (X_N) averaging 1.5 members and were assumed to migrate via a variance gamma process (M_{nomad}), as estimated from a GPS-collared nomad (Fig. 10.10). Nomads were assumed to contact their local pride according to the average rate of pride-nomad contacts per pride (C_N).

When a pride contacted another pride or nomadic coalition, only a subset of the pride was involved in the interaction (G), and the number of lions involved depended on the size of that pride. When nomads contacted prides, all members of the

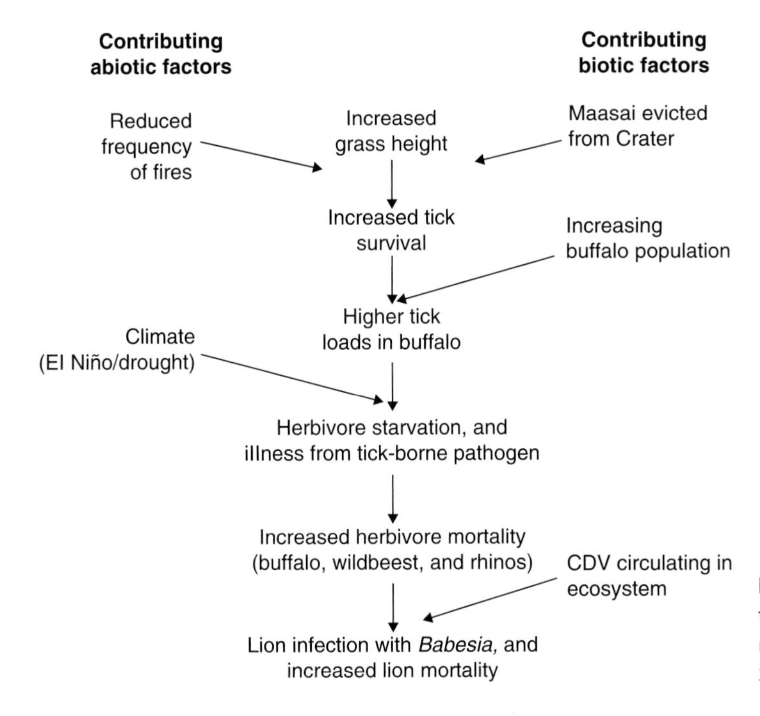

Figure 10.9 Abiotic and biotic factors in the Ngorongoro Crater which led to mortality in Crater lions (Fyumagwa *et al.* 2007; Munson *et al.* 2008).

coalition were assumed to be present. Inter-group contacts of resident males were incorporated into the pride contact patterns.

Epidemiological model

In this network model, prides move through each susceptible, exposed, infectious, and recovered class as a unit; prides contact other prides as a function of their distance within territory adjacency networks; male coalitions transmit disease between their residential prides; and nomads migrate and contact prides according to empirically estimated rates. CDV was introduced into this network and was transmitted among prides according to incubating/infectious parameters estimated from the domestic dog literature. Simulations were run across a range of transmissibility values (probability that an infection is passed during a contact between a susceptible and infectious individual). Model output was compared to three characteristics of the 1994 outbreak: (1) 17/18 study prides were infected; (2) infection spread in a discontinuous pattern through the study area (Fig. 10.8c); and (3) CDV took 35 weeks to spread 100 km to the Maasai Mara National Reserve (MMNR, Kenya; Cleaveland *et al.* 2007; Craft *et al.* 2008).

Table 10.2 Lion demographics.

Demographic parameters	Estimated quantities	Distributions
A: area of ecosystem	10,000 km^2	N/A
N_P: number of prides in ecosystem	180	$U(150, 200)$
X_P: pride sizes (number of females and cubs over 3 months old)	$X_P \sim$ Gamma (k, θ) with $k = 2.226$ and $\theta = 4.707$ Mean pride size = 10.48	θ: N (4.707, 1.243) k: N (2.226, 0.636)
η: fraction of prides sharing males	0.882	η: N (0.882, 0.078)
ς: rate at which territorial male coalitions switch prides	0.25 switches/day	ς: N (0.25, 0.070)
M_{adj}: territory adjacency model	$\ln\left(\frac{p_{adj(AB)}}{1-p_{adj(AB)}}\right) = 1.483 - 0.386. {}^*S_{AB}$ (S_{AB} = the number of prides located in the joint radius of A and B) Mean number adjacent prides = 7.36	Intercept $\sim N$ (1.483, 0.225) Slope $\sim N$ (−0.386, 0.041)
ψ: proportion of prides recently 'split off'	0.063	$\psi \sim N$ (0.063, 0.021)
N_N: number of nomads	180	U (150–200)
X_N: nomad group sizes	$X_N \sim$ Log-normal (μ, σ) with $\mu = 0.292$ and $\sigma = 0.446$ Mean group size = 1.51	$\mu \sim N$ (0.292, 0.065) $\sigma \sim N$ (0.446, 0.046)
M_{nomad}: nomad movement model Horizontal (x) and vertical (y) displacements in km per 2 weeks are given by gamma distributions	$Disp_x \sim Gamma_x$ (k_x, θ_x) with $k_x = 0.382$ and $\theta_x = 2.85$ $Disp_y \sim Gamma_y$ (k_y, θ_y) with $k_y = 0.714$ and $\theta_y = 1.743$	$k_x \sim N$ (0.382, 0.029) $\theta_x \sim N$ (2.85, 0.02) $k_y \sim N$ (0.714, 0.029) $\theta_y \sim N$ (1.743, 0.019)

Source: From Craft *et al.* 2009.

Table 10.3 Contact parameters.

Contact parameters		
C_P: rate of pride–pride contacts per pride	4.55 contacts/2 weeks	C_P: $N(4.55,0.573)$
$M_{contact}$: contact weighting model	$\ln\left(\frac{w_{c(A,B)}}{1-w_{c(A,B)}}\right) = \alpha + \beta_d d_t(A,B) + \begin{cases} -\beta_s & \text{if recently split} \\ \beta_s & \text{otherwise} \end{cases}$ where $W_{c(A,\ B)}$ is the weighting factor for the contact rate between A and B and d_t (A, B) is the territory distance between the prides.	
	$\alpha = 3.265$ $\beta_d = 1.698$ $\beta_s = 0.696$	α : $N(3.265,0.371)$ β_d : $N(1.698,0.264)$ β_s : $N(0.696,0.220)$
C_N : per pride rate of pride–nomad contacts	7.136 contacts/2 weeks	C_N : $N(7.136,1.018)$
G : pride group size during contact $G' = \log(G+1)$	$G' \sim N(\mu_{G'}, \sigma_{G'})$ with $\mu_{G'} = 0.447 + 0.014 \cdot X_P$, $\sigma_{G'} = 0.232$ Mean group size $= 3.65$	$\mu_{G'}$ intercept $\sim N(0.447,0.057)$ $\mu_{G'}$ slope $\sim N(014,0.004)$ $\sigma \sim N(0.232,0.022)$

Source: From Craft *et al.* 2009.

Nomads

Although nomads are numerous and travel long distances, their impacts on model CDV disease dynamics were surprisingly low. In fact, for extensive outbreaks with 95% of prides infected, nomads only accounted for 10% of all transmissions, whereas the vast majority of transmissions were pride-to-pride (neighbours, 53.1%; second degree neighbours, 27.8%; and third degree neighbours, 8.3%) and prides four prides away or greater, and shared males only accounted for less than 1% of transmissions. To assess the effects of nomads on CDV prevalence, spatial spread, and velocity, simulations were run where nomads migrated at an unrealistically fast rate and were removed altogether. In the simulations (regardless of the presence or migration rate of nomads), it was possible to infect at least 95% of prides, as seen in 1994. Accelerating the migration rate of nomads only slightly increased overall CDV prevalence among prides and removing nomads from the simulations slightly decreased overall prevalence (Fig. 10.11). The spatial spread of CDV, driven by pride-to-pride transmission, was wave-like throughout the ecosystem and when we either increased the nomad migration rate or removed nomads from the simulations, the overall spread in the population remained wave-like, however was correlated at longer network distances with the high nomad migration rate (Fig. 10.12; for the correlation methods see the supplementary material in Craft *et al.* 2009). Finally, the model results produced a wave of CDV that travelled at a velocity consistent with the observed velocity. However, because there was no difference in the velocity of the epidemic when nomads were removed from the simulations, this again showed that nomads were not driving the spatial spread (Fig. 10.13).

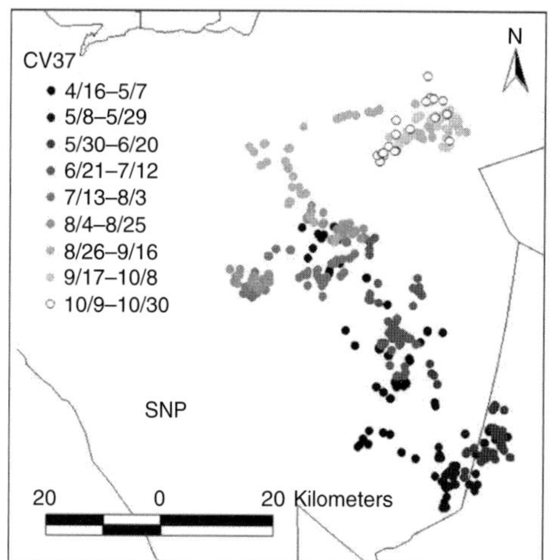

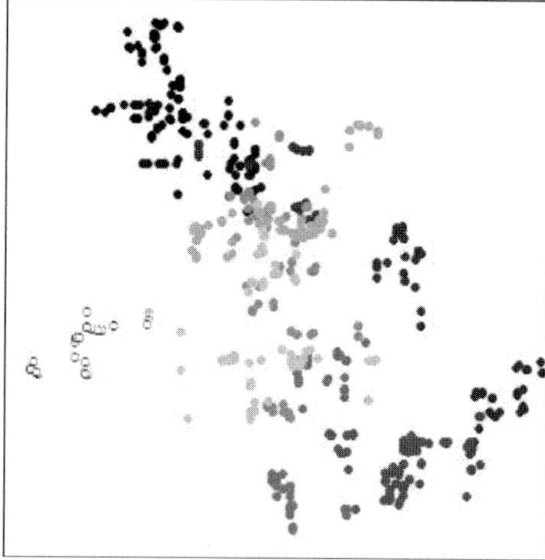

Figure 10.10 Movement patterns of a Serengeti National Park (SNP) nomadic lion in 2006 (left), where spatiotemporal locations are represented by shades of gray (dark are early locations, whereas white are late locations and the month/day of locations are indicated in the legend) versus simulated nomad (right) with grey shades representing the same temporal scale (6 months). The time steps on the simulated nomad exactly mirror the time steps on the actual.

Did lions maintain the 1994 CDV outbreak themselves?

Model results showed that the observed 1994 CDV spatial spread pattern and velocity were likely to occur at low transmissibilities, while the observed prevalence was likely at higher transmissibilities, and only a few selected simulations exhibited both the observed prevalence and velocity (Fig. 10.14). The results from the model suggest that epidemics could not have been as large and as slow as the observed 1993–94 outbreak; hence, the lion-to-lion transmission model lacked a critical component of the actual transmission dynamics. Lions could not maintain distemper on their own, and the missing piece of transmission was presumably multiple introductions of disease from other wild carnivore species, such as spotted hyenas (*Crocuta crocuta*) and jackals (*Canis* spp.).

It is reasonable that other carnivores were involved in the fatal Serengeti outbreak, as all families in the order Carnivora are susceptible to CDV (Williams 2001), and lions frequently interact with hyenas and jackals at carcasses (Cleaveland *et al.* 2008). A multi-host explanation for the observed CDV dynamics is also consistent with (a) a genetic analysis of a single CDV variant found in lions, hyenas, bat-eared foxes (*Otocyon megalotis*), and domestic dogs at the time of the epidemic (Haas *et al.* 1996; Roelke-Parker *et al.* 1996; Carpenter *et al.* 1998); (b) observations of a few sick carnivores at the time of the epidemic (but no known effects on hyena or jackal populations; Roelke-Parker *et al.* 1996); (c) serological reconstruction of an epidemic in hyenas, dogs, and lions (Fig. 10.8b; Kock *et al.* 1998; Harrison *et al.* 2004); and (d) the concept of morbilliviruses requiring a much larger critical community size than 3000 lions (Bartlett 1960; Grenfell *et al.* 2001). In other words, the 1994 CDV epidemic observed in lions was likely fuelled by multiple carnivore species.

Multi-host dynamics

If lions could not produce the observed CDV outbreak, and other wild carnivores were feasibly involved in transmission to the lion population, could a multi-host spatial model account for the patchy pattern of CDV spread seen in lions in 1994 (Fig. 10.8c)? To test this hypothesis, a stochastic susceptible-infected-recovered multi-host model was constructed which allowed transmission between a highly territorial species, like lions, and 1–2 more gregarious hosts, such as hyenas and jackals (Craft *et al.* 2008). Social structure of each species was explicitly modelled by

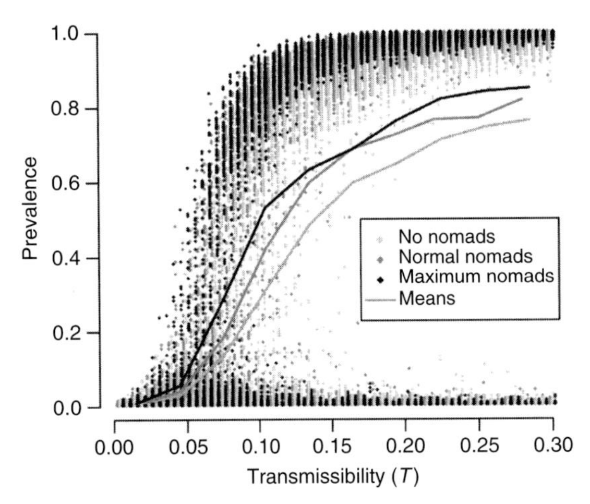

Figure 10.11 Prevalence across a range of transmissibilities for simulations with realistic movement patterns of nomads ('normal nomads') with maximum nomad migration, and no nomads in the simulation. The mean for prevalence at each transmissibility was plotted for the overall model ecosystem (180 prides, solid lines).

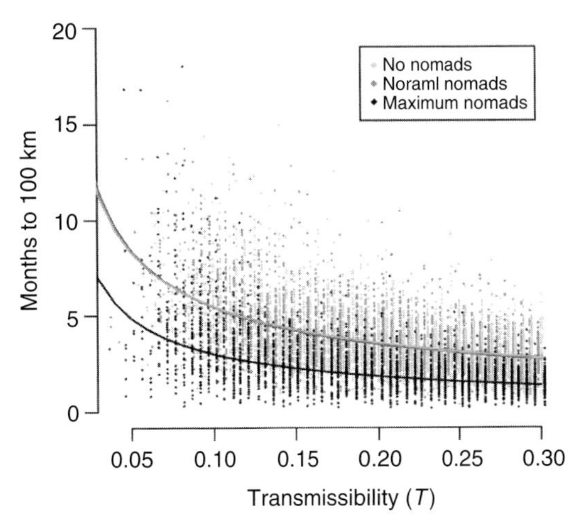

Figure 10.13 Velocity across a range of transmissibilities. Each point represents the time until the disease reached 100 km from the first infected pride for a single simulated epidemic starting at a randomly chosen pride in the subset. The lines show the least squares linear regression on log–log transformed values. The no nomads and normal nomads lines are on top of each other, while the maximum nomad line is below.

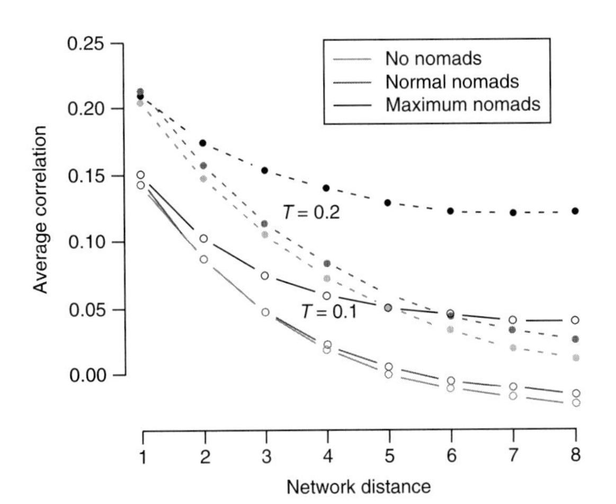

Figure 10.12 Network correlograms for simulated epidemics. In simulated epidemics, the average correlation in the timing of infectious periods between randomly chosen prides decreases with increasing network distance. This is plotted for transmissibility values of $T = 0.1$ and $T = 0.2$.

varying within- and between-group transmission rates (e.g. isolated vs well-connected territorial structures), while interspecific transmission with sympatric carnivores occurred at both low and high rates. According to model results, when other gregarious species were coupled with lions at low transmissibility, the erratic and discontinuous patterns of CDV spatial spread were similar to those seen in lions in 1994 (Craft *et al.* 2008). Based on this simplified model, it is difficult to identify which carnivore species were likely involved in repeat transmission into the lion population but, rather that low interspecific contact rates, could have accounted for the high prevalence and erratic spatial spread of CDV seen in 1994 in the lion population.

The results of both the network and the multi-host models, in combination with the observational and viral work, suggest that lions are a non-maintenance population for CDV, and because lions cannot independently maintain chains of CDV transmission, CDV control efforts should focus on other carnivores besides lions. Domestic dogs are a likely maintenance population for CDV, but whether other wild carnivores are part of this maintenance population remains unknown.

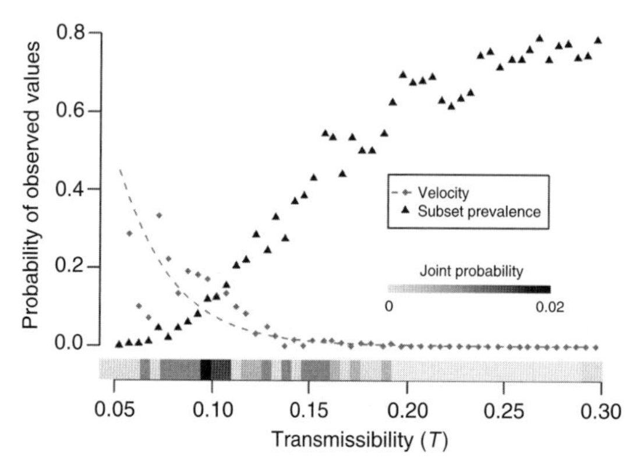

Figure 10.14 Probability of observed epidemic values across a range of transmissiblities. The probability of the observed velocity is calculated as the fraction of simulations that took at least 35 weeks to reach 100 km. The probability of the observed prevalence is calculated as the fraction of simulations that infected at least 17 of the 18 prides in the study area (subset). The joint probability is calculated as the fraction of simulations that exhibited both the observed velocity and prevalence. (From Craft *et al.* 2009.)

Conclusions

Even within large, well-protected areas like the Serengeti, species like lions can be threatened by infectious disease (Cleaveland *et al.* 2007). These diseases can originate from outside the protected area, and outbreaks can be triggered by climatic factors. As we expect more climatic extremes from global climate change, this could have unexpected effects on disease dynamics in wild animal populations (Munson *et al.* 2008). Disease dynamics are complex and understanding them requires coordinated and integrated ecosystem-level approaches (Cleaveland *et al.* 2008). In order to conserve free-ranging lions, and wild felids in general, we need to effectively integrate veterinary epidemiology into carnivore conservation and management, and focus our efforts on long-term, integrative, cross-species, cross-pathogen research (Cleaveland *et al.* 2007; Haydon, D.T. 2008).

Which management approach should be adopted to protect wild felids from infectious disease threats? It is logistically infeasible to protect all cats from all diseases—some diseases are non-pathogenic and resources are limited. For a start, it is important to understand the potential impacts of disease on long-term population viability (Driciru *et al.* 2006). Ironically, this does not necessarily mean the total elimination of a pathogen from a system. Studies have shown that depending on reservoir dynamics and resource availability, instead of attempting to eliminate a disease, prevention of the largest outbreaks that would decrease population numbers below a viable threshold may be more practical (Vial *et al.* 2006; Cleaveland *et al.* 2007). So how do we prioritize which diseases, and in which situations, to focus our efforts? Maybe we should focus interventions on diseases that are of anthropogenic origin (i.e. viruses associated with humans and their domestic dogs like rabies, CDV, and parvovirus), and focus concerted effort on small, fragmented populations that might not recover from a decline in population size due to disease. Specifically, what lessons can we learn from the Serengeti Lion Project's disease studies?

First, studies of disease dynamics in Serengeti lions show that endemic diseases like gastrointestinal macroparasites, FIV, and FHV can persist in low-density or small populations, such as the small population of Crater lions and the low-density lions on the Serengeti plains. On the other hand, epidemic diseases either need a large number of susceptibles in order to persist (FCV), or the ability to infect a suite of hosts (CDV, FPV/CPV, and FCoV). Ecological studies in Serengeti lions also illustrate that co-infection can either lessen or increase virulence, as seen in trypanosome and CDV/*Babesia* infections, respectively.

Secondly, disease status should be considered in lion relocations, with different viruses present in different populations, as seen when comparing the nearby lion populations of Ngorongoro Crater and Serengeti (Hofmann-Lehmann *et al.* 1996). In addition, FIV, FHV, and rarely FCV and FPV/CPV infections can persist in seropositive hosts and asymptomatic carriers can continue to transmit, or shed, the virus (Driciru *et al.* 2006; Gaskell *et al.* 2006). A translocation could turn into a conservation disaster if a shedding individual was introduced into a totally susceptible population.

Finally, we have learned that in the Serengeti some diseases are harder to control than others. This is likely related to the concept of R_0 (the number of secondary infections produced by one infectious individual in a completely naive population). Through the domestic dog vaccination campaign, Hampson et al. 2009 demonstrate that because the R_0 for rabies is surprisingly low (around 1.2), the elimination of canine rabies is logistically feasible. On the other hand, despite extensive dog vaccinations, Serengeti lions were still exposed to CDV in 2006 (Munson *et al.* 2008). If CDV is similar to other morbilliviruses like measles with its high R_0 (Lloyd-Smith *et al.* 2005), then CDV is more contagious than rabies. If we want to eliminate CDV to protect lions and other carnivores, we would likely need to increase vaccination coverage of domestic dogs (and perhaps other carnivores). However, it may be that the total elimination of CDV from the ecosystem may not be practical, and emphasis should instead be placed on protecting small, fragmented populations, like wild dogs, from CDV. Alternatively, if we wanted to protect an isolated population of lions from excess mortality from CDV/*Babesia* co-infection, instead of focusing on the CDV, we could reduce lion tick load by keeping levels of ticks to a minimum in the ecosystem, as the Ngorongoro Crater authorities are currently doing with controlled burns (Fyumagwa *et al.* 2007).

Acknowledgements

Thanks to Lauren Meyers and Erik Volz for crucial collaboration on the CDV network model; to Sarah Cleaveland, Clarence Lehman, Craig Packer, and an anonymous reviewer for extensive comments on the manuscript; and to Harriet Auty, Andy Dobson, Katie Hampson, Tiziana Lembo, and Jennifer Troyer for useful discussion on diseases in the Serengeti. M.E.C. was supported by NSF grants (DEB-0225453, DEB-0343960, BE-0308486, EF-0225453, DEB-0749097, and DEB-0710070) with additional funding from Lincoln Park Zoo Field Conservation Funds; the University of Minnesota's Graduate School; Department of Ecology, Evolution and Behavior; and the Office of International Programs.

CHAPTER 11

African lions on the edge: reserve boundaries as 'attractive sinks'

Andrew J. Loveridge, Graham Hemson, Zeke Davidson, and David W. Macdonald

A pride of lions crossing the Hwange National Park (HNP) boundary fence. © J.E. Hunt.

Introduction

Lion conservation status in Africa and key drivers of lion population decline

Lions (*Panthera leo*) are among those large carnivore species that have experienced significant range collapse. Indeed, their range loss over the past 500 years has been estimated at 85% (Morrison *et al.* 2007). Cardillo *et al.* (2004) warn that the species could move from its current International Union for Conservation of Nature (IUCN) red list status of 'vulnerable' to a status of 'threatened' if active conservation measures are not taken. Key drivers of decline have been loss of native prey and habitat due to human agricultural expansion, and extirpation of populations in conflict with pastoralists over livestock loss. Population declines have accelerated with the increasing availability of modern firearms and progressively easier access to agricultural poisons. Although lion populations are challenging to census accurately in many areas of Africa (Loveridge *et al.* 2001), lion populations in Africa are currently thought to be between 23,000 and 39,400 lions across the entire continent (Chardonnet 2002; Bauer and van der Merwe 2004), with the largest populations centred on large, protected areas.

Lions have been extirpated from a number of small reserves in recent times, including the temporary extirpation of lions from Amboseli National Park, Kenya, in the 1990s (Frank *et al.* 2006) and near extirpation of lions from Liuwa Plains National Park, Zambia, due to conflict with local pastoralists (Purchase *et al.* 2007). In order to conserve species that both range widely and come into conflict with people, it is important that processes occurring on reserve boundaries are well understood to provide the foundation for focused conservation efforts and mitigation of deleterious impacts.

Anthropogenic processes that have detrimental impacts on carnivores at reserve edges

Edge effects are processes that, due to the juxtaposition of two contrasting environments, have impacts on ecosystems. Boundaries of ecosystems, reserves, or fragmented habitat patches are exposed to disturbance that is not experienced by the interior of these areas. The ratio of border length to total area is smaller for large habitat patches, thus edge effects tend to be less pronounced in large, intact ecosystems than in small habitat fragments, where perturbations may extend through the entire patch (Woodroffe and Ginsberg 1998; Laurance 2000; Borgerhoff Mulder and Coppolillo 2005).

Large obligate carnivores, particularly the large felids, do not readily coexist with people in savannah landscapes where livestock production is the primary rural industry (Weber and Rabinowitz 1996; Linnell *et al.* 1999; Woodroffe 2000), and population declines are often due to conflicts over livestock (Frank and Woodroffe 2001; Polisar *et al.* 2003). Protected areas often offer the best prospects for the preservation of populations of many large felids. Nevertheless, at the boundaries of protected areas, carnivores have an increased probability of encountering humans, domestic ungulates, or anthropogenic features such as roads. These areas are therefore hotspots of conflict between large felids and people. High levels of mortality on the edges of reserves can compromise protected populations and lead to local extinctions (Woodroffe and Ginsburg 1998; Brashares 2003). This is particularly true for felids in reserves that are small in relation to the home range size of the species (Harcourt *et al.* 2001; Parks and Harcourt 2002). Brashares *et al.* (2001) found that extinction of carnivores in West African reserves was correlated with both human population density in areas adjacent to reserves and with reserve size (and thus the relative impact of edge effects on the entire protected area). This suggests that if protected areas are to be effective and their integrity maintained over long periods of time, boundary areas need to be especially safeguarded.

There are a number of anthropogenic processes that impact wild carnivores at the edges of reserves, and the type of protected area boundary may determine levels of conflict between carnivores and people. Some boundaries are unfenced, or demarcated with fences that are not wildlife proof. If these boundaries are directly adjacent to agricultural land, there is significant potential for conflict between people and wildlife from the protected area. In some regions, buffer zones between fully protected areas and farmland act to insulate the protected area from potentially destructive anthropogenic activities that might otherwise take place directly on the boundary. Utilization of carnivore and prey species in buffer areas, coupled with habitat conversion, creates a density gradient that may serve to limit levels of conflict on buffer zone/farmland edges. Protected-area boundaries that are appropriately fenced, or are marked by geographical barriers to movement such as rivers, are often termed 'hard' boundaries. These are often effective in reducing levels of conflict (Hermann *et al.* 2002; Nyhus and Tilson 2004b). However, fenced boundaries are expensive to maintain (approximately US$40/km/year for the Kgalagadi Transfrontier Park; Funston 2002) and few barriers are completely impermeable to carnivore movement (Anderson 1981). Furthermore, fencing limits the potential benefits that local communities can gain from utilization of wildlife species moving out of protected areas.

The extent of road networks and infrastructure has been used as a surrogate for intensity of human activity (Sanderson *et al.* 2002a). Road density and levels of use also have direct impacts on wildlife through increasing the numbers of animals killed accidentally by vehicles, as well as providing access to wild areas for settlers, poachers, and hunters (Wilkie *et al.* 2000). For example, Kerley *et al.* (2002) demonstrated that Amur tigers (*P. tigris*) living in areas of high road density had lower survival and reproductive rates than did tigers inhabiting pristine areas. Encroachment of settlers or subsistence hunters into wilderness areas often results in depletion of prey species; this may have catastrophic impacts on carnivore populations relying on these prey animals (Karanth and Stith 1999; Nyhus and Tilson 2004b). Illegal killing and poaching have been shown to reduce the population densities of carnivores in boundary areas of reserves, and this can subsequently reduce the effectiveness of reserves to protect biodiversity. For example, the illegal hunting of badgers (*Meles meles*) reduced badger population densities in peripheral areas of Dõnana National Park, Spain. This reduced the effectiveness of the protected area by 37% (Revilla *et al.* 2001). Similar processes may have caused the reduction in density of lions in the peripheral areas of the Serengeti National Park, Tanzania, in the 1990s (Packer 1990).

African savannah and grassland ecosystems are often utilized by pastoralist communities and their domesticated animals. Domestic stock within carnivore home ranges increases encounter rates and may result in carnivores becoming 'problem animals' when they kill livestock. Livestock depredation can lead to retaliatory killing of carnivores by farmers or herders, which elevates levels of mortality and may lead to population declines (Loveridge *et al.*, Chapter 6, this volume). For example, in Laikipia, Kenya, Woodroffe and Frank (2005) found that retaliatory killing of lions caused the lion population to decline by around 4% per annum. In addition, encounters with domestic species can expose wild animals to disease pathogens (Haydon *et al.* 2006; Munsen *et al.* Chapter 9, this volume). For example, lion and other carnivore populations in the Serengeti National Park suffered significant declines due to an epidemic of canine distemper virus thought to have been contracted from domestic dogs from villages surrounding the park (Packer *et al.* 1999; Cleaveland *et al.* 2000).

Finally, utilization of carnivores in areas adjacent to reserves can impact population density and dynamics. Wide-ranging species, in particular, may be vulnerable to harvests if they move outside protected areas. Forbes and Theberge (1996) recorded strikingly elevated mortality (from trapping and hunting) when wolves (*Canis lupus*) followed seasonal migrations of prey beyond the boundaries of Algonquin National Park, Canada. Similarly, Canada lynx (*Lynx canadensis*) were vulnerable to over-harvest when their ranges expanded beyond protected area boundaries during slumps in snowshoe hare populations (Bailey *et al.* 1986). In another case, the edge effect caused by fur trapping in the vicinity of Riding Mountain National Park, Canada, led to high mortality rates for lynx, whose ranges were only partly encompassed by the protected area, leading to temporarily extirpation of the species from the park (Carbyn and Patriquin 1983). Trophy hunting and commercial trapping have the potential to cause population declines if poorly regulated (Harris *et al.* 2002; Whitman *et al.* 2004; Loveridge *et al.* 2007c). Furthermore, the demographics, dynamics, and behaviour of populations in protected areas can be adversely affected if over-harvesting occurs on reserve boundaries (Yamazaki 1996; Loveridge *et al.* 2007c).

There has been considerable debate as to whether the hunting of lions can be carried out on a sustainable basis. Creel and Creel (1997) suggest that hunting in Selous Game Reserve, Tanzania, is only sustainable with an offtake of 4% adult males, while Greene *et al.* (1998), using a model taking into account the effects of male infanticide, suggested that removal of 10% of adult male lions will result in a stable harvest. More recently, a model parameterized with data from the Serengeti suggested that removal of males of 6 years and older will not affect the persistence of a lion population, though removal of younger males or females is predicted to result in population declines (Whitman *et al.* 2004). It is clear from all of these studies that uncontrolled and unsustainable trophy hunting on the boundaries of protected areas has the potential to have detrimental impacts on protected populations (Loveridge *et al.* 2009a).

Purpose of this chapter

Since lions are a wide-ranging species reliant on protected areas for their conservation, and a species that conflicts strongly with people, they may be particularly vulnerable to edge effects. In this chapter, we present data collected during two studies of African lions undertaken in southern Africa. Our purpose in combining data from two discrete lion populations, each facing different threats, is to form a synthesis of the causes of conflict and the trajectories of lion populations compromised by different human activities on protected area edges. One study took place in and around Hwange National Park (HNP), Zimbabwe, from 1999 to the present, and focused primarily on the impact of trophy hunting and other anthropogenic mortality on the lion population. The second study was undertaken in Makgadikgadi Pans National Park (MPNP), Botswana, from 1999 to 2003 and focused on the behavioural ecology of lions in a seasonally variable environment, and in particular on the ecology of livestock-raiding by lions, and retaliatory killing by local farmers.

The two study sites were situated in central southern Africa. MPNP is approximately 135 km south-west of HNP. Both fall within the central and northern Kalahari ecosystem (Fig. 11.1). The region has one wet and one dry season per annum. HNP receives a mean of 650 mm, MPNP 450 mm of rain falling between November and April. All of MPNP and around two-thirds of HNP fall within the Makgadikgadi Basin, the remnants of paleo-lake Makgadikgadi (Potts *et al.* 1996).

Hwange National Park case study

Hwange National Park study area

HNP is 14,900 km^2, situated in north-western Zimbabwe (between latitudes 18° 30′ and 19° 50′ and longitudes 25° 45′ and 27° 30′). Park vegetation consists largely of woodland and scrubland (predominantly *Baikiaea plurijuga, Terminalia sericea, Combretum* spp., and *Coleospermum mopane*); less than 10% of the entire area is open or bushed grassland (Rogers 1993). Around

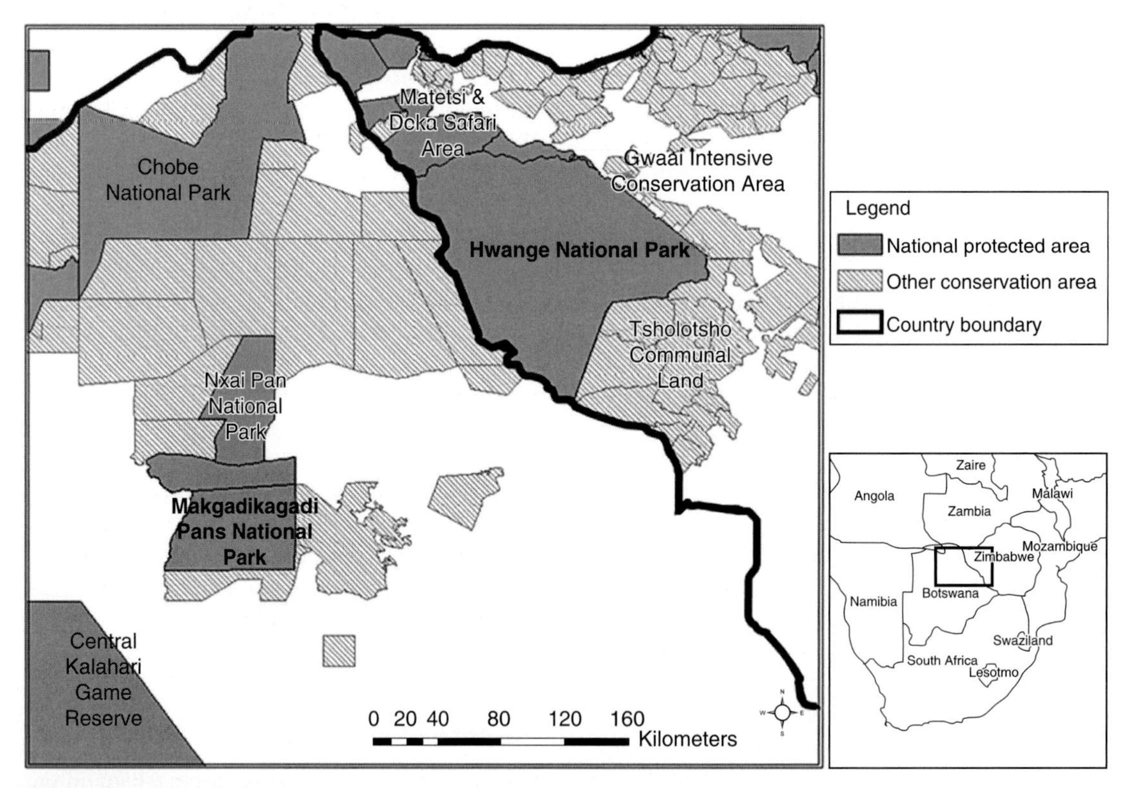

Figure 11.1 Map showing our study sites: Hwange National Park (HNP) and Makgadikagadi Pans National Park (MPNP), as well as other places mentioned in the text. Inset map shows the location of main map in southern Africa.

two-thirds of the park is situated on deep aeolian Kalahari sands, the remainder on Karoo sediments and Batoka basalt soils (Rogers 1993; Potts *et al.* 1996). The study site was in the northern area of the Park and parts of the surrounding land and covered around 7000 km^2, of which 6000 km^2 was situated in HNP. Although no sport hunting occurs within HNP, the Park is surrounded by safari areas and hunting concessions. Concession areas form either buffers between the park and areas of human settlement (forestry land, Deka and Matetsi safari areas, and Gwaai intensive conservation area [ICA]), or are communal areas (Tsholotsho communal land) where subsistence agriculture is the predominant activity and wildlife is utilized under the Communal Areas Management Plan for Indigenous Resources (CAMPFIRE) scheme and generally has lower densities of wildlife (Dunham 2002). The southern boundary of the Park is formed by the Zimbabwe–Botswana border (Fig. 11.1). Livestock is kept in areas east of the park and in the communal land to the north of the Gwaai conservancy buffer zone.

Based on spoor counts, demographic and radio-tracking data, we estimated that lion density was 2.6 lions/100 km^2 in the study site (Loveridge *et al.* 2007c). However, lower densities are expected in the drier southern areas of the park. Common large ungulate species include buffalo (*Syncerus caffer*), giraffe (*Giraffa camelopardalis*), zebra (*Equus burchelli*), kudu (*Tragelaphus strepsiceros*), impala (*Aepyceros melampus*), sable (*Hippotragus niger*), roan antelope (*H. equinus*), and warthog (*Phacochoerus africanus*). Elephants (*Loxodonta africana*) occur at high densities, particularly in the northern sectors of the park (Dunham 2002). In HNP, lions prey on all these species (Loveridge *et al.* 2006).

Trophy hunting of lions occurred on all hunting areas surrounding HNP from 1999 to 2003 and included Matetsi Safari Area hunting units 3–5, Deka Safari Area (1708 km^2), Gwaai ICA south of the Bulawayo to Victoria Falls road (881 km^2), Sikumi and Ngamo forest lands (1542 km^2), and Tsholotsho CAMPFIRE area (the five wildlife management areas closest to HNP; 4536 km^2; Fig. 11.1). During 2004, hunting of lions was suspended on private land (making up 10.2% of land surrounding the park). A moratorium on lion hunting was put in place from 2005 to 2008 owing to concerns over sustainability of quotas over the previous 5 years (Fig. 11.2).

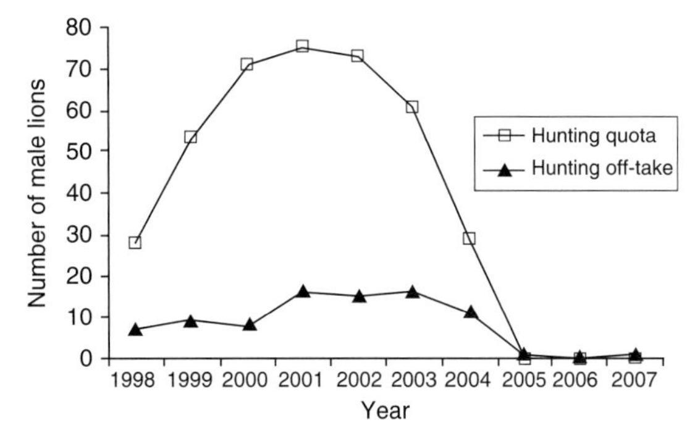

Figure 11.2 Trophy hunting quotas and trophy hunting offtake for male lions in the hunting concessions adjacent to Hwange National Park (HNP).

Methodology of the Hwange study

Capture and marking of study animals

Study animals were immobilized in order to fit radio-collars using standard methodologies for the species (Hemson 2003; Fahlman *et al.* 2005; Loveridge *et al.* 2007c). In HNP, we tagged 96 lions (47 females and 49 males). Lions were fitted with standard Very High Frequency (VHF) or Global Positioning System (GPS) radio-collars and a numbered ear tag. We attempted to radio-collar all adult (>4 years old) males within the study-site and at least one adult female from each pride. From 2000, we also caught subadult males (3–4 years old) and where practicable radio-collared them. For the purposes of this chapter, prides are defined as female groups and their offspring, and pride size as the number of adult females in the group. Male groups are referred to as coalitions where more than one male associated together.

Collection and analysis of radio-tracking data

We radio-tracked from a vehicle and from mid-2001, we used an aircraft to locate radio-collared lions. GPS collars were programmed to take 13 fixes per night, hourly from 18:00 through until 06:00 (night fixes) and then one fix at 09:00 and another at 16:00 (day fixes) and programmed to download data remotely (Hemson 2002). Later versions could be downloaded 'on demand' using a remote control. We located lions as often as possible (every day, once per month). In addition to our own sightings, we used detailed records of lion sightings recorded by safari guides from tourist camps in and around the protected area.

To minimize autocorrelation, home ranges for GPS-collared lions were calculated from one night fix every 24 h, whereas all VHF radio tracking fixes were used if at least 24 h apart. Home ranges were calculated using 90% kernels with smoothing factor h_{ref} (Hemson *et al.* 2005) in Ranges 7 (Kenward *et al.* 2003). Seasonal ranges were defined as 'early dry' (March–June), 'late dry' (July–October), and 'wet' (November–February), coinciding with varying levels of natural water availability and concomitantly prey abundance (Loveridge *et al.* 2009b).

To compare survival of individuals on the boundary of the park with those in the interior of the park, marked lions were assigned to either edge or core groups, depending on whether they occupied home ranges on or straddling the park boundary (edge prides) or home ranges in the centre of the park (core prides). Core pride ranges were a mean distance of 15.6 ± 6.7 km, (range 3.8–36.2 km) from the park boundary. We used a Kaplan–Meier survival analysis to determine the survival (number of days tagged) of tagged individuals living at the edge or in the core of HNP. This procedure allows for staggered entry of subjects into the study and for truncation of data (e.g. when animals can no longer be found because of radio-collar failure; Pollock *et al.* 1989). Individuals whose home ranges moved from the park interior to the boundary ($n = 11$) were included in the 'edge' group for this analysis (the period the animal had spent in the 'core' area was excluded). Tagged animals that were lost to the study (e.g. through radio failure), but whose fates could not be reliably determined, were 'censored' from the analysis at the day they were last found. Equality of survival distributions was tested with a log-rank (Mantel–Cox) test.

Collection and analysis of demographic and survival data

Whenever possible we made detailed observations on the composition of lion groups, although due to the dense nature of the woodland in the park this was not always possible. The effects of proximity to the Park boundary on lion prides were examined. Those prides whose ranges drifted from the park core to the boundary during the study ($n = 2$) were included in the 'edge' sample; no prides moved from the edge to the core. Only prides that we had observed for more than 1 year were used in analyses. The mean size of each pride was calculated as the mean number of females in the pride over the period the pride was monitored. A *t*-test was used to compare the mean sizes of prides in the core and at the edge of the park. A Kaplan–Meier survival analysis (SPSS 14.0 for Windows) was used to compare survival rates (years survived) of prides from the edge and the core of the park.

In the study, pride data were collected on the presence and survival of cubs. In order to test whether cubs were more likely to survive close to or away from the boundary, we classed cubs according to whether they belonged to prides on the park boundary or in the park interior (edge and core prides defined as above). We compared cub survival before and after trophy hunting was suspended. We compared the survival of cubs up until 1 year between edge and core prides and before and after hunting was suspended using a chi-square test, adjusted with Yate's correction for continuity. Those prides for which there were sufficient data (i.e. cohorts—synchronous litters within the same pride—we had monitored for more than 2 years) were also compared in terms of survival of cubs to 2 years of age.

Utilization and problem animal control data

Parks and Wildlife Management Authority (PWMA) records were used to assess the numbers of lions on trophy hunting quotas and numbers shot each year during the study by sport hunters or as problem animals around HNP. However, because hunting records were not always complete (especially for Gwaai ICA), we augmented these where possible with field data (study animals shot and reports of lions hunted). We encouraged hunting operators to report lions shot and to return collars and ear tags of study animals. In addition, photographic safari operators, hunters, and PWMA staff reported incidences of other lion mortality in the study area. Where these reports could be confirmed, they were included in our records. For the purposes of calculating quotas and offtake levels, we included in analysis only those areas directly adjacent to HNP (Fig. 11.2).

Edge effects and anthropogenic mortality of lions in Hwange National Park

Between 1999 and 2007, we tagged 47 females (24 on the edge and 23 in the core of the park) and 49 males (11 in the core and 38 on the edge). We monitored 37 prides, 22 on the boundary, and 15 in the core of the park.

Sources of mortality for HNP lions

In total, we recorded 90 mortalities of both marked and unmarked animals. Categories of mortalities were trophy hunting, illegal killing, problem animal control (PAC), vehicular collisions, natural causes, and deaths due to uncertain causes. Thirty-five lions were trophy hunted and, in addition, at least nine juveniles and subadults died as an indirect result of trophy hunting when new pride males killed cubs whose sires had been trophy hunted. This is almost certainly an underestimate as we have only presented data for those deaths where a carcass was seen, and have not included individuals that disappeared but whose deaths were not confirmed. Thirty-one lions died through being illegally snared, trapped, or poisoned (Fig. 11.3) and a further five were shot as problem animals (four on PAC licence and one by a safari guide in self-defence). Seven lions were killed by trains on the railway line that demarcates the northern boundary of the Park. Seven lions died of natural causes (including an aging female killed by spotted hyenas, *Crocuta crocuta*) a female killed by a neighbouring pride, a male killed in a fight with other males, and three females killed while hunting (by a zebra, a giraffe, and a buffalo, respectively). The causes of five deaths were uncertain, but four of these occurred in the core of the park and were therefore likely to have been caused naturally. A total of 43 marked animals died over the study period. The single largest source of mortality was trophy hunting, with 27 individuals (57% of marked animals) shot (15 adult males, 5 adult females, and 6 subadult males and 1 sub adult female). Nine marked animals (19%) were snared, trapped, or poisoned (one adult male, eight adult females). Seven marked lions died from various other causes, including intraspecific fighting (two females), train collisions (one female), and unknown causes (four females).

Impact of trophy hunting

We were able to assess the impacts of trophy hunting retrospectively by monitoring the changes in

Figure 11.3 An anti-poaching patrol with the remains of a snared subadult lion, Hwange National Park (HNP). Thirty-one of 90 recorded lion mortalities around the park were due to illegal wire snares, traps, or poisoning. © J.E. Hunt.

population demographics during the period in which trophy hunting was suspended (2005–07), and comparing this to the period 1999–2004 when trophy hunting took place in the surrounding hunting concessions.

Impact of trophy hunting on population demographics

Between 1999 and 2004, 68% of tagged male lions were trophy hunted, with 30% of these being subadult (<4 years of age; Loveridge *et al.* 2007c). The number of adult males in the population declined between 1999 and 2003, when the population was hunted heavily (Loveridge *et al.* 2007c). Male numbers increased markedly in the study area after trophy hunting was suspended from between 7 and 10 (1999–2004) to between 16 and 32 (2005–07) and differed significantly before and after hunting was suspended (Wilcoxon: $W = 10.0, z = -2.121$, $P = 0.034$; Fig. 11.4). Similarly, the number of male coalitions in the study area increased from between 4 and 7 to between 10 and 14 once trophy hunting was suspended, with the number of coalitions in the study area differing significantly before and after hunting was suspended (Wilcoxon: $W = 10.0$, $z = -2.160$, $P = 0.031$; Fig. 11.4). Coalitions that formed after 2005 were larger than coalitions that formed before the moratorium came into effect in 2005 (mean coalition size before moratorium: 1.6 ± 0.83, $n = 15$; after moratorium: 2.4 ± 0.9, $n = 12$; Mann–Whitney test: $U = 42.0$, $z = -2.539$, $P = 0.01$). The proportion of adult males (adult male:adult female sex ratio) in the study population was negatively correlated with the trophy hunters' 'harvest' of males in surrounding hunting concessions (Spearman's correlation = −0.689, $P = 0.04$, $n = 9$; Fig. 11.5).

Impact of trophy hunting on spatial movements

Between 2004 and 2007, mean annual home range size of male lions decreased from 647.3 ± 274.4 to 324.4 ± 189.8 km^2 ($R^2 = 0.79$, $F_{1,3} = 49.03$, $P = 0.03$; Fig. 11.6a and b), while that of female lions decreased from 447.1 ± 266.8 to 293.3 ± 209.6 km^2 ($R^2 = 0.79$, $F_{1,3} = 48.05$, $P = 0.02$). Similarly, variation in monthly home range size of both sexes declined and the difference in home range size between sexes was reduced from 50% to 14% over the same period. Overlap of annual home range decreased for males from 60.9 ± 2.4% in 2003 to 5.6 ± 3.9% in 2007 ($F_{1,4} = 8.11$, $R^2 = 0.43$, $P = 0.029$), but increased for females from 10.6 ± 14.3% in 2003 to 32.3 ± 32.4% in 2007 ($F_{1,4} = 6.01$, $R^2 = 0.09$, $P = 0.017$).

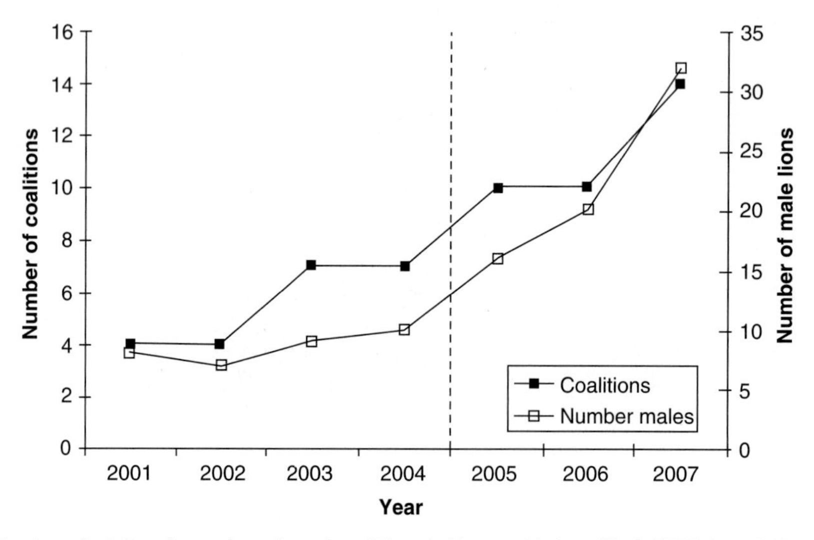

Figure 11.4 Number of adult males and number of coalitions in Hwange National Park (HNP) from 2001 to 2007. Dotted vertical line separates period trophy hunting occurred (2001–04) from period trophy hunting was suspended (2005–07).

Impact of trophy hunting on cub survival

Infanticide caused by incoming males was frequently observed when territory-holding males were removed (Loveridge *et al.* 2007c). Cub survival appeared to improve once trophy hunting was suspended, possibly due to the reduced turnover of territorial males. Of the 240 cubs in 42 cohorts whose fates we were able to follow, 144 (in 28 cohorts) were born between 1999

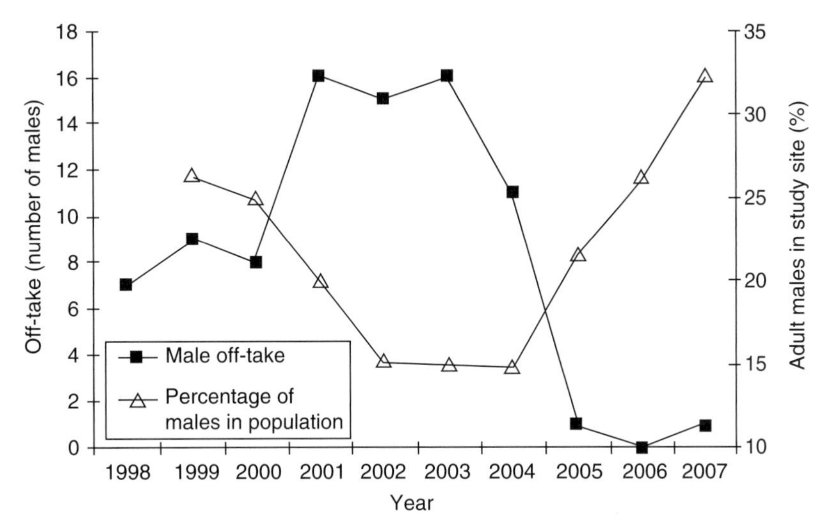

Figure 11.5 Percentage of adult males in adult proportion of population plotted against the harvest of adult males. (Modified from Loveridge *et al.* 2007c.)

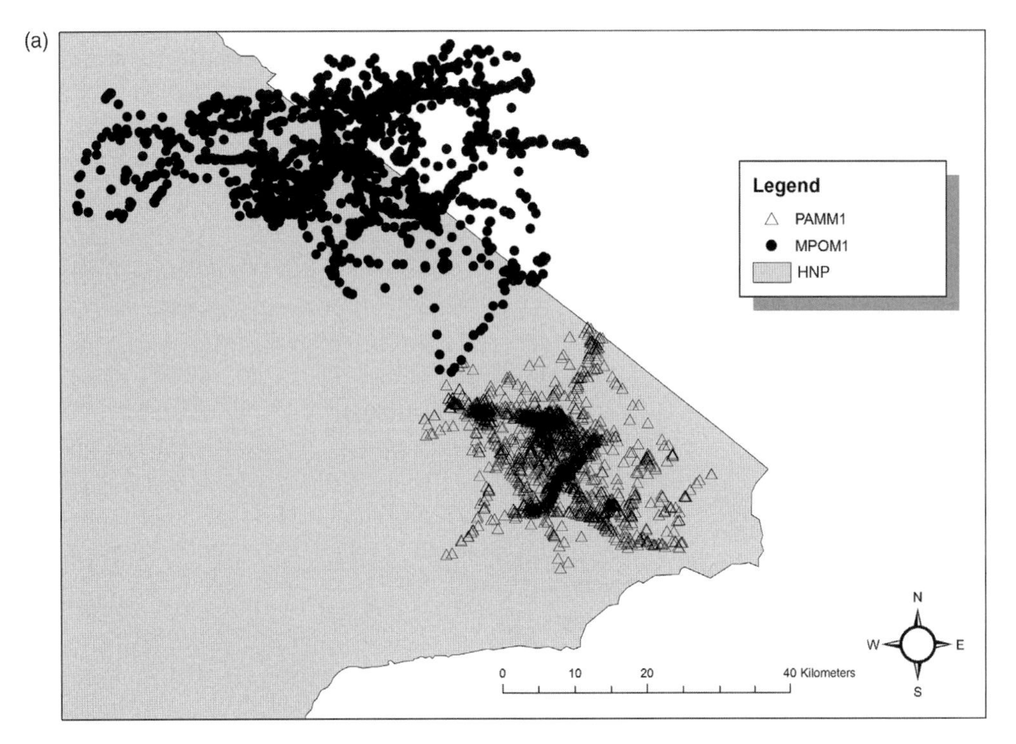

Figure 11.6a Map of the eastern side of the Hwange National Park (HNP) study area showing male lion ranges (represented by GPS locations) (a) between August 2002 and July 2003, during the period male lions were trophy hunted outside the Park.

(b)

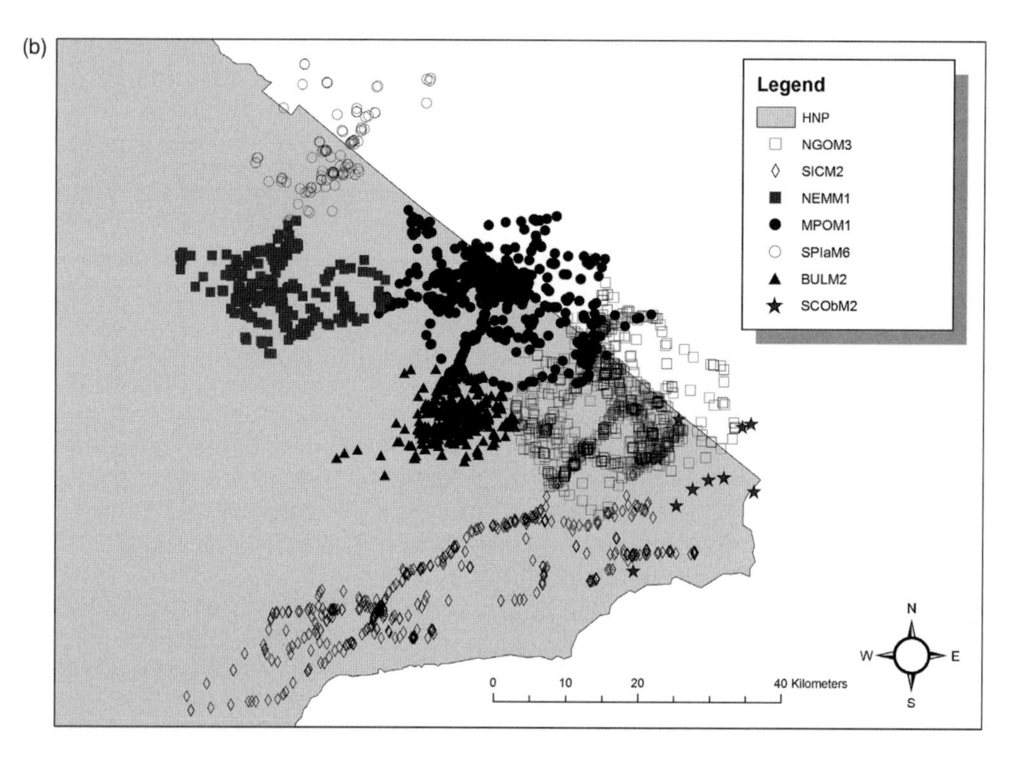

Figure 11.6b Map of the eastern side of the Hwange National Park (HNP) study area showing male lion ranges (represented by GPS locations) between November 2007 and November 2008, during the period when lion trophy hunting was suspended. Numbers of male lions increased in the study area after hunting was suspended. Note the smaller, more numerous ranges of the unhunted population (Fig. 11.6b) compared to the hunted population (Fig. 11.6a).

and the end of 2004; years when trophy hunting occurred. Ninety-six (in 14 cohorts) were born between 2005 and 2007, after trophy hunting was suspended. The chances of survival differed between periods. Cubs born during 1999–2004 were less likely to survive to 1 year ($X^2 = 5.069$, d.f. $= 1$, $P = 0.024$) or 2 years ($X^2 = 7.534$, d.f. $= 1$, $P = 0.006$). From 1999 to 2004, 95 (66.0%) survived to 1 year, and 57 (41.3%, $n = 138$) to 2 years. After 2005, of 96 cubs observed, 77 (80.2%) survived to 1 year, 36 (64.3%, $n = 56$) survived to 2 years.

Effect of proximity to the park boundary on pride size and reproductive success

Pride size

Pride size differed between prides on the park boundary and prides in the park interior (Fig. 11.7). Using the number of adult females as a measure of pride size, we found that prides on the park boundary were smaller (with a mean of 1.9 ± 0.94 females per group) than prides at the core of the park (mean 4.2 ± 1.5 females, $t = -5.733$, d.f. $= 35$, $P = 0.00$; equal variances assumed).

Cub survival

In total we were able to follow the fates of 240 cubs in 13 edge prides and 10 core prides. There were, in total, 110 cubs in 24 cohorts in edge prides and 130 cubs in 18 cohorts in core prides. More cubs survived to 1 year in core prides than they did in edge prides ($X^2 = 53.05$, d.f. $= 1$, $P < 0.005$); 53 (48.2%) of 110 edge cubs survived to 1 year, while 119 (91.5%) of 130 cubs in core prides survived. Similarly, more cubs survived to 2 years of age in core compared to edge prides, with 19 (19.2%) of 99 edge cubs and 74

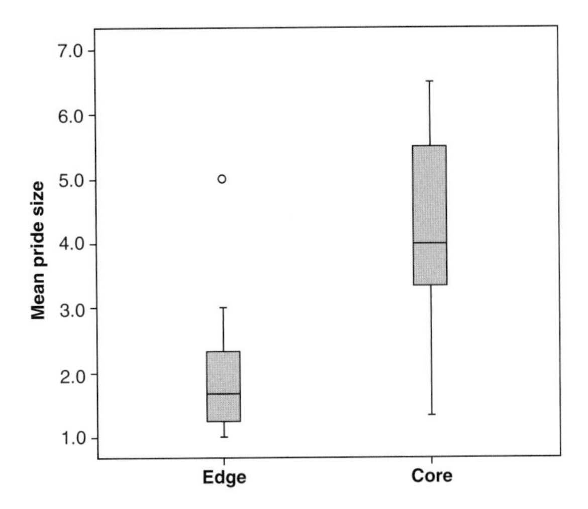

Figure 11.7 Box plots of mean sizes of lion prides in Hwange National Park, grouped according to whether the pride was located adjacent to or straddling the park boundary (edge), or in the park interior (core). Length of box represents 50% interquartile range, horizontal line the median, bars cover lowest and highest values, and circles represent outliers.

(77.9%) of 95 core cubs surviving ($X^2 = 64.61$, d.f. $= 1$, $P < 0.005$).

Survival of prides and individual lions in the core and on the edge of the park

The survival rates of prides differed between edge and core areas (Fig. 11.8a). Prides away from the park boundary were extant for a longer period than were those close to the park boundary, which were more likely to go extinct (log-rank [Mantel–Cox] test: $X^2 = 4.497$, d.f. $= 1$, $P = 0.03$); 10 of 22 boundary prides went extinct over the 9 years of the study, while only 1 of the 15 core prides did so. Mean survival time on the edge was 5.4 ± 0.6 years (95% CI = 4.2–6.6), while survival time in the core area was 7.4 ± 0.52 years (95% CI = 6.4–8.5).

Similarly, the survival times of tagged females in the core of the park were longer (mean days survived 1806.2 ± 253.1 SE; log-rank [Mantel–Cox] test: $X^2 = 5.328$, d.f. $= 1$, $P = 0.02$) than those in prides at the boundary (mean 970.8 ± 194.9 SE days; Fig. 11.8b). On the edge, 17 of 24 tagged females died, 16 from anthropogenic causes ranging from wire snares and trophy hunting to vehicular accidents. Only 6 of 23 tagged females died in the core of the park, all of natural causes.

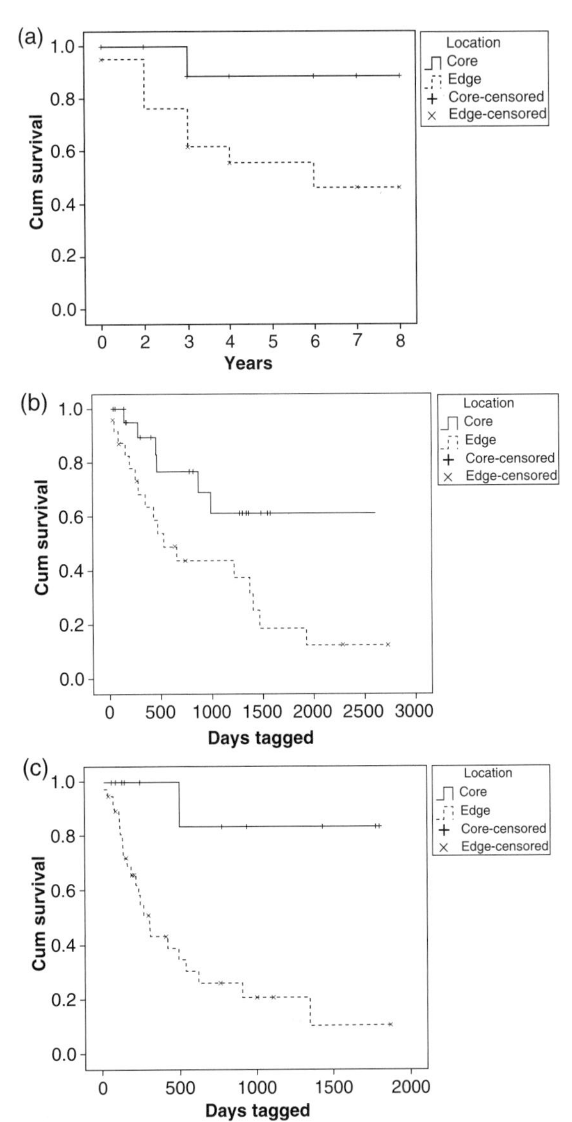

Figure 11.8 Cumulative survival of (a) prides; (b) tagged lionesses; and (c) tagged male lions on the edge (dotted line) and in the core (solid lines) of the Hwange National Park (HNP).

Due to the fact that male ranges were extensive (90% local convex hull, kernels: mean 534 km^2, range 101–1156 km^2; Loveridge *et al.* 2009b), all male ranges, except those in the very centre of the park, came into contact with the park boundary. For male lions (11 in the core, 38 on the edge), the difference in survival between the edge and core of the park was particularly marked because of the high mortality due to trophy hunting (log-rank [Mantel–Cox] test: $X^2 = 8.669$, d.f. = 1, $P = 0.003$; Fig. 11.8c). Males on the edge survived a mean of 567.9 ± 116.2 SE days, while those in the core survived 1576.0 ± 197.2 SE days, with 24 dying on the edge and only 1 in the core. Of the 24 edge males dying, 22 were trophy-hunted, 1 was killed in a wire snare, and 1 died of natural causes.

Edge effects and anthropogenic mortality of lions in HNP

The removal of territorial individuals, and the resulting gaps in the territorial structure of populations, may cause individuals from surrounding home ranges to be attracted to the depopulated area by reduced competition for resources (Loveridge *et al.* 2007c). In the case of male lions, one main resource may be access to mating opportunities within unattended prides. Variations on this 'vacuum effect', and similar perturbations, are also evident in other mammalian species. For example, badger groups removed in bovine tuberculosis control operations (Cheeseman *et al.* 1993; Macdonald *et al.* 2006; Woodroffe *et al.* 2006), lynx populations depleted by fur trapping (Bailey *et al.* 1986), and experimentally depopulated brush-tailed possum (*Trichosurus vulpecula*) home ranges (Efford *et al.* 2000) were all recolonized through immigration by immediate neighbours. From 1999 to 2004, adult males were removed from the HNP population by sport hunters on the border of the park. Each removal of a male or coalition created vacancies within the territorial structure close to the border of the park. Male groups from nearer the core of the park successively filled these territorial vacuums. For lions, creation of territorial vacuums on the boundaries of protected areas results in attraction of males to these areas from further inside the protected area. In effect, this exposes males from well within the core of a protected area to sport hunting or other mortality on the park boundary. For this reason, high mortality in boundary areas may weaken the potential of even very large protected areas to guard populations of wide-ranging carnivores.

Although both sexes were trophy hunted, males were more heavily affected because they are preferred as trophies (Grobbelaar and Masulani 2003); quotas of up to 60 male lions were granted annually between 1999 and 2003 (Loveridge *et al.* 2007c). While, in lions, male mortality is naturally high (because subadult males disperse and adult males fight for access to females), excessive offtake of males from a population can influence population viability, particularly if young and subadult males are killed (Whitman *et al.* 2004). Removal of many adult male lions from the HNP population greatly reduced the density of males and the number of coalitions in the park. Surviving males were able to range widely (some in excess of 1000 km^2) without encountering rivals. This inflated home range size in turn increased the probability of lions encountering the park boundary, and hence also increased the risk of mortality (through trophy hunting or by other causes). When trophy hunting was suspended, both male density and the number of coalitions increased in the study area, and male ranging was curtailed because male movement became increasingly constrained by encounters with rival males or coalitions. This resulted in reduced home range size, reduced overlap between ranges, and reduced variability in seasonal ranges. Changes in ranging behaviour observed in females may be the result of adjustments to the changes in male home range settlement and avoidance of non-pride males. Smaller ranges reduced the probability of an individual lion encountering the park boundary and decreased the vulnerability of a large proportion of the population.

Trophy hunting also reduced both coalition size and the length of tenure of males. The size of coalitions has a direct effect on reproductive success in male lions (Packer and Pusey 1982, 1987), and the length of tenure in a pride determines whether males are able to raise offspring to maturity. In the case of HNP, we found that males who lost coalition partners to trophy hunting often lost tenure to stronger coalitions ($n = 3$; Loveridge *et al.* 2007c; Andrew J. Loveridge and Zeke Davidson, personal observation). Coalitions formed after trophy hunting was suspended were larger and had prolonged tenure of female prides. The increased length of male tenure may in part account for the increase in cub survival that we observed after trophy hunting was suspended. Factors other than reduced territorial male turnover (such as increasing the availability of prey and improved female survival) may also account for increased cub survival. However, prey density did not change over the study period (Loveridge *et al.* 2009b) and edge prides were exposed to similar levels of anthropogenic threat (and mortality) both before and after trophy hunting was suspended.

In HNP, lionesses were exposed to the risk of mortality through a number of unnatural causes: trophy hunting, snares, poisoning, and through railway accidents. These risks were ubiquitous on the park boundary, but rare in the park core, resulting in high levels of mortality in boundary prides and higher likelihood of prides going extinct. As a result of high adult mortality, prides on the edge were, on average, composed of fewer than two females. These prides were much smaller than those in the park core, which averaged more than four females per pride. In Serengeti National Park, optimal pride size was between 3 and 10 females. Prides of three or more enjoyed greater reproductive success than did smaller prides (Packer *et al.* 1988). Moreover, larger prides are able to gain and defend territories with high-quality resources and to protect cubs better against infanticidal males. In HNP, high mortality on the boundary pushed edge prides below this threshold pride size, which may account for the lower cub survival in these prides. In addition, because adult male mortality was high on the boundary, boundary prides were exposed to higher territorial male turnover. This resulted in higher rates of infanticide. Female lion social groups can only expand if female offspring survive and join the pride; unrelated females rarely, if ever, form groups. Edge prides are therefore further disadvantaged if, because of reduced reproductive success, few daughters survive to boost pride size. Thus, edge effects not only had direct impacts on population demography through mortality of adults and reduced viability of edge prides, but also had indirect effects through the reduced reproductive success of edge prides.

Makgadikgadi Pans National Park case study

Makgadikgadi Pans National Park study area

MPNP is 4900 km^2 and lies between latitude 20° and 21° and longitude 20° and 26° in Central Botswana. Nxai Pan National Park adjoins MPNP to the north. To the south and east of MPNP lie the Makgadikgadi Salt Pans. These flat, dry, alkaline expanses become seasonally inundated, forming Botswana's largest wetland area (Thomas and Shaw 1991). The vegetation consists of grasslands associated with scattered pans (seasonal waterholes), grasslands on well-drained saline sands, and tree savannah dominated by *Acacia* spp., *Lonchocarpus* spp., *Boschia* spp., and *Terminalia* spp. Riverine vegetation occurs along the Boteti River bordering the west of the park (see Hemson [2003] for detailed habitat descriptions).

The wildlife of the area is more diverse than is typical for the Kalahari. Mesic species occur along the Boteti River and xeric species occur nearer the salt pans. Most significantly, the area supports a migratory population of Burchell's zebra (*E. burchelli*) and blue wildebeest (*Connochaetes taurinus*). These two species move *en masse* from west to east in the wet season, returning to the west in the dry season as the availability of potable surface water and fresh grazing declines in the east (Thomas and Shaw 1991; Kgathi and Kalikawe 1993). Migratory movements of zebra and wildebeest create periods of local wild prey abundance and scarcity in different parts of the MPNP on a seasonal basis. The park is surrounded on approximately 75% of its boundary by wildlife management areas predominately used for livestock-rearing by local people (over 150 cattle-posts and 4 villages; Fig. 11.9). The predominant source of income in the area is from livestock (Meynell and Parry 2002). Overall, the area supports less than one person per square kilometre, well below the national average of three persons per square kilometre (Central Statistics Office 2001). Most of the people affected directly by livestock predation live in cattle-posts, which consist of 2–10 huts and associated livestock pens or kraals and a well or borehole. The structures of kraals vary and can consist of piled thorn bush walls, poles, and wire or pole stockades,

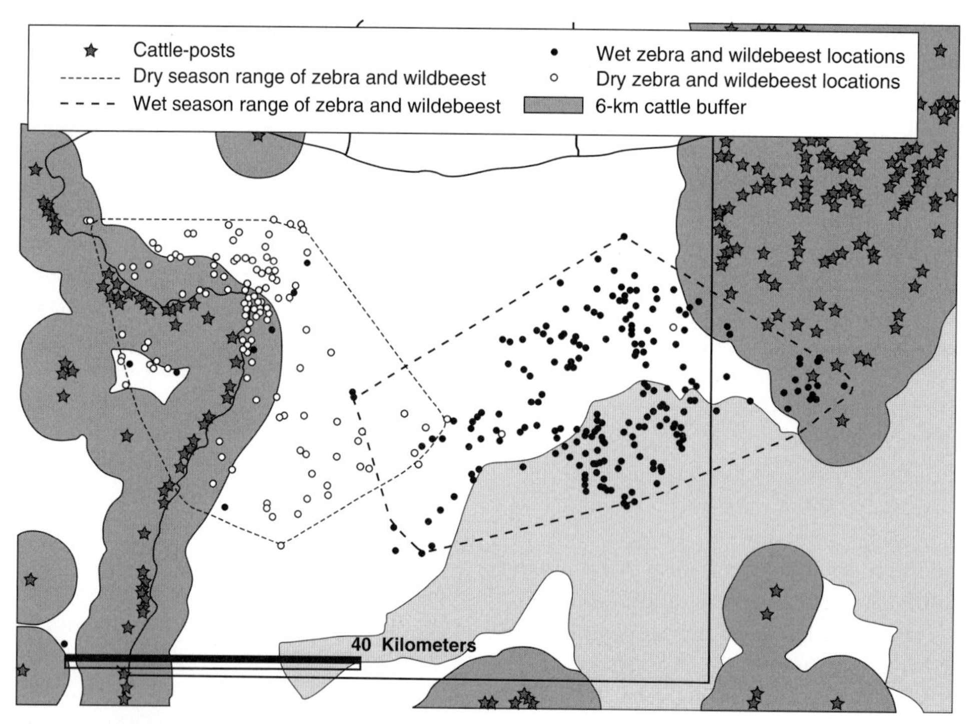

Figure 11.9 The study area, including the cattle-posts and 6-km cattle buffer migrant prey home ranges and Ntwetwe Salt Pan (grey shaded area).

depending upon the local availability of materials. People also employ herders, light fires next to their kraals, and use guard dogs. They also attempt to chase predators away with torches, by banging pots and pans, or by discharging firearms in attempts to limit stock loss.

A calling station survey in 1999 (see Karanth *et al.*, Chapter 7, this volume) indicated that 39 (28–59; 95% confidence limits) adult and subadult lions lived in the area. Subsequent data from spoor, radio, and GPS-collar tracking enabled us to narrow our estimate to 32–41 adults and subadults, giving a density of 0.5–0.7/100 km^2.

The study area encompassed the park and 3500 km^2 of adjacent land. The Makgadikgadi study area was split into eastern and western zones (Fig. 11.9), each of 4200 km^2. The two zones encompassed the dry and wet season ranges of the zebra and wildebeest. Prey species were categorized as either wild (resident and migratory) or livestock (Table 11.1), and population estimates were made using data from the Botswana Department of Wildlife and National Parks aerial survey and Botswana Aerial Surveys Information System (Wint 2000). We defined the periods during which zebra were detectable in a zone as 'good' periods (roughly equivalent to the dry season in the west and the wet season in the east) and the periods during which zebra were not detected in a zone as 'lean' periods (Fig. 11.9).

Methodology used in the Makgadikgadi study

Prey and livestock movements in MPNP

The areas used by the migratory zebra and wildebeest during good and lean periods were calculated using radio-tracking data from an ongoing ecological study of zebra ($n = 22$) and wildebeest ($n = 7$) in the Makgadikgadi (C.J. Brookes, unpublished data). Data were stratified into lean and good periods, and areas used by migratory ungulates were estimated using 95%

Table 11.1 Estimated numbers of animals and biomass in different prey classes in the eastern and western study zones during different periods of prey abundance, mean mass per individual, mean mass of species killed in kill incidences, and proportions of biomass estimates by class within area and prey abundance period.

Zone	Period	Prey class	Estimated number of animals	Estimated biomass in tons (kg/1000)	Proportion of biomass	Mean mass per animal (kg)	Proportion of kill incidences	Mean mass of species per kill incident (kg)
West	Good	Migratory	18,888	5,293	0.56	280	0.86	278
		Resident	3448	482	0.05	140	0.05	137
		Livestock	11,459	3,593	0.38	314	0.09	292
	Lean	Migratory	0	0	0	—	0	—
		Resident	4815	665	0.16	138	0.14	169
		Livestock	11,459	3,593	0.84	314	0.86	349
East	Good	Migratory	18,888	5,293	0.51	280	0.70	279
		Resident	2463	267	0.03	109	0.03	110
		Livestock	17,799	4,807	0.36	270	0.26	299
	Lean	Migratory	0	0	0	—	0	—
		Resident	1391	154	0.03	111	0.16	167
		Livestock	17,799	4,807	0.97	270	0.84	344

Notes: Biomass estimates for prey species were calculated using the mean mass of adult females of the species from Skinner and Smithers (1990) and Mason (1988). Smaller prey species (< 20 kg) not detected in aerial surveys were excluded from the analyses of prey abundance and lion predation. These items were found to represent only 0.2% of the biomass observed killed by lions (Hemson 2003).

minimum convex polygons around all location data, using the harmonic mean as a point from which to exclude outliers (Fig. 11.9). Overlaps between these polygons and individual lion home ranges were calculated as the percentage of the latter that overlapped with the former, and used as an index of the availability of these key species to lions during different periods (Ranges 7; Kenward *et al.* 2003).

The area utilized by cattle was estimated by creating a buffer around all mapped cattle-posts using data from nine cows, from different cattle-posts, fitted with GPS collars (Televilt, GPS-Simplex). The collars took one fix per hour for a total of 177 cow days (mean 20 days) and data were gathered during both periods (72 days lean and 105 good) and during wet and dry seasons. There was no evidence that the distances moved from cattle-posts differed between seasons. We assumed that the movements of these collared cattle are representative of cattle movements to and from cattle-posts and an approximation of these same patterns for livestock in general.

For each of the 3847 GPS locations, the distance to the closest cattle-post at which that cow had spent the night (cows occasionally spent the night in cattle-posts other than their own) was calculated using ArcView 3.2 GIS (ESRI 1992) and the Animal Movement Extension (Hooge *et al.* 1999). These distances were analysed by hour to describe the daily movements of the cattle. The distance within which 95% of all points were encompassed was used as the outer limit of the intensive cattle area, and used to create the 'cattle buffer' by merging radii from each cattle-post by the use of ArcView 3.2 (ESRI 1992). The density of stock within buffers of different radii from cattle-posts was calculated by dividing the percentage of locations for a given hour that were within each 1-km wide radius from the cattle-post minus the next inner radius to give densities with a series of rings at 0–1, 1–2, 2–3, 3–4 km, etc. Locations within 250 m of a cattle-post were conservatively assumed to be inside the kraal (livestock enclosure).

Capture and marking of study animals and analysis of lion radio-tracking data

In MPNP, similar methods and equipment were used to capture and radio-collar lions as those used in the HNP study. In MPNP, we captured 30 lions (8 adult males, 18 adult females, 3 subadult males, and 1 subadult female). We collared all adult male coalitions in the study and collared at least one female in each pride; we also opportunistically collared larger subadult males (near adult) when encountered.

Lion home range estimates were calculated using concave clusters (Kenward *et al.* 2003) for good and lean periods. In MPNP, lions that moved so that their ranges retained an overlap with migratory ungulate areas all year round were called 'rovers' and those that did not were labelled 'residents'. The amount of overlap between seasonal ranges of each individual is presented as percentages of the lean period range overlapped by the good period range. Seasonal changes in overlap of lion home ranges with the area used by livestock were calculated as a percentage of good period home range overlapping with the cattle buffer, expressed as a percentage of the lean period range overlapping the buffer. This enabled us to accommodate the considerable variation in home range sizes of different lions (18.1–1274 km^2), and provided a robust index of the proportion of time lions spent in different areas, irrespective of the absolute overlap.

In MPNP, the distance to the nearest cattle-post from each lion location was calculated. The mean of these distances was calculated for each lion, and additional means were calculated for each season and for each hour for which data were available. Mean straight line distances moved per hour were also calculated between successive points in a time series for each lion using the Animal Movement Extension in ArcView 3.2. The mean proportion of locations for the lean period for stock killer males and females in successive 1-km radii from cattle-posts was calculated for three periods: late evening (19:00–23:00), morning (00:00–05:00), and daytime (06:00–18:00), to test whether foraging strategies varied through the 24-h cycle.

We used a Kaplan–Meier survival analysis to determine the survival (number of days tagged) of livestock-raiding and non-raiding lions in MPNP. This procedure allows for staggered entry of subjects into the study and for truncation of data (e.g. when animals can no longer be found because of radio-collar failure; Pollock *et al.* 1989).

Collection of lion-kill data in MPNP

Lion predation data were collected between January 1999 and March 2003. Data were collected during 370 h of continuous observation of foraging lions. Ad hoc data on lion kills were also collected when radio-collared lions located for a study of movement and range use were found feeding or through opportunistic location of feeding lions. Carcasses that we could not confirm as lion prey beyond reasonable doubt (including those that may have been scavenged) were excluded from this analysis. All kills were categorized by species. Kill incidences were separated into eastern and western zones by location, and further into 'lean' and 'good' periods.

Utilization and problem animal control data

No lions were trophy hunted around the MPNP study site during the study period. Lion trophy hunting was banned in Botswana between 2001 and 2004, because of concerns for the viability of lion populations. Despite this ban, lions were killed both legally and illegally in response to livestock losses. We gathered data from PAC reports and directly from observations of radio-collared animals and associated pride mates.

Ecology of livestock-raiding by lions in MPNP

Home ranges and prey preferences of 'residents' and 'rovers'

The first major rains of the year resulted in the zebra and wildebeest leaving the Boteti River (where there is permanent water) for the summer grazing areas further east, returning only when surface water evaporated.

The most abundant class of prey available to lions during good periods was migratory prey (zebra and wildebeest). Livestock were the most abundant ungulates during migrant scarce lean periods and the next most abundant after migrants during good periods (Table 11.1), although their distribution was limited to edge areas of the park (95% of all cattle locations were within 6 km of cattle-posts; Fig. 11.9). Resident prey was least abundant at all times. Larger prey species made up a larger proportion (both by numbers and biomass) of the wild migrant and livestock classes than did the resident wild prey class.

The response of MPNP lions to changes in wild prey abundance depended on their proximity to the boundary of the national park and the linked availability of livestock as potential prey. Lions living on the boundary had more stable home ranges than did those in the core of the park. All lions with ranges on the edge of the park were known stock-raiders. These lionesses' home ranges did not change in size significantly between the seasons (Mann–Whitney test: $P = 0.19$, $n = 3$, range 350–979 km^2). During both periods of migrant abundance and scarcity, the seasonal ranges of boundary, livestock-raiding lionesses overlapped more than did those of non-raiding lionesses (Mann–Whitney test: $P = 0.0282$, $n = 3$ and 6; Figs. 11.10 a and b).

We recorded 187 successful hunts in which 241 animals were killed by lions (Table 11.2). Multiple killing events were most frequent for livestock, with 1.7 times as many animals killed per incident (predominately goats, of which there were 4.9 times more animals killed than successful hunts) compared to the one kill per successful hunt for resident wild prey and 1.06 kills per hunt for migratory prey.

Lions in different areas responded to changes in relative prey abundance by utilizing the most abundant prey type in their area most frequently. During good periods when migrant prey were abundant they dominated the diet of all lions, making up 79% and 83% of the kill incidences for residents and rovers, respectively. However, when migrants were scarce, livestock became the most important prey class for resident lions. While rovers only rarely ventured into areas without migrants, and thus only infrequently experienced migrant prey scarcity, they responded differently to this scarcity and killed resident prey more frequently than when migrants were present.

These two dietary patterns had implications for the way lions used space. Rovers had larger home ranges (mean 536 km^2, $n = 3$, SD = 235) than did resident lionesses (mean 112.4 km^2, $n = 6$, SD = 98) during migrant scarce periods (Mann–Whitney test: $n = 6$ and 3, $P = 0.0282$), but not during migrant abundant periods (mean 206.4 km^2, SD = 202; Mann–Whitney test: $n = 6$ and 3, $P = 0.1556$). During migrant abundant periods, residents used larger home ranges than they did during lean periods (Wilcoxon signed rank test: $P = 0.018$, estimated median = +100.8%, $n = 6$) and home ranges overlapped less with the livestock areas during both the day ($P = 0.018$, $n = 6$, estimated median = −28.89%) and night ($P = 0.050$, $n = 4$, estimated median = −26.00%) than they did during lean periods.

Patterns of lion movement in relation to cattle-posts

The pattern of movements of resident lions relative to cattle-posts during lean periods paralleled those of cattle more closely than during good periods (Fig. 11.11a and b). In contrast, the movements of rovers did not mirror those of cattle (Fig. 11.11c). Seventy-eight per cent of all GPS locations of residents were within 6 km of a cattle-post during lean periods. Residents made the closest approaches to cattle-posts during lean periods between 22:00 and 04:00 with a modal distance of 1–2 km. In the evenings, residents were most frequently located between 3 and 4 km from a cattle-post. By 06:00, fewer than 1% of resident locations were within 1 km of a cattle-post, and during the day they were most often 2–3 km away. Some rovers occasionally approached close to a cattle-post between 22:00 and 04:00. However, we could not correlate these to livestock predation incidents. We observed a resident lioness kill a zebra foal at a cattle-post, and by following spoor it was clear that lions used water sources associated with cattle-posts, indicating that livestock predation was not the sole activity at or near these sites.

The mean distance of 45 livestock kill incidents by resident lions from the nearest cattle-post was 3.5 km (0.3–6.9 km) during lean periods and 7.6 km (3–24.5 km, the latter probably being a long-term stray) during good periods. The mean distance of four resident prey kill incidents by residents during lean periods was 5.3

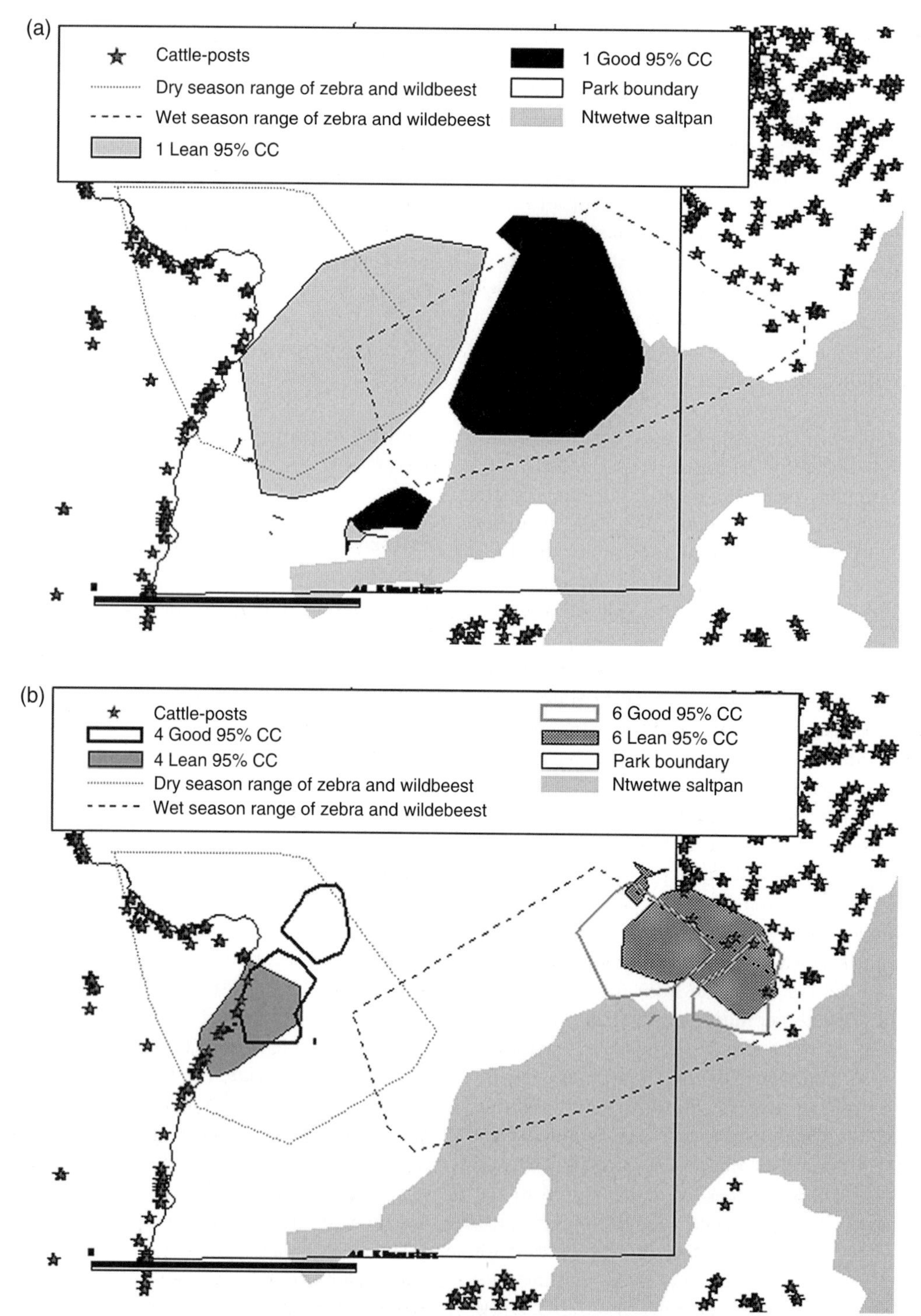

Figure 11.10 Two maps showing the seasonal shift in home ranges of (a) a rover lioness (ID = 1); (b) two resident lionesses (4 and 6), as estimated using 95% concave clustered polygons and the wet and dry season home ranges of the migratory prey. Females 4 and 6 shift range towards cattle-posts during periods of migrant scarcity, whereas one entirely relocates to the east to follow the migratory ungulates.

Table 11.2 Percentage kill incidents by prey class and period for 'residents' (lions that did not follow migratory prey) and 'rovers' (lions whose ranges tracked seasonal movements of migratory prey species).

	Good periods			Lean periods	
	Migratory prey	Resident prey	Livestock	Resident prey	Livestock
Residents	79 (n = 64)	1 (n = 1)	20 (n = 16)	13 (n=5)	87 (n=34)
Rovers	83 (n=48)	16 (n=9)	2 (n=1)	66 (n=6)	33 (n=3)

km (range 3.2–10.5 km). Most stray livestock were found between 4 and 7 km from cattle-posts at night (Fig. 11.9).

Survival of stock-raiders and non-stock-raiders

In MPNP, of 30 tagged lions (11 males and 19 females), 20 were known stock-raiders, while 8 were never found to kill stock. For two male lions, tagged at the end of the study, it was unknown whether they killed livestock. Male lions had extensive ranges and all came into contact with the park boundary. As lion survival rates did not differ between sexes (log-rank test: chi-square = 0.166, d.f. = 1, P = 0.684; males 1196.6 ± 211.6 days, 95% CI = 781.9–1611.3; females 1286.8 ± 124.4 days, 95% CI = 1043.0–1530.6), we analysed MPNP males and females together. Seven (two males and five females) out of 20 known stock-raiders were killed as stock-raiders, and one stock-raiding territorial male disappeared (and is presumed to have been killed). None of the eight non-raiders (all females) died. However, there was no statistical difference in survival between stock-raiders and non-stock-raiders (chi-square = 3.533, d.f. = 1, P = 0.06). Nevertheless, the difference in survival approached statistical significance, suggesting that stock-raiders may be at a greater risk of being killed than non-raiders (Fig. 11.12). Comparisons of survival times between raiders and non-raiders could not be computed because no non-raiders were known to have died during the study.

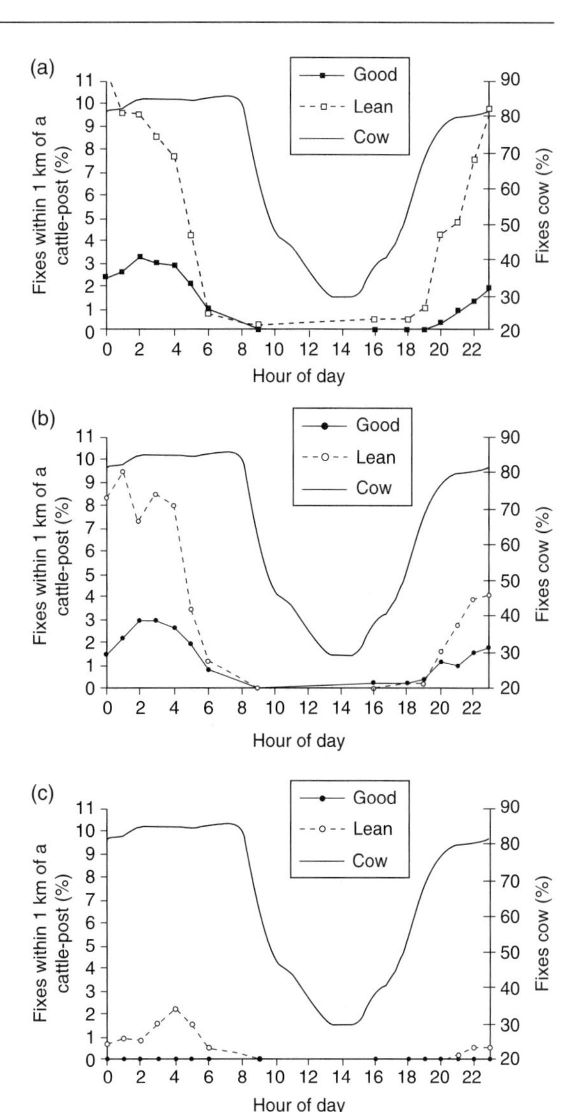

Figure 11.11 Percentage of GPS-collar fixes at different hours of the day within 1 km of cattle-posts by season compared with percentage of cattle fixes within 1 km of cattle-posts (solid grey line and right hand y-axis) for (a) stock-raiding females; (b) males; and (c) non-stock killer females.

Edge effects and livestock-raiding

In the Makgadikgadi, the lions that shared space with livestock killed livestock. While some individuals may have had a proclivity for livestock-raiding, our data suggest that any individual variation observed might

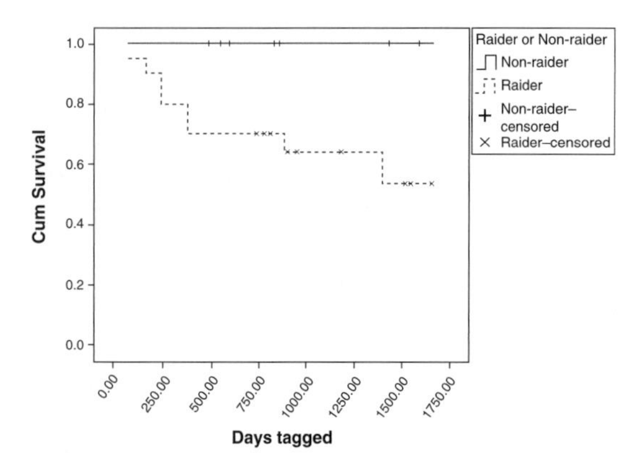

Figure 11.12 Cumulative survival of livestock-raiding ('Raider') and non-livestock-raiding ('Non-raiders') lions around Makgadigadi National Park (MPNP).

also be explained by seasonal variation in relative availability of livestock and wild prey in an individual's home range. While the degree to which any given population may turn to livestock as a food source may vary, it is plausible that livestock-raiding may occur wherever large felids come into contact with unguarded livestock during periods of wild prey scarcity (Polisar *et al.* 2003; Patterson *et al.* 2004). Seasonal patterns in livestock-raiding observed elsewhere may also be driven by changes in wild prey availability (Bauer 2003; Hoogesteijn and Hoogesteijn 2008). In some instances, high wild prey availability appears to reduce frequency of livestock predation (Cozza *et al.* 1996; Meriggi and Lovari 1996), but in others has been identified as a predictor of higher rates of livestock loss (Stahl *et al.* 2001b; Treves *et al.* 2004), perhaps because reservoirs of wild prey may sustain predator populations in mixed-use landscapes.

Problem animals were in no sense a small proportion of individuals in our study; livestock availability relative to wild prey availability appeared to drive livestock predation, and exposure of a large proportion of the lion population to high relative livestock availabilities created a large sub-population of stock-raiders. Makgadikgadi lions that regularly shared space with livestock used behaviours that increased their chances of encountering livestock, and reduced the chances of encountering people by approaching cattle-posts most closely at night. These lions frequently killed livestock. This in turn influenced the causes of mortality of these lions through trapping, shooting, and poisoning. In contrast, lions that configured their ranges to ensure access to migrant prey rarely killed livestock and were not subject to anthropogenic mortality.

Home range size for a carnivore of a given size is often determined by the availability of limiting resources such as prey (Gehrt and Fritzell 1998; Ferguson *et al.* 1999; Grigione *et al.* 2002). If home range size is an index of habitat quality, then findings from MPNP suggest that areas in which livestock and migrant prey are found are areas of better quality than are areas with only migrant prey. Livestock are probably a predictable resource that may buffer resident lions against the unpredictability in the dispersion of wild prey, particularly migratory species. This 'buffering' enables resident lionesses to use smaller areas, whereas rovers move over larger areas to ensure access to less reliably distributed wild prey. While this situation may not to be replicated in areas without migratory prey, border areas of many protected areas are likely to have higher numbers of ungulates (a combined biomass of livestock and wild prey), raising the possibility that these areas are 'better' lion habitat in terms of prey, than are areas with only wild prey. Such areas of stable prey availability may be more attractive to lions. However, when livestock are present there are also considerable consequences. The lifestyles of livestock-raiding lions had implications for their survival; all 10 (7 tagged and 3 untagged) of the adult lions that died in the study were stock-raiding residents, and at least 8 of these were killed by people in retaliations for livestock loss.

Outcomes and conclusions from the two study sites

Burgeoning human populations in many parts of the world make preservation of large carnivorous animals a major challenge for conservationists. Many large carnivores already face significant threats and most species come into conflict with people through competition for space and resources. Carnivores that predate on domestic stock may be particularly at risk. The extensive ranges of many large carnivores increase the likelihood of contact with protected area edges and with humans or anthropogenic landscape features such as roads (Woodroffe 2000; Mattson 2004). Lions studied in HNP and MPNP have some of the largest home ranges of any population of lions studied (but see Stander 2005a), making them good candidates for demonstrating edge effects in large carnivores. In our studies, wide-ranging behaviour was particularly observed in male lions, which appeared to increase their ranges freely into areas from which other males had been removed. In Hwange, males ranged extensively (101–1156 km^2 annually) when male lion density was low during the period that lions were heavily hunted. All mature males in Makgadikgadi had large ranges (206–1074 km^2) that overlapped the park boundary, despite the protected area being almost 5000 km^2. All adult males were also stock-raiders, perhaps as a result of this overlap.

Figure 11.13 Livestock owners at a cattle-post adjacent to Makgadigadi Pans National Park (MPNP) skin a lion recently killed in retaliation for livestock-raiding. © G. Hemson.

Reserve edges as 'attractive sinks'

Reserve edges may be attractive to carnivores, despite the threats present in these areas. One attractant for large carnivores is the presence of domestic ungulates. Encounters with humans under these circumstances are liable to be antagonistic, and are likely to result in mortality of carnivores, as livestock raiding often results in retributive killing by herders (Loveridge *et al.*, Chapter 6, this volume; Fig. 11.13). In Makgadikgadi, lions, particularly those living near the park edge, killed livestock beyond the park boundary, appearing to utilize livestock as an alternative food resource when the availability of natural prey was low. Thus, perversely, livestock may provide an important food subsidy for these lions during lean periods, allowing a greater density of lions to survive than the available natural prey might otherwise support. However, livestock-raiding is a risky behaviour, and lions engaging in it were more vulnerable to being killed by herders than those that did not kill livestock. In fact, all mortalities recorded in the Makgadikgadi study were due to retaliatory or pre-emptory killing by livestock herders.

Carnivore population density is often reduced by anthropogenic mortality at the interface of protected and non-protected areas (Seidensticker *et al.* 1990). Reduced levels of intraspecific competition and competition-free access to resources (e.g. mates, food resources, and denning sites) may attract carnivores from the better-protected cores of reserves into edge areas, thus exposing them to higher levels of threat. This is typified by trophy hunting of male lions in hunting concessions on the boundary of HNP. Trophy hunting reduced the density of males in boundary areas and created territorial vacuums (Loveridge *et al.* 2007c). These vacuums, and the untended female prides within them, were attractive, competition-free areas for male lions. Males from the park centre could expand their ranges into these vacant territories without resistance. However, because of the proximity to hunting concessions and the increased likelihood of being shot, boundary territories were risky for male lions. Hwange lions demonstrate how offtake at the edge can compromise the survival of viable populations, even within very large protected areas, by creating territorial voids into which core animals are drawn, effectively exposing a much greater proportion of the park's population to mortality on the boundary (Loveridge *et al.* 2007c).

Solutions and management

It is clear that large carnivore populations protected within large national parks can be profoundly affected by edge effects. Reducing mortality on park boundaries may be crucial to the long-term survival of many large felid populations, particularly those whose home ranges are large relative to protected-area size.

Hunting areas adjacent to fully protected areas may act as population sinks for hunted species and it is becoming increasingly clear that management of wide-ranging carnivore species needs to occur across extensive multiuse landscapes (Laundre and Clark 2003; Stoner *et al.* 2006). While trophy hunting can benefit conservation by setting aside habitat, raising revenue, and providing incentives which promote wildlife conservation (Loveridge *et al.* 2007c), care needs to be taken to ensure that trophy hunting is sustainable through careful monitoring of quotas, trophy quality, and population demographics (Loveridge *et al.* 2009a). Recent work suggests that harvest of male lions over the age of 6 years has little impact on population viability (Whitman *et al.* 2004). In practical terms, this requires accurate age estimation by hunters in the field; it may also require careful monitoring and aging of trophies, validation of trophy age estimates, and the means to penalize hunters who shoot under-age specimens. Although blanket measures may not always be desirable, the moratorium on trophy hunting lions around HNP has seen a marked recovery of this population, suggesting that previous quotas were unfeasibly high, but also demonstrating that lion populations, when well protected, are able to recover rapidly from perturbation.

Where livestock is kept in close proximity to protected carnivore populations, use of traditional or modern livestock-guarding practices can limit livestock loss and hence promote tolerance of carnivores by local people (Loveridge *et al.*, Chapter 6, this volume). Protective activities undertaken by livestock owners and maintenance of adequate levels of natural prey are often key determinants of whether or not livestock are lost to predation (Ogada *et al.* 2003; Frank *et al.* 2005). In addition, compensation for livestock losses and management of natural resources for the benefit of local communities may increase the tolerance local people have for carnivores and hence limit retaliatory killing.

One solution to manage human–carnivore conflict is by separation of wild carnivores and livestock by fencing protected areas. The cost of one such 240-km fence erected along the western boundary of the MPNP in 2004 was approximately US$3 million (Hemson 2003). In comparison, the total loss to herders in the surrounding area through wildlife depredation was approximately US$50,000 per annum (with an average of US$168 per household due to lion depredation; Hemson 2003). While the fence has provided the additional benefits of improved disease management and meat export quality to the cattle industry, it must be viewed as an extremely expensive conflict-management option. Personal communications from people working in the vicinity suggest that the boundary fence is already quite permeable to lions and other carnivores but that the numbers of large carnivores have increased inside the park. This suggests that the fence may have done little to limit lion movements or livestock depredation, but that its value may have been in limiting access to the protected area by humans and livestock and this may in turn provide some protection for lions.

Acknowledgements

We are grateful for the assistance of numerous field assistants and volunteers in undertaking fieldwork for these projects. We were supported by the Darwin Initiative for Biodiversity Grant 162/09/015, Disney Wildlife Conservation Foundation grant, SAVE Foundation, Regina B. Frankenburg Foundation, Rufford Foundation, Eppley Foundation, Panthera Foundation, Rivington and Joan Winant, the Mantis Collection, Installite Contracting, and Lilian Jean Kaplan Foundation. Finally, we owe our thanks to several photographic safari operators in Botswana and Zimbabwe for logistical support and access to concession land. We thank Dr A. Gosler, J. Loveridge, and A. Cotterill for their comments on the manuscript.

CHAPTER 12

Tiger range collapse and recovery at the base of the Himalayas

John Seidensticker, Eric Dinerstein, Surendra P. Goyal, Bhim Gurung, Abishek Harihar, A.J.T. Johnsingh, Anil Manandhar, Charles W. McDougal, Bivash Pandav, Mahendra Shrestha, J.L. David Smith, Melvin Sunquist, and Eric Wikramanayake

An adult tigress needs nearly 3000 kg of meat-on-the-hoof each year; feeding cubs requires an additional 50%. Tigers meet this demand by killing large, hoofed mammals—large deer, wild pigs, and wild cattle. © John Seidensticker.

The tiger's (*Panthera tigris*; Fig. 12.1) threatened status became widely appreciated 40 years ago. Since then, the human population has doubled in South and Southeast Asia (UN 2007), contributing to massive land-use change across most of the tiger's once extensive range. All signs point to an impending range collapse: (1) tigers now occupy only 7% of their historic range (Sanderson *et al.* 2006); (2) tiger geographic range has declined by 40% in only 10 years (Dinerstein *et al.* 2007); and (3) tiger numbers in India are down to ~1410 individuals >1.5 years of age (Jhala *et al.* 2008), about half the number reported in the 2001–02 national estimate (Government of India 2005). More comprehensive surveys and studies across the shrinking range confirm that we are learning more about fewer tigers.

The overall gloomy trend masks positive results in the recovery of tiger populations in a few selected landscapes. In this chapter, we report on one such effort centred in the Terai zone of southern Nepal and adjacent northern India (Fig. 12.2). These remaining forests and tall-grass savannas skirt the southern foothills of the Himalayas and contain the highest densities of tigers on earth. After providing a brief primer on Terai ecology, we summarize advances in tiger biology—many originating from

Figure 12.1 Smithsonian-Nepal Tiger Ecology Project Tigress 101 (January 1974), the first tiger captured and fitted with a radio transmitter collar. She was subsequently tracked by foot, from trained elephants, and from tree platforms (from which this photo was taken), as she lived and reared her cubs in the riverine savanna/grassland and *sal* forest in eastern Chitwan National Park, Nepal. Her territory abutted farmland with villages nearby but she did not venture beyond the natural vegetation cover in the park. She produced at least four litters. Her first territory was eventually taken over by one of her daughters and she subsequently established a new one in lower quality habitat. Tigress 101 lived to be at least 12 years old before she was poisoned by villagers. © John Seidensticker.

studies conducted in Terai reserves—that inform conservation efforts in the Terai and elsewhere. The results of these field studies have expanded the conservation focus from protecting increasingly isolated core reserves to designing conservation landscapes that can sustain wild tigers as meta-populations. Finally, we show how socio-economic interventions, in the form of community-managed forestry programmes, have provided the underpinning of landscape-scale conservation. The ambitious, overriding goal of this restoration programme, called the Terai Arc Landscape (TAL; Joshi *et al.* 2002), is to facilitate recovery of wild tiger and prey populations and to ensure that species and habitat conservation becomes mainstreamed into the rural development agenda.

Nearly 10 years since its inception, the TAL experiment illustrates how critical tiger dispersal and breeding habitat can be restored, while enhancing local livelihoods. The importance of bottom-up habitat restoration efforts, such as TAL, complement top-down, trans-national strategies and enforcement to combat the illegal trade in tiger parts, key elements required to spark a range-wide recovery of tigers.

Tigers in the Terai

Terai: an ecological primer

Unlike the mid-elevation forests of the Himalayas, the Terai forests at the foot of the mountains on the northern edge of the Ganges Plain were sparsely settled until the early 1950s. 'The Terai . . . was commonly invoked as a region almost defined by death. This tract was considered so deadly as to be impassable for Indians and Europeans alike . . . ' in the early nineteenth century (Arnold 2006, p.49). A deadly strain of malaria precluded extensive farming of the flood plain. The Nepalese Terai was considered a biological barrier to military campaigns emanating from colonial India and remained largely intact until the late 1950s and early 1960s. Widespread use of dichlorodiphenyltrichloroethane (DDT) eradicated malaria and opened the Terai as an agricultural frontier (Rose 1971).

Fanning out from the southernmost Himalayan foothills, the Terai incorporates the gravelly Bhabar and sandy areas formed from the alluvium deposited from the high mountains and the adjacent, parallel Sivalik range (in India) and Churia Hills (in Nepal; Fig. 12.3a). We refer to the low hills and flood plains collectively as the Terai. The hot, humid, monsoon (June–September) is marked by heavy rains and flooding, the latter being the major structuring force in this ecological community. Winter temperatures are occasionally near freezing in the misty, early mornings. The water table, a critical variable in determining distributions of tigers, lies 5–37 m deep. A high water table creates marshy areas that are drained by numerous southeasterly flowing rivers (Shukla and Bora 2003), and these areas attract tigers and their prey. An 'inner' Terai zone is

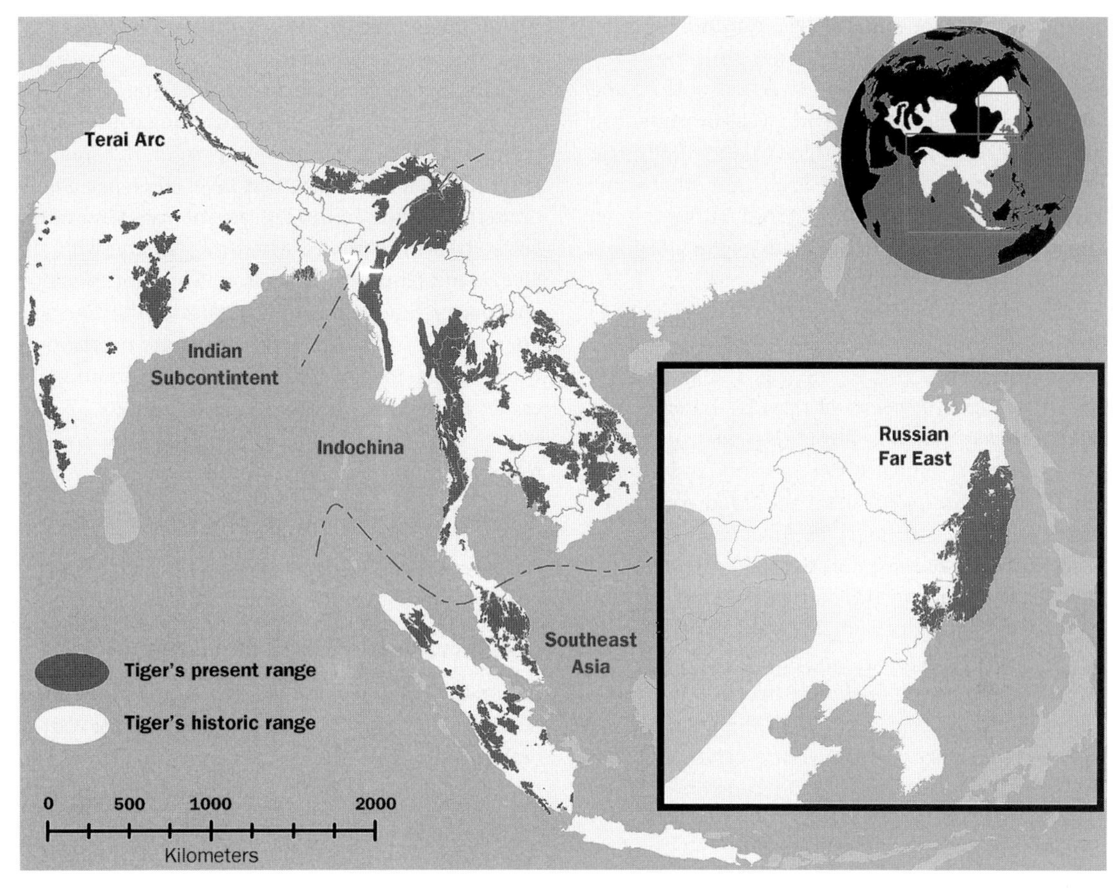

Figure 12.2 Historic (*c.* 1850) and present distribution of tigers. Tigers inhabit only 7% of their historic range today. The Terai Arc Tiger Conservation Landscape (TAL) is located in northern India and Nepal in the forests at the base of the Himalayas. Map courtesy of World Wildlife Fund-US, Wildlife Conservation Society, and Smithsonian Institution.

composed of several broad river or 'dun' valleys—the most famous being Dehra Dun in India and Chitwan in Nepal—lying between the Sivalik/Churia Hills and the outer range of the Himalayas (Kimura 1999).

Some of the world's tallest grass species occur on these flood plains and create the world's tallest grasslands, a globally rare phenomenon (Lehmkuhl 1989, Wikramanayake *et al.* 2002). Today, most of the fertile areas have been converted to agriculture, with sugar cane and rice grown widely.

A quintessential tiger landscape

In the first half of the twentieth century, chroniclers of the great hunts described an abundant tiger population in the Terai. These royal 'shikars' were staged when the threat of malaria abated during the 2-month winter season. For example, the ruling Rana family of Nepal and their guests shot 41 tigers during the 1933 winter hunt and 120 tigers during the 2-month 1937–38 season (December–January) in the Chitwan Valley (Smythies 1942). Tiger enthusiasts learned of the Terai and the adjacent Himalayan foothills from the books of F. W. Champion, who pioneered the use of flash camera traps in Asia. His classic accounts, *The Jungle in Sunlight and Shadow* (1925) and *With a Camera in Tiger-Land* (1927), offered the first outsider's glimpse of this remarkable jungle. The western Terai foothills came to life in Jim Corbett's hunting stories (1946). After World War II, forest departments of India and Nepal (after Nepal opened to outsiders in 1954) encouraged tiger hunting to generate revenue; this practice continued until the enactment of protective legislation in India in 1972 and Nepal in 1973. Before the ban, wealthy American hunters stalked tigers in the Terai and in 1968, near India's border with western Nepal, shot what some consider the largest example on record. This specimen is on permanent display in the Smithsonian's National Museum of Natural History.

Today, the 11 major protected areas (Table 12.1) established along the Terai provide snapshots of the former glory of this entire zone (Fig. 12.3b). Healthy populations of five species of cervids—barking deer (*Muntiacus muntjak*), hog deer (*Axis porcinus*), spotted deer (*Axis axis*), sambar (*Rusa unicolor*), and swamp deer (*Rucervus duvauceli*) share the flood plain with wild boar (*Sus scrofa*), gaur (*Bos frontalis*), wild water buffalo (*Bubalus bubalis*; now extirpated), nilgai (*Boselaphus tragocamelus*), and blackbuck (*Antilope cervicapra*). The steep slopes of the Sivaliks and Churias support serow (*Capricornis thar*) and goral (*Naemorhedus goral*). All serve as prey for the tiger and sympatric predators, dhole (*Cuon alpinus*) and leopard (*Panthera pardus*). Tigers also take the young calves of Asian elephants (*Elephas maximus*) and greater one-horned rhinoceros (*Rhinoceros unicornis*).

Tiger populations on the Chitwan flood plain could quickly recover from the intensive hunts of the 1930s because the grasslands and forests thriving on rich alluvium supported extraordinary prey densities. The large mammal assemblage, even with the loss of large herds of wild water buffalo and swamp deer across the lowlands, rivalled the ungulate biomass of the most productive African rift valleys (Seidensticker 1976b). Consequently, tiger densities in the protected flood plains in the Terai rank among the highest recorded (>15 tigers over 1 year of age/100 km^2; Carbone *et al.* 2001).

The flagship national parks of India and Nepal anchor the TAL from west to east. Corbett National Park was established as India's first national park in 1936 and Chitwan as Nepal's first park in 1973, both of which are now World Heritage Sites. The other core areas in the Terai were gazetted after 1972, although the Nepalese core-protected areas served as royal hunting reserves in the previous decades. All of the protected areas contained villages that have since been relocated.

There are extensive blocks of recovering, selectively logged *sal* forests in the ~1000 km long conservation corridor of the TAL, extending from the Yamuna River in India to the Bagmati River in Nepal. Development pressures in the TAL are intense. Most of the larger rivers are dammed where they emerge from the Sivalik and Churia Hills to store water for irrigation and power generation for the cities of the Gangetic Plain. Five of 12 Terai forest divisions overlapping with the TAL in India hold large-scale plantations of exotic softwoods (Johnsingh and Negi 2003). The highly productive, tall-grass savannah and riverine forests that previously characterized the sandy areas of the Terai have been largely reduced to those included within the major protected areas.

Conservation science challenge

Restoration of tiger populations across their range poses one of the greatest challenges in conservation biology. The few remaining populations survive in mostly fragmented forest landscapes and recovery entails creating large spaces with abundant prey that are free of intensive human disturbance (Seidensticker *et al.* 1999). In the TAL specifically, the 11 reserves nest within a human-dominated landscape of extreme poverty and many landless people. Thus, tiger conservation faces the added challenge of addressing land tenure issues, local rights, and the legitimate economic aspirations of the rural poor (Tharoor 2007). At the other end of the economic spectrum, a rising Asian middle class has created economic conditions and markets that drive rampant poaching and illegal trade in tiger parts (Dinerstein *et al.* 2007; Damania *et al.* 2008).

Conservation efforts in any tiger landscape, whether in the TAL, the Russian Far East (RFE), or the Sundarbans in the Gangetic delta of India and Bangladesh, must address two fundamental questions: what controls tiger population numbers, and what factors affect the probability of population persistence? These questions concern the tiger's ecology and the mix of ecology and human values. Years of experience tells us

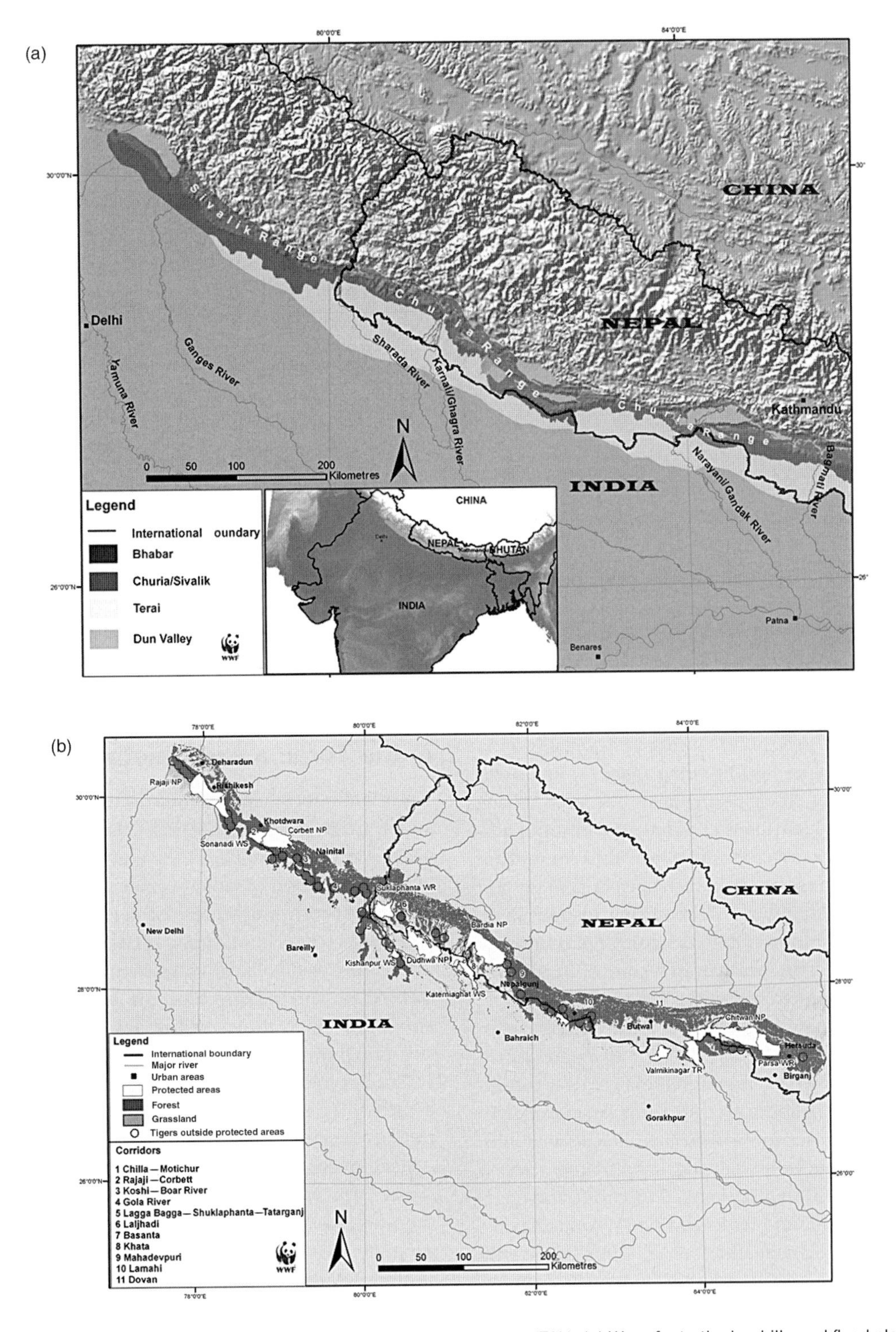

Figure 12.3 Features of the Terai Arc Tiger Conservation Landscape (TAL). (a) We refer to the low hills and flood plains at the base of the outermost ranges of the Himalayas collectively as the Terai. The Terai incorporates the gravelly *Bhabar*, sandy river flood plains, the parallel Sivalik range (in India) and Churia Hills (in Nepal), and the 'inner' Terai zone composed of several broad river or 'dun' valleys lying between the Sivalik/Churia Hills and the outer range of the Himalayas. (b) TAL land cover, protected areas, locations of tigers found living outside the protected areas, and key bottlenecks that must be maintained as connecting corridors to enable the dispersal of tigers and other large mammals across the landscape.

Table 12.1 Protected areas in the Terai Arc Landscape (TAL).

Major protected areas: east to west	Date established	Size (km^2) (additional buffer)
Parsa Wildlife Reserve, Nepal[a]	1984	499 (298)
Chitwan National Park, Nepal[a]	1973	932 (750)
Valmiki National Park, India[a]	1989	336 (348)
Sohelwa Wildlife Sanctuary, India	1988	453
Bardia National Park, Nepal[a]	1977	968 (327)
Katarniaghat Wildlife Sanctuary, India	1976	551
Dudhwa National Park, India[a]	1958	490 (124)
Kishanpur Wildlife Sanctuary, India	1972	227
Suklaphanta Wildife Reserve, Nepal[a]	1975	305 (243)
Corbett Tiger Reserve incorporating Corbett National Park and Sonanadi Wildlife Sanctuary, India[a]	1936	1288
Rajaji National Park, India[a]	1983	821
Total		6870

[a] Core tiger-breeding areas.

that tiger populations can be restored by (1) stopping the killing of tigers; (2) reducing anthropogenic impacts on the core tiger-breeding areas; (3) increasing the number of core-breeding areas; (4) reconnecting the cores through forest corridors; (5) ensuring adequate prey populations in cores and connectors; (6) creating the institutional arrangements to support these interventions; and (7) practising adaptive management and innovation in designing field conservation efforts. These measures in whole, or in part, are being implemented within the TAL today.

The proximate challenges of maintaining tigers, even at the local level of protected areas, are daunting. Insular reserves, the strongholds of tiger conservation efforts in South Asia, are rarely large enough to ensure the conservation of ecologically, demographically, and genetically secure tiger populations that are resilient to disturbance events (Woodroffe and Ginsberg 1998; Soulé and Terborgh 1999; Wikramanayake *et al.* 2004; Ranganathan *et al.* 2008. Not only must tigers be able to disperse among the reserves, but conservation interventions must also seek to expand tiger populations in all core areas and encourage breeding between core areas through habitat recovery. These principles underpin the TAL programme.

Science-based tiger conservation in the Terai: a short history

Tiger field research has its roots in the Terai jungles, and in Chitwan National Park in particular. Science-based work on tigers began there in 1972, catalysed by the innovation of radio-telemetry. Following collared or known individuals, Seidensticker (1976a), McDougal (1977), Sunquist (1981), and Smith (1993) followed collared or known individuals and focused on quantifying tiger life history. Baseline knowledge on mating, rearing, foraging, dispersal, and refuging (after Eisenberg [1981]) emerged. In addition, ecological responses to differences in prey density, prey species composition, and habitat associations were revealed.

Field investigations need to adapt to local circumstances. The first scientists in Chitwan assumed their site would be a remote, pristine wilderness in which to study free-ranging tigers. Their naiveté ended abruptly during the reconnaissance phase: Chitwan's tigers inhabited a small island (~1000 km^2) in the middle of a human-dominated landscape (Seidensticker *et al.* 1999). This situation proved typical across many South Asian wildlands, where dynamic land-use change was already underway. Accordingly, field research refocused on the numerical and behavioural responses of tigers, leopards, and their large ungulate prey at the park–village interface as a critical step in understanding the persistence and extirpation of large mammals in Asian landscapes.

In 1972, McDougal and colleagues began the longest running longitudinal study of tigers that continues to this day. McDougal's study linked closely with the Smithsonian-Nepal Tiger Ecology Project founded in 1973 by H. Mishra, J. Seidensticker, R. Simons, and K. M. Tamang, and was followed in subsequent years by M. Sunquist, J. L. D. Smith, and E. Dinerstein (2003). Dinerstein (1979a, 1979b, 1980) first investigated tiger ecology on the flood plains of Nepal's Karnali River in 1975, work that was continued by P. Wegge and associates (2004) beginning in 1992. D. Smith, M. Shrestha, B. Gurung, and E. Wikramanayake (Smith *et al.* 1999; Wikramanayake *et al.* 2004; Gurung *et al.* 2006b) investigated tiger occupancy along the length of the Nepalese Terai to describe tiger meta-population structure.

Less progress was made in understanding tiger ecology in the Indian Terai until the Wildlife Institute of India (WII) was formed in Dehra Dun, adjacent to Rajaji National Park, in 1982. A. J. T. Johnsingh and associates documented tiger occupancy and breaks in forest connectivity along the length of the Indian Terai (Johnsingh and Negi 2003; Johnsingh *et al.* 2004). A. Harihar, A. J. Kurien, B. Pandav, and S. P. Goyal (Harihar *et al.* 2007) from WII documented the recovery of tiger and prey populations in the Chilla Range within Rajaji National Park following the voluntary resettlement of graziers out of the areas in 2002.

Deconstructing the resilience of the Terai Arc Landscape

As noted by one of the earliest Terai researchers, being a tiger biologist requires one to become an ungulate biologist (M. Sunquist, personal communication), because ungulate distributions hold the key to understanding tiger ecology. Through a series of studies (Chitwan: Seidensticker 1976b; Mishra and Wemmer 1987; Suklaphanta: Schaaf 1978; Bardia: Dinerstein 1979a, 1979b, 1980) a better comprehension of population dynamics, habitat preference, and life history of tiger prey emerged. The impacts of human disturbance on the forests in the western Sivaliks and the manifestation of these impacts on ungulate densities and, ultimately, on tiger densities were reported by Harihar *et al.* (2007). Dinerstein (1987) took Sunquist's argument one step further: to be a tiger or ungulate biologist, we needed to be plant ecologists, because understanding successional patterns and the natural disturbance factors that shape the vegetation is key to tiger and prey population dynamics, habitat preference, and recovery. The resilience of the Terai ecosystem, as it turns out, is linked to how dominant plant species respond to habitat disturbance through annual fires and monsoon flooding.

To this end, we discuss critical elements of ecosystem resilience and the potential to conserve a viable tiger meta-population in the TAL. First, we briefly describe natural disturbance regimes and vegetation dynamics. Second, we illustrate how the distribution of plant communities and prey distributions influence tiger behavioural systems and life history characteristics (work done primarily in Chitwan). Third, we assess tiger range collapse in the Terai and the potential for recovery. Finally, we document responses by tigers to newly available habitats regenerated by interventions through community forestry programmes in the Nepalese TAL.

Plant succession in Terai forests and grasslands

Plant succession and vegetation responses to disturbance regimes in the Terai were initially modelled by Dinerstein (1987). The major vegetation types along the riverbanks are tall grass, riverine forest, and derived savanna grasslands. Away from the rivers, *sal* (*Shorea robusta*) forest is dominant, often in pure stands or mixed with other tree species. Tall-grass areas and early riverine forests are generally flooded during the monsoon. Moist-mixed riverine forest is more common where flooding is less severe but where the soil remains waterlogged during the monsoon. Secondary open mixed forest is common along rivers, where soils are deeper and better drained than in riverine forest. This type often abuts mature *sal* forest with shrub and grass understories that grows in dryer soils away from the rivers. Open *sal* forest occurs on the dry, steep, south-facing slopes of the Churia and Sivalik Hills.

Key to restoring prey and tiger populations is conservation of the earliest successional stage of the flood plain ecosystem. The riverbeds are dominated by near mono-specific stands of the grass *Saccharum spontaneum*, a wild sugar cane, and support the highest ungulate biomass and highest density of breeding tigers in Asia. This dominant association is adapted to annual, severe disturbance events: monsoon floods. *Saccharum spontaneum* sprouts new shoots soon after inundation by floods, and also after cutting, grazing, or burning. Because it remains green all year long, it is a major food plant for large ungulates.

Without human interference, the successional sequence is from short grass to tall grass and from diverse successional forests to continuous mature forests. The shift from short grass and intermediate-sized *Imperata cylindrica* to tall elephant grasses results in lower deer densities, particularly spotted deer because the grasses become less palatable and unreachable as they mature. Areas of the tall-grass *Narenga porphyracoma* and *Themeda* spp. support the lowest ungulate biomass among grassland types.

Burning, cutting, and domestic livestock grazing alter community structure of *sal* forest, early riverine forest, moist-mixed riverine forest, and secondary open mixed hardwood forest. The result is a derived savannah grassland type favourable to deer and other ungulates. *Sal* is fire resistant and its seedlings are much less palatable to deer and other ungulates than other tree species, which may explain its emergence as the dominant vegetation type. But where *sal* forest has been cleared, derived savannahs can be maintained through annual burning. The intermediate-sized *Imperata* grass, used locally for thatch, dominates old village sites for a number of years depending on the grazing pressure and water table. Where there is a high water table and low grazing pressure, *Imperata* grassland is eventually replaced by tall grasses such as *Narenga porphyracoma*. Heavy grazing pressures by domestic livestock combine with annual fires to arrest grassland succession; when protected from these pressures, these tall grasslands recover rapidly.

Spotted deer mostly frequent derived savannah grasslands and early riverine and moist, mixed riverine forests, whereas swamp deer and hog deer prefer short or tall grasslands. Sambar frequent mature *sal* forest and riverine forest. Nilgai and blackbuck survive only in short grasslands and are quickly killed by tigers in tall grass. Wild boar, rhinoceros, and water buffalo also prefer *Saccharum spontaneum* grassland and secondary forest to mature *sal* forest.

The predominant invasive species that reduce the carrying capacity of Terai habitats are *Lantana camara, Chromolaena odorata, Cannabis sativa*, and *Mikania micrantha. Mikania* first appeared in Chitwan in the late 1990s and has taken over large areas of riverine forest and grasslands; wild ungulates avoid it. *Lantana*, with little forage value, is found across the Terai, especially in sandy soils near river banks heavily overgrazed by domestic livestock. *Cannabis* grows naturally in many areas but has also escaped from cultivated lands and encroached into the grassland of some protected areas (e.g. Corbett Tiger Reserve), where managers expend great energy and resources trying to remove it (J. Seidensticker, personal observation). The threat of invasion of *Saccharum spontaneum* grasslands by exotic species combined with land clearing and overgrazing by domestic livestock suggest that the most critical habitat for tiger restoration in the TAL is also the most threatened.

Variation in the tiger's behavioural and life history traits in response to resource distribution

We summarize the modal characteristics of the tiger behavioural systems and life history strategy in Table 12.2, based largely on data from Chitwan.

Feeding ecology

Tigers are the largest obligate meat-eaters in Asian wildlands, with population density limited by available large ungulate biomass. Therefore, prey density is the principal driver of tiger density in systems largely unaffected by poaching of either tigers or prey. Karanth *et al.* (2004c) developed a model that

Table 12.2 Modal tiger life history and behavioural systems.

Body size	Males (235 kg, 200–261) 169% > females (140 kg, 116–164)	Sunquist and Sunquist (2002)
General activity	Usually active and moving between sunset and sunrise although they sometime hunt during the day	Sunquist (1981)
Social system	Dispersed; Eisenberg social complexity rating: 2 lowest out of 20 Females defend stable home ranges or territories and are intolerant of conspecifics except dependent offspring and during sexual receptivity; females do relinquish portions of their exclusive home ranges or territories to mature daughters, after which there is no persistent association. Female territory size tracks available prey biomass Breeding males defend territories that overlap those of several breeding females. Size is dependent on the assertiveness of adjacent males The transient portion of the population is composed of post-dispersal, non-breeding adult animals that move through in their search for open territories	Sunquist (1981); Smith *et al.* (1987)
Mating	Polygynous; little or no synchrony of reproduction among females; individual males establish and maintain territories over a number of years; male defence of a mate is transitory but he excludes other adult breeding males from his territory, which is large enough to overlap a number of female territories. This restricts a female's chance to breed with males other than the resident territorial male, unless her territory is overlapped by two males. Females maintain individual stable territories that may be subsequently subdivided among female offspring	Sunquist (1981); Smith *et al.* (1987); Smith and McDougal (1991); Barlow *et al.* (2009)
Rearing	Gestation is 103 days; lactation is 90 days. Single adult female lives on a stable territory and raises her young alone; no direct male care, but territorial males exclude other males, resulting in increased cub survival; females mobility is restricted during the first months of her cubs' life; the females begins to move with cubs when they are 2 months of age	Eisenberg (1981); Sunquist (1981); Smith (1993)
Dispersal	Tigers disperse at 23 months (19–28) of age, 1.6 months after a new litter is born. Males disperse further than females and settle in poorer habitat. There is a tendency towards female-biased philopatry by offspring, resulting in female kin clustering with a mean degree of relatedness between females of 0.35	Smith *et al.* (1987); Smith and McDougal (1991); Smith (1993)

(Continued)

Table 12.2 (Continued)

Foraging	Low seasonal flux in the food base that may be highly predictable in some habitats and unpredictable in others; solitary staking and ambush hunting style; high defence of resources: territorial males and female hunt alone (or with their larger offspring) in their exclusive areas; the resident territorial male areas overlap those of the females. No cooperative hunting between adults; females with large offspring may show some cooperation when hunting very large prey; food sharing between females and cubs, even non-reproducing subadults sharing her territory; males dominate over the females at kills	Sunquist (1981); Seidensticker and McDougal (1993)
Refuging	Both males and females 'lie up', resting during the day in thick cover, sometimes in pools of water during very hot periods; females conceal their very young cubs in sheltered areas but move them to new sites frequently	Seidensticker (1976a); Sunquist (1981)
Anti-predatory behaviour	Individualistic; cubs are vulnerable to newly established territorial male tigers, perhaps dhole packs, and fires set by people to burn them out	Sunquist (1981); Smith and McDougal (1991)

predicts tiger densities based on prey densities. Their key finding was that prey densities in alluvium, such as in the Kaziranga National Park in the Brahmaputra River Valley of India, supported among the highest tiger densities recorded (16.76 > 1 year of age/100 km^2). Kaziranga is similar in vegetation to the marshy areas of the Terai flood plains.

Tiger prey items span a wide range in body mass (20 to ~1000 kg) but, on average, tigers kill ungulates in the ~100 kg size class (Seidensticker 1976a, Seidensticker and McDougal 1993). Although tigers kill prey as encountered when hunting, they selectively seek and kill large-bodied ungulate prey, thereby gaining access to considerable portions of the ungulate biomass attributed to a relatively few individuals. The average kill rate is ~50 ungulates/adult tiger/year, while a tigress with large cubs kills 60–73 ungulates/year. They remove about ~10% of the standing ungulate biomass annually (Sunquist 1981). However, tigers are frequently deprived of kills by humans. This kleptoparasitism is a significant factor (e.g. 10 tiger kills were taken by people in an 8-month period in Sunquist's study area) that determines kill rates and food availability both inside and outside protected areas (Sunquist 1981; A. J. T. Johnsingh and E. Dinerstein, personal observation).

Across South Asia, ungulate biomass varies with water availability, soil type, vegetation cover, accessibility to crops, urbanization, and protection (Eisenberg and Seidensticker 1976). Wild ungulate biomass in Chitwan, and across the Terai, follows a gradient, with highest densities and biomass in riverine forest and tall/short-grass mosaic grazing areas on newly recovering flooded sites (67 individuals/1 km^2, 2800 kg biomass/1 km^2). This contrasts to lower densities and biomass found in *sal* forest-clad hills but near permanent water (18 individuals/1 km^2, 1300 kg biomass/km^2; Seidensticker 1976b). Radio-collared tigers on the flood plain avoided open farmlands, and only occasionally visited adjacent *sal* forests (Sunquist 1981; Smith 1993). In Chitwan, the numerical response to this habitat gradient for resident adult-breeding tigers was dramatic: the density of tigers living in tall-grass savannah/gallery forest was 7/100 km^2, while in mixed riverine/tall-grassland savannah and low elevation *sal* forest it was

5/100 km^2, and dropped to 2.7/100 km^2 in upland *sal* forest.

Tiger sociality

Briefly, the resident-breeding tiger population in Chitwan lived in exclusive home ranges (or territories), while the transients were those unable to establish territories (Table 12.2). Tiger territories averaged 23 (13–51) km^2 for females and 68 (33–151) km^2 for males and were defended both directly and indirectly against other same-sex adults (Sunquist 1981; Smith *et al.* 1989), but with substantial overlap between opposite sexes. Territory size for males were as large as 151 km^2 to as small as 33 km^2 over time, depending on the number of resident males that were living on the park's flood plain (Smith *et al.* 1999). Sunquist (1981) found that a single male territory overlapped those of as many as five breeding female territories and Smith (1993) found one resident male territory overlapped those of seven breeding females. Typically, one male's range overlapped three or four female territories (Table 12.2). Breeding males attempted to mate with as many females as possible, and performed no parental activities.

The area occupied by breeding adults also contained non-breeding adults, which formed ~20% of the population. In 1976, for example, Sunquist (1981) and Smith (1993) found that in their flood plain study area in Chitwan there were 15 resident adults, 14 pre-dispersal offspring, and 3 post-dispersal transient tigers. Karanth *et al.* (2006) estimated the transient portion of the tiger population to be 18% in Nagarahole National Park, India. These were usually young (2–4 years of age), post-dispersal individuals, or uncommonly, old tigers, usually females, displaced from their territories.

Breeding system and tiger lifetime reproductive success

Lifetime reproductive success of multiple females in the population has yet to be documented. But the tigress Chuchchi in Chitwan produced, over 15 years, 5 litters and a total of 16 cubs, 11 of which survived to dispersal age (McDougal 1991). Chuchchi's three daughters settled locally, all producing cubs; one in the territory of an adjacent female following her death, a second daughter claimed the northern portion of her mother's range, and the third dispossessed her then 14-year-old mother of the southern portion of her range. Chuchchi then roamed widely outside her regular home range, although the actual extent of these movements was uncharted. Eventually, she was killed by a young male, the son of an unrelated neighbouring tigress, when she returned to her former territory. Three sequential territorial males whose ranges overlapped hers sired her five litters and four others may have bred or attempted to breed with her but those litters perished (McDougal 1991). Sunquist (1981) estimated optimal lifetime production for a tigress living in this habitat as 13–18 cubs, with about half that number reaching adulthood. In the RFE, optimal lifetime production for a tigress was 12.1 cubs but only 6.5–7.3 cubs reached 1 year of age (D. Miquelle, personal communication).

Impact of prey density on tiger density and reproductive success

Over the past decade, monitoring programmes have demonstrated that naturally occurring tiger densities vary by a factor of 30, from <0.5/100 km^2 (tigers 1 year of age and older) in the temperate forests of the RFE to 15+/100 km^2 in the flood plain tall-grassland savannah and riverine forests of the Terai. The number of potential prey animals varies from <2/km^2 (<400 kg/km^2) in the RFE to >60 individuals/km^2 (2800 kg/km^2) in the high productivity habitats in India and Nepal. Low prey density translates to lower encounter rates with prey, greater search time, and higher energy expenditures per kill and ultimately, less total energy in the ecosystem to support top carnivores (Sunquist and Sunquist 2001).

In the harsh environmental conditions of the RFE, mean litter size was 2.4, but decreased to 1.3 at 1 year (Kerley *et al.* 2003). In contrast, in a high prey density area like Chitwan, mean litter size was 2.98 (variance 0.48), decreasing to 2.45 (variance 0.62) by year one, and 2.22 (variance 0.51) by year two (Table 12.3).

Table 12.3 Reproductive parameters of tigers living in high- (Chitwan) and low-density Russian Far East (RFE) habitats.

Study site	Age at first reproduction females (years)	Mean litter size	Inter-birth interval (months)	Mean dispersal age (months)	Litters per lifetime
RFE	3.5–4.5	2.5	21.8	18.8	5
Chitwan	3.4	2.98	21.6	23	4–5

Sources: RFE—Kerley *et al*. (2003) and Dale Miquelle (personal communication, 2008); Chitwan—Smith and McDougal (1991) and Smith (1993).

The mean lifetime production of breeding offspring by female tigers in Chitwan was 2.0 (variance 3.26; Smith and McDougal 1991).

A major source of cub mortality is infanticide, which occurs when a resident-breeding male is usurped by another (Smith and McDougal 1991). Also, when young cubs are born during the dry season, their refuge sites in tall grass are subjected to human-caused fires that contribute to cub mortality. Finally, when a tigress is poached, cubs younger than ~15 months perish when left on their own (Smith 1993).

Karanth and Stith (1999) created a life-stage-based stochastic model to show that prey depletion has a strong negative impact on tiger populations by decreasing cub survival, reducing the carrying capacity for breeding adults, and decreasing population size. The exact ages of tigers in any wild tiger population is unknown, so they used a model with four life-stages for both sexes: breeder, transient, juvenile, and cub. Breeding female tigers were assumed to have stable, exclusive home ranges and their densities positively correlated with densities of ungulates. To reflect the size of actual Indian subcontinent tiger reserves, they set the carrying capacity (K) for tigresses from a high of 24 to a low of 3. Based on this range of tigress-carrying capacities, they modelled the possible effects of tiger poaching and prey depletions by depressing survival rates of the affected tiger's demographic stage. From this model, they found that the persistence of tiger populations is a function of the high reproductive potential of tigers.

More recent work by Damania *et al.* (2008) shows that even in high-density prey areas, modest-sized tiger populations (~30 breeding individuals) are fragile, unstable, and vulnerable to extinction and can be expected to crash with a 2% mortality rate; larger populations (>80 breeding adults) can withstand a 10% mortality rate. Prey-deprived tiger populations collapse rapidly with a very small increase in poaching because the carrying capacity for breeding females is depressed, cub survival drops, and tiger numbers decline rapidly, mainly due to additional effects of lowered cub survival. Thus, prey depletion may significantly constrain tiger recovery.

The pattern is clear: where prey densities are low, as in the RFE, the result is low tiger reproductive output and a greatly diminished resilience in the tiger meta-population. The resilience of the Terai landscape to sustain a tiger meta-population is greatly reduced when and where prey populations are depleted through poaching. At present, resilience in the Terai landscape is now dependent on the few secure areas where prey densities are high enough for a tigress to support litters. The recovery of prey populations is a prerequisite for tiger population recovery.

Dispersal system

Understanding tiger dispersal was greatly facilitated by the use of light aircraft to follow radio-collared tigers. In Chitwan, Smith (1993) found that subadults dispersed when they reached 19–28 months of age, ~45 days after the birth of their mother's new litter. Males dispersed on average 33 km (greatest distance = 71 km) and females ~10 km (greatest distance = 43 km). Tigers avoided open farmland; indeed, none of 7000 locations of radio-collared tigers recorded by Smith (1993) occurred in cultivation or villages. Young adult females remained within their mother's home range in better quality habitats;

a total of 16 of 25 females tracked settled near their mothers (Smith 1993). These young females took another year to establish exclusive breeding territories with intensively marked boundaries. One young female that left her natal area shifted localities 13 times over 8 months before settling 33 km west of her natal area. In contrast, only 6 of 32 (19%) subadult males settled near or within their natal area and only 40% of males survived to become established breeders. Dispersing males occupied marginal habitats and temporary post-dispersal territories ranging from 17 to 70 km^2 until able to evict established adult males from prime habitats (Smith 1993).

Intra-guild interactions

Tigers, leopards, and dholes typically coexisted in the riverine forest and tall-grass vegetation of the Terai. Dholes now occur in Chitwan but are absent from the other TAL reserves. Tigers are behaviourally dominant to leopards and dholes (Karanth and Sunquist 2000). The two large cats differ in modal body size by a factor of about four and in average size of prey killed (leopards ~25 kg and tigers ~100 kg). However, they prey on different size classes of the same species. Leopards differed from tigers in their activity periods and in their microhabitat use (Seidensticker 1976a).

High tiger density appears to depress leopard density in Chitwan and in Rajaji (McDougal 1988; Harihar *et al.* 2007). For example, when the density of resident-breeding tigers occupying prime tall-grass savannah/riverine forest in western Chitwan increased to 8/100 km^2 in 1987, six leopards were killed by tigers over a 21-month period (McDougal 1988). Leopards avoided areas where tiger density was high, instead residing at the peripheries of the park between the high-density tiger areas and croplands. Where tigers have been extirpated in the tall-grass Indian Terai, leopards have recolonized (Johnsingh *et al.* 2004). India's Nagarahole National Park offers a contrast to Chitwan, where both leopard and tiger abundance was facilitated by abundant prey in a wider range of size classes (Karanth and Sunquist 2000). Wild ungulate prey outside the park is often quite low, so both tigers and leopards subsist on domestic ungulates—goats and calves of cattle and domestic water buffalo—and for leopards, domestic dogs (Seidensticker *et al.* 1990).

Tiger genetics and small population size

Smith and McDougal (1991) modelled the rate of inbreeding in the Chitwan population and postulated that it increased by 2% per generation. Kenney *et al.* (1995) warned that even if a relatively low level of poaching continued over time, the probability of extinction increases sigmoidally because as the duration of poaching increases, the probability of extinction increases rapidly. At lower levels of poaching pressure, a small incremental increase has little effect, but there is a critical zone of poaching pressure in which a small incremental increase greatly increases the probability of extinction in a population the size of that living in Chitwan—and this is one of the largest tiger populations on the Indian subcontinent. At any particular level of poaching pressure, the probability of extinction varies with the duration of poaching. But even if a tiger population survives a period of poaching and fully recovers to pre-poaching population levels, loss of genetic variability due to the rapid population decline during poaching creates a bottleneck effect. This, in turn, creates conditions where the coefficient of inbreeding will increase rapidly in tiger subpopulations.

Tiger numbers in the Terai are depressed because tigers are subjected to depressed prey populations, increased mortality imposed by poaching, and substantial habitat fragmentation, degradation, and loss. The risks to tiger subpopulation viability because of the loss of genetic diversity were initially thought to be of concern but not to pose the substantial and immediate additive threat that the modelling by Kenney *et al.* (1995) suggests.

Insights from field studies for the design and management of the Terai Arc Landscape

The TAL design builds on insights derived from over three decades of field work on tigers and their prey.

The TAL has three key goals: (1) to restore connectivity for tigers across the 11 protected areas embedded in a 1000 km long landscape; (2) to increase the breeding population of tigers to at least 500 by 2020; and (3) to harmonize location of tiger-breeding areas and dispersal corridors between reserves with the rural development agenda.

Restoring connectivity for tigers

The key insight gleaned from our studies of tiger demography and conservation genetics is that small reserves alone are inadequate to restore tiger populations; resilience and recovery will require a landscape-scale approach. Yet, to reassemble the TAL, we must first understand the extent, locations, and rate of its fracturing.

Half a century ago, before malaria was eradicated in the TAL, tigers probably formed a single interbreeding population. About 15 years after eradication, the first Director of Project Tiger interviewed forest officers in the Terai in order to map the tiger population (Sankhala 1977). His map, created in 1972, identified three subpopulations: (1) the westernmost component extending from the Yamuna River in the west—spanning Rajaji, Corbett, and Dudhwa National Parks in India and the Suklaphanta Wildlife Sanctuary, Nepal—to the Karnali River in the east, forming the western boundary of Bardia National Park, Nepal; (2) a spatially distinct population occupying the Sivalik/Churia Hills along the India–Nepal border between the Karnali and the Narayani (Gandak in India) rivers; and (3) a segregated population in the Chitwan Valley and the adjacent Parsa Wildlife Reserve in Nepal and Valmiki National Park in India (Fig. 12.3b). Although Sankhala's map depicts three distinct populations, contiguous forest cover in 1972 probably allowed tigers to disperse between them.

The tiger's vulnerability to even moderate habitat fragmentation and dispersal barriers created by roads and other infrastructure became apparent during the 1980s through the work of Seidensticker (1986) and Smith (1993) and this vulnerability was confirmed by Kerley *et al.* (2002) working in the RFE. The rapid expansion of agriculture and subsequent clearing of forest reduced connectivity, as did the growth of villages and towns along the Terai rivers. Road networks in rural areas grew from dirt tracks crossable by tigers to asphalt highways with truck and car traffic. Hydroelectric and irrigation projects, which are substantial barriers for tigers, appeared on all the major rivers emerging from the Himalayas, excepting the Karnali. Improved irrigation along the base of the mountains attracted new settlers from the hills and the Gangetic Plain. Human population density in the Terai grew dramatically to reach very high levels for a rural landscape (~500 people/km^2 in India, about half that in Nepal).

Taken together, these pressures resulted in massive land-use change and increasing insularization of tiger reserves. Geographic information system (GIS) tools allowed Dinerstein *et al.* (1997) to assess the impact on connectivity and meta-population function. They hypothesized, based on forest cover, infrastructure, and natural barriers that by 1995 tigers had been further divided into at least seven subpopulations in the Terai, which they called Tiger Conservation Units. Smith *et al.* (1999) hypothesized four subpopulations of tigers living along the Nepalese Terai, based on land-cover maps used to assess the extent and quality of tiger habitat, differential dispersal ability through a given habitat, and ground surveys. Relying on similar land-cover data and occupancy surveys, Johnsingh *et al.* (2004) hypothesized only two tiger-breeding units in the Nepal Terai and three in the Indian Terai from the western Nepal–India border to the Yamuna River in the far west. They did not report on occurrence of tiger breeding outside reserves, but found that tiger occurrence was 'patchy and variable, influenced by prey availability, cover and disturbance levels' (Johnsingh *et al.* 2004, p. 54). They also identified 18 sites where the habitat was either highly degraded or constricted, threatening connectivity and restricting tiger movement across the TAL. In the most recent assessment of landscape connectivity, Sanderson *et al.* (2006) delineated seven tiger subpopulations in the TAL utilizing the most recent land-cover data available, human influence data, tiger dispersal distance, area of tiger presence, and minimum core/area stepping stone size (Table 12.4).

These studies depicted the 'landscape pathology' of this once healthy, intact tiger landscape. The study by Wikramanayake *et al.* (2004) provided a scientific

Table 12.4 Terai Tiger Conservation Landscape (TCL) spatial and land-cover parameters. Protected areas and buffer zones (Table 12.1) comprise ~35% of the total area identified as TCLs.

TCL name	Total area (km^2)	Habitat area (km^2)	Largest patch (km^2)	Habitat as percentage of landscape area
Chitwan	4055	1257	560	31
Bardia South	499	206	83	41
Bardia	6777	3272	740	48
Suklaphanta	1144	467	300	41
Corbett	5996	1758	250	29
Rajaji	1044	301	172	29
Yamuna	322	120	82	37
Total	19,837	7,381		37

Sources: Dinerstein *et al*. (2006) and Sanderson *et al*. (2006).

framework for landscape recovery for tigers and became the blueprint for the TAL. Their model identified potential dispersal corridors and strategic transit habitat for tigers, and made recommendations for land management and restoration beyond reserve boundaries to enhance dispersal. Relying on Landsat satellite images, they used an ArcView-based cost-distance model to establish a grid (1 ha) and scored each grid with a 'dispersal cost' based on habitat parameters and the probability of tigers using those habitats. Combining these metrics, they identified and classified corridors into one of three classes reflecting the probability of use depending on the dispersal cost. The corridor analysis was expanded to include strategically placed 'stepping stones' and community forests that would restore and improve dispersal value of corridors (Fig. 12.4).

Adoption of the model by the Nepal government and donor agencies allowed field workers to target corridor restoration in critical bottlenecks. Initial results have been quite promising. Within 5 years of the forest restoration efforts, tigers were detected in four of the five corridors, including a tigress with cubs in the Basanta forest corridor that connects the Churia Hills in Nepal with the Dudhwa Tiger Reserve in India, indicating that tigers may actually be resident in these linkages. The Khata corridor is a second transnational corridor where tigers have been confirmed following restoration efforts. This corridor connects Nepal's Bardia National Park with Katerniaghat Wildlife Sanctuary in India (Fig. 12.4; Wikramanayake *et al.* 2010). So far in the Nepalese Terai, six bottlenecks to tiger dispersal were identified for restoration and 200 local Community User Groups re-established 220 km^2 of forest corridors. Gurung *et al.* (2006b, personal observation) found evidence of tigers breeding outside protected areas in eight locations in the Nepalese Terai. This latter finding helps validate a major assumption of the TAL, that breeding populations can exist outside the 11 gazetted tiger reserves.

The least-cost model and, most importantly, its successful implementation show that GIS-based models can be useful tools to gain government support and to identify tiger conservation landscapes and corridors for recovery. A major issue that remains, however, is monitoring occupancy and meta-population structure across a 1000 km conservation landscape.

Increasing the breeding population of tigers to at least 500 by 2020

Population recovery in the TAL is anchored by seven core protected areas or contiguous protected areas (Table 12.1) supporting abundant prey that are relatively free from human disturbance. In these refugia, tigresses rear cubs that will eventually disperse to other areas through stepping stone or continuous habitat corridors. These subpopulations with breeding tigers are from west to east: Rajaji, Corbett, Suklaphanta, Dudhwa, Bardia, Katerniaghat, and Chitwan/Parsa/Valmiki (Fig 12.3b). The forested

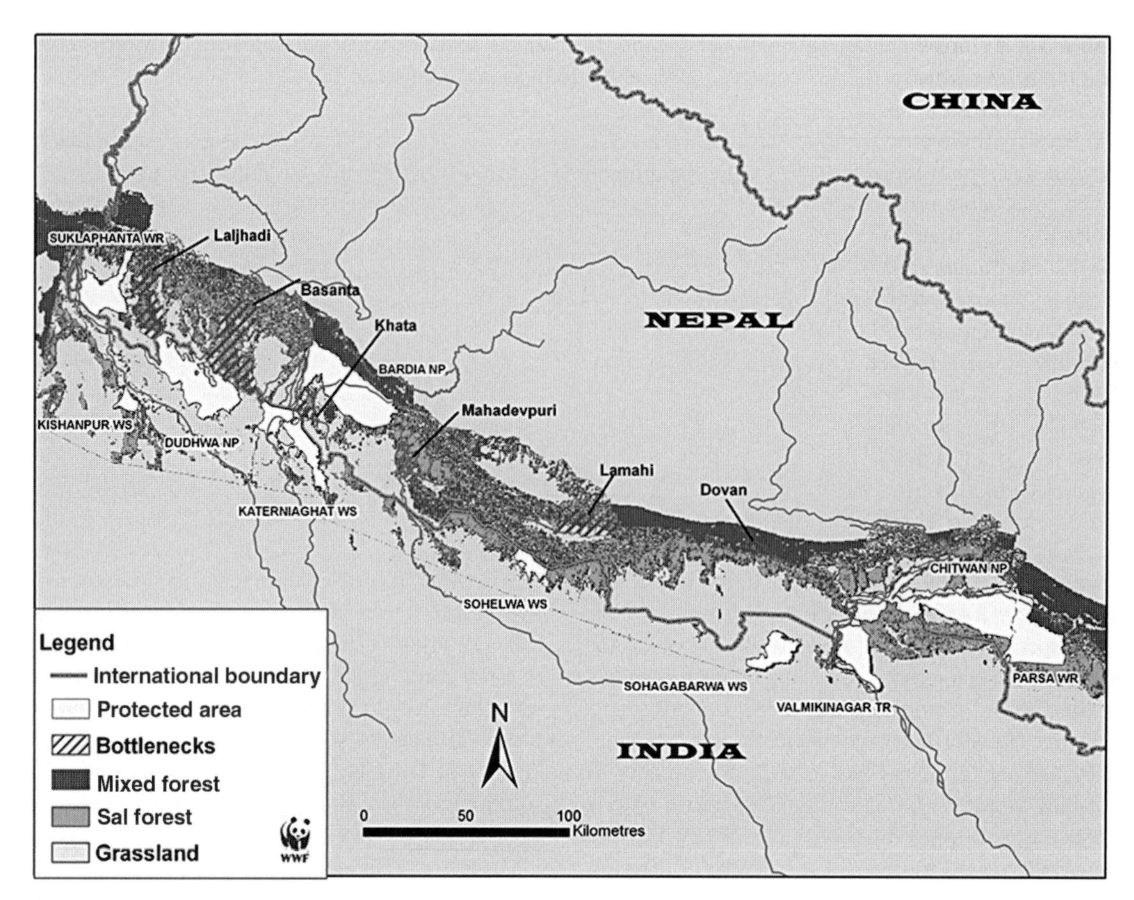

Figure 12.4 Western Nepal Terai corridors and connections: community forestry projects used to harmonize tiger-breeding areas and dispersal corridors between reserves with the rural development agenda. Community forests have resulted in a net increase in forest cover in Nepal, earned local community support for conservation, and expanded conservation efforts far beyond protected areas.

areas between these core areas act largely as dispersal routes rather than as tiger-breeding habitat. To achieve the 2020 goal of 500 breeding tigers will obviously require expanding the number of breeders beyond these core areas to buffer zones and other potential breeding areas nested in the landscape.

Three local issues could prevent the TAL programme from reaching its numeric goal. First, heightened exposure to organized poachers, reduced wild-prey biomass, and human–tiger conflict in non-reserve areas could continue to depress tiger numbers.

Second, the remaining habitat outside reserves offers an unpromising ratio for recovery of tigers: ~85% of remaining vegetation in the Terai is *sal* forest that offers poor forage quality for tiger prey, constituting what one tiger researcher has termed a 'green desert' (B. Pandav, personal communication). These flat *sal* forests and hill *sal* forests also typically lack permanent surface water during the 8-month dry season. Dinerstein (1987) found that deer seldom use habitats >2 km from available surface water, thus these *sal* forests are occupied by tigers only in the monsoon.

The third constraint is human encroachment. The Nepalese Terai attracts many landless people both from the hills to the north and from across the Indian border. Encroachment into the Basanta Corridor (Fig 12.4) has become a serious issue, in part generated by the landless poor, and compounded by the resettlement of former combatants from the Maoist insurgency. Conservationists have about a 10-year window in which to secure the pathways through major dispersal bottlenecks and potential tiger-breeding areas along alluvial stretches in the corridors, after which such interventions will be too costly or impossible because of settlement pressures.

Balancing these proximate threats is an ecological maxim: the Terai is a forgiving landscape. If tigers are given reasonable protection and wild prey is sufficient, they and their most preferred habitat rebound rapidly. *Saccharum spontaneum* grasslands can recover from intense overgrazing by domestic stock in <2 years to support tigers again. We cite two episodes of recovery. First, tigers and prey increased in number when protection was effectively established for Chitwan, when it became a national park in 1973. The number of adult tigers occupying prime tall-grass savannah riverine forest in the western end of the park doubled from 4 to 8/100 km^2 between 1976 and 1987 (McDougal 1988), with the same increase in tiger density thought to have occurred through the rest of the park's flood plain.

Second, tigers were absent from the Baghmara and Kumrose areas in the Chitwan Buffer Zone at the eastern end of the park in 1993. A locally managed regeneration project began in 1994 to restore heavily overgrazed areas adjacent to high-density tiger habitats. By 1996, five tigers, including a tigress and cub, were regularly using these formerly degraded areas. Significantly, poaching incidents that had occurred near the regeneration areas ceased after the locally managed buffer zones began to recover in 1994, largely because tigers generated tourist revenue for local residents, providing an incentive to protect rather than poison tigers (Dinerstein *et al.* 1999, Dinerstein 2003).

A change in attitude regarding voluntary resettlement and equitable land transfer programmes has opened up new possibilities for landscape management. Once viewed as pariah projects, new resettlement programmes, initiated by local villagers who want to shift from remote areas closer to towns and population centres, are underway. Two examples are worth noting.

Tigers and prey recovered after pastoralists were voluntarily resettled from the Chilla Range (148 km^2), a series of three small dun valleys at the western end of the TAL in Rajaji National Park, in 2002. By 2006, the valley supported >70 individual ungulates/km^2 and the resident tiger density had climbed from 0 to 2.5/100 km^2, including breeding females. This prey density could potentially support $\sim$14 tigers/100 km^2. The source population for this recovery is presumably Corbett Tiger Reserve, 40 km to the east, which is connected through a forested corridor (Harihar *et al.* 2007). The resettlement of the pastoralists created space for a new tiger source population and also 'unblocked' an adjacent 2-km riverine forest corridor and stepping stone islands where the Ganges River divides the Rajaji National Park. This allowed the first known attempt by tigers to cross the Ganges in the two decades that A. J. T. Johnsingh monitored this crossing (Seidensticker and Lumpkin 2006).

The recently completed land transfer of the Padampur enclave in eastern Chitwan offers another ideal opportunity to study a citizen-initiated land transfer in the Terai. About 14,000 villagers were shifted over a 10-year period to New Padampur, a site $\sim$20 km from the former park enclave and much closer to the town of Narayanghat. Similarly, the increase in critical habitat for tigers and endangered greater one-horned rhinoceros in eastern Chitwan is being studied by K. Thapa and N. Dhakal (personal communication). These examples are beginning to show that win–win situations are possible in areas where benefits to tiger conservation and improving rural livelihoods may be achieved simultaneously.

Harmonizing location of tiger-breeding areas and dispersal corridors between reserves with the rural development agenda

A wise field biologist once said, 'Conservation is ten percent science and 90 percent negotiation' (R. Cowling, personal communication). The past decade of addressing the human dimensions of the TAL can best be characterized as a protracted, yet fruitful, negotiation. Based on the best science available, conservationists have advocated for tigers and people. Seidensticker (2002) and Dinerstein (2003) describe how Chitwan's managers were sensitized to addressing the needs of local people who lost access to the resources now protected by the parks. Bouts of poaching made it clear that local people would have to become part of the protection of those resources.

They would have to experience the park not as a no-trespassing zone that belongs to outsiders and tourists but as a source of non-consumptively generated revenues. In 1993, new legislation in Nepal permitted managers and local communities to negotiate an arrangement whereby local communities received a substantial share of the tourist revenues generated by the park. These revenues were invested through newly empowered local committees to improve livelihoods through better training, education, and health facilities.

That legislation conserved the buffer zones of the Nepalese tiger reserves, but not the corridors between them. The second law, enacted in 1978, allowed village user group committees to take control of government forests and the land to be 'handed over' after completion of a management plan; this act changed the course of conservation in the Nepalese Terai (Dinerstein 2003). This second act allowed Nepalese communities far from protected areas to benefit directly from community forestry. Most of the people in the Terai are still dependent on forest resources to augment daily livelihoods and to survive times of food insecurity with food, fodder, fuel wood, and medicinal plants harvested from the forest. Therefore community-based management and *usufruct* rights to these forests are welcomed. Community forests have actually resulted in a net increase in forest cover in Nepal, earned local community support for conservation, and expanded conservation efforts far beyond protected areas (Wikramanayake *et al.* 2010; Fig. 12.4). These programmes have been maintained even through the recently ended 10-year insurgency in the Nepalese Terai.

Future of the TAL and its tiger population

At the base of the Himalayas, one of the important chapters in the recovery of wild tigers is currently being written. How it will end depends on continually adjusting to the changing situation, being opportunistic, and incorporating new knowledge.

The first piece of new information is derived from the science: the ecological recovery of the TAL requires restoring and expanding vital *Saccharum spontaneum* grasslands and associated riverine forest, the prime breeding sites for tigers. These habitats must be strategically targeted even if it means diverting attention from *sal* forest corridors. Securing these vital habitats along river courses must be a conservation imperative if the TAL is to reach its ambitious target of 500 breeding tigers.

The second aspect of conservation in the TAL is a critical need for innovations to finance habitat restoration. These can range from carbon credit programmes to introduction of hybrid, stall-fed cattle, and/or methane gas digesters. Additionally, new types of trust funds to meet the recurrent costs of restoration or some other types of user payments must be explored.

As the density of tigers living in prime habitat in protected areas increases, an increasing number of subordinate, post-dispersal tigers move out and begin to use the margins of protected areas. Conflict with the people using these habitats is inevitable. Gurung *et al.* (2006a) reported that a total of 37 tigers have killed 88 people around Chitwan in the past 30 years, with an average of 1.5/year from 1978, when the first human death was recorded, to 1998, and 8.25/year from 1999 to 2006. Half of the victims were cutting grass, which required them to bend down or sit, dramatically altering the physical profile presented to a tiger (Seidensticker and McDougal 1993). Ten of the 17 tigers removed were subordinates living in marginal habitats; 9 of these were judged to be impaired due to fights with other tigers, old age, or gunshot wounds. Appropriate conservation schemes are required to address this confounding problem.

The illegal demand for tiger parts and products could overwhelm local efforts in the TAL. This market originates far beyond the boundaries of the parks and even beyond India and Nepal, and could undermine tiger protection measures everywhere. This generic problem can only be countered with international cooperation and interventions (Dinerstein *et al.* 2007).

Coda

The Terai wildlife sanctuaries are pearls set in a thin green necklace strung along the base of the Indian and Nepal Himalayas. The conservation story of these wildlands informs the region and the world that we can sustain tigers and other large mammals in the face of massive land-use change and an intensifying human footprint. The science we report here is but the first phase, the engineering segment, in the conservation system required to sustain these habitats and their wildlife. The second, larger phase is implementing a compelling vision intertwining society and this ecosystem, realized through the life work and dedication of many conservation practitioners too numerous to thank here for their efforts.

Attempting the impossible is often foolhardy but it is the occupation of visionaries. All of us draw inspiration from the examples of two such visionaries, the late Dr T.M. Maskey, the first warden of Nepal's Chitwan National Park, and Billy Arjan Singh, who remains unbowed in his fight to create and sustain India's Dudhwa National Park. The zeal for conserving the foundations of the Terai-Arc trace back to their unequalled dedication. And so we dedicate this chapter to them.

CHAPTER 13

Amur tiger: a case study of living on the edge

Dale G. Miquelle, John M. Goodrich, Evgeny N. Smirnov, Phillip A. Stephens, Olga Yu Zaumyslova, Guillaume Chapron, Linda Kerley, Andre A. Murzin, Maurice G. Hornocker, and Howard B. Quigley

An adult male. © Dale Miquelle.

Introduction

Although the stereotypic image of tigers usually places them in a steamy Indian jungle, in fact the species occupies diverse habitats across a wide range of latitudes. The Siberian or Amur tiger (*Panthera tigris altaica*) is the northernmost, and probably faces the most severe suite of environmental constraints. Although tigers in general have moderate to low levels of genetic diversity (Wentzel *et al.* 1999; Luo *et al.* 2004), and the Amur tiger appears to be particularly homogenous (Luo *et al.* 2004; Russello *et al.* 2004), its ability to adapt to the hostile environments of north-east Asia demonstrate the behavioural and physiological plasticity of this species. Understanding how a population survives at the edge of a species' distribution provides a lens through which to understand the ecological constraints imposed not only on that population but the species as a whole. Because Amur tigers are already stressed by extreme environmental conditions, we should expect them to be more sensitive to human disturbances. Thus, understanding the impact of humans on tigers at the northern fringe of their range provides a basis for conserving not only Amur tigers, but all subspecies.

We hypothesize that the primary constraints on tiger distribution and abundance are (1) prey distribution and abundance; and (2) mortality and habitat loss caused by people, factors considered critical for other large felids as well (e.g. Woodroffe 2000, Carbone and Gittleman 2002, Dinerstein *et al.* 2007, and Hayward *et al.* 2007). In this chapter we investigate how the abundance and distribution of prey influences density, home range size, reproductive parameters, and survival rates of Amur tigers. By comparing characteristics of Amur tigers with those of more southern populations, especially Bengal tigers (*P. t. tigris*), we reveal how varying ecological constraints influence conservation strategies.

Description of Amur tiger range

The present geographic range of Amur tigers is primarily confined to the Russian Far East provinces (Krais) of Primorye and Khabarovsk (Fig. 13.1). There are probably less than 400 adult and subadult tigers in Russia (Miquelle *et al.* 2007), with about 95% occurring in the foothills and

Figure 13.1 The Sikhote-Alin and East Manchurian (Changbaishan) Mountain ecosystems, and current distribution of tigers in Northeast Asia, based on preferred habitat and track locations from winter surveys in 2005 in the Russian Far East (Miquelle *et al.* 2007), and in northeast China, based on verified reports and surveys in 1998 and 1999 (Jiang 2005; Yu 2005).

main ranges of the Sikhote-Alin Mountains. There is also an isolated population of 20 or so tigers distributed within the East Manchurian (or Changbaishan) mountain system of eastern Jilin and southeastern Heilongjiang Provinces of China and Southwest Primorski Krai, Russia. Today, most tigers in the East Manchurian Mountains occur on the Russian side, but the majority of available habitat is on the Chinese side. Historically, the largest tracts and most productive tiger habitat (and therefore the largest numbers of Amur tigers), no doubt were found in Northeast China. When tigers nearly disappeared from the Russian Far East in the 1940s (Kaplanov 1948; Miquelle *et al.*, in press), dispersal from China probably contributed to their recovery in Russia (Heptner and Sludskii 1992). However, the development of Northeast China, starting in the late nineteenth century and increasing dramatically in the 1940s, led to habitat loss and intensive hunting (Ma 2005) so that by the 1990s the few tigers still reported in Northeast China were probably mostly dispersers from Russia (Yu 2005).

In both Sikhote-Alin and East Manchurian forest ecosystems, peak altitudes are generally 500–800 m above sea level, few reaching 1000 m or more. Collectively, these two Tiger Conservation Landscapes (TCLs; Sanderson *et al.* 2006) represent a zone where East Asian coniferous–deciduous forests and the boreal forests merge, resulting in a mosaic of forest types that vary with latitude, elevation, topography, proximity to the Sea of Japan, and past history.

Over 70% of Primorye and southern Khabarovsk is forest-covered and, despite more extensive clearing, large tracts of forest also remain in Northeast China (Fig. 13.2). The original dominant forest was composed of a mixture of Korean pine (*Pinus koraiensis*) and deciduous tree species including birch (*Betula* spp.), basswood (*Tilia* spp.), aspen (*Populus* spp.) and in the north and at higher elevations, conifers such as spruce (*Picea* spp.), fir (*Abies* spp.), and larch (*Larix* spp.). Most of these forests have been selectively logged in the past, and human activities, in association with fire, have resulted in conversion of many low elevation forests to secondary oak (*Quercus mongolica*) and birch (*B. costata, B. platyphylla*, and others) forests.

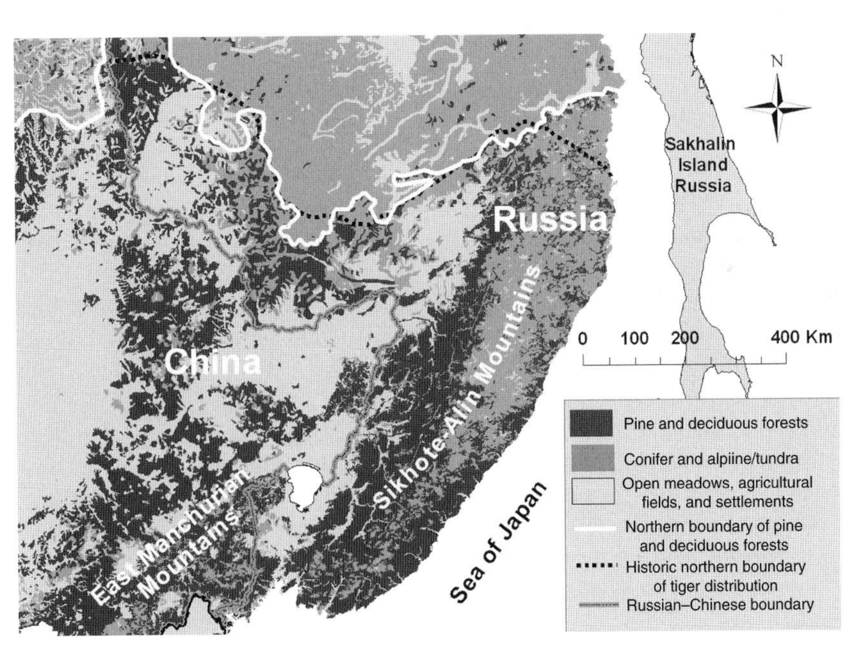

Figure 13.2 Northernmost distribution of tigers, reconstructed from Heptner and Sludski (1992), and distribution of the mixed-pine (*Pinus koraiensis* and *Pinus sylvestris*) broad-leaved and pure deciduous forests, based on data from two sources: a vegetation map of the Amur Basin developed under V. H. Sochavi (Soviet Academy of Sciences, 1968) and a map of habitat types of Sikhote-Alin by V. V. Ermoshin, A. A. Murzin, and V. V. Aramilev (Institute of Geography, Far Eastern Branch of the Russian Academy of Sciences) interpreted from Landsat 5 satellite imagery.

We have intensively studied Amur tigers from 1992 to the present in the Sikhote-Alin Zapovednik (Reserve), an IUCN Category I protected area which straddles the Sikhote-Alin Mountains (Fig. 13.1) and is the largest protected area containing tigers in north-east Asia.

The climate in this region is monsoonal, with 75–85% precipitation (650–800 mm in central Sikhote-Alin) occurring between April and November. The January monthly mean temperature is −22.6°C on the inland side of the central Sikhote-Alin Mountains, but the Sea of Japan moderates temperatures (and snow depths) on the coastal side of the range, when the January monthly mean is −12.4°C. The frost-free period (which influences plant productivity) varies between 105 and 120 days/year. Snow depth varies greatly by locale and year, but averages 22.6 ± 2.9 cm in mid-February (60-year mean ± 95.6°C) in the inland central Sikhote-Alin and only 13.7 ± 3.5 cm on the coast (data from Primorye Weather Bureau). However, heavy snowfalls, most common in March, can have dramatic impacts on ungulate and tiger populations.

Food habits and prey preferences of Amur tigers

By necessity tigers have developed the capacity to prey on a wide range of species. However, there is doubtless a suite of species and body weight ranges that tigers prefer, which likely maximize the difference between energy expended and gained in acquiring prey, tempered by the risk associated with capture (Hayward and Kerley 2005). Tigers are likely to forage optimally when taking the largest prey that can safely be killed (Pyke *et al.* 1977; Hayward and Kerley 2005), often ungulates their own size or larger (Sunquist and Sunquist 1989). The list of primary prey across Asia usually includes a suite of large and medium-sized cervids, as well as suids and bovids (Table 13.1).

The ungulate complex of north-east Asia is represented by seven species, with red deer (*Cervus elaphus*; for body weights see Table 13.2), wild boar (*Sus scrofa*), and Siberian roe deer (*Capreolus pygargus*) the most common: all are found throughout the Sikhote-Alin and much of the eastern Manchurian Mountains, but are rare in higher altitude spruce–fir forests. Red deer become rarer in the southern portions of this region, where they are replaced by sika deer (*C. nippon*) which reach their northern limits in central (inland) and northern (along the coast) Primorye. Roe deer are often associated with forest 'edge' habitats, but also occur in forests throughout the region. Manchurian moose (*Alces alces cameloides*; 281 kg; Bromley and Kucherenko 1983) reach the southern limits of their distribution in central Sikhote-Alin Mountains, and are sparsely distributed in the inland boreal forests, but more abundant further north. Musk deer (*Moschus moschiferus*) are most commonly associated with conifer forests, but can be found in a wide variety of forest habitats. Goral (*Nemorhaedus caudatus*; 32 kg; Voloshina and Myslenkov 1994), a rare wild goat, are mostly limited to coastal cliff habitat. Unique to north-east Asia is the absence of wild bovids in the spectrum of available prey for tigers (Table 13.1). Although bovids do not always represent a large percentage of the items selected, because of their size they can provide a significant proportion of the total biomass consumed by tigers (Karanth 1993).

Representing both potential competitors and prey, brown bear (*Ursus arctos*) and Asiatic black bear (*U. thibetanus*) are common in north-east Asia.

Given a relatively broad range of potential prey, it is not surprising that Amur tigers have been reported eating everything from eagles to seals to brown bears. However, of 720 kills made by tigers found in and around Sikhote-Alin Zapovednik between 1962 and 2003, 94% were ungulates, with red deer the primary prey item (54%) followed by wild boar (28%; Miquelle *et al.* 2005c). Roe deer accounted for only 6% of the diet. Sika deer and musk deer accounted for 3% and 1% of the kills, respectively, while bears (both Himalayan and brown bears) accounted for <3% of the kills.

Data from six other sites across tiger range in Russia confirm that red deer and wild boar are the two key prey species (63–92% of kills, collectively) and that combined with the two medium-sized cervids (sika and roe deer), these four ungulates comprise 81–94% of kills (Miquelle *et al.* 1996). Kill data sometimes under-represent smaller prey items but

Table 13.1 Summary of ungulate assemblages in tiger range.

Type	Chitwan, Nepal[1]	Kanha, India[2]	Nagarahole, India[3]	Sunderbans, Bangladesh[4]	Huai Kha Khaeng, Thailand[5]	Taman Negara, Malaysia[6]	Java[7]	Sumatra[8]	Sikhote-Alin Zapovednik, Russia[9]
Large deer[10]	Present	Present	Present	Formerly	Present	Present	Present	Present	Present
Medium-sized deer[11]	Present	Present	Present	Present	Formerly	Absent	Absent	Absent	Present
Small deer[12]	Present	Present	Present	Present	Present	Present	Present	Present	Present
Wild pigs[13]	Present	Present	Present	Present	Present	Present	Present	Present	Present
Large bovids[14]	Present	Present	Present	Formerly	Present	Present	Present	Present	Absent

[1] Sunquist (1981); [2] Schaller (1967); [3] Karanth (1993); [4] Heindrichs (1975); [5] Seidensticker (1986); [6] Kawanishi (2002); [7] Rabinowitz (1989); [8] O'Brien et al. (2003) and Seidensticker and Suyono (1980); [9] this book.

[10] Large deer-*Alces alces, Cerrus. elaphus, C. unicolor, C. duvauceli, C. schomburgki, C. timorensis*.

[11] Medium-sized deer-*C. nippon, Capreolus capreolus, Axis axis, A. porcinus, C. eldi.*

[12] Small deer-*Muntiacus spp. Moschus moschiferus, Tragulus spp.*

[13] Wild suids-*Sus scrofa, S. verrucosus, S. barbatus*.

[14] Large bovids-*Bos gaurus, B. frontalis, B. javanicus, Babulus babulis*.

Table 13.2 Jacobs' index as a comparison of preference by tigers for five ungulate species, based on proportion of kills and proportion of available ungulate prey numbers (from average prey density) for nine 3-year periods between 1966 and 2002 in Sikhote-Alin Zapovednik, Russia.

Species	Jacobs index	95% CI	Proportion of prey numbers (availability)	Proportion of kills by tigers	Adult female body mass (kg)
Wild boar	0.749	0.094	0.054	0.292	86.0[a]
Red deer	0.263	0.139	0.453	0.589	149[b]
Sika deer	−0.571	0.295	0.035	0.024	74[b]
Roe deer	−0.737	0.104	0.341	0.070	35[b]
Musk deer	−0.862	0.117	0.117	0.013	12[c]

Notes: Jacobs' index (Jacobs 1974) is estimated as $D = (r - p)/(r + p - 2rp)$, where r is the proportion of total kills made up by a species and p is the proportional availability of the prey species. D ranges from +1 to −1, where +1 indicates maximum preference and −1 maximum avoidance.

[a] Bromley and Kucherenko (1983); [b] Dalnikin (1999); [c] Prikhodko (2003).

analyses of scats in Sikhote-Alin Zapovednik also confirmed that these four ungulates were the dominant food items (86% of diet items in scat, compared to 90% of kills), although other prey, including both species of bears and smaller prey items such as badgers (*Meles leucurus*), appeared more commonly in summer scats.

Given their relative size and abundance, as well as the risks associated with capturing them, we would predict that red deer and wild boar should be not only the most common, but also the most preferred prey items for Amur tigers. To test this, we used kill data from Sikhote-Alin Zapovednik for the years 1962–72 (from Gromov and Matyushkin 1974) and 1980–2003 (from Miquelle *et al.* 1996, 2005c) to estimate percentage of occurrence of various prey in the diet, and used yearly surveys of ungulates based on winter track counts (1966–2003) converted to numbers (Stephens *et al.* 2006) to provide an estimate of proportional abundance. We focused on that portion of the Zapovednik east of the Sikhote-Alin Divide to estimate prey densities, because that is where nearly all tiger kills were reported and because prey densities and proportions vary greatly east and west of the divide (Stephens *et al.* 2006). Because of small sample sizes in some years, we estimated proportions of each species killed during 3-year intervals, and calculated the mean prey abundance for these same 3-year intervals. Because surveys are conducted in winter when some prey species are not active (e.g. bears and badgers), we focused on the five most common prey species: red deer, wild boar, roe deer, musk deer, and sika deer (moose are virtually absent within this study area). Jacobs' index was used as an index of preference (Hayward and Kerley 2005).

The results (Table 13.2) reveal that red deer and wild boar, the two largest ungulates, are the most preferred, with the smaller cervids—musk, roe, and sika deer—clearly not preferred. However, given that wild boar are smaller than red deer (Table 13.2), it would be expected, based on body mass alone, that they would be less preferred, when in fact, the opposite is true (Table 13.2). Although wild boar, with their sharp tusks, are probably more dangerous to hunt, they may be easier to stalk than red deer (Yudakov and Nikolaev 1977). Groups of wild boar, with heads down and while foraging noisily in the leaf litter, are much easier to approach than red deer, and are likely slower when attempting escape. This greater vulnerability may explain their preferred status. Roe deer, though common, may be too small to be worth hunting. Sika deer numbers have expanded dramatically in the coastal area of Sikhote-Alin Zapovednik, and have largely replaced red deer as the most common cervid in the southern Sikhote-Alin Mountains. However, this trend, which may be related to global climate changes (Zaumyslova 2000) may not be beneficial for tigers. Although sika deer appear capable of reaching higher densities than red deer or wild boar (Gaponov 1991; Gaponov *et al.* 2006) they are half the size of red deer, and so tiger-foraging efficiency may decrease where they are a primary prey. Despite this concern, tiger populations in southern Sikhote-Alin appear to be surviving and reproducing despite the changing suite of available prey (Miquelle *et al.* 2007).

What limits tiger distribution to the north?

The historical northern limit of permanent populations of tigers in Russia (from Heptner and Sludski [1992]) coincides strikingly (despite imprecise knowledge of the tiger's historical range) with the northern limits of the pine-mixed deciduous forest complex of the Russian Far East (Fig. 13.2). This congruence suggests that these forests represent a proxy for the northern limits of the tigers' primary prey: wild boar and red deer (Fig. 13.3a and b). Although clearly neither species is wholly dependent on these forest complexes (both extend to completely different forest ecosystems in Europe), their northernmost limits in the Russian Far East extend only slightly beyond these mixed pine and deciduous forests. Indeed, the zones of highest densities occur, with a few exceptions for red deer, in the southernmost regions of the Russian Far East (Fig. 13.3a and b), wholly within the limits of this forest complex. Although red deer (and to a lesser extent wild boar) occur beyond those limits, their densities are likely too low to sustain a tiger population. Although other factors such as extreme cold and deep snows probably play a role in defining the northern limits of tiger distribution, the coincidence of prey distribution, habitat types, and tiger distribution suggests clear linkages. Thus, although individual dispersing tigers have been reported more than 1000 km north and west of their historical range (Heptner and Sludski 1992; Kirilyuk and Puzansky 2000), stable populations are restricted to more southern regions where more productive forest systems support prey densities high enough to allow for successful reproduction to occur (see below).

The clear linkages between habitat types, prey distributions, and tiger distribution provide insight into where tigers historically occurred in nearby China (Fig. 13.2), which coincides well with existing information on their former distribution there (Ma 2005; Yu 2005). This information, in conjunction with recent data on where suitable habitat still remains, provides a mechanism for defining potential tiger recovery zones in Northeast China (Miquelle and Zhang 2005).

Tiger density and home range size in relationship to prey density

In the absence of human disturbance, available prey biomass is expected to be the primary determinant of carnivore density (Carbone and Gittleman 2002; Karanth *et al.* 2004c), a prediction supported by many field studies (van Orsdol *et al.* 1985; Stander *et al.* 1997c; Eberhardt and Peterson 1999; Fuller and Sievert 2001; Hayward *et al.* 2007).

In Sikhote-Alin Zapovednik, estimates of ungulate and tiger numbers have been collected annually since 1966 (Smirnov and Miquelle 1999; Stephens *et al.* 2006). Using average female body mass as a measure of body size in ungulate populations (with males weighing more, and subadults and young weighing less), we converted estimates of numbers of ungulates into total ungulate biomass for the Zapovednik (excluding moose and goral, which represent a minor part of the total biomass, and only 1.4% of the tigers' diet). A regression analysis revealed a strong linear relationship between the total available prey biomass and tiger density (Fig. 13.4). Because the most preferred prey—wild boar and red deer—are also the most abundant, this relationship is similar whether we include all ungulate biomass, or only that of the preferred species.

(a)

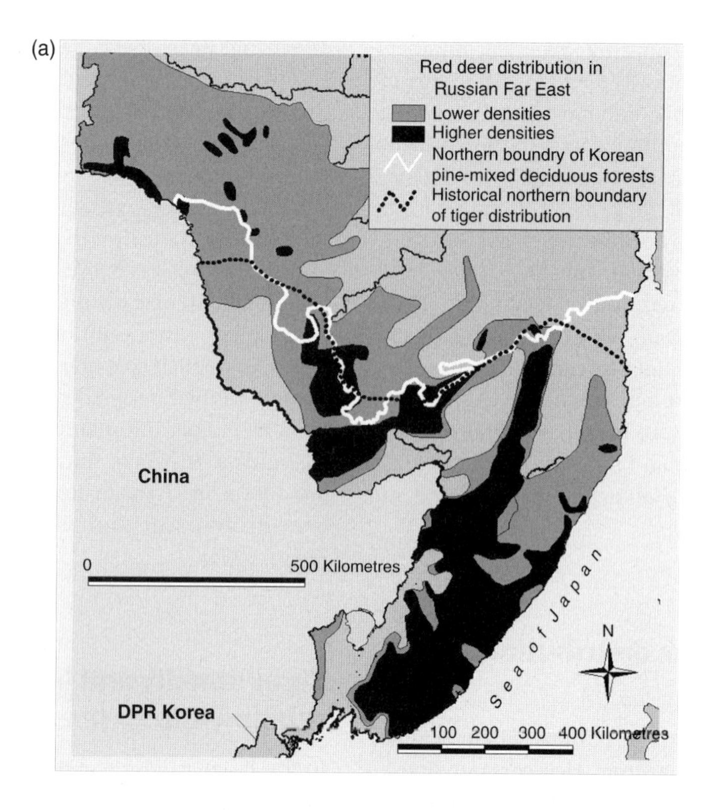

(b)

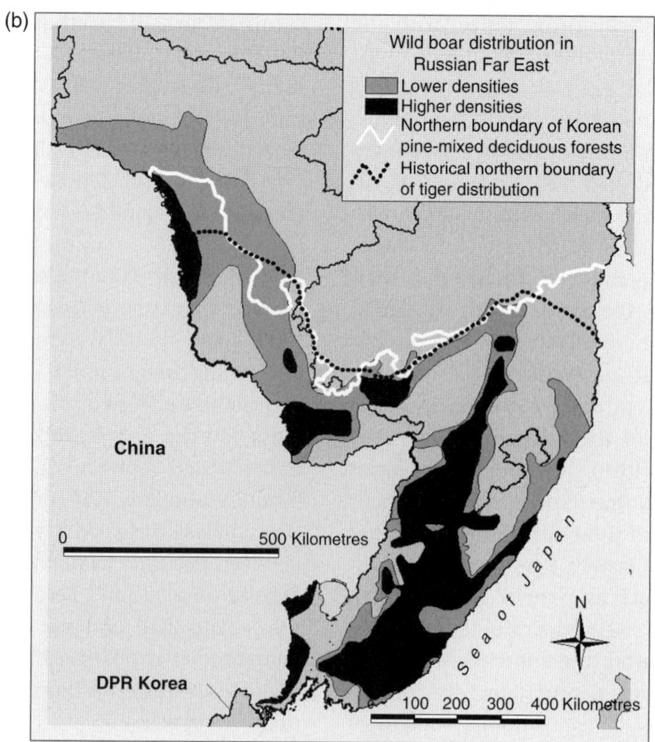

Figure 13.3 Distribution of (a) red deer; and (b) wild boar in the Russian Far East (reconstructed from Bromley and Kucherenko 1983) in relation to northernmost distribution of tigers and northern limits of the pine-mixed deciduous forest complex (from Fig. 13.2).

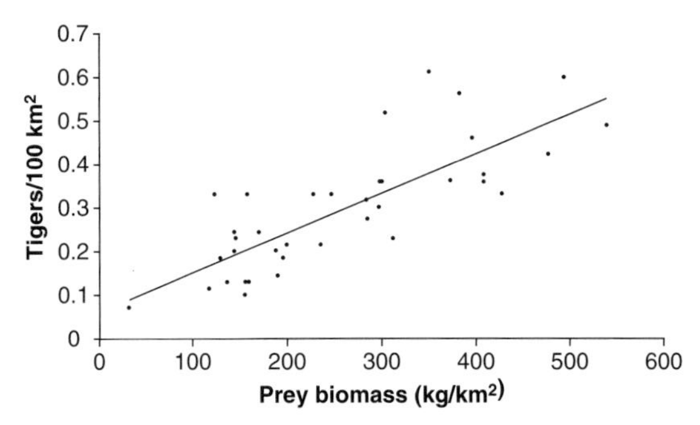

Figure 13.4 The relationship between prey biomass (kg/km^2) and tiger population density (adult tigers/100 km^2) in Sikhote-Alin Zapovednik, Russian Far East ($Y = 0.0009X + 0.623$, $r^2 = 0.63$, and $P < 0.0001$), based on data collected from 1966 through 2002. Ungulate estimates are based on winter track counts converted to animal abundance using the Formozov formula (Mirutenko 1986; Stephens *et al.* 2006), and tiger numbers are based on an expert assessment based on interpretation of relative track size, track age, and distance between tracks (Smirnov and Miquelle 1999; Miquelle *et al.* 2006).

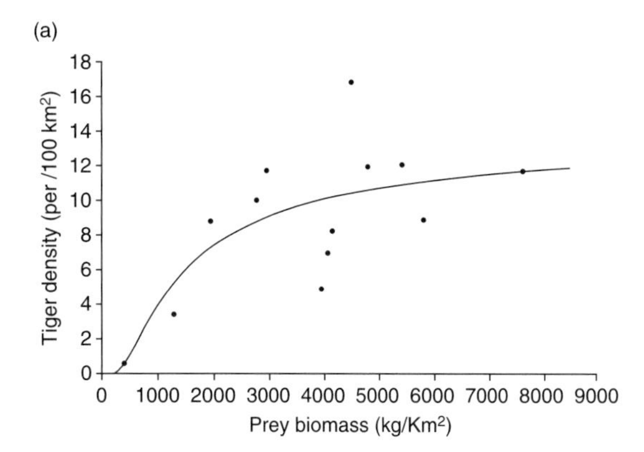

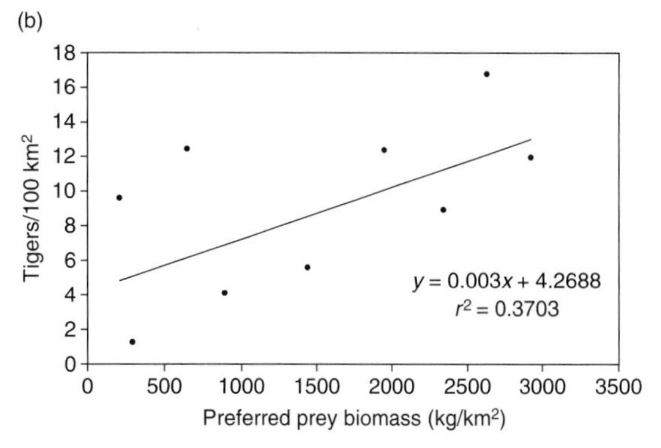

Figure 13.5 The numerical response of tigers (animals/100 km^2) to changes in (a) total prey biomass ($Y = 13.683e^{-1229.52/X}$) and $r^2 = 0.53$); and (b) preferred prey biomass ($y = 0.003x + 4.2688$, $r^2 = 0.37$, and $P = 0.08$). (Data from Schaller 1967; Tamang 1982; Thapar 1986; Karanth 1991; Stoen and Wegge 1996; Smirnov and Miquelle 1999; Karanth and Nichols 2000; Stephens *et al.* 2006.) The curvilinear relationship in (a) is represented by a Michaelis–Menton function of the form $Y = ae^{b/X}$, where Y is tiger density, X is prey density, and a and b are constants.

However, when we compared a greater range of both tiger and prey densities from 13 sites across Asia (Miquelle *et al.* 2005c), a curvilinear relationship was revealed (Fig. 13.5). This Type II functional response has not been reported for any other large carnivores, but accords with Karanth *et al.*'s unrestricted model (2004c) of tiger density and prey numbers. (The fact that they used prey numbers, rather than biomass, does confound the comparison, but it seems unlikely that conversion to biomass would dramatically change the relationship.)

There are several possible explanations for why this relationship is not linear. The sample size was small ($n = 13$), and methods and accuracy varied. There may also be errors in estimating prey density, tiger density, or both. However, the parallel with Karanth *et al.*'s finding (2004c), which relied upon independent data and a more rigorous design, suggests that measurement errors are not the explanation. Alternatively, the curvilinear fit might indicate over-representation of non-preferred species. To assess this, for 9 of the 13 sites for which it was possible, we recalculated available biomass for the range of species between 86 and 550 kg, with the lower limit representing the weight of female wild boar (a preferred prey of Amur tigers), and the upper limit representing the approximate weight of adult female gaur. The relationship based solely on prey within what is presumed to be the preferred weight range is linear (Fig. 13.5b), similar to other large carnivores (Hayward *et al.* 2007), but is not strong and marginally non-significant ($r^2 = 0.37$ and $P = 0.08$). Whether this relatively weak linear relationship (Fig. 13.5b) is a result of insufficient data is as yet unclear, but suggests that a curvilinear relationship (Fig. 13.5a)

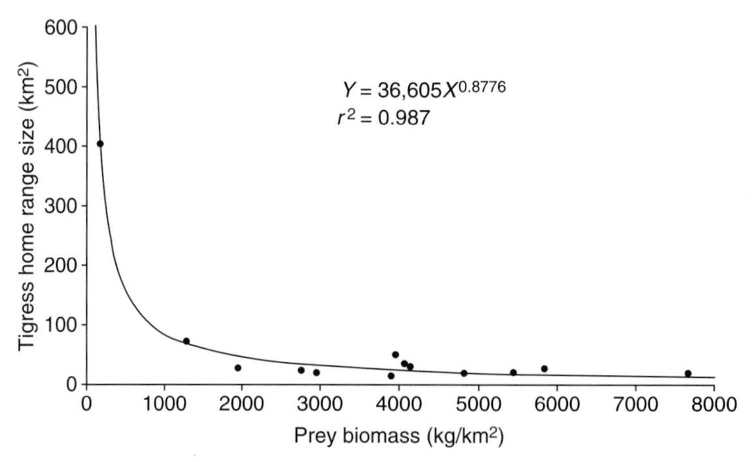

Figure 13.6 A hypothetical relationship between prey biomass and home range size of adult resident female tigers, $Y = aX^{(b/X)}$, where Y is female home range size, X is prey biomass, a = 18.41, and b = 205.506. This relationship assumes that all tiger populations contain territorial females, and there are on average 2.5 individuals per female home range, including an adult resident female, a third of a male (assuming males cover home ranges of 3 females), 0.7 cubs (e.g. Amur tigers on average produce 1.4 cubs/year, but only half survive to 12 months), and 0.5 transients per home range.

may be more appropriate for tigers, providing a tantalizing indication that other factors (perhaps social regulation, see below) may be influencing tiger densities when prey biomass is extremely high.

Wherever they have been studied, tigers have demonstrated a spacing system in which females defend territories that overlap little with neighbouring females, and males defend territories that include one to nine tigresses (Smith *et al.* 1987; Goodrich *et al.* 2005). Females are expected to maintain home ranges sufficiently large to ensure adequate resources for rearing young, but as small as possible to avoid unnecessary defence costs (Smith *et al.* 1987; Sandell 1989). Consequently, female home range size is predicted to be inversely related to prey density (Sandell 1989), but there are too few data to test this for tigers. However, assuming there are 2.5 animals/female home range in the average tiger population (assuming, in addition to the adult female, 0.3 adult males/female home range, 0.7 cubs/year/home range, and 0.5 transients), it is possible to derive a prediction for female tiger territory size using estimates of prey biomass and tiger densities from Fig. 13.5a. This model (Fig. 13.6) provides reasonable estimates where measurements of home range exist, correctly predicting 445 km^2 for tigresses in Sikhote-Alin Zapovednik Russia (vs the actual estimate of 440 km^2; Goodrich *et al.* 2005) but slightly overestimating home range size for Chitwan National Park, Nepal (41 km^2 vs the actual estimate of 20.7 km^2; Smith *et al.* 1987). Across the range of tigers in Asia, female home range size appears to calibrate so as to provide a similar prey biomass per home range. Thus, although home range size of resident female Amur tigers in Sikhote-Alin Zapovednik is, on average, 22 times larger than that of Bengal tigers in Chitwan, total prey biomass per female home range is approximately equal (actually 1.2 times greater in Sikhote-Alin than Chitwan), using data from Tamang (1982) and Stephens *et al.* (2006) to calculate prey biomass.

These results suggest that prey density is a driving force in determining home range size in female tigers, as predicted for solitary carnivores (Sandell 1989), and consequently is of major importance in considering area requirements for viable populations of tigers. The curvilinear relationship in Fig. 13.6 also suggests that as prey biomass becomes very large, there may be a minimum home range size for adult females (10–20 km^2), below which social regulation may start to influence tiger densities.

At the other end of the spectrum, Figs 13.4 and 13.5a suggest that tigers in Sikhote-Alin Zapovednik are close to the lower limits of prey densities at which tigers can exist. Without the constraints of rearing cubs, home ranges can continue to increase to ensure sufficient prey can be found, as demonstrated by male Amur tigers which maintain home ranges on average in excess of 1000 km^2 (Goodrich *et al.* 2005). We suggest that, at 440 km^2, female Amur tigers within our study area are nearing the limits of home range size in which reproduction can successfully occur. While prey densities decline further to the north where tigers still occur (and presumably home ranges increase in size), the logistics of capturing ample prey in the vicinity of immobile young cubs, finding sufficient food to feed subadult cubs still dependent on their mother (amounting to 2–4 times her individual energetic needs), and the energetic demands of covering such large areas may make significantly larger home ranges non-viable for breeding females.

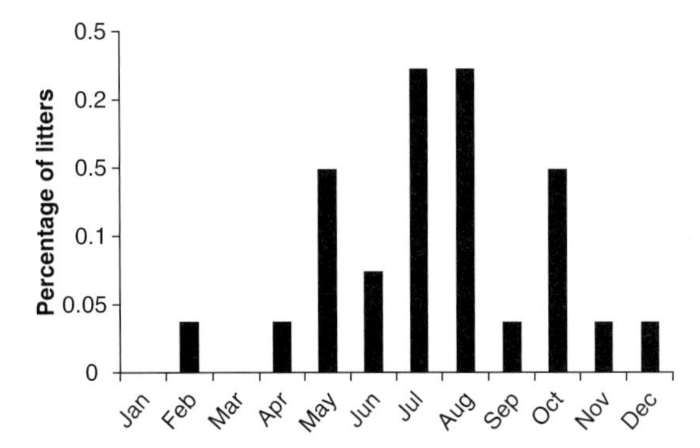

Figure 13.7 The percentage of 27 Amur tiger litters born in each month to 13 tigresses in and around Sikhote-Alin Zapovednik, 1992–2004 (Kerley *et al.* 2003, 2005).

Reproduction, survival rates, and population persistence

All northern ungulates demonstrate a birthing pulse in late spring and early summer ensuring that the most nutritious forage will be available to females to compensate for the energetic demands of lactation, and providing the longest period of quality forage for growth of neonates before facing the rigours of a northern winter. Surprisingly, Amur tigers do not demonstrate a similar birthing pulse in spring that might take advantage of vulnerable young ungulates for feeding young cubs. In fact, while we have observed tigers giving birth in all but 2 months of the year, the peak birthing season is July and August (Fig. 13.7). Cougars (*Puma concolor*) in northern environments demonstrate a very similar pattern (Cougar Management Guidelines Working Group 2005). The period of greatest energy demands for female felids raising cubs is likely not when caring for newborns (even though there is a high energy cost of lactation over the first few months), but when females must provide food for nearly full-grown subadults, which still rely on their mother, and require as much or more energy (in the case of male cubs) than she does. Our data suggest that Amur tiger cubs disperse from their natal home ranges on average at 19 months, and that they are moving largely independently of their mother by 18 months of age (Kerley *et al.* 2003). This suggests that the most demanding period for females would be when cubs are 15–18 months old. Male cubs at this time are already larger than their mothers, but seem to be largely dependent on them for food. For females giving birth in July or August, the peak demands of feeding a family would occur between October and February of the next year. As demonstrated above (Table 13.2), tigers prefer larger prey, but this preference is likely to be even stronger for females with large cubs. Small packets of prey (e.g. newborn ungulates) would not provide sufficient food to feed a family, even if kills were made daily: we have observed that females with two to three yearling cubs can consume an adult red deer in 24–36 h. By late autumn and early winter, calves of the year are sufficiently large to provide reasonable volumes of meat, and are still relatively easy to catch. Also, hunting ungulates is probably easier for felids in winter, when fresh tracks in snow provide clues to finding prey (Yudakov and Nikolaev 1987), deep snow can inhibit escape by prey, and soft snow acts as a muffler for silent stalking. Thus, the shift in birthing season from spring and early summer, characteristic of Bengal tigers (Smith and McDougal 1991; Smith 1993) and captive Amur tigers (Seal *et al.* 1987), to late summer may be a strategy for females to ensure maximum prey biomass is available for their offspring prior to independence and dispersal, when energetic demands of the family unit are greatest.

Given their hostile environment, it might be expected that reproduction and survival rates of Amur tigers would be lower than those in more southerly populations. However, the scanty available data

Table 13.3 Comparison of reproduction parameters of Amur tigers in Sikhote-Alin Zapovednik, Russia (Kerley *et al.* 2003), and Bengal tigers in Chitawan National Park, India (Smith and McDougal 1991; Smith 1993).

Parameter	Bengal tigers			Amur tigers		
	n	Mean	95% CI	*n*	Mean	95% CI
Age of first reproduction	5	3.5	—	6	3.7	0.6
Interbirth interval[a]	7	21.6	—	7	21.4	4.4
Litter size at first observation[b]	49	2.98	0.19	22	2.5	1.1
Litter size at 1 year	38	2.45	0.25	25	1.3	0.5

[a] For Amur tigers, for litters that survived >2 months; for Bengal tigers, for litters that survived >2 weeks.

[b] Age at first observation in Nepal was 2–3 months, and in Russia averaged 4.1 + 1.3 months.

suggest that most reproductive and survival parameters of Amur and Bengal tigers are similar. Age of first reproduction, interbirth interval, and litter size at birth (or slightly thereafter) are all similar in Bengal tigers in Chitwan National Park, Nepal, and Amur tigers in Sikhote-Alin Zapovednik (these being the only two complete datasets that exist for tigers in the wild; Table 13.3). The only significant difference is a nearly 50% reduction in litter size after 1 year in Russia compared to Nepal. This difference is partially due to the fact that 43% of cub mortality in our study was directly or indirectly human-caused (Goodrich *et al.* 2008), and such sources of cub mortality were probably rare in Nepal when studies were conducted. But this difference may also be associated with greater energy demands on female Amur tigers forced to travel over larger home ranges in search of dispersed prey and greater vulnerability of cubs to predation and other mortality factors when mothers are away. In either case, the difference in cub production appears to have marginal impact on population growth rates in model simulations (Karanth and Stith 1999) and appears to be much less important than adult female survival rates in determining population persistence (Chapron *et al.* 2008a).

Two additional key parameters in determining population persistence in large felid populations appear to be age of first reproduction and interbirth interval (Chapron *et al.* 2008a). These two parameters greatly influence potential population growth rates, and because these parameters are greater in tigers than other large solitary felids such as leopards (*P. pardus*) or cougars, potential growth rates are lower (Fig. 13.8). Consequently, for tiger populations to persist, survival rates of adult females must be higher than for these other solitary felids, and must average over 84% (Fig. 13.8). In this regard, the only two estimates of survival rates for tigers are not encouraging: Goodrich *et al.* (2008) reported adult female survival rates of 0.81 ± 0.10 (SE) over a 12-year period based on mortalities of radio-collared animals in Russia, and Karanth *et al.* (2006) reported 0.77 ± 0.05 (SE) survival rates for an entire population of tigers in Nagarhole National Park, India, using mark-recapture analyses of camera-trap data. However, as noted above, the data from Russia included a period of intense poaching, in the absence of which, female survival rates were higher than the 84% threshold suggested by Chapron *et al.* (2008a) for population persistence, and the estimate from India includes emigration (i.e. emigration and death could not be differentiated) and combined survival rates of transients and adults of both sexes. Undoubtedly, survival rates of adult resident females are considerably higher than the averaged population value in Nagarhole. Thus, despite dramatic differences in many facets of their biology, key demographic parameters and the constraints on population persistence appear to be remarkably similar across tiger subspecies in very different environmental settings.

Conservation implications

Our studies of the northernmost population of tigers demonstrate the capacity of tigers to adapt to extreme environmental gradients. Variation in prey, tiger, and human densities, as well as human impacts

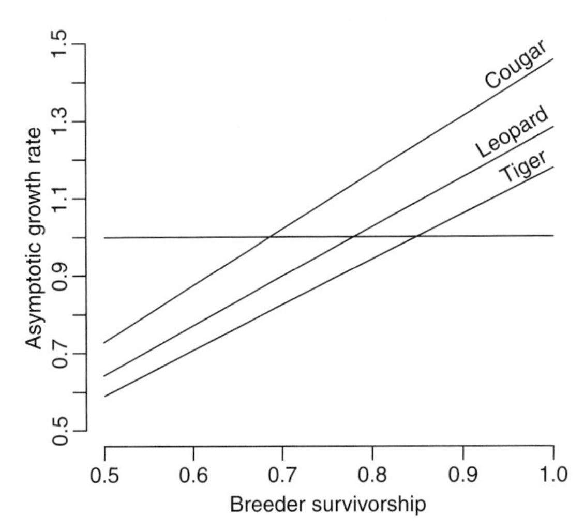

Figure 13.8 Asymptotic growth rates (lambda) of cougars, leopards, and tigers as a function of adult breeding female survival rates and species-specific demographic characteristics (age of first reproduction and interbirth interval), derived from a branching process model designed specifically for solitary felids (Chapron *et al.* 2008a).

are all parameters that influence appropriate conservation strategies for tigers. Here we consider two key components of tiger conservation: first, how human impacts that directly affect mortality influence viability of tiger populations, and secondly, how to consider land-use planning (habitat conservation) for tigers given the range of conditions under which they live. These components address what we believe to be the two key threats to survival of tigers: poaching and habitat loss. Comparing conservation approaches in the Indian subcontinent versus north-east Asia is useful in that we consider the two ends of the spectrum of human–tiger relationships: one (India) where islands of habitat with tigers are surrounded by a sea of humanity, and the second (Russia) where isolated 'islands' of human settlements are surrounded by vast forested habitats of tigers.

Anthropogenic impacts as mortality agents

Although little information exists, the vast majority of tiger mortalities appear to be human-caused (Schaller 1967; Miquelle *et al.* 2005a; Goodrich *et al.* 2008). In Russia, the most unbiased estimate indicates that 83% of mortalities were human-caused, with poaching the dominant factor (Goodrich *et al.* 2008). A similar situation probably prevails across much of Asia, and direct human-caused mortality is likely to be the crucial determinant of the tiger's future. While relatively high mortality rates of cubs and subadults can be tolerated, mortality rates exceeding 16% for resident, breeding females is likely to lead to population declines, and ultimately to extinction (Chapron *et al.* 2008a). Age of first reproduction and interbirth intervals—variables that appear to be more or less physiologically and/or ecologically constrained—are high in comparison to other large felids, and consequently restrict the tiger's ability to respond to high mortality rates by increasing birth rates. These findings are paralleled in lions, the other large felid of approximately the same size, which appear to be able to sustain only limited levels of human-caused mortalities (Whitman *et al.* 2004). Tiger populations do not appear to be nearly as 'resilient' to human perturbations as has previously been thought (Sunquist *et al.* 1999).

Prey recovery efforts are essential for recovery of tiger populations across much of the remaining suitable habitat in Asia (Karanth and Stith 1999; Karanth *et al.* 2006), and it is clear that, in the absence of high rates of human-caused mortality, tiger density is directly related to prey abundance (Figs 13.4 and 13.5a; Karanth *et al.* 2004c). However, because humans are the primary mortality agent of tigers, reduction of human-caused mortality is an essential prerequisite to their conservation. Existing tiger conservation strategies focusing solely on landscape planning (Miquelle *et al.* 1999a; Wikramanayake *et al.* 2004) or prey recovery (Karanth and Stith 1999) alone may not be sufficient if mortality rates of adult breeding females exceed 16%. Despite evidence of some variation in reproductive success between Amur tigers and other populations, the key parameters that limit responses by tigers (age of first reproduction and interbirth intervals) appear similar across their range (Table 13.3). The importance of reducing human-caused mortality on adult females appears to be universal for tiger populations across Asia. Localized extinctions of tiger populations in India (Dinerstein *et al.* 2007) demonstrate that poaching can eliminate tiger populations, despite high prey densities and the potential for high tiger densities. Hence, control of poaching and other sources of human-caused mortality must be at the forefront of tiger conservation initiatives across their range.

Habitat loss and fragmentation

Like tigers elsewhere, Amur tigers prey on medium-sized and large cervids and wild boar, although wild cattle, an important component of the diet in some parts of Asia, are absent in the north. Because plant productivity declines with latitude, carrying capacity for these preferred prey in northern temperate forests is reduced by an order of magnitude in comparison to southern areas. Prey biomass estimates in high-quality habitats in India range from 2000 to nearly 7500 kg/km^2, whereas in the Russian Far East, quality habitat supports a prey biomass of less than 600 kg/km^2 (Fig. 13.4). Consequently, female Amur tigers maintain home ranges that are more than 20 times larger than those of Bengal tigers to find sufficient food to raise young (Fig. 13.6), and their birthing season has shifted to increase hunting success when demands on females with young are greatest. Given these constraints, tiger densities in the Russian Far East rarely exceed one animal/100 km^2, whereas some parts of India can boast of more than 16 tigers/100 km^2.

The conservation implications of these ecological differences are vast. Within the Indian subcontinent, tiger conservation is largely predicated on retaining viable populations within individual protected areas (e.g. Thapar 1999). Such a tactic would be impossible in the north. At 4000 km^2, Sikhote-Alin Zapovednik is the largest protected area in Russia within tiger range, yet harbours fewer than 30 animals, at least half of which regularly use areas outside its boundaries. A similarly sized reserve in India could contain 640 tigers in an ideal setting. Because Amur tigers require vast home ranges, no single protected area can retain a viable population of tigers. Yet protected areas do provide a haven for tigers and their prey. On unprotected lands adjacent (and presumably similar) to zapovedniks, tigers occur at 50% of the densities observed in zapovedniks because prey densities are lower and anti-poaching efforts are generally weaker (Miquelle *et al.* 2005b). Thus the reserve network provides source populations. Therefore, a successful tiger conservation strategy in north-east Asia will depend upon creating and effectively managing a core network of protected areas that are connected to, and interspersed with, multiple-use lands where tiger conservation is integrated with sustainable use of natural resources by humans (Miquelle *et al.* 1999b, 2005a; Miquelle and Zhang 2005).

Although there are more tigers in India than in any other country, its burgeoning human population has fragmented remaining tiger populations into small, isolated pockets of habitat. The total amount of land available for tiger conservation is quite similar in north-east Asia (Russia and Northeast China) and the Indian subcontinent—271,298 versus 227,569 km^2—but in north-east Asia there exist two TCLs (Sikhote-Alin and east Manchurian mountains) that average 135,649 km^2, while in the Indian subcontinent there are 40 TCLs that average only 2100 km^2 in size (Sanderson *et al.* 2006). Of course, the potential number of tigers that can be held in these landscapes is dramatically different. Ranganathan *et al.* (2008) suggest that with effective management both inside protected areas and in adjacent matrices, over 6000 tigers could be retained on the Indian subcontinent, but in a more likely scenario where adjacent land matrices are hostile for tigers, only around 3600 tigers would survive (provided protected areas are well managed). In north-east Asia, tiger densities could increase if effective ungulate management is practiced outside protected areas, and poaching is eliminated within protected areas. However, densities above 1.0 tigers/100 km^2 are likely to occur only in the most productive habitats, and numbers exceeding 1000 Amur tigers are unlikely.

At least in the short term, the focus on protected areas has been successful in retaining tigers in the Indian landscape, although this strategy is a risky one, as the probability of extinction for any one of those isolated populations is relatively high. This is because their survival is dependent upon strong government regulation and compliant people—neither of which is likely to be continuous. Ranganathan *et al.* (2008) suggest that survival of tigers in the vast majority of these landscapes will be dependent on how the environment surrounding protected areas is managed: under less favourable conditions in the surrounding areas, the estimated median population size for 129 of the 150 protected areas is only eight tigers. As this seems to be the scenario developing in India in the near future, we are likely to witness many localized extinctions of the kind experienced in Sariska National Park (Dinerstein *et al.* 2007). But even with more effective landscape management in adjacent territories, which would increase median population size to 14 (Ranganathan *et al.* 2008), the long-term prospects of these small populations is bleak, as genetic drift and inbreeding depression would likely become major concerns, and active manipulation of populations would be necessary to retain the genetic integrity of the Bengal subspecies. Ranganathan *et al.* (2008) suggest that there are 21 protected area clusters that are sufficiently large to retain viable populations of tigers even in a hostile environment, and it is these areas that are probably the last hope for tigers in the Indian subcontinent.

In contrast to the Indian subcontinent, tiger habitat in the Russian Far East is largely unfragmented, although the risk of future fragmentation exists (Carroll and Miquelle 2006). Nonetheless, despite the fact that Amur tigers require home ranges exceeding 400 km^2, at approximately 430–500 tigers (Miquelle *et al.* 2007), Russia retains what appears to be the largest existent single population of tigers in the world in a single unbroken tract of 192,000 km^2 of forests, of which approximately 128,000 km^2 (66%) is potential tiger habitat (Miquelle *et al.* 1999a). Although future development and timber harvest is unavoidable (but can be managed to reduce impact), a declining human population across Russia and specifically an exodus of people from the forest villages within tiger habitat provide hope that this landscape may not undergo serious fragmentation in the near future.

One similarity that arises from this comparison of conservation strategies is the urgent need in both India and north-east Asia to improve management outside protected areas to increase the effective amount of habitat for tigers across the greater landscape matrix. In India, the task is on the one hand easier because the scale (total amount of land surrounding protected areas) is much smaller; improved management on relatively small parcels of land can greatly increase total population sizes in isolated TCLs. In Russia, multiple-use scenarios that are compatible with tiger conservation must be initiated across a vast forested landscape, making the task a more complicated one. However, in India, the demands of a swelling population of 1.1 billion people make the task there more daunting, while in Russia the total human population surrounding tiger habitat is less than 2 million, with the vast majority concentrated in a few cities and peripheral agricultural lands. Hence, the scale of the landscape is vast in Russia, but the scale of problems associated with the human population is much greater in India.

In Northeast China, a surprising amount of habitat still remains (Fig. 13.2), presently largely devoid of tigers. While more fragmented than nearby Russia, with a denser human population exerting pressures on the landscape, there nonetheless still exists the potential for recovery of tiger numbers in Northeast China across a broad swath of habitat along the provincial boundaries of Jilin and Heilongjiang. This process is underway, with the recovery of tigers in the Hunchun Tiger and Leopard Reserve along the Sino-Russian border in Jilin Province (Fig. 13.1). Although it will take time, there exists a great opportunity to increase substantially the total population of Amur tigers, and develop a viable population, separate from the existing Sikhote-Alin population, in the East Manchurian Mountains (Miquelle *et al.*, in press). Thus, although threats to survival of the Amur tiger continue to loom, those threats are understood and rectifiable, if there is sufficient political will and constraint of human demands on the landscape.

Based on the wealth of legend and lore that has grown up around the tiger in both Russia and Northeast China, it is clear that not only the natural landscape, but also the human landscape will be the poorer if this population disappears. Yet there is sufficient reason for optimism that this northernmost representative of *P. tigris* will continue to roam the forests of north-east Asia, interacting with natural prey and exposed to the full spectrum of natural forces that it has survived in for so long, if humans can provide the minimum requirements of space, undisturbed habitats, and freedom from direct human persecution. It is a simple recipe, if we have the collective societal will to make it happen.

CHAPTER 14

An adaptive management approach to trophy hunting of leopards (*Panthera pardus*): a case study from KwaZulu-Natal, South Africa

Guy A. Balme, Luke T. B. Hunter, Pete Goodman, Hayden Ferguson, John Craigie, and Rob Slotow

A radio-collared female leopard in the Phinda-Mkhuze ecosystem, South Africa. © Brett Pearson.

In principle, the trophy hunting of large carnivores has substantial potential to foster their conservation (Lindsey *et al.* 2007; Loveridge *et al.* 2007b). Predators may be tolerated to a greater extent when they augment human livelihoods, and hunting may be a means by which private landowners and local communities can generate revenue from the presence of carnivores (Leader-Williams and Hutton 2005). If the profits realized from harvesting a few individuals are sufficient incentive for people to tolerate the larger population, the goals of trophy hunting and conservation are compatible. However, poorly managed trophy hunting may produce a variety of deleterious effects on carnivore populations. Excessive and sustained harvesting of a species can lead to its local extinction, or reduce numbers to such an extent that a population is no longer viable in the long term (Treves and Karanth 2003). Similarly, even when quotas for trophies appear conservative, hunting may constitute additive rather than compensatory mortality, particularly given that other anthropogenic sources of mortality are rarely included in the calculation of quotas (Caro *et al.* 1998). Additionally, uncontrolled trophy hunting may erode the genetic diversity of a species by consistently selecting the fittest individuals in a population that would normally survive to propagate the species, particularly adult breeding males and females (Harris *et al.* 2002; Coltman *et al.* 2003). Finally, hunting can alter demographic patterns in carnivore populations, resulting in impacts on non-hunted cohorts (Tuyttens and Macdonald 2000). Accordingly, for trophy hunting to be considered a legitimate tool in the management and conservation of carnivores, it is essential that hunting practices avoid detrimental impacts on populations and are demonstrated to be biologically sustainable.

The African leopard is one of the most sought-after big game trophies (Turnbull-Kemp 1967; Grobbelaar and Masulani 2003). Although resilient in the face of human pressure, leopards have been eradicated from vast tracts of their former range due mainly to loss of habitat, depletion of natural prey, and direct persecution by people (Nowell and Jackson 1996). Leopards once occurred throughout most of the African continent with the exception of the hyper-arid interiors of the Sahara and Namib deserts (Fig. 14.1). They are now virtually extinct in North Africa, have disappeared from most of the West African coastal belt, and continue to decline outside of protected areas in large parts of East and southern Africa (Hunter *et al.*, in press). Ray *et al.* (2005) estimate that leopards have been eradicated from 36.7% of their historic African range. Notwithstanding these recent declines, leopards are often deemed to warrant low conservation priority. Their wide geographic range and ability to persist in regions where other large carnivores have been extirpated has given rise to a widespread assumption that their conservation status is assured. Although listed on appendix I of the Convention for the International Trade in Endangered Species (CITES) of flora and fauna, 12 African countries are permitted to export a quota of leopard skins procured through trophy hunting (CITES 2007). In 2008, the quota for all states combined was 2648 leopard skins/year, with an additional 10 skins from the Democratic Republic of Congo and Gabon obtained from sources other than hunting (Table 14.1).

Trophy hunting is regulated and accounts for a smaller proportion of leopard deaths than other human-mediated mortality such as problem animal control and illegal persecution; for example, an average of 445 leopards were legally hunted per year in the 1980s compared to an estimated minimum of 770 killed by other anthropogenic sources in South

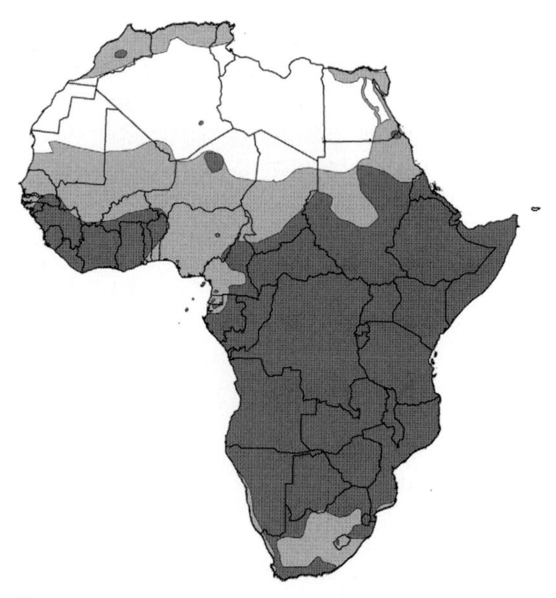

Figure 14.1 Current (dark grey) and historical leopard distribution. (Modified from Ray *et al.* 2005, with permission.)

Table 14.1 The annual CITES quotas for the 12 African countries permitted to export leopard skins procured through trophy hunting in 2008.

Country	Quota
Botswana	130
Central African Republic	40
Ethiopia	500
Kenya	80
Malawi	50
Mozambique	120
Namibia	250
South Africa	150
Tanzania	500
Zambia	300
Zimbabwe	500
Uganda	28

Africa, Botswana, Zimbabwe, and Zambia combined (Martin and de Meulenaer 1988). Nevertheless, there is remarkably little scientific input on the allocation of harvest quotas and the implementation of hunting practices. Population estimates most widely quoted by countries proposing to increase or introduce quotas are derived from an overly simplistic modelling exercise that correlated leopard numbers with rainfall (Martin and de Meulenaer 1988). This model was widely criticized for omitting critical factors such as anthropogenic mortality and prey availability, and for relying upon questionable assumptions, for example, that leopards occur at maximum densities in all available habitats. Accordingly, the final estimate of 714,000 leopards in sub-Saharan Africa was considered an impossible overestimate (Jackson 1989; Norton 1990; Bailey 2005). Despite recent advances in survey techniques that furnish accurate estimates of leopard numbers at moderate cost (Henschel 2001; Balme *et al.*, 2009a), few authorities employ these techniques in routine management activities such as setting hunting quotas.

Despite the lack of data, the demand for leopard hunting in Africa appears to be growing. The annual quotas of four countries (Namibia, South Africa, Tanzania, and Mozambique) have increased since 2002, and Uganda recently introduced trophy hunting of the species, raising the combined quotas of these countries from 485 to 1048 (CITES 2007). Further potential for over-exploitation exists where regulations are abused or ignored. In Tanzania, females comprised 28.6% of 77 trophies shot between 1995 and 1998, even though only males are legally hunted there (Spong *et al.* 2000a). In Zimbabwe, excess skins from illegal hunts are reportedly smuggled into Zambia and Mozambique to be exported under the quotas of those countries (I. A. Caldwell, personal communication). As an explicit condition of CITES certification, African countries that allow leopard hunting commit to do so in a way that is compatible with national and sub-national conservation objectives for the species. However, there remains (1) no rigorous data on the numbers, densities, or population trends of leopards anywhere they are hunted; (2) few empirical data on the impact of hunting on leopard populations; and, finally, (3) no national, provincial, or local regulatory framework for harvesting leopards established by an assessment and consideration of numbers 1 and 2.

A variety of management strategies have been proposed to regulate trophy hunting of comparable carnivore species that have similar lacunae in knowledge (Logan and Sweanor 1998; Laundrė and Clark 2003). In lieu of detailed demographic data, any sensible approach must incorporate an understanding of the basic population dynamics of the target species. In particular, the source–sink metapopulation structure of large felid populations has been used as a proxy to partition areas into sub-populations, each with distinct management objectives (Cougar Management Guidelines Working Group 2005). In this context, source areas are closed to hunting and function as inviolate refuges where natural population dynamics can occur without being affected by human influence. They serve as robust biological savings accounts that ensure long-term population persistence and provide dispersing animals for numeric and genetic augmentation of neighbouring sinks and other, more distant sources (Weaver *et al.* 1996). Sink areas, in contrast, are those sub-populations where harvesting is permitted. They are generally, but not always, less suitable for carnivores due mainly to the presence of people, and concomitant ecological changes such as reduced prey availability and habitat quality (Cougar Management Guidelines Working Group 2005).

We adopted a similar approach to develop a novel protocol for the trophy hunting of leopards in KwaZulu-Natal (KZN), South Africa. We previously demonstrated that poorly regulated harvesting contributed to the low reproductive output of leopards in the Phinda-Mkhuze ecosystem, one of the largest leopard populations in the province (Balme and Hunter 2004; Balme *et al.*, 2009b). In that context, we developed a new protocol for the trophy hunting of leopards in KZN, intended to be compatible with both the conservation objectives for the species as defined by the statutory conservation authority, Ezemvelo KwaZulu-Natal Wildlife (EKZNW), and the desire of private landowners and the hunting industry to utilize leopards. The chief objectives of the new hunting protocol were to (1) mitigate the detrimental effects of leopard hunting in KZN; (2) enhance both the sustainability and economic productivity of leopard hunting in KZN; and (3) ensure compliance with international and national norms and standards for leopard hunting (Department of Environmental Affairs and Tourism, South Africa 2006). In this chapter, we review the historical background of leopard hunting in KZN, summarize the biological impacts of hunting on a local leopard population, and present the new protocol for trophy hunting of the species in the province.

Trophy hunting of leopards in KwaZulu-Natal

KZN province is situated on the eastern side of South Africa (26° 45′–31° 10′ S and 28° 45′–32° 50′ E; Fig. 14.1) in the biologically rich transition zone between the Indian Ocean and the Drakensberg escarpment. Although it is only the sixth largest (92,100 km^2) of South Africa's nine provinces, it contains the most people (estimated at 9.5 million in August 2001; Statistics South Africa 2005). Altitude ranges from sea level in the east to over 3450 m above sea level in the west. Vegetation types vary from tropical dune forests near the coast to moist lowland and upland grasslands, a variety of subtropical forests, and semi-arid savannas further inland, all of which contain the megafauna typical of these habitats in Africa, though many now occur only in protected areas (Goodman 2003). Approximately 58% of the land area in KZN is used for stock farming (including of wild game), 17% for crops, 8% for commercial forestry, and roughly 7% is set aside for conservation (South African National Biodiversity Institute 2005).

The most suitable leopard habitat in KZN is found in the mesic northeastern parts of the province on commercial game ranches, cattle farms, and within state-managed protected areas. Leopards also occur on communally owned Zulu lands but generally at lower densities, chiefly due to habitat degradation and lack of natural prey (Balme and Hunter 2004). We refer to both private and communal land as 'properties'. Legally, leopards are categorized as 'specially protected game' and fall under the mandate of EKZNW on public, private, and communal lands. They are strictly protected inside provincial public parks (IUCN Protected Areas Management Category II), and all trophy hunting of the species currently takes place on privately owned land. Otherwise, leopards cannot legally be destroyed, unless they represent a threat to life or property, and then only if the affected party has applied for and been granted a 'destruction permit' by EKZNW. Despite these restrictions, leopards are killed opportunistically and illegally on private and communal lands in the province by livestock owners and game ranchers (Balme *et al.* 2009b).

From 1996 onwards, the annual quota of CITES permits allocated for trophy hunting in KZN has varied between 5 and 10 leopards (Burgener *et al.* 2005). Since 2003, it has been stable at 5 animals/year. By comparison, in 2006, Limpopo Province was allocated 50 permits, North West Province received 20 permits, and the adjoining Mpumalanga Province received 10 permits (Burgener *et al.* 2005). Prior to 2006, CITES permits were allocated in KZN using a random draw process whereby the first five applicants drawn received a permit and the remaining applicants were placed sequentially on a waiting list. If a leopard was not successfully hunted within 2 weeks, the permit defaulted to the next person on the waiting list, and so on.

For such a widespread species, removing five animals from the provincial population does not seem excessive. However, hunting effort was not evenly distributed throughout KZN. Only 5 of the 22 conservation districts that fall within potential leopard range in the province were awarded CITES permits between 2000 and 2005 (Table 14.2). The distribution of tags across these five districts was also uneven, with most (79%) allocated to the neighbouring Nyalazi and Mkhuze districts (Table 14.2), and specifically to private land surrounding the Phinda-Mkhuze Game Reserves, which constitute a single contiguous leopard population. During the same period, Nyalazi had more than double the amount of leopards shot on CITES hunts than any other district (Table 14.2). Sixteen properties in the province were allocated CITES permits between 2000 and 2005; however, leopards were successfully hunted in only seven of them. The number of leopards shot per property during this period ranged from 1 to 4 (mean = 2.29 ± 1.21 SE). Additionally, and contrary to national stipulations (Department of Environmental Affairs and Tourism, South Africa 2006), leopards were frequently hunted on the same property in consecutive years. Three privately owned game farms hunted leopards for 2 successive years and another game farm hunted leopards for 4 consecutive years. None of these properties was larger than 25 km^2, which is slightly less than the mean home range size of female leopards resident in protected areas in the region (mean = 27.57 ± 5.72 SE km^2; G.A. Balme, University of KwaZulu-Natal, unpublished data).

This high offtake contributed to a significant local effect on the leopard population. The average annual mortality rate (AMR) of this population between 2002 and 2005 was 0.401 ± 0.070 SE (Balme *et al.* 2009b), more than double that recorded for the species in similar habitat under protection (Bailey 2005). Male leopards suffered particularly high mortality (AMR = 0.538 ± 0.023 SE), due partly to their higher vulnerability to trophy hunting (Balme *et al.* 2009b). Males are more desirable to trophy hunters because of their larger size, and males also utilize larger home ranges and cover greater daily distances than females (Mizutani and Jewel 1998; Bailey 2005), increasing their chances of moving into areas where they can be hunted. In some felid and ursid species, unnaturally high turnover among males can result in elevated levels of infanticide (Swenson et al. 1997; Wielgus and Bunnell 2000; Logan and Sweanor 2001; Whitman *et al.* 2004). Although infanticide is a natural event in leopards (Ilany 1986; Scott and Scott 2003), high levels of male turnover promotes a situation in which females cannot successfully raise cubs because of constant incursions by new males. The consequences of this in the Phinda-Mkhuze population included low survival rates among cubs (AMR = 0.574 ± 0.089 SE), delayed age at first parturition (45.33 ± 1.76 SE months), reduced conception rates (19%), and low annual litter production (0.643 ± 0.127 litters/female/year; Balme *et al.*, 2009b). The combined effects of high mortality and depressed reproduction led to a negative population growth rate during the study period (λ = 0.978). It is important to note that human hunting was not the only source of leopard mortality and many losses in the

Table 14.2 The number of CITES permits allocated to, and the relative success of, hunts within conservation districts in KwaZulu-Natal (KZN) province, South Africa, from 2000 to 2005.

	Number of permits allocated		Successful hunts		
Conservation district	*n*	%	*n*	%	Hunt success per district (%)
Dundee	1	2	0	—	0
Mdletshe Tribal Authority	1	2	0	—	0
Mkhuze	15	32	4	25	27
Nyalazi	22	47	9[a]	56	41
Vryheid	8	17	3	19	38
Total	47		16		Mean = 34

[a] In addition to this total, an adult female leopard was critically wounded in a hunt in the Nyalazi district in 2003 but not recovered; she is unlikely to have survived.

population were the result of illegal persecution rather than legal trophy hunting (Balme *et al.*, 2009b). Nevertheless, the concentration of trophy hunting effort on a single population contributed to higher AMRs than previously documented for the species, prompting EKZNW to ratify the new protocol towards the end of 2005.

Protocol

Our protocol was based on five key recommendations intended to achieve objectives 1–3 listed earlier. These recommendations, and the rationale behind them, were as follows:

1. Cap the number of CITES permits allocated in KZN to five each year. When South Africa's annual CITES quota of 75 leopards was doubled in October 2004, 10 permits were made available for trophy hunting in KZN. A population and habitat viability analysis (PHVA) which evaluated the possible effects of increasing the leopard harvest suggested that the KZN population had a much greater likelihood of suffering significant impacts from hunting than did more robust populations elsewhere in South Africa (Daly *et al.* 2005). This was primarily because the level of illegal killing of leopards by humans was estimated to be among the highest in the country. According to the PHVA, the risk of extinction rose from 11% to 62% and the mean population size declined from an estimated 393 animals to 217 with a quota of 10 leopards (Fig. 14.2; Daly *et al.* 2005). Thus, despite the relatively large estimated population size and carrying capacity, the removal of just five additional leopards per year put this population at substantially greater risk (Daly *et al.* 2005).
2. Allocate applications for leopard hunts to individual properties rather than hunting outfitters. In the past, any number of hunting outfitters could apply to hunt leopards on the same property, with potential for consecutive hunts during the year and multiple animals being removed from a single farm in a year. Under the new protocol, landowners applied for a hunt to take place on a particular property (this applies whether the land is privately or communally owned). If successful, the property, not the property owner or hunting outfitter, was allocated the leopard hunt. It then fell to the property owner to negotiate the best deal from an outfitter and inform EKZNW of the name of the outfitter and timing of the proposed hunt so that the CITES permit could be issued.
3. Ensure a more even distribution of CITES permits across the province. As discussed earlier, CITES permits were formerly allocated unevenly across the province where it was both feasible to hunt leopards and where landowners sought to host hunts. To distribute hunting opportunities more evenly, we demarcated five leopard-hunting zones (LHZ), each with a population considered extensive and robust enough to sustain hunting (Fig. 14.3). Each LHZ covered an area of at least 600 km^2 and lay adjacent to a protected area of at least 290 km^2 in which hunting was prohibited. Based on density estimates from mark–recapture statistics applied to camera trapping we conducted in the main land use types in which leopards occur in KZN (11.11 leopards/100 km^2 in protected areas, 7.17 leopards/100 km^2 in game ranches, and 2.49 leopards/100 km^2 in non-protected areas; Balme *et al.* 2009b), we estimated that each LHZ contained at least 25 adults and was adjacent to a protected source population with at least 32 adults. Each LHZ qualified for a single permit (i.e. only one animal can be hunted each year) and had its own individual draw and waiting list. A CITES permit allocated to one LHZ could not be used in another LHZ, regardless of demand, or of the success of hunts. If a leopard was not shot on the first allocated property within 1 month (rather than the previously stipulated 2 weeks), the permit was moved to the next property on the waiting list of that particular LHZ, and so on.
4. Link the likelihood of obtaining a tag to the size of the property on which the hunt will take place. Properties within an LHZ were still drawn at random; however, the chance of a property being selected was weighted according to its size. This was based on the assumption that, in the same area, larger properties are likely to sustain more

leopards. Therefore, for every 1.0 km^2 of land per property, that property qualified for a single 'ticket' in the draw for that LHZ. Thus, a 5 km^2 property was eligible for 5 tickets, a 25 km^2 property was eligible for 25 tickets, and so on. In any one draw, the probability of any property being selected increased with its size. Once a property had been allocated a CITES tag, all of its tickets were removed before the next draw which established the first property on the waiting list, and so on.

5. Restrict the trophy hunting of leopards in KZN to adult males. South Africa is one of the few countries that have historically allowed the hunting of female leopards. The PHVA demonstrated that harvesting only males resulted in lower extinction rates for all provinces but was most beneficial in KZN, where the risk of extinction decreased by almost half (Daly *et al.* 2005). Therefore, it was proposed that only adult male leopards over 3 years old should be legally hunted in KZN. At this age, male leopards are easily distinguishable from females; the majority of 3-year-old males weigh over 60 kg (mean $= 64.50 \pm 3.7$ SE kg and $n = 10$; G.A. Balme and L.T.B. Hunter, Panthera, unpublished data) which is considerably heavier than females (mean $= 35 \pm 3.5$ SE kg and $n = 7$) and they show numerous obvious physical differences (Fig. 14.4). All hunters taking leopards in South Africa are required to be accompanied by a professional hunter (PH) who is responsible for reporting the hunt to EKZNW. An EKZNW officer is required to inspect the trophy within 24 h of the hunt and, provided all legal stipulations are met, issue a CITES tag so that the trophy can be exported. It is the responsibility of the PH conducting the hunt to judge whether the animal to be shot is an adult male. A printed guide is provided by EKZNW (Balme and Hunter, in press) to PHs and hunting clients to assist them in sexing and aging animals so that the likelihood of shooting females is reduced.

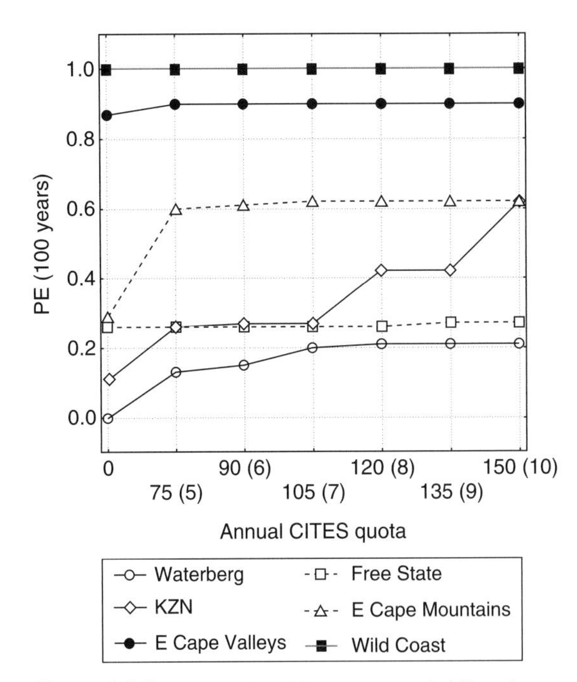

Figure 14.2 Effects of CITES quota on probability of extinction of leopard populations in South Africa. (From Daly *et al.* 2005, used with permission.) KwaZulu-Natal quota in parentheses.

Discussion

This protocol represents the first rigorous effort to ensure that trophy hunting of leopards in South Africa is sustainable. Prior to this, and presently elsewhere in South Africa, CITES permits were allocated on a relatively *ad hoc* basis that resulted in a concentration of hunting effort on a single leopard population. Our protocol addressed this problem by distributing hunts over a much larger area and is based on the meta-population management approach used to manage puma (*Puma concolor*) and grizzly bear (*Ursus arctos horribilis*) populations in North America (Servheen *et al.* 1999; Laundrė and Clark 2003). All LHZs in KZN adjoin large protected areas that potentially act as source populations to replace animals shot on surrounding farms and game ranches. The implicit assumption is that each LHZ is capable of sustaining the removal of one leopard a year to trophy hunting without threatening the demographic or genetic viability of that sub-population or the larger regional population. In other parts of South Africa, source populations that are not found on formally protected land may need to be identified. In this context, there is considerable scope for the private sector to contribute; for

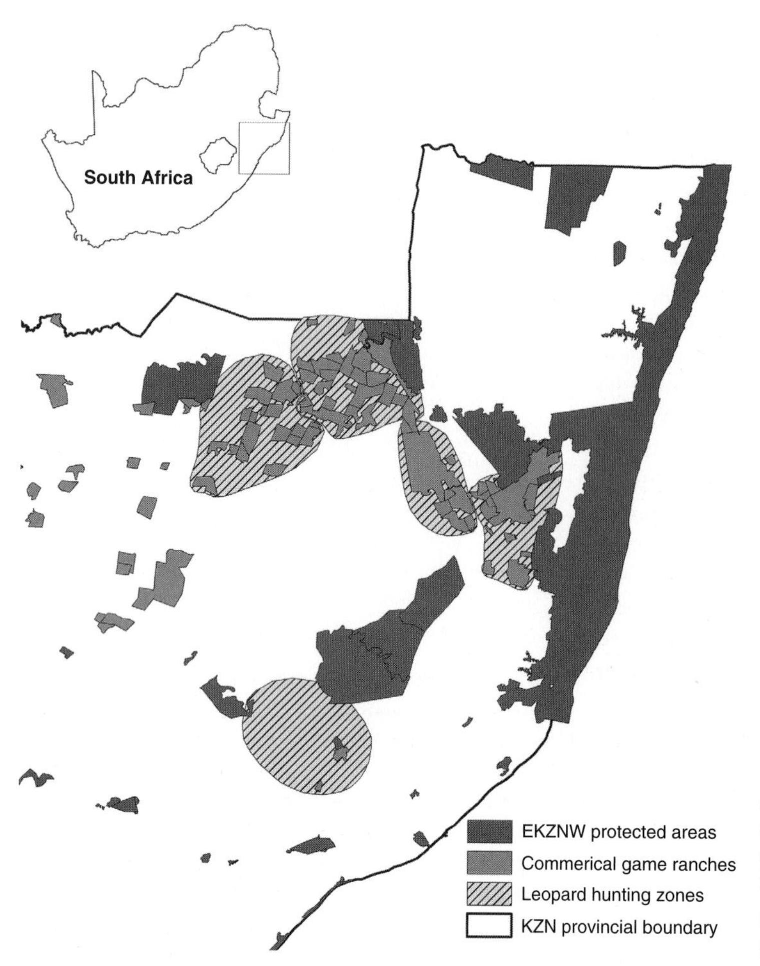

Figure 14.3 The location of designated leopard-hunting zones (LHZ) in relation to commercial game ranches and Ezemvelo KwaZulu-Natal Wildlife (EKZNW) protected areas and conservation districts in KwaZulu-Natal (KZN) Province, South Africa.

example, by the consolidation of commercial game ranches and private reserves that do not rely on consumptive utilization of leopards into 'conservancies', in which multiple, private landowners develop agreements to manage and conserve their combined areas as a single unit (the 'conservancy model'; Hunter *et al.* 2003). However, it is critical that source areas are large enough that losses in adjacent sinks do not have detrimental effects on the source population. Determining the size of a source area depends on various factors such as prey density, habitat quality, and harvest rates in neighbouring sink populations. Beier (1993) and Logan and Sweanor (2001) suggested that a minimum size of 1000–2200 km^2 was needed to sustain a viable puma population for 100 years. In KZN, only two protected areas (Hluhluwe-Imfolozi Game Reserve and the Mkhuze/Mun-Ya-Wana/Makhasa/St Lucia Complex) adjoining a LHZ were this size, but leopards reached higher densities in these areas than pumas in the above examples and all sub-populations were likely large enough to withstand the removal of a single animal by hunting.

Although sink areas in our protocol comprised non-protected farm land, commercial game ranches, and communal lands, leopard population growth rate need not be negative and, if the management objective is to maintain hunting, harvest levels must not exceed immigration rates (Cougar Management Guidelines Working Group 2005). Indeed, provided other anthropogenic sources of leopard mortality are curtailed, populations occupying game-ranching areas have the potential for positive growth given the widespread availability of natural prey and habitat (Lindsey *et al.* 2005). This may benefit ranch

(a)

(b)

Figure 14.4 Extract from a guide for PHs to assist in the accurate identification of suitable leopards for hunting (G.A. Balme and L.T.B. Hunter, Panthera, unpublished report): a) adult female leopard highlighting narrower chest area and proportionally smaller head; and b) adult (> 3-years-old) male leopard showing developed neck region with prominent dewlap and much larger head.

owners seeking to profit from hunting leopards but will only be realized when removal of leopards for other reasons (primarily by illegal shooting and trapping in retribution for real and perceived losses of livestock) are reduced. Most importantly, we do not accept that a sink is an area where uncontrolled removal of leopards is permissible. Private landowners who hunt leopards and benefit financially from their presence must foster populations on their lands if both hunting and leopards are to persist in those areas. This may be assisted by our recommendation of linking leopard hunting to the size of properties. As well as rewarding larger landowners with an increased opportunity of hosting a hunt, this may encourage smaller landowners with an interest in hunting leopards to increase the size of their land or consolidate neighbouring properties, and therefore foster the growth of larger contiguous areas of suitable habitat for leopards.

While we believe we have been conservative in planning LHZs, we have not addressed a prevailing problem anywhere that leopard quotas are assigned, which is the quantification of additional leopard deaths due to people. Caro *et al.* (1998) warned that high levels of traditional hunting by resident communities and illegal poaching probably resulted in unsustainable trophy hunting quotas for leopards in Tanzania, though that country subsequently doubled its quota to 500 in 2002. In KZN and South Africa generally, leopards are illegally killed primarily by pastoralists and farmers, though there is some evidence for organized poaching; for example, skins of at least 58 individuals were seized in the Mkhuze district in July 2004, apparently destined for international markets (Hunter *et al.*, in press). During a 6-year period, 52.6% of all anthropogenic leopard mortality ($n = 19$) recorded in the Nyalazi district was illegal (Balme *et al.*, 2009b). This is undoubtedly a minimum estimate given that most illegal killings are concealed and even accidental deaths of leopards such as road deaths are rarely reported. The ongoing sustainability of leopard hunting in each LHZ will depend on increased quantification and mitigation of the numbers of individuals killed illegally.

In addition to illegal persecution, significant numbers of leopards are legally destroyed every year through problem animal control usually undertaken by the landowner. In principle, landowners wishing to destroy a damage-causing leopard must demonstrate that a particular individual is preying upon livestock or creating some other form of damage and that steps have been taken to bring about a non-lethal solution to the problem (Ferguson 2006). In practice, destruction permits are regularly awarded on the basis of little evidence and there is no mechanism to ensure that landowners remove the individual responsible for the damage (G.A. Balme and L.T.B. Hunter, unpublished data). Although recent policy changes by EKZNW attempt to address this (Ferguson 2006), the number of leopards killed as problem animals (3.42 ± 1.21 SE; EKZNW, unpublished data) generally exceeds the numbers hunted legally every year and often occurs on the same properties as trophy hunting. Ideally, damage-causing leopards could be hunted under the CITES quota (Daly *et al.* 2005) but this is not a viable option in KZN. Trophy hunts are planned months and sometimes years in advance, while stock-killing incidents are largely unpredictable. With only five hunts taking place during the course of a year, it is unlikely that the two events will occur concurrently. Additionally, for this to work, the allocation of CITES permits would have to be linked to complaints, rather than by the existing lottery process. Such a system might foster incentives for false claims about damage-causing leopards to increase the chances of receiving a CITES tag. Similarly, landowners with a genuine problem of losing livestock would be rewarded, potentially promoting lax husbandry. It is possible that this idea may be more applicable elsewhere such as in the Limpopo Province where there is both a larger quota and greater conflict between leopards and farmers (Daly *et al.* 2005). In Namibia, a 'Hunters' Hotline' was established to link farmers that were losing livestock to leopards with the Professional Hunters Association (Stein 2008). While this partially overcame the problem of sourcing clients available at short notice, it did not encourage farmers to improve their management techniques. Furthermore, a questionnaire survey revealed that only 12% of farmers were willing to use the hotline (Stein 2008). Most farmers thought it was easier and more beneficial to destroy the problem leopard themselves. In Namibia, a destruction permit can be obtained after a damage-causing animal has been killed.

Our protocol was not created to disadvantage hunters or prevent the consumptive utilization of leopards. In fact, the new system may improve the overall economic productivity of hunting as more professional hunters and landowners stand to benefit from the limited CITES permits available. Previously, the industry was monopolized by a small number of outfitters operating in the Nyalazi and Mkhuze districts. By distributing permits evenly across the province, it gives a larger number of outfitters the opportunity of hosting a leopard hunt. Additionally, extending the validity of a permit from 2 weeks to 1 month greatly improves the chance of any individual hunt being successful without increasing the harvest quota. Individual outfitters therefore stand to gain as their clients are typically charged a daily rate and there is now potential for longer visits. Similarly, there is increased potential for more property owners to benefit financially over time from having leopards on their land. The prospect of receiving revenue from a trophy hunt every few years may mitigate damage caused by leopards and encourage landowners to allow the continued presence of leopards on their farms where formerly they stood no chance to host a hunt. Thirty-nine percent of landowners interviewed in the Nyalazi district (n = 18) supported the new protocol, 11% felt that the allocation of CITES permits should be linked to the control of damage-causing leopards, 11% preferred the old process, and 39% showed no interest in hunting leopards (Balme *et al.*, 2009b). More widely, if applied correctly, this framework may ultimately permit increased quotas and hunting in more areas of the province. If landowners benefiting from hunts reduce the persecution of leopards and foster population growth on their lands, the potential exists to increase quotas and expand or add LHZs to new areas.

Under our protocol, the potential also exists for a wider cross-section of the community to benefit from leopard hunts. Historically, leopard hunting has been restricted largely to private farms and game ranches and has rarely taken place on communally owned Zulu land. Hunting outfitters had little incentive to develop wider relationships as they were guaranteed access to a small number of private landowners mostly within the Nyalazi and Mkhuze districts. Under the new system, outfitters who negotiate with local tribal authorities (TAs) to conduct hunts will increase their chances of securing access because communal lands are generally much larger than individual private holdings and therefore stand a greater likelihood of being drawn in the lottery. At least two TAs, the Mdletshe TA and the Mthembeni TA, already have a system in place for such negotiations and received a CITES permit for a leopard hunt in 2001 and 2007, respectively. Other communities have developed structures for hunting other trophy species that could easily be extended to include leopards. For example, the Makhasa TA hosted a black rhino (*Diceros bicornis*) hunt in 2006 that generated almost US $11,000 for the community (EKZNW, unpublished data). Communal areas in KZN comprise approximately 30,000 km^2, much of it with considerable potential to be important leopard habitat (McIntosh *et al.*

1996). This will only likely be realized when communities are encouraged to change current land-use practices that limit the viability of the land to harbour leopards and other wildlife. With its low operating costs and high profit margin, trophy hunting may provide one such incentive (Lewis and Jackson 2005).

The implementation of our protocol is only the start of an adaptive process that will rely upon careful monitoring of leopard numbers and losses to people. This currently takes place in only one LHZ as part of a long-term ecological study on the species (Hunter *et al.* 2003; Balme and Hunter 2004; Balme *et al.* 2009a, b). Data from this study suggest that the new protocol has been successful in facilitating the recovery of the Phinda-Mkhuze leopard population (Balme *et al.*, 2009b). Since its implementation in 2006, estimated population growth has increased by 16% ($\lambda = 1.136$), annual mortality rate has dropped to 0.134 ± 0.016 SE, and the reproductive output of the population has improved (Balme *et al.*, 2009b). Specifically, mean age at first parturition has decreased (33.67 ± 1.85 SE months), conception rates have improved (39%), and annual litter production has increased (0.709 ± 0.135 SE litters/female/year; Balme *et al.*, 2009b). Most notably, all cubs ($n = 14$) born to radio-collared females survived to independence after the change in hunting protocol. As a result of lowered harvest rates and increased social stability, population density increased from 7.17 ± 1.12 SE leopards per 100 km^2 in 2005 to 11.21 ± 2.11 SE leopards per 100 km^2 in 2009 (Balme *et al.*, 2009b).

Although it is unlikely that such intensive monitoring could be extended to other LHZs, the process may be assisted by recent improvements in survey methodologies for cryptic species. Camera trapping is one such method that furnishes accurate density estimates for leopards with modest financial resources and relatively little technical expertise (Henschel and Ray 2003; Balme *et al.*, 2009a). The monitoring of trophy quality and composition will also help gauge whether there are any undesirable demographic shifts within a population (Anderson and Lindzey 2005). The new reporting requirement for trophy age and sex will improve monitoring of trophy quality across all LHZs in KZN. In Botswana and the Niassa Province in Mozambique, assessments of trophy quality are used to promote sustainable leopard hunting (Begg and Begg 2007; Funston, personal communication). Quotas are assigned independently to regions, depending on the sex and age of animals taken in the previous hunting season. These systems are self-regulating and encourage ecologically sound hunting by penalizing hunters that shoot females or under-age leopards. Although no similar penalties are currently in place for hunters in KZN and the hunting of males only (recommendation 5) is simply advised as best practice, more stringent control is likely in the near future; for example, a policy is being drafted that will prevent PHs receiving a CITES permit for a period of 3 years if they hunt female or subadult (<3 years old) leopards (EKZNW, unpublished data). Evaluating and incorporating the socio-economic attitudes of human communities coexisting with leopards will also be crucial in successfully implementing this protocol. Aside from anecdotal information (e.g. Hunter *et al.* 2003), there are no data available on the attitudes of local communities towards leopards and their management, nor on the underlying factors influencing attitudes. Practical solutions exist for conducting widespread surveys in the region (Lindsey *et al.* 2005) and should be adopted more widely in parallel with the implementation of the new protocol in all LHZs.

Our protocol has potential to act as a standard for hunting the species in other provinces of South Africa, and further afield in Africa. Currently, no other national or regional authority assigns quotas based on a detailed understanding of the species' ecology and the potential impacts of hunting as we have attempted here. Implementing this approach on a national scale will require that local conservation authorities clarify leopard population status and distribution within each province of South Africa. As a first step, we recommend the delineation of leopard sub-populations that could act as source or sink areas utilizing GIS coverages that include habitat and land-use type, records of leopard presence, patterns of problem-leopard complaints, historical utilization patterns of leopards, and the demand for CITES tags (Cougar Management Guidelines Working Group 2005). Although parameters will vary, this approach is relevant to most countries that permit sport hunting of leopards. Similarly, our approach may have value for other large, wide-ranging felids that are sport-hunted, for example, cheetahs (*Acinonyx jubatus*) in Namibia which are hunted exclusively on private land (CITES 1992), pumas in Latin American populations, where sport hunting is growing (Ojasti 1997), and Eurasian lynx (*Lynx lynx*) in Norway, where quotas are assigned on the basis of depredation complaints (Andrèn *et al.* 2006). Except for a handful of sites where lions have been reintroduced (Hunter *et al.* 2007c), the leopard is the apex predator in many landscapes in KZN and, as such, fulfils a vital role in functioning ecosystems. We believe the implementation of this protocol will be a key factor in maintaining leopards in those landscapes by balancing the ecological requirements of the species with the needs of multiple stakeholders.

Acknowledgements

We are grateful to all at EKZNW who helped to develop and implement this protocol, and to the various professional hunters, game-ranch managers, farmers, and landowners who have agreed to adopt it in the field. We are thankful to CC Africa/&Beyond and the Wetland Authority for allowing us to conduct the research on leopards in their reserves. We are especially grateful to Kevin Pretorius (CC Africa/ &Beyond) for inviting us to initiate research into the effects of hunting on leopards in this region, and for continued logistic and material support. This study was supported by Albert and Didy Hartog, Panthera, Christian Sperka and the Wildlife Conservation Society. G.A. Balme was supported while writing this chapter by a Panthera Kaplan Award and received a NRF bursary from grant #FA2004050400038. Nicole Williams is thanked for her considerable effort in proofreading and formatting the manuscript. We thank David Macdonald, Michael J. Chamberlain, Cynthia Jacobson, and five anonymous reviewers whose detailed comments improved the chapter. Our thanks to Brett Pearson for providing the photograph of Ntombi and to S. Nijhawan (Panthera), who prepared the maps.

CHAPTER 15

Cheetahs and ranchers in Namibia: a case study

Laurie Marker, Amy J. Dickman, M. Gus L. Mills, and David W. Macdonald

Namibian cheetahs have the largest home ranges yet recorded, with 90% living outside protected areas, resulting in cheetah conservation that requires land use planning on a scale rarely seen in conservation. © A J Loveridge.

Introduction

Cheetahs (*Acinonyx jubatus*) were once one of the most widely distributed of all land mammals, but over recent decades their numbers have plummeted by almost 90%, from around 100,000 in 1900 to fewer than 12,000 in 1990 (Marker 1998; IUCN/SSC 2007a, b; Morrison *et al.* 2007). Over roughly the same period their geographic range, once incorporating at least 44 countries spanning the Middle East, the Indian subcontinent, and the length of Africa, has shrunk to 29 countries, with most of the extirpation having occurred over the past 50 years (Marker 1998). Even these dismal figures cloak the reality: 50% of countries still containing cheetahs no longer have viable populations (Marker 2002), with many surviving cheetahs now dispersed in small, fragmented populations (Marker 1998). Cheetahs have fared worst in the northern part of their historic range. Formerly abundant throughout the Middle East and India, the critically endangered Asiatic cheetah (*Acinonyx jubatus venaticus*) survives in small, fragmented pockets, as do those in north, west, and central Africa (Nowell and Jackson 1996; Marker 1998).

The reasons for this precipitous decline vary with locality but include combinations of habitat loss and fragmentation, natural prey base depletion, and persecution due to conflict with humans (Nowell and Jackson 1996; Marker 2002). Considering the general vulnerability of the remaining populations, the most secure cheetah populations are those that are large and less fragmented. These are confined to sub-Saharan Africa, particularly Tanzania and Kenya and most especially Namibia, which supports an estimated 3000 individuals, or perhaps 25% of the world's cheetahs (Marker 1998; Marker *et al.* 2007). Numerically at least, the cheetahs of Namibia are therefore the most important in the world, making it vitally important to understand the threats they face and potential options for alleviating them. Moreover, an estimated 95% of remaining Namibian cheetahs now range across human-dominated land (Marker 2002), a situation likely to become more common for wild felids as wilderness shrinks ever further. The purpose of this case study is to explore the conflict between cheetahs and people in this landscape and, considering what is known of cheetah biology, to consider practical solutions.

Study area

The farmlands of north-central Namibia form the single most important stronghold for cheetahs in southern Africa, so this is where we based our study (Fig. 15.1). Our core research area is primarily on farmland about the Waterberg Plateau (insert in Fig. 15.1), a 100-km long elevation that rises 1870 m above sea level. Farms at around 1670 m above sea

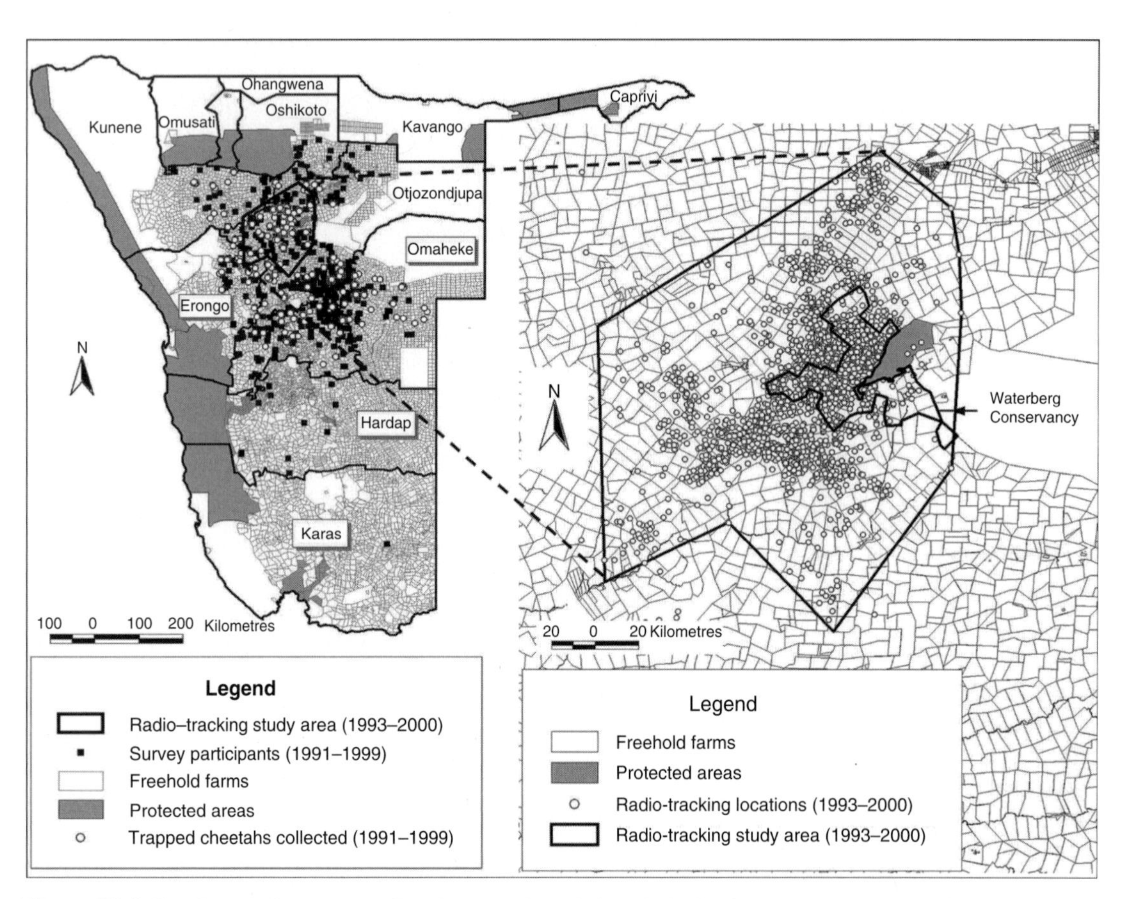

Figure 15.1 The Cheetah Conservation Fund's research study area, including farms where questionnaire surveys were conducted (1991–99), farms where trapped cheetahs were collected by CCF (CCF; 1991–2007), and CCF's core study area (insert), including the Waterberg Conservancy and fixes from CCF's cheetah radio-tracking study (1993–2000).

level surround the plateau and form the whole study area encompassing districts of Windhoek, Okahandja, Omaruru, Otjiwarongo, and Grootfontein in the regions of Omahake, Khomas, and Otjozondjupa (between 19° 30′–23° 30′ S and 16°–19° E). Most commercial farms in this area are between 5000 and 20,000 ha (averaging 8000 ha). The farms comprise largely thornbush savannah, consisting of grassland with clumped trees and shrubs, stocked with cattle and small livestock mingling with wildlife (Marker-Kraus *et al.* 1996).

Namibia has three seasons: a hot, dry season from September to December, a hot, wet season from January to April, and a cold, dry season from May to August (Marker-Kraus *et al.* 1996). Annual rainfall is highly variable, with the majority of rain falling between November and April. The mean annual rainfall in the Waterberg study area over a 40-year period was 123.4 mm (SD = 27.8) for the hot, dry season, 348.6 mm (SD = 58.3) for the hot, wet season, and 2.8 mm (SD = 7.4) for the cold, dry season.

Human–cheetah conflict in Namibia

Namibia boasts an impressive 112,000 km^2 of protected areas, constituting 21% of the country, yet an estimated 95% of its 3000-strong cheetah population occurs in the 275,000 km^2 of private farmland (Marker 2002). This farmland is characterized not only by domestic stock, with over 1040 head of livestock on a typical 8000 ha farm (Marker 2002), but also by abundant wild prey, with an average biomass of 400 kg/km^2. Wild ungulates, 88% of which are free-ranging rather than confined to game-fenced areas, occur in a 1:2 ratio to livestock in the study area (Marker 2002). Cheetahs on the farmland obtain access to artificial waterpoints (142 waterpoints on average per 1000 km^2; Marker-Kraus *et al.* 1996) and are relatively free from intra-guild competition from lions (*Panthera leo*) and spotted hyenas (*Crocuta crocuta*; Laurenson 1994; Durant 2000a) in the area, following extirpation of these species from the farmlands in the 1900s (Marker 2002).

However, despite these seemingly favourable ecological conditions, farmlands still involve risk to cheetahs, as a Namibian Nature Ordinance in 1975 legally entitled farmers to kill them if they are considered to pose a threat to their livelihoods or stock (Nowell and Jackson 1996). This entitlement has traditionally been taken up with gusto, so much so that farmers have been cited as halving Namibia's cheetah population between 1975 and 1987 (Morsbach 1987). Moreover, the figure of 6700 cheetahs reported as 'removed' (usually killed) during the 1980s is probably a substantial underestimate (CITES 1992). Since 1992, Namibia has a legal quota for trophy hunting or exporting cheetahs, allowing the removal of 150 animals per year (CITES 1992); however, this quota is dwarfed by the magnitude of offtake due to conflict with humans. Here, and in the face of a 90% global decline in cheetahs over the past century, we report on our efforts to understand and resolve this conflict that imperils their remaining Namibian stronghold. This interdisciplinary project, which combines wildlife conservation and human development, has grown under the aegis of the Cheetah Conservation Fund (CCF), a Namibian non-governmental organization (NGO) established by one of us (L. Marker) in 1990 specifically to foster sustainable coexistence between farmers and wild predators.

Actions spring from perceptions, and in the case of predators perceptions may distort reality (e.g., Baker and Macdonald 2000 and Macdonald and Sillero-Zubiri 2004b). The first step, therefore, was to survey the Namibian farmers to quantify evidence of their losses to cheetah and their retributive actions, to evaluate the quality of that evidence, and, importantly, to document and understand their attitudes and perceptions.

A 1991–93 survey of 241 farmers in the north-central commercial farmlands (the stronghold of both national and regional cheetah ranges) revealed alarming results; an average of 19 cheetahs annually were removed (1–250) per farm per year (Marker *et al.* 2003c). Of the 6818 cheetahs reported removed, 83% of the cheetahs were either shot on sight or live-caught in traps where they were then shot or placed in captivity after trapping them (14.5%; Marker-Kraus *et al.* 1996). Due to the sociality of cheetahs, entire social groups can be removed in one trapping effort. However, these figures are skewed by a few farmers, as over half (55.3%) of the farmers replied that they had removed fewer than 10 cheetahs each year (Marker 2002; Marker *et al.* 2003c). The number of cheetahs removed did increase to an average of 29 cheetahs per year from each farm where cheetahs were considered particularly problematic (Marker *et al.* 2003c). Worryingly, removals were not clearly driven by current stock loss or perceived problematic status; 60% of farmers who did not consider cheetahs problematic still performed substantial removals (presumably as a form of prophylaxis), eliminating on average 14 cheetahs every year (Marker *et al.* 2003c).

The surveys revealed a striking relationship between the presence of so-called play-trees on farms and the level of cheetah removal from those farms, despite there being no link between play-tree presence and perceived cheetah problems (Marker *et al.* 2003c). These trees are commonly visited by cheetahs, probably functioning as foci for scent-marking (Fig. 15.2). The predominant method of capture was to place one or more cage traps of approximately 2 m by 0.75 m at the base of play-trees, exploiting the relative sociality of cheetahs to remove entire social groups at one time (Marker *et al.* 2003b). With an average home range size for these cats of 1651 km^2, and a mean home range overlap of 16% (Marker *et al.* 2008a), a single play-tree could draw in up to six groups of cheetahs ranging across several thousand square kilometres of farmland. Our radio-tracking data showed that a typical cheetah's range encompassed 21 farms, thereby exposing each cheetah to the management interventions of 21 different farmers (Marker *et al.* 2008a), making obvious the potential impact of either a renegade farmer or a renegade cheetah (as a parallel, killing lions on one 180 km^2 Kenyan ranch caused perturbations over an area of 2000 km^2; Woodroffe and Frank 2005).

Figure 15.2 Cheetahs trapped at a tree used for scent-marking, known locally as a 'play-tree'. Setting up several adjacent traps allows entire social groups to be removed at once. © L. Marker.

The level of removal was undeniably alarming, but the wholesale cheetah capture on the farmlands had one small, but scientifically important, silver lining: it allowed us to examine trapped cheetahs and thereby gain invaluable data on their biology, ecology, and conservation threats. The most salient results from this research are detailed below, and these data are being used to develop conservation strategies for reducing cheetah removal from the farmlands.

Biology and ecology of cheetahs on Namibian farmlands

Demography and social structure

Between 1991 and 2000, we examined 412 captured cheetahs in 228 social groups, of which 41.3% (n = 170) were adult males, and 51.2% of these were caught in a coalition, revealing a mean coalition size of 2.3 males, although it is possible that some coalition males may have avoided capture (Marker *et al.* 2003a). Adult females represented 14.6% (n = 60) of our examined cheetahs, of which 27 were females with cubs (n = 81) with a mean litter size of 3.0. In addition, 19.2% (n = 79) were orphan cubs representing 44 litters with a mean of 1.7 cubs per litter (Marker *et al.* 2003a).

Some farmers allowed us to tag and release cheetahs back onto their farms rather than shooting them or placing them in captivity. Of these, 19 litters from 10 radio-collared females revealed a mean litter size of 3.2 post-emergence (defined as when the cubs leave the den and begin to follow the mother, at approximately 6 weeks of age) and a mean interbirth interval of 24 months, about the time when a litter was raised to independence (defined at 18 months when the cubs leave their mother; Caro 1994; Marker *et al.* 2003a).

The majority of captured cheetahs were adults, with young cheetahs (<30–48 months old) comprising 23% of captures and prime-aged adults (>48–96 months old) making up a further 25% (Marker *et al.* 2003a). Among adults, there was a strong bias towards the capture of adult males, with a male/female capture ratio of 2.9:1, despite a 1:1 sex ratio for cubs aged 12 months or less (Marker *et al.* 2003a). This is likely to be because territorial males in particular frequent play-trees (Marker 2002).

Mortality rates were calculated from 67 recorded wild cheetah deaths, including 45 marked cheetah and 22 that had never been handled, which represented multiple ages and sexes. All cheetahs handled were assigned age classifications that were based on both experience with captive cheetahs and on information from previous studies (Marker and Dickman 2003). From the data collected, ages at mortality were derived from estimated age at capture, which is explained in some detail in Marker *et al.* (2003a). Although these data are subject to some biases, it provides the best data currently available for Namibian farmland cheetah and allowed for the calculation of age-specific mortality rates, which ranged from 20 to 28% for the first 3 years of life before dropping to a mean of 2% between 3 and 5 years of age (Marker *et al.* 2003a). Marked male cheetah (both radio-collared and tagged) provided us an opportunity to calculate male mortality rates, which peaked at 82% for prime adults of 5–6 years old, but the few marked cheetahs that reached 6 years old ($n = 4$) all lived for a further 4 years (Marker *et al.* 2003a). Causes of death could be established for 63 wild cheetahs (Fig. 15.3): 40 of these were males, a significant bias as males are more easily trapped and shot. However, as males are caught more often in coalitions, the number of capture events shows almost equal female to male coalition groups (Marker *et al.* 2003b). In our sample of tagged and collared cheetah, over three-quarters of deaths (79%) were human-caused: The single most important cause of death was shooting on livestock farms, accounting for 36% of reported mortality (Marker *et al.* 2003b). Although human-caused mortality was most likely over-represented, due to the ease of finding cheetah killed by farmers compared to those dying of disease, tagged and radio-collared animals provided good insight into mortality. We found that shooting in order to protect livestock or game was responsible for 48% of wild cheetah mortality, far more than the 11% caused by trophy hunting (Marker *et al.* 2003b).

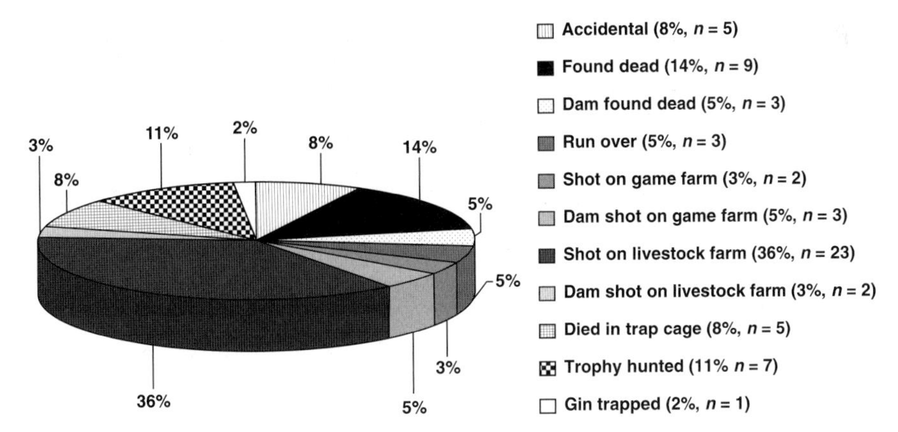

Figure 15.3 Causes of death of 63 wild cheetahs reported to the Cheetah Conservation Fund as dying on the Namibian farmlands during the study (Marker *et al.* 2003b).

The bias towards the removal of adults revealed here, and the concomitant peak of mortality between 5 and 6 years of age, is of particular concern, as it has been shown to impact cheetah populations far more than the removal of cubs (Crooks *et al.* 1998).

Spatial ecology and habitat selection

Over time, CCF has developed a friendly relationship with Namibian farmers that allowed us to tag and release or radio-collar trapped cats. We were able to release 173 captured cheetahs back onto the farmlands between 1991 and 1999 (Marker *et al.* 2003b). Between 1993 and 1999, we fitted 41 adult cheetahs (26 males and 15 females) with VHF radio-tracking collars with an external antenna (280 g, life expectancy of >36 months, Advanced Telemetry Systems, Minnesota) before release (Marker *et al.* 2008a). Cheetahs were released in their original social groups, with only one individual radio-collared per social group. Following release, radio-collared cheetahs were tracked weekly from a Cessna 172 aeroplane, using dual antenna procedure, with the animal's location determined using a portable global positioning system (Marker *et al.* 2008a). This project allowed us to gather vital information regarding the spatial ecology of Namibian cheetahs and better understand how they were utilizing their farmland habitat.

Release sites depended on the cooperation of local farmers and ranged from 0 to 600 km (with a mean of 114.3 km, SD = 165.1) from the site of capture (Marker *et al.* 2008a). Overall there were no statistically significant differences in the overall home range size between translocated cheetahs and resident cheetahs (Marker 2000; Marker *et al.* 2007).

Final home range analysis was restricted to cheetahs with enough fixes to reach an asymptotic level ($\geq$30), as determined using Ranges V (Kenward and Hodder 1996). Twenty-seven cheetahs (19 males and 8 females) had sufficient fixes ($\geq$30 and reached an asymptote) to calculate overall home range size, which averaged 1651 km^2 (SD = 1594 km^2) for a 95% kernel and 1716 km^2 (SD = 1610 km^2) for a 95% minimum convex polygon (MCP; Marker *et al.* 2008a). Females tended to have larger overall ranges than either single or coalition males (Table 15.1), but these differences between sexes or social categories were not statistically significant (Marker *et al.* 2008a). Mean home range overlap averaged 15.8% (SD = 17.0) across all collared cheetahs in the study area (Marker *et al.* 2008a).

Thirty-four per cent ($n = 14$) of the cheetahs were released outside of their original home range (>100 km). Although distance moved had some effect on spatial ecology, releasing a cheetah further from the capture site was linked to larger core (50% kernel) home ranges for females and to ranging further in the first year for single males. Cheetah relocated away from their capture site did not range over larger areas in the first year than in subsequent years. Overall home ranges could be established for six of the translocated cheetahs and showed no statistically significant differences in the overall home range size between translocated and non-translocated cheetah ($z = -1.960$, $P = 0.050$; Marker 2002).

There was no significant difference between wet and dry season home ranges for any of the sex and social categories (Marker *et al.* 2008a). However, the sizes of the core home ranges (50% kernel ranges) for males in a coalition were significantly smaller than those of females (Table 15.1), but there were no detectable differences between other social categories or seasons. Average core home ranges were 13.9% (SD = 5.3) of overall range size for all sex and social categories. Core areas comprised, however, a significantly smaller percentage of single males' overall home range in the wet season (11.3%, SD = 5.0%) than in the dry season (14.5%, SD = 2.9), but no significant differences were detected for other social categories (Marker *et al.* 2008a).

Habitat type was recorded for every fix obtained and revealed that overall, radio-collared cheetahs appeared to utilize thick bush (>75% canopy cover), medium bush (30–75% canopy cover), and sparse bush (<30% canopy cover) approximately in proportion to their availability (Marker *et al.* 2008a). However, when examined by social category it emerged that female cheetahs avoided thickly bushed areas, significantly preferring medium bush. Conversely, coalition males appeared to select for areas of thick bush, while utilizing medium bush areas far less than expected. At the resolution of our analyses, single males utilized all habitat types in approximate proportion to availability (Marker *et al.* 2008a).

Table 15.1 Information regarding home range sizes of all social categories of cheetahs radio-tracked on the Namibian farmlands during this study, from 1993 to 1999 (Marker *et al*. 2008a).

Cheetah social categories		Distance (km) from capture site to release site	Body mass (kg)	Number of months tracked	Total number of fixes	Flights located (%)	Overall home range size		Core home range size (50% kernel)	Wet season home range size		Dry season home range size	
							95% kernel	95% MCP		Number of fixes	95% kernel	Number of fixes	95% kernel
Single males (n = 15)	Mean	80	46.2	14.7	63.5	84.3	1490.3	1697.9	219.7	35.8	1134.8	37.1	2225.8
	SD	107.8	4.7	7.7	39.3	11.2	1202.9	1700.1	239.3	14	957.8	21.5	2051.8
Coalition males (n = 11)	Mean	43.2	46	16.5	75	85.7	1344.3	1608.4	207.5	57.1	1013.2	65.7	2377.9
	SD	70.8	4.9	16.6	75.3	17	1358.2	1091	295.1	26.3	852.7	39.2	3027.1
All males (n = 26)	Mean	64.4	46.1	15.5	68.3	84.9	1436.5	1664.9	215.2	43.6	1090	46.1	2273.8
	SD	94.1	4.7	12	56.3	13.7	1226.5	1471.4	253.1	21.5	898.1	30.4	2314.5
All females (n = 15)	Mean	200.7	35.4	19.1	68.3	90.9	2160.7	1836.3	397.8	53.5	1405.1	56.8	1942.6
	SD	222.4	5.2	23.9	87.3	10.3	2269.4	2010.3	578	44.4	1032.4	49.7	2272.3
All cheetahs (n = 41)	Mean	114.3	42.4	16.8	68.3	87.1	1651.1	1715.7	269.3	47	1198.7	49.3	2175.7
	SD	165.1	7.1	17.1	68.1	12.7	1594.2	1610.4	376.2	30.9	940.4	36.4	2263.3

Note: SD, standard deviation.

The most striking result found from the radio-tracking was the large size of home ranges of Namibian farmland cheetahs. Even the ranges for coalition males here (the smallest in our study was 1344 km^2) dwarfed those found elsewhere: They were seven times larger than the 188 km^2 95% kernel range calculated for a Kruger National Park single male (Broomhall 2001), with a 12-fold difference in magnitude when 95% MCP ranges were compared (Marker *et al.* 2008a). Single males and females showed a similar pattern, with range sizes in Namibia outstripping those in Kruger National Park 6-fold for single male 95% kernels, 8-fold for single male 95% MCPs, 10-fold for female 95% kernels, and 11-fold for female 95% MCPs (Broomhall 2001; Broomhall *et al.* 2003; Marker *et al.* 2008a).

The Kruger ranges were based on small sample sizes ($n = 6$), but even when comparing with a long-term dataset on cheetah spatial ecology from the Serengeti (Caro 1994), the minimum polygons found in Namibia were 2.2 times larger than Serengeti ranges for both females and single males (Marker *et al.* 2008a). At first glance the difference between other males is striking—1608 km^2 in Namibia and only 37 km^2 on the Serengeti plains, a 43-fold difference—but the Serengeti data are for territorial males and we did not have the behavioural information needed to discern whether the coalitions radio-tracked on the farmlands were territorial or not. Nonetheless, it remains obvious that Namibian cheetahs range over vast areas on an annual basis, which is of conservation concern as it is likely to put them at increased risk of anthropogenic conflict, as discussed above. The causes of these extensive movements, for example social perturbation, remain to be demonstrated.

Feeding ecology and prey selection

Understanding the feeding ecology of Namibian cheetahs, and in particular gauging the importance of livestock in their diet, is fundamental to developing effective strategies for their conservation on farmlands. In 2002, we performed nine feeding trials on captive cheetahs at the CCF research station in order to determine correction factors for differential digestion of prey species (Marker *et al.* 2003e), an important step for accurately performing scat analysis on wild carnivores (Floyd *et al.* 1978). The correction factor was calculated as the correlation between the weight of the prey species consumed per field collectable scat and the average weight of the prey species (subadult weights were used; Marker *et al.* 2003e). Analysis followed a least squares regression plot (Floyd *et al.* 1978; Weaver 1993), which yielded an equation where y is the average weight in kilograms of prey consumed, per collectable scat and x is the average weight of individual species. By multiplying y by the frequency of occurrences (n) of each prey species in the sample, it was possible to obtain a total weight consumed of each species and calculate the ratio of biomass consumed between prey species. The total weight of each species consumed was then divided by the average estimated weight in order to compute the number of individuals consumed, and the ratios were computed relative to a 16 kg, or kudu-calf weight, size (Marker *et al.* 2003e). The following correction factor equation, $y = 0.0098x + 0.325$, can be used to provide information on the relative contribution of different species of prey reported as a part of the cheetah's diet (Marker *et al.* 2003e).

We used this correction factor to analyse the contents of 78 wild cheetah scats (37 males, 9 females, and 32 unknown sex; Marker *et al.* 2003e). The most common prey species detected were kudu (*Tragelaphus strepsiceros*) and eland (*Taurotragus oryx*), while remains of domestic stock were found in 6.4% of scats that contained identifiable prey remains (Table 15.2). Larger antelope (eland; red hartebeest, *Alcelaphus buselaphus*; oryx, *Oryx gazelle*; and kudu) hair was found in higher amounts in male cheetah scat (Fig. 15.4; Marker *et al.* 2003e). Sixty-nine per cent of scats from male cheetahs contained these species, compared to only 9% of those from females, which more frequently contained evidence of smaller antelope such as duiker (*Sylvicapra grimmia*) or steenbok (*Raphicerus campestris*) (Fig. 15.4; Marker *et al.* 2003e). Domestic stock remains were found only in scats from male cheetahs, but due to the sample size of scats from females this analysis only provides an indication of feeding behaviour.

These data, revealing that 4.3% of scats contained evidence of domestic calves, while 2.1% contained sheep remains (Table 15.2), allowed rates of livestock consumption by cheetahs to be calculated using the correction factor obtained through the feeding trials (Marker *et al.* 2003e). Assuming a minimum density

Table 15.2 Contents of 78 wild cheetah scats collected on various Namibian farms, including the number and per cent of samples that had various species remains, and the per cent of samples that had identifiable prey remains (Marker *et al*. 2003e).

	Total scats collected	All scats collected (%)	Scats containing identifiable prey remains (%)
Total scats collected	78	100	—
Number with identifiable remains	60	76.9	—
Cheetah hair only	13	16.7	—
Kudu	10	12.8	21.3
Eland	7	9	14.9
Kudu/eland	12	15.4	25.5
Steenbok	4	5.1	8.5
Gemsbok	2	2.6	4.3
Red hartebeest	3	3.8	6.4
Hare	3	3.8	6.4
Bird	2	2.6	4.3
Domestic calf	2	2.6	4.3
Sheep	1	1.3	2.1
Warthog	1	1.3	2.1

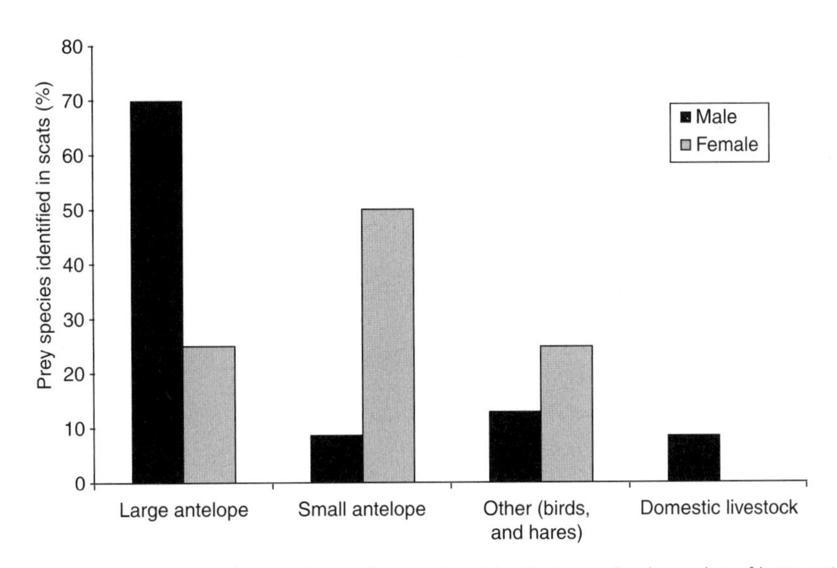

Figure 15.4 The percentage of scats from male and female cheetahs that contained remains of large antelope species (e.g., eland, red hartebeest, oryx, or kudu), small antelope species (e.g., steenbok or duiker), other species such as hares and birds, and domestic stock (Marker *et al*. 2003e).

of 2.5/1000 cheetahs/1000 km^2 on the farmlands and an average farm size of 8000 ha (Marker 2002; Marker *et al.* 2003e), we estimated a minimum rate of livestock depredation by cheetahs to be 0.01 calves and 0.004 sheep/km^2, or 0.76 calves and 0.32 sheep annually on an average-sized farm of 8000 ha (Marker *et al.* 2003e). However, these inferences are highly dependent on density estimates. If the highest reported estimates of cheetah density on the Namibian farmlands are used (34 cheetahs/1000 km^2; Stander 2001b), the rates of livestock depredation due to cheetahs could be as high as 10.3 calves and 4.4 sheep/farm/year (Marker *et al.* 2003e).

Morphometrics and genetics

Handling large numbers of wild-caught cheetahs enabled us to collect data on the morphology of Namibian cheetahs and to compare these to data from other populations. We measured 241 wild cheetahs between 1991 and 1999, using a standardized methodology to assess a suite of 16 morphometric variables, from muzzle girth to tail length, and placed each cheetah in an age category as described by Marker and Dickman (2003). For both sexes, full adult body mass was reached only after cheetahs reached prime adult age >48 months of age (Marker and Dickman 2003). We found significant sexual dimorphism, with males larger for every variable measured. For instance, prime adult males had a mean mass of 45.6 kg (SD = 6.0), while females weighed in at a mean of 37.2 kg (SD = 5.1). Cheetahs we examined had been caught in seven different geographical regions; however, no statistically significant differences were detected in prime adult body mass between regions, for either males or females (males: $F = 3.034$, d.f. $= 1$, $P = 0.084$; females: $F = 0.294$, d.f. $= 1$, $P = 0.591$; Marker and Dickman 2003).

Under anaesthesia, physical examinations were performed on all cheetahs ($n = 241$), in which all were given one of three physical condition scores, depending on their condition, musculature, body fat, injuries, and external parasites (Marker and Dickman 2003). Sixty-three per cent ($n = 151$) of the cheetahs scored were classed as being in excellent condition (Marker and Dickman 2003). No significant relationships were found between physical condition and sex of the cheetah, the region it was captured in, or the season of its capture. Overall, age was not related to condition, although individuals older than 96 months were in significantly poorer condition than younger animals, while being held in captivity for ≥30 days prior to examination was also linked to poorer condition (Marker and Dickman 2003).

Cheetahs measured here tended to be heavier than those weighed in captivity. However, there can be marked variation even within a subspecies: males measured here were similar to fellow Namibian cheetahs measured in Etosha (McLaughlin 1970), which averaged 44.1 kg but were 15% smaller than those from the Kalahari Gemsbok National Park (Labuschagne 1979) and 17% smaller than those measured in South Africa (McLaughlin 1970). Marker and Dickman (2003) set out a clear and standardized methodology for collecting morphometric data from cheetahs, which, if adopted more widely, would facilitate comparisons of morphological variability between different cheetah populations.

Morphological examination of wild Namibian cheetahs did reveal some unforeseen and interesting results: of 208 cheetahs whose mouths were examined between 1992 and 1999, 40.9% showed evidence of focal palatine erosion (FPE), an abnormality where the first lower molar perforates the upper palate (Fitch and Fagan 1982). This condition had been observed in captive cheetahs and was attributed to a 'soft' captive diet (Phillips *et al.* 1993), and had never before been recorded in wild cheetahs. Other physical abnormalities were also noted: 9.1% of cheetahs examined were missing one of the two upper premolars, 11.5% were missing both, and 31.7% had crowded lower incisors (Marker and Dickman 2004a). Missing premolars and crowded incisors were both associated with more severe FPE, indicating a possible link between the traits.

The relatively high incidence of these traits reported here is novel in a wild population and shows that a soft captive diet, although it may exacerbate the condition, is not the causal factor. It is possible that it is associated with an environmental factor specific to Namibian cheetahs, and/or it may have a genetic component. Cheetahs have been shown to have extremely limited genetic heterozygosity (O'Brien *et al.* 1983), and very low genetic variation has been linked to dental abnormalities in populations of highly inbred white tigers (Marker-Kraus

1997), as well as to physical abnormalities in Florida panthers (Johnson *et al.* 2001c).

Irrespective of any link to FPE or dental abnormalities, the genetic structure of Namibian cheetahs was of great interest to us, and examining wild-caught cheetahs provided a valuable opportunity to collect blood and conduct genetic analyses. Between 1991 and 2000, total genomic DNA was extracted from 313 cheetahs from 7 different geographic regions of the country (Marker 2002; Marker *et al.*, 2008a). Variation among 38 polymorphic microsatellite loci was assessed between 89 unrelated cheetahs, and an additional 224 individuals were assessed to investigate questions of gene flow, geographical patterns of genetic variation, and questions relating to cheetah behavioural ecology.

The mean expected heterozygosity (H_e) was similar across all regions, ranging from 0.640 to 0.708, and was comparable to that found in the Serengeti (H_e = 0.599; O'Brien *et al.* 1987c; Marker 2002; Marker *et al.*, 2008a). Despite cheetahs being sampled from seven different regions, F_{st} values were low, averaging 0.021 (± 0.02). These values are a measure of the level of genetic divergence within groups, and the low levels reveal a panmictic population, with a lack of differentiation and high recent gene flow between populations in different regions. This high rate of recent gene flow may be linked at least in part to the high levels of perturbations suffered by the Namibian cheetah population over the past 30 years, as exploited populations may exhibit higher dispersal levels (Johnson *et al.* 2001b; Marker 2002; Marker *et al.* 2008a). This evidence for a panmictic population is important for developing conservation strategies, as it shows that the entire Namibian population can be managed as a single unit and that, if necessary, cheetahs can be translocated from region to region without significantly altering rates of gene flow.

Overall health and disease

Opportunistic biomedical sample collection on wild-caught cheetahs has provided baseline information on the health and reproductive fitness of Namibia's cheetahs while contributing to solving some of the questions surrounding health and reproductive problems in captive cheetahs. For instance, captive cheetahs tend to exhibit unusually severe inflammatory responses to common feline viruses, which may be linked to the low genetic heterozygosity mentioned above, as well as to the stresses of captivity (Marker 2002; Munson *et al.* 2004a). As shown above, cheetahs on the Namibian farmlands have huge ranges that encompass many different farms with some adjacent to towns and cities, putting them in close proximity to domestic dogs and cats (Marker 2002). These domestic animals are often unvaccinated (Schneider 1994) and may be important vectors of diseases such as canine distemper, which has affected wild cheetahs in the Serengeti (Roelke-Parker *et al.* 1996). To examine the extent to which Namibian cheetahs have been exposed to such viruses, we analysed sera collected from 81 cheetahs of different ages and different regions of the country between 1992 and 1998, none of which exhibited any symptoms of viral disease (Munson *et al.* 2004a). We evaluated these samples for antibodies against canine distemper virus (CDV), feline corona virus (feline infectious peritonitis virus; FCoV/FIP), feline herpesvirus 1 (FHV1), feline panleukopenia virus (FPV), feline immunodeficiency virus (FIV), and feline calicivirus (FCV) and for feline leukemia virus (FeLV) antigens.

CDV, FCoV/FIP, FHV1, FPV, and FCV antibodies were detected in 24%, 29%, 12%, 48%, and 65% of the wild population, respectively, but neither FIV antibodies nor FeLV antigens were present in any cheetah tested. CDV antibodies were detected in cheetahs of all ages sampled between 1995 and 1998, suggesting the occurrence of an epidemic in Namibia during the time when CDV swept through other parts of sub-Saharan Africa (Munson *et al.* 2004a). The data gathered here reveal that wild Namibian cheetahs have been infected with feline and canine viruses that are known to cause serious clinical disease in captive cheetahs. These results should be borne in mind when considering translocations: stress from capture and transport could increase susceptibility to viral infections, and infected cheetahs from one location could threaten immunologically naive animals in another. Given that translocation is commonly used in management strategies for cheetahs in Namibia, these risks mean that all cheetahs should undergo strict quarantine before translocation is conducted.

To summarize, opportunistic access to captured cheetahs on the farmlands has provided informative data on their biology and ecology. Some of these results, such as the vast ranges of cheetahs encompassing many farms, and the worryingly high mortality and removal of adults, emphasize the need for immediate and effective strategies to be developed and implemented on the farmlands if the Namibian cheetah population is to be conserved. Possible ways of accomplishing this are discussed below.

Developing conflict resolution strategies

To develop effective conflict resolution strategies aimed at the long-term conservation of farmland cheetah populations, it is crucial to understand the reasons behind their widespread removal. Of 376 cheetahs where we were able to gauge reasons for capture, 91% (n = 373) were trapped as they were perceived to pose a threat to livestock or farmed game (Marker *et al.* 2003b). The majority of these (198 animals) were trapped to protect livestock (Fig. 15.5), while the remaining 145 were captured on game farms (Marker *et al.* 2003b). However, although livestock farmers removed a greater number of cheetahs, the ratio of livestock/game farms in the study was 3:1 (Marker *et al.* 2003b), meaning that game farmers actually removed proportionally more cheetahs. This difference may be linked to the increase in game prices (more than 50% in the 10-year period prior to the survey; Erb 2004). In addition, many game farms are stocked with rare and valuable species such as roan (*Hippotragus equinus*) and sable (*Hippotragus niger*) antelope, and these prices increased from US$6500 in 1990 to US$13000 in 2000 (Erb 2004).

It is clear from these figures that the farmers' perceptions of threat to either their livestock or farmed game are the driving forces behind cheetah removals and that effectively addressing these perceptions will be a prerequisite to successful conflict resolution. Possible mitigation strategies on both livestock and game farms are discussed below.

Conflict mitigation on livestock farms

Our initial farm survey clearly showed that farmers strongly believe that cheetahs pose a significant threat to their livestock, but further research has shown that the reality of the threat is probably far less than is commonly perceived (Marker *et al.* 2003c). With over 70% of the country's game species found on farmlands, where livestock and wildlife coexist, cheetah have greater access to livestock than wild game; however, scat analysis from wild cheetahs revealed a strong preference for wild prey, with domestic stock being detected in only 6.4% of scats (Marker *et al.* 2003e). Moreover, there is scant evidence that many of the cheetahs trapped because

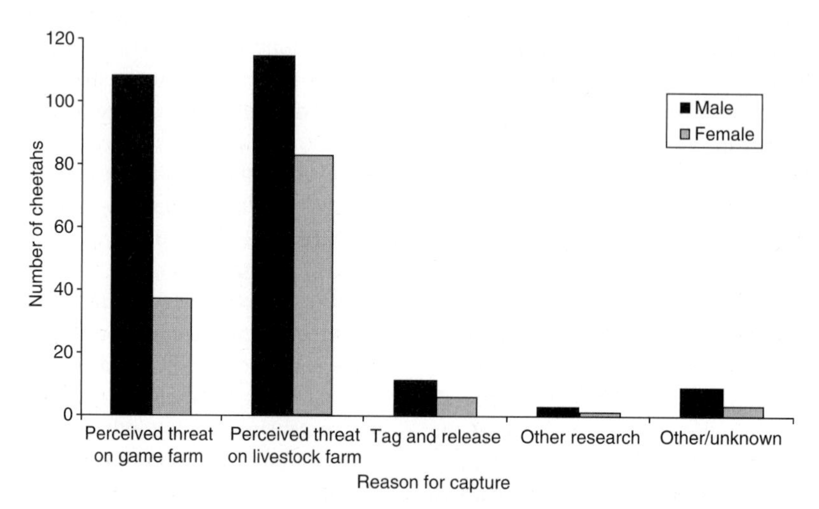

Figure 15.5 Reported reason for capture of 376 cheetahs examined between 1991 and 1996 (Marker *et al.* 2003b).

they were regarded as 'problems' were actually causing any depredation at all. We investigated the circumstances surrounding all 198 cases where cheetahs were captured due to a perceived threat on a livestock farm and looked for any evidence that the animal concerned was indeed a problem—this evidence included current predation, being caught on a livestock kill, and capture in or attempting to enter an enclosed livestock kraal near human habitation (Marker *et al.* 2003b). Such evidence was found in only six cases (3%), and in four of these cases the cheetahs exhibited physical problems or injuries that would hinder normal hunting: three had injured feet or broken teeth, while one was emaciated and in very poor condition (Marker *et al.* 2003b). Overall, 11.4% of 376 cheetahs examined ($n = 43$) had such physical problems. We found a statistically significant relationship between the occurrence of such problems and the likelihood of there being direct evidence of a cheetah being implicated in livestock depredation, indicating that cheetahs with a health or physical problem were more likely to become a problem animal (Marker *et al.* 2003b).

However, even if our data suggest that cheetahs, particularly normal, healthy individuals, are unlikely to pose much risk to livestock, the important fact remains that farmers still consider them as problematic and are therefore removing them by the hundreds. As has been seen in many previous studies of human–predator interactions, it is crucial to address people's perceptions, erroneous or not, as they are what ultimately drives removals (Rasmussen 1999; Sillero-Zubiri and Laurenson 2001). Therefore, the first step in reducing conflict here is to acknowledge farmers' concerns and work with them to minimize the perceived threat as effectively as possible.

Diverse techniques have been explored to alleviate predation on stock (Sillero-Zubiri *et al.* 2007), but the one that seemed likely to have most relevance to the Namibian situation was the use of specialized livestock guarding dogs. The approach is well known and is developed from the mediaeval precedent using large, specially bred dogs to protect transhumance sheep flocks from wolves (*Canis lupus*) in Europe (Coppinger and Coppinger 1980; Landry 1999). This form of husbandry has been revived through the use of dogs, as well as guarding animals such as donkeys, mules, and llamas (Marker-Kraus *et al.* 1996), in order to discourage predation by coyotes (*Canis latrans*), wolves, and bears (*Ursus arctos*), as well as theft by people (Andelt 1995; Marker *et al.* 2005a).

The objectives of the guarding dog programme were to evaluate how effective these dogs could be in terms of reducing livestock losses on the Namibian farmlands, with the ultimate aim of developing an appropriate method for diminishing the level of conflict with farmland predators and helping to foster a sustainable coexistence of predators and people.

In 1994, 10 Kangal/Anatolian Shepherds were imported to Namibia from the Birinci kennels in the United States: 145 dogs were subsequently bred on selected farms in Namibia in working environments and 143 were placed as working guardians (Fig. 15.6). Two additional males were later imported for breeding purposes. All the dogs bred in-country were kept with their mother for 6–8 weeks, in a kraal with livestock in the vicinity of the puppies (Marker *et al.* 2005a). Until 2003, all livestock guarding dogs were provided to farmers free of charge, with CCF bearing all the costs for breeding, raising, vaccinating, and neutering young dogs. We neutered all dogs placed as guardians to avoid interbreeding with local dogs, and we found that neutering made no difference in the effectiveness of guarding dogs (Marker *et al.* 2005a). After 2003, commercial farmers were asked to pay the costs incurred while raising the puppy to placement age and for its neutering, although all costs were still covered for owners on communal farms. In 2003, the cost for commercial farmers usually came to N$800 (approximately US $130) for both male and female puppies, including neutering, which still made them very cheap compared to the sale price of such dogs in South Africa, where livestock guarding dogs routinely fetch around N$4000 (US$600)

All potential owners completed a questionnaire in order to assess levels of livestock depredation on farms prior to guarding dog placement. The performance of the dogs was evaluated on the basis of 334 follow-up questionnaires conducted on 117 farmers who had received guarding dogs between August 1995 and January 2002 (Marker *et al.* 2005a). These surveys included questions related to the dogs' attentiveness, trustworthiness, and protectiveness—three

Figure 15.6 A Kangal breed of Anatolian Shepherd livestock guarding dog working independently with its charges (goats and sheep) on Namibian farmland. © L. Marker.

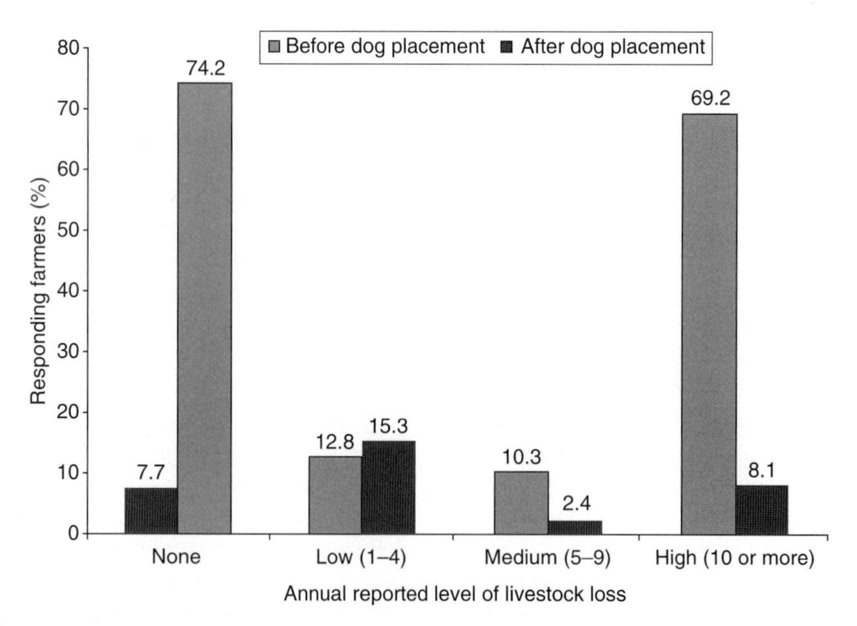

Figure 15.7 Annual levels of livestock loss reported by farmers before getting a livestock guarding dog (light grey), and levels of loss reported once they had owned a livestock guarding dog for more than a year (dark grey; Marker *et al.* 2005a).

key areas for effective livestock guardians (Coppinger and Coppinger 1980)—as well as ones related to farmer satisfaction with the programme.

Livestock guarding dogs had a substantial impact on the level of livestock depredation suffered: 69% of farmers lost more than 10 head of stock per year to all predators (i.e., jackal, cheetah, and leopard) before getting a dog, whereas 74% of them reported no depredation at all once they had owned a guarding dog for a year (Fig. 15.7; Marker *et al.* 2005a). Perhaps even more importantly, the farmers recognized the benefits of the programme. Sixty-nine per cent reported their dog's behaviour as good or excellent, 70% regarded the dog as an economic asset, and 93% said they would recommend the programme (Marker *et al.* 2005a).

However, the programme was not flawless. Although dogs scored highly on attentiveness (mean score of 0.88 on a 0–1 scale) and protectiveness (mean = 0.71), they showed relatively low trustworthiness around stock, with a mean score of 0.44 (Marker *et al.* 2005a). Ninety-five per cent of farmers reported behavioural problems with their dogs at some point: The most common issue was dogs chasing game, reported in 46% of cases, while a quarter of farmers reported dogs harassing or killing livestock. This was a particular problem for dogs under a year old (Marker *et al.* 2005a). However, the majority of problems (61%) were solved with the corrective training recommended by CCF.

Over a third of the dogs (34%, $n = 49$) that entered the programme were reported to have died while they were still actively working as guardians (Marker *et al.* 2005b). On average, dogs had a working lifespan of 4.31 (SD = 0.31) years. Females lived slightly, but not significantly, longer, at 4.65 (SD = 0.45) years compared to 4.16 (SD = 0.40) years for males (Marker *et al.* 2005b). Accidents were the single most common cause of working dog deaths ($n = 22$), followed by culling by the owner ($n = 12$), presumably because the farmer was dissatisfied (all incidents of culling occurred on commercial ranches, while all cases of disease and poisoning occurred on communal farms). The mortality rate reported here is lower than the 48% reported for guarding dogs used in the United States (Lorenz *et al.* 1986). The effectiveness of the dogs at reducing losses and the satisfaction of the farmers suggest that this is a valuable strategy for reducing livestock depredation (Marker *et al.* 2005b).

Conflict mitigation on game farms

Although resolving conflict on livestock farms is a critical aspect of farmland cheetah conservation, it in no way addresses the entire problem. As mentioned above, game farmers, who made up only 25% of farms reporting captures, were responsible for trapping 42% of the cheetahs whose sites of capture were known (Marker *et al.* 2003b). Our radio-tracking data showed extensive spatial movements of cheetah throughout all farms, without concentration on game farms. Therefore, reducing this disproportionately high level of removal on game farms is of critical importance for long-term cheetah conservation. However, addressing the perceived threat to game species is a tougher proposition. Scat analysis on the farmlands showed that cheetahs were hunting a significantly higher percentage of natural prey species such as kudu or gemsbok (Marker *et al.* 2003e), and hunting such species is not aberrant predator behaviour. In addition, our surveys with the farmers corroborated these studies (Marker *et al.* 2003c).

Trophy hunting of game species is a huge industry in Namibia, bringing in US$285 million/year (Terra Daily 2008), with trophy prices of US$1800 for an eland, US$1000 for a kudu, and US$600 for a gemsbok or red hartebeest—even a duiker or steenbok can command US$250 (Ndandi Namibia Safaris 2008). These prices exclude additional revenue associated with a trophy hunt such as guide time and taxidermy, so rearing animals for hunting can be a very lucrative business compared to livestock, where a domestic cow might command only US$200 at market (Marker *et al.* 2003a). However, with kudu and eland comprising at least 61% of farmland cheetah diet on both game farms and livestock ranches (Table 15.2), supporting a healthy cheetah population on game farms alone clearly carries a substantial price tag for these farmers.

Traditionally, the predominant line of defence for Namibian game farmers has been electric fencing, supplemented by killing any cheetah that transgress these boundaries (Marker 2002). Farmers report electrification to be 70–80% effective, but the price of fence maintenance is frequently prohibitive (Marker-Kraus *et al.* 1996). One of the main problems with fencing comes from warthogs (*Phacochoerus aethiopicus*), aardvarks (*Orycteropus afer*), and porcupines (*Hystrix africaeaustralis*), all of which dig holes under fences and hence provide access for cheetahs and other predators (Schumann *et al.* 2006). A possible mitigation of this problem is the use of 'swing gates', which, based on the principle of cat flaps, allow these animals to traverse fences without creating access for cheetahs. To examine the possible efficacy of this technique, we installed 14 swing gates, which consisted of a 45 cm by 30 cm metal frame covered with galvanized fencing (mesh size 75 mm), along a 4800 m section of game fencing known to be a common access point for cheetahs into the game-fenced area (Schumann *et al.* 2006). Heat-sensitive cameras and spoor tracking were used to examine which species used the gates during three phases of the project—before gate installation, when gates were installed but tied open, and when gates were closed.

Before gate installation, various species used open holes to breach the fence; warthogs were recorded most commonly (in 58% of cases), but black-backed jackals (*Canis mesomelas*; 22%), aardvarks (17%), porcupines (2%), cheetahs (0.5%), and leopards (*Panthera pardus*: 0.5%) all crossed the barrier this way. After holes were filled and gates were installed and tied open, warthogs used the gates most frequently (85% of cases), and steenbok (5%), baboons (*Papio ursinus*: 5%), jackals (2.5%), and aardvarks (2.5%) all passed through the open gates. After gate closure, only warthogs (73%), porcupines (18%), and aardvarks (9%) were recorded as using them to pass through the fence (Schumann *et al.* 2006). Significantly, although leopard, cheetah, jackal, caracal (*Felis caracal*), and African wildcat (*Felis nigripes*) were all photographed passing by the gates after installation, no predators were recorded passing through the game fence once swing gates had been installed (Schumann *et al.* 2006).

This technique, where materials cost US$40.32/km using six gates per kilometre, was 18 times cheaper than electrifying a fence, which costs a minimum of US$752/km for materials alone (Schumann *et al.* 2006). Moreover, labour costs were US$4.32/km for installing 1 km of swing gates, whereas 12 times more workers are needed to install 1 km of electric fencing, entailing labour costs of US$51.84/km (Schumann *et al.* 2006). Electric fences also require checking twice a week, as well as the application of herbicides to prevent plants short-circuiting the fence, all of which add more economic costs to this strategy (Schumann *et al.* 2006).

However, despite the operational and economic promise of using swing gates in fences, their success raises the paradox that, if widely applied, they will deny cheetah access to the landscape on which they depend. Therefore, this should not be seen as a long-term solution for cheetah–game farmer conflict, but rather a useful non-lethal method for reducing predator access to small game-fenced areas and alleviating intense conflict in the shorter term, while other, longer-term coexistence strategies can be developed and implemented.

One of the most promising of these longer-term strategies is the conservancy concept, where land owners or occupiers pool and manage their resources collectively, with the aim of sustainable long-term management and conservation of the land (Marker-Kraus *et al.* 1996). The first freehold commercial Namibian conservancy was established in 1992, followed by the first four communal conservancies, gazetted in 1998. As of 2006, there were 64 conservancies (Fig. 15.8), covering over 19% of Namibia's surface area (over 12 million ha) and supporting more than 130,000 people (NACSO 2006). In addition, many conservancies border on other conservation areas, thus increasing the conservation management areas and creating more connectivity and open systems, and therefore allowing animals to move freely and extensively. These conservancies also provide for increased tourism concessions, which directly benefit the conservancy members (NACSO 2006). The conservancy model can have various benefits. A survey of 1192 households across seven conservancies in Namibia showed that conservancy membership had a positive impact on household welfare (Bandyopadhyay *et al.* 2004), with direct income and benefits to conservancy members estimated at over US$2.2 million in 2004 (USAID 2005).

The management of conservancies involves stewardship over wildlife and its habitat, and this has been demonstrably associated with population increases for endangered species such as the black

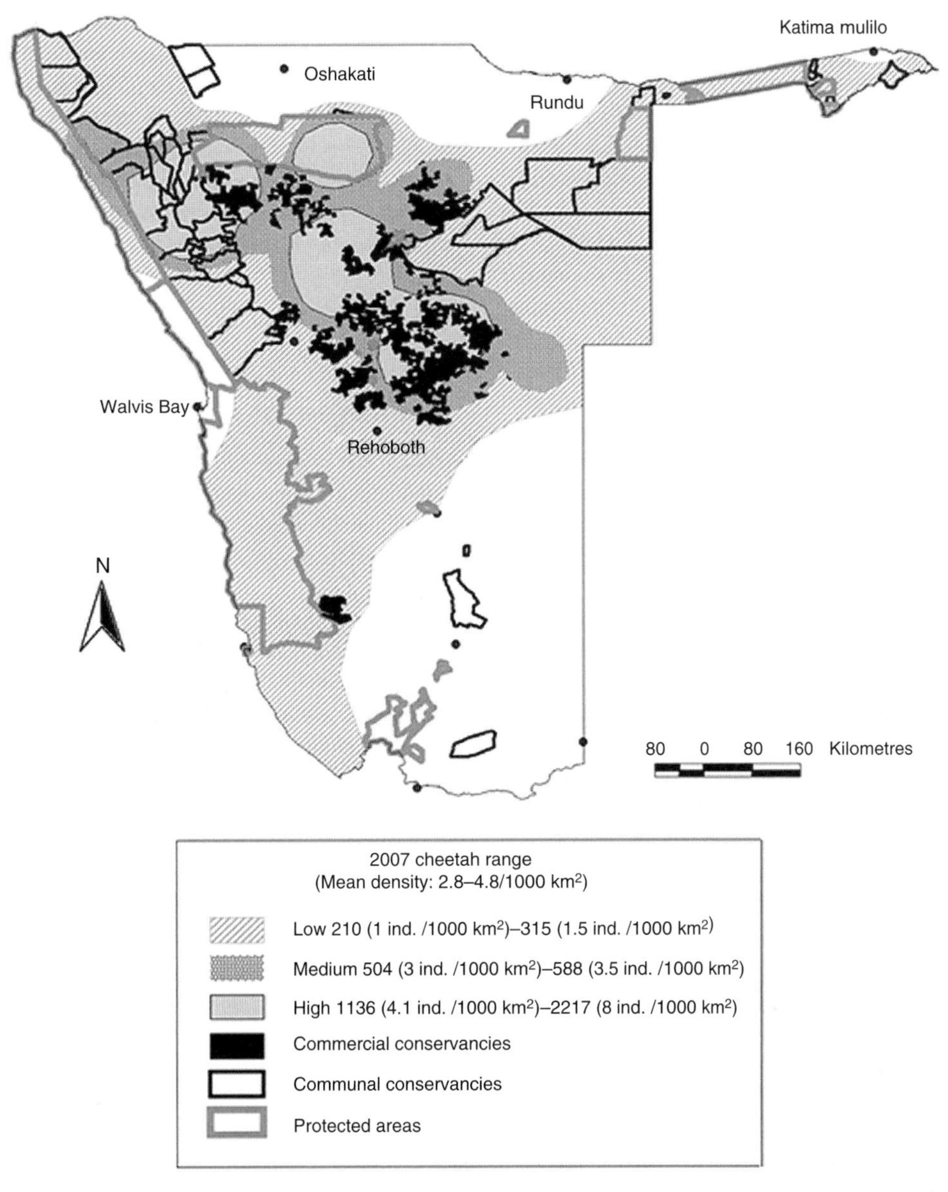

Figure 15.8 Distribution and density (high, medium, and low) estimates of cheetah in Namibia based on sightings and observations, overlaid with communal and commercial (freehold) conservancies (Marker *et al.* 2007).

rhino (*Diceros bicornis*), whose population has doubled over the past 20 years, and Hartmann's mountain zebra (*Equus zebra hartmannea*), whose population has increased by over 30% from approximately 9500 in 2001 to over 12,500 in 2005 (USAID 2005; NACSO 2006). Annual harvestable value of wildlife to landholders has increased dramatically since conservancy formation: in north-west Namibia this figure increased from almost nothing in 1994 to nearly US$5 million in 2007 (NACSO 2008). Whereas individual landholdings, with their local preoccupations, may feel compelled to act against predators such as cheetah, the perceived sharing of risk within the conservancy model (with its larger area and multiple landholders) is associated with more benign, landscape-scale, tolerance of predators (ABSA 2003). Due to the increase and management of wildlife in these conservancies, there has been a marked increase in density and a shift in range use of cheetahs in the past 10 years (Fig. 15.8; Marker *et al.* 2007), in particular with their movement into the north-western communal conservancies where cheetah sightings have increased by 12-fold in just 4 years (NACSO 2006). This has also increased human/livestock and predator conflict. However, with conservancy management systems in place, conflict can be better mitigated, thus reducing livestock losses and allowing for other benefits to communities for having predators. In the Nyae Nyae communal conservancy, in the north-east of Namibia, the cost of keeping leopards was calculated and the average annual loss per adult conservancy member caused by leopard predation on livestock was US$0.03, while annual benefits per adult were US$6.00 through ecotourism (Stander 2001a).

Trophy hunting of cheetahs as an enhancement of the species

The cheetah is recognized as a sustainable resource in Namibia, and Namibia is one of three countries where a quota exists for the trophy hunting of cheetahs (CITES 1992). It has been proposed that trophy hunting offers an economic incentive for landowners to increase tolerance towards cheetahs.

In an effort to support long-term conservation strategies for the cheetah, the Namibian Professional Hunters Association (NAPHA) developed a special committee called RASPECO (Rare Species Committee). In 1994, this committee developed guidelines and programmes that support sustainable utilization of rare species, such as the cheetah, for the enhancement of the species and to offer an economic incentive for landowners to increase tolerance towards cheetahs. It was envisioned that through the development of this programme the indiscriminate catching and killing of cheetahs would stop, thus securing a future for the animal in Namibia.

As part of this programme, NAPHA members were asked to participate by signing a 'COMPACT for the Management of Cheetah' on their farms, ensuring that all trade in cheetahs is sustainable and legal and that hunting quotas were enforced. NAPHA members represent about 15% of the 3000-plus landholders in Namibia. As of November 1998, more than 190 hunters had signed, comprising 1.5 million ha and representing more than 70% of the land where NAPHA members hunt (US Federal Register 2000).

Incentives for conservation

Despite the promise of many of these strategies outlined above, merely lessening the costs of predation will never be enough to encourage the long-term conservation of cheetah populations on Namibian farmlands. For this to happen, farmers must value cheetahs in some capacity and feel that they benefit from them—one obvious example would be their capacity, as with other large cats, to generate revenue through ecotourism. Grounds for optimism in this regard include the often cited example of lions in Amboseli National Park, which are estimated to generate US$0.5 million each, annually, through tourism (Western and Henry 1979; Emerton 1998). Unfortunately, the cheetah's secretive habits, combined with the predominance of thick bush, make this option unpromising on Namibian farmlands.

An alternative approach pioneered by CCF is the 'Cheetah-Country Beef' initiative. Developed from the model of 'dolphin-friendly tuna', where the ethical concerns of consumers drive industrial standards that are benign to biodiversity, this scheme provides an incentive for farmers to abide by guidelines for predator-friendly farming, paramount among which is not to kill cheetah indiscriminately (Marker 2002). Qualifying farmers will be able to market their beef under the imprimatur of 'Cheetah-Country Beef', attracting premium prices. Our business plan shows an increasing amount of beef to be certified over the first 5 years as farmers join the scheme due to the price premium provided to the farmers. The premium has been calculated to be, on average, £0.77/pound or an increase of £70,000–347,000 in profit to the group of participating farmers (Fig. 15.9; Bell 2006). As of yet, this initiative has not been

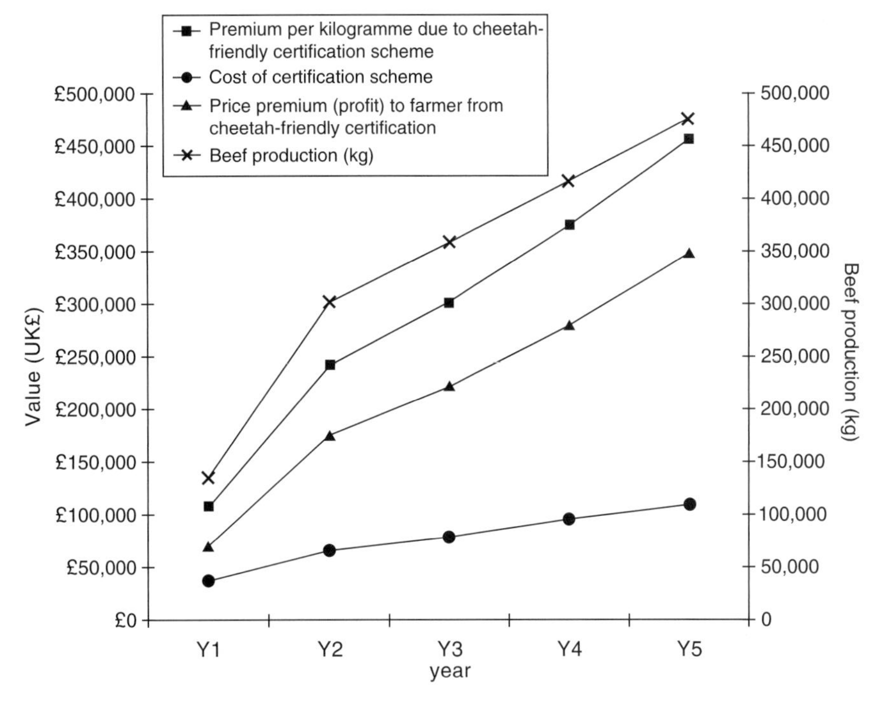

Figure 15.9 Using the minimum recommended price premium provided to the farmers during a 5-year period, an increase of income to the farmers would be generated from the cheetah-friendly certification process, which would increase the number of cheetah-friendly farmers, thus increasing the beef production (kg) and decreasing the cost of certification (Bell 2006).

launched; however, potential European sales are currently being negotiated.

CCF has also pioneered the 'Bushblok' programme. Since the 1970s, due to increasing aridity, livestock overgrazing, fire suppression, and the extirpation of megaherbivores (Marker 2002), the north-central mixed woodlands known as the Namibian farmland savannahs have been encroached by dense, *Acacia*-dominated thornbush. This thornbush has encroached over 14 million ha of farmland, primarily since the severe 1980 drought (Marker-Kraus *et al.* 1996) and has reduced beef production on the affected land by up to 30% or 34,000 t/year, which is worth about N$102 million (US$25 million; Quan *et al.* 1994). It also has deprived cheetahs of a favoured habitat (Muntifering *et al.* 2006) and increased their risk of eye injuries (Marker-Kraus *et al.* 1996; Bauer 1998). If these trends continue they are further likely to disadvantage both farmers and cheetahs and exacerbate conflict between them. Adopting well-proven technology, CCF initiated a scheme where encroaching bush is selectively harvested, compacted into 'Bushblok' fuel logs, and sold overseas. The profits will then be used to subsidize the prices of fuel logs to local communities. The project not only generates foreign currency, increases productivity, and improves habitat, it also reduces local deforestation by providing alternative sources of firewood. Each 1000 tons of finished Bushblok represents 12 jobs for harvesters in the field. A further 12 find employment at the processing factory. While still in the start-up phase, eventually profits from each 1000 tons will allow sufficient subsidy to internal sales to replace unsustainable firewood harvest.

One of the greatest stimuli to valuing wildlife is education. In addition to running workshops on predator ecology, conservation, and management, CCF has published two guidebooks for farmers on predator and livestock management, which have been distributed to over 10,000 Namibians (Marker-Kraus *et al.* 1996; Schumann 2003). Looking ahead, CCF has pushed for inclusion of predator-related topics in the national curriculum, and between 1991 and 2006, CCF's education staff worked with over 220,000 students at schools and those visiting the research centre and have developed a range of educational materials for teachers to use in local schools. In addition, over 750 subsistence, emerging and resettled farmers, and conservancy members have been trained in week-long courses, and over 100 young professional conservation scientists and educators from 15 countries have been trained in month-long training courses. Course evaluations from the 2006 integrated livestock, wildlife, and predator management courses showed that 69% of the trainees rated the overall courses as very good and 31% as good. Participants were asked to rank the most important topics taught. When looking at the overall selection of topics, the economics of livestock management ranked as the most important topic, while the value of predators ranked second, and veterinary care and herd health ranked third. Rangeland management and predator track identification tied for fifth place. The majority (88%) of the participants said identifying livestock losses to predators was new to them, and more than 80% said both biodiversity and raising and training livestock guarding dogs was new information (Schumann and Fabiano 2006).

Conclusions

Through employing a wide variety of approaches, CCF has made important strides in reducing the costs and improving the benefits of human–cheetah coexistence on the Namibian farmlands. The fact that almost three-quarters of farmers with livestock guarding dogs reported no depredation a year after dog placement proves that relatively simple conflict resolution strategies can be effective on livestock farms, while innovations such as swing gates can assist game farmers in reducing the threat of predators on their farms as well. On a larger scale, incentive programmes to encourage conservation-friendly land management, such as the conservancy movement and the certification of 'Cheetah-Friendly' products, have been well received and have the potential to reap substantial benefits to the communities concerned. Education programmes have reached hundreds of thousands of schoolchildren, encouraging future generations of farmers to consider conservation as something that can be beneficial to them, rather than a hindrance.

Many of these benefits are intangible and difficult to measure, but numerous indicators suggest that these programmes are having an important effect in terms of cheetah conservation. Annual follow-up surveys conducted with the 241 farmers initially interviewed in 1991–93 revealed that annual removals of cheetahs on farms where they were considered problematic had dropped from an average of 29.1 (SD = 47.8) in 1991–93 to 3.5 (SD = 2.6) by 1999 (Marker *et al.* 2003a). Removals on farms where cheetahs were not considered problematic also dropped significantly from a mean of 14.3 (SD = 26.9) per year per farm in 1991–93 to 2.0 (SD = 3.1) by 1999 (Marker *et al.* 2003a). Over the course of the study, the overall numbers of cheetahs removed annually per farm declined by almost 90%, from a mean of 19.8 (SD = 35.9) in 1991 to 2.1 (SD = 2.6) in 1999 (Marker *et al.* 2003a). Farmers were also more likely to allow the release of trapped cheetahs back onto the farmland during later years of the study, although this difference was not statistically significant (Marker *et al.* 2003b). Indeed, in the latter years of the study, the level of cheetah removals was significantly correlated with actual stock losses rather than being conducted as a purely preventative measure, as seemed to be the case in the initial surveys (Marker *et al.* 2003a). Research on these farmers' attitudes also revealed an increased tolerance for cheetahs on their land between 1991 and 1999, with fewer cheetahs removed per head of livestock lost (Marker *et al.* 2003a).

Overall, this case study from the Namibian farmlands highlights the complexity of trying to conserve large wild felids long term on private land, and where dealing with the resulting conflict requires a broad, multifaceted, and flexible approach. However, progress can be made. Data suggest that local landowners in Namibia have become increasingly tolerant of cheetahs on their land since CCF has worked there, and the number of cheetah removals has dropped substantially over time (Marker *et al.* 2003a). Although many challenges undoubtedly lie ahead, both in Namibia and elsewhere, especially as land tenure systems change and environmental challenges become more acute, this programme has shown that there are tools that can be used to help people and wild felids coexist on private land, hopefully to the long-term benefit of both.

CHAPTER 16

Past, present, and future of cheetahs in Tanzania: their behavioural ecology and conservation

Sarah M. Durant, Amy J. Dickman, Tom Maddox, Margaret N. Waweru, Tim Caro, and Nathalie Pettorelli

A cheetah mother and three of her five cubs on the Serengeti plains in Tanzania. © Sarah Durant.

Introduction

The cheetah (*Acinonyx jubatus*) is famed for its speed, which has been recorded at 103 km/h (Sharp 1997), making it the fastest creature on land. Despite a hunting strategy reliant on high speed, cheetahs are habitat generalists, occupying deserts, grassland savannahs, and thick bush (Myers 1975). However, despite this, the cheetah presents major challenges for conservationists in the twenty-first century. Formerly widely distributed across Africa and southwest Asia, the species has experienced dramatic reductions in numbers and geographic range in recent decades (Ray *et al.* 2005). All large cats need large areas to survive; yet, cheetahs range more widely, and hence need larger areas, than most other species. In this chapter, we draw together findings from a 30-year study of free-ranging cheetahs in the Serengeti National Park in Tanzania, targeted studies on human–cheetah conflict outside the protected areas, and an analysis of cheetah distribution to explore issues affecting the conservation of cheetahs within Tanzania. Finally, we discuss an approach for conserving the species across its range.

The long-term study of cheetahs on the Serengeti plains was initiated by George and Lory Frame in 1974–77, continued by TC over 1980–90, and by SMD from 1991. The Serengeti Cheetah Project's historical study area covers some 2200 km^2 of short- and long-grass plains and the woodland borders in the south-east of the Serengeti National Park. The project locates cheetahs within this area and uses a card index system and, more recently, computer software to identify individuals by unique spot patterns on their face and body. Demographic information including data on reproduction, survival, and ranging patterns has been collected on each individual cheetah in the study area throughout the study. The long-term project has more recently spawned a number of more short-term projects in different areas in Tanzania focused on human–cheetah conflict and establishing the distribution of the species across the country.

Cheetah ecology and behaviour

Mating system and ranging patterns

The long-term study of individual cheetahs on the Serengeti plains has shown that cheetahs exhibit a social system not yet observed in any other cat species. Cheetah females are solitary, but do not maintain territories, and remain tolerant of other females, having large, overlapping home ranges (25% average lifetime overlap for unrelated females and 50% average lifetime overlap for mothers and daughters; Caro 1994; Laver 2005). Cheetah males can be territorial (29% of 87 males; Caro and Collins 1987b) and can also be social. Fifty-nine per cent of male cheetahs in the Serengeti form groups, forming permanent coalitions of two or three individuals (40% and 19% out of 110 males, respectively; Caro and Collins 1986), usually brothers (83% of 22 coalitions were comprised of brothers; Caro and Collins 1986), which stay together for life (Caro and Durant 1991). Males in groups are more likely than single males to take and retain territories against male intruders (67% of adult male coalitions vs <5% of singletons held territories; Caro and Collins 1987b). In the Serengeti, radio-collaring has shown that male territory size averaged 48 km^2 (n = 9, SE = 7.6), while females ranged over 833 km^2 (n = 19, SE = 85.1), respectively, over 2–4 years (Caro 1994). Annual ranges for females are lower, averaging 596 km^2 (n = 10, SE = 66; Laver 2005). Estimates of home range for non-territorial males are only available from non-radio-collared individuals (located during routine searching) measuring 777 km^2 (n = 9, SE = 153.0), similar to that for females. This system, where males are social and hold small territories, and females are solitary, moving across several male territories each year, has not been reported for any other mammal species (Caro and Kelly 2000; Gottelli *et al.* 2007).

Recent genetic analysis has demonstrated that female cheetahs are highly promiscuous, with high levels of multiple paternity within litters: Of 23 litters with more than 1 cub, 43% had 2 or 3 fathers. There is no evidence of mate fidelity; of eight females where there were data from two to three litters, none returned to the same male (Gottelli *et al.* 2007). It has been hypothesized that territoriality in male cheetahs provides them with increased access to females. This is supported by observations that males' territories are centred on areas with relatively high densities of female cheetahs (Caro 1994). Territories tend to occupy areas that may be particularly attractive to females, containing important resources such as cover or water (Caro and Collins 1987a). Scarcity of suitable territory sites may lead to high male–male competition and hence selection for unrelated as well as related males to form groups to gain better access to territories, leading to the formation of permanent male coalitions (Caro 1994). This is supported by the fact that males in coalitions are more likely to take and retain territories than are single males (see above), and that territorial males are in better health and condition than non-territorial males (Caro *et al.* 1989). Genetic data are currently insufficient to test whether territoriality results in clear reproductive benefits for males.

This complex and unusual system can be confusing for the inexperienced observer and the details given above could only be uncovered by long-term study. Cheetahs appear strangely tolerant of each other; female cheetahs are often found a few hundred meters from other females, concentrated around transient (such as prey and water) and permanent resources (such as trees and kopjes) on male territories, which in turn tend to attract non-territorial males and males from adjacent territories, leading to a clustering of cheetahs in ephemeral hot spots (Durant 1998). The picture is further confused by the fact that territorial males make temporary forays off their territories when prey are scarce (Caro 1994). In general, in the study area, once a cheetah is found by an observer, another cheetah is likely to be nearby (Durant 1998). Non-territorial males, however, still risk being killed by resident territorial males (Caro 1994), but potential reproductive benefits from the abundance of females found on male territory sites may outweigh the risks.

Demography

Cheetahs give birth to their first litter at 2 years after a 3-month gestation (Caro 1994). The cubs are kept in a lair for the first 2 months of their life, during which time their mother leaves to hunt every morning and returns at dusk (Laurenson 1993). Cheetah cub mortality can be high, especially during the denning period: On the Serengeti plains and woodland borders, in 26 out of the 36 litters monitored by Laurenson (1994), all cubs died before emergence from the lair. Cubs died mostly (73%) because they were killed by lions (*Panthera leo*) or spotted hyenas (*Crocuta crocuta*). Their mother cannot defend them against these much larger predators (Laurenson 1994). Cubs also died from abandonment if their mother was unable to find food, and a single litter each died from exposure and fire. In the Serengeti, mortality of cubs from birth to independence at 14 months was recorded at 95% (Laurenson 1994). However, Laurenson's study was conducted during a period of high lion density, when the cheetah population was in decline and mortality is almost certainly lower during periods of lower lion density (Laurenson 1995c; Durant *et al.* 2004). Female age is also important: a female becomes more successful at rearing cubs to independence as she gets older (from 2 to 6 years) and then success rate declines after year 7 (Durant *et al.* 2004).

If cubs survive, they will stay with their mother for an average of 18 months, after which they will roam with their littermates for a further 6 months (Caro 1994). At this time, females split from their siblings and go on to produce their first litter, while surviving males will stay together for life. Males appear to benefit from this period of adolescent sociality because males with sisters have higher survival than males on their own or accompanied only by their brothers (Durant *et al.* 2004), which could be explained by greater access to food caught by their more accomplished sisters (Caro 1994). Subsequently, coalitions of brothers may reap reproductive benefits through increased access to territories. There are no clear overall survivorship benefits or costs to adolescent sociality for females accompanied by brothers or sisters (Durant *et al.* 2004), despite possible group-related antipredator benefits, and it is not uncommon for females to separate early during this period, particularly in large litters. During adolescence, cheetahs are especially social, frequently approaching unknown cheetahs. Single males may meet and join up with unrelated males to form a coalition; males in coalitions may acquire additional coalition members and mixed groups may occasionally acquire extra female or male adolescents. Such associations may last from a few hours to months, or even years, in the case of male coalitions. Mean mortality during adolescence from year 1 to 2 is 0.32 for females and 0.61 for males (Durant *et al.* 2004), probably because males begin suffering the impacts of male–male aggression as they approach adulthood.

Longevity in the wild may reach 14 years 5 months for females and 11 years 10 months for males (S.M. Durant, unpublished data), however females have never been recorded as reproducing beyond 12 years (Durant *et al.* 2004). Mean annual adult mortality is 0.15 for females and 0.31 for males with variances in these rates of 0.125 and 0.210, respectively (Durant *et al.* 2004). In Namibia, juvenile mortality is much lower than in Serengeti, probably due to the lack of other large predators in the agricultural landscapes in Namibia. Adult survival is similar in the two areas (Marker *et al.* 2003a; Durant *et al.* 2004; Marker *et al.*, Chapter 15, this volume). Causes of adult mortality in the Serengeti are difficult to discern due to the rapid disappearance of carcasses in this scavenger-rich ecosystem. However, both female and male adult cheetahs have been killed by lions; a female died from internal injuries received while bringing down an adult male Grant's gazelle (*Gazella granti*); and a male died from encephalitis (although to date, no mass mortality due to a disease epidemic has been recorded in this ecosystem). Male cheetahs are killed by other males, probably during territorial disputes (Caro 1994).

Food

Cheetahs are predominantly diurnal, and hunting activity is highest during the early morning and late afternoon (Schaller 1968, 1972), although hunting can occur throughout the day and at night (Caro 1994). Normally, a fast chase follows a concealed stalk. Because of their unrivalled speed and acceleration, cheetahs can hunt successfully even if they start a chase at a much greater distance than bulkier and heavier large cats, such as lions and leopards. They take a wide variety of prey, depending on habitat and geographic location, but prefer medium-sized prey: the size of a Thomson's gazelle (*Gazella thomsonii*), springbok (*Antidorcas marsupialis*), or juvenile impala (*Aepyceros melampus*; Mills 1984; Caro 1994; Broomhall *et al.* 2003; Mills *et al.* 2004). Prey recorded during systematic observations of kills on the Serengeti plains were 67% Thomson's gazelle, 14% hares (*Lepus* spp.), 7% Grant's gazelle, and 7% wildebeest (*Connochaetes taurinus*; Hunter *et al.* 2007a). Other recorded prey species (<5%) included impala, hartebeest (*Alcelaphus buselaphus*), Burchell's zebra foals (*Equus burchelli*), dik-dik (*Madoqua kirki*), bat-eared fox (*Otocyon megalotis*), warthog (*Phacochoerus aethiopicus*), reedbuck (*Redunca redunca*), and common eland calves (*Taurotragus oryx*; Hunter *et al.* 2007a). Prey selection has been shown using discrete choice models to vary with sex of the hunting cheetah, as well as with the seasonal availability of prey (Cooper *et al.* 2007). In woodlands, cheetahs take a greater diversity of prey, with impala being the dominant species (S.M. Durant, unpublished data).

Thomson's gazelle are partly migratory, following a more restricted movement pattern than the wildebeest migration, for which the Serengeti is famous. The gazelle move out to the phosphorus-rich grasslands in the south and east of the park when the rains start in November, and move north and west into the woodlands in the dry season around June (Durant *et al.* 1988). They generally move south slightly ahead of the wildebeest at the start of the rains, and move north and west off the plains a month or two behind the wildebeest at the beginning of the dry season, as their small mouths are better able to access small shoots of grass left behind by the broad muzzle of the wildebeest (Fryxell *et al.* 2004). In the Serengeti, nearly all cheetahs follow the Thomson's gazelle migration on and off the plains, but a few individuals remain out on the short grass plains throughout the dry season, picking off resident prey such as hares and the more drought-tolerant Grant's gazelle, or in the woodlands during the wet season persisting on resident prey (Durant *et al.* 1988). Territorial male cheetahs may make brief forays off their territories, in search of food when migratory prey is not present (Caro and Collins 1987a). However in general, male cheetahs are faithful to their territories: out of four territorial male groups, comprised of three coalitions and one singleton, only the singleton shifted territories between years (Laver 2005).

Cheetah hunting success on the Serengeti plains is generally high. On average 53% ($n = 368$) of prey are caught and killed once a cheetah starts its chase and 36% ($n = 589$) of prey are captured following the first stalk. These figures hide a wide variation in kill rates, depending on prey age (FitzGibbon and Fanshawe 1989; Caro 1994), sex (FitzGibbon 1990b), and species (FitzGibbon 1990a; Caro 1994). Generally, small prey such as gazelle fawns are associated with a high hunting success rate from first stalk (80%, $n = 100$), while large prey such as adult Thomson's gazelle are associated with a much lower success rate of 17% ($n = 355$). Hunting success varies with the reproductive state, as females suffering high demands of lactation can increase their hunting success on adult gazelle from 3% when they had no cubs, to 13% when they were denning, to 20% when accompanied by their cubs (Laurenson 1995a). Success also varies with the herd size of gazelle: 48% of gazelle kills are of solitary individuals, although 98% of gazelle are found in groups (FitzGibbon 1990b). The decision of a cheetah to hunt or not depends on a number of factors, including the prey species in the vicinity, the presence of lions and spotted hyenas, and whether a female has dependent cubs (Durant 1998; Cooper *et al.* 2007).

Competition

In the Serengeti, large carnivores have a strong impact on cheetah cub survival. For example, lions have an overall negative impact on survival of adult females, except when the population of cheetahs is high (relative to average density), suggesting that the picture is further complicated by additional, unmeasured, parameters (Durant *et al.* 2004; Fig. 16.1). On account of its smaller size, a cheetah will nearly always lose in direct competition with a lion or spotted hyena (only in a few rare instances have cheetahs been recorded as resisting a hyena approach when on a kill); 11% of kills are lost in total, 9% to spotted hyenas and 2% to lions (Hunter *et al.* 2007a).

However, there are less obvious effects of spotted hyenas and lions on cheetahs, through their indirect impact on space use by cheetahs. Predator avoidance is a key strategy employed by cheetahs to reduce predation on cubs and to avoid kleptoparasitism of their kills (Durant 1998, 2000a; Hunter *et al.* 2007). Cheetahs tend to avoid both lions and hyenas, and are found in areas of relatively low prey density, probably because high densities of prey attract the other large carnivores (Durant 1998, 2000a, b). However, both prey and predators move around the ecosystem, in part due to variation in rainfall patterns, and hence areas attractive to cheetahs are transitory in both time and space, leading to the formation of temporary hot spots of cheetahs in areas low in predators and occupied by low densities of prey, with cheetahs perhaps subsisting on lower prey densities through high hunting success. While such hot spots are transitory, they are more likely to occur on male territories, where there is an increased availability of cover (Caro 1994).

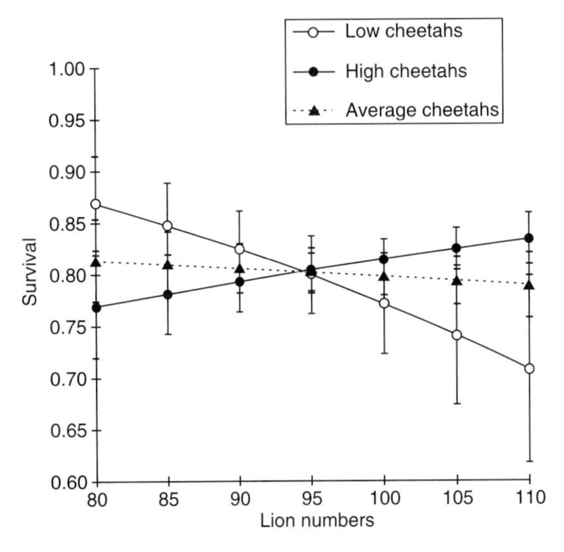

Figure 16.1 Relationship between adult female survival and cheetah and lion numbers. Results are predicted from a logistic regression model, with survival as the dependent variable (from Durant *et al.* 2004). Low cheetah numbers correspond to the lowest number of adults, average to the average number of adults, and high cheetah numbers to the highest number of adults observed in the study area during the long-term study. Lion numbers reflect the range of population size of adult female lions resident in the study area over this study.

The low competitive ability of the cheetah and its consequent predator/competitor avoidance strategy leads it to depend on high mobility to avoid other large predators (Durant 1998) and may have contributed to the evolution of the unusual combination of ranging patterns and semi-sociality observed in the species. Females with dependent cubs need to be mobile to avoid other predators. Males, which are less vulnerable because they are larger, live in small groups with associated antipredator benefits (Caro 2005) and are able to roam across smaller areas, which they try to monopolize and from which they exclude other males. Male cheetahs are also able to move more rapidly away from danger, as they are not accompanied by dependent cubs.

Individual choices, survival, and reproductive success

Reproductive success of female cheetahs varies markedly between individuals. Over 5 consecutive years, the reproductive success of 38 well-known females varied from 0 to 11 cubs surviving to 12 months (Pettorelli and Durant 2007a). Fifty-one per cent of females never managed to produce surviving cubs, while 45% of cubs originated from just 5 out of 63 matrilines (Kelly 2001). In consequence, many females never manage to produce surviving cubs, while a few successful ones are able to found cheetah 'dynasties' lasting over several generations. Individual differences are partly driven by longevity: females that live longer are able to produce more cubs over their lifetime, and their female offspring live longer (Pettorelli and Durant 2007b). As yet, it is not clear what drives these differences, but given high cub mortality from predation, predator avoidance may play a role. Previous results have linked both long life and reproductive success to successful avoidance of lions, perhaps because reproductively successful females acquire avoidance traits from their mothers (Durant 2000b).

Unusually for mammals, female reproductive success is very skewed, with a very low number of females contributing to subsequent generations (Kelly 2001), but male reproductive success may be more egalitarian, with a number of different males contributing to subsequent generations (Gottelli *et al.* 2007). Out of 15 cubs from 10 litters where paternity could be assigned, there were 10 different fathers, with one male siring a maximum of three cubs. Many fathers of cubs in the study area originate from outside the study area, while females show little mate fidelity either within litters (i.e. within a single oestrus), or between litters (Gottelli *et al.* 2007). This suggests that more males than females are likely to contribute to subsequent generations of cheetahs.

Conservation

Cheetahs used to be widespread across Africa and southern Asia into India. However, there are no cheetahs left in Asia except for a small population in Iran, and only a few populations in North and West Africa. Most cheetahs surviving today are concentrated in sub-Saharan Africa. The first status survey for cheetahs was in the early 1970s (Myers 1975). Later surveys of selected countries were conducted in the 1980s (Gros 1996, 1998, 2002; Gros and Rejmanek 1999), and a summary of current knowledge of global status was collated in 1998 (Marker 1998) and has recently been updated for eastern and southern Africa (IUCN/SSC 2007a, b, c). Information on status and distribution of this species largely comes from interviews because cheetahs are shy and rarely seen. Furthermore, the ranging patterns and behaviour of the species incline it to cluster at hot spots, making estimating numbers additionally problematic (Durant *et al.* 2007). The Serengeti volcanic grasslands in northern Tanzania, around which the Serengeti National Park is sited, are an important ecoregion for cheetahs. However, cheetahs are found in a variety of habitats in East Africa, particularly the *Acacia-Commiphora* ecoregions, and also in miombo, Sudanian savannah, and Ethiopian grasslands and woodlands (IUCNSSC 2007a). This section explores results from the long-term Serengeti study and from studies elsewhere in Tanzania in the context of cheetah conservation in the region.

Population viability

Cheetahs always live at low population densities, seldom above two adult cheetahs per 100 km^2. In the Serengeti, where densities of predators are relatively high, the cheetah population estimate is 210 adults, about 20% that of leopards, <10% that of lions, and <5% that of spotted hyenas (Caro 1994), because cheetahs are ultimately limited by the presence of other predators, rather than prey. The implications are serious, in that viable populations of cheetahs need perhaps an order of magnitude more area for their conservation than the already large areas needed by other large carnivore species.

A key benefit of a long-term study is that it enables quantification of demographic parameters of a population, not only mean survival and reproductive rates, but, most importantly, the variance in those rates. The variability in survival and reproduction are key to understanding the impact that fluctuations in demography have on population dynamics, and hence the risk of extinction (Lande *et al.* 2003). Cheetahs in the Serengeti not only suffer from unusually high cub mortality (Laurenson 1994), but also from unusually high variance in their demographic parameters (see above; Kelly and Durant 2000). This is partly driven by strong family effects, in that individuals in family groups are more likely to suffer similar fates (Pettorelli and Durant 2007a). Cubs from the same litter have more similar first-year survival than unrelated cubs, doubling the variance of the long-term reproductive success of females compared with an assumption of complete independence of fates between cubs (Pettorelli and Durant 2007a). Such effects increase overall stochasticity in demography and make small populations even more vulnerable to extinction.

In the long term, cheetah populations will depend not only on demographic viability, but also on genetic viability, as small population size makes them vulnerable to genetic problems due to losses of genetic variability (Lande *et al.* 2003). Although a small number of females contribute to future generations (Kelly 2001), a higher number of males are able to contribute than expected (Gottelli *et al.* 2007). This suggests that losses in genetic diversity from small populations may not be as high as first thought. Cheetahs are thought to h ave relatively low levels of genetic diversity, and some authors have suggested that low diversity could cause difficulties for cheetahs in reproduction and make them susceptible to disease (O'Brien *et al.* 1983, 1985b, 1986). However, these results have been disputed (Caughley 1994; Merola 1994). Results from the Serengeti study show that cheetahs in the wild do not have breeding problems and show no discernible signs of deleterious impacts of low genetic diversity (Caro and Laurenson 1994; Durant *et al.* 2007). Furthermore, disease has not had a serious impact on the population since the study began, despite a canine distemper epidemic that swept through the lion population in the Serengeti ecosystem in 1994 (Roelke-Parker *et al.* 1996).

Examination of the viability of the study population under three different scenarios (low, average, and high lion densities) showed the clear potential impact lions can have on cheetah viability (Kelly and Durant 2000). If lion density were to stay at the highest densities recorded in the study area, then the cheetah population would not be viable and would depend on immigration from surrounding areas. While this would not have posed a problem for cheetahs when they were widespread across Africa and able to move easily between source and sink, today their habitat is becoming increasingly fragmented, and their movement restricted. If cheetah populations are to remain viable, they need to interconnect to areas with low predation risk (e.g. because habitats are unsuitable for lions, because of the impacts of humans on lions, or because habitats make it easier for cheetahs to avoid lions), which are able to provide source populations to supplement temporary or permanent sinks. This demonstrates the importance to cheetah conservation of maintaining heterogeneity across an entire landscape.

Living with people

The Serengeti cheetah population is principally located in a national park without resident people. However it is impacted by tourism. Cheetahs are an important attraction for visitors to the Serengeti (Shemkunde 2004) and hence visitor fees provide indirect benefits to cheetah conservation. However, once cheetahs are seen they can be monopolized by vehicles for hours at a time, which may stay too close, disrupting hunting activity and even separating individuals. Such activities have resulted in cub deaths. The impacts of tourism can be managed through off-road driving restrictions, particularly around cheetah denning areas where cheetahs are most vulnerable, effective policing (although this is difficult across large areas), and through the education of drivers and tourists, such as through a campaign initiated in the Serengeti using posters and leaflets (Durant *et al.* 2007).

Extensive ranging patterns mean that cheetahs are likely to range outside even the largest protected areas. This takes cheetahs into contact with people beyond park boundaries, where they are vulnerable to human–cheetah conflict. In southern Africa, cheetahs are persecuted by farmers, who see them as a threat to their livestock and to valuable game species (Marker *et al.* 2003b). Human–cheetah conflict has been investigated in three areas in Tanzania. Levels of conflict were low in Loliondo Game Controlled Area, on the borders of the Serengeti, but were higher in the more populated Ngorongoro Conservation Area (Maddox 2003), and very high along the southern boundary of Ruaha National Park in central Tanzania (Dickman 2005; Fig. 16.2). The first two areas were occupied by stable low-to-medium density human populations (5 and 11 persons per square kilometre, respectively) dominated by traditional pastoralism (Maasai) with low levels of immigration, while the latter area was occupied by high density agro-pastoralist populations (32 persons per square kilometre) from a number of different ethnic groups, with high levels of immigration and high levels of underlying conflict.

The study around Ruaha has revealed that although depredation of goats is one of the key drivers of negative attitudes, lack of income from conservation and little recognition of conservation-related benefits were both linked to increased hostility towards cheetahs and other predators, while underlying factors such as ethnicity affected the degree of reported conflict (Dickman 2005, 2008). Despite

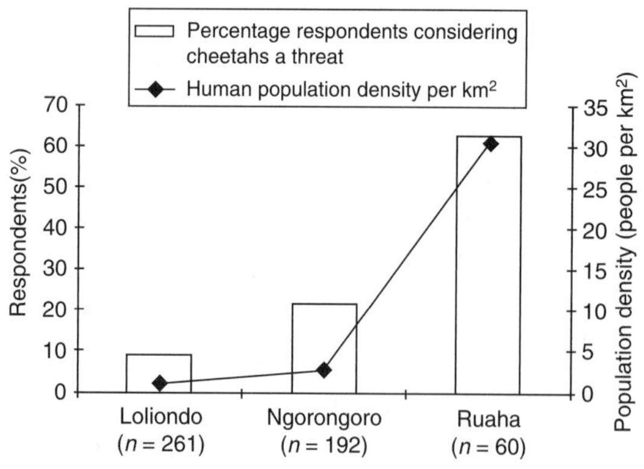

Figure 16.2 Percentage of respondents in Loliondo Game Controlled Area (Maddox 2003), Ngorongoro Conservation Area (Maddox 2003), and areas bordering Ruaha National Park (Dickman 2005, 2008) who considered cheetahs as a threat to livestock or humans. Approximate population densities for each region are also shown, as well as the total number of people interviewed in each study area.

their traditional reputation of harmonious coexistence with wildlife, the Maasai actually reported the highest conflict score with cheetahs, averaging 1.6 compared to 1.2 for other ethnic groups. Understanding the complex factors behind human–cheetah conflict will be critical for developing the most effective mitigation strategies to assist in the long-term conservation strategies for this species outside the boundaries of protected areas.

Landscape conservation

While it is important to maintain cheetahs as a component of functioning ecosystems, it is a sad fact that, due to the low density and wide-ranging patterns of cheetahs, most protected areas are too small to support viable populations. No single population in East Africa is known to harbour more than 710 adult cheetahs, that is, the population occupying the Tsavo/Serengeti/Mara landscape (IUCNSSC 2007a). Most populations extend beyond the boundaries of protected areas, and hence are vulnerable to increasing fragmentation due to accelerating economic development in the region (IUCNSSC 2007a). It is unlikely that protected areas will be increased in size in the future, and hence, if cheetahs are to be conserved in the long term, their conservation needs to take place across an international landscape of protected areas and human use areas, allowing sufficient dispersal between areas to maintain demographic and genetic viability. In 2007, 11.0% (103,685 km^2) of Tanzania's total area was estimated to hold resident cheetah populations; however, over half, 52% (54,211 km^2), of this range was outside protected areas (IUCN/SCC 2007a; Fig. 16.3).

In 2006, the World Bank ranked Tanzania as 18th lowest in its GNI (Gross National Income) and 4th lowest in its Purchasing Power Parity in the world (World Bank 2007). Tanzania had a human population of 39.5 million in 2007 in an area of nearly 1 million km^2 and a per capita income of US $350 per year (World Bank 2007). The population is growing at 2.6% per year against a sub-Saharan Africa rate of 2.3% (World Bank 2007). Nevertheless, Tanzania had positive economic growth of around 6% per year over 2004–06 (World Bank 2007). Wildlife tourism, particularly photographic tourism, is economically important to Tanzania, contributing US$824 million in 2005, 25% of the total foreign revenue (Bank of Tanzania 2007; World Bank 2007). This makes Tanzania's biodiversity, including large cats that are sought after by visitors, economically important for the country's development. However, despite the importance of wildlife resources for the economy, conservation is, by necessity, low on the list of the country's priorities. More pressing needs, such as basic education and health, inevitably take precedence. Land-use change is unavoidable when confronting the urgent needs of human development. However, if such changes are carefully planned, it is possible to prevent the loss of corridors between protected areas, ameliorating the overall impacts of land-use change, and safeguarding biodiversity. Hence a commitment to a strong land-use planning strategy would maintain wildlife tourism, while increasing Tanzania's capacity to support its human population.

Towards a range-wide conservation strategy

Global estimates of cheetah numbers were 14,000 in the 1970s (Myers 1975) and 'less than 15,000' more

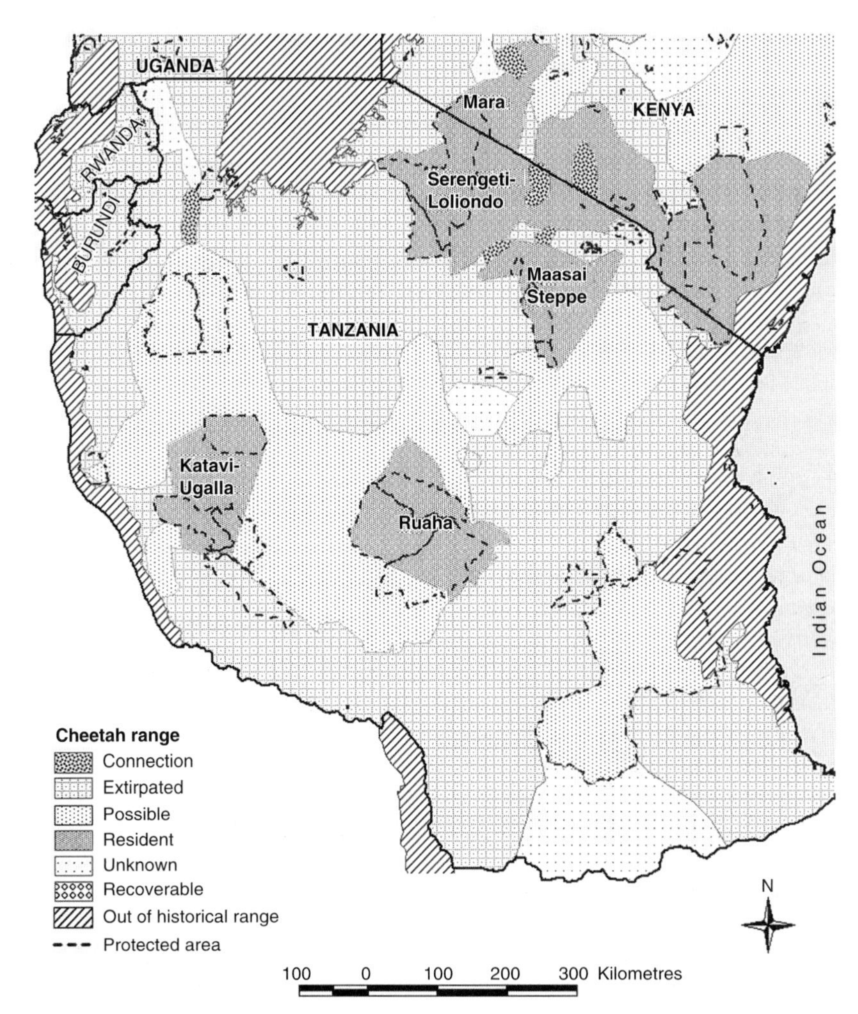

Figure 16.3 Distribution of cheetahs in Tanzania in 2007, as mapped by participants at the recent eastern African regional cheetah and wild dog conservation strategy workshop (IUCNSSC 2007a). Protected areas shown in this map include national parks, game reserves, and conservation areas, and are all within IUCN Categories I–IV.

recently (Marker 2002), but these are very approximate. The species is listed as vulnerable under the IUCN red list (IUCN 2006c). The distribution of the species has contracted markedly from its historical range (IUCN/SCC 2007a, b), and so it is likely that the 1970s estimate is an underestimate or the later estimate an overestimate. These declines are largely attributed to habitat loss and fragmentation (IUCN/SCC 2007a, b, c). The disappearance of the species from most of Asia was due to over-exploitation of its prey, agriculture, and the capture and use of cheetahs for hunting (Caro 2000). Today, in sub-Saharan Africa, persecution, due to perceived or actual conflict with livestock or game ranching, also plays a strong role in the decline of the species (Myers 1975; Marker *et al.* 2003b, c). Conservation efforts are further hampered by a lack of capacity within cheetah range states (Durant *et al.* 2007).

The wide-ranging behaviour of cheetahs coupled with increasing fragmentation requires large-scale land-use planning for this species. This depends on concerted and coordinated efforts by range states guided by a clear strategic framework. These needs led to the launch of a range-wide conservation planning process for cheetahs in 2007. In this process, cheetah conservationists joined forces with conservationists of another wide-ranging African carnivore—the African wild dog. Although superficially dissimilar, both cheetahs and wild dogs have similar ecological needs and the planning process was able to make use of the synergies between the two species.

The range-wide conservation planning process uses a series of regional workshops, bringing together specialists on the species' biology and conservation managers from governmental and non-governmental conservation organizations, to summarize current information and to develop a regional conservation strategy. These are followed by national workshops to develop national action plans in order to implement the strategy. The aim is to ensure that all current cheetah range is covered by a strategic regional plan and national action plan, with the full support of national wildlife authorities and conservation agencies, the implementation of which should secure cheetah populations into the future (IUCN/SCC 2007a, b).

Conclusions

Conservation of any species is best informed by high-quality science, and the cheetah is no exception. The study in the Serengeti has generated a wealth of information that has been used to increase our understanding of the species and inform its conservation, and more such studies are needed across its range. However, cheetah conservation depends ultimately on people, and their future can only be secured if we are able to reduce the key threats of habitat loss and fragmentation. Their large range and low density means that this requires land-use planning on a scale rarely seen before in terrestrial conservation. This would be difficult to achieve in developed regions, let alone in sub-Saharan Africa, where many people face difficulties securing food for their own families. Nonetheless, economic development progressing hand in hand with cheetah conservation, to the benefit of both people and biodiversity, is a feasible goal, if an appropriately designed land-use planning strategy can be successfully implemented. Conservationists, armed with sound scientific knowledge of cheetahs and their habitats, can help raise political will to achieve this goal.

Acknowledgements

We would like to thank TANAPA, TAWIRI, and the Tanzania Commission for Science and Technology for providing permission for the long-term study. Many individuals and organisations have assisted in the study over the years and the recent range-wide planning process, and we thank them for their support. Financial support has been provided by a large number of organizations, principally the Howard G. Buffett Foundation, Wildlife Conservation Society, Frankfurt Zoological Society and National Geographic Society. Finally, we thank the communities in and around Ruaha, LGCA, and NCA for their willingness to help with our surveys and their universal hospitality.

CHAPTER 17

Jaguars, livestock, and people in Brazil: realities and perceptions behind the conflict

Sandra M. C. Cavalcanti, Silvio Marchini,* Alexandra Zimmermann, Eric M. Gese, and David W. Macdonald*

Jaguar on a riverbank, Paraguay river, Pantanal, Brazil. © Edson Grandisoli.

Introduction

The jaguar (*Panthera onca*) is the largest predator in the Neotropics and is arguably the most charismatic species for conservation in Central and South America. Regrettably, the jaguar is also the carnivore that is least compatible with humans in twenty-first-century Brazil. This fundamental incompatibility is due to the jaguar's need for abundant, large prey, as well as extensive, undisturbed habitat. Humans (also large, top predators) have competed directly with jaguars for food (i.e. native and domestic ungulates) for as long as they have coexisted (Jorgenson and Redford 1993), and lately threaten them directly and indirectly through deforestation and habitat fragmentation. Moreover, jaguar predation on livestock (particularly cattle) (Fig. 17.1) provokes retaliatory persecution by humans (Hoogesteijn and Mondolfi 1992).

Persecution looms as the *coup de grace* to jaguar populations outside protected areas (Nowell and Jackson 1996) and, due to their wide-ranging movements, threatens jaguars within protected areas as well (Woodroffe and Ginsberg 1998). Although disentangling the contributions of persecution and habitat loss may be difficult, jaguar distribution

* Sandra Cavalcanti and Silvio Marchini are joint first authors.

Figure 17.1 A jaguar stands over a young calf that it has killed on a ranch in the southern Pantanal. © Pantanal Jaguar Project archive.

(Fig. 17.2) and abundance have declined drastically in recent decades (Hoogesteijn and Mondolfi 1992). Our case study focuses on Brazil, where the jaguar is a threatened species (Machado *et al.* 2005), although internationally it is classed as near threatened (i.e. it may be threatened with extinction in the near future; IUCN 2007).

Efforts to protect jaguars by curbing persecution by humans have been based on what might be termed a 'bio-rational' understanding of the problem. Insofar as the root of the problem is livestock-raiding then, so this rational goes, if we can find ways to effectively reduce jaguar predation on cattle (e.g. by the use of electric fences, aversive conditioning, and translocation), then persecution by cattle ranchers should subsequently decline (Cavalcanti 2003; Hoogesteijn 2003). Preventative actions combined with monetary compensation to ameliorate the financial costs of lost livestock are aimed at alleviating the economic burden on ranchers who coexist with jaguars, on the assumption this will reduce the motivation for ranchers to kill them.

Although this bio-rational thinking may be plausible to a scientifically trained conservationist, we hypothesized that human persecution of jaguars may be less related to livestock depredation than previously believed, and the economic justification for killing jaguars may be equally unclear. We argue that the ultimate motivator of retaliatory persecution may not be the actual impact of jaguars on human safety or livestock, but rather the cultural and social perceptions of the potential threat that

Figure 17.2 Current (black) and historical (dark grey) jaguar distribution range.

jaguars pose when attacking humans and killing livestock. In conflicts between people and carnivores, the perceived impacts often exceed the actual evidence (Conover 2001; Chavez and Gese 2005, 2006; Chavez *et al.* 2005; Sillero-Zubiri *et al.* 2007). In addition, factors not directly related to the impacts that jaguars have on human livelihoods (e.g. perceived social status of jaguar hunters in the community and thrill of the chase) may also be involved in the persecution of jaguars, making evaluations for the economic rationale for killing jaguars even more unclear. Such imprecise linkages between reality and perception could prove perilous to a threatened species, adding a potentially lethal element to already significant risks posed by retributive killing, and rendering irrelevant many biologically based conservation actions and mitigation measures.

To explore this 'perception blight', this chapter addresses the realities and perceptions behind the conflicts involving jaguars, livestock, and cattle ranchers in Brazil. We use the Pantanal region to quantify the importance of cattle in jaguar ecology, and techniques adapted from the social sciences to examine the ranchers' perceptions about jaguar depredation on cattle and other perceptions about jaguars and jaguar hunting that may be relevant in dealing with conflicts between ranchers and jaguars. We then investigate how these social and cultural perceptions may translate into the persecution of jaguars. Finally, we discuss how information on the ecological, economic, social, and cultural dimensions of a human–carnivore conflict can be fruitfully integrated into a strategy that encompasses both individuals and populations (of both jaguars and humans) in an attempt to promote coexistence between jaguars, livestock, and people.

Jaguars, livestock, and people

The jaguar occurs from the south-western United States to northern Argentina, across an area of 11.6 million km^2 and occupies a diverse array of habitats, including xeric shrublands, dry forests, montane grasslands, moist lowland forest, wet savannahs, and mangroves (Zeller 2007). Even though 36% of jaguar distribution overlaps protected areas (Zimmermann and Wilson, in preparation), studies indicate that very few of these areas offer true protection from human influences for both jaguars and their prey. Indeed, the edges of protected areas often become hot spots for human–wildlife conflict (Woodroffe and Ginsberg 1998; Loveridge *et al.*, Chapter 11, this volume). The human geography outside these protected areas is varied so jaguars may coexist with people holding a range of different perceptions and levels of tolerance for wildlife. Outside protected areas, the most common land-use form is livestock ranches, followed by logging areas, forest matrix lands, agricultural areas, and other forms of land use (Zeller 2007). On a continental scale, jaguars occur mostly in areas with a low Human Footprint Index (HFI; 95% of jaguar range is in areas of <35 HFI), and low cattle densities (96% in areas with $\leq$7.5 cattle/km^2; Zimmermann and Wilson, in preparation). Nevertheless, hunting of prey used by jaguars and direct human persecution of jaguars (most often in retaliation for livestock depredation) are, according to 130 jaguar experts, the most serious threats to the survival of the jaguar (Zeller 2007).

Conflicts between humans and jaguars occur in many different socio-economic and cultural contexts and vary in their severity, but appear to be most extensive in regions with large cattle ranches, where human densities are low, cattle densities are moderate, and small areas of wilderness containing natural prey still persist. There are several such vast rangelands in South America, most notably the Pantanal, Llanos, Beni, and Chaco regions of Argentina, Brazil, Bolivia, Colombia, Paraguay, and Venezuela. The best studied of the above regions of Brazil, the Pantanal, is the focus of our chapter.

Conflicts between ranchers and jaguars over livestock are widespread and have been documented throughout jaguar range (e.g. Belize: Rabinowitz 1986; Brazil; Crawshaw and Quigley 1991; Dalponte 2002; Conforti and Azevedo 2003; Michalski *et al.* 2006a; Azevedo and Murray 2007b; and Palmeira *et al.* 2008; Costa Rica: Saenz and Carrillo 2002; Argentina: Schiaffino *et al.* 2002; Venezuela: Scognamillo *et al.* 2002; and Polisar *et al.* 2003). Nevertheless, several ecological, socio-economic, cultural, and historical aspects of the relationships between people and jaguars have made Brazil particularly important for jaguar research and conservation.

Conflict in Brazil

Brazil covers 40% of the land area of Latin America. Even though estimates of jaguar abundance are as scarce for Brazil (Almeida 1986; Quigley and Crawshaw 1992; Soisalo and Cavalcanti 2006) as for other parts of their range (cf. Wallace *et al.* 2003; Maffei *et al.* 2004a; and Silver *et al.* 2004), Brazil does contain the two largest population strongholds for jaguars (Sanderson *et al.* 2002b): the wetlands of the Pantanal (140,000 km^2) and the rainforests of Amazonia (3,400,000 km^2). The southern Pantanal of Brazil has the highest density of jaguars recorded (estimates range from 6.7 to 11.7 individuals/100 km^2; Soisalo and Cavalcanti 2006). The Pantanal is also home to the largest jaguars, with the weight of males averaging 100 kg (females are typically 10–20% smaller than males) and the largest males reaching 158 kg (Seymour 1989). Jaguars were widely distributed throughout Brazil until 1500, but have since been extirpated from entire regions (Sanderson *et al.* 2002b; Fig. 17.2). Some jaguars still remain in fragments of the Atlantic forest and the Cerrado, but large jaguar populations are present only in Amazonia and the Pantanal, where human population density has historically been low.

Brazil is also home to the world's largest commercial cattle herd (>200 million head) and is the world leader in beef exports (Nepstad *et al.* 2006). Due to ecological and historical reasons, there is overlap between areas where beef production flourishes and jaguars survive, namely, the Pantanal and the agricultural frontier of southern Amazonia (Thornton *et al.* 2002). Cattle ranchers have a long tradition of killing jaguars (Hoogesteijn and Mondolfi 1992) in retaliation for livestock losses.

Cattle ranching also threatens jaguars indirectly, insofar as it is the major driver for the high and rapid level of deforestation in Amazonia, being the primary reason for >66% of habitat loss in the region (Nepstad *et al.* 2006). Between 1987 and 2006, an average of 18,000 km^2 of prime jaguar habitat was lost in this region every year, mostly from the Amazonian agricultural frontier (PRODES 2007). In the past two decades, Brazil has lost larger areas of jaguar habitat than any other country.

In 1967, the Brazilian Wildlife Protection Act prohibited commercial exploitation of wildlife and wildlife products derived from their capture, pursuit, or destruction. The Convention on International Trade in Endangered Species (CITES) of 1973 made it illegal to trade jaguar skins or parts for commercial gain. The CITES listing, in combination with the Brazilian legislation and anti-fur campaigns, brought about a sharp decline in the fur trade, helping to reduce the pressure on jaguar populations in the wild. However, jaguar persecution continues (Crawshaw 2002; Michalski *et al.* 2006a), now very rarely for the illegal trade, but more because of their perceived threat to people and their livelihoods. The indiscriminate killing of jaguars is one of the most serious threats to their survival across all of Latin America (Zeller 2007).

Jaguars, livestock, and people have coexisted in Brazil for many decades across a wide range of ecological, cultural, and socio-economic settings. From small family-run farms in the dry Caatinga to commercial large-scale ranches in the wetlands of the Pantanal, from old traditional cattle ranches in the Atlantic rainforest to recent settlements on the Amazon agricultural frontier, Brazil is the perfect test tube in which to explore the interacting chemistry of jaguars, livestock, and people.

Pantanal

The Pantanal is located in the geographic centre of South America and spans the borders of Brazil, Bolivia, and Paraguay (Fig. 17.3). With a highly seasonal climate, the Pantanal receives an average of >1.2 m of rainfall annually, which causes vast amounts of areas to be flooded and a subsequent flush of green grasses to be available for both native and domestic ungulates. The Pantanal is characterized by savannahs interspersed with isolated islands of secondary forest, which are an important refuge for both predators and prey. Gallery forests border temporary and permanent rivers and provide long corridors for wildlife movement.

Almost a third of published scientific articles on jaguar biology and conservation concern Brazil. While these topics have been addressed in the Brazilian Amazon (Oliveira 2002b; Michalski *et al.* 2006a), Cerrado (Silveira and Jacomo 2002; Palmeira *et al.* 2008), and Atlantic Rainforest (Garla *et al.* 2001; Leite

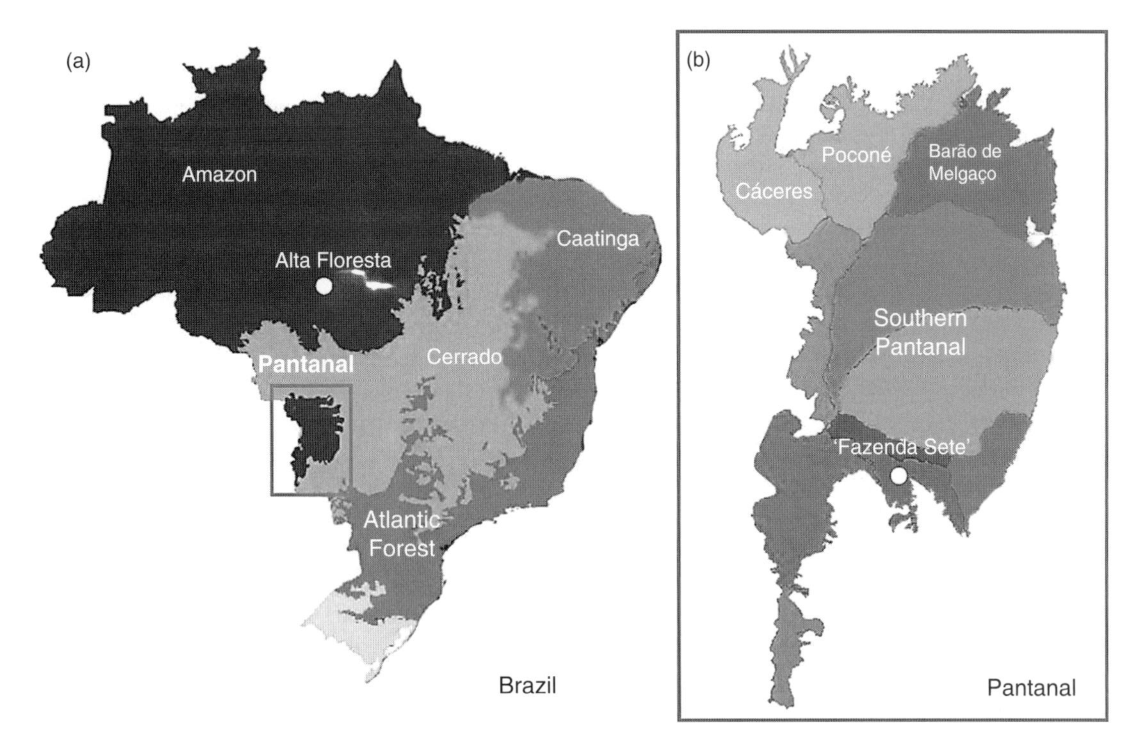

Figure 17.3 (a) Map of Brazil showing major biomes, the Amazon study site (Alta Floresta), and the Pantanal (highlighted by the box); and (b) map of the Pantanal showing its subregions. This study was conducted in the three subregions of northern Pantanal, namely Cáceres, Poconé, and Barão de Melgaço, and a ranch in southern Pantanal ('Fazenda Sete').

et al. 2002; Conforti and Azevedo 2003; Crawshaw *et al.* 2004; Cullen *et al.* 2005), the Pantanal accounts for the greatest portion of publications about jaguars (e.g. Schaller and Vasconcelos 1978; Schaller 1979, 1983; Schaller and Crawshaw 1980; Crawshaw and Quigley 1984, 1991; Crawshaw 1987, 2002; Quigley 1987; Quigley and Crawshaw 1992, 2002; Dalponte 2002; Zimmermann *et al.* 2005a; Soisalo and Cavalcanti 2006; and Azevedo and Murray 2007a, b, and Cavalcanti and Gese 2009).

In this landscape mosaic, cattle have been ranched for >200 years (Wilcox 1999). The Pantanal consists almost entirely of large cattle ranches (e.g. average ranch size 12,950 ha, SE = 22,444 ha; Zimmermann *et al.* 2005a). Cattle are raised extensively in the region, with an average cattle density of 16 head/km^2 (Mourão *et al.* 2002). People and jaguars, however, have coexisted uneasily. Jaguars have long been blamed for killing cattle and, in the past, ranch owners employed men solely to hunt jaguars. The extent of retaliation by ranchers was considerable. For example, in the early 1980s, 68 jaguars were killed over 8 years on one ranch alone (P. Crawshaw, as cited in IUCN/SSC 1986). Whether as a result of governmental legislation or the economic crisis in cattle ranching caused by the severe flood of the 1970s, the rate at which jaguars are killed appears to have declined and jaguar abundance in the Pantanal appears to be increasing (Crawshaw 2002). Nonetheless, as ranchers own 95% of this vast region, the future of jaguars in the Pantanal is inextricably linked to the ranchers' perceptions and attitudes towards them.

Assessing the realities and perceptions behind the conflict

In this chapter, we will attempt to weave together findings from several studies conducted by the authors between 2000 and 2008 which explored human–jaguar conflict in the Pantanal and the Amazon from various perspectives: jaguar predation rates on a cattle ranch, perceptions and attitudes of ranchers towards jaguars and livestock losses, and the various factors that may shape human beliefs and behaviour towards jaguars.

To document the realities of jaguar predation on livestock and native prey, the Pantanal Jaguar Project quantified kill rates, composition of prey killed, characteristics of prey killed, and patterns of predation on a ranch in the southern Pantanal ('Fazenda Sete' in Fig. 17.3; Cavalcanti 2008). In addition, Global Positioning System (GPS) telemetry provided information on jaguar movements (Cavalcanti and Gese 2009) and facilitated analysis of habitat use and spatial patterns of predation (on both domestic and native species) in relation to the type and distribution of vegetation and other landscape attributes (Cavalcanti *et al.*, in preparation). Ten jaguars were equipped with GPS radio-collars (Televilt, Sweden), which recorded their locations at 2-h intervals, enabling us to identify kill sites and thereby to find and document 438 carcasses of prey (including the identity of the predator, the date and approximate time of death, the period for which the predator stayed by the carcass, and the vegetation cover at the kill site; Cavalcanti 2008; Cavalcanti *et al.*, in preparation).

Meanwhile, the Coexistence Project studied the factors determining people's perceptions of jaguars and how these perceptions translated into human persecution of jaguars. Interviews with ranchers were used to document the following: (1) socio-demographic variables; (2) description of the property; (3) respondents' knowledge about jaguars and depredation problems; and perceptions of the jaguars' impact on (4) livestock; and (5) human safety; together with perceptions of (6) an increase in jaguar abundance; (7) degeneration of economic situation; (8) the social acceptability/desirability of persecuting jaguars, including the importance of traditional jaguar hunting; (9) the ease or difficulty of this persecution; (10) attitudes towards both jaguars and persecution; and (11) intention to persecute jaguars. Answers in either a binary yes/no or in 3- or 5-point scale formats enabled us to construct measurement scales (0 to 10 for knowledge and perceptions and −10 to 10 for attitude) and combine responses into an additive score for each variable (the higher the score the greater the knowledge or perception and more positive the attitude). In order to assess the degree to which the findings from the Pantanal can be extrapolated to other regions or whether attitudes are culturally specific to human–jaguar conflicts, we replicated this study on an agricultural frontier area in southern Amazonia (municipality of Alta Floresta). Like the Pantanal, the Amazon site hosts relatively high densities of both jaguars and livestock, but as a recently established agricultural frontier it differs in many social and cultural aspects from the Pantanal. Unlike the Pantanal, habitat loss is a major threat to jaguars on the Amazon frontier. This study involved 45 ranchers in two sub-regions of the northern Pantanal (Cáceres and Poconé) and 106 ranchers in Amazonia (Fig. 17.3; Marchini and Macdonald, in preparation-a).

We also examined the attitudes and conservation values of 50 ranchers from an earlier study in the three subregions of the northern Pantanal, namely, Cáceres, Poconé, and Barão de Melgaço (Fig. 17.3). In this study, we investigated the associations between attitudes and socio-economic variables such as rancher age, ranch size, cattle herd size and density, reported cattle losses, and level of involvement in tourism. Attitudes were explored using a series of suggested statements regarding jaguars and conservation, and responses were recorded on a five-point Likert scale so that they could be combined into an additive score, and the relationships between the combined score and potential explanatory variables could be analysed (Zimmermann *et al.* 2005a).

Realities of jaguar foraging ecology

Radio-tracking (Cavalcanti 2008) revealed that native species comprised 68.3% of the prey killed, with the remainder being cattle (31.7%). For individual jaguars, the number of cattle killed varied widely among cats (Fig. 17.4). Individuals also differed in the species diversity of their diets; although collectively the 10 jaguars killed 24 prey species, some killed few prey species, while others killed many (Table 17.1). Jaguars killed predominantly ungulates, but they also killed and consumed other predators, such as maned wolves (*Chrysocyon brachyurus*), crab-eating foxes (*Cerdocyon thous*), coati (*Nasua nasua*), and raccoons (*Procyon cancrivorous*).

Based on kills reported by ranch hands, Crawshaw and Quigley (2002) calculated that cattle comprised 46% of jaguar kills in the southern Pantanal. In their data, small prey was probably under-represented as these may be killed and consumed in secluded sites (see also Schaller 1979). This bias might also affect our findings, which included a small proportion of the biomass killed and consumed (e.g. birds; caiman lizard, *Dracaena paraguagensis*; coati; small anaconda *Eunectes notaeus*; and armadillo, *Euphractos sexcinctus and Dasypus novencinctus*). Homing in on radio-collared jaguars, Crawshaw and Quigley (2002) found 17 prey items of which 29% were cattle and 41% were white-lipped peccaries (*Tayassu pecari*)—a close match to our overall finding of cattle accounting for 31.7% of jaguar kills, but varying seasonally between 19.2% and 48.9%, respectively, for the wettest and driest periods of the 4-year field study (Cavalcanti 2008).

Calves (<1 year old, <174 kg) accounted for 69% of cattle killed by jaguars (Cavalcanti 2008), which is higher than Crawshaw and Quigley (2002) reported (43%) in their study in the same area in the southern Pantanal; perhaps again due to bias in carcass detection or annual variation. These findings from the Pantanal are broadly consistent with those reported elsewhere. In Venezuela, jaguars attacked young cattle (weaned calves and heifers 1–2 years of age) more often than they did adults (Hoogesteijn *et al.* 1993; Farrell 1999; Scognamillo *et al.* 2002). In north-east Argentina, cattle between 1 and 3 years comprised the majority of jaguar kills (Perovic 2002). Younger calves of 3–9 months of age comprised the majority of jaguar kills in northern Goiás, central-western Brazil (Palmeira *et al.* 2008). Azevedo and Murray (2007a) found that in the southern Pantanal predation risk was higher among calves up to 12 months of age.

Although jaguars can kill mature bulls (Hoogesteijn *et al.* 1993), we documented no jaguar attacks on an adult bull, and only one instance of jaguars scavenging on a bull carcass. Contrary to the beliefs of ranchers, the GPS data indicated that jaguars scavenged a proportion of their prey (we found six instances, involving three individuals, of feeding substantially from cattle that had died from other causes; see also Lopez-Gonzalez and Piña 2002). Therefore, scavenging complicates the interpretation of diet analyses based on undigested remains in jaguar faeces.

At 19 kill sites located by GPS-tracking, the remains of two different prey species were found (Cavalcanti 2008). We deduced this might have occurred when a jaguar killed a species scavenging from the original kill, and in 79% of these occasions this was a plausible explanation (e.g. one of the carcasses was a potential scavenger, such as feral hog, Susscrofa; peccary; armadillo; raccoon; or caiman, *Cayman crocodylus yacare*). This 'scavenger-trap' hypothesis seemed inappropriate for the remaining 21% of

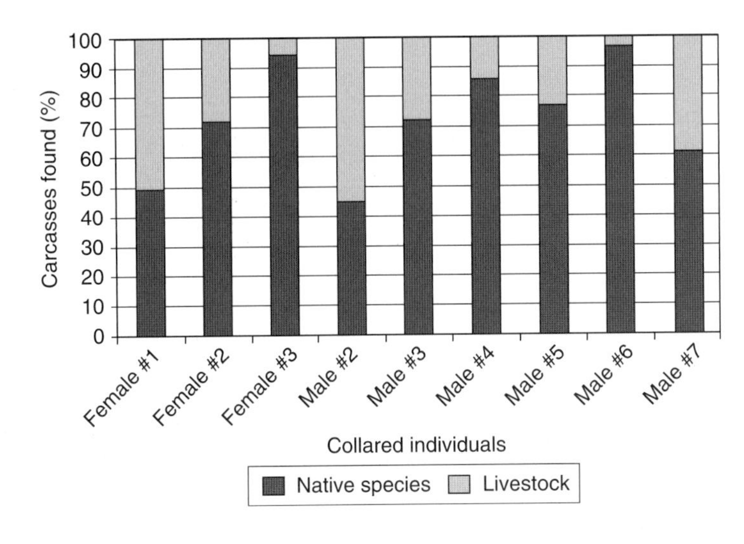

Figure 17.4 Distribution of native prey species and livestock killed by collared jaguars, November 2001 to April 2004, southern Pantanal, Brazil. (From Cavalcanti 2008.)

Table 17.1 Distribution of prey species (n [percentage of kills]) detected at kill sites for 10 individual jaguars, November 2001 to April 2004, southern Pantanal, Brazil.

Prey	Adult female				Adult male					
	#1 (*n* = 80)	#2 (*n* = 123)	#3 (*n* = 22)	#4 (*n* = 5)	#1 (*n* = 47)	#2 (*n* = 36)	#3 (*n* = 18)	#4 (*n* = 40)	#5 (*n* = 36)	Subadult male #6 (*n* = 27)
Tapir, *Tapirus terrestris*	0	0	0	0	0	0	0	1 (2.50)	1 (2.78)	0
Birds[a]	0	1 (0.81)	0	0	1 (2.13)	1 (2.78)	0	0	0	0
Calf *Bos taurus*	30 (37.5)	18 (14.6)	0	3 (60.0)	24 (51.1)	3 (8.33)	2 (11.1)	7 (17.5)	1 (2.78)	6 (22.2)
Capybara *Hydrochoerus hydrochaeris*	4 (5.0)	1 (0.81)	0	0	0	0	0	1 (2.50)	0	3 (11.1)
Marsh deer, *Blastocerus dichotomus*	3 (3.75)	2 (1.63)	1 (4.54)	0	4 (8.51)	1 (2.78)	0	2 (5.0)	0	3 (11.1)
Maned wolf, *Chrysocyon brachyurus*	2 (2.50)	1 (0.81)	0	0	0	0	0	0	0	0
Land turtle, *Geochelone carbonaria*	1 (1.25)	0	0	0	0	0	0	0	0	0
Caiman, *Cayman crocodylus yacare*	10 (12.5)	52 (42.3)	9 (40.9)	1 (20.0)	4 (8.51)	8 (22.2)	7 (38.9)	3 (7.50)	8 (22.2)	5 (18.5)
Crab-eating fox, *Cerdocyon thous*	0	1 (0.81)	0	0	0	0	0	1 (2.50)	1 (2.78)	0
Raccoon, *Procyon cancrivorus*	0	0	1 (4.54)	0	0	1 (2.78)	0	0	1 (2.78)	0
Feral hog, *Sus scrofa*	3 (3.75)	4 (3.25)	1 (4.54)	0	6 (12.8)	1 (2.78)	1 (5.55)	2 (5.0)	1 (2.78)	0
Coati, *Nasua nasua*	0	0	1 (4.54)	0	1 (2.13)	1 (2.78)	0	0	0	2 (7.40)
Peccary[b]	7 (8.75)	23 (18.7)	5 (22.0)	0	4 (8.51)	11 (30.5)	6 (33.3)	14 (35.0)	20 (55.6)	2 (7.40)
Anaconda, *Eunectes noctaeus*	0	0	0	0	0	0	0	0	0	1 (3.70)
Giant anteater, *Myrmecophaga tridactyla*	7 (8.75)	2 (1.63)	0	0	1 (2.13)	3 (8.33)	0	0	1 (2.78)	0
Lesser anteater, *Tamandua tetra-dactyla*	1 (1.25)	1 (0.81)	0	0	0	3 (8.33)	0	2 (5.0)	0	0
Armadillo[c]	2 (2.50)	0	3 (13.6)	0	0	0	0	1 (2.50)	0	0
Adult cattle, *B. taurus*	9 (11.2)	16 (13.0)	1 (4.54)	1 (20.0)	2 (4.25)	2 (5.55)	2 (11.1)	5 (12.5)	0	4 (14.8)
Brocket deer[d]	1 (1.25)	0	0	0	0	1 (2.78)	0	1 (2.50)	2 (5.55)	1 (3.70)
Caiman lizard, *Bracaena paraguayensis*	0	1 (0.81)	0	0	0	0	0	0	0	0

[a] Includes an egret (*Egretta alba*), a jabiru stork (*Jabyru mycteria*), and a boat-billed heron (*Cochlearius cochlearius*).

[b] Although collared peccaries (*Tayassu tajacu*) were present, the vast majority killed by jaguars were white-lipped peccaries (T. pecari).

[c] Includes two species of armadillos present in the study area, *Euphractos sexcinctus* (*n* = 4) and *Dasypus novencinctus* (*n* = 1).

[d] Includes both species, *Mazama americana and Mazama ouazoubira*.

Source: From Cavalcanti (2008).

double kills, insofar as neither of the victims was a scavenger (e.g. calf; brocket deer, *Mazama* spp.; giant anteater, *Myrmecophaga tridactyla*; and lesser anteater, *Tamandua tetra dactyla*).

Jaguars are often considered nocturnal predators. However, we found the time of day at which jaguars killed was evenly distributed throughout the 24-h period, even when examining individual prey species (Cavalcanti 2008). Jaguars appear to be adaptable to the movement and activity patterns of various prey species and readily exploit these species when they are active or vulnerable to predation.

When examining the seasonality of predation patterns by jaguars, we found the average number of cattle, caiman, and peccaries (the three major prey species) killed by radio-collared jaguars each season indicated a peak of predation on cattle in the dry seasons of each year (Fig. 17.5; Cavalcanti 2008). The frequency of predation on caiman appeared to be constant throughout all months of 2002, while predation appeared to peak during the wet seasons (February–March) of 2003 and 2004. There may be an inverse relationship between predation on cattle and caiman; as water levels recede in the Pantanal, caiman move with these levels and predation declines; conversely, as water levels recede cattle are moved into these areas for grazing and predation on cattle increases. The fluctuation of water levels is the major driver in this ecosystem, dictating the availability and vulnerability of prey species, including cattle. The frequency of predation on peccaries also appeared to be constant throughout 2002, then increased in 2003 and 2004. Seasonally, the mean number of peccaries killed each month appeared to be lowest during the wet seasons (February–March; Fig. 17.5). However, despite an apparent tendency for the number of cattle killed each month to have declined over the 4-year study, statistically the actual seasonal predation rates on cattle did not decline between 2002 and 2004 (Fig. 17.6). Conversely, while the data suggest an increase in the number of caiman killed each month, the observed seasonal predation rates on caiman did not increase statistically over the seasons. Predation rates on peccaries did increase significantly between the wet season of 2001–02 and the dry season of 2004 (Fig. 17.6). The increase in jaguar predation rates on peccaries during the study occurred during a period of relatively high

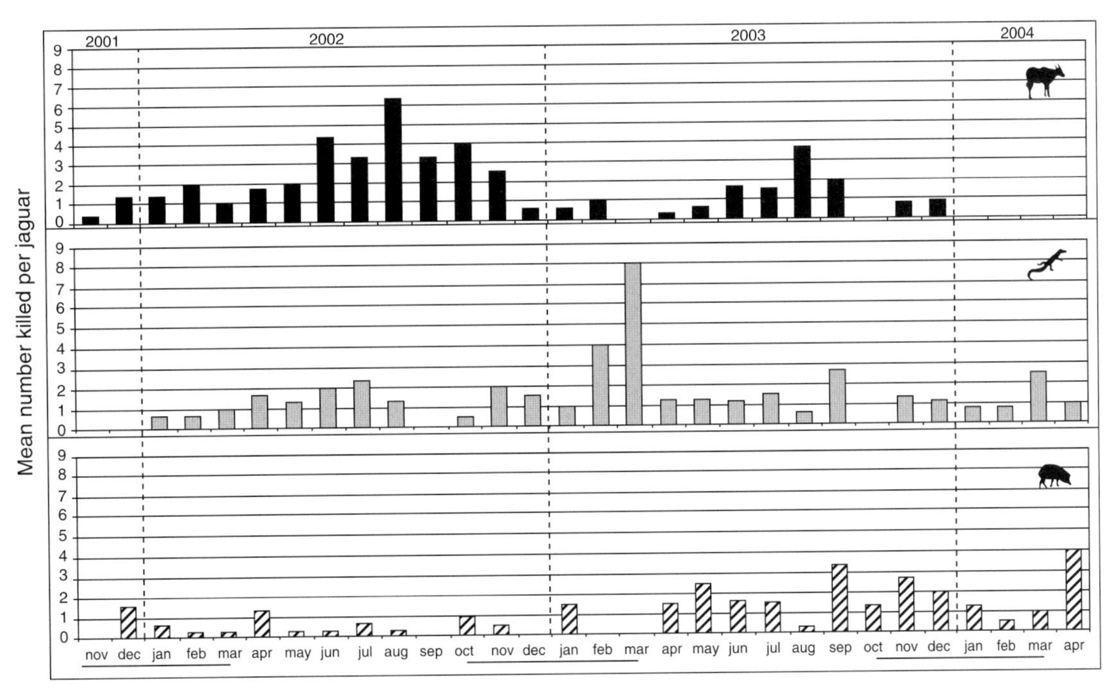

Figure 17.5 Distribution of the mean number of cattle, caiman, and peccary killed per month by collared jaguars, November 2001 to April 2004, southern Pantanal, Brazil. (From Cavalcanti 2008.)

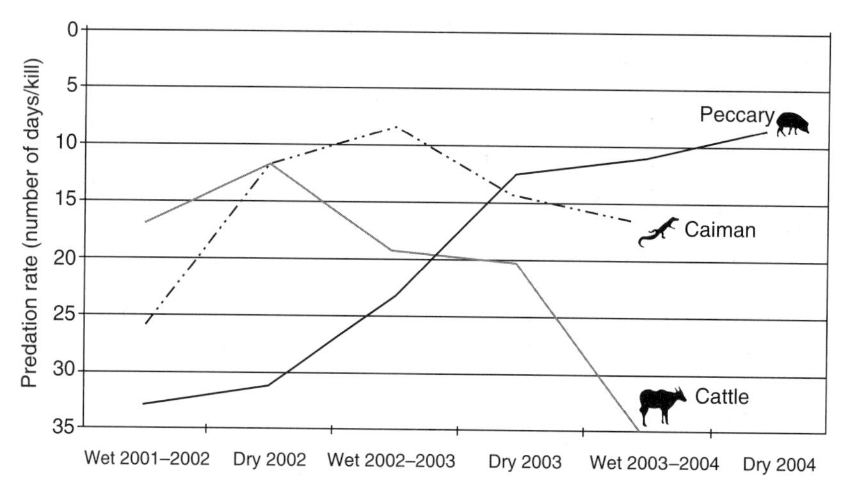

Figure 17.6 Seasonal variation in jaguar predation rates of caiman, peccary, and domestic cattle, November 2001 to April 2004, southern Pantanal, Brazil. (From Cavalcanti 2008.)

peccary densities (9.63 individuals/km^2; Keuroghlian 2003). This increased predation rate on peccaries during the study suggests that the availability of alternative prey could reduce jaguar predation rates on cattle and could serve as a buffer species.

Because jaguars are ambush predators, an obvious prediction would be that kills were associated with dense vegetation. Cavalcanti *et al.* (in preparation) found that while the 10 GPS-collared jaguars used forests and shrublands preferentially, kills were made in habitats in proportion to their availability. Cattle, caiman, and peccaries killed by jaguars (n = 327) were distributed in the various habitat classes according to their availability, except during the dry season, when caiman and peccaries were mainly killed in shrublands and forests, respectively. Male and female jaguars consistently selected shrublands during both wet and dry seasons. Although there was little evidence that particular species were killed in particular habitats, there was a tendency for cattle to be killed further than expected from water.

Some authors have hypothesized that jaguar predation on cattle is a function of the distribution, availability, or proximity to forest habitat or forest edges (Rabinowitz 1986; Hoogesteijn *et al.* 1993; Michalski *et al.* 2006a; Palmeira *et al.* 2008). Hoogesteijn *et al.*'s comparison (1993) of three ranches in Venezuela led to the conclusion that jaguars killed cattle closer to forested areas. Rabinowitz (1986) reported jaguars readily killed domestic livestock that entered forested areas, but not when cattle were in open pastures. Quigley (1987) reported cattle were killed only in gallery forests and forest patches, although some might have been dragged there from the forest edge. This differs from the findings of Cavalcanti *et al.* (in preparation) reported above, who found that during the wet season, cattle were killed by jaguars significantly closer to forest edges than in the dry season. During the wet season, cattle were able to forage in chest-deep water, but they needed dry ground on which to spend the night. Therefore, they might spend more time closer to forests, which are typically associated with higher and drier ground. Several authors have suggested keeping cattle herds away from forested areas as a strategy to minimize jaguar attacks (Rabinowitz 1986; Hoogesteijn *et al.* 1993; Michalski *et al.* 2006a; Palmeira *et al.* 2008), but at least in the Pantanal, we recorded jaguar attacks on cattle in other habitats as well (Cavalcanti *et al.*, in preparation).

Individual variation in jaguar diets: do 'problem animals' exist?

Since jaguars differed individually in their diet (Cavalcanti 2008), we examined whether some jaguars contributed more than others to the levels of domestic stock losses (cf. Linnell *et al.* 1999). There was no straightforward answer. While prey remains of individual jaguars indicated that cattle comprised >50% of the diet for some individual jaguars, for others it did not exceed 5%. Nevertheless, each of the 10 radio-collared jaguars killed cattle. Whether or not killing the predominant cattle-killers would ameliorate the problem (e.g. as suggested by Rabinowitz 1986 and Hoogesteijn and Mondolfi 1992) depends on the causes of this individual variation (i.e. causes may include availability and vulnerability of prey, preference for particular prey species, or cultural learning from their mother). However, we also found that for some individuals that had >50% of their kills comprised of cattle in 2002 (a dry year), these same jaguars exhibited an appreciable decline in cattle kills in 2003 (a wet year). Again, water levels, and the consequent movement of both caiman and cattle, likely played an important role in the availability of these two key prey species within individual jaguar home ranges and therefore influenced encounter rates (Cavalcanti 2008).

Previous analyses of the variation in the level of livestock depredation suggest that males are more likely to kill cattle than are females (e.g. Rabinowitz 1986, Stander 1990, and Chellam and Johnsingh 1993). However, results from the 4 years of study found no differences between males and females in the level of predation on cattle (Table 17.1; Cavalcanti 2008). It is conceivable that jaguars, especially females, may kill cattle in excess of their needs, which might be considered a mechanism for teaching their young to hunt (A. Silva, V. Correia, A.T. Neto, and B. Fiori, personal communication). Surplus killing is almost universal amongst the *Carnivora* (Kruuk 1972), so it is unexpected that it has not been reported for jaguars. In general, the time interval between kills, and the time spent at each kill, was related to prey size (Cavalcanti 2008). After killing and consuming a small prey item, a jaguar generally killed again in a shorter time (3.0 days before making another kill) as compared to when they killed larger prey (5.1 days). Similarly, the length of time jaguars stayed at a carcass site significantly increased with increasing body mass of prey; 16.0 h were spent on small prey, increasing to 27.9 h on larger prey (Cavalcanti 2008). Some authors have speculated that livestock depredation is more prevalent among subadult than adult felids (Rabinowitz 1986; Stander 1990; Saberwal *et al.* 1994), whereas others conclude adults are more likely to kill cattle than younger individuals (Bowns 1985; Esterhuizen and Norton 1985). In our study, stock killing occurred at a rather constant rate among individuals. On average, jaguars killed one calf every 13.3 ± 15.5 (SD) days, while adult cows were killed at a lower rate of 25.5 ± 18.4 (SD) days between kills, although these rates varied annually (Cavalcanti 2008). The level of rainfall in any given year appeared to be the most influential factor affecting individual jaguar predation rates on cattle by determining the availability of cattle on the landscape (Cavalcanti 2008). During wet years, native prey were also available to jaguars and cattle were less vulnerable to predation. Conversely, during dry years, cattle were more dispersed over the landscape, exposed to more jaguar territories, thereby increasing encounter rates between individual jaguars and domestic prey. In addition, the poorer condition of cattle influenced their vulnerability to predation during the dry years. Husbandry practices are also likely to have a large influence on jaguar predation. In the Pantanal, calves were generally born over several months, prolonging the time period over which vulnerability to jaguar predation was increased. In addition, pregnancy rates of cows are generally well below optimal, often between 60% and 75%. Native ungulates usually swamp predators by having a short birth pulse, thereby decreasing the length of time that young are exposed or vulnerable to predation. Shortening the birth pulse and increasing the number of pregnant cows within a cattle operation could, in theory, reduce overall predation losses within individual jaguar territories, where calving grounds are located by swamping an individual cat with far more prey than can be killed; assuming that satiation of the predator causes an asymptote in the kill rate.

A common hypothesis in terms of large cat predation on livestock is that it is prevalent among wounded or sick predators, and this idea has been mooted for jaguars (Rabinowitz 1986; Fox and Chundawat 1988; Hoogesteijn *et al.* 1993). Indeed, two studies in Venezuela both revealed that the majority of the jaguars (75% and 53%) killed as part of predation control had previously sustained severe wounds, precluding them from hunting normally (Hoogesteijn *et al.* 1993), although the condition of jaguars not killed could not be confirmed. In our study, all radio-collared jaguars were in excellent physical condition at the time of capture (Cavalcanti 2008), as were those documented by Schaller and Crawshaw (1980) and Hopkins (1989). The oldest individual radio-tracked (a male estimated to be >11 years old) had two missing canines (a broken lower canine on his first capture and a further broken upper canine on his second capture) and killed white-lipped peccaries, feral hogs, and marsh deer at a similar rate (7.1 ± 5.6 [SD] days between kills) as all the other radio-tracked jaguars (no significant difference between the jaguar kill rates; Cavalcanti 2008).

Perceptions about depredation and persecution

Depredation problems caused by jaguars have been reported by 82% of the landowners in the northern Pantanal (Zimmermann *et al.* 2005a; Marchini and Macdonald, in preparation-a). Not surprisingly, jaguars were considered the most detrimental species to human livelihoods by 73% of 110 ranchers and ranch hands interviewed in both the southern and northern Pantanal (Marchini 2003). Reported losses to jaguars ranged from 0% to 11% of their livestock holdings, with greater proportional losses among smaller ranches and smaller herds ($r = -0.590$ and -0.716, both $P < 0.001$; Zimmermann *et al.* 2005a) with losses averaging between 2.1% (Marchini and Macdonald, in preparation-a) and 2.3% (Zimmermann *et al.* 2005a) of their livestock holdings. In absolute terms, the greatest reported loss was 80 calves in 1 year from a herd of 2000 head on a 13,200 ha ranch. Given the average price of a calf in the region (approximately US$228 in 2008), this case translated into a monetary loss of US$18,240 (Marchini and Macdonald in preparation-a). Over one-third of the respondents (38%) ranked jaguars as a greater problem affecting their income from cattle than floods, droughts, rustling, and disease (Zimmermann *et al.* 2005a).

Most ranchers (62%) reported that jaguar attacks show no seasonal pattern (Zimmermann *et al.* 2005a). As for variation between years, 24% of the respondents believed the frequency of attacks within their ranches was increasing, 35% believed it was declining, and 41% stated it was not changing (Marchini and Macdonald, in preparation-a). Most ranchers (72%) believed that jaguars varied in their dietary preferences and thus believed that 'problem jaguars' were the ones killing cattle (Marchini and Macdonald, in preparation-a).

These findings suggest that perceptions of jaguar depredation might sometimes exceed reality, as ecological studies addressing jaguar depredation in the Pantanal and elsewhere revealed lower losses of livestock holdings (0.83% in two ranches of northern Pantanal, 0.3% in one ranch of southern Pantanal, and 1.26% in southern Amazonia; Dalponte 2002; Azevedo and Murray 2007b; Michalski *et al.* 2006a, respectively). However, when predation rates were estimated from GPS-collared jaguars (Cavalcanti 2008), the ranch foreman reported the ranch lost on average 70 head of livestock out of 6000 head annually to jaguar predation (1.2% of livestock holdings). During a dry year (2002), a jaguar killed an average of 2.1 calves/month and 0.6 adult cows/month, for a total of 2.7 head of cattle/month. Extrapolating this kill rate to half (not all jaguars had equal access to cattle), the estimated resident (80%) population of jaguars on the ranch (6.7 jaguars/100 km^2; Soisalo and Cavalcanti 2006) would generate an estimated loss of about 390 head of livestock. Conversely, during wet years (2003), jaguars killed an average of 0.5 calves/month and 0.3 adult cows/month, for a total kill rate of 0.8 head of cattle/month. Again extrapolating to half, the resident jaguar population on the ranch generated an estimated 118 head of livestock lost. During a wet year (2003), the perceived losses (70 head; 1.2% of cattle) and the estimated losses from jaguar predation rates (118 head; 1.9% of cattle) were similar. Conversely, during a dry year (2002), predation rates indicated that over five times more cattle were lost (390 head; 6.5% of cattle) to jaguar predation than the ranch foreman perceived (70 head). Therefore, the level of rainfall influences the number of cattle lost annually by determining the access cattle have to the landscape and the number of jaguars to which they are exposed, in addition to increasing their vulnerability due to poor body condition. In addition, we generally found that cattle killed by the radio-collared jaguars were found by the ranch hands only rarely and unreported losses are likely higher than previously believed. Ranch hands easily found cattle kills in open fields and pastures, while missing most kills in the dense cover of shrublands and forests.

We emphasize that these extrapolations of predation rates are from only one study and may not be representative of all ranches in the Pantanal. However, it does raise the issue that accurate and unbiased documentation of jaguar kill rates on livestock and native prey are needed, to lend credibility to claims on both sides of the argument regarding losses sustained by livestock operations. In a study examining wolf (*Canis lupus*) predation on livestock in central Idaho, USA, researchers reported that ranchers found only one in eight of the actual kills documented (Oakleaf *et al.* 2003). During the years of wolf reintroduction into the United States, government personnel consistently agreed that a rapid response time and accurate documentation of actual losses were critical to any compensation programme proposed for ranchers, and lack of data can often lead to heated debate about the actual level of losses sustained by a ranching operation. Some ranchers were very diligent in keeping track of losses, while others were less accurate and blamed predators for more losses than actually occurred.

In addition to the perceptions about the level of jaguar depredation on livestock, other beliefs and perceptions about jaguars and jaguar hunting may be relevant in dealing with conflicts between ranchers and jaguars. Many ranchers (30%) held the perception that jaguar abundance was currently increasing (4% perceived it as decreasing; Marchini 2003). In the Pantanal subregion of Cáceres, 80% believed jaguar abundance was increasing and there was a widespread perception that jaguar numbers were now abnormally, and unbearably, high (Marchini and Macdonald, in preparation-a).

A few ranchers (15%) in the subregion of Poconé believed that jaguars caused cattle mortality even without preying on them (Marchini and Macdonald, in preparation-a). The rationale was that jaguars scared cattle out of the 'capões' (dry forest patches where cattle find refuge during floods), from which the cattle then ran to flooded areas, where they drowned or got stuck in the mud and starved to death. This belief in 'indirect predator-induced mortality' of livestock needs further investigation.

Some people perceived jaguars as a threat to human safety. Many respondents (53%) believed that jaguars attacked people even when not provoked (Marchini and Macdonald, in preparation-a). A rural school in Cáceres closed its doors in 2008 because the pupils refused to attend classes after several sightings of jaguars in the vicinity. This episode occurred prior to an incident on 24 June 2008 when a young fisherman was killed by a jaguar while sleeping in his tent on a riverbank of the Paraguay River, in the subregion of Cáceres. This was the first officially documented, unprovoked, fatal attack of a jaguar on a human in Brazil, and was widely covered by the national media. Prior to this incident, attacks were almost invariably associated with hunting situations in which the jaguar was cornered or injured. Jaguars are also known to attack in order to defend their cubs or the carcass upon which they are feeding. The impact of the above event on people's perceptions of the risk that jaguars pose to human safety is currently being assessed.

Many ranchers unashamedly admit that killing jaguars is socially acceptable. Only 15% of the respondents believed their neighbours or family would disapprove of them killing jaguars (Marchini and Macdonald, in preparation-a). For many, killing jaguars is considered one of the traditions of the Pantaneiro culture. Additionally, there is the general view that all aspects of the Pantaneiro culture should be cherished and preserved. Indeed, a prevailing opinion is that hunting jaguars is an act of bravery and a test of dexterity among cowboys. Shooting a jaguar enhances a cowboy's reputation. Even when ranch owners have specifically banned jaguar hunting, some ranch hands may continue to kill jaguars (S. Cavalcanti, personal observation).

The extent to which a rancher perceives the difficulty of killing a problem jaguar may affect the likelihood of actually pursuing this option. The general approach is to use dogs to find and pursue the jaguar. Either the jaguar climbs a tree or turns at bay on the ground, whereupon the hunters arrive and kill it. In the Pantanal, hiring a professional hunter who owns a pack of trained dogs can be relatively easy and affordable (sometimes a cow is offered in exchange for the service), but in other regions, the difficulty and cost of hiring a hunter may discourage small ranchers from killing jaguars. Several small landowners on the Amazon agricultural frontier, for instance, told us they had never killed a jaguar but

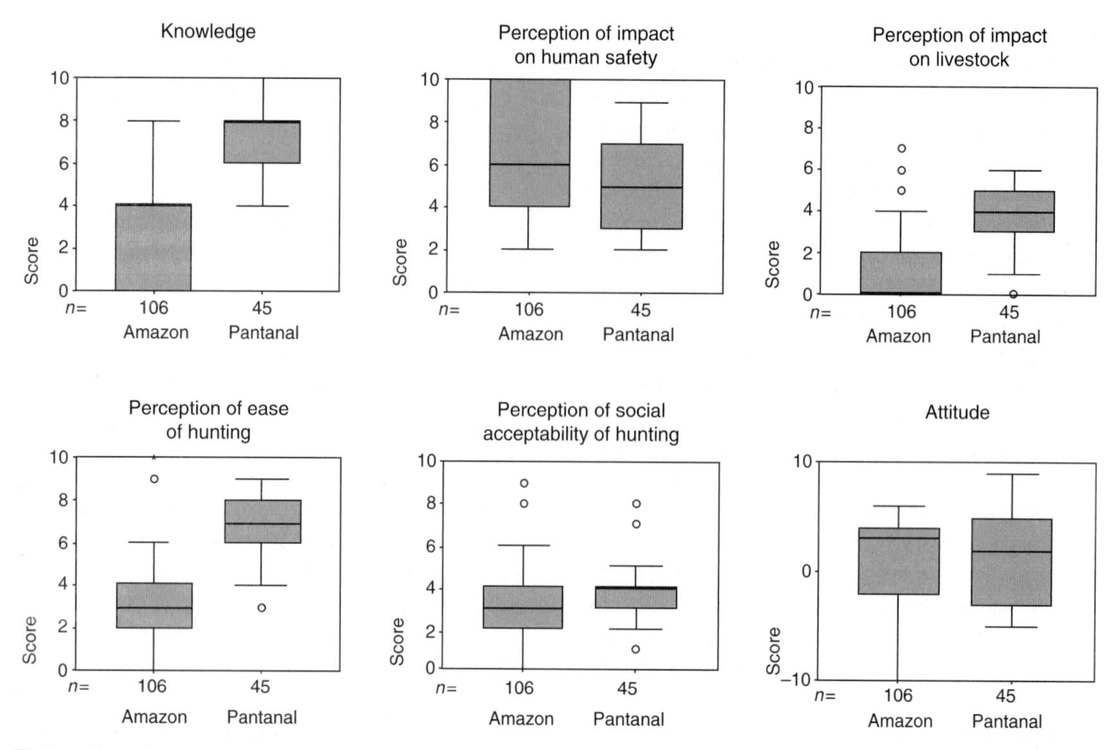

Figure 17.7 Graphs showing the measurement scales, distribution of average score values, and differences between Amazon and Pantanal. The box indicates the median, 25% and 75% quartiles, and whiskers are the largest values that are not outliers, while circles mark outliers.

would have killed the cat if they had had the capacity to do so (Marchini and Macdonald, in preparation-a).

Surprisingly, jaguars also elicit positive feelings among ranchers. All our respondents considered the jaguar a 'beautiful' or 'very beautiful' animal (Marchini and Macdonald, in preparation-a), and 16% would choose the species to be the symbol of the Pantanal (only the jabiru stork, *Jabyru mycteno* the official symbol of the region, ranks higher; Marchini 2003). Although we met ranchers who confessed hatred of jaguars, irrespective of their behaviour, the average attitude score value of ranchers in the region, for questions assessing an individual's like or dislike for jaguars (e.g. 'How would you feel if all jaguars disappeared?') and unfavourability or favourability towards jaguar persecution (e.g. 'Would killing any jaguar that shows up in your property this year improve your livelihood?'), was positive (Fig. 17.7; Marchini and Macdonald, in preparation-a).

Finally, the economic decline in the region may exacerbate the conflict between ranchers and jaguars. In recent decades, growing competition within the cattle industry, higher taxes, and generational land splitting has made cattle ranching less profitable in the Pantanal (Swarts 2000). Indeed, 95% of the ranchers believed their economic situation is worse now than in the past (Marchini and Macdonald, in preparation-a). A decline in the profit margin from cattle ranching may decrease their tolerance of jaguar depredation on their cattle. The growth of ecotourism in the region has brought the hope of better days for some ranchers (and conservationists as well), although ecotourism alone seems unlikely to be a universal solution.

Variation in perceptions and its determinants

In order to understand how and why the above perceptions vary, we examined correlations between perceptions and socio-economic and demographic variables (Table 17.2). Details of these analyses are in Marchini and Macdonald (in preparation-a).

The perceived impact of jaguars on livestock, which was measured using questions about recent and past depredation events, the magnitude of the loses (from none to very large) on his ranch, as well as neighbouring and relatives' ranches, and the current trends in the jaguar depredation problem (decreasing, unchanged, and increasing), was positively correlated with the perception of increasing jaguar abundance ($r = 0.41$, $P < 0.02$) and declining economic situation ($r = 0.47$, $P < 0.04$). There was also a negative correlation between their attitude towards jaguars ($r = -0.61$, $P < 0.0001$) and the number of years attending school ($r = -0.49$, $P < 0.0001$): ranchers had stronger negative attitudes towards jaguars in relation to fewer years in school (education varied greatly among respondents, from 33% being illiterate to 22% with higher education). Attitudes towards jaguars was also negatively correlated with the respondents' perception of the deterioration in the economic situation ($r = -0.57$, $P < 0.04$) and positively correlated with years in school ($r = 0.36$, $P < 0.0001$), which was negatively correlated with age ($r = -0.50$, $P < 0.001$). The rationale for using different questions to assess the perceptions of the impacts on livestock is that a rancher's evaluation of these impacts is not based solely on his recent losses to jaguars. Different ranchers, depending on their background and socio-economic situation, see the same loss as small or large. The perceived impacts of jaguars on human safety were measured with questions regarding (a) the potential for unprovoked attacks on humans by jaguars, as well as any perceived man-eating habits among jaguars; (b) first- or second-hand reports of jaguar attacks on people (fatal or not); and (c) the magnitude of threats to human safety posed by jaguars and the innate fear of jaguars (none to very large). All of these perceptions were positively correlated, with a perceived increase in jaguar abundance ($r = 0.45$, $P < 0.02$) and negatively correlated with the respondent's knowledge of jaguar ecology and depredation problems ($r = -0.53$, $P < 0.0001$). In summary, if a person's perception of the jaguars' impact on livestock and human safety determines retaliatory persecution, then perceptions of increasing jaguar abundance and declining economic situation, attitudes towards jaguars, years in school, age and knowledge of jaguar ecology, and depredation problems may all play a role in conflicts between humans and jaguars in the Pantanal.

Some differences between the perceptions in the Pantanal versus the Amazonia region are relevant to this discussion (Fig. 17.7; Marchini and Macdonald, in preparation-a). The perception of the jaguars' impact on livestock was stronger in the Pantanal than in the Amazon ($t = -9.966$, $P < 0.0001$, d.f. $= 149$), whereas the perceived threat to human safety was higher in the Amazon than in the Pantanal ($t = 2.919$, $P = 0.004$, d.f. $= 149$). Even though attitude to jaguars was similar in the two regions ($t = -1.112$, $P = 0.268$, d.f. $= 149$), in the Pantanal, attitude was correlated with the perceived impact on livestock (see above), whereas in the Amazon, it was correlated with the perceived impact on human safety ($r = -0.39$, $P < 0.01$). We also found differences in the perception of social acceptability of jaguar hunting; assessed by whether the respondent felt his family and neighbours would approve or disapprove if he killed jaguars. The acceptability of killing jaguars was higher in the Pantanal than in Amazonia ($t = -2.962$, $P = 0.004$, d.f. $= 149$) and so was the perceived ease of persecuting jaguars ($t = -13.044$, $P < 0.0001$, d.f. $= 149$). In addition, people in the Pantanal were more knowledgeable about jaguars and depredation problems than were people on the Amazon frontier ($t = -7.684$, $P < 0.0001$, d.f. $= 149$). For instance, whereas 89% of the respondents in the Pantanal could tell the difference between jaguar and puma tracks, only 7% of the respondents in the Amazon were correct in their identification skills.

From perceptions to persecution

From a conservation standpoint, what ultimately matters in conflicts between people and jaguars is the level of persecution and its impact on a carnivore population. To investigate the relationship between perceptions, attitudes, and persecution, we used a hierarchical cognitive model based on the correlations mentioned above and adapted from the 'Theory of Planned Behaviour' (TPB, Ajzen 1985). This is an influential theory in social psychology attempting to

Table 17.2 Zero-order correlations between perception, knowledge, and attitude scores, and demographic and socio-economic variables in the Pantanal.

No.		1	2	3	4	5	6	7	8	9	10	11
1	Age	1										
2	Years in school	−.50**	1									
3	Property size	.28	.08	1								
4	Knowledge about jaguars and depredation	−.15	−.28	−.17	1							
5	Perception of increase in jaguar abundance	.05	−.12	−.08	.18	1						
6	Perception of decline in economic situation	.43	−.25	.38	.08	.26	1					
7	Perception of impact on human safety	−.08	.27	−.11	−.53**	.45*	.10	1				
8	Perception of impact on livestock	.18	−.49**	−.14	.16	.41**	.47**	.19	1			
9	Perception of ease of hunting jaguars	−.14	.09	−.12	.22	.13	−.11	.13	−.15	1		
10	Perception of social acceptability/ desirability of jaguar hunting	−.10	.19	−.04	−.25	.39	.29	.07	−.18	−.07	1	
11	Attitude to jaguars and jaguar hunting	−.16	.36*	.14	−.06	−.33	−.57**	−.04	−.61**	.04	−.05	1

** Correlation is significant at the 0.01 level (two-tailed).
* Correlation is significant at the 0.05 level (two-tailed).

predict a person's behaviour (see also Macdonald *et al.*, Chapter 29, this volume). In the vocabulary of the TPB, a person's behaviour is explained by behavioural intention, which is preceded by 'attitude towards the behaviour'. Intention also depends upon 'subjective norms', which is a person's perception of the social acceptability or desirability of the action in question, and 'perceived behavioural control', which is the actor's perception of the ease or difficulty of performing the specific action. Background factors such as age, education, wealth, occupation, culture, and knowledge may influence these attitudes and perceptions, but are not incorporated in the causal model (Ajzen 1985). In our model, jaguar persecution is represented by the intention to persecute jaguars, which is preceded by attitude towards persecution, norms regarding this behaviour, and perceived behavioural control over it. Given the central importance of the perceptions of jaguars' impact on livestock and on human safety in the conflicts between people and jaguars, we expanded our TPB model to address explicitly those perceptions as potential determinants of attitude towards persecution (Fig. 17.8). This approach allowed us to assess the relative importance of the different components of the causal chain of jaguar persecution so that more effective interventions could be devised to decrease persecution.

Marchini and Macdonald (in preparation-b) assessed the intention to persecute jaguars via the question: 'Would you kill any jaguar that shows up in your property?' The answer to this question was expressed in the form of a dichotomy: a person either intended to persecute or not. Evidence of recent persecution of jaguars was found in 27 ranches (8 in the Pantanal and 19 in Amazonia), which facilitated validation of this measurement. Most (81%) of the people who had killed jaguars in the previous 2 years said that they intended to persecute any jaguar that showed up on their ranch, whereas 20% of the people who had not killed any jaguar expressed the intention to persecute ($\chi^2 = 35.301$, d.f. = 1, $P < 0.001$). This seeming association between declared intentions and actions strengthens our conclusion that we should take seriously the statements other ranchers made to us about their intentions to kill jaguars. Almost 60% of the landowners in the Pantanal declared their intention to kill any jaguar that showed up on their land, whereas in the Amazon about 20% of the landowners did so.

Regression analysis revealed that attitudes and subjective norms significantly explained the variation in the intention to persecute jaguars in the Pantanal: more favourable attitudes towards jaguar persecution and a greater perception of the social acceptability of jaguar hunting were associated with a greater intention to kill jaguars ($\beta = -0.259$, $P = 0.01$ and $\beta = 0.497$, $P = 0.024$, respectively; $-2\log$ likelihood = 46.722). Several ranchers in Marchini and Macdonald's sample (in preparation-b) also expressed the view that killing jaguars was appropriate on the grounds that it was a tradition passed from generation to generation. The important influence of these social norms was unsurprising considering that many ranchers in northern Pantanal were

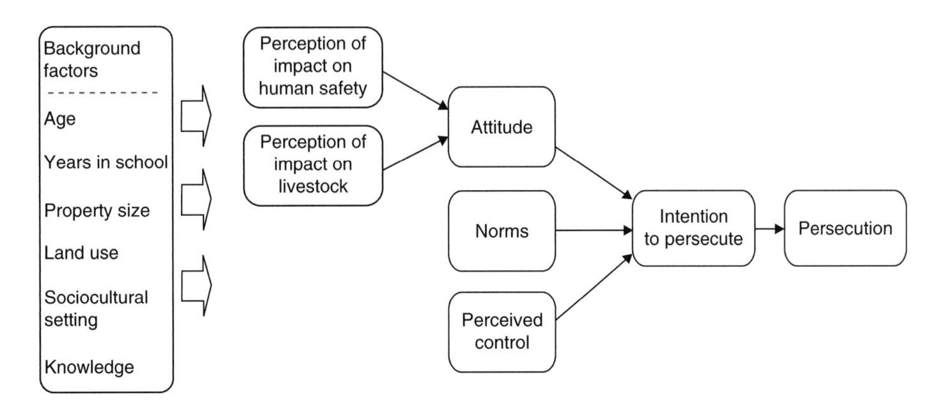

Figure 17.8 Hierarchical cognitive model of jaguar persecution adapted from the Theory of Planned Behaviour (Ajzen 1985).

interrelated, with a network of family bonds linking ranches. Attitude, in turn, was correlated with the perceived impact of jaguars on livestock (R^2 adjusted $= 0.364$, $F = 26.138$, $P < 0.0001$). Cattle ranching is an icon of the Pantaneiro culture. The few traditional families that together own a substantial portion of the lands in the northern Pantanal have been raising cattle in the region for generations, and this has been the only viable economic activity. The economic and cultural centrality of cattle ranching in the region doubtless affects the high correlation between perceptions of the jaguars' impact on livestock and the attitudes towards persecution. Although cattle ranching in the Pantanal is generally undertaken at such a large scale that the loss of a few cattle is unlikely to seriously impact the ranchers' livelihoods, for the majority of ranchers, such losses are unacceptable and may be higher than actually realized given detection rates of kills.

In contrast, in the Amazon, the intention to persecute jaguars was significantly explained by attitude to persecution and perceived ease or difficulty of persecuting ($\beta = -0.481$, $P < 0.0001$ and $\beta = 0.663$, $P = 0.011$, respectively; $-2\log$ likelihood $= 66.831$). Indeed, a significant proportion of the landowners, and particularly those with smaller properties, favoured the idea of killing jaguars, but did not intend to engage in this activity because they believed that they lacked the means (or were not brave enough, as they told us) to do so. In the Amazon sample, the perceived social acceptability or desirability of persecution did not significantly affect the intention to persecute jaguars, which might reflect the reality that in this frontier area people typically have little interaction, or shared background, with their neighbours. However, their attitudes were heavily associated with the perceived risk of jaguars on human safety (R^2 adjusted $= 0.488$, $F = 18.385$, $P < 0.0001$). A fear of jaguars is common among the frontiersmen, who were largely immigrants with little experience of jaguars and the forest.

Finding solutions for the future of jaguar–human coexistence

Direct persecution of jaguars by people, combined with the hunting of prey used by jaguars, is the most significant threat to the long-term survival of jaguars throughout their range (Sanderson *et al.* 2002b; Zeller 2007). Most persecution is directed at jaguars living near or within areas of livestock raising. Jaguars kill livestock and this creates a conflict with ranchers from an economic perspective. Several aspects of jaguar ecology and behaviour elucidated by our studies have direct implications for this economic aspect of jaguar conservation. The obvious, and traditional, response has been attempting to curtail jaguar depredation on livestock through preventive measures. A radical, but evidence-based, alternative would be for all stakeholders to recognize the reality that cattle are routinely a component of jaguar diet in the region. Under the Biodiversity Impacts Compensation Scheme (BICS) model proposed by Macdonald (2000; elaborated with respect to carnivore conflict by Macdonald and Sillero-Zubiri 2004b), the approach would be to refine management interventions to reduce negative impacts (stock losses), and then find other mechanisms to offset irreducible damage; in this case, alternative mitigation measures to make the residual stock losses to jaguars bearable. Additionally, while kills of domestic stock may be related to a lack of natural prey (Saberwal *et al.* 1994; Vos 2000), in that predators have no alternative choice of food, this chicken-and-egg logic can be reversed insofar as domestic stock adds to the carrying capacity of the environment for predators. Schaller (1972) found that the more abundant a preferred species was, the more likely it was to fall prey to lions. By extension, in the Pantanal, cattle are both the most abundant and the most vulnerable prey, so some level of jaguar predation is an inevitable and a natural part of ranching, like drought or soil fertility (Soisalo and Cavalcanti 2006). By analogy, there are limits to the feasibility of mitigating such environmental effects on agriculture, and limits to what society deems an acceptable cost of environmental intervention. For example, the latter is clearly illustrated in Europe by payments to farmers for custody of nature under the Common Agricultural Policy (Dutton *et al.* 2008). To the extent that irreducible damage by jaguars to cattle ranchers must be offset (rather than tolerated as an inevitable consequence of farming in jaguar country), solutions might lie in financial instruments such as tax benefits, favourable credits, or a regional increase in beef prices. The significance of livestock losses to jaguars will be proportionally diminished by ranchers improving other aspects of rudimentary herd husbandry that currently account for more losses than does jaguar predation (Hoogesteijn *et al.* 1993). That said, the quest for efficiency will eventually bring the farmer into head-on collision with those losses to jaguars that are unavoidable, and society will need to decide who is to bear these costs.

Recently, there has been an effort in the Pantanal to alleviate jaguar–livestock conflict in the form of a compensation programme (Silveira *et al.* 2006). Such programmes have been explored worldwide (Saberwal *et al.* 1994; Wagner *et al.* 1997; Vos 2000; Naughton-Treves *et al.* 2003; Swenson and Andrén 2005) but their effectiveness is debated (Nyhus *et al.* 2003, 2005; Bulte and Rondeau 2005; Maclennan *et al.*, 2009). Unverifiable losses, fraudulent claims, overly bureaucratic procedures and associated time lags in payment, payments below market values, lack of sustainable funding, high administrative costs, and moral hazard are some of the drawbacks associated with compensation programmes (Bulte and Rondeau 2005; Nyhus *et al.* 2005; Zabel and Holm-Müller 2008). Ideally, such schemes would be closely monitored, but in the Pantanal, this is challenging because retaliatory, illegal killing of jaguars is often clandestine.

An alternative to compensation involves 'performance payments' (Nyhus *et al.* 2005; Zabel and Holm-Müller 2008). By analogy with agri-environment schemes elsewhere, payments would be conditional on some measure of effective jaguar conservation in an area (Ferraro and Kiss 2002; Zabel and Holm-Müller 2008). As with all environmental payments (and compensation schemes), it would be essential to have effective monitoring, robust regulation, and care to avoid unintended consequences.

However, the results of our studies demonstrated that the problem goes beyond the economics and into the realms of culture—depredation on stock and retributive killing turn out to be more loosely linked than is often supposed. Although prejudices against jaguars are deeply ingrained within the culture of cattle ranching, attitudes can change over generations. Wolves were eradicated from the Rocky Mountain region of the United States by the 1930s, but are now making a dramatic comeback after reintroduction efforts in 1995. It may have taken decades, but policies towards wolves slowly changed over time as ecological studies and social attitudes reflected an increasing appreciation for the role that top predators play in ecosystem dynamics. In the case of the Pantanal, given that cowboys are ultimately the ones whose behaviour will directly impact jaguar conservation, one priority would be to make them stakeholders in jaguar conservation, and this could be a potent ingredient of any performance-related scheme. Examples from the Amazon and Africa illustrate the potential of community-based resource management in wildlife conservation (Lewis *et al.* 1990; Castello 2004; Frost and Bond 2008). It will require ingenuity to formulate, and then regulate, a scheme that delivers benefits to both landowners and local communities from successful custody of 'their' jaguars. For example, mechanisms might be sought to channel payments both to landowners and into wider community benefits (e.g. education, health, and economic development) to encourage, ideally in ways that even foster, peer pressure against those persons killing jaguars.

Our synthesis reveals that while jaguars do indeed kill livestock in the Pantanal, this is not the only, nor perhaps even the most important reason, why people kill jaguars. Therefore, in many cases, jaguar conservation may need to be approached in many different ways. As described here, in our case studies from the Pantanal and the Amazon, the motivations for killing jaguars include not only traditions and social rewards, but also the fear and misconceptions of the threat that jaguars pose to humans, the social incentives for persecution, as well as the economic viability of ranching. These insights may lead us towards approaches to decrease persecution that rely on gradual changes in the values, attitudes, and social norms concerning jaguars and jaguar persecution and that are tailored for the specific region. For example, whereas in the Pantanal communication campaigns to influence the social norms concerning jaguar hunting may significantly contribute to decrease in persecution, in the Amazon education to increase knowledge and improve perceptions about jaguars' threat to human safety might be more effective. Although the Pantanal is very important for jaguar conservation in the long term (Sanderson *et al.* 2002b), it would be unwise to generalize too readily from this particular situation to other parts of the jaguar's range. Nonetheless, conditions in the Pantanal are similar to those in, for example, the tropical-wet savannahs of the Venezuelan Llanos and the Bolivian Beni, so there is scope for an international analysis of cross-regional patterns in jaguar conflict (Zimmermann and Macdonald, in preparation).

Unquestionably, practical conservation must be underpinned by sound science. The illuminating power of data to allow for informed discussions and dispel misconceptions is illustrated by the findings we report on jaguar predatory behaviour in the Pantanal. However, the tensions between people and wildlife are so complicated that while ecological science is necessary as a foundation for solutions, it is not sufficient to deliver them. Drawing on methodologies from the social sciences, we have shown that the link between jaguar depredation on cattle and retaliatory persecution is only part of the story. To change peoples' actions will thus require a more far-reaching involvement that examines and understands their perceptions and traditions.

Acknowledgements

The Pantanal Jaguar Project was supported by the Wildlife Conservation Society; the U.S. Department of Agriculture; National Wildlife Research Centre at Utah State University, Logan, Utah; the National Scientific and Technological Development Council in Brazil; the Conservation, Food, and Health Foundation; the Mamirauá Civil Society; and Brazilian Foundation for Sustainable Development. The People and Jaguars Coexistence Project, in collaboration with the Wildlife Conservation Research Unit (Oxford University), has been supported by Cristalino Ecological Foundation, Instituto HSBC Solidariedade, Anglo American Brazil, Rainforest Concern, Cleveland Metroparks Zoo, North of England Zoological Society (Chester Zoo), Woodland Park Zoo, O Boticário Foundation, Fauna & Flora International, and Kevin Duncan. The global survey of human–jaguar conflicts project is funded by North of England Zoological Society (Chester Zoo) in collaboration with Wildlife Conservation Research Unit (Oxford University). Input from Paul Sparks.

CHAPTER 18

The ecology of jaguars in the Cockscomb Basin Wildlife Sanctuary, Belize

Bart J. Harmsen, Rebecca J. Foster, Scott C. Silver, Linde E.T. Ostro, and C. Patrick Doncaster

Male jaguar in the Cockscomb Basin, Belize (2003). © Bart Harmsen, Wildlife Conservation Society.

The cryptic habits of Neotropical felids present a substantial challenge for biologists wishing to study their behaviour and ecology. Their characteristically low population densities across generally inhospitable habitats require ingenious methodologies to obtain information about the presence and activity of individuals that may entirely elude direct observation. The jaguar (*Panthera onca*) has proved to be one of the more difficult large felids to study successfully, evidenced in our still limited stock of knowledge about its ranging and feeding behaviours and interactions with other cats.

This case study summarizes our knowledge of the jaguar population in and around the Cockscomb Basin Wildlife Sanctuary (CBWS), in Belize, Central America. It outlines the beginnings of research on what has become one of the best-studied jaguar populations to date. Here we present data on population density, demographics, territoriality, temporal and spatial interactions with pumas (*Puma concolor*), and feeding ecology of jaguars in a tropical forest habitat.

Physical description of Cockscomb Basin Wildlife Sanctuary

The CBWS encompasses 425 km^2 of evergreen and semi-evergreen broadleaf tropical forest on the eastern slope of the Maya Mountains in south-central Belize in Central America, 15 km from the Caribbean Sea (Fig. 18.1). The sanctuary includes the watersheds of two major rivers, divided by a series of hills (< 1000 m a.s.l.), which separate the sanctuary into its east and west basins. The two basins are surrounded by mountains, with elevations ranging from 50 to 1120 m (Kamstra 1987). The east basin is mostly flat, with 75% of the area lying below 200 m (Kamstra 1987), and has been the focal area of our research. The climate follows a seasonal wet and dry pattern, with a distinct dry season from January to May. The average annual precipitation is about 2700 mm, and average temperature is 25–28°C (Silver 1997).

The forests of the east basin consist primarily of various stages of disturbed and secondary growth. It was selectively logged from 1888 until 1961, when a hurricane levelled 90% of the canopy (cited in Kamstra 1987). After this, major logging operations ceased, although legal selective logging continued until the mid-1980s, when the flora and fauna came under formal protection. To the east of the CBWS, the land stretching to the coast forms a mosaic of community and private lands, including an unprotected broadleaf forest buffer, shrublands, pine savannah, small-scale tourist resorts, traditional milpa, cattle pastures, shrimp farms, and fruit plantations (Foster 2008). The farms and villages suffer intermittent livestock predation by jaguars and retaliatory lethal control of jaguars is common (Foster 2008).

The CBWS forms part of the Maya Mountain forest block, ~5000 km^2 of contiguous forest in southern Belize made up of 14 protected areas and forest reserves (Fig. 18.1). This forest block connects with the remaining forested areas of Guatemala and, through unprotected forest patches, to northern Belize and the Mexican Yucatan. Thus, the CBWS forms a critical link in the Mesoamerican Biological Corridor, at the heart of a protected area complex that may contain one of the largest jaguar populations remaining in Central America (Sanderson *et al.* 2002b).

Jaguar research in the Cockscomb Basin Wildlife Sanctuary

With its large contiguous expanse of relatively undisturbed and protected moist tropical forest, the CBWS offers a study site for jaguars that, if not unique, is extremely rare in Mesoamerica. This is a habitat in which jaguars have not been extensively studied elsewhere, and one that is relatively free from the anthropogenic pressures of hunting and habitat destruction that threaten jaguar populations throughout much of their range. The reserve is relatively easy to access, has a well-mapped and extensive system of foot trails, the cooperation of most of the communities surrounding the park, and the flora and fauna have been the subject of extensive studies (Rabinowitz and Nottingham 1986; Kamstra 1987; Silver 1997; Ostro 1998; Silver *et al.* 2004; Harmsen 2006; Foster 2008; Harmsen *et al.* 2009, in press b; Foster *et al.* in press b).

The jaguars of the Cockscomb Basin were first studied in the 1980s. Rabinowitz and Nottingham (1986, 1989) used telemetry to analyse jaguar ranging behaviour, conducted an analysis of jaguar diet, and surveyed the mammalian community. They concluded that the area contained one of the highest densities of jaguar in Central America. In association with the Belize Audubon Society (BAS), they recommended to the Belizean government that it be protected. The area was declared a jaguar reserve in 1986, and is currently managed by the BAS.

In 2000, the Cockscomb Jaguar Research Program began to explore the use of remotely triggered camera traps as a means of studying jaguar populations. What started as an opportunistic deployment of cameras, to find out whether jaguars could be photographed reliably in rainforest, has eventually developed into long-term population monitoring of this well-protected jaguar population. Unlike pumas, individual jaguars can be reliably identified by their spot pattern, as revealed by a pair of photographs taken by cameras straddling the trail at each camera station (Silver *et al.* 2004). Jaguars can also be sexed more easily than pumas, with their posture while walking tending to reveal the male's testes and female's absence of testes (Harmsen 2006). Once a jaguar had been sexed, it could be linked to its individual identification such that further photographs did not require a view beneath the tail.

The CBWS has few vehicular roads in the interior, so cameras cannot be checked daily. The remoteness and thick vegetation requires that all camera traps are distributed and maintained on foot, which limits

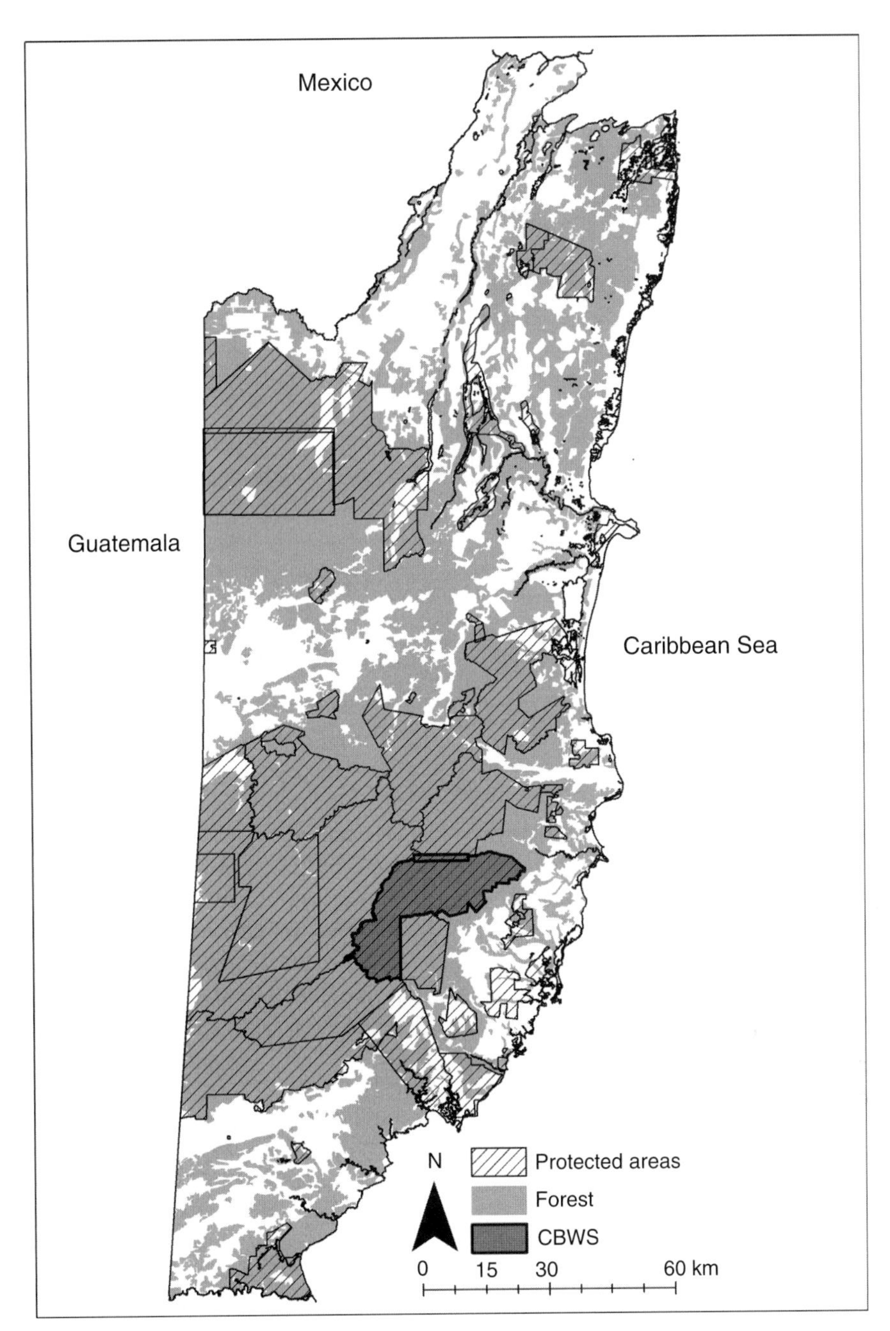

Figure 18.1 The Cockscomb Basin Wildlife Sanctuary (CBWS) in relation to other terrestrial protected areas and forest cover in Belize, Central America (from Meerman and Sabido (2001)).

options for changing camera locations during a survey. Consequently, a relatively static grid of camera stations (Fig. 18.2) was designed, compared, for example, to the more flexible rotation of camera locations used by Karanth and Nichols (1998) for studying Indian tigers (*Panthera tigris*) in an area that was accessible by vehicle. Maintaining the same camera sites first required optimizing their

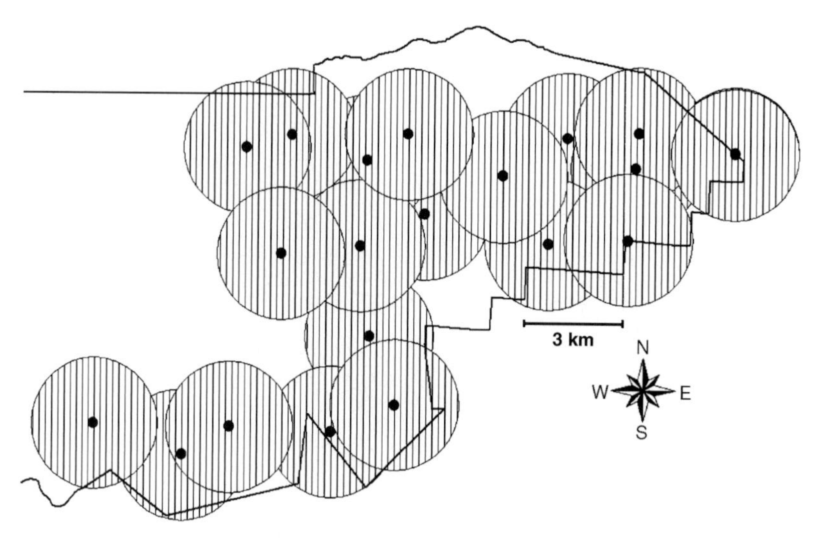

Figure 18.2 Camera-trap array used for dry-season surveys within the western half of the Cockscomb Basin Wildlife Sanctuary (CBWS; outlined). The black dots represent the cameras, and the circles represent a 10-km^2 area surrounding each camera. (From Harmsen 2006.)

positions; thereafter a static grid had the advantage of controlling variability in the probability of capture from station to station between surveys.

In 2002, Silver *et al.* (2004) carried out the first jaguar abundance survey, examining photographs to estimate abundance through a capture–recapture analysis of individual jaguars. We adapted a methodology pioneered by Karanth and Nichols (1998) for tigers in India (Silver *et al.* 2004; Karanth *et al.*, Chapter 7, this volume), adjusting the camera-trapping grid and modifying camera placement to maximize jaguar captures (in this chapter, capture is used synonymously with photograph). An array of 20 camera stations, each consisting of two heat- and motion-sensitive cameras (Cam Trak South Ltd), were used to assess jaguar population abundance across an area of at least 236 km^2 (Silver *et al.* 2004). This sample area assumed a minimum jaguar home range of 10 km^2 (based on Rabinowitz and Nottingham 1986), and was determined by drawing overlapping circular buffers of 10 km^2 around each camera station and calculating the total area encompassed by the merged perimeter (Fig. 18.2). Camera stations were placed no more than 3 km apart in a pattern that ensured that no area greater than 10 km^2 within the array lacked a camera (Silver *et al.* 2004). The survey was conducted over a period of 59 days between February and April 2002, and surveys lasting 60–90 days have subsequently been repeated once yearly during the dry season. Due to logistical issues, the number of camera stations was reduced to 19 after the initial survey. Abundance was estimated with capture–recapture closed population models using the software program CAPTURE (Rexstad and Burnham 1991). To estimate densities for each study site, we divided the abundance calculated above by the effective sample area. The effective sample area was determined by adding to each station a circle with a radius equal to half the mean maximum distance among multiple captures of individual jaguars during the sample period (Wilson and Anderson 1985) and then calculating the area bounded by the merged perimeter.

From 2003 to 2007, the established stations were operated throughout the year, and additional camera traps were placed in a variety of arrays within CBWS across a total of 119 locations (Harmsen 2006). Between 2004 and 2006, Foster (2008) established 105 camera locations in the lands neighbouring the eastern border of the CBWS, expanding the total study area to ~524 km^2 (based on 10 km^2 circular buffers centred on each station). She conducted four large-scale surveys (22–46 stations with effective sampling areas of up to 532 km^2) spanning the protected forest, unprotected forest buffer, and unprotected fragmented landscape. With one of the largest and most intensive camera-trap datasets of any large felid, we were able to analyse the photographic data to tackle questions not just about jaguar density, but also about their demography, ranging behaviour, territoriality, habitat use, and activity patterns. The following sections outline the methods for doing this and some of our key results emerging from the ongoing work.

Jaguar density in CBWS

Between 80 and 90 individual jaguars were photographed from 2002 to 2007 in and around the CBWS (Harmsen 2006; Foster 2008). A range is given because the dataset includes some single-side photographs of individuals, due to camera failure. Jaguar density estimates from the CBWS range from 3.5 to 11.0 individuals/100 km^2 during this period (Table 18.1). These densities are at the high end of densities found elsewhere in the geographical range of jaguars (Table 18.2; Maffei *et al.*, in press). The lower estimates from the first 2 years of trapping in the CBWS reflected lower numbers of individuals captured than in subsequent years. Moving some camera locations even small distances onto paths more often frequented by jaguars led to the capture of more individuals and more captures of each individual, suggesting that camera placement is important in the accuracy of population estimates (Harmsen 2006). Because optimal camera placement maximizes the number of captures used to obtain the abundance estimate, the resulting estimate is likely to be more accurate as well as being higher. From 2004 to 2007, cameras retained their positions on the same trees, and population estimates remained higher during this period than in the previous years before positions were fixed (Table 18.1).

Harmsen (2006) also carried out a survey within a single wet season, and revealed no evidence of seasonal variation in density estimates (Table 18.1). Although capture rates of jaguars at individual cameras were highly variable between locations, comparisons between seasons indicated no strong seasonality (general linear model [GLM] with season and location as factors, season: $F_{2,60} = 2.82$, $P = 0.09$; location: $F_{9,60} = 2.73$, $P < 0.05$; interaction season × location: $F_{18,60} = 1.20$, $P = 0.29$; Harmsen 2006).

Outside the CBWS, Foster (2008) detected a decline in jaguar density with distance from the contiguous forest block. Lethal control in retaliation for livestock predation is common on the lands neighbouring the CBWS and is the most likely cause for the observed reduction in jaguar density with distance from

Table 18.1 Population density of jaguars in Cockscomb Basin Wildlife Sanctuary (CBWS) taken from annual dry-season jaguar population surveys 2002–2007. Last line shows the single wet-season survey in 2003.

Year	Capture rates	Sample area (km^2)—using accumulated HMMDM[a]	Density/100 km^2	SE of density
2002	0.04	322	4.3	1.2
2003	0.06	289	3.5	0.7
2004	0.03	319	11.0	3.1
2005	0.04	319	6.6	3.1
2006	0.07	319	6.3	1.8
2007	0.08	319	6.9	1.3
2003[b]	0.05	196	6.1	1.7

[a] HMMDM is half the mean of the maximum distance (in metres) moved by any individual animal that was photographed in more than one location during a survey. It reflects the degree to which jaguars range during the survey period. The accumulated HMMDM is the value pooled across all individuals from all surveys.

[b] Survey undertaken in wet season with reduced number of camera stations.

Source: From Harmsen (2006).

Table 18.2 Jaguar density in different regions of the neotropics based on telemetry and/or camera-trap data. Repeated surveys at the same site are shown as separate estimates; alternate methods of estimation using the same data are shown as ranges.

Country	Habitat	Density/100 km^2 (SE)	Method	Study
Mexico	Dry tropical deciduous forest	1.7	Telemetry	1
	Semi-deciduous and seasonally flooded forest	4.5–6.7	Telemetry	2
Guatemala	Subtropical moist forest with seasonally inundated bajo forest	~0.7–1.7	Camera	3
Belize	Subtropical moist forest with seasonally inundated bajo forest	8.8 (2.3), 11.3 (2.7)	Camera	4a
	Tropical evergreen seasonal broadleaf secondary forest	5.3 (1.8)	Camera	4b
	Tropical evergreen and semi-evergreen broadleaf lowland secondary forest	8.8 (3.7), 18.3 (5.2), 4.8 (1.0), 11.5 (5.5)	Camera	5
	Deciduous semi-evergreen seasonal forest with pine	7.5 (2.7)	Camera	6a
Costa Rica	Tropical lowland primary rainforest	7.0 (2.4)	Camera and telemetry	7
Brazil	Atlantic forest—semi-deciduous rainforest	2.2 (1.3)	Camera and telemetry	8
		3.7	Telemetry	9
	Pantanal—seasonally inundated alluvial plains: pasture, grassland, cerrado, gallery forest, deciduous, and semi-deciduous forest	4.0	Telemetry	10
		7	Telemetry	11
		1.6	Telemetry	12
		5.7 (0.8)–10.3 (1.5)	Camera and telemetry	13
		5.8 (1.0)–11.7 (1.9)		
Bolivia	Open canopy forest with palms	1.7 (0.8), 2.8 (1.8)	Camera	14, 6b
	Chaco alluvial plain forest	5.1 (2.1), 5.4 (1.8)	Camera	15a, 6c
		2.1	Camera	16a
	Chaco-chiquitano transitional dry forest with scrub	11.1	Camera	16b
		4.0 (1.3)	Camera	6d
		2.3 (0.9), 2.6 (0.8), 3.1 (1.0)	Camera	15b

Notes: 1—Chamela-Cuixmala Biosphere Reserve, Jalisco, Núñez *et al.* (2002); 2—Calakmul Biosphere Reserve, Campeche, Ceballos *et al.* (2002); 3—Mirador-Rio Azul National Park, Novack (2003); 4—(a) Gallon Jug Private Estate; and (b) Fireburn Private Reserve, Miller (2006); 5—CBWS, Harmsen (2006); 6—(a) Chiquibul Forest Reserve; (b) Madidi National Park, Tuichi Valley; (c) and (d) Kaa-Iya National Park, Gran Chaco, Silver *et al.* (2004); 7—Corcovado National Park, Salom-Pérez *et al.* (2007); 8—Morro do Diabo State Park, Cullen *et al.* (2005); 9—Iguaçu National Park, Crawshaw *et al.* (2004); 10—Acurizal and Bela Vista Ranches, Pantanal, Schaller, and Crawshaw (1980); 11—San Francisco Ranch, Pantanal, Azevedo and Murray (2007a); 12—Pantanal, Quigley, and Crawshaw (1992); 13—Fazenda Sete Ranch, Pantanal, Soisalo, and Cavalcanti (2006); 14—Madidi National Park, Tuichi Valley, Wallace *et al.* (2003); 15—(a) and (b) Kaa-Iya National Park, Gran Chaco, Maffei *et al.* (2004b); 16—(a) Kaa-Iya National Park, Gran Chaco; and (b) Kaa-Iya National Park, Gran Chaco, Maffei *et al.* (2004a).

Source: From Foster (2008).

contiguous forest and proximity to human development (Foster 2008).

Jaguar habitat use

The high jaguar densities observed in the CBWS suggest that local conditions are particularly favourable for these cats, with abundant prey, optimal habitat features, and minimal competition with other predators or humans. Analysis of capture rates across multiple camera locations can be used to reflect site usage by jaguars across the landscape and thus make inferences about habitat preferences (Foster *et al.*, in press a; Harmsen *et al.*, in press b). Within the dense homogenous forest of the CBWS, Harmsen *et al.* (in press b) found no association between the capture rates of male jaguars and altitude, slope, or the distance to waterways. The only habitat variables significantly correlated with male jaguar capture rates were those associated with trails, specifically trail width, seasonality, and age. Males were captured more frequently on wider, older, permanent trails than on narrow, fresh, or seasonal trails, presumably because the latter provided reliable and easily accessible travel routes through the otherwise dense secondary vegetation (Harmsen *et al.*, in press b). In contrast, Foster *et al.* (in press a) found that capture rates of female jaguars within the forest did not vary with trail variables, but increased with proximity to flowing water. They suggest that the waterways may provide alternative travel routes through dense forest which allow females with young to avoid trails dominated by males. Harmsen *et al.* (2009) suggested that the lack of territoriality and high flux in male dominance documented in the CBWS could contribute to higher levels of infanticide. As for many other felids, infanticide has been reported in jaguars (Soares *et al.* 2006) and is expected to be more common in areas with unstable male territories.

Jaguar territoriality

Rabinowitz and Nottingham (1986) found that the home ranges of five radio-collared male jaguars within Cockscomb encompassed between 28 and 40 km^2. Levels of overlap between the males were high, with one male's range overlapping others by 80%. Multiple males used the same logging roads, leading Rabinowitz and Nottingham (1986) to conclude that these roads were a limiting resource in the dense jungle environment. Despite the high level of range overlap, they never observed jaguars in the same place at the same time. Each would stay for periods of 4–14 days within a small area averaging 2.5 km^2 (SD = 0.6 km^2, $n = 16$), before moving a larger distance again. However, jaguars did react to the disappearance of conspecific males in the area with spatial shifts in range use, indicating that territoriality played a part in the ranging behaviour of male jaguars in this area.

Camera traps can be used to examine territoriality and ranging behaviour of jaguar populations, with the advantage over radio-tracking that they can sample the full membership of the area, offsetting the disadvantage of less intensive information on any one individual (Harmsen *et al.* 2009). The network of stations in our study encompassed the same logging roads (now mostly overgrown) that featured in Rabinowitz and Nottingham's study (1986). While the camera-trap data do not track the movements of individuals over time, they do provide timed observations of every jaguar passing the fixed camera location, thereby giving a more complete picture of spatial overlap of every individual in the area. Harmsen *et al.* (2009) collated data from 42 established camera stations over 108 location-months, where a location-month is the pooled data from a single camera-trap location for 1 calendar month. These revealed a mode of two male jaguars passing the same point within 1 month (56 of 108 sample months, see Fig. 18.3), though three males per month was not uncommon (21 of 108; Harmsen *et al.* 2009). Although the sample size of 15 females was much lower, the most common occurrence was for a single female to be photographed per month per location (9 of 15 samples, see Fig. 18.3). The frequency of multiple males passing by a camera station within a single month suggests that, at least for parts of their range, male jaguars do not maintain exclusive territories in the CBWS (Harmsen *et al.* 2009), confirming the pattern suggested by the telemetry data of Rabinowitz and Nottingham (1986). Extensive overlap between adult male jaguar ranges has also been observed in the Pantanal (Cavalcanti 2008), and

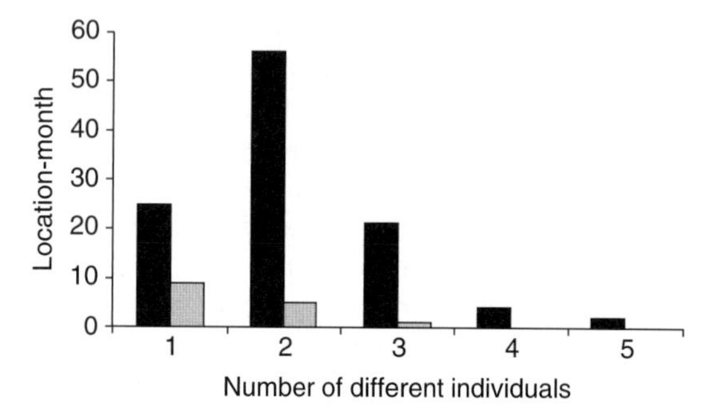

Figure 18.3 Frequency distribution of individual jaguars photographed per 'location-month' (pooled data per camera-trap location per calendar month). Only location-months with >1 capture of either sex are counted. Black bars show males (n = 108 location-months) and grey bars show females (n = 15 location-months). (From Harmsen *et al.* 2009.)

contrasts with the characteristic felid social system of exclusive territories within the sexes (Sunquist and Sunquist 2002). While there was no clear evidence to support exclusive territoriality, we do not discount the possible exclusive use of core areas, or that jaguars use trails to demark territory boundaries. Nevertheless, neither possibility completely explains the overall spatio-temporal spread of males across all 119 trap stations used to date, one-fifth of which were positioned off established trails (Harmsen *et al.*, in press b).

If male jaguars maintain exclusive territories, we would expect to find territory holders having a markedly higher capture frequency than conspecifics captured at the same location. However, Harmsen *et al.* (2009) found that the most frequently captured jaguar at any one location had only one or two more photographs taken than the next most frequently photographed individual, with a mode of one excess capture. Also, the identity of this most frequently captured individual changed from month to month. Of the 21 camera stations capturing at least three different individuals, half had a new 'resident' (i.e. most photographed) adult in consecutive months, and 20% of the camera stations switched back to individuals that had previously held a highest capture score per month. Only two locations had a consistent record of a single male dominating the location for over a year. The findings of Harmsen *et al.* (2009) suggest that territoriality is weakly expressed in a low consistency of site usage, with just 2 of the 21 locations indicating the possibility of an exclusively used core area.

Radio-tracking by Rabinowitz and Nottingham (1986) revealed how the death of a male jaguar in Cockscomb resulted in range shifts in the two neighbouring males. Harmsen (2006) observed a similar pattern over a 4-year period in Cockscomb using camera traps. Three adult male jaguars were captured in an area continuously camera-trapped for 2 years. During this time, one of the males was captured on average weekly, the other two less frequently, approximately once every 2 months. Following the abrupt disappearance of the resident male, capture rates of the two remaining males increased by a factor of 5 and 8.5, respectively.

Harmsen *et al.* (2009) further explored the data for evidence of male jaguars avoiding each other in time. If males use temporal avoidance as a strategy, we would expect them to move through areas sequentially, with new males only moving into an area once they perceive previous males have left, resulting in a delay between different males being recorded. If this strategy was in effect, we would expect a shorter time interval between captures of the same male than between captures of different males. In fact, the mean interval between photographs at each location of same males (8.6 days, $n = 151$) was longer than the interval between different males (7.2 days, $n = 248$; $P < 0.05$; Harmsen *et al.* 2009), suggesting that temporal avoidance was not a mechanism used by males. Other camera-trap studies on jaguars and felids have not explored capture data at this level, yet it can be informative about behaviour (e.g. see also the section below on jaguar–puma interactions), given a sufficiently dense network of trap stations.

Capture patterns of male and female jaguars

The mean sex ratio of individual jaguars photographed on a year-round basis from 2002 to 2007 within the CBWS was skewed, at three males per female. During annual dry season surveys this ratio ranged from 4:1 to 15:1 per survey (Table 18.3;

Table 18.3 Ratio of male to female jaguars photographed during annual camera-trap surveys.

Year	Male/female
2002	9.0:1
2003	8.0:1
2004	4.7:1
2005	4.7:1
2006	15.0:1
2007	4.4:1

Source: From Harmsen (2006).

Harmsen 2006). Although this may reflect the true situation, similarly skewed sex ratios from camera traps have been reported from other camera surveys in Central America, and in other parts of the jaguars' range (Foster 2008; Maffei *et al.*, in press).

The skewed capture rate has a number of possible explanations arising from methodological sampling biases (Harmsen 2006; Foster 2008). In the CBWS, male jaguars range more widely than females, and so are sampled over a larger area by the camera array (Harmsen 2006; Foster 2008). Cameras that are spaced further apart than the average female home range diameter may fail to detect all females within the population (Foster 2008). These biases have been addressed by Foster by conducting single-sex density estimates based on sex-specific sampling areas (Foster 2008). Males are also more active and have higher overlap in home ranges than females (Rabinowitz and Nottingham 1986; Harmsen *et al.* 2009). Both these behaviours will increase male capture probability over that of females (Harmsen 2006). Inter-sexual differences in habitat use and the avoidance of males by females (Foster *et al.* in press a) may also contribute to the observed male-biased sex ratio, as males dominate the wider trails which are also preferentially used by researchers as camera-trap locations. Harmsen *et al.* (in press b) attempted to address this issue by placing cameras extensively off-trail but they detected no female jaguars in these locations. It is not yet apparent where to place cameras in order to maximize female captures. The sex ratio of captures outside the reserve, in a more open and fragmented landscape is much closer to 1:1, potentially because the more even distribution of roads and paths may make it more difficult for males to dominate them (Foster 2008).

Two adult jaguars were captured in the same photograph on two occasions, but for only one of these events could we confirm that the pair was a male and female (Harmsen 2006). This particular male and female were photographed together 12 times across a period of 2 days, and mating calls were heard from the direction of the camera, which was situated close to the research station (Harmsen 2006). Males do not appear to seek out females outside of these short-term mating events, and communication between the sexes probably takes place mostly at a distance using marking behaviour. Nevertheless, Harmsen (2006) showed that whenever two jaguars were captured at the same location within a 24-h period they were more likely to be of opposite sex than both males, despite the overall lower capture rate of females than males (7% of 248 male–male events, and 14% of 67 male–female or female–male events were same-day captures: $X_1^2 = 4.20$, $P < 0.05$).

Jaguar diet

Jaguar diet is highly variable, with more than 85 prey species taken across their geographic range, and between 8 and 24 prey species documented in the diet at any one site (in Foster 2008). Peccaries, deer, large caviomorph rodents, armadillos, and coatis are the main prey taken (Oliveira 2002a). Studies from across the jaguar's geographical distribution show their willingness to take a range of prey items, from cattle weighing over 200 kg down to small rodents, with only peccary and deer (white-tailed, *Odocoileus virginianus*, or red brocket, *Mazama americana*) universally taken—or available to be taken—across all regions (e.g. Rabinowitz and Nottingham 1986; Emmons 1987; Taber *et al.* 1997; Núñez *et al.* 2000; Crawshaw and Quigley 2002; Leite and Galvão 2002; Scognamillo *et al.* 2003; Novack *et al.* 2005). Both of these preys are also popular game species throughout the jaguar's geographical distribution, highlighting the need for no-hunting zones outside as well as inside nature reserves, which can act as refuge sources for key prey to vulnerable predators.

Dietary studies of the Cockscomb jaguar population have been undertaken periodically over the past two (Table 18.4) decades. Rabinowitz and Nottingham (1986) first examined jaguar diet using scat analysis before the area was formally protected (Table 18.4). They characterized jaguars as opportunistic predators with a diet that largely reflected the relative abundance of prey species. The nine-banded armadillo (*Dasypus novemcinctus*) made up 54% of prey items taken and they speculated that this reflected intense hunting pressure by humans on other popular game species. Although the CBWS suffers occasional incursions by poachers at the edges today (R. Foster and B. Harmsen, personal observation), it is likely that prey populations have improved since hunting within the study area became illegal. In a 4-year study of feeding ecology conducted 20 years after the establishment of the CBWS, Foster *et al.* (in press b) found that armadillos remained the most common species in jaguar diet, still comprising half the prey items taken in a sample of 204 jaguar scats. At the same time, the occurrence of peccaries in the diet increased by a factor of 4, suggesting that protection may have aided peccary recovery. The continued high predation on armadillos despite this increase in use of larger ungulates led Foster *et al.* (2008) to hypothesize that armadillos in CBWS are abundant, are exploited opportunistically by jaguars, and may play an important role in maintaining the high jaguar density observed in CBWS. Although jaguars may adapt to a wide range of prey, including relatively small species such as armadillos, models of the energetics of reproduction predicted that female breeding is probably limited by the availability of larger prey such as peccaries and deer (Foster 2008). In the unprotected fragmented lands neighbouring the CBWS, jaguars rarely ate large wild ungulates; instead, a diet of relatively small prey was supplemented with larger domestic animals, primarily cattle (Foster *et al.* in press b).

Interaction between jaguars and pumas

Jaguars are sympatric with pumas across their entire range and are of a similar size and weight class: head-body lengths are 1.10–1.85 m for jaguars and 0.86–1.54 m for pumas, with adult body weights of 30–158 kg for jaguars and 24–120 kg for pumas (Emmons and Feer 1997; Reid 1997). Where they coexist, however, jaguars tend to be larger than pumas (Iriarte *et al.* 1990). The jaguars and pumas inhabiting Belize are among the smallest in their geographic ranges (Iriarte *et al.* 1990). Studies to identify mechanisms of coexistence of these similar-size predators have been conducted in and around the CBWS, investigating spatial and temporal partitioning and differences in food habits (Harmsen 2006; Foster 2008; Harmsen *et al.* 2009; Foster *et al.*, in press b).

The ratio of jaguars to pumas within the CBWS is unknown due to the difficulty of identifying individual pumas from camera-trap photographs. Although it is sometimes possible to identify individual pumas for abundance estimation (e.g. Kelly *et al.* 2008), this technique relies on sufficient variation in scars and skin parasites within the study population to consistently distinguish all individuals through time. Although pumas in the CBWS could not be individually identified to a level which allowed reliable density estimation, total captures of jaguars and pumas from 2003 to 2005 were similar (619 and 576, respectively; Harmsen 2006). If jaguars and pumas have similar chances of capture across the survey grid, similar total captures would imply similar abundances. However, Harmsen *et al.* (in press b) found that short-term recognition of pumas within the CBWS was possible over small areas and short time periods. This revealed fewer pumas than jaguars, and high capture rates of a few individual pumas versus relatively few captures of numerous jaguar individuals. Harmsen *et al.* (in press b) found that pumas followed trails more consistently than jaguars, and had fewer excursions off-trail, while jaguars used the surrounding forest matrix more than pumas. Individual pumas were more likely to be captured at multiple locations along the trail within short time periods, compared to individual jaguars, which were more often captured only once. Individual pumas were captured at significantly more consecutive cameras along the trail than jaguars, indicating a higher tendency for pumas to remain walking on trail than jaguars. In contrast, when individuals of either species were captured at the start and end of the trail, jaguars were less likely to be captured on the middle cameras than were pumas, indicating a higher tendency for jaguars to leave the trail than pumas (Harmsen *et al.*, in press b). The differential trail use means that individual pumas have a greater probability of being photographed on trails than do individual jaguars, thereby biasing a comparison of relative capture rates towards pumas for any camera network in which the majority of camera traps are on established trails. Given the overall similarity in capture rates of jaguars and pumas, this indicates that jaguars are probably more abundant within the CBWS than pumas. Differences in trail usage by pumas and jaguars suggest a need for caution in trying to index abundance from relative capture rates of these two cat species for camera grids which mainly use trail-based camera locations. This finding also indicates that jaguars prefer moving off trail through the forest matrix compared to pumas. Harmsen *et al.* (in press b) suggested that the short-limbed, stocky jaguar is better suited to travelling through thick vegetation compared to the leaner, long-limbed puma, whose agility is lost in dense undergrowth.

Harmsen *et al.* (2009) found that jaguars and pumas inhabiting the CBWS had similar 24-h activity patterns, both being primarily nocturnal (Fig. 18.4). Jaguar and puma capture frequencies were positively correlated at each camera location ($r = 0.65$ between $\log_{10}$ captures, $P < 0.01$, $n = 93$; Harmsen *et al.* 2009), indicating that both species use the same areas within CBWS in roughly equal proportions (Fig. 18.5a). However, the two species appear not to use the same areas at the same time. Their capture rates were never simultaneously high at the same location within the same month (Fig. 18.5b; Pearson correlation

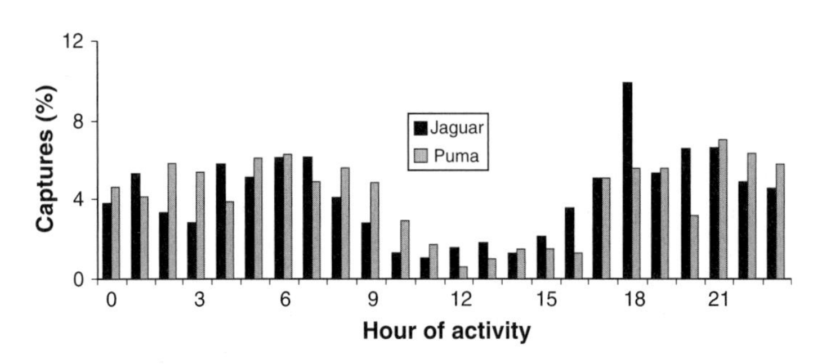

Figure 18.4 Percentage of jaguar and puma captures per hour (jaguar $n = 397$, puma $n = 413$), based on camera-trap data collected from 2003 to 2005. (From Harmsen *et al.* 2009.)

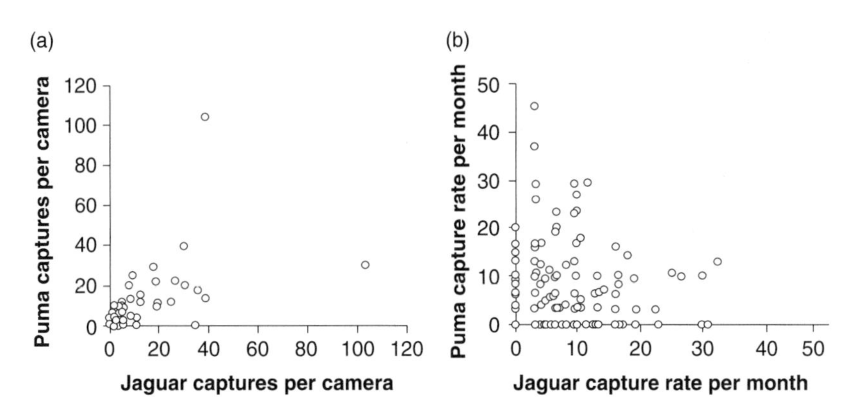

Figure 18.5 Relationship between jaguar and puma captures in the CBWS from 2003 to 2005: (a) Jaguar and puma captures at each camera location; and (b) jaguar and puma capture rate per 'camera-month' (a single camera location over the period of 1 month). (From Harmsen *et al.* 2009.)

between $\log_{10}$ non-zero captures, $r = 0.09$, $P = 0.17$, $n = 258$; Harmsen *et al.* 2009). Further investigation of the time intervals between consecutive captures of jaguar and puma at the same locations showed significantly longer interspecific intervals between jaguar and puma captures than the intraspecific intervals between jaguar–jaguar and puma–puma captures ($P < 0.01$; Harmsen *et al.* 2009). The mean interval between sequential jaguar–puma photographs was 8.5 days ($n = 166$), and between sequential puma–jaguar photographs was 10 days ($n = 173$). But the mean interval between jaguar–jaguar photographs was 7.6 days ($n = 278$), and between puma–puma photographs was 7.5 ($n = 258$), suggesting that jaguars and pumas actively avoid one another more than their conspecifics (Harmsen *et al.* 2009). Scognamillo *et al.* (2003) likewise never found jaguars and pumas in close proximity in the Venezuelan Llanos, despite similar habitat use and activity schedules. Their result was not calibrated against intraspecific avoidance, however, which is a prerequisite for estimating interspecific competition.

Although Harmsen *et al.* (2009) found evidence of interspecific avoidance between jaguars and pumas, it is unknown whether this behaviour is mutual or one-sided. The more powerful jaguar may be expected to dominate the somewhat lean, agile puma, a supposition that is supported by anecdotal evidence of jaguars killing and eating pumas (B. Miller, personal communication). Under such a scenario, pumas would be expected to avoid direct confrontation with jaguars. Harmsen *et al.* (in press a) studied the scrape-marking behaviour of both cat species in sympatry and found that pumas mark more frequently than jaguars on forest trails. They suggest this may be a mechanism by which pumas communicate discretely with other conspecifics while temporally avoiding jaguars.

Because sympatric jaguars and pumas often use the same habitats and follow similar activity patterns (e.g. Núñez *et al.* 2002; Scognamillo *et al.* 2003; and Harmsen *et al.* 2009, in press b), it is reasonable to hypothesize that in such areas coexistence may be facilitated by selecting different prey species. Foster *et al.* (in press b) compared the food habits of jaguars and pumas in sympatry using the largest sample of scats to date (362 jaguars and 135 pumas). The approximate 3:1 ratio of jaguar to puma scats supports the conclusions of Harmsen *et al.* (in press b) and Foster *et al.* (in press b) that jaguars are more abundant in the area than pumas.

Within the CBWS, Foster *et al.* (in press b) found that the main size classes of prey taken by jaguars and pumas were similar, but they had significantly lower dietary overlap than expected by chance, with each cat species relying heavily on a single different small prey. Jaguars mainly ate armadillos, while pumas predominately ate pacas (*Agouti paca*). Both cats also took larger prey, mainly red brocket deer by pumas and white-lipped peccaries (*Dictolyes pecari*) by jaguars (Foster *et al.*, in press b). Although pumas did eat some white-lipped peccaries, they were mainly juveniles, while jaguars tended to take the adults. Foster *et al.* (in press b) suggest that jaguars, equipped with powerful jaws and short muscular limbs, are better adapted than pumas to hunt and kill dangerous species like white-lipped peccaries and armoured species such as armadillos. In contrast, the longer limbed puma is perhaps better suited to catching agile prey such as pacas and brocket deer. They suggest that the time spent by pumas walking on trails (Harmsen *et al.*, in press b) may limit predation on armadillos because when in the undergrowth these species may be less easy to detect than larger species such as pacas. In contrast, Foster *et al.* (in press b) suggest that jaguar encounter rates with armadillos are likely to be higher than for pumas due to the jaguars' habit of walking off trail (Harmsen *et al.*, in press b).

Camera and scat data provided evidence that, unlike jaguars, pumas were virtually absent in fragmented landscape neighbouring the CBWS (Foster 2008; Foster *et al.*, in press b). Foster *et al.* (in press b) suggest that the scarcity of pumas, compared to jaguars, in the unprotected lands outside the CBWS may reflect competition with humans for pacas and deer, both prized game species in the region, and a reluctance to hunt livestock. In contrast, they propose that jaguars are flexible in their use of the human-influenced landscape due to their ability to exploit armadillos, a less-valued and potentially more abundant game species in Belize, and occasionally hunt domestic animals when larger ungulates are scarce (Foster *et al.*, in press b).

Future directions for jaguar research in the CBWS

While camera trapping is a technique that has been widely employed for assessing population abundance, we have also found it invaluable for gaining wider insight into the behavioural ecology of the jaguar. Camera data from continued jaguar surveys will allow us to address questions about home range tenure, and population fluctuations in the CBWS jaguar population. Foster (2008) highlighted the need for field data on demographic rates and dispersal to improve population viability analyses of jaguars. Long-term population data from the CBWS will soon be sufficient to estimate jaguar survival and recruitment. Annual surveys monitoring the population have continued into 2008, though funding for future surveys is uncertain. Continuation of this research is crucial, as the data will ultimately provide the first reliable field-based estimates of jaguar vital rates. Expanding the size of the survey areas will help us better define the size of jaguar home ranges, their ranging behaviour, and home range use, to complement the more intensive, though less extensive, data from radio-tracking which we hope to start collecting in the near future. The properties of jaguar dispersal are virtually unknown from any part of its geographical range, and they require elucidation. Density estimates from camera-trap studies combined with movement data obtained from radio-collared members of the population could contribute to a better understanding of jaguar dispersal. Valid assessments of the probability of long-term jaguar persistence require more information about dispersal, and ranging behaviour in general, not only in Belize, but also in other range countries (Foster 2008).

Around the world, many camera-trap datasets have been collected for abundance estimates of elusive large cats; some of these could be deployed productively to study interactions within and between felid species, especially shared use of space in relation to population density and diet. Camera trapping provides a useful complement to telemetry as a method of studying social interactions, by virtue of it more fully sampling the complete array of players within an area. In future work, we will be using radio-telemetry in combination with camera trapping to fill important gaps in our understanding of how jaguars move through their environment daily and seasonally, and the mechanisms that allow jaguar to coexist with each other and with pumas. There remain many questions concerning jaguar life history and social interactions that will be solvable only through simultaneous application of multiple techniques.

Jaguar density estimation using camera traps will greatly benefit from improved data on ranging behaviour, particularly of females (Foster 2008). Since females appear to avoid trails dominated by males (Foster *et al.*, in press b), there is an urgent need to optimize methods to detect female jaguars off-trail within forests (Foster 2008; Harmsen *et al.*, in press b). Camera traps are not suitable for capture off-trail as they depend on trails as funnels to direct the target species past the camera (Harmsen *et al.*, in press b). An alternative method that is currently being explored is the genetic analysis of scats which are collected off-trail using trained detector dogs. A recent pilot study in the CBWS has demonstrated a high rate of scat detection off-trail by the dogs, but methods to genotype scat samples to sex and to the individual are still being refined (A. Devlin, unpublished data; C. Pomilla, personal communication). Once refined, these genetic methods will also allow an investigation of sexual differences in prey selection, and energetics models which predict that breeding females require a higher frequency of large prey in their diet than males (Foster 2008) can be formally tested. Estimates of wild prey abundance will allow us to assess whether there are sufficient prey available for female reproduction, and will also improve our understanding of the extent to which differences between jaguar and puma feeding habits influence or facilitate their coexistence. Because low visibility in dense secondary forest may hinder techniques such as distance sampling (Foster 2008) to assess prey availability, pilot studies in the CBWS are assessing the relative abundance of pacas and armadillos using a number of independent methods that can be cross-validated (Bies and Harmsen, unpublished data) and there are also plans to begin monitoring the larger prey species. Genetic data from scats collected in and around the CBWS are also contributing to a database for the analyses of the genetic diversity of jaguars across their entire range (C. Pomilla, unpublished data).

Basic natural history data are sometimes undervalued in conservation programmes which, understandably, want to solve practical problems with a high degree of urgency. While the natural histories of other endangered big cats such as lions (*Panthera leo*) and tigers are well known (e.g. Sunquist and Sunquist 2002), we are only just beginning to accumulate such data on the jaguar. Our knowledge of jaguar ecology in the CBWS serves as a platform from which we can answer critically important questions about how these animals survive in the moist forests of Central America.

CHAPTER 19

Snow leopards: conflict and conservation

Rodney M. Jackson, Charudutt Mishra, Thomas M. McCarthy, and Som B. Ale

Adult male snow leopard photographed in February 2003 using a TrailMaster 1550 film camera trap at an elevation of nearly 13,000 feet in Hemis National Park, Ladakh, J&K State, India. © Snow Leopard Conservancy.

Introduction

The endangered snow leopard (*Panthera uncia*) is perhaps the world's most elusive and charismatic large felid. The wild population is roughly estimated at 4500–7500, spread across a range of 1.2–1.6 million km^2 in 12 or 13 countries of South and Central Asia, namely Afghanistan, Bhutan, China, India, Kyrgyzstan, Kazakhstan, Nepal, Mongolia, Pakistan, Russia, Tajikistan, Uzbekistan, and possibly also Myanmar (Fox 1994; Nowell and Jackson 1996; Fig. 19.1). Snow leopards suffer from similar conservation challenges as other large felids, namely persistence at naturally low densities, extensive home ranges, dependence upon prey whose populations are low and/or mostly declining, and high vulnerability to poaching and other anthropogenic threats. They inhabit mountainous rangelands at elevations of 3000 to over 5000 m in the Himalaya and Tibetan Plateau, but as low as 600 m in Russia and Mongolia (Sunquist and Sunquist 2002). Snow leopard habitat is among the least productive of the world's rangelands due to low temperatures, high aridity, and harsh climatic conditions, with an average peak graminoid biomass of 170 kg ha^{-1} (asymmetric 95% CI 128–228 kg ha^{-1}; Mishra 2001). Consequently, prey populations are also relatively low, ranging from 6.6–10.2 blue sheep (*Pseudois nayaur*) km^{-2} in productive habitat in Nepal (Oli 1994) to 0.9 ibex (*Capra ibex*) km^{-2}

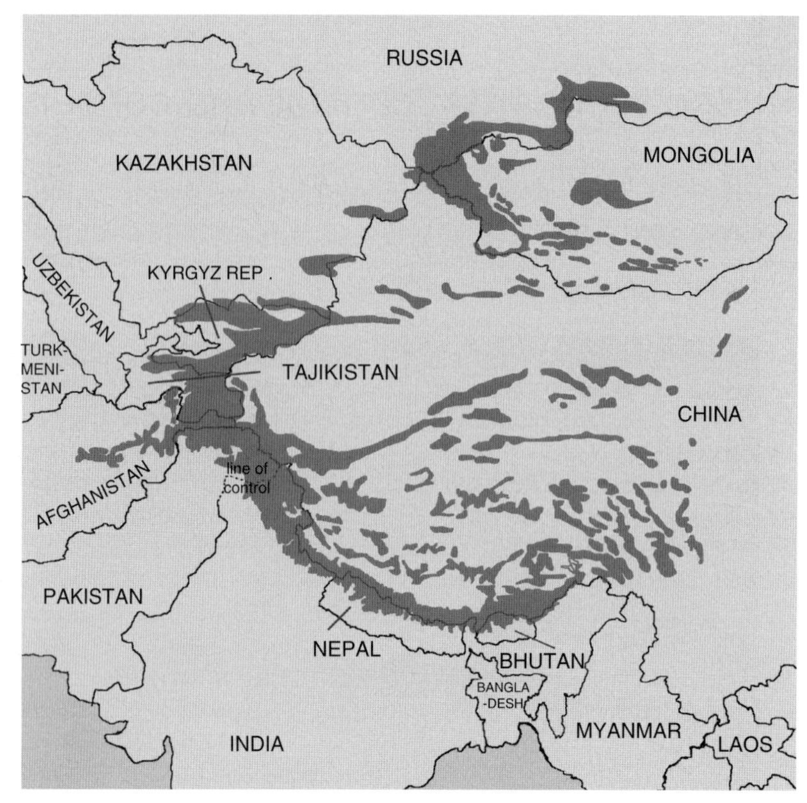

Figure 19.1 Distributional range of snow leopards. (Adapted from Fox 1994.)

in marginal habitat of Mongolia (McCarthy *et al.* 2005).

Traditional pastoralism and agro-pastoralism are the predominant land uses and sources of local livelihood in snow leopard habitat, with seven range countries having over 25% of land area under permanent pasture, >50% of their human population involved in agro-pastoralism, >40% living below national poverty levels, and average per capita annual incomes of US $250–400 (Mishra *et al.* 2003a). Although relatively few humans live in snow leopard habitat, their use of the land is pervasive, resulting in ever-increasing human–wildlife conflict, even within protected areas. Snow leopards, and other sympatric wildlife range across the larger landscape beyond protected areas, albeit at low densities. Since few (if any) protected areas are free of human influence, the snow leopards' survival hinges upon an uneasy coexistence with subsistence pastoralists and farmers eking out their living from the same harsh environment.

In this chapter, we review pertinent aspects of snow leopard behaviour and ecology, followed by an assessment of threats arising from the above-mentioned livestock–predator–wild prey paradigm. We conclude with a discussion of alternative conservation measures currently being applied in efforts to reverse this species' declining population trend.

Snow leopard: an overview of adaptation for mountain living

The snow leopard is a large felid, with males weighing 45–55 kg and females 35–40 kg, a shoulder height *c.* 60 cm and head-body length of 1.8–2.3 m (Hemmer 1972). With its smoky-grey pelage tinged with yellow and patterned with dark grey, open rosettes and black spots, the snow leopard is especially well camouflaged for life among bare rocks or patchy snow (Figure 19.2). It has a well-developed chest, short forelimbs

Figure 19.2 Snow leopard investigating a scent-sprayed rock situated on a corridor frequently used by a other individuals. © Snow Leopard Conservancy.

with sizable paws, long hind-limbs, and a noticeably long tail (75–90% of its head and body length), giving it an amazing agility for negotiating steep terrain or narrow cliff ledges (Sunquist and Sunquist 2002). Adaptations for cold include an enlarged nasal cavity, long body hair with dense, woolly underfur (belly fur up to 12 cm in length), and a thick tail that can be wrapped around the body for added warmth while at rest. Mating occurs between January and mid-March, a period of intensified social marking and vocalization (Ahlborn and Jackson 1988). In captivity, oestrous lasts 2–12 days, with a cycle of 15–39 days (Nowell and Jackson 1996). Age at sexual maturity is 2–3 years (Sunquist and Sunquist 2002). There is no information on longevity in the wild. Litter size is usually two to three and exceptionally as large as seven. Dispersal is said to occur at 18–22 months of age, and sibling groups may remain together briefly at independence (Jackson 1996). This may explain reported sightings of as many as five snow leopards in a group (Hemmer 1972).

Snow leopards favour steep, rugged terrain well broken by cliffs, ridges, gullies, and rocky outcrops (Jackson and Ahlborn 1989; Sunquist and Sunquist 2002); however, in Mongolia and Tibet, they occupy relatively flat or rolling terrain when sufficient cover is available (Schaller 1998; McCarthy 2000). Home range size varies from 12 to 39 km^2 in productive habitat in Nepal (Jackson and Ahlborn 1989) to 500 km^2 or more in Mongolia with its open terrain and lower ungulate density (McCarthy *et al.* 2005). Densities range from <0.1 to 10 or more individuals per 100 km^2, but current knowledge is insufficient for generating a reliable range-wide population estimate.

The four telemetry studies to date reveal largely overlapping male and female home ranges, but with use of a particular area usually separated temporally. In Nepal, 42–60% of home range locations for four cats occurred within 14–23% of their respective home areas: these commonly used core areas intersected the most favourable local topography, habitat, and prey base (Jackson and Ahlborn 1989). Solitary and typically crepuscular, snow leopards remain within a small area for a week before shifting to another part of their home area. Mountain ridges, cliff edges, and well-defined drainage lines serve as common travel routes and sites for the deposition of signs, including scrapes, scats, and scent marks (Ahlborn and Jackson 1988). Core areas are often used by more than one snow leopard and are marked significantly more frequently than non-core sites, suggesting that such marking may help space individuals and thereby facilitate more efficient use of sparse resources (Jackson and Ahlborn 1989). In Nepal's rugged habitat, leopards moved up to 7 km daily (straight-line distances), but averaged *c.* 1 km (Jackson and Ahlborn 1989); whereas in Mongolia, their daily movements were considerably greater at 12 km, with one female covering 28 km within a single day (McCarthy *et al.* 2005).

Ultimately such differences may reflect variation in prey availability. An opportunistic predator, the snow leopard is capable of killing prey up to three times its own weight, so that only adult camel (*Camelus bactrianus*), kiang (*Equus* spp.), and wild yak (*Bos grunniens*) are excluded as potential prey (Schaller 1998). The snow leopard's distribution coincides closely with its principal prey, the bharal (blue sheep) and ibex, ungulates with mean body weights of 55 and 76 kg, respectively (Mishra *et al.* 2002). These caprids reportedly contribute more than 45% and up to 66% of the snow leopard's diet, with livestock providing as much as 40–58%, though generally more on the order of 15–30% (Schaller *et al.* 1988; Oli *et al.* 1993; Chundawat and Rawat 1994; Jackson 1996; Bagchi and Mishra 2006). Important supplementary prey items (approximately 12–25% of diet) are marmot (*Marmota* spp.), pika (*Ochotona* spp.), hares (*Lepus* spp.), small rodents, and game birds. Annual prey requirements are estimated at 20–30 adult bharal, with radio-tracking indicating a large kill every 10–15 days (Jackson and Ahlborn 1984; Jackson 1996). Unless disturbed, a snow leopard may remain on its kill for up to a week (Fox and Chundawat 1988).

Status and threats to snow leopards

Snow leopards were first listed as endangered under the 1972 IUCN red list data. The 1975 enactment of the Convention on International Trade in Endangered Species (CITES) of Wild Flora and Fauna prohibited all international commercial trade of Appendix 1 species (which includes snow leopards). All snow leopard range states except Tajikistan are party to CITES, with Bhutan and Kyrgyzstan joining recently. The snow leopard has been designated as a 'concerted action' species under the Convention on Conservation of Migratory Species of Wild Animals, to which five range states are party (India, Mongolia, Pakistan, Tajikistan, and Uzbekistan) and thus obliged to conserve and restore its habitat.

Although officially fully protected in all 12 range countries, such protection is rarely enforced because of lack of awareness, insufficient political will to uphold regulations, shortage of funds and trained personnel, and the low priority some governments afford to biodiversity conservation (McCarthy and Chapron 2003; Theile 2003). Five countries (India, Mongolia, Nepal, Pakistan, and Russia) have developed National Snow Leopard Action Plans. However, their implementation has been often hindered by insufficient awareness amongst decision-makers, inadequate enabling legislation, and, most importantly, scarce funding.

With subsistence agro-pastoralism extensively practiced across snow leopard range, the widespread conflict over livestock depredation attributed to this felid (as well as the wolf, *Canis lupus*) comes as little surprise (Nowell and Jackson 1996; Mishra *et al.* 2003a). Depredation rates vary widely from under 1% in parts of Mongolia or western China (Schaller *et al.* 1987, 1994) to over 12% of livestock holdings in hot spots in Nepal (Jackson *et al.* 1996) and India (Mishra 1997; Bhatnagar *et al.* 1999), but they typically average 1–3% (Oli *et al.* 1994; Namgail *et al.* 2007b). These and other studies confirm depredation to be highly site-specific, with losses varying greatly between successive years and even between nearby settlements. Herders are especially angered by events of surplus killing when a snow leopard enters a corral and up to 50 or more of the confined sheep and goats are killed in a single instance (Jackson and Wangchuk 2001); in the Hemis National Park, India, such events (14% of all incidents) accounted for 38% of all livestock lost (Bhatnagar *et al.* 1999) and probably led to most retributive action against snow leopards.

Annual economic losses associated with depredation events range from about US$50 to nearly US $300 per household, a significant sum given per capita annual incomes of US$250–400 (Oli *et al.* 1994; Jackson *et al.* 1996; Mishra 1997; Bhatnagar *et al.* 1999; Ikeda 2004; Namgail *et al.* 2007b). Typically <10% of households suffer disproportionate loss, usually from corralled sheep and goats kills, or when unguarded, but high-valued yaks and horses are killed on the open range (Jackson *et al.* 1996; Ikeda 2004). Complacent guarding, poorly constructed night-time pens, favourable stalking cover, and insufficient wild prey are cited as the primary factors contributing to livestock depredation.

Snow leopards are capable of killing nearly all types of domestic animals, and while herders take measures to reduce the risk of depredation (Mishra *et al.* 2003b), these are often insufficient. Livestock numbers and biomass may be an order of magnitude higher than wild ungulates. In Nepal, for example, livestock biomass may reach 1700 kg km^{-2} (Jackson *et al.* 1996) compared to 330 kg km^{-2} for bharal in the same season (Oli 1994), so the probability of encountering domestic animals is typically higher. Bagchi and Mishra (2006) reported higher livestock (58%) in snow leopard diet in an area with more livestock (29.7 heads km^{-2}) and fewer wild ungulates (2.1–3.1 bharal km^{-2}) in comparison to an adjoining area with less livestock (13.9 km^{-2}) but more wild ungulates (4.5–7.8 ibex km^{-2}), where livestock formed 40% of its diet. These data highlight the importance of livestock as prey for some snow leopard populations, and the potential role that local communities may unintentionally play in sustaining them. Superstition can also be a powerful force, for some Buddhist herders believe depredation results from deities they or the community angered.

The relative abundance of livestock and wild ungulate prey is considered a reasonable predictor of livestock depredation risk (Bagchi and Mishra 2006). Other indicators include the distance to snow leopard travel lanes (e.g. ridges and cliffs) and broken terrain ('depredation hot spots'), along with lax guarding by herders, in part resulting from insufficient manpower or funds for hiring communal shepherds (Jackson *et al.* 1996). Although livestock losses that herders attribute to wild predators tend to be higher than actual carnivore-caused mortality, it is the 'perceived' level of depredation that most defines people's negative attitudes and subsequent reaction towards wild predators (Mishra 1997; Woodroffe *et al.* 2005). As shown by Bagchi and Mishra (2006), negative attitudes may have a strong economic basis even in culturally similar areas: thus, communities with more access to alternative income displayed greater tolerance towards snow leopards, despite losing on average 1.1 livestock per family annually. A nearby community heavily dependent upon animal husbandry, however, held more negative feelings despite losing fewer livestock (0.6 animals).

The snow leopard's habitat is frequently overstocked with livestock, in some cases with herd densities compromising animal production itself (Mishra *et al.* 2001). The resultant competition for forage, along with human disturbance, is cited as causes for natural prey population declines (Mishra *et al.* 2002, 2004). As livestock numbers increase, so natural prey populations tend to decrease, as illustrated in a study where bharal density was 63% lower (2.6 bharal km^{-2}) in a rangeland supporting 30% more livestock than an otherwise comparable area sustaining 7.1 bharal km^{-2} (Mishra *et al.* 2004). Declining prey populations are a serious and often chronic threat to snow leopards, along with the cascading effect of escalating livestock depredation rates leading to intensified herder retribution.

Wild ungulates present protein sources for some impoverished local communities, thereby encouraging further prey depletion (Mishra and Fitzherbert 2004; Mishra *et al.* 2006). Similarly, excessive sport hunting has resulted in a decline of argali (*Ovis ammon*) in Mongolia (Fox 1994; Theile 2003). By contrast, Pakistan's community-based ungulate trophy-hunting programme has generated substantial economic benefit, with 80% of the >US$ 25,000 trophy fee being distributed to the local community, where it constituted US$150 per household for one settlement. Concurrently, the number of markhor (*Capra falconeri*) increased due to stringent community-imposed curb on poaching. However, one negative outcome is that some beneficiary communities have demanded monetary compensation when snow leopards prey upon this rare ungulate (Hussain 2003b).

Poaching and trading in the snow leopard's exquisite fur and highly valued bones or body parts (used in traditional Asian medicine) is another significant threat, fuelled by illegal international markets that operate in many range countries (Theile 2003). Koshkarev (1994) and Koshkarev and Vyrypaev (2000) estimated a three- to fourfold decrease in the snow leopard population of Kyrgyzstan and Tajikistan during the 1990s, with poachers taking up to 120 animals in a single year. With its emerging economic power and high demand for rare wildlife, China is judged a key catalyst for the recent poaching surge in neighbouring Mongolia (Wingard and Zahler 2006).

The primary trade centres for snow leopard pelts and body parts are thought to be Afghanistan, Pakistan, China, and the border towns of Mongolia (Theile 2003). The fur trade that thrived in Afghanistan during the 1970s (Rodenburg 1977) re-emerged following the fall of the Taliban government and influx of international aid workers and soldiers (Mishra 2003; Mishra and Fitzherbert 2004).

As human populations expand and new roads penetrate into previously inaccessible mountains, so snow leopard populations become increasingly subjected to poaching and retaliation for livestock depredation (Hussain 2003b; Theile 2003; Mishra *et al.* 2006). Up to a third of the snow leopard's range falls along politically sensitive international borders (Fig. 19.1) complicating trans-boundary conservation initiatives. In fact, over the past 50 years there have been several wars, along with low-intensity factional or international conflicts that continue in countries like Afghanistan.

Conserving the snow leopard: the needs

As highlighted above, extant populations of snow leopards face a multitude of conservation challenges. These include direct internal threats from local people, external threats like wars or lack of international cooperation, and indirect threats related to underdeveloped economies, illegal markets for fur and bones, or lack of public awareness. The importance of each factor varies across landscapes and between countries, thus necessitating locally appropriate conservation action (McCarthy and Chapron 2003). In Fig. 19.3, we present a general conceptual threat-based model (Salafsky and Margoluis 1999) for snow leopard conservation that summarizes these threats, identifies desirable conditions, and highlights the necessary interventions for achieving conservation targets.

Clearly, managing human–snow leopard conflict is critical, requiring reduction of livestock depredation levels through better animal husbandry and wild prey restoration, as well as devising sustainable means for offsetting or sharing economic losses. Incentive programmes for garnering local community support for snow leopard conservation are imperative; it is largely when tangible economic returns are realized that communities are willing, indeed able, to assume their role as conservation partners and become effective stewards of their surrounding environment (Western *et al.* 1994; Jackson and Wangchuk 2001; Mishra *et al.* 2003a). Additionally, effective snow leopard conservation hinges upon better, more inclusive national conservation policies, developing greater awareness through conservation education programmes targeting local and national levels, and by fostering international cooperation and financial assistance.

The establishment of protected areas is obviously important. Even by the mid-1990s there were 109 protected areas encompassing 276,000 km^2 within snow leopard range (Green and Zhimbiev 1997), although most are too small to support more than a few resident or breeding individuals. The total area under protected status has since increased substantially, especially with additions in China (which supports some 60% of snow leopard range) and Mongolia, but mega-reserves like the vast Changtang complex of Tibet harbour few cats (G.B. Schaller, personal communication). The Global Environment Facility and other multilateral organizations have funded trans-boundary protected area conservation projects, like the Trans-Altai project along the borders of Russia, Mongolia, Kazakhstan, and China, or the World Bank initiative involving Uzbekistan, Kyrgyzstan, and Tajikistan. Nonetheless, most protected areas still function as paper parks, with poorly regulated resource management. Barren rock and ice landscapes constitute a high proportion of land types present (Allnutt *et al.* 2005), with important wildlife populations, including snow leopard populations, lying outside the network (Jackson and Ahlborn 1991; Mishra *et al.* 2010). One notable exception is the Sagarmatha National Park in Nepal, where snow leopards were extirpated in the 1970s, but have recently recolonized the area following the recovery of wild ungulate populations (Ale *et al.* 2007).

Over the past 15 years, conservation organizations have set up small-scale, threat-based snow leopard conservation initiatives, working directly with local communities and governments. These serve as useful pointers for designing more effective range-wide programmes. In the following section, we review and synthesize some of these initiatives within context of the threat-based conservation model presented above.

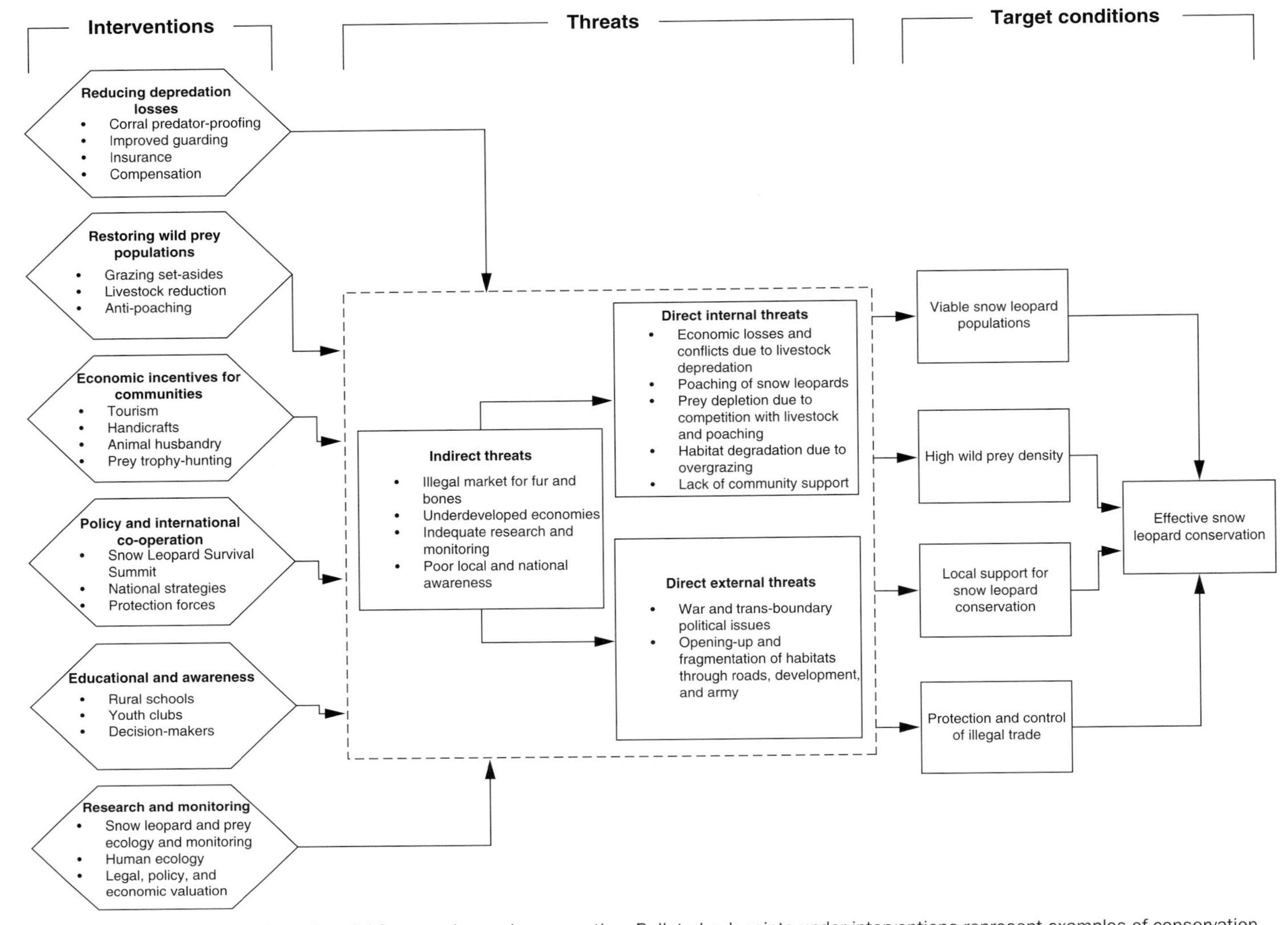

Figure 19.3 A conceptual threat-based model for snow leopard conservation. Bulleted sub-points under interventions represent examples of conservation interventions discussed in this chapter.

Threat-based interventions for snow leopard conservation

Addressing human–snow leopard conflicts

Addressing human–snow leopard conflict requires reducing livestock depredation and offsetting costs (real or perceived) of resulting losses. Better anti-predator livestock management is an important means of reducing livestock depredation by wild carnivores. In Ladakh, India, this has been achieved using predator-proof corrals, significantly reducing livestock losses to snow leopards (Jackson and Wangchuk 2001). This simple and effective conservation partnership provides predator-proofing material such as wooden doors, poles, and chain-link fences to local communities for construction and improvement of traditional livestock corrals. Such improvements ($n = 16$) have virtually eliminated multiple killing of sheep and goats in Hemis National Park, thereby removing long-held animosity towards the species. Livestock has declined from over 30% of the snow leopard's diet (Chundawat and Rawat 1994) to around 11% currently (U.R. Sharma, unpublished data). In India's Spiti Valley, where village livestock are herded communally by a few herders, vigilant herding to reduce depredation losses has been encouraged through an incentive system which provides cash rewards to herders for better shepherding and fewer depredation losses in the pastures (Mishra *et al.* 2003a).

Relatively low wild ungulate and high livestock abundance lead to increased livestock depredation by snow leopards and sympatric carnivores. In one high-conflict area in India, wild ungulate recovery has been facilitated by creating grazing set-asides on village land through a community-based conservation incentives programme (Mishra *et al.* 2003a). These livestock-controlled areas or village wildlife reserves, where grazing and resource extraction are curtailed, allowed intensified use by snow leopards in response to wild prey recovery from *c.* 4 blue sheep km^{-2} to more than 9 km^{-2} (C. Mishra, unpublished data). Together with improved herding practices, livestock depredation losses have been reduced from 12% to 4% of holdings annually (Mishra 1997), although the contribution of livestock to the snow leopard's diet declined less drastically from 58% to 40%. Within a few years following the natural recolonization of snow leopards to the Sagarmatha National Park, the number of Himalayan tahr (*Hemitragus jemlahicus*) declined by nearly 70%, which researchers attribute to heavy predation of kids by this carnivore (S. Lovari, personal communication). This suggests that a possible outcome of better nature protection and more wild prey could lead to increased snow leopard numbers or use of an area, intensified predation upon wild prey, and possibly also greater livestock depredation losses over the long-run. This underscores the need for multi-pronged conflict management programmes; in this case, wild prey recovery must be accompanied by better livestock protection.

Livestock are inherently vulnerable to depredation as they possess reduced anti-predatory abilities, so loss will continue to occur even under vigilant herding practices or high wild prey densities (Madhusudan and Mishra 2003). Therefore, addressing human–carnivore conflict by offsetting or sharing depredation-related economic impacts is necessary (Woodroffe *et al.* 2005; Treves *et al.* 2006). Compensation, an accepted predator conservation strategy in the Western United States, has been extended to Asian species like snow leopard and tiger (*Panthera tigris*). Although effective in some instances (Nyhus *et al.* 2003), compensation programmes in developing countries face numerous challenges (Loveridge *et al.*, Chapter 6, this volume). For instance, low compensation amounts (offsetting only 3–35% of the financial losses), false claims, corrupt disbursement officials, difficulty in authenticating claims, and bureaucratic apathy often associated with state-run programmes (Mishra 1997; Jackson and Wangchuk 2004) have resulted in increased resentment towards protected areas harbouring snow leopard populations.

Given these difficulties with compensation programmes, conservationists have helped communities establish locally managed insurance programmes for snow leopard depredation. In Pakistan's Baltistan region, Hussain (2000) established a community-managed insurance programme, which currently covers 3 settlements and 151 families, and is supported through premiums contributed by participating families along with contributions from local tourism. A similar programme, started in the Kibber village of

Spiti Valley in India (Mishra *et al.* 2003a), was supported with conservation funds (60% over 5 years) in addition to premiums contributed by the participating families (remaining 40%) and became financially self-sustaining within 5 years. This programme currently insures livestock worth US$60,000, with an average contribution of US$20 per year by each participating household (C. Mishra, unpublished data). It has been expanded to cover nine villages (*c.* 200 herding families) in Spiti and Ladakh. Such programmes have led to greater tolerance of snow leopards amongst local communities. With local management by village-level committees and family-paid premiums, local ownership is strengthened, as is internal peer pressure against corruption or false claims.

Incentive programmes for snow leopard conservation

Until local people have a meaningful stake in conservation, the long-term outlook for snow leopards will not be encouraging. Hence, conservationists often employ economic incentive programmes to gain local community support and involvement in snow leopard conservation. One such programme, Snow Leopard Enterprises, generates income for herder–artisans in Mongolia through handicraft development and sales, in exchange for community support of snow leopard conservation (Mishra *et al.* 2003a). Artisans receive training and simple tools to develop culturally appropriate woollen products which are marketable in the west. In exchange, communities sign conservation contracts stipulating a moratorium on snow leopard and wild ungulate poaching. Contract compliance brings herders a 20% bonus over the agreed price of their products. The bonus is split between artisans and a community conservation fund. A single contract violation loses the entire community's bonus. Resultant peer pressure results in collaboration to stop poaching by locals or outsiders. Compliance is monitored by nearby protected area rangers and law enforcement agencies. Currently 29 communities and over 400 herder families benefit from increased household incomes of nearly 40%. Until recently, there had been no known cases of snow leopards being killed at any project site, while pre-1997 poaching levels were as high as 0.5 cats per year per site (Snow Leopard Trust, unpublished data). The bonus had only been withheld twice and both resulted for poaching of ibex. In the beginning of 2009, however, there was a tragic loss of a snow leopard radio-collared a few months earlier by the Snow Leopard Trust in the Tost Mountains, after it got trapped in a snare set up for wolves and was shot by a herder. Wolf trapping is legally permitted in Mongolia. This incident resulted in a loss of bonus, as well as ongoing prosecution of the herder by the local administration. Over US $90,000 worth of Mongolian handicrafts was sold in 2007 and further growth is expected. Proceeds from sales now cover all programme costs, freeing donor funds for expansion. Success of this project in Mongolia has led to expansion in five Kyrgyzstan and three Pakistan villages.

The largely self-sustaining, 'Himalayan Homestays' incentive programme in India builds upon existing tourism and trekking to enhance livelihoods for local people and garner support for snow leopard conservation (Jackson and Wangchuk 2004). This UNESCO-supported initiative provides villagers training in running homestays and nature guiding, with bookings facilitated by local travel agents. Individual households operate through women's groups and revenue accrues from 'bed and breakfast' stays in village homes (rotated among households), catering and handicraft sales at tented cafes on trekking routes, and nature guiding. The first traditional homestays were established in Hemis National Park, India's premier snow leopard protected area. Now >100 families in 20 communities in Ladakh, Zanskar, and Spiti participate. Homestay operators (*c.* 40) in prime snow leopard habitat earn US$100–650 (average US$230) during the 4-month tourist season (Snow Leopard Conservancy, unpublished data). Tourist visitation increased from 37 in 2001 to >700 by 2006. Client satisfaction exceeds 85% and tourists welcome the cultural interaction. Another *c.* US$400 in sales from cafes is shared among 4–8 families. Approximately 10–15% of homestay profits go into a village conservation fund which has supported tree planting, garbage management, and recently the establishment of a village wildlife reserve for the threatened Tibetan argali (*Ovis ammon hodgsonii*). One community constructed predator-proof corrals, another paid a full-time herder to guard livestock in high summer pastures, and a third insured large-bodied, high-valued livestock like yak through a national livestock insurance programme.

Livestock losses to disease (>50% of many herds) far exceed losses to snow leopards (<0.5% of average herd) in Chitral, Pakistan (Snow Leopard Trust, unpublished data). In response, a pilot livestock vaccination programme was established in a single village in 2003, and in exchange for tolerating depredation, 1452 village livestock were vaccinated against common diseases. Disease losses then declined by *c.* 90%. Participants agreed to cease snow leopard persecution, reduce their livestock holdings, and improve fodder handling methods to increase forage availability for wild herbivores. The programme creates economic incentives by increasing livestock survival and productivity, with sales of excess animals bringing each family *c.* US$400 per annum. The programme was expanded to eight villages in early 2008, with the goal of being self-sustaining and free of donor support within 5 years.

International cooperation, policy, and law enforcement

Given its widespread occurrence along international borders, the snow leopard is a species acutely in need of concerted international conservation efforts and cooperation, especially as funds are difficult to generate within most range countries. The establishment of the International Snow Leopard Trust 25 years ago represented an early step towards catalysing international cooperation among conservation partners in the range states, thus stimulating awareness, information exchange, on-the-ground research, and the establishment of pilot conservation projects. The Snow Leopard Trust and the more recently established Snow Leopard Conservancy represent two international organizations wholly dedicated to range-wide conservation of snow leopards and their habitat.

The Snow Leopard Trust played a key role in organizing international symposia, generally convened in range countries. Jackson and Fox (1997) reviewed the role these meetings played in setting research and conservation priorities. In 2002, the Snow Leopard Survival Summit brought together knowledgeable individuals from around the world to discuss and prioritize emerging threats to the snow leopard on a regional basis, and to identify and recommend 'best-practice' conservation, research, and policy actions appropriate to alleviating such threats. In addition to developing the Snow Leopard Survival Strategy (McCarthy and Chapron 2003), this summit led to the establishment of the Snow Leopard Network (SLN), a global alliance of institutions and individuals involved in snow leopard conservation. The SLN facilitates information sharing, supports advocacy through position papers and statements, and operates a small grants programme for snow leopard research and conservation.

The top-down 'fences and guns' approach to conservation is especially prone to failure in snow leopard habitat, given its remoteness, harshness, and subsistence livelihoods of local people, along with the fact that most habitat is located outside of protected areas. Governments are increasingly realizing the importance of more participatory approaches, and are thus developing more inclusive national snow leopard conservation strategies which identify a central role for local communities, as well as encouraging sound science, in underpinning the species' conservation.

Government agencies are primarily responsible for patrolling and preventing poaching, although an independent collaborative effort between a conservation organization Naturschutzbund Deutschland (NABU) and government to stop leopard poaching, called the Gruppe Bars, was launched in Kyrgyzstan in 1999. The specially equipped team, recruited in part from Ministry of Environment law enforcement officers, sought to halt illegal trade in snow leopards and other endangered species through a country-wide network of informants, undercover operations, and awareness programmes. Within 3 years, this effort led to the confiscation of a live snow leopard, 14 pelts, 162 firearms, numerous traps (including 79 snares specifically designed for snow leopards), numerous other wildlife trophies, and the arrest of 110 poachers (Dexel 2003).

Recently, following renewed use of real felid furs for lining their traditional ceremonial cloaks, and at the behest of His Holiness the XIV Dalai Lama, some Tibetans publicly burned skins and reverted back to using imitation fur products. There is unfortunately no easy solution to the problem of illegal national and international wildlife trade. Since users of traditional Chinese medicine value bones from tigers, as well as other large felids used as substitutes, efforts at law enforcement and the strengthening of surveillance networks should target these species concurrently (Theile 2003; Nowell 2007).

Education and awareness

Conservationists have also undertaken educational initiatives to promote awareness of the precarious status of snow leopard populations and the need to conserve them, especially by local communities. For example, in Spiti Valley, this is being achieved through classroom and outdoor activities aimed at sensitizing school children to conservation issues (Trivedi *et al.* 2006). Appropriate educational tool kits for teachers and products for children and others, such as books and posters, have been developed and are being used to promote awareness in 10 rural schools under the pilot phase.

Youth clubs have been formed in several villages to help teachers implement educational and awareness activities, including those aimed at conflict resolution. In Ladakh and Zanskar, locally relevant educational materials and a teachers' handbook have been developed based on the region's wild biodiversity, threats, and conservation actions. Classroom workshops are led by conservationists, while the State Education Department is making these materials available to teachers throughout the district (Snow Leopard Conservancy, unpublished data). Similarly, in Nepal, a series of reading booklets about snow leopards and their role in the environment were produced in Nepali and English, with one booklet being translated into three other languages, including Tibetan Braille. These books have been used in 33 schools in Dolpo, Mugu, Mustang, and Manang, all important snow leopard areas.

Research and monitoring

The first ground-based very-high frequency (VHF) radio-tracking study was launched in relatively productive habitat in Nepal in the 1980s (Jackson 1996), followed in the 1990s by research in low-productivity habitats in Mongolia using a combination of VHF and satellite tracking (McCarthy 2000). These seminal studies have contributed much of our current understanding of snow leopards. They brought to light the high variation in home range size, highlighting the need to be cautious when extrapolating data to larger areas (McCarthy *et al.* 2005). Studies by Chundawat (1992) and Oli (1991) further contributed to understanding this felid's prey and dietary requirements. This was followed by applied research on wildlife and human ecology, which helped understand the direct and cascading effects of livestock grazing on human–snow leopard conflicts (Mishra 2001). Further applied research focused on human–snow leopard conflict (Oli *et al.* 1994; Mishra 1997; Bhatnagar *et al.* 1999; Hussain 2003b; Namgail *et al.* 2007b), human ecology (Mishra 2000; Mishra *et al.* 2003b; Namgail *et al.* 2007a), and the relationships between livestock, wild prey, and the snow leopard (Mishra *et al.* 2002, 2004; Bagchi *et al.* 2004; Bagchi and Mishra 2006). More recently, a snow leopard satellite-tracking study was launched in Pakistan (McCarthy *et al.* 2007). The realization that snow leopards must be conserved in a landscape with pervasive human use has encouraged multidisciplinary research with high emphasis on human ecology, as well as the use of social science tools (Jackson and Wangchuk 2004).

Development of techniques to monitor snow leopards (or their prey populations) has remained a challenge given the combination of harsh topography, severe climate, and the felid's cryptic nature. Until recently, snow leopard monitoring focused on quantifying indirect signs such as scats, scrapes, and other social markings that the cat leaves in relatively predictable micro-habitats (Jackson and Hunter 1996). More recently, camera-trapping and the use of capture–recapture statistical frameworks for estimating snow leopard populations has been found effective (Jackson *et al.* 2006; McCarthy *et al.*, 2008). Population surveys incorporating non-invasive genetic sampling have been successfully used for examining population status in diverse species (Schwartz *et al.* 2007). Preliminary work suggests molecular tools may be useful for monitoring snow leopards through species and individual identification of field-collected scats (Conradi 2006; Waits *et al.* 2007; Janečka *et al.* 2008).

Table 19.1 Show leopard conservation interventions: guidelines and comparisons.

Conservation action	Key activities	Cost, technical, and logistical factors	Potential pitfalls	Monitoring needs
Grazing management	Promote grazing practices that reduce impacts on wildlife	Low cost (excluding set-aside payments); moderate technical requirement	▪ Determining existing grazing patterns or land tenure disputes ▪ Grazing plans designed without input from community likely to fail	▪ Pasture quality and indicators developed by local herders ▪ Numbers and productivity of wild and domestic ungulates for grazing plan compliance
Wildlife-based ecotourism	Establish tourism that provides financial benefits to local people and creates incentives to protect natural resources	Moderate to high (may require substantial skills training and infrastructure development; marketing critical)	▪ Political instability, security, and health issues of importance to clientele ▪ Viewable wildlife often wary of humans ▪ Short season and leakage of revenue ▪ Financial benefits not equitably distributed ▪ Maintenance of prices and servicing standards may be difficult to achieve	▪ Numbers and trends of wildlife ▪ Quality of tourist attractions ▪ Level of economic benefit of ecotourism to local people ▪ Local attitudes towards wildlife and tourists ▪ Strong incentives for compliance
Cottage industry	Provide income to residents of snow leopard habitat through handicraft sales linked with wildlife conservation	Moderate to high cost (getting products to high-value markets, skills training, maintaining standards and marketing outreach)	▪ Semi-skilled artisans (products may not consistently meet market standards) ▪ Strong international competition ▪ Inconsistent participation after training investment ▪ Market saturation requires continued new or unique product development	▪ Numbers and trends of wildlife for anti-poaching compliance ▪ Other indicators determined collaboratively by community (compliance incentives) ▪ Number of participants benefiting ▪ Financial impact at household and community levels ▪ Public attitudes to snow leopards
Community-managed prey species trophy-hunting	Establish sustainable trophy hunting to provide return to local people as an incentive to protect ungulates and snow leopards	Moderate (externally driven planning and decision-making; high technical demands)	▪ Corruption at national and local level ▪ Lack of awareness of law among foreign outfitters/clientele ▪ Insufficient hunting fee revenues reach local level (lack of incentive to protect) ▪ Poor monitoring of trophy species ▪ Perverse incentive to persecute snow leopards	▪ Numbers and trends of wildlife ▪ Harvest statistics (hunting effort, trophy size, etc.) ▪ Numbers of local people or communities gaining benefit ▪ Financial impact at household/ community levels

Animal husbandry	Provide training in animal husbandry and veterinary care to improve monetary return at lower stock levels or to offset depredation costs	Low to moderate (linked with government veterinary extension capacity)	▪ Long-term commitment of community, government, or NGO may be difficult to maintain ▪ Low skill level for effective veterinary training program ▪ Limited acceptance of fewer high-quality animals versus large unproductive herds	▪ Numbers of livestock and financial returns ▪ Livestock health, incidences of disease, and other mortality ▪ Stocking density and carrying capacity of pastures ▪ Attitudes towards depredation/predators
Livestock insurance	Establish locally managed subscription-based insurance scheme to offset depredation economic losses	Moderate over long term but potential high start-up costs	▪ Initial investment into capital fund can be high ▪ Validation of claims can be difficult and contentious ▪ Fails to address root cause of depredation	▪ Numbers of livestock and financial returns ▪ Livestock health and incidences of depredation ▪ Attitudes towards depredation and targeted predator species
Education outreach	Raise public awareness for snow leopard conservation	Low to moderate (hinges on collaboration with local school teachers and education departments)	▪ Low levels of education and literacy ▪ Linguistic, cultural, or ethnicity barriers ▪ Limited capacity of education system ▪ Dissemination in remote areas difficult	▪ Baseline surveys to determine current levels of awareness ▪ Monitoring to evaluate program effectiveness
Applied research	Investigate snow leopard and prey ecology, behavior, etc., including ecosystem and landscape dynamics	Moderate to high (dependent upon outside researchers and institutions)	▪ Research topics often not of interest to protected area managers ▪ Tendency to exclude communities from research (i.e. information 'mining' only)	▪ Ensure project targets priority topics and management issues ▪ Dissemination to general public and decision-makers

Conclusions

Conserving snow leopards is largely contingent upon the cooperation and goodwill of stakeholders, starting with local communities and extending to the international conservation community. Trade in snow leopard pelts and body parts are best arrested in tandem with actions addressing the tiger trade. Snow leopard conservation hinges on the equitable involvement and decision-making of local communities: there is no substitute to community-based incentive programmes, despite the challenges and costs involved (see Table 19.1 for a summary of conservation actions, costs, and some considerations for implementation). However, robust connections must exist between the rationale for providing such incentives, benefits they bring, and the community's vested responsibility for protecting snow leopards, its prey, and habitat. These include clearly articulating each stakeholder's conservation responsibilities, arrangements for reciprocal financing or in-kind contributions, and providing the project with efficient tools to enable participatory planning and action along with supporting collaborative compliance monitoring and project evaluation (Jackson and Wangchuk 2001). Substantial local support can be secured by valuing traditional ecological knowledge while building institutional capacity for adaptive management (Berkes *et al.* 2000).

Incentive programmes tend to be heavily subsidized, limited to relatively small areas or supported only over the short-term as part of larger Integrated Conservation Development Programs (Sanjayan *et al.* 1997; Mishra *et al.* 2003a; Wells and McShane 2004). Project transaction costs and human resources are high, since developing the necessary skills for undertaking relatively complex, competitive market-based enterprises like handicrafts production, traditional homestays, and nature guiding is time-consuming. Creating a self-sustaining market for such goods or services and implementing monitoring activities for ensuring compliance with species or general biodiversity conservation goals add to costs. Many donors fail to appreciate that significant returns on community-based programmes may not be forthcoming for 5–10 years, while implementing agencies are hard-pressed to demonstrate tangible results within the typical 2–5-year time frame expected by donors.

Central and South Asia's mountain areas are renowned for their varied climates, geography, biodiversity, and cultures that change dramatically over relatively short horizontal or vertical distances. Consequently, even closely spaced valleys typically exhibit different ecosystems, economies, and ethnic groups whose values, livelihood strategies, and influences have evolved to address a specific set of environmental, social, and political factors (Bishop 1990; Mishra *et al.* 2003b). Generally, each conservation initiative must address a different set of threats and conditions, along with the particular aspirations of the targeted community. The small community-based conservation and conflict management efforts described above were effective largely because of their well-informed and participatory focus. A critical, though often overlooked, element enabling successful conservation models involves site-specific scientific knowledge of wildlife and human ecology. This calls for both biological and social science expertise. Monitoring human socio-economy and land use, as well as the responses of snow leopards and prey species to conservation efforts, allows for adaptive management. While more experimental conservation models need to be assessed, conservationists must also work with national governments to formulate appropriate policy so that these inclusive, science-based conservation approaches can be scaled up and included in national conservation frameworks.

The real long-term challenge lies with moving communities beyond their harsh and insecure subsistence livelihood into more economically viable and environmentally friendly activities. Ultimately, local people must be encouraged to perceive snow leopards and other large carnivores as being worth 'more alive than dead'. They must also assume greater responsibility for protecting their herds from predators. This will enable an enduring coexistence between predators and humans across snow leopard range.

CHAPTER 20

Pumas and people: lessons in the landscape of tolerance from a widely distributed felid

Tucker Murphy and David W. Macdonald

Camera-trap photograph of a puma living in close proximity to human settlement in the Araucanía Lake District of Chile. © Nicolás Gálvez, Darwin Initiative.

Amigo del cristiano, cougar, catamount, chim blea, puma, panther, pangi, mountain lion, mountain screamer, deer tiger, ghost cat—these names represent some of the multiple attempts people have made to encapsulate and describe this elusive and solitary felid over the time they have shared the American landscape. In total, 42 different names for *Puma concolor* have been identified in English (Barnes 1960) and a similar number exist in both Spanish and Native American languages (Young and Goldman 1946). Perhaps these varied monikers hint at the ambivalence people have experienced in defining this cryptic and crepuscular cat; pumas have been perceived as the friend of humans by some cultures, and persecuted for the threat they pose to livestock and people by others (Anonymous 1892; Young and Goldman 1946; Barnes 1960). Certainly, these numerous names reflect the extensive range of pumas, which has brought them into contact with diverse cultures. Their range stretches from the Straits of Magellan to the Canadian Yukon and is the broadest of any New World terrestrial mammal, except humans. Within this range, pumas vary in body size, and are immensely adaptable, such that they occupy almost every type of biogeographic zone from tropical rainforest to desert (Young and Goldman 1946). As a result, pumas are an ideal model organism to examine how culture and context shape our relationship with large felids. In this case study, we investigate how the exigencies of depredation, as well as the threat pumas pose to humans, are managed and perceived at study sites in both North America (Montana) and South America (Chile).

Biological characteristics

Pumas have not always occupied such an extensive range. Since they evolved in North America from a common ancestor with the American jaguarundi (*Herpailurus yaguaroundi*) and the African cheetah (*Acinonyx jubatus*) 5–8 million years ago (Janczewski *et al.* 1995; Johnson and O'Brien 1997; Werdelin *et al.*, Chapter 2, this volume), pumas' evolutionary history has been characterized by colonization, extinction, and recolonization events. Approximately 3 million years ago, when the Panamanian land bridge formed, pumas crossed from North to South America during the Great American Interchange (Marshall *et al.* 1982). Subsequently, the North American population of pumas was extirpated in the Pleistocene mass extinction event (10,000–12,000 years ago) and replaced by a founding population of South American pumas. According to mitochondrial gene sequence and microsatellite loci evidence, the current North American population of pumas derives from a core of genetic diversity in eastern South America (Culver *et al.* 2000).

Figure 20.1 Map showing pumas' extensive range and depredation studies. A and B indicate our two study sites, in Montana and Chile, respectively. All other numbers mark sites referred to in the text, where puma depredation has been studied. 0, California (Torres *et al.* 1996); 1, south-east Arizona and north-west Mexico, respectively (Cunningham *et al.* 1995; Rosas-Rosas *et al.* 2008); 2, Venezuelan Llanos (Polisar *et al.* 2003; Scognamillo *et al.* 2003); 3, central-western Brazil (Palmeira *et al.* 2008); 4, Bolivian Altiplano (Pacheco *et al.* 2004); 5, southern Brazil (Mazzolli *et al.* 2002); 6, Andean foothills of Patagonia (Franklin *et al.* 1999).

Throughout their range (Fig. 20.1), puma body size and diet vary with latitude. They are the fourth largest felid, and are generally comparable in size to snow leopards and African leopards (Logan and Sweanor 2001); however, pumas closer to the equator have a lower average body weight (<35 kg between latitudes of 0° N/S to 20° N/S), than those at the far north and south of their distribution (>50 kg at latitudes >40° N/S; Kurtén 1973; Iriarte *et al.* 1990). Populations of pumas in temperate habitats favour larger prey (>30 kg at latitudes >40° N/S; e.g. ungulates in North America) from a smaller number of taxa (Iriarte *et al.* 1990). In contrast, puma populations in tropical habitats subsist on smaller, more varied prey (<15 kg between a latitude of 0° N/S to 20° N/S; Iriarte *et al.* 1990). Since mean body mass of prey has been positively correlated with puma body mass, prey size is believed to have imposed selective pressures on puma size. Difference in diet is largely dictated by prey availability and vulnerability, as well as habitat characteristics. Preference of smaller prey in the tropics may also reflect resource partitioning with larger sympatric jaguars (*Panthera onca*; Scognamillo *et al.* 2003).

Conflict

Regardless of latitude, pumas prey on domestic animals or livestock wherever they range adjacent to or within puma habitat. Studies from areas as varied as the Sonoran Desert (Cunningham *et al.* 1995; Rosas-Rosas *et al.* 2008), the Bolivian Altiplano (Pacheco *et al.* 2004), the Venezuelan Llanos (Polisar *et al.* 2003; Scognamillo *et al.* 2003) and the grasslands, and Andean foothills of Patagonia (Franklin *et al.* 1999) have all reported a high incidence of livestock loss. Sheep or goats are often the primary victims of depredation (e.g. southern Brazil, Mazzolli *et al.* 2002; southern Chile, Rau and Jimenez 2002), though pumas will opportunistically attack other livestock, hobby animals and pets (California, Torres *et al.* 1996). Occasionally, depredation of adult cattle weighing up to 130 kg has been recorded (Cougar Management Guidelines Working Group 2005). However, pumas generally favour smaller prey and cattle losses are greatest when calves are born in puma habitat (e.g. calves 0–6 months old accounted for 100% of puma depredation on kills checked in southern Brazil, Palmeira *et al.* 2008; see also Cunningham *et al.* 1995; Crawshaw and Quigley 2002). Depredation may also be greater in areas with low abundance of native prey (Crawshaw and Quigley 2002; Polisar *et al.* 2003; Loveridge *et al.*, Chapter 6, this volume).

Several trends combine to create a global challenge for big cats in general and pumas in particular. Conservation efforts are intensifying, sometimes leading to increasing numbers of pumas. However, numbers of humans and their environmental footprint are also increasing almost everywhere, so these modern pumas almost always live alongside people. As a result, opportunities for puma depredation on livestock and domestic animals may increase. In fact, over the past 30 years depredation incidents in the western United States have multiplied (Padley 1997). Though the root cause is unknown, factors such as declining deer numbers, elimination of bounties, increasing puma numbers and changes in land use have been suggested (Cougar Management Guidelines Working Group 2005).

Often the outcome of these depredations is conflict between pumas and people. Conservationists consider conflict as one of the three major threats facing felids globally; the other two linked concerns being a diminishing prey base and habitat modification (Mazzolli *et al.* 2002; see also Macdonald *et al.*, Chapter 29, this volume). Conflict was at the root of the extirpation of the Zanzibar leopard (*Panthera pardus adersi*) and Barbary lion (*Panthera leo leo*) and it has led to the elimination of tigers (*Panthera tigris*) from most of China, the lion (*Panthera leo*) from North Africa and south-west Asia, and the puma from the entire eastern half of the United States and parts of Argentina (Young and Goldman 1946; Nowell and Jackson 1996). The likelihood of conflict between humans and felids may be increased by the threat these large cats pose to public safety and human life. Puma attacks on humans occur rarely; for instance Beier (1991) documented nine fatal attacks and 44 non-fatal attacks in North America from 1890–1990. However, when they do occur they receive substantial publicity and accounts in the popular press may be prone to hyperbole (e.g. Baron 2004).

Human values

Methods of resolving environmental problems have often been weighted towards ecological approaches (Riley 1998). In the context of human–felid conflict, ecological approaches usually centre on reducing depredation through mitigation efforts, such as improving livestock husbandry (Mazzolli *et al.* 2002; Rosas-Rosas *et al.* 2008). However, human values also play a considerable role in conflict. In *The Strategy of Conflict*, game theorist and economist Thomas Schelling (1960) contends that situations of pure conflict in which two human antagonists are completely opposed are rare, and arise only in wars of complete extermination. As a result, 'winning' in human conflict does not have the strictly competitive meaning of winning relative to one's adversary; rather, it means gaining relative to one's value system (Schelling 1960).

The importance of understanding human values in situations of human–carnivore conflict is increasingly reflected in the literature of conservation biology (see Cavalcanti *et al.*, Chapter 17, this volume). Macdonald and Sillero-Zubiri (2004b) point out that 'prejudices cannot be changed until they are first identified and second—if possible—debunked'. Decker and Chase (1997) add further support to this idea and argue that a human–wildlife problem could only be considered solved when stakeholders believe it to be so. Of course, in the past, common ground has not always been found between humans and carnivores, and 'wars of complete extermination' have occurred (e.g. the Falkland Island wolf, *Dusicyon australis*). Nevertheless, a step towards devising solutions is to understand such animosity, particularly considering the role of perception—as distinct from evidence—in the deaths of many felids and carnivores in general (Cavalcanti *et al.*, Chapter 17, this volume). Studies of the human dimension of conflict have also revealed that stakeholders' tolerance of felids may be increased through education, even in the absence of a reduction in overall depredation rates (Marker *et al.* 2003c).

Management in North and South America

At the intersection between conflict and human values is puma management. In its essence management is an attempt to influence puma behaviour and populations in a way that accords with human desires. However, the Cougar Management Guidelines Working Group (2005) point out that given pumas' secretive nature, their low population densities, their impact on wild and domestic prey, and the threat they pose to human safety 'public attitudes about them differ widely'. As a result, management strategies also vary considerably. Certainly, broad differences exist in the management of pumas between North and South America.

In North America, puma hunting is legal except in the eastern United States, where pumas have been extirpated everywhere but Florida and are listed under the Endangered Species Act (Cougar Management Guidelines Working Group 2005). Wildlife managers in the United States often issue depredation permits or employ lethal means to resolve human–puma conflict. In theory, management decisions in the Unites States are made democratically at a regional (or state) level and diverse stakeholder groups, often with limited direct contact with pumas, may influence decisions about the management and control of their populations. In practice, since appointed game commissions make most regulatory decisions, they may be biased towards protecting the harvest of revenue-generating species such as deer and elk. The interests of agricultural communities, which may be less favourable towards predators, also tend to be strongly represented on game commissions (H. Quigley, personal communication 2008). Nevertheless, throughout the United States, groups as varied as hunters (of both pumas and their ungulate prey), livestock owners, conservationists, and homeowners have a stake in decisions about puma management.

Throughout South and Central America killing pumas is prohibited (except in Peru, and in El Salvador, where pumas have been nearly extirpated; Lopez-Gonzalez 1999; Logan and Sweanor 2001). Protection of pumas in South America is often mandated at a national level and entails very little stakeholder input (C. Bonacic, personal communication 2008). Therefore only those stakeholders in direct contact with pumas are capable of influencing their future populations by illegally killing them. However, if stakeholders retaliate against pumas they are at risk of suffering legal consequences.

Stakeholders in North and South America

Thus far, the most comprehensive studies of the human dimension of human–puma conflict have been limited to North America (Kellert *et al.* 1996; Riley 1998). We sought to expand the scope of understanding by investigating human dimensions in Chile. To highlight some of the broad differences in relevant stakeholder groups in North America and South America we compared their involvement and beliefs in Chile to published results for Montana (Riley 1998).

In Montana, Riley (1998) used a self-administered questionnaire to assess the human dimension of puma management and conflict. In total, 1378 questionnaires were delivered to randomly selected households throughout Montana. The questionnaire was concerned mainly with stakeholder involvement with pumas, their beliefs about pumas, and their assessment of the risks posed by pumas. Since management decisions are made democratically, questionnaires were not targeted at any sector of stakeholder. In the Araucanía Lake Region of Chile stakeholders were asked the same set of questions. Those stakeholders in closest contact with pumas, who posed the greatest direct threat to them, and consequently held sway over their future, were targeted for interviews. This group was comprised mainly of owners of small livestock holdings.

We employed chi-square tests to determine whether a significant difference existed between the interactions or beliefs of Montanan stakeholders and Chilean stakeholders. In cases where expected values were <10 and did not meet the restrictions of standard chi-square test, we applied Yate's correction for continuity. Since we were only interested in understanding broad distinctions between the two study areas and different methodologies necessarily applied to each area (possibly compromising some assumptions of the chi-square test), significance was treated as a rough indicator of differences. All tests were performed in SPSS (version 16.0 for Macintosh).

The most common form of involvement among Montanan stakeholders with pumas was to have read or heard of encounters between pumas, pets, livestock, or people, or to have read or heard of their control by management authorities (Table 20.1a). In contrast, most Chilean stakeholders claimed to have observed a puma or knew a friend or neighbour who had had contact with one. Among Chilean livestock owners, even second-hand information on pumas was rarely derived from newspapers or reading; instead it was passed by word of mouth. Overall, more Chilean than Montanan stakeholders responded positively to the presence of pumas in their country. However, they were more negative about the presence of pumas near their houses (Table 20.1b). Chilean stakeholders also believed that encounters between pumas and people showed a decreasing trend, whereas, Montanan stakeholders perceived an increasing trend. Each group exhibited significantly different beliefs about whether risks were well understood by experts, with Chileans believing they were not and Montanans believing that they were (Table 20.1c).

Table 20.1a Table comparing the involvement of Montanan (Riley 1998) and Chilean stakeholders with pumas. Results listed as not available (NA) were either considered not relevant or not translatable to a particular study area.

Involvement	Chile (%)	Montana (%)	Statistic
Observed a puma in the wild	75	36.3	$\chi^2 = 29.976$ $P \ll 0.0001$
Read or heard of a puma being killed by authorities	0	70.1	$\chi^2 = 120.805$ $P \ll 0.0001$
Read or heard of a puma being killed in the area	43	NA	NA
Read or heard of a puma being liberated by authorities	55	NA	NA
Had a pet threatened or attacked by a puma	15	2.0	$\chi^2 = 29.21$ $P \ll 0.0001$
Had livestock threatened or attacked by pumas	65	2.0	$\chi^2 = 29.21$ $P \ll 0.0001$
Have been personally threatened by a puma	2	3.3	$\chi^2 = 0.085$ $P = 0.7703$
Read or heard of pets being threatened/attacked by a puma	40	68.4	$\chi^2 = 17.321$ $P \ll 0.0001$
Read or heard of people being threatened/attacked by a puma	42	65.5	$\chi^2 = 13.696$ $P = 0.0002$
Know a friend/neighbour who had an encounter with a puma	94	26.9	$\chi^2 = 113.897$ $P \ll 0.0001$

Note: NA not available.

Table 20.1b Table showing the response of Chileans and Montanans (Riley 1998) to belief statements. Attitudes in Chile and Montana mirrored one another to a large extent, except in regard to the presence of pumas within the country or state and the presence of pumas near a stakeholder's house. Results listed as not available (NA) were either considered not relevant or not translatable to a particular study area.

Belief statements	Study Site	Disagree (%)	Neither (%)	Agree (%)	Statistic
The presence of pumas is a sign of a healthy environment?	Chile	17	8	75	$\chi^2 = 3.2094$
	Montana	14.2	16.7	69.1	$P = 0.2010$
Pumas help to maintain prey populations in balance with their environment?	Chile	10	5	85	$\chi^2 = 3.966$
	Montana	13.0	12.7	74.3	$P = 0.1377$
The presence of pumas in Chile/Montana increases my overall quality of life?	Chile	10	14	66	$\chi^2 = 29.929$
	Montana	29.5	30.6	39.9	$P \ll 0.0001$
The presence of pumas near my home increases my overall quality of life?	Chile	75	12	13	$\chi^2 = 14.32$
	Montana	49.9	28.4	21.7	$P = 0.0008$
Pumas should have the right to exist wherever they may occur?	Chile	NA	NA	NA	NA
	Montana	44.8	11	44.2	
I enjoy having pumas...	Chile	24	11	65	$\chi^2 = 1.6900$
	Montana	17.9	14.6	67.5	$P = 0.4295$
But I do worry about the problems pumas may cause	Chile	27	9	64	$\chi^2 = 3.5447$
	Montana	33.6	14.6	51.8	$P = 0.1699$

Note: NA, not available.

Case studies

Since the stakeholder groups able to exert influence over puma populations in North and South America differ widely in their involvement with this felid, to further unravel human values in relation to pumas and people in each place we drew on different methods and data sources. We will now follow the thread of these different data sources through each study area before weaving them together to suggest more comprehensive conclusions about the relevance of management and education in mitigating conflict.

Case 1: second-hand sources in Montana

In Montana, Riley (1998) has shown that the perception of current puma populations, attitudes towards pumas, and beliefs about risks predict stakeholder desires about future puma populations. Since the majority of stakeholders in Montana (and in North America) may influence management decisions despite having only read or heard of pumas, we sought to understand the information sources, such as newspapers, that may shape their perceptions, attitudes, and beliefs about risks. The anthropologist Lévi-Strauss (1987) contends that our perceptions and the values we place on nature are shaped by the stories we tell. Therefore the stories chosen by newspaper editors, presumably because they will sell, are also likely to be the stories that strike the most powerful chords in the human psyche. Since newspaper articles convey important cultural values and events, sociologists have used newspaper archives as representations of how animals become problems within a culture (Jerolmack 2008). Furthermore, with the shift of human populations from rural to

Table 20.1c Table showing the comparison of the risk beliefs of stakeholders in Chile and Montana. Respondents indicated the number between the two words that best respresented their opinion. Riley's original 5-point scale (1998) was combined into a 3-point scale in order to be more comprehensible during oral interviews with stakeholders in Chile.

Risk statement			1&2	3	4&5		Statistic
Encounters between pumas and people are....	Chile (%)	**Decreasing**	65	11	23	**Increasing**	$\chi^2 = 61.78$
	Montana (%)		12.01	17.9	69.9		$P \ll 0.0001$
You are... to live with the risk associated with pumas	Chile (%)	**Unable**	22	10	68	**Able**	$\chi^2 = 4.38$
	Montana (%)		21.4	21.0	57.6		$P = 0.1117$
The risks from pumas...well understood by experts	Chile (%)	**Are not**	42	30	27	**Are**	$\chi^2 = 10.302$
	Montana (%)		28.7	21.9	49.5		$P = 0.0058$

urban environments, the role of the media (and scientists) in conveying information about a species is becoming increasingly influential.

However, common narratives do not always correlate directly with actual puma behaviour (Kellert *et al.* 1996). For example, early European and Asian settlers in North America suffered from a lack of direct and accurate knowledge about this secretive species, which they had never encountered in the Old World. Instead, their beliefs about pumas were often rooted in stories and observations related to African lions (Young and Goldman 1946; Bolgiano 1995). Perhaps no better example exists of narratives departing from biological reality than the American President and political father of conservation, Theodore Roosevelt calling pumas 'The big horse-killing cat, the destroyer of the deer, the lord of stealthy murder—with a heart both craven and cruel.' Later, without any apparent irony, he went on to claim that, 'No American beast has been the subject of so much loose writing or of such wild fables as the cougar' (Bolgiano 1995; Kellert *et al.* 1996).

To investigate how common narratives portray puma biology and behaviour we examined the representation of pumas in newspaper archives of Montana. Do certain narratives appear more frequently than others? Are particular narratives deemed more important than others? In what direction might these narratives skew human perceptions?

In order to deepen our understanding of how newspapers portray pumas, we also compared their representation to that of other predators. Past studies suggest that where pumas are sympatric with other predators such as wolves in North America (Kellert *et al.* 1996) or jaguars in Central and South America, they are often ignored or considered culturally less important (Conforti and Azevedo 2003; S. Marchini, personal communication 2007). Kellert *et al.* (1996) argued that, 'Perhaps strong attitudes towards mountain lions tended to develop only when wolves were not common enough to be culturally relevant.' As evidence they cited the fact that many of the cultures that venerated pumas lived beyond the historic range of grey wolves (Mech 1981). Of the Cherokee, Creek, Chickasaw, Seminole, Zuni, and several cultures in South America that revered pumas or considered them friendly to humans (Young and Goldman 1946; Wright 1959; Barnes 1960; McMullen 1984) only the Zuni also shared their territory with wolves (Kellert *et al.* 1996). In Montana newspapers, we compared the categories of narratives written about pumas to those written about wolves. We also investigated whether pumas became more culturally relevant over a short time period when wolves were removed from the ecosystem.

Study area

Montana, the fourth largest state in the United States (381,156 km^2), is illustrative of some of the broader principles of North American puma management. Here, in western North America (Fig. 20.1) pumas and other carnivores (such as wolves and coyotes) were historically killed under a bounty system in which hunters received payment for skins brought to authorities (Cougar Management Guidelines Working Group 2005). Since 1971, pumas have been classified as 'game species', to be hunted under a quota system. Although, puma populations are notoriously difficult to census and collecting demographic information requires intensive long-term studies (e.g., Logan and Sweanor 2001), based on (less reliable) removal records, there is speculation that puma populations in western North America have increased from historically low levels in the 1900s (Padley 1997; Riley 1998; Cougar Management Guidelines Working Group 2005). If an increase in pumas did occur, it may plausibly be attributed to an increase in the abundance of ungulate prey, which have benefited from favourable habitat conditions and conservative management (Riley 1998; Cougar Management Guidelines Working Group 2005).

In frequency and biomass, white-tailed deer (*Odocoileus virginianus*), mule deer (*Odocoileus hemionus*), and elk (*Cervus elaphus*) comprise the majority of pumas' prey in Montana (Murphy 1983; Hornocker *et al.* 1992; Williams 1992). However, their prey choice may vary with location. One study on the eastern front of the Rockies found that pumas' diet consisted of 43% deer, 27% elk, and 18% bighorn sheep (*Ovis canadensis*; Williams 1992). In another study in northern Yellowstone National Park, elk were preferred over deer (Hornocker *et al.* 1992).

Depredation of domestic sheep by pumas in Montana is low. From 1989 to 2007, on average, pumas accounted for less than 2% (542) of sheep killed by predators, whereas coyotes accounted for approximately 73% (19,726) of sheep killed and wolves for 0.6% (170; US Department of Agriculture National Agricultural Statistics Service 2008a). Since sheep holdings in Montana are typically comprised of range flocks (>500 sheep) that graze on native and unimproved farmland, puma depredation events are unlikely to have a large proportional impact on the holdings of individual livestock owners. Domestic sheep numbers in Montana have also been on the decline (from 3,462,000 in 1940 to 265,000 in 2008; US Department of Agriculture National Agricultural Statistics Service 2008b).

Similar to pumas' depredation of sheep, their depredation of cattle is also low compared to other predators. In a 2001 study, 71.9% of calves (2300 heads) and 33.3% of cattle (200 heads) lost to depredation were attributed to coyotes, whereas <3% (100 heads) of calf losses were attributed to pumas. Losses of adult cattle to pumas and wolves experienced by livestock owners were so low as to go unrecorded by the US Department of Agriculture (US Department of Agriculture National Agricultural Statistics Service 2001). In contrast to domestic sheep, cattle numbers have been on the rise in Montana (from 1,148,000 in 1940 to approximately 2,600,000 in 2008; US Department of Agriculture National Agricultural Statistics Service 2008c).

Since 1890, there have also been one fatal and at least five non-fatal puma attacks on humans recorded in Montana. All of these attacks occurred within the last 20 years (Beier 1991; Etling 2004; Cramer 2007).

Though debate may exist over changes in puma populations, human populations in Montana unquestionably have been on the rise. During the 1990s, the mountain West was the fastest-growing region of the United States, with a population growth rate of 25.4%. Montana still has one of the lowest population densities (2.51 people km^{-2} in 2006) in the United States; however, as it undergoes rapid urbanization the nature and localization of these populations has shifted. Cities in Montana are growing much faster than the rural population. Simultaneously, a demographic movement out of the prairie and into the mountainous regions is occurring as people are attracted to natural amenities (Riebsame *et al.* 1997; Hansen *et al.* 2002). One consequence of these changes is that natural buffers around national parks, wilderness areas, and nature reserves may be shrinking (Hansen *et al.* 2002). Another consequence of increased urbanization and shifts in demography may be an alteration in the cultural landscape and associated shifts in beliefs and attitudes towards the environment and land-use practices (Riebsame 1997; Riley 1998; Rasker *et al.* 2006).

Methods

We collected archives from a group of editorially independent newspapers in the Montane region of Montana: the *Missoulian, Daily Inter Lake, Glendive Independent, Glendive Times/Ranger-Review, Yellowstone Monitor,* and the *Whitefish Pilot.* Each newspaper served a community in Montana ranging from approximately 5000 to 65,000 inhabitants and generally reflects local issues over national concerns. The *Missoulian* had the largest circulation of any paper sampled (32,378) and is currently Montana's third largest newspaper. The span of articles included in our analysis ranged from 1883 to 2007.

In order to develop a concept of the relative importance of particular themes to stakeholders, we classified newspaper articles into different narrative categories based on common subject matter and storylines. We then quantified the total frequency of particular narrative categories. We used frequency of appearance of a story on the front page as a proxy for the importance of specific narratives. Although this frequency may vary on any particular day based on other newsworthy events, over time we expect the narratives considered to be of greater consequence to appear more frequently on the front page. We compared the frequency of particular narratives on the front page to the frequency of articles in less prominent sections of the newspaper. The frequency of front-page puma narrative categories was also compared to the frequency of front-page wolf narrative categories. Since the frequency of articles on the inside of the newspaper exhibited right skew, we $\log_{10}$-transformed these data, so that they better approximated a normal distribution. The difference in frequencies was analysed using a paired *t*-test (SPSS version 16.0 for Macintosh).

To test the hypothesis that pumas become more culturally relevant when wolves are not present, we divided articles on pumas and wolves into groups, consisting of a period corresponding to pre-extirpation of wolves (1883–1930), the period post-extirpation but prior to reintroduction of wolves (1948–95), and the period following the reintroduction of wolves (1995–2007). The period post-extirpation but prior to reintroduction corresponded with a time in which wolves were almost non-existent in Montana, though pumas were still present (Riley 1998). Within this period, we compared the number of articles about pumas to the number of articles about wolves. In our analysis, we also included the number of articles about another canid predator responsible for the majority of sheep depredation in Montana, the coyote (*Canis latrans*).

Results

Over the entire time period analysed (1883–2007) many more articles were written about wolves (1085) than either pumas (274) or coyotes (76, Fig. 20.2a). Although, fewer articles appeared about pumas than wolves, within a subgroup for which newspaper section and page number were recorded (225) approximately 30% of the total number of articles written about pumas appeared on the front page of the newspaper. This result contrasted with the total proportion of wolf articles appearing on the front page, which was less than 20% ($n = 921$) and proportion of front page coyote articles, which was 3% ($n = 70$ Fig. 20.2b).

No difference was observed in the frequency of articles about pumas compared to wolves between the time periods of pre-extirpation and post-reintroduction ($\chi^2 = 0.99$, $P = 0.32$). The difference between the frequency of articles pre-extirpation and the period when wolves were absent was also not significant ($\chi^2 = 0.39$, $P = 0.53$). However, a significant difference was observed between the period when wolf populations were low (or non-existent) and post-reintroduction (Fig. 20.2c, $\chi^2 = 12.86$, $P = 0.0001$).

By narrative category the most prevalent articles on all pages of the newspaper relate to 'pumas killed in conflict', 'hunting and management', and 'sightings' (Fig. 20.3a). However, a comparison of the frequency of articles on front page of newspapers to articles on inside pages reveals that some of the most prevalent narrative categories are deemed significantly less important than other, much rarer, categories (Fig. 20.3b, Paired *t*-test: $t_{12} = 13.562$, $P = 0.000$, $n = 12$). Narratives of 'conflict' and of puma attacks on humans dominate the front page, whereas stories of pumas killed in conflict (either by stakeholders or wildlife managers) tend to appear on the inside pages of the newspaper. Puma narratives on the front page also differed considerably in frequency and type from narratives about wolves on the front page (Fig. 20.3c).

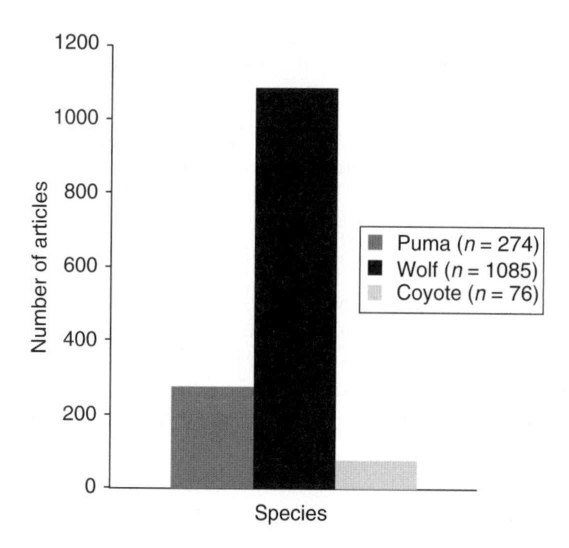

Figure 20.2a Total number of articles appearing in six Montana newspapers (1883–2007) about three different predator species that come into conflict with humans: pumas (*Puma concolor*), wolves (*Canis lupus*), and coyotes (*Canis latrans*).

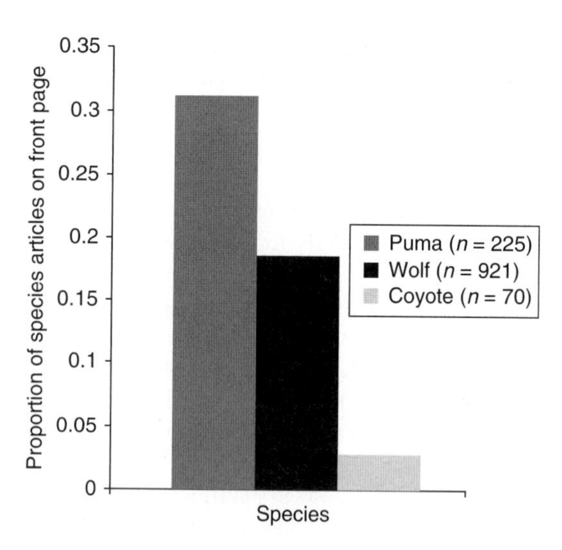

Figure 20.2b Proportion of total articles about a particular species that appeared on the front page of newspapers. Of the total articles written about a species, a much higher proportion of puma articles ran on the front page of the newspaper than articles about either wolves or coyotes.

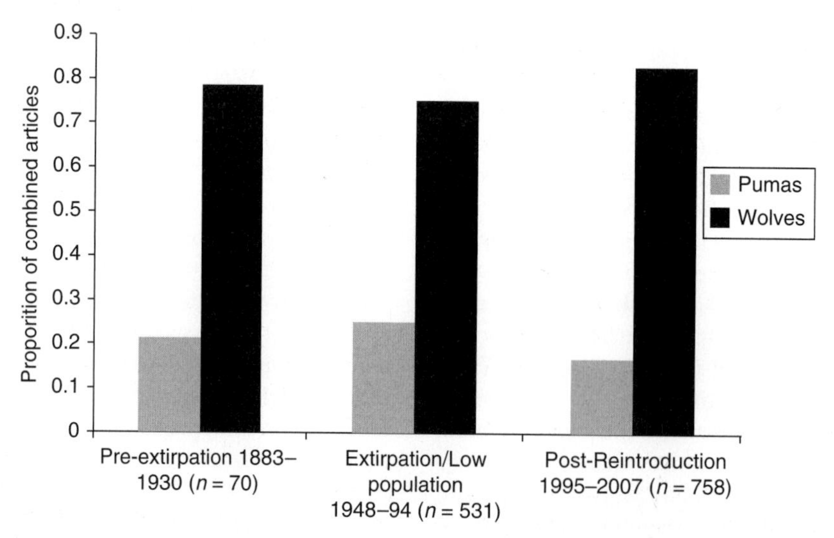

Figure 20.2c Frequency of articles about pumas in comparison to the frequency of articles about wolves over different time periods, which correspond to changes in wolf populations.

Discussion

Do second-hand sources give preference to particular narratives?

On the front page of the newspaper dramatic narratives of conflict—such as stories 'human encounters with pumas' and 'puma attacks on humans'—appear with significantly more frequency than articles concerning hunting and management. Articles about pumas killed in conflict are also relegated to the later pages of the newspaper. Thus, the narratives

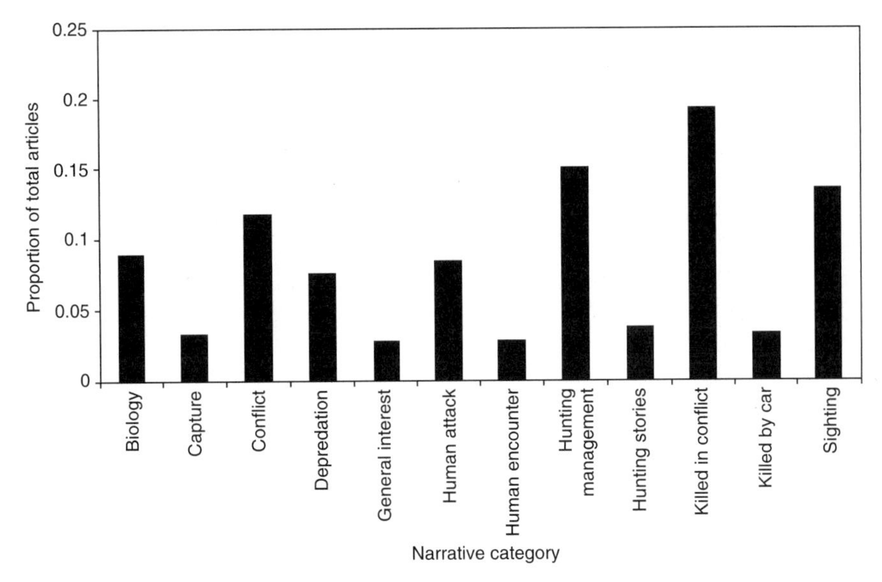

Figure 20.3a Comparison between the frequency of narratives on the front page of the newspaper and the frequency of narratives on the inside pages of the newspaper. Articles were collected from the Montana archives of newspapers, for which page number was identified ($n = 214$).

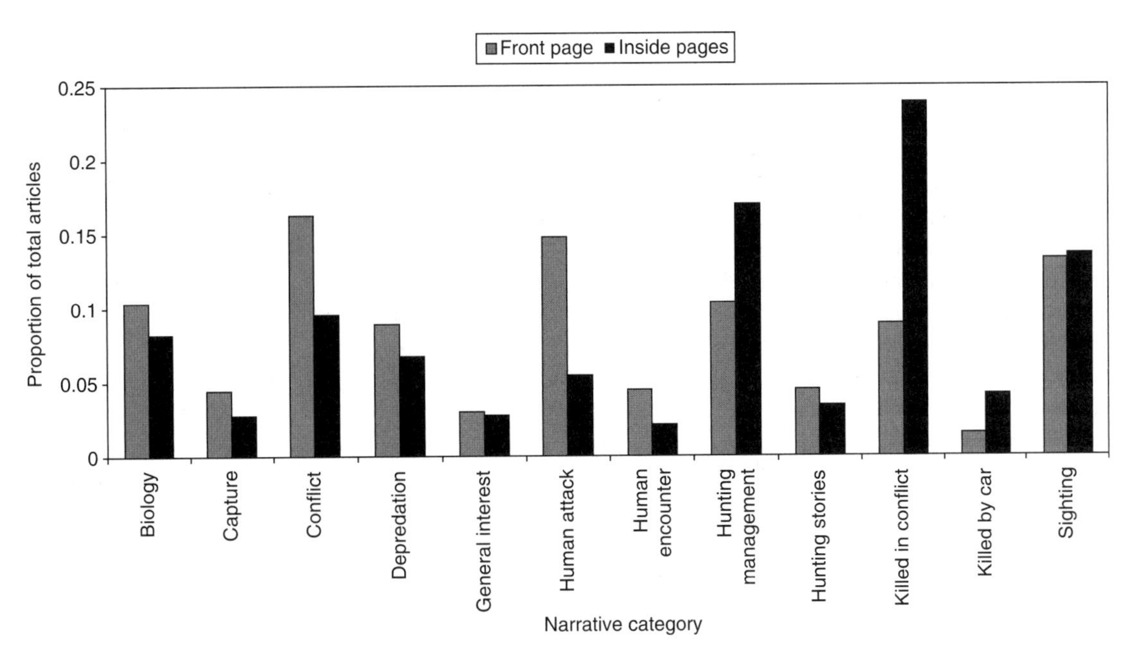

Figure 20.3b Chart showing the proportion of articles in Montana newspapers about pumas by narrative category ($n = 214$).

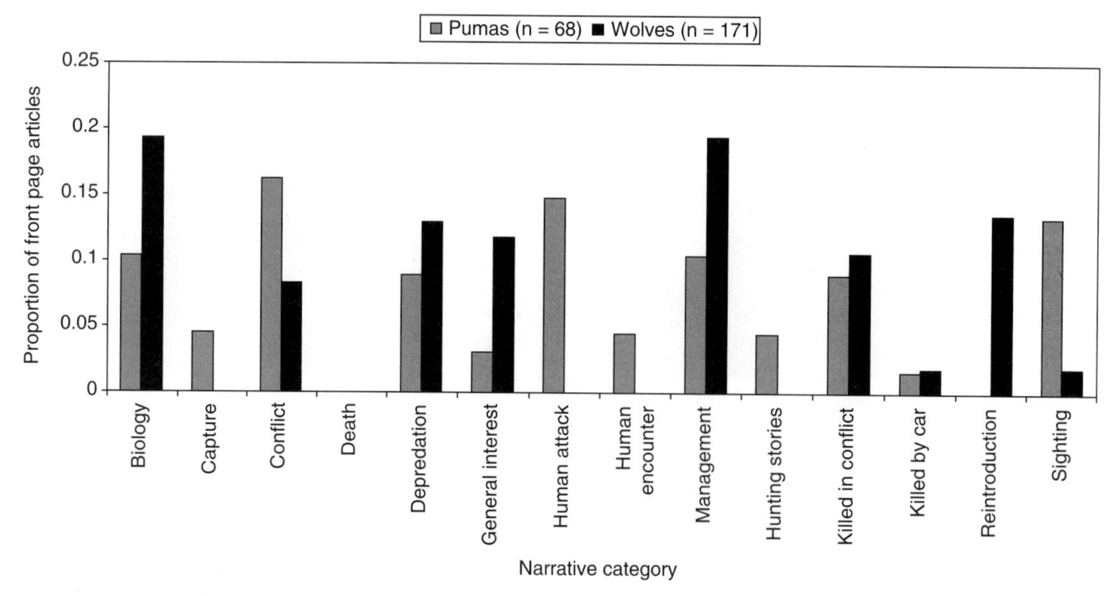

Figure 20.3c Comparison between the puma narratives (n = 68) on the front page of the newspaper and wolf narratives (n = 171) on the front page of the newspaper. Puma narratives garnering front-page attention were dominated by conflict, human attacks, and sightings. On the other hand, wolf narratives were dominated by stories about wolf biology and management. Only one article relating to coyotes appeared on the front page (not shown here); it was about conflict.

of human–puma interaction given priority in newspapers are often fear-based stories of conflict, rather than the much commoner occurrence of puma control by wildlife authorities.

Unsurprisingly, issues of management and reintroduction were deemed culturally more important for wolves than pumas. More revealingly, a much larger proportion of wolf stories on the front page were written about their biology or for general interest than articles about pumas (e.g. 'Wolf-watcher keeps ghostly vigil over orphaned pups;' 'Wolves, not bears, the rave at Yellowstone,' and 'Soul of the wolf'). The movement of wolves was also tracked on the front page, whereas 'sightings' of typically elusive pumas dominated front-page coverage. The fact that puma sightings draw attention may be of particular concern given that they are notoriously unreliable (Hamilton 2007; S. Riley, personal communication 2008). Beier and Barret (1993) describe three cases in which groups of two to five experienced witnesses with prolonged, repeated, daylight views of housecats and coyotes misidentified these animals as pumas. In both, the frequency of articles and narrative themes, newspapers selling stories of wildlife present a skewed portrayal of pumas and may contribute to a distorted perception of pumas among North American stakeholders.

Do pumas only gain cultural relevance when wolves are not present?

Even during the years after wolves were mostly extirpated from the state of Montana and prior to their reintroduction (1948–94), Montana newspapers contained a much higher frequency of articles about wolves than pumas. Admittedly, during part of this period, much of the attention wolves attracted related to the raging debate over wolf reintroduction. In 1993, the Environmental Impact Statement for the reintroduction of wolves generated 160,000 public comments from 40 countries, one of the largest volumes of public comment received on a planned federal action up to that time (Bangs *et al.* 2005). The attention wolves attract may be attributed, in part, to their potency as a symbol of wilderness as well their perceived threat to ranchers (Kellert *et al.* 1996). However, the very fact that the potential benefits and costs of reintroducing wolves drew more attention than the existing threat of pumas, which were responsible for one fatal and two non-fatal attacks on humans in

Montana in the early 1990s, also reflects the extent to which pumas are often ignored. Since the frequency of puma articles in relation to wolf articles was significantly greater in the period of extirpation compared to the period of post-reintroduction, though not pre-extirpation, the hypothesis that pumas would assume greater cultural relevance when wolves were not present was only weakly supported.

Why do wolves attract such disproportionate attention in relation to pumas, even when they are not physically present? Perhaps the explanation rests, in part, with humans' ability to relate to this visible and social canid over more cryptic and solitary pumas. Human understanding and empathy for social canids is also evident in the domestication of dogs, which occurred as early as 130,000 years ago, long before the domestication of cats, approximately 9500 years ago (Wayne and Ostrander 1999; Kruuk 2002; Driscoll *et al.* 2007). However, not only was the number of articles written about wolves far greater than those written about pumas, it also far exceeded the number of articles written about another highly visible social canid species: coyotes. Although coyotes in Montana account for a disproportionate amount of cattle and sheep depredation in relation to other predators, they receive relatively little media attention compared to wolves. The fact that wolves culturally eclipse both pumas and coyotes may reflect wolves' long and layered history with humans in Old World landscapes. The relationship between wolves and the European settler of North America extends back thousands of years, whereas they have only had several hundred years of experience sharing landscapes with pumas and coyotes (Lopez 1978).

Though articles on pumas appear in newspapers far less frequently than those on wolves, a high proportion of puma interactions with humans that do appear are considered important. Perhaps, the prominence of puma articles simply reflects the tendency of newspapers to give importance to rare ('newsworthy') events. However, it may also imply that because of human cultural and psychological tendencies a much higher threshold exists for articles on pumas to appear in the newspaper than those on wolves; therefore the majority of puma interactions reported are those with the most dramatic narratives.

Case 2: stakeholder tolerance in Chile

In Chile, only the views of stakeholders living in closest direct contact with pumas and suffering the highest costs of coexistence are relevant. Although second-hand information continues to play a role among this group it is mainly transmitted by word of mouth and is thus difficult to quantify. Most of the relevant Chilean stakeholders also claimed to have had first-hand experience with pumas. Since these stakeholders may influence population levels through their actions, it is essential to understand their tolerance for the costs of coexisting with pumas.

Decker (1994) argued that tolerance levels depend upon perceptions of what is at stake. What is at stake in turn is influenced by a host of other variables such as attitudes, beliefs, knowledge and experiences with wildlife, the aesthetic value of the referent species, the economic condition of the stakeholder, the economic dependency of the stakeholder on the resource being threatened, and stakeholders' attitudes towards management (Decker and Purdy 1988; Craven *et al.* 1992).

Our study aimed to understand the role of some of these variables in determining tolerance among Chilean livestock owners. We examined factors that our discussions with local people had indicated to be important in the Araucanía Lake Region of Chile. We investigated whether the tolerance of livestock losses was simply related to livestock holdings and other economic measures; or, whether variables such as stakeholder age or particular beliefs produce a good model for tolerance.

Study area

Though in theory the protection of pumas may be considered an important step for their conservation, the reality is more complicated, as is illustrated at our study site in Pucón, Chile. One of the southernmost communes in the Cautín Province of the Araucanía Lake Region, Pucón covers an area of 1429 km^2 that due to past (and continuing) volcanic activity ranges widely in elevations (from 200 m to 3747 m). Historically, livestock has been the main industry in this area; as a result, puma depredation has always been a concern. Prior to the introduction of the *Ley de Caza*

(Hunting Law) that provided protection to pumas, livestock-owner response to depredation was simple. Darwin (1839) recorded that pumas in Chile were 'generally driven up bushes or trees, and then either shot or baited to death by dogs'. Though a minimum puma density of 2.5/100 km^2 has been estimated further south in the distinct ecoregion of Torres del Paine National Park (Fig. 20.1, site 6; Franklin *et al.* 1999), very little is known about puma populations and densities within the Araucanía Lake Region.

Puma diet in the Araucanía Lake Region of Chile, however, is better studied. Based on 62 scats collected opportunistically over 14 years, Rau and Jimenez (2002) have shown that 44.3% of pumas' diet consists of non-native European hares (*Lepus europaeus*), 20.2% rodents, and 19.0% native pudu (*Pudu puda*; the world's smallest deer with an average adult weight of 8.5 kg; Iriarte *et al.* 1990). Also present, though less prevalent in puma diet were birds (5.1%) and marsupials (1.3%; Rau and Jimenez 2002). Based on this study, non-native wild boar (*Sus scrofa*) and carnivores including foxes (*Lycalopex culpaeus* and *Lycalopex griseus*), and introduced feral dogs (*Canis familiaris*), were not found in puma diet, though they are found throughout the Araucanía Lake Region. The high incidence of European hares in puma diet is a widespread phenomenon in southern South America (Yañéz *et al.* 1986; Iriarte *et al.* 1990; Franklin *et al.* 1999) and Rau and Jimenez (2002) hypothesize that it results from either an increase in hare abundance or the almost complete extinction of pudus.

Domestic livestock comprised 8.9% of pumas' diet, including 5.1% goats (*Capra hircus*), 2.5% sheep (*Ovis aries*), and 1.3% cattle (*Bos taurus*; Rau and Jimenez 2002). Because their study area was dominated by uncleared forest, Rau and Jimenez (2002) suggest that livestock as a percentage of pumas' diet may have been under-represented in their study compared to other areas of the Araucanía Lake Region (e.g. Muñoz-Pedreros *et al.* 1995). Other studies have found that areas of uncleared forest contain a greater abundance of natural prey and have lower rates of livestock depredation (e.g. Crawshaw and Quigley 2002; Polisar *et al.* 2003). However, any depredation, no matter how small, may be of considerable concern to individual livestock owners, given that commercial farms are rare and most livestock owners have smaller holdings on which they practice traditional techniques of husbandry. In contrast to Montana, a single puma depredation event may result in the loss of a considerable proportion of an individual owner's stock.

Records of puma attacks on people in South America are sparse and have not been systematically compiled, as they have in North America. Darwin (1839) reported hearing of two men and one woman killed by pumas. In modern times only one fatal attack has been documented in Chile: a fisherman was killed by a puma, in 1998, in Torres del Paine National Park (Franklin *et al.* 1999).

Currently, a shift is occurring from agriculture and animal husbandry to tourism, logging, and salmon farming in the Araucanía Lake Region. This shift is part of a larger trend in which natural habitat is converted into a more urban landscape that is threatening biodiversity throughout Chile and the Americas (Pauchard *et al.* 2006). Of the total population of 14,532 people in Pucón, approximately 8023 are urban dwellers and 6509 rural dwellers (45% of the population). Pucón's population density, of 10.16 km^2, is higher than that of Montana.

Most of this population is concentrated in privately owned lowland forest areas, which have been cleared and fragmented and are under threat from logging, environmental degradation, and pressure from heavy tourism during certain parts of the year. National Parks and National Protected Areas (SNASPE) in the Araucanía Lake Region contain 80% of the remaining stands of araucaría (monkey-puzzle) trees and are located primarily above the winter snowline (approximately 1100 m above sea level, though as low as 800 m during harsh winters). As a result, these areas may provide insufficient winter range for pumas (and their prey). However, there is an absence of evidence to underpin the creation of new protected areas.

In the face of potential habitat limitations, the protection of pumas may increase conflict and lead to stronger negative perceptions of pumas among local stakeholders. Not only do livestock owners bear the economic costs of puma depredation, but also if they retaliate against pumas they are at risk of suffering legal consequences such as a heavy fine or prison term. If, on the other hand, stakeholders report puma depredation events to the Agriculture and

Livestock Service (SAG), the outcome for both human and puma interest is less certain. Upon receiving a depredation report, SAG's standard protocol has been to attempt to capture and translocate all pumas in the area. From 1991–2001, SAG captured 23 pumas in the Araucanía Lake Region and released them into protected areas. Of this group, 12 pumas were released into Conguillío National Park, an area with 256 km^2 of potential habitat, 8 pumas were released into Villarrica National Park (460 km^2) and 1 puma each was released into the National Reserves, Malleco (139 km^2), Malalcahuello (57 km^2), and Alto Bio-Bio (39.55 km^2,Fig. 20.4).

Nothing is known of the outcome of these expensive and time-consuming translocations. One hypothesis is that translocating so many pumas to protected areas with limited habitat is tantamount to killing them. A study in the San Andres Mountains of New Mexico showed that the survival rates of translocated pumas was lower than that of pumas in reference populations and that a translocated puma may travel as much as 494 km to return to the site of original capture (Ruth *et al.* 1998). Studies have also revealed that conflicts with carnivores recur in the same locations even after removal of a few individuals (Evans 1983). Currently, SAG feels it has few options in the management of puma–human conflict, and lacks the economic resources to study it, an issue to which our project is intended to contribute data and discussion.

Methods

We used questionnaires to assess stakeholder beliefs and perceptions of the extent of conflict. We collected baseline data on perceptions of past puma depredation, seasonality of depredation events, and methods employed to prevent livestock losses. Based on local market value, the following cost per head estimates were applied in US dollars: horses \$320, cattle \$230, sheep \$50, goat \$50, swine \$60, and fowl \$5. We also assessed stakeholder opinions of current management policies of the Agricultural and Wildlife Services (SAG). Tolerance was measured following the index established by Romañach *et al.* (2007), asking respondents how many sheep they were willing to lose per year without killing the predator responsible. Variables such as number of livestock, sheep holdings as a percentage of total capital base, depredation history, stakeholders' age, and beliefs were tested for correlation with the index of tolerance. Since the dependent variable (tolerance) was not normally distributed, we used a nonparametric correlation test (Spearman's Rho).

A regression model was developed to explain tolerance. Because few livestock owners were willing to accept the loss of more than one sheep, we used a binary logistic regression. To meet the model's requirements categorical variables were transformed into dummy variables. All variables were selected *a priori* and tested individually in the logistic regression before being included in the model. The variables tested in the model were stakeholders' age, current number of sheep, whether stakeholders had suffered puma depredation in the past 10 years, whether stakeholders enjoy having pumas, and finally, whether stakeholders worried about the problems that pumas might cause. The best model was selected based on fit. Spearman's Rho correlations were also used to determine whether particular kinds of involvement appeared linked to belief statements. All analysis was carried out in SPSS (version 16.0 for Microsoft).

Results

Overall, the sheep-holdings of stakeholders in the Araucanía Lake Region ranged from 3 to 120. On average, stakeholders owned 17 sheep (*Ovis ares*, SE = 2.25), 4 goats (*Capra hircus*, SE = 1.25), 8 cows (*Bos taurus*, SE = 1.18), 1 horse (*Equus caballus*, SE = 0.15), 23 fowl (SE = 2.95), and 1 pig (*Sus scrofa domestica*, SE = 0.29). Of these holdings, approximately 8% of sheep and lambs, 7% of goats, and less than 0.5% of cattle and calves were reported lost to pumas. Approximately 4% of sheep and lambs and less than 1% of cattle were also reported lost to feral dogs. Less than 0.5% of sheep and lamb losses were attributed to foxes (Fig. 20.5a).

Almost 50% of stakeholders were willing to accept the loss of ≥ 1 sheep but <4 to pumas per year without seeking retribution against pumas (Fig. 20.5b). The mean number of sheep stakeholders were willing to lose per year was under one (0.79, SE = 0.122). However, 5% of stakeholders were willing to lose three sheep per annum. The percent of total sheep

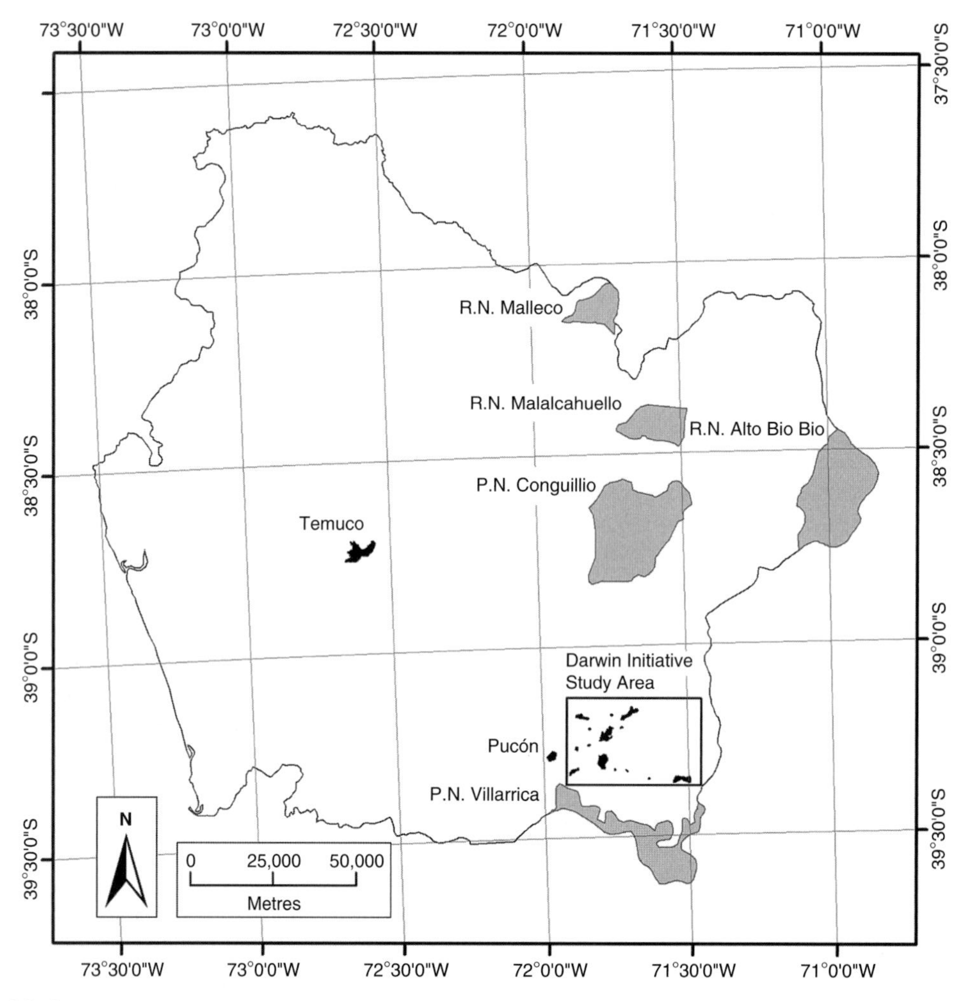

Figure 20.4 Map of study area in Chile, showing the location and size of surrounding national parks (P.N.) and national reserves (R.N.) where pumas have been translocated in the past. © Robert Petitpas.

holdings such a loss represented varied. The highest proportion of total holdings a stakeholder was willing to tolerate losing was 100%; however, this was an exceptional case in which total holdings were small ($n = 2$). On average, tolerant stakeholders were willing to lose approximately 15% (SE = 2.5) of their total sheep holdings to pumas per annum, without seeking retribution. Though many stakeholders tolerated the loss of one sheep per year, all of those interviewed believed that pumas may attack more than one sheep in a single depredation event. They offered predominantly two explanations for this behaviour: pumas killed numerous sheep to drink their blood or in order to teach their young to hunt (Fig. 20.5c).

Tolerance was strongly correlated with stakeholders' age (Fig. 20.6a; Spearman's correlation: −0.331, $N = 58$, $P = 0.011$), the extent that livestock owners worry about the problems that pumas might cause (Spearman's correlation: −0.366, $N = 58$, $P = 0.005$), and whether they enjoyed the existence of pumas (Spearman's correlation: 0.280, $N = 58$, $P = 0.034$). A binary regression model containing these three independent variables provided the best fit for the dependent variable of how many sheep losses livestock owners would tolerate in a year

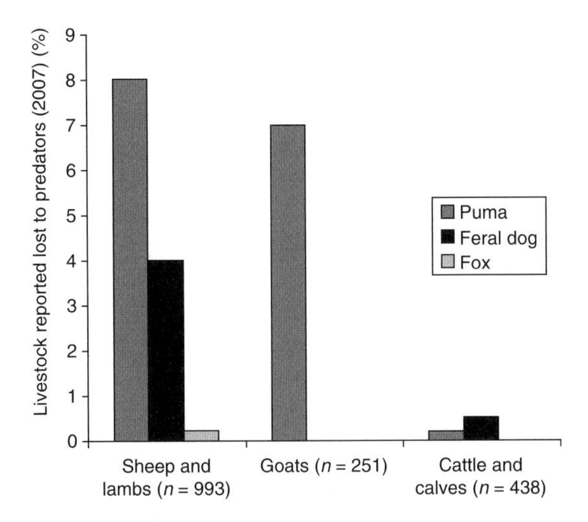

Figure 20.5a Percent of livestock reported lost to different species during 2007. Livestock owners reported considerably fewer attacks on cattle than on sheep. Though reported losses give an indication of the total percent of livestock lost to each predator, depredations by pumas and feral dogs are difficult to distinguish and may be conflated.

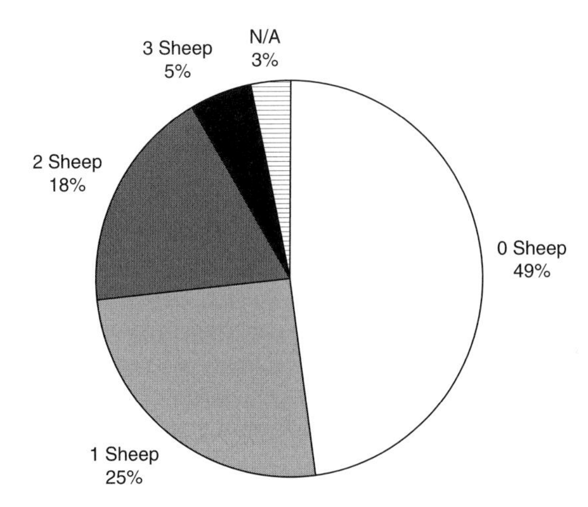

Figure 20.5b Number of sheep losses livestock owners are willing to accept per year without seeking retribution against pumas. Almost 50% of stakeholders are willing to accept ≥1 sheep but <4 lost to pumas per year. The mean number of sheep stakeholders were willing to lose per year was under 1 (0.79 ± 0.122). This measure was used as a proxy for tolerance.

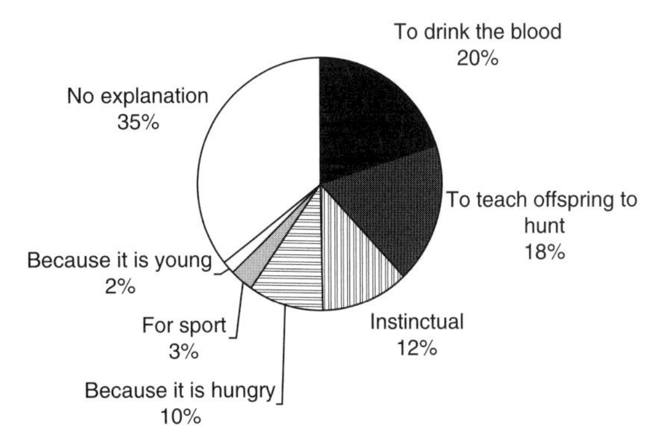

Figure 20.5c One hundred percent of livestock owners (n = 60) believe that pumas attack >1 sheep in a single depredation event, and this chart shows the most common reasons they proposed to explain this behaviour. Of stakeholders who offered an explanation, the majority believed that pumas attack >1 sheep to drink their blood or in order to teach their young to hunt.

(Binary regression: $F_{1,58}$, χ^2 = 12.926, Nagelkerke $R^2 = 0.2665$ $P = 0.009$).

Whether stakeholders had previously suffered puma depredation had little bearing on tolerance (Binary regression: $F_{1,58}$, χ^2 = 0.140, Nagelkerke R^2 = 0.003, P = 0.708). The number of sheep owned by a stakeholder did not explain tolerance (Binary regression: $F_{1,58}$, χ^2 = 1.010, Nagelkerke $R^2 = 0.023$, $P = 0.315$). The total economic value of sheep-holdings as percentage of the total value of livestock holdings also was not linked to tolerance (Binary regression: $F_{1,58}$, $\chi^2 = 0.019$, Nagelkerke $R^2 = 0.000$, $P = 0.890$).

The greatest proportion of Chilean stakeholders liked having pumas, but worried about the problems they might cause (Fig. 20.6b). Stakeholders' tendency to worry about the problems pumas may cause was negatively correlated with whether they believe the existence of pumas near their house increased their quality of life (Spearman's correlation: −0.328, N = 60, P = 0.010). Also, whether a stakeholder worried about the problems pumas may cause was positively correlated with whether a stakeholder had read or heard of a person being attacked by a puma (Spearman's correlation: 0.311, N = 60, P = 0.040). A positive correlation existed between stakeholder enjoyment of having pumas and their belief that the presence of pumas is a sign of a

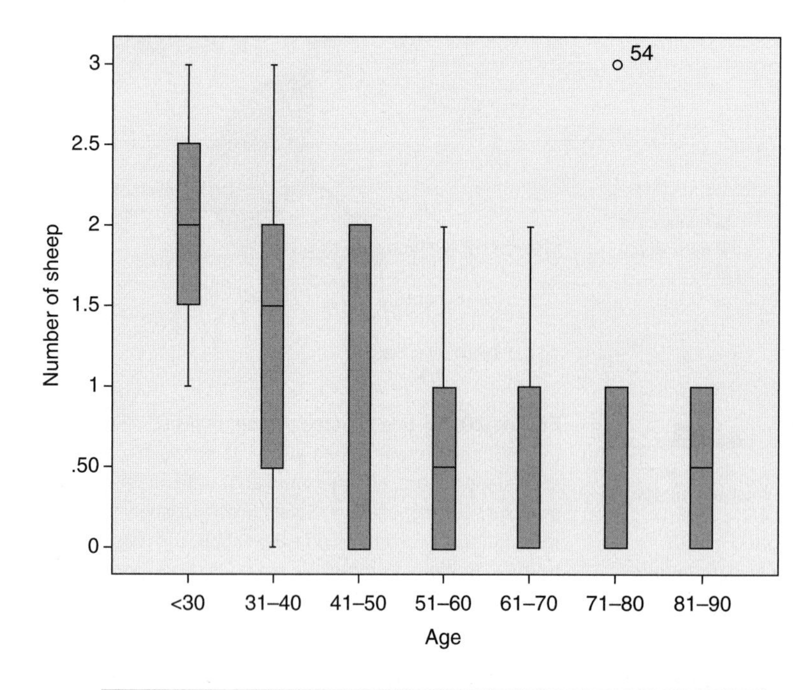

Figure 20.6a Association between age and the number of sheep livestock stakeholders were willing to lose without seeking retribution against pumas. Stakeholders >40 years old tolerated fewer livestock losses per year than those in lower age brackets.

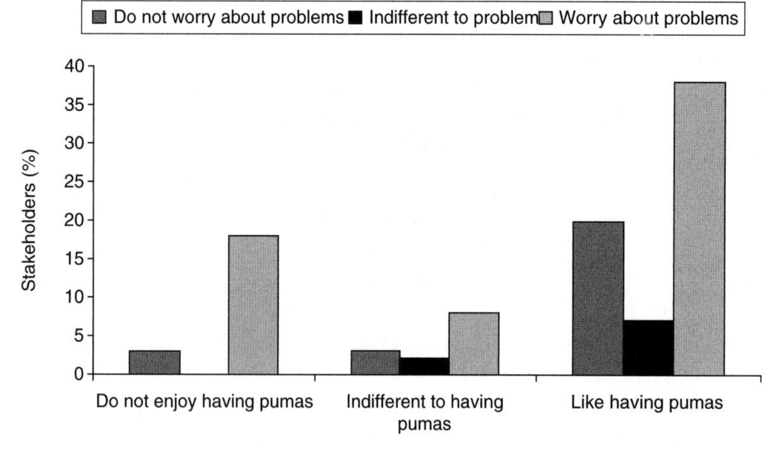

Figure 20.6b Relationship between liking of the presence pumas and worrying about the problems they might cause. The greatest proportion of Chilean stakeholders enjoys having pumas, but worries about the problems they might cause.

healthy environment (Spearman's correlation: 0.464, $N = 60$, $P = 0.000$).

Only 5% of livestock owners interviewed in Chile were willing to call the Agricultural and Wildlife Services (SAG) in the event of depredation. The most common explanation (58% of stakeholders) offered for not calling SAG was 'no vale la pena' (it was not worth the trouble). Forty-three percent of stakeholders said that they had heard of a puma being killed in the area.

Discussion

What economic conditions and social beliefs increase tolerance of the costs of coexisting with pumas?

The total number of sheep owned by stakeholders did not have bearing on their tolerance of losses. Nor did the economic value of sheep holdings as a percentage of the value of stakeholders' total holdings affect tolerance. This trend differed from Romañach *et al.*'s findings (2007) that stakeholders with large holdings were willing to tolerate the loss of as many as six sheep; while stakeholders with smaller holdings rarely tolerated the loss of more than one sheep. However, variation in sheep holdings among stakeholders in Chile was much lower than between commercial farmers and community members in Kenya (Romañach *et al.* 2007).

The relationship between tolerance and age, with older stakeholders exhibiting less tolerance for pumas, is consistent with the findings of other studies (Fig. 20.6a). For example, in a study of jaguar–human conflict in the Pantanal of Brazil, Zimmerman *et al.* (2005a) found that respondents >60 years old held more negative views towards jaguars than younger respondents. Williams *et al.* (2002b) observed a similar trend for stakeholders in relation to wolves. They suggest that '[n]egative attitudes associated with age are probably a cohort effect, and we should not expect the aging populations in the United States and Europe to lead to more negative wolf attitudes'. Whether values are different in younger age cohorts of Chilean stakeholders as the result of more diverse educational opportunities or different priorities is difficult to determine (and our study was not designed to investigate this relationship). Increasing opportunities in urban areas, and in tourism may also lead younger generations to place less value on livestock holdings. Another possible explanation for age-related differences in tolerance is that older stakeholders have a different baseline for conflict. Most stakeholders claim that puma encounters were higher in the past than they are currently. Perhaps older stakeholders recollect a higher rate of encounter and therefore perceive that pumas pose a considerable threat to their livestock and livelihood.

Beliefs about pumas appear to play a greater role in whether livestock holders are willing to tolerate losses than economic concerns. We found that both, whether stakeholders enjoy having pumas *and* worry about the problems they may cause, were significant factors in our model explaining tolerance of livestock losses. Since most stakeholders enjoy having pumas and like the existence of pumas in their country but not near their house, these worries need to be addressed. Often Chilean stakeholders were ambivalent about their beliefs in relation to pumas (Fig. 20.6b) and justified their continued tolerance of pumas with more philosophical statements, such as 'Dios los dejó en la Tierra' (God put them on Earth) or 'Siempre han existido' (They've always been here).

Conservation conclusions

What role can management play in increasing tolerance of pumas?

Macdonald and Sillero-Zubiri (2004b) observed that some problems associated with human–wildlife conflict cannot just be talked out of existence. In situations where reserves are unlikely be expanded to provide better biological 'savings accounts' and stakeholders must coexist with pumas, good management will likely be the tipping point for tolerance. To be effective, management cannot just take into account the ecological requirements of a species. It must be calibrated contextually and developed in collaboration with stakeholders.

Management of conflict should be directed towards the problems that have been identified as most critical based on surveys of stakeholders. For example, of the livestock owners interviewed in Chile none was willing to accept the loss of more than three sheep a year (worth approximately US $150 and representing 18% of average ovine capital and approximately 7% of average total capital base). Yet all stakeholders in Chile reported that pumas are capable of killing more than three sheep in a single night, and believed that they often do. This behaviour, known as surplus killing (Kruuk 1972), is close to universal among carnivores (and is associated with food caching and other means of using the opportunistic additional kills; Macdonald 1976). Among felids it has been reported in lynx (*Lynx lynx*; Odden *et al.* 2002) and leopards (*Panthera pardus*; Sangay and Vernes 2008).

Not only could surplus killing decimate livestock holdings within a single night, but Chilean stakeholders also struggled to understand these events and created intensely negative narratives to explain their occurrence. Most commonly, the reason stakeholders offered for surplus killing was that pumas drank the blood of the sheep or were teaching their young to hunt (Fig. 20.6a). These explanations portray pumas as either bloodthirsty creatures (similar to the mythical 'chupacabra') or animals that use the livelihood of livestock owners in order to train their offspring how to attack and kill prey. A key factor in improving perceptions and increasing the likelihood of tolerance of pumas will be to focus on reducing the amount of surplus killing through management practices. In order to achieve this goal, it is crucial to investigate and verify cases of surplus killing, especially considering that feral dogs may also be responsible for these events, though pumas receive the blame (Nicolás Gálvez, personal communication 2008). A further important area of investigation will be to determine whether management practices have an effect on the number of stock killed per depredation event.

Simultaneously, better management has the potential to improve stakeholders' perceptions of wildlife authorities. Most Chileans have heard that wildlife managers release pumas near livestock holdings (Table 20.1a). As a result, they believe that wildlife authorities do not understand the risk associated with living in proximity to pumas and are unwilling to call authorities in the event of depredation. Therefore, better management in Chile may involve the allocation of funds currently used in translocation to test other solutions (e.g. guard dogs, covered corrals or, as a last resort, lethal control).

How can education shape future human values?

Better management of pumas may limit livestock losses and improve the efficiency of husbandry practices; however, as long as pumas and people inhabit the same areas, some level of depredation will likely occur. The risk pumas pose to human safety can also be reduced by providing empirically based information about appropriate behaviour in the case of an encounter (e.g. Beier 1991); however, it will never be eliminated completely. Therefore, conservation biologists have no choice but to seek to understand and promote beliefs that lead to a landscape of tolerance.

To achieve tolerance, compensation schemes have been employed by conservationists throughout the world with mixed success (Nyhus *et al.* 2005). However, this approach may not be suitable for livestock owners in the Araucanía Lake Region with limited holdings, since we found that beliefs played a greater role in the tolerance of minor losses than economics. In their study of conflict between jaguars and livestock owners in Brazil, Cavalcanti *et al.* (Chapter 17, this volume) demonstrated a similar disjunction between economics and killing jaguars: the intent to kill was only loosely linked with depredation on livestock. Though both studies approach the issue of conflict from different perspectives, they illustrate the importance of the stakeholder's value system. Whether stakeholders intend to kill felids regardless of losses or whether they tolerate losses without seeking retribution, they are hoping to 'win' relative to their own value system, not relative to pumas or jaguars.

Perhaps, the best recourse available to influence culture and value systems is education. Since stakeholder involvement with pumas may vary greatly, education campaigns must be targeted towards the particular concerns and preoccupations of different groups. Macdonald and Sillero-Zubiri (2004b) also suggest that the education that is appropriate in the developed world may differ from what is appropriate in the developing world.

For those living in close proximity to pumas beliefs are influential, as evidenced by the fact that in Chile those who enjoyed having pumas in the area were more tolerant of livestock losses. Also, whether stakeholders enjoyed having pumas was positively correlated with the belief that the presence of pumas was a sign of a healthy environment. However, for those who must coexist in close proximity to pumas highlighting shared benefits or positive qualities of pumas is certainly not enough; combating worries and negative perceptions must also be a part of education. Accurate information about the risks of puma attack as well as measures for preventing livestock predation may help allay fears. To ensure that depredation is not misattributed to pumas, it is essential that stakeholders in Chile are provided with information about distinguishing puma kills from feral dog kills. Thus, wildlife managers must also be good communicators, whose role is to provide practical education that empowers stakeholders to improve husbandry and identify the animal responsible for depredation events.

Education directed at more urbanized groups, less directly involved with pumas, must combat the most common fears and misperceptions associated with this species. Willingly or not, conservation biologist and wildlife managers have no choice but to deal with the media. Therefore, knowing the cultural relevance of a species and human narrative tendencies related to it is essential. In the face of such tendencies and concerns for human safety, managers are responsible for creating messages to reassure the public. Presenting scientific facts, data, or analysis in the form of anecdote may increase the appeal of the message (Hamilton 2007). For example, in Montana one approach towards education may be to publish stories of human–puma interactions in which attacks did not occur or examples of 'sightings' of pumas that were proven to be housecats, dogs, or other animals. Here, again, managers should function as educators. If they are to have success in communicating their message they must be viewed by the public as both a reliable source for information about pumas and the first recourse in dealing with problems related to pumas.

In the future, this form of education and the presentation of scientific fact in an appealing and engaging manner will likely become increasingly relevant as the nature of human–puma conflict comes to reflect areas such as Montana. The trend towards greater urbanization is already occurring; in 2007, for the first time in history, the majority of people in the world were living in towns and cities, rather than rural areas (UN-Habitat 2006). In this changing landscape, only by developing a deeper understanding of human values and tendencies may conservationists foster the management strategies and cultural values under which the existence of pumas is both a tenable and tolerable option for stakeholders.

Acknowledgements

We are very grateful to Nicolás Gálvez and Cristian Bonacic of Fauna Australis and the Pontificia Universidad Católica de Chile, Jerry Laker of the Macaulay Institute, and the Darwin Initiative for their support of the fieldwork necessary to complete this chapter. Robert Petitpas kindly prepared a map of our study site in the Araucanía Lake Region of Chile. We would also like to thank livestock owners for generously consenting to be interviewed. Richard Ladle, Howard Quigley, and Seth Riley all contributed valuable criticism and comments on this manuscript. Funding from the Darwin Initiative and Rhodes Trust made this work possible.

CHAPTER 21

Long-term research on the Florida panther (*Puma concolor coryi*): historical findings and future obstacles to population persistence

Dave Onorato, Chris Belden, Mark Cunningham, Darrell Land, Roy McBride, and Melody Roelke

A male Florida panther investigates a motion-sensitive camera trap. © Florida Fish and Wildlife Conservation Commission.

The diverse names given to *Puma concolor*, including mountain lion, cougar, catamount, painter, puma, and panther (Guggisberg 1975), attest to the historically widespread distribution of this carnivore throughout large regions of the Americas (Currier 1983). The ability to inhabit a variety of biomes allowed one subspecies of puma, the Florida panther (*P. c. coryi*) (Fig. 21.1), to extend from the south-central plains and flatwoods of western Louisiana through the south-eastern states of Arkansas, Mississippi, Alabama, Georgia, parts of Tennessee, and South Carolina to the swamplands of southern Florida (see inset in Fig. 21.2; Young and Goldman 1946). This adaptability allowed the panther to persist, despite the constant persecution of large carnivores by European settlers or the alteration of wildlands that accompanied the colonization of the south-eastern United States. Herein we first present a timeline of the panther's precipitous decline towards extinction. Subsequently, we discuss knowledge accrued during more than 30 years of research on Florida panthers, including findings regarding demography, correlates of inbreeding depression, morbidity and mortality factors, impacts of habitat

Figure 21.1 A female Florida panther photographed with young kittens using a motion-sensitive camera trap. © Florida Fish and Wildlife Conservation Commission.

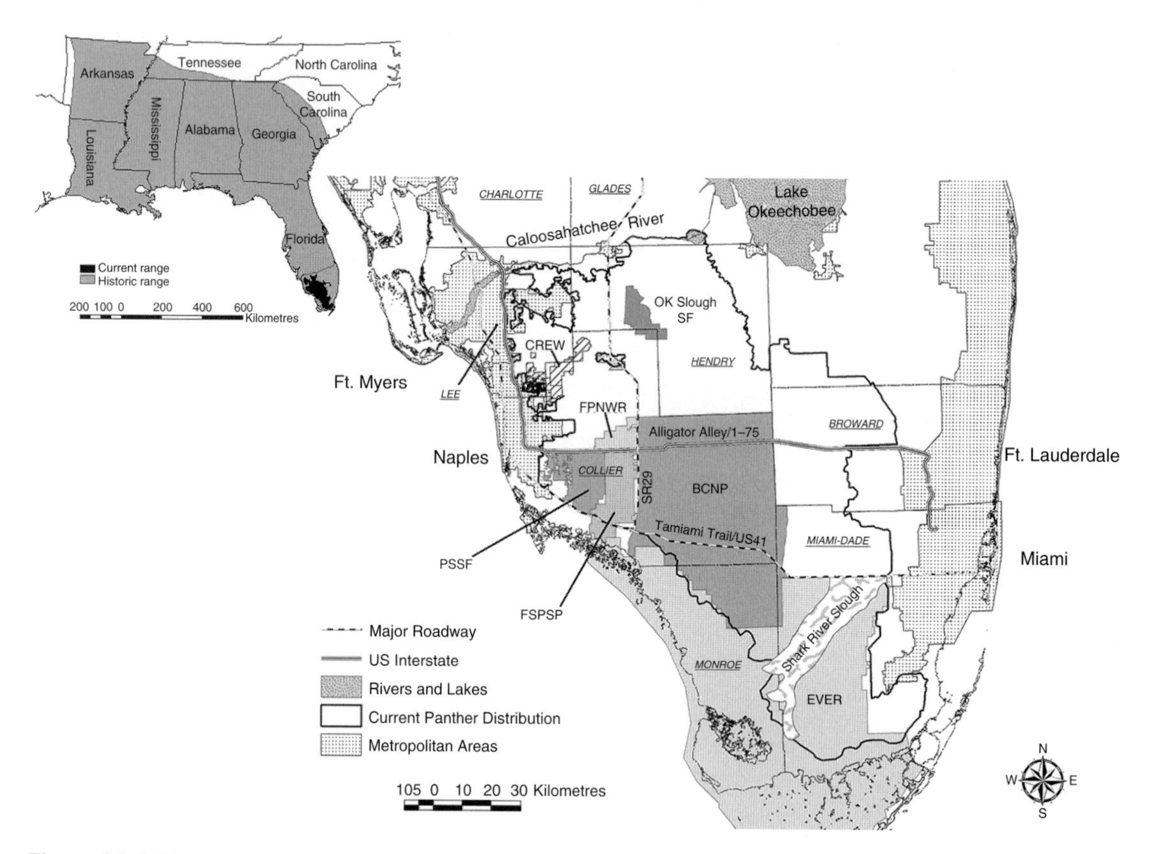

Figure 21.2 The location of varied roadways, counties, metropolitan areas, public lands, and other geographical locations referred to in the text. Florida counties are italicized and underlined; acronyms are defined in the text. Note the Caloosahatchee River and Shark River Slough; features of the landscape that may serve as semipermeable barriers to migration and dispersal events. The inset map depicts the historic and current breeding range of the Florida panther (*Puma concolor coryi*) in the south-eastern United States.

loss, and genetic introgression. Combined, these findings help predict the likelihood of achieving recovery of the only remaining population of *P. concolor* east of the Mississippi River.

A history of the Florida panther

The decline of the Florida panther coincided with the settlement of the south-eastern United States during the early 1800s. Early European explorers reported the widespread presence of *P. concolor* along the eastern seaboard as early as the sixteenth century (Guggisberg 1975). The first colonies established on the east coast in the seventeenth century led to the peopling of this region and the eventual formation of the United States. Panthers, and carnivores in general, were viewed by colonizing Europeans as competitors for game and a potential threat to livestock and humans. Panthers were considered a menace throughout the southern frontier, making them a frequent topic of folklore in Florida (Williams 1976). In 1832, 13 years prior to achieving statehood, a bounty was established in Florida to be paid by the county courts for the scalps or hides of panthers. Townshend (1875) noted that settlers attempted to destroy all predators, including bobcats (*Lynx rufus*), red wolves (*Canis rufus*), and panthers, and that they were becoming increasingly scarce. By 1887, Florida lawmakers had authorized a US$5 statewide bounty for panther scalps. By some calculations of relative worth, this bounty would be equivalent to US$112 in 2007 prices (Officer and Williamson 2008), a significant amount for many settlers in the late nineteenth century. By the late 1800s, it was apparent that unregulated take was beginning to have an impact on panthers in the south-east. For instance, Beverly (1874) commented in a popular article in *Forest and Stream* that the panther was so rarely encountered that it was regarded as mythical by many hunters.

By the beginning of the twentieth century, the Florida panther had escaped complete extirpation only within several large wilderness areas, mainly within Florida. These natural areas were considered inhospitable to early homesteaders, bounty hunters, and poachers (Tinsley 1970). Cory (1896) recorded that panthers were common in the wilder areas of both coasts of Florida and that Indian tribes reported numerous panthers within the Big Cypress region (encompassing areas currently included in the Florida Panther National Wildlife Refuge [FPNWR], Fakahatchee Strand Preserve State Park [FSPSP], and western portions of Big Cypress National Preserve [BCNP]; see Fig. 21.2), an area that would eventually encompass much of the last remaining habitat for the endangered felid.

By the late 1920s, the panther survived only in central and south Florida (Young and Goldman 1946; Tinsley 1970; Alvarez 1993). The proliferation of ranching, agriculture, and development in the south-east continued to gradually eliminate panthers within an increasingly fragmented population. Furthermore, a white-tailed deer (*Odocoileus virginianus*) eradication programme that began in Florida during the 1930s to control the cattle-fever tick resulted in the extermination of over 10,000 deer between 1939 and 1941. The resulting decrease in deer density coincided with a reported increase in panther livestock depredation. In one case, two panthers shot near Naples, Florida, in 1939 were reported to have killed 20–30 head of cattle (Alvarez 1993). Panther depredations helped solidify the mindset of those who found panthers to be a nuisance. The combination of these factors likely hastened the decline of the panther.

By the 1930s, many believed the Florida panther was extinct; but in 1935, Dave Newell, a Florida sportsman, with Arizona houndsmen Vince and Ernest Lee tracked and killed eight panthers during a 5-week hunt in the Big Cypress region in Collier County (Newell 1935). Another successful hunt into the swamps of the Fakahatchee Strand in 1946 (Tinsley 1970) established that a few panthers continued to persist in southern Florida.

In recognition of this decline, the Florida panther was afforded partial protection as a game animal by the Florida Game and Freshwater Fish Commission (FWC)[1] in 1950. Panthers could be legally hunted

[1] The Florida Game and Freshwater Fish Commission, the Marine Fisheries Commission, and several sections of the Florida Department of Environmental Protection were combined into the Florida Fish and Wildlife Conservation Commission (FWC) in 1999. We use the current acronym (FWC) of this agency throughout the text to avoid confusion.

only during deer season. Panthers, deemed nuisance animals because of livestock depredation, could be hunted by special permit at any time. In 1958, the panther was removed from the native-game animal list and given complete legal protection by FWC.

The core of the remaining wilderness in south Florida was penetrated with the construction of the Tamiami Trail (US Highway 41) in 1928 to connect Tampa with Miami via Naples (Fig. 21.2). Nevertheless, the majority of panther habitat north and south of the Naples–Miami section of the Tamiami Trail remained difficult to access. The construction of Alligator Alley (Fig. 21.2) in 1966–67 further facilitated public access to remaining patches of panther habitat in remote areas of the Big Cypress region. Other major access roads were built from Alligator Alley into the heart of the Big Cypress region, permitting recreational use of these areas by hunters and off-road vehicle enthusiasts. Concurrently, a vast system of canals was constructed to drain the wetlands that constituted large parts of southern Florida, allowing development of areas once considered uninhabitable for humans. The combination of easier access, a tropical climate, and inexpensive real estate was the catalyst for an exponential increase in the human population that continues in Florida today and is projected to continue into the foreseeable future. As expected, this chain of events resulted in the dilution of the remote character of the Big Cypress region and assisted in the degradation of panther habitat.

Following the passing of the Endangered Species Preservation Act (Public Law 89–669), the U. S. Department of Interior listed the Florida panther as 'Endangered' on 11 March 1967 (US Federal Register 1967). The Endangered Species Act (ESA) of 1973 (Public Law 93–205) established authority for protection of species listed as 'Threatened' or 'Endangered'. This designation by the federal government raised awareness of the plight of the Florida panther and initiated research and management aimed at averting an apparently imminent extinction.

Early research

When the Florida panther was listed under the ESA, little was known of its natural history. In February and March of 1973, R. McBride and R. M. Nowak, funded by the World Wildlife Fund, initiated a survey for panthers in Florida (Nowak and McBride 1974). A female panther was captured by McBride near Gopher Gully in Glades County (Fig. 21.2). This animal was approximately 10 years old, appeared to be in poor physical condition, and was infested with ticks. Panther signs were also documented in the Fakahatchee Strand (FSPSP in Fig. 21.2) in southern Collier County (Nowak and McBride 1974).

When FWC began research on the Florida panther in 1976, it was questionable whether a reproductively viable population of panthers still occurred in Florida. There were reported sightings in other areas, but indisputable records of the species were confined to south Florida. To understand the status of the remaining population of panthers, FWC set up a 'Florida Panther Record Clearinghouse' to collect and analyse panther reports (Belden 1978a). The clearinghouse was advertised in an article in *Florida Wildlife Magazine* (Belden 1977) and in local newspapers. Cooperators, including government employees and segments of the general public, received the clearinghouse record form, tips on identifying panther tracks (Belden 1978b), and directions for making plaster casts.

While compiling records in the clearinghouse, FWC biologists searched for field sign in habitat remnants historically used by panthers. From October 1976 to June 1990, biologists spent 4427 h searching for tracks, scrapes, kills, or other sign on >15,200 km of dirt roads, trails, fire lanes, levees, and tree islands at 52 sampling sites. Only 16 sites yielded panther signs (3 sightings, 118 track sets, 29 scats, and 64 scrapes), all of them in the Big Cypress region, the Everglades region, and sporadically along the St Johns river watershed in north-eastern Florida (Belden *et al.* 1991).

Research involving the capture and very high frequency (VHF) radio-collaring of panthers by FWC began in 1981. The objectives of this research were to understand the basic biology and habitat requirements of the species. Studies have included estimation of home range sizes, resource selection, morphological descriptions, diet, causes of mortality, and reproduction (Table 21.1). Panther prey studies, including deer population dynamics, herd health, reproduction, and mortality were also completed (McCown *et al.* 1991; Schortemeyer *et al.* 1991; Land *et al.* 1993). Telemetry data and tracking surveys showed that breeding populations of panthers persisted only in Collier, Hendry, and Miami-Dade counties (Fig. 21.2).

Table 21.1 Findings related to demography and habitat selection of Florida panthers (*Puma concolor coryi*) resulting from research initiated in 1981.

	Value	Citation
Home range	435–650 km^2 male, 193–396 km^2 female	Belden *et al.* (1988); Maehr *et al.* (1991); Comiskey *et al.* (2002)
Resource selection studies (highest-ranking habitat)	Upland-hardwood, hardwood swamp, pinelands, cypress swamp	Kautz *et al.* (2006)
	Upland forest, wetland forest	Land *et al.* (2008)
Food habits	White-tailed deer, wild hogs, marsh rabbits, raccoons, armadillos	Maehr *et al.* (1990); Dalrymple and Bass (1996)
Primary cause of mortality[a]	Intraspecific aggression	FWC (unpublished data)
Average litter size	2.51 kittens (range = 1–4, mode = 3)	FWC (unpublished data)
Dispersal	Mean maximum distance: 68 km (males); 20 km (females)	Maehr *et al.* (2002b)

[a] In radio-collared panthers.

Threats to panthers identified by early research

Genetic issues and correlates of inbreeding documented pre-introgression

The critically small size of the Florida panther population in the 1980s and early 1990s made it extremely vulnerable to demographic stochasticity and extinction. Chance events such as highway mortalities, poaching, or declines in prey populations have the potential to devastate small, isolated populations. In addition, disease is a constant threat to endangered populations. For example, a disease outbreak of canine distemper led to the virtual extinction of the remnant population of wild black-footed ferret (*Mustela nigripes*) in Wyoming in 1987 (Williams *et al.* 1988). Decreased genetic diversity, increased neonatal mortality, abnormal reproductive traits, congenital defects, and lowered disease resistance usually accompany inbreeding in mammals (Roberts 1971; Ralls *et al.* 1979; O'Brien *et al.* 1983; O'Brien *et al.* 1985b). The Florida panthers examined in early research seemed to be experiencing many of these problems (Roelke *et al.* 1993b).

Inbreeding depression and associated low levels of genetic variation have been shown to have an effect on demographic parameters such as reproductive success and survival in small mammals (Crnokrak and Roff 1999) and captive populations of brown bears (*Ursus arctos*; Laikre *et al.* 1996) and wolves (*Canis lupus*; Laikre and Ryman 1991). Florida panthers that persisted in the 1980s and early 1990s exhibited some of the lowest levels of genetic variation of any wild felid populations, with an observed heterozygosity level (H_o) of 0.16, which was much lower than levels seen in more contiguous populations of *P. concolor* in western North America (H_o = 0.33; Driscoll *et al.* 2002). Inbreeding and loss of genetic variation, along with development and habitat loss in southern Florida, threatened the persistence of the Florida panther. Ameliorating these threats became the goal of panther research and management.

Several correlates of inbreeding have been described for Florida panthers. These include notable phenotypic traits such as the kinked tail (Fig. 21.3) and mid-dorsal pelage whorls (Wilkins *et al.* 1997). The importance of these morphological irregularities on survivorship or the evolutionary potential of panthers is probably trivial. Other evidence of inbreeding detrimental to individual fitness is atrial septal defects (ASD) and cryptorchidism. ASDs result

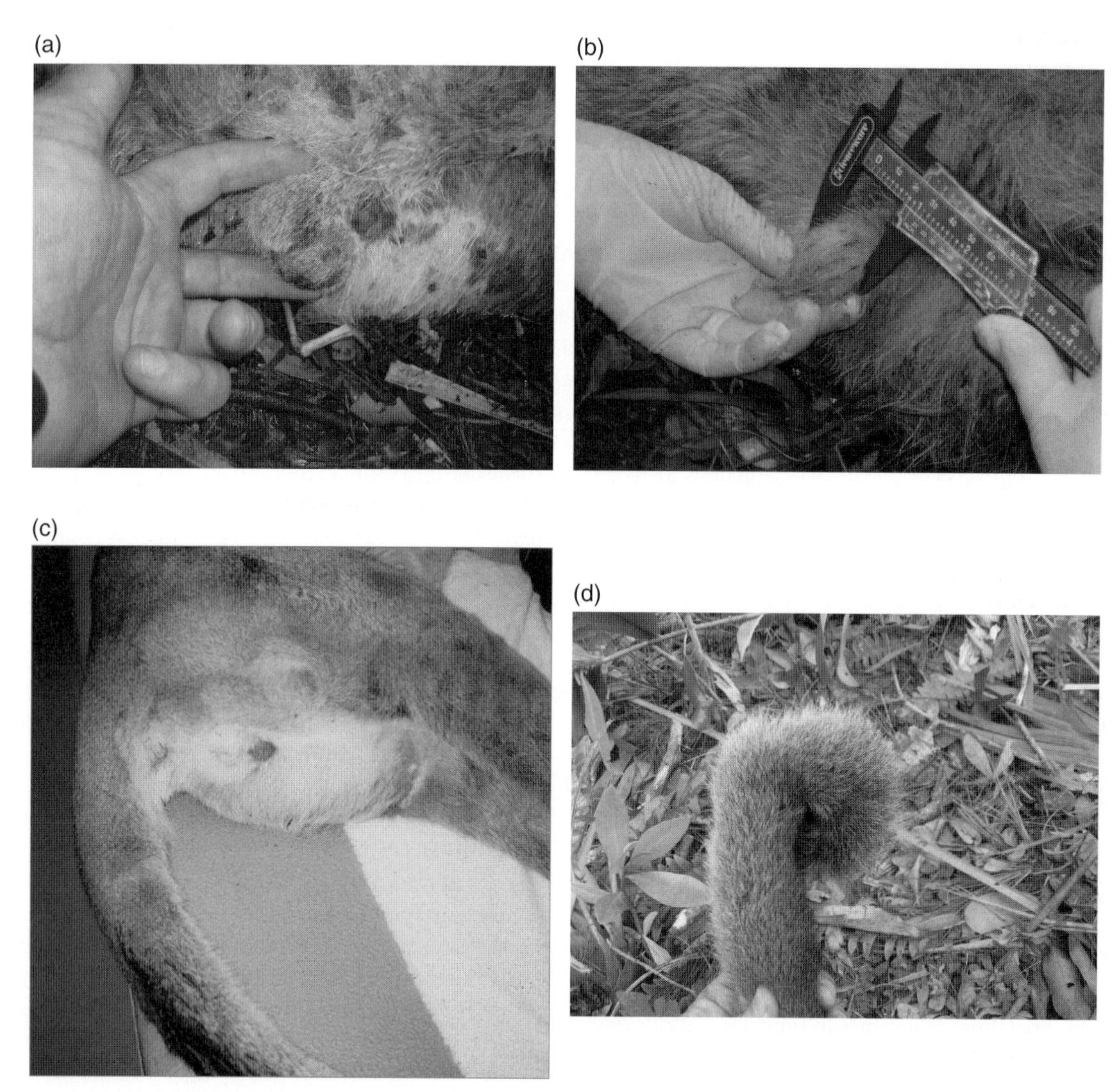

Figure 21.3 Purported correlates of inbreeding in the Florida panther. These include (a) a comparison of normally developed testicles; (b) unilateral cryptorchidism; (c) bilateral cryptorchidism; and (d) the kinked tail. © Florida Fish and Wildlife Conservation Commission.

when the opening between two atria fails to close normally during birth and can lead to cyanosis and heart failure. Although small ASDs (<5 mm diameter) do not typically impact health, large ASDs (e.g. 25 mm) have been linked to the death of several panthers (Roelke *et al.* 1993b; Cunningham *et al.* 1999).

Cryptorchidism, or the failure of one or both testes to descend into the scrotum (Fig. 21.3), results in reduced fertility and has been linked to certain genes in domestic and laboratory animals (Overbeek *et al.* 2001; Williams *et al.* 2007). It is suspected of being coupled with inbreeding depression in panthers (O'Brien *et al.* 1990; Barone *et al.* 1994; Mansfield and Land 2002). Cryptorchidism was first identified among older, pure Florida, male panthers captured during the early years of the FWC study. The remnant population that persisted in the 1980s included several cryptorchid yet reproductively active males. Field data have not documented bilaterally cryptorchid males to breed successfully and one captive bilaterally cryptorchid male was aspermic on electroejaculation. Unilaterally cryptorchid individuals have viable sperm on electroejaculation and can impregnate females (Mansfield and Land 2002), although they are likely to be at a reproductive disadvantage in comparison to competitors with normally descended testes. Felines in general have low sperm concentrations in their semen, with a high percentage of deformed spermatozoa, perhaps due to genetic and hormonal etiologies (Roelke *et al.* 1993b; Jayaprakash *et al.* 2001). Florida panthers, in particular, have been shown to have a high percentage of malformed spermatozoa (93.5%; Barone *et al.* 1994) compared to samples from captive felines, including the cheetah (*Acinonyx jubatus*; 64.6%), leopard (*Panthera pardus*; 79.5%), and tiger (*Panthera tigris*; 37.5%) (Wildt *et al.* 1988).

Impacts of environmental contaminants

Environmental contaminants were, and continue to be, a possible threat to the panther. Methyl mercury, dichlorodiphenyldichloroethylene (DDE, a breakdown product of the insecticide DDT), and polychlorinated biphenyls (PCBs) have been identified in panther tissue. These contaminants bioaccumulate in the aquatic food chain and reach their greatest concentrations at higher trophic levels. Although panthers feed mostly on ungulates such as white-tailed deer and feral swine (*Sus scrofa*), smaller prey from the aquatic food chain represent on average 5–19% of the biomass consumed by free-ranging panthers (Maehr *et al.* 1990; Dalrymple and Bass 1996). Raccoons (*Procyon lotor*) are a likely dietary source of mercury in panthers, and American alligators (*Alligator mississippiensis*) and river otters (*Lutra canadensis*) also may contribute to the presence of this contaminant. Roelke *et al.* (1991) found mercury concentrations in the liver of raccoons to average 24 ppm in Shark River Slough, and suggested that the geographic distribution of these levels mirrored that found in panthers. More recent findings of Barron *et al.* (2004) from simulation analyses predict that mercury exposure risks to panthers have decreased since 1992.

High levels of DDE and PCBs have been found in some panthers. An adult male panther that died of unknown causes in Everglades National Park (EVER) had hair mercury concentrations of 150 ppm and DDE and PCBs in fat of 100 ppm each. High mercury, DDE, and PCB concentrations have been found most consistently in panthers from the Everglades ecosystem and Fakahatchee Strand, whereas lower levels were generally found in individuals utilizing upland habitat associated with the core portion of their range in south-western Florida (Roelke *et al.* 1991; FWC, unpublished data). Panthers with high levels of one contaminant were more likely to have other contaminants as well.

The effect of these contaminants on panthers is unknown. In domestic cats, liver mercury levels of 37–145 ppm were associated with high mortality (Harada and Smith 1975). Individual panthers that have died of unknown causes had liver mercury concentrations as high as 110 ppm (Roelke *et al.* 1991). Mercury levels as high as 40 ppm were found in hair of neonatal kittens (Shindle *et al.* 2003). However, mercury or other contaminants have not definitively been identified as the cause of death for any panther. Subclinical effects of mercury, PCBs, DDE, and other toxins may work additively or synergistically with other stressors to reduce fitness and impair reproduction. Further research concerning environmental contaminants in Florida panthers is ongoing.

Implementation of genetic introgression and preliminary findings

Controversy and project implementation

By the early 1990s, documented annual counts indicated that the remaining population of panthers in Florida was ≤ 30 individuals (Fig. 21.4; McBride *et al.* 2008), was threatened by genetic depletion, and was on the brink of extinction. It was clear that drastic measures were needed. A workshop convened a diverse group of scientists to develop a plan to ameliorate genetic impoverishment in the Florida panther population (Seal 1994). Several management regimes were proposed, including: (1) initiating a captive breeding programme involving pure Florida panthers and Texas pumas to allow for the subsequent

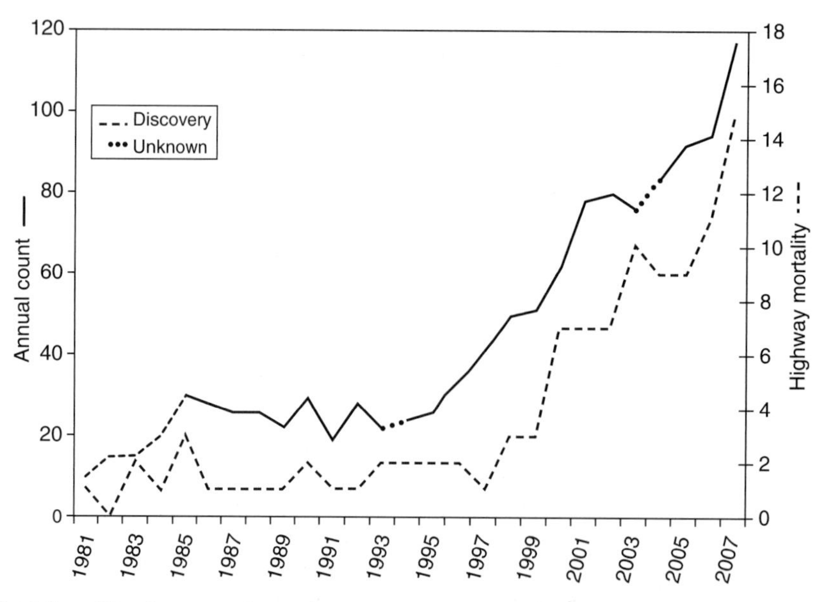

Figure 21.4 Depiction of the documented annual count of Florida panthers and annual highway mortality 1981–2007. The small dashed line in the documented annual counts from 1981 to 1985 reflects an increasing number of panthers added to the sample as they were discovered. The annual documented count represents the total number of panthers documented by physical evidence during each calendar year, except in 1994 and 2004, when censuses were incomplete. Highway mortalities closely track fluctuations in the documented annual count. The subsequent increase in the documented annual counts and highway mortalities of panthers since 1995 coincided with the implementation of genetic introgression. (Taken from McBride *et al.* 2008.)

translocation of intercrossed (*P. c. coryi* × *P. c. stanleyana* or [*P. c. coryi* × *P. c. stanleyana*] × *P. c. coryi*), captive-bred and raised individuals into south Florida; (2) introducing genetic material from captive stock that was purported to contain both pure panther and South American puma lineages; (3) translocating wild female, Texas pumas (*P. c. stanleyana*) into south Florida; or (4) no action at all (USFWS 1994).

Alternative (3) was a plan to release eight female Texas pumas into habitat occupied by pure Florida panthers to mimic historical gene flow between these two subspecies of *Puma*. The goal was to diversify the genetic composition of the panther population with Texas genes to a level at which approximately 20% of the average genotype of Florida panthers was of Texas origin. This proportion was estimated to be sufficient to ameliorate the negative impacts of inbreeding depression, at least temporarily (Seal 1994; Land and Lacy 2000). Upon achieving this goal, the introduced Texas cougars would be removed from the wild population.

The implementation of a genetic introgression programme was not without its critics. Maehr and Caddick (1995) expressed concerns about the potential for outbreeding depression and the lack of documented cases of inbreeding depression in wild populations. Although these concerns were plausible, few examples of outbreeding depression have been validated in the wild, as noted by Maehr and Caddick (1995), and reversal of inbreeding depression in the wild through introgression has subsequently been documented in species such as greater prairie chickens (*Tympanuchus cupido pinnatus*; Westemeier *et al.* 1998). After much debate, it was decided that to rectify the genetic impoverishment of the Florida panther, wild Texas pumas would be released directly into the wild Florida population (USFWS 1994). Most of those involved with the development of the final environmental assessment judged the potential benefits of genetic introgression through alternative (3) (e.g. increased genetic variation, population growth, and population expansion) to outweigh any risks.

In 1994, eight female pumas were captured by R. McBride on private ranchlands in south-western Texas. These pumas were radio-collared, quarantined, and underwent thorough health evaluations before being introduced into south Florida (Seal 1994). Between March and July of 1995, the FWC, with the assistance of the National Park Service (NPS) and United States Fish and Wildlife Service (USFWS), released the eight Texas female pumas in EVER (2 females), BCNP south of Alligator Alley (2), FSPSP (2), and BCNP north of Alligator Alley (2; areas depicted in Fig. 21.2). Although there were some long-distance movements immediately after release, most females established home ranges in south Florida within a few months. Three females died (highway mortality, illegal kill, and unknown) and five successfully reproduced. A minimum of 20 F1 descendents were produced and in 2003, the three Texas female pumas that continued to persist in the wild were removed to captivity.

Preliminary assessment of the impacts of genetic introgression on purported correlates of inbreeding

Once implemented, the genetic introgression project allowed panther researchers the unique opportunity to evaluate the artificial restoration of gene flow between Florida panthers and pumas from Texas. Although the effects of increasing genetic variation in inbred populations have been extensively tested in domesticated or laboratory populations of animals, there have been few opportunities to follow this process through multiple generations in wild populations (Lacy 1997). Genetic introgression provided an unprecedented opportunity to assess the progression of this management intervention through comparisons of historic data collected 14 years pre-introgression and 13 years of data collected during field monitoring and laboratory research post-introgression. Among these comparisons was an assessment of the prevalence of purported correlates of inbreeding depression in pre- vs post-introgression time periods.

The presence of an ASD is difficult to diagnose *in vivo*, and a definitive diagnosis is typically made at necropsy. All panther carcasses suitable for necropsy, regardless of the cause of death, are examined for the presence of ASDs. Fig. 21.5a depicts the linear relationship between the proportion of animals with ASDs and birth year cohorts from 1975 through 2005 (Cochran-Armitage trend test, $Z = 2.15$, $P = 0.033$, $n = 130$). A logistic regression model of the linear association between birth year and the log-odds of being born with ASD indicated that these odds decreased by 33.2% every 5 years (95% profile likelihood CI: −54.1% to −3.3%). It should be noted that although ASDs are potentially detrimental to the health and survival of individual panthers, these defects have never been extremely common. These findings indicate that the prevalence of ASD has been steadily decreasing in panthers born since 1975 and that a further decline associated with genetic introgression cannot be statistically corroborated at this time.

A high proportion of cryptorchid (both unilaterally and bilaterally) males were born in the two cohorts just prior to 1995 (Fig. 21.5b). Some research has concluded that documented consanguineous matings (FWC, unpublished data; Roelke *et al.* 1993b) may have resulted in elevated levels of cryptorchidism. In fact, the repeated documentation of cryptorchid panthers within the population during the 1980s and early 1990s was an important rational for genetic introgression.

The occurrence of cryptorchidism in Florida panthers has decreased sharply since the introduction of the Texas pumas. Statistically, the proportion of individuals born cryptorchid within the five cohort groupings in Fig. 21.5b were different (generalized linear mixed model, F-test $= 4.56$, $P = 0.002$, $n = 133$). Post hoc pairwise comparisons delineate the 2001–05 grouping as having a significantly lower proportion of cryptorchid panthers compared to panthers estimated to be born between 1986–1990 (Tukey's pairwise comparisons $t = 3.89$, adjusted $P = 0.001$) and 1991–1995 (Tukey's pairwise comparisons $t = 3.20$, adjusted $P = 0.015$). We conclude that genetic introgression is correlated with a decline in cryptorchid individuals. Furthermore, cryptorchidism has not been observed in male panthers born from Texas pumas or female F1 intergrades. However, cryptorchidism has not disappeared from the population and continues to be periodically documented in adult male panthers as recently as 2008, perhaps due to remaining pure panthers or backcrosses between them and descendents of the Texas pumas.

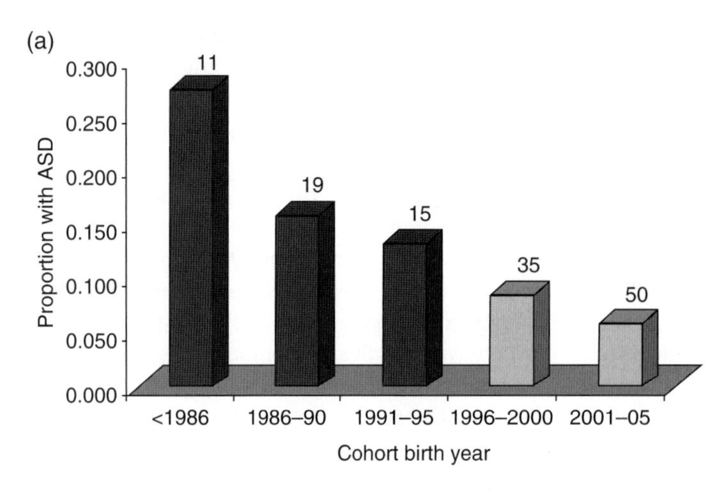

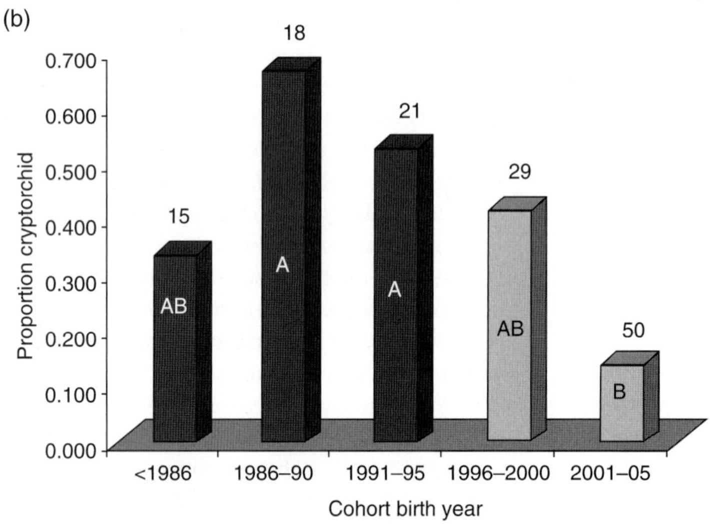

Figure 21.5 Proportion of Florida panthers born in specific year cohorts that were (a) ultimately diagnosed as having ASD; or (b) were unilaterally or bilaterally cryptorchid. Grey bars represent panther cohorts that include animals that were born after the implementation of the introgression project in 1995. Values above the bars represent the total number of panthers that were examined for ASD or cryptorchidism within respective cohorts. Letters within bars in (b) relate to post hoc pairwise statistical comparisons described in the text.

Population expansion and attempts to recolonize former range

The most recent (2007) documented population count for Florida panthers delineated the presence of 117 adult and subadult panthers in south Florida (Fig. 21.4; McBride *et al.* 2008). Coinciding with the introduction of Texas pumas and the growth of the population has been an expansion into habitats where collared panthers were not regularly found in the past. Areas include a large portion of BCNP, private lands to the north of BCNP, portions of Picayune Strand State Forest (PSSF), the Corkscrew Regional Ecosystem Watershed (CREW), and sporadically to the north of the Caloosahatchee river (Fig. 21.2). Panthers historically were present in EVER, but this sub-population was always partially isolated from the core population in south-western Florida by the semipermeable barrier resulting from hydrocycles within the Shark River Slough (Fig. 21.2). By 1995, the panther population in EVER could be described as ephemeral. Cooperators from FWC and NPS agreed to release two female Texas pumas to reinvigorate the population in the Park. Currently (2008), the Park supports a minimum of three reproductive-age females and at least one male panther (transient or resident), as verified by the birth of two litters in 2006. Recently (2006–07), three unmarked adult panthers (two females and one male) have been killed on roadways just to the east of EVER. This indicates some reproduction in EVER was not documented by our sampling; a consequence of both the multitude of logistical constraints when conducting capture work in the remote sections of the Park and the less intense capture effort in EVER. Nevertheless, EVER has not supported a large population of panthers in recent history and it is unlikely that it currently supports >10 panthers. The population in EVER will likely always be tenuous, perhaps ephemeral, and dependent upon migration and dispersal events of panthers residing to the north-west in BCNP. Nevertheless, maintaining the small subpopulation of panthers in EVER has management implications (discussed later in this chapter) for the perseverance and viability of panthers from the standpoint of meta-population theory and the susceptibility of endangered populations to cataclysmic events.

All panthers monitored north of the Caloosahatchee river (Fig. 21.2) since 1988 have been male and the dispersal of panthers towards the north is associated with the population growth that has occurred since genetic introgression. Dispersal movements have the potential to allow panthers to recolonize historic habitat outside of the confines of southern Florida. Success of this range expansion to the north will of course rely upon female dispersal, a topic that will encompass future research and management objectives (see below).

In concluding our discussion of genetic introgression, we must outline some caveats related to our preliminary findings. Making a direct correlation between the increase in the documented population count for the panther population (McBride *et al.* 2008) and genetic introgression is premature prior to completing final genetic analyses. Regardless, it is clear that the population today (2008) has improved reproductive output and population size in comparison to the smaller population in the early 1990s.

Current challenges to the persistence of panthers in Florida

Loss of habitat, highway mortality, and impact of increased panther–human interactions

The core panther habitat, defined as the area where panther reproduction occurs, is 919,000 ha (Kautz *et al.* 2006). Over 70% of this habitat is on publicly owned lands that are managed for multiple uses including public hunting, watershed restoration, nature viewing, and endangered species conservation. Most of these public lands are contiguous and include BCNP, EVER, FSPSP, FPNWR, and the PSSF (Fig. 21.2). Private lands used by panthers support a variety of economic activities such as agriculture (citrus, vegetable, and cattle production), forest products, and hunting leases.

Privately owned panther habitat occurs mainly in three Florida counties: Collier, Hendry, and Lee. The human population in these three counties has increased by 25.2%, 11.7%, and 29.6%, respectively from 2000 to 2006, with the addition of 197,977 new residents (US Census Bureau 2007). Collier and Lee counties are projected to increase from a combined populace of just over 900,000 to more than 1,600,000 by 2030 (Office of Economic and Demographic Research Florida Legislature 2008). This growth is putting pressure on the panther's remaining habitat as the human population footprint expands inland from the western Florida coastline. Development of panther habitat not only renders it unsuitable for panther use, but it can also lead to loss of functionality of adjacent habitat. For example, habitat corridors can become constricted by surrounding development, resulting in fragmented habitat that is unable to support panthers. Although <30% of core panther habitat is privately owned, habitat loss in this part of the panther's range could reverse the increases in population since the mid-1990s by severing connections with publicly owned lands. Kautz *et al.* (2006) reviewed several population viability models constructed for Florida panthers and found that populations of 80–100 panthers are stable over 100 years. However, if panther numbers fall to 60–70 animals, the population is barely viable for the same period, while a population of <50 animals faces a high probability of extinction. Even if the current (2007) documented panther count of 117 (McBride *et al.* 2008) remains stable for the next century, the loss of 30% of their habitat would compromise the long-term prospects for this population.

Unfortunately, one of the best indicators of the presence of panthers has been the discovery of individuals killed by vehicles (Fig. 21.4). This source of mortality resulted in the loss of 0–3 panthers per year between 1981 and 1995 ($\bar{X} = 1.3$, SE $= 0.21$). Although the number of highway mortalities appears low during that period, a loss of two panthers could have represented 10% of the 20–30 panthers that were thought to remain in south Florida. Ironically, the loss of panthers on Florida's roads helped raise awareness about the plight of panthers and their tenuous status in Florida.

The increase in the human population of southern Florida has resulted in higher levels of traffic on existing roads, as well as the construction of new roads, contributing to a greater risk for panthers when crossing roadways. Highway mortality continues to be a concern. Between 1996 and May of 2008, 85 panther highway mortalities were recorded (FWC, unpublished data; FWC 2006). Some of this increased mortality is due to an increase in the numbers of panthers: more panthers and dispersing animals moving away from tracts of public lands should result in higher vehicle mortality. Increases in traffic, particularly at night, may increase the risk to panthers. The FWC and its partners at the state and federal level have been proactive in attempting to mitigate vehicle mortality on roadways in south Florida by the construction of wildlife crossings, establishing speed zones, installing rumble strips, and public information campaigns. The first wildlife crossings were installed when Alligator Alley in Collier County was converted into I-75; a four-lane, high-speed interstate. These structures are dry-ground underpasses that allow wildlife to pass safely beneath the highway. Twenty-four wildlife crossings and 12 other underpasses modified for panther use were constructed within a 40-mile stretch of I-75. A continuous barrier fence directs animals to the crossings. Seven wildlife crossings have been installed under other south Florida highways. These crossings are used not only by panthers but also other wildlife, including white-tailed deer, bobcats, American black bears (*Ursus americanus*), American alligators, and many more (Lotz *et al.* 1997).

Although effective, wildlife crossings are expensive, with current construction cost estimates of US $4 million (2008) each, with associated fencing costs estimated at US$85/meter. Wildlife crossings are easier to integrate during the planning stages for new road construction or major road improvement projects. It is more costly to retrofit crossings into existing roads as stand-alone projects, and there is less money available for retrofitting. Florida panther numbers have increased in spite of the increase in panther highway mortality; this source of mortality cannot be totally eliminated, but panther managers are seeking construction of new crossings as funding opportunities arise.

The rapid development of south-western Florida results not only in loss of panther habitat, but it also brings humans closer to panthers. Consequently, human–panther interactions have recently increased, including panther sightings, encounters, and depredation of livestock and pets. Panthers are predators, and although there are no documented attacks on humans in Florida, attacks on people by western pumas have increased over the past two decades. From 1890–1990, there were 9 reported fatal attacks by western pumas and 44 non-fatal attacks, resulting in 10 human deaths (one attack involved 2 deaths) and 48 injuries (Beier 1991). From 1991–2003, western pumas killed 10 people in 73 attacks (Chester 2006). Given this trend, in 2004 the FWC, USFWS, and NPS established a Florida Panther Interagency Response Team to manage human–panther interactions while promoting public safety and assuring the continued existence and recovery of this endangered animal. This team of experts and agency representatives developed a Florida Panther Response Plan (USFWS 2006) that provides guidance to agencies dealing with human–panther interactions. Proactive public outreach is a part of the response plan, and these include developing information brochures, hosting town hall meetings, updating agency web sites, and posting signs at public use areas in panther habitat.

Increased likelihood of infectious disease transmission between panthers and domestic animals at the urban–wildland interface

Many problems result from the transformation of panther habitat into a densely populated, developed landscape. One is an increased likelihood that panthers on the outskirts of core panther range will rely on marginal and disturbed habitat along the urban–wildland interface. A greater opportunity for contact with humans and their domestic animals along this interface not only increases the incidence of panther–human interactions, but also the threat of infectious disease transmission between domestic animals and panthers.

Expansion of the urban–wildland interface may have contributed to an outbreak of feline leukemia virus (FeLV) in panthers between 2001 and 2004. Feline leukemia virus is a retrovirus found in about 4% of feral domestic cats in Florida (Lee *et al.* 2002). It is the most common infectious cause of mortality in this species; however, reports of FeLV in free-ranging, non-domestic felids have been quite rare. Florida panthers have been monitored for FeLV since 1978. Results were negative until the winter of 2002–2003, when two panthers tested positive for FeLV antigen (Cunningham *et al.* 2008). Additionally, retrospective ELISA antibody and PCR tests on archived serum and tissues showed evidence of exposure and latent FeLV infections, indicating that the introduction of the virus into the population may have occurred in 2001 (Brown *et al.* 2008; Cunningham *et al.* 2008). Three more panthers tested positive for FeLV in 2004, including a male thought to have been infected during an aggressive encounter with an infected radio-collared male. All infected panthers died within 5 months of their first positive sample collection. Causes of death were FeLV-related (anemia, septicemia; $n = 3$) or resulted from intraspecific aggression ($n = 2$). The source of the outbreak is believed to be an infected domestic cat (Brown *et al.* 2008). Once introduced, the virus is thought to have spread panther-to-panther (Cunningham *et al.* 2008).

A vaccination programme initiated in 2003 may have helped end the epizootic, although many factors likely contributed. Panthers in the Okaloacoochee Slough area (OK Slough State Forest; Fig. 21.2) and between this slough and the remainder of the population to the south were initially targeted for vaccination. As of 2007, it is estimated that approximately 40% of the living population has received at least one inoculation against FeLV, primarily through vaccination at capture (Cunningham *et al.* 2008). The last case of FeLV was diagnosed in July 2004, indicating the outbreak is probably over. However, the potential for reintroduction remains, as indicated by the recent finding of a domestic cat in the stomach of a necropsied panther killed by a car along the urban–wildland interface proximal to EVER (Fig. 21.6).

Another domestic cat disease that has been diagnosed in panthers is notoedric mange. *Notoedres cati* is a sarcoptiform mite that causes intense pruritis, inflammation, and secondary infections in domestic cats, non-domestic felids, raccoons, and other wildlife species, including pumas (Uzal *et al.* 2007). In 1992, an infestation was diagnosed in a neonatal kitten from the FPNWR (Maehr *et al.* 1995). Subsequently, in 1993, the kitten's dam was observed with clinical signs of notoedric mange and successfully treated. Notoedric mange is a potentially epizootic disease, especially in dense felid populations. Combined, these cases suggest that a panther population restricted to the confines of southern Florida will be continually challenged by infectious disease epizootics along the urban–wildland interface.

Implications for recovery and future research

Defining recovery and the hurdles associated with population expansion

Even with the growth of the population since the mid-1990s, the Florida panther remains one of the most endangered felids in the world. The extant breeding population of panthers is currently restricted to the historic range in southern Florida, south of the Caloosahatchee river (Fig. 21.2). Being a single population, the panther is at risk of catastrophes such as disease outbreak or sea level rise. A major criterion for achieving recovery and delisting is to attain three reproductively viable populations of panthers with 240 adult and subadult individuals each (USFWS 2008b). This population size is derived from a population viability analysis (PVA) that suggested populations with these panther numbers have a high probability of persistence, low extinction probability, an ability to retain 90% of their observed heterozygosity, and tolerate some habitat loss (Root 2004). Results of PVAs or any modelling exercise should be interpreted cautiously, since they rely on the quality of datasets and parameters selected for analysis. Achieving multiple populations of 240 panthers

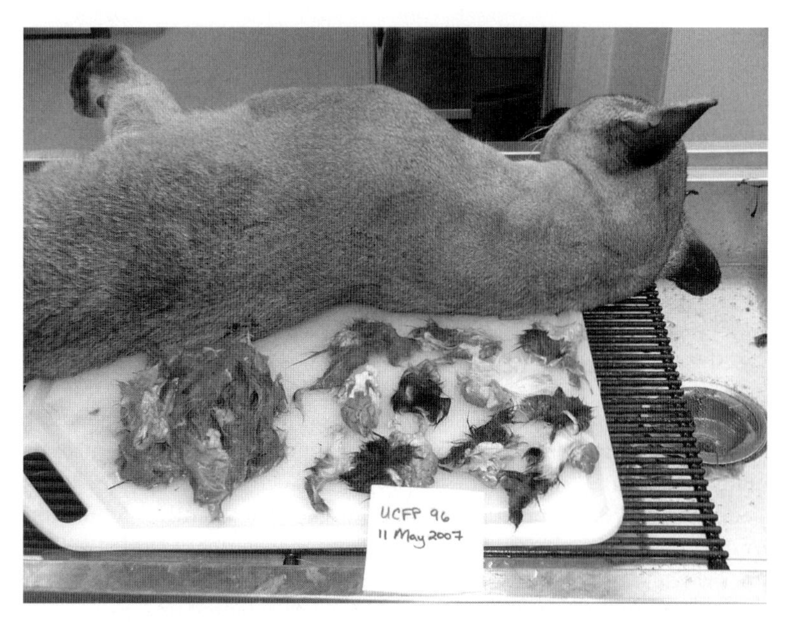

Figure 21.6 Remains of domestic cats (*Felis sylvestris catus*) found during a necropsy of a road-killed panther near EVER. Presence of domestic felids in the diet of panthers attests to the increased risk for disease transmission between the two species, especially along the urban–wildland interface. © Florida Fish and Wildlife Conservation Commission.

underscores the daunting challenge for researchers and wildlife managers.

A recent analysis of potential panther reintroduction sites within their former range revealed that sufficient habitat exists (Thatcher *et al.* 2006). Models were developed from a combination of available habitat and surveys of expert opinion. Results ranked Okefenokee National Wildlife Refuge (located in south-eastern Georgia), Ozark National Forest (western Arkansas), and Felsenthal National Wildlife Refuge (south-central Arkansas) with the highest likelihood of supporting panther populations. These combined the best expert model scores (46.9, 30.4, and 19.2, respectively) and effective habitat area (4112, 5270, and 8161 km^2, respectively; Thatcher *et al.* 2006). Unfortunately, socio-political obstacles such as opposition by some residents, public perceptions of large carnivores, and local political stances, currently make the probability of reintroduction into those sites unlikely. Nevertheless, public and political attitudes can change over time. A similar scenario has been witnessed in western Texas, where black bears were once eradicated, only to be subsequently tolerated and accepted after natural recolonization in the late 1980s (Onorato and Hellgren 2001).

A feasibility study (Belden and Hagedorn 1993; Belden and McCown 1996) revealed that panthers could survive in northern Florida, but would likely succeed only in large re-establishment areas ($\geq$2590 km^2) with low human density ($\leq$3.5 occupied housing units km^{-2}) and a sufficient population of white-tailed deer ($\geq$1 deer per 36 ha). The ever-expanding human population in Florida reduces the likelihood of proceeding with such an introduction. Combined, these findings demonstrate the challenges associated with animal reintroductions, especially large carnivores. History has revealed that these projects are frequently fraught with impediments, and have notably low success rates (Earnhardt 1999; Breitenmoser *et al.* 2001).

Natural recolonization can permit better opportunities for conserving endangered species. Characteristics of this process are described for several carnivores, including otters (Blundell *et al.* 2002), wolves (Valiere *et al.* 2003), brown bears (Waits *et al.* 2000) and black bears (Onorato *et al.* 2004; Hellgren *et al.* 2005). Natural recolonization often gives an endangered species more protection than that given to reintroduced populations, which the USFWS can designate as 'experimental'. This difference was evident in management and conservation plans for the non-essential experimental population of grey wolves in south and central Idaho versus the population naturally recolonizing the Idaho panhandle from Canada (US Federal Register 2005).

The successful recolonization of former habitat north of the Caloosahatchee river by a breeding population of panthers will depend upon the dispersal of females from core panther habitat into habitats to the north across the river. Recent (2007) survey data indicate that there are no female panthers occupying habitat immediately north of the river (R. McBride, personal communication) and historical survey records (1972–2007) do not document any female panthers north of the Caloosahatchee river since 1972 (R. McBride, personal communication). Although unlikely, the eventual dispersal by a female panther cannot be ruled out. The presence of adult and subadult male panthers indicates that panthers can survive on public and private lands north of the river.

Three radio-collared male panthers dispersed across the river between 1997 and 2000 (Maehr *et al.* 2002b). The movement path of one of these dispersing males has been described as 'frustrated' due to the circular movement pattern of his travels north of the Caloosahatchee river. Eventually, this male ranged >200 km before his return to habitat just north of the river prior to radio-collar failure (Maehr *et al.* 2002b). Recently (March 2007), another male monitored north of the Caloosahatchee river since 2004 was killed on Interstate 4 near Orlando, approximately 150 km north of his home range (FWC, unpublished data). Clearly, his home range north of the Caloosahatchee river provided sufficient habitat and prey. The question remains as to what factor caused these two male panthers to make extensive movements further to the north. We believe that the ultimate cause of frustrated movements was the inability of these panthers to locate breeding females within their home ranges north of the river, as opposed to the lack of sufficient habitat.

Recolonization of the panther's former range north of the Caloosahatchee river will not accomplish the USFWS recovery goal for additional populations of 240 animals (USFWS 2008b). Nevertheless, it could catalyse the expansion of the panther population from the confines of southern Florida and increase the current panther population. There are indications that the remaining panther habitat south of the Caloosahatchee may be saturated. Mortality from intraspecific aggression is the major cause of death in radio-collared Florida panthers (Fig. 21.7), a behaviour that has been attributed to population density in the Iberian lynx (*Lynx pardinus*; Ferreras *et al.* 2004). Logan and Sweanor (2001) assessed whether intraspecific aggression was correlated with puma density in undisturbed and contiguous mountain ranges in New Mexico. Their results showed no link between density and aggressive encounters between conspecifics, although the relevance of their findings to the unnaturally fragmented habitat of south Florida may be problematic. Regardless of how population density impacts the Florida panther, aiding natural dispersal by promoting habitat restoration and preserving corridors or relocating female panthers north of the Caloosahatchee River has the potential to help alleviate density-dependent stressors.

The presence of a reproductive population north of the Caloosahatchee River could benefit panthers with respect to meta-population dynamics. A subpopulation north of the river would not only increase the population size, but would decrease the probability that a catastrophe (e.g. disease outbreak) would significantly impact the population. Although the panther population in southern Florida cannot be defined as a traditional 'meta-population' (Hanski and Simberloff 1997), it can be considered as such under recent meta-population concepts that

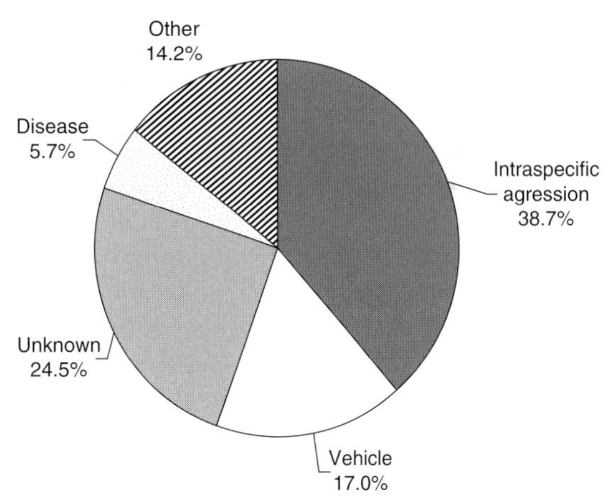

Figure 21.7 Causes of mortality in radio-collared Florida panthers between 1982 and 2007. Diseases that have resulted in the death of panthers include pseudorabies, rabies, and feline leukaemia. Uncommon sources of mortality (e.g. poaching, euthanasia, and liver failure) were included in the 'other' category.

accommodate the definitive characteristics associated with current panther population dynamics (Akçakaya *et al.* 2007). For example, the small, ephemeral 'sub-population' in EVER inhabits an area that only periodically experiences gene flow from across the Shark River Slough separating EVER from BCNP. While this natural barrier has the potential to increase the likelihood of inbreeding and reduce gene flow to EVER, it can also serve as a partial obstacle to epidemics that may affect the core population in south-western Florida. Similarly, a sub-population of panthers north of the Caloosahatchee River could serve as a buffer for the main population to the south through benefits afforded by this semi-permeable barrier between two sub-populations. Initiating or promoting recolonization north of the Caloosahatchee River will be complex and controversial. Nevertheless, this option now holds the most promise for assuring the continued persistence of the panther, as it could increase the population to a larger size than has been reached in decades.

Future research objectives

As research on the Florida panther prepares to enter its fifth decade, researchers are taking a retrospective look at the wealth of data that has been collected on this population while also delineating future research goals to assist in the recovery of the panther. Two current research priorities are a collaborative project to develop a PVA for panthers; and a comprehensive resource selection study based on nocturnal and diurnal locations collected by GPS radio-collars.

The conclusions of four previous PVAs developed for Florida panthers (Seal and Lacy 1989; Seal and Lacy 1992; Maehr *et al.* 2002a; Root 2004) have ranged from '100% chance of persistence for the next 100 years' to '100% chance of extinction in the next century'. A possible reason for the disparity may be that a 'consensus' approach was used to develop input values for some models, a technique that can result in the selection of values derived from opinionated discussions instead of the available data (Beier *et al.* 2003). We will complete a PVA that will use available data, consider the limits of these data, and avoid the use of subjective estimates of parameters. This process will involve parameter estimation and population modelling professionals, panther biologists, geneticists, and interested stakeholders.

The FWC has been deploying GPS radio-collars on panthers since 2002. Data collected by these GPS collars can be used to assess resource selection, movement rates, and predation patterns of panthers at a precision level and scale not possible with conventional VHF-telemetry datasets. A comprehensive GPS dataset inclusive of locations collected across the diel period from panthers throughout their current range could answer some remaining questions about panther habitat use. Combined, these two projects should assist in orienting research, management, and conservation initiatives for the immediate future.

Conclusion

Long-term projects in conservation biology and wildlife management are the exception as opposed to the rule. Much ground-breaking research has been conducted during the past 30 years on the Florida panther. Although some of this research has been controversial, all researchers that have been associated with varied aspects of panther conservation and management have contributed to the continued persistence of this endangered carnivore. The current (2008) population size is more amenable to recovery than was the dwindling population of the 1980s. That said, all FWC research and management activities focused on Florida panthers inevitably depend on the support of the people of Florida, who continue to do so with contributions to the Florida Panther Research and Management Trust Fund (FPRMTF). Floridians compiled over US$1,700,000 for the FPRMTF in the 2006/07 fiscal year. As we enter the twenty-first century, the panthers of southern Florida face many pressures, so public sentiment is of the utmost importance. Public opinion can drive policy. As researchers and managers, we must provide the most accurate and reliable information to the public and their elected representatives, in the hope of preserving and eventually recovering one of the last remaining icons of wilderness in the southern United States.

Acknowledgements

FWC's panther research and management activities are funded entirely by the FPRMTF. This trust fund is supported principally by the sale of Florida panther license plates. The FWC panther research and management group thank the citizens of Florida who continue to support our efforts to conserve and manage panthers through contributions to this fund. We would like to acknowledge the following individuals and agencies for their continued collaboration on varied aspects of panther research and management: Marc Criffield, Paul Kubilis, Mark Lotz, Rocky McBride, Rowdy McBride, Cougar McBride, David Shindle, Alexander Smith, Mario Alvarado, John Benson, Sonny Bass, Deborah Jansen, Annette Johnson, Lori Oberhofer, and Steve Schulze. We also extend our appreciation to the staffs at BCNP, EVER, FPNWR, FSPSP, PSSF, OK Slough State Forest, and the Big Cypress Seminole Indian Reservation for their continued cooperation and support of our research and management efforts.

CHAPTER 22

Reversing cryptic extinction: the history, present, and future of the Scottish wildcat

David W. Macdonald, Nobuyuki Yamaguchi, Andrew C. Kitchener, Mike Daniels, Kerry Kilshaw, and Carlos Driscoll

Scottish wildcat. © E.A. Macdonald.

Introduction

Following the demise of the lynx (*Lynx lynx*) by early medieval times (Hetherington *et al.* 2005), the Scottish wildcat (*Felis silvestris grampia*) is the only remaining indigenous felid in Britain. In a country that has also lost the wolf (*Canis lupus*) (1743) and bear (*Ursus arctos*) (*c.*1000–2000 years ago) (Kitchener and Bonsall 1999), the wildcat survives as an icon of Highland wilderness. Wildcats have been used in heraldry since the thirteenth century, including the Clan Chattan ('Clan of the wildcat'), among others (whose clan motto was, wisely, 'touch not the cat without a glove') (Clan Chattan Association 2008). This evocative, charismatic species is included in Schedules 5 and 6 of Britain's Wildlife and Countryside Act 1981 (Macdonald and Tattersall 2001), the highest legislative protection for a wild mammal in the United Kingdom. It also appears in Annex IVa of the European Habitats Directive 1992 and is identified as a European Protected Species in the Conservation (Natural Habitats, &c.) Regulations 1994, as amended (SNH 2008). More recently, it has been listed in the 2007 revision of the UK Biodiversity Action Plan (Biodiversity Reporting and Information Group 2007), and is an acknowledged priority of the relevant national authority, Scottish Natural Heritage, in their Five-Year Species Action Framework under criterion 1a ('Native species that are critically endangered in Scotland [or elsewhere, or demonstrating significant decline], or for which Scotland is a stronghold [including species that are only found there i.e. endemic] and there is a continuing threat to the species in the immediate future'; SNH 2007).

Research has revealed not only the precariousness of the Scottish wildcat's situation, and the urgency of restorative action (e.g. Macdonald *et al.* 2004b), but also that the predicament of this species illustrates, in microcosm, fundamental questions in both the science and ideology of species' conservation. These questions concern the definition of species and the consequences of their legal classification, the nature and consequences of introgression, the biological and social considerations in species restoration, the role of captive breeding, and questions spanning conservation genetics and ethics regarding the nature of extinction.

The initial decline of the Scottish wildcat has been attributed partially to habitat loss, in particular loss of forested areas, and partially to direct killing through hunting for its fur and persecution as a predator (Kitchener 1995; Macdonald *et al.* 2004b). The development of sporting estates in Scotland from the mid-nineteenth century increased persecution, leading to a further severe decline (Langley and Yalden 1977; Tapper 1992; Macdonald *et al.* 2004b). Carnivores often come into conflict with people (Macdonald and Sillero-Zubiri 2004a; Loveridge *et al.*, Chapter 6, this volume) and the Scottish wildcat is no exception. As predators on grouse moors they were, and still are, at risk from snares set legally for red foxes (*Vulpes vulpes*) and are often incidentally shot through 'lamping' (using a powerful lamp to pick out the eyeshine [reflection from the *tapetum lucidum*] to allow shooting at night) aimed primarily at foxes and feral domestic cats. In addition to accidental and/or deliberate mortality resulting from predator control, the Scottish wildcat is also threatened by habitat loss and fragmentation (Easterbee *et al.* 1991; Stahl and Artois 1994). The urbanized and industrialized Central Belt in Scotland is thought to act as a physical barrier to the southwards movement of wild cats (Easterbee *et al.* 1991), despite there being a large amount of suitable habitat in southern Scotland.

However, a more insidious threat facing the Scottish wildcat is that posed by the domestic cat (Macdonald *et al.* 2004b). Two important issues have been identified; the first is that wildcats suffer from a vulnerability to feline diseases transmitted by feral cats (Stahl and Artois 1994; Munsen *et al.*, Chapter 9, this volume). In Scotland, studies have shown that the wildcat population is commonly infected with the major viruses of the domestic cat, including the feline leukaemia virus (FeLV), which is readily transmitted among young cats via infected body fluids, such as during fighting or mating, and is almost always fatal (McOrist *et al.* 1991; FAB 2008) and may increase the mortality of wildcats (Fromont *et al.* 2000; Macdonald *et al.* 2004b). Other diseases found in the Scottish wildcat population include the feline calicivirus (FCV), which can cause 'cat flu' and affect the joints. Cats infected with both FeLV and FCV may develop pneumonia. In addition to FCV, the feline coronavirus (FCoV) and the feline foamy virus (FFV) have also been found in the Scottish wildcat population (Daniels *et al.* 1999). It is not known what effect these diseases may have on the wildcat population.

Secondly, and more importantly, indigenous wildcats are closely related to, and hybridize extensively with, introduced domestic cats, and the introgression between the two may result in cryptic extinction of the wildcat (Kitchener 1995; Rhymer and Simberloff 1996; Beaumont *et al.* 2001). As well as the fear that widespread hybridization could lead to genetic extinction of 'pure' wildcat populations (Suminski 1962), it can also lead to difficulties in identification of 'pure' wildcats (Daniels *et al.* 1998). This difficulty in identification not only makes surveys and population estimates difficult (Macdonald *et al.* 2004b), but also leads to legal difficulties because, in the United Kingdom, the Scottish wildcat is legally protected and the feral cat can be legally controlled (Daniels *et al.* 1998).

Therefore, the conservation scientist, striving to protect wildcats and thereby to conserve biodiversity, faces the question of whether it is possible to distinguish 'pure' wildcat specimens from hybrids. To biologists, accustomed to dealing with natural continua, and with judgments made on the basis of probabilities rather than certainties, these problems, while challenging, are not insuperable. However, in this particular case conservationists face a further cause for concern, namely that a recent proposal to add the wildcat/domestic cat hybrid to the Wildlife and Countryside Act 1981 was rejected (FQR 2002), so although wildcats are protected under the Wildlife and Countryside Act 1981, their hybrids are excluded. Consequently, while the likelihood that a wild-living cat that harbours a few genes of domestic ancestry might have little bearing on its biodiversity or evolutionary value, it could have a defining impact on whether it were to enjoy protection under the law as currently phrased. If the law is to achieve the purpose for which it was obviously drafted, that is, protection of the Scottish wildcat, then biologists must agree a way of defining and diagnosing those specimens which can be said, beyond reasonable doubt, not to be hybrids. Alternatively, legislation could be adapted to consider a definition based on geographical area as opposed to physical appearance. The logic behind this approach would be that wildcats were protected for where they lived and consequently how they behaved; that is, what they are (Daniels 1997; Daniels *et al.* 1998). However, there are many practical difficulties with this approach, including the risk that wildcats illegally killed could be moved outside the protected area to pretend that they had been killed legally.

Unless the latter approach is adopted by Scottish and UK legislation, we are stuck with working out the best way to protect cats based on what they look like; that is, physically defining the animal. We are far from being the first to be alert to the problem of defining the Scottish wildcat; Darwin (1868) raised the issue and illustrations in a succession of published natural histories make clear how long-standing is the uncertainty (Fig. 22.1).

Much of the debate on definition concerns taxonomy and systematics—topics some might think of as dustily academic and remote from the practicalities of conservation, whereas, as explained above, it is currently legally relevant as the basis for defining the Scottish wildcat. Another reason for being interested in the taxonomy and systematics of the Scottish wildcat is in order to shed light on the phylogenetic relationship between the wildcat and the domestic cat, the fundamentally interesting process of domestication, and how taxonomists deal with that.

Authorities differ in whether they regard Scottish wildcats and domestic cats as separate at the level of species or subspecies, and these alternatives have different legal consequences. For example, the offspring of a wildcat and a domestic cat would, if these two are different species, be defined as hybrids, which are offered no protection under current British law. While it is clear that wild-living cats remain in the Scottish Highlands, it is unclear to which extent some of them are the same animal as the Scottish wildcat of prehistory, and the task of distinguishing which, if any, of them is free from hybridization is biologically difficult. Our aim is to develop and implement a plan for the conservation and restoration of the Scottish wildcat, acknowledging that this involves partly separate issues at the levels of protecting individuals and of restoring viable populations. To underpin that plan, our research over almost two decades has addressed many scientific, ethical, logical, and logistical issues. In this chapter we discuss these issues, highlighting their wider relevance to wildlife conservation and suggesting the way ahead.

What is the wildcat?

The wildcat is a small cat belonging to the 'domestic cat lineage', which diverged from other felid branches ~6 million years ago (Ma) (Johnson *et al.* 2006b; Werdelin *et al.*, Chapter 2, this volume). The lineage consists of five closely related small cats that are naturally distributed across Africa and the Palaearctic: jungle cat (*F. chaus*), black-footed cat (*F. nigripes*), sand cat (*F. margarita*), Chinese steppe cat (*F. bieti*), and wildcat (*F. silvestris*).

The wildcat is one of the most widely distributed small cats, and its range extends from western Europe to western China and throughout the African continent (Stahl and Artois 1994; Nowell and Jackson 1996) (see Fig. 22.1). Traditional taxonomy divides the wildcat into three groups (European, Asian, and African wildcats) based on their morpho-geographical characteristics (Pocock 1951; Roberts 1951; Haltenorth 1953; Guggisberg 1975; Heptner and Sludskii 1992). However, a recent taxonomic revision based on molecular data indicates some changes, as discussed in the following text. Morphologically, the European wildcat typically has a bushy tail with a broadly rounded black tip, more-or-less conspicuous continuous stripes, and its distribution includes all but the most northerly parts of Europe, and Asia west

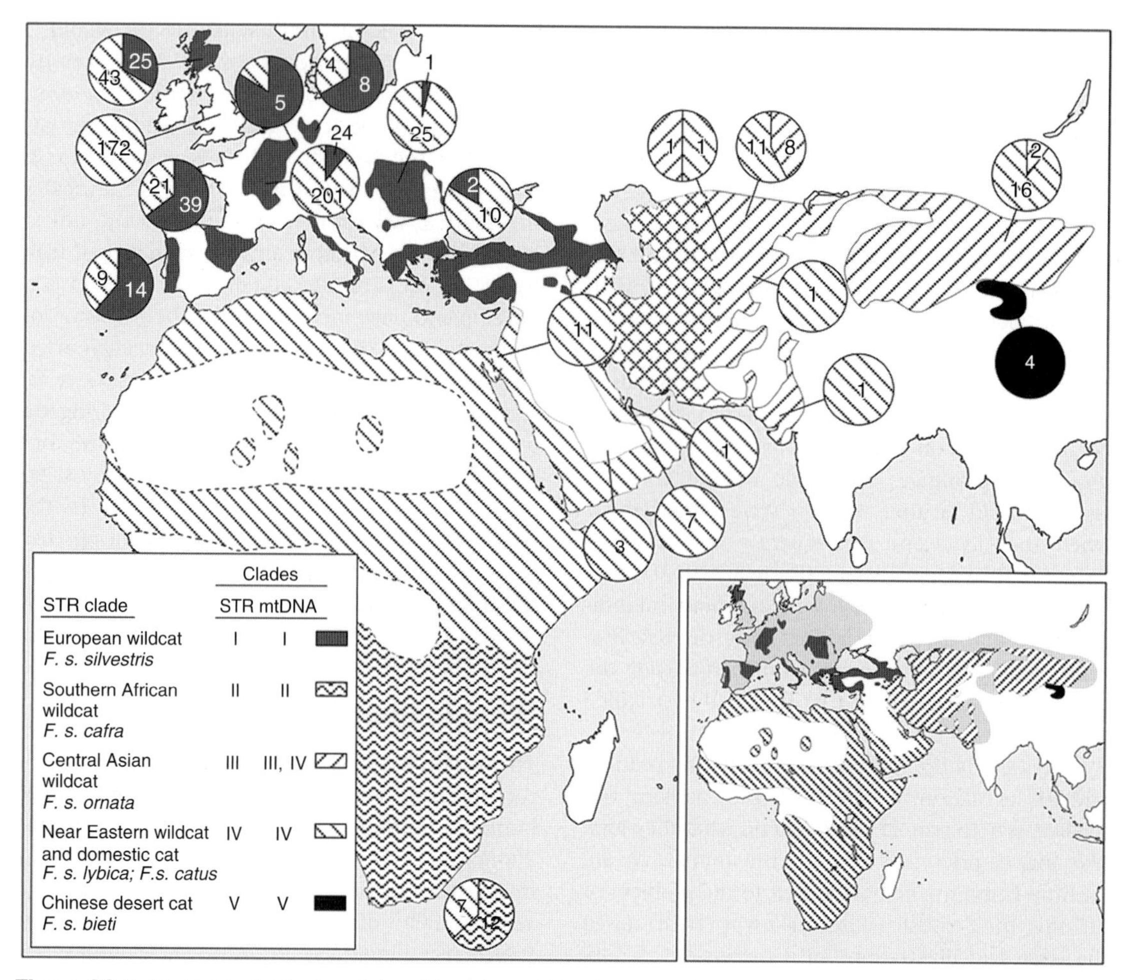

Figure 22.1 Worldwide distribution of the wildcat: its traditional classification and intraspecific phylogeny. (From Driscoll *et al.* 2007.) The wildcat is customarily divided into European, Asian, and African wildcats (see the smaller map). The main map shows the intraspecific phylogeny of the wildcat based on both mtDNA and nuclear STR (or microsatellite). Mitochondrial DNA haplotype frequencies are indicated in pie charts with number of specimens carrying each mtDNA haplotype clade.

of the Urals and the Caspian Sea. The African wildcat has a tapering tail and inconspicuous body stripes, while the otherwise similar Asian wildcat has a distinct spotted pelage. The African wildcat is found throughout Africa and Arabia, except in very arid deserts and tropical forests. Populations of wildcats on the Mediterranean islands (including Crete, Corsica, Sardinia, and the Balearic Islands) are often considered to belong to the North African wildcat lineage, but are probably early feral domestics (Vigne 1992; Kitchener 1998). Following the molecular work of Driscoll *et al.* (2007), it is noteworthy that previous references to the African wildcat should be split between the North African *F. s. lybica* and the sub-Saharan or southern African *F. s. cafra*. The Asian wildcat (*F. s. ornata*) is found in the arid regions and the steppes of western and central Asia, from Iran to India and as far east as Mongolia and western China. The ubiquitous domestic cat has traditionally been considered to have stemmed from the 'African wildcat' by *c.* 4000 years ago, probably in the Near East, and is now found virtually throughout the world (Fitzgerald 1988; Serpell 2000). However, there is no substantial evidence for Egypt or Egyptian culture playing any significant role in the initial domestication of the cat. In fact, the linked domestications of grain and cats had likely been accomplished in the Fertile Crescent some 4000–6000 years before even predynastic Egypt (Zeuner 1963; Clutton-Brock 1999).

To shed further light on the intraspecific phylogeny of the wildcat, we examined the skull morphology of wild-living cats from Britain and Europe (218 skulls), Central Asia (103), and southern Africa (145) (Yamaguchi *et al.* 2004c). All our analyses, based on 39 skull variables, placed British and European wildcats close together, and distinct from the Central Asian and southern African wildcats, which were also similar to each other (Fig. 22.2). British and European wildcats were characterized by proportionally shorter cheek tooth rows (Mean [Pm_3—M_1/ greatest length of skull] $\times 10^2$: British 21.60, SE = 0.13; European 21.82, SE = 0.12; Central Asian 23.02, SE = 0.10; and southern African 22.80, SE = 0.09) with smaller teeth (Pm_4 breadth/greatest length of skull) $\times 10^2$: British 3.24, SE = 0.02; European 3.21, SE = 0.02; Central Asian 3.74, SE = 0.02; and southern African 3.46, SE = 0.02) compared to those of Central Asian and southern African wildcats. Compared to the magnitude (fivefold on a measure called the Mahalanobis distance) of the differences between forest wildcats (= European wildcats) and steppe wildcats (= southern African and Asian wildcats), we could detect no substantial difference between the British wildcat and pooled continental European wildcats based on their skull morphology in spite of *c.* 7000–9000 radiocarbon years of separation. Those southern African wildcats were later found to form the *'cafra'* clade, distinct from the *'lybica'* clade (the direct ancestor of the domestic cat) based on the latest molecular phylogenetic research (Driscoll *et al.* 2007; see below). However, Driscoll *et al.* (2007) also found that they are nonetheless closer to each other than are they to forest wildcats, and therefore the foregoing arguments still hold.

Furthermore, to investigate the relationships between domestic cats, their indigenous wild progenitors and related species of the genus *Felis*, we developed a collection of tissue specimens from 749 individuals sampled across the range of the wildcat (Driscoll *et al.* 2007). The samples included wildcats and feral domestic cats on three continents, in addition to 54 domestic cats of various pedigree breeds, 12 sand cats, and 6 Chinese steppe cats. We extracted DNA from each specimen, sequenced 2604 bp of mitochondrial (mtDNA) *ND5* and *ND6* genes, and genotyped 36 variable short tandem repeat (STR) or microsatellite loci for each cat. The results offer a new understanding of the wildcat's intraspecific phylogeography. In particular, the results suggest that the Chinese steppe cat, which has been customarily

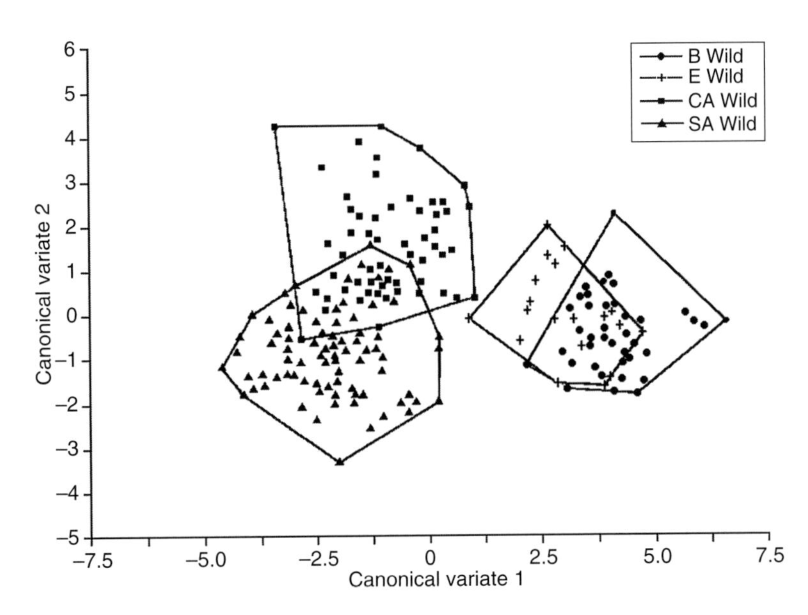

Figure 22.2 Morphological similarities among skulls of British, European, Central Asian, and southern African wildcats. (From Yamaguchi *et al.* 2004b.) The graph shows bivariate plot of the first two canonical variates based on six principal components extracted from 39 variables concerning skull morphology.

treated as an independent species, is phylogenetically so close to the wildcat that they could be considered the same species (Fig. 22.3) (the extent to which *ornata* and *bieti* are sympatric, or separated by altitude, merits research). These results also indicate that the wildcat of southern Africa should be recognized as *F. s. cafra* rather than *F. s. lybica*. Most evocatively, these results support the fragmentary archaeological evidence that the wildcat was first domesticated in the Fertile Crescent of the Near East, probably about 9–10,000 years ago, in what was evidently a single widespread event.

A mitochondrial clade, composed of both contemporary domestic cats and *F. s. lybica* sharing five deep matrilines, provides evidence of the incorporation of regional native wildcat matrilines, which are expected to significantly predate domestication, into the domestic cat gene pool as genes for domestication spread. This happened as those individual proto-domestic wildcats, fit for 'urban' life, incorporated mitochondria horizontally from geographically proximal wildcat lineages, perhaps aided by human selection for tameness and the transportation of these proto-domestic individuals. A similar progression is proposed in horses (Jansen *et al.* 2002).

So, rather than suppose multiple independent domestications, the most parsimonious explanation for the diversity and depth of the mitochondrial genetic variation found in modern domestic cats is that it stems from the spread of an initial domesticate which interbred with local wildcats. In sum, the evidence points to a single domestication 'event', although this 'event' was really a long process over a large area.

Evolutionary history of the wildcat

The modern wildcat is probably descended from Martelli's wildcat (*F. s. lunensis*), which is known from Europe and may date back as early as the late Pliocene *c.* 2 Ma (Kurtén 1965b, 1968; Kitchener 1991). Fossil remains suggest that the transition from Martelli's wildcat to the modern wildcat may have occurred during the middle Pleistocene, possibly by *c.* 0.45–0.35 Ma (Kurtén 1965b; Yamaguchi *et al.* 2004b). In comparison to this long and contiguous fossil record in Europe, wildcat fossils are recorded only from the late Pleistocene onwards (less than *c.* 130,000 years ago) in Africa (both north and south of the Sahara) and the Middle East, although we are unaware of any published references concerning fossil records of the wildcat from Asia (Yamaguchi *et al.* 2004b). If we interpret this lack of fossil record as the absence of the wildcat both in Africa and the Middle East, it is possible that the wildcat may have expanded its range suddenly, rapidly and, in evolutionary terms, very recently during the late Pleistocene. Our re-evaluation of Kurtén (1965a, b) suggests that late Pleistocene Palestine, less than *c.* 50,000 years ago, was inhabited by wildcats characterized by their proportionally larger cheek teeth, compared to those of both late Pleistocene and present-day wildcats in central and northern Europe and Britain. One of the explanations for the proportionally larger cheek teeth found in present-day Central Asian and southern African wildcats (compared to those of European and British wildcats) may be that these two share a common ancestor possessing that character in the geologically recent past. This suggests that Asian and African wildcats are both derived from a large-toothed wildcat such as the one that inhabited the Near East during the late Pleistocene (Yamaguchi *et al.* 2004b). This speculation, based on morphological analysis, is tentatively supported by our subsequent molecular work (Driscoll *et al.* 2007; also see Fig. 22.3).

The evolution of the modern wildcat probably consisted of at least three different range expansions. Firstly, by the late Pleistocene there was an exodus from Europe, which had been the centre for wildcat evolution for nearly 2 million years. The initial range expansion may have involved (some) earlier wildcat lineage(s) suggested by the relict Chinese steppe cat population in north-western China, which may have hybridized with expansions of later steppe wildcats during one or more glacial maxima (see Figs. 22.1 and 22.3) (a detailed morphological examination of the Chinese steppe cat would be helpful to shed further light on this hypothesis). Secondly, virtually all those early wild cats have been replaced by the wave of wildcats from the steppe lineage that appeared, relative to the forest wildcat lineage, more recently. However, as Asian and African wildcats possess consistently

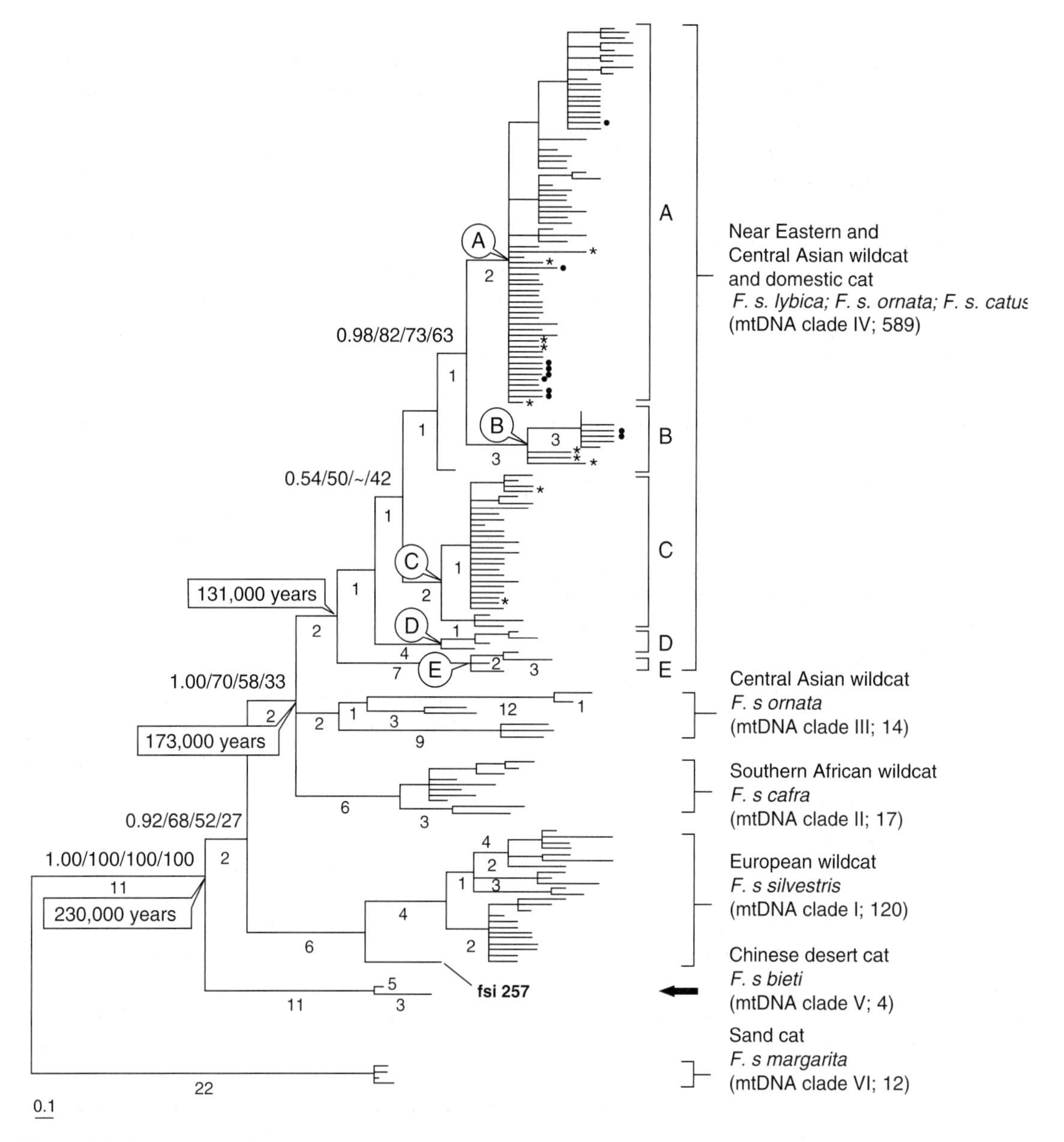

Figure 22.3 Phylogenetic tree (maximum likelihood) of mtDNA sequence (minimum evolution/neighbor joining phylogram of 2604 bp of the ND5 and ND6 gene) of 176 haplotypes discerned from 742 cats sampled across range of the wildcat. (From Driscoll *et al.* 2007.) Confidence/bootstrap values (Bayesian/maximum parsimony/maximum likelihood/minimum evolution) are adjacent to nodes. Bootstrap values are based on 1000 iterations. Numerical clade designations and number of individuals are indicated in parentheses following corresponding common name. Letters A through E designate lineages within mtDNA IV. Confidence/bootstrap values for these nodes are as follows: A, 1.00/87/71/54; B, 1.00/82/80/80; C, 0.97/63/59/42; D, 1.00/98/99/88; E, 1.00/100/100/82 (as above).

distinct coat patterns (e.g. Pocock 1951), this stage may have involved a series of expansions and contractions possibly affected by the late Pleistocene glacial–interglacial cycles or it may have resulted from the separation of these populations on two different continents (Yamaguchi *et al.* 2004b). This speculation is also tentatively supported by the more complex mtDNA haplotype structure in the western Asian wildcat populations in comparison to those in Europe and southern Africa (Driscoll *et al.* 2007; also see Fig. 22.1). Whenever and wherever they may have emerged initially, the domestic cat wave has, with human assistance, swept the earth, with the outcome (widely problematical for conservation) that domestic cats are among the most successful mammals on the planet.

Wildcat biology, taxonomy, and the law: an enigma wrapped in a riddle inside a conundrum

The latest phylogeographical analysis suggests that the wildcat (*F. silvestris*) consists of five subspecific groups, including the three traditional subspecies: the European or forest wildcat (*F. s. silvestris*), the Asian wildcat (*F. s. ornata*), and the African wildcat (*F. s. lybica*), with the additional recent recognition of the southern African wildcat (*F.s.cafra*) and the incorporation of the Chinese steppe cat into the species (*F. s. bieti*) (see Werdelin *et al.*, Chapter 2, this volume). However, there are more taxonomy-oriented views, and one of them may be that this clade represents three diagnosable phylogenetic species—*F. bieti, F. silvestris*, and *F. lybica* (including *ornata* and *cafra* as further subspecies)—which have radiated relatively recently. Although derived from the wildcat only recently, for practical reasons the domestic cat is often given a separate species status (*F. catus*), though some authorities assign it status as a subspecies of the wildcat as *F. s. catus* (e.g. Randi and Ragni 1991; Masuda *et al.* 1996; Driscoll *et al.* 2007). This leads back to a minefield with regard to whether or not the domestic cat is a species. In terms of biological processes and phylogeny, whether domestic cats are treated as a subspecies of *F. silvestris* or a separate species (*F. catus*) might seem arbitrary. However, these taxonomic niceties are of the highest operational importance because the current legislation intended to protect the wildcat is framed in terms that can be effective only if the wildcat is recognized as a separate species and, moreover, the legislation offers no protection to hybrids, which may nevertheless carry important wildcat genes. Further compounding the difficulty in finding a working solution, the *F. s. catus* nomenclatural construction is currently illegal under the ICZN's rules, since a species' name must be the one first published (i.e. the first published name has priority over later published names, which are then agreed to refer to conspecific taxa). Because the name *Felis catus* Linnaeus, 1758 predates *Felis silvestris* Schreber, 1777, the species name would have to revert to *Felis catus* (as it was in much of the nineteenth century) (and hence *F. c. silvestris, F. c. lybica*, and *F. c. ornata*), if they were regarded as conspecific. However, the usage of *F. silvestris* as the scientific name for the wildcat is conserved by the International Commission on Zoological Nomenclature, along with 16 other specific names concerning similar circumstances (ICZN 2003). (This strange case, where the colloquial name is actually more precise and informative than the scientific name, must set a precedent.)

In the rarefied debate between taxonomic precedence and legal precision, if the domestic cat is classified as a subspecies of *F. silvestris*, then arguably it would enjoy just as much legal protection as does the wildcat. In that case, perversely, law framed with the intention of protecting wildcats would also protect one of the greatest threats to them, feral domestic cats! A worse complication could arise if it was decided that the species' name of both the animals in question was officially *Felis catus*, the name generally reserved for the domestic cat (and not, after all, *Felis silvestris*, the name used in the Act for wildcats), thus awkwardly making wildcats an official target of eradication. Similar legislative difficulties concerning wildcat conservation exist internationally and elsewhere in its range countries. The Convention on International Trade in Endangered Species (CITES) lists all felid species, except the domestic cat, on Appendices I or II, thereby restricting or regulating their international commercial trade (CITES 2008). It is feared that classifying the wildcat as conspecific with domestic cats would effectively delist it from

CITES, opening up a potential 'free for all' in international trade for the wildcat. That said, since the collapse of the Soviet Union there is essentially no trade in the wildcat (Nowell and Jackson 1992), which had previously been traded for fur. Further enlivening this theatre of the absurd, legislation passed within the last decade in the United States, European Union (EU), and Switzerland effectively offers the domestic cat more protection from economically motivated trade than the wildcat, by banning the importation, exportation, manufacture, or sale of domestic cat fur (e.g. Federal Trade Commission 2000; Defra 2007), while allowing all the same for the wildcat. A further unhappy quirk is that no CITES import permit is required of any country for Appendix II animals, such as *F. silvestris* (CITES 2008), so if, for example, China were to export cat jackets (whether, wildcats, desert cats, or domestic cats) as *F. silvestris* spp., no CITES license would be required to import them, though export permits are required and individual national laws may have some bearing.

Both formal taxonomy and legal definitions laudably aim to reflect accurately the messy biological reality that evolution presents; however, these aspirations are increasingly overwhelmed by the recent ever-accelerating accumulation of molecular phylogenetic findings using not only modern DNA, but also ancient DNA, techniques. Two views might be characterized, respectively associated with two views of what constitute species, the Biological Species Concept (BSC) and the Phylogenetic Species Concept (PSC), which we will explain later in the chapter. As a brief introduction, one view is that domestication is a speciation process, and this process may have gone so far as to qualify the domestic cat as a species. This view, which draws on the PSC, might be convenient under current British law, by making it clear that *F. catus* is not protected, whereas *F. silvestris* is. However, this consolation could be short-lived in so far as we have a unique and fixed diagnostic mtDNA molecular marker that defines Scottish wildcats that, under the PSC, could require the Scottish wildcat to be renamed *Felis grampia*, a species not mentioned under the law. This would clearly be absurd. Similarly, if, for example, *lybica, silvestris, bieti*, and *catus* were all elevated to species status, how would they fare under law currently specific to *silvestris*? Another view is that it is not appropriate to force domesticated forms into a scheme intended to interpret natural selection, because doing so opens the door to absurdities, by extension, of considering each breed of cat a potential species under the PSC. The latter view would hold domestics apart and designate their taxonomy differently to other animals, since they did not evolve under natural selection. This view would have domestics and wildcats as the same species under the BSC (which is the concept enshrined in most conservation laws, although inconveniently at odds with the current wording of the British Act protecting wildcats).

Most readers will probably be relieved to know that these debates (BSC vs. PSC) are so far beyond the scope of this chapter to resolve that we will soon leave them (although excellent chapters, especially 5 and 10, in Wheeler and Platnick [2000], shed light on the debate). As biologists, we recognize the difficulty of trying to force biological reality into the somewhat artificial concept of formal species, and then to compound the problem by further squeezing the outcome into the absolutely artificial legal categorizations required for practical conservation. But as conservationists we are obliged to wade into the morass to retrieve the best available pragmatic outcome. So, putting aside legal imperatives for the time being, we will briefly review the taxonomic arguments.

Taxonomy is dynamic, as species concepts and consensus change, yet while species concepts are crucial to many biodiversity issues (much of conservation, biodiversity studies, ecology, and related legislation concern this taxonomic level), even biologists have failed to agree on a single species definition (Mallet 2007). The most widely accepted concept is the BSC (Mayr 1970), in which species are defined as 'groups of interbreeding natural populations who are reproductively isolated from other such groups', for example, potentially interbreeding populations. Although generally accepted immediately after being developed by Mayr and Dobzhansky (Dobzhansky 1937; Mayr 1942, 1970), and despite being the most commonly used working definition, the BSC suffers criticism regarding its practical application. For example, the application of the BSC may be problematic because of interspecific hybridization between clearly delimited species (Skelton 1993) due to its heavy emphasis on reproductive compatibility. The BSC has been challenged by many other concepts (see Mallet 2007). For example, the PSC (Hennig 1966; Cracraft 1989) defines species as 'an irreducible (basal) cluster of organisms, diagnosably distinct from other such clusters, and within which there is a parental pattern of ancestry and descent' (Cracraft 1989). Unlike the BSC, the PSC almost always recognizes extensively hybridizing, diagnosably distinct populations as separate species (Cracraft 1997). A third and similar contender is the concept of Evolutionary Significant Units (ESUs; Simpson 1951; Ryder 1986), which conceives of 'a lineage (ancestral-descendant sequence of populations) evolving separately from others and with its own unitary evolutionary role and tendencies'. General applicability is a primary benefit of the ESU due to its 'fuzzy' meaning: the term can apply to species, subspecies, populations, stocks, or races. The ESU is the facultative unitary definition used in application of the US Endangered Species Act, though the Act is based on the principle of the BSC.

Each of these concepts differs in specific ways and so has specific consequences for the case of the Scottish wildcat. For instance, under the BSC wildcats and domestic cats are the same species because they often and freely interbreed without human intervention producing viable offspring that then go on to mate with either of the parent groups. Thus, by the very fact that they 'hybridize', domestic cats and wildcats are the same species. The BSC provides for recognition of subspecies, mating between which is properly termed 'intercrossing' (whereas hybridizing refers to the mating of two species). Difficulties with applying the BSC occur when 'good' species produce fertile offspring.

In contrast, the PSC does not recognize subspecies (or at least not as much as does BSC), but holds that a group is a species if distinguishable ('Where they exist, well-defined, diagnosable "subspecies" should simply be called species'; Wheeler and Platnick 2000, p. 134), and so this concept views domestication as a speciation process. Because domestic cats are diagnosably distinct from wildcats on such characters as cranial volume, the PSC recognizes the two as distinct species. A question therefore arises, that if the domestic cat is a distinct species under the PSC, rather than a subspecies, then are the other 'subspecies' also 'good species'? With the coming of molecular taxonomy, adherence to this conceptual view would lead to a proliferation of species since researchers can find diagnostic characters in almost every taxon they investigate. For instance, Asiatic and African lions would be distinct species because they differ in a fixed and diagnostic manner at a single genetic locus. As ESUs, European wildcats and domestics are considered as separate units due to their distinct histories and behavioural divergence, without specifying the threshold of divergence or distinction between them. Universal application of the ESU concept is hampered by this very flexibility, which therefore requires a consensus.

Against what may be intellectually diverting machinations of implementing Britain's Wildlife and Countryside Act 1981, the biological realities for the Scottish wildcat remain sombre. It is worth mentioning that in a broader context conservationists have started to recognize merit in attaching priority to hybrids under certain circumstances. For example, in 2000, the Parties to CITES adopted a resolution which specifically makes it possible to include hybrids under the protection of the CITES Appendices, if they form distinct and stable populations in the wild, or if they have a recent lineage (going back four generations) that includes a species in CITES Appendix I or II; the wildcat is in Appendix II. Similar proposals were advanced in the United States (US Federal Register 1996) with a view to aiding the recovery of listed species (eliminating intercross progeny where they interfere with conservation efforts for a listed species, but fostering intercrossing if this was a means of preserving the remaining genetic material of a listed species). However, important as these advances in conservation thinking may be, for the foreseeable future the law can offer no protection to hybrid Scottish wildcats, and relies on expert opinion agreeing on what constitutes the wildcat as a species. That being said there remains scope for testing and developing this legal definition, for example, the 'area-based approach', under existing quinquennial review mechanisms.

In short, the problems of definition are not fundamentally biological, but procedural and legal; if the current law does not recognize the relevance of subspecies, then it should be changed to do so, and thus reflect evolutionary reality. In the meantime, if calling domestic cats *F. catus* enables the purpose of the law to be fulfilled in Britain, then so be it.

Wild cats in Britain

Palaeogeological, morphological, and genetic evidence suggest that the present Scottish wildcat is the descendant of continental European ancestors (fossil evidence at Thatcham in Berkshire suggests that the wildcat was already in Britain by *c*. 9600–10,050 radiocarbon years ago; Coard and Chamberlain 1999), who were isolated in Britain as the sea level rose, after the last glacial maximum by ~7000–9000 radiocarbon years ago (Yalden 1982; Yamaguchi *et al*. 2004b; Driscoll *et al*. 2007).

By 1800, the distribution of the wildcat was already restricted to northern England, Wales (where it was scarce) and Scotland (Langley and Yalden 1977), and, as previously mentioned, its decline has been attributed partly to habitat loss (reduction of forest cover) and partly to hunting for its fur and sport, and persecution. With the rise of the sporting estates, particularly in Scotland from the mid-nineteenth century, and of the gamekeeper, the decline of the wildcat (alongside many other carnivores) continued, and by *c*. 1850 it had almost disappeared from west central Wales and was only 'hanging on' in Northumberland (Langley and Yalden 1977; Tapper 1992; Kitchener 1998). By 1880, the wildcat survived only in Scotland, and by 1915 it was believed to be restricted to the north-west of the Highlands (Langley and Yalden 1977). Decline of the wildcat is thought to have abated during the First World War with the associated decline of gamekeepers (Tapper 1992), and in response to a significant reduction in persecution, apparently wildcat numbers rapidly increased. Recolonization was aided by reforestation by the Forestry Commission (established in 1919) and reportedly by the 1950s the wildcat had recolonized most of the area north of the central belt of Scotland (Jenkins 1962; Kitchener 1995).

In Scotland, a distribution map based largely on a questionnaire survey carried out in the 1980s (1983–87) was produced by Easterbee *et al*. (1991) based on 499 units of 10 km^2, and more than 400 people supplying information. Easterbee *et al*. (1991) suggested that wildcats were restricted to the area north of a line between Edinburgh and Glasgow, and the main populations appeared to occur in the east, north-east, and south-west. In the 1990s (1992–94), Balharry and Daniels (1998) carried out a further survey based on carcass location and live-trapping. Their results suggested that the distribution of wild-living cats, including some wildcat, was more restricted than that suggested by Easterbee *et al*. (1991) and confined more to the east of the country. However, these results are likely to be biased by distribution of traffic and keepers. Harris *et al*. (1995), extrapolating from distribution estimates and radio-tracking data, suggested a total pre-breeding population in Scotland of 3500 wildcats. Estimating total population based on home range data in Daniels (1997) for the same distribution given by Harris *et al*. (1995) produces an estimate of 4200. However, these results, as well as those of the historical records, should be interpreted with caution, as the authors themselves highlighted the fact that many respondents (and in fact researchers as well) were unsure how to distinguish between wildcats, feral domestic cats, and their hybrids. Therefore, only an unknown proportion of these wild-living cats may have met the criteria we will propose below as diagnostic of wildcats. By the same token, many other 'hybrids' containing wildcat genes may have been discounted.

Scottish wild cats—the last British wildcat with an identity problem

Although originally classified as a separate Scottish species, *Felis grampia*, by Miller in 1907 (and later reclassified as a subspecies *Felis silvestris grampia*, Miller 1912a; Pocock 1951), the modern view is that, despite their possible identifiable genotype (Pierpaoli *et al*. 2003; Driscoll *et al*. 2007), it is currently unclear if there is any taxonomic distinction between the Scottish wildcat and the wildcats in continental Europe, or between any of the European wildcats (Randi and Ragni 1991; Wilson and Reeder 1993), although there is a clear cline in pelage markings from distinctly striped wildcats in the west through to virtually unstriped wildcats in the east (A.C. Kitchener, unpublished data). However, the source of a major logical difficulty concerning the identity of the contemporary Scottish wildcat is that domestic cats had been brought to Britain long before zoologists started to study them.

It is traditionally assumed that the domestic cat was spread by the Romans, reaching Britain *c.* 2000 years ago (Clutton-Brock 1987; Serpell 2000). However, archaeological evidence suggests that the domestic cat may have been in Britain far longer, with their bones recovered from Iron Age (3200–2586 years ago) and Late Iron Age (2400–2100 years ago) sites (Smith *et al.* 1994; O'Connor 2007) (domestic cat bones have also been recovered from Portchester castle, Hampshire and Gussage All Saints, Dorset). The type specimen of *Felis grampia* kept at the Natural History Museum, London (register number BMNH 4.1.25.3) was collected in 1904, at least 2000 years after the two forms had been in potential sympatry. In this context, any historical account of the wildcat in Britain, as well as anywhere else, must be interpreted with caution, owing to the difficulty of distinguishing hybrids and the absence of evidence regarding both the spread of feral domestic cats and the extent of historical introgression (Daniels *et al.* 1998; Daniels and Corbett 2003). This difficulty is emphasized by the writings of early naturalists. For example, in 1790 Hodgeson and Bewick wrote 'The domestic Cat, if suffered to escape into the woods, becomes wild, and live on small birds and such other game as it can find there; it likewise breeds with the wild one.' In the absence of wildcat specimens prior to the arrival of the domestic cat, all 'wildcat' specimens including the type specimen cannot logically evade the possibility of previous introgression.

Is there something out there?

Do Scottish wildcats actually exist? A range of morphological and genetic characteristics has been proposed to discriminate between wildcats, domestic cats, and their apparent hybrids, including pelage patterns, body measurements, gut lengths, skull characters, and both mtDNA and nuclear microsatellites (see Macdonald *et al.* 2004b for an overview). However, previous authors have attempted to identify contemporary specimens of pre-Iron Age wildcats (tacitly assuming that individuals matching this description still exist), classifying cats with some but not all of the characteristics assigned to wildcats as 'hybrids' (e.g. Suminski 1962; Schauenberg 1969, 1977; Corbett 1979; Ragni and Randi 1986; French *et al.* 1988; Randi and Ragni 1991; Jefferies 1991; Kitchener and Easterbee 1992; Fernandez *et al.* 1992; Hubbard *et al.* 1992; Randi *et al.* 2001). Considering the possible contact with domestic cats for 2000 years (or more), as well as the lack of information on pre-Iron Age wildcats, such an assumption should not be made casually; nonetheless although hybrids are pervasive (see below), introgression has not erased all evidence of the autochthonous wildcat.

We employed a different approach by quantifying the variation found in the contemporary wild-living cat population, and then interrogating that variation to identify statistically distinguishable categories. Interestingly, although we did not assume the existence of any group (such as wildcat, domestic cat, or hybrid), our analyses have repeatedly detected a consistent pattern among the wild-living cats in Scotland. The pattern is that some wild-living cats are more similar (others less similar) to the domestic cat, on the bases of the examined characters (Fig. 22.4). The emergence of the same pattern again and again has strongly encouraged us to believe that there are still some wild-living cats (those furthest from domestic) in Scotland that cannot be dismissed as mere feral domestic cats.

Defining the Scottish wildcat

Under the circumstance, where no reference characteristics (of the pre-Iron Age wildcat) are known, the most straightforward answer may be to define as the wildcat those whose characteristics are furthest from those of the domestic cat (Reig *et al.* 2001). However, each of the different methods, both morphological (French *et al.* 1988; Daniels *et al.* 1998; Reig *et al.* 2001; Yamaguchi *et al.* 2004c; Kitchener *et al.* 2005) and genetic (Beaumont *et al.* 2001; Daniels *et al.* 2001), identifies a somewhat different non-domestic cat grouping. For example, on the basis of data from eight or nine nuclear DNA microsatellite loci, Beaumont *et al.* (2001) and Daniels *et al.* (2001) found

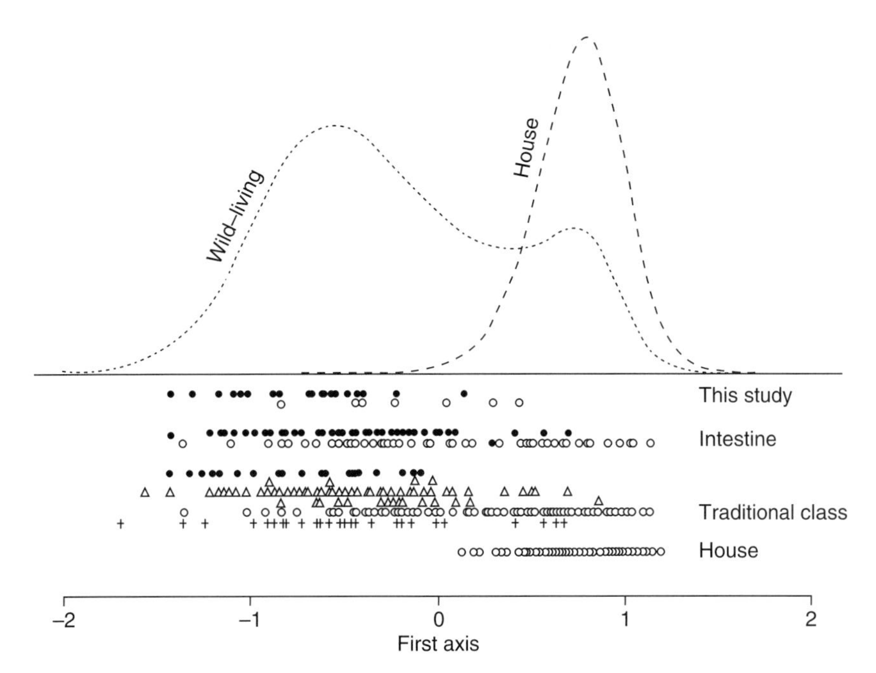

Figure 22.4 Estimated density curves for the score of individual cats along the first axis of principle components analysis of allele-sharing differences. (From Daniels *et al.* 2001.) Scores of individual cats are plotted beneath density curves: cats from Daniels *et al.* (2001) labelled 'this study' are tabby (back circles) or non-tabby (white circles); cats from Daniels *et al.* (1998), here labelled as 'intestine', are Group-1 (black circles) and Group-2 (white circles); 'traditional class' after Beaumont *et al.* (2001) are wild cats (black circles), hybrids (white triangles), domestics (white circles), and unclassified (crosses), based on the traditional pelage. House cats ('house') are followed; Beaumont *et al.* (2001).

many wild-living cats were clearly genetically differentiated from pet domestic cats (thus suggesting the existence of two genetic groups: 'non-domestic' and 'domestic') (Fig. 22.5). However, when some of the animals were classified as either 'domestic' or 'non-domestic' based on other characteristics, including tabby and non-tabby coat patterns (Daniels *et al.* 2001), the sizes of limb bones and intestines (defined as Group-1 [less domestic] and Group-2 [more domestic]; Daniels *et al.* 1998), and five pelage characteristics traditionally associated with wildcats (Beaumont *et al.* 2001), it became clear that there was no dyad which produced the identical classification (Fig. 22.5). This would be expected in a heavily introgressed population. Of 64 cats in museums that were classified as wildcat on the basis of 39 skull parameters, 27% were classified as either hybrid or domestic on the basis of traditional pelage, whereas of cats classified as wildcats on the basis of traditional wildcat pelage, all save 8% displayed wildcat skull characteristics (Yamaguchi *et al.* 2004c). In short, 'wildcats' with traditional pelage largely have furthest-from-domestic characteristics in terms of their skulls (Yamaguchi *et al.* 2004c), pelage markings (Kitchener *et al.* 2005), gut and limb bone lengths (Daniels *et al.* 1998, 2001), and genetic characteristics (Beaumont *et al.* 2001; Daniels *et al.* 2001).

The incongruities between different systems for classifying cats are schematized in Fig. 22.5. Cats at the centre qualify as furthest-from-domestic by all criteria; they have traditional pelage markings, and their genes and skulls are at an extreme from the domestic values, and they are also partly Group-1 cats, as defined by Daniels *et al.* (1998). One might define the wildcat as those that are furthest-from-domestic on all criteria (the inner circle in Fig. 22.5) and diagnosed by their 'traditional' pelage.

At this point it would be helpful to be precise about 'traditional' wildcat pelage patterns—recognizing that there are no specimens enabling us to

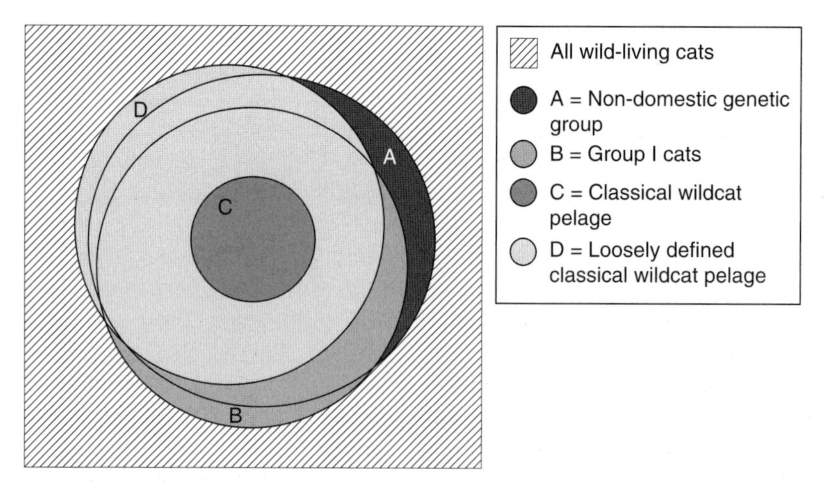

Figure 22.5 Possible proportions of qualified wildcats among all wild-living cats in Scotland defined by various criteria. The rectangle represents all wild-living cats, circle A represents non-domestic genetic group (Beaumont *et al.* 2001), circle B Group-1 cats (Daniels *et al.* 1998), and circle C traditional wildcat pelage (Kitchener *et al.* 2005). The circle D represents the relaxed definition based on the pelage. They overlap, but no dyad is identical.

check what the pelage characters of Scottish wildcats were like before possible introgression. That being said, assuming that the 'traditional' (i.e. Victorian) portrayal of a wildcat's pelage has merit, we have devised an easy scoring system to identify cats with and without 'traditional' wildcat pelage markings (Kitchener *et al.* 2005). The analysis indicated that we had to examine only seven pelage characters (7PS: extent of dorsal line, shape of tail tip, distinctness of tail bands, broken stripes on flanks and hindquarters, spots on flanks and hindquarters, stripes on nape, and stripes on shoulder: Fig. 22.6). We found that, on the basis of the variation in the total scores for the 7PS (each scored either 1, 2, or 3), it is possible to define the Scottish wildcat and, under current knowledge, such cats likely possess little or no recent domestic-cat ancestry (Beaumont *et al.* 2001, Kitchener *et al.* 2005). We propose that any cat that has a 7PS score of 19 or greater, and no scores of 1, should be regarded as a wildcat, unless there is evidence to the contrary. This diagnosis can be supported by assessing the eight additional pelage characters (white on chin, stripes on cheek, dark spots on underside, white on flank, white on back, colour of tail tip, stripes on hind leg, and colour of the back of ear), for which a wildcat should not have a score of 1 for any of these characters.

At a molecular level, Driscoll *et al.* (2007) revealed that clade-IV (see Fig. 22.3) is a large clade of related mtDNA haplotypes that comprise wildcats in the Near Eastern Asian/North African region, cosmopolitan domestic cats across the world, including Asia, Europe, the United States, and representatives of 33 domestic cat breeds. That Near East Asian (clade-IV) wildcats and domestic cats share a mtDNA kinship implies that domestication probably occurred in the same region as clade-IV wildcats live today (Fig. 22.1). Furthermore, since clade-I (see Fig. 22.3) includes the majority of wildcats in Europe, then putative wildcats in Europe carrying clade-IV mtDNA probably derive from hybridization of indigenous wildcats with domestics in the region. However, at a practical level for most purposes the 7PS system provides a practical diagnosis for the Scottish wildcat.

Inconvenient truth

The Scottish wildcat appears to be highly introgressed with the domestic cat. It is, therefore, difficult to combine the aspirations to find a legally operable basis of defining, and thus protecting, individual specimens and also provide a means of

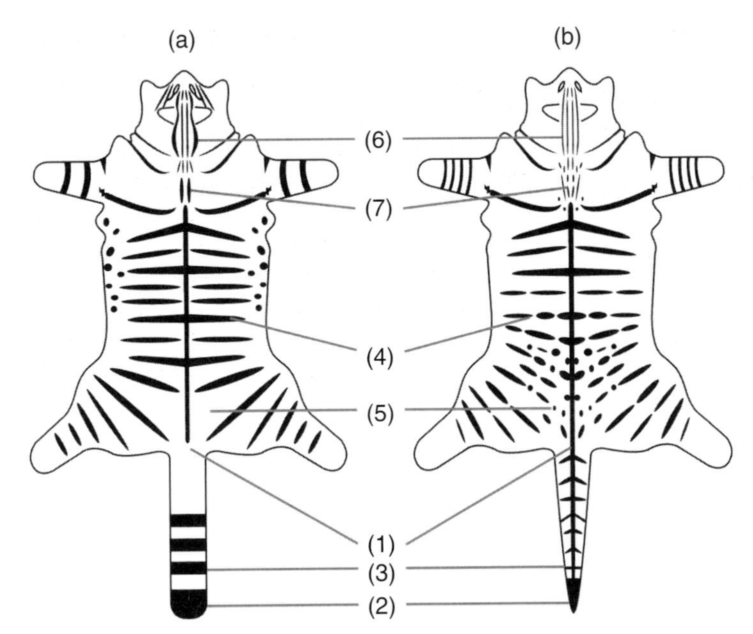

Figure 22.6 Seven pelage characters useful to diagnose a free-ranging tabby cat (b) from a Scottish wildcat (a) (Kitchener *et al.* 2005). Numbers correspond with (1) extent of dorsal line; (2) shape of tail tip; (3) distinctness of tail bands; (4) broken stripes on flanks and hindquarters; (5) spots on flanks and hindquarters; (6) stripes on nape; and (7) stripes on shoulder.

conserving the potentially valuable biodiversity within the introgressed population. Indeed, the inconvenient reality is that these two goals cannot practically (and within the existing legal framework) be achieved in one step alone. In terms of internally consistent criteria for the protection of individual specimens under existing law, there is a risk that a stringent definition based on the 7PS may define the Scottish wildcat to the brink of extinction, although this may be the reality of the situation.

Among the wild-living cats collected throughout Scotland in the early 1990s, *c.* 60% (out of *c.* 200) belonged to the non-domestic genetic group (Beaumont *et al.* 2001), *c.* 45% (out of 152) met the Group-1 criterion (Daniels *et al.* 1998), and *c.* 12% (out of 187) possessed traditional wildcat pelage (Daniels *et al.* 1998). In the early 1990s, Harris *et al.* (1995) estimated a total of 3500 wildcats in Scotland. For the sake of illustration we will suppose that they were wild-living cats similar in variation to the wild-living cats analysed in the studies reported above. If so, and asking how many of these 3500 would meet the definition according to the non-domestic genetic group, Group-1 criterion, skull characters, or traditional pelage, the answers would be ~2100, 1600, 500, and 400, respectively. By any definition it seems that at least *c.* 40% (and possibly as much as *c.* 90%) of wild-living cats in Scotland fail to qualify and, therefore, represent a divergence away from the wildcat (the flip side of the logic that defines a wildcat as furthest from a domestic cat). Clearly, these estimates of the Scottish wildcat population rest on many untested assumptions, and arguably assume the worst case; we certainly do not present these figures as precise. However, they illustrate our grave concern that the conservation status of the Scottish wildcat may be much more precarious than generally supposed. Furthermore, introgression may have worsened the situation in the intervening decade since Harris *et al.*'s estimate, although we have little information on the progress and dynamics of hybridization. In comparison to the Scottish situation, an estimated *c.* 90% of wild-living cats apparently possess traditional wildcat pelage in Bulgaria (Spassov *et al.* 1997), a situation we can only envy.

Discussing conservation of the Scottish wildcat based on a stringent definition based on 7PS might have tactical advantage in attracting attention to its plight, but could be counterproductive if it led to disadvantaging all other less strictly defined specimens. The more strictly the Scottish wildcat is defined, the rarer it will be, and therefore, paradoxically, the more important to its conservation is likely to become the biodiversity locked up in the bodies of cats that *just* failed to qualify. If it were possible to conserve the Scottish wildcat simply by protecting those specimens meeting strictly defined criteria, then the task of conservation would be greatly simplified. However, in practice, this strikes us as implausibly optimistic. Moreover, it remains to be seen whether cats just failing to qualify for a more strictly defined wildcat have any conservation value; does their continuing existence within a population enhance the survival of the wildcat, or do they contribute to continued genetic dilution? A greater understanding of the progress and dynamics of introgression is required in the context of a population genetics approach in order to answer this question.

Practical solution: a relaxed definition

We are mindful that for the definition of the Scottish wildcat to be useful, it must be operable not only in museums and zoos, but also in courts of law; for prosecution cases, access to pelage, skull, gut, and tissues for DNA analysis may all be feasible. However, an extreme interpretation of the stringent definition based on traditional pelage could lead to the conclusion that *c.* 90% of all the wild-living cats should be removed from Scotland. Such an operation may not only be unrealistic, but also unnecessary. Many individuals falling outside the stringent definition nonetheless might have characteristics different from those of the domestic grouping, and their morphology and genes may still owe much to the ancestral pre-Iron Age wildcat population. For example, in Fig. 22.7, it is clear that quite a few cats whose pelage diagnoses them as 'wild/hybrid', 'hybrid', 'hybrid/domestic', and even 'domestic' actually possess q values that are comparable to those of the pelage-based 'wildcats' (Beaumont *et al.* 2001). Q values quantify the mean admixture proportion that is the probability that the genetic character of a cat (based on nine nuclear microsatellites) is furthest from the domestic cat, and hence perhaps closest to its pre-Iron Age ancestors. Fig. 22.7 reveals that of all the cats studied, the individual with the highest q value had domestic pelage (bottom right corner in Fig. 22.7) (Beaumont *et al.* 2001)! Therefore, if as seems likely, wildcats meeting the stringent definition are rare, those that just fail to qualify may have features (both morphological and genetic) that could usefully contribute to the restoration of the Scottish wildcat population (Macdonald *et al.* 2004b). In this context, the stringent definition may be accepted for protection of individual cats, but a separate workable definition may be necessary, and indeed unavoidable, for the conservation of the Scottish wildcat as a population. In practice, this may involve making comparisons of mtDNA between wild-living Scottish and British domestics and worldwide domestics (which are likely to be the same), and also to Continental wildcats which are a fair comparison and proxy for pre-Iron Age British wildcats. By comparing different European populations with varying proportions of scientifically defined wildcats, we could determine which pelage and skull characters are most likely to have been present in pre-Iron Age wildcats. It would also be possible to extract a DNA from fossil remains throughout Europe to test whether there is a genetic contiguity between wildcats over the past 10–20,000 years.

If we include all cats that do not score 1 for any of the 7PS, an estimated 50% of the Scottish wild-living cats qualify for this relaxed definition, dramatically increasing the genetic and morphological resource potentially useful to restore the Scottish wildcat population (Macdonald *et al.* 2004b; Kitchener *et al.* 2005). This assumes that natural selection is acting on the wild-living cat population, thus selecting for wildcat characteristics. In practical terms, when considering management at the individual level, this means that any cat with a distinct striped tabby coat pattern and a thick-ringed tail with a black blunt tip (which never occurs in the pelage group closest to the domestic cat) is effectively a wildcat. Considering management at a population level, an alternative could be to define a wildcat by where it lives as opposed to what it looks like, but this approach requires first identifying the wildcat hot spot areas and then reconciling this level of population management with the individual management of conspicuous feral or hybrid cats therein. Right now, nobody knows what blend of individual versus population-based approach will work best, and experimental approaches should be devised to test this.

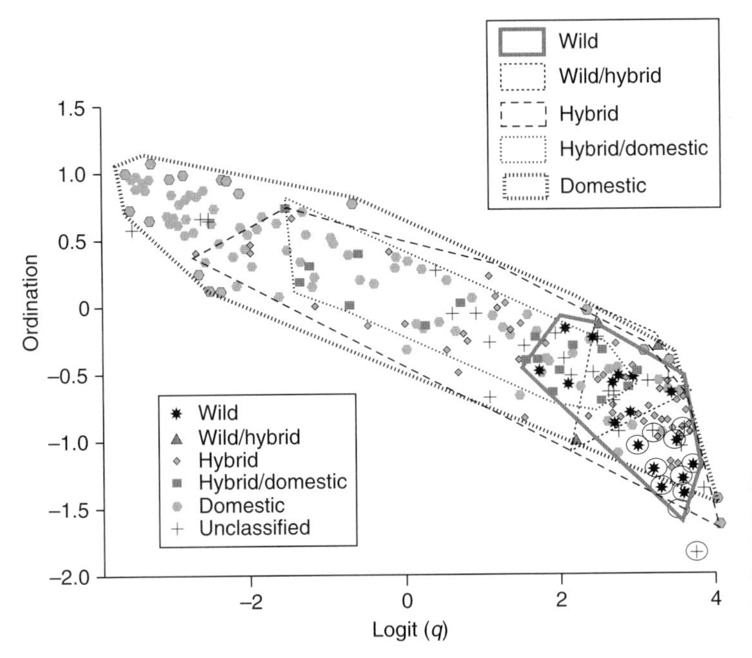

Figure 22.7 Pelage-based classification and the q value. (From Beaumont *et al.* 2001.) Note that all 'wild' cats possess highest q values, but other categories, even 'domestic', include many cats with similar q values.

We may not have much time

All the indications are that saving the Scottish wildcat from cryptic extinction requires immediate action. Beaumont *et al.* (2001) report that among cats possessing traditional wildcat pelage, all eight individuals collected before 1970 have q values greater than 0.95. On the other hand, only 6 out of 20 individuals collected after 1970 have q values greater than 0.95. In the same vein, in the original analyses by French *et al.* (1988), they showed that the statistical description of wildcat skulls collected after 1953 (between 1953–63 and 1975–78) had shifted away from the characteristics of those collected between 1901 and 1948 and towards the characteristics of domestic cat skulls, but there appeared to be a partial reversal to the skull morphology of pre-1950s wildcats in the later sample. Interpretation is complicated because we do not know how conventional evolutionary forces such as natural selection, as distinct from hybridization, may have been affecting skull morphology during history. Admittedly, this result takes no account of whatever variation might have occurred in pre-Iron Age Scottish wildcats (as, e.g., in their pelage), or the direction in which they might have evolved even without introgression, but it flags the ominous reality that something was changing (and seemingly worsening) to the nature of the Scottish wildcat during the twentieth century.

Conservation actions to restore the Scottish wildcat

Until recently, the focus of research has been on defining and diagnosing wildcats. Ample is now known to initiate conservation action, and although some issues remain uncomfortable, posterity will doubtless judge harshly any dithering in the face of imminent national extinction.

Direct persecution remains a serious threat to the Scottish wildcat. Regardless of (or perhaps because of) the current legal minefield, hundreds of wild-living cats are killed annually, deliberately or incidentally, in the grouse moors and upland edges of pheasant shooting. Any potential benefits in terms of disease reduction or introgression abatement gained by removing domestic, feral domestic, or hybrid cats may therefore be outweighed by the death of many of the last remaining wildcats (however defined). Enforcing a precautionary principle of banning all cat-killing in areas where wildcats are most likely to occur could be of immediate benefit and could also provide breathing room where further research into the impact of introgression, as outlined below, could take place.

The goal

How can we identify wildcat populations of relative purity and those where introgression is widespread? Yamaguchi *et al.* (2004c) used skull morphology to illustrate possible consequences of defining a wildcat population in which ~80% of wild-living cats possess the traditional pelage. On the other hand, genetic analyses based on eight to nine microsatellite loci (Beaumont *et al.* 2001; Daniels *et al.* 2001) can be interpreted to suggest that if *c.* 20% of the population possesses traditional wildcat pelage, it may qualify as a 'non-domestic' population. Confirming wildcat identity under the stringent definition necessitates a closer examination of the cats than is generally feasible in the field, whereas the relaxed definition we propose could be easily applied in the field and may include approximately four times as many cats. This is appealing because misdiagnosis is more likely to 'do no harm'. Therefore, we propose an initial goal of restoring Scottish wildcats to a point where at least 80% would qualify for the relaxed definition.

Monitoring the distribution of phenotypes

The proportion of qualifying wildcats doubtless varies between populations of wild-living cats. In 2007, SNH initiated a new field survey to evaluate this variation. Daniels *et al.* (1998) raise the possibility that the distribution of Group-1 (furthest-from-domestic) cats can be related to specific environmental features, such as the suitability of land for forestry, on which basis they partitioned Scotland into high, medium, and low probability zones (Fig. 22.8). A new survey, employing the 7PS pelage characters, may provide new data that allows us to recognize areas with higher population densities and proportions of wildcats.

Interestingly, although few ecological differences have so far been detected between different groups of wild-living cats, females of Group-2 (domestic) cats showed less seasonality in oestrus, suggesting that some ecological differences may exist between wildcats and feral domestic cats (Daniels and Macdonald 2002; Daniels *et al.* 2002). Group-1 cats (wildcats and 'wild-living' cats) generally avoid high-mountain areas above 600 m, exposed coasts, and fertile lowlands with extensive agriculture (Easterbee *et al.* 1991; SNH 2008). Daniels *et al.* (1998) also established that Group-1 cats were more likely to be found in cold areas and areas deemed to be poor for forestry, preferring to live in the margins of mountains and moorlands with rough grazing, often combined with forest and some arable land (Easterbee *et al.* 1991; SNH 2008). This led to the proposition that wildcats or Group-1 cats may be more ecologically adapted to these conditions. Conversely, in rural areas, domestic feral cats (e.g. Group-2 cats) are often found in their greatest densities near farms and villages, where they have access to good shelter and extra food provisioned by human caretakers (see Kilshaw *et al.* 2008). A

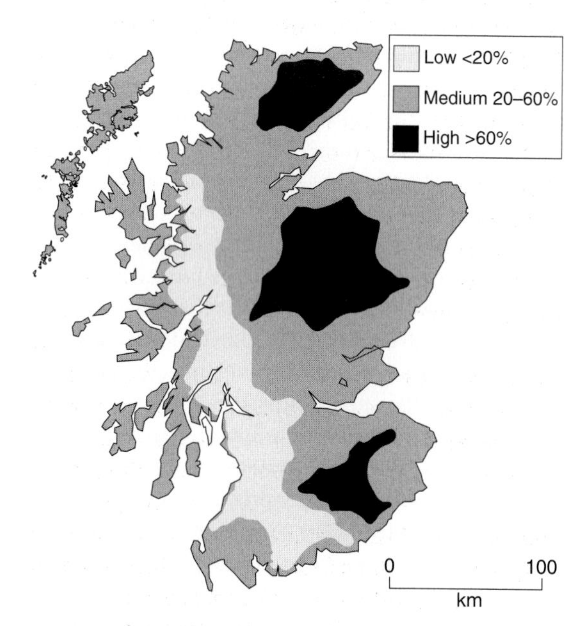

Figure 22.8 Regional zonation. (Adapted from Balharry and Daniels 1998.) The map shows areas of Scotland in which there is a high, medium, and low probability of Group-1 cats occurring, based on the occurrence of environmental variables associated with these cats.

study carried out on solitary feral cats in Avonmouth, the United Kingdom, indicated that 'high-risk areas' or areas where feral cats were often found included derelict land, dockyards, factory premises, hospitals, institutions, large industrial sites, military sites, refuse tips, and trading or industrial estates; other areas included farms, particularly those producing eggs, and around farms (Page and Bennett 1994). None of these site types are particularly prominent in the Scottish Highlands. Whether or not such preferences really exist, anthropic environments clearly play to the strengths of domestic cats; restoring native habitats would at least help level the playing field and aid wildcat recovery.

There are two schools of thought regarding environmental effects on the genetic composition of wild-living cats. One prediction is that both genotypes and phenotypes of wild-living cats in Scotland will change through natural selection until they more closely mirror those that probably prevailed in pre-Iron Age Scottish wildcats. However, we do not know how long this process will take, perhaps thousands of generations, and while it continues the existence of domestic phenotypes may both complicate the implementation of protection and may detract from support for, or satisfaction with, Scottish wildcat conservation. Furthermore, a pre-Iron Age Scottish environment is presumably required for the selection pressures appropriate to reshaping a pre-Iron Age Scottish wildcat. An alternate prediction to the contrary is that, although their fur may be less dense and their build lighter, domestic cats also carry genes for an openness to sociality, lack of fear of people, disease resistance, and perhaps a greater ecological flexibility so important in an anthropogenic world. Thus, the domestic might actually outperform the wildcat in the twenty-first century version of 'the wild'. A stronger and more provocative version of this prediction holds that modern domestic cats will out-compete wildcats regardless of the environment, whether anthropic or truly wild, simply due to their greater ecological flexibility, opportunistic feeding strategies, and ability to adapt to change. Under either scenario therefore, there may be a case for selectively removing some genotypes/phenotypes to accelerate and enhance restoration.

Therefore, a two-pronged remedial approach suggests itself. The first prong would concentrate conservation efforts on areas having suitable wildcat habitat (and possibly 'wild-living' hybrids and low-density feral cats) that are located as far as possible from human population centres (with associated higher feral cat population densities). The second prong would be directed towards improving the situation of wildcats in less suitable habitat (and therefore places where feral cats are less abundant, and perhaps less adapted), which currently has extremely low population densities of or no feral cats.

Our team is currently undertaking a scoping study to explore implementation of these pathways, and associated ideas for captive breeding and restocking. For example, the Alladale Wilderness Reserve in the Scottish Highlands encompasses a mixture of moorland, mixed woodland adjoining mountainous areas, and pasture. The Reserve is currently free of feral cats and although it is mainly 'medium' to 'poor' Scottish wildcat habitat, there are areas of 'good' habitat present and much potential for improvement through woodland regeneration that is forecast to produce abundant rodents associated with young growth.

It is therefore worth investigating whether, with some level of habitat improvement, the Scottish wildcat could be reintroduced into areas such as the vicinity of Alladale, where there is scope for a 'pure' bred wildcat population to develop without the threat of hybridization with, and disease transmission from, feral cats which is currently driving existing wildcat populations to extinction.

Code of practice for wildcat conservation

Although the Wildlife and Countryside Act 1981 legally protects only strictly defined wildcats, there is a need for a Code of Practice for Wildcat Conservation which, although lacking legal teeth, should be supported by a compelling education programme. This might be based on zonation, perhaps established on the basis of ecological or population characteristics in which levels of protection would be much more stringent. If zonation is successful, the most stringently protected zones might be progressively expanded.

The idea of designating areas with variable levels of protection for the conservation of particular species is not without precedent (similar zonations have been proposed for wolves in the United States by the IUCN Wolf Specialist Group under the adoption of the 'Wolf Manifesto'; Anonymous 1974). Furthermore, in the United Kingdom a legislative mechanism for designating such areas currently exists under the EC Habitats Directive 1992 (Council Directive 92/43/EEC on the Conservation of Natural Habitats of Wild Fauna and Flora), which requires member states to define 'Special Areas for Conservation' (SAC) for all Annex I and II species, which was transposed into British national law through the Conservation (Natural Habitats, &c.) Regulations 1994 (as amended). For example, 168 SAC sites have been designated for the otter, *Lutra lutra*, in the United Kingdom, with a further three possible sites being considered; the largest of these 168 otter sites is an extensive (*c.* 143,538 ha) area of the Caithness and Sutherland peatlands (JNCC 2008). Approximately 230,000 ha have been proposed as SACs within the Eastern Highland area outlined by Daniels *et al.* (1998) as most similar to the habitat used by Group-1 (non-domestic) cats. Frustratingly, however, this option is not open for the wildcat, which is listed on Appendix IV of the EC Habitats Directive. Nonetheless, an area-based approach to wildcat conservation and restoration can fruitfully be informed by the model of SACs focusing on conservation action. While not legally enforceable, a Code of Practice for Wildcat Conservation could focus on Special Areas for Wildcat Conservation (SAWCs) based on optimal wildcat habitat (see Daniels 1997; Macdonald *et al.* 2004b), and with a suitably vigorous education campaign and creative involvement of all stakeholders (including gamekeepers), these could prove a powerful tool for achieving the recovery of local wildcat populations.

The wildcat is used to market Scotland as a tourist destination. Recent estimates for the annual income generated by other 'charismatic' animals in Scotland (dolphins [£5 million], eagles [£3.5 million], and otters [£2.3 million]) suggest that the wildcat could have a similar value (D. Bryden, personal communication). These values should be enhanced in the education of all sectors of society, focusing on gamekeepers, landowners, domestic-cat owners, and school children. We see nature as part of nationhood, and see merit in people attaching value, both cultural and commercial, to their wildlife as an element of their national heritage. In that sense, we advocate an educational message that the wildcat is a part of the British heritage.

Controlling the densities of feral cats: an uncomfortable decision

None of the proposed conservation action to restore the Scottish wildcat could succeed if the flow of domestic-cat genes into wild-living populations cannot be staunched. All the discussions of the possible merit of removing (de facto, shooting or trapping) some feral cats and hybrids necessitate that we are clear about an ethical issue. Clearly, killing a cat is sad, even if on balance it is judged to be necessary. Equally clearly, it is not an action to be taken lightly. However, if we have to kill cats failing to meet the current definition of the wildcat, it would be morally dubious to kill those individuals while taking no steps to prevent their replacement from domestic stock.

In countries where feral cats are a threat to native wildlife, various means to reduce contact between wildlife and domestic cats are taken. For example, in parts of Australia local government has been encouraged to introduce 'curfews', requiring owners to confine cats between certain hours (Paton 1993). In South Africa, Stuart and Stuart (1991) advocate neutering programmes, limiting the number of cats per household and licensing, as measures for protecting both the wildcat and indigenous wildlife. In Germany, domestic cats may be killed if they are more than 300 m away from human habitation (H. Heinrich, personal communication). All these measures are designed to reduce the numbers of cats moving into the wild either directly (killing) or indirectly (through reduced levels of reproduction). Animal welfare is a top priority in terms of both direct and indirect control methods; neutering, together with reduction of vital resources (e.g. secure rubbish), are recommended for ethical, welfare, and public safety reasons (Gunther and Terkel 2002). There is a need for a wide consultation on the usefulness and acceptability of such measures in Scotland, and this should involve especially veterinarians, the farming community, and game managers, along with the conservation community.

Kilshaw *et al.* (2008) examined the feasibility of carrying out feral cat management in the Cairngorms National Park to benefit of the Scottish wildcat. The study proposed a combination of intensive neutering of feral cats, subsidized neutering of local domestic pet cats, education of domestic cat owners, changes in game keeping to prevent accidental mortality of wildcats, and education of gamekeepers to ensure correct identification of wildcats (Kitchener *et al.* 2005) to be carried out in and around areas of suitable wildcat habitat (see Daniels 1997; Macdonald *et al.* 2004b). The study highlighted the importance of public and landowner support to ensure the successful implementation and long-term success of feral cat management with the long-term aim of reducing the level of hybridization and transmission of potentially fatal felid diseases from feral cats to the Scottish wildcat. Significantly, the study stressed the importance of support by local pet owners, as irresponsible cat ownership is the first step in the establishment of a feral cat problem, and training stakeholders in the correct identification of the Scottish wildcat from both hybrids and tabby domestic cats. The results from this study have recently been implemented in a new initiative, the Highland Tiger Project (www.highlandtiger.com). This project is currently based in the Cairngorms National Park and aims to encourage responsible cat ownership and increase public awareness of the Scottish wildcat.

Captive breeding

Kitchener (1995, 2002) raises the possibility of captive breeding and reintroduction. While research, monitoring, and conservation action are getting underway, there is a real risk that we may lose the Scottish wildcat through cryptic extinction. Captive breeding offers a vital insurance policy for the future conservation action of the wildcat, just in case the situation in the wild worsens suddenly for whatever reason. There have been five attempts to reintroduce wildcats in Europe, only one of which claims success (Stahl and Artois 1994; Breitenmoser *et al.* 2001; see also Kitchener 2002), although it remains too early to judge others. However, reintroductions of other felids, particularly lynx, can succeed if planned appropriately and sufficient stock is reintroduced (Breitenmoser *et al.* 2001).

The British and Irish Association of Zoos and Aquariums (BIAZA) studbook for the Scottish wildcat is vested with Neville Buck of Port Lympne Wild Animal Park, and we are currently scoping the feasibility of an experimental release programme in Sutherland. However, the current captive population contains no more than nine animals that conform to the strict definition, which may be too few to establish a genetically sustainable population to ensure the survival of the species. Initial results from the scoping study indicate that although reintroduction is theoretically possible, long-term viability of the population can only be assured through long-term management, including regular supplementation of new individuals into the population (Kilshaw and Macdonald 2010). This, supported by previous reintroductions of the European wildcat which have identified that a large breeding base of captive individuals is required to ensure success not only to prevent the risk of genetic degeneration, but also because of the high mortality rates (70–80%) of individuals in the first few weeks of release (Stahl and Artois 1994). Therefore, in addition to building of the captive population through careful management, further animals will be required to be taken from the wild to enhance the potential of the founder population. Although there may be some ethical qualms about doing this with such a critically endangered felid, a successful precedent has been established with the world's most critically endangered cat, the Iberian lynx, *L. pardinus*, which may number less than 200 animals in the wild (Guzmán *et al.* 2004; Wildt *et al.*, Chapter 8, this volume). By taking a non-viable third kitten from litters of three, a captive population is slowly being established. It could be better to take action now with perhaps 400 ± wildcats, rather than waiting until the situation worsens. Through combined conservation action of education, removal of feral domestic cats, restoration of habitat and reintroduction, we will judge our efforts to save the wildcat to have been successful when we can no longer call it the Scottish wildcat, because it has been restored to suitable habitats throughout the length and breadth of Britain.

CHAPTER 23

The changing impact of predation as a source of conflict between hunters and reintroduced lynx in Switzerland

Urs Breitenmoser, Andreas Ryser, Anja Molinari-Jobin, Fridolin Zimmermann, Heinrich Haller, Paolo Molinari, and Christine Breitenmoser-Würsten

Eurasian lynx feeding on a roe deer in the Swiss Alps. Camera-trapping photo by KORA.

Introduction

The persecution of predators as a consequence of their conflicts with human interests is, together with the reduction of their living space, the main reason for the decline of carnivore populations (Loveridge *et al.* Chapter 6, this volume). The source of conflict is always predation: predation on humans, their livestock, or their game (Kruuk 2002). The least understood, and most difficult to assess, of these conflicts is the competition over game. This conflict is very topical in Europe, where presently brown bear (*Ursus arctos*), wolf (*Canis lupus*), and Eurasian lynx (*Lynx lynx*) are making a comeback. In the nineteenth century, both large herbivores and large carnivores were heavily reduced or completely eradicated in most regions of Europe. The recovery of the wild ungulate populations began around 1900 and was actively supported by hunters and conservationists. The returning large carnivores now interface with an abundant and heavily managed ungulate population and a wildlife management system not designed for the presence of large predators.

The Eurasian lynx, Europe's largest felid, was reduced to a few remnant populations in Scandinavia and Eastern Europe by 1900. The population nuclei in the Pyrenees and the Alps, the species' last stronghold in Western Europe, became extinct around 1940. The remnant populations in northern and eastern Europe have slowly started to recuperate since 1950, and reintroductions in Central and Western Europe began in 1970 (Breitenmoser and Breitenmoser-Würsten 2008). The recovery is, however, impeded by a serious conflict with traditional hunting. Experts across all range countries have ranked illegal killing prompted by conflict with hunters as the primary threat to lynx (von Arx *et al.* 2004). Červeny *et al.* (2002) concluded from a confidential survey of hunters in the western Czech Republic that illegal killing was the main source of mortality in this reintroduced lynx population and that legal protection had virtually no effect. Conflicts occur throughout the species' range, but seemingly most intensely in the wake of active reintroduction, as in Switzerland in the early 1970s. The perceived impact of lynx on roe deer (*Capreolus capreolus*) and chamois (*Rupicapra rupicapra*), their main prey, has been a constant source of conflict between hunters, wildlife managers, and conservationists. Periods of increasing lynx abundance and simultaneous declines in roe deer and chamois resulted in repeated spates of illegal killing. Today, 35 years after the first releases, the reintegration of the lynx into the ecosystem of the Alps and the Jura Mountains, far less into a balanced management system, remains incomplete. Attacks on domestic livestock by lynx have turned out to be a relatively minor problem, made tractable by prevention and compensation. Competition with hunters has emerged as the most important factor for the lynx re-establishment.

From 1971 to 1976, 24 lynx, imported from the Carpathian Mountains of Slovakia, were released at eight different sites in the Swiss Alps and in the Jura Mountains without further monitoring (Breitenmoser and Baettig 1992; Breitenmoser *et al.* 1998; Fig. 23.1). Soon after the return of the first lynx, hunters alleged that a considerable reduction of roe deer and chamois had occurred, a claim strongly opposed by wildlife biologists and conservationists at the time (e.g. Kurt 1991). Our early field studies, which began more than 10 years after the releases, revealed that the predation impact of lynx in the aftermath of the recolonization had been at least locally considerable, even leading to temporary local extinction of roe deer (Haller 1992). We attributed this to a 'front-wave effect', postulating that the prey, which had not experienced large predators for more than a century, and therefore was naive and extremely vulnerable to predation, would readapt to the presence of the large carnivore and the predator–prey system would stabilize after this initial phase (Breitenmoser and Haller 1993). Subsequent experience, however, demonstrated that the populations of the lynx and its main prey fluctuate widely and repeatedly, obviously as a consequence of independent factors such as winter mortality of roe deer, with predation reinforcing the amplitude of the fluctuations.

Understanding the significance of predation and dynamics of predator–prey systems is crucial for the conservation of carnivore populations, particularly when the predators are competing for prey with humans and conflicts need to be mitigated through management. Here, we summarize our observations on predation and predator–prey dynamics in the reintroduced Eurasian lynx populations in Switzerland and address the following questions: (1) Can we

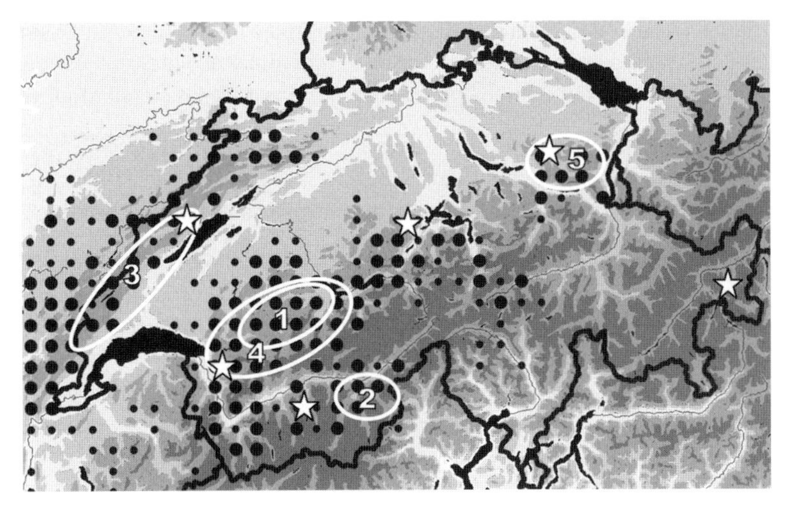

Figure 23.1 Lynx distribution and study areas in the Swiss Alps and the Jura Mountains. Stars indicate release sites initiating the populations. All releases with the exception of the lynx reintroduced in North-eastern Switzerland 2001–06 (5) took place in the early 1970s. Present distribution according to a 10 × 10 km grid. Big black dots indicate permanently, small dots sporadically occupied cells (von Arx *et al.* 2004). Study areas: 1, 4 = north-western Alps I and II; 2 = Central Alps; 3 = Jura Mountains; 5 = North-eastern Switzerland.

adhere to the 'front-wave hypotheses' in the light of subsequent experience? (2) What was the impact of predation on ungulate populations compared to other anthropogenic or natural effects? (3) What are the consequences of the recovery of a top predator for wildlife management? Although assembled retrospectively, and thus suffering from some methodological inconsistencies, our five field studies, spanning 25 years of the lynx's re-establishment and its relationships with prey and hunters in Switzerland, illuminate the range of conditions and fluctuations in this relatively simple predator–prey system within a complex environment.

Studies and study areas

Lynx reintroduction started in 1971 in the Swiss Alps, and in 1974 in the Jura Mountains (Breitenmoser 1983; Breitenmoser and Baettig 1992). Officially, only six animals were released in the Alps and four in the Jura, but at least 19 additional lynx were set free without authorization in different parts of the mountain ranges (Breitenmoser and Breitenmoser-Würsten 2008; Fig. 23.1). The two nuclei increased, spread, and incorporated some isolated animals from the unofficial releases. However, surveillance, monitoring, and field research only started in March 1983 (NWAlps I in Table 23.1), when we radio-tagged the first two lynx in the north-western Alps (Fig. 23.1). In 1985, we established a study area at the edge of the expanding population in the Central Alps (CAlps in Table 23.1), and in 1988, in the Jura Mountains (Jura in Table 23.1). A third reintroduction was initiated in 2001 in the foothills of the Alps in North-eastern Switzerland (NE-CH in Table 23.1), where lynx originating from the Alpine and the Juran population were released.

The Swiss Alps and the Jura Mountains are mixed landscapes with pastures and montane and subalpine forests, covering 28% of the north-western Alps and up to 57% of the Jura Mountains. The timberline in the Alps is between 1800 and 2300 m, way below the highest peaks of 4000 m. Crests in the Jura reach 1400–1700 m, below the natural timberline. The climate along the north slope of the Alps and in the Jura is moderate to subalpine; however, it varies strongly between years. Summers can be hot and dry to cool and rainy. Winter used to bring 1–2 m of snow lasting for 4 months, but recent winters have been mild, with lasting snow only at higher altitudes. The central Alps

Table 23.1 Lynx field studies in Switzerland and prey taken by radio-tagged animals. Livestock includes 87% sheep, 8% goats, and 5% other domestic species. For more differentiated lists, see Breitenmoser and Breitenmoser-Würsten (2008) and publications mentioned in the text. Locations of study areas in Fig. 23.1.

Study	Years	Lynx	Prey items						
			Roe deer	Chamois	Livestock	Red fox	Hares	Others	Total
NWAlps I	1983–88	11	63	38	4	0	9	8	122
CAlps	1985–88	6	20	48	8	0	2	2	80
Jura	1988–98	29	428	133	3	37	13	8	622
NWAlps II	1997–2001	44	203	81	80	16	19	24	423
NE-CH	2001–07	10	172	55	1	1	5	3	237
Total		100	886	355	96	54	48	45	1484
Per cent			60	24	6	4	3	3	100

are more continental than the other study areas (Fig. 23.1), but conditions also vary strongly between years. The main valleys are densely settled and industrialized; the mountain economy is based on livestock farming (predominately cattle, but also sheep and goats), forestry, and large-scale tourism. Local hunters can hunt everywhere with the exception of some small game sanctuaries. The most important game is roe deer and chamois, and also, in some areas, red deer (*Cervus elaphus*).

This synthesis summarizes our observation of radio-tracked lynx in five studies (Table 23.1), typifying different predator and prey situations.

The first 11 lynx radio-tagged between 1983 and 88 in the north-western Alps (1 in Fig. 23.1) gave insights into the land tenure system, movement patterns, and prey choice (Haller and Breitenmoser 1986; Breitenmoser and Haller 1987). More than 10 years after lynx had recolonized the study area, predator and prey populations were stable and predation impact was lower than we had assumed from anecdotal observations in previous years, indicating that prey had already readapted to the presence of the large cat and the predator–prey system had reached equilibrium (Breitenmoser and Haller 1987).

In the Central Alps (2 in Fig. 23.1), we followed six lynx at the edge of the expanding Alpine population from 1985 to 1988, in a situation where the regional wildlife authority claimed a serious drop in roe deer and chamois abundance. Haller (1992) documented intensive hunting by lynx in a game sanctuary, leading to the temporary disappearance of roe deer and a decrease of the local chamois population from 800 to 300–400 in the years 1980–88. This strong predation impact contrasted with our findings in the north-western Alps and seemed to confirm the expected 'front-wave effect' (Breitenmoser and Haller 1993).

Research in the reintroduced population in the Jura Mountains (3 in Fig. 23.1) started 14 years after the releases and lasted from 1988 to 1998. We studied spatial organization and demography (Breitenmoser *et al.* 1993; Breitenmoser-Würsten *et al.* 2007a, b) and food ecology of lynx (Jobin *et al.* 2000; Molinari-Jobin *et al.* 2002, 2004, 2007). Lynx showed seasonal preferences for different prey categories, but the overall predation impact over the years remained rather stable.

After the mid-1990s, attacks on livestock in the north-western Alps increased. We started a second study in 1997 (4 in Fig. 23.1; NWAlps II in Table 23.1), more than 25 years after the re-establishment of the population. We found higher lynx abundance and stronger predation impact than in any previous study. The yearly roe deer hunting bag in the study area decreased from 1065 in 1994 to 117 in 1999, triggering a violent controversy and illegal killing of lynx (Breitenmoser and Breitenmoser-Würsten 2008).

The consequence of these events was a new management concept for lynx (Buwal 2004), allowing the removal of stock raiders and the reduction of 'over-abundant' lynx populations through translocations or culling. From 2001 to 2008, 13 lynx (six males, seven females; eight from the north-western Alps and five from the Jura) were translocated to the north-eastern Swiss Alps (5 in Fig. 23.1 and Table 23.1). It is too early to assess the success of these translocations, but they allowed a study of prey choice and predation patterns during the initial phase of a reintroduction (Ryser *et al.* 2004).

Prey spectrum and prey selection

The Eurasian lynx is a medium-sized cat of 17 kg to 25 kg for females and males, respectively. Their preferred prey is roe deer wherever the two species coexist, with other wild and domestic ungulates as the most important alternative prey, followed by hares (*Lepus* sp.) and other medium-sized mammals or large birds (Breitenmoser and Breitenmoser-Würsten 2008; Table 23.1). Creating an objective list of dietary items from a variety of records is problematic, as detection methods and the chance of finding a specific prey item depends on its type, size, kill site, and handling by the predator (Karanth *et al.* Chapter 7, this volume). Wild ungulates and even prey as small as foxes, hares, or capercaillies were found by locating lynx at the kill and feeding sites at night and searching the spot next day by snow tracking or tracking with dogs. Shepherds or farmers often reported sheep and goats killed to the game wardens (for compensation) and removed the carcasses before lynx had finished eating them. Table 23.1 lists only domestic kills, which we were able to attribute to one of the radio-tagged lynx. Smaller prey items, such as rodents, were only detected by chance. Scat analyses (Breitenmoser and Haller 1987; unpublished data) have, however, shown that they are insignificant. Prey items found ranged from mouse to red deer (combined as 'others' in Table 23.1), but 'kill series' (intensive monitoring of individual lynx for a period of 60–90 nights) revealed that roe deer (886 out of 1448 prey items = 60%; Table 23.1), and chamois (24%) form the staple food of lynx in Switzerland, livestock, red foxes, and hares occurring 6%, 4%, and 3% of the time, respectively. These species are marginal as lynx prey (Molinari-Jobin *et al.* 2007), but can be locally and temporarily important for the survival of individual lynx.

The proportion of the main prey, roe deer, and the preferred alternative prey, chamois, in the diet varied between the studies depending on their local and temporal abundance, availability, and vulnerability to predation in certain habitats. The highest proportion of roe deer was observed in North-eastern Switzerland (73%; Table 23.1) and in the Jura Mountains (69%). In the Central Alps, however, lynx killed more chamois (60%) than roe deer (25%) as a consequence of the depressed roe deer population during the study period and the high availability of chamois (Haller 1992). Most of the chamois population lives in the extended alpine zone of the inner Alps, building a source population for the forest-living chamois. In the forested habitats, where lynx hunt, chamois are more vulnerable to predation than roe deer. In the Jura Mountains, where no alpine zone exists, chamois are, in proportion to their relative abundance, killed more often by lynx than are roe deer (Jobin *et al.* 2000). In the north-western Alps, the proportion of both roe deer (51–48%) and chamois (31–19%) were equally reduced in the second study, but all other prey (mainly livestock) had significantly increased in the diet from 17% to 33% ($P < 0.001$). During the NWAlps II study, the populations of the main prey (roe deer) and the alternative prey (chamois) were decreasing, the latter as a result of an epidemic of infectious keratoconjunctivitis (Degiorgis *et al.* 2000), forcing lynx to hunt less preferred prey, for example, livestock. Livestock had a percentage occurrence of 19%, the highest proportion observed over all the studies (Table 23.1). Lynx showed a functional response not only to longer-term prey population changes, but also to seasonal availability of prey categories. In the Jura Mountains, where the prey populations and the overall predation impact remained rather stable over the years, they killed mainly fawns in summer, chamois males during their rut, and roe deer females in winter (Molinari-Jobin *et al.* 2002, 2004). Within home ranges shared by females and males, we found a total of 618 roe deer and 254 chamois preyed upon by lynx, with females killing significantly more roe deer than chamois (74:26%), whereas males killed more chamois (45:55%; $P < 0.001$). The roe deer abundance in the lynx areas was probably at least twice that of chamois, indicating that male lynx preferentially targeted chamois, perhaps reducing intersexual competition.

Kill rhythm and consumption

To estimate the impact of predation, it is important to assess mean and variance of consumption. In Switzerland, a Eurasian lynx kills and consumes roughly an ungulate per week. We estimated that solitary lynx in the north-western Alps killed 55 ungulates per year, or 66 if dependent cubs were included (Breitenmoser and Haller 1987). Daily consumption in the Jura Mountains was 3.8 kg (±1.3 kg, $n = 27$) for males, 3.1 kg (±1.4 kg, $n = 27$) for solitary females, 4.8 kg (± 1.1 kg, $n = 17$) for females with one cub, and 7.1 kg (± 1.4 kg, $n = 16$) for females with two cubs, from October to March. Kill series revealed that lynx utilized a fawn for 2–3 nights, an adult roe deer or chamois for 4–5, and sometimes up to 7 nights. The average search time for male lynx was 2.5 days, for females 1.5–2 nights. The extrapolated yearly consumption was 56 ungulates for an adult male lynx, 59 for a solitary female, 72 for a female with two cubs, and 57 for a subadult animal (Molinari-Jobin *et al.* 2002). Lynx ate on average 70–80% of all consumable parts of a carcass. Domestic sheep were less completely consumed ($n = 191$): 46% were consumed over several nights, 44% in 1 night, and 13% were killed but not eaten. Apart from situations where lynx were heavily disturbed, we observed exceptional wasting or surplus killing of wild prey only twice. In the Central Alps, a lynx killed four chamois on the same day (Haller 1992). In December 2007, a lynx immigrated into the Swiss National Park and killed within a few weeks at least five chamois and three red deer (the latter taken in very fast succession within 300 m). The lynx consumed about 18 kg of a total biomass of >250 kg; three kills remained untouched (H. Haller, personal communication). In common with examples of spatially concentrated predation (see below), the two cases occurred in sanctuaries with very high densities of naive ungulates. Other reports of surplus killing from the 1970s and 1980s were related to artificial feeding of ungulates, for example, a lynx in the Lötschental (Central Alps) killed at least 18 roe deer in spring 1985 in the vicinity of feeding stations (Haller 1992).

Lynx home ranges and distribution of kills

An important feature of predation is the decision of the predator where to hunt, hence the distribution of kill sites. Lynx live solitarily in large home ranges, shared only with mates and eventually with the cubs of the year. Home range sizes observed in Switzerland were 72–221 km^2 for resident females and 135–363 km^2 for resident males (Table 23.2). Lynx typically hunt across their entire home ranges (Fig.

Table 23.2 Home ranges (in km^2) of resident female and male lynx in Switzerland. Home ranges were calculated as minimum convex polygons of peripheral radio fixes excluding excursions (for the translocated lynx in NE-CH the initial trip from the release site to the place of residence) and large marginal areas of unsuited habitat (e.g. lakes). Means of home ranges differ significantly (Mann-Whitney U-test): NWAlps I > NWAlps II (females: $U = 86$, $P = 0.002$; males: $U = 21$, $P = 0.048$); NWAlps I + Jura > NWAlps II + CAlps + NE-CH (females: $U = 343.5$, $P < 0.001$; males: $U = 94.0$, $P = 0.001$). All values taken from the publications mentioned in the text and compiled in Breitenmoser and Breitenmoser-Würsten (2008).

Study	Females			Males		
	n	Mean	SDEV	*n*	Mean	SDEV
NWAlps I	4	221	139	2	275	450
CAlps	3	72	42	1	135	—
Jura	12	161	66	5	357	181
NWAlps II	14	93	43	11	181	79
NE-CH	3	100	20	3	152	40

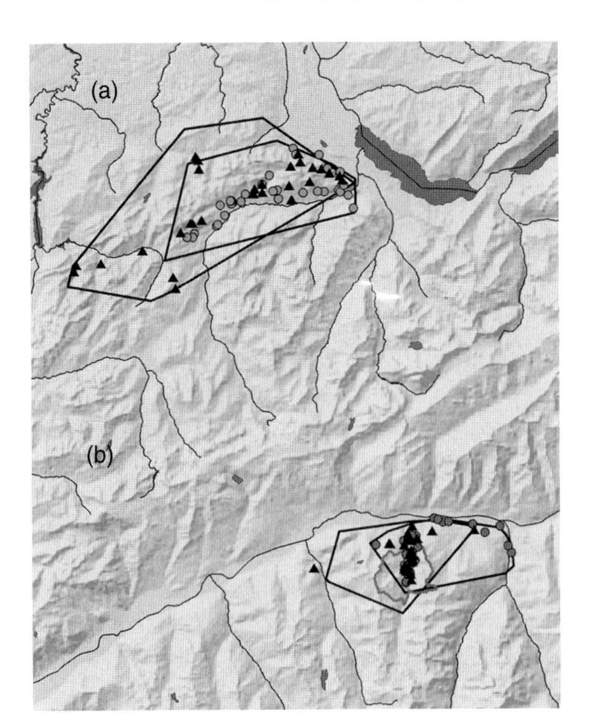

Figure 23.2 Distribution of lynx kills in two pairs of overlapping home ranges (polygons) of resident female and male lynx in the north-western Alps (a) F_2, 224 km^2, 29 kills; M_2, 428 km^2, 26 kills and Central Alps; (b) F_9, 119 km^2, 29 kills; M_4, 134 km^2, 34 kills. The kill distribution in the north-western Alps illustrates prey distribution in an established situation. The two lynx at the expanding front of the population in the Central Alps had smaller home ranges and hunted in a concentrated area within a game sanctuary (irregular polygon). Distances between consecutive kills were larger ($U = 615.5$, $P = 0.048$) in the north-western Alps (5.67 km, $n = 45$) than in the Central Alps (2.65 km, $n = 35$). (From Breitenmoser and Haller 1987; Haller 1992; and Breitenmoser and Haller 1993.)

23.2a) and rarely make two successive kills in the same sector. The average distance between consecutive kills in kill series was 12.68 ± 8.57 km ($n = 56$) for males, 6.54 ± 5.50 km ($n = 28$) for single females, and 2.62 ± 2.05 km ($n = 88$) for females with cubs. Distributing hunting effort over the entire range may help to keep prey vigilance low and retain the element of surprise needed for a successful hunt (Breitenmoser and Haller 1993).

There are, however, exceptions to this general pattern: a male lynx from the Jura Mountains translocated to North-eastern Switzerland killed 16 roe deer over a 4-month period in a forest of 4 km^2 in the vicinity of Zurich (Ryser *et al.* 2004). In a valley in the Central Alps, Haller (1992) observed highly localized predation by two lynx (Fig. 23.2b). Forty-eight out of 63 kills were made in an area of 8.5 km^2 in the centre of their overlapping home ranges (Breitenmoser and Haller 1993). Both cases of localized predation occurred in game sanctuaries in newly colonized areas with highly abundant and naive prey. None of the other translocated lynx showed a similar behaviour. However, an adult female lynx in the Jura Mountains once remained in a forest of 16 km^2 for 11 weeks and killed there 12 roe deer and 4 red foxes in an area where blown-down trees had led to an increase in roe deer density. Although mass predation can be an effect of the 'front-wave', this is not always the case and it may also be observed in an established situation under specific circumstances such as clumped prey.

Impact on ungulate populations

We have so far observed three different predator–prey situations: (1) changing predation impact during the recolonization of areas with naive and abundant prey; (2) low-to-moderate impact after the postulated readaptation; and (3) a dynamic predator–prey phase with high impact in the north-western Alps after a 15-year period of stability. We use ungulate mortality (hunting bag, road kills, and natural mortality) as an indicator of long-term population dynamics. They are the only data consistently available from all cantons for the past 40 years. Natural mortalities fluctuate from year to year, and hunting quotas are fixed according to the (inconsistent) population estimations, but the long-term changes and the comparison of different causes of mortality is a good indicator of population trends.

Situation 1: newly colonized area

We missed the recolonization phase in the north-western Alps, but retrospective analysis of the data from the canton of Obwald, where the first eight lynx were released between 1971 and 73, suggests that the returning predator had, with a delay of

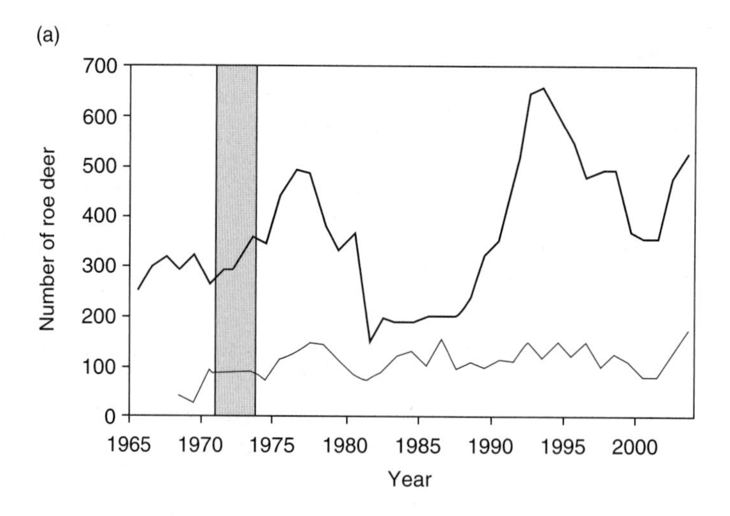

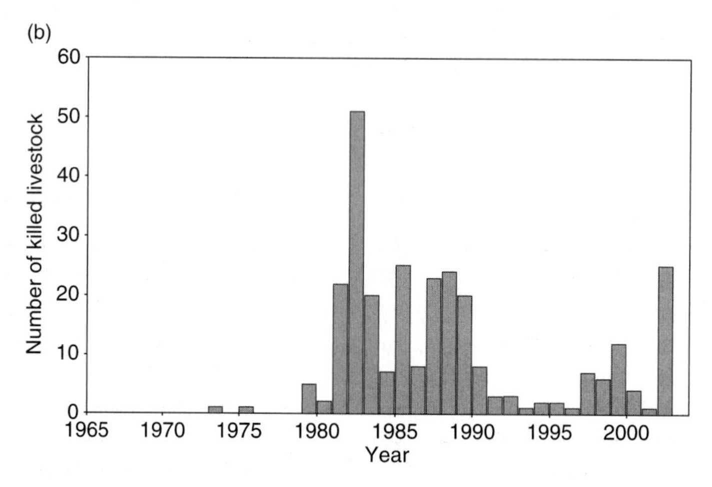

Figure 23.3 Roe deer mortality after the release of (a) lynx; and (b) sheep killed by lynx in the canton of Obwald, central Switzerland. The releases (grey bar in [a]) marked the start of the reintroduction and of lynx predation. First, the hunting bag (thick, black line) and the 'other known mortality' (thin, black line) increased. Only after a few years, the two parameters indicated a sudden drop of the roe deer population, which was at the time attributed to the increasing impact of lynx predation and coincided with a first wave of attacks on sheep (b). After about 1980, the lynx population decreased and remained low, whereas the roe deer population increased once more.

3–5 years, a considerable impact on the roe deer population (Fig. 23.3). The total known roe deer mortality dropped from 632 in 1977 to 223 in 1981, but then increased again to 799 until 1992. Attacks on livestock were high from 1981 to 1989. This may have been a consequence of low roe deer availability. There was no systematic lynx monitoring, but according to game warden reports, lynx abundance decreased in the early 1980s and remained low.

This observation gave rise to the 'front-wave hypothesis', which we confirmed in the Central Alps. From the data compiled by Haller (1992), we estimate that lynx in the central part of the game sanctuary killed 7–8 ungulates per km^2 per year, almost exclusively chamois (Fig. 23.2). The study covered the last 2 years of a period of high lynx presence from 1982 to 1987. During this period, the roe deer, kept year-round in this rough alpine valley by artificial winter feeding, had disappeared, and the chamois population had dropped from about 800 in 1980 to 300–400 in 1987. At the end of this period, the resident lynx expanded their home ranges and started to hunt outside the game sanctuary (Fig. 23.2), in what we interpreted as a numerical response to the reduced prey availability (Breitenmoser and Haller 1993).

With the translocations of lynx to North-eastern Switzerland, we observed the first phase of a reintroduction. Two of the 11 lynx translocated from 2001 to 2007 left the region. All other animals (for as long as they were alive and monitored) all established a 'classical' land tenure system with equally spaced females and overlapping male home ranges of moderate size (100 and 152 km^2, respectively; Table 23.2). Prey consisted of 73% roe deer and 23% chamois, with a high average consumption rate of 79% of the carcasses. Within the home ranges established from 2001 to 2003, lynx predation was 10% of the estimated roe deer and chamois population, which we considered moderate for a newly settled area. Lynx mortality was high and population growth was very slow (Ryser *et al.* 2004). The reduced monitoring effort after 2003 did not allow a quantitative assessment of the ongoing impact of predation. However, to date we have not observed a 'front-wave effect'.

Situation 2: re-established predator–prey balance

In the first study in the north-western Alps and the Jura Mountains, we observed the postulated established predator–prey relationship, where the prey regained vigilance and this forced lynx to search and hunt over larger areas. This limited predator density, as indicated by significantly larger home ranges of resident lynx compared to the other three studies (Table 23.2). Lynx predation in the NWAlps I was 0.38 roe deer and 0.24 chamois per km^2 per year, respectively, or 6–9% of the roe and 2–3% of the chamois population as estimated by the wildlife service. Over this time period lynx removed less than a quarter of the roe deer and chamois removed by human hunters (Table 23.3).

The study in the Jura Mountains started 14 years after the lynx reintroduction. During the 10-year study, roe deer and chamois populations steadily increased (roe deer hunting bag increased by about 60%; Molinari-Jobin *et al.* 2007). However, we did not detect a clear numerical response in lynx. Number and home range size of resident females in the study area remained stable, but we observed a shortage of males from 1992 to 1994 after two resident males had been poached, resulting in low recruitment (Breitenmoser-Würsten *et al.* 2007a, b). The average adult lynx density in the study area of 710 km^2 was 0.7–0.8 lynx per 100 km^2 and the yearly consumption 0.50 roe deer per km^2 and 0.12 chamois per km^2. Lynx took about 9% and 11% of the estimated roe deer and chamois populations, respectively, or 60% and 116% of the hunting bag (Table 23.3).

Situation 3: high impact of predators on prey in the north-western Alps

In the mid-1990s, lynx monitoring revealed an increase in the population in the north-western Alps. Decreasing roe deer mortalities (Fig. 23.4b) and increasing attacks on livestock (Fig. 23.4a) indicated another shift in the predator–prey balance. Mean home range size of resident females and males had decreased by 60–70% compared to the first study (Table 23.2), the resident lynx density in the north-western Alps was 1.4–1.5 per 100 km^2 (Table 23.3) or 1.9–2.1 per 100 km^2 including non-territorial sub-adult 'floaters' (Breitenmoser-Würsten *et al.* 2001). The average yearly predation in the study area was now estimated to be 1.20 roe deer per km^2 and 0.48 chamois per km^2; the predation impact on roe deer had therefore tripled and on chamois doubled compared to the 1980s (Table 23.3). Lynx took 36–39% of the local roe deer population as estimated by the wildlife service; predation was clearly the highest single mortality factor and almost four times the average hunting bag 1997/98 of the study area (Table 23.3). A comparably high predation impact had, until this point, only been observed by Okarma *et al.* (1997) in eastern Poland, where lynx consumed 21–36% of the local roe deer population.

Table 23.3 Lynx predation on roe deer and chamois in the north-western Alps I and II and in the Jura Mountains compared to the official population estimations and other known losses (hunting bag and animals perished) in the reference areas. Wildlife statistics for NW Alps II are averages for 1997/98. 'Predation relative' expresses the observed offtake by lynx as a percentage of the official population estimation for the study area.

	NWAlps I 1983–85		Jura 1988–98		NWAlps II 1997–2001	
	n	%	*n*	%	*n*	%
Reference area (km^2)	154		710		330	
Adult lynx/100 km^2	0.4–0.6		0.7–0.8		1.4–1.5	
Roe deer						
Hunting bag	250	*53*	586	*50*	100	*24*
Deer perished	166	*35*	234	*20*	145	*37*
Lynx predation	58	*12*	354	*30*	395	*39*
Total known mortality	474	*100*	1174	*100*	640	*100*
Roe deer population	700		4890		1100	
Predation relative		*6-9*		$\leq$*9*		*36-39*
Predation impact	Small		Moderate		Strong	
Chamois						
Hunting bag	240	*62*	75	*42*	91	*20*
Chamois perished	112	*29*	17	*10*	221	*47*
Lynx predation	37	*10*	87	*49*	157	*33*
Total mortality	389	*100*	179	*100*	469	*100*
Chamois population	700		1060		2600	
Predation relative		*2-3*		$\leq$*11*		*6*
Predation impact	Small		Moderate		Moderate	

Sources: Data compiled from Breitenmoser and Haller (1987); Molinari-Jobin *et al*. (2002); and Breitenmoser and Breitenmoser-Würsten (2008).

The increase of the lynx population was a numeric reaction to rising roe deer abundance, which was, in turn, probably a consequence of low winter mortality during a series of mild winters from 1989 to 1994 (Breitenmoser and Breitenmoser-Würsten 2008). Interpreting the mortality record (Fig. 23.4b), the roe deer population increased until 1994 and then declined until 1999. The lynx population peaked in 1999; a delay explained through a functional response evidenced by the higher proportion of chamois, smaller wildlife species, and an exceptionally high portion of 19% of livestock in the cats' diet (Table 23.1). Some of the radio-tagged lynx showed no response, while others almost completely switched to chamois or sheep (functional response type III).

Discussion

Wild ungulate populations in Europe have increased markedly over the past 50–100 years. In Switzerland, four out of five indigenous ungulate species were virtually extinct in the nineteenth century (roe deer; red deer; ibex, *Capra ibex*; and wild boar, *Sus scrofa*) but have recovered to probably higher densities than ever achieved in historic times (Breitenmoser and Breitenmoser-Würsten 2008). Nowadays they provide hunting opportunities and hence income for public or private owners. On the other hand, they cause browsing damage and economic losses to forestry and agriculture. For example, the prevention of browsing damage in Bavaria, Germany, is estimated to cost 52 Euro per hectare yearly

(F. Leitenbacher, personal communication). Potential game densities in the diverse cultivated landscape of Switzerland are high, but ungulate populations in the mountains are mainly limited by winter mortality. Winter conditions vary considerably between years (with a general tendency to milder winters) and as a result ungulate populations have the potential to fluctuate considerably. Such fluctuations will likely be observed more often as ungulate populations reach the carrying capacity of their habitats after decades of increase. But these changes are neither understood nor accepted. The aim of European wildlife management is maximizing harvest and minimizing damage by stabilizing ungulate populations below carrying capacity where recruitment is strong. This is an adaptively managed process, often based on pressure from one or the other sector rather than on clearly defined goals and population monitoring, and therefore measures are implemented with a certain delay.

The return of the lynx has complicated the system, both ecologically and socio-politically. We have studied a relatively simple predator–prey system in a highly complex environment with a highly diverse landscape, changeable climate, and a variety of anthropogenic factors (hunting, forestry, livestock husbandry, and tourism). The lynx was the only top mammalian predator; roe deer were its main prey, with chamois as an effective substitute prey species. Early, but not well-documented experience (Fig. 23.3) has generated the 'front-wave hypothesis', which we found supported in the Central Alps (Fig. 23.2b). In the light of subsequent experience, we need to restrict this hypothesis to situations where prey is not only naive, but also abundant. Spatially concentrated and surplus killing occurred repeatedly when lynx arrived in protected areas, but not in the open hunting areas of the release sites of North-eastern Switzerland. On the other hand, lynx abundance and the impact of predation can also temporarily increase in areas already occupied for a long time.

In the early 1990s, lynx showed a clear numerical response to an increase of roe deer in the north-western Alps (Fig. 23.4). Between 1994 (roe deer peak) and 1999 (lynx peak), the increasing attacks on livestock indicated a reduced availability of wild ungulates (for an alternative explanation for sheep raiding in the French Jura Mountains, see Stahl *et al.* 2001a, b). The elevated utilization of domestic sheep and goats, red fox, hares, and small mammals allowed the continued increase of the predator population and its impact on roe deer and chamois. Hunters attributed the observed decline of the roe deer population, which was very obvious, although not as strong as the drop in the hunting bag from 1994 to1999 in Fig. 23.4 suggests, to lynx predation. It is impossible to disentangle the effect of predation, hunting, and increased mortality due to harsher winters on the decline of the roe deer population in the mid-1990s (Fig. 23.4b). But in these years, lynx predation was indeed the most important cause of mortality for roe deer in the north-western Alps (Table 23.3).

The use of alternative prey suggested a limited functional response of the predator (Figs. 23.3 and 23.4), but the reduced availability of roe deer and chamois eventually led to a numerical response of lynx. The decline of the lynx population after 1999 was marked. However, we cannot tell to what extent this was a natural phenomenon and how much anthropogenic losses contributed to population decline. The peak in the lynx population triggered a series of illegal protest killings. Lynx paws were mailed to the wildlife authorities, a shot lynx was disposed of in a shopping mall, and several lynx were poisoned. Our observations suggest that the delayed numeric response of the predator to changing prey abundance, acting in parallel to the similarly delayed management response, is enhancing rather than mitigating fluctuations of ungulate populations and feeds the controversy.

Debates were sparked off by lynx attacks on livestock, but the underlying and more serious conflict was the competition with hunters for game. Wildlife in Switzerland belongs to the cantons (states), who sell hunting licenses to resident hunters. Some hunters irreconcilably oppose the presence of (large) carnivores, but many claim to tolerate lynx if there are not 'too many'. 'Small' or 'moderate' impacts by lynx on game species (slightly below the hunting bag; for example, NWAlps I and Jura in Table 23.3) were associated with limited conflict, although occasional illegal killings occurred. Whilst we lack reliable census data for forest-living ungulates, the search effort of hunters increased after 1995 (Fig. 23.4) with declining roe deer abundance (and maybe rising vigilance) to a point where they were no longer able to redeem the licenses they had acquired. When lynx predation clearly exceeded the hunting bag (e.g. NWAlps II in Table 23.3), the conflict erupted into systematic persecution of lynx.

From 1997 to 2003, 17–22 lynx killings became known to the wildlife authorities in the NW Alps (cantons of Bern, Fribourg, and Vaud; Ceza *et al.* 2001; P. Juesy, personal communication). Only one case resulted in conviction of the offender. Available information indicates that hunters are responsible for the illegal killing (Ceza *et al.* 2001). However,

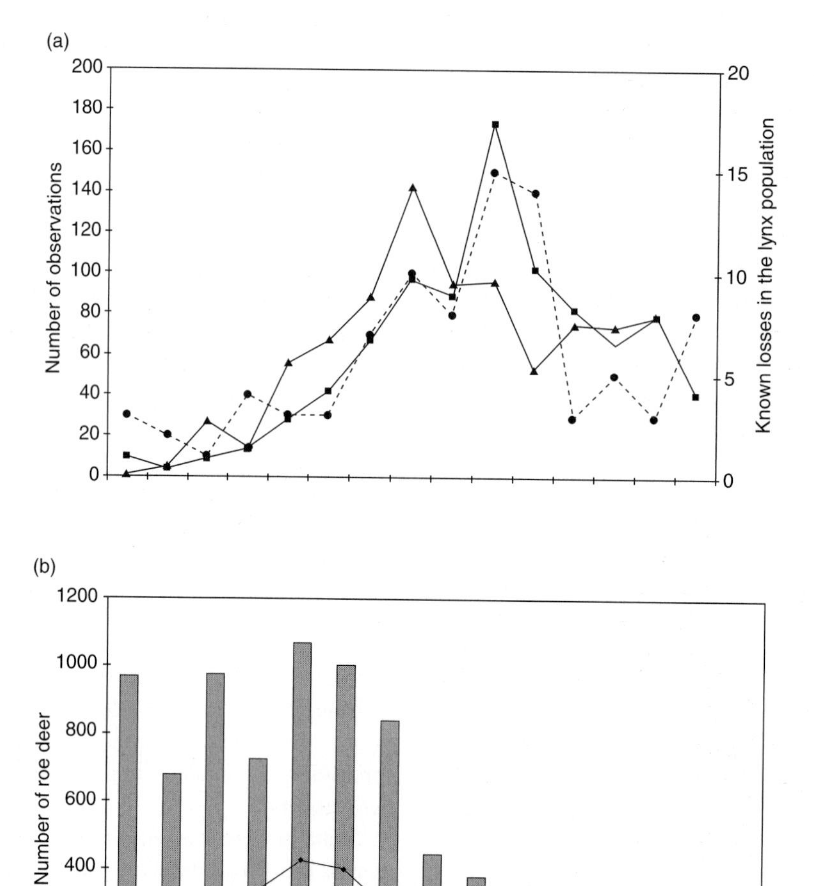

Figure 23.4 Development of (a) lynx monitoring indicators: triangles, chance observations; squares, attacks on livestock; dots, known lynx mortality; and hunting bag; and (b) roe deer mortality: bars, hunting bag; curve, other known mortality excluding lynx predation in the north-western Alps (Zimmermann *et al.* 2005c). The numerical response of lynx on first the increase and then the decrease of its main prey and the strong predation impact of lynx in the years 1997–2000 suggest a typical predator–prey system. However, changing winter mortality and hunting management of roe deer and human-caused losses of lynx contributed to the observed population fluctuations.

the conflict goes beyond the anger of hunters over a competitor for game. Seventy-five per cent of the Swiss accepted the presence of lynx in an inquiry in 1997 (Wild-Eck and Zimmermann 2001). When the above-mentioned series of illegal killings became public in 2001, the approval increased in the city of Zurich to 88%, but it decreased in the north-western Alps to 47% (Hunziker *et al.* 2001). People declared their solidarity with hunters and sheep breeders (both part of the local society), resulting in several political motions at the regional and federal level aiming to repeal the legal protection of lynx.

In this case, and in previous similar, but less well-documented cases, the social capacity to tolerate conflict was clearly below the ecological carrying capacity of the species concerned. Coexistence seemed possible at densities of about one lynx per 100 km^2 (residents and floaters); the density observed in the north-western Alps during peak years was about two lynx per 100 km^2. The Swiss lynx management plan, first enacted in 2000 and revised in 2004 (Buwal 2004), aims to mitigate this conflict through reducing local abundance and expanding the distribution of the species through translocation projects, such as the one in North-eastern Switzerland. The problem, however, is that the requirements to be met for control of a local lynx population as defined in the management plan (increasing lynx abundance, decreasing prey abundance, and the absence of browsing damage over several years) only allow for such a delayed management response that they more likely enhance than mitigate natural mechanisms and fluctuations.

Conclusions

The real challenge of reintroducing lynx into the cultivated landscapes of Europe is not biological or ecological. Ecological conditions for large carnivores in many parts of Europe are today better than in historic times (Breitenmoser and Breitenmoser-Würsten 2008). The potential carrying capacity for wild ungulates and their predators is high in the cultivated landscapes of Switzerland with a mosaic of forests and pastures. While the ungulate populations are controlled for economic reasons (to prevent browsing damage), the lynx population can grow to its ecological carrying capacity, which is higher than local societies are willing to tolerate. The real challenge is reintegration of the predator into existing land use and wildlife management systems and human perception of nature and landscape in general, in order to gain the support of local people. Living with large carnivores in a modern, cultivated landscape requires a compromise between traditional land users and the conservation community. On one hand, hunters, wildlife, and forest managers have to accept that wildlife populations fluctuate and that predation can, under certain conditions, impede harvest or management goals. On the other hand, conservationists have to understand that maintaining the present traditional land use (hunting, livestock husbandry, and forestry) requires not only the (temporary) control of herbivore, but also of carnivore populations.

CHAPTER 24

Iberian lynx: the uncertain future of a critically endangered cat

Pablo Ferreras, Alejandro Rodríguez, Francisco Palomares, and Miguel Delibes

Adult Iberian lynx in the wild at Doñana National Park. © Hector Garrido (Estación Biológica de Doñana, CSIC).

The Iberian lynx (*Lynx pardinus*) is at risk of being the first member of the Felidae to go extinct since the last glaciation. Over the past 20 years its status has plummeted successively from vulnerable, to endangered then critically endangered. Despite regional, national, and international efforts, their numbers have fallen by at least 80%, from 1000 to fewer than 200, between 1980 and 2008. How did this situation develop and why have conservation initiatives failed to remedy it in the educated and developed Western Europe? In this chapter, we describe the slide of the Iberian lynx into an extinction vortex and, unhappily, face up to the enormity of the challenge to save it. Crucial factors are that: (1) conservationists, being insufficiently aware of this species' plight, reacted too late; (2) extreme vulnerability of the species is now largely due to uncontrollable natural causes; which (3) have been exacerbated, intentionally or unknowingly, directly or indirectly, by human activities. To escape the resulting vortex, passive conservation measures are necessary but not sufficient. If the Iberian lynx is to survive, active, indeed assertive, evidence-based conservation must include the restoration of habitat (including wild rabbit populations), reintroductions, supplementary feeding, and captive breeding. Even then, a recovery is not guaranteed, and much now depends on chance.

Need for conservation knowledge: changes in human attitudes towards the Iberian lynx

With most species undescribed, it is obvious that ignorance handicaps biodiversity conservation (Wilson 2001). It is not surprising that many invertebrate species in remote wilderness areas go unstudied, but it is startling when one of the most charismatic carnivores in Europe, the Iberian lynx, slips towards extinction unnoticed.

Part of the complacency stemmed from uncertainty over the lynx's taxonomic status. Miller (1912b) recognized it as a species, but some later authors demoted it to a subspecies of the Eurasian lynx (*Lynx lynx*; Ellerman and Morrison-Scott 1951; Corbet 1978). Others underestimated its rarity by overestimating its distribution, which they believed extended throughout southern Europe and perhaps Asia Minor (e.g. van den Brink 1970). In the past 20 years, it has been confirmed to exist only in the Iberian Peninsula, where its status as a distinct species has been confirmed by fossil, morphological, ecological, and genetic evidence (Werdelin 1981; García-Perea 1992; Beltrán *et al.* 1996; Johnson *et al.* 2004).

The Iberian lynx has not engendered the blend of fear and fascination prompted by the wolf (*Canis lupus*) and the bear (*Ursus arctos*), so culturally it attracted only passing mention in the annals of game hunters. By the time Cabrera (1914) undertook the first serious study of Iberian mammals, it was considered vermin, in the same category as the common and resilient red fox (*Vulpes vulpes*), and persecuted as a menace to small game and poultry. This situation prevailed until 1973 and 1974, when it was protected by Spanish and Portuguese regulations, respectively. Until the end of the twentieth century, the role of the Iberian lynx as a top predator, and its influence on smaller carnivores and rabbits, were unknown (Palomares *et al.* 1995, 1996).

Valverde's pioneering observations (1957) on the Iberian lynx's activity and hunting behaviour were rigorous but fragmentary, laying a foundation for a spate of research in the 1980s, initially in the Doñana National Park (south-west Spain) and in the 1990s in the Sierra Morena mountains (south-central Spain; see Fig. 24.2; Delibes *et al.* 2000, unpublished data).

Knowledge of the species brought a radical shift in perception. The Iberian lynx was transformed from vermin to Iberian icon. There remains a significant risk that this change came too late.

With the recent available information, and based on four factors—habitat specificity, geographic range size, body size (which provides a rough estimate of population density), and an index of human persecution, called 'active threat'—the Iberian lynx was ranked as the most vulnerable wildcat in the world (Nowell and Jackson 1996). We first describe the natural traits (range size, habitat specialization, and body size) that determine the natural vulnerability of the Iberian lynx. Then we reconstruct the path that brought it to the verge of extinction, list major issues that should guide conservation action, and prioritize conservation-oriented research lines for the future.

Natural vulnerability of the Iberian lynx

Evolution and a small geographic range

Palaeontological evidence identifies the Pliocene–Early Pleistocene form *L. issiodorensis* as the common ancestor of the Iberian, Eurasian, and Canada lynx (Kurtén 1968). This is consistent with a comparison of sequence variation in felid mitochondrial DNA (mtDNA), suggesting that these three congeners diverged into monophyletic lineages during the Early Pleistocene (1.53–1.68 Ma; Johnson *et al.* 2004). It is thought that the smaller Iberian lynx was the product of evolution in isolation within glacial refuges of the southern Iberian Peninsula (Ferrer and Negro 2004; Johnson *et al.* 2004), preceded by a gradual trend in body-size reduction documented in the fossil record (through *L. i. valdarnensis* and *L. pardinus spelaeus*; Werdelin 1981). Apparently *L. issiodorensis* hunted ungulates of the size of a roe deer (*Capreolus capreolus*), and the lineage towards smaller forms could have avoided competition from other felids by exploiting smaller prey, such as Pleistocene leporids (Werdelin 1981). These were larger than the modern European rabbit (*Oryctolagus cuniculus*), currently the staple prey of the Iberian lynx, which originated on the Iberian Peninsula (López Martínez 2008). The oldest European rabbit fossil is 0.6 Ma (López

Martínez 2008), whereas mtDNA analyses date its divergence as a distinct species at about 2.0 Ma (Biju-Duval *et al.* 1991; Branco *et al.* 2000), slightly earlier than the Iberian lynx.

The distribution of Late Pleistocene Iberian lynx fossils indicates that the range of the Iberian lynx reached about 650,000 km^2, that is, 11 to 21 times smaller than the current range of other lynx species (Palomares and Rodríguez 2004). The fossil record identifies the northernmost boundary of *L. pardinus spelaeus* as southern France (Rodríguez and Delibes 2002), matching the distribution of *Oryctolagus cuniculus* during the glacial periods of the Pleistocene (López Martínez 2008). One potential drawback of body-size reduction in Iberian lynx ancestors might have been a lower efficiency in hunting larger prey during interglacial periods. The expansion of the European rabbit to northern France and the Iberian peninsula during the Upper Pleistocene (0.1–0.01 Ma; López Martínez 2008) was apparently followed, at least in part of this area, by a parallel expansion of the Iberian lynx, as shown by recent fossil and subfossil records in eastern and southern France during the Upper Pleistocene and the Early Holocene (0.035–0.005 Ma; Callou and Gargominy 2008). A subsequent return and confinement of the Iberian lynx to the Iberian Peninsula could be due to a later contraction of the range of the European rabbit. Other factors, such as unsuitable hunting habitat in the temperate forests of central Europe during interglacial periods have been suggested (Ferrer and Negro 2004), but the ultimate reason for the ever-shrinking range of the Iberian lynx is far from clear.

Subfossil remains show that the Iberian lynx was widely distributed across most of the Iberian Peninsula, excluding a strip in the north and north-west (Fig. 24.1). By 1950, it was found only in 10% of its historical range, in the south-west of the Iberian Peninsula (Fig. 24.2), although some uncertain, relict populations were still reported in the north and east (Cabrera 1914; Rodríguez and Delibes 1990). Around 1980, the species was confined to 10 highly fragmented populations representing less than 3% of Iberia (Rodríguez and Delibes 1990; Castro and Palma 1996). Total population size was about 1100 individuals, and medium-to-high lynx density (>0.13 individuals per km^2) was found only in 17% of the lynx range (Rodríguez and Delibes 1992). A similarly low percentage of locations with medium-to-high lynx density was found using independent data for different periods between 1950 and 1988 (Rodríguez and Delibes 2002). By 2002, and according to Guzmán *et al.* (2004), less than 200 individuals remained (with less than 50 breeding females), in two small, isolated (230 km apart) populations (Sierra Morena and Doñana, Fig. 24.2). Although the total population could be slightly underestimated (according to own simultaneous data from Doñana; Román *et al.* 2005), it is unquestionable that in 50 years the Iberian lynx range has been reduced by 99%, from 58,000 km^2 in 1950 to 350 km^2 in 2000 (Rodríguez and Delibes 1990, Guzmán *et al.* 2004).

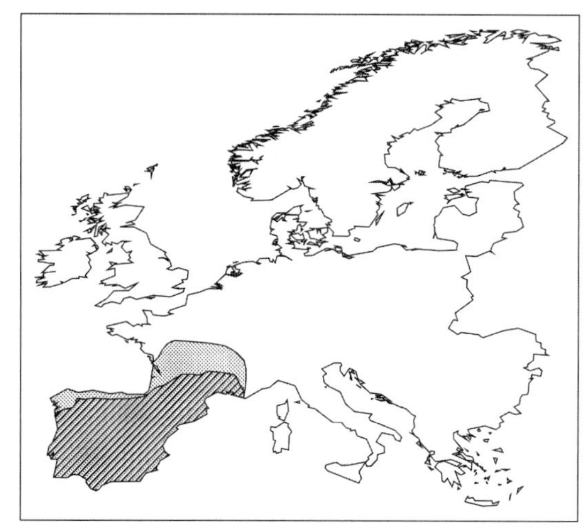

Figure 24.1 Hatched area: estimated maximum historical range of the Iberian lynx according to Rodríguez and Delibes (2002). Dotted area: natural distribution of the European rabbit (Iberian lynx's staple prey) during the maximum glacial period, before human-driven expansion. (After López Martínez 2008.)

A rigorous food and habitat specialist

Regarded as one of the strictest habitat and prey specialists among felids, the Iberian lynx is restricted to Mediterranean shrubland (Palomares *et al.* 2000). This habitat dependence is related to its specialization on European rabbits, whose natural distribution and origin are restricted to SW Europe (Fig. 24.1). One rabbit provides roughly the daily energy

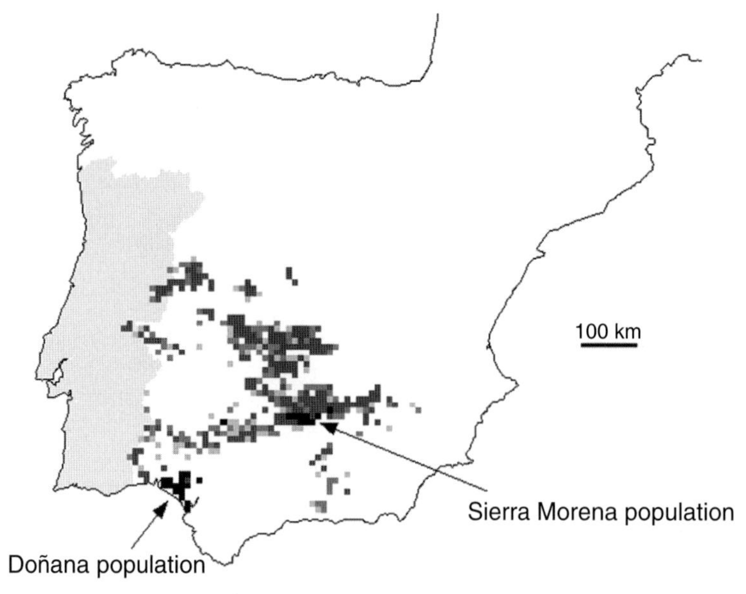

Figure 24.2 Contraction of the Iberian lynx geographic range in Spain since 1950, represented on a 10-km UTM grid. Cell shading from light grey to black corresponds to the distribution estimated for 1950 (lightest grey), 1975, 1985, and 2002 (black). The main difference between the 2002 map and the estimated distribution for 2008 is that the small relict of central Sierra Morena has most likely gone extinct. Comparable data were not available for Portugal (shaded area). (Modified from Rodríguez and Delibes 2002 and Rodríguez 2007.)

required by an adult Iberian lynx (673–912 Kcal; females and males, respectively), constituting what is known as a 'ration prey' (Aldama *et al.* 1991). Although Iberian lynx hunt other prey such as ungulates, small mammals, ducks, and other birds, these represent a small fraction of its diet. Studies have shown that rabbits account for 80–99% of the Iberian lynx's diet, suggesting it may be unable to switch to alternative prey (Delibes 1980; Aldama and Delibes 1990; Beltrán and Delibes 1991; Fedriani *et al.* 1999; Palomares *et al.* 2001; Gil-Sánchez *et al.* 2006; Zapata *et al.* 2007). The lowest rabbit densities within stable Iberian lynx breeding territories average between 1 and 4.5 individuals per ha in autumn and spring (lowest and highest rabbit abundance, respectively; Palomares *et al.* 2001). This trophic specialization helps explain why Iberian lynx densities (average population density < 0.21 individuals per km^2; Rodríguez and Delibes 1992) are lower than those reported for lynx species of similar size, which may hunt a higher number of prey species (e.g. up to 1 individual per km^2 for *L. rufus*; McCord and Cardoza 1982).

Historic and recent distribution of Iberian lynx is associated with landscapes rich in Mediterranean scrubland (Rodríguez and Delibes 1992; Fernández *et al.* 2006), usually supporting high rabbit densities. Studies in Doñana (Palomares *et al.* 2000, Palomares 2001) indicate that Iberian lynx show a preference for mid-density scrubland (58% ground cover), with frequent occurrence of tall shrub species (20% ground cover), used as diurnal refuges and as secondary dens by females (Fernández and Palomares 2000), low tree cover, and absence of pine and eucalyptus plantations (Palomares *et al.* 1991). Suitable breeding habitat also includes places for giving birth, such as hollow trees (Fernández and Palomares 2000; Fernández *et al.* 2002) or rock cavities (Fernández *et al.* 2006). The availability of water within home ranges might be another limiting factor as lynx visited a water point three times per day on average ($n = 97$, 24-h tracking periods), as compared to 1.3 times per day to any other points within the home range (Palomares *et al.* 2001). Home range size of resident lynx was inversely related to rabbit density ($R^2 = 0.986$, $n = 5$

territories), density of scrub-pasture ecotones and mean coverage of tall shrubs ($R^2 = 0.65$; Palomares 2001; Fernandez *et al.* 2003), probably because interspersion of woody cover and pastures not only offers food and refuge for rabbits, but also a good hunting habitat structure for lynx (Fernández *et al.* 2003).

Dispersing lynx have slightly different needs. Compared with areas used by resident lynx, dispersing lynx tolerate lower scrub density (45% vs 56%) and a higher proportion of trees (32% vs 14%; Palomares *et al.* 2000). The vegetation structure of areas used for dispersal clearly differs from areas rarely or never used (Palomares *et al.* 2000). Rabbit abundance is also higher in areas used by resident individuals (mean = 78 pellets per m^2) than in dispersal areas (mean = 35 pellets per m^2), which, in turn, is slightly higher than in unused sites (mean = 32 pellets per m^2).

Obviously, specialization generates dependence and therefore vulnerability in our rapidly changing world (Davies *et al.* 2004). The strong dependence on a single prey and a specific habitat makes the Iberian lynx an obligate user of such resources, and inevitably links its persistence to their fate. A sharp decrease in rabbit abundance and reduction of other components of habitat quality, as a result of human-caused transformations or fragmentation, have contributed to the recent contraction of the Iberian lynx's geographic range (Delibes *et al.* 2000; Rodríguez and Delibes 2002, see below).

Size and social system as determining factors of density

The Iberian lynx is a medium-sized felid, weighing approximately half as much as its Eurasian congener *L. lynx*, and about the size of *L. canadensis* and *L. rufus*. On average, adult male Iberian lynx weigh 12.8 kg (22 kg *L. lynx*) and females 9.3 kg (18 kg *L. lynx*) (Beltrán and Delibes 1993; Nowell and Jackson 1996).

It is estimated that in order to maintain a healthy to moderate-sized population (e.g. 10–15 breeding females) of Iberian lynx, about 150 km^2 of more or less continuous preserved habitat are needed. Rabbit density primarily determines lynx density, through its negative effect upon home range size (Palomares *et al.* 2001). However, like most solitary felids, lynx have exclusive territories (Ferreras *et al.* 1997). This might impose a limit in the potential for compression of lynx territories as a response to increasing rabbit density. The highest adult Iberian lynx density was reported in some places within Doñana as 0.77 adults per km^2 (Palomares *et al.* 2001). Male territories are on average 1.2 times larger (male ranges: 9.3 ± 1.2 km^2; female ranges: 7.8 ± 2.0 km^2; Ferreras *et al.* 1997; Palomares *et al.* 2001) and significantly overlap those of females. In Doñana, high competition forces males to have small territories, resulting in a mating system closely resembling monogamy (Ferreras *et al.* 1997). Possibly for this reason, fatal injuries as a result of territorial disputes are sometimes observed in wild-living adults (Fedriani *et al.* 1995; Ferreras *et al.* 1997). In comparison, a polygynous mating system could occur in Sierra Morena, where densities of lynx are lower (0.185 adults per km^2; Consejería de Medio Ambiente de Andalucía 2006) and extensive tracts of suitable habitat are available (Fernández *et al.* 2006). Recent genetic data indicate that males can sire multiple litters per year, which would suggest a polygynous mating system (authors, unpublished data).

High dispersal rates of both males and females (80% of subadults have dispersed by 2 years of age, $n = 41$) are also precipitated by competition for territories in Doñana (Ferreras *et al.* 2004). Isolation of the Doñana population by natural and anthropogenic barriers may have resulted in shorter natal dispersal distances (maximum distance recorded is 38 km, $n = 27$) than those reported for other similar-size solitary felids (Ferreras *et al.* 2004).

Evidence from field studies (data from females radio-tracked since they were young; Palomares *et al.* 2005) and captive breeding indicate that the age of first reproduction is ~3 years (Wildt *et al.*, Chapter 8, this volume). Breeding rate (probability of an adult resident female giving birth in a given year) was estimated as 83% ($n = 29$ observations of female-years) in the northern Doñana population (see Fig. 24.2; Palomares *et al.* 2005), which supports one of the highest rabbit densities (5.6 rabbits per ha within lynx home ranges on average; Palomares *et al.* 2001) across the Iberian lynx's range. Lower breeding rates are expected in areas with lower rabbit densities. Although Palomares *et al.* (2005) did not detect any correlation between the abundance of

European rabbits and lynx-breeding parameters (total number of cubs born, and number of breeding females) in the northern Doñana subpopulation, at a larger spatial scale, a declining trend in the probability of breeding in the whole Doñana population (from 54% to 31%, $n = 13$ monitored territories between 2002 and 2004) coincided with a sustained decrease in rabbit numbers (from 43.3 ± 3.8 to 11.6 ± 1.7 rabbit pellets per m^2; Fernández *et al* 2007) due mainly to the effects of rabbit haemorrhagic disease (see below). Breeding performance is not only closely associated with landscape variables which influence rabbit abundance (effect of the landscape index—based on distance to water points, density of pastureland–shrubland ecotones, and mean shrub coverage—on the number of breeding records within a territory, $R^2 = 0.49$, $P = 0.021$; Fernández *et al.* 2007) but also with less well-known factors, such as individual quality and experience, or probability of finding mates (Palomares *et al.* 2005).

Average litter size is close to three (Palomares *et al.* 2005), and plasticity in this trait is less apparent (63% litters consisted of three cubs, range 2–4, $n = 16$ wild-born litters) than in other lynx species such as the Canada lynx (*L. canadensis*) which, at peak prey densities, can have an average litter size of up to five (Mowat *et al.* 1996). Iberian lynx cubs suffer moderate mortality: 25% die before they reach 3 months ($n = 63$). Consequently, at 3 months most litters (64%, $n = 11$) are comprised of only two cubs (Palomares *et al.* 2005). Only 57% of cubs ($n = 42$ known-fate cubs) reach the age of dispersal (Palomares *et al.* 2005) and 48% of these survive and settle in a territory (Ferreras *et al.* 2004). As a result, under a third of the lynx born eventually reproduce. Based on this knowledge, in Doñana, where some litters have three or more cubs, the smallest have been removed, and incorporated into the captive breeding programme (Wildt *et al.*, Chapter 8, this volume).

Human-mediated threats and global change

The natural vulnerability of Iberian lynx favoured a process of decline that was exacerbated by human-mediated threats. Some of them (e.g. direct persecution) stem from human attitudes and can be changed in a relatively short time, but others (e.g. habitat destruction and transmission of diseases by feral domestic carnivores) are global trends that are very difficult to reverse at the local scale. Three significant human threats are distinguished: (1) direct persecution and other non-natural causes of mortality; (2) habitat deterioration; and (3) indirect effects from processes such as climate change or exposure to pathogens.

Direct persecution and other causes of non-natural mortality

During the nineteenth and early twentieth centuries, Iberian lynx populations were exploited for fur. Brehm (1880) reported that between 200 and 300 lynx pelts arrived annually to Madrid from the surrounding areas. From this and other fragmentary information, trapping and hunting appear to have been an important source of mortality for the Iberian lynx at that time (Rodríguez and Delibes 2004).

During the second half of the twentieth century, the Iberian lynx continued to suffer high non-natural mortality (Rodríguez and Delibes 2004). Between 1950 and 1988, information on 1258 non-natural lynx deaths was collected in central and southern Spain (averaging 31.5 losses per year). Trapping was responsible for 62% of reported deaths during this period, followed by shooting (26%). Road casualties resulted in 1% of deaths. Given the low population size of the Iberian lynx in 1980, this level of non-natural mortality may have contributed significantly to its rapid decline, but did not substantially affect the spatial pattern of relative abundance across its range. Indeed, mortality was higher in relatively dense lynx local populations (Rodríguez and Delibes 2004).

One possible explanation for the lack of a strong negative relationship between non-natural mortality levels and local density is that Iberian lynx were not specifically targeted. Predator control was directed towards carnivores that killed livestock or towards those abundant enough to deplete populations of game species, primarily red fox. Traps set to exploit furbearers were aimed at common species able to maintain a pelt market, but the Iberian lynx was rare in most places (Rodríguez and Delibes 2002). In addition, where abundant, rabbits were culled for meat, pelts, and to reduce agricultural damage, by specialized trappers who set hundreds of leg-hold traps, where lynx were also regularly captured. A parsimonious explanation is that lynx losses could simply be in proportion to their abundance, rather than due to any targeted persecution. However, the spatial pattern of mortality also seemed related to land use, and high levels of non-natural mortality, resulting from predator control, were recorded where small game hunting was a valuable resource. Predator persecution was 4.5 times higher on estates where rabbits were a game resource than on estates where rabbits were not hunted (Rodríguez and Delibes 2004). In that study, relative mortality was estimated as the ratio between death reports and total reports (deaths and sightings) in 10-km cells. The mortality ratio was independent of population size in that it varied little in both large and small populations (~0.37). Therefore, the potential edge effects in small lynx populations did not result in increased mortality. The mean mortality ratio remained fairly stable around 0.65 until 1970–74, and then showed a steady decline to 0.13 in 1985–89. In Sierra Morena, which has harboured the largest lynx population at least since 1950, the onset of this decline in mortality ratio (from 0.66 to 0.42 in 5 years) coincided with legal protection of the Iberian lynx in 1973. In other regions, a reduction in non-natural mortality ratios started in 1960, long before legal protection. However, in an independent study García-Perea (2000) inspected 44 lynx carcasses and found that the proportion of deaths from traps and snares increased after 1973 from 47% to 76%, indicating that legal protection alone was not effective in preventing the killing of Iberian lynx. It was unclear whether these 44 specimens constituted a random sample regarding their cause of death. One factor contributing to the reduction of non-natural mortality was a change in land use, from small game to big game, prompted by a higher economic profitability of big game (Rodríguez and Delibes 2002, 2004; see below). Road mortality has been reported to increase (e.g. González-Oreja 1998), although these observations may result from the fact that, except where radio-tracking is used, road mortality is more detectable than deaths caused by other factors.

In Doñana, survival analysis from radio-tracking data showed that human activities caused up to 75% of annual mortality rates during the 1980s (Ferreras *et al.* 1992). Illegal trapping (with leg-hold traps and neck-snares) and road accidents accounted for 58% of lynx losses in this period. In the 1980s and early 1990s most resident lynx in Doñana lived inside a National Park (see Fig. 24.3) and exhibited high survival rates (annual rate 0.9; Ferreras *et al.* 1992; Gaona *et al.* 1998), whereas dispersing lynx had much higher mortality rates. For example, the annual survival rate of dispersing individuals 12–24 months old was 0.48, compared with 0.92 for non-dispersing individuals of the same age (Ferreras *et al.* 2004). Habitat saturation led dispersing lynx outside the National Park where they were frequently illegally trapped (73%, $n = 11$) or killed by cars (9%; Ferreras *et al.* 2004). The situation changed after the arrival of rabbit haemorrhagic disease in 1990, which caused rabbit density to drop drastically within the National Park (Moreno *et al.* 2007; see below). Because of this, a large proportion of resident lynx moved outside the park (Román *et al.* 2005). Consequently, a larger fraction of resident lynx are now exposed to human activity and their mortality rates are expected to increase. Moreover, during this period over 100 km of paved roads have been built in the area surrounding Doñana (Fig. 24.3), causing a substantial proportion of lynx deaths (21%, $n = 24$, in the 1980s and 48%, $n = 60$, between 1992 and 2007), despite measures taken to reduce the impact of vehicles (reduced speed limits, underpasses associated with channelling fencing; Rosell *et al.* 2003). Park surroundings include large coastal developments, intensive greenhouse crops, a variety of tourist attractions and, therefore, intense traffic. This movement of resident lynx outside the park represents a significant threat to the persistence of the Doñana meta-population (Gaona *et al.* 1998; Ferreras *et al.* 2001).

Declines of many carnivores are usually attributed to deliberate persecution (Woodroffe 2001). However, the case of the Iberian lynx is remarkable in that apparently it bore high levels of non-natural mortality without being directly targeted. Instead, they were hunted opportunistically and indiscriminately caught in traps set mostly for other animals, and increasingly involved in traffic accidents. Non-

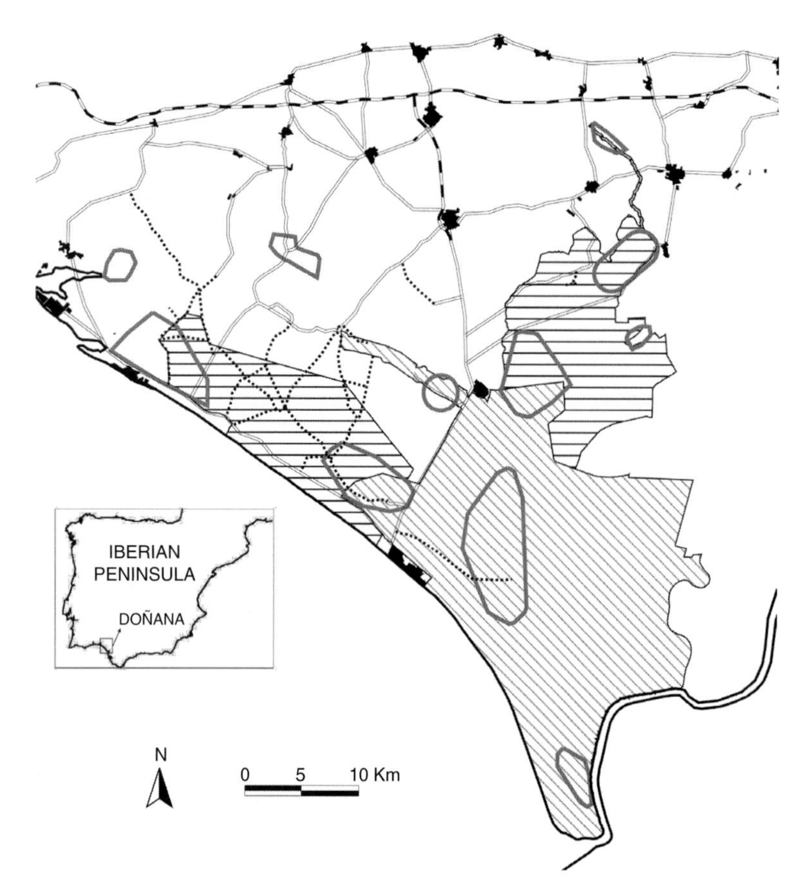

Figure 24.3 Spatial structure of the Doñana Iberian lynx metapopulation. Thick grey lines define local breeding populations; double and dashed lines represent highways; double lines represent roads; dotted lines represent consolidated dirt roads; diagonally striped area represents Doñana National Park; horizontally striped areas represent buffer protected area ('Natural Park'); and black areas represent towns.

natural mortality remains today as a serious threat to the few remaining lynx populations (Delibes *et al.* 2000; Ferreras *et al.* 2001, 2004), especially because most populations are very small (Rodríguez and Delibes 1992; Guzmán *et al.* 2004).

Habitat degradation

Lynx persecution and predator control can be regulated, but habitat destruction is more difficult to reverse, as it results from socio-economic demands that may conflict with lynx conservation. Intensification of agriculture and forestry, human migration to cities, and expansion of the transport network have all altered the Iberian lynx habitat.

Early settlers of the Iberian Peninsula generated cultural landscapes that might have favoured species linked to successional stages of Mediterranean forest, such as the European rabbit (Flux 1994). In this way, humans may have initially benefited lynx by generating mosaics that improved habitat and prey (Rodríguez and Delibes 2002). Transformation of these heterogeneous cultural landscapes into landscapes dominated either by open land (e.g. extensive agriculture in large tracts of the central plains of Iberia) or by homogeneously dense woody cover (e.g. blocks of scrubland in big game estates; see below) probably entailed a loss of habitat quality for the Iberian lynx (Rodríguez and Delibes 1992; Delibes *et al.* 2000). For instance, agricultural expansion during the first half of the twentieth century caused the loss of lynx habitat, although we lack data on the impact this had. During the second half of the century, native vegetation was extensively replaced by conifer and eucalyptus plantations. These may be used by dispersing lynx but are not suitable for resident individuals, since they are typically poor in rabbits (Palomares *et al.* 2000). Suitable habitat was therefore progressively reduced to isolated blocks or patches within a homogeneous forested or deforested matrix.

According to simulation models, the reduction of carrying capacity caused by habitat loss greatly increases the probability of extinction of lynx meta-populations, especially if habitat loss occurs in local source populations. For instance, a reduction of the carrying capacity of the Doñana meta-population from nine to seven territories increases the risk of extinction in 100 years from 45% to 80% (Gaona *et al.* 1998; Ferreras *et al.* 2001).

Open habitats, such as croplands, are avoided even by dispersing lynx (Palomares *et al.* 2000), and large tracts of open land efficiently isolate lynx populations (Rodríguez and Delibes 1992). Consequently, a composite index of isolation by distance that considered immigration from several nearby populations was a good predictor of population extinction (Rodríguez and Delibes 2003).

Another form of habitat degradation is the loss of natural linear structures, such as riparian vegetation, which can act as corridors for dispersing lynx (Palomares *et al.* 2000). The loss of such elements connecting local populations increases the probability of extinction of lynx meta-populations (Gaona *et al.* 1998). Moreover, simplification of the matrix structure could also limit the movements of dispersing lynx, reducing the connectivity between local populations, determining their extinction and, therefore, reducing meta-population persistence (Revilla *et al.* 2004). Simulation models indicate that maximizing connectivity between local populations would decrease the risk of extinction of the Doñana meta-population within 100 years from 45% to 2% (Gaona *et al.* 1998; Ferreras *et al.* 2001). Increasing connectivity would require low-cost measures such as the restoration of riparian vegetation or other linear elements lost, in most cases, to aggressive agricultural practices. These measures would conform to government land-use policy and would require effective law enforcement.

In SW Spain, the rise of the big game business started in 1970. The buoyant Spanish economy allowed an increase in the number of resident hunters, from 140,000 to 1,300,000 between 1946 and 1987 (López-Ontiveros 1991). Big game became a profitable business and most estates were fenced to build up dense populations of game ungulates. This was followed by a progressive transformation of landscapes, favouring thick and continuous patches of scrubland where big game can live undisturbed. A major outcome was that patches of scrubland interspersed with grassland (favourable for rabbits and lynx), formerly maintained by traditional management, were replaced by large blocks of scrubland, where rabbits do not attain high densities (Rodríguez and Delibes 2002).

In addition to habitat loss and transformation, the economic development of Spain during the second half of the twentieth century led to considerable expansion of the transportation network across lynx habitat (Rosell *et al.* 2003). The first high-speed railway built in Spain (from Madrid, in the centre, to Seville, in the south) fragmented lynx populations in a pre-extinction phase, both in Toledo Mountains and Sierra Morena, perhaps contributing to the extinction of local populations. Several motorways were constructed across habitat occupied by remnant lynx populations in the 1980s. Roads also preceded further human developments (agriculture, urban land, ecotourism, hunting, and forestry), bringing humans into contact with the least-disturbed areas of lynx habitat and making lynx populations progressively more vulnerable to such deterministic factors (Rosell *et al.* 2003).

Other processes related to global change

The possible impact of other global processes, such as the expansion of pathogens, genetically modified organisms (GMOs), proliferation of feral domestic animals, and climate change, on the recovery of the Iberian lynx is discussed below.

The Myxoma virus, which is common and less virulent in South American lagomorphs, was introduced to Australia in order to control invasive European rabbits around 1950, and to France by a farmer to reduce agricultural damage to his property in 1952. Very quickly, from France it spread throughout Europe, killing 99% of Britain's rabbits and presumably a similar proportion in Spain and Portugal. The lasting effects of the disease on rabbit densities are still obvious today. The expected response of a specialist predator such as the Iberian lynx was a sudden and generalized decline in numbers characterized by multiple local extinctions. However, a significant increase in the rate of range contraction did not occur until 20 years after the introduction of the myxomatosis. In other words, lynx populations were able to survive in most areas, although probably at lower density. These lower densities made the Iberian lynx more vulnerable to other environmental factors (Rodríguez and Delibes 2002).

In 1984, the Haemorrhagic Disease Virus (HDV) was discovered on domestic rabbit farms in China. It was detected in Italy in 1987 and soon across the rest of Europe (although it was a different strain, which apparently originated from previously harmless European strains; Forrester *et al.* 2006), and was introduced to Australia and New Zealand for rabbit control purposes. In the early 1990s, the virus killed 70–90% of domestic rabbits and 50–60% of wild rabbits in Spain (Villafuerte *et al.* 1994). Numbers in many areas, including those where the Iberian lynx occurred, have not yet recovered. For instance, in Doñana the estimated mean maximum monthly rabbit density declined from 8.6 rabbits per ha in June 1989 to about 1.5 rabbits per ha thereafter (Moreno *et al.* 2007). Some researchers (Angulo and Cooke 2002) have predicted that new exotic viruses will affect rabbit populations even if they acquire resistance to myxomatosis and HDV. At present, scientists from Australia, where rabbits are considered a pest, are working to modify the Myxoma virus to reduce rabbit fertility through a transmissible immunocontraception (Angulo and Cooke 2002). It is easy to imagine the devastating effects of this modified virus on rabbit-dependent predators, if it eventually arrives in the Iberian Peninsula.

In addition to impacting rabbit populations, pathogens may also directly affect the Iberian lynx. Worldwide, natural ecosystems are suffering from increasing human perturbation and, consequently, invasive feral domestic animals, such as cats and dogs, can transmit diseases to wild carnivore populations (Roelke-Parker *et al.* 1996; Randall *et al.* 2006). The Mediterranean ecosystem is no exception. A seroprevalence study in the 1990s revealed a low frequency of antibodies for most contagious diseases in the two lynx populations (Roelke *et al.* 2008; but see Luaces *et al.* 2008). However, more recently (2006/7), a virulent and sudden epidemic of feline leukaemia, probably transmitted by feral domestic cats, killed a significant proportion of the remaining lynx in Doñana (Meli *et al.* 2009; Millán *et al.* 2009).

We do not know how climate change will affect the Iberian lynx, either directly or indirectly, through the distribution or abundance of rabbits or the structure of Mediterranean scrubland. Predictions suggest that global warming will be particularly severe in SW Europe, with a significant increase in summer temperatures, although effects on patterns of precipitation are less clear (Diffenbaugh *et al.* 2007). At the very least, these conditions will considerably enhance the risk of forest fires, with potentially catastrophic consequences if they occur in the currently restricted lynx areas.

Preventing the arrival of rabbit viruses, eliminating feral cats and dogs, or reducing the frequency and dimensions of forest fires, which currently show an increasing trend, are daunting tasks, making the recovery of lynx even more uncertain despite conservation efforts.

The Iberian lynx population decline towards an extinction vortex

Hypotheses about the relative role of different factors (environmental vs anthropogenic and stochastic vs deterministic) responsible for the decline of the Iberian lynx have been derived from patterns of distribution and abundance within its geographic range, as well as the timing, magnitude, and location of range contraction during the past 50 years (Rodríguez and Delibes 2002). Analyses of such patterns revealed that peripheral populations had lower densities than central populations. In central areas of the lynx geographic range, there was a marked trend of increasing lynx density along mountain ranges towards the east, following a gradient of decreasing precipitation and productivity, a gradual shifting of the dominant land use from livestocking to game, and an increase in rabbit density (Rodríguez and Delibes 2002).

While the spatial pattern suggests a close relationship between lynx abundance and rabbit abundance, the timing of extinction peaks indicates that more than one causal factor was probably involved in the reduction of rabbit abundance, and that lynx populations resisted a sudden and sustained shortage in prey abundance for almost 20 years (Rodríguez and Delibes 2002). However, many lynx populations succumbed to further destabilizing factors that operated mainly in the 1970s. The structural homogenization of Mediterranean forest landscapes followed the massive emigration of the rural population towards urban areas, and extensive areas of native forest were converted to forestry. The third concurrent factor was the rise of big game hunting, leading to habitat management unfavourable for rabbits. Non-natural mortality associated with diffuse persecution, as a side-effect of predator control, helped to weaken or to drive lynx populations to extinction, mostly on hunting estates with small game (Rodríguez and Delibes 2004).

These deterministic factors apparently operated with uniform intensity across most of the lynx range, which is consistent with the observed density-dependent probability of persistence of local populations. Local extinction in larger, denser, self-sustainable populations probably resulted from the sole action of deterministic factors (Rodríguez and Delibes 2002).

Simulations of the process of range contraction under a random distribution of local extinctions showed that populations occupying areas <100 km^2 went extinct at frequencies higher than larger ones (Rodríguez and Delibes 2003). Indeed, only lynx populations living in areas <500 km^2 went extinct during the contraction process. Deterministic disturbance alone could not explain the high levels of extinction observed in areas occupied by small populations. Simulations also showed that isolated populations occurred at frequencies lower than expected, suggesting that they went extinct at higher rates than larger populations. The positive relationship between population isolation and extinction rates suggests that demographic instability in small populations was often prompted by reduced immigration and the subsequent disruption of meta-population equilibrium (Rodríguez and Delibes 2003).

Reduced to only two very small, isolated populations (Fig. 24.2), the Iberian lynx is currently within an extinction vortex (Gilpin and Soulé 1986), and is exposed to a high risk of extinction in the short term. In addition to deterministic factors, lynx genetic diversity is very low in Sierra Morena and extremely low in Doñana (Johnson *et al.* 2004). Recent studies suggest that reduced genetic variability affects individual fitness, probably through depressed immunocompetence (Peña *et al.* 2006). Demographic stochasticity also limits the ability of lynx populations to recover. In Doñana, most of the animals that died during the past few years were males. As of the summer of 2007, there is evidence of a lack of adult males, with several females established in scattered territories far away from the closest reproductive male. This Allee effect is characteristic of small populations and a clear symptom of the underlying extinction vortex. Furthermore, as lynx populations occupy only two small areas, there is an additional permanent risk derived from environmental stochasticity, in the form of catastrophic rains, droughts, or fires (e.g. Palomares 2003). In summary, inbreeding depression, Allee effect, environmental catastrophes, and other stochastic factors seem to be the most important forces in a typically positive feedback vortex.

Scenarios for conservation

It is not possible to escape from an extinction vortex using only passive conservation measures and waiting for the self-recovery of populations. To reverse Iberian lynx demographic trends, proactive management is needed to counteract the detrimental effects of prey scarcity and random fluctuations associated with small numbers.

General aims of active management should be: (1) to improve and recover rabbit populations across the historic range of the Iberian lynx; (2) to eliminate causes of non-natural mortality and habitat loss; (3) to improve reproduction and survival of present lynx populations; (4) to increase its area of occupancy (Gaston 1991) by natural recolonization of territories surrounding those already occupied, and creation of new lynx populations through reintroduction; (5) to ensure a healthy and productive captive lynx population; (6) to counteract the negative effects of loss of genetic diversity; and (7) to minimize disease transmission from wild reservoirs and domestic animals.

Currently, management consists of attempting recovery of natural rabbit populations, supplementary feeding of lynx, habitat management for both lynx and rabbits, captive breeding, and measures to prevent non-natural mortality.

Techniques commonly used to foster recovery of rabbit populations include restocking with wild rabbits (Moreno *et al.* 2004) and managing vegetation (Moreno and Villafuerte 1995). Wild rabbits are normally released into new areas after checking their genetic clade (Branco *et al.* 2002), and a quarantine period (Calvete *et al.* 2005). Vegetation is managed to ensure appropriate interspersion of scrubland and open areas, and artificial warrens are built to facilitate refuge and reproduction of translocated rabbits (Román *et al.* 2006). Although satisfactory results have been obtained on a small scale (Moreno *et al.* 2004), there are still no clear guidelines on how to augment rabbit numbers over areas large enough to support a self-sustaining lynx population. The home range size of an adult lynx is between 500 and 1500 ha, and the minimum area to rehabilitate rabbit populations needs to incorporate several lynx territories. Another indirect measure of enhancing rabbit populations is to encourage sustainable rabbit hunting in areas where lynx occur.

While wild rabbit populations recover, lynx need to be supplemented with food in order to, at least, maintain present numbers. Supplementary feeding is therefore a major short- and midterm management practice for lynx conservation. To accomplish this goal, small enclosures supplemented with domestic rabbits are now being tested in Doñana, with quite promising results. All 29 lynx with such enclosures within their home ranges used them. They needed, on average, 14 weeks to make regular use of the supplemented food (López-Bao *et al.* 2008). Simple enclosures of 4 × 4 m and 1.3 m tall were enough to successfully feed lynx. Lynx had access to enclosures by jumping, and rabbits could refuge in shelters to elicit hunting behaviour in visiting lynx. Lynx also used enclosures designed for wild rabbit restocking as an additional food source.

Although supplementary feeding appears to be a useful management option to maintain and augment lynx numbers in the near future, several aspects need to be addressed before it can be used with confidence. Supplementary feeding means that lynx have access to a predictable and stable source of food. This can potentially alter lynx behaviour, which is largely determined by the spatial and temporal distribution of food resources. If food is found in a predictable manner, lynx could: (1) reduce daily movements and, as a consequence, home range size; (2) share feeding stations, increase contact between individuals, decrease intraspecific territorial aggression, and increase the risk of disease transmission; (3) extend time spent in maternal territories, thereby delaying natal dispersal; and (4) modify their hunting techniques, thereby compromising successful learning of young lynx to hunt rabbits in the wild once they leave areas with supplementary feeding. On the other hand, supplementary feeding should: (1) allow lynx to keep territories in supplemented areas irrespective of the density and temporal fluctuations of wild rabbits; (2) decrease lynx home range size, allowing increased lynx densities; (3) increase reproductive success and productivity; (4) allow the settlement of dispersing lynx in areas with lower wild rabbit densities by increasing the area of available habitat; and (5) attract lynx away from human-impacted areas, thus reducing human-caused mortality. All these issues need to be scientifically addressed in order to make supplementary feeding an effective and secure conservation tool for Iberian lynx.

A captive breeding programme, started in 2003, aims to have at least 60 captive lynx by 2009, distributed among five different facilities (Wildt *et al.*, Chapter 8, this volume). The programme aims to maintain 80% of the present genetic variability through careful selection of founders and appropriate cross-breeding. In 2010, individuals should be available for reintroduction in the field. However, at present, the main challenge is to learn how to breed lynx in captivity and to generate effective protocols for raising healthy and valuable lynx for potential reintroductions. Many aspects need to be investigated in the near future regarding reproduction, physiology, diet, diseases, behaviour, and husbandry of captive lynx before these goals can be successfully accomplished.

Finally, to prevent road casualties, underpasses are being built in areas where lynx occur, and mitigation measures to decrease traffic speed are being implemented. For instance, six underpasses have been built at points where lynx have repeatedly been killed by cars (although whether lynx use underpasses has not been properly quantified), and traffic-calming measures have been implemented, including reducing the maximum speed limit, and the installation of speed humps, speed cameras, and road signs announcing entry into lynx areas. Recently, the regional government of Andalusia launched a programme to address the problem of road deaths in Doñana. Legal protection has little effect if not followed by appropriate monitoring and law enforcement, accompanied by information campaigns to increase the awareness of people living or hunting in lynx areas (Delibes *et al.* 2000). Measures to prevent illegal hunting should continue to be implemented by managers. In addition, infectious diseases have recently been detected as a potential mortality source for Iberian lynx, therefore the numbers of domestic carnivores (cats and dogs) in lynx areas must be lowered to minimize disease transmission (Millán *et al.*, 2009).

There have been some complaints in the newspapers about the high public investment required for conserving each Iberian lynx. Although there is no estimate of how much society is willing to invest for conserving lynx, there are some indirect indicators of public acceptance for lynx conservation programmes. For instance, in 2002 the Andalusian authorities launched a public petition on the Internet in support of the conservation of the Iberian lynx. This petition has been signed by 66,000 people, including private individuals, companies, and public institutions. In November 2007, the governments of Portugal and several Spanish regions from where lynx recently disappeared have endorsed this petition.

Meta-population management should be urgently considered to prevent Iberian lynx extinction. The Doñana lynx population needs to be demographically and genetically augmented. Short term, demographic management could include relocation of lynx from sinks to sources within Doñana, or translocation of lynx from the larger population of Sierra Morena to the source of the Doñana meta-population. Spatially explicit models clearly indicate that if present source areas lose one more territory, lynx extinction in Doñana will occur within the next 15–20 years. However, an increase in carrying capacity to 23 territories within source areas, effective connectivity between source areas, and translocation of 10 lynx in the next 5 years would decrease the probability of lynx extinction over the next 100 years in Doñana to less than 5% (Revilla *et al.* 2007). At present, some areas in the source nuclei lack reproductive lynx (males or females) because no lynx are available to occupy these territories. Doñana lynx have a lower genetic variability than those of Sierra Morena (Johnson *et al.* 2004; J.A. Godoy, personal communication), and mating is now occurring between closely related individuals (parent–offspring, siblings). Translocations of lynx from Sierra Morena to Doñana may therefore reduce the risks of inbreeding in the Doñana population. However, this may potentially increase the risk of introducing parasites (Mathews *et al.* 2006), such as the blood parasite *Cytauxzoon* sp., which at present has been only detected in Sierra Morena (Millán *et al.* 2007). Taking measures to prevent this risk, one adult male from Sierra Morena has been successfully translocated to northern Doñana in winter 2007 and the first mixed litters were born in the wild in spring 2008.

The recently declared feline leukaemia epidemic (2007) in the Doñana population has prompted the design and implementation of an urgent control plan. All wild lynx of the Doñana population are being captured and tested for the disease. Sero-negative animals are vaccinated, whereas sero-positive lynx are removed from the wild and kept in captivity. Feral dogs and cats are being culled in lynx areas, as vectors and reservoirs of this and other lynx diseases.

The establishment of new populations should be encouraged through reintroductions into areas where lynx have recently disappeared and where habitat and rabbit abundance are optimal for lynx (or at least amenable to restoration), reducing the risk of simultaneous extinction of remaining populations, expanding the current geographic range, and increasing the probability of survival in the wild over a longer timescale.

Conclusions: new lines of research and the fate of the Iberian lynx

Basic research on the biology and ecology of Iberian lynx has had an important role in guiding conservation and implementing recovery plans. Studies on social and spatial organization, reproduction, activity patterns, or habitat use, have been important in understanding lynx requirements. However, the focus of research should shift to: (1) learning how to restore the ecological conditions that allow lynx settlement and reproduction; (2) learning more about new, and sometimes subtle, threats, such as risks associated with inbreeding, or a higher susceptibility to infectious diseases of wild populations; and (3) evaluating the response of free-ranging lynx to management measures, and optimizing procedures leading to successful captive breeding, restocking, and reintroduction.

The Iberian lynx provides an opportunity to learn about extinction in a changing world, where nature conservation is an increasing public demand. However, lynx conservation conflicts with other public interests (Loveridge *et al.*, Chapter 6, this volume), such as the demands of the urban society for public welfare, often involving aggressive land-use changes, or high-speed transportation infrastructures, which tend to exclude more naturally sensitive species. Moreover, in a globalized world, such species are especially affected by unexpected processes, including disease and large-scale changes (such as land-use changes).

The Iberian lynx can act as a model for the conservation of other, little-known felid species. Some of these felid species may not yet be recognized as highly endangered, and the threats they face may be unknown. These species may have already entered into an extinction vortex which will need a significant amount of human and material effort to be alleviated. Only time will tell whether the conservation measures for the Iberian lynx will succeed and whether the extinction process will be reversed, but many useful lessons for the conservation of other felids will be drawn along the way.

Acknowledgements

The information contained in this chapter has been collected as a result of several research projects, funded mostly with public funds from Spanish regional and national administrations, as well as several EU LIFE projects. E. Revilla, J. Calzada, J.A. Godoy, A. Vargas, J. Román, G. Ruiz, J.J. Aldama, N. Fernández, J.V. López-Bao, J.C. Rivilla, A. Píriz, S. Conradi. and M.A. Simón, among other scientists and managers, have contributed to the current knowledge on the Iberian lynx reflected in these pages. A. Rodríguez was supported by Consejería de Innovación, Ciencia y Empresa, Junta de Andalucía, through a research grant during the preparation of this chapter.

CHAPTER 25

Cyclical dynamics and behaviour of Canada lynx in northern Canada

Mark O'Donoghue, Brian G. Slough, Kim G. Poole, Stan Boutin, Elizabeth J. Hofer, Garth Mowat, and Charles J. Krebs

The large paws and long legs of Canada lynx facilitate travel over the deep powdery snows of the boreal forest. © Mark O'Donoghue.

Introduction

Canada lynx (*Lynx canadensis*) inhabit boreal and montane coniferous forests throughout northern North America. These medium-sized cats (9–12 kg) are famed for their regular '10-year' cycles in abundance that follow those of their main prey, snowshoe hares (*Lepus americanus*). Predation by lynx and a suite of other mammalian and avian predators is thought to interact with hare–vegetation dynamics to generate the hare cycle (Krebs *et al.* 1995). Records of fur harvest kept by the Hudson's Bay Company and Statistics Canada since the early 1800s have provided a time series of unusual length that has been analysed extensively by ecologists interested in population regulation, predator–prey interactions, and geographical structuring of population cycles (Elton and Nicholson 1942; Keith 1963; Royama 1992; Ranta *et al.* 1997; Stenseth *et al.* 1997, 1998; Gamarra and Solé 2000).

Throughout most of their range in the contiguous northern boreal forests, populations of lynx reach cyclical peaks in abundance every 8–11 years, followed by steep crashes in numbers when snowshoe hares become scarce. These patterns are reflected in levels of fur harvest of lynx (Fig. 25.1). Peak densities may be up to 17 times those at cyclical lows, and they are roughly synchronous, within 2–3 years, across boreal North America. The drastic fluctuations in numbers of lynx are accompanied by corresponding changes in their demography, movements, foraging behaviour, and social interactions, so any discussion of Canada lynx must consider the long-term cyclical patterns in virtually all aspects of their ecology.

Despite the historical prominence of analyses of the lynx cycle, there have been surprisingly few field studies of lynx in North America, and most of those have been short term (2–3 years; e.g. Saunders [1963b] in Newfoundland; Parker [1981] and Parker *et al.* [1983] in Nova Scotia) and from the southern part of the species' range (e.g. Mech [1980] in Minnesota; Carbyn and Patriquin [1983] in southern Manitoba; Koehler [1990] in Washington; Apps [2000] in southern British Columbia; Squires and Laurion [2000] and Squires and Ruggiero [2007] in Montana and Wyoming). Even the pioneering studies of Lloyd Keith and colleagues in central Alberta (Nellis and Keith 1968; Nellis *et al.* 1972; Brand *et al.* 1976) were carried out in a landscape that was about one-third cleared agricultural land, which is atypical of the intact boreal forests characteristic of most of the species' range. Much of current understanding of the demographics and changes in diet associated with the lynx population cycle has come from analyses of carcasses collected from trappers (e.g. Saunders 1963a; Van Zyll de Jong 1966; Brand and Keith 1979; O'Connor 1986; Quinn and Thompson 1987).

Increasing prices of lynx pelts in the early and mid-1980s (they more than tripled in five years to a high of $720 Canadian in 1984–85) caused concerns that lynx might be over-harvested in accessible areas. This led to the initiation of several longer-term (6–10 year) field studies in the core of the geographical range of lynx in north-western North America. This chapter summarizes the results of three studies in the Yukon Territory and the Northwest Territories (hereafter 'NWT') in northern Canada. Along with the central Alberta research further south, these represent the first field studies that span the whole 10-year lynx cycle.

The three studies

All three studies were carried out in relatively intact boreal forest habitats with either low or very light trapping activity. Two were located in the southern

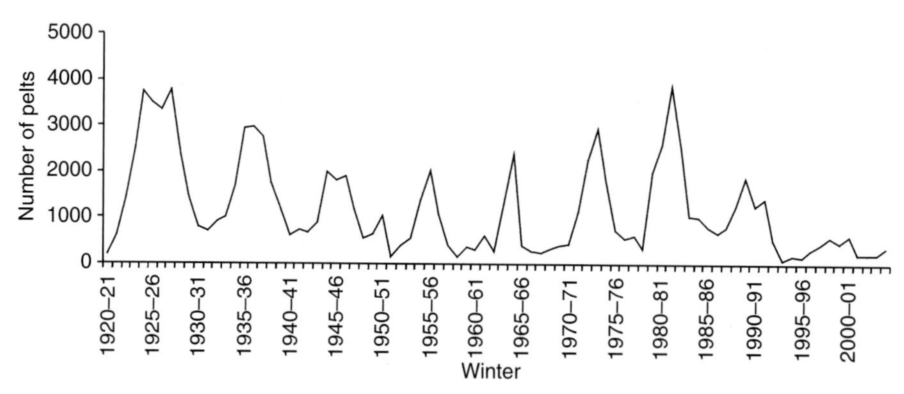

Figure 25.1 Reported number of lynx harvested in the Yukon Territory, Canada, from 1920–21 to 2004–05. (Data from Yukon Department of Environment.)

Yukon Territory (Kluane Lake study area, 350 km^2, 60° 57′ N, 138° 12′ W; Snafu Lake study area, 301 km^2, 60° 15′ N, 135° 20′ W; Fig. 25.2) and one in the south-western NWT (135 km^2, 61° 35′ N, 116° 45′ W; Fig. 25.2). The climate in all three areas is cold continental, with mean January temperatures from −20°C to −28°C, mean July temperatures from +10°C to +16°C, and continuous snow cover typically from October to May.

The Kluane Lake study (hereafter 'Kluane') was conducted in a broad glacial valley bounded to the north and south by alpine tundra. Elevations in the study area range from about 830 m to 1170 m. The dominant vegetation was mature white spruce (*Picea glauca*) forest with scattered aspen (*Populus tremuloides*) stands and, especially at higher elevations and in riparian habitats, a dense but patchy shrub layer of willows (*Salix* spp.), bog birch (*Betula glandulosa*), and soapberry (*Shepherdia canadensis*). Lynx abundance, survival, activity patterns, movements, and social organization were studied by live-trapping, radio-collaring, and monitoring 56 lynx from 1987 through 1996. Regular monitoring of lynx tracks in the snow along a 25-km transect (run an average of about 1100 km per winter) and intensive snow-tracking (following 2500 km of lynx trails) during winters from 1987–88 through 1996–97 were used to supplement our data on abundance and to collect information on recruitment, foraging behaviour, and habitat use. This research was a part of a long-term study of the structure of the vertebrate food web in the boreal forest, and the abundance and dynamics of prey species and other predators were monitored regularly by intensive live-trapping, mark–recapture, and telemetry (Krebs *et al.* 2001).

The Snafu Lake study area (hereafter 'Snafu') was on a plateau characterized by weathered mountains and dissected by creek valleys; elevations range from 800 to 1950 m. Most of the area (73%) had burned in 1958 and was vegetated with a matrix of dense young lodgepole pine (*Pinus contorta*), white spruce, aspen, and willows; only about 9% of the area was mature forest, and these stands varied greatly in size and were scattered through the area; 14% was alpine tundra and lakes. One-hundred-and-three lynx were live-trapped and radio-collared and these were intensively

Figure 25.2 Locations of study areas in northern Canada. Shaded area is the geographical range of Canada lynx (after Anderson and Lovallo 2003), and the heavy lines are the Canada–United States border.

monitored from 1986 through 1994 to gather data on abundance, recruitment, survival, movements, and social organization; another 232 lynx were ear-tagged only, and these provided additional information on survival and movements. Carcasses of lynx trapped near the study area were collected during these same years to assess reproduction and population composition. Abundance of hares was assessed in each habitat type using pellet counts (Krebs *et al.* 1987).

The NWT study area was in flat terrain averaging 200–230 m in elevation. Most of the study area was forested with black spruce (*Picea mariana*), white spruce, jack pine (*Pinus banksiana*), aspen, and balsam poplar (*Populus balsamifera*), with sedge meadows and willows in lacustrine depressions. Numbers, survival, recruitment, movements, and social organization were studied by monitoring 71 lynx live-trapped and radio-collared from 1989 through 1995. Abundance of hares was assessed by habitat type using pellet and track counts.

Snowshoe hares reached peak abundances at similar times on all three study areas (1989–90 at Kluane and NWT, 1 year later at Snafu). The studies complemented each other well, as the focus of the Snafu and NWT studies was on demography, movements, and social organization, while the Kluane study concentrated on foraging behaviour and predator–prey interactions.

Cyclical dynamics

Abundance

Lynx populations showed strong numerical responses to cyclical changes in abundance of hares, each with a lag of 1 year (Fig. 25.3). At Kluane, the density of hares peaked in 1989–90 at about two hares per hectare and then declined over the next two winters to reach a low of 0.1 per hectare by 1992–93; the amplitude of the hare cycle was 26–44-fold (Boutin *et al.* 1995). Lynx responded with a 7.5-fold fluctuation in densities, peaking at 17 per 100 km^2 in 1990–91 and then rapidly declining over the next three winters to 2.6 per 100 km^2 (O'Donoghue *et al.* 1997). At Snafu, hare densities were considerably higher, peaking at 7.5 hares per hectare in 1990–91 and then crashing

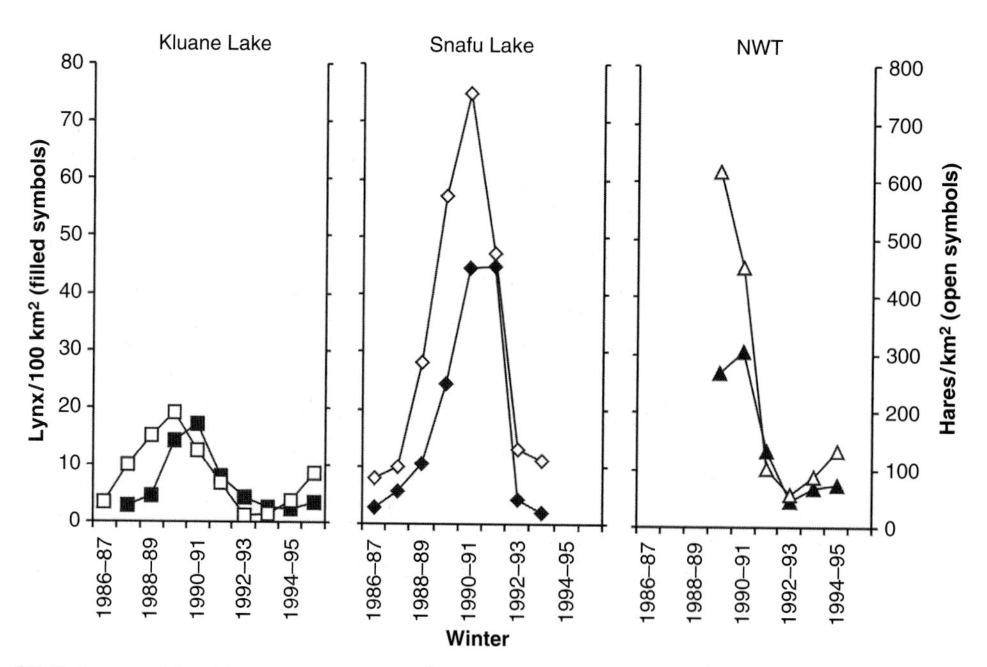

Figure 25.3 Estimated densities of lynx (per 100 km^2) and snowshoe hares (per km^2) during winter at Kluane (December estimates; from O'Donoghue *et al.* 1997, 2001), Snafu (March estimates; from Slough and Mowat 1996), and NWT (November estimates; from Poole 1994) study areas.

to 1.3 per hectare over the next 2 years (Slough and Mowat 1996). Lynx densities were likewise higher at Snafu, increasing 17-fold to a high of 45 per 100 km^2 in 1990–91 and 1991–92 and then rapidly declining to four per 100 km^2 by the first winter of low hare abundance. Densities of hares at NWT peaked in 1989–90, the year the study started, at six per hectare, and then declined during the next 2 years to one per hectare (Poole 1994). Lynx densities reached 30 per 100 km^2 in 1990–91 and then declined sevenfold to 4.4 per 100 km^2 over the next two winters. The peak densities measured at Snafu and NWT were the highest yet recorded for Canada lynx (previous estimates include: 10 per 100 km^2 in Alberta, Brand *et al.* 1976; 20 per 100 km^2 in Nova Scotia, Parker *et al.* 1983).

The patterns of decline of lynx populations were similar in all three areas. During the first winters of declining abundance of hares, lynx were able to find ample food in areas with localized concentrations or greater functional availability of hares, and their populations continued to grow. One to two years later though, numbers of hares had reached very low abundance and lynx populations likewise crashed. In Alberta, the pattern of the numerical response of lynx to the hare cycle was similar, with lynx densities peaking a year after those of hares (Brand *et al.* 1976).

Reproduction and juvenile survival

Modelling analyses have concluded that two main demographic changes drive population declines of lynx from cyclical peaks: collapses in recruitment of young and declines in adult survival (Brand and Keith 1979; Steury and Murray 2004). Results of our three studies showed consistent demographic responses by lynx to declining hare numbers.

Almost all successful production of kits occurred when snowshoe hares were relatively abundant, during the later increase, peak, and early decline phases of the hare cycle. Consistent with other studies (Brand and Keith 1979; O'Connor 1986), carcass collections around Snafu showed that most (71–93%) adult females were pregnant at all phases of the hare cycle except during the low, when only about a third had recent placental scars (Slough and Mowat 1996). Likewise, most (48–100%) female kits also became pregnant in their first winter during years when hares were most abundant. The rate of viable litter production, as measured by closely monitoring radio-collared females during early summer at Snafu and NWT, was more variable. Almost all adult collared females gave birth to litters, in late May, when hares were abundant, with litter sizes averaging 3.5 to 5.3 (Poole 1994; Mowat *et al.* 1996; Slough and Mowat 1996; Fig. 25.4); year-old kits also gave birth to viable litters at the cyclical peak. By the second year after the peak in numbers of hares though, few to no females showed any evidence of having kits at den sites. Based on movements of collared females, most litters born were likely lost at or very soon after birth at this time (Mowat and Slough 1998). Birth rates were not directly measured at Kluane but, as in the other two studies, few family groups were present in winter by the second year after the hare peak (O'Donoghue *et al.* 1997; Fig. 25.4).

Survival rates of kits through their first winter varied with the hare cycle as well: survival rates were fairly high (0.64–0.83 at Snafu, Slough and Mowat 1996; 0.50–0.82 at NWT, Poole 1994) when hares were abundant, but low in other years (0.00–0.22 at Snafu; 0.00 at NWT). Kits of yearling females survived quite poorly (0.08–0.26) in all years. Litter sizes observed during winter averaged 2.0–4.4 on the three study areas during winters with high densities of hares, and 0.8–2.0 when hare numbers were low (Fig. 25.4).

Similar patterns of declining litter sizes and rates of pregnancy, yearling reproduction, and juvenile survival during lynx population crashes have been noted in other field studies and carcass collections (Nellis *et al.* 1972; Brand *et al.* 1976; Brand and Keith 1979; Parker *et al.* 1983; O'Connor 1986; Staples 1995). Compared to other felids (Eisenberg 1986), lynx have high reproductive potential: litter sizes as high as eight have been observed (litters with six kits remaining are not unusual in winters of peak hare densities) and females may reproduce during their first year. They can thus take advantage of the rapid increases in abundance of their main prey species by producing numerous young, some of which will survive to carry their populations through the next cyclical low.

Adult survival and mortality

Survival of adult and yearling lynx was generally quite high (67–100%) at all phases of the hare cycle except in the first 1–2 years following the cyclical

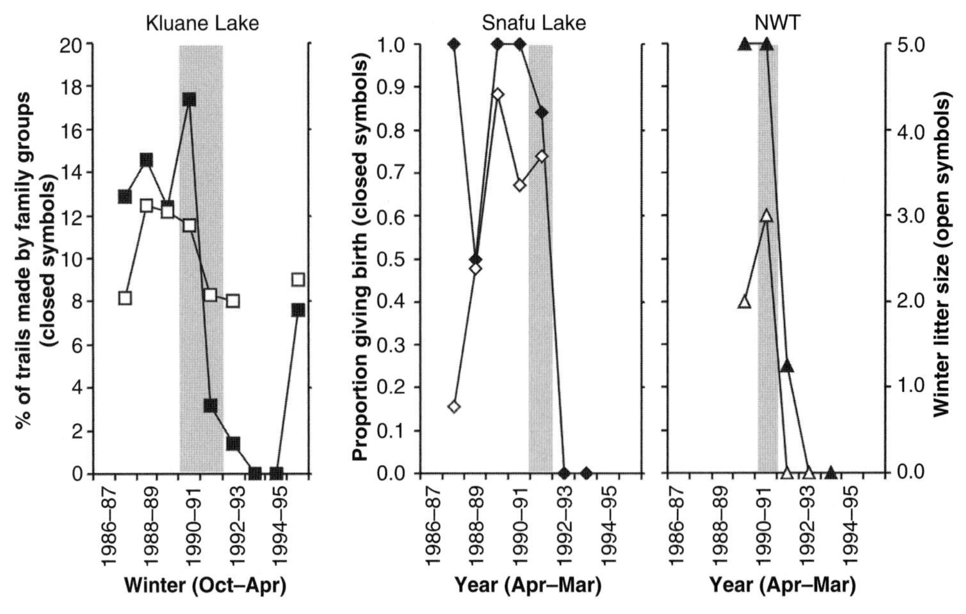

Figure 25.4 Indices of litter production (percentage of lynx trails made by family groups at Kluane and proportion of adult-collared females successfully giving birth to litters at Snafu and NWT) and winter litter sizes (mean litter sizes for whole winter at Kluane, estimated adult litter sizes in March at Snafu, and midwinter adult litter sizes at NWT) in northern Canadian study areas. Shaded areas of grey are winters of major decline in abundance of hares. (From Poole 1994; Slough and Mowat 1996; O'Donoghue *et al.* 1997, 2001.)

declines in hare abundance (Fig. 25.5). Most radio-collared lynx did not survive the first year of low hare numbers: only 11% ($n = 9$) at Kluane, 40% ($n = 30$) at Snafu, and 27% ($n = 16$) at NWT. Trapping was the main cause of mortality of collared lynx in all three study areas through the increase, peak, and decline phases of the cycle (Poole 1994; Slough and Mowat 1996; O'Donoghue *et al.* 1997), while natural causes—including starvation (11 confirmed cases) and predation by wolves (*Canis lupus*; 1 confirmed), wolverines (*Gulo gulo*; 3 confirmed, 2 suspected), and other lynx (5 confirmed)—led to most mortalities during cyclical lows. The drastic decline in prey abundance associated with the hare decline therefore led to poor body condition and higher susceptibility to predation; intra-guild predation on lynx by coyotes (*Canis latrans*) and other lynx has also been reported elsewhere (Elsey 1954; O'Donoghue *et al.* 1995; Apps 2000). Lynx that lived into the third year of low hare numbers had apparently learned to subsist in these suboptimal conditions and had high survival rates (Fig. 25.5).

Emigration

Both adult and young lynx often dispersed, especially during the decline and early low phases of the hare cycle (Fig. 25.5). Emigration rates of radio-collared animals were generally highest (means 41% at Kluane, 62% at Snafu, and 67% at NWT) 1–3 years after the cyclical peak. Early in this period, dispersing lynx were still in good enough condition to travel long distances from their original home ranges, and 42 animals travelled straight-line distances of over 100 km in the three studies; 16 of these moved over 500 km, and one lynx moved 1100 km (Slough and Mowat 1996; O'Donoghue *et al.* 1997; Poole 1997). There was no evidence of different rates or distances of emigration by age or sex at NWT (Poole 1997). At Snafu, however, of 30 male kits born on the study area, all dispersed as kits or yearlings, while 8 of 21 females stayed until they were 2–4 years old and then emigrated as adults after hare numbers had crashed; 70% of the immigrants to this study area were males (Slough and Mowat 1996). Survival rates of dispersers did not differ from those of residents at NWT (Poole

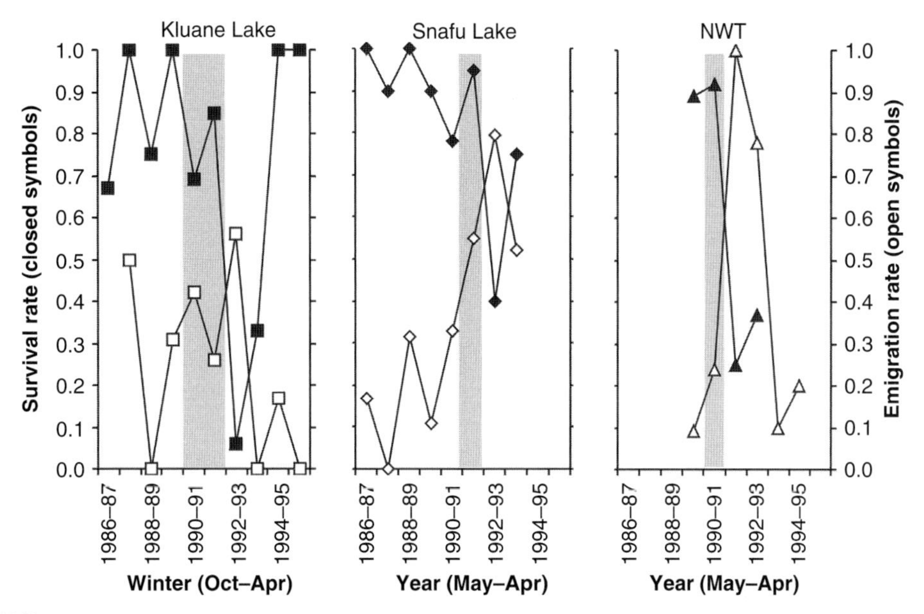

Figure 25.5 Survival and emigration rates of adult and yearling lynx at Kluane (October through April rates; from O'Donoghue *et al.* 2001), Snafu (mid-May through mid-May rates; from Slough and Mowat 1996) and NWT (June through May rates; from Poole 1994, 1997) study areas. Shaded areas of grey are winters of major decline in abundance of hares.

1997), where maximal effort was put into following animals once they left their original home ranges.

Long-distance movements of lynx during cyclical crashes of hares, previously reported as unusual events (Nellis and Wetmore 1969; Mech 1977), are apparently typical; three tagged lynx travelled more than 250 km from Kluane during the previous population decline in the early 1980s (Ward and Krebs 1985). Given the distances covered, there are few data about the eventual fates of dispersing lynx except those that are trapped, but at least eight dispersers were documented as successfully establishing new home ranges in the NWT study (Poole 1997). Male lynx are apparently more vulnerable to trapping than are females (Bailey *et al.* 1986; Quinn and Thompson 1987), perhaps related to their wider-ranging movements, so evidence of sex-biased dispersal that is based on trapping data may be biased.

Foraging behaviour and habitat use

Diet

The close tie between lynx and snowshoe hares is apparent in all published analyses of the diets of lynx: hares are the dominant prey killed and consumed by lynx throughout their range, in all seasons, and during all phases of the hare cycle. Lynx kill a variety of alternative prey, mostly squirrels, mice and voles, and birds, more so when hares are scarce and during the snow-free seasons when the availability and diversity of prey is higher. Lynx occasionally kill larger animals such as caribou (*Rangifer tarandus*), thinhorn sheep (*Ovis dalli*), and white-tailed deer (*Odocoileus virginianus*) or scavenge on their carcasses (Saunders 1963a; Stephenson *et al.* 1991; Staples 1995; Squires and Ruggiero 2007).

We determined diets of lynx only at Kluane (Fig. 25.6). Hares made up more than 90% of the estimated prey biomass consumed by lynx during winter, except during the three winters of lowest hare abundance when they accounted for 54–67% of prey. However, the large drop in available prey biomass associated with the cyclical decline (we estimated a fourfold decrease) led to major changes in the hunting behaviour of lynx. During the three winters of the low in hare numbers, lynx hunted and killed more red squirrels (*Tamiasciurus hudsonicus*; 51–79% of all kills) than any other prey; in these winters, squirrels represented an estimated 20–44%

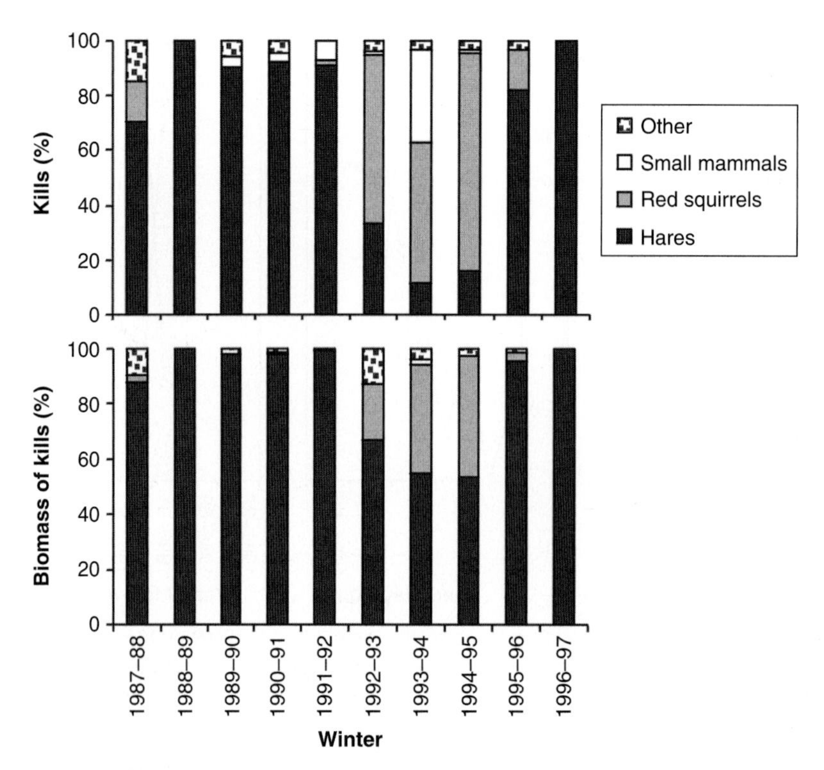

Figure 25.6 Winter diets of lynx at Kluane by percentage and percentage biomass of kills (n = 572), as determined from snow-tracking. (From O'Donoghue *et al.* 1998b, 2001.)

of prey biomass consumed. Lynx also killed voles (mostly *Microtus* spp.), flying squirrels (*Glaucomys sabrinus*), muskrats (*Ondatra zibethicus*), a weasel (*Mustela erminea*), a red fox (*Vulpes vulpes*), spruce grouse (*Falcipennis canadensis*), and, in summer, Arctic ground squirrels (*Spermophilus parryii*), but none of these species made substantial contributions to the total prey biomass consumed.

The change in hunting behaviour of lynx during the cyclical low was large enough to suggest prey switching from hares to squirrels (O'Donoghue *et al.* 1998a), *sensu* Murdoch (1969). Squirrels made up a greater proportion (1.5 to 5 times more) of the diet of lynx at the low than was predicted from the null model based on prey preferences shown over the whole cycle—which was unexpected for a specialist predator. Squirrels were the main alternative prey in studies in Alaska (Staples 1995), British Columbia (Apps 2000), and Washington (Koehler 1990), but the Kluane results represent the highest use reported to date.

Functional responses

Cyclical fluctuations in numbers of hares led to a fivefold change in kill rates by lynx at Kluane (Fig. 25.7). In the cycle in hare numbers from 1987–88 through 1994–95, we observed the highest kill rates (1.2 hares per day) during the early decline phase and the lowest at the cyclical low (0.3 hares per day); kill rates were phase-dependent, being higher at comparable hare densities during the decline phase than in the increase (O'Donoghue *et al.* 1998b, 2001). Estimated kill rates were higher though (1.4–1.6 hares per day) during the subsequent increase in hare abundance in 1995–96 and 1996–97. Low numbers of hares were associated with longer chases (annual means varied from 2 to 15 m for successful chases and 6 to 23 m for

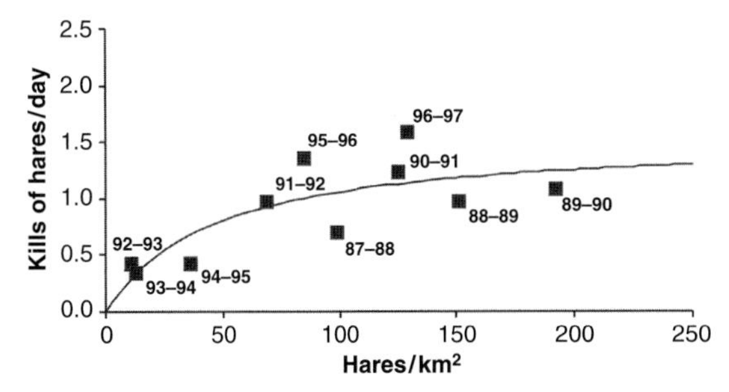

Figure 25.7 Functional response of lynx to changes in densities of snowshoe hares at Kluane from 1987–88 (labelled '87–88') through 1996–97, fitted with a Type-2 curve (Holling 1959). (From O'Donoghue *et al.* 1998b, 2001.)

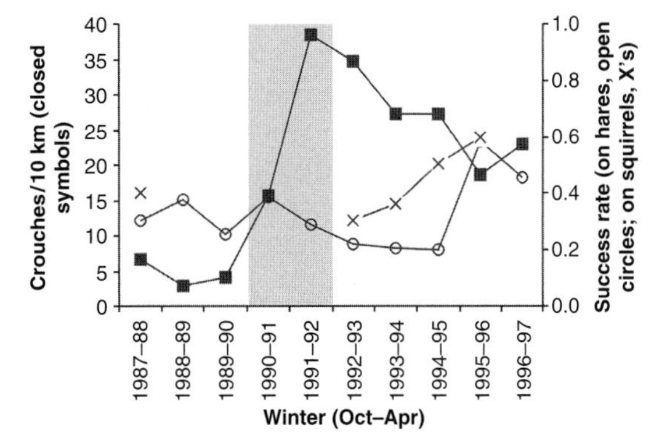

Figure 25.8 Frequency of crouches or 'ambush beds' along the trails of lynx and hunting success (% of chases that were successful) of lynx chasing hares and red squirrels at Kluane. Shaded areas are winters of major decline in abundance of hares. (From O'Donoghue *et al.* 1998b, 2001.)

unsuccessful chases) and lower success rates at catching hares (hunting success varied from 20% to 39% per winter). However, there was little variation in the travel rates (about 1.0 km/hr), percent time spent active (39–44%), and amount eaten per kill (78–95% based on prey remains) by lynx among cyclical phases (O'Donoghue *et al.* 1998b, 2001).

The estimated kill rates at Kluane were higher than those calculated in Alberta (0.2–0.8 hares per day over the whole hare cycle; Nellis and Keith 1968; Brand *et al.* 1976), Nova Scotia (about 1 hare per day during a winter of high hare abundance; Parker 1981), and Newfoundland (about 0.5 hares per day, abundance of hares not noted; Saunders 1963a), and greater than needed to fulfil the estimated energetic requirements of lynx (400–600 g/day; Nellis *et al.* 1972). The only other functional response calculated for lynx was in Alberta, where Keith *et al.* (1977) calculated a threefold increase in kill rates by lynx with increasing densities of hares, peaking at about 0.75 hares per day.

Hunting tactics

Lynx typically hunt either by slowly stalking prey and then bounding after flushed animals or by waiting in 'ambush beds' near well-used trails or middens of hares and red squirrels. Use of these different tactics changed significantly as abundance of hares declined and lynx switched increasingly to hunting red squirrels at Kluane. Between 1989–90 and 1991–92 (peak to late decline phases), use of ambush beds increased tenfold, and lynx initiated a progressively higher proportion of chases of hares from beds (Fig. 25.8; O'Donoghue *et al.* 1998a). Hunting success from ambushes was about the same as from stalks at this time, but use of beds may have helped conserve energy during a period of relative food shortage. From 1990–91 to 1994–95 (decline through low phases of the hare cycle), hunting success by lynx chasing hares declined from 39% to 20%, while hunting success chasing red squirrels increased from 31% to 51% during the last 3 years of this period (Fig. 25.8). These results suggest that either lynx surviving through the cyclical low became increasingly skilled at hunting red squirrels or that only those lynx that were good at hunting alternative prey survived. Hunting tactics were also affected by vegetative cover: during the cyclical increase, lynx typically stalked hares in open spruce forest and ambushed when using denser cover (Murray *et al.* 1995). Use of ambush beds by lynx has been noted in Newfoundland (Saunders 1963a, b), Alberta (Nellis and Keith 1968; Brand *et al.* 1976), and Montana (Squires and Ruggiero 2007), but few ambush sites were found while snow-tracking in Nova Scotia (Parker 1981).

Scavenging and caching

Lynx seldom scavenged at any phase of the hare cycle at Kluane compared to sympatric coyotes, and cached entire carcasses of only 1.6% of radio-collared hares they killed (O'Donoghue *et al.* 1998b). Caches of uneaten portions of hare carcasses were typically made by pulling snow over them. We monitored seven caches made by lynx at hare kill sites, and lynx returned to six of them within 2 days of making the kills and consumed all edible parts remaining. At NWT, a radio-collared lynx was observed feeding on a bison (*Bison bison*) carcass for at least 4 days. Most studies of lynx have reported low levels of scavenging. In Alberta, however, carcasses of livestock and wild ungulates were the most important alternative food source for lynx (Nellis and Keith 1968; Nellis *et al.* 1972; Brand *et al.* 1976).

Influences of snow

The large paws of lynx are well adapted for travelling through the deep powdery snows of the boreal forest and allow them to utilize habitats less favourable to similar-sized predators such as coyotes and bobcats (*Lynx rufus*) (Parker 1981; Murray and Boutin 1991). Snow conditions may still affect hunting behaviour and success of lynx, however. At Kluane, lynx used higher elevation areas with deeper snow than did coyotes but, during the cyclical increase, selected routes of travel with shallower snow than immediately adjacent to their trails (Murray and Boutin 1991). Snow tended to be harder and shallower at sites of successful chases than at failed attempts during this period, but this relationship did not hold through the cyclical low. Lynx also frequently (10–28% of their travel distance each winter) travelled on the trails of hares, which would lead to higher encounter rates with prey and also provide packed snow on which to walk and initiate chases (O'Donoghue *et al.* 1998a). In Alberta, Nellis and Keith (1968) suggested that softer snow cover in the last year of their study led to lower hunting success by lynx. Snow was softer at sites of failed chases compared to sites where kills were made in April in Nova Scotia (Parker 1981), while an earlier study adjacent to Snafu found no relationship between snow characteristics and hunting success during a cyclical increase (Major 1989).

Habitat selection

Lynx selected denser forested habitats (characterized by abundant hares) in all three study areas, and generally avoided large, open areas. In the mostly mature spruce forests of the Kluane study area, lynx selected habitats according to hare abundance and used progressively denser habitats from the peak of the hare cycle through the cyclical decline (Murray *et al.* 1994; O'Donoghue *et al.* 1998a, 2001). This paralleled changes in habitat use by hares, although hares used denser habitats than lynx in all winters. At Snafu, lynx consistently selected regenerating (28–34 years old) pine habitats and riparian willow thickets where hares were abundant (Mowat and Slough 2003). Likewise, dense 20–60-year-old coniferous and deciduous habitats regenerating after fires were selected by lynx in all years at NWT (Poole *et al.* 1996).

Hunting success by lynx did not vary by habitat type at Kluane, where most forests were mature (Murray *et al.* 1994, 1995), whereas other studies in Alaska (Staples 1995), Nova Scotia (Parker 1981), and an earlier study near the Snafu study area (Major 1989) found that a higher proportion of chases were successful in mature forest habitats than in the dense young forests where hares were most abundant. Hares are apparently more successful in escaping lynx in shrub thickets than in open cover, and lynx frequently hunted along the edges of the densest cover.

Selection for young regenerating forests and dense shrub habitats by lynx and hares is widespread (Parker *et al.* 1983; Staples 1995; Hodges 2000; McKelvey *et al.* 2000c), but lynx denning habitat may differ among areas. At Snafu, most (85%) lynx dens were under downed trees in the regenerating burned area, while some were also under dense shrubs or low canopies of trees (Slough 1999). In their southern montane range, lynx consistently select mature forest types for denning (Aubry *et al.* 2000). Regardless of the surrounding habitat type, the most important factors in den site selection seem to be dense overhead and lateral cover.

Social behaviour

Home ranges

A typical spatial organization among felids is one of intrasexually exclusive territories among adults, with male home ranges broadly overlapping those of females (Kleiman and Eisenberg 1973; Eisenberg 1986; see Macdonald *et al.*, Chapter 5, this volume). Territories are maintained mostly by passive means—scent-marking, scratching, and some vocalizations. Juveniles are typically tolerated within their natal home ranges until they reach sexual maturity at which time they disperse, although female offspring sometimes settle in parts of or next to their mothers' home ranges. This model is consistent with our observations of home range use by lynx in our three study areas during most cyclical phases.

Mean annual home range sizes of adult male lynx ranged from 17 to 47 km^2 in size and those of females from 11 to 33 km^2 during the cyclical increase, peak, and decline phases of the hare cycle in the three study areas (Fig. 25.9). While home range sizes were not directly related to densities of hares, they tended to increase in size from the first year of the cyclical decline (when lynx densities were the highest) through the low. Male home ranges were generally larger than those of females at Kluane and Snafu, but the two were similar in size at NWT (Poole 1994; Slough and Mowat 1996; O'Donoghue *et al.* 2001). Overlap of home ranges of adult males with those of other males was variable among years and individuals. At both Kluane and NWT, most adult males maintained home ranges that were largely exclusive of other males, but in each area there were several males that did not fit this pattern, their home ranges overlapping considerably with those of several other males (Poole 1995; O'Donoghue *et al.* 2001). With the exception of these animals, overlap of home ranges (95% minimum convex polygons) among males was 7% at Kluane and 2% at NWT during the peak year in lynx numbers. Overlap among home ranges of adult females was low (11% at peak lynx densities) at Kluane, and among the home ranges of most females at NWT except for three pairs of females that extensively overlapped (overlap for all females 43%). At Snafu, overlap of home ranges (90% clusters) was 21% among males and 17% among females over all years of the study. Home ranges of males overlapped broadly with those of females in all study areas, and each male range typically included those of one to two females.

Once hare populations had crashed, the spatial organization of lynx broke down in all three study areas. High emigration and mortality rates of resident lynx were coupled with increased sizes of home ranges of animals that survived, and the establishment of larger home ranges by animals new to

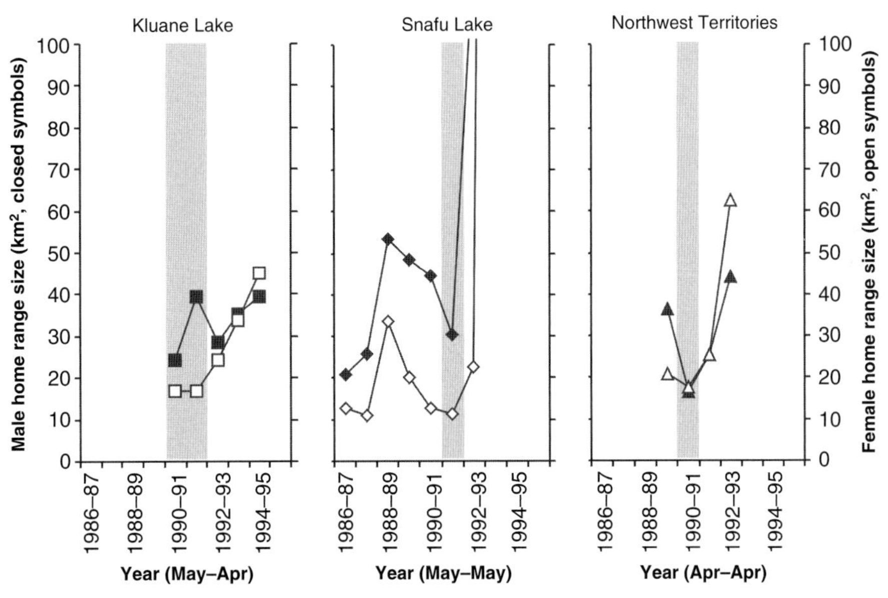

Figure 25.9 Mean home range sizes (95% minimum convex polygons; Mohr 1947) of lynx in northern Canadian study areas (from Poole 1994; Slough and Mowat 1996; O'Donoghue *et al.* 2001). At Snafu, male home ranges averaged 119 km^2 in 1992–93 and 266 km^2 in 1993–94, and female home ranges averaged 507 km^2 in 1993–94. Shaded areas are winters of major decline in abundance of hares. © Mark O'Donoghue.

the areas (Fig. 25.9). The few remaining lynx at Snafu covered home ranges of several hundred square kilometres by the second year of the cyclical low. Some lynx may also become nomadic during cyclical lows, as suggested at NWT by Poole (1994) and at Kluane during the previous hare cycle by Ward and Krebs (1985), but distinguishing between nomadic movements and activities in very large home ranges such as those observed at Snafu is difficult with the limited sample sizes typical of these studies. At Kluane, lynx captured during and after the second year of the cyclical low established new, apparently stable home ranges (O'Donoghue *et al.* 2001).

Estimates of home range sizes of lynx from other boreal studies were similar to those we observed (16–21 km^2 in Newfoundland, Saunders 1963b; 11–50 km^2 in Alberta, Brand *et al.* 1976; and 12–32 km^2 in Nova Scotia, Parker *et al.* 1983), but most studies in the southern parts of the lynx's range (where densities of hares are typically lower than in the north) have measured home range sizes of several hundred square kilometres (Mech 1980; Carbyn and Patriquin 1983; Apps 2000; Squires and Laurion 2000), similar to those observed at Snafu at the cyclical low.

Female–kit relationships

Lynx kits remain within the home ranges of their mothers throughout most of their first year of life. Kits were ear-tagged at their dens at both Snafu and NWT, so most of our data on family interactions come from those two studies. Young kits stayed near their den sites for 6–8 weeks until they were weaned and could travel with their mothers; females sometimes relocated kits during that time at Snafu (Slough 1999). Adult females localized their movements around their den sites while kits were very young, but gradually increased their movements and reoccupied their whole home ranges by late summer. Kits travelled as a group and hunted with their mother until February or March, when the family bond weakened and kits began exploratory movements on their own (Poole 1995). By late March to early April, when kits were about ten months old, they regularly travelled independently and some began to disperse from their natal home ranges (Mowat *et al.* 1996; Slough and Mowat 1996). Peak periods of natal dispersal at NWT were in April/May and October/November (Poole 1997).

We observed some confirmed and suspected longer-term associations between female lynx and their female offspring. At Snafu, female kits were occasionally with their mothers in May after their mothers had given birth to new litters and, in one instance, a yearling gave birth to kits of her own within 800 m of her mother's den site (Mowat and Slough 1998). Mothers and female offspring were also located together at other times during the study even when both had kits. The home ranges of six yearling females and one 2-year-old female that did not disperse from Snafu had higher overlap (32% and 30% of home ranges, respectively) with the home ranges of their mothers than the average overlap among all females (17%). While no known kits settled within the home ranges of their mothers at NWT, the home ranges of three pairs of females (aged 4 and 3, 2 and 1, and 9 and 1) overlapped extensively, but it is unknown if they were related. One group of three adults observed during the cyclical low at Kluane included two females suspected of being a mother and her daughter (O'Donoghue *et al.* 1998a). Extensive overlap in home ranges between pairs of females was also noted in Minnesota (Mech 1980) and in Manitoba (Carbyn and Patriquin 1983). In Manitoba, one group of two females, with one and two kits respectively, stayed together for a month in January before one of the females was trapped.

Group hunting

Family groups (mothers with their kits) of lynx hunted together as a unit during most of the winter until the kits were 9–10 months old. While travelling, kits often followed in the tracks of their mothers, but typically fanned out when moving through areas of concentrated hare activity; based on our snow-tracking at Kluane, hares apparently flushed by one lynx were sometimes killed by another. This same hunting pattern was observed at Snafu (Mowat and Slough 1998) as well as elsewhere (Saunders 1963a; Parker 1981). While Brand *et al.* (1976) only noted tracks of kits at 7 of 17 kills made by family groups in Alberta, we found that kits shared in feeding on most kills.

While adult lynx generally did not travel with other adults outside of the breeding season (March/April) during the cyclical increase, peak, and early decline, we observed several groups of two to three adult lynx hunting together during the late decline and low (1991–92 through 1993–94) at Kluane (O'Donoghue *et al.* 1998a), and we saw tracks of some of these groups repeatedly (Breitenmoser *et al.* 1993b). We did not know the relatedness of most of these animals. The Kluane snow-tracking data suggest that group hunting benefited lynx in at least one winter when hare numbers were low (O'Donoghue *et al.* 1998a). In earlier winters, per-individual kill rates of mothers with kits were lower as group size increased (by a mean of about 0.1 hares killed per day per kit in the group), likely due to the hunting inexperience of kits, especially in early winter. Parker *et al.* (1983) noticed a similar pattern in Nova Scotia, and found that hunting success of family groups increased as the winter progressed (14–16% January–February and 20–26% March–April), which they attributed to increasing experience of kits. At Kluane, however, per-individual kill rates of adult lynx hunting in groups were about equal in one winter and more than double the next year, compared to those of lynx hunting alone at the cyclical low. Cooperative hunting among adults is seldom seen in most felids (Kleiman and Eisenberg 1973; see Macdonald *et al.*, Chapter 5, this volume), and has been reported for lynx only once (Barash 1971). While pairs of adult lynx have been seen together outside of the breeding season (e.g. Parker *et al.* 1983), it is still the exception rather than the rule; our Kluane data suggest that hunting in adult groups may confer advantages of higher food intake when numbers of hares are low.

Conclusions and questions

We make the following generalizations from our three studies:

1. Lynx show 7–17-fold fluctuations in abundance with a 1-year lag in response to the 10-year snowshoe hare cycle. Lynx numbers are higher, on average, where dense early successional forests support higher densities of hares.
2. Cyclical declines in lynx abundance are associated with a collapse of recruitment, high dispersal rates early in the decline, and low *in situ* adult survival later in the decline.
3. Long-distance movements by lynx are typical during cyclical declines, and some of these dispersers successfully establish new home ranges.
4. Lynx feed mostly on snowshoe hares at all phases of the cycle but, at least in some areas, red squirrels are important alternative prey through the cyclical lows.
5. Lynx show clear functional responses to changing abundance of hares, with kill rates that may vary fivefold. Maximal kill rates of 1.2–1.6 hares per day are higher than the estimated energetic needs of lynx.
6. Lynx may adjust their hunting tactics between stalking and ambushing as abundance, vulnerability, and habitat use by hares and use of alternative prey change over the hare cycle.
7. Lynx selectively use denser forest types, especially early successional habitats where snowshoe hares are most abundant, at all phases of the cycle.
8. The core home ranges of adult lynx are mostly intrasexually exclusive and broadly overlapping between sexes, but there is evidence that females sometimes share home ranges and remain associated with female offspring.
9. The sizes of male home ranges are typically larger than those of females. When numbers of snowshoe hares crash, the existing spatial organization breaks down, and sizes of lynx home ranges may increase greatly (two- to fourfold increases in average home ranges at Kluane and NWT and 9–45-fold increases at Snafu).
10. Besides groups of mothers with kits, lynx are primarily solitary hunters, but temporary adult groups may form during the cyclical decline and low hare abundance.

Many questions about the ecology of lynx remain, and some have important implications for lynx conservation. These are highlighted as research priorities below.

Patterns and effects of dispersal

There are many uncertainties about the patterns and effects of dispersal of lynx. Data from Snafu suggest that males emigrate at higher rates than females, and Rueness *et al.* (2003b) found lower rates of divergence in mitochondrial DNA than nuclear DNA in eastern Canada, which could be caused by male-biased dispersal. However, dispersal rates from NWT were not sex-biased, and immigrants into a study area in northern Minnesota in the 1970s had an even sex ratio (Mech 1980). There is also little information about how often dispersers are successful (i.e. successfully reproduce in a new area), as few studies have had the resources to follow animals over hundreds or even thousands of kilometres. Low genetic diversity (global $F_{st} = 0.033$ in western North America) among lynx populations over vast areas of their contiguous range (Schwartz *et al.* 2002; Rueness *et al.* 2003b) suggest fairly high gene flow, although there is some evidence of genetic population structuring related to geographical or ecological barriers (Rueness *et al.* 2003b). Spatial models have demonstrated that even fairly low levels of dispersal among independently cycling lynx populations could lead to patterns of phase synchronization similar to those observed in North America (Ranta *et al.* 1997; Blasius *et al.* 1999). Analyses of time series of lynx harvest data and field observations have also shown that patterns of lynx occurrence in the northern contiguous United States are most likely caused by influx of lynx from cycling populations to the north in Canada (Mech 1980; McKelvey *et al.* 2000a), again suggesting high rates of movement of lynx related to the hare cycle. Cyclical influxes of dispersers may lead to higher rates of successful colonization than similar rates of constant dispersal (Blasius *et al.* 1999; McKelvey *et al.* 2000b). Research is needed to evaluate both the importance of immigration for maintaining southern populations and the effects of emigration on the success of reintroduction programmes in disjunct southern montane areas (Steury and Murray 2004). In addition, the ongoing expansion of human activities in all regions of lynx distribution makes the study of dispersal of lynx a conservation priority.

Habitat patchiness and fragmentation

While patterns of habitat use by lynx and snowshoe hares seem consistent in northern boreal forests, where both species select early successional habitats, they remain poorly understood in the southern montane part of their geographical ranges (Ruggiero *et al.* 2000). The pattern of habitat distribution on the landscape and the degree of natural patchiness or human-caused fragmentation may have important effects on hare dynamics and population cycling (Dolbeer and Clark 1975; Wolff 1980), and therefore on lynx. Given the patchy nature of montane landscapes, lower densities of hares and lynx, and higher human densities in the southern areas, understanding habitat relationships of hares and lynx will be important for successful conservation. Even in northern boreal forests, where lynx populations are presently secure, increased industrial development and changing landscapes due to climate change will likely increase the degree of forest fragmentation in the coming years.

Effects of trapping

Along with marten (*Martes americana*), lynx are the mainstay for many northern trappers. Levels of fur harvest have varied greatly over the past two centuries, with some peak harvests 5–10-fold different than others (Mowat *et al.* 2000). Analyses of time series suggested that changes in trapping pressure could have led to the sudden shifts in the cyclical amplitude observed since the early 1700s (Gamarra and Solé 2000). Low and declining harvests from the 1920s to 1950 prompted speculation that this was partially due to over-harvest of lynx by trappers (De Vos and Matel 1952), but the evidence is equivocal. Although boreal lynx populations can sustain fairly high rates of harvest during late cyclical increases, peaks, and declines (Quinn and Thompson 1987), cases of local over-harvest have occurred (Bailey *et al.* 1986). The collapse in recruitment of lynx when hares are scarce means sustainable harvest levels are very low at this cyclical phase. Trapping low-density isolated populations may remove most or all animals (Carbyn and Patriquin 1983). We do not know, however, whether trapping has any long-term effects on the amplitude and peak densities attained by lynx populations during cyclical fluctuations. At present, trapping pressure is low over much of the range of lynx because of low fur prices, high fuel prices, shifts to wage-based economies and inaccessibility of many areas in the north, and harvest restrictions in the south. The fur market can change rapidly, however, and accessibility will likely increase in the northern range of lynx, so we need to better understand the long-term effects of trapping on lynx populations. Restriction of harvest during cyclical lows (Brand and Keith 1979) and the establishment of large untrapped refugia (Slough and Mowat 1996) have both been proposed as management strategies for minimizing the risk of over-harvest of lynx.

Importance of alternative prey

Snowshoe hares are by far the most important prey of lynx and cyclical fluctuations in the abundance of hares have major effects on lynx densities, demography, and behaviour. Canada lynx do not live in areas with no hares and they do not reproduce when densities of hares are low. Most studies of lynx in the southern portions of their range have shown that lynx consume more alternative prey than in northern regions (Aubry *et al.* 2000; but see Squires and Ruggiero 2007). However, the Kluane data indicate that even in northern areas, other prey species, in this case red squirrels, may be heavily used. Lynx that survived through the cyclical low at Kluane did so largely by becoming skilled at catching squirrels; this may be true for many other areas of mature forest in the north as well. Lynx living in areas comprised mostly of early successional forests, like those at Snafu, would typically have higher densities of hares but lower densities of red squirrels available as alternative prey than those living in mature forests. The extent to which lynx require alternative prey and the densities of those prey species required to sustain viable lynx populations are unknown and have important implications for the persistence of lynx populations in the southern part of their range.

Patterns of cyclical population turnover

Breitenmoser *et al.* (1993b) developed a 'core population hypothesis' to describe the life history strategy of lynx, which are faced with large but predictable fluctuations in prey abundance. They proposed that a core population of resident lynx occupy large and stable home ranges throughout the hare cycle. These residents maximize their reproductive success by expending energy on reproduction only during favourable periods, defending territories from unrelated like-sexed lynx at all times, and tolerating related offspring within their territories at certain times of the cycle. Specific predictions are: (1) resident lynx should occupy the same territories throughout the cycle. These will be shared with related females whenever prey densities are adequate to support them and with male offspring during the cyclical decline only, when competition for mates is relaxed; (2) survival of residents should be higher than that of transients during the cyclical declines; (3) female offspring should stay and reproduce within their mothers' territories, resulting in clusters of increasingly overlapping home ranges of related females; and (4) male offspring should disperse during cyclical increases, but some may remain resident in parental home ranges during declines.

Thus far, the data offer mixed support for this hypothesis, and there is a lack of information on the relatedness of most lynx in these studies to test some predictions. Between the cyclical peak in the early 1980s and the peak 10 years later, eight lynx were known to have persisted at Snafu, and one adult male lived in approximately the same area at Kluane (Breitenmoser *et al.* 1993b). In all three studies summarized here (peak 1989–91), home range sizes of lynx remained mostly constant through the cyclical decline of hares, but the social organization largely broke down during the first years of the cyclical low and home range sizes increased. All known residents at Kluane and NWT either died or dispersed when hare numbers collapsed, resulting in complete population turnover, while two of the three known survivors at Snafu had been born there. Contrary to the second prediction, no difference in survival rates was found for lynx that remained resident or those that emigrated from NWT. We did see evidence that some pairs of related females shared home ranges in all three study areas, but have no data to indicate that this resulted in clusters of related females. Some yearling males also did remain within their natal home ranges during the cyclical declines, but subsequent emigration and mortality rates were so high that we did not see evidence of males permanently settling in them. As concluded by Breitenmoser *et al.* (1993b), evaluation of this hypothesis will require long-term data on social organization, relatedness, and fitness of lynx in an untrapped area.

Cyclical changes and regional structuring of the lynx–hare interaction

Time series analyses have suggested that the hare–lynx interaction is qualitatively different during the cyclical increase than in the decline phase (Stenseth *et al.* 1998). Many factors, including the age structure of the lynx population, body conditions of both hares and lynx, foraging behaviour of both predator and prey, kill rates of lynx, and the abundance of alternative prey and other predators of hares are different between the cyclical phases and could affect the lynx–hare interaction. Based on results of the Kluane study showing higher kill rates by lynx during the cyclical decline, Stenseth *et al.* (1998) proposed that the phase-dependency in the functional response of lynx to hares was a likely cause of differences between phases observed throughout the north. Analyses of lynx fur harvest returns also suggested that the lynx–hare interaction was structurally similar within broad geographical regions that corresponded to large climatic zones in northern North America (Stenseth *et al.* 1999), and that these corresponded with genetic structuring of lynx populations (Rueness *et al.* 2003b). This led to the hypothesis that climatic factors may affect the hare–lynx interaction, possibly by differences in snow conditions (most likely, snow hardness) affecting the sinking depth and thus the hunting success of lynx (Stenseth *et al.* 2004b). Studies of mechanisms and factors affecting the hunting success of lynx at different cyclical phases and in different climatic zones would be required to test this hypothesis. Such studies would also have broad implications for the potential effects of climate change on the lynx–hare cycle, which has been an enduring icon of the boreal forest.

CHAPTER 26

Black-footed cats (*Felis nigripes*) and African wildcats (*Felis silvestris*): a comparison of two small felids from South African arid lands

Alexander Sliwa, Marna Herbst, and M. Gus L. Mills

(a) Female black-footed cat looking up from a grooming bout; (b) Female African wild cat scanning her surroundings for prey movements. © Alex Sliwa and Marna Herbst.

Some of the leading causes for the decline of felid populations are habitat loss, habitat degradation, and persecution. Africa's two smallest cat species, the black-footed cat (BFC) (*Felis nigripes*) and the African wildcat (AWC) (*Felis silvestris*) occur in southern Africa's grasslands and semi deserts and are affected by all these causes of decline. Additionally, the AWC is threatened by hybridization with domestic cats (*Felis catus*) (Smithers 1983; Nowell and Jackson 1996; Macdonald *et al.*, Chapter 22, this volume). Our objectives were to: (1) explore the origins of, and morphological differences between, the two species; (2) compare their life histories and ecological parameters; (3) compare ecological factors that impact species abundance and distribution; and (4) identify gaps in research knowledge, particularly with relevance to conservation management of the species. While variation in diet, home range size, resting site use, and activity patterns were present between the two species, we could not discern significant differences in these parameters, or in population threats. We propose that collaborative research and concerted action planning will maximize the efficiency of financial resources to develop applied conservation solutions for both species.

Introduction

The BFC, also called the small-spotted cat, is the smallest cat species in Africa and among the smallest in the world. Endemic to the arid grassland, dwarf shrub, and savannah of the Karoo and Kalahari in the western parts of southern Africa (Fig. 26.1; Smithers 1983), it has the most restricted distribution of any African cat species (Nowell and Jackson 1996). It shares much of its habitat with the widespread AWC, which ranges throughout most of the African continent (Fig. 26.1; Smithers 1983; Nowell and Jackson 1996). Erratic rainfall affects the food resources in the Kalahari study area described here and throughout the distribution range of the BFC (Leistner 1967; Nel *et al.* 1984; van Rooyen 1984).

Although the species differ markedly both in coat patterns and size (Fig. 26.2) there is considerable confusion by the general public, and thus in their distribution records in southern Africa (A. Sliwa, personal observation). However, the contemporary distribution of the two species suggests that the BFC is sensitive to habitat and climatic variables, while the AWC has a very broad ecological niche, inhabiting almost all African habitats, with the exception of the tropical rainforests and true deserts. Within the northern portion of the AWC's distribution, the sand cat (*Felis margarita*) inhabits the driest parts of the Sahara (Sunquist and Sunquist 2002). The sand cat is a small cat similar in several morphological adaptations to the BFC (Huang *et al.* 2002). AWCs inhabiting truly arid habitats are also smaller in stature and mass, for example, the *gordoni* wild cats of the Eastern Arabian peninsula average only 77–78% in head–body length and 51–53% in mass, compared to the kalahari (*F.s. gordoni* head body [HB] length: ♂♂ 50.3 to 65 cm; ♀♀ 47 to 60 cm. Mass: ♂♂ 2.7 to 5.1 kg, ♀♀ 2.0 to 3.9 kg; unpublished data measurements on *Felis silvestris gordoni* by Breeding Centre for Endangered Arabian Wildlife, Sharjah, United Arab Emirates; Phelan and Sliwa 2005; Kalahari *silvestris*—Herbst & Mills in press, unpublished data).

All African *Felis* species have been little studied (Nowell and Jackson 1996), thus no clear limitations for their ecological separation have been defined. In this chapter, we summarize what is known about the behaviour and ecology of BFCs and AWCs from two intensive field studies in South Africa and make suggestions for future research and conservation measures. The results of the only small African cat species provide the basis for comparing them in this chapter. A study of both species in sympalry is still lacking; however, the present study areas are only 500 km apart in relatively similar habitat in the Northern Cape Province, Republic of South Africa (Fig. 26. 3a, and b).

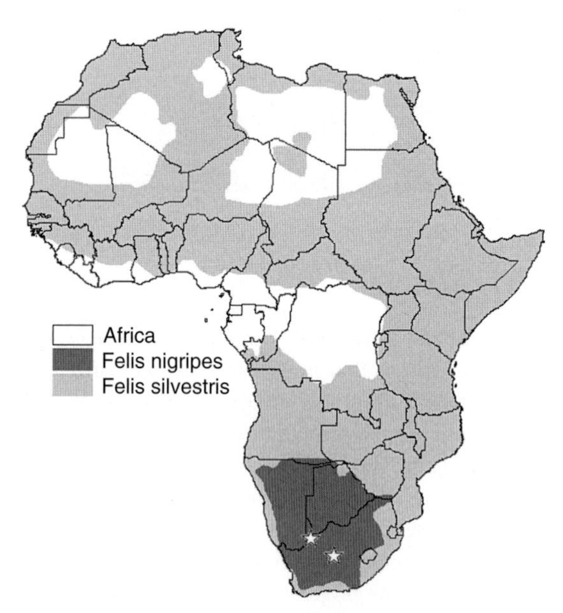

Figure 26.1 Distribution of the AWC, *Felis silvestris* and BFC, *Felis nigripes* in Africa. The two stars mark the location of the study areas.

Origin and size

The two cat species belong to the Old World domestic cat lineage (Johnson and O'Brien 1997; Werdelin *et al.*, Chapter 2, this volume), however, the BFC is thought to have diverged from the other *Felis* species about 3 million years ago (Johnson *et al.* 2006b). The wildcat (*Felis silvestris*) of Europe, Africa, and Asia has been the subject of continuous taxonomic debate. Nowell and Jackson (1996) divided wildcats into four groups: (1) the *silvestris* group comprising the heavily furred forest cats of Europe and the Caucasus; (2) the *ornata* group including the light-bodied

(a)

(b)

Figure 26.2a and b (a) AWC female, © M. Herbst; and (b) BFC male. © A. Sliwa.

steppe cats of Asia; (3) the *lybica* group comprising the long-legged AWCs of Africa and the near East; and (4) the domestic cat, *Felis catus*.[1] Genetic analysis confirms that these four groups of 'wild cats' are phylogenetically very close to each other (Driscoll *et al.* 2007; Pocock 1907; Macdonald *et al.*, Chapter 22, this volume), and that interbreeding with domestic cats may severely threaten the status of true wildcats. This process is accelerated by habitat loss and increased contact with human settlement and associated domestic cats (Macdonald *et al.* 2004b, Yamaguchi *et al.* 2004a, b; Macdonald *et al.*, Chapter 22, this volume).

BFCs were shorter (♂♂ = 45/♀♀ = 40 cm HB) and smaller in mass (♂♂ = 1.9/♀♀ = 1.3 kg) than AWCs (♂♂ = 65/♀♀ = 60 cm HB; ♂♂ = 5.1/♀♀ = 3.9 kg) in the respective study areas close to Kimberley and Twee Rivieren, South Africa (Sliwa 2004; Herbst and Mills in press, unpublished data), the difference in body mass being almost threefold. Smaller size allows the BFC to conceal itself better in very short vegetation and find refuge in burrows of fossorial mammals, most commonly those of springhares (*Pedetes capensis*), but also in those of the Cape ground squirrel (*Xerus inauris*), South African

[1] For the current nomenclature of wildcats, and discussion of controversy over the domestic cat classification, see Macdonald *et al.* (Chapter 22, this volume).

porcupine (*Hystrix africaeaustralis*), and aardvark (*Orycteropus afer*). In parts of its distribution, the BFC utilizes abandoned hollow termitaria (Smithers 1983; Olbricht and Sliwa 1997). In contrast, the Kalahari AWCs spent most of the day resting under dense bushes and vegetation (85%), holes and caves (11%), and open shade (4%) ($n = 304$; observations of cats resting or sleeping before an activity period; Herbst unpublished data).

Study areas

The BFC study took place between December 1992 and September 1998 on the 114-km^2 game farm 'Benfontein' (28° 50′ S; 24° 50′ E), owned by De Beers Consolidated Mines Ltd, 10 km south-east of Kimberley (Fig. 26.3a). This area lies at the centre of the known distribution of BFCs (Nowell and Jackson 1996). The study area encompassed 60 km^2 with a variety of arid vegetation communities (Sliwa 1996, 2004, 2006), including the elements of three major biomes: Kalahari thornveld, pure grassveld, and Nama Karoo, which meet in the Kimberley area (Acocks 1988). An ephemeral pan and its specialized plant communities in the north dominate the farm, but in the south the vegetation changes into grassveld and finally Kalahari thornveld with deeper sandier soils on higher ground. Grass length ranges from $\leq$5 cm close to the pan to >100 cm in the Kalahari thornveld, where scattered camelthorn trees (*Acacia erioloba*) are interspersed in an open savannah. The climate is 'semi-arid continental' (Schulze and McGee 1978), with cool, dry winters (mean $T = 8$°C in July) and hot summers (23°C in January). Annual rainfall was 431 ± 127 SE mm for the past 50 years (Weather Bureau, Department of Environmental Affairs, Pretoria) and occurs mainly in spring and summer. For analysis, the year was divided into three seasons of 4 months each: winter—May–August; summer—November–February; autumn/spring—March–April and September–October. Populations of wild bovids, springbok (*Antidorcas marsupialis*), blesbok (*Damaliscus dorcas*), and black wildebeest (*Connochaetes gnou*) are harvested at irregular intervals by sport hunting and are culled for meat, but aside from this, human activity in the study area is minimal. In the south-eastern quarter of the farm, varying numbers of cattle (*Bos taurus*) are grazed.

The AWC study was conducted from March 2003 to December 2006 in the Kgalagadi Transfrontier Park (KTP). The main study area was along the southern part of the Nossob riverbed and surrounding dune areas (26° 28′ 17.7″ S, 20° 36′ 45.2″ E) (Fig. 26.3b). The KTP, incorporating the Kalahari Gemsbok National Park (South Africa) and the neighbouring Gemsbok National Park (Botswana), is a 37,000 km^2 area in the semi-arid southern Kalahari system, which forms part of the south-west arid biotic zone (Eloff 1984). The KTP is a wilderness area with minimum human impact; only limited tourism activities are present on two main roads in the riverbeds of the park. Herds of springbok, blue wildebeest (*Connochaetes taurinus*), red hartebeest (*Alcelaphus buselaphus*) and gemsbok (*Oryx gazella*) are dominant and large predators such as lion (*Panthera leo*), leopard (*Panthera pardus*), spotted hyena (*Crocuta crocuta*), brown hyena (*Hyaena brunnea*), and cheetah (*Acinonyx jubatus*), and smaller carnivores such as caracal (*Caracal caracal*), black-backed jackal (*Canis mesomelas*), Cape fox (*Vulpes chama*), honey badger (*Mellivora capensis*), small-spotted genet (*Genetta genetta*), and various raptor species are common in the KTP.

The vegetation of the Kalahari is described by Acocks (1988) as the western form of the Kalahari thornveld comprising an extremely open scrub savannah. Four main habitat types were identified and described as: (1) the dry riverbed and immediate surroundings; (2) the adjacent *Rhigozum* veld; (3) the sandy dune areas; and (4) the calcrete ridges and limestone plains. For more detailed descriptions of the vegetation, see Bothma and De Graaff (1973). The study site is characterized by low, irregular rainfall (Mills and Retief 1984), varying between 200 and 250 mm annually. Three seasons are recognized in the KTP: (1) a hot–wet season (HW) ranging from January to April, with mean monthly temperatures equal to or greater than 20°C, with 70% of the annual rainfall falling during this period; (2) the cold–dry season (CD) ranging May–August with mean monthly temperatures below 20°C and scarce rainfall; and (3) the hot–dry season (HD) ranging September–December with monthly temperatures ~20°C and rainfall generally not more than 20% of the annual rainfall (Mills and Retief 1984).

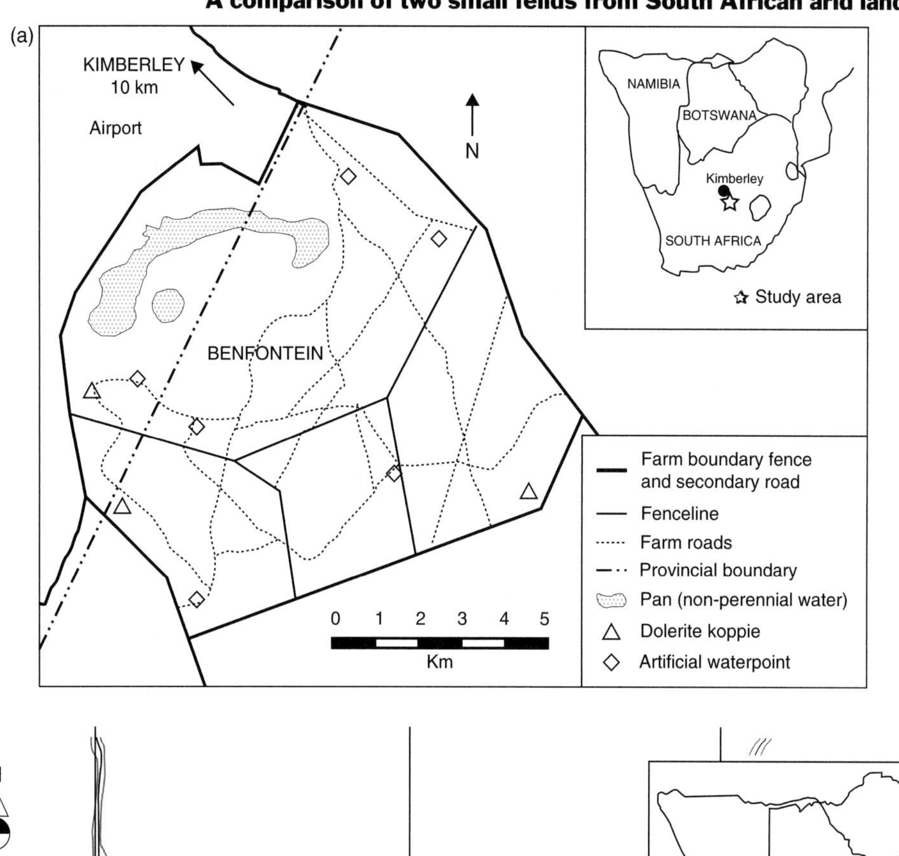

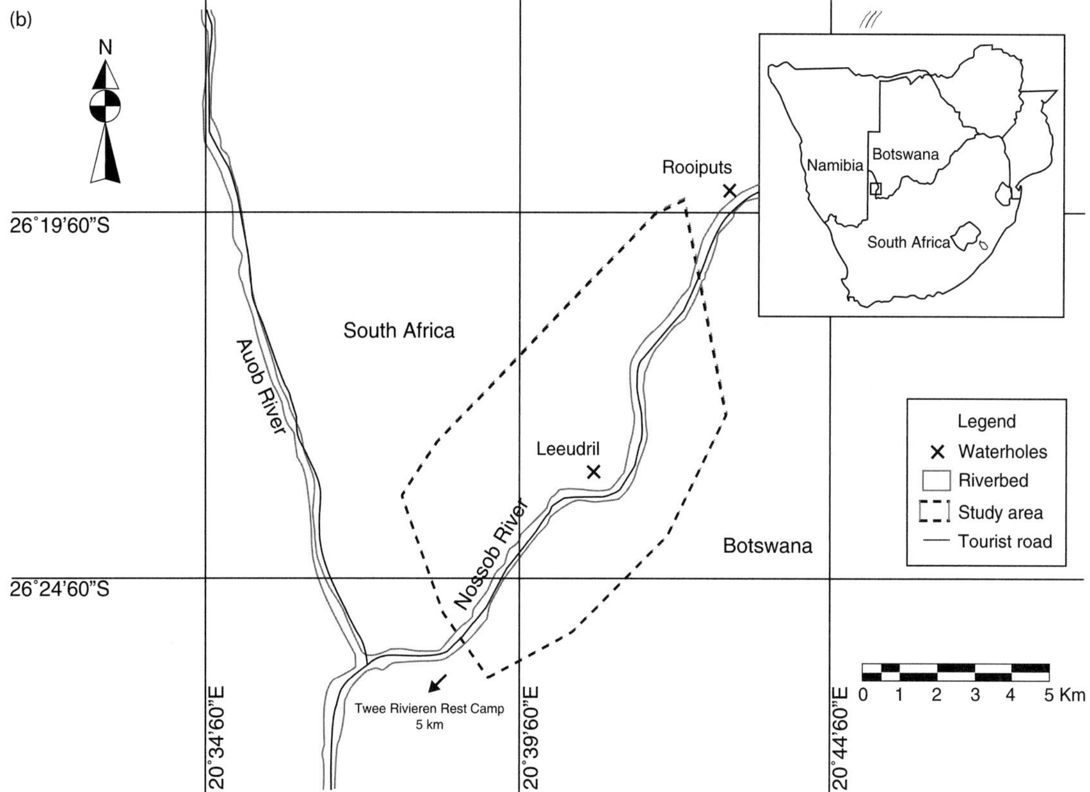

Figure 26.3a and b (a) Study area for BFCs, the game farm 'Benfontein', on the border of the Northern Cape and Free State provinces, South Africa. To the north-west of the boundary fence, marked by a thick black line, is Kimberley airport. The pan (solid grey) in the northern part of the study area, the road system, and some special features are shown. (b) Map of study area for AWCs around the Leeudril waterhole, indicating the riverbed and associated vegetation in the KTP. The Nossob river forms the unfenced border between South Africa and Botswana.

Methods

BFCs were detected with a spotlight at night, then were either followed to a hole and dug out by hand, or caught with a net while hiding on the ground. They were also trapped in specially made wire-cage traps, 30 × 30 × 100 cm, baited with dead birds. BFCs were anaesthetized by intramuscular injection of 20 mg/kg ketamine-hydrochloride and 10 mg/kg acetyl-promazine in order to fit custom-built radio-collars. All radio-transmitters (AVM Instrument Co., Livermore, CA, USA) operated in the 148–150 MHz frequency range. The radio-collars weighed 50 g and had a battery life of 6–8 months. Cats were weighed to the nearest 50 g, measured, and aged based on a combination of tooth wear, body mass, reproductive condition, and subsequent territorial behaviour. A cat was classified as adult when it had permanent dentition with slight discolouring or chipping and adult body size and mass or females had used nipples. It was classified as subadult if it was independent, had clean, white, unchipped teeth and, in females, unused nipples with <1 kg body mass. Resident adult males spray marked on a regular basis while non-resident males and subadult males did not. Twenty-one BFCs (six adult males, nine adult females, two subadult females—one became a resident adult, and four subadult males—three of which became resident adults during the study) were captured a total of 50 times. Twenty cats were radio-collared but three collars either stopped transmitting or were dropped after 2–6 days (Sliwa 2004). The remaining 17 individuals were each radio-tracked discontinuously over a period of 418 ± 355 days (mean ± SD; range: 16–1254 days).

AWCs were either caught in cage traps (10 cats), or immobilized while free-ranging by using a dart gun (18 cats). Cage traps (50 × 50 × 150 cm) were baited with chicken pieces. A crush plate enabled a hand injection to be administered. AWCs were then immobilized with 25 mg/ml Zoletil (2.5 mg/kg Tiletamine hydrochloride with Benzodiazephine derivative Zolazepam) in order to fit them with radio-collars. Radio-collars weighing 80–85 g from African Wildlife Tracking CC. were used, with a battery life of ~18 months. Trapping cats with cage traps did not prove to be very efficient with 1.4% success rate (1244 cages set over a trap period of 301 nights). Darting free-ranging cats was more effective. It was possible to approach cats with a vehicle at night and temporarily deprive them of sight with a spotlight. Qualified SA National Parks wildlife veterinarians used a CO_2 rifle (Dan-inject JM Standard model) with a standard dart syringe (10.5 mm; 1.5 ml capacity) and fitted with a stopper to reduce penetration. Cats were only darted when a clear shot was possible from a distance of 10 m. Eighteen cats were successfully darted with a combination of drugs (Butorphanol:Medetomidine and Zoletil: Medetomidine) and antagonists (Naltrexone for Butorphanol, and Antipamezole for Medetomidine—Zoletil does not have an antidote; Herbst and Mills, in preparation). In all cases, a small skin sample, taken from a nick in the ear, was collected for DNA analysis and, if relevant, a radio-collar was fitted. Eight AWCs, consisting of three adult females and four adult males and one young male were radio-collared.

Cats in both studies were observed directly from a four-wheel-drive vehicle after an initial habituation period of 1–3 weeks. At night, cats were observed with a low-powered handheld spotlight and focal animals were closely followed at a distance of 10–100 m. The beam of the spotlight was kept slightly behind the cat to avoid illuminating the prey or the cat. When a prey item was caught the observer attempted to identify it to the species level, where possible, and its average mass was taken from the literature and museum mammal collections for later diet analysis (Sliwa 2006; Herbst *et al.*, unpublished data; Tables 26.1–26.3). Details of the focal cat's behaviour, together with the location and length of the dominant vegetation since the last observation, were recorded onto an audio recorder whenever the cat changed direction or behaviour, or after 15 min. Single fixes were also recorded sporadically. The BFC study included 12 observation periods, each lasting a mean of 50 ± 29 SE days. A total of 17,450 fixes was obtained while following BFCs over a distance of 2000 km for 3125 h, including 1600 h of direct observation (Sliwa 2004). For AWCs, 10,979 fixes and 1538 h of direct observations were recorded (Herbst and Mills in press, unpublished data).

Life history and ecology comparisons

Social organization and spatial system

Both species are solitary. A maximum of 10 adult BFCs were radio-collared simultaneously in summer 1998 in the 60-km^2 study area, with no further cats sighted, giving an estimated density of 17 adults/100 km^2 (Sliwa 2004). During 2005/6, a total of 10 AWCs were radio-collared on the 53-km^2 study area and three non-radio-collared cats were regularly sighted, giving a minimum estimate of 25 cats/100 km^2. Mean annual home range sizes, using the 100% minimum convex polygon method (MCP) (Mohr 1947), was 20.7 ± 3.1 SE km^2 for five male BFCs, and 10 ± 2.5 SE km^2 for seven adult females (Sliwa 2004). Mean annual home range (100% MCP) was 9.8 ± 3.4 SE km^2 for four male AWCs, and 6.1 ± 1.1 SE km^2 for three females. This suggests that despite their smaller size, BFCs have home ranges 64–111% larger than AWCs between the studies, although this difference could have been due to prey resources.

Resident adult male BFCs' ranges overlapped with up to four different females. Intra-sexual overlap was slight for adult males (2.9%), but considerable for females (40.4%) (Fig. 26.4a). Home ranges were relatively stable, with mean shifts in range centres from one season to the next of 835 ± 414 m. In addition, the extent of overlap of seasonal ranges of the same individuals was 68 ± 11% (Sliwa 2004). Resident adult male AWCs' range overlapped with up to four different females. Intra-sexual overlap between adult females was 39.8%, but only 5.8% between adult

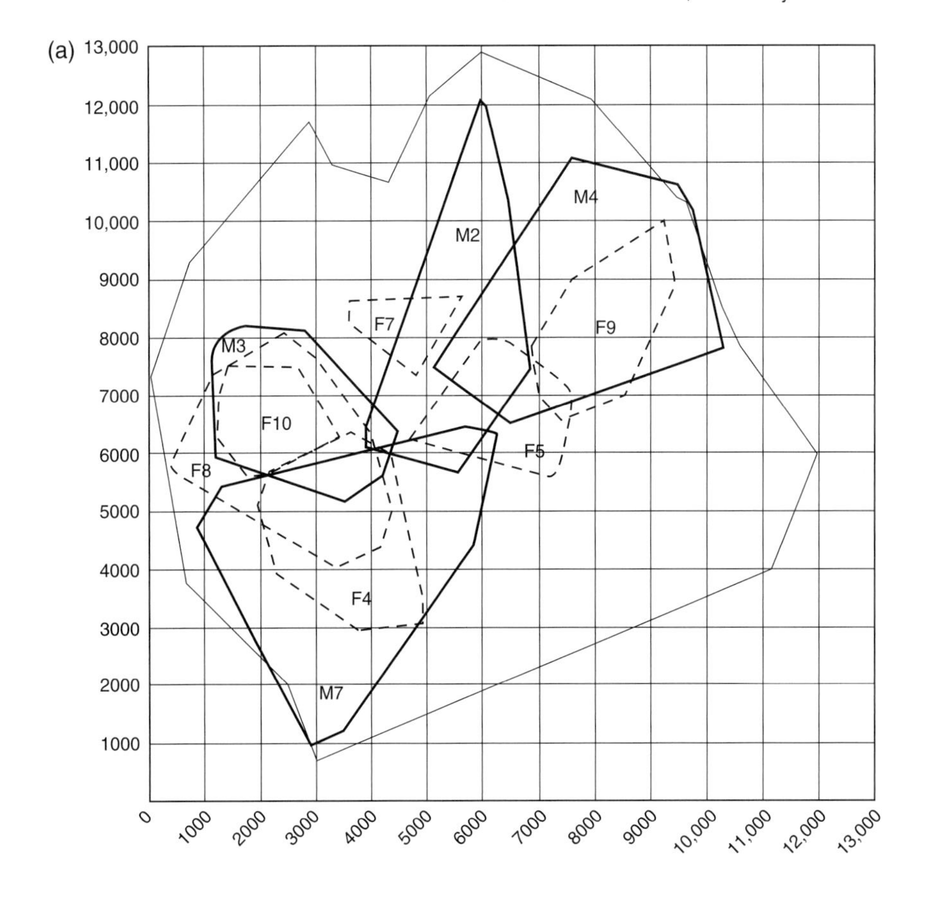

Figure 26.4a 100% MCP home ranges calculated from all records for 10 seasonal ranges of BFCs tracked during the summer or non-mating season 1998 (January, February, and March) on a 1-km^2 grid. Outline of the boundary fence of 'Benfontein' game farm given (males, thick solid lines; females, thin broken lines).

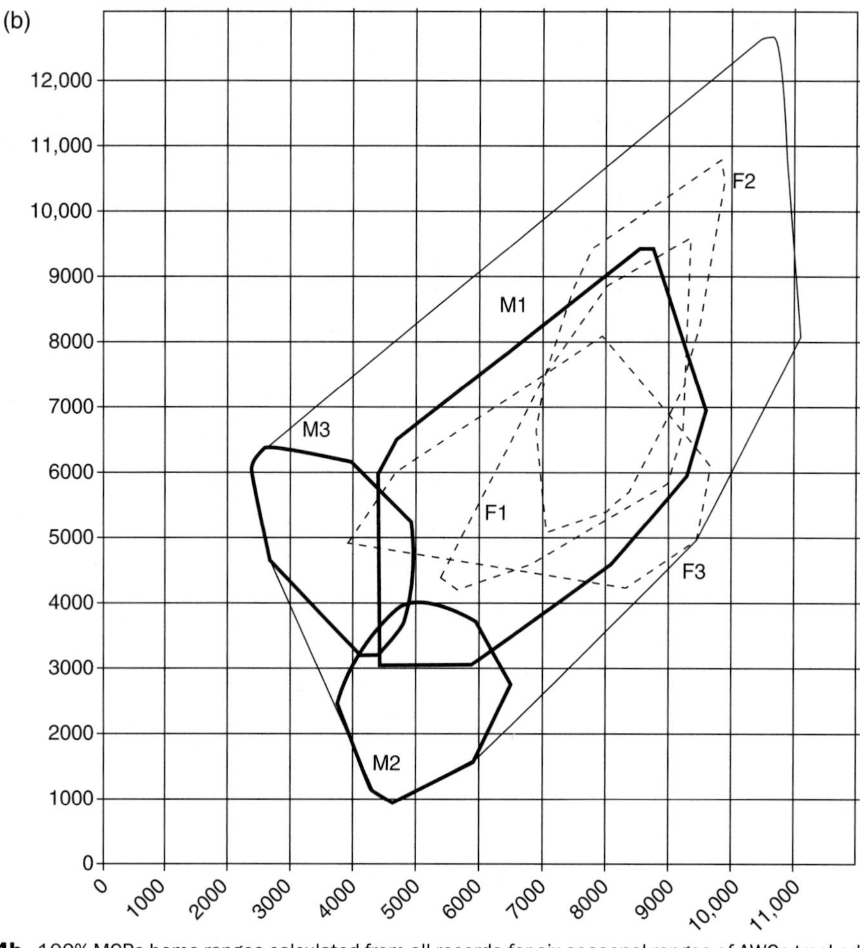

Figure 26.4b 100% MCPs home ranges calculated from all records for six seasonal ranges of AWCs tracked during 2004 and 2005 on a 1-km^2 grid. Outline of overall area of study site given (males, thick solid lines; females, thin broken lines).

male cats (Fig. 26.4b). However, when a subadult male was included in the analysis the overlap increased to 9.7%. The social organization is thus very similar between the two species and both adhered to the 'classical' felid system (Kitchener 1991; Sunquist and Sunquist 2002).

Communication

Female BFC marking frequency varied from no sprays/night to up to 268/night. Males' marking frequencies ranged widely from 0 to 598 sprays/night during the mating season. Adult resident males spray mark regularly (mean = 18 sprays/km), in contrast to non-resident and subadult males who mark only rarely (~1 spray/km; Sliwa 2004; Sliwa, unpublished data). Females left an average of 6.5 ± 10.7 marks/km (range = 0–44). Females exhibit urine scent-marking patterns depending on their current reproductive state. The highest spray-marking frequency (36 sprays/h) of one female occurred 1.5 months before conception of her litter, dropping to a lower frequency (<1 spray/h) during pregnancy, while being entirely absent when she reared young (Molteno *et al.* 1998). Urine marks were deployed in proportion to intensity of use (Molteno *et al.* 1998). The primary function of urine spraying in females is likely advertisement of reproductive condition and may play an additional role in social spacing (Sunquist 1981).

Female AWCs showed urine spray-marking patterns that were related to their current reproductive status. In all cases, where females increased spray marking ($n = 10$), they either had kittens ($n = 5$) or they were in the presence of a male cat ($n = 5$). Spraying varied from zero to 50 sprays per observation period (observation period = 8 h or more of continuous following), giving an estimated 3.6 ± 8.7 sprays/km. The primary function of spray marking for females is probably to advertise their reproductive status to male cats, however, unlike with BFCs, spray marking was still performed by females raising young. Male AWCs show much less spatial and seasonal variation in spray marking than females and spraying ranged from 0 to 183 sprays per observation period and an estimated 13.6 ± 23.5 sprays/km moved. BFCs are seasonal breeders while AWCs seem to be more opportunistic, with females already coming into oestrus while they are still suckling kittens. Thus, the variation in spray marking between the two species might be explained by a difference in the mating system.

Male BFCs have a surprisingly loud call, reminiscent of that of a large domestic tomcat, with calling bouts spaced 10–30 min apart ($n = 19$) between July and December, coinciding with the mating season. They usually called after sniffing a urine spray mark, often after demonstrating flehmen. Female 'loud' calling, similar to that of males, was heard only once, when two competing males moved away from her (Sliwa, unpublished data). The 'loud' call probably supplements spray marking, serving both as spacing and attracting mechanisms during the brief female oestrus, lasting for only 36 h (Leyhausen and Tonkin 1966; Sliwa, unpublished data). The tonal frequency is an octave lower than in the larger-bodied *Felis* species (Peters *et al.*, 2009). In addition, BFCs utter softer vocalizations while communicating between mother and kittens and during mating between the male and female (Sliwa, unpublished data).

Both male and female AWCs have a loud call, which is mostly evident when male cats are calling females and vice versa. The 'prau' call described by Dards (1983) and Leyhausen (1979) is a short, relatively high-pitched cry, with a rapid rise in frequency and may be repeated. As with BFCs, 'loud' calling by the male AWC usually follows after sniffing a urine spray mark followed by flehmen behaviour. The purpose of these calls is possibly to attract, advertise receptiveness, or establish spacing between cats (Kitchener 1991b). On two occasions, male cats uttered a surprising loud whining–singing sound while courting a female. Females may, in addition, call loudly to kittens after returning from a hunt to locate them in dense vegetation. Upon reunion, much softer vocalizations and rubbing between the mother and her kittens occur. Softer vocalizations were also evident during mating between male and female cats. In summary, both species are not very vocal in general, and use vocalizations in a similar context, although only in brief periods throughout the year.

Reproduction and mating behaviour

Wild BFCs mate between late July and March, leaving only 4 months where no mating occurs. The main mating period starts at the end of winter, in July and August (7 of 11 [64%] matings) resulting in litters born in September/October (Sliwa, unpublished data). One or more males follow the female, which is receptive for only 36 h (Leyhausen and Tonkin 1966; Sliwa, unpublished data) and copulate up to 10 times (Sliwa, unpublished data; $n = 3$ mating sequences). After a 63–68-day gestation period (Schürer 1988; Olbricht and Sliwa 1997) an average of two kittens (1–4) are born inside a springhare burrow or hollow termitarium (Smithers 1983; Olbricht and Sliwa 1997). On the day of parturition, females only leave the maternal den for several hours. However, after 4 days they will have resumed their normal routine of hunting throughout the night, only returning at dawn to suckle the kittens (Olbricht and Sliwa 1997), leaving the kittens for up to 10 h continuously per night. After their first week, kittens are moved frequently, perhaps to reduce the risk of predation. In their second month, they start to eat solid food and are weaned at 2 months (Olbricht and Sliwa 1995). The mother carries prey back to them, both while at the den and later when kittens are left in patches of long grass waiting for her return. Older kittens are presented with live prey that they learn to hunt and kill, as observed in cheetahs (Caro 1994). Kittens become independent at about 5 months, when their milk dentition is replaced by permanent dentition. Up to two litters may be raised by a female in a year. One female had litters in

February and then 8 months later in October 1994 (Olbricht and Sliwa 1997; Molteno *et al.* 1998).

For the AWC, no clear seasonality in breeding was evident. However, from all litters ($n = 15$) observed during the study period, eight were conceived during the HD seasons, four during the HW seasons, and three during the CD. At the beginning of the study (2003) food availability was low and no litters were conceived for a 14-month period. However, after an increase in rodent numbers each female produced up to four litters in a 12-month period. An average of 3 kittens (1–5) per litter was born, with kittens being born in dense vegetation, holes in the ground, or small crevices in calcrete ridges. Kittens were moved frequently to new dens. They emerged from the den ($n = 5$) after 7–10 days, not wandering further than a few metres. The mother spent most of her time at the den and made short hunting trips around the den area. As kittens developed, the mother stayed away for extended periods, leaving the kittens in dense vegetation or in close proximity to trees. Initially, she hunted for herself and returned to the den to suckle the kittens. However, as kittens approached 5 weeks of age she carried live prey back to them. The kittens played and practised their hunting skills on the stunned prey and either ate it or left the dead prey for the mother, who ate it or covered the remains. Kittens remained with the mother for 2–4 months after which they dispersed.

Males spent on average 1.7 ± 0.5 SE days ($n = 6$) with a receptive female while chasing, playing, and courting. Mating involves grabbing the female by the scruff of neck and the female lunging after stimulation (Smithers 1983; Sunquist and Sunquist 2002). Male cats did not assist in the rearing of kittens, although they twice visited females with kittens.

Social interactions

For both species of cats, very few intraspecific interactions were observed. Adult BFCs of opposite sex met rarely (two incidences) outside the mating season, resulting in a brief nose-to-nose sniff of each other. Agonistic interaction was observed only once between males during the mating season, where the resident cornered and threatened the transient while vocalizing; however, no physical contact took place. A subadult male encountered an adult female on two occasions, and travelled for 300 and 160 m with her while attempting to play. Because a subadult male is unlikely to approach a strange female, we tentatively assume this interaction was between a mother and an offspring. No such visits were recorded while a female was attending to kittens. A radio-marked subadult male played with another subadult cat on one occasion (Sliwa, unpublished data). In the AWC older kittens did return to the den ($n = 3$), especially when litters were born shortly after each other, sometimes within a 3-month period. These older kittens played with the younger siblings (observed in two different litters, in one of which the older kitten returned for three consecutive nights) and joined the mother on hunting forays. On these occasions the mother did not provide prey to the older kitten, which hunted its own prey and did not return to the den with its mother. No provisioning of food to younger siblings was observed.

AWCs were solitary except for the short periods (2–4 months) when females cared for kittens or during the brief mating periods, when males trailed receptive females (1–2 days). Twice male cats visited dens with kittens. The mother remained with the kittens, pulling her ears back and uttering a soft, hissing sound after which the male left. Often in encounters ($n = 10$), AWCs may stare at each other for several minutes at a distance without any interactions, after which they walk away from each other. Two males were observed fighting, spitting, scratching, and caterwauling, after which they ran away from each other. On four occasions, the dominant male cat in the study area stalked up to smaller subadult male cats and chased them away.

Interspecific interactions

On five occasions black-backed jackals circled cornered adult BFCs. Each time the BFC attacked, succeeding in driving the jackal away (Olbricht and Sliwa 1997; A. Sliwa, personal observation). However, kittens and inexperienced subadult cats are more likely to be in danger of predation, particularly when two jackals are involved. In the three cases this was observed, both jackals attempted to bite the cat in the back, making it more difficult even for an adult cat to stand its ground, although no incident of killing was directly observed. Black-backed jackals also stole hares (*Lepus* sp) from AWCs on two of the six occasions they were seen to catch one, having been attracted by the noise of the chase through vegetation (as opposed to the sounds of the prey—on only one of the six did a hare cry out loud). Afterwards, the cat took cover in thick vegetation. Although larger mammals such as hares contribute a large amount of food for an AWC, the pirating of kills (kleptoparasitism) by jackals probably contributes to the cats' preference for hunting smaller rodents.

On three occasions in the Kimberley study site, BFCs, on sensing an AWC, squatted low until the AWC passed without detecting them. Recently, two radio-marked adult BFCs were reported killed by a caracal and one by black-backed jackal (2007, B. Wilson and J. Kamler, personal communication). There were numerous interactions between BFCs with other species, resulting in the investigation of the other species or vice versa with no specific outcome, for example, aardwolf (*Proteles cristatus*), South African hedgehog (*Atelerix frontalis*), springhare, springbok, and even ostrich. Once, a male BFC stole a *Tatera* gerbil kill from a striped polecat (*Ictonyx striatus*), by driving it away. Also a marsh owl (*Asio capensis*) trailed a hunting BFC on three consecutive nights and captured small birds flushed by it (Sliwa 1994).

On five occasions, AWCs avoided larger predators (leopards, lions, cheetahs, and caracals) by running or hiding from them in dense vegetation. There have been records of caracals and leopards killing and consuming AWCs in the study area (M. Herbst and M.G.L. Mills, personal observation). AWCs chased away Cape foxes and small-spotted genets in rare encounters. A giant eagle owl (*Bubo lacteus*) twice tried to grab a large adult male AWC on his back, while the cat was crossing a clearing in the riverbed. The owl was unable to lift the cat and the cat then ran into thick vegetation.

Activity cycle and movement patterns

BFCs were strictly crepuscular and nocturnal, with cats leaving and returning to their dens within 30 min of sunset and sunrise (Olbricht and Sliwa 1997). Occasionally, though, during particularly cold and wet conditions they were seen basking close to their den during daylight. Their activity period varied with the length of the night, according to the season, from 10–14 h. They were active throughout the night, once they left the den at dusk until they returned to a den at dawn, travelling an average of 662 ± 89 m/h (Fig. 26.5). Part of their activity involved sitting outside rodent burrows, for between 30 and

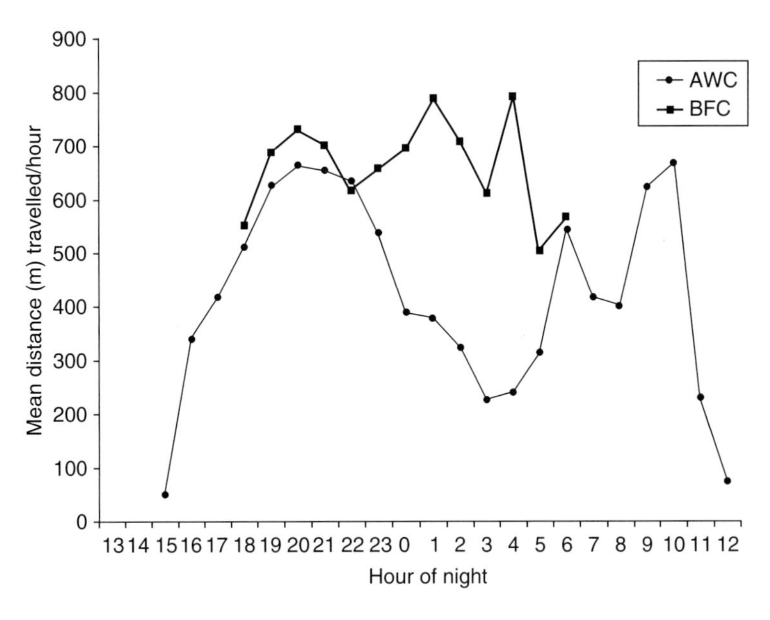

Figure 26.5 Activity as a function of average distances moved during each hour of the day/night for BFC (BFC; *n* = 10,85 nights) and AWC (AWC; *n* = 8,91 nights).

120 min and (judging by the constant movement of their ears) poising to pounce. On frequent occasions, these longer stationary periods resulted in a successful pounce. BFCs used predominately grassy habitats and were never observed to enter rocky or more densely wooded habitats.

AWCs were not as nocturnal as generally believed (Smithers 1983; Sunquist and Sunquist 2002) and their activity patterns depends on season and food availability. Typically, they became active as the sun was setting, with a peak activity time between 20:00 and 22:00, followed by a decrease in activity in the predawn hours. At dawn, there was an increase in activity and they remained active until late in the mornings, especially in winter months (Fig. 26.5). There are periods when cats lie down in front of rodent burrows, waiting for prey to appear. Although the cats may close their eyes, their heads are up and their ears constantly move, remaining alert to the sounds around them (from 344 observations 27% resulted in successful kills, 9% were unsuccessful catching attempts, and in 64% no attempts were made). Sometimes the cat would eventually lower the head and spread out laterally, resting and remaining in that position for several hours before continuing to hunt again. In contrast to BFCs, they do not have a shelter to which they return during the day. AWCs rest in thick vegetation (47%), in the shade of *Rhigozum* bushes (33%) or holes in the ground, trees, small crevices (16%), or just in the open (4%).

Average distances travelled per night by 10 BFCs (5♂♂/5♀♀) during 85 nights, where they were continuously observed for their entire activity period, was 8.42 ± 2.09 SE km (4.42–14.61 km). For eight AWCs (5♂♂/3♀♀) on 94 nights, the distance was 5.1 ± 3.35 km (1.07–17.37 km). So BFCs travelled about 65% further per night than AWCs, and this difference could have been influenced by prey abundance.

Diet

During the BFC study, 1725 prey items were consumed by 17 habituated cats (Sliwa 2006). Average prey size was 24.1 ± 47.4 g (SD). Males fed on significantly larger prey than did the females (8 ♂♂ average = 27.9 ± 53.2 g, $n = 795$ items; 9 ♀♀ = 20.8 ± 41.5 g, $n = 930$; Mann Whitney U-test: $U = 349244$, $P = 0.042$). Fifty-four prey species (Tables 26.1 and 26.2)

Table 26.1 Non-mammalian prey species (for mammals see Table 26.2) captured by *F. nigripes* on Benfontein Farm, on the border of the Northern Cape and Free State provinces, South Africa (their frequency of consumption and average mass).

Scientific name	Species identified	Number caught	Average individual body mass (g)	Mass consumed
Solpuga sp.	Solifuge	1	1.0	1
Opisthophthalmus glabrifrons	Shiny burrowing scorpion	7	1.0	7
Hodotermes mossambicus	Harvester termite (alates)	~390 (*5 incidences*)	0.15	58.5
Planipennia	Lacewings, ant lions	34	0.5	17
Saltatoria	locusts, and grasshoppers	93	1.5	139.5
Lepidoptera	large moths + beetles	26	1.0	26
Total invertebrates	>10 species	~551	~	**249**
Reptiles + frogs				
Lamprophis fuliginosus	Brown house snake	7	5–80	154
Lycophidion capense	Cape wolf snake	1	50	50
Mabuya capensis	Cape skink	1	4	4

Pachydactylus capensis	Cape gecko	3	3	9
Pachydactylus mariquensis	Marico gecko	2	3	6
Pyxicephalus adspersus	Giant bullfrog	1	400	250
Pseudaspis cana	Mole snake (juvenile)	1	110	110
Tompterna cryptotis	Tremolo sand frog	1	5	5
Total Reptiles/ amphibians	8 species	17		**588**
Birds				
Anthropoides paradisea	Blue crane (chick)	1	130	130
Anthus cinnamomeus	Grassveld pipit	5	25	125
Calandrella cinerea	Red-capped lark	29	26	754
Cercomela sinuata	Sickle-winged chat	1	18.5	18.5
Chersomanes albofasciata	Spike-heeled lark	128	26	3,328
Cisticola aridula	Desert cisticola	17	10	170
Columba guinea	Speckled pigeon	1	347	300
Eremopterix verticalis	Grey-backed finchlark	4	18	72
Eupodotis afraoides	White-quilled bustard	5	670[a]	2,110
Francolinus levaillantoides	Orange River francolin (scavenged)	1	370	—
Galerida magnirostris	Thick-billed lark	1	30	30
Malcorus pectoralis	Rufous-eared warbler	2	10	20
Mirafra apiata	Clapper lark	51	32	1,632
Mirafra sabota	Sabota lark	1	25	25
Mirafra africanoides	Fawn-coloured lark	1	20	26
Myrmecocichla formicivora	Southern anteating chat	9	48	432
Oenanthe pileata	Capped wheatear	1	28	28
Pterocles namaqua	Namaqua sandgrouse	1	180[a]	150
Rhinoptilus africanus	Double-banded courser	8	89	712
Telophorus zeylonus	Bokmakierie	1	65	65
Turnix sylvatica	Kurrichane button-quail	5	42	210
Unidentified small birds		13	20	260
Eggs of respective:	Black bustard, coursers, and larks	2 + 3 + 6	40, 10, 2.5	125
Nestlings of larks		5	~10	50
Total birds:	21 species	302		**10,773**

[a] For calculation: 20% of overall mass was subtracted to account for mass of feathers and bones remaining.

Table 26.2 Mammals consumed by BFCs. Average mass of mammals were taken from Skinner and Smithers (1990) and the collection of the McGregor Museum, Kimberley. *Antidorcas, Cynictis, Lepus, Pronolagus*, and *Xerus* were included in Fig. 26.6a as 'larger mammals'. All the other mammal taxa were pooled into 'smaller mammals'.

Scientific name	English name	Number consumed	Average mass of one (g)	Mass consumed
Antidorcas marsupialis[a]	Springbok (only scavenged)	1	3000[a]	1100
Crocidura sp.	Reddish-grey musk shrew	17	9	153
Cynictis penicillata	Yellow mongoose	2	830[a]	900
Dendromus melanotis	Grey climbing mouse	75	9	675
Desmodillus auricularis	Cape short-tailed gerbil	5	52	260
Gerbillurus paeba	Hairy-footed gerbil	152	26	3952
Lepus capensis[1]	Brown hare	13	1500[a]	4330
Malacothrix typica	Large-eared mouse	595	16	9520
Mus minutoides	African Pygmy mouse	276	7	1932
Pronolagus rupestris[a]	Smith's red rock rabbit (juvenile)	1	1600[a]	200
Saccostomus campestris	Pouched mouse	2	46	92
Tatera leucogaster	Bushveld gerbil	87	71	6177
Xerus inauris[a]	Ground squirrel	2	600[a]	520
Unidentified rodent		16	10	160
Total mammals	14 species	1246		**29,971**

[a] Individual consumption masses were estimated by weighing the carcass after feeding.

were classified by their average mass into different size classes for mammals, birds, amphibians/reptiles, and for invertebrates. Smaller mammals (5–100 g) constituted the most important prey class (54%), followed by birds (26%) and then larger mammals (>100 g; 17%) (Fig. 26.6a). Males and females took prey size classes at significantly different proportions, most notably for small birds (♀♀ = 21% vs. ♂♂ = 13%) and larger mammals (♀♀ = 9% vs. ♂♂ = 25%) (Sliwa 2006).

During the AWC study, 2553 prey items were observed being caught, of which 81% could be identified to one of five food categories (invertebrates, reptiles, birds, small mammals [<500 g], larger mammals [>500 g]), and comprising 26 species (Table 26.3). Nineteen per cent of the food items were classified as unknown as they were too small and consumed too quickly to be identified, thus data could be biased towards larger food items. From the HD season of 2003 to the CD season of 2004 (September 2003–August 2004), 97% of these total unknown food items were recorded when rodent numbers were lowest and invertebrate consumption was highest. Excluding unknowns, mammals made up to 82% of the cumulative prey biomass consumed (73% small mammals and 9% larger mammals), followed by birds (10%) and reptiles (6%) (Fig. 26.6b). The most frequently captured prey items were small mammals (44%), followed by reptiles (23%). Small mammals almost exclusively consisted of murids, with only one recorded macroscelid preyed upon (Bushveld elephant shrew, *Elephantulus intufi*). During 1538 h of observations on eight habituated AWCs, a total of 85.8 kg of prey items were consumed, with small mammals contributing to 61.9 kg of the

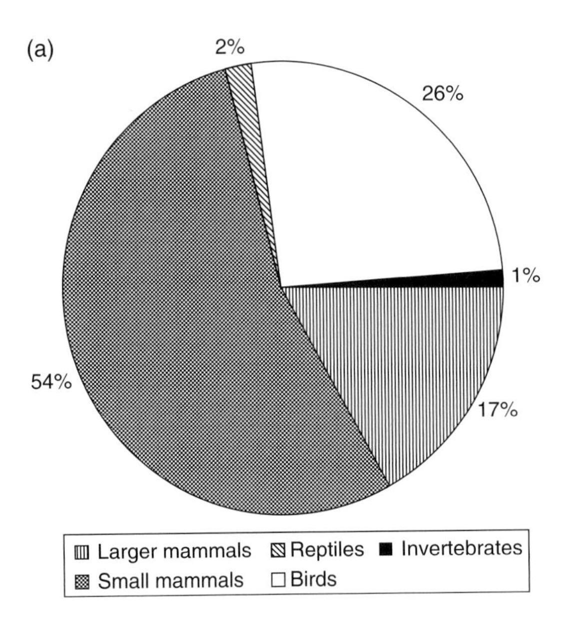

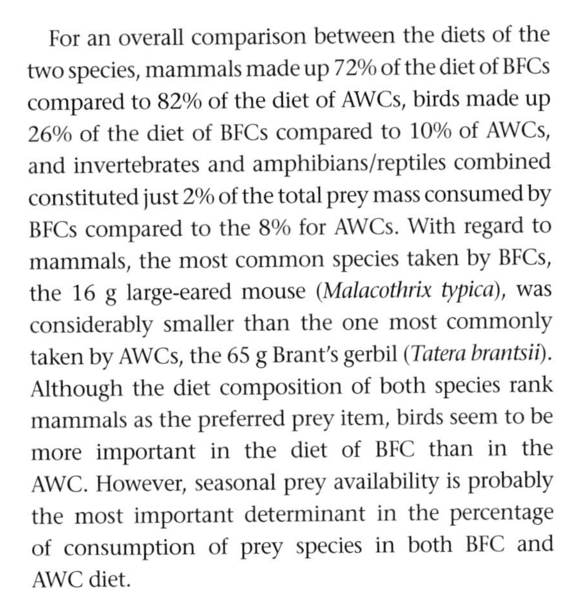
For an overall comparison between the diets of the two species, mammals made up 72% of the diet of BFCs compared to 82% of the diet of AWCs, birds made up 26% of the diet of BFCs compared to 10% of AWCs, and invertebrates and amphibians/reptiles combined constituted just 2% of the total prey mass consumed by BFCs compared to the 8% for AWCs. With regard to mammals, the most common species taken by BFCs, the 16 g large-eared mouse (*Malacothrix typica*), was considerably smaller than the one most commonly taken by AWCs, the 65 g Brant's gerbil (*Tatera brantsii*). Although the diet composition of both species rank mammals as the preferred prey item, birds seem to be more important in the diet of BFC than in the AWC. However, seasonal prey availability is probably the most important determinant in the percentage of consumption of prey species in both BFC and AWC diet.

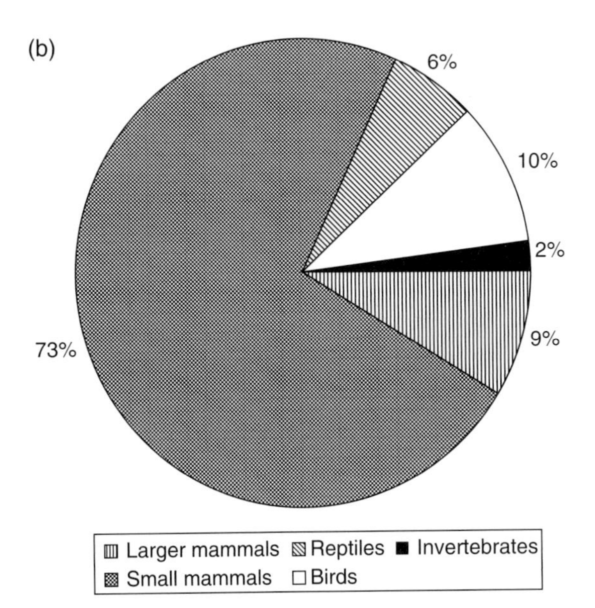

Figure 26.6a and b Prey composition from direct observations expressed as percentage of total biomass consumed by BFCs (a) and AWCs (b), pooled for five prey classes and for both sexes combined.

diet. There were no significant differences in the prey size of AWC sexes and both preferred small mammals. AWC females consumed more birds than males (Herbst and Mills in press).

Seasonal variation in the diet

For the three 4-month seasons of the year recognized in the BFC study, ectothermic prey items were unavailable during winter, when larger birds and mammals (>100 g) were mainly consumed. Small rodents like the large-eared mouse (*Malacothrix typica*, 595 captures) were particularly important (34.5% of all captures, 23% of total prey mass) for females during the spring and early summer when they were suckling kittens. Male BFCs showed less seasonal variation than females in prey size classes consumed (Sliwa 2006). This sex-specific difference in prey size consumption may ultimately help to reduce intraspecific competition. Despite this difference, the largest part of the diet (57%) of both sexes was made up by small-sized prey ((♀♀ = 66% vs ♂♂ = 49%) (Sliwa 2006; Fig. 26.7a).

Small mammals and reptiles were prey items most commonly consumed by AWCs, and combined, these contributed to more than 57% of the prey numbers eaten in each season. Small mammals contributed more than 65% to the cumulative biomass consumed by AWCs over all seasons. During the study, reptiles showed significant seasonal variation, being most common in the HW season (18% of the biomass of the diet of AWC), to less than 1% during the cold months, when reptiles are known to

Table 26.3 Prey items captured by AWCs in KTP during 2003 to 2006, documented from direct observations. Prey items presented in prey categories and in order of decreasing cumulative mass (g) of prey items consumed by AWCs.

Species identified	Scientific name	Number caught	Average individual body mass (g)	Mass consumed (g)	% occurrence
Larger mammals					
Spring hare	*Pedetes capensis*	3	2000	6000	
Hare sp.	*Lepus* sp.	2	2000	4000	
Ground squirrel	*Xerus inauris*	1	625	625	
Subtotal		6	4625	10,625	0.24
Small mammals					
Rodents (unidentified)		1100	50	55,000	
Brant's gerbil	*Tatera brantsii*	50	65	3250	
Brant's whistling rat	*Parotomys brantsii*	28	80	2240	
Striped mouse	*Rhabdomys pumilio*	19	32	608	
Damara mole rat	*Fukomys damarensis*	3	131	393	
Hairy-footed gerbil	*Gerbillurus paeba*	11	26	286	
Short-tailed gerbil	*Desmodillus auricularis*	2	46	92	
Pygmy mouse	*Mus indictus*	6	5	30	
Bushveld elephant shrew	*Elephantulus intufi*	1	42	42	
Subtotal		1220	477	61,941	47.79
Birds					
Lark sp.		50	60	3000	
Namaqua sandgrouse	*Pterocles namaqua*	8	300	2400	
Cape turtle dove	*Streptopelia capicola*	9	150	1350	
Spotted thick-knee	*Burhinus capensis*	1	320	320	
Namaqua dove	*Oena capensis*	1	42	42	
Subtotal		69	872	7112	2.70
Reptiles					
Common barking gecko	*Ptenopus garrulous*	488	5	2440	
Sand snake	*Psammophis* sp.	5	200	1000	

Giant ground gecko	*Chondrodactylus angulifer*	34	23	782	
Ground agama	*Agama aculeata*	13	25	325	
Kalahari tree skink	*Mabuya occidentalis*	5	10	50	
Subtotal		545	263	4597	21.35
Invertebrates					
Locusts	Order Orthoptera	47	4	188	
Moths	Order Lepidoptera	80	2	160	
Insects (unidentified)		73	2	146	
Formicidae	Order Hymenoptera	5	2	10	
Ant lion	Order Neuroptera	3	2	6	
Beetle	Order Coleoptera	2	2	4	
Scorpion	*Opistophthalmus wahlbergii*	5	5	25	
Solifugidae		4	2	8	
Unknown		494	2	988	
Subtotal		713	23	1535	27.93
Total		**2553**	**6260**	**85,810**	**100**

hibernate (Branch 1998). The percentage biomass contributed by birds also indicates significant seasonal variation (HD months = 17%, CD months = 1.6%). Because the categories 'Insects', 'Unknown', and 'Other' contributed less than 1.5% to the total prey biomass consumed, these categories were omitted from the analyses. Although the dietary composition for both sexes differed significantly between seasons, small mammals contributed most to the total prey biomass eaten over all seasons (♂♂ = 70% and ♀♀ = 57%; Fig. 26.7b).

Biomass consumed per distance moved

In order to compare the energy needs for both species, we calculated the biomass consumed per night and the distance moved. The average prey mass consumed per night for BFCs was 237 ± 105 g (67–611 g) and for AWCs it was 401 ± 358 g (2–2250 g). The latter consumed an average of 107.9 ± 133.8 g/km travelled (range 0.94–979.9 g), while BFCs consumed only 30.3 ± 17.1 g/km (6.5–110.2 g). This translates to an average of 13.7 ± 17.2 (range 1–113) prey items captured by AWCs per night, compared with 12.4 ± 5.3 (range 2–26) captured by BFCs. While the number of prey items caught per night is similar for the species, the difference in biomass consumed per kilometre travelled is 3.5-fold. When this is calculated for the two species per kg body mass, it is 18.9 g/km/kg of cat for BFCs (mean = 1.6 kg body mass for sexes pooled) and 24 g/km/kg for AWCs (mean 4.5 kg), a 38.6% higher prey mass consumption per km and kg body mass.

During HD and HW seasons, AWCs consumed more biomass per kilometre than during the CD season (HD = 130.3 ± 177 g/km, HW = 107.8 ± 105.6 g/km, and CD = 75.8 ± 48.4 g/km). However, during the winter (i.e. CD) months the cats travel further per observation period (8 h or more of continuous observations) (HD = 4.2 ± 2.5 km, HW = 4.8 ± 4.2 km, and CD = 6.5 ± 3.4 km) and they are active over a longer period, including early afternoons and

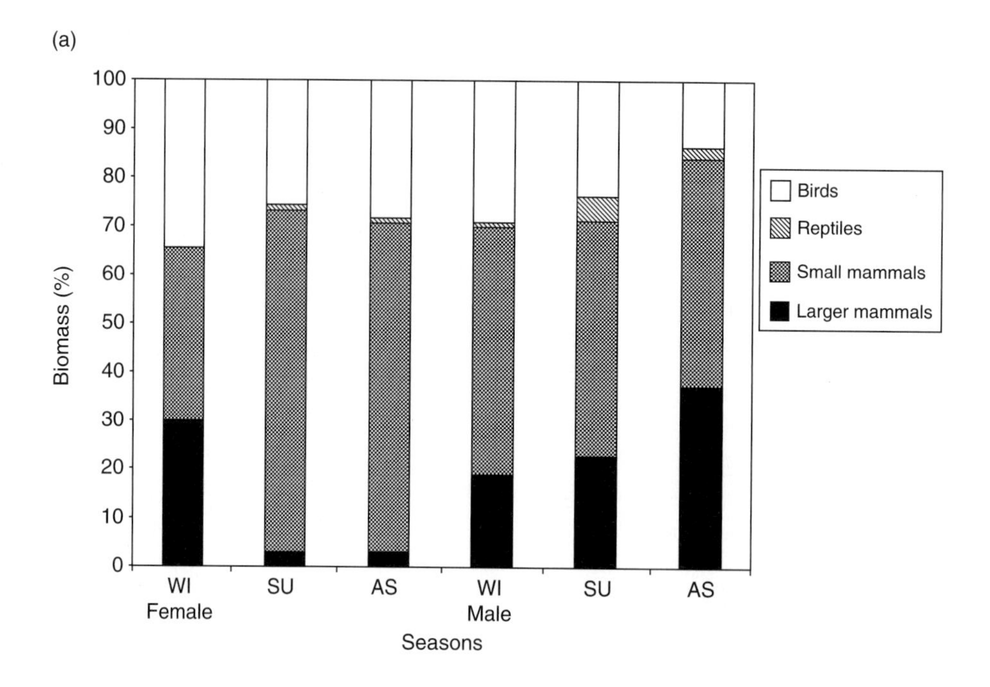

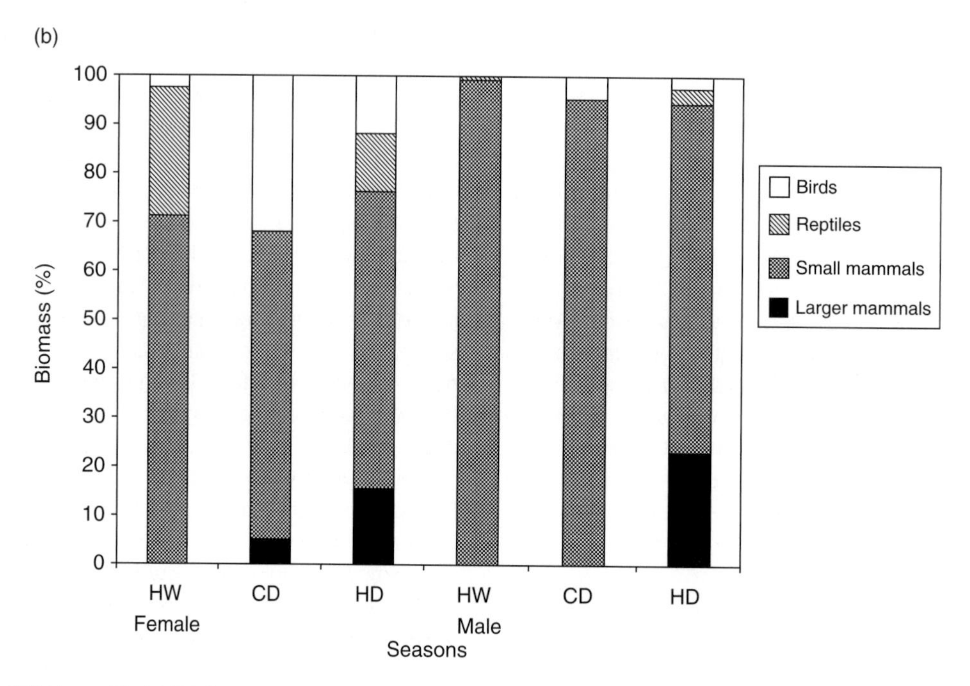

Figure 26.7a and b (a) Total prey mass consumed in the four prey categories with percentages larger than 1.5% by male and female BFCs across different seasons from visual observations (WI, winter; SU, summer; AS, autumn/spring). Invertebrates were not considered. (b) Total prey mass consumed in the four prey categories with percentages larger than 1.5% by male and female AWCs across different seasons from visual observations (HW, hot–wet; CD, cold–dry; HD, hot–dry). Invertebrates were not considered.

late mornings. BFCs consumed similar biomass per kilometre during all seasons (summer = 31.1 ± 15.7 g/km, winter = 29.5 ± 22.3 g/km, and autumn/spring = 30.2 ± 14 g/km). BFCs travelled similar distances in all seasons (summer = 8.9 ± 2.1 km, winter = 8.7 ± 2.3 km, and autumn/spring = 7.8 ± 1.8 km).

Conclusions and recommendations

We compared BFCs with AWCs to see if body size differences might also explain differences in the life history and ecology of these two small felid species (Table 26.4). Observed differences between the species may also reflect the environmental differences of the study areas and also the rainfall patterns of the study periods. Some possible but as yet not fully tested hypotheses are discussed below.

Both cats are mainly nocturnal; however, AWCs are more flexible and hunt during daylight. BFCs have a set activity period from dusk till dawn and return to rest mostly within dens during daylight. This may reduce predation risk by diurnal raptors as well as persistent mobbing of BFCs by passerine birds, the latter seen often when they travel at dusk and dawn (Olbricht and Sliwa 1997). Although AWCs in the Kalahari face similar predation risks with even larger predators present, their larger body size may be more advantageous for hunting in daylight hours, possibly being less susceptible to diurnal raptor predation. Alternatively, a more diurnal activity regime might reduce interspecific competition since the AWC is part of a carnivore guild of various smaller- and similar-sized carnivores in the Kalahari.

In many predator studies, prey abundances and availabilities have been found to be crucial factors in facilitating and determining distributions and coexistence (Creel and Creel 1996; Durant 1998; Karanth and Sunquist 2000). BFCs and AWCs fed mainly on mammalian prey between 5 and 100 g. There was, however, a difference between the most frequently captured prey species. BFCs hunted mostly large-eared mice (*Malacothrix typica*; mean = 16 g) (Sliwa 2006), whereas AWCs took Brant's gerbil (*Tatera brantsi*; 65 g). Expressed as prey mass per unit kilogram of cat, BFCs took 10 g of prey and AWCs 14.4 g of prey. Both species consumed a similar number of prey items each night, therefore the fact that the most commonly available rodent eaten by AWCs was larger than that eaten by BFCs resulted in a larger biomass consumed by AWCs. However, when comparing the percentage biomass consumed per unit (kg) body mass of cat, BFCs consumed 14.9% of their body mass per night compared to 8.9% for AWCs. This is probably due to the higher metabolism related to smaller body size in BFCs, and the need to cover longer distances per night to capture enough prey to sustain their energy needs. Despite the strong seasonal variation in biomass consumed per distance moved by AWCs (76–130 g/km), even the lowest prey mass consumed in the cold winter season by AWCs was 2.5 times that consumed by BFCs (30 g/km) (Table 26.4).

The distribution of BFCs may be influenced by the availability and abundance of certain prey species and prey sizes (i.e. the large-eared mouse is absent in the Kalahari ecosystem and AWC study site). One could describe the BFC as a habitat specialist that shows a preference for grassland and avoids woody or rocky areas. It moves further per night, while consuming less biomass per distance than the AWC. This is reflected in the larger annual home ranges and distances travelled per night of BFCs. Alternatively, these differences could have been due to differences in prey abundance between sites.

BFCs consumed more birds (26%) in comparison to AWCs (10%), probably resulting from their smaller size and agility, and being able to conceal themselves better in short vegetation. Thus, a greater abundance of small birds in a habitat may favour BFCs over AWCs. Although we could not compare bird abundance on each site, the body size of a cat may be negatively correlated with hunting success of small birds, as demonstrated in the differential hunting success by BFC sexes. However, AWCs may not need to supplement their diet with birds and the larger-sized and more abundant rodents might be sufficient for their dietary requirements. During seasons with low rodent numbers in the Kalahari AWCs changed their diet accordingly and took more invertebrates and reptiles during the warmer seasons than BFCs and, to a lesser extent, birds. Seasonal variation in the Kalahari contributed largely to differences in AWC diet and the biomass

Table 26.4 A summary of the ecological and life history traits of AWCs and BFCs.

	Study site	Years data collected	Adult cats radio-collared	Head-body size (cm)	Weight (kg)	Resting places	Estimated densities (cats/ 100 km^2)	Home range (MCP 100%)	Intra-sexual overlap	Urine spray marking/km	Average litter size (range)	Maximum litters per year	Activity	Distance travelled per night (km)	Biomass (g) per distance consumed
F. silvestris	KTP, Northern Cape, South Africa and Botswana	2003–06	♂ = 5 ♀ = 3	♂ = 65 ♀ = 60	♂ = 5.1 ♀ = 3.9	No fixed resting place—dense vegetation or in trees	25	♂ = 9.8 km^2 ♀ = 6.1 km^2	♂ = 5.8% ♀ = 39.8%	♂ = 13.6 ♀ = 3.6	3 (1–5)	4	Mainly nocturnal, although active mornings and afternoons	5.1 ± 3.35	401 ± 358/night 108 ± 134/km
F. nigripes	Benfontein, Kimberley, Northern Cape/Free State, South Africa	1992–98	♂ = 8 ♀ = 10	♂ = 45 ♀ = 40	♂ = 1.9 ♀ = 1.3	Den sites in holes or hollow termitaria	17	♂ = 20.7 km^2 ♀ = 10 km^2	♂ = 2.9% ♀ = 40.4%	♂ = 12.6 ♀ = 6.5	2 (1–4)	2	Nocturnal	8.42 ± 2.09	237 ± 105/night 30 ± 17/km

Source: *Felis nigripes* (Sliwa 1994, 2004, 2006); *Felis silvestris* (Herbst, unpublished data).

consumed per night, with less seasonal variation in the BFC study.

The AWC was possibly better able to respond reproductively to temporary food restrictions and superabundances than the BFC, although a climatic variation between sites confounds this data. AWCs have larger litter sizes and may raise up to four litters per year, while reproduction can fail entirely in years with low-prey abundance. Data for comparisons from the BFC is still lacking. AWC mothers take short hunting trips around the den, while female BFCs may need to travel longer distances to capture sufficient prey for their dependent offspring.

Research gaps in relation to conservation management

Both species were influenced by the presence of competitors and predators. A high density of mesocarnivores like jackals and caracal would both result in harassment, pirating of kills, and even intra-guild predation. This has been observed in other predator guilds (Palomares and Caro 1999) specifically between foxes (*Vulpes macrotis, V. velox*, and *V. vulpes*) and coyotes (*Canis latrans*) (Moehrenschlager and List 1996) and for large felids between tiger (*Panthera tigris*) and leopard (Seidensticker 1976a), and among cheetah, lion, and leopard (Caro 1994), but recently also proposed for smaller felid guilds in tropical America comprised of ocelot and oncilla (*Leopardus pardalis, L. tigrinus*,) (Oliveira *et al.*, Chapter 27, this volume). In South African farming communities where livestock depredation occurs, densities of jackals and caracals are regulated through predator control. In the protected area of the southern Kalahari, there is little interference from human activities and predator numbers are mainly regulated by available food resources (Mills 1990).

Increasing human impact, through population growth and changes in land-use patterns (small holdings farming, irrigation, and overgrazing), may also affect the two cat species differently. The BFC avoids human contact (Olbricht and Sliwa 1997; Sliwa 2004), while a male AWC radio-monitored in the same study area stayed close to permanent water and human habitation (Sliwa, unpublished data). However, if species like jackals and caracals are removed from small-stock farming areas, this may be to the advantage of small cats, especially the BFC. The AWC seems to have a higher tolerance to human-modified habitats, and may profit from increasing rodent populations associated with farming; however, it may also be threatened in its genetic integrity through hybridization (Nowell and Jackson 1996; Yamaguchi *et al.* 2004c) and disease transfer (Mendelssohn 1989; Macdonald *et al.* 2004b) from domestic cats associated with man.

Studies of smaller African felids are in their infancy, especially within their carnivore guild. A number of key questions arise from our comparative research: (1) What are the maximum levels of habitat loss, degradation, and fragmentation both species could tolerate? (2) What influences the distribution of the BFC—when does competition pressure from potential predators and competitors become too high, leading to its exclusion from certain areas? (3) Is conservation management for both species similar or mutually exclusive? (4) Could AWCs negatively affect BFC numbers, especially given this behaviour among other felid species?

There is an urgent need for comparative studies of small felids in order to address specific conservation questions. We trust that our studies will both contribute to the basic understanding of BFC and AWC ecology, as well as provide the baseline data for future research and conservation measures for small African felid studies.

Acknowledgements

The BFC study was funded by: Endangered Wildlife Trust, South Africa; San Diego Zoological Society, Chicago Zoological Society, Columbus Zoo, John Ball Zoo Society, Project Survival in the USA; International Society for Endangered Cats and Mountain View Farms in Canada; People's Trust for Endangered Species in the United Kingdom; and Wuppertal Zoological Garden, Germany. We thank De Beers Consolidated Mines for permission to work on Benfontein Farm. Beryl Wilson, Arne Lawrenz, Gershom Aitchison, Enrico Oosthuysen, Gregory and Nicola Gibbs, and Andrew Molteno helped with capturing and tracking cats.

The AWC study was funded by the Endangered Wildlife Trust's Carnivore Conservation Group,

Elizabeth Wakeman Henderson Charitable Foundation, and the Kaplan Award Program from the Wildlife Conservation Society. We are grateful to South African National Parks, Department of Wildlife and National Parks, Botswana (KTP), and the Mammal Research Institute, University of Pretoria.

We thank Chris and Mathilde Stuart for data on AWC distribution on the African continent (Fig. 26.1) and the two editors of this volume, as well as Mike Daniels, Nobuyuki Yamaguchi, and one anonymous reviewer who greatly improved the quality of our manuscript.

CHAPTER 27

Ocelot ecology and its effect on the small-felid guild in the lowland neotropics

Tadeu G. de Oliveira, Marcos A. Tortato, Leandro Silveira, Carlos Benhur Kasper, Fábio D. Mazim, Mauro Lucherini, Anah T. Jácomo, José Bonifácio G. Soares, Rosane V. Marques, and Melvin Sunquist

Ocelot *Leopardus pardalis* from eastern Amazonia. © 'Projeto Gatos do Mato—Brasil'.

Introduction

The ocelot (*Leopardus pardalis*, 6.6–18.6 kg) is the third largest of the 10 felid species found in the Neotropical region (Sunquist and Sunquist 2002; Oliveira and Cassaro 2005). Ocelots extend from south Texas through to northern Argentina, and have been the subject of several studies on feeding ecology, home range, activity patterns and density (e.g. Ludlow and Sunquist 1987; Emmons 1988; Crawshaw 1995; Jacob 2002; Di Bitetti *et al.* 2006; Moreno *et al.* 2006). The ocelot appears to be strongly associated with dense habitat cover and preys predominantly on small rodents (see Sunquist and Sunquist 2002). However, the extent to which environmental or biological variables constrain ocelot presence and distribution remains poorly understood. Throughout their broad geographic range ocelots live in sympatry with the smaller margays (*Leopardus wiedii*, 2.3–4.9 kg), little spotted cats (*Leopardus tigrinus*, 1.7–3.5 kg), jaguarundis (*Puma yagouaroundi*, 3.0–7.6 kg) and, to a lesser extent, with pampas cats (*Leopardus colocolo*, 2.5–3.8 kg) and

Geoffroy's cats (*Leopardus geoffroyi*, 2.8–7.4 kg) (Oliveira and Cassaro 2005; Lucherini *et al.* 2006). They also coexist with pumas and jaguars through most of their range. There is considerable overlap among these sympatric small felids, not only in geographic range but also in habitat use, feeding ecology, activity patterns, and body size, suggesting the potential for interspecific competition (Oliveira 1994; Nowell and Jackson 1996; Sunquist and Sunquist 2002). Additionally, competitive intraguild interactions are now acknowledged as important in influencing animal communities (Dayan *et al.* 1990; Creel and Creel 1996; Donadio and Buskirk 2006; St-Pierre *et al.* 2006). In this regard, the mesopredator and niche release theories are of special interest (Crooks and Soulé 1999; Moreno *et al.* 2006). The mesopredator release theory predicts that with the decline of top predators, mesocarnivore numbers would rise (Crooks and Soulé 1999), whereas the disappearance of a potential competitor would broaden the niche of its subordinate competitors (Moreno *et al.* 2006). However, it is unknown to what extent these interactions influence felid dynamics in the neotropics.

Here we tackle these issues by, first, characterizing ocelot ecology across its broad geographic range in an effort to understand the extent to which environmental variables influence ocelot demography and ecology; we then compare ocelot ecology to that of the other sympatric small cats with a view to revealing their ecological differences. We also assess the potential for competition between ocelot and other felids, combined with testing the competitive niche release hypothesis. Finally, we consider the potential role that ocelots might play as mesopredators in the dynamics of the small-felid assemblage of the lowland tropical neotropics.

Methods

To describe ocelot ecology and the environmental variables that might influence it, we compiled information from 35 sources detailing home range, habitat use, diet, and density across the species' full geographic range, including sites from the northernmost (south Texas) and southernmost (north-eastern Argentina and southern Brazil) extremities of its range (Fig. 27.1). In our analyses we consider small felids as all those smaller than the ocelot; sympatric species to include the ocelot, jaguarundi, margay, and little spotted cat; and mesopredators as all small- to medium-sized lowland felids (i.e. ocelot, jaguarundi, Geoffroy's cat, pampas cat, margay, and little spotted cat).

Habitat was grouped into rainforest (Panama, Peru, Brazil-Pará [PA], Paraná [PR]), semi-arid habitats, which include scrub, chaparral, and dry forest (Texas and Mexico), and open flood plains/savannahs (Venezuela, Brazil-Mato Grosso do Sul-MS, and Goiás/Emas-GO) according to ecoregional similarities (Olson *et al.* 2001). Dense cover is considered here as either closed canopy or dense understorey, as described in their original studies. Body mass of adult male and female ocelots are combined for each site.

Differences in ocelot home range size and habitat use among sites and between sexes in different areas were compared using mean values. For comparing home ranges, whenever possible we used minimum convex polygons (MCP) incorporating 95% of fixes (using sources listed in Table 27.1). For habitat use and preference (but not for home range-estimates) for Emas National Park (ENP) (Brazil-GO) we only used 95% fixed kernel estimates. Preference/avoidance for each habitat type was measured using Ivlev's index of selectivity or preference (1961) (−1, complete avoidance, to +1, complete preference), on second and third order selection (Aebischer *et al.* 1993), as described in Manfredi *et al.* (2006) and Rocha (2006). Habitat categories for Emas include grassland savannah, scrub/woodland savannah, gallery forest, flooded grassland, and pasture/agriculture.

Ocelot diet was assessed by the analysis of droppings from study sites across the species range. Dietary results are presented as percent occurrence (number of times an item was found as a percentage of total number of prey items in droppings). We quantified diet using both percent occurrence and biomass contribution. For biomass calculations only mammalian prey were included. Prey were classed as very small mammals (<100 g: marsupials, rodents, others), small mammals (0.1–0.699 kg: marsupials, rodents), medium-sized mammals (0.7–1.5 kg: marsupials, rodents, rabbits, primates), large mammals (1.51–10 kg: rodents, xenarthrans, primates, carnivores, others), very large mammals (>10 kg: ungulates), birds, squamates (lizards and snakes), and other vertebrates. Maximum prey mass is defined as the mass of the

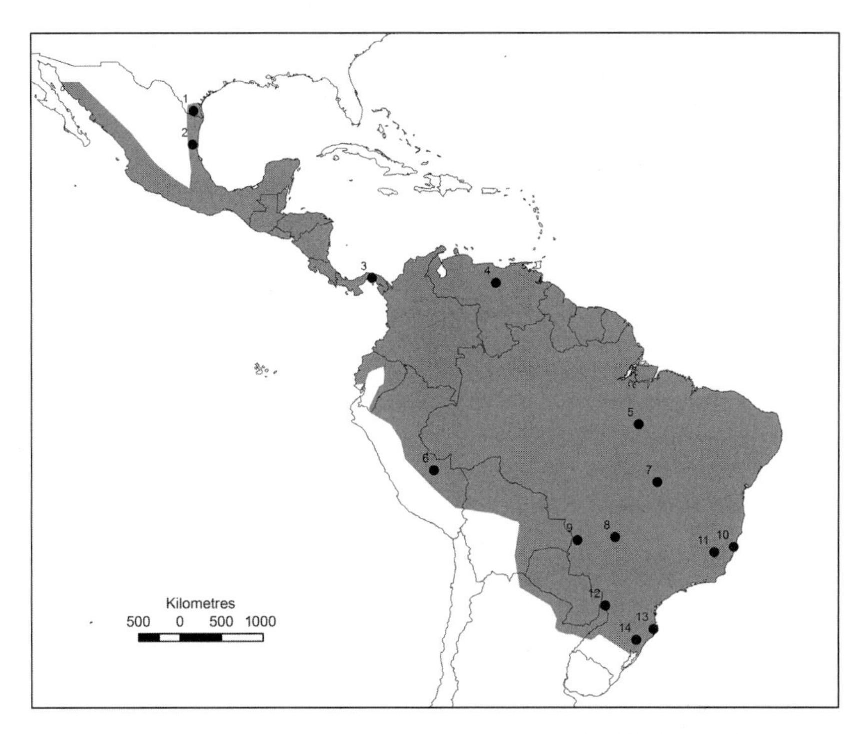

Figure 27.1 Ocelot distribution range showing the locations of study sites listed in the text. 1, S-Texas/United States (Lower Rio Grande Valley); 2, Tamaulipas/Mexico (Los Ebanos Ranch); 3, Barro Colorado Island/Panama; 4, Central Llanos-Guarico/Venezuela (Hato Masaguaral/Hato Flores Morada); 5, Carajás-Pará/Brazil; 6, Cocha Cashu Biological Station/Peru; 7, Tocantins/Brazil (Lajeado Dam); 8, ENP-GO/Brazil; 9, S-Pantanal-MS/Brazil; 10, Linhares-ES/Brazil (Vale do Rio Doce Natural Reserve); 11, Caratinga Biological Station-MG/Brazil; 12, Iguaçu National Park-PR/Brazil; 13, Serra do Tabuleiro State Park-SC/Brazil; 14, São Francisco de Paula National Forest-RS/Brazil. (Modified from Oliveira and Cassaro 2005.)

three largest species taken by each felid. Diets of sympatric species are compared using mean mammalian prey mass and niche overlap. To test for competitive niche release, we compared ocelot diet in areas with and without jaguars following Moreno *et al.*'s calculations (2006), and little spotted cat diet in the presence and absence of ocelots.

Mean prey body mass is often used as a metric in the assessment of trophic separation among carnivores (e.g. Oliveira 1994, 2002a; Ray and Sunquist 2001). The mean body mass of mammalian prey (MWMP) is determined as the arithmetic mean of the body mass of all mammalian prey consumed (geometric means tend to underestimate the mean mass of prey Walker *et al.* 2007). Body masses of prey species are from original studies, or from summaries in Fonseca *et al.* (1996) or Eisenberg and Redford (1999). All prey items were assumed to be adult size, unless otherwise noted. However, to avoid over-representation, prey with a body mass larger than 1.5 times the felid's body mass were assumed to be juvenile and assigned 40% of the adult mass. Canine diameter, a metric used to evaluate carnivore segregation and character displacement (Dayan *et al.* 1990), was assessed for Brazilian specimens of sympatric species. Mean values were correlated with felid body mass, MWMP, and maximum prey mass.

Diet niche overlap is calculated using Pianka's index: $O_{jk} = \sum P_{ij}P_{ik}\sqrt{(\sum P_{ij}^2 \sum P_{ij}^2)}$, where P_{ij} and P_{ik} are the proportions of the *i*th resource used by the *j*th and the *k*th species, respectively (Pianka 1973). Values range from 0 (no similarity) to 1 (complete similarity). Calculations are made at the lowest taxonomic category of prey possible.

Table 27.1 Average home range (km^2) of adult lowland Neotropical felids in various habitats.

Species	Male	*n*	Female	*n*	Habitat	Location	Source
Ocelot	12.3	5	7.0	3	Thorn scrub/oak forest	S. Texas, United States	Tewes (1986)
	6.3	3	2.9	3	Thorn scrub/oak forest	S. Texas, Unites States	Laack (1991)
	8.1	2	9.6	2	Low tropical forest/pasture	Taumalipas, Mexico	Caso (1994)
	31.2	1	14.7	1	Central American rainforest	Cockscomb, Belize	Konecny (1989)
	30.8	2	30	3	Central American rainforest	Ciquibul, Belize	Dillon (2005)
	10.6	2	3.4	6	Gallery forest/flood plains	Llanos, Venezuela	Ludlow and Sunquist (1987)
	7.0	2	1.8	3	Amazon rainforest	Cocha Cashu, Peru	Emmons (1988)
	19.2	1			Savannah	Tocantins, Brazil	Trovati (2004)
	90.5	1	75	1	Savannah	ENP, Brazil	Silveira (unpublished data)
	5.4	1	2.6	2	Deciduous-gallery forests/flood plains	Pantanal, Brazil	Rocha (2006)
			1.3	3	Deciduous-gallery forests/flood plains	Pantanal, Brazil	Crawshaw and Quigley (1989)
	11.7	4	7.2	4	Tropical forest	Morro do Diabo, Brazil	Jacob (2002)
	38.8	6	17.4	5	Subtropical forest	Iguaçu, Brazil/Argentina	Crawshaw (1995)
Jaguarundi	9.6	8	8.9	5	Low tropical forest/pasture	Taumalipas, Mexico	Caso (1994, personal communication)
	94.1	2	20.1	1	Central American rainforest	Cockscomb, Belize	Konecny (1989)
	25.3	1	18.0	1	Savannah	Tocantins, Brazil	Trovati (2004)
			40.2	1	Savannah	Emas Nat. Park, Brazil	Silveira (unpublished data)
	8.5	1	1.4	1	Forest/savannah/eucalyptus plantation	São Paulo, Brazil	Michalski *et al.* (2006b)
	17.6	1	6.8	1	Subtropical forest	Iguaçu, Brazil/Argentina	Crawshaw (1995)
	23.4	3			Subtropical forest fragments/ agriculture	Taquari, Brazil	Oliveira *et al.* (2008)
Margay	4.0	3	0.9	1	Low tropical forest/pasture	Mexico	Caso (personal communication)
	11.0	1			Central American rainforest	Cockscomb, Belize	Konecny (1989)
	15.9	1			Subtropical forest	Iguaçu, Brazil	Crawshaw (1995)
			20	1	Subtropical forest fragments/ agriculture	Taquari, Brazil	Oliveira *et al.* (2008)

Little spotted cat	4.8	2			Savannah	Tocantins, Brazil	Trovati (2004)
	17.1	1	0.9	1	Savannah	Serra da Mesa, Brazil	Rodrigues and Marinho-Filho (1999)
			25.0	1	Savannah	ENP, Brazil	Silveira (unpublished data)
	8.0	2	2.0	1	Subtropical forest fragments/ agriculture	Taquari, Brazil	Oliveira *et al*. (2008)
Pampas cat	27.7	4			Savannah	ENP, Brazil	Silveira (unpublished data)
Geoffroy's cat	3.1	2			Swamp forest/wet grassland/ agriculture	Arroio Grande, Brazil	Oliveira *et al*. (2008)
	5.0	2	1.9	2	Wet pampas grassland	Campos del Tuyú, Argentina	Manfredi *et al*. (2006
	8.9	2	5.1	1	Pampas grassland/agriculture	Tornquist Park, Argentina	Manfredi *et al*. (in preparation)
	2.2	4	1.1	2	Argentine Monte	Laguna de Chasicó	Manfredi *et al*. (in preparation)
	2.0	3	0.3	1	Scrubland	Lihue Calel National Park, Argentina	Pereira *et al*. (2006)
	9.2	5	3.7	2	Patagonian woodland/steppe	Torres del Paine, Chile	Johnson and Franklin (1991)

Estimates of felid abundance are based on records of tracks, photographs, or live captures. At each site felid species were ranked based on abundance indices, which varied from 1 (most abundant) to 5 (least abundant) (Oliveira *et al.* 2008, submitted).

Normality tests were conducted prior to determining the use of parametric or non-parametric statistics. Data in the form of percentages are converted into arcsin ($y_t = \arcsin \sqrt{y}$) for statistical analyses (Zar 1999).

Ocelot home range and habitat use

Home range

Variation in home range size is expected to reflect differences in resource availability between areas (e.g. Sandell 1989; Gompper and Gittleman 1991). Ocelot home range size varies considerably between sites and between sexes across the continent (1.3–90.5 km^2, Table 27.1). The smallest ranges for males and females are in Brazil's Pantanal flood plains (5.4 km^2 and 1.3 km^2, respectively) and the largest in the savannahs of ENP (90.5 km^2 and 75 km^2, respectively) also in Brazil (Crawshaw and Quigley 1989; Rocha 2006; Silveira, unpublished data). Both are open areas, but the former is seasonally flooded and the latter is dry. Although sample sizes for Emas included only one male and one female, both were non-dispersing adults. At all sites, mean home range sizes of adult males (mean ± standard deviation, 22.6 km^2 ± 24.2, range 5.4–90.5 km^2, n = 30 animals from 12 sites) are significantly larger than those of adult females (14.4 km^2 ± 20.8, range 1.3–75 km^2, n = 36 individuals from 12 sites) (t = 3.386, d.f. = 10, P = 0.007) in accordance with Crawshaw (1995).

Relationships between home range, body size, and environmental variables

Ocelot body mass (6.6–18.6 kg, n = 112) differs among regions (F = 5.835, P < 0.001). However, these differences do not correlate with latitude (r = −0.385, P = 0.306, Fig. 27.2), but varied with habitat types grouped as rainforests (mean = 11.1 kg ± 2.2, n = 56), open savannahs/flood plains (mean = 10.0 kg ± 2.4, n = 29), and semi-arid habitats (scrub, chaparral, and dry forest; mean = 8.7 kg ± 1.4, n = 27) (H = 21.72, d.f. = 2, P < 0.001). Pairwise comparisons show that rainforest ocelots are significantly larger than those of savannahs/flood plains (t = −2.085, d. f. = 83, P = 0.040) and scrub/chaparral/dry forest (t = 5.004, d.f. = 81, P < 0.001), but those of semi-arid habitats and savannahs/flood plains do not differ significantly (t = 664.5, P = 0.087). Ocelot body mass did not differ among evergreen forest sites (F = 1.741, P = 0.170). The pattern in ocelots contrasts with that for jaguar, which are larger in flood plains than in rainforests, due to larger size and greater abundance of large prey (>15 kg) in flood plains (Hoogesteijn and Mondolfi 1996). A possible, but untested, explanation for the greater mass of ocelots in rainforest than elsewhere is that larger ocelot prey may be more readily available in rainforests (e.g. Oliveira 1994; Nowell and Jackson 1996; Sunquist and Sunquist 2002). Agoutis (*Dasyprocta* spp.), pacas (*Agouti paca*), armadillos (Dasypodidae), large marsupials, monkeys, and sloths (*Bradypus, Choloepus*), which are important prey for ocelots, generally reach their highest densities in rainforests (see Glanz 1982; Emmons 1984).

When data from Emas are excluded due to their extremely large values, there is an intriguing, but not significant, tendency for average ocelot body mass and mean home range to be positively correlated (r = 0.767, P = 0.075, n = 6, Fig. 27.3). There are no differences in mean ocelot home range size by habitat type across all habitat categories (thorn scrub/dry forests, flood plains, savannahs, and forests) (F = 2.438, P = 0.157), suggesting that variations among areas are related to availability of prey and not to habitat structure *per se*. No correlation exists between mean annual precipitation and mean adult ocelot home range per site (r^2 = 0.079, P = 0.500); either for males (r^2 = 0.128, P = 0.384) or females (r^2 = 0.022, P = 0.729). In fact, in the Texas chaparral, where average annual rainfall is only 640 mm and habitat is considered suboptimal for the species, ocelot home range sizes are only 25% of those in wetter (1500 to 1700 mm/year) and presumably prime rainforest habitats of Belize and southern Brazil (Tewes 1986; Laack 1991; Crawshaw 1995; Dillon 2005). While precipitation in Texas is rather low, rodents and rabbits are abundant (Tewes 1986). These facts suggest, unexpectedly, that abiotic variables have little detectable direct influence on ocelots' home range size and, consequently, their density.

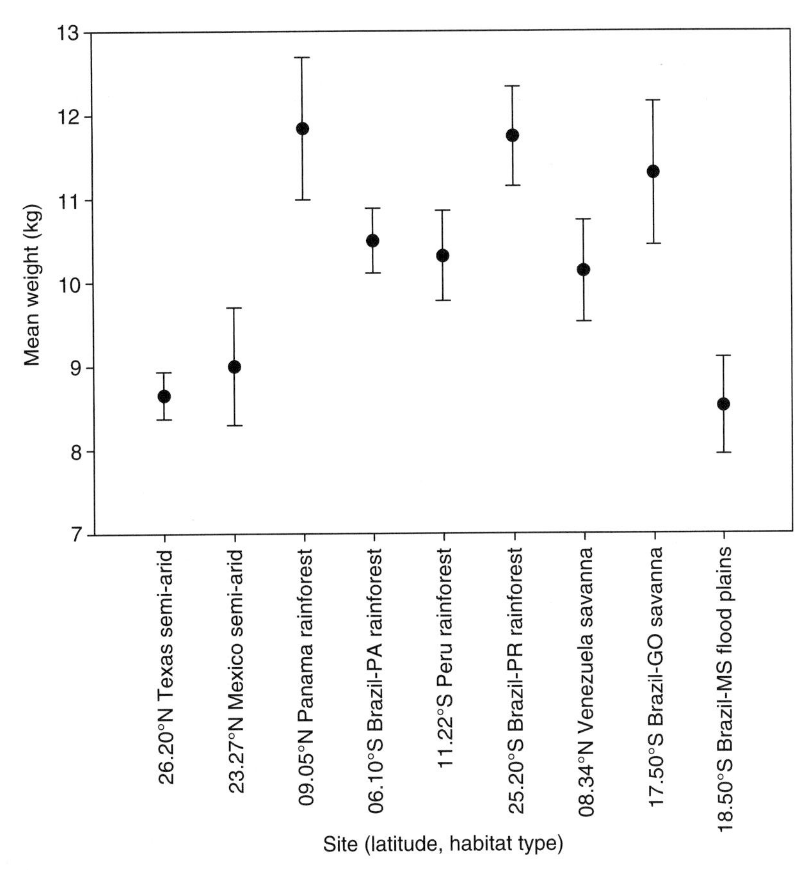

Figure 27.2 Ocelot mass (mean ± standard error) in the Americas, grouped by habitat similarity and latitude. (Data sources: Tewes 1986; Ludlow and Sunquist 1987; Emmons 1988; Laack 1991; Crawshaw 1995; Rocha 2006; Companhia Vale do Rio Doce, unpublished data; Crawshaw and Quigley, unpublished data; Moreno and Kays, unpublished data; Silveira, unpublished data.)

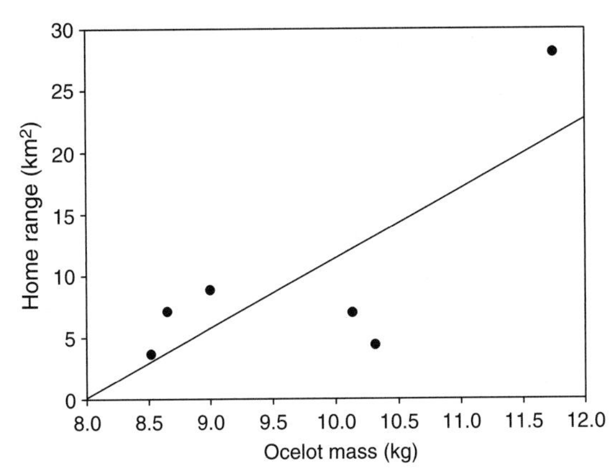

Figure 27.3 Correlation of mean ocelot body size and mean home range in the Americas.

Habitat use

Several studies highlight the importance of dense habitat cover for conservation of ocelots (e.g. Tewes 1986; Ludlow and Sunquist 1987). In South Texas, 51% of ocelot locations were in dense habitat cover, which comprised only 1% of the study area (Harveson *et al.* 2004). In northern Mexico, 97.6% were in dense brush (i.e. dense woody cover) (Caso 1994). The strong selection of ocelots for dense cover in south Texas could also be related to avoidance of the similar-sized and sympatric bobcat (*Lynx rufus*) (Tewes 1986; Fisher 1998). In the open savannahs of ENP, ocelots favoured areas of dense forest and savannah, but they also used pastoral/agricultural land, and avoided grassland savannah and flooded grasslands. Forest cover

Table 27.2 Summary of habitat use versus availability of sympatric felids at ENP (Brazil).

	Home range within each habitat type (usage within home range, %)				
	Grassland savannah	Scrub savannah	Gallery forest	Flooded grassland	Pasture/ agriculture
Availability	42.50	23.22	1.83	2.61	29.83
Ocelot M1	16.67 (0)	41.06 (55.17)	5.42 (6.90)	0.90 (0)	35.95 (37.93)
Ocelot M2	8.86 (2.70)	28.22 (29.73)	4.01 (27.03)	1.32 (5.41)	57.58 (35.14)
Ocelot F1	6.91 (4.58)	37.39 (45.04)	5.65 (4.58)	1.35 (0)	48.70 (45.80)
Jaguarundi F1	22.25 (22.37)	38.38 (42.11)	2.33 (1.32)	0.45 (0)	36.59 (34.21)
Little spotted cat F1	30.80 (29.41)	40.75 (39.71)	1.30 (0)	0.20 (2.94)	29.95 (27.94)
Pampas cat M1	100.00 (100.00)	0	0	0	0
Pampas cat M2	59.70 (36.74)	3.84 (23.53)	0 (0)	0.04 (0)	36.43 (36.74)
Pampas cat M3	54.49 (58.33)	12.58 (8.33)	2.70 (16.67)	0.16 (0)	30.07 (16.67)

comprised only 1.8% of the total area (Table 27.2). Ocelots varied individually in their habitat use in Emas but, on average, scrub/woodland savannah and pasture/agriculture were used in proportion to their availabilities (Ivlev's third order values: 0.09 and −0.08, respectively), grassland savannah and flooded grassland were avoided (−0.6, −0.5, respectively), and forest was positively selected (0.3). In the Pantanal flood plains, too, ocelot home ranges were configured to encompass forests and scrub/woodland savannahs, and under-represented open grassland savannahs (Rocha 2006).

One suggestion is that, due to their frequent association with dense cover, ocelots might occupy a narrower range of microhabitats than expected from their wide geographic distribution (Emmons 1988). However, throughout Brazil, ocelots commonly use heavily disturbed areas in the Amazon, Cerrado, Atlantic Forest, and Pantanal biomes (Oliveira, personal observation; R. Boulhosa, personal communication; E. Payan, personal communication; Silveira, unpublished data). Indeed, they have been found in habitats varying from heavily logged and fragmented, to early and late successional forests, the outskirts of major cities and towns, disturbed scrub/woodland savannah and mosaic of forest (disturbed or not), and savannah with eucalyptus plantations and agricultural areas. In short, despite being associated with dense cover, ocelots show a higher level of habitat plasticity than previously thought.

Comparative ecology of ocelots and other sympatric cats

Comparative home range

The limited data on adult home range sizes of the smaller, sympatric lowland felids also show considerable variation (Table 27.1). Jaguarundi (mean body mass 5 kg) averages 24.2 ± 20.4 SD km^2, margay (3.3 kg) 12.3 ± 7.6 SD km^2, and little spotted cat (2.4 kg) 10.9 ± 9.6 SD km^2. The expected home range sizes based on their body mass and metabolic needs (Lindstedt *et al.* 1986) would be 8.9 km^2, 5.8 km^2, and 4.2 km^2, respectively. For ocelots (11 kg) the average and expected home range sizes are similar, at 20.0 ± 25.2 km^2 and 20.1 km^2, respectively. Thus, the average home range sizes of the smaller lowland felids are 2.5 (2.1–2.7) times larger than what would be expected based on body size. Pairwise comparisons of mean home range values of male and female of the smaller species show that male ranges are significantly larger than those of females ($W = -36.00$, $T+ = 0.00$, $T- = -36.00$, $P = 0.008$), as was noted for ocelots.

There is considerable inter- and intraspecific variation in home range size among the sympatric felids, resulting in extensive overlap in size and no significant differences among mean adult home ranges ($F = 0.749$, $P = 0.529$, Table 27.1).

Home range size in solitary male carnivores is, on average, 2.5 times that of females, and this value is also significantly larger than expected based on energy

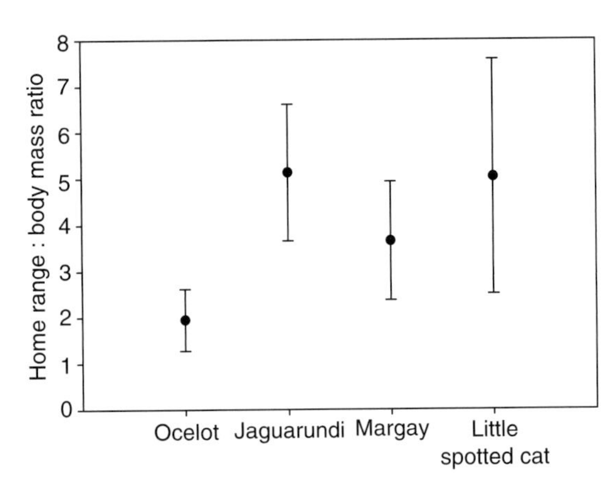

Figure 27.4 Home range size:body mass ratio (mean ± standard error) of sympatric lowland felids. (Data sources: Table 27.1.)

requirements, suggesting that access to several females plays an important role in determining male territory size (Sandell 1989). Thus, the larger home ranges of male lowland Neotropical felids should be linked to the distribution of females, rather than to the availability of food resources (Ludlow and Sunquist 1987; Manfredi *et al.* 2006), a pattern typical in the Felidae.

Average home range:felid body mass ratio among the sympatric assemblage is smallest for ocelots (1.9 km^2 per kg of body mass) and larger for jaguarundis, little spotted cats, and margays (5.1, 5.0, 3.7 km^2 per kg of body mass, respectively; Fig. 27.4). Hypotheses to explain the larger than expected home range sizes of the smaller species include avoidance of larger guild-members through the use of areas of lower prey density, living along the periphery of the dominant species range, or use of different habitats where they live in sympatry (Creel and Creel 1996; Durant 1998; Palomares *et al.* 1998). Through such mechanisms, larger, dominant species may attain higher densities and have smaller home range sizes than do some smaller, subordinate species due to avoidance of the former by the latter (Creel and Creel 1996; Woodroffe and Ginsberg 2005).

Comparative habitat use

Although most felids use open habitats, all lowland species, except pampas cats, are closely associated with forest cover. At localities with a mosaic of open and closed environments, including agricultural areas, the use of closed habitats by ocelots, jaguarundis, margays, and little spotted cats was similar, and on average 74% (40–98%) of the location records, despite the fact that dense cover sometimes comprised only 11% of the total vegetation (Ludlow and Sunquist 1987; Caso 1994; Harveson *et al.* 2004; Trovati 2004; Michalski *et al.* 2006b; Oliveira *et al.* 2008; Silveira, unpublished data). Margays seem even more strongly associated with dense cover than are ocelots (Oliveira 1998b, personal observation). Even jaguarundis, which are typically associated with open habitats (Oliveira 1998a), use forested habitats part of the time. It seems that only pampas cats are a predominantly open-habitat specialist, but they too can be found in forests (Oliveira 1994; Sunquist and Sunquist 2002).

In the savannahs of Emas, pampas cats prefer open grassland savannahs, and also use pastoral/agricultural lands, but avoid forest and scrub/woodland savannah. One female jaguarundi favoured scrub savannah and used forest and pastoral/agricultural lands slightly more than its availability, whereas one female little spotted cat used forest and pastoral/agricultural lands less than its availability; both species avoided the open grasslands. They used scrub/woodland savannah, open grassland, and pastoral/agricultural lands in proportion to availability. Unexpectedly, jaguarundis and little spotted cats both avoided forest (Ivlev's = −0.3 and −1, respectively), and differed only in that flooded grassland was strongly avoided by the former and selected by the latter. Otherwise, all species strongly avoided flooded grasslands (Table 27.2). The spatial separation among small felids in Emas could be due to avoidance of ocelots. In contrast to Emas, 92.3% and 70% of locations of two male little spotted cats in the savannah biome of Tocantins State (central Brazil) were in forests, and only 7.7% and 30% in savannah. In contrast, jaguarundis in Tocantins were located 17% of the time in forests and 83% in savannahs (Trovati 2004). Additionally, in an agriculture-forest patch mosaic, both species were strongly associated with forest cover (Oliveira *et al.* 2008).

There appear to be parallels within the Neotropical small felid guild to the differential habitat use by jackal species in areas of sympatry and allopatry. In that case, the dominant species narrows the number of habitats used by the subordinate species, which avoids the more aggressive competitor (Loveridge and Macdonald 2002; Macdonald *et al.* 2004c).

In agricultural landscapes in Brazil, radio-tracked ocelots, jaguarundis, margays, and little spotted cats only rarely used open fields, but preferred field borders near forest cover (Oliveira *et al.* 2008; Silveira, unpublished data). These edge habitats may offer both high rodent abundance and cover. While Geoffroy's cats tended to spend more time in densely vegetated habitat (Manfredi *et al.* 2006; Lucherini *et al.*, unpublished data) they may also use open agricultural fields (Oliveira *et al.* 2008). Occasional use of open areas is not restricted to jaguarundis, pampas cats, and Geoffroy's cats. Faecal analysis of little spotted cat diet in secondary Atlantic rainforest revealed prey from both forest and field/forest edges, but not from agricultural fields (Facure-Giaretta 2002). In summary, ocelots, margays, little spotted cats, and Geoffroy's cats favour dense cover, jaguarundis use both open and closed habitats, whereas pampas cats use predominantly open habitats.

Ocelot foraging ecology, niche release, and the sympatry issue

Ocelot feeding ecology

Ocelot diet varies throughout its range (Fig. 27.5). Small mammals (rodents and marsupials <600 g) are generally the numerically pre-eminent prey (61.3% ± 28), but their biomass contribution is less (10.7% ± 11.1). Conversely, large rodents (>1 kg), armadillos, ungulates, and possibly sloths and monkeys, contribute much more in terms of biomass (Table 27.3). Large rodents, especially paca and agouti, are important in terms both of biomass (27.1% ± 15.3), and the frequency with which they are eaten (11.4% ± 8.1) (e.g. Bianchi 2001; Aliaga-Rossel *et al.* 2006; Moreno *et al.* 2006), and occurred in ocelot diets at 11 of 13 study sites. This revises the earlier perception that small rodents are the major prey of ocelots (reviewed by Oliveira 1994; Nowell and Jackson 1996; Sunquist and Sunquist 2002). At some sites (e.g. Panama and Caratinga-Brazil) sloths and monkeys also make a considerable contribution (18.8–26.9%) (Moreno *et al.* 2006; Bianchi and Mendes 2007). Ungulates, especially deer (*Mazama* spp.), are infrequently taken (3.7% ± 3.4), but when they are, can contribute substantially to biomass intake (29% ± 26.7).

The MWMP taken by ocelots, while highly variable, averages 1.5 ± 1.1 kg, and is highest in Atlantic rainforest (Linhares, Espirito Santo State, in Southeast Brazil) at 3.3 kg, and lowest in the Llanos flood plains of Venezuela, at 0.3 kg (Ludlow and Sunquist 1987; Bianchi 2001) (Table 27.4). In Linhares this reflects the prevalence of armadillos and

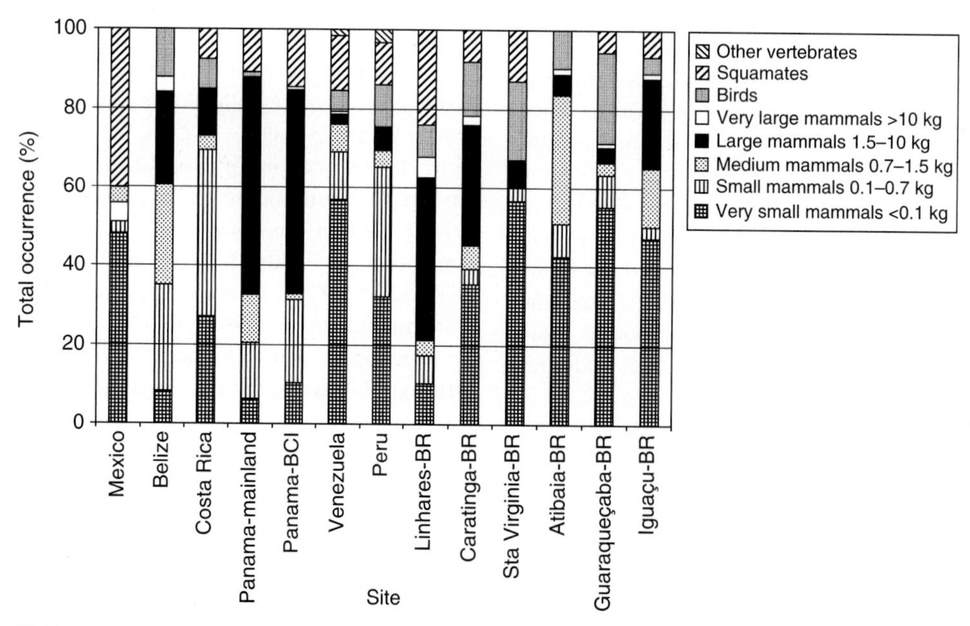

Figure 27.5 Ocelot diet (total occurrence, %) throughout the Americas. (Data sources: Table 27.3.)

Table 27.3 Contribution of main prey items by percentage of occurrence (#) and biomass to the diet of ocelots throughout the Americas (values in percentage). Percentage of occurrence is derived from the number of occurrences of each diet item as a percentage of all occurrences of all diet items.

Site	Large rodents		Armadillos		Sloths		Monkeys		Ungulates		Small mammals	
	#	Biomass	#	Biomass	#	Biomass	#	Biomass	#	Biomass	#	Biomass
Mexico[1]									9.43	90.86	90.57	9.14
Belize[2]	9.23	33.00	15.38	24.06					4.62	21.45	40.00	5.00
Costa Rica[3]	9.09	47.56									81.82	25.68
Panama-mainland[4]	17.31	18.15	11.54	14.52	26.92	41.40					15.38	1.52
Panama-BCI[4]	25.00	26.96	2.71	3.43	23.37	40.83	7.07	6.56	1.09	3.19	35.88	3.59
Venezuela[5]	1.88	12.34	0.94	8.59					0.94	33.81	87.32	23.25
Peru[6]	8.33	54.25					1.67	2.37			85.83	34.9
Linhares-BR[7]	23.81	38.53	27.38	26.59			4.76	2.78	8.33	20.74	19.05	1.11
Caratinga-BR[7]	15.29	20.68	8.24	12.52			18.82	61.94			49.41	1.09
Sta Virginia-BR[8]			5.26	43.78	5.26	46.78					84.21	9.43
Atibaia-BR[9]	1.82	9.34	3.64	13.63					1.82	22.04	56.36	3.96
Guaraqueçaba-BR[10]	3.23	8.30					3.23	26.02	1.61	28.74	95.16	17.15
Iguaçu-BR[11]	10.94	28.68	10.94	29.18					1.56	11.37	56.25	2.74
Mean	11.45	27.07 ±	9.56 ±	19.59 ±	18.52 ±	43.00 ±	7.11 ±	19.93 ±	3.68 ±	29.03 ±	61.33 ±	10.66 ±
Standard deviation		15.27 8.11	8.17	12.45	11.62	3.28	6.84	25.41	3.42	26.73	28.01	11.08
Paired t-test	−4.114		−2.927						−3.901		10.834	
d.f.	10		8						7		12	
P	0.002		0.019						0.006		<0.001	

Notes: [1] Villa Meza *et al.* 2002; [2] Konecny 1989; [3] Chinchilla 1997; [4] Moreno *et al.* 2006; [5] Ludlow and Sunquist 1987; [6] Emmons 1987; [7] Bianchi 2001; [8] Wang 2002; [9] Facure-Giaretta 2002; [10] Vidolin 2004; [11] Crawshaw 1995.

BR, Brazil.

Table 27.4 Primary ocelot prey by percentage of occurrence (vertebrates) and biomass contribution (mammals), and mean weight of mammalian prey (MWMP) throughout the Americas, all weights in grams (g).

Habitat/location	Main vertebrate prey (occurrence, %)			Main mammalian prey (biomass)			MWMP	n
	Species	Prey mass	All items (%)	Species	Prey mass	Mammals (%)	(g)	
Tropical dry forest/ Mexico[1]	1. *Ctenosaura pectinata*	800	36.4	1. *Odocoileus virginianus*[12]	15,000	90.9	1465.74	51
	2. *Liomys pictus*	50	32.3	2. *Liomys pictus*	50	2.1		
	3. *Marmosa canescens*	30	9.1	3. *Sigmodon mascotensis*	180	0.7		
Tropical rainforest/ Belize[2]	1. *Didelphis marsupialis*	1000	25.7	1. *Agouti paca*	8000	33.0	2238.00	49
	2. *Philander opossum*	400	20.3	2. *Dasypus novemcinctus*	3500	24.1		
	3. *Dasypus novemcinctus*	3500	13.5	2. *Mazama americana*[12]	10,400	21.4		
Tropical rainforest/Costa Rica[3]	1. *Proechimys semispinosus*	375	34.6	1. *Dasyprocta punctata*	4000	47.6	764,55	23
	2. *Heteromys desmarestianus*	75	15.4	2. *Proechimys semispinosus*	375	20.1		
	3. *Tylomys watsoni*	240	7.7	3. *Potos flavus*	2500	14.9		
	3. *Dasyprocta punctata*	4000	7.7					
	3. *Penelope purpurescens*	1700	7.7					
	3. *Iguana iguana*	3000	7.7					
Tropical rainforest/ mainland-Panama[4]	1. *Bradypus variegatus*	4300	18.8	1. *Bradypus variegatus*	4300	37.6	2861.15	49
	2. *Dasyprocta punctata*	3000	13.0	2. *Dasyprocta punctata*	3000	18.1		
	3. *Dasypus novemcinctus*	3600	8.7	3. *Nasua nasua*	4500	15.1		
	3. *Iguana iguana*	3000	8.7					
Tropical rainforest/BCI-Panama[4]	1. *Dasyprocta punctata*	3000	16.2	1. *Dasyprocta punctata*	3000	24.0	2852.38	190
	2. *Proechimys semispinosus*	375	15.4	2. *Choloepus hoffmanni*	5700	22.8		
	3. *Bradypus variegatus*	4300	8.5	3. *Bradypus variegatus*	4300	18.0		
Llanos gallery forest/ floodplains/ Venezuela[5]	1. *Zygodontomys brevicauda*	50	37.6	1. *Odocoileus virginianus*[12]	16000	21.5	349.91	160
	2. *Sigmomys alstoni*	40	18.8	2. *Sylvilagus floridanus*	800	14.0		
	3. *Holochilus brasiliensis*	300	14.6	3. *Holochilus brasiliensis*	300	12.5		

Amazon rainforest/Peru[6]	1. *Proechimys* spp.	280	31.6	1. *Proechimys* spp.	280	26.6	491.58	62
	2. *Oryzomys* spp.	70	21.5	2. *Dasyprocta variegata*	4000	20.3		
	3. *Birds*	—	10.7	3. *Agouti paca*	8000	13.6		
Tropical rainforest/ Linhares—Brazil[7]	1. *Dasypus* spp.	3200	15.5	1. *Agouti paca*	8000	31.8	3294.94	77
	2. *Tupinambis merianae*	1200	8.8	2. *Dasypus* spp.	3200	26.6		
	3. *Agouti paca*	8000	7.4	3. *Tayassu tajacu*[12]	7600	11.0		
Tropical rainforest/ Caratinga—Brazil[7]	1. *Calomys* sp.	25.5	12.9	1. *Alouatta guariba*	5650	37.9	2104.64	60
	2. Birds	—	12.1	2. *Brachyteles hypoxanthus*	13,500	22.6		
	3. *Alouatta guariba*	5650	9.7	3. *Agouti paca*	8000	13.4		
Tropical rainforest/Sta. Virgínia—Brazil[8]	1. *Akodon* sp.	28.7	23.3	1. *Bradypus variegatus*	3900	46.8	438,76	17
	2. *Monodelphis* sp.	48	20.0	2. *Dasypus novemcinctus*	3650	43.8		
	3. Colubrid snakes	—	13.3	3. *Akodon* sp.	28.7	2.4		
Tropical rainforest/ Atibaia—Brazil[9]	1. *Sphiggurus villosus*	1325	19.7	1. *Sphiggurus villosus*	1325	30.3	953.74	34
	2. *Oligoryzomys nigripes*	16.7	14.8	2. *Mazama americana*[12]	11,560	22.0		
	3. *Didelphis* spp.	1500	9.8	3. *Didelphis* spp.	1500	17.2		
	3. *Akodon* sp.	28.7	9.8					
	3. Birds	—	9.8					
Tropical rainforest/ Guaraqueçaba—Brazil[10]	1. Cricetidae	50	49.5	1. *Mazama* spp.	9000	28.7	505.16	60
	2. Birds	—	21.5	2. *Cebus/Alouatta*	4075	26.0		
	3. *Tupinambis merianae*	1200	5.4	3. *Tamandua tetradactyla*	5200	16.6		
Subtropical rainforest/ Iguaçu—Brazil[11]	1. Small rodents	50	37.8	1. *Dasypus novemcinctus*	3300	29.2	1236.86	56
	2. *Didelphis aurita*	1500	8.5	2. *Dasyprocta azarae*	3200	24.3		
	2. *Dasypus novemcinctus*	3300	8.5	3. *Didelphis aurita*	1500	13.3		
	4. *Dasyprocta azarae*	3200	7.3					

[1] Villa Meza *et al*. 2002; [2] Konecny 1989; [3] Chinchilla 1997; [4] Moreno *et al*. 2006; [5] Ludlow and Sunquist 1987; [6] Emmons 1987; [7] Bianchi 2001; [8] Wang 2002; [9] Facure-Giaretta 2002; [10] Vidolin 2004; [11] Crawshaw 1995; [12] 40% of body mass.

n, sample size (number of scats).

pacas (average body mass of 3.2 kg and 8 kg, respectively), which comprise 40.5% of mammalian items and 23% of total vertebrates taken (Bianchi 2001). Conversely, the mean body mass of all vertebrate prey taken in the Llanos increases to 0.6 kg when iguanas (*Iguana iguana, c.* 2.7–3.0 kg) are included (Ludlow and Sunquist 1987). The figures presented here double the previously reported (Oliveira 1994) mean size of ocelot mammalian prey, due to including a larger number of studies (e.g. Crawshaw 1995; Bianchi 2001; Moreno *et al.* 2006).

Emmons (1987) predicted that the average adult ocelot would eat 0.56–0.84 kg of meat per day, and the figures reported in this review either fall within this range or above it. Given that the average carnivore biomass is typically 1–3% of the biomass of their prey (Vézina 1985), the prey biomass predicted to sustain one ocelot for a year would range from 366 to 1100 kg, or 33–100 kg of prey per kilogram of ocelot, which is below the ratio reported for the larger jaguars and pumas (Emmons 1987). At all sites where MWMP was >1 kg, prey mass of at least one of the main vertebrate species was ≥0.8 kg. At 10 of 13 sites (76.9%), at least one main prey of ocelots weighed ≥0.8 kg (Table 27.4). However, there was no statistically significant relationship between MWMP with the body mass of the three main (in terms of biomass consumed) mammalian prey ($r^2 = 0.344$, $P = 0.262$), but a significant positive correlation existed between MWMP and the body mass of the three most numerically prominent vertebrate prey ($r^2 = 0.904$, $P = 0.006$).

Carnivore density is known to be constrained by metabolic needs and prey abundance (Carbone and Gittleman 2002). Sunquist (1992) argued that it would be virtually impossible for a lactating female to raise cubs on a diet of small mice (*c.* 50 g, e.g. *Oryzomys, Zygodontomys* spp.). Emmons (1988) reported that a lactating female ocelot increased her activity from 12–14 h/day to almost 23 h/day, and still was unsuccessful in rearing her kittens. This prompts the hypothesis that the limiting factor for ocelot persistence may be the availability of larger sized prey (≥0.8 kg), such as agouti and armadillos, to complement the intake of small mammals.

Dietary comparisons within the Neotropical felid assemblage

The ocelot's fundamental niche is much broader than those of the smaller species, with prey ranging from small mice to collared-peccary (*Tayassu tajacu*), brocket (*Mazama* spp.), and white-tailed deer (*Odocoileus virginianus*) (Konecny 1989; Bianchi 2001; Villa Meza *et al.* 2002). Jaguarundis, margays, and little spotted cats prey predominantly on small mammals (<1 kg) (Fig. 27.6), but, like ocelots, they also prey on birds, squamates, and occasionally larger mammals. This raises the possibility that exploitative competition occurs between ocelots and smaller cats as well as among the smaller felids, as has been noted for larger carnivores (Woodroffe and Ginsberg 2005). However, among the array of prey, predators tend to specialize on those that they can capture most easily, effectively, and profitably (Sinclair *et al.* 2003).

The ocelot is the only felid within its guild that consistently takes prey weighing ≥1 kg. The smallest prey size for all sympatric felid species is in the range of 10–20 g (bats, small mice, and marsupials). However, maximum prey masses diverge considerably, but are positively correlated with mean felid mass ($r = 0.884$, $P = 0.0001$, see also Oliveira 1994, 2002a; Oliveira and Paula, unpublished data). Competition within carnivore guilds is probably asymmetrical. Larger species, whilst supported by larger prey not available to their smaller competitors, may deplete the latter's food supply by opportunistic predation on small prey (Woodroffe and Ginsberg 2005). The diets of sympatric felids (Fig. 27.5 and Fig. 27.6) suggest competitive pressure from the larger ocelots upon the smaller jaguarundis, margays, and little spotted cats.

Canine diameter, a useful measure of ecological segregation among carnivores (Dayan *et al.* 1990), differs significantly between sympatric Neotropical small felids, excepting margays and jaguarundis (Oliveira, unpublished data). Mean upper canine diameters correlate with felid mean body mass ($r = 0.980$, $P = 0.0199$, Fig. 27.7) and average MWMP ($r = 0.987$, $P = 0.0125$). Conversely, there is no correlation between canine diameter and maximum prey mass. Thus, canine diameter seems to reflect the modal, rather than the maximal, size of prey.

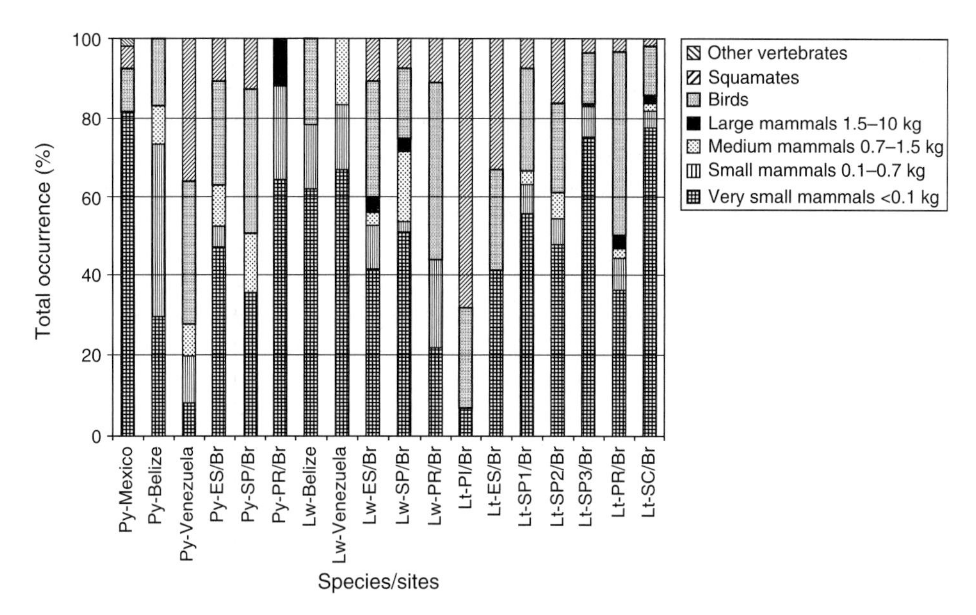

Figure 27.6 Jaguarundi (Py), margay (Lw), and little spotted cat (Lt) main prey items (total occurrence, %) in the Americas. (Data sources: Py: Bisbal 1986; Mondolfi 1986; Konecny 1989; Facure and Giaretta 1996; Felipe 2003; Guerrero *et al.* 2002; Oliveira 2002a; Rocha-Mendes 2005; Lw: Mondolfi 1986; Konecny 1989; Wang 2002; Felipe 2003; Rocha-Mendes 2005; Lt: Olmos 1993; Facure and Giaretta 1996; Facure-Giaretta 2002; Nakano-Oliveira 2002; Wang 2002; Rocha-Mendes 2005; M. Tortato, unpublished data.)

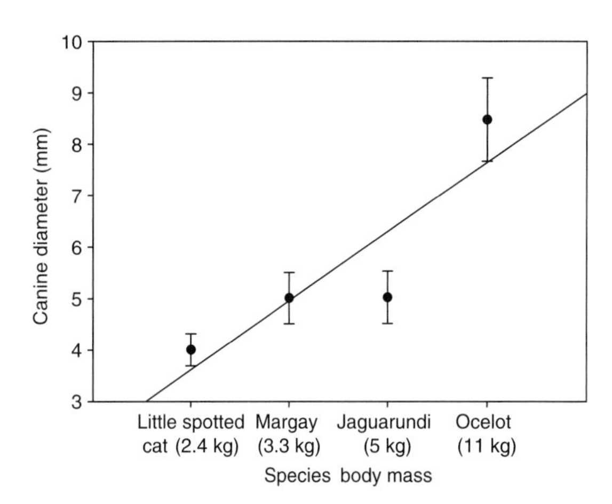

Figure 27.7 Correlation of mean upper canine diameter (±SD) and mean body mass for little spotted cats (n = 28), margays (n = 34), jaguarundis (n = 24), and ocelots (n = 47) in Brazil (r = 0.980, P = 0.0199). (Data source: Oliveira, unpublished data.)

Food niche overlap values between ocelots and margays average 0.41 (0.68–0.19, $n = 3$), between ocelots and little spotted cats it is 0.66 (0.81–0.45, $n = 3$), and between ocelots and jaguarundis it is 0.49. Given the differences in body size and canine dimensions, these values are high. The highest overlap for ocelots is with the smallest species, the little spotted cat, but small sample sizes (and scarcity of studies of sympatric felids) may distort these comparisons.

A pairwise comparison of the food niches of the smaller species shows an average overlap of 0.62 (±0.253, range 0.18–0.91) (Konecny 1989; Olmos 1993; Facure and Giaretta 1996; Facure-Giaretta 2002; Nakano-Oliveira 2002; Wang 2002; Felipe 2003; Rocha-Mendes 2005). Therefore, the available datasets provide no evidence of any significant difference in overlap in food niche between ocelots and other sympatric felids ($t = -0.757$, d.f. $= 14$, $P = 0.461$) or among the smaller species ($F = 0.841$, $P = 0.484$).

The obvious potential for competition amongst these felids may be offset by the considerable difference in MWMP between ocelots and the smaller species, and despite their overlap on more abundant prey, each may be selecting specific prey (Pimm 1991). Dietary overlaps varying from very high to very low have been reported for Neotropical felids (Oliveira 2002a; Moreno *et al.* 2006). In the high

Andes dietary overlap of pampas cats and Andean cats (*Leopardus jacobita*) was very high (82%), but they nonetheless emphasized different principal prey (Walker *et al.* 2007; see Marino *et al.*, chapter 28, this volume). Furthermore, high overlap in one niche dimension might be compensated by low overlap in other axes, such as space use and time of activity, or even seasonal temporal variation in hunting of prey (Fig. 27.8), thus facilitating the sympatry (Konecny 1989, Oliveira *et al.* 2008; Silveira, unpublished data).

MWMP differs significantly between sympatric Neotropical felids ($H = 20.362$, d.f. $= 3$, $P < 0.001$; Table 27.4, Fig. 27.9). Pairwise comparisons are also significantly different ($P < 0.05$) for every combination except for margay–little spotted cat and margay–jaguarundi. Margays are intermediate between the other two species, in so far as they are closer in size and appearance to little spotted cats, but closer to jaguarundis in feeding morphology (e.g. canine diameter, jaw length; Kiltie 1984; Oliveira and Cassaro 2005; Oliveira, unpublished data). Not surprisingly, the largest difference between the smallest and largest MWMP is for the ocelot (9.4-fold), whereas the range of prey sizes taken by jaguarundi, margay, and little spotted cats is 2.3-, 3.9-, 4.9-fold, respectively. Mean MWMP represents 13.7% of ocelot mass (3.8–30%), 6.2% of jaguarundi mass (4–9.1%), 7.3% of margay mass (3.1–12.2%), and 6.3% of little spotted cat mass (2.4–11.9%).

Assessing the potential for competitive release

While ungulates, armadillos, and large rodents are taken by pumas and jaguars (see Oliveira 2002a), and are favoured by humans as well (Jorgenson and Redford 1993; Oliveira 1994), these prey are also taken by ocelots (Table 27.3). Thus, there is potential for niche overlap between ocelots and big cats. Alternatively, ocelots could be taking larger prey only in areas where jaguars are absent (Moreno *et al.* 2006). To test for a jaguar dietary release on mesopredators, as suggested by Moreno *et al.* (2006), we compared ocelot MWMP from areas with robust jaguar populations ($n = 8$) to those areas with marginal occurrence or absence ($n = 5$) of jaguars, and found no dietary differences ($t = -0.951$, d.f. $= 11$, $P = 0.362$). In fact, the highest MWMP was from an area with a substantial jaguar population (Bianchi 2001; Garla *et al.* 2001). MWMP for ocelots was $11.8 \pm 11.0\%$ of the jaguar's and $19.6 \pm 13.4\%$ of the puma's mean prey body mass where all three were sympatric (Fig. 27.10; Emmons 1987; Crawshaw 1995; Chinchilla 1997; Brito 2000; Bianchi 2001; Garla *et al.* 2001). Therefore, it seems unlikely, on the basis of their MWMP, that ocelot compete substantially with the big cats.

To test for dietary release in the absence of ocelots, we compared the MWMP of little spotted cats from areas with moderate to robust ocelot populations ($n = 2$) to those where ocelots were rare or absent ($n = 3$), and found no dietary differences ($t = -1.627$, d.f. $= 3$, $P = 0.202$). Although sample sizes were small, this suggests that, just as jaguars do not constrain ocelot diets, ocelots do not appear to be constraining the diets of little spotted cats. Furthermore, the average MWMP of little spotted cats is 9.7% of the ocelot's (references in Table 27.3 and Fig. 27.6), suggesting considerable scope for resource partitioning between them.

Factors affecting ocelot density and dynamics

A species' abundance is influenced by proximate factors such as prey density, habitat, presence of potential competitors and predators, and ultimately by environmental variables, especially those that influence prey productivity.

Abundance rank

The larger ocelot seems to be the most versatile felid within the lowland, small-medium Neotropical assemblage. In Brazil, regardless of habitat type ($n = 24$), ocelots were the most abundant cat in 84.2% of all areas examined (Fig. 27.11). Of the smaller sympatric species, margays usually ranked second in abundance in rainforests and last in savannahs, where jaguarundis were most abundant. Little spotted cats usually ranked second or first only in areas where ocelots were absent or rare (Oliveira *et al.*

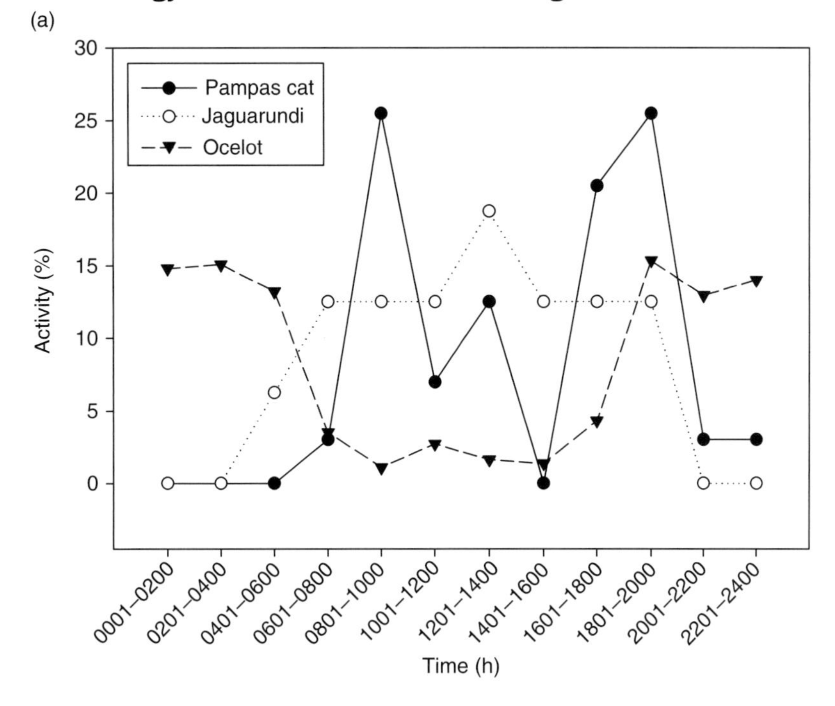

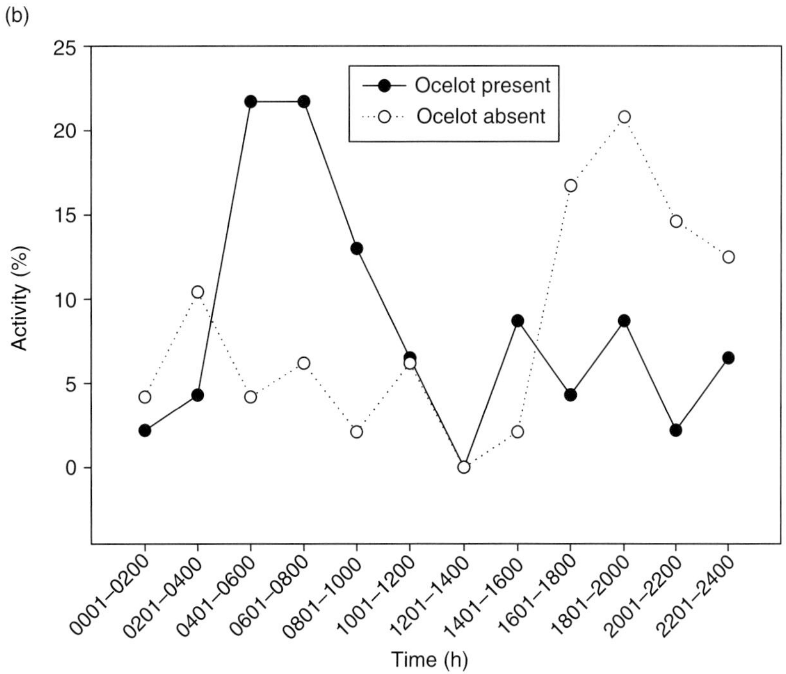

Figure 27.8 Activity patterns of (a) ocelots, jaguarundis, and pampas cats in the savannahs of ENP, central Brazil; and (b) little spotted cats in mixed broadleaf–pine forests in the presence (São Francisco de Paula National Forest) and absence (Serra do Tabuleiro State Park) of ocelots in southern Brazil, determined by photographic records. In Emas ocelots are nocturnal, whereas jaguarundis and pampas cats are predominantly diurnal. Differences in little spotted cat activity between areas in Fig. 27.8b could be related to avoidance of predominantly nocturnal ocelots (Data sources: Oliveira *et al.* 2008; Silveira, unpublished data.)

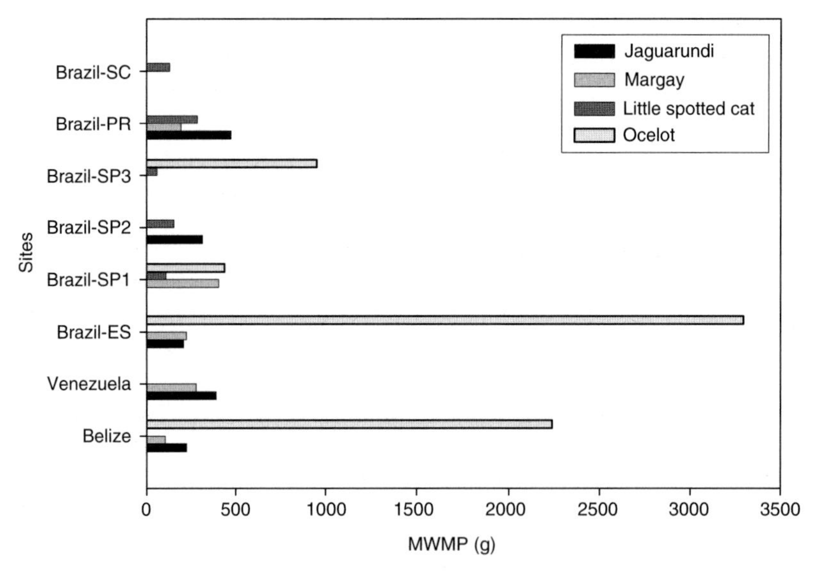

Figure 27.9 Mean weight of mammalian prey (MWMP, g) of jaguarundis, margays, little spotted cats, and of ocelots in areas of sympatric occurrence with the smaller species. (Data sources: same as Figs. 27.5 and 27.6.)

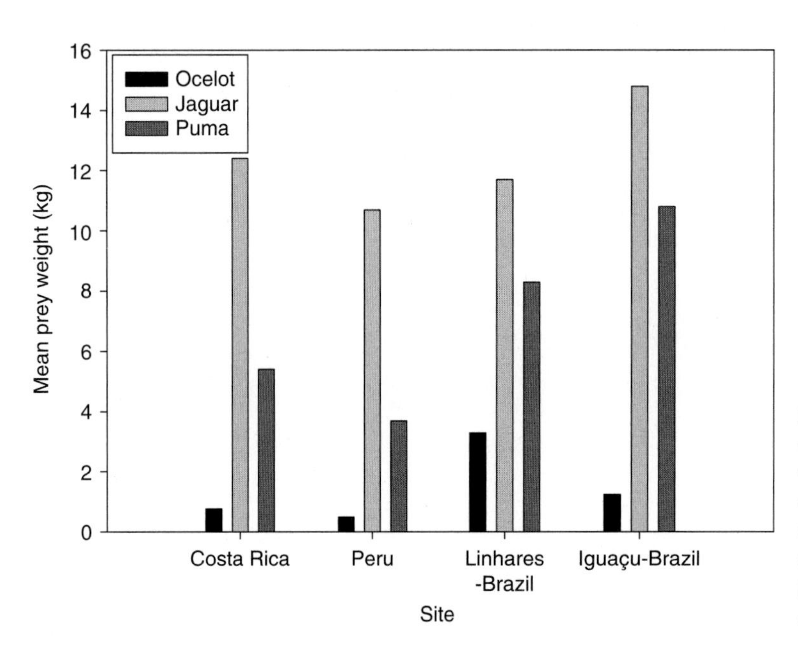

Figure 27.10 Mean weight of ocelot mammalian prey, and mean weight of jaguar and puma vertebrate prey in areas of sympatry. (Data source: Emmons 1987; Crawshaw 1995; Chinchilla 1987; Brito 2000; Bianchi 2001; Garla *et al.* 2001.)

2008, submitted). In the temperate Southern Cone, where the ocelot is absent, Geoffroy's cats might take its ecological role. Geoffroy's cat abundance-ranked first in four locations where this species was sympatric with other small cats and ocelots were absent (Lucherini and Luengos Vidal 2003; Cuellar *et al.* 2006; Lucherini *et al.*, unpublished data; Pereira *et al.*, unpublished data).

The effect of environmental variables

Maffei *et al.* (2005) reported a positive but insignificant correlation ($r^2 = 0.332$) between ocelot density and precipitation. Expanding their dataset and grouping estimates from the same vegetation type of the same region led to a significant and stronger relationship between these two variables ($r^2 = 0.415$, $P = 0.032$, $n = 11$). However, if the data were not grouped in this way, the correlation disappeared ($r^2 = 0.114$, $P = 0.145$, $n = 20$), as did the correlation between precipitation and home range size. Ocelot density and home range size did not correlate with habitat types (grouped in forest, transitional, and open formations) ($F = 0.309$, $P = 0.738$).

The influence of prey

Prey biomass is known to influence carnivore space use and density (Litvaitis *et al.* 1986). Carbone and Gittleman (2002) highlighted the remarkable consistency in average population density of carnivores in general (despite their ecological variations) to that of prey biomass.

Analysis of food habits clearly indicates that larger prey ($\geq$0.8 kg) is especially important for ocelots. The only study on predator–prey relations for a small Neotropical felid that combines data on predator density and prey density or biomass is of Geoffroy's cats in Argentina. The cat's density declined from 0.29 km^{-2} to 0.03 km^{-2} following the numerical decline of its main prey from 0.56 km^{-2} to 0.06 km^{-2} (Pereira *et al.* 2006).

A test of MWMP against ocelot density at five sites (using Konecny [1989] MWMP for Dillon [2005] density estimation; Emmons 1987, 1988; Ludlow and Sunquist 1987; Crawshaw 1995; Moreno *et al.* 2006), revealed no correlation ($r = 0.247$, $P = 0.688$). Although the dataset is very limited, it does suggest that mean prey mass is not a predictor of ocelot density, although the density of main prey by percentage of occurrence might be.

Ocelot density and comparisons with the smaller species

Ocelot density varies considerably between areas, from 0.08 to 1.0 individuals per km^2, and averaged 0.31 $\pm$ 0.22 ($n = 22$) (Ludlow and Sunquist 1987; Emmons 1988; Jacob 2002; Trolle and Kéry 2003, 2005; Dillon 2005; Maffei *et al.* 2005; Cuellar *et al.* 2006; Di Bitetti *et al.* 2006; Rocha 2006; Oliveira *et al.* 2008, submitted; Moreno and Kays, unpublished data). Ocelot densities were higher than those of the smaller jaguarundi, Geoffroy's cat, margay, and

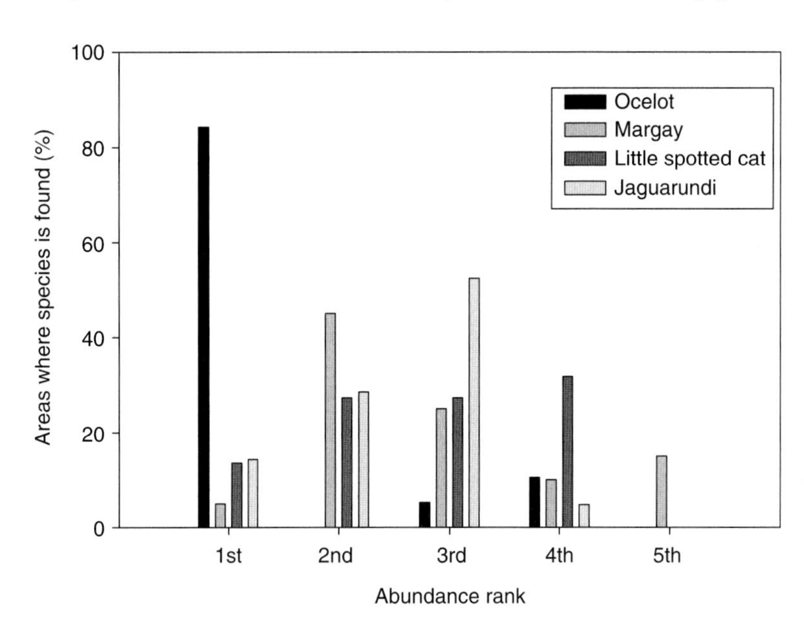

Figure 27.11 Lowland sympatric felid species ranked according to their order of abundance (1, most abundant; 5, least abundant) at each site where they occur in forests (n = 12 areas), savannahs (n = 5), transitional areas/mosaic (n = 5), semi-arid scrub (n = 1), and Pantanal flood plains (n = 1) in Brazil. (Adapted from Oliveira *et al.* 2008.)

little spotted cat combined (mean, 0.187 km^{-2} ± 0.159 km^{-2}, n = 22; Cuellar *et al.* 2006; Oliveira *et al.* 2008, submitted; Caso, unpublished data). Thus, density estimates of ocelots and those of the smaller species are significantly different (t = 2.161, d.f. = 42, P = 0.036). Density estimates of the smaller Neotropical felids (Oliveira *et al.* 2008, submitted) would suggest that they should, according to Yu and Dobson (2000), be categorized as 'widespread, but everywhere small population size'.

Carbone and Gittleman (2002) report strong negative correlation in average population density of carnivores in relation to carnivore mass and strong positive correlation with prey biomass, despite variation in species' ecology. However, in Brazil, felid density estimates are not correlated with body mass (Oliveira *et al.*, submitted). Based on their body size, the small cats should have smaller home ranges and higher population densities than ocelots. In fact, the smaller felids have larger than expected home ranges and lower than expected population densities. Their expected population densities, based on Carbone and Gittleman (2002), would be 0.91 km^{-2} for little spotted cats, 0.69 km^{-2} for margays, 0.48 km^{-2} for jaguarundis, and 0.54 km^{-2} for Geoffroy's cats, which are, on average, 3.5 times larger than their observed mean density. Conversely, the expected density of ocelots (0.24 km^{-2}) is 1.3 times less than its observed mean. This suggests that the smaller sympatric felids may deviate from expected densities due to the effect of the larger ocelots.

Interestingly, the smaller felids reach higher densities only in areas where ocelots are either absent or in low numbers. Ocelot densities, on the other hand, are all from areas where both larger and smaller felid species are found. Additionally, even the highest densities of jaguarundis, margays, little spotted cats, and Geoffroy's cats (0.2–0.4 individuals per km^2) are lower than that of the ocelot (0.5–1.0 individuals per km^2) (Emmons 1988; Cuellar *et al.* 2006; Oliveira *et al.* 2008, submitted; Caso, unpublished data; Moreno and Kays, unpublished data). The density of smaller cats in areas where ocelot density was <0.1 km^{-2} or absent was significantly higher than where ocelot density was >0.1 km^{-2} (t = −5.564, d.f. = 19, P < 0.001). Interspecific predation and competition, major factors influencing carnivore population density (Fuller and Sievert 2001; Donadio and Buskirk 2006), seem to strongly influence the population dynamics of lowland Neotropical felids. The larger and dominant species, contrary to expectation, attain higher densities than smaller subordinate species, which range far more widely than the larger guild members, presumably to avoid intraguild predation (Palomares and Caro 1999; Woodroffe and Ginsberg 2005). Dominance rank is highly correlated with body mass and has been detected in other assemblages (Brown and Maurer 1986; Buskirk 1999; French and Smith 2005). Body-size differences associated with interference competition normally vary by a factor of 1.5–4, but could be as high as 7 (Buskirk 1999). The frequency and intensity of carnivore interspecific killing reaches its maximum when the larger species is 2–5.4 times larger than the other guild members, as at this size range there is a higher potential for high diet overlap and, in this way, a higher benefit for the larger species to eliminate the smaller potential competitors (Donadio and Buskirk 2006). The ocelot is 2.2–4.6 times larger than the jaguarundi, margay, and little spotted cat. Dietary overlap of ocelots with these smaller sympatric species is also high (as shown above) and, as dietary overlap seems to be a factor likely to motivate interspecific killing (Donadio and Buskirk 2006), we speculate that intraguild killing, or its potential (Woodroffe and Ginsberg 2005), might be the mechanism by which ocelots affect small cat dynamics in the neotropics.

Ocelot density does not appear to be impacted by the presence of the larger puma and jaguar (Oliveira *et al.*, unpublished data), as predicted by the mesopredator release theory of Crooks and Soulé (1999). Conversely, ocelot density appears to impact negatively the numbers of smaller little spotted cat, margay, jaguarundi, and Geoffroy's cat (Fig. 27.12) in what has been called the 'ocelot effect' (or the 'pardalis effect'; Oliveira *et al.* 2008, unpublished data). If true, then this dominant mid-sized carnivore might determine the dynamics of the mesopredator community in the neotropics, rather than the top predators, as predicted by the mesopredator release theory. In this case, the prediction is that as ocelot numbers decrease small felid numbers should rise due to reduced intraguild predation by ocelots (Fig. 27.13). This is what has been observed over the past 8 years (2000–07) at a site in southern Brazil (Oliveira *et al.* 2008).

The emerging pattern within the lowland felid guild is that the density of ocelots is not influenced by that

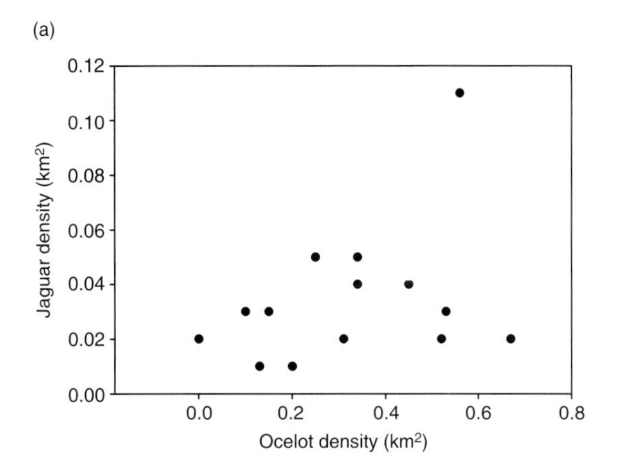

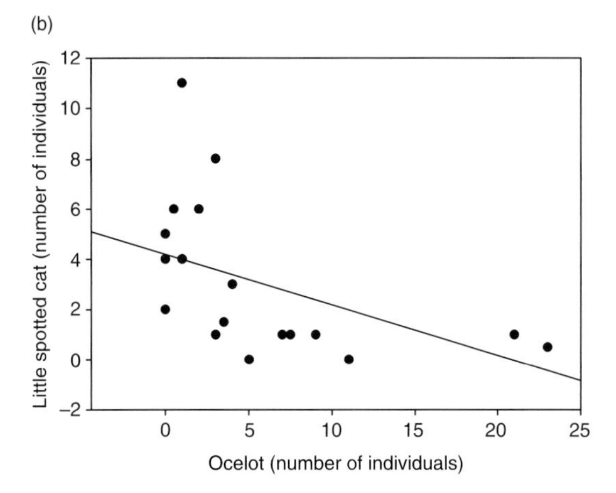

Figure 27.12 Demographic interactions of ocelots with (a) potential predators, jaguars; and (b) potential competitors, little spotted cats. (Modified from Oliveira et al. 2008, unpublished data.)

of other felids, but rather by availability of prey and adequate habitat. In turn, the density of the smaller sympatric jaguarundis, Geoffroy's cats, margays, and little spotted cats seem to be influenced first by ocelot numbers and then by other environmental factors typically associated with habitat and prey base (Oliveira *et al.* 2008, unpublished data). Carnivore assemblages elsewhere show similar trends (see Macdonald *et al.*, Chapter 1, this volume). Cheetahs (*Acynonyx jubatus*) and African wild dogs (*Lycaon pictus*) avoid areas with high prey densities because these are areas where their intraguild competitors and predators such as lions (*Panthera leo*) and spotted hyenas (*Crocuta crocuta*) reach high population densities (Laurenson *et al.* 1995; Creel and Creel 1996; Mills and Gorman 1997; Palomares and Caro 1999). A similar relationship between the density of ocelots and the smaller felid species is expected and would have conservation implications for the latter species. Even the largest protected areas, that typically harbour ocelot population densities >0.1 km^2, may not hold viable populations of the smaller species, whose conservation would be dependent upon their protection outside reserves, in ever dwindling natural habitats (Oliveira *et al.* 2008).

Concluding remarks

Ocelots are generalists, highly adaptable, and the dominant felid in the lowland mesopredator assemblage of the neotropics. According to first principles, we had expected environmental variables to play a key role in determining ocelot numbers through primary productivity, and thus the influence of prey abundance on their home range and population density. However, although rainfall was correlated with ocelot density, it was not correlated with home range size, so these relationships cannot be assumed and merit further exploration.

Among the sympatric small cats, small rodents were the prey whose undigested remains were most commonly found in droppings; however, ocelots were the only species that consistently took larger prey (>1 kg), such as agoutis, pacas, armadillos, sloths, and monkeys. Small rodents contributed relatively little to the ocelot's intake of biomass, and the abundance of larger prey emerges as the likely limiting factor on their persistence. Furthermore, we detected no evidence of competitive dietary niche release within the lowland Neotropical felid assemblage, at least in so far as the absence of larger species had no observable effect on the diet of the smaller species. Body mass of main vertebrate prey (as defined by occurrence in diet) emerged as a useful predictor of the average mass of prey taken by ocelots. The patterns of ocelot habitat use revealed by our review indicate a greater degree of plasticity than often thought. Ocelots have been associated not only with dense cover (from pristine to highly disturbed), but also use adjacent open areas, including boundary areas of agricultural fields. This versatility

Figure 27.13 The 'ocelot effect' (or the '*pardalis* effect') occurs when the dominant mid-sized carnivore (ocelots) impact the dynamics of the mesopredator community in tropical America. As ocelot numbers decline, smaller felids numbers increase due to reduced intraguild predation (pictures not to scale). © 'Projeto Gatos do Mato—Brasil'.

may explain why ocelots are generally the most abundant species in most of the habitats in which they occur.

Our synthesis of available data also suggests that ocelot numbers have not been detectably influenced by either the larger jaguar and puma (potential predators) or by the smaller species (potential competitors). Density estimates of ocelots are significantly higher than those of the smaller jaguarundis, Geoffroy's cats, margays, and little spotted cats. The pattern that emerges is apparently one of coexistence between ocelots and the smaller species, but the lower densities of the smaller felids may reflect intraguild predation by ocelots, or the threat of it. If these premises hold true, then the 'ocelot effect' may be a key factor shaping the dynamics of the small-felid community of the lowland neotropics.

Acknowledgements

Project 'Gatos do Mato—Brasil' (Wild Cats of Brazil Project) was funded by Brazil's National Environmental Fund (FNMA conv. 001/04), with additional support from Fundação O Boticário de Proteção à Natureza, Conservation International-Brazil, FATMA, FAPEMA, ISEC-Canada, and Instituto Pró-Carnívoros, which we acknowledge. We also thank all 12 partner institutions and all project members. Emas National Park Carnivore Project was funded by Memphis Zoo, FNMA, Earthwatch Institute, and Jaguar Conservation Fund (JCF). The invaluable comments of Renata Leite-Pitman, Andrew Loveridge, David Macdonald, three anonymous reviewers, and the help of Natália Torres (JCF), as well as the unpublished data provided by several researchers, are deeply appreciated.

CHAPTER 28

Highland cats: ecology and conservation of the rare and elusive Andean cat

Jorgelina Marino, Mauro Lucherini, M. Lilian Villalba, Magdalena Bennett, Daniel Cossíos, Agustín Iriarte, Pablo G. Perovic, and Claudio Sillero-Zubiri

Andean cat in Salar de Surire, north Chile. © Jim Sanderson.

Introduction

The concept of rarity is central in conservation, as is the linked topic of prioritization, and both issues are complicated (Dickman *et al.* 2006b; Mace *et al.* 2006). Rarity can be defined numerically or probabilistically (e.g. IUCN red list criteria), and while generally viewed as the undesirable consequence of a failure in conservation, some species are naturally rare, possibly as the result of being ecologically specialized (e.g. Ethiopian wolves, *Canis simensis*; Sillero-Zubiri *et al.* 2004). A rare animal population is usually defined as one with low numbers of individuals, but large populations can appear to be rare if animals show elusive behaviour and survey procedures prove ineffective (i.e. an error in estimation), or if animals are sparsely distributed over a vast area and little distinction is made between terms like density and abundance. Whatever definition of rarity is involved, rarity poses problems when monitoring populations (Thompson 2004) and for conservation planning (Mace *et al.* 2006).

The 4–4.5 kg Andean cat (*Leopardus jacobita*) is arguably one of the rarer and certainly among the least known of the small cats (Nowell and Jackson 1996; Yensen and Seymour 2000; García-Perea 2002; Villalba *et al.* 2004; Fig. 28.1). In this case study we evaluate the extent to which the Andean cat's rarity is actual or a perception, whether natural or due to anthropogenic factors, and what important conclusions arise for its conservation. The species' rarity is primarily evident from the paucity of published and unpublished records (Villalba *et al.* 2004). The Andean cat was first described in 1865 from a skin collected near the Bolivia–Argentina border (Cornalia 1865). Until recently, information was limited to three skulls and 14 skins in museums (Johnson *et al.* 1998). Originally classified as *Felis*, the Andean cat was moved into a new genus *Oreailurus* on the basis of skull morphology (Cabrera 1940). The molecular phylogeny derived from these few specimens (Johnson *et al.* 1998) gives support to the proposal that Andean cats, together with the Pampas cat (*L. colocolo)* and several small- to medium-sized South American felids, belong to the ocelot lineage (*L. pardalis*; Herrington 1986; see Werdelin *et al.*, Chapter 2, this volume).

Andean cat populations are threatened by a suite of anthropogenic factors, including human persecution, local elimination of major prey species (notably chinchillas *Chinchilla lanigera* and *C. brevicaudata*), and alteration of natural habitats by livestock grazing, mining, salt extraction, and tourism (Villalba *et al.* 2004). The nature and extent of such negative impacts, however, are largely speculative. Hunting by the Aymara and Quechua peoples living in the High Andes, who have traditionally used skins of small wild cats for ceremonial and spiritual purposes, is locally important in some areas, while in others negative attitudes are rooted in the belief that these small cats predate on livestock or confer bad luck (see Villalba *et al.* 2004). For example, in north Argentina, 46% of 35 local inhabitants who reported they were able to recognize the Andean cat said they would hunt and kill small wild cats (Lucherini and Merino 2008). In spite of limited data, in 2002 Andean cats were reclassified from 'Vulnerable' to 'Endangered', after surveys indicated smaller and more fragmented populations than had previously been suspected (Acosta *et al.* 2008).

Over the past decade, field biologists working under the Andean Cat Alliance (AGA in Spanish—www.gatoandino.org), have contributed to advancing our knowledge on the ecology and distribution of this species. This review examines the still scarce and fragmentary information for the Andean cat in light of the paradigm of rarity, and incorporates new field studies from around the triple frontier (TF) between Argentina, Bolivia, and Chile, where protected areas tessellate across international borders. This represents the epicentre of many recent sightings of Andean cats and the focus of an international network of research teams sharing common methodologies under the AGA banner. We discuss the role of the Andean cat as a flagship species for the monitoring and conservation of High Andean vertebrate diversity

Figure 28.1 Andean cat, Salar de Surire, Chile. Distinctive characteristics of Andean cats include the brown-yellowish blotches along vertical lines at the sides of the body, giving the appearance of continuous stripes, and the six to nine wide dark brown/black rings along the tail. © Jim Sanderson.

within this TF area, and emphasize the role of transfrontier collaborations.

Ecology of a highland specialist: rarity through specialization?

One explanation for rarity is that rare species are specialized, thus only occurring at low density or across small geographical ranges. The emergence of a more definitive distribution map of Andean cats has been a major step (Fig. 28.2), the result of extensive searches across their suspected range and of the progressive refinement of field techniques (Perovic *et al.* 2003; Villalba *et al.* 2004; Cossíos *et al.* 2007b; Lucherini *et al.* 2008; Napolitano *et al.* 2008). In the next section we discuss whether inadequate techniques may account, at least partially, for the species' rarity. New records confirm that Andean cats are endemic to the Central Andes of Argentina, Bolivia, Chile, and Peru,

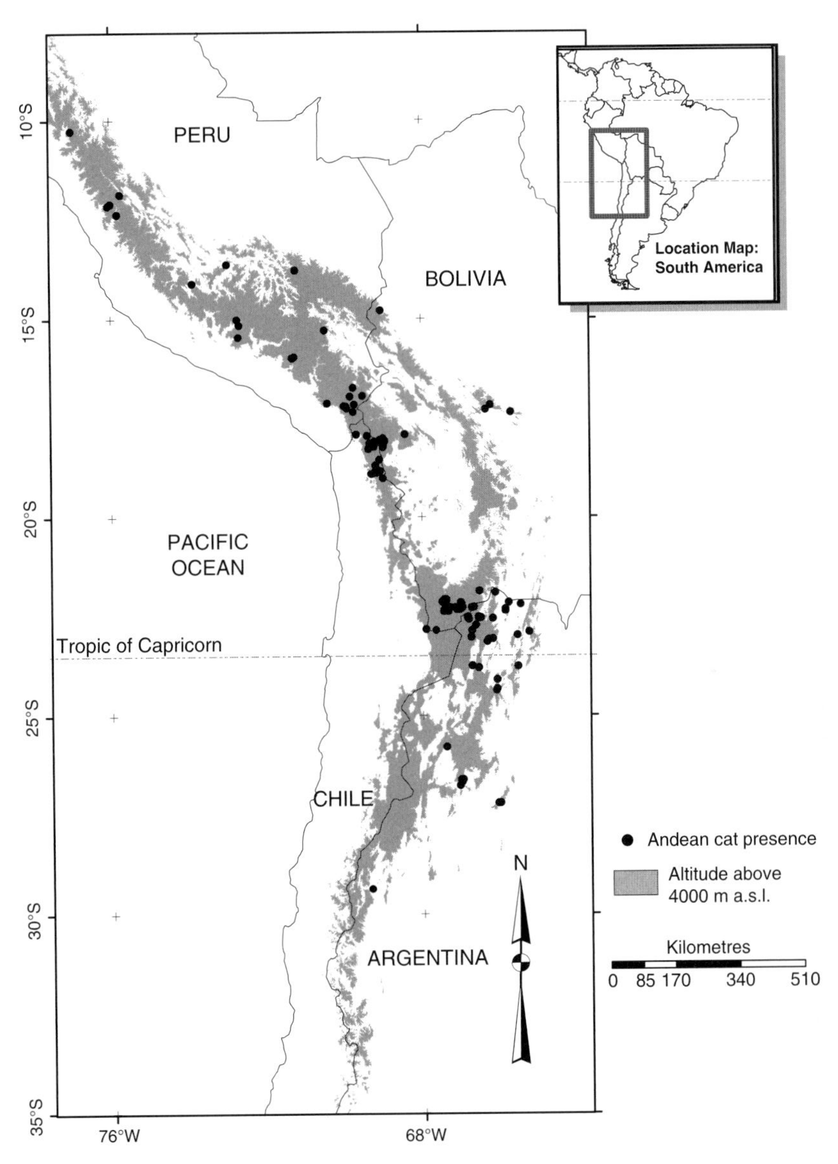

Figure 28.2 Distribution of Andean cats, showing all known records.

and largely restricted to high altitudes between 3000 to 5500 m a.s.l. (mainly above 4000 m), but as low as 1800 m in the most southerly location (Mendoza, Argentina; Sorli *et al.* 2006; Fig. 28.2).

This distributional range largely corresponds with the High Andes biogeographical region, from above 4200 m in tropical areas but at lower elevations towards the South (Cabrera and Willink 1973), where diminishing rainfall and shorter rainy seasons also result in greater aridity (Brush 1989). Compared to other members of the carnivore guild, including the similar-sized Pampas cat, the culpeo fox (*Pseudalopex culpaeus*) and the much larger puma (*Puma concolor*), only Andean cats are restricted to this biogeographical zone (Redford and Eisenberg 1992; Rau *et al.* 1998). In this severe climate, with low precipitation (<100–800 mm annually), frequent frosts, intense solar radiation, low temperatures (annual mean between 0° and 4°C) and large variation in daily temperature (Villalba *et al.* 2004), vegetation is sparse and mainly consisting of grasses (*Festuca* and *Stipa* spp.), some creeping herbs, and bushes in more sheltered sites. A varied community of high-altitude rodents, mainly herbivores, constitutes the main prey of carnivores in the High Andes. Andean cats are consistently the most specialized High Andes carnivore in terms of their diet (Table 28.1). Even when not the most frequent food item (up to 70% of the food items in faeces from the TF in Bolivia), the mountain vizcacha (*Lagidium viscacia*) contributed the greatest biomass to the Andean cat's diet in three study areas, in large part because its adult body mass is about 25 times that of other common prey such as cricetine mice. The Pampas cat consumes mostly cricetines, tuco-tucos (*Ctenomys* spp.) and/or birds in variable proportions, while culpeo foxes have more generalist diets, including all of these prey items, as well as many invertebrates and birds, European hares (*Lepus europaeus*), and carrion of the highland specialist vicuñas (*Vicugna vicugna*), guanacos (*Lama guanicoe*) and livestock (llamas, *Lama glama*). The consistency of dietary items represented in Table 28.1 indicates that the apparent trophic specialization of Andean cats is not a mere artefact from lack of study.

Nowell and Jackson (1996) observed that Andean cat distribution largely coincides with the historic distribution of chinchillas, a suspected key prey before persecution over the last century eliminated them from most of their range (Iriarte and Jaksic 1986). Andean cat distribution also largely overlaps with that of its present main prey, the mountain vizcacha, a gregarious, medium-sized (2–2.5 kg) chinchillid that also occurs in the Patagonian steppes (Walker *et al.* 2000; Table 28.1). These strict specialists of rocky habitats have patchy distributions and by using this habitat to escape predators, they are arguably more difficult to catch than many other prey species (Walker *et al.* 2003, 2007). More behavioural observations are needed to support the hypothesis that the characteristic long, thick, and cylindrical tail of Andean cats (66–75% of head and body length) is an adaptation to hunting chinchillas and mountain vizcachas in cliffs and boulders (Nowell and Jackson 1996), by helping with balance (Hickman 1979).

This dependence on mountain vizcachas helps explain why Andean cat distributions are generally associated to rocky formations, although the use of caves for shelter may also contribute to the cat's habitat specificity. Direct observations of Andean cats often occur around rocky formations, in or near a cave, including those of females with kittens (e.g. two sightings in Argentina, Lucherini *et al.* 2004; one in Bolivia, Villalba 2002). The same association is found for Andean cat faeces (e.g. all 10 samples collected in Sud Lipez, Bolivia, L. Villalba and N. Bernal, unpublished data), which are commonly found in latrines in rocky areas (e.g. five of six samples from Bolivia, O. Torrico, unpublished data). Latrines appear to be largely species-specific (e.g. samples of faeces from five latrines in Chile were either exclusively from Pampas or Andean cats; Napolitano *et al.* 2008) and to be regularly used (of 21 latrines found in rocky areas in Khastor, Bolivia, 19 had both fresh and old droppings, and 22 to 23 days after being cleared, 5 had new droppings; L. Villalba *et al.*, unpublished data). Observations of the single Andean cat radio-collared to date, a female (L. Villalba, unpublished data) confirm its use of rocky caves for daytime resting, for consuming prey (remains of mountain vizcachas were found in one cave) and presumably also for reproduction (observations of the radio-collared female with kittens).

Clearly, Andean cats are rare in the sense that they are endemic, restricted in distribution, and rocky outcrop specialists. It remains unclear how their specialized diet and the patchy vizcacha distribution relate to their energetic and spatial requirements, but

Table 28.1 Diet of carnivores in areas of the High Andes, expressed as percent occurrence of food items in faeces (number of times an item occurred as a percentage of the total of prey items recorded).

		Argentina			Chile		Bolivia		
		AC	PC	CF	AC	PC	AC	PC	CF
Rodents		**68.0**	**79.3**	**41.6**	**82.0**	**71.0**	**100.0**	**83.7**	**74.6**
Phyllotis spp	50 g	1.3	21.7	14.8	27.6	31.5	0.0	5.1	10.5
Other cricetine	30–50 g	38.7	28.4	13.8	4.0	8.5	26.7	40.4	29.9
Abrocoma cinerea	250 g	0.0	3.6	1.9	0.0	1.0	0.0	10.0	1.5
Ctenomys spp.	200 g	0.0	20.2	7.9	6.3	1.0	0.0	2.0	9.0
Lagidium viscacia	1.6 kg	28.0	5.3	3.2	44.1	29.0	73.3	26.3	23.9
Birds		**10.7**	**9.0**	**16.8**	**18.0**	**27.5**	**0.0**	**15.1**	**13.4**
Flamingo spp.	4.5 kg				0.0	23			
Tinamou spp.	250 g				18.0	4.5			
Reptiles		**5.3**	**4.8**	**3.8**	**0.0**	**0.0**			
Insects		**4.0**	**1.3**	**11.3**	**0.0**	**1.5**	**0.0**	**0.0**	**4.5**
Large mammals		**2.7**	**1.7**	**15.5**	**0.0**	**0.0**	**0.0**	**0.0**	**7.5**
Other[a]		**9.3**	**4.0**	**11.0**	**0.0**	**0.0**	**0.0**	**1.0**	**0.0**
Number of faeces		57	504	399	33	75	18	45	43

Notes: AC, Andean cat; PC, Pampas cat; CF, culpeo fox.

Bolivia (Viscarra 2008): Locality at ~4200 m, north of Eduardo Avaroa National Reserve between 22–23° S latitude; Argentina (Walker *et al*. 2007) and Chile (Napolitano *et al*. 2008): The samples are from larger areas across broader elevation ranges (between 3200 to ~4800 m in each case and 18° and 29–22° latitude, respectively), although 87% of the Andean cat faeces in Argentina are from the TF.

[a] Other prey, includes unidentified mammals, marsupials, Caviidae, *Lepus europeaus* (in culpeo faeces from an area in Argentina) and greater chinchillas *Chinchilla brevicaudata* (in three culpeo samples from Argentina).

the only known home range (that of 'Sombrita', the solitary female radio-tracked in Bolivia for a period of 8 months) covered 65.5 km^2 with a centre of activity coinciding with a series of rocky formations (95% Minimum Convex Polygon, Kernel Centre, from 66 locations between April and December 2004; L. Villalba, unpublished data). The large home range does not seem to reflect dispersal, as there was only one location clearly separated that was excluded as an outlier. The Andean cat's use of steep, rocky, or broken terrain, cliffs and ridgelines would place them close to their main prey and to the only type of cover available at such altitudes, while using latrines may facilitate mutual avoidance and coexistence within high-quality patches.

Elusive highland dwellers: rarity through low detection?

The paucity of Andean cat data most likely results from a combination of elusiveness and low numbers, along with the logistical constraints of surveying the remote High Andes. Our extensive review of published and unpublished records (Table 28.2) demonstrates that the cats are notoriously difficult to see (only 17 direct sightings recorded in the four range countries between 1990 and 2006). This is predictably related to the cat's elusive behaviour, which includes the use of caves, and mainly crepuscular-nocturnal habits (Table 28.3). An additional complication is that Andean cats look very similar to Pampas cats (Cossíos *et al.* 2007a), so reliable reports are rare. Specifically designed interviews have been used with the Aymara and Quechua people inhabiting these mountains, but the skins found in their possession have provided the best proof of the presence of Andean cats in some areas. With advances of faecal DNA extraction over the past decade (see Culver *et al.*, Chapter 4, this volume), the reliance on faeces has now become the cornerstone of Andean cat research (Table 28.2). Techniques used to identify Andean cat DNA from epithelial cells on droppings include restriction enzyme digestion patterns (Cossíos and Angers 2006), sequencing of mitochondrial genes (Napolitano *et al.* 2008), and the mtDNA control region (Cossíos *et al.* 2007b). In recent years, camera-trap surveys are becoming a common technique among Andean cat researchers (Table 28.2), whereas attempts to capture and radio-track Andean cats remain limited to two occasions (Table 28.2), one successful, in areas where Andean cat presence had been confirmed and there was evidence of their permanence.

As the techniques available to study felids in general and Andean cats in particular have improved (see Karanth *et al.*, Chapter 7, this volume) data on this species have increased (Table 28.2). Between 1990 and 2006, surveys across the four range countries contributed 46 skins and 49 faeces (confirmed using DNA), and between 2004 and 2006 camera traps captured 21 photographs (averaging 7.0 records per annum, compared to 2.0 faeces, and 3.1 skins from DNA identification; these rates do not account for variation in sampling effort or the spatial extent of study areas). These detection rates, however, are still low for the level of effort required. For example, overall capture frequency across seven sites in three study areas of north Argentina was 0.49 photos/100 trap days for Andean cats and 1.27 photos/100 trap days for Pampas cats; and this study used lures and located cameras close to places where other signs had been detected (Lucherini *et al.* 2008). The burden of evidence is that Andean cats do indeed live at low population densities, and seemingly below the levels of Pampas cats, which consistently showed higher detection rates across study areas and methodologies (Table 28.2). It would seem fair to conclude that Andean cats are widely distributed across their known range, but also that numbers are generally low, and that this is not an artefact of inadequate field techniques or of lack of studies.

The ghost of intraguild competition: rarity through competition?

Intraguild competition has been widely demonstrated among the Carnivora in general and has been highlighted as an important factor in canid ecology by Macdonald and Sillero-Zubiri (2004a). It is also emerging as a factor in felid communities (see Oliveira *et al.*, Chapter 27, this volume) and could possibly contribute to the rarity of Andean cats

Table 28.2 Records of Andean cat presence provided by various techniques, with some indication of rates of detection. When possible a ratio of Pampas to Andean cat (PC/AC) detection is shown.

Region	Rate of AC detection	PC/AC	Source
Direct sightings			
Bolivia (Sud Lípez)	2 sightings (October 2001)		Villalba (2002); N. Bernal (unpublished)
Argentina (3 regions)	10 (25 PCs) (1990–2001)	2.5	Perovic *et al*. (2003)
Argentina (4 regions)	2, in one region (1999–2004)		Lucherini *et al*. (2004)
Argentina (Mendoza)	1 (2006—new record for the region)		Sorli *et al*. (2006)
Peru (10 regions)	No sighting (5 PCs) (2001–06)		Cossíos *et al*. (2007b)
Chile (Region I)	1 (1999—first photographic record)		Sanderson (1999)
Chile (Region I)	2 (2006)		A. Iriarte (unpublished)
Interviews			
Bolivia (4 regions)	46 of 137 interviewees (1999–2004)		Villalba *et al*. (unpublished)
Argentina (4 regions)	22 of 40 (1999–2004)		Lucherini *et al*. (2008)
Peru (7 regions)	339 of 696 (2002–06)		Cossíos *et al*. (2007b)
Chile (5 regions)	115 of 280 (1998–2006)		A. Iriarte *et al*. (unpublished)
Skins			
Bolivia (6 regions)	10 of 37 wildcat skins (1998–99)	3.7	L. Villalba and N. Bernal (unpublished)
Argentina (3 regions)	12 of 39 (1990–2001)	3.3	Perovic *et al* (2003)
Argentina (4 regions)	3 of 11 (1999–2005)	3.7	Lucherini *et al*. (2008)
Peru (10 regions)	19 of 86 (2002–06)	4.5	Cossíos *et al*. (2007b)
Chile (Region I)	2 of 8 (1998, 1 month)	4.0	Sanderson (1999)
Chile (Region I)	10 of 36 (1999–2002)	3.6	A. Iriarte (unpublished)
Faeces			
Bolivia (Sud Lípez)	7 of 34 faeces (2004)		L. Villalba and N. Bernal (unpublished)
Bolivia (Sud Lípez)	7 of 17 faeces (2007)		O. Torrico (unpublished)
Bolivia (Sud Lípez)	17 of 46 faeces (2007)	2.7	Viscarra (2008)
Argentina (4 regions)	12 of 246 cat faeces (2000–06)	20.5	M. Lucherini *et al*. (2008)
Argentina (Jujuy)	7 of 105 cat faeces (2005–06)	15.0	M. Lucherini *et al*. (unpublshed)
Peru (4 regions)	11 of 194 cat faces (2002–06)	17.6	Cossíos *et al*. (2007b)
Chile (Region I)	33 of 108 cat faeces (January–April 2004)	3.3	Napolitano *et al*. (2008)
Camera traps			
Bolivia (Sud Lípez)	First photographic record		Villalba (2002)
Bolivia (Sud Lípez)	0 of 13 photos of carnivores (April 2003)		L. Villalba *et al*. (unpublished)
Bolivia (Sajama)	First photographic record		Barbry and Gallardo (2006)
Argentina (3 regions)	12 in 2434 trap/days (2004–06)	2.6	Lucherini *et al*. (2008)
Chile (Region I)	7 of 98 photos of carnivores (2005–06)		A. Iriarte *et al*. (unpublished)
Capture			
Chile (Region I)	0 after 591 trap days (1998)		Sanderson (1999)
Bolivia (Sud Lípez)	1 (adult female); 1 Pampas cat		Delgado *et al*. (2004)

(Lucherini and Luengos Vidal 2003; Walker *et al.* 2007). While Andean cats are mostly limited to high elevations, they do not segregate from Pampas cats into relatively disjunct distributions along an altitudinal axis. Two studies, one in north Chile (Napolitano *et al.* 2008) and one in north Argentina (Perovic *et al.* 2003) showed substantially overlapping distributions (by at least 1000 m, from 3500 to 4500 m in Argentina and from 3700 m to 4500 m in Chile), but logistic regressions indicated that in Chile the probability of finding Pampas cat faeces decreased with altitude, while the opposite trend was observed for Andean cats; and in Argentina the mean elevation of Andean cat records (4236 $\pm$ SE 140 m, $n = 24$) was significantly higher than that of Pampas cats (3567 $\pm$ SE 67, $n = 50$). A recent faecal record from Vilama, Argentina, confirmed the presence of Pampas cats at 5000 m (M. Lucherini, unpublished data).

Coexistence between Andean and Pampas cats is also evident at the local level. Faeces from both species were recorded in a same rocky outcrop (Argentina, Perovic *et al.* 2003), as close as 6 m apart (Chile, Napolitano *et al.* 2008), in the same standardized 400 m transects on two occasions (Bolivia, Viscarra 2008) and in two shared latrines (Bolivia, O. Torrico, unpublished data; Argentina, M. Lucherini *et al.*, unpublished data). Photos of both felids were captured in 6 out of 25 camera-trap stations in Argentina (J. Reppucci *et al.*, unpublished data). Such localized sympatry is also apparent between the culpeo and these two cat species, from signs along standard transects (Table 28.4) and from footprints in latrines (12 of 19 latrines in Bolivia were visited by both culpeos and small cats; L. Villalba, unpublished data).

Theory predicts that the intensity of competition between these sympatric carnivores will equal the product of their relative abundance and the shared use of a resource (Begon *et al.* 1990). The apparently higher abundance of culpeos and Pampas cats (Lucherini and Luengos Vidal 2003; Perovic *et al.* 2003; Walker *et al.* 2007; Table 28.2) certainly raises the possibility that they are exerting intense resource competition on the Andean cat. In such cases, coexistence could be facilitated by niche segregation, particularly if species are of similar size, morphology, and food habits (Farrell *et al.* 2000), as is the case with these two small felids.

Regarding food partitioning, in north Argentina the cats' standardized niche breadth (Colwell and Futuyma 1971) was similar (Andean Cat [AC] $B_{sta} = 0.19$ and Pampas Cat [PC] $B_{sta} = 0.15$), but greater for culpeo fox ($B_{sta} = 0.62$), a far more generalist species (Walker *et al.* 2007). Here the Andean cat had a diet specialized towards mountain vizcacha, while the Pampas cat specialized on tuco-tucos and cricetines (Table 28.1). In Chile, the overlap in food niche (measured by the Pianka index, Pianka 1973) was extensive between the two cats (0.82), but the trophic niche of Pampas cats was wider ($B_{sta} = 0.35$

Table 28.3 Activity patterns of a female Andean cat (AC) in Bolivia, expressed as percentage of the locations when the animal was active in each period (averages over May–June 2004, it may include some activity without significant movement; L. Villalba, unpublished data) and from camera traps in Argentina, expressed as percentage of photos of Andean cat (AC), Pampas cat (PC), and culpeo fox (CF) taken in given time periods (M. Lucherini *et al.*, unpublished data).

	Radio-collared female (Khastor, Bolivia)		Camera traps (Loma Blanca, Argentina)			
	Period	AC active (%)	Period	AC, % ($n = 36$)	PC, % ($n = 56$)	CF, % ($n = 32$)
Night-time	19:45–4:30	54.1	Night-time	66.7	62.5	56.3
Dawn	5:30–7:30	13.0	1 h before and after sunrise	5.6	7.1	18.8
Daytime	7:45–16:30	20.5	Daytime	25.0	28.6	6.3
Dusk	16:45–19:30	48.5	1 h before and after sunset	2.8	1.8	18.8

compared to 0.25; Napolitano *et al.* 2008), feeding more frequently on birds (especially flamingos) and small rodents. In the TF area in Bolivia, a recent study showed greatly overlapping diets between the two cats (0.82)—and between Pampas or Andean cat and the culpeo (0.88 and 0.80, respectively)—but a much more specialized diet in Andean cats (B_{sta}: 0.1, compared with 0.4 for culpeo and Pampas cat), which consumed mainly mountain vizcachas while Pampas cats preyed more frequently on small cricetines and birds. While the three carnivore species share a common pool of prey species in the High Andes, culpeos are consistently more generalists and Pampas cats tend to specialize to various degrees in relation to local conditions, showing greater flexibility than the Andean cats and relying much less on the mountain vizcacha in terms of biomass consumed.

With regards to temporal segregation in areas of sympatry, the two cats appear to have similar activity patterns (according to data from camera traps in one localized area, although telemetry would also indicate high dusk activity in the radio-collared Andean cat; Table 28.3) and both consume nocturnal and diurnal prey (diurnal small, cursorial rodents, and diurnal/crepuscular, sometimes nocturnal, mountain vizcacha), whereas the culpeo appears to be more strictly nocturnal (Table 28.3). There is an indication of spatial segregation as well; although both cat species use rocky caves (e.g. 9 of 18 Pampas cat faeces and 2 of 5 AC faeces were found in caves in one study area in Bolivia; Viscarra 2008), Andean cats spend more time foraging in steeper, rocky areas frequented by mountain vizcachas (Napolitano *et al.* 2008). Such spatial segregation appears more evident between culpeos and the two cats, for example in the frequency of their signs in open versus rocky habitats, and in the use of caves and latrines (Table 28.4).

Evidence of resource partitioning is still inconclusive, but it does provide a tantalizing indication of some segregation between culpeos and the small cats, and between the two cats, mainly along the trophic axis. The more specialized, restricted Andean cat could be a more efficient resource exploiter and still have smaller local populations if it uses rarer or lower quality resources (Glazier and Eckert 2002), in this case the mountain vizcachas, a large prey item with very patchy distribution (Walker *et al.* 2003). Adaptations to cold, high-altitude or arid environments could also give the Andean cat a competitive edge over the more widespread and generalist Pampas cat in overlapping areas, including its long fur and thick, long tail, and large auditory bullae—a structural specialization that provides habitat-specific survival value to animals inhabiting arid environments such as some gerbillines (Lay 1972) and sand cats (Huang *et al.* 2002), because of the attenuation of sound in deserts.

A simple ecological model

In the absence of detailed ecological studies, simple models can help us uncover the relationships between environmental conditions and the Andean

Table 28.4 Summary of results from searches of carnivore signs along standard transects (Cossíos *et al.* 2007a) in two study areas in the TF.

	Percent of transects with signs of:						Faeces		Foot-prints	Signs in caves (%)	Signs in open areas (%)	Signs in rocky formations (%)
	n	C	F	P	C+F		*n*	Latrines (%)	*n*			
Vilama (ARG)	37	35	59	11	24	F	39	0	—	13	51	38
						C	44	57	—	64	14	86
Avaroa (BOL)	110	30	35	2	15	F	18	5	51	3	81	16
						C	62	35	10	40	23	77

Notes: C, small cats (faeces of Andean or Pampas cats); F, culpeo fox; P, puma. Latrines here defined as a discrete accumulation of >5 faeces (maximum numbers observed: 40 and 45 faeces of AC and 11 and 5 faeces of foxes in Argentina and Bolivia, respectively).

cat's habitat requirements (see Boyce *et al.* 2006). Habitat models that use presence–absence data and logistic regressions are increasingly applied for this purpose and to quantify habitat availability (e.g. Massolo and Meriggi 1998; Palomares *et al.* 2001). The principle is to contrast used versus unused habitat units or locations with a set of explanatory variables to determine habitat suitability. The resulting 'Resource Selection Function' offers a quantitative characterization of resource use and, when combined with geographic information system (GIS), it can accommodate virtually any type of resource and spatial data (Boyce and McDonald 1999).

A key assumption of habitat models is that the factors limiting the distribution and abundance of the studied organism are known, and that data are available on key resource variables. With this objective in mind we developed a simple biological model (Fig. 28.3) that postulates two key, possibly limiting, factors for Andean cats, namely: (1) the presence of

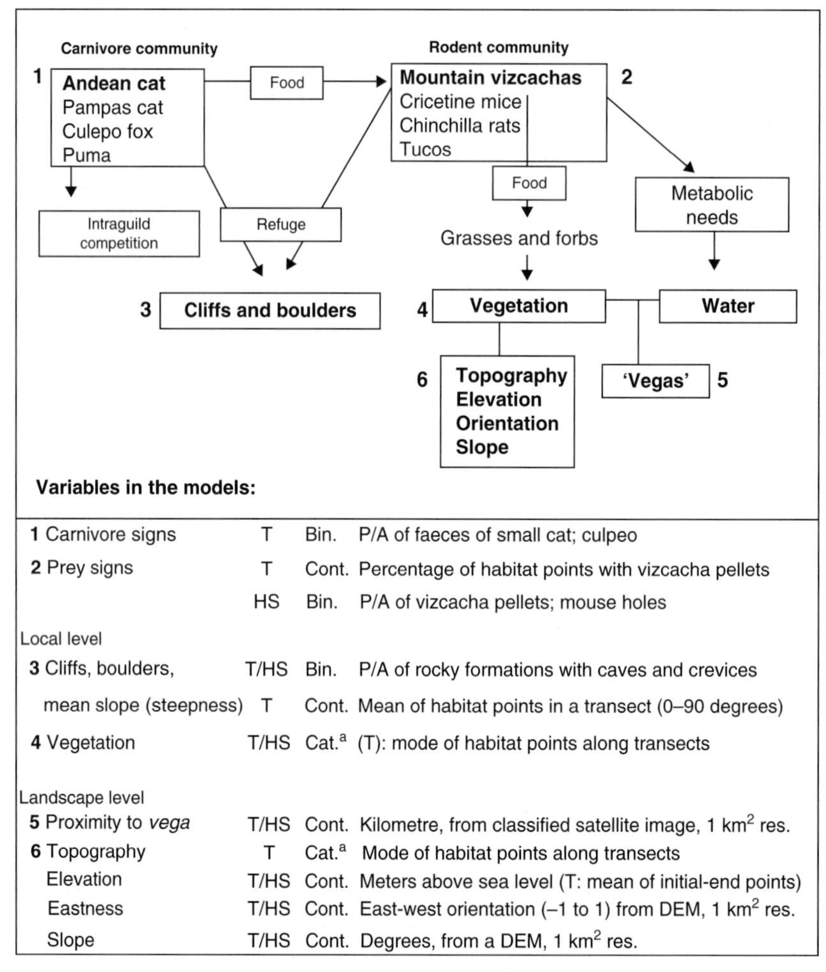

1 Carnivore signs	T	Bin.	P/A of faeces of small cat; culpeo
2 Prey signs	T	Cont.	Percentage of habitat points with vizcacha pellets
	HS	Bin.	P/A of vizcacha pellets; mouse holes
Local level			
3 Cliffs, boulders,	T/HS	Bin.	P/A of rocky formations with caves and crevices
mean slope (steepness)	T	Cont.	Mean of habitat points in a transect (0–90 degrees)
4 Vegetation	T/HS	Cat.[a]	(T): mode of habitat points along transects
Landscape level			
5 Proximity to *vega*	T/HS	Cont.	Kilometre, from classified satellite image, 1 km^2 res.
6 Topography	T	Cat.[a]	Mode of habitat points along transects
Elevation	T/HS	Cont.	Meters above sea level (T: mean of initial-end points)
Eastness	T/HS	Cont.	East-west orientation (–1 to 1) from DEM, 1 km^2 res.
Slope	T/HS	Cont.	Degrees, from a DEM, 1 km^2 res.

Figure 28.3 Biological model relating components of the High Andes ecosystem expected to affect the presence or survival of carnivores, particularly Andean cats, and their main prey, the mountain vizcacha. T, data collected along standard transects; HS, data collected at habitat samples; Bin., binary; Cont., continuous; Cat., categories. [a].(Vegetation: grassland, heaths, bare, 'vega', and other; Topography: peak, mountain slope, hill, plateau, broad valley, narrow rocky valley, boulder, and cliff). Landsat Thematic Mapper Plus source: US Geological Survey; DEM source: Shuttle Radar Topography Mission. Landsat TM + supervised classification with Idrisi software (Clark Labs, Clark University, Worcester, United States).

rocky formations (boulders, ridges, and cliffs); and (2) the availability of mountain vizcachas, themselves specialists of rocky habitats with crevices and tunnels to nest and escape predators (Walker *et al.* 2000). This generalist herbivore consumes grasses and to a lesser extent shrubs and herbs (Galende *et al.* 1998; Cortes *et al.* 2002) and appears to require a source of free water for its survival (Tirado *et al.* 2007). The distribution of plant food resources in these highlands are expected to be determined by slope, orientation, and elevation, with the characteristic high-altitude marshes known as 'vegas' or 'bofedales' as the pockets of productivity *par excellence*, as well as key sources of water for wildlife and livestock alike. These systems of bogs and shallow pools are continuously refreshed by melting ice or subterranean water and concentrate most of the green biomass in an otherwise barren landscape (Squeo *et al.* 2006). Data on these key habitat variables were collected across selected study sites in the TF using a standardized methodology (Cossíos *et al.* 2007a). The TF region contains systems of 'vegas' along drainage lines and valleys, some grasslands and shrublands (mainly 'tholas' *Parastrephia* and *Lepidophyllum* spp.), and extensive areas of exposed sandy or rocky terrain, devoid of substantial plant cover. The high peaks demarking international boundaries also delimit major basins to each side of the frontier, with permanent and seasonal lakes, such as the complex of lagoons in the Vilama Ramsar Site or the extensive salt pans of Chalviri and Tara at the centres of Avaroa and Flamencos Reserves, respectively (Fig. 28.4). The most humid and productive site is Vilama, possibly due to its generally easterly orientation.

We collected 755 habitat samples (5-m radius) from across the TF region, consisting of records of presence/absence of mountain vizcacha pellets (used to index abundance in the Patagonian steppes by Walker I 2000) and environmental variables. We used this dataset for 'vizcacha models' to predict the probability of its presence from combinations of key environmental variables (using logistic regressions and logit link functions; Manly *et al.* 1993). For a second dataset we tallied carnivore signs and associated environmental variables from 147 transects (20 m wide and 400 m long, with 'habitat points' encompassing a 5-m radius at intervals of 20 m along each transect) in north Avaroa (Bolivia) and Vilama (Argentina). As faeces of the two small cats are indistinguishable to the eye, we constructed 'cat models' to predict the probability of detecting signs of small cats from combinations of key environmental variables (using Poisson regressions with log link functions, with numbers of signs as the dependent variable and length of transect as the offset variable; Manly I 1993). As suggested by their diets and use of habitat, rocky formations and vizcachas are also expected to affect the abundance of the sympatric Pampas cat in the High Andes. To model both counts of cat signs and vizcacha presence, we compared a small number of candidate models, developed *a priori* using the variables which were measured on the basis of their likely biological relevance (Fig. 28.3). Thus, our aim for model selection was the testing and comparison of a few specific hypotheses, rather than trawling through the large number of model configurations theoretically possible among these predictor variables. For this reason (and because predictors were intercorrelated), sequential logistic regression models were used for both models (i.e. variables considered the most limiting, taking account of the plausible sequence of cause and effect, were entered first and the variance explained subtracted at each step). We measured the goodness of fit of the candidate models using the deviance (= −2*log-likelihood). In Table 28.5 we present models which provided the best fit from a reduced combination of relevant variables. (In more complex models additional variables were not significant, nor did they improve model fit.)

The simplest 'cat model', with vizcachas as the sole explanatory variable, provided a good fit to the data, although the model improved with the addition of other local variables, reflecting the availability of broken, rocky terrain (Table 28.5). There was no evidence of landscape-level variables or of the presence of culpeo signs affecting the probability of locating small cat signs. We ran separate 'cat models' for each study area (Vilama, Avaroa, and Tara/Tatio) because of a significant interaction with abundance of vizcachas within transects, which likely reflected variation in climate and productivity between these basins. Simple 'vizcacha models' also provided good fit to the data, with vegetation (classes ranging from dense vegetation to barren terrain), distance to the nearest 'vega', and presence of rocky formations as

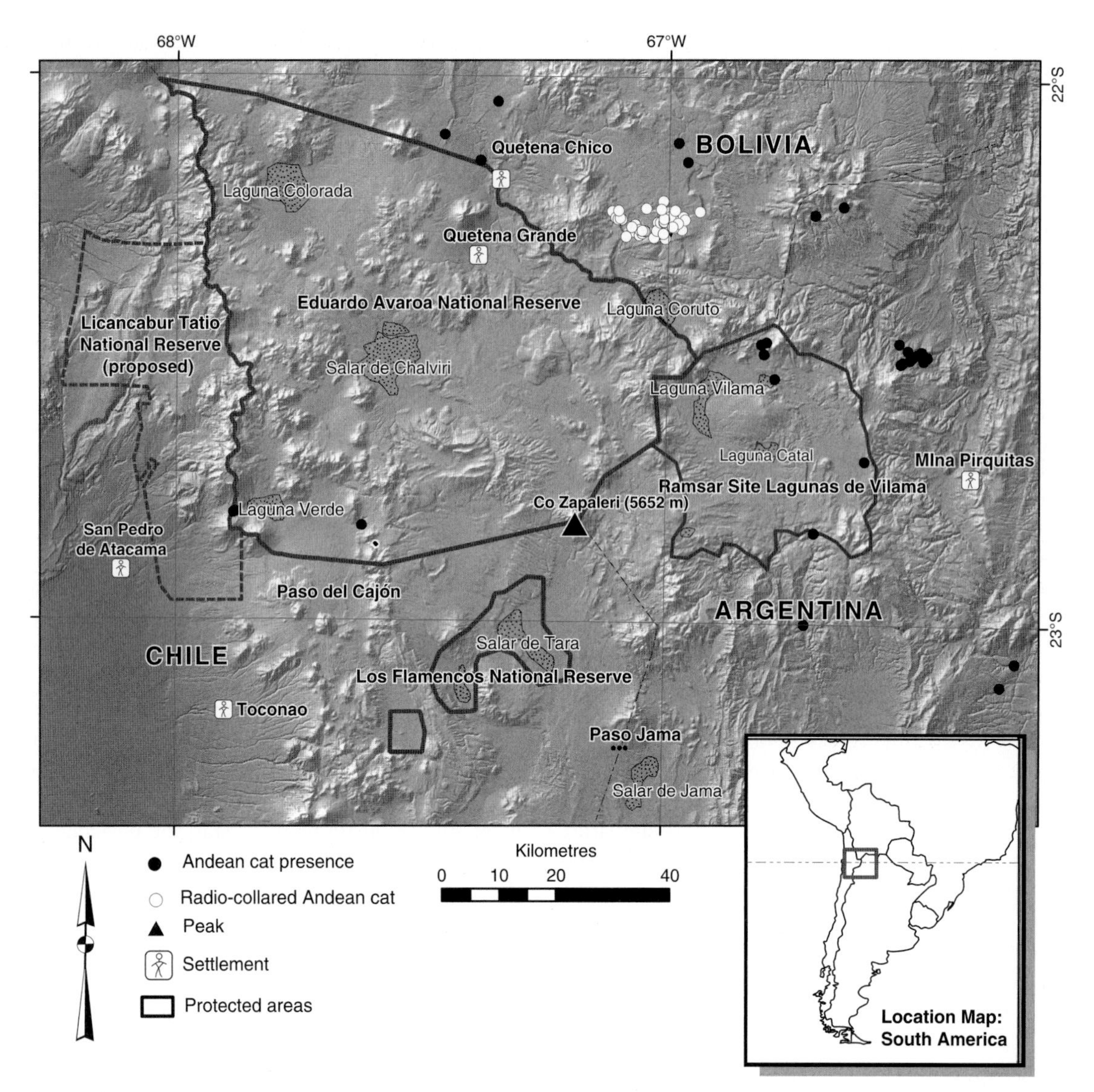

Figure 28.4 Triple Frontier study area, centred on the Cerro Zapaleri peak, showing adjacent protected areas in each country and confirmed Andean cat records.
Source: Shuttle Radar Topography Mission Digital Chart of the world; ESRI Andean Cat Alliance database.

useful explanatory variables (Table 28.5). A significant interaction between study area and proximity to 'vegas' required separated models; only in Tara/Tatio, the most arid, were samples with vizcachas significantly closer to 'vegas' than those without pellets (Table 28.6). This variable depicts a higher density of 'vegas' in Vilama, particularly in comparison to Tara/Tatio (see Table 28.6), with restricted systems of 'vegas' within a dry and barren matrix (Fig. 28.5). More complex models with landscape-level variables did not produce appreciably better models (Table 28.5); the significant effect of 'eastness' in Avaroa and Tara/Tatio may indicate the importance of the influx of humidity from forested lowland to the east in these comparatively dry areas.

Overall, the statistical models support our hypotheses of a few key resources limiting the distribution of small cats and their main prey in this

Table 28.5 Models of diverse complexity and their goodness of fit, as expressed by the deviance (= 2*log-likelihood).

Variables in the model	Goodness of fit: Deviance		
	Vilama (Argentina)	Avaroa (Bolivia)	Tara/Tatio (Chile)
Cat model	*n* = 36 T	*n* = 111 T	—
Vizcacha	69.3 (d.f. 34)	208.5 (d.f. 109)	
Vizcacha + **rocky formations**	40.2 (d.f. 33)		
Vizcacha + **means slope steepness**		170 (d.f. 108)	
Vizcacha model	*n* = 473 HS	*n* = 148 HS	*n* = 134 HS
Simplest biological model			
Vegetation + dist. to 'vega' + **rocky formations**	294.5 (d.f. 292)		
Vegetation + dist. to 'vega' + **rocky formations**		168.0 (d.f. 130)	
Vegetation + **dist. to 'vega'** + **rocky formations**			55.1 (d.f. 130)
Plus landscape level variables			
+ **slope**[a] + eastness + northness	468.1 (d.f. 426)		
+ slope + **eastness** + northness		169.3 (d.f. 139)	54.5 (d.f. 109)

Notes: Variables in bold had a significant effect ($P > 0.05$).
[a] Nearly significant (chi-square 311, $P = 0.078$).

region, as in our initial model (Fig. 28.3). Small cats utilize rocky areas, where the main prey is present, while vizcachas select more vegetated areas. The water and plant biomass concentrated in 'vegas' are limiting resources for vizcachas only in the more arid regions in the Chilean side of the TF. Larger-scale selection is not apparent for cats or the prey, although vizcachas appear to select eastern, more humid orientations in the TF areas of Chile and Bolivia.

Table 28.6 Comparison of the mean distances to a 'vega' of habitat samples with and without vizcacha pellets in each study area.

	Mean distance to 'vega' (km)
Vilama	Absences (155) = 0.56 (SD 0.03) Presences (318) = 0.66 (SD 0.03) Overall = 0.63 (SD 0.49) range 0–2.7
Avaroa	Absences (49) = 2.83 (SD 0.32) Presences (99) = 2.78 (SD 0.23) Overall = 2.80 (SD 2.28) range 0–8.1
Tara/Tatio	Absences (15) = 3.47 (SD 0.35) Presences (99) = 0.40 (SD 0.28) Overall = 3.13 (SD 3.73) range 0–10.9

Note: In Tara/Tatio only, samples with vizcachas were significantly closer to 'vegas' than those without.

Conservation of Andean cats and their role as a High Andes flagship

It would seem possible to hypothesize that Andean cats are naturally rare and that the best hope for the continued survival of these cats and their prey is the remoteness of their habitat, but there is also an argument to be made for their active protection. The Andean cat is a species of conservation concern because of its low numbers, restricted distribution, and relatively low levels of genetic variation in some populations (Napolitano *et al.* 2008). So far it has been impossible to assess if, or to what extent, Andean cat numbers have declined due to human activities, either in the TF region or elsewhere. Further methodological refinements would eventually allow monitoring population changes in areas affected by human activities. Nowell and Jackson (1996) argue that the usual human-induced causes of rarity, namely habitat loss/modification and direct persecution, can only be partially responsible for the observed low densities, as there have been no significant changes in land use of the High Andes over the past 2000 years. Human densities in the High Andes have, if anything, decreased in recent times, and the negative effects from grazing by domestic camelids, sheep, and goats upon rodents are localized, rather than widespread. Still, hunting pressure resulting from the traditional use of small cat skins appears to be locally important in some areas. Over the last century, intensive hunting for chinchillas, presumably a major prey for the Andean cat, led to their extirpation (Iriarte and Jaksic 1986). Finally, expanding mining operations could impact water resources in the High Andes, and unregulated off-road tourism may alter vegetation cover and soil structure, with unknown effects on biological diversity.

For their long-term persistence Andean cat populations will require large areas, encompassing sufficient habitat patches with caves and mountain vizcachas, and protected from intense competition with other sympatric carnivores, or where concentrations of locally abundant prey species may facilitate coexistence, albeit at low density. Walker *et al.* (2000) suggest that Andean cats may be better equipped to hunt vizcacha in rocky formations, but the prospect of a properly controlled experimental test of trophic competition is remote. With improved techniques to measure relative abundances, however, insights into competitive interactions within the Andean carnivore guild could be gained by comparing communities of varying composition. Improving the long-term prospects of these carnivores may now depend more heavily on maintaining and extending both the range and number of vizcacha colonies, along with ensuring connectivity between protected areas, and coaxing local people to reduce hunting pressure through education campaigns (Lucherini and Merino 2008). The vizcacha, by itself, might serve as an important indicator of potential areas occupied by Andean and Pampas cats, and is thus in its own right a keystone species of important conservation concern. The Resource Selection Functions presented here could be used to produce maps of prey suitability in the TF, from which to identify areas required to sustain local carnivore populations, and the degree of connectivity between them. Identifying 'productivity' corridors may be relevant to the long-term survival of Andean cats in this arid region, as arguably the population is continuous between adjacent protected areas, and substantial portions of the TF are dry and of low productivity (Fig. 28.5). Andean cat records on either side of the Argentina–Bolivia border are as close as 20 km apart, and a record in Bolivia is in the vicinity of Chile (O. Torrico, unpublished data, Fig. 28.5), suggesting Andean cat presence in the Chilean side of the TF, suspected but yet to be confirmed. Important areas of the TF in Bolivia and Chile present a pattern of vast dry extensions with a few more humid patches and corridors (Fig. 28.5). Notably, only 2% of the 1610 km^2 of the Vilama Ramsar site are 2 km or further away from a 'vega', compared to 26% of the 420 km^2 Flamencos National Reserve (Chile), 32% of the 6750 km^2 Eduardo Avaroa National Reserve (Bolivia), and 11% of the proposed 1770 km^2 Licancabur-Tatio.

Being mostly restricted to the High Andean ecosystem, a priority biome in Latin America because of its high level of endemism (Dinerstein *et al.* 1995), the Andean cat has potential as a flagship for its conservation. The other sympatric carnivore species invariably range across various biogeographic zones. Although rare and elusive, Andean cats also have charisma and religious significance among the Aymara and Quechua peoples, for whom they symbolize prosperity and abundance (see Villaba *et al.* [2004]). In its flagship role, the Andean cat has so

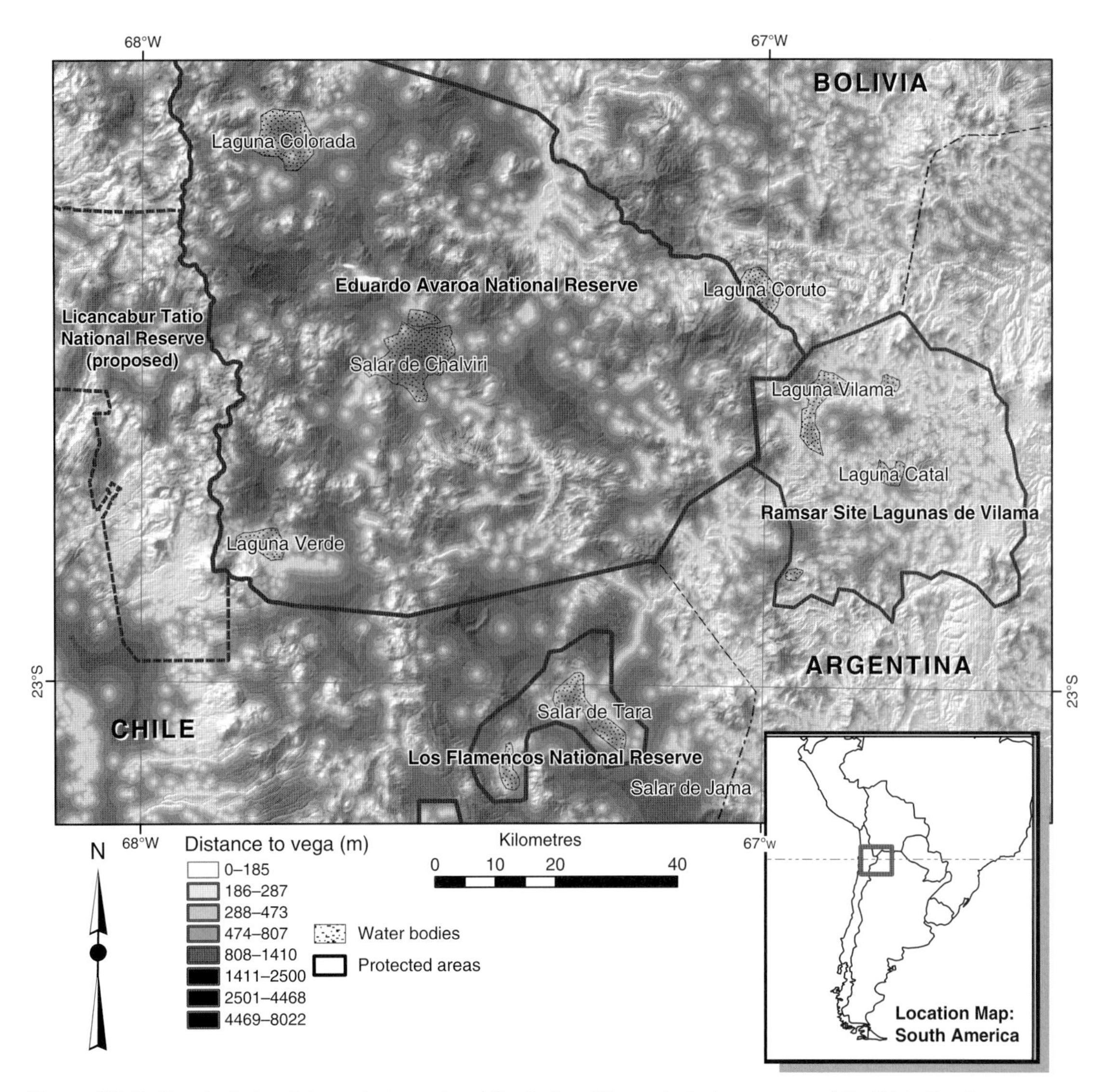

Figure 28.5 Map depicting distance to 'vegas' and the limits of the protected areas around the Triple Frontier. Source: Shuttle Radar Topography Mission Digital Chart of the World; ESRI.

far helped raise awareness of the need to understand and protect High Andes biodiversity, mainly among researchers, donors, and international conservation agencies. The synergy created by researchers and conservationists working under the Andean Cat Alliance across the species range, has already resulted in the production of an Action Plan for its conservation (Villalba *et al.* 2004), a field manual with standardized methodologies for Andean carnivores and their prey (Cossíos *et al.* 2007a), and common educational strategies to increase awareness of the need to protect Andean ecosystems among local communities (Merino 2006). However, there is a need to improve the image of Andean cats among communities with more generalized negative attitudes (Lucherini and Merino 2008).

Presently, the Andean Cat Alliance is working with the authorities of protected areas in the TF to develop coordinated activities to address common threats to the High Andes biodiversity, such as the illegal hunting of vicuñas, livestock grazing and movements, unregulated off-road tourism, as well as highlighting the needs for wildlife monitoring. This initiative includes a precursor of international collaboration in the region, the High Andes Flamingo Conservation Group (GCFA in Spanish—www.flamencosandinos.org), which works in the TF on critically important wetlands for the conservation of three species of flamingos (*Phoenicopterus chilensis, P. jamesi; P. andinus;* Caziani *et al.* 2001). GCFA and AGA have joined forces in this common endeavour. Potentially, Andean cats and flamingos might also help foster compliance with environmental impact legislation from the mining industry, particularly regarding the use of water, and best-practice guidelines for the burgeoning off-road tourism market.

PART III

A Conservation Perspective

CHAPTER 29

Felid futures: crossing disciplines, borders, and generations

David W. Macdonald, Andrew J. Loveridge, and Alan Rabinowitz

A puma jumping a cravass. © Andy Rouse.

'In our world' ***said Eustace*** **'a star is a huge ball of flaming gas.' 'Even in your world'** ***replied Ramandoo*** **'that is not what a star is, but only what it is made of.'**

——C.S. Lewis, *The Voyage of the Dawn Treader*

Introduction

Conservationists, in our experience, increasingly find it necessary to navigate the corridors of government as they do the backwaters of the wildernesses they seek to protect. Decision-makers are as prone to fashionable catchphrases as are those in any other sector, and two such phrases that are current in British government-speak extol the merits of 'joined-up thinking' and lament the inescapability of 'hard choices'. These two phrases capture the essence of the view of felid conservation that we present in this chapter, thereby illustrating that the languages of conservation and politics have become inseparable.

Consider, first, three things each of which is irredeemably joined up. First, the animals, together with the plants and their environment. No single individual can be understood, or conserved, other than in the context of its society, population, or metapopulation. No one species can be disentangled from the community of its predators, prey, and competitors, or from the ecosystem of which it is a working part. This joined-upness, the linkage, or conjugality of nature, demands joined-up thinking about conservation, and leads progressively from protected areas to corridors through landscapes to transboundary parks, and thus inevitably to land-use policy and politics. Second, the research disciplines and perspectives. Biology, and associated natural sciences, are necessary, but no longer sufficient, as the intellectual framework for conservation which now blends knowledge and techniques from the social sciences, economics, and thus politics. This rupture of the biological silo is embodied in the increasing migration of courses on conservation from university departments of Biology to those of Geography. This move rightly acknowledges the joined-up essentials of interdisciplinarity (but should not become an excuse for forgetting that conservation's taproots and priorities lie essentially in biodiversity). Third, human enterprises. A consequence of biodiversity conservation's welcome emergence on the world stage, in the late twentieth century, is that it must now make its case alongside those of other human enterprises, most notably development and economic growth (Sachs *et al.* 2009). This twenty-first century juxtaposition, and the daily advance of globalization, makes almost every strategic conservation decision a political one. Comparing the government documents describing the purposes of England's first statutory body for conservation in 1947 (the Nature Conservancy) to its current successor, launched in 2005 (Natural England), neatly illustrates the progressively central nature of the human dimension in biodiversity conservation. In 1947, five functions were ascribed to protected areas, of which the last and most briefly mentioned was 'amenity' for people. In 2005, the strapline for Natural England was 'for people, places and nature'—people conspicuously listed first of these beneficiaries of nature.

Macdonald *et al.* (2007) elaborate upon the entwining of these three, already complicated, dimensions of biodiversity conservation—the species, the disciplines, and the human enterprises—arguing for 'alignment' between them—alignment being a concept used in the business community to describe the process of harnessing the disparate individual and institutional drivers within an enterprise to a common purpose. Macdonald and Sillero-Zubiri (2004c) develop related ideas in the context of conserving wild canids—with which there are parallels to the wild felids that are the topic of this chapter. When, below, we turn to joined-up thinking about felid conservation, it is thus in a context where wildlife conservation is a dynamic, highly technical, and rapidly changing field. It embraces problems spanning the challenges faced by particular species—some charismatic, others obscure, some imperilled, others pestilential—to the grand global linkages of the twenty-first century between biodiversity, livelihoods, food security, health, and climate change.

Whether the focus is on species or ecosystems, on protection or sustainable use, on wildlife or on people, there is a need to join up the science that informs the policies that deliver the solutions. Since we three authors are old enough to remember when conservation was largely the domain of people wearing muddy boots, many of whom followed their calling as part of an attempt to avoid such humdrum daily matters as politics, there is some irony that today's conservationist functions in a deeply political arena, and one that becomes more so with each irreversible bound towards globalization (Sachs *et al.* 2009).

Before leaving this introduction to joined-upness, let us briefly note that the obvious response to what we have said so far is 'of course!' The reason that there is so much interconnection within, and between, the three spheres—nature, disciplines, and human enterprises—is that the world is in reality full of connectedness. Thinkers have struggled to manage this messiness by cleaving the ball of string into manageable chunks, but the divisions never really existed, and with the new globalized perspective the challenge is now to fit the connections together again. Calling for joined-upness is not so much a new idea, as remembering where we started.

Turning to hard choices, it should be obvious that in a world of finite resources and burgeoning human consumption, the joined-upness of nature, disciplines, and human enterprises makes hard all but the most technical micro-decisions. Drawing our examples from wild felids, we will illustrate how joined-up thinking confronts hard choices. There are several reasons why felids make particularly good paradigms for this discussion. They are top predators, which range widely and place heavy demands on resources. Although in matters of human–wildlife conflict (HWC) perceptions often run far adrift of reality, the sometimes awkward truth is that conflict between felids and people can be genuine and indeed grievous—there is a real problem to be faced. Finally, felids are among the most enigmatic, beautiful, and fascinating creatures, and among the most powerful human emblems. That this makes them paradigmatic, emblematic, and charismatic is partly why their conservation presents challenges that are both practically and intellectually difficult. If we cannot save felids, what can we save, and if we can save felids, we have saved much else besides. In this chapter, however, we begin with two discouraging examples intended to shock the reader into realizing in just how dramatic a fashion even the most charismatic felids are already slipping through our fingers.

Lions, tigers, and natural benchmarks

Two non-exclusive goals for conservationists could be preventing extinction and conserving naturalness (Czech 2004; Mace *et al.* 2007). Extinction is more easily diagnosed than naturalness, and is resilient to the Shifting Baseline Syndrome (Pauly 1995), the phenomenon whereby each generation resets its perception of the benchmark (e.g. the biomass of fish stocks that is considered 'sufficient') in the face of continuing erosion of environmental capital. Furthermore, humans universally prioritize the preservation of rare things (Kontoleon and Swanson 2003; Courchamp *et al.* 2006) so it is tactically advantageous that many arguments have been based on risks of extinction. Nonetheless, to provide a less transitory baseline, while being alert to the reality that humans are themselves part of nature, Yamaguchi *et al.* (in preparation, explained in Macdonald *et al.* 2007) explore a concept of naturalness defined as the point immediately before anthropogenic influence is detectable (either empirically or through deduction) for the first time in the history of the species (this modifying an earlier definition by Angermeier 2000).

What might constitute a natural benchmark against which to judge the contemporary status of large emblematic felids? We will illustrate possible answers for tigers (*Panthera tigris*) in China and lions (*P. leo*) everywhere, taking, respectively, perspectives over 2000 and 10,000 years. We provide these two examples while fully aware that the attempt to understand naturalness must not be confused with a delusion that it can be restored—this will often be impossible considering the current extent of anthropogenic impacts on the earth (although we do echo Manning *et al.* 2006 in thinking that conservationists may use 'stretch goals and backcasting' tactics to guide ambitious restoration projects). Rather, we provide a sobering perspective on what has gone, a backcloth against which to consider the urgency of protecting what remains, and a stimulus to restore where it is practicable to do so.

Only a century ago as many as 100,000 tigers may have roamed the forests of Asia, but today fewer than 5000 may survive in the wild, mostly in small isolated fragments that encompass only 7% of their historic range (Dinerstein *et al.* 2007; Seidensticker *et al.*, Chapter 12, this volume.). A compendious search of archaeological remains enabled a team of Chinese researchers to compile a map of the historic

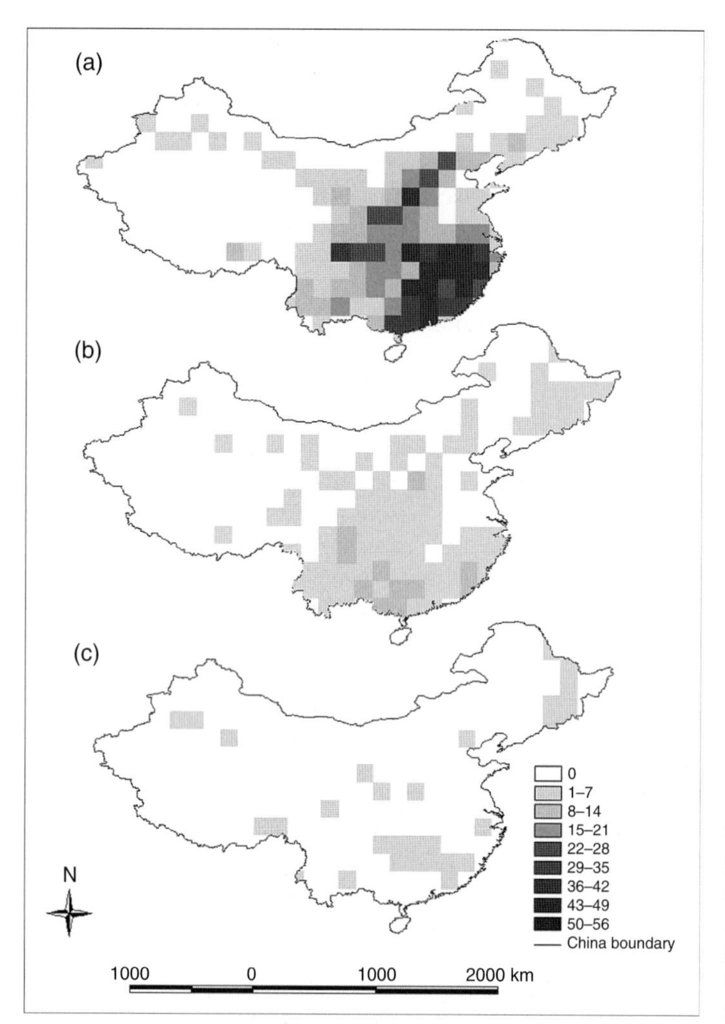

Figure 29.1 Historic records of tiger presence in China depicted a record of densities in each of three periods: (a) 1640–1911 in Qing Dynasty; (b) 1911–49; and (c) 1949–90 in the People's Republic of China. (Endi Zhang, Aili Kang, and Eve Li, personal communication.)

distribution of tigers in China (Fig. 29.1). The currently restricted whereabouts of the 37–50 (IUCN 2008) or so tigers surviving in contemporary China provides a poignant contrast between that which has already been lost and that which remains to the saved. This is a range contraction in China of about 95%.

Taking an even longer perspective, Yamaguchi *et al.* (2004a, in preparation) suggest that the lion's well-documented fossil (Kurtén and Anderson 1980) and bio-molecular history (Barnett *et al.* 2006a) make it a good candidate for provoking debate on naturalness. Classified only as IUCN Vulnerable (Bauer *et al.* 2008) the lion's accessible prominence in tourist areas may have created illusory complacency about its future. However, in 2002, a shocking estimate put the wild lion population of Africa at *c.* 18,000–47,000 (Bauer and van der Merwe 2002; Chardonnet 2002). Lions have been extirpated from around 85% of their historical range in West and Central Africa and 30% of their range in East and southern Africa (Fig. 29.2). Alarmingly, these figures use as their benchmark the lion range obtained 150 years ago, a time which already follows radical–historical contraction (IUCN 2006b). Much of this loss has been shockingly recent (see Fig. 29.3); e.g. in 1894 Frederick Selous wrote 'where but half a century ago every species of wild game native to that part of the world, from the ponderous elephant to the graceful springbuck, was to be met with in such surprising numbers..., one may now travel for days without seeing a single wild animal. In British Bechuanaland the elephant and the rhinoceros are as extinct as the mammoth in England, . . . and in many districts where fifty years ago the magnificent music of the lion's roar was the traveller's constant lullaby, no sound now disturbs the silence of the night.' (Selous 1894). To put this in an even deeper context, fossils reveal that following its exodus from Africa *c.* 500 thousand years ago (Kyr), Pleistocene lions colonized most parts of Eurasia, crossing the Beringian subcontinent to Alaska not much later than 300 Kyr, and finally reaching the southern parts of North America *c.* 130–120 Kyr. However, towards the end of the Pleistocene, the lion's enormous geographical range began what was to become a massive contraction. They disappeared from the northern Holarctic by *c.*10 Kyr, and from most of the North African–south-western Eurasian region by the end of the nineteenth century (Yamaguchi *et al.* 2004a). Thus, the range of the lion shrank by an estimated 96–98% to around 1.5–3.0 million km^2 (Fig. 29.4a and b). Depending on whether the range from which lions disappeared had an average carrying capacity comparable to that of contemporary lions in the Kalahari or the Serengeti, an estimated *c.* 1.42 or 4.8 million lions would have lived in the area from which they were lost. No earlier trend anticipates or explains the sudden contraction of their range and population, which coincides ominously with increased anthropogenic activities (Yamaguchi *et al.* 2004a). Barnett *et al.* (2006b) analysed ancient mitochondrial DNA (mtDNA) to suggest that, during the late Pleistocene, the lion consisted of three main phylogenetic groups: the modern lion distributed in Africa and south-western Eurasia, the cave lion (customarily given a Latin name *P. l. spelaea*) from Europe across northern Eurasia to Alaska, and the American lion (customarily called *P. atrox*) in the southern part of North America (see Fig. 29.3). Thus, assuming that

Figure 29.2 The end of the trail: following a dramatic range contraction 10 Kyr ago, possibly associated with the spread of humans, the modern lion persisted throughout Africa and South-west Asia until the nineteenth century, but by the mid-twentieth century, all but one lion population outside sub-Saharan Africa had been wiped out, with a sole population surviving in India. Lion trails in the snow of the Atlas Mountains or, as in this case, desert valleys of North Africa, came to an end shockingly recently. This image, from a postcard bought in Paris by Mr John Edwards, was reputed to have been taken in 1940 from an aeroplane flying from Casablanca (Morocco) on its way to attack Dakar (Senegal), then in the hands of the Vichy government. The last lions disappeared from the then Palestine during the eleventh- and twelfth-century Crusades, from Turkey in 1870, Iraq 1918, and Iran 1942 (Nowell and Jackson 1996). In North Africa, the last recorded lion was killed in Tunisia in 1891 near Babouch in the north-west, and in Algeria in 1893 near Batna, *c.* 100 km south of Constantine (although there are reports that the last Algerian lion may have been shot in 1943 and Karradine Kesting [personal communication] reports a lion seen there in 1956). The last lion in Morocco was apparently recorded in 1942 on the northern side of the Tizi-n-Tichka pass in the Atlas Mountains, near the road between Marrakech and Ouarzazat. (From Yamaguchi and Haddane 2002.)

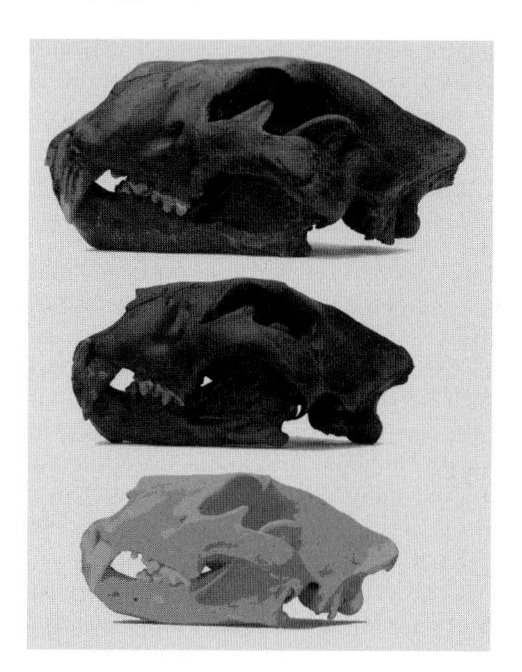

Figure 29.3 Skulls of the lion guild of 10 Kyr, revealing the miniature dimensions of the modern form: *top*: American lion (*atrox* group) from California; *middle*: the so-called European cave lion (*spelaea* group), although this specimen is from Alaska; and *bottom*: the modern African lion. (See Barnett *et al.* 2009.) © N. Yamaguchi.

the cave lion and the American lion possessed intra-taxon mtDNA diversities comparable to those in the modern lion, the lion has lost more than 60% of its mtDNA diversity.

However, size is flexible on an evolutionary time-scale, and may be particularly affected by competition within the predatory guild as well as by the relative sizes of predator and main prey. For example, the European lion (*P. l. fossilis*) seems to have been at its biggest when it first appeared in Europe during the mid-Pleistocene, and became smaller, and known as the cave lion, in the Late Pleistocene. Similarly, before modern lions arose in Africa 74–203 Kyr ago, very big lions inhabited some parts of Africa during the Early–mid-Pleistocene, including the Elandsfontein (South Africa) lion, which may be as big as the Middle Pleistocene lion (Yamaguchi *et al.* 2004a). In so far as Late-Pleistocene lions are sometimes classified as *spelaea* if larger and *leo* if smaller, it is hard to know whether contemporaneous cave and modern lions differed in size.

A provocative speculation, emphasizing the insight offered by a natural benchmark, is therefore that the earliest detectable human influence on the lion dates to before *c.* 10 Kyr. So, perhaps, had mankind not intervened, the current 'natural state' of the lion might have been that which prevailed just before the 10 Kyr mark. This perspective suggests that human impacts have caused lion populations to decline to an extent that mirrors the well-publicized anthropogenic devastation of the North Sea fish stock, for which it is estimated that *c.* 97% of the 'natural' biomass of large fish weighing 4–16 kg has been lost to fisheries (Jennings and Blanchard 2004).

Most natural benchmarks can probably never be regained (that is not the point, however), but they can help guard conservation against a succession of dangerously migratory goals. At least each movement of the goal should be recorded, and reread each time a further shift is considered. Furthermore, for those who argue that current human population, and the projected human footprint by 2050, is not a fundamental problem for a sustainable environment, it may be helpful to take stock of what has already been destroyed.

Joining up the cats (through science)

Wild felids—especially the big ones—are the miners' canaries of terrestrial biodiversity, and barometers of the success, or failure, of conservation. They need to be studied in joined-up ways, not only as part of complicated predator–prey communities, but within a framework of interspecific rivalries among them, and between them and other predators. This is similarly true for canids (Macdonald and Sillero-Zubiri 2004c); Oliveira *et al.* (Chapter 27, this volume) and Durant *et al.* (Chapter 16, this volume), for example, powerfully illustrate this. Among the biggest academic questions for which felids provide a potent model are the roles of community assembly, intra-guild competition, and meso-predator phenomena in felid communities. That the populations of which these communities consist are increasingly fragmented, and their metapopulations disrupted, have direct bearing on their conservation. Michalski and Peres' finding (2005) that all four felid species (of which the

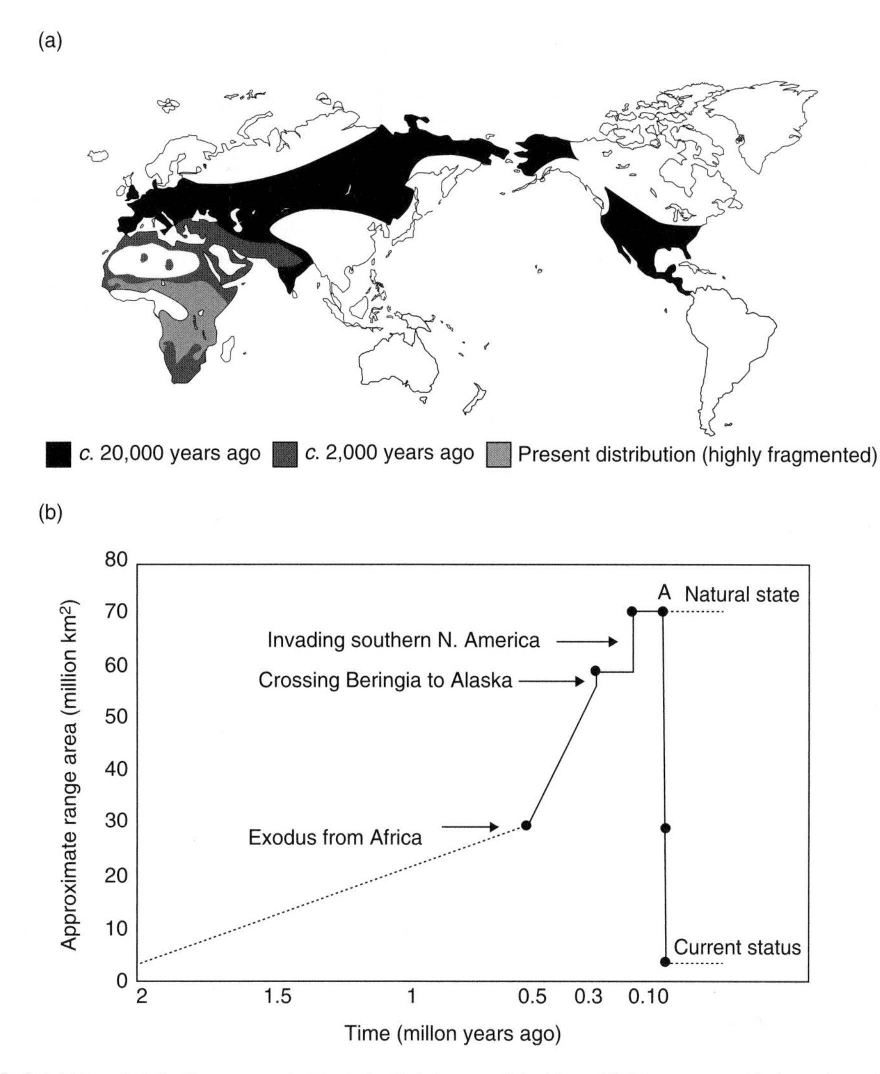

Figure 29.4 (a) Lion distribution ranges in the Late Pleistocene (black), *c.* 2000 years ago (dark grey), and present (light grey). (b) The constant range expansion of the lion during the Late Pleistocene, followed by the sudden contraction at the end of the Pleistocene throughout the Holocene, starting around the point A. (From Yamaguchi *et al.*, in preparation.)

smallest was the jaguarondi, *Herpailurus yaguarondi*) were found only in the larger fragments of a Brazilian forest landscape (more adaptable canids and mustelids were less sensitive to patch size) illustrates felids' potential to serve as umbrella species (i.e. 'miners' canaries'). In many places, the top priority is to preserve healthy source populations of felids, and if they can be secured, the next step is to create genetic corridors to link populations on a continental scale. Several extraordinary initiatives are seeking to create such linkages—here, as examples, we mention the two initiatives already underway, involving the jaguar, *P. onca*; the tiger; a third one just starting on the southern African lions (see also Loveridge *et al.*, Chapter 11, this volume); and finally, the aspirational case of the Balkan lynx, *Lynx lynx balcanicus*.

Genetic corridors

Corridors are envisioned to act as a surrogate for an intact landscape by allowing the species in question to exchange, naturally, genetic material among now-isolated population fragments. Such constructions should slow inbreeding of isolated subpopulations and prevent their differentiation while maintaining a full level of genetic variability characteristic of each subspecies. In some cases, a more proactive approach may be required. For instance, if corridors are geographically impracticable, individuals, or their genes in the form of sperm, may be transferred from reserve to reserve to create 'virtual corridors' in order to maintain a genetically integral population. This sort of management mimics the functioning of metapopulations in nature and, using this as a metaphor, has been termed metapopulation management in the context of an attempt to conserve African wild dogs, *Lycaon pictus*, in South Africa (Davies-Mostert *et al.* 2009).

Jaguar corridor initiative

Ranging from the south-western United States through northern Argentina, the jaguar's habitat has shrunk by more than 40% since the early 1900s (Sanderson *et al.* 2002b). Such loss, compounded by continuing persecution (Cavalcanti *et al.*, Chapter 17, this volume), has caused jaguar numbers to decline in many parts of its range. Nonetheless, much jaguar habitat remains, supporting core populations embedded in a matrix of human-modified landscapes. Thus, in 2004 the Jaguar Corridor Initiative was launched with the idea of 'connecting the dots', by extending jaguar conservation and management beyond core-protected areas and into the human landscape, creating a genetic corridor linking existing jaguar populations throughout the entire species' range (Rabinowitz 2006; White 2009; see Fig. 29.5).

The primary catalyst to this initiative was the discovery that there was no detectable racial variation between jaguar populations; indeed, there appeared to have been high levels of gene flow across the species' range (Eizirik *et al.* 2001; Ruiz-Garcia *et al.* 2006). However, to maintain this gene flow obviously requires keeping open the channels for dispersal, thereby preventing the shrinkage of effective population size, increased genetic drift, and increased inbreeding—all factors adding to extinction risk of jaguar populations and, cumulatively, of the species (Frankham 2005).

Clearly, despite high rates of deforestation in Latin America, there are still pathways outside protected areas that enable jaguars to move between populations. The Jaguar Corridor Initiative—the first to identify and protect existing jaguar populations and the dispersal pathways between them at such a large geographic scale—seeks to capitalize on this situation in its development of a range-wide conservation strategy focusing on the species as a whole, not just high-priority populations.

As a first step, a geographic information system (GIS)-based Least Cost Corridor Analysis (LCCA) (e.g. Walker and Craighead 1997; Adriaensen *et al.* 2003) was performed for each country within the Mesoamerican region (with the exception of El Salvador). Informed by a consensus of expert opinion on the optimal landscape characteristics for jaguar dispersal, and the locations of known jaguar populations, the LCCA identified likely pathways of movement between populations on which travel would be the least restricted and the most secure, thereby providing a swath of cells that offers a dispersing individual the best chance of success (Singleton *et al.* 2002; Beier *et al.* 2006). The next step, validation, involves biologists exploring the corridors, ground truthing jaguar presence, and documenting prey availability, human land-use patterns, human perceptions of jaguars, and government or industry plans for development. Once a proposal is developed for the maintenance or restoration of a corridor, it is presented to government agencies and local partners for approval and incorporation into national or regional land-use planning strategies.

Since the functioning of these corridors, and the protection of the jaguars travelling them, rests heavily on the goodwill of local people, the initiative focuses on the mitigation of conflict between jaguars and ranchers and continual engagement with farmers, park managers, indigenous groups, school teachers, mayors, and non-governmental organizations (NGOs). As of 2006, the Comisión Centroamericana de Ambiente y Desarrollo (CCAD; the Central American Commission for Environment and Development) gave unanimous approval of the Jaguar Corridor Initiative at the ministerial level. The Jaguar Corridor Initiative is now an official and crucial component of the larger Mesoamerican Biological Corridor Program started in 1998 to link all the protected areas in Central America.

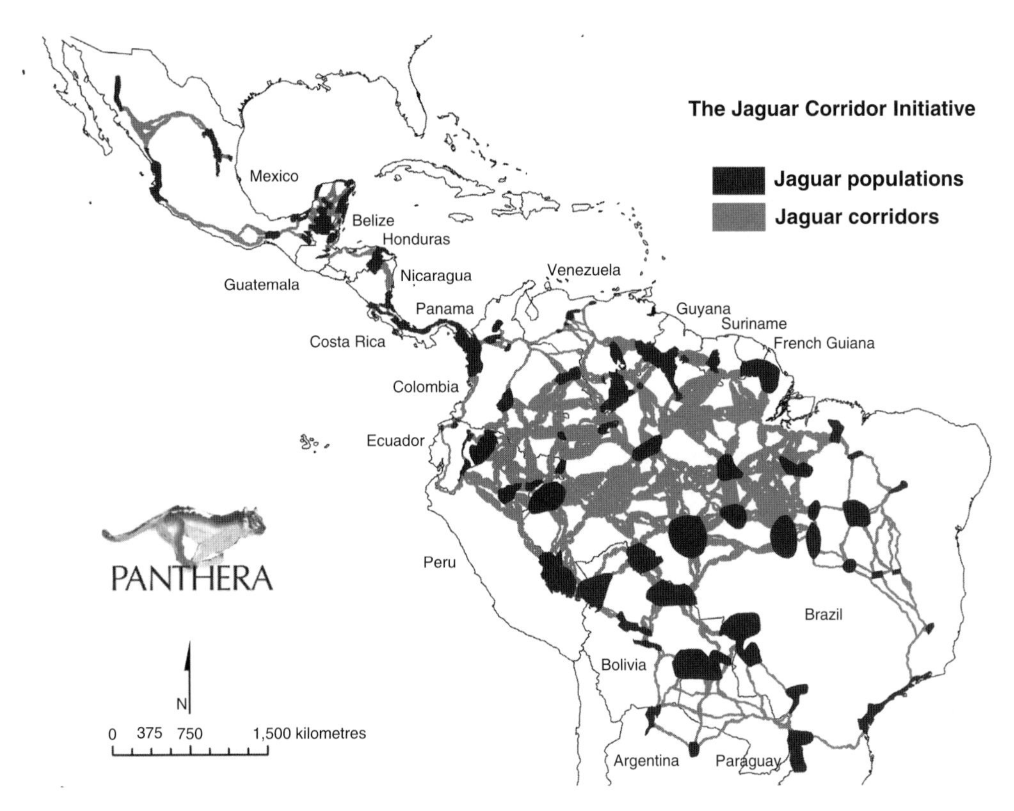

Figure 29.5 Panthera's uniquely ambitious proposal for corridors linking jaguar populations throughout their range.

The Jaguar Corridor Initiative is now extending beyond Mesoamerica, southwards through Columbia, the Guiana Shield, and Brazil. A major focus in South America has been in the Brazilian Pantanal, of which 95% is under the management of 2500 private ranches that support up to 8 million cattle, where conflict with jaguars is intense (Cavalcanti *et al.*, Chapter 17, this volume). The Panthera Pantanal Project aims to create, as a demonstration to be replicated elsewhere, an extensive protected jaguar corridor within which profitable cattle ranching is integrated with jaguar conservation. By 2009 the programme encompassed two contiguous ranches, totalling 700 km^2 and 5000 cattle, and is developing public–private partnerships to ensure genetic connectivity and tolerance of jaguars on a landscape scale. Mindful of the link between biodiversity conservation and community development, the project is developing an education initiative with local schools and a health-care programme in partnership with Mt Sinai Hospital of New York.

Tiger corridor initiative (TCI)

Believing that tigers could recover within their remaining habitat if critical threats were ameliorated, a science-based programme called Tigers Forever (TF) was initiated in 2005 by two NGOs, Panthera and the Wildlife Conservation Society (Tigers Forever 2009). Focusing first on critical tiger sites in India—Myanmar, Thailand, Cambodia, Lao PDR, Indonesia, Malaysia, Russia, and China—the target, bravely declared, is to increase tiger numbers by at least 50% at these sites in 10 years, and then, by 2017 to foster an increase in tiger numbers in the landscapes surrounding these critical sites. Conserving tiger populations requires not only ingenious and coordinated efforts to mitigate critical threats, but also vigilant surveillance and long-term population monitoring to measure success or failure. Without these, even the best tiger habitats are at risk of becoming 'empty forests', following the dismaying precedent of India's Sariska National Park (a Project Tiger Reserve created in 1978 and once heralded as a prime tiger-watching destination, which was declared devoid of tigers in 2005). The threats persist: trade in skins and other body parts continues to threaten tigers, with a resurgence in the fashion of wearing tiger furs in Tibet, and an expanding trade in tiger parts for medicinal purposes (Loveridge *et al.*, Chapter 6, this volume).

Experience with the Jaguar Corridor Initiative makes clear that saving or restoring isolated tiger populations will not guarantee the species' survival. Focusing again on range-wide protection, with functioning genetic corridors within human-dominated landscapes beyond protected areas, Panthera launched the TCI in 2008. The goal is to identify and create safe passages between protected core populations, based on regional, bi-national, and national tiger corridors. Modelled on the Jaguar Corridor Initiative, the programme is predicated first on GIS modelling, then ground truthing and data collection, before proposing potential tiger corridors to regional and national governments. In the case of tigers, however, their populations are sufficiently fragmented that this planning involves several corridors, the largest of which is the multinational Himalayan–Indo-Malayan corridor, potentially connecting tiger populations from Nepal into Bhutan and northern India through Myanmar and other South-east Asian countries. The scheme has particular cultural resonance in the Kingdom of Bhutan, where the tiger is a significant religious symbol (Guru Rinpoche is said to have brought Buddhism to Bhutan while riding on the back of a flying tiger), and which supports 115–150 tigers (McDougal and Tshering 1998; Wang and Macdonald 2009a), fully protected by the Forest and Nature Conservation Act (1995). Bhutan's small human population (<1 million), extensive natural habitats (*c.* 72% forest cover), and progressive attitudes to conservation, enhance its importance for tiger conservation; however, road building and industrialization inevitably threaten fragmentation of tiger habitat. Showing uncommon foresight, the Bhutanese government has launched pre-emptive landscape-scale conservation, designating 26% of the country as protected areas (tigers occur in at least six of these; Sangay and Wangchuk 2005) and in 1999 an additional 9% was designated as constituting biological corridors—the combination called the Bhutan Biological Conservation Complex (B2C2). Twelve biological corridors collectively cover 3640 km^2, and vary in length from 16 to 76 km (Nature Conservation Division 2004)—the corridors themselves are sufficiently extensive that breeding tigers have been found in three of them.

Other important genetic corridors include a bi-national China–Russia Tiger Corridor, connecting the Siberian tiger (*P. t. altaica*) populations of Sikhote Alin, Russia, and Heilongjiang and Jilin Provinces, China; a national tiger corridor through the island of Sumatra and another in peninsular Malaysia, and several possible Indian corridors, most importantly in the Western Ghats of Karnataka.

One consideration in corridor design is that female tigers are territorial which, in so far as corridors are long and thin, may restrict gene transfer between refugia. This could be more evident in mtDNA than in the nuclear genes, but the latter too will be restricted (this concept underlies the explanation; (see section on 'Caspian tigers') for the facultative monomorphism of the Caspian (*P. t. virgata*) and Amur tigers (*P. t. altaica*) as a result of funnelling through the natural Gansu Corridor (Driscoll *et al.* 2009).

African lion corridor

The year 2009 saw the 10th anniversary of intensive research on lions, and their prey, in Hwange National Park (HNP), Zimbabwe, by Oxford's WildCRU team (Loveridge *et al.* 2006; Loveridge *et al.* 2007c; Valeix *et al.* 2009a; Valeix *et al.* 2009b; Loveridge *et al.* 2009b). The work of this team places the lions in the context of adjoining land-uses, such as trophy hunting in adjacent safari areas and conflict with subsistence farmers, but also in a wider regional context linking, for example, with shared lessons from the same team's work in Botswana (Loveridge *et al.*, Chapter 11, this volume). Hwange, where the lion work is now being replicated on leopards (*P. pardus*), lies in a constellation of protected areas in Zimbabwe, Botswana, Zambia, Namibia, and Angola and has the potential to be one of the hubs of a network of corridors between major source

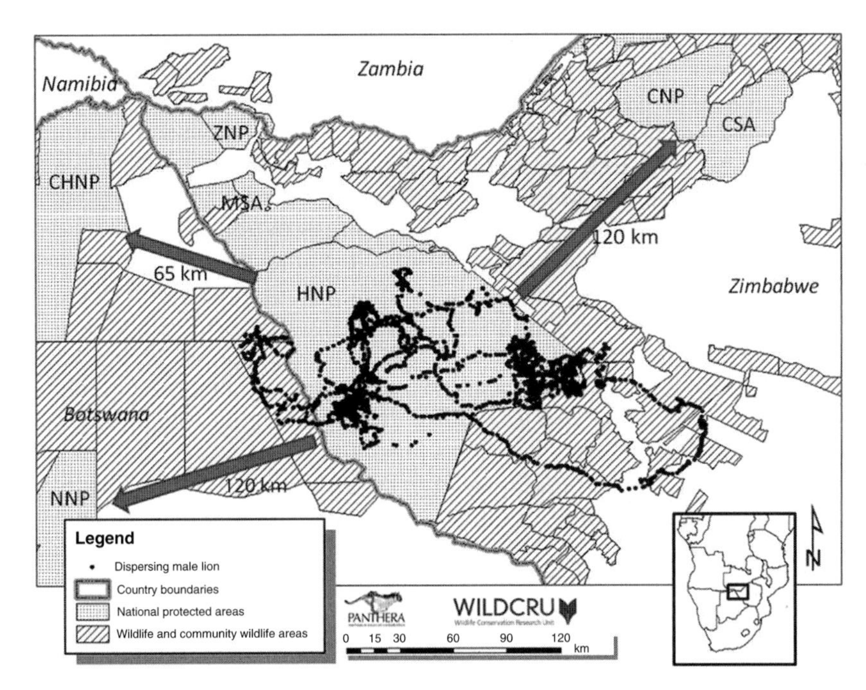

Figure 29.6 Map of corridor areas leading from HNP (indicated by arrows and distances) that potentially allow dispersal of lions between protected area complexes in the region. These are shown in the context of data collected over a period of 19 months (October 2007–February 2009) from a dispersing 4-year old male lion collared with a GPS-satellite collar. It is clear from this individual's movement patterns that significant potential exists for gene flow between protected populations in the region, provided corridors are identified and measures taken to ensure they remain intact in the face of increasing human population pressure. CHNP, Chobe National Park; NNP, Nxai Pan National Park/Makgadikgadi Pans National Park complex; HNP, Hwange National Park; MSA, Matetsi Safari Area; ZNP, Zambezi National Park; CNP, Chizarira National Park; CSA, Chirisa Safari Area. Inset map shows location in southern Africa.

populations (Fig. 29.6). This area falls within the planned Kavango-Zambezi Transfrontier Area, the governmental memorandum of understanding for which was signed in December 2006. This transfrontier conservation area will amalgamate conservation areas in the five countries. Although no formal management agreements are in place, potential corridors already exist between HNP and nearby protected areas that provide refugia for lion subpopulations. The first of these links HNP to Chizarira National Park and Chirisa Safari Area, via the Gwaai Valley Conservancy, Mzola Forest, and existing Community Wildlife Areas in the Hwange and Binga districts. The second links HNP with Chobe National Park via existing protected areas to the north-east and the third may link HNP with Nxai Pan and Makgadikgadi Pans National Parks. All these protected areas are well within the potential dispersal distances of lions (Fig. 29.6), with the furthest being no more than 120 km away. Nevertheless, these corridors need to be ground truthed and feasibility assessments completed before planning can be put in place to ensure that habitat remains intact and that livelihoods are protected in corridor areas inhabited by people. The key questions lie beyond biology—can human–lion conflict be mitigated, can incentives and mechanisms be found to ensure community participation, and can the necessary integration be achieved between such constituencies as local communities, rural authorities, private landowners, regional conservation agencies, and international donor agencies? If this project succeeds then, at a broader scale, it can sit at the hub of the spider's web of linkages planned throughout Africa within Panthera's Project Leonardo.

This lion corridor raises a question of vocabulary. The word 'corridor' conjures up an image of a ribbon of natural habitat, functionally analogous to, but scaled up from, hedgerows ramifying between woodlands in British farmland. However, while this vision may sometimes be apt—a strip of usable habitat, ideally several home range diameters in width, connecting the subpopulations of an endangered felid—a more general image may be expressed in terms of the topography of a landscape of risk. For lions venturing from the protected hub at Hwange, the web of possible routes to other protected areas leads through variously risky landscapes. In addition to ensuring that strands of the web traverse habitats through which lions are inclined to travel, the task is to ensure that the risks of their being killed en route are reduced in enough places that the odds of completing a safe journey allow sufficient lions to make it. In this sense, a corridor may be considered not so much as a ribbon of habitat but rather, probabilistically, as valleys through the peaks of risk that together offer a fair hope of safe passage. Aligning felid behaviour and the geometry of landscapes of risk is a scholarly challenge to those thinking about the future of conservation corridors.

The process of joining up fragmented felid habitat will almost always involve restoring damaged environments. Indeed, while much of the story of twentieth-century conservation was about preventing further deterioration, we foresee a shift now towards restoration.

Balkan lynx

Following the last Ice Age, European lynx (*L. lynx*) probably spread north from a southern refugium, perhaps on the Italian peninsula or in eastern France, through the Alps and across land bridges to Scandinavia. Such a journey is impossible in modern, environmentally fragmented Europe; so conservationists have transported lynx from northern Europe (phylogenetically indistinguishable from the extinct Alpine form [Gugolz *et al.* 2008; Breitenmoser *et al.*, Chapter 23, this volume]) to repopulate the Alps artificially. The Caucasian and Balkan populations are remnants of separate Ice Age refugium. Until the nineteenth century, lynx in the Balkans were distributed continuously from Slovenia in the north-west across the entire Dinaric range, Albanian mountains, south to the southern Pindos Mountains in Greece, and east to the Bulgarian–Turkish border (Miric 1978). However, they were eradicated from the north southwards until, today, <100 survive in the mountains of south-eastern Balkans, mainly in the border areas between Macedonia and Albania. The next closest populations are in the Carpathians to the north-east and the Dinarics to the north-west (these latter being descendants of individuals reintroduced from the Carpathians in the 1970s). Currently, there is no evidence that the Balkan remnants are connected with the Carpathian and Dinaric populations (the latter reaching south to the Neretva river valley in Bosnia and Herzegovina, some hundreds of kilometres from the closest Balkan lynx in the Albanian Alps; von Arx *et al.* 2004). However, since the mountains spanning the Dinaric–Pindos range are continuous between them, the two subspecies (*L. l. carpathicus* and *L. l balcanicus*) would be likely to meet, were their numbers to increase (Fig. 29.7). This prospect raises the bold but exhilarating thought of creating a network of corridors between their scattered mountain refugia. Preliminary scoping suggests that routes, which would have minimal economic impact, already exist for these corridors. Although data are few on lynx presence in the south-eastern mountains of Bosnia and Herzegovina and the adjoining parts of their range in Montenegro, it seems likely that the Neretva river valley has been a barrier to the further expansion of the Dinaric lynx population. Lynx signs are occasionally reported in the adjacent mountains in western Serbia and along the border between Montenegro and Bosnia and Herzegovina; but whether these lynx are the advance guards of an expanding Dinaric population or are remnants of the indigenous Balkan, is unknown. Nonetheless, for the future, the topology of the Dinaric Mountains could foster the management of a corridor to join these populations via a dispersal corridor along continuous mountain ridges and slopes, with no major human or natural barriers (A. Trajce, personal communication).

Restoration

Conservation and the scholarship that guides it, moves in phases; and recent decades have seen a

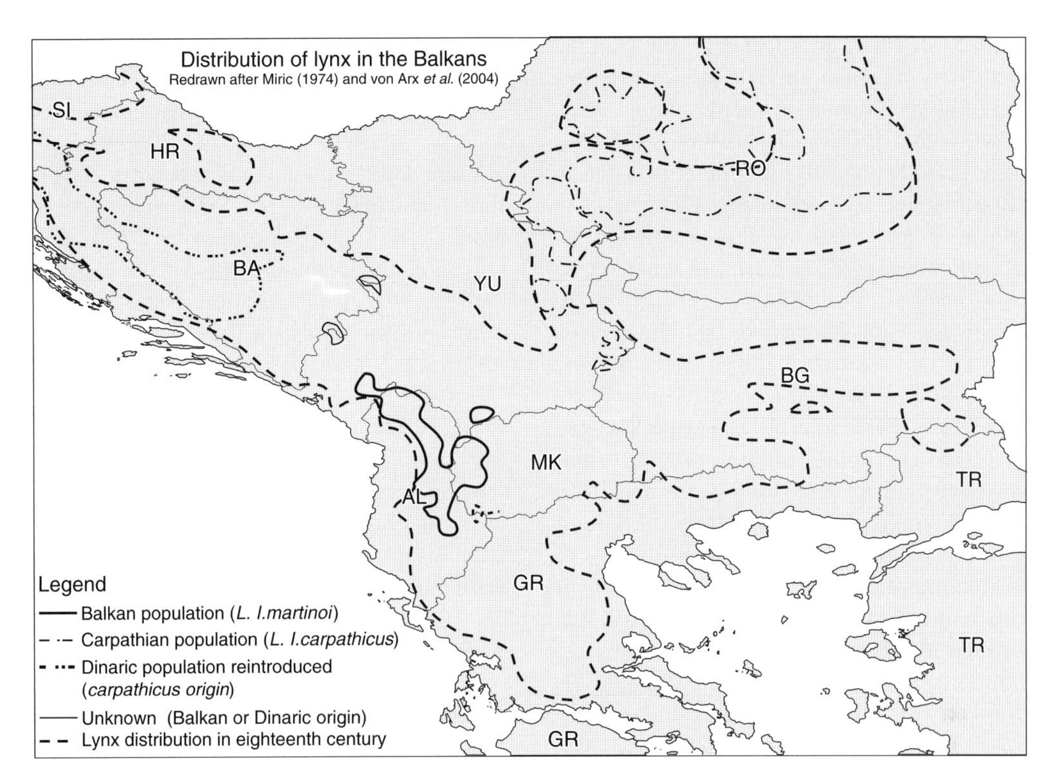

Figure 29.7 Comparison of the distribution of the lynx in the Balkans in the eighteenth century and in present day. (A. Trajec, personal communication.)

burgeoning of behavioural, ecological, and genetical knowledge, and excitement at the prospects of community-based initiatives. The greatest focus has been on halting the loss of biodiversity, formalized in Millennium Development Goal 7 (incorporating the Malahide declaration and added to the Convention on Biodiversity in 2002), that aims 'to achieve by 2010 a significant reduction of the current rate of biodiversity loss at the global, regional and national levels as a contribution to poverty alleviation and to the benefit of all life on Earth' (Sachs *et al.* 2009). For the coming decades, we foresee, and advocate, an increasing emphasis on restoration. As Macdonald (2009) emphasizes, the challenge is not so much to recreate the past—an unlikely goal for a world with 6.7 billion people and rising—but to design the future through the creation of new habitats fit for purpose in the twenty-first century. Arguably, felids' role as umbrella species, and their charisma, makes them particularly suitable flagships for, and barometers of, the restoration movement. To be realistic, because large felids can be dangerous, while easily advocated from the safety of an armchair, restoring felids is an ambitious goal; and one that carries very serious responsibilities.

One important tool in restoration is reintroduction, an approach with a history of good intentions and poor outcomes (the latter mainly due to inadequate adherence to the sensible IUCN guidelines that have been in place since 1998; IUCN 1998). Three key issues are relevant, of which the first is the complexity of genetic matching between the former extinct populations and the surviving populations from which candidates for reintroduction may be selected (we illustrate this below with the examples of the Barbary lion, *P. l. leo* and Caspian tiger). Second, Johnsingh and Madhusudan (2009) argue that despite the perilous status of the tiger in India, attempting to reintroduce them now would be a nugatory exercise because the threats that endangered them in the first place are undiminished—reminiscent of filling a leaky bucket (but see Caspian tiger example below). Third, instances of poorly planned reintroductions, ranging from Eurasian lynx to African lions (Breitenmoser *et al.* 2001; Linnell *et al.* 2009; Slotow and Hunter, 2009) offer lessons for the future. These include the importance of economic instruments such as the conferral of user rights to landowners, thereby increasing the economic value of wildlife (Lindsey *et al.* 2009), or, as for cheetahs (*Acinonyx jubatus*), developing a compensation programme for damages caused by reintroduced individuals (Marnewick *et al.* 2009). Hayward and Somers (2009) have devoted a book to the reintroduction of large carnivores, which confirms that success depends on considerations ranging from genetics (e.g. Frankham 2009) to economics (Lindsey *et al.*, 2009) and social factors (e.g. Al-Kharousi 2006), and considering especially the capacity of human communities to cope with, and benefit from, the reintroduction (Reading and Clarke 1996; Macdonald *et al.* 2002).

The modernity of some notable extinctions is shocking—the Bali tiger (*P. t. balica*) and Barbary lion driven extinct in 1939 and 1942, respectively. There is a chilling sense of shared responsibility in the awful realization that two of us were already committed naturalists when, in 1968, the last Caspian tiger died, and were already professional biologists at the demise of the Javan tiger (*P. t. sondaica*) in 1979. What prospects are there of righting these wrongs? On the principle of last-out-first-in, we consider the prospects of restoring two of these recent losses. Then, taking a British perspective and mindful of the unbecoming tendency for conservationists from the developed world to propose reintroductions comfortably far from their own backyard, we discuss the reintroduction of lynx to Scotland.

Barbary lion

The recent historical proximity of lions to Europe is caught in the delicious observation in 1864 by Ormsby (1864) that 'a man who has dined on Monday in London can, if he likes, by making the best use of express trains and quick steamers, put himself in a position to be dined on by a lion in Africa on the following Friday evening'. The lion Ormsby had in mind was the North African Barbary lion or Atlas lion, renowned for the males' dark and copious manes. Although virtually wiped out by the early decades of the twentieth century (the last one was killed in 1942 in Morocco), the Berber people of the Atlas Mountains had previously made frequent gifts of these lions to the Sultans and Kings of Morocco. Yamaguchi and Haddane (2002) report how the appearance of surviving specimens in the King's collection so closely resembled that of the Barbary lions that the Moroccan authorities considered them as candidates for reintroduction of the animals into the Atlas Mountains (Fig. 29.8).

However, is the Barbary lion sufficiently distinct to be a priority for conservation? Recent molecular work revealed that Barbary lions were phylogenetically distinct (Barnett *et al.* 2006a). Unfortunately, the same study suggests that the mtDNA of the captives was not a match to them, although it is distinguishable from that of eastern–southern African lions (Barnett *et al.* 2006a). Indeed, the Barbary lion is phylogenetically closer to the Asiatic lion (*P. l. persica*) than to sub-Saharan African lions (Barnett *et al.* 2006b), and the Asiatic lion may be even considered a close sister population to the Barbary lion, at least based on mtDNA-based phylogeographic analysis. Assuming no pure Barbary lions survive in captivity, then the next best option as a step towards restoring a functioning predator community in the Atlas Mountains might be to introduce the Asiatic lion to North Africa. The Asiatic lion develops an enormous mane in cooler climates (Barnett *et al.* 2007) and to judge by the observations in Algeria of a nineteenth-century French hunter, North African lions lived in very cold places—Gérard, following lion track in the snow wrote 'notwithstanding this uncomfortable carpet, however, and this temperature so little congenial to those of her species [if we may believe our scientific naturalists], a lioness had settled in the country' (Gérard 1856).

Let us indulge in a dream. Between the Middle and High Atlas lies a rocky mountainous area where green oaks dominate a landscape in which the endangered Barbary leopard, (*P. p. panthera*), may still survive. The area supports minimal human activity, but is home to Barbary macaques (*Macaca sylvanus*), Barbary sheep (*Ammotragus lervia*), and wild boar (*Sus scrofa*). To restore the original Moroccan fauna, the Cuvier's gazelles (*Gazella cuvieri*) might be reintroduced, as might be the Barbary red deer (*Cervus*

(a)

(b)

Figure 29.8 Postcards dated 1907 picture Barbary Lions as they slip towards extinction, in (a) the Sultan of Morocco's menagerie at Fez; and (b) in a Parisian menagerie, where the tawdriness of their confinement is made more lurid by the painted backdrop of an expansive oasis scene. © J. Edward's collection.

elaphus barbarus) which has already been reintroduced elsewhere in Morocco from Tunisia. There has been talk that a fenced enclave, perhaps as large as 100 km^2, could hold a managed group of lions (loosely similar to the fenced reserves for African wild dogs that Davies-Mostert *et al.* [2009] describe). Eventually, perhaps scientific planning and changing attitudes would allow the fence to be breached. This wondrous dream is probably, for the time being, just that! Nonetheless it provokes a very modern, pragmatic question about the role of fenced areas in carnivore conservation. Sandom *et al.* (in press) point out that while fenced areas, intended to keep carnivores and people apart, are often spoken about with enthusiastic waving of the arms, the pithy questions thereby engendered: that is, what level of biological 'naturalness' does an enclosure of a given size provide; and what level of management intervention does such an enclosure necessitate?—are not commonly addressed; and the philosophically tricky question of the value attributed to areas along the continuum from cage-to-enclosure-to-park-to wilderness is not, either. In the cost–benefit analysis of such issues, it is almost invariably hoped that the quasi-natural circumstances of the fenced, but still charismatic, carnivores will generate revenue and employment to sustain the enterprise. Indeed, the aforementioned Atlas restoration area could be accessible by main road from tourist destinations such as Fès and Marrakech, and would revitalize Ormsby's easy link between London and lions. Against such optimism, discussions about restoring these lions rested on the implementation of a phased plan—identification of surviving Barbary lions (or at least their genes), their selective breeding, the ecological assessment of the reintroduction site, its protection and restoration, releasing the lions, and monitoring the outcomes while promoting ecotourism. Science got off to a promising start, largely confirming the Barbary lion as unique among modern lions, phylogenetically (and probably ecologically; Yamaguchi and Haddane 2002; Barnett *et al.* 2006a). However, sadly, scientists have yet to clear the second hurdle as it appears that the Moroccan king's lions do not appear to be 'pure' Barbary lion on their mother's side, and no alternative source has yet been uncovered.

Caspian tigers

The Caspian tiger (Fig 29.9) recently inhabited Central Asian forest and riverine systems from Xinjiang to Anatolia in a range seemingly disjunct from that of other tigers. The Caspian tiger's obscure biogeographical origin and uncertain phylogenetic placement in the tiger family tree has puzzled naturalists for over a century. In 2006, a molecular investigation of tigers suggested the surprising existence of a new subspecies in peninsular Malaysia, *P. t. jacksoni*, prompting questions as to the taxonomy of the extinct Caspian tiger. To discover what were, in genetic terms, its nearest neighbours, Driscoll *et al.* (2009) generated a composite 1257 bp mtDNA haplotype from museum specimens of 20 wild-caught Caspian tigers, sampled across their historical range (Fig. 29.10). This revealed that Caspian tigers carry a major mtDNA haplotype that differs by a single residue from the monomorphic haplotype found in contemporary Amur (Siberian) tigers, which inhabit the Russian Far East. A somewhat arcane consequence of this finding is that in accordance with international rules on taxonomic nomenclature, Caspian and Amur tigers (*P. t. virgata*, Illiger, 1815 and *P. t. altaica*, Temminck, 1844, respectively) should be synonymized under *P. t. virgata*, Illiger, 1815. Thus, Caspian tigers are not extinct, they just live in the Russian Far East; according to these taxonomic niceties it is Amur tigers that never existed! Returning to reality, a phylogeographical analysis of *P. t. virgata* in the context of the five living tiger subspecies provides evidence that Caspian tigers colonized Central Asia less than 10,000 years ago via a westerly migration though the narrow Gansu Corridor (Silk Road), followed by a more recent easterly return across northern Asia to establish the Amur tiger population living today in the Russian Far East. Therefore, the severe genetic depletion characteristic of contemporary Amur tigers likely reflects these founder migrations, pre-dating any anthropogenic influence.

That Caspian and Amur tigers are very closely related was perhaps not a surprise to some. That the Amur population is derived from the Caspian may have been more of a revelation; but the finding that the historic Caspian population was genetically depauperate was unexpected. The conservation import of this last finding is that the mitochondrial monomorphism seen in today's Amur tigers is a natural feature of the population (being derived from the Caspian) and not the result of human actions. Of course, this does not absolve humans from the tragic loss of population and habitat that Amur tigers have suffered; it simply means that modern conservationists need not obsess about the comparative lack of genetic variation in Amur tigers, since this has been the case for at least 10 Kyr.

Were any tigers from the former Caspian range currently extant then, in the jargon of tiger conservation, they would probably be designated a Management Unit (MU) separate from, but conspecific with, the Amur population, with which they would form an Evolutionarily Significant Unit (ESU). Such abstract considerations aside, it is clear that the deeper level of divergence shown between *virgata/altaica* and the other designated subspecies requires *virgata/altaica* to be distinctly maintained as an ESU in captivity and in the wild.

The most wide-reaching ramification of Driscoll *et al.*'s findings (2009) is that the entire tiger range of Central Asia is open to reintroductions from existing Amur stock. Because tiger conservation has adopted a 'landscape approach' (WWF 2002), which seeks to preserve all trophic levels naturally present in the environment, repatriation of appropriate tigers to their historic natural range could provide a conservation framework for vast areas of natural habitat. Having mentioned this cat, we will provocatively let it go among the pigeons by musing on where Caspian tigers might be repatriated. Conversations with Ali Aghili and Carlos Driscoll focus on Iran and, for example, the Golestan National Park, where the last tiger was shot in 1959; further west there is potential habitat in the Atrak valleys, Gorgon peninsula, and Miankaleh protected area. There is also prime habitat in Turkey, and one could even imagine corridors linking Azerbaijan, Armenia, and Iran. Tigers were hunted to extinction in the former Soviet Tigrovaya

(a)

(b)

Figure 29.9 Caspian tiger (a) in Berlin Zoo c. 1900; and (b) killed in northern Iran, 1940s.

('tiger park') in Tadjikistan, so perhaps sufficient prey remain there to support them; while in Kazakhstan the swamps formed of the Illi river as it fans into Lake Balkhash could be tiger parkland.

The possible benefits of this vision of the future are varied. They encompass providing an outlet for the 'surplus' tigers in Chinese tiger farms to restoring ecological balance and biodiversity in Central Asian forests. This would involve filling a vacant top trophic-level niche and endowing parks with flagship, revenue-generating tigers. Moreover, because small populations are subject to stochastic extinction when restricted to a single site, translocated populations could provide a geographically removed 'second homeland' as insurance against catastrophe from epidemics, local declines in prey, or even a local socio-political change in wildlife protection. In this way tiger populations established in, for example, Turkey, Iran, or the Urals could genetically insulate the parent Russian population from extinction.

Anyway, these molecular revelations may offer a new interpretation of the journeys of the tigers depicted on numerous petroglyphs and artefacts found in the Yinshan, Helanshan, and Mazongshan

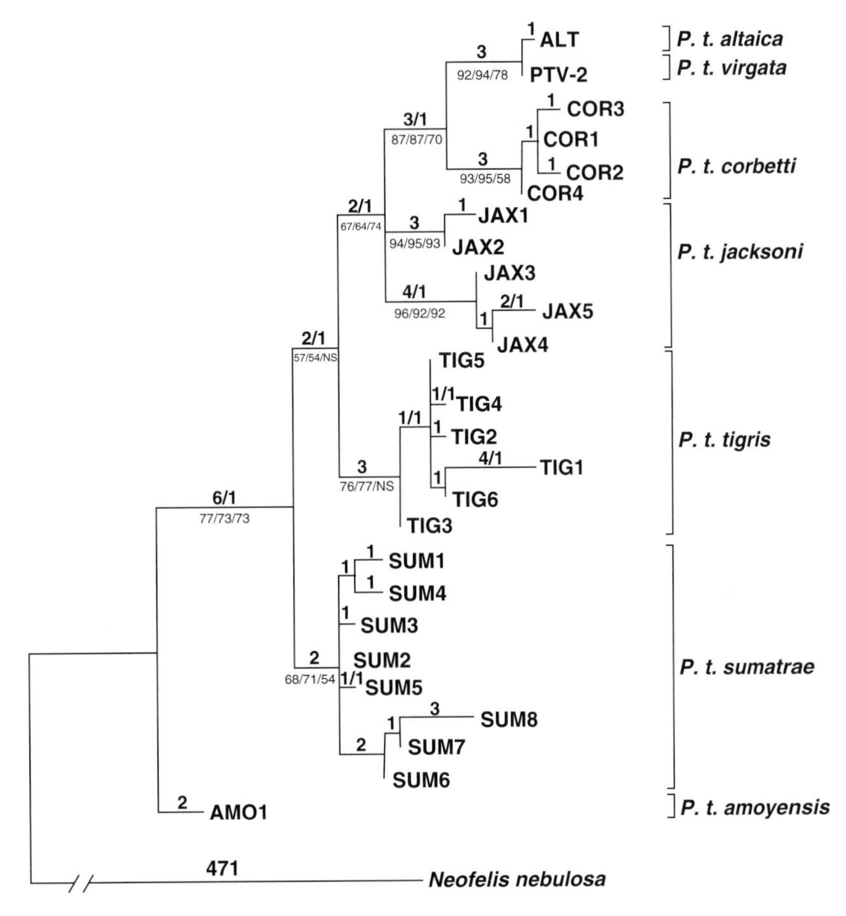

Figure 29.10 The first rooted tree of all extant tiger subspecies shows that the perilously rare South China tiger (*P. t. amoyensis*) is basal and clearly distinct, while the living Amur tigers (*P. t. altaica*) are the most derived. There is a close sister relationship between the living Amur tigers and extinct Caspian tigers (*P. t. virgata*), from which Amur tigers are derived. The difference between Amur and Caspian tigers is less than that found within other recognized subspecific tiger groups. Driscoll *et al.*'s conclusion is that *P. t. virgata* and *P. t. altaica* are essentially the same. Tiger subspecies: *altaica*, Amur; *virgata*, Caspian; *corbetti*, Indo-Chinese; *jacksoni*, Malay; *tigris*, Bengal; *sumatrae*, Sumatran; *amoyensis*, South China; *Neofelis nebulosa*, clouded leopard. (From Driscoll *et al.* 2009.)

Mountains along the eastern leg of the Silk Road, threading through the Gansu Corridor from the Ordos Plateau, to the Tien-Shan Mountains. Tantalizing depictions of Caspian tigers' span history and in Central Asia date back to the Bronze Age (2500–1500 BC). Such images include a bronze axe depicting a Caspian tiger attacking a boar. This axe was found among the Oxus Treasure initially acquired by fourth–fifth-century BC merchants somewhere along the River Oxus (now Amu Darya) that runs north-westward from the Afghan border to the Aral sea. Alexander the Great (356–323 BC) killed one midway along the Silk Road near Samarkand in contemporary Uzbekistan. A buckle from second-century Tadjikistan depicts a tiger attacking a Bactrian camel. A silver plate from the third and fourth-century Sassanian dynasty from Oxus depicts a hunter prince (probably Peroz Kushahshah, who reigned from 245 to 270 AD) spearing a tiger (Tanabe 2001). In 2001, thousands of gold figurines of tigers were found in a tumulus in the Altai-Sajan Mountains dating to the seventh-century-BC Pazyrak culture, in Arzhan in the extreme south of Siberia. In 2005, a team from the Northwest University of China explored China's eastern Tien-Shan Mountains in the county of Barkol (translated from Turkish as Tiger Lake) and found a petroglyph dated to the late Bronze or Early Iron Age that appears to depict a tiger fighting a leopard (Li Tao, personal communication) (e.g., see Fig. 29.11a, b, and c). This vivid insight that the ancients appreciated intra-guild hostility may also constitute evidence that the eastern Tien-Shan Mountains provided a supportive conduit for the ancestral Caspian tiger expansion from the Hexi Corridor, Gansu to the west and north. Furthermore, tigers were formerly known in the middle and west of Inner Mongolia, whence an unbroken corridor of suitable habitat leads to the Caspian sea and Transcaucasia in the west and to Lake Baikal and the Russian Far East to the east (Fig. 29.11d).

(a)

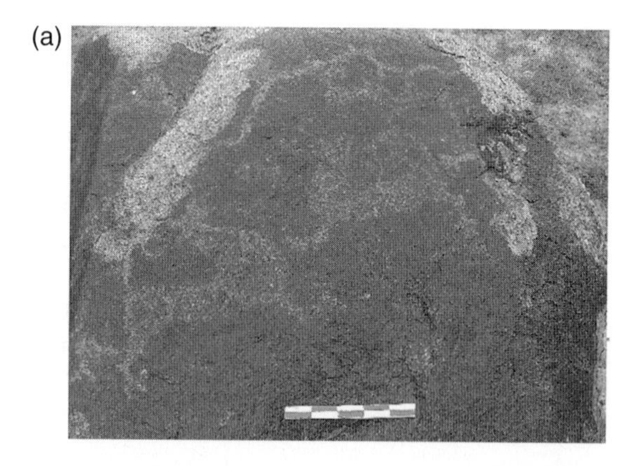

(b)

(c)

Figure 29.11a–c Archaeological images of tigers along the Silk Roads. Late Bronze or Early Iron Age petroglyphs from (a) Tien-Shan Mountains depicting spotted (leopard) and striped (tiger) felids fighting. (Photo courtesy of J. Wang and J. Ma.); (b) Yinshan Mountain, Inner Mongolian Autonomous Region depicting tigers. (Photo courtesy of Li Tao.); (c) Sassanian Persian silver plate depicting royal prince hunting tiger (The Japanese private collection; Photo courtesy of Shogakukan Ltd.)

(d)

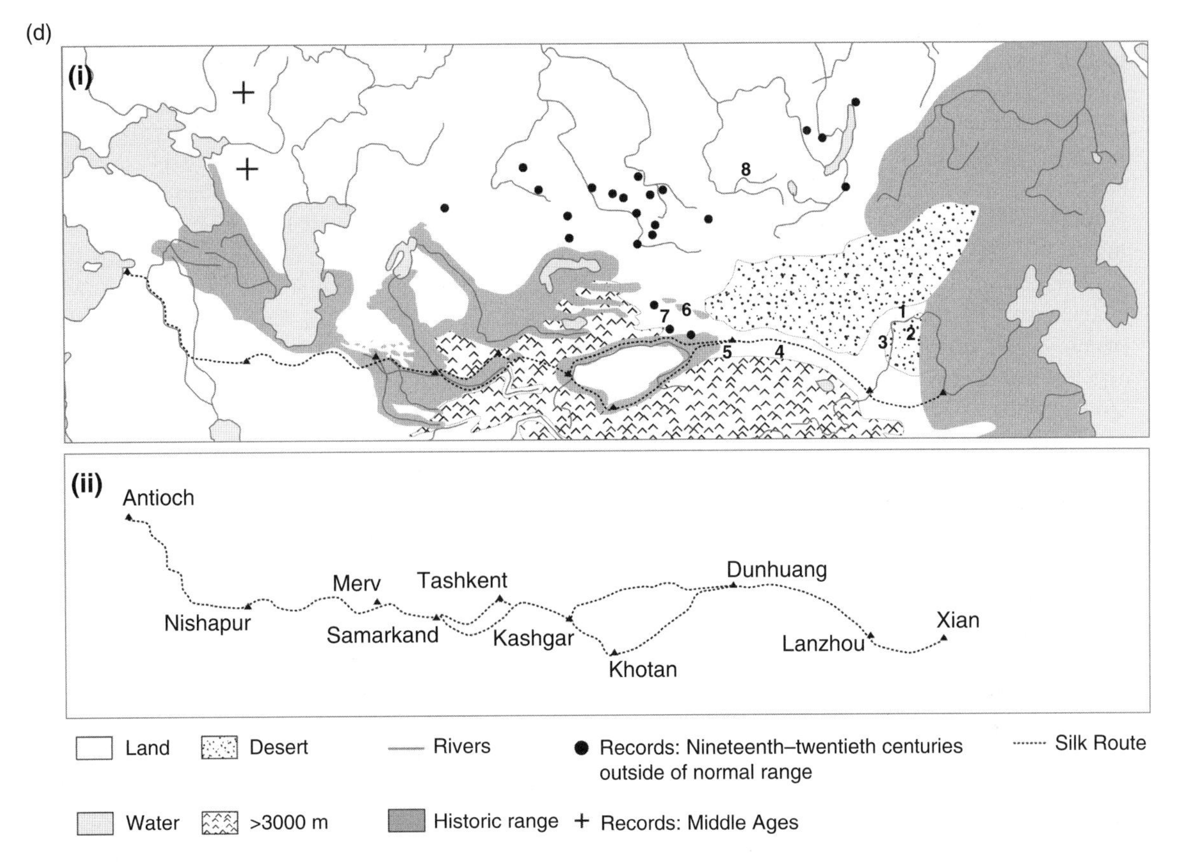

Figure 29.11d Map of tiger sightings (Mazak 1981, 1983; Heptner and Sludskii 1972), annotated with archaeological sites. (Information provided by Li Tao). (i) Physical map of Central Asia showing the historical Silk Route and the range of tigers c. 1800, prior to major anthropogenic habitat contraction (1–3). Numbers indicate archaeological sites where either tiger petroglyphs (PT) or tiger artefacts (AT) are found. (1) PT, Yinshan Mountain, Inner Mongolia; (2) AT, Ordos Plateau, Inner Mongolia, Shannxi Province; (3) PT, Helenshan Mountain, Ningxia Region, Gansu Province; (4) PT, Mazongshan Mountain, Gansu Province; (5) PT, Tien-Shan Mountains, Hami, Xinjiang Region; (6) PT, Tien-Shan Mountains, Barkol, Xinjiang Region; (7) PT, Tien-Shan Mountains, Teksun, Xinjiang Region; (8) Arzan 2, Altai-Sajan Mountain, Tuva Republic, Russia. (Li Tao, personal communication.) (ii) Trace of Silk Road, indicating major trading cities.

Reintroducing lynx to Scotland

Carbon-dated bones reveal that at least 12,650 years ago lynx occurred in Britain, and some bones in Yorkshire dated as recently as 1550–1842 ^{14}C years ago demonstrate that lynx survived in medieval Britain (Hedges *et al.* 1994). A downturn in their fortunes may have begun between 10,000 and 4000 years ago and has been linked to climate change, associated loss of forest habitat and prey and, latterly, persecution. Smout *et al.* (2004) suggest that by 2000 years ago, and before major human populations caused substantial deforestation, only 25% of land cover in Scotland was forested. By the thirteenth century roe deer (*Capreolus capreolus*) had become scarce in Britain, and in the Central Lowlands and Southern Uplands of Scotland both red (*C. elaphus*) and roe deer (probably an important prey of British lynx) were extinct by the late seventeenth century (Ritchie 1920).

There would be an argument under the EU Habitats Directive for reintroducing lynx to Scotland, if people had played a hand in its demise. Furthermore, the condition that previous impediments to their survival have been reversed is certainly met in so far as nowadays, roe deer are so numerous, and still increasing, that they are commonly regarded as a pest. If lynx had the capacity to limit the current high populations of deer, that too could motivate their reintroduction. Neither is an especially compelling case (Hetherington 2006), but recreating a large carnivore community could delight some, although it may be seen by others as foolish meddling. Hetherington *et al.* (2008) modelled the feasibility of such a reintroduction. They assumed home range sizes characteristic of the Swiss Jura (females 45 km^2, males 74 km^2 encompassing female home ranges), and assessed the extent and connectivity of woodland in Scotland. This process identified 20,678 km^2 of potential habitat in 30 patches and two main networks of 15,000 km^2 in the Highlands, and 5330 km^2 in the Southern Uplands, with an additional contiguous 810 km^2 in England (the two main networks being partitioned by major motorways within Scotland's Central Belt).

On the continent, while lynx prefer roe deer they will prey on ungulates up to the size of a red deer hind, taking also lagomorphs, small mammals, and ground-nesting birds (Breitenmoser *et al.*, Chapter 23, this volume). Hetherington *et al.* (2008) extrapolated from known densities of deer in woodland (e.g. 14.7 roe deer km^{-2} and 5–15 to 0.3–31.1 red deer km^{-2}), to reach an overall ungulate density of 12.2 km^{-2} in the Highlands and 6.5 km^{-2} in the Southern Uplands, comparable to levels in the Swiss Jura, where lynx thrive. Then, using a relationship between lynx density and log ungulate biomass km^{-2} they estimated that the Highlands and the Southern Uplands could sustain 2.63 and 0.83 lynx/100 km^{-2}, respectively. This equates to ~400 lynx in the Highlands and 50 in the Southern Uplands (including the Kielder Forest in England). A Population Viability Analysis suggested such a Highland population would be viable, whereas the more southerly one might not be.

What might be the pros and cons of such a reintroduction? Lynx kill sheep; for example, in Norway, where the latter are grazed in forests and the density of roe deer is low (Odden *et al.* 2002). In the French Jura, lynx kill 100–400 sheep annually; however, 70% of attacks occur in only 1.5% of the area inhabited by lynx, and of those flocks experiencing damage, most suffer only one to two attacks per year. In the Swiss Alps, 80% of farmers who lost livestock during the years 1979–99 lost fewer than three and 95% of attacks were within 360 m of the forest edge (Stahl *et al.* 2001a; Stahl *et al.* 2002). Although the abundance of roe deer in Scotland might minimize depredation on sheep, it would nonetheless occur (and since deer populations are widely considered to be damagingly high, any reduction in their numbers would be a bonus; although perhaps a source of competition with deer stalkers). Indeed, Scottish sheep are different to, and managed differently to, Continental ones, and might thus be much more susceptible to predation. There might also be some predation on grouse; Hetherington concludes, however, that that would be small. Conservationists might fear for impacts upon capercaillies (*Tetrao urogallus*), wildcats (*Felis silvestris*), and pine martens (*Martes martes*).

This example, perhaps more vividly than most, illustrates the central importance of consumer choice in conservation. Lynx occurred in Scotland before, and conditions are now such that they could be there again, apparently (although, realistically, a first step might be a large-fenced reserve [Sandom *et al.*, in press]). They could cause problems to some, delight to others. They would cause financial loss to some, while doubtless generating revenue for others (e.g. through ecotourism); and it would surely not be beyond the wit of a nation such as Scotland to balance these costs and benefits with suitable economic instruments (e.g. compensation for loss of stock). The question is simply whether society wants lynx, and for those who think the answer should be yes, the road ahead lies in ordering their priorities and honing their advocacy. The answer has some significance further afield because this is a case of a developed country considering the denizens of its own backyard; should the British public prove unenthusiastic about having lynx in its midst, this would not enhance the British conservationists' case that South Americans, Africans, and Asians should welcome, respectively, jaguars, lions, and tigers into theirs.

Transboundary parks

According to the Convention on Biological Diversity (1992), in situ conservation of viable populations of species is fundamental to the maintenance of global biodiversity. Currently, ~11.5% of the Earth's land surface consists of protected areas. A growing trend and significant element in conservation has been the development of areas managed jointly for conservation and resource management between adjacent countries (Mittermeier *et al.* 2005). These make up to 16.8% of protected areas and are termed, variously, transboundary conservation areas (TBCAs) or transfrontier conservation areas (TFCAs); transboundary protected areas (TBPAs) or occasionally 'Peace Parks'. The first TBCA, Waterton-Glacier International Peace Park, was established in 1932 between Canada and the United States. Since then 188 transboundary complexes have come into existence, including 818 protected areas in 112 countries (Besançon and Savy 2005). TBCAs not only function at a scale appropriate to ecological processes, but may reduce international tensions. For instance, the establishment in 1998 of the Cordillera del Condor between Peru and Ecuador was part of an initiative to facilitate a settlement to the long-running and hotly disputed boundary disagreement between the countries (Mittermeier *et al.* 2005).

Few TBCAs have been developed specifically to promote conservation of wild felids (one exception being the establishment of the Manas TBCA between Assam, India, and Bhutan, the initial planning for which was originated by Project Tiger in 1974; Gupta 2005). Nevertheless, their scale is perfectly adapted to conserving top predators. For instance, La Amistad Biosphere Reserve (between Panama and Costa Rica) and Maya Tropical Forest (Belize, Mexico, and Guatemala) provide significant corridors for the movement of jaguars, with the former also protecting six species of wildcat (Nations *et al.* 2005; Ramirez 2005). In southern Africa extensive TBCAs have been set up in recent decades, providing strongholds for declining large cat species such as African lions and leopards. Examples include the 36,000 km^2 Great Limpopo Transfrontier Park (GLTP) between South Africa, Mozambique, and Zimbabwe and the ambitiously planned Kavango-Zambezi Transfrontier Park (KAZA-TFCA) which, when formally established, will cover 129,000 km^2 and include protected areas in Angola, Botswana, Namibia, Zambia, and Zimbabwe (Braack 2005; Hanks 2005). These mega-parks accommodate biological process such as migration and dispersal that are too extensive for protection within increasingly fragmented habitat in national parks, and they are especially relevant to big felids (and small ones, Marino *et al.*, Chapter 28, this volume).

Impediments to joining up and the quest for linkage

Joining up, through space and time, emerges as the priority for radical felid conservation. Genetic corridors link fragmented sub-populations in space; restoration offers the opportunity of creating a present to link with the past; and transfrontier parks link jurisdictions. All these operations function at a landscape scale, and all the landscapes are inhabited by people. This focuses the conservationist's attention on human livelihoods; and, all too often, impediments to joined-upness deriving from them.

The dream of joining up sub-populations demands understanding of dispersal—an aspect of felid biology that is notoriously difficult to study. The need to disperse almost presupposes a secure breeding source population; an attribute all too scarce among many big cats. Tigers in the Western Ghats, India, provide a promising example. In more than 1.5 million km^2 of potential tiger habitat in south and south-east Asia, proximate threats such as depletion of ungulate prey and persecution for the Traditional Chinese Medicines (TCM) market, are impacting source populations to extents that overshadow the ultimate threats of habitat loss and fragmentation. Before the source populations can be linked, they must first be secured. However, confirming scientifically that local source populations are producing surplus tigers that are capable of dispersing along corridors is difficult. Happily, Karanth *et al.*'s long-term population study of tigers (2006) in Nagarahole National Park, Western Ghats, India has devised a methodology to quantify tiger population dynamics, which has now been applied to over 22,000 km^{2}of tiger habitat in Western Ghats. This metapopulation thus provides a framework within which a collaboration between Karanth and the WildCRU team is now studying tiger dispersal as a foundation for 'joining up' this landscape.

The impediments to joined-upness—linkage—in felid conservation take all the varied forms of Diamond's Evil Quartet, and more; but can loosely be categorized as habitat loss, persecution, over-exploitation and the ragbag category of conflicts. While the latter will be our main focus in this section, the case studies of this volume (Chapters 10–28) painfully illustrate the other issues. Thus loss of habitat as well as poaching threatens the tigers of the Terai arc in southern Nepal and northern India (Seidensticker *et al.*, Chapter 12, this volume); habitat loss now threatens the Florida panther (*P. c. cougar*), having previously been nearly eradicated by hunting (Onorato *et al.*, Chapter 21, this volume). Persecution exacerbated the plight of the Iberian lynx, *L. pardinus* (Ferreras *et al.*, Chapter 24, this volume), and persecution of jaguars in the Pantanal goes far beyond an immediate response to stock-killing (Cavalcanti *et al.*, Chapter 17, this volume), as does that of cheetahs in Namibia (Marker *et al.*, Chapter 15, this volume). Eurasian lynx not only face conflict through killing livestock, but are also persecuted by hunters who compete with them for ungulate prey species, roe deer and chamois, *Rupicapra* sp. (Breitenmoser *et al.*, Chapter 23, this volume). Over-exploitation, for example for body parts or fur has worsened the predicament of the snow leopard, *P. uncia* (Jackson *et al.*, Chapter 19, this volume), as it did the Iberian lynx in the nineteenth century (Ferreras *et al.*, Chapter 24, this volume) and threatens the extinction of the Amur tiger (Miquelle *et al.*, Chapter 13, this volume).

Although scarcely a welcome accolade, no vertebrate family may rank higher than the Felidae in the hierarchy of HWC. Building on the foundation laid by Loveridge *et al.* (Chapter 6, this volume), we will explore why this is so, and what can be done about it. First, which among the multifarious strained relationships between people and wildlife qualify as conflict? Alexandra Zimmermann (personal communication) offers a thought-provoking definition that HWC is the situation that arises when the behaviour of a (otherwise non-pestilential) wild animal species poses a direct and recurring (real or perceived) threat to the livelihood and/or safety of a person or a community, in response to which, persecution ensues. This definition excludes, for example, hunting (whether for sport or for sustenance); disease control; road traffic kills; and turns crucially on the usage of the word pest and the *intention* of the human actor. Actually, any distinction between a lion killing a cow, a fox killing a hen, or a rat killing a pheasant exists only in terms of people's predispositions to think more readily of particular taxa as 'pests' (one of us, DWM, recently caused a stir by delivering a plenary talk on lions, tigers, and cheetah at an international conference on pest biology that was otherwise dominated by lectures on rodents). Furthermore, a definition so affected by intention courts anomalies when different people do the same thing for different reasons (while none of the horsemen who engaged in fox-hunting in Britain were seeking sustenance, some of the time some of them may have believed they were controlling foxes in response to conflict; while others, or even the same ones on different occasions, may have been in pursuit of sport). However, such niceties aside, Zimmerman's definition gives a sense of what HWC means in terms of the strained relationship between wild felids and people. Like all generalities, however, it provokes points for debate: for example, what sympathy to grant 'persecution' for cultural/religious/spiritual reasons—for example, some rural Zambians kill species of birds and reptiles believed to be manifestations of Christian demons (G. Wegner, personal communication); and many Mozambicans believe that sorcerers can magically convert either sticks of wood, or even themselves, into 'people–lions', which can then be used to kill the sorcerer's enemies, so lions are commonly killed in order to thwart such sorcery (West 2001). Similarly, it is traditionally believed in Benin that vengeful deities have the power to make leopards attack people (Ben-Amos 1976); leopards have also long been associated with witchcraft in what was the Gold Coast of Africa, as witches were thought to ride on their backs in order to attend secret nocturnal meetings (Field 1955).

Herbert Spencer's maxim and cultural relativism

Demonic birds and reptiles and bewitched twigs or sorcerers masquerading as lions are clearly fiction—that is, propositions entirely unsupported by empirical evidence. Nonetheless, as liberal youths we might have defended peoples' right to believe what they wished, (rightly) relishing the cultural richness of such myths and folklore. However, in the face of a crisis in biodiversity conservation this luxurious liberalism, intended as a tolerant generosity of spirit, is revealed on mature scrutiny as more akin to condescension. As Herbert Spencer opined (going to the heart of what might be called conservation's usufructuary dilemma), 'Every man is free to do that which he wills, provided he infringes not the equal freedom of any other man' (Spencer 1900)—which poses a conundrum for conservation, however: as the *same* lion may be one man's trophy; another's photo-opportunity; another's bloody nuisance and, very likely, the embodiment of another's enemy. While the foundations of each view are open to moral scrutiny and its consequences open to quantification and utilitarian cost–benefit analysis, the fourth proposition differs from the first three in being most palpably based on manifest nonsense. Of course, nothing can be proven to be false—we cannot *prove* that the lion whose attack you fear is not your human enemy in disguise; or that a leopard is not secretly transporting witches on its back—but it can be shown to be so improbable that it is sensible to treat it as nonsense. To be tediously even-handed, we should note that some people tolerate felids for reasons as nonsensical as those for which others kill them: Nepali Buddhists protect snow leopards as the 'mountain god's dog' (Ale 1998). However, tolerating nonsense, whether out of a polite and kindly desire not to hurt another's feelings, or a well-founded fear of reprisals for political incorrectness, surely fails Spencer's test when it comes at the cost of endangering species and causing suffering to individuals.

Nonetheless, the prevalence of myth, superstition, and solipsism make it very hard for practical conservationists to argue straightforwardly from evidence born of hypothesis testing. Indeed, to do so risks vilification by the many proponents of 'cultural relativism', who hold that all cultural systems are equally valid. This view gained sufficient weight that, in 1947, anthropologists lobbied the United Nations for the Universal Declaration of Human Rights to accommodate all cultural perspectives; the attempt failed, but these ideas have had an influence far beyond the bounds of professional anthropology (Lukes 2008). Consequently, many Western conservationists have been fearful to gainsay superstitions in less developed countries (LDCs) lest they be accused of cultural or moral imperialism. Of course, superstition's pernicious effect on wildlife is not confined to the LDCs—an example, albeit historical, closer to home is the fifteenth-century legend which fuelled persecution of wildcats now extinct in the north of England (Macdonald *et al.*, Chapter 22, this volume). A memorial in the church at Barnburgh in South Yorkshire, claims that the 'British tiger' (the wildcat) attacked one Sir Percival Cresacre in the woods at High Melton and that 'a furious fight followed which ended at the church porch where both Sir Percival and the cat were found dead the following morning' (Lovegrove and Lovegrove 2007). Indeed, the supernatural phenomena that underpin the world's most sophisticated religions are no less implausible than the bewitched 'people–lion' hypothesis (Dawkins 2006; Hitchens 2007). Beliefs, of course, can be turned to good or ill, within the dizzying vortex of doing the right thing for the wrong reasons and vice versa: Hazzah (2006) shows that religious Maasai are more likely to kill lions, while Dickman's study (2008) into the corollaries of human–felid conflict in Ruaha (see below) revealed that people who adhered to external religions were more hostile towards felids. However, there is a groundswell of interfaith ecumenism promoting initiatives to identify scriptural imperatives to advance environmental conservation: such as the Judeo-Christian Stewardship Conservation Ethic, which was developed by people who interpret biblical stories such as Noah's Ark as showing that God intended the Earth to be populated with the panoply of animal species (Callicott 1994; Borgerhoff Mulder and Coppolillo 2005). In Britain, the Archbishop of Canterbury, Rowan Williams (Ebor Lecture, 29 March 2009) having correctly stated that 'Ecological questions are increasingly . . . defined as issues of justice' (he had in mind justice between rich and poor and between generations), went on to equate unintelligent behaviour towards the environment with ungodliness; while there may be no more evidence for this than for transmogrified lions, conservationists (as distinct from conservation science) will doubtless take succor wherever it is offered. The substantive point for felid conservation is that people may kill felids for reasons that can be questioned by science (e.g. to what extent does lynx predation disadvantage roe deer hunters [Breitenmoser *et al.*, Chapter 23, this volume]), or for reasons on which science has no comment. Of the latter, it pleases both wealthy American hunters and impoverished Maasai warriors to kill lions (in both cases probably because of their [different] cultural and [shared] biological backgrounds). It would be prejudiced to judge one as being more or less correct ethically than the other (although one might have readier access to knowledge on the consequences of their actions than the other; while a Utilitarian might observe that their relative degrees of wealth would affect any cost–benefit analysis of the outcomes). However, somewhere towards the edge of the continuum of daft motives for doing things come fascinating, culturally rich, and quaint but nonetheless patently nonsensical beliefs—and transmogrified lions bring into focus the need for conservation to consider whether it should pussyfoot around moral relativism when extinction is at stake (others will ask the same question with regard to animal welfare). A source of inspiration lies in the field of Human Rights, where Lukes (2008), using the example of cultural variation in the acceptability of wife-beating, argues that a break with relativism rests on redrawing of the boundaries between 'acceptable' and 'unacceptable' action based on global consensus-building. When it comes to assessing the moral cost of actions involving both endangered species and suffering to individual humans, conservationists currently lack a global consensus as well developed as the yet imperfect framework discussed by Lukes on human rights.

From local lessons to global good practice

Among the factors predisposing felids to conflict with people, the foremost is livestock depredation, but this is exacerbated by the depletion of natural prey (livestock provide an alluring alternative), habitat disturbance (e.g. increase in edge effects of protected areas which often increase animal/human encounter probabilities and create population sinks; Woodroffe and Ginsberg 1998), and livestock management practices (levels of protection implemented, as well as general condition of livestock; Hoogesteijn 2003; Ogada *et al.* 2003). At the level of individuals, a predator's condition, including health, age, and territory has also been correlated with depredation; but, as Zimmermann *et al.* (in press) point out, has not been proven to predispose carnivores to prey on livestock (Rabinowitz 1986; Linnell *et al.* 1999; Miquelle *et al.* 1999b; Wydeven *et al.* 2004; Marker *et al.*, Chapter 15, this volume).

A systematic review by Inskip and Zimmermann (2009) of all 37 species showed that nine felids are particularly prone to conflict: caracal (*Caracal caracal*), cheetah, Eurasian lynx, jaguar, leopard, lion, puma (*Puma concolor*), snow leopard, and tiger. According to Inskip and Zimmermann's criteria, the severity of conflict differed significantly among felid weight groups, being severer for large cats than for medium or small ones (the latter two not differing significantly; Fig. 29.12). Indeed, of the middleweights, only the 17 kg caracal and 23 kg Eurasian lynx nudge into the top division of conflictors, otherwise monopolized by species over 50 kg.

The factor which pushes felids up the HWC ranking and above canids, the other main contenders (Macdonald and Sillero-Zubiri 2004c), is their tendency to attack people (a risk that is generally highest around protected areas, where communities are often already strained by other aspects of HWC; see below). Inskip and Zimmermann (2009) associate attacks on humans with felids with an average body mass greater than 50 kg—principally leopards, lions, and tigers; attacks by pumas and jaguars are comparatively rare (Beier 1991; Perovic and Herrán 1998; Quigley and Herrero 2005; Altrichter *et al.* 2006), and there are no reports of snow leopards or cheetahs attacking humans in the wild.

HWC is particularly vivid when the wild protagonist is endangered or where the conflict seriously threatens human life or livelihood. Bhutan's Jigme Singye Wangchuck National park, which is home to tigers, leopards, and 5000–6000 humans, illustrates both (Wang and Macdonald 2006). In 2000, 274 households in six different geogs (sub-districts) within the park were questioned on their family circumstances, their farming enterprises, their experiences with wildlife, and use of the park. Of the farming households, 12% supplemented their income by collecting timber forest products. Of all households surveyed, 21.2% reported a total loss of 2.3% of their domestic animals to wild predators over 12 months ($n = 76$). This loss equated to an average annual financial loss equal to 17% (US$44.72) of a family's total per capita cash income (average $250 per annum). The financial cost of total reported losses amounted to US$12,252, of which leopard and tiger kills accounted for 82% (US$10,047). Annual mean livestock loss per household (of those that reported loss) was 1.29 head of stock, equating to more than two-thirds of their annual cash income of $250.

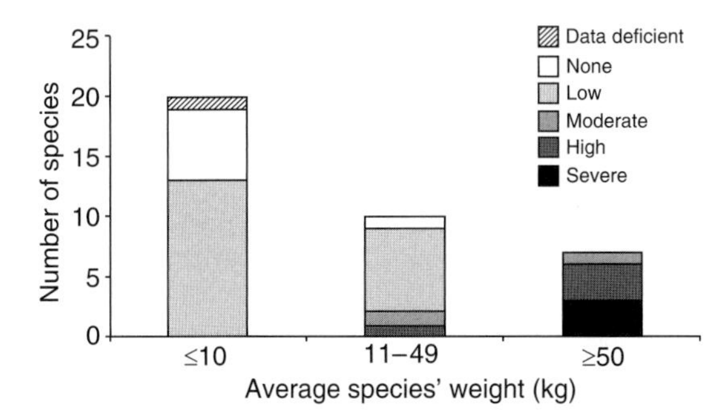

Figure 29.12 Body mass and the scale of conflict. The number of species in each weight category for which there is no evidence of conflict (white), or evidence of low (light grey), medium (mid-grey), high (dark grey), or severe (black) conflict. Hatched portions represent those species classed as data deficient. (From Inskip and Zimmermann 2009; see also Loveridge *et al.*, Chapter 6, this volume.)

Lax herding, inadequate guarding practices, and overgrazing contributed to livestock losses. Approximately, 60% of the households lacked proper stables for corralling their livestock at night; additionally, there was a significant correlation between the number of livestock lost and the distance between the household and the grazing pasture.

For conservation, things have been looking up: the Jigme Singye Wangchuck Park was gazetted in 1993, and the Forest and Nature Conservation Act implemented in 1995. However, farmers perceive a cost: 52% of them blamed their hardships on the park and the Act. Wang and Macdonald (2009b) confirm that tigers and leopards do feed on livestock. The top five problems listed by farmers, in order of decreasing severity, were: (1) increased predation; (2) lack of fodder; (3) increased incidence of disease; (4) insufficient pasture; and (5) reduced milk quality. In the short term these grievances might be defused by stock intensification programmes, including pasture improvement, and financial compensation. In the longer term, compromise might lie in a stock insurance scheme, and there might be merit in relaxing the resource-use restrictions in the Forest and Nature Conservation Act of 1995. Certainly, farmers should be involved in solving the conflicts, and should bear some responsibility—improving their own herding and guarding practices, for instance; and building adequate corrals (ideally they would also share any profits generated by tolerating predators).

Felid pests, give and take, and 'holistic' innovation

Perhaps the derogatory term 'vermin' has inured some people to the startling numbers of predators routinely killed in Europe and North America. Over 97,000 coyotes (*Canis latrans*) are killed annually in North America (Gompper 2002), and ~400,000 red foxes (*Vulpes vulpes*) in the United Kingdom as part of routine agricultural and game-rearing enterprises (the efficacy of these culls in reducing stock-losses is not always obvious; e.g. Berger 2005; Macdonald and Johnson 1996; Macdonald *et al.* 2003 point out that more money may be lost to the rabbits [*Oryctolagus cuniculus*] that the foxes would have eaten than is saved by killing the foxes). Perhaps the contempt bred of familiarity with such heavy tolls renders less shocking the thousands of black-backed jackals (*C. mesomelas*) killed by Namibian and South African farmers? Perhaps it is prejudice born of their larger size and startling beauty that makes it more jarring to realize that caracal and leopards are killed as pests on South African farms (Fig. 29.13a and b). For example, there are records of 3377 caracal (along with 24,589 black-backed jackals) killed between 1966 and 1992 in the Free State Province of South Africa (Oranje Jag Hunting records 1952–92), and an average of 2100 caracals were killed annually between 1931 and 1952 in the Cape Province (Moolman 1986). In the past five years, licenses to catch 80 leopards have been issued in the Western Cape Province (D.L. Hignett, personal communication). Bounties have been paid on South African predators since 1650, and continued to be paid on leopards into the 1960s (Norton 1986). The Nature Conservation Ordinance (No. 19) of 1974 removed leopards from the 'vermin' list; subsequently, a permit has been required to trap and kill them. In short, farming and felids have traditionally been regarded as incompatible (Esterhuizen and Norton 1985; Stuart *et al.* 1985)—although evidence that killing predators is efficacious in reducing agricultural losses is no more conclusive in Africa than it is in Europe, according to Butler (2000). Indeed, caracals are thought to be expanding their range, one suggestion being that they gain advantage from the widespread persecution of jackals (Pringle and Pringle 1979).

In line with the situation of cheetahs of Namibia (Marker *et al.*, Chapter 15, this volume), these rangelands amount to an estimated 80% or more of Africa (Mackinnon and Mackinnon 1986) and a high proportion of leopard and caracal habitat, so these tallies are clearly relevant to the species' future. Indeed, only 8–13% of the potential range of leopards in Africa lies within protected areas, and in the Republic of South Africa only Kruger National Park and Kgalagadi Transfrontier Park may be sufficiently large to maintain genetically viable leopard populations (Boitani *et al.* 1999; Daly *et al.* 2005; Marker and Dickman 2005). Proliferating small game reserves may offer sanctuary and prey for leopards, but they are also the focus of competition with hunters (Hunter and Balme 2004). Against this background, the spectre of the Cape lion (*P. l. melanochaitus*), extinguished in 1858, castes a cautionary shadow over the leopard's destiny (Friedman and Daly 2004).

In this context, the former Cape Province of South Africa is an illuminating case. There, between 1977 and 1982, an average of 690 leopard-related stock deaths were reported annually (Esterhuizen and Norton 1985), and between 1977 and 1980 at least 110 leopards were killed legally (Stuart *et al.* 1985). These tallies raise not only conservation issues, but also welfare ones, since both leopards and caracals are commonly killed in gin traps (J. McManus, personal communication, see Fig. 29.14a). Data are fragmentary, but the Eastern Cape Department of Environmental Affairs and Tourism reports at least 28 leopards killed in a 300 km^2 district within the Eastern Cape region during the period 2005–08 (data collated by Landmark Trust). There, McManus (2009) estimated 1.3 leopards/100 km^2 (0.9 females and 0.4 males). Against this background, McManus (2009) reports on exciting pilot studies initiated by the Landmark Foundation (www.landmarkfoundation.org.za). Having first captured, then fitted with GPS collars, 12 potentially problematic leopards (as of 4 April 2009), the Foundation released the animals with the promise of compensating landowners for any stock they might kill. In exchange, landowners undertook to forego lethal control and to adapt their husbandry to reduce predation risk (Fig. 29.14.). By May 2009, only one farmer had claimed compensation (of R50,000 [$6000] for eight calves killed in 1 year); having received the compensation the farmer caught the leopard in a gin trap and allegedly stabbed it. However, three case studies illustrate more encouraging outcomes.

(a)

(b)

Figure 29.13 (a) A bag of six caracal. (Photo R. Harrison-White.) (b) A leopard hunt in 1958 illustrates intergenerational tradition: the 14-year-old boy on the right now runs the family farm, where he grew up to be committed to leopard control. © T. Barry.

First, one farmer, with 4500 head of small livestock on 120 km^2, had, between 1984 and 2004, lost an annual average of 130 head of livestock, valued at R78,000 ($9400). During that period he had commissioned four annual culls, using hunting and gin traps, with an average annual tally of eight caracals and/or jackals, at a yearly cost of R24,000 ($2900) (conversion rate of R8.29 to US$1). In 2005, this farmer undertook a trial with two Anatolian dogs (see Marker *et al.* 2005a). Over the next 4 years his losses averaged 35 livestock per year. The profit-and-loss accounts for these two contrasting periods associate the sheep-guarding dogs with an annual saving of about R63,000 ($7600; Table 29.1a). The second case study involves a farmer in Graaff Rienet, in the Eastern Cape, who was accustomed to an average annual expenditure of R24,000 ($2900) on lethal control of predators, and nonetheless reported losses of 450 sheep. Having purchased eight alpacas (R10,000 each; $2900) and put protective collars on 400 sheep (R28 each; $3), a total investment of R91,200 ($11,000), his losses the following year fell to 150 head of stock (Table 29.1b). The third study involved monitoring the stock losses of ten farmers in the Baviaanskloof during the period October 2007–September 2008; having equipped 12,000 of their sheep with 'Dead Stop' protective collars (see Fig. 29.14b). Previously, these farmers had reported losses of 1350 head during the year (11.3% of 12,000 stock) valued at R675,000 ($81,130; with a by-catch of three leopards killed in gin traps ostensibly set for caracal or jackal). During the trial, only 50 head (0.42% of the flock) were lost. In this case, the profit-and-loss account suggests a saving of R414,000 ($49,760) in the first year and R750,000 ($90,144) in the second as the initial costs were recovered by the end of the first year (Table 29.1c).

All three case studies suggest an economic benefit of non-lethal control to farmers, which is generally considered a prerequisite to increased tolerance towards carnivores (Hunter and Balme 2004; Lindsey *et al.* 2005). McManus (2009) studied leopards in the Baviaanskloof, and found that under this heavy mortality regime they ranged very extensively. This might be attributed to low densities of wild prey, but it also raises the possibility of a perturbation effect (Tuyttens *et al.* 2000) and is reminiscent of the spatial responses of heavily hunted male lions (Loveridge *et al.* 2007c, Chapter 11, this volume). Her work with the Landmark Trust suggests that a 'holistic' approach, underpinned by science, offers an effective and economic alternative to the killing of leopards and caracals that has historically been considered to be the only option (Stuart *et al.* 1985). This combines innovative compensation, modern devices, and mediaeval guard dogs (Marker *et al.*, Chapter 15, this volume).

(a)

(b)

Figure 29.14 Lethal and non-lethal approaches to controlling leopard predation on sheep. (a) Injury sustained by a pregnant female leopard caught in a gin trap in August 2006 on farmland in the Eastern Cape, who chewed off three toes to free herself. After 11 days of veterinary care she escaped, and her characteristic spoor has been seen intermittently in the ensuing 3 years; (b) Sheep fitted with re-usable 'Dead Stop' protective collar, costing R20 each, made of epoxy-coated metal mesh. © J. McManus.

Joined up human enterprise

Not so long ago it was possible to observe the functioning of innumerable traditional systems by means of which local, often tribal, peoples appeared to

Table 29.1. a) Comparison of the costs of lethal control of predators and the use of Anatolian dogs on a 10,000-ha sheep and angora goat farm (after McManus 2009); b) a comparison of the costs of lethal control of predators and the use of a combination of Alpacas and protective collars on a 5000-ha sheep farm; and c) a comparison of the costs of lethal control of predators and the use of protective collars on a collection of 10 sheep farms within one valley.

a)	Cost of lethal control (per year)	Cost of non-lethal control (Anatolian dogs) (per year)	Difference (amount saved by implementing non-lethal instead of lethal control) per year
Costs (Rand/year)	R24,000	R18,000	−R6,000
Average losses per year			
No. animals	211	35	+176
Value*	R105,500	R17,500	+R88,000
Total cost per year	R129,500	R35,500	+R94,000

b)	Lethal control (per year)	Cost of non-lethal control (per year; 8 alpacas and 400 collars)	Difference (amount saved by implementing non-lethal instead of lethal control)
Costs (Rand/year)	R24,000	0 after initial cost	+R24,000 (after year 1)
Initial costs (one-off)		R91,200	
Average losses per year			
No. animals	450	150	+300
Value*	R225,000	R75,000	+R150,000
Total cost year 1	R249,000	R166,200	+R82,800
Total cost year 2	R249,000	R75,000	+R174,000

c)	Cost of lethal control (per year)	Cost of non-lethal control (protective collars) (per year)	Difference (amount saved by implementing non-lethal instead of lethal control)
Costs (Rand/year)	Unknown (estimated R100,000)	0 after initial cost	+R100,000 (after year 1)
Initial costs (one-off)		R236,000	
Average losses per year			
No. animals	1350	50	+1300
Value*	R675,000	R25,000	+R650,000
Total cost year 1	R775,000	R361,000	+R414,000
Total cost following years	R775,000	R25,000	+R750,000

* Approximately R8.32: US$1.

manage a balanced coexistence with wildlife. Notwithstanding any delusional romance regarding a probably mythical noble savage, it nonetheless seemed possible to hope that sustainable systems could be carried into the future as islands of respectful engagement between rural peoples and nature. Today, in the face of a rampant human population crisis, globalization, and the Internet, and cheaply available means of predator extermination, this hope seems quaintly naive. The agglomeration of traditional societies into nation states, and the expansion of the market economy, increasingly dissolves traditional forms of natural resource management (Gatzweiler 2006). In responding to this new reality, felid conservationists are already deploying a diversity of social science tools. These include monetary compensation schemes (Jackson *et al.*, Chapter 19, this volume; Loveridge *et al.*, Chapter 6, this volume), non-consumptive use through tourism (Durant *et al.*, Chapter 16, this volume), and extractive use through sport hunting (Balme *et al.*, Chapter 14, this volume; Loveridge *et al.*, Chapter 11, this volume). Engagement with local people and communication and education approaches are key tools in the resolution of HWC (Marker *et al.*, Chapter 15, this volume).

Conservation's audacious dream still rests on the hoped-for win–win between biodiversity conservation and poverty alleviation (Sachs *et al.* 2009), but as human populations rise and issues of equity become starker, any comfortable middleground between conservation and development may diminish. The history of conservation's attitudes to poverty can be categorized on a continuum from ecocentric to anthropocentric. Ecocentric approaches put conservation first, investing nature with intrinsic value, and therefore treating its conservation as a separate and legitimate goal from that of poverty reduction. At the other end of the spectrum, anthropocentric approaches prioritize human development and poverty reduction, and consider the conservation of wildlife important mainly because of the dependence of people (which, in a development context, are marginalized) upon it (Adams *et al.* 2004; Wegner and Pascual 2008). A parallel continuum exists between reliance on regulation and incentivization. Low-flexibility regulatory instruments of the 'command and control' or 'fence and fine' ilk seek to protect wildlife from human interference. They can be relatively straightforward to understand, and easy to enforce, but can be expensive and inefficient. Higher-flexibility mechanisms, on the other hand, promote the sustainable management of resources through economic incentives and moral persuasion. Wegner and Pascual (2008) conclude that an increase in flexibility also increases efficiency and can result in more harmonious and sustainable outcomes, but with that promise comes a complexity (and need for effective governance) that can undermine effectiveness, particularly in developing countries.

The ecocentric–anthropocentric continuum

As explained above, thus far the pendulum of conservation's history has swung hard from an ecocentric, protectionist approach, relying heavily on regulation, towards a community-based one (Macdonald *et al.* 2007; Swallow *et al.* 2007). Nonetheless, even in the heyday of protectionism, conservationists were rarely blind to human poverty (Adams *et al.* 2004; Wegner and Pascual 2008). Some considered poverty an impediment to conservation, and thus took the pragmatic view that it needed to be alleviated; others saw it as a moral obligation that conservation should not worsen the predicament of the disadvantaged, therefore arguing that they be compensated for opportunities lost; for example, as a result of their having been excluded from protected areas. From these traditional protectionist perspectives poverty alleviation may have been more a prerequisite or corollary of conservation than a goal in itself. Early protectionism—sometimes characterized, perhaps with a derisory nuance, as command-and-control conservation—was doubtless a product of its era. It faced practical and moral problems, such as how to fence large areas and how to ameliorate the injustices of resettling or excluding people. A subsequent era sought solutions in the alignment of concern for people and nature, within the unifying concept of sustainable development (WCED 1987) which has matured into the Ecosystem Service Framework and informed the design of Integrated

Conservation and Development Projects (ICDPs) and Payments for Ecosystems (PES) schemes (e.g. Engel *et al.* 2008).

As one pendulum has swung from protectionism to community involvement, another has swung from the belief that central governments should manage conservation to bottom-up direction by local communities. Considering the spatial scale and ethical complexity of conserving felids, Alexander Pope might have had this type of unhelpful dichotomy in mind when in 1732 he concluded that all 'Forms of Government let fools contest; whatever is best administered is best.' Certainly, in an analysis that concludes that both extremes are flawed, but each has its comparative advantage; Barrett *et al.* (2001) argue that both are required, and they cheerfully conclude, 'Inept, corrupt, or dysfunctional institutions are not easily reformed, so we do not mean to suggest that this is an easy task, just that it is necessary.' Because the task is difficult, the global beneficiaries of felid conservation should engage with local stakeholders to deliver it. Whether the unit of governance is government or community, conservation requires (but in many of the countries with custody over wild felids, is not always blessed with) competent, trustworthy, and authoritative institutions capable of awarding and withdrawing incentives, monitoring outcomes, and adapting to change.

The motivation for the alignment of conservation with human well-being is at one and the same time ethical and operational—being grounded in the conviction that it is not only immoral and often politically inconceivable, but also ineffective and often counterproductive, to pursue wildlife conservation while neglecting the well-being of the human constituencies it affects. Thus while it might be more comfortable for self-righteous conservationists to rail against the threats posed by avaricious multinationals, or even a selfish bourgeoisie, the more challenging priority is how the avant-garde conservationist reconciles poverty alleviation and biodiversity conservation.

'Consider the goat' wrote Norton-Griffiths (2007) in a paper that serves as a primer for the economics of free-market conservation 'and suppose you were not allowed to use it in any way at all (no marketing, no slaughter, no use of milk, meat or skin), and that if you were discovered to be using it you would at worst be shot dead, or at best imprisoned—indeed imagine that all you could do with your goat was hope a minibus of tourists would drive by and photograph it, from which you might obtain a meagre reward. How many goats do you imagine would remain in Kenya?'

This wry parable provokes the question of whether felids can pay their way in the general context of the cost-effectiveness of conservation—the balancing of the net benefits arising from using land for conservation against the opportunity cost of doing so. The opportunity costs of conservation to a landowner are the forgone benefits of development (Norton-Griffiths and Southey 1995). In short, does a stakeholder gain more from retaining those lions, leopards, jaguars, and cheetah than it costs him? The algebra is slipperier than it at first appears; because, while it is often clear who is bearing the opportunity costs of conservation, it can be less obvious how payment of these costs should be shouldered by various beneficiaries. This is because the benefits of conservation can be local (e.g. tourist revenues from people coming to photograph the lion, devalued by the number of cows it eats; additionally, the veterinary bills incurred by diseases transmitted by its prey, and the cost of the grass they eat, etc.); national (e.g. the reputational, and thus perhaps commercial, gain the country gets from being proud home to, for example, clouded leopards); or global (e.g. the delight that the world community takes in knowing that jaguars prowl the Amazon, and even the carbon sequestration achieved by forest retention to accommodate them). The apparent cost-effectiveness would depend on which of these three categories of benefit is considered, and whether the relevant beneficiaries are prepared to stump up their share of the payments (Norton-Griffiths 1996). Local stakeholders may be unwilling selflessly to carry the cost of the nation's enjoyment of, or even its profit from, the biodiversity in their backyards. Nations may want more than a glow of selfless satisfaction, and a congratulatory pat on the back from the world community, if they are to freeze their development to safeguard the biodiversity conserved within their borders. In economic terms, a rational decision on whether to put up with big (or little) cats depends, at first glance, on the balance between the marginal benefits of conservation and those of development (at second glance, complexities emerge; e.g. industrialization might draw people away from the countryside, conceivably relieving conflict). Almost all the pressures of the twenty-first century are tipping this balance against conservation, making it increasingly difficult for felids or other wildlife to pay their way; thus, if the overarching influence is that of market economics, it is increasingly rational for landowners to get rid of them. This is why Pearce and Moran (1994) concluded that the biggest threat to conservation is the failure to establish pricing mechanisms which take account of the true externalities associated with the loss or conservation of biodiversity (Pearce *et al.* 2007). It is also why, in the postscript to this chapter, we emphasize the importance of valuing nature beyond its price (see below). Swanson (1994) argues that the main threat to wild species is no longer over-exploitation, but the fact that land will be allocated to a species only in proportion to its ability to generate more competitive returns than would be expected from the best alternative use of the land and the mining of the existing stock of natural resources on that land.

Norton-Griffiths (1996, 1998) provides a worked example of opportunity costs for the Maasai pastoralists living around the Maasai Mara Nature Reserve. This has direct relevance to felid conservation because big cats eat the Maasai cattle, in addition to which the felids' prey eats the grass potentially eaten by the cattle, and communicates disease to them. In Kenya, where annually human population grows at >3%, cultivation at >8%, and wildlife decreases at >3% per annum, over 70% of wildlife lives beyond protected areas on rangelands. Even the US$500 million annual revenue of tourism has been insufficient to prevent the loss of 60–70% of Kenya's wildlife since 1977 (although the loss has been less from areas visited by tourists than in areas rarely visited [32% vs 50% lost, respectively]).

The government of Kenya retains ownership of the lions, leopards, cheetah (and all wildlife) on the Maasai group ranches, and regards supplying them as a public good. From the Maasai perspective, however, these big cats are a public *bad*, adding to their production costs (pestilential lions eat their cows, make it dangerous for their children to wander around at night; see Maclennan *et al.*, 2009) and creating massive opportunity costs by blocking alternative development of 'their' land (Fig. 29.15.). Norton-Griffiths (1996, 1998) estimates an annual discrepancy of $26.8 million between the current annual returns from conservation to the Maasai landowners of $1.3 million and the potential net returns of $28.8 million if their land were fully developed. If one considers that the Maasai have a basic right to benefit from the development potential of their land, the question arises of how this is to be balanced with the right of the Kenyan government to save lions (and all that falls under that aegis) for the benefit of all Kenyans and the world at large. This usufructuary dilemma is central to conservation (see below). One answer would be to compensate the Maasai to the tune of $26.8 million annually, as payment for favouring conservation over the production of cereals. Where could this money come from? Your favoured answer will depend on your political inclinations regarding free markets versus state intervention. One possibility would be to charge tourists (Norton-Griffiths estimates $80 per visitor per day would do it), another would be to reverse the 1977 ban on hunting and all forms of consumptive use (which had hitherto generated annual revenue of $24 million [at 1977 values]). Then there are the payments for ecosystem services

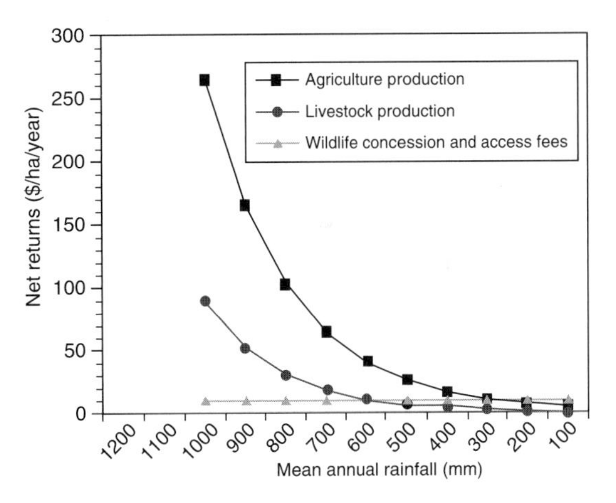

Figure 29.15 The dynamic relationship between differential net returns to landowners from agriculture, livestock, and wildlife 'production' varies with local rainfall. The average returns to wildlife of $10 ha^{-1} year^{-1} are competitive with agriculture only in very dry areas of <300 mm annual rainfall, and with livestock below 600 mm annual rainfall. Net returns to livestock are calculated with wildlife—they would be 66% higher if wildlife were eliminated! (From Norton-Griffiths 2007.)

(see below) which might be funded privately, as is lion production around Amboseli (see below), or nationally as is the case with British farmers whose encouragement of wildlife is funded by European taxpayers (Dutton *et al.* 2008).

Each model has its merits. Even the entirely free market can deliver conditions conducive to wildlife: in southern Africa privately owned and managed conservation areas are roughly 20% (60,000 km^2) the size of state areas (300,000 km^2). In Namibia, thanks to the Conservation Ordinance 1967 (which privatized the ownership of wildlife on privately owned land), wildlife on private ranches has increased by 70%. On the other hand, while it may be possible to market lions or cheetahs, the flat-headed cat may be a harder sell. Whatever system of incentives and regulation is devised, the goal is to create a bottom line which persuades landowners and users to become partners in conservation rather than its opponents.

Wherever you personally stand on the merits of the free market, it is inescapable that a country's conservation, and wider environmental policy, cannot be isolated from its economic development policy. The poorest are likely to depend heavily on nature, while the more developed may depend on agriculture—both, however, are likely to be at odds with predators in general and big felids in particular. Disadvantaged countries, and impoverished individuals, depend on natural assets, including wildlife, more than do the rich (World Resources Institute 2005). As Barrett (2008) says, the rural poor earn their living by applying their labour to the fruits of nature, so as their numbers (and consumption) increase there is less to go around and its marginal value to them is eroded; raising the spectre of a double whammy in which natural resources and human well-being both collapse. Further, poverty inclines the poor towards over-exploitation because they are uninsured, cannot secure credit and thus have no alternative but to under-invest in conserving the nature on which they depend; the risks being worsened by weak institutions for both conservation and development.

Weak institutions and the associated failure to clarify property rights, famously led Hardin (1968) to coin the expression 'tragedy of the commons' in the context of the pastoralists in East Africa. Although there is little empirical evidence that the tragedy of the commons actually applies to these pastoralists (Lybbert *et al.* 2004), it is easy to see how the idea could be extended from property rights to describe how larger herds degrade rangelands more severely leading to a poverty trap of escalating poverty and degradation. This is a situation which increasing human population exacerbates, taking wildlife (especially the sort that preys on cows) with it. However, wildlife, and especially big cats (made both susceptible and problematic by their predatory habits) are likely to be at odds with both the poor and the less poor. Kerapeletswe and Lovett (2001) document how the poorest 20% of Botswanans secure more than half their income from wild products; while of their environmentally derived income, about 90% comes from collecting wild or natural materials. These people are likely to over-harvest natural resources and be in interference–competition with big cats. For the richest 20% closer to 10% of their income comes from wild products, with more than 75% of their environmentally derived income coming from livestock grazing. These people are likely to be in direct scramble–competition with big cats.

Two not entirely mutually exclusive views characterize the causes of poverty: one, championed by De Soto (2000) blames dysfunctional institutions; while the other, advocated by Sachs (2005), blames shortage of money (and other productive assets) to overcome the market imperfections that prevent the poor from accumulating capital and thus grasping new technology. The basics of poverty economics are helpfully introduced by Carter and Barrett (2006) in terms which make obvious the ways in which poverty alleviation and biodiversity conservation are entwined. In most of the low-income world, asset poverty is a primary, intractable causal factor behind biodiversity loss. One reason is peoples' inability or unwillingness to invest in the future, given their immediate survival needs. Another reason is that vulnerable non-poor will commonly go to great lengths to protect the assets on which their future livelihoods depend; if that means killing wildlife, so be it. A third reason is that those with meagre assets commonly suffer very low returns to their labour when mixed with their (limited) land or capital, thus they commonly enjoy higher returns when they exploit common pool resources: wildlife and their habitats. Whatever the mechanism that induces poor people in particular to under-invest in wildlife conservation, poverty reduction is obviously central to wildlife conservation on pragmatic grounds (let alone moral ones).

The key point for lessons for conservation in general, and that of felids in particular, is that there exists a poverty line, or a variant of it that Carter and Barrett call the Micawber threshold (a charmingly Dickensian allusion to perpetually insolvent debtors) below which the poor are economically incapable of bettering their situation without significant outside help; people in this situation have no incentive to save or preserve anything, including wildlife and especially big cats, which are damaging and dangerous. If people below the poverty line dominate the stakeholders affecting the felids with which you are concerned, and if the endowment upon them of property rights to environmental resources does not bring them above the Micawber line, they will not preserve those resources effectively. In this way projects could fail, money and hope be squandered, and the distribution of property rights could backfire. Furthermore, those above the Micawber line, perhaps agriculturalists living alongside felids, when faced with economic setbacks, will wish to defend their own assets. At this point their tolerance of losses (perhaps inflicted by a felid) is likely to evaporate and they are likely to turn to exploiting any other resource which is publicly available, which could include the land supporting the felids. In referring to these people, have in mind that what they would do is exactly what you too would do in their circumstances. It seems therefore, that to protect felids (singled out here as umbrellas for wildlife in general), it is essential not only to protect their human neighbours from poverty and where possible to increase the value of wildlife to them, but also to regulate the impact on wildlife of those travelling the road to affluence.

Making wildlife profitable is made harder because, especially in developing countries, it is often treated as an economic externality—available under open access to whomever grabs it—and this lack of ownership or tenure creates conditions for over-exploitation. In short, if society wants lions rather than maize farms, tigers rather than rice paddies, clouded leopards rather than timber and palm oil, then in the face of escalating human needs mechanisms will have to be found that makes it profitable for people, especially poor people, to tolerate them (Alexander 2000). How this is to be achieved, by what blend of ingenious economics and biological wisdom, is the great question of this decade for conservation (an important sub-question is how much of the heavy lifting will be left to non-use or perhaps ecosystem payments, as opposed to direct use, such as hunting and tourism). The special role in the answer for felids may be that their charisma, and the world's willingness to pay for it, may provide a lever whereby land supporting felids becomes the most valuable biodiversity land in the world.

Ultimately, the balance of opportunity costs needs to be reversed. Currently, the opportunity cost of conserving nature is often so high that it is prohibitively expensive. We need to foster circumstances under which it is the opportunity cost of destroying nature that is prohibitively high—either that can arise because nature becomes valuable, or because people have something better to do than to destroy it (i.e. in economic terms, it is no longer profitable for them to draw down the biodiversity asset)—the more productive opportunities people have as alternatives to activities that damage biodiversity, the greater the hope for big cats and all that shelter in the shade of their umbrella.

How many felids do we need?

If conservation is to align with development and all aspects of environmental policy, not to mention other issues of global significance, then felid conservationists need to address their own version of Norton-Griffiths' provocative question, 'How many wildebeest do you need?' (He mused that 300,000 of the current 1.5 million might satisfy the tourist experience of mass migrations, for example). Considering that 24 of 37 felids are of some conservation concern (IUCN 2008), or worse, the general answer—as many as we can get—is not trite. For felids, if the question of how many we need is answered solely in terms of their direct market value and the ecosystem services needed as minimal life-support for humanity, then the future will be bleak. A minimalist answer which nonetheless recognizes their existence value might, more technically, be as many as ensures that no felid species is threatened; or which reduces extinction risk below, for example, 1%, in the next 100 years. When the question is phrased in terms of habitat protection, Smith *et al.*'s observation (1998) that when the proportion of pristine to degraded habitat fell to 50% tigers failed to reproduce, and at 30% they disappeared, offers some insight. Foster (2008) found that jaguar population densities were similar in areas with 100% and 30% forest cover, but the latter may have been an ecological sink.

Addressing this question brings us face to face with joined-upness extending into the future—planning felid conservation for today's world is plainly silly, when it is already clear that on the time-line of 50–100 years during which today's plans must roll out, the impacts of globalization and geopolitics, climate change, population growth, and altered age structures, desperate problems in food security, and a faltering reverence for GDP as the yardstick of success, will have set a very different stage for the next act of the felid play.

Life, livelihoods, ghouls, and nuisance: human relations with large felids

To add context to abstract considerations of the 'how' by which conservation and communities can coexist, we present four vignettes. Using felid examples, these illustrate, respectively, some socio-economic factors relevant to (a) the preservation of human life; (b) the impact of culture on attitudes to conflict; and (c) and (d) the operational realities of compensation schemes.

Preservation of human life: the case of Selous's lions

Large felids are dangerous to people who, one would suspect, have accordingly loathed them since their first ancestral encounters. Kruuk (2002) suggests that the 'special' status of large carnivores in the human psyche may therefore be inherited, and Macdonald (2009) observes that the notion of reintroducing these animals is distinctly postmodern. Puncture marks on early hominid skulls incriminate leopards as long-standing hazards: one of the best-known examples of leopard predation comes from a juvenile *Paranthropus robustus* fossil (SK 54) from Swartkrans, South Africa, which appears to have been dragged by the head to its final resting place (Brain 1969). The sabre-toothed cat *Megantereon* may also have preyed upon our hominid ancestors (Lee-Thorp *et al.* 2000), and jaguars are implicated in 6% of deaths among Ache Indians in Paraguay (Hill and Hurtado 1996). Hart and Sussman (2009) describe humanity's roots as prey, especially to lions and tigers; perhaps explaining Barrett's discovery (1999) that children of only 3–4 years of age display a sophisticated understanding of death as a result of predation.

The uncomfortable reality is that many calls to retain or restore large felids come from people whose ancestors long ago removed them from their own backyards (Macdonald *et al.* 2000a). Loveridge *et al.* (Chapter 6, this volume) review, among others, the 294 human deaths to tigers between 1984 and 2001 in the Indian Sundarbans, the 100 plus people attacked yearly by leopards in India (Athreya *et al.* 2004), and the recent unprecedented killing by a jaguar in Brazil (De Paula *et al.* 2008); nowhere, however, are the motives to adopt a 'not in my backyard' stance more poignantly understandable than in Tanzania, where between 1990 and 2004 almost 563 people were killed and over 300 more injured by lions (Packer *et al.* 2005b).

This blight is concentrated in the south-eastern part of the country, where lion attacks have been increasing over the past 20 years. This may be linked, indirectly through habitat destruction and prey depletion, to the 2.9% annual growth rate which took the human population of Tanzania from 23.1 to 34.6 million during the period 1998–2004 (Packer *et al.* 2005b). According to Packer *et al.* (2005b), lions are known to follow bush pigs (*Potamochoerus larvatus*) into the midst of human settlements and agricultural areas. Once led there in pursuit of pigs, they learn that humans can be prey and begin actively to hunt people. Lions will enter the centre of villages and attack people right outside their homes or even break into huts on agricultural fields and attack sleeping people. This selection of human prey seems to be unusual, in so far as many documented cases of carnivore attacks on humans follow chance encounters or involve sick/injured animals (but see Loveridge *et al.*, Chapter 6, this volume).

For both humanitarian and economic reasons, this problem must be solved. Current research points to a number of village-level interventions that could prevent attacks. As bush pigs are a major attractant to lions, excluding bush pigs from agricultural field with fencing, trenches, or repellents could be one way to prevent lions from entering agricultural fields and coming into contact with people. Other methods include fencing around peoples' homes that surrounds yards and outhouses, stronger house and hut construction, and centralization of village resources such as water and firewood (Kushnir *et al.*, in press). Indeed, Kushnir (personal communication) suggests that a holistic approach to village planning, both lion-sensitive and development-oriented, could include planting fuelwood species near villages to reduce deforestation; designating some areas for agriculture while prohibiting burning elsewhere to reduce habitat destruction; and perhaps establishing village wildlife reserves in order to generate revenue from tourism and hunting and thereby improving conservation. In addition, she argues that mitigation should focus on the root of the problem, which may plausibly be argued to be that habitat destruction leads to prey depletion, exacerbated by over-hunting by people of the lion's prey; which in turn leads to an increased frequency of man-eating by lions. The solution thus lies ultimately in conservation of the lion's habitat and prey, which would be fostered by an education campaign showing the link between poaching wildlife and lion attacks on people (Kushnir, personal communication). The scenario is thus that bush pigs invade agricultural land, and lions follow them there, hence encountering people—thus, if the bush pigs could be excluded, the lions would be less inclined to enter villages and would not encounter people, and thus would be less tempted to eat them. However, it remains to be proven whether the tendency for lions to follow bush pigs arises from a shortage of other, preferred, prey. Villagers suffering lion attacks report that they see fewer natural prey animals than do those free from lion attacks.

Why should these villagers tolerate any risk to their lives from lions? One answer might lie in the wider perspective of the value of lions nationally. Tanzania is a critical country for lion conservation, being home to 25–50% of all African lions (Chardonnet 2002; Bauer and Van Der Merwe 2004). Lions are not only top predators in the natural ecosystem, but are also of economic importance to Tanzania, as wildlife tourism brings in a large proportion of foreign revenue (Wade *et al.* 2001). Figures of the following sort quickly date. However, in 2004, Baldus (2004) estimated that lion hunting in Tanzania brought in about UD$6–7 million per annum to the hunting industry and $2.4 million per annum to the Tanzanian Wildlife Division. In 2008, the costs include a Trophy fee per lion of $4900; a permit fee of $1250 (for more than 10 days); a conservation fee of $3150 (for the hunter); an observer fee of $2100; and a trophy handling fee of $500, totalling $11,900. Added to these government charges would be the cost of bait: Zebra ($1200), Buffalo ($1900), or Wildebeest ($650), and then fees to hunting companies for food, accommodation, and the provision of a professional hunter. A cheap safari might cost $50,000, but costs could commonly reach $100–150,000.

Against this background an analogy—beguiling but false—might be that tolerating a human death or two annually is as an occupational hazard, akin to the risks a miner accepts in order to gain a living from a natural resource. The analogy is flawed because these villagers enjoy no income whatsoever from lions, and indeed many live nowhere near protected areas, so to suppose that their selflessness extends to tolerating the risk of being eaten for the sake of the national economy is fatuous. What is relevant is the tolerance of wildlife displayed by many African communities, at least in terms of levels of stock losses (e.g. Loveridge *et al.*, Chapter 11, this volume) that would not be countenanced by Western farmers. This raises the thought that the well-meaning preoccupations of Western biologists may introduce a European agitation regarding conflict to people who would otherwise tolerate it as a natural phenomenon in much the same way they do droughts and disease. However, while tolerating some loss of cattle is one thing, human deaths are another, and it would be wrong for conservationists to deny the risks that these treasured felids can pose to those living alongside them.

Impact of culture on attitudes to conflict: the case of Ruaha's big cats

As Cavalcanti *et al.* (Chapter 17, this volume) emphasize, hostility towards large felids is not always straightforwardly linked to peoples' experiences, or fears, of depredation. In Ruaha, Tanzania, Dickman (2008) studied the sociology of conflict in a context in which people initially reported losing a mean of 12.4 stock animals per month (which equated to 13.8% of average herd size) and depredation accounted for around 10% of those losses on average (1.15 livestock per month, or 1.2% of herd size); with disease accounting for over 60%. However, long-term monitoring showed that losses to depredation had been markedly overestimated, with losses to depredation actually closer to 0.26% of total herd size in an average month over a year-long period.

Not surprisingly, Dickman found levels of retaliatory lion-killing extremely hard to pin down—only 7% of people ($n = 19/268$) openly admitted to having killed predators, with six admitting to killing lions; anecdotal evidence, however, convinced her that poisoning and snaring is much more widespread than these initial figures would indicate. Against this background, Dickman (2008) found that a person's ethnicity was pivotal to their attitudes towards wildlife, due to the beliefs, traditions, values, and cultural ideologies linked to that ethnic group. There, a statistically highly significant relationship existed between a person's ethnicity and their level of hostility towards wild cats and other large carnivores. Having controlled, statistically, for their current involvement with livestock, people from traditionally pastoralist ethnic groups such as the Maasai and Barabaig were less tolerant of big cats and other predators than were other groups.

Furthermore, people who adhered to organized religions were significantly more hostile towards carnivores than were those who were not religious or who held animistic beliefs. This trend emerged particularly strongly among respondents from traditional pastoralist backgrounds, with both Maasai and Barabaig pastoralists who had converted to Christianity showing more antagonism towards large carnivores than did their kinsmen retaining traditional beliefs. In some cases, these attitudinal associations were understandable. For example, Muslim respondents were more hostile towards large cats and other carnivores than were most other groups, at least in part because they could not eat the meat of depredated livestock as it was not Halal. Conversely, several of the Maasai with traditional beliefs viewed occasional depredation less negatively, a blessing in disguise providing an otherwise rare opportunity to eat meat. A small level of depredation by lions and leopards was actually viewed by some people as advantageous, as they believed that it would lead to an increase in the number of young animals born the next season, but this belief did not hold for attacks by hyenas (*Crocuta crocuta*) and other large carnivores.

Hostility towards predators can be exacerbated as a result of social tensions between human communities. Thus, in her study around Ruaha National Park, Dickman (2008) recorded that a key reason given by respondents from the Hehe tribe to explain their intense dislike of carnivores was their belief that the rival Gogo tribe trained predators to kill stock belonging to Hehe households, so that the Gogo could eat meat without killing their own stock (and see cultural relativism). Similarly, some Maasai perceived that their traditional rivals, the Barabaig, benefited from lion presence on the grounds that if the Barabaig killed a lion they were rewarded with cattle from fellow Barabaig, whereas the Maasai themselves felt excluded from such benefits. Interestingly, although this perceived inequality was cited several times by disenchanted Maasai, none of the Barabaig interviewees mentioned any such benefit to them of having lions in the area.

Certain trends were specific to one ethnic group but not others. For instance, tolerance towards felids and other large carnivores increased with wealth (measured by number of cattle owned), but only for Hehe households. Similarly, it was also only among Hehe that it emerged that the heads of households were significantly more antagonistic towards large felids and other carnivores than were other members of the household, presumably because household heads bore the brunt of responsibility for any losses. It appears that some ethnic groups (in this case the Hehe) and people within the ethnic groups (in this case, people other than the heads of households) show more flexibility in their views towards large felids than do others—important information when devising conflict-mitigation strategies.

Stakeholders may subscribe to very different views. In Ruaha, Dickman's interviews revealed that beneath peoples' stated general support for the park's presence there was resentment of lost access to grazing land

and water (Dickman 2008). There was also a perception of heavy-handedness by local wildlife authorities, one villager having been killed as a suspected poacher. While such extremes are rare, over 10% of people who had experienced any interactions with personnel from the nearby park reported it as negative (commonly being chased away while grazing close to protected area boundaries). Meanwhile, only 1.5% of people ($n = 4$) reported receiving any income from tourism. The park did conduct some outreach but, worryingly, any contact with the park, whether negative or positive, seemed to inflame conflict in so far as people who had had park contact were significantly more hostile towards felids and other carnivores than were those who had not. This suggests that deep-seated hostility towards the park was an important factor in shaping human–felid conflict.

Overall, while depredation was cited as the main reason for antagonism towards large felids and other carnivores, being significantly linked to the intensity of reported conflict when examined alone, it was not retained as one of the most important predictors of hostility in the final model, suggesting that other factors are actually more important in driving conflict. The final model, including ethnicity, adherence to an external religion, conflict with other species, and direct sighting of predators, was significant and accounted for 26% of variation in reported conflict intensity (Dickman 2008). In short, attitudes towards large felids are complex and arise from a variety of social, cultural, and economic drivers as much, or more, than from personal experiences of depredation. The evidence we have unearthed on this human facet of felid conservation is fragmented, but sufficient to suggest a wide-ranging research agenda, and to make clear that effective mitigation measures will have to be multifaceted, addressing these numerous drivers of conflict, rather than focusing on reducing stock losses to predators alone.

Do communities exist? The case of Makgadikgadi's lions

One possibility is that a community might tolerate predators because the losses incurred by some sectors are more than offset by the gains in others. Hemson *et al.* (2009), revealed some painful realities about the meaning of 'community' when they explored this regarding the 32–41 adult and subadult lions in the 4900-km^2 Makgadikgadi Pans National Park in Botswana (Hemson *et al.* 2009). Botswana has impressive conservation credentials, with over 18% of its land designated as protected areas, and a further 21% as wildlife management areas (WMA) and controlled hunting areas (CHA; Lawson and Mafela 1990). The hope is that communities in WMAs will benefit from sport hunting, game ranching, veldt products, and photographic tourism, thereby fostering synergy between conservation and development. Botswana is also the only member of the Southern African Development Community (SADC) with a state-funded compensation system. Nonetheless, Hemson *et al.* (in press) found many people were actively hostile to lions. Indeed, in their study area a stock-killing episode in 2000 resulted in eight lions being killed in retaliation during the ensuing year. This emphasizes the numerical reality that however many hundreds of people in a community might be persuaded to value lions, it takes only one person with some poison or traps to obliterate entire prides.

Hemson *et al.* (2009) explored the question of whether the advantages the lions brought to tourism offset the losses caused by their depredations on cattle. The study area was populated by about 4200 adults (55% of Botswana are under 19 years old); residents from 86 cattleposts (84 had stock at the time), 55 tourism employees, and 76 villagers were surveyed. The cattle-posts were 0.2–22 km (average 5.1 km) from the national park, and were home to an average of 3.9 people and 63.5 cows. Negative attitudes towards the national park and lions were most frequent in cattle-post owners, followed by villagers and then tourism employees. At first, it seemed paradoxical that reported losses to lions did not correlate with quality of preventative measures at specific cattle-posts (e.g. the size or type of their livestock enclosures; numbers of dogs, or herders). It emerged that this was because herders took a relaxed attitude to husbandry, with 81% stating that they did not fetch their stock home in the evening (Hemson *et al.* [2009] estimated that 13% of the average herd stayed in the park overnight). Fifty-two herdsmen (59%) claimed to have lost livestock to lions in the previous year; of which 117 kills out of 375 were reported outside the cattle-post, and 22 within kraals (107 kills were at

Table 29.2 The value of livestock owned, sold, and lost to theft, disease, and lions per annum and the distribution of these losses across the community.

Value of	Mean (StDev)	Median (range)	Of 86 cattleposts
Livestock owned	US$14,119 (21,617)	US$7061 (0–634,650)	6 no livestock
Stock sold	US$716 (1555)	US$176 (0–8815)	31 no sales
Losses to disease	US$482 (1523)	US$47 (0–13,113)	67 no losses
Losses to theft	US$172 (430)	US$0 (0–2435)	42 no losses
Losses to lions	US$283 (592)	US$95 (0–4528)	34 no losses

Source: Adapted from Hemson *et al*. (in press).

night, 30 by day). The mean value of livestock lost to lions was estimated at US$ 646.0 per annum for each cattle-post reporting losses. Thus, across the 84 cattle-posts, the total annual loss of lions totalled ~US$ 33,643 per annum (for context, sales totalled $62,000 and rustling cost $14,000; Table 29.2).

It was only among employees of safari camps (and then only marginally) that the majority said they liked living near the national park. Of 132 interviewees who expressed a view on why they did not like living near the park, 101 cited stock losses to predators. Few people felt that they or their community benefited from tourism. Herdsmen were not motivated to improve their stockmanship and only 15% said they wanted to do so.

While most respondents throughout the community took the view that herdsmen were responsible for livestock, the general feeling was that government or state wildlife authorities were responsible for controlling livestock predation. When asked what the government could do to reduce livestock loss, the reply most frequently obtained (61% of 173 respondents) was that it could erect a fence between their grazing areas and the national park (although the herdsmen use the park as a grazing area); 16% suggested it could kill lions (and indeed, our photos of snares set in existing fences illustrate a determination to turn words into action). In short, the general perception was that the benefit derived from lions accrued to somebody else (government or tour operators), and that somebody else—the government—should therefore assume responsibility for the big cats. Tour operators felt they already gave enough money to the government, so they too put the onus on the latter (Fig. 29.16).

That compensation, at 80% of market value, could not be claimed, as the fact of cattle being killed within the park was evidently insufficient incentive for retrieving stock at night. Consequently, 33% of incidents reported by cattle-post respondents had not been reported for compensation and 21% of those that were, were unsuccessful. Thus, Hemson *et al.* (in press) estimated that each cattle-post recoups less than half of the value of stock that is lost to lions, which the herdsmen estimated as an annual average loss to lions for each cattle-post of US $168. Revenue from tourism in the study area was estimated as US$ 150,000 per annum tax paid to government; US$ 55,000 resource rental paid to the Central District Council; and a lease fee of US$ 6600 per annum paid to the local community (together with salaries roughly equal to US$ 105,600 per annum). If, as seems doubtful, salaries as well as the lease fee are counted as community benefits, then overall income might be said to exceed the crude estimate of the costs of living with lions; these benefits, however, do not necessarily accrue to the same individuals who suffer the losses. The rather unpromising summary is that the community wanted fewer lions, the right to kill them, more compensation, and the national park fenced. They did not appear willing to take action, other than killing lions, to address the problem of livestock loss. Rather, they viewed the government as solely responsible for the costs incurred by the maintenance of wildlife.

Hemson *et al.* (in press) conclude that in this case the 'community' existed only as the emergent property of individuals whose priorities were, understandably, the livelihoods of their own families; to whom, therefore, the fact that lions 'paid for' a school or

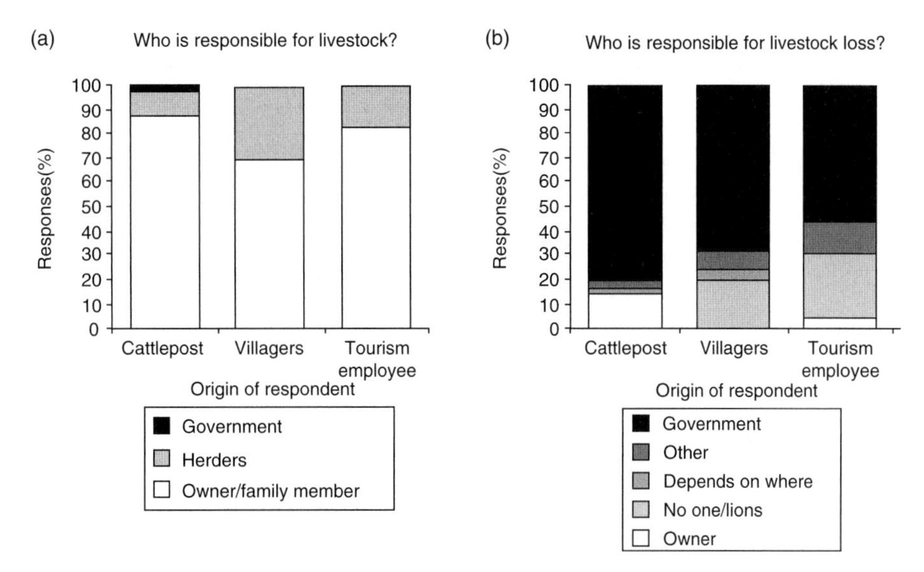

Figure 29.16 Proportions of respondents from each category and responses to the question of (a) 'Who is responsible for livestock?'; and (b) 'Who is responsible for livestock loss?' (Adapted from Hemson *et al.*, in press.)

hospital elsewhere in the region was irrelevant. Since only one disgruntled person would be sufficient to devastate the lion population, it is crucial that compensation reach the people who carry the costs.

Operating a compensation scheme—the case of the lions of Amboseli

The almost universal difficulties of designing, funding, and implementing compensation schemes are exemplified in East Africa, where human and livestock populations are dense (human population of the area has tripled over the past 20 years; Campbell *et al.* 2003). Protected areas are unfenced and corruption and patronage are rampant (Kenya remains in the upper quartile of most corrupt countries in the world according to Transparency International 2008).

Maclennan *et al.* (2009) provide a case study from the Amboseli ecosystem of southern Maasailand, Kenya, where a privately funded livestock compensation scheme has contributed towards a reduction in the rate of lion-killing. Amboseli, Kenya's third most visited national park, nestles amidst a constellation of unfenced ranches, across which roams wildlife—including predators that daily kill livestock. Lions are crucial to Amboseli's US$3.5 million tourist revenue. In 2001/2, mainly in retaliation for livestock losses, and partly for traditional reasons, the Maasai community owning the 1200 km^2 Mbirikani ranch speared 24 lions. Fearing their local annihilation, in 2003 a local NGO drew on American donors to establish the Mbirikani Predator Compensation Fund (MPCF). At a cost of US$44,201 per year, this has conserved an average of 13 lions per year, using the following techniques during 2004–2006: programme rules and national laws were enforced with the help of partner organizations, such as a local game scout; also a 'Lion Guardian' programme was setup, which employs young would-be lion killers to monitor lions instead and to assist in preventing depredation. So, wider society—in this case the NGO representing the ideological interests of conservation and the financial interests of ecotourism—valued lions sufficiently to pay herdsmen not to kill them. This, in contrast to the hitherto more familiar 'polluter pays' principle of environmental regulation, is a case of the developing genre of 'beneficiary pays' programmes, collectively known as Payments for Ecosystem Services—in this case, the continued presence of lions (see below).

A fixed rate is paid per head of each species of livestock (US$192.86 per cow, US$28.57 per goat, and US$85.71 per donkey), and each claim against the scheme is verified by a MPCF representative. The team examined 1694 stock-killings—claims averaged 52 per month. Eight monitors were employed to verify claims, and if anybody in the community kills a carnivore, the entire community loses its payments. Nearly 2.3% of livestock were lost to predators annually. Owners were compensated at full market value, but most were penalized by 50–70% for avoidably sloppy husbandry, so the US$33,000 per annum paid out was less than half the value of the predated stock. Fraud is the Achilles heel of many compensation schemes, and particularly challenging to quantify. However, the fact that claims for livestock depredated each month were not correlated with livestock market prices at a nearby town provides reassurance that corruption was not an obvious problem within the MPCF.

The original motivation of the MPCF was solely to offset peoples' livestock losses to lions, in return for a commitment from the community not to kill lions (the community lost ~0.1% of their herd or 36 animals per year to lions). The community refused the initial offer and demanded compensation for losses to *all* large predators, threatening to kill all large carnivores if compensation were not forthcoming. In addition to lion-induced losses, they lose 2.21% of their herd to other large carnivores per year. On average, US$33,116 is paid for livestock claims per year (staff wages and administrative expenses cost a further US$11,085, and payments per case vary between 27% and 94% of market price). Carnivores incur ~US$70,000 worth of stock losses every year. Not surprisingly, many tensions bedevilled the first years of the MPCF, which had to be shut down on several occasions, for up to several months at a time, in response to fraud or non-adherence to the rules of the programme.

Interestingly, most kills of livestock (43% of the claims) were made by hyenas (lions accounted for only 7% of stock-killing), while 55% of the animals claimed for were strays left untended at night in the bush. Indeed, the project demonstrated that bringing in all livestock at night, to a well-secured, hyena-proof boma could reduce the stock losses, the costs of the programme and animosity towards large carnivores. Frustratingly, however, even as the MPCF matured, between 2003 and 2006 there was no reduction in people's tendency to claim for stray animals.

Taking into account the costs of compensation (US$ 44,201 per annum) and scientific monitoring (US$36,498 per annum), keeping the average lion alive cost US$ 6124 a year. Has the investment of $80,699 per annum over 3.5 years worked? It's too early to say, but in the interim a short-term metric of success might be the number of lions killed on the ranch, and this has surely diminished. However, lion numbers on Mbirikani did not increase, perhaps because people on neighbouring properties continued to kill them. Nonetheless, without the scheme there probably would not have been any lions left; it is now being extended to the neighbouring properties too.

Joined-up disciplines in the quest for solutions

So, when its your own backyard, it can be seriously difficult living with wildlife, especially the sort that can eat not only your livelihood, but also your family. Incentivizing people to do so raises a mix of questions in psychology, sociology, and economics—perilous topics for an innocent naturalist. What is to be done? In an essay entitled *Theory into practice without bluster*, Macdonald and Sillero-Zubiri (2004b; fig. 23.3, p. 368) posed the same general question with regard to canids, and suggested a framework for thinking about this elaborated from Macdonald's Biodiversity Impacts Compensation Scheme (BICS 2001). This envisages an iterative process flowing from problem to solution. BICS develops the notion that problems can be partitioned between reducible and irreducible elements. The balance between these will shift as currently intractable elements are rendered reducible by new innovation (unfortunately, it is also possible for currently reducible problems to become irreducible when, for example, they cross some ecological threshold). Macdonald and Sillero-Zubiri (2004b) suggest that most HWC problems fall within only a small number of categories, most interventions following a similar and predictable trajectory (iteratively cycling between planning, implementation, and evaluation). The priority at any one time is to mitigate the reducible problems, thereby minimizing the current level of conflict. Conflict can be partitioned between that which is bearable (more or less willingly) by the afflicted stakeholders, and that which is unbearable. The extent to which stakeholders will bear a conservation cost (such as predation), will be affected by their tolerance, in turn affected by the value they attribute to the problematic felid, which in turn may be greatly affected by education. Two options are relevant to the unbearable component in current conflict: either to control (generally to kill) the problematic predator, or to compensate the aggrieved stakeholder.

At the micro level of alleviating or mitigating the problem there are three components: (a) the efficacy of the techniques; (b) the competence of the diagnostician and technician; and (c) the competence of the recipient. The stable of techniques to mitigate HWC that is applicable to felids has not grown recently beyond the mediaeval armoury of better husbandry, barricades, and repellents. Nonetheless, research continues with other taxa (e.g. on aversive conditioning; Baker *et al.* 2007), with some inspirational innovation, such as the use of chillies or bees to repel elephants (Parker and Osborn 2006; King *et al.* 2007) and the beginnings of a 'holistic' approach to felid pests (e.g. leopards, see below). An important question is whether the many, penetrating lessons learnt at local scales, regarding felids from snow leopards to pumas, from caracal to jaguars, can be 'scaled up' into useful generalizations to underpin a practical decision-support system to deliver best practice globally. For example, Zimmermann and Macdonald (in preparation) are undertaking a meta-analysis of >100 cases of jaguar conflict across the entirety of its 19-state range. With this broad canvas in mind, a group of 30 felid specialists gathered at Oxford's WildCRU in 2007 to thrash out how to join up lessons learnt locally into a transferrable decision-support system for human–felid conflict. Their deliberations suggest that a feedback loop starting with local experiences (including short-term interventions, long-term management, and preventative actions) could refine general principles expressed in guidance systems (technical guidelines and information networks) which in turn would be applied locally (Zimmermann *et al.*, in press).

The competence of the specialists who diagnose the problem and propose the solution raises a twenty-first century conundrum. Because of the emerging awareness of the interdisciplinarity of conservation, enmeshed with development, the specialists need competence in topics as diverse as predator–prey dynamics and community micro-finance and insurance. Current training rarely aims for such diversity of ability in any one person, and career structures tend to penalize those who attempt it; alternatively, project financing rarely permits the assembly of stable teams combining the necessary blend of skills. There is thus an awkward need for a new sort of specialist, and a new *modus operandi*, adapted to this new perception of reality, and perhaps structurally more closely analogous to a medical team (i.e. a unit combining the varied skills of paramedic, anaesthetist, surgeon, and physiotherapist, etc.) than has been typical in the environment sector hitherto (Macdonald *et al.* 2007). As we look towards retaining and restoring felids, especially bigger ones, this new breed of individual conservationist (or new structure of conservation team), will need to combine all the skills of biologists, geographers, sociologists and economists to predict where and when HWC and other threats to the persistence of felids will foment, and pre-emptively to alleviate them. Doubtless these plans will be laid at the level of entire landscapes, sometimes spanning international borders, and wrestling with issues spanning predator–prey demography to community land rights.

Finally, the competence and motivation of the recipient of advice is clearly crucial to the success of its implementation. Conservation conversations, if not the published literature, are replete with stories where dependency or carelessness causes soluble problems to remain unsolved. Ultimately, the energy and competence of conservationists is born of the value they attribute to nature, and that appreciation of value needs to be nurtured, as does the motivation to deliver results.

Social science—the unfulfilled promise?

Two related facts have been obvious to conservation biologists for years. First, the increasing centrality of the human dimension to conservation makes it essential to understand and manage nature in the context of the human enterprise, and thus conservation science must integrate the natural and social sciences. Second, despite the longed-for synergistic win-win between conservation and development, the increasing human population (currently growing by 1.5 million each week) inevitably throws these two grand purposes into conflict, ensuring the politicization of conservation decisions. Both these issues are introduced in Macdonald *et al.* (2007), and both have caused a hopeful generation of conservation biologists to turn expectantly to social scientists for enlightenment and salvation. Perhaps expectations were unrealistic—after all, in the interdisciplinary reality symbolized by that messily entangled ball of string mentioned in our introduction, no one discipline can untie all the knots—but our sense is that this hopeful generation feels that the promise of social science is as yet unfulfilled.

Certainly, there appear to be many more biologists learning about social science than vice versa. Of course, one might charge that biologists fall at the interdisciplinary fence through inadequate understanding of the theories that explain human behaviour, the tools to influence it, and the research methods to assess it (Campbell 2005; Lélé and Norgaard 2005). Nonetheless, the need for joined-up thinking with which we began this chapter is so obvious, and our conviction so deep that the success of felid conservation lies in the joined-up hinterland of natural and social sciences, and their unified input to politics, that we will now attempt our own (biologist's) perspective of what social sciences can and must contribute in future. Conservation actions are one aspect of human behaviour, and successful conservation requires that the underlying political, economic, social, and cultural contexts that shape behaviours be not only understood but also influenced (Mascia *et al.* 2003; Fox *et al.* 2006). In Chapter 1 (Macdonald *et al.*, this volume) we report that the authors of this book identify felid–human conflict as the paramount issue, and this is developed in Chapter 6 (Loveridge *et al.*, this volume). Here, we introduce some techniques offered by social science to add to the conservationist's toolkit and then briefly mention some cases where they have been applied to felids.

Techniques

People hold various, sometimes contradictory, views about humanity's relationship with the natural world. Differences of opinion regarding human–felid conflict will probably grow in frequency and intensity, not least at the interface between conservation and development. We turn, optimistically, to the social sciences to seek better understanding of such disagreements and perhaps even to learn how better to mediate them (Herda-Rapp and Goedeke 2005), although this mediation will be achieved in an arena far beyond that of academia.

Possible categories for grouping social science tools include: governance, capacity building and organizational development, adaptive management, demographic and gender issues, socio-economic assessments and alternative livelihoods, participatory approaches to conservation and communication, and education initiatives. The approach most commonly borrowed by conservationists in the context of HWC is an assessment of attitudes (illustrations in Murphy and Macdonald [Chapter 20, this volume] and Cavalcanti *et al.* [Chapter 17, this volume]). Unfortunately, in a review of 19 papers in which conservationists had employed social science techniques, Browne-Nuñez and Jonker (2008) found only one that met their criteria for professionalism, which included describing a theoretical framework to explain the relationship of attitudes to behaviour and addressing the reliability and validity of the findings. It may be, therefore, that while biologists have been busy in learning social science tools, they have not been thorough in their execution.

One theoretical structure to explain the relationship of attitudes to behaviour is the Theory of Planned Behaviour (TPB; Ajzen and Fishbein 1980), which holds that behaviour is not determined by attitudes alone, but also by social norms and perceptions of control over the behaviour. Drawing on psychology and neuroscience, this is the terrain occupied by cognitive psychologists—a profession whose goals may have more in common with those of conservationists than is the case for branches of social science preoccupied only with understanding human behaviour, in so far as both cognitive psychologists and conservationists are interested in the practicalities of bringing about changes in human behaviour.

Marchini and Macdonald (in preparation-b; see Cavalcanti *et al.* Chapter 17, this volume) used TPB to study ranchers in the Pantanal and Amazonia, confirming that peoples' mindsets vary from place to place. This is pivotal to conservation in so far as actions grow from perceptions which manifest as opinion, as captured in Herbert Spencer's pithy epithet 'Opinion is ultimately determined by the feelings, and not by the intellect' (Spencer 1851). Marchini and Macdonald's report (submitted) illustrates the importance of feelings, or more precisely emotions, elicited by big cats; that is, 88% of 1033, 10–19 year olds in the Pantanal and Amazonia used emotion-laden adjectives in response to a question about jaguars: fear and anger were prevalent, followed by awe. Indeed, Manfredo (2008) places emotions at the heart of human attraction to, and conflict over, wildlife. This component of human–wildlife interactions has been neglected, probably because conservationists emphasize rationality as a means of combating emotional arguments; and because, for academics, emotions are both difficult to measure and conceptually complex. Emotion and cognition involve different parts of the brain, but interact in decision-making.

Returning to opinion, those of the Pantaneiros, typically from long-established, land-owning dynasties, were different from the Amazonians who were mostly immigrants with small properties and scant experience of jaguars or forest. Pantaneiros, especially those with least education and most economic pessimism, regarded jaguars as a menace to their stock; whereas the Amazonians were more worried about jaguars attacking them than they were their cows (Marchini and Macdonald in preparation-b.). While 60% of Pantaneiros expressed a continued willingness to persecute jaguars, only 20% of Amazonians did so. Their mindsets differed strikingly: Pantaneiros were heavily influenced by their perception of what was acceptable to their peers—who tended to view killing jaguars as a legitimate tradition—while Amazonians were oblivious to peer endorsement, and believed they lacked the capacity to kill jaguars (although they quite liked the idea), and were too scared to try. Intention to kill jaguars turned out to be less directly linked to stock losses, and more to do with social acceptance that killing jaguars was part of the natural order of things. As a result, the project shifted attention from the largely nugatory task of persuading jaguars not to eat cattle, to teaching children the value of jaguars, and thereby to exert moral pressure on their parents. Ultimately, it is helpful to appreciate the distinction between that which is universal in HWCs and that which is site- or culture-specific (Manfredo and Dayer 2001).

Postscript: what is next?

Hard choices

Trends in world population and human consumption, the worsening crisis in food security and escalating demands for all environmental services, and the space requirements of felids as top predators, combine to make scanning the horizon dispiriting. Three things are obvious. First, things will get worse before they (might) get better. Second, to survive the short term we need radical plans, which involve developing financial instruments, and finding means of alleviating poverty. Third, in thinking about conservation, there is an urgent need to combine day-to-day fire fighting with a new and radical long view. All of this has implications for the status of protectionism as a conservation tool.

Short of famine, disease, war, or other interventions of a distressingly Old Testament genre, reducing the burgeoning human population, it is obvious that self-preservation and the precedence of some social and economic imperatives are set to degrade ecosystems even further, thereby devastating biodiversity. It is also obvious that felids, as top predators precariously aloft food pyramids, will be among the hardest hit. The only potential bulwark against this unfolding tragedy is that society places value on these animals and the areas that sustain them. This value can take two forms. It may be financial; in which case when fully itemized it must exceed the value of alternative, unsustainable uses of the species and their habitat. It may also be measured in terms only dubiously monetizable, as conceptualized in existence value (Pearce *et al.* 2007).

In both cases, education is crucial. For direct or indirect financial (or use or non-use) value, science and education enable citizens to comprehend that the clouded leopard (*Neofelis nebulosa*), or at least the forest that supports it, may be worth more to them in a full economic analysis than would be the palm oil that appears superficially more valuable. For non-monetizable value, treasuring the leopard depends solely on education to the effect that its beauty transfixes and its lifestyle enthrals. Unexpectedly, this is why pure research can be among the most powerful applied weapons, because its findings, packaged in television documentaries and the like, enhance existence value. Interestingly, joy derived from nature is emerging as a quantifiable ecosystem service in so far as evidence accumulates that health, immunocompetence and psychological well-being benefit from contact with nature, which may thus diminish health service costs. So, in terms of Fig. 29.16, if ways are not found of making it advantageous for citizens to protect felids, or any other aspect of biodiversity, the tendency to convert wilderness to whatever use is most profitable will inevitably prevail, when push comes to shove. Although wilderness is already greatly degraded (see above), it seems certain that it will now get worse because, in the face of the current burgeoning demands of many economic sectors, it is already inconceivable that financial mechanisms can be found to persuade people to hold even the current line.

Rohan (2000) draws a distinction between the term *value* when used as a verb (assigned on the basis of meaning, goodness, and worth) and as a noun (a stable and enduring cognitive measure), such that the process of valuing (verb) something is based on the enduring values (noun) that a person holds. In the United States, with increasing affluence and education and urbanization, values (noun) are reputedly shifting from materialism (focused on physical security and economic well-being, also use and management of wildlife for human benefit) towards post-materialism (focused on quality of life, self-expression and self-esteem, and mutualistic orientations towards wildlife; Inglehart 1990; Manfredo *et al.* 2003). One might thus suspect that while the educated upper class in, for example, Brazil might hold post-materialist values in regards to jaguars, the great majority of the Brazilians, including those who share the land with the big cats, destroy jaguar habitat, and persecute jaguars, are more concerned with their own physical security and economic well-being than with the status of jaguar populations. In this sense, the future of felids may depend not only on the effectiveness of economic instruments that make people value (verb) them more, but also on the speed with which the luxury of post-materialist values (noun) can be spread through the 'lower strata' of society as affluence and education increase (i.e. we want to educate people to coexist with big cats). Escape from poverty not only begins a journey towards affluence and education that meets a humanitarian imperative, but may initiate an attitudinal shift which appreciates nature as having higher intrinsic value, a shift that may be accelerated if the escape has been funded by economic instruments raising the financial value assigned to wildlife.

Direct values

Why might a sovereign state accede to an external request to modify or curb a given practice? The answer lies in the transnational consequences of that practice. Consider the farming of tigers in order to supply their bones to TCM, in which case the answers separately concern welfare and conservation issues (see below). Failure to keep captive tigers in accordance with high standards of animal welfare (e.g. as required for domestic stock in the United Kingdom), would surely attract international opprobrium; with reputational and financial consequences that might influence Chinese policy. Even if acceptable welfare standards were achieved, the mind-your-own-business riposte fails to the extent that TCM creates either a demand (for wild tigers) in consumers (e.g. in China) that damages the interests (viability of tiger populations) of non-consuming producer-countries (e.g. India), or to the extent that tigers are regarded as a global resource, requiring their stewardship by all for the greater good. People, or states, might change their behaviour for fear of losing (or gaining) either reputation or money, and in general the former is likely to affect the latter. The balance in this quadrant of reputational and financial gains and losses dictates the direction of policy and behaviour change.

Market forces—threat or solution?

Farming versus harvesting

Recent reassessments of tiger numbers make disquieting reading. In 2007 India, which contains by a large margin the most tigers, more than halved the estimate of its tiger population; from over 3000 to less than 1400 (Chundawat *et al.* 2008). Given that far more tigers exist in captivity than the wild (estimates stretching between 5000–14,000 in the United States, and 5000 in Chinese farms), what contribution might this captive stock make to tiger conservation?

This question touches on the more general issue of the impacts of supply and demand in conservation thinking—prompting Norton-Griffiths (1995) to ask provocatively whether bans motivated by conservation can be the kiss of death: that is, if a ban increases the supplier's fixed costs, leading to inflated price, and thus, perversely, an increased incentive to secure the product: hurry while stocks last—might it not lead to *accelerated* slaughter of the target felid? Better, perhaps, in general to tackle the demand side—perhaps by educating consumers but perhaps also by economic devices such as flooding the market. Hence the question arises as to whether captive tigers could take the pressure off wild tigers in supplying the TCM market. The main driver of tiger destruction is illegal trade in their parts for TCM, with bones ranging between US$1250–3750/kg (a skin may be worth US $16,000; EIA-WPSI 2006; Nowell and Xu 2007). In a controversial letter to the *New York Times*, Barun Mitra argued 'tiger-breeding facilities will ensure a supply of wildlife at an affordable price, and so eliminate the incentive for poachers and, consequently, the danger for those tigers left in the wild' (Mitra 2006)—although this proposition prompted outrage from conservationists (Save the Tiger Fund 2009), it can be recast as an hypothesis, at the nub of which is the general—and interesting—question of whether wildlife *farming* is substitutable for wildlife *harvests*. Norton-Griffiths (1995) had anticipated this dilemma a decade earlier by pondering notional 'free range' and 'battery-farmed' markets in tiger penises.

A 1998 survey estimated that more than 27 million tiger-bone products were being traded annually, and, despite the fact that 88% of Shanghai residents knew tigers were endangered, 51.6% of the conurbation's residents were prepared to use these products (Save the Tiger Fund 2009). Anticipating a lucrative market, China and other East Asian countries built captive breeding centres for tigers in the 1980s. They had not yet developed the infrastructure or competence to rear tigers effectively when, in 1993, the trade in medicines containing tiger parts was banned throughout China. Having invested substantial private money into their facilities and more than 5000 captive tigers (Gratwicke *et al.* 2008a), these tiger farms now subsist as loss-making tourist attractions; hunger for the profits that renewed trade in tiger parts could bring still smoulders. The argument espoused by Barun Mitra is that swamping the market with sustainable, farmed, tiger products would eradicate the demand for wild tigers. The 35 NGOs who have formed the International Tiger Coalition to oppose vehemently on conservation grounds any reopening of trade has flatly rejected this substitutability hypothesis, alluring as it is. Opposition has also come from the European Union and the United States. In 2007, the 56th meeting of the CITES secretariat rejected requests to lift the ban.

To test the substitutability hypothesis requires information on both supply and demand. There is no estimate of what the total demand for tiger products from both wild and farmed sources would be if the sale of farmed tiger products were once more allowed in China, nor is the substitutability of the wild and farmed products in the market known. Trade in other species offers contradictory evidence on substitutability: there is evidence that crocodile-farming has protected wild stocks (Thorbjarnarson 1999), whereas farmed bear bile, sold legally and cheaply in China, appears not to assuage demand for high-priced bear bile of wild provenance. Discouragingly, a recent survey found that 71% of Chinese respondents would prefer wild tiger parts (Gratwicke *et al.* 2008b). The degree to which consumers are willing to substitute farmed for wild tiger products at various prices and levels of supply is known to economists as the *cross-price elasticity*; right now there is no measure of this vital parameter. However, in the absence of data on real trade, there would be a way to estimate cross-price elasticity, using 'stated preference interviews' during which members of the public state which products they would prefer to buy. This is a priority because without an estimate of demand, it is hard to say whether farmed stock could potentially meet it. Current estimates, based on total sales from hospitals in China in 1992, are probably a woeful underestimate.

Assuming that tiger farmers could meet total demand, the way to countermand any preference for wild tiger products would be to sell the farmed produce at an irresistible discount; but could it be produced sufficiently cheaply to make this realistic? Barun Mitra (op. cit.) estimated that the cost of raising a tiger is *c.* $4000 (Lapointe *et al.* 2007). However, figures from a farm in Guilin suggest the costs would be much higher. Guilin Xiongsen Bear and Tiger Garden (2007) stated that it has a US$4.9 million annual shortfall remaining after receipts from tourism. The park claims that, if the ban were lifted, it could produce 600 cubs per year. If so, it would need to make US$8160 per tiger to cover current costs (and even more to clear its accrued debts).

Would these production costs be sufficiently low ruinously to undercut the poacher? Well, sufficient poison and bait to kill a tiger costs about $20. The costs of trafficking are unknown, but probably not high (see Loveridge *et al.*, Chapter 6, this volume). The balance of inadequate evidence is that farmed tiger products could not undercut the wild products nor entirely replace demand for them whereas, in terms of enforcement, black markets may be beyond control. Therefore, it seems unlikely that swamping the market with captive tiger products would destroy the market in wild tiger parts. Of course, a partisan Chinese voice might ask why China should spend millions on wildlife enforcement and forsake a lucrative internal market (and indeed a perceived medicinal benefit) for the sake of (largely) Indian tigers. The answer would lie neither in science nor commerce, but in the strength of advocacy expressed by the voice of the international community that values wild tigers.

If, for the poacher, the value of tiger parts remains greater in cash than in the wild, then policing to ensure the big cats' protection will remain essential. Policing the trade in tiger parts may become harder if one sector of it is legal. Given the likely disparity in prices between wild and farmed products poachers would have two non-exclusive and seemingly attractive options. Either they could supply a premium illegal market of wild tiger parts, or they could launder the sale of these parts through a legal market. In short, farming tigers would not negate the need to police the illegal trade; indeed, it seems likely to make the task harder and more expensive. Nonetheless, in so far as consistency is a goal, there might be irony in the fact that conservation places so much importance on preserving seemingly 'useless' species in the hope that pharmaceutical prospecting will one day attribute great value to them, yet is far from rejoicing that millions of Chinese attribute just such value to tigers. The critical point, however, is that the harvest of tigers is currently so grotesquely mismanaged that it threatens their extinction.

A second question concerning captive tigers is whether they could contribute to preserving the species' genetic diversity. Already there are more tigers in the United States alone than there are wild tigers worldwide, and the captive population has the potential to be immeasurably larger (both in census and genetic effective size) than any wild population could be. A provocative thought is that from a genetic perspective it makes no difference whether the tiger lives in a cage or in a forest, so long as there is no artificial selection or biasing towards more fecund individuals; the captive population could be an invaluable genetic storehouse, but there is no consensus on how best to make use of this potential.

Hunting and the 'value' of hunted wildlife: how do charisma, size, and rarity determine demand?

The 'value' of products derived from wild tigers is clearly reflected in the high price consumers are willing to pay. From an economic perspective, price provides an unambiguous index of quality and desirability. Two factors that might affect price and that are worth evaluating from a conservation perspective are rarity and size.

Considering first rarity, the rate at which a consumer's (in this case a collector's) perception of value changes as rarity increases or decreases might be critical for its conservation. If, as stocks dwindle, the costs of harvesting a rare species exceed its value such that harvesting becomes unprofitable, a respite from harvesting might be provided during which its numbers recover. Alternatively, if perceived value increases with rarity faster than the costs of harvesting, an endangered species might get sucked into a vicious vortex whereby exploitation remains profitable even up to the point of extinction (Courchamp *et al.* 2006). Indeed, Sadovy and Chung (2003) describe the plight of a fish, the giant yellow croaker, *Bahaba taipingensis*, which fell victim to such a vortex. Discovered only in

1937, their swim bladders were so valued by TCM that soon 50 t were being landed annually. By 2000, over 100 boats could catch only 10 specimens; the single fish caught in 2008 fetched $130,000. The croaker's value was inflated so spectacularly by rarity that its exploitation remained economic, even as it approached extinction.

A second consideration is size (which we know is associated with rarity to some extent for most taxa). While bigger might be thought to be 'better', bigger species may be more demographically fragile in so far as they are generally slower to reproduce and therefore more vulnerable to exploitation (Peters 1983).

To assess the impact of both rarity and size on species hunted for sport in Africa, Johnson *et al.* (submitted) surveyed the prices of both felids and bovids in brochures selling hunting safaris. Rarity had no detectable bearing on the prices of felids, but it did increase the price of bovids (price increased relatively slowly with increasing rarity, much less than in proportion with our rarity index: it was proportional to numbers available to the power of +0.25). Felids, however, were priced entirely in proportion with their size (slope on log–log scale of +1.0; Fig. 29.17). The cost of hunting a lion compared with that of a lioness was, in statistical terms, completely explained by the cats' relative size difference.

In bovids, by contrast, large species were proportionately less expensive to hunt compared with small species (the slope on log–log scale was *c*.0.40). The explanation for this difference may lie in the fact that the size of bovids' horns, rather than the

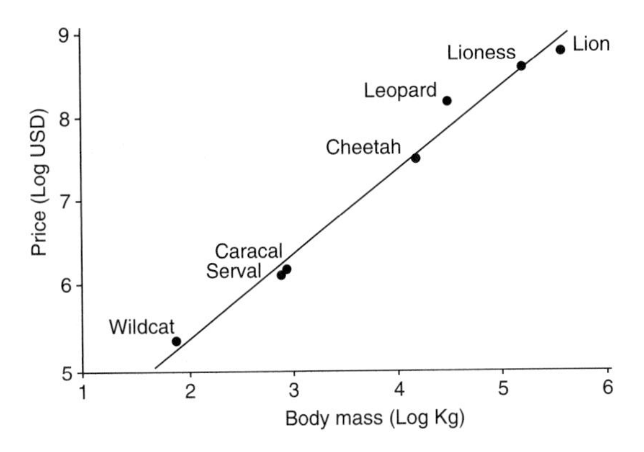

Figure 29.17 The average price of felids offered for sale in brochures advertising hunting in Africa. (Taken from Johnson *et al.*, submitted.)

animals' bulk, determines their value; horns being relatively smaller in larger bovids. In short, for felids, big means pricey; to the extent that more than 500 lions are shot for trophies annually, each at about US $50,000, all in economies where both hard currency and lions are scarce (Loveridge *et al.* 2007c; Chapter 6, this volume; see Table 29.3.).

Indirect values—swaps and trading

Arguably the most topical challenge to felid conservation is devising financial instruments that will persuade people to protect big cats, both as emblems and umbrellas of wilderness. In this book there are examples of conventional mechanisms such as compensation, insurance schemes, the development of alternative livelihoods, brand price premium (e.g. cheetah-friendly meat), and the more avant-garde Payments for Wildlife Services (PWS). Exciting possibilities are opened up by the precedent of carbon trading under the Clean Development Mechanism (CDM) (e.g. Coomes *et al.* 2008), provoking the question of how to design biodiversity offsets, as envisaged under the BICS (see above; Macdonald 2001b). At the time of writing, Reduced Emissions by Deforestation and Degradation (REDD) schemes are currently trailblazers for linking biodiversity benefit to the carbon benefit of avoiding deforestation; this field, however, is currently so aflame with controversy that its parameters are certain to have shifted by the time readers reach this paragraph (Sukhdev 2008). By compensating states for avoiding carbon emissions as a result of not destroying forests they would otherwise have felled, REDD may thereby protect the biodiversity—including that of felids—therein (Parker *et al.* 2008). These schemes pose regulatory challenges (e.g. demonstrating additionality and avoiding leakage [Kanninen *et al.* 2007; Myers 2007]) and they could backfire; but there is real hope that they represent a new dawn across the landscape of biodiversity conservation. Put simply, a variant of REDD might strengthen the protection of Borneo's forests *and* the clouded leopards therein.

Lessons from elementary economics suggest some requirements for the functioning of market-based instruments for the conservation of wildlife, including access to relevant national and international

Table 29.3 Number of lion trophies explored from different African countries between 2000 and 2008.

Country	2000	2001	2002	2003	2004	2005	2006	2007	2008
Botswana	30	8	1	0	0	13	22	26	0
Ethiopia	0	2	2	3	1	0	1	1	0
Mozambique	29	13	10	15	15	24	18	17	18
Tanzania	314	226	225	213	139	209	220	92	0
South Africa	144	136	147	162	174	208	242	371	0
Zambia	38	23	3	38	43	68	68	63	0
Zimbabwe	61	75	81	90	93	88	61	37	0

Source: CITES trade statistics, derived from CITES Trade Database (UNEP-WCMC 2009). Note that these data include the secondary export of specimens. There is often a time lag between lions being hunted, trophies being exported, and member countries submitting data, so numbers for recent years should be treated with caution.

markets and adequate terms of trade (Aylward *et al.* 1998), as well as adequate investment in physical infrastructures (roads, lodges, etc.; Adams and Infield 2003). A particularly important requirement is the establishment of secure property and user rights—because owners will be more inclined to invest in conservation or sustainable use if they are confident of personally accruing the resulting rewards (Bulte *et al.* 2003). Norton-Griffiths (1998) compared wildlife losses during the years 1977–94 in five districts in Kenya to reveal that in those where ecotourist revenue went largely to the tourist business and the government, some 29–65% of wildlife had been lost; whereas wildlife had held its own in national parks in which revenues were shared among group ranches; and wildlife increased by 12% on private land on which owners reaped all the benefits derived there from. As Norton-Griffiths wryly points out, livestock are owned and have value, and in consequence Kenya is stuffed full of livestock; wildlife is not owned and accordingly has little or no value to those on whose land it is found. Since wild animals have no value, they are being eradicated wholesale.

Managing property rights may be too expensive a task (and require too much trust) for governments in some developing countries (Swanson 1995), so one suggestion has been that the state could entrust communities with the responsibility to preserve (and sustainably to use) nature in return for payments under Integrated Conservation and Development Projects (ICDPs). However, not least because ICDPs often fail to compete with more remunerative but unsustainable practices, Wegner and Pascual (2008), conclude that they tend to supplement rather than substitute existing livelihoods (see Barrett *et al.* 2001 and the section above on 'The ecocentric–anthropocentric continuum'). Strengthening of property rights, education, and awareness will all count for little if tolerating felids or their habitat and prey causes landowners a net cost (Engel *et al.* 2008).

Economic incentives (or 'market-based instruments') can take variously ingenious forms (e.g. reviewed by Wegner and Pascual 2008). For example, landowners may be compensated for the opportunity costs incurred in conserving problematic predators such as felids. The objective is to make the lion population worth more in the bush than would be the profits (or savings) in the bank, yielding interest, from its destruction. Variants of this idea described in this book (see above and Jackson *et al.*, Chapter 19, this volume) are being explored for lions in Kenya (Maclennan *et al.*, 2009) and the snow leopard in India (Mishra *et al.* 2003a) under the guise of PWS. PWS involves direct transfer of value from those who benefit from wildlife conservation (they like lions sufficiently to pay for them) to compensate those who bear the opportunity costs locally (they are paid enough to tolerate lions).

Supping with the devil

An awkward coincidence is that many areas of special conservation value are politically unsavoury or even war zones, and many species urgently needing conservation live under regimes censured by the international community for their human rights records. The tigers of northern Myanmar or the lions of Zimbabwe surely need conservation, but both live under governments that much of the international community would ostracize. Indeed, many of the Western liberals who would rightly care deeply for the conservation of, for example, Chinese tigers, Iranian leopards, Pakistani snow leopards, or Venezuelan jaguars might falter before fulsome engagement with the governments of these and many other range states. The juxtaposition of conservation urgency and dismal human rights is sufficiently widespread that it raises the question of whether conservation scientists should avoid, even boycott, projects in countries whose human rights records they view as problematic. Similar, if perhaps less ethically painful, questions might be posed about engaging with governments that tolerate or practice corruption (see Smith *et al.* 2003).

One of us faced a particularly vivid instance of this question when in 2006, after more than a decade of fieldwork, the world's largest tiger reserve was established in the Hukaung Valley of northern Myanmar (Rabinowitz 2008). Joy at the creation of this reserve, an area nearly half the size of Bhutan, was tempered by criticism in the media of working in a dictatorship internationally deplored for human rights abuses. Nervousness over this tension might explain why funding for this park, from either government or non-profit sectors, has been unexpectedly slow to materialize. Indeed, an article in *Nature*, entitled *Conservation in Myanmar: Under the Gun*, concluded '[Rabinowitz] seems to prioritize the needs of wildlife above those of a brutally repressed population'—seemingly positioning an irreconcilable dichotomy between caring for tigers and people. But the reality is that some charitable donors give for people-causes and others for animals; that is one of the charitable sector's strengths. Furthermore, biologist's skills can most effectively be directed at biological problems, and their inputs to conservation, and the money associated brings resources and employment to rural dwellers in ways that are surely humanitarian. In order to retain our funding, others of us have been required to dispel donor government fears that research to conserve lions in Zimbabwe would not be construed as supporting a regime denounced by Britain. In short, engagement should not be confused with approval.

How should these dilemmas be addressed? One solution may be to draw on the generally apt analogy between conservationists tending the environment and physicians tending the body. In a medical crisis, organizations like the International Red Cross work tirelessly for the well-being of imperilled people irrespective of the odium with which their governments may be viewed. Indeed, for aid agencies to withhold care to the sick on the basis of their politics, or their rulers', would be regarded as abhorrent. Without emergency medical intervention, imperilled people die. So too, without conservation intervention, imperilled species will go extinct. If this analogy is valid, then conservationists should, like doctors, work where they are needed; whether that be a terrorist stronghold or under a vindictive regime. Time is short to save these tigers, snow leopards, or lions, and it is likely to have run out long before these species find themselves under comfortable democracies. Of course, from the vantage point of an armchair an alluring distinction may seem to exist between the need to work in a country and the need to collaborate with its government. In practice, though, this may offer only illusory comfort, as it is generally impossible to work somewhere without government approval, or indeed active involvement, especially if one seeks truly lasting solutions. The pressures on biodiversity may be worst when democracy is at its lowest ebb; in the future, the hoped-for more acceptable government may be unable to restore all that has been lost if conservationists withhold their balm until the regime has international approval. Indeed, the anti-engagement school may be fickle when it comes to judging democracy and environmental records: some environmentalists argue that attacks on the environmental legislative framework by the first Bush regime went relatively unremarked because environmental abuse transacted in the corridors of a democracy may be easier to get away with than when perpetrated by peasants with axes.

The generality of modern conservation is that biodiversity provides essential services to humanity, and that wildlife conservation and human development have to advance together and, where possible, to mutual advantage (a goal which can be particularly challenging in the case of big cats, which rural people may prefer to be without). At a finer scale, an implied need to prioritize either wildlife or repressed people forces a false dichotomy. Neither medicine nor conservation is a luxury, to be withheld where conditions are not quite right. Medics help people, conservationists help nature (and thus people), and both should do so where they are needed, rather than where it is politically amenable.

A pact across generations

In undertaking a journey, it is well to keep in mind the destination. Taking the long-term view, where might we hope to reach? One answer might be a human population enjoying a healthy, equitably high and sustainable standard of living, alongside functioning ecosystems populated with natural levels of biodiversity. In the context of felids, these ecosystems would be sufficiently extensive and interconnected to sustain robustly viable populations of big cats and full communities of smaller ones (the jaguar and tiger corridor initiatives described above blaze this trail). This vision of human well-being doubtlessly relies on advanced technology to sustain people in cleverly engineered 'green' towns supported by sophisticated communications networks. It relies on human ingenuity and rationality rising to Hamlet's challenge because, when it comes to saving the environment, and thus ourselves, there *are* indeed more things in heaven and earth to be considered, than have so far been dreamt of in our philosophy. It relies on valuing the spiritual in Nature.

Optimizing the solutions will require the father and mother of all cost–benefit analyses. If each person is to enjoy a satisfyingly high standard of living, then the only way this can be achieved while leaving sufficient space to deliver the associated vision for the natural world (itself a prerequisite for the desired well-being of the people), will be for there to be only a fraction of the projected number of people. Political compulsion and the Old Testament mechanisms of famine, pestilence, and strife are similarly horrific (although sadly not unlikely), so the only optimistic road to this goal is one of phased population reductions over many generations. This would require an intergenerational pact for which there is no precedent in human history.

This destination, or even the much more modest one of somehow accommodating the 5 billion people alongside which large felids are likely to coexist by 2020 (Population World Net 2009), places into sharper focus the question of which staging posts should be planned along the journey (to avoid the current probability that there will soon be insufficient wilderness left to reseed the functioning ecosystems alongside which we hope our descendants will thrive). It seems likely that protectionism, so long unfashionable, will become essential. However, neo-modern protectionism would have to live alongside development, during which phase the imperative of protecting the environment for future (hopefully less populous) generations, cannot trump concern for the basic well-being of existing people. The solution will hopefully include large-scale restoration, and would doubtless include zonation that allowed for various levels of use, and would have to be paid for.

To draw finally on Herbert Spencer: 'No one can be perfectly free till all are free; no one can be perfectly moral till all are moral; no one can be perfectly happy till all are happy' (Spencer 1851). This philosophical absolute is perhaps too grand for the earthily practical conservation biologist, but it is a pointer towards a journey that spans disciplines, links biodiversity conservation, sustainability, and development, and makes the leap from science to practice. It is a big jump, but there is no choice.

Epilogue

So, the journey to felid conservation is challenging, and in this chapter we have sought to suggest some directions it might take, and pitfalls it should avoid. If we lose our way, will the ubiquitous domestic cat be the epilogue to 40 million years of felid evolution? Musing on this dismal thought, at first we wondered what percentage of the lion, ocelot

(*Leopardus pardalis*) and manul (*Otocolobus manul*) genome might be saved as a genetic ghost in such a domesticated felid epitaph. The answer, as Carlos Driscoll penetratingly explained, is none. While all such mutations involve variation, and most variation is random, the differences that make a lion a lion and a domestic cat a domestic cat are not random. Doubtless any meaningful sequence of leonine DNA and the homologous cat sequence will vary by some unknown extent, and nobody can know which parts of that variation would survive evolution; so in that sense nothing that constitutes essential lion (or ocelot or manul) exists in the cat. If we lose wild felids, then perhaps the domestic cat's mannerisms—the pounce, the growl, the face-rub—will remind us of felids past, but that will be all. Better, then, that we set about the enormous task of conserving them, which in the long run probably can happen only if we make it part of saving ourselves as well.

Acknowledgements

We thank the following for their important input to this chapter: Amy Dickman, Chris Barrett, Carlos Driscoll, Adam Dutton, Kerry Kilshaw, Paul Johnson, Eve Li, Harunnah Lyimo, Silvio Marchini, Seamus Maclennan, Jeannine McManus, Laurent Mermet, Ros Shaw, Aleksander Trajco, Giulia Wegner, Claudio Sillero, Mark Stanley-Price, Marion Valeix, Nobby Yamaguchi, and Alexandra Zimmermann. We are also grateful for the selfless help of Stephan Bodini and his editorial eagle eye, to LI Teo for his archaeological insights, and to John Edwards for access to his marvellous archive of old felid pictures.

© D. W. Macdonald.

References

Abbadi, M. (1993) The sand cat in Israel. *Cat News*, **18**, 15–16.

ABSA. (2003) *Game Ranch Profitability in Southern Africa*. Absa Group Economic Research, Rivonia, South Africa.

Abu-Baker, M., Nassar, K., Rifai, L., Qarqaz, M., Al-Melhim, W., and Amr, Z. (2003) On the current status and distribution of the jungle cat, *Felis chaus*, in Jordan (Mammalia: Carnivora). *Zoology in the Middle East*, **30**, 5–10.

Ackerman, B. B., Lindzey, F. G., and Hemker, T. P. (1984) Cougar food habits in southern Utah. *Journal of Wildlife Management*, **48**, 147–55.

Ackley, C. D., Yamamoto, J. K., Levy, N., Pedersen, N. C., and Cooper, M. D. (1990) Immunological abnormalities in pathogen-free cats experimentally infected with feline immunodeficiency virus. *Journal of Virology*, **64**, 5652–5.

Acocks, J. P. H. (1988) Veld types of South Africa. *Memoirs of the Botanical Survey of South Africa*, **57**, 1–146.

Acosta, G., Cossios, D., Lucherini, M., and Villalba, L, M. (2008) *Leopardus jacobita*. In: IUCN 2009. IUCN Red List of Threatened Species. Version 2009.1. <www.iucnredlist.org>. Accessed online 18 September 2009.

Acosta-Jamett, G. and Simonetti, J. A. (2007) Conservation of *Oncifelis guigna* in fragmented forests of central Chile. In *Felid Biology and Conservation Conference 17th–20th September 2007: Programme and Abstracts*, p. 63. Wildlife Conservation Research Unit, Oxford, UK.

Acosta-Jamett, G., Simonetti, J. A., Bustamante, R. O., and Dunstone, N. (2003) Metapopulation approach to assess survival of *Oncifelis guigna* in fragmented forests of central Chila: a theoretical model. *Mastozoologia Neotropical*, **10**, 217–29.

Adams, D. B. (1979) The cheetah: native American. *Science*, **205**, 1155–8.

Adams, W. M. and Infield, M. (2003) Who is on the gorilla's payroll? Claims on tourist revenue from a Ugandan national park. *World Development*, **31**, 177–90.

Adams, W. M., Aveling, R., Brockington, D., *et al.* (2004) Biodiversity conservation and the eradication of poverty. *Science*, **306**, 1146–9.

Addie, D. D. and Jarrett, O. (2006) Feline coronavirus infections. In C. E. Greene (ed.), *Infectious Diseases of the Dog and Cat*, pp. 88–102. W.B. Saunders, Philadelphia, PA.

Adriaensen, F., Chardon, J. P., de Blust, G., *et al.* (2003) The application of 'least-cost' modelling as a functional landscape model. *Landscape and Urban Planning*, **64**, 233–47.

Aebischer, N. J., Robertson, P. A., and Kenward, R. E. (1993) Compositional analysis of habitat use from animal radio tracking data. *Ecology*, **74**, 1313–25.

Aghili, A., Masoud, R., Murdoch, J. D., and Mallon, D. P. (2008) First record of Pallas's cat in northwest Iran. *Cat News*, **49**, 8–9.

Ahearn, S. C., Smith, J. L. D., Joshi, A. R., and Ding, J. (2001) TIGMOD: an individual-based spatially explicit model for simulating tiger/human interaction in multiple use forests. *Ecological Modeling*, **140**, 81–97.

Ahlborn, G. G. and Jackson, R. (1988) Marking in free-ranging snow leopards in west Nepal: a preliminary assessment. In H. Freeman (ed.), *Proceedings of the Fifth International Snow Leopard Symposium*, pp. 25–49. Wildlife Institute of India and International Snow Leopard Trust, Seattle, WA.

Ajzen, I. (1985) From intentions to actions: a theory of planned behavior. In J. Kuhl and J. Beckman (eds.), *Action-Control: From Cognition to Behavior*, pp. 11–39. Springer, Heidelberg, Germany.

Ajzen, I. and Fishbein, M. (1980) *Understanding Attitudes and Predicting Social Behavior.* Prentice-Hall, Englewood Cliffs, NJ.

Akçakaya, H. R., Mills, G., and Doncaster, C. P. (2007) The role of metapopulations in conservation. In D. W. Macdonald and K. Service (eds.), *Key Topics in Conservation Biology*, pp. 64–84. Blackwell Publishing, Oxford, UK.

Akersten, W. A. (1985) Canine function in *Smilodon* (Mammalia: Felidae; Machairodontinae). *Contributions in Science*, **356**, 1–22.

Al-Kharousi, Y. H. (2006) Poaching of re-introduced Arabian oryx in Oman, will accession to CITES help? *Re-Introduction News*, **25**, 23–5.

Aldama, J. J. and Delibes, M. (1990) Some preliminary results on rabbit energy utilization by the Spanish lynx. *Doñana Acta Vertebrata*, **17**, 116–21.

Aldama, J. J., Beltrán, J. F., and Delibes, M. (1991) Energy expenditure and prey requirements of free-ranging Iberian lynx in Southwestern Spain. *Journal of Wildlife Management*, **55**, 635–41.

Ale, S. (1998) Culture and conservation: the snow leopard in Nepal. *International Snow Leopard Trust Newsletter*, **16**, 10.

Ale, S. B., Yonzon, P., and Thapa, K. (2007) Recovery of snow leopard *Uncia uncia* in Sagarmatha (Mount Everest) National Park, Nepal. *Oryx*, **41**, 89–92.

Alexander, K. A. and Appel, M. J. (1994) African wild dogs (*Lycaon pictus*) endangered by a canine distemper epizootic among domestic dogs near the Masai Mara National Reserve, Kenya. *Journal of Wildlife Diseases*, **30**, 481–5.

Alexander, K. A., Kat, P. W., Munson, L. A., Kalake, A., and Appel, M. J. G. (1996) Canine distemper-related mortality among wild dogs (*Lycaon pictus*) in Chobe National Park, Botswana. *Journal of Zoo and Wildlife Medicine*, **27**, 426–7.

Alexander, K. A., Kat, P. W., House, J., *et al.* (1995) African horse sickness and African carnivores. *Veterinary Microbiology*, **47**, 133–40.

Alexander, R. M., Bennett, M. B., and Ker, R. F. (1986) Mechanical properties and function of the paw pads of some mammals. *Journal of Zoology*, **209**, 405–19.

Alexander, S. E. (2000) Resident attitudes towards conservation and black howler monkey in Belize: the Community Baboon Sanctuary. *Environmental Conservation*, **27**, 341–50.

Aliaga-Rossel, E., Moreno, R. S., Kays, R. W., and Giacalone, J. (2006) Ocelot (*Leopardus pardalis*) predation on agouti (*Dasyprocta punctata*). *Biotropica*, **38**, 691–4.

Allendorf, F. W., Leary, R. F., Spruell, P., and Wenburg, J. K. (2001) The problem with hybrids: setting conservation guidelines. *Trends in Ecology & Evolution*, **16**, 613–22.

Allnutt, T. F., Wikramanayake, E. D., Dinerstein, E., Loucks, C. J., Jackson, R., and Carpenter, C. (2005) Protected areas in the Himalaya. In U. R. Sharma and P. Yonzon (eds.), *People and Protected Areas of South Asia*, pp. 116–21. IUCN World Commission on Protected Areas South Asia and Resources, Himalaya, Kathmandu, Nepal.

Almeida, A. (1986) A survey and estimate of jaguar populations in some areas of Mato Grosso. In B. des Clers (ed.), *Wildlife Management in the Neotropical Moist Forest*, pp. 80–9. International Council for the Conservation of Game, Paris, France and Manaus, Brazil.

Altrichter, M., Boaglio, G., and Perovic, P. (2006) The decline of jaguars *Panthera onca* in the Argentine Chaco. *Oryx*, **40**, 302–9.

Alvarez, K. (1993) *Twilight of the Panther: Biology, Bureaucracy and Failure in an Endangered Species Program*. Myakka River Publishing, Sarasota, FL.

American Association of Zoo Veterinarians (2008) <www.aazv.org>.

Amlaner, C. J. and Macdonald, D. W. (eds.) (1980) *A Handbook on Biotelemetry and Radio Tracking*. Pergamon Press, Oxford, UK.

Andelt, W. F. (1995) Livestock guarding dogs, llamas and donkeys for reducing livestock losses to predators. *Colorado University Cooperative Extension Bulletin*, Number 1.218.

Anderson, A. E., Bowden, D. C., and Kantter, D. M. (1992a) The puma on Uncompahgre Plateau, Colorado. *Colorado Division of Wildlife Technical Bulletin*, **40**, 1–116.

Anderson, A. J., Greiner, E. C., Atkinson, C. T., and Roelke, M. E. (1992b) Sarcocysts in the Florida Bobcat (*Felis rufus floridanus*). *Journal of Wildlife Diseases*, **24**, 116–20.

Anderson, C. R. and Lindzey, F. G. (2005) Experimental evaluation of population trend and harvest composition in a Wyoming cougar population. *Wildlife Society Bulletin*, **33**, 179–88.

Anderson, C. R. and Lindzey, F. G. (2003) Estimating cougar predation rates from GPS location clusters. *Journal of Wildlife Management*, **67**, 307–16.

Anderson, C. R., Lindzey, F. G., and McDonald, D. B. (2004) Genetic structure of cougar populations across the Wyoming Basin: metapopulation or megapopulation. *Journal of Mammalogy*, **85**, 1207–14.

Anderson, E. M. (1987) *Bobcat behavioral ecology in relation to resource use in southeastern Colorado*. PhD dissertation, Colorado State University, Fort Collins, CO.

Anderson, E. M. and Lovallo, E. M. (2003) Bobcat and lynx. In G. A. Feldhamer, B. C. Thompson, and J. A. Chapman (eds.), *Wild Mammals of North America: Biology, Management, and Conservation*, pp. 758–86. Johns Hopkins University Press, Baltimore, MD.

Anderson, J. L. (1981) The re-establishment and management of a lion *Panthera leo* population in Zululand, South Africa. *Biological Conservation*, **19**, 107–17.

Anderson, R. M. and May, R. M. (1991) *Infectious Diseases of Humans: Dynamics and Control*. Oxford University Press, Oxford, UK.

Anderson, R. M. and May, R. M. (1979) Population biology of infectious diseases. 1. *Nature*, **280**, 361–7.

Andersson, K. (2004) Elbow-joint morphology as a guide to forearm function and foraging behaviour in mammalian carnivores. *Zoological Journal of the Linnean Society*, **142**, 91–104.

Andheria, A. P., Karanth, K. U., and Kumar, N. S. (2007) Diet and prey profiles of three sympatric large carnivores in Bandipur Tiger Reserve, India. *Journal of Zoology*, **273**, 169–75.

Andrén, H., Linnell, J. D. C., Liberg, O., *et al.* (2006) Survival rates and causes of mortality in Eurasian lynx (*Lynx lynx*) in multi-use landscapes. *Biological Conservation*, **131**, 23–32.

Andrews, C. W. (1914) On the Lower Miocene vertebrates from British East Africa, collected by Dr. Felix Oswald. *Quarterly Journal of the Geological Society of London*, **70**, 163–86.

Angermeier, P. L. (2000) The natural imperative for biological conservation. *Conservation Biology*, **14**, 373–81.

Angulo, E. and Cooke, B. (2002) First synthesize new viruses then regulate their release? The case of the wild rabbit. *Molecular Ecology*, **11**, 2703–9.

Angulo, E., Deves, A., Jalmes, M. S., and Courchamp, F. (2009) Fatal attraction: rare species in the spotlight. *Proceedings of the Royal Society of London Series B*, **276**, 1331–7.

Anon. (2005a) Fishing cat found dead in central India. *Cat News*, **43**, 33.

Anon. (2005b) Mountain lions in US mid-west and Canada. *Cat News*, **42**, 34.

Anon. (1996) The mystery of the Formosan clouded leopard. *Cat News*, **24**, 16.

Anon. (1992) Cat News. *Cat News*, **16**, 26.

Anon. (1989) Snow leopards outwit hunters. *Cat News*, **10**, 8.

Anon. (1974) Manifesto on wolf conservation. *IUCN Bulletin*, **5**, 17–18.

Anon. (1892) Man's friend, the puma. *The New York Times*. Ochs Sulzberger, New York.

Antonevich, A.L., Naidenko, S.V., Bergara, J. *et al.* (2009). A comparative note on early sibling aggression in two related species: the Iberian and the Eurasian lynx. In A. Vargas, C. Breitenmoser, U. Breitenmoser, (eds.), *Iberian Lynx Ex-situ Conservation: An Interdisciplinary Approach*, pp. 156–63. Fundación Biodiversidad, Madrid, Spain.

Antón, M. and Galobart, A. (1999) Neck function and predatory behavior in the scimitar-toothed cat *Homotherium latidens* (Owen). *Journal of Vertebrate Paleontology*, **19**, 771–84.

Antón, M., Morales, J., Salesa, M. J., and Turner, A. (2004) First known complete skulls of the scimitar-toothed cat *Machairodus aphanistus* (Felidae, Carnivora) from the Spanish late Miocene site of Batallones-1. *Journal of Vertebrate Paleontology*, **24**, 957–69.

Antunes, A., Troyer, J. L., Roelke, M. E., *et al.* (2008) The evolutionary dynamics of the lion *Panthera leo* revealed by host and viral population genomics. *PLoS Genetics*, **4(11)**, **e1000251**.

Anyonge, W. (1996) Locomotor behaviour in Plio-Pleistocene sabre-tooth cats: A biomechanical analysis. *Journal of Zoology*, **238**, 395–413.

Appel, M. J. G. (1987) Canine distemper virus. In M. J. G. Appel (ed.), *Virus Infections of Carnivores*, pp. 132–59. Elsevier Science, New York.

Appel, M. J. G. and Summers, B. A. (1995) Pathogenicity of morbilliviruses for terrestrial carnivores. *Veterinary Microbiology*, **44**, 187–91.

Appel, M. J., Yates, R. A., Foley, G. L., *et al.* (1994) Canine distemper epizootic in lions, tigers, and leopards in North America. *Journal of Veterinary Diagnostic Investigation*, **6**, 277–88.

Apps, C. D. (2000) Space use, diet, demographics, and topographic associations of lynx in the southern Canadian Rocky Mountains: A study. In L. F. Ruggiero, K. B. Aubry, S. W. Buskirk, *et al.* (eds.), *Ecology and Conservation of Lynx in the United States*, pp. 351–72. University Press of Colorado, Boulder, CO.

Aranaz, A., De Juan, L., Montero, N., *et al.* (2004) Bovine tuberculosis (*Mycobacterium bovis*) in wildlife in Spain. *Journal of Clinical Microbiology*, **42**, 2602–8.

Arizona Game and Fish Department (2006) Mexican wolf blue range reintroduction project. Interagency Field Team Annual Report for 2006. <www.azgfd.gov/w_c/wolf/reports.shtml>. Accessed online 15 Nov 2008.

Armesto, J. J., Rozzi, R., Smith-Ramírez, C., and Arroyo, M. T. K. (1998) Conservation targets in South American temperate forests. *Science*, **282**, 1271–2.

Arnold, D. (2006) *The Tropics and the Travelling Gaze: India, Landscape and Science 1800–1856*. University of Washington Press, Seattle, WA.

Artois, M. and Remond, M. (1994) Viral diseases as a threat to free-living wild cats (*Felis silvestris*) in Continental Europe. *Veterinary Record*, **134**, 651–2.

Arzate, C. N. M., Martinez, A. R., Sierra, R. G., and Lopez-Gonzalez, C. A. (2007) Spatial ecology and abundance of Mexican bobcats in northwestern Mexico to assess its conservation status. In *Felid Biology and Conservation Conference 17–20 September 2007: Programme and Abstracts*, p. 119. Wildlife Conservation Research Unit, Oxford, UK.

Association of Zoos & Aquariums (2008) <www.aza.org>. Accessed online 10 Sept 2008.

Athreya, V. (2006) Is relocation a viable management option for unwanted animals? The case of the leopard in India. *Conservation and Society*, **4**, 419–23.

Athreya, V., Thakur, S., Choudhary, S., and Anirudh (2008) Human-leopard conflict in the Pune District. Wildlife Protection Society of India. <www.wpsi-india.org/projects/human_leopard_conflict.php>. Accessed online 4 Mar 2008.

Athreya, V., Thakur, S.S., Chaudhuri, S., and Belsare, A.V. (2004) A study of man-leopard conflict in the Junnar Forest Division, Pune District, Maharashtra. Report to Maharastra State Forest Department and the Wildlife Protection Society of India, New Delhi, India.

Aubry, K. B., Koehler, G. M., Squires, J. R., *et al.* (2000) Ecology of Canada lynx in southern boreal forests. In L. F. Ruggiero, K. B. Aubry, S. W. Buskirk, *et al.* (eds.) *Ecology and Conservation of Lynx in the United States*, pp. 373–96. University Press of Colorado, Boulder, CO.

Austin, S. C. and Tewes, M. E. (1999) Ecology of the clouded leopard in Khao Yai National Park, Thailand. *Cat News*, **31**, 17–18.

Austin, S. C., Tewes, M. E., Grassman Jr, L. I., and Silvy, N. J. (2007) Ecology and conservation of the leopard cat *Prionailurus bengalensis* and clouded leopard *Neofelis nebulosa* in Khao Yai National Park, Thailand. *Acta Zoologica Sinica*, **53**, 1–14.

Avenant, N. L. and Nel, J. A. J. (2002) Among habitat variation in prey availability and use by caracal *Felis caracal*. *Mammal Biology*, **67**, 18–33.

Avenant, N. L. and Nel, J. A. J. (1998) Home range use, activity, and density of caracal in relation to prey density. *African Journal of Ecology*, **36**, 347–59.

Averbeck, G. A., Bjork, K. E., Packer, C., and Herbst, L. (1990) Prevalence of hematozoans in lions (*Panthera leo*) and cheetah (*Acinonyx jubatus*) in Serengeti National Park and Ngorongoro Crater, Tanzania. *Journal of Wildlife Diseases*, **26**, 392–4.

Avise, J. C. (2000) *Phylogeography: The History and Formation of Species*. Harvard University Press, Cambridge, MA.

Avise, J. C., Arnold, J., Ball, R. M., *et al.* (1987) Intraspecific phylogeography: the mitochondrial DNA bridge between genetics and systematics. *Annual Review of Ecology and Systematics*, **18**, 489–522.

AWAGR. (2001) Status and needs for conservation of lions in West and Central Africa—An information exchange workshop. West African Working Group Report, Limbe, Cameroon.

Azevedo, F. C. C., de and Murray, D. L. (2007a) Spatial organization and food habits of jaguars (*Panthera onca*) in a floodplain forest. *Biological Conservation*, **137**, 391–402.

Azevedo, F. C. C., de and Murray, D. L. (2007b) Evaluation of potential factors predisposing livestock to predation by jaguars. *Journal of Wildlife Management*, **71**, 2379–86.

Azlan, M. J. and Sanderson, J. G. (2007) Geographic distribution and conservation status of the bay cat *Catopuma badia*, a Bornean endemic. *Oryx*, **41**, 394–7.

Azlan, M. J., Lading, E., and Munau (2003) Bornean bay cat photograph and sightings. *Cat News*, **39**, 2.

Bagchi, S. and Mishra, C. (2006) Living with large carnivores: predation on livestock by the snow leopard (*Uncia uncia*). *Journal of Zoology*, **268**, 217–24.

Bagchi, S., Mishra, C., and Bhatnagar, Y. V. (2004) Conflicts between traditional pastoralism and conservation of Himalayan ibex (*Capra sibirica*) in the trans-Himalayan mountains. *Animal Conservation*, **7**, 121–8.

Bailey, T. N. (2005) *The African Leopard: Ecology and Behaviour of a Solitary Felid*, 2nd edn. The Blackburn Press, Caldwell, NJ.

Bailey, T. N. (1993) *The African Leopard: Ecology and Behavior of a Solitary Felid*, 1st edn. Columbia University Press, New York.

Bailey, T. N. (1981) Factors of bobcat social organization and some management implications. In J. A. Chapman and D. Purseley (eds.), *Proceedings of the Worldwide Furbearer Conference*, pp. 984–1000. Frostburg, MD.

Bailey, T. N. (1974) Social organisation in a bobcat population. *Journal of Wildlife Management*, **38**, 435–46.

Bailey, T. N. (1972) *Ecology of bobcats with special reference to social organisation*. PhD dissertation, University of Idaho, Moscow, ID.

Bailey, T. N., Bangs, E. E., Portner, M. F., Malloy, J. C., and McAvinchey, R. J. (1986) An apparent overexploited lynx population on the Kenai peninsula, Alaska. *Journal of Wildlife Management*, **50**, 279–90.

Baker, J. R. (1968) Trypanosomes of wild mammals in the neighbourhood of the Serengeti National Park. *Symposium of the Zoological Society of London*, **24**, 147–58.

Baker, S., Dutton, A., and Macdonald, D. W. (2006) *The Wildlife Trade in East and Southeast Asia: An Overview and Analysis of the Trade with Specific Reference to Animal Welfare*. WSPA-Wildlife Conservation Research Unit, Oxford, UK.

Baker, S. E. and Macdonald, D. W. (2000) Foxes and fox-hunting on farms in Wiltshire: a case study. *Journal of Rural Studies*, **16**, 185–201.

Baker, S. E., Johnson, P. J., Slater, D., Watkins, R. W., and Macdonald, D. W. (2007) Learned food aversion with and without an odour cue for protecting untreated baits from wild mammal foraging. *Applied Animal Behaviour Science*, **102**, 410–28.

Baldus, R. D. (2006) A man-eating lion (*Panthera leo*) from Tanzania with a toothache. *European Journal of Wildlife Research*, **52**, 59–62.

Baldus, R. D. (2004) Lion conservation in Tanzania leads to serious human-lion conflicts. *Tanzania Wildlife Discussion Paper*, **41**, 1–63.

Balharry, E. and Daniels, M. J. (1998) Wild-living cats in Scotland. Research, Survey and Monitoring Report No. 23. Scottish Natural Heritage, Inverness, Scotland.

Ballesio, R. (1963) Monographie d'un machairodus du gisement Villafranchien de Senèze: *Homotherium crenatidens* Fabrini. *Travaux du Laboratoire de Géologie de la Faculté des Sciences de Lyon*, **9**, 1–129.

Ballou, J., Gilpin, M. E., and Foose, T. J. (eds.) (1995) *Population Management for Survival and Recovery*. Columbia University Press, New York.

Balme, G. A. and Hunter, L. T. B. (unpubl.) A guide to aging and sexing leopards in southern Africa. Unpublished report. Wildlife Conservation Society, New York.

Balme, G. A. and Hunter, L. T. B. (2004) Mortality in a protected leopard population, Phinda Private Game Reserve, South Africa: a population in decline? *Ecological Journal*, **6**, 5–11.

Balme, G. A., Hunter, L. T. B., and Slotow, R. (2009a) Evaluating methods for counting cryptic carnivores. *Journal of Wildlife Management*, **73**, 433–441.

Balme, G. A., Hunter, L. T. B., and Slotow, R. (2009b) The impact of conservation interventions on the dynamics and persistence of a persecuted leopard population. *Biological Conservation*, **142**, 2681–90.

Balme, G. A., Hunter, L. T. B., and Slotow, R. (2007) Feeding habitat selection by hunting leopards *Panthera pardus* in a woodland savanna: prey catchability versus abundance. *Animal Behaviour*, **74**, 589–98.

Bandyopadhyay, S., Humavindu, M. N., Shyamsundar, P., and Wang, L. (2004) *Do Households Gain from Community-Based Natural Resource Management? An Evaluation of Community Conservancies in Namibia*. World Bank, Washington DC.

Bangs, E. E., Fontaine, J. A., Jimenez, M. D., *et al.* (2005) Managing wolf–human conflict in the northwestern United States. In R. Woodroffe, S. Thirgood, and A. Rabinowitz (eds.), *People and Wildlife: Conflict or Coexistence?*, pp. 340–56. Cambridge University Press, Cambridge, UK.

Bank of Tanzania (2007) *Tanzania Tourism Sector Survey: The 2005 International Visitors' Exit Survey Report*. Bank of Tanzania, Dar es Salaam, Tanzania.

Banks, D. and Wright, B. (2007) *Use and Availability of Asian Big Cats Skins at the Litang Horse Festival, Sichuan, August 2007*. Environmental Investigation Agency and Wildlife Protection Society of India, London, UK and New Delhi, India.

Barash, D. P. (1971) Cooperative hunting in the lynx. *Journal of Mammalogy*, **52**, 480.

Barbry, T. and Gallardo, G. (2006) First camera trap photos of the Andean cat in the Sajama National Park and Natural Area of Integrated Management, Bolivia. *Cat News*, **44**, 23.

Bard, J. B. (1981) A model for generating aspects of zebra and other mammalian coat patterns. *Journal of Theoretical Biology*, **93**, 363–85.

Barker, I. K. and Parrish, C. R. (2001) Parvovirus infections. In E. S. Williams and I. K. Barker (eds.), *Infectious Diseases of Wild Mammals*, pp. 131–46. Iowa State University Press, Ames, IA.

Barlow, A. C. D., McDougal, C., Smith, J. L. D. *et al.* (2009) Temporal variation in tiger (*Panthera tigris*) populations and its implications for monitoring. *Journal of Mammalogy*, **90**, 472–478.

Barnes, C. T. (1960) *The Cougar or Mountain Lion*. Ralton Company, Salt Lake City, UT.

Barnett, R., Shapiro, B., Barnes, I., *et al.* (2009) Phylogeography of lions (*Panthera leo*) reveals three distinct taxa and a late Pleistocene reduction in genetic diversity. *Molecular Ecology*, **18**, 1668–77.

Barnett, R., Yamaguchi, N., Shapiro, B., and Nijman, V. (2007) Using ancient DNA techniques to identify the origin of unprovenanced museum specimens, as illustrated by the identification of a 19th century lion from Amsterdam. *Contributions to Zoology*, **76**, 87–94.

Barnett, R., Yamaguchi, N., Barnes, I., and Cooper, A. (2006a) Lost populations and preserving genetic diversity in the lion *Panthera leo*: implications for its ex situ conservation. *Conservation Genetics*, **7**, 507–14.

Barnett, R., Yamaguchi, N., Barnes, I., and Cooper, A. (2006b) The origin, current diversity and future conservation of the modern lion (*Panthera leo*). *Proceedings of the Royal Society of London Series B*, **273**, 1–7.

Barnett, R., Barnes, I., Phillips, M. J., *et al.* (2005) Evolution of the extinct sabre-tooths and the American cheetah-like cat. *Current Biology*, **15**, R589–90.

Barney, E. C., Mintzes, J. J., and Yen, C. F. (2005) Assessing knowledge, attitudes and behavior toward charismatic megafauna: the case of dolphins. *Journal of Environmental Education*, **36**, 41–55.

Baron, D. (2004) *The Beast in the Garden: A Modern Parable of Man and Nature*. W.W. Norton & Company Ltd, New York.

Barone, M. A., Roelke, M. E., Howard, J. G., Brown, J. L., Anderson, A. E., and Wildt, D. E. (1994) Reproductive characteristics of male Florida panthers: comparative studies from Florida, Texas, Colorado, Latin America, and North American zoos. *Journal of Mammalogy*, **75**, 150–62.

Barrett, C. B. (2008) Poverty traps and resource dynamics in smallholder agrarian systems. In A. Ruijs and R. Dellink (eds.), *Economics of Poverty, Environment and Natural Resource Use*, pp. 17–40. Springer, Dordrecht, the Netherlands.

Barrett, C.B., Brandon, K., Gibson, C., and Gjertsen, H. (2001) Conserving tropical biodiversity amid weak institutions. *Bioscience*, **51**, 497–502.

Barrett, H. C. (1999) *Human cognitive adaptations to predators and prey*. PhD dissertation, University of California, Santa Barbara, CA.

Barron, M. G., Duvall, S. E., and Barron, K. J. (2004) Retrospective and current risks of mercury to panthers in the Florida Everglades. *Ecotoxicology*, **13**, 223–9.

Barry, J. C. (1987) Large carnivores (Canidae, Hyaenidae, Felidae) from Laetoli. In M. D. Leakey and J. M. Harris (eds.), *Laetoli: A Pliocene Site in Northern Tanzania*, pp. 235–58. Clarendon Press, Oxford, UK.

Bartlett, M. S. (1960) The critical community size for measles in the United States. *Journal of the Royal Statistical Society Series A—General*, **123**, 37–44.

Baskin, J. A. (1981) *Barbourofelis* (Nimravides) and (Nimravides), with a description of two new species from the late Miocene of Florida. *Journal of Mammalogy*, **62**, 122–39.

Bauer, G. A. (1998) Cheetah—running blind. In B. L. Penzhorn (ed.), *Proceedings of a Symposium on Cheetahs as Game Ranch Animals*, pp. 106–8. Wildlife Group of South African Veterinary Association, Onderstepoort, South Africa.

Bauer, H. (2008) Synthesis of threats, distribution and status of the lion from the two lion conservation strategies. In B. Croes, R. Buij, H. de Iongh, and H. Bauer (eds.), *Management and Conservation of Large Carnivores in West and Central Africa*, pp. 13–28. Institute of Environmental Sciences (CML), Leiden University, Leiden, the Netherlands.

Bauer, H. (2003) *Lion conservation in west and central Africa*. PhD dissertation, Leiden University, Leiden, the Netherlands.

Bauer, H. and de Iongh, H. H. (2005) Lion (*Panthera leo*) home ranges and livestock conflicts in Waza National Park, Cameroon. *African Journal of Ecology*, **43**, 208–14.

Bauer, H., de Iongh, H. H., Princee, F., and Ngantou, D. (2001). Status and needs for conservation of lions in West and Central Africa: An information exchange workshop report. Conservation Breeding Specialist Group (IUCN/SSC), Apple Valley, MN, USA. 114pp.

Bauer, H. and Nowell, K. (2004) West African lion population classified as regionally Endangered. *Cat News*, **41**, 35–6.

Bauer, H., and van der Merwe, S. (2004) Inventory of free-ranging lions *Panthera leo* in Africa. *Oryx*, **38**, 26–31.

Bauer, H. and van der Merwe, S. (2002) *The African Lion Database*. IUCN Species Survival Commission, African Lion Working Group. <www.african-lion.org/ALD_2002.pdf>. Accessed online 18 Dec 2008.

Bauer, H., Nowell, K., and Packer, C. (2008) *Panthera leo*. 2008 IUCN Red List of Threatened Species. <www.iucnredlist.org>. Accessed online 18 Dec 2008.

Baxby, D., Ashton, D. G., Jones, D. M., and Thomsett, L. R. (1982) An outbreak of cowpox in captive cheetahs: virological and epidemiological studies. *Journal of Hygiene*, **89**, 365–72.

Beaumont, G. de, (1990) Contribution a l'étude du genre *Nimravides* Kitts (Mammalia, Carnivora, Felidae). L'espèce *N. pedionomus* (MacDonald). *Archives des Sciences, Genève*, **43**, 125–57.

Beaumont, G. de, (1978) Notes complementaires sur quelques félidés (Carnivores). *Archives des Sciences, Genève*, **31**, 219–27.

Beaumont, G. de, (1964) Remarques sur la classification des Felidae. *Eclogae Geologicae Helvetiae*, **57**, 837–45.

Beaumont, M., Barratt, E. M., Gottelli, D., *et al.* (2001) Genetic diversity and introgression in the Scottish wildcat. *Molecular Ecology*, **10**, 319–36.

Beaver, B. V. (1992) *Feline Behaviour: A Guide for Veterinarians*. WB Saunders Company, Sydney, Australia.

Beck, B. B., Rappaport, M. R., Stanley Price, M. R., and Wilson, A. C. (1994) Reintroduction of captive-born animals. In G. M. Mace, P. J. Olney, and A. Feistner (eds.), *Creative Conservation: Interactive Management of Wild and Captive Animals*, pp. 265–86. Chapman and Hall, London, UK.

Beekman, S. P. A., Kemp, B., Louwmanand, H. C. M., and Colenbrander, B. (1999) Analyses of factors influencing the birth weight and neonatal growth rate of cheetah (*Acinonyx jubatus*) cubs. *Zoo Biology*, **18**, 129–39.

Begg, C. M. and Begg, K. S. (2007) Trophy monitoring in Niassa National Reserve, Mozambique: lion, leopard, buffalo, hippo and crocodile. Unpublished internal report for SRN, Maputo, Mozambique.

Begg, C., Begg, K., and Muemedi, O. (2007) *Preliminary Data on Human–Carnivore Conflict in Niassa National Reserve, Mozambique, Particularly Fatalities due to Lion, Spotted Hyaena and Crocodile*. Sociedade para a Gestao e Desenvolvimento da Reserva do Niassa, Maputo, Mozambique.

Begon, M., Harper, J. L., and Townsend, C. R. (1990) *Ecology: Individuals, Populations and Communities*. Blackwell Scientific Publications, Oxford, UK.

Beier, P. (1993) Determining minimum habitat areas and habitat corridors for cougars. *Conservation Biology*, **7**, 94–108.

Beier, P. (1991) Cougar attacks on humans in the United States and Canada. *Wildlife Society Bulletin*, **19**, 403–12.

Beier, P. and Barrett, R. H. (1993) *The Cougar in the Santa Ana Mountain Range, California*. Orange County Cooperative Mountain Lion Study, Berkeley, CA.

Beier, P. and Cunningham, S. C. (1996) Power of track surveys to detect changes in cougar populations. *Wildlife Society Bulletin*, **24**, 540–8.

Beier, P., Penrod, K. L., Luke, C., Spencer, W. D., and Cabañero, C. (2006) South coast missing linkages: restoring connectivity to wildlands in the largest metropolitan area in the USA. In K. R. Crooks and M. A. Sanjayan (eds.), *Connectivity Conservation*, pp. 555–86. Cambridge University Press, Cambridge, UK.

Beier, P., Vaughn, M. R., Conroy, M. J., and Quigley, H. (2003) *An Analysis of Scientific Literature Related to the Florida Panther*. Florida Fish and Wildlife Commission, Tallahassee, FL.

Beier, P., Choate, D. C., and Barrett, R. H. (1995) Movement patterns of mountain lions during different behaviors. *Journal of Mammalogy*, **76**, 1056–70.

Beissinger, S. R. and Westphal, M. I. (1998) On the use of demographic models of population viability in endangered species management. *Journal of Wildlife Management*, **62**, 821–41.

Belden, R. C. (1978a) Florida panther investigation—a 1978 progress report. In R. R. Odum and L. Landers (eds.), *Proceedings of the Rare and Endangered Wildlife Symposium*, pp. 123–33. Georgia Department of Natural Resources, Atlanta, GA.

Belden, R. C. (1978b) How to recognize panther tracks. *Proceedings of the Annual Conference of Southeastern Fish and Wildlife Agencies*, **32**, 112–15.

Belden, R. C. (1977) If you see a panther. *Florida Wildlife*, **31**, 31–4.

Belden, R. C. and McCown, J. W. (1996) *Florida Panther Reintroduction Feasibility Study*. Florida Game and Freshwater Fish Commission, Tallahassee, FL.

Belden, R. C. and Hagedorn, B. W. (1993) Feasibility of translocating panthers into Northern Florida. *Journal of Wildlife Management*, **57**, 388–97.

Belden, R. C., Frankenberger, W. B., and Roof, J. C. (1991) *Florida Panther Distribution*. Florida Game and Freshwater Fish Commission, Tallahassee, FL.

Belden, R. C., Frankenberger, W. B., McBride, R. T., and Schwikert, S. T. (1988) Panther habitat use in southern Florida. *Journal of Wildlife Management*, **52**, 660–3.

Bell, D. (2006) *Cheetah Country Beef Business Plan*. MeatCo, Windhoek.

Bell, R. H. V. (1982) The effect of soil nutrient availability on the community structure in African ecosystems. In B. J. Huntley and B. H. Walker (eds.), *Ecology of Tropical Savannas*, pp. 193–216. Springer-Verlag, Berlin, Germany.

Belousova, A. V. (1993) Small Felidae of Eastern Europe, Central Asia and Far East. Survey of the state of populations. *Lutreola*, **2**, 16–21.

Beltrán, J. F. and Delibes, M. (1993) Physical characteristics of Iberian lynxes (*Lynx pardinus*) from Doñana, southwestern Spain. *Journal of Mammalogy*, **74**, 852–62.

Beltrán, J. F. and Delibes, M. (1991) Ecología trófica del lince Ibírico en Doñana durante un periodo seco. *Doñana Acta Vertebrata*, **18**, 113–22.

Beltran, J. F. and Tewes, M. E. (1995) Immobilization of ocelots and bobcats with ketamine hydrochloride and xylazine hydrochloride. *Journal of Wildlife Diseases*, **31**, 43–8.

Beltrán, J. F., Rice, J. E., and Honeycutt, R. L. (1996) Taxonomy of the Iberian lynx. *Nature*, **379**, 407–8.

Ben-Amos, P. (1976) Men and animals in Benin art. *Man*, **11**, 243–52.

Benson, J. F., Chamberlain, M. J., and Leopold, B. D. (2006) Regulation of space use in a solitary felid: population density or prey availability. *Animal Behaviour*, **71**, 685–93.

Berger, K. M. (2005) Carnivore–livestock conflicts: effects of subsidized predator control and economic correlates on the sheep industry. *Conservation Biology*, **20**, 751–61.

Bergström, R. and Skarpe, C. (1999) The abundance of large wild herbivores in a semi-arid savanna in relation to seasons, pans and livestock. *African Journal of Ecology*, **37**, 1–12.

Berkes, F., Colding, J., and Folke, C. (2000) Rediscovery of traditional ecological knowledge as adaptive management. *Ecological Applications*, **10**, 1251–62.

Berry, W. L., Jardine, J. E., and Espie, I. W. (1997) Pulmonary cryptococcoma and cryptococcal meningoencephalomyelitis in a king cheetah (*Acinonyx jubatus*). *Journal of Zoo and Wildlife Medicine*, **28**, 485–90.

Bertram, B. C. R. (1979) Serengeti predators and their social systems. In A. R. E. Sinclair and P. Arcese (eds.), *Serengeti II: Dynamics, Management and Conservation of an Ecosystem*, pp. 215–25. University of Chicago Press, Chicago, IL and London, UK.

Bertram, B. C. R. (1976) Kin selection in lions and in evolution. In P. P. G. Bateson and R. A. Hinde (eds.), *Growing Points in Ethology*, pp. 281–301. Cambridge University Press, Cambridge, UK.

Bertram, B. C. R. (1975a) Social factors influencing reproduction in wild lions. *Journal of Zoology, London*, **177**, 463–82.

Bertram, B. C. R. (1975b) Weights and measures of lions. *East Africa Wildlife Journal*, **13**, 141–3.

Bertrand, A. S., Kenn, S., Gallant, D., Tremblay, E., Vasseur, L., and Wissink, R. (2006) MtDNA analyses on hair samples confirm Cougar, *Puma concolor*, presence in southern New Brunswick, eastern Canada. *Canadian Field-Naturalist*, **120**, 438–42.

Besançon, C. and Savy, C. (2005) Global list of internationally adjoining protected areas and other transboundary initiatives. In R. Mittermeier, C. F. Kormos, C. G. Mittermeier, P. Robles Gil, T. Sandwith, and C. Besançon (eds.), *Transboundary Conservation: A New Vision for Protected Areas*. CEMEX/Conservation International, Arlington, VA.

Beverly, F. (1874) The Florida panther. *Forest and Stream*, **3**, 290.

Bezuijen, M. R. (2003) The flat-headed cat in the Merang river region of south Sumatra. *Cat News*, **38**, 26–7.

Bezuijen, M. R. (2000) The occurrence of the flat-headed cat *Prionailurus planiceps* in south-east Sumatra. *Oryx*, **34**, 222–6.

Bhagavatula, J. and Singh, L. (2006) Genotyping faecal samples of Bengal tiger *Panthera tigris tigris* for population estimation: a pilot study. *BMC Genetics*, **7**, 48.

Bhatnagar, Y. V., Mathur, V.B., and McCarthy, T. (2003). A Regional Perspective for Snow Leopard Conservation in the Indian Trans-Himalaya. In Bhatnagar, Y.V. and Sathyakumar, S. (eds.). *WII ENVIS Bulletin*. Wildlife Institute of India, Dehradun.

Bhatnagar, Y. V., Wangchuk, R., and Jackson, R. (1999) *A Survey of Depredation and Related Human-Wildlife Conflicts in Hemis National Park Ladakh Jammu and Kashmir*. International Snow Leopard Trust, Seattle, WA.

Bhattacharyya, T. (1989) *Report on the Survey on the Status and Distribution of Fishing Cat (*Felis viverina*) in Howrah District of West Bengal*. Indian Society for Wildlife Research, Calcutta, India.

Bianchi, R. C. (2001) *Estudo comparativo da dieta da jaguatirica* Leopardus pardalis *(Linnaeus, 1751) na Reserva Natural da Vale do Rio Doce-ES e na Estação Biológica de Caratinga-MG*. MSc dissertation, Universidade Federal do Espírito Santo, Vitória, Brazil.

Bianchi, R. C. and Mendes, S. L. (2007) Ocelot (*Leopardus pardalis*) predation on primates in Caratinga Biological Station, Southeast Brazil. *American Journal of Primatology*, **69**, 1173–8.

Biek, R., Akamine, N., Schwartz, M. K., Ruth, T. K., Murphy, K. M., and Poss, M. (2006a) Genetic consequences of sex-biased dispersal in a solitary carnivore: Yellowstone cougars. *Biology Letters*, **2**, 312–15.

Biek, R., Drummond, A. J., and Poss, M. (2006b) A virus reveals population structure and demographic history of its carnivore host. *Science*, **311**, 538–41.

Biek, R., Zarnke, R. L., Gillin, C., Wild, M., Squires, J. R., and Poss, M. (2002) Serologic survey for viral and bacterial infections in western populations of Canada lynx (*Lynx canadensis*). *Journal of Wildlife Diseases*, **38**, 840–5.

Biggins, D. E., Godbey, J. L., Hanebury, L. R., *et al.* (1998) The effect of rearing methods on survival of reintroduced black-footed ferrets. *Journal of Wildlife Management*, **62**, 643–53.

Biju-Duval, C., Ennafaa, H., Dennebouy, N., *et al.* (1991) Mitochondrial DNA evolution in lagomorphs—origin of systematic heteroplasmy and organization of diversity in European rabbits. *Journal of Molecular Evolution*, **33**, 92–102.

Biknevicius, A. R. and Ruff, C. B. (1992) The structure of the mandibular corpus and its relationship to feeding behaviours in extant carnivorans. *Journal of Zoology*, **228**, 479–507.

Biknevicius, A. R. and Van Valkenburgh, B. (1996) Design for killing: craniodental adaptations of predators. In J. L. Gittleman (ed.), *Carnivore Behavior, Ecology and Evolution*, pp. 393–427. Cornell University Press, Ithaca, NY.

Biknevicius, A. R., Van Valkenburgh, B., and Walker, J. (1996) Incisor size and shape: implications for feeding behaviors in saber-toothed 'cats'. *Journal of Vertebrate Paleontology*, **16**, 510–21.

Bininda-Emonds, O. R. P., Decker-Flum, D. M., and Gittleman, J. L. (2001) The utility of chemical signals as phylogenetic characters: an example from the Felidae. *Biological Journal of the Linean Society*, **72**, 1–15.

Bininda-Emonds, O. R. P., Gittleman, J. L., and Purvis, A. (1999) Building large trees by combining phylogenetic information: a complete phylogeny of the extant Carnivora (Mammalia). *Biological Reviews*, **74**, 143–75.

Biodiversity Reporting and Information Group (2007) Report on the species and habitat review. Report to the UK Biodiversity Partnership. <www.ukbap.org.uk/bapgrouppage.aspx?id = 112>. Accessed online 15 Dec 2008.

Biro, Z., Szemethy, L., and Heltai, M. (2004) Home range sizes of wildcats (*Felis silvestris*) and feral domestic cats (*Felis silvestris f. catus*) in a hilly region of Hungary. *Mammalian Biology*, **69**, 302–10.

Bisbal, F. J. (1986) Food habits of some neotropical carnivores in Venezuela (Mammalia, Carnivora). *Mammalia*, **50**, 329–39.

Bisceglia, S. B. C., Pereira, J. A., Teta, P., and Quintana, R. D. (2007) Food habits of Geoffroy's cat (*Leopardus geoffroyi*) in the central monte desert of Argentina. *Journal of Arid Environments*, **72**, 1120–6.

Bishop, B. C. (1990) Karnali under stress: livelihood strategies and seasonal rhythms in a changing Nepal Himalaya. Geography Research Papers Nos. 228–9, University of Chicago Press, Chicago, IL.

Bissett, C. and Bernard, R. T. F. (2007) Habitat selection and feeding ecology of the cheetah (*Acinonyx jubatus*) in thicket vegetation: is the cheetah a savanna specialist? *Journal of Zoology*, **271**, 310–17.

Bjork, K. E., Averbeck, G. A., and Stromberg, B. E. (2000) Parasites and parasite stages of free-ranging wild lions (*Panthera leo*) of northern Tanzania. *Journal of Zoo and Wildlife Medicine*, **31**, 56–61.

Björklund, M. (2003) The risk of inbreeding due to haitat loss in the lion (*Panthera leo*). *Conservation Genetics*, **4**, 515–23.

Blasius, B., Huppert, A., and Stone, L. (1999) Complex dynamics and phase synchronization in spatially extended ecological systems. *Nature*, **399**, 354–9.

Blundell, G. M., Ben David, M., Groves, P., Bowyer, R. T., and Geffen, E. (2002) Characteristics of sex-biased dispersal and gene flow in coastal river otters: implications for natural recolonization of extirpated populations. *Molecular Ecology*, **11**, 289–303.

Blythe, L. L., Schmitz, J. A., Roelke, M., and Skinner, S. (1983) Chronic encephalomyelitis caused by canine distemper virus in a Bengal tiger. *Journal of the American Veterinary Medical Association*, **183**, 1159–62.

Bohlin, B. (1940) Food habits of the machaerodonts, with special regard to *Smilodon*. *Bulletin of the Geological Institutions of the University of Uppsala*, **28**, 156–74.

Boitani, L., Corsi, F., and De Baise, A. (1999) *A Databank for the Conservation and Management of African Mammals*. Institution of Ecological Application, Rome, Italy.

Boitani, L. and Fuller, T. K. (eds.) (2000) *Research Techniques in Animal Ecology: Controversies and Consequences*. Columbia University Press, New York.

Bolgiano, C. (1995) *Mountain Lion: An Unnatural History of Pumas and People*. Stackpole Books, Mechanicsburg, PA.

Bolton, L. A. and Munson, L. (1999) Glomerulosclerosis in captive cheetahs (*Acinonyx jubatus*). *Veterinary Pathology*, **36**, 14–22.

Bolton, L. A., Lobetti, R. G., Evezard, D. N., *et al.* (1999) Cryptococcosis in captive cheetah (*Acinonyx jubatus*): two cases. *Journal of the South African Veterinary Association*, **70**, 35–9.

Bonacic, C., Galvez, N., Amar, F., Laker, J., Murphy, T., and Macdonald, D. W. (2007) Puma and farmer interactions: a multi-scale approach in three eco-regions of the Chilean Andes. In *Felid Biology and Conservation Conference 17–20 September 2007: Programme and Abstracts*, pp. 96–7. Wildlife Conservation Research Unit, Oxford, UK.

Bond, I., Child, B., de la Harpe, D., Jones, B., Barnes, J., and Anderson, H. (2004) Private land contribution to conservation in South Africa. In B. Child (ed.), *Parks in Transition. Biodiversity, Rural Development and the Bottom Line*, pp. 29–61. Earthscan, London, UK.

Bookhout, T. A. (ed.) (1994) *Research and Management Techniques for Wildlife and Habitats*. Wildlife Society, Bethseda, MD.

Boomgaard, P. (2001) *Frontiers of Fear: Tigers and People in the Malay World, 1600–1950*. Yale University Press, New Haven, CT.

Borgerhoff Mulder, M. and Coppolillo, P. (2005) *Conservation. Linking Ecology, Economics and Culture*. Princeton University Press, Princeton, NJ.

Bornstein, S., Morner, T., and Samuel, W. M. (2001) *Sarcoptes scabiei* and Sarcoptic mange. In W. M. Samuel, M. J. Pybus, and A. A. Kocan (eds.) *Parasitic Diseases of Wild Mammals*, pp. 107–20. Iowa State University Press, Ames, IA.

Bornstein, S., Roken, B., and Lindberg, R. (1997) An experimental infection of a lynx (*Felis lynx*) with *Sarcoptes scabiei* var. *vulpes*. *16th International Conference of the World Association for Advancement of Veterinary Parasitology*, Sun City, South Africa.

Bosman, A. M., Venter, E. H., and Penzhorn, B. L. (2007) Occurrence of *Babesia felis* and *Babesia leo* in various wild felid species and domestic cats in Southern Africa, based on reverse line blot analysis. *Veterinary Parasitology*, **144**, 33–8.

Bothma, J. d. P. (1998) *Carnivore Ecology in Arid Lands*. Springer, Berlin, Germany.

Bothma, J. d. P. and De Graaff, G. (1973) A habitat map of the Kalahari Gemsbok National Park. *Koedoe*, **16**, 181–8.

Bothma, J. d. P., Knight, M. H., le Riche, E. A. N., and Van Hensbergen, H. J. (1997) Range size of southern Kalahari leopards. *South African Journal of Wildlife Research*, **27**, 94–9.

Bothma, J. d. P. and le Riche, E. A. N. (1986) Prey preference and hunting efficiency of the Kalahari desert leopard. In S. D. Miller and D. D. Everett (eds.), *Cats of the World: Biology, Conservation, and Management*, pp. 389–414. National Wildlife Federation Institute for Wildlife Research, Washington DC.

Boutin, S., Krebs, C. J., Boonstra, R., *et al.* (1995) Population changes of the vertebrate community during a snowshoe hare cycle in Canada's boreal forest. *Oikos*, **74**, 69–80.

Bowland, J. M. (1990) *Diet, home range and movement patterns of serval on farmland in Natal.* MS dissertation, University of Natal, Pietermaritzburg, South Africa.

Bowns, J. E. (1985) Predation-depredation. In J. Roberson and F. G. Lindzey (eds.), *Proceedings of the Second Mountain Lion Workshop*, pp. 204–15. Utah Division of Wildlife Resources, Salt Lake City, UT.

Boyce, M. S. and McDonald, L. L. (1999) Relating populations to habitats using resource selection functions. *Trends in Ecology & Evolution*, **14**, 268–72.

Boyce, M. S., Rushton, S. P., and Lynam, T. (2006) Does modelling have a role in conservation? In D. W. Macdonald and K. Service (eds.), *Key Topics in Conservation Biology*, pp. 134–44. Blackwell Publishing Ltd, Oxford, UK.

Braack, L. (2005) The Great Limpopo Transfrontier Park: a benchmark for international conservation. In R. Mittermeier, C. F. Kormos, C. G. Mittermeier, P. Robles Gil, T. Sandwith, and C. Besançon (eds.), *Transboundary Conservation: A New Vision for Protected Areas*, pp. 85–98. CEMEX/Conservation International, Arlington, VA.

Bradbury, J. W. and Vehrencamp, S. L. (1976) Social organization and foraging in emballonurid bats I. Field studies. *Behavioural Ecology and Sociobiology*, **1**, 337–81.

Bradshaw, J. W. S. and Cameron-Beaumont, C. (2000) The signalling repertoire of the domestic cat and its undomesticated relatives. In D. C. Tuner and P. Bateson (eds.), *The Domestic Cat, the Biology of Its Behaviour*, Vol. 2, pp. 67–93. Cambridge University Press, Cambridge, UK.

Bradshaw, J. W. S., Goodwin, D., Legrand-Defretin, V., and Nott, H. M. R. (1996) Food selection by the domestic cat, an obligate carnivore. *Comparative Biochemistry and Physiology—A Physiology*, **114**, 205–9.

Bragin, A. P. (1986) *Population Characteristics and Social-Spatial Patterns of the Tiger (*Panthera tigris*) on the Eastern Macroslope of the Sikhate-Alin Mountain Range, USSR.* Pacific Institute of Geography, Vladivostok, Russia.

Brain, C. K. (1981) *The Hunters or the Hunted? An Introduction to African Cave Taphonomy.* University of Chicago Press, Chicago, IL.

Brain, C. K. (1969) The probable role of leopards as predators of the Swartkrans Australopithicines. *The South African Archaeological Bulletin*, **24**, 170–1.

Branch, B. (1998) *Field Guide to Snakes and Other Reptiles of Southern Africa*, 3rd edn. Struik Publishers Ltd, Cape Town, South Africa.

Branch, L. C. (1995) Observations of predation by pumas and Geoffroy's cats on the plains vizcacha in semi-arid scrub of central Argentinia. *Mammalia*, **59**, 152–6.

Branco, M., Monnerot, M., Ferrand, N., and Templeton, A. R. (2002) Postglacial dispersal of the European rabbit (*Oryctolagus cuniculus*) on the Iberian Peninsula reconstructed from nested clade and mismatch analyses of mitochondrial DNA genetic variation. *Evolution*, **56**, 792–803.

Branco, M., Ferrand, N., and Monnerot, M. (2000) Phylogeography of the European rabbit (*Oryctolagus cuniculus*) in the Iberian Peninsula inferred from RFLP analysis of the cytochrome b gene. *Heredity*, **85**, 307–17.

Brand, C. J. and Keith, L. B. (1979) Lynx demography during a snowshoe hare decline in Alberta. *Journal of Wildlife Management*, **43**, 827–49.

Brand, C. J., Keith, L. B., and Fischer, C. A. (1976) Lynx responses to changing snowshoe hare densities in central Alberta. *Journal of Wildlife Management*, **40**, 416–28.

Brashares, J. S. (2003) Ecological, behavioural, and life-history correlates of mammal extinctions in West Africa. *Conservation Biology*, **17**, 733–43.

Brashares, J. S., Arcese, P., and Sam, M. K. (2001) Human demography and reserve size predict wildlife extinction in West Africa. *Proceedings of the Royal Society of London Series B*, **268**, 2473–8.

Braun, B., Dehnhard, M., Voigt, C., *et al.* (2007) Pregnancy diagnosis in Iberian lynx with witness relaxin test. In M. East and H. Hofer (eds.), *Contributions to the 6th International Wildlife Research Conference in Behaviour, Physiology and Genetics*, p. 48. Leibniz Institute for Zoo and Wildlife Research, Berlin, Germany.

Braun, C. L. and Smirnov, S. N. (1993) Why is water blue? *Journal of Chemical Education*, **70**, 612–13.

Breazile, J. E. (1987) Physiological basis and consequences of distress in animals. *Journal of the American Veterinary Medical Association*, **191**, 1212–15.

Breeden, S. (1989) The happy fisher. *BBC Wildlife Magazine*, **7**, 238–41.

Brehm, A. E. (1880) Los linces—*Lynx*. In *La Creación. Historia Natural. Vol. I, Mamíferos*, pp. 254–69. Montaner y Simón, Barcelona, Spain.

Breitenmoser-Würsten, C., Vandel, J. M., Zimmermann, F., and Breitenmoser, U. (2007a) Demography of lynx *Lynx lynx* in the Jura Mountains. *Wildlife Biology*, **13**, 381–92.

Breitenmoser-Würsten, C., Zimmermann, F., Stahl, P., *et al.* (2007b) Spatial and social stability of a Eurasian lynx (*Lynx lynx*) population: an assessment of 10 years of observation in the Jura Mountains. *Wildlife Biology*, **13**, 365–80.

Breitenmoser-Würsten, C., Zimmermann, F., Ryser, A., *et al.* (2001) Untersuchungen zur Luchspopulation in den Nordwestalpen der Schweiz 1997–2000. *KORA Bericht*, **9**, 1–88.

Breitenmoser, U. (1983) Zur Wiedereinbuergerung und Ausbreitung des Luchses *Lynx lynx* in der Schweiz. *Schweizerische Zeitschrift für das Forstwesen*, **134**, 207–22.

Breitenmoser, U. and Baettig, M. (1992) Wiederansiedlung und Ausbreitung des Luchses (*Lynx lynx*) im Schweizer Jura. *Revue suisse de Zoologie*, **99**, 163–76.

Breitenmoser, U. and Breitenmoser-Würsten, C. (2008) *Der Luchs—ein Grossraubtier in der Kulturlandschaft*. Salm Verlag, Bern, Switzerland.

Breitenmoser, U. and Haller, H. (1993) Patterns of predation by reintroduced European lynx in the Swiss Alps. *Journal of Wildlife Management*, **57**, 135–44.

Breitenmoser, U. and Haller, H. (1987) Zur Nahrungsökologie des Luchses Lynx lynx in den schweizerischen Nordalpen. *Zeitschrift für Säugetierkunde*, **52**, 168–91.

Breitenmoser, U., Breitenmoser-Würsten, C., Morschel, F., Zazanashvili, N., and Sylven, M. (2007) General conditions for the conservation of the leopard in the Caucasus. *Cat News Special Issue 2*, **2**, 34–9.

Breitenmoser, U., Mallon, D., and Breitenmoser-Würsten, C. (2006a) A framework for the conservation of the Arabian leopard. *Cat News Special Issue*, **1**, 44–7.

Breitenmoser, U., von Ark, M., and Breitenmoser-Würsten, C. (2006b) Lessons from the reintroduction of the Eurasian lynx in Central and Western Europe. In *Iberian Lynx Ex-Situ Conservation Seminar Series: Book of Proceedings*, pp. 151–2. Fundación Biodiversidad, Sevilla and Doñana, Spain.

Breitenmoser, U., Breitenmoser-Würsten, C., Carbyn, L. N., and Funk, S. M. (2001) Assessment of carnivore reintroductions. In J. L. Gittleman, S. M. Funk, D. Macdonald and R. K. Wayne (eds.), *Carnivore Conservation*, pp. 241–81. Cambridge University Press, Cambridge, UK.

Breitenmoser, U., Breitenmoser-Würsten, C., Okarma, H., Kaphegyi, T., Kaphegyi-Wallmann, U., and Müller, U. M. (2000) *Action Plan for the Conservation of the Eurasian Lynx (*Lynx lynx*) in Europe*. Council of Europe, Strasbourg, France.

Breitenmoser, U., Breitenmoser-Würsten, C., and Capt, S. (1998) Re-introduction and present status of the lynx in Switzerland. *Hystrix*, **10**, 17–30.

Breitenmoser, U., Kavczensky, P., Dotterer, M., *et al.* (1993a) Spatial organization and recruitment of lynx (*Lynx lynx*) in a re-introduced population in the Swiss Jura Mountains. *Journal of Zoology*, **231**, 449–64.

Breitenmoser, U., Slough, B. G., and Breitenmoser-Würsten, C. (1993b) Predators of cyclic prey: is the Canada lynx victim or profiteer of the snowshoe hare cycle? *Oikos*, **66**, 551–4.

Briones, V., de Juan, L., Sanchez, C., *et al.* (2000) Bovine tuberculosis and the endangered Iberian lynx. *Emerging Infectious Diseases*, **6**, 189–91.

Brito, B. F. A., de (2000) *Ecologia alimentar da onça-parda,* Puma concolor, *na Mata Atlântica de Linhares, Espírito Santo, Brasil. MS dissertation, Universidade de Brasília, Brasília, Brazil.*

Bromley, G. F. and Kucherenko, S. P. (1983) *Ungulates of the Southern Part of the USSR*. Nauka, Moscow, Russia.

Bronson, E., Bush, M., Viner, T., Murray, S., Wisely, S. M., and Deem, S. L. (2007) Mortality of captive black-footed ferrets (*Mustela nigripes*) at Smithsonian's National Zoological Park, 1989–2004. *Journal of Zoo and Wildlife Medicine*, **38**, 169–76.

Broomhall, L. S. (2001) *Cheetah* Acinonyx jubatus *in the Kruger National Park: a comparison with other studies across the grassland-woodland gradient in African savannas*. MSc dissertation, University of Pretoria, Pretoria, South Africa.

Broomhall, L. S., Mills, M. G. L., and Du Toit, J. T. (2003) Home range and habitat use by cheetahs (*Acinonyx jubatus*) in the Kruger National Park. *Journal of Zoology*, **261**, 119–28.

Brown, D. E. and Lopez-Gonzales, C. A. (2001) *Borderland Jaguars: Tigres de la Frontera*. University of Utah Press, Salt Lake City, UT.

Brown, E. W., Yuhki, N., Packer, C., and O'Brien, S. J. (1994) A lion lentivirus related to feline immunodeficiency virus: epidemiologic and phylogenetic aspects. *Journal of Virology*, **68**, 5953–68.

Brown, J. H., and Maurer, B. A. (1986) Body size, ecological dominance and Cope's rule. *Nature*, **324**, 248–50.

Brown, J. L. (2006) Comparative endocrinology of domestic and nondomestic felids. *Theriogenology*, **66**, 25–36.

Brown, J. L., Graham, L. H., Wu, J. M., Collins, D., and Swanson, W. F. (2002) Reproductive endocrine responses to photoperiod and exogenous gonadotropins in the Pallas's cat (*Otocolobus manul*). *Zoo Biology*, **21**, 347–64.

Brown, J. L., Terio, K. A., and Graham, L. H. (1996) Fecal androgen metabolite analysis for non invasive monitoring of testicular steroidogenic activity in felids. *Zoo Biology*, **15**, 425–34.

Brown, M. A., Cunningham, M. W., Roca, A. L., Troyer, J. L., Johnson, W. E., and O'Brien, S. J. (2008) Genetic characterization of feline leukemia virus from Florida panthers. *Emerging Infectious Diseases*, **14**, 252–9.

Brown, M., Lappin, M. R., Brown, J. L., Munkhtsog, B., and Swanson, W. F. (2005) Exploring the ecological basis for extreme susceptibility of Pallas's cats (*Otocolobus manul*) to fatal toxoplasmosis. *Journal of Wildlife Diseases*, **41**, 691–700.

Brown, M., Lappin, M. R., Brown, J. L., Munkhtsog, B. M., and Swanson, W. F. (2003) Pallas's cat in Mongolia. *Cat News*, **38**, 28–9.

Brown, R. E. and Macdonald, D. W. (eds.) (1985) *Social Odours in Mammals*. Oxford University Press, Oxford, UK.

Browne-Nuñez, C. and Jonker, S. A. (2008) Attitudes toward wildlife and conservation across Africa: a review of survey research. *Human Dimensions of Wildlife*, **13**, 47–70.

Bruning Fann, C. S., Schmitt, S. M., Fierke, J. S., *et al.* (2001) Bovine tuberculosis in free-ranging carnivores from Michigan. *Journal of Wildlife Diseases*, **37**, 58–64.

Brush, S. B. (1989) El ambiente natural y humano de los Andes centrales. In *Informe sobre los Conocimientos*

Actuales de los Ecosistemas Andinos, p. 53. UNESCO/PUMA —MAB, Montevideo, Uruguay.

Bryant, H. N., Russell, A. P., Laroiya, R., and Powell, G. L. (1996) Claw retraction and protraction in the Carnivora: skeletal microvaration in the phalanges of the Felidae. *Journal of Morphology*, **229**, 289–308.

Bryman, A. (2004) *Social Research Methods*. Oxford University Press, Oxford, UK.

Buckley-Beason, V. A., Johnson, W. E., Nash, W. G., *et al*. (2006) Molecular evidence for species-level distinctions in clouded leopards. *Current Biology*, **16**, 2371–6.

Buddle, B. M., Skinner, M. A., and Chambers, M. A. (2000) Immunological approaches to the control of tuberculosis in wildlife reservoirs. *Veterinary Immunology and Immunopathology*, **74**, 1–16.

Bull, M. E., Kennedy Stoskopf, S., Levine, J. F., Tompkins, W. A. F., Loomis, M., and Gebhard, D. G. (2003) Evaluation of T lymphocytes in captive African lions (*Panthera leo*) infected with feline immunodeficiency virus. *American Journal of Veterinary Research*, **64**, 1293–300.

Bulmer, M. G. (1974) A statistical analysis of the 10 year cycle in Canada. *Journal of Animal Ecology*, **43**, 701–17.

Bulte, E. and Damania, R. (2005) Wildlife farming and conservation: an economic assessment. *Conservation Biology*, **19**, 1222–33.

Bulte, E. H. and Rondeau, D. (2005) Why compensating wildlife damages may be bad for conservation. *Journal of Wildlife Management*, **69**, 14–19.

Bulte, E. H., van Kooten, G. C., and Swanson, T. (2003) *Economic Incentives and Wildlife Conservation*. CITES, Geneva, Switzerland. <www.cites.org/eng/prog/economics/CITES-draft6-final.pdf>. Accessed online 15 Dec 2008.

Bunnefeld, N., Linnell, J. D. C., Odden, J., Van Duijn, M. A. J., and Andersen, R. (2006) Risk taking by Eurasian lynx (*Lynx lynx*) in a human-dominated landscape: effects of sex and reproductive status. *Journal of Zoology*, **270**, 31–9.

Burgener, M. A., Greyling, A., and Rumsey, T. (2005) *Status Quo Report on the Policy, Legislative and Regulatory Environment Applicable to Commercial and Recreational Hunting in South Africa*. The Wildlife Trade Monitoring Network (TRAFFIC) East/Southern Africa, Johannesburg, South Africa.

Burger, J., Rosendahl, W., Loreille, O., *et al*. (2004) Molecular phylogeny of the extinct cave lion *Panthera leo spelaea*. *Molecular Phylogenetics and Evolution*, **30**, 841–9.

Burnham, K. P., Anderson, D. R., and Laake, J. K. (1980) Estimation of density from line-transect sampling of biological populations. *Wildlife Monographs*, **72**, 1–202.

Burns, R. J., Zemilcka, D. E., and Savarie, P. J. (1996) Effectiveness of large livestock protection collars against depredating coyotes. *Wildlife Society Bulletin*, **24**, 123–7.

Burton, A. M., Navarro Perez, S., and Chavez Tovar, C. (2003) Bobcat ranging behavior in relation to small mammal abundance on Colima Volcano, Mexico. *Anales del Instituto de Biología, Universidad Nacional Autónoma de México, Serie Zoología*, **74**, 67–82.

Buskirk, S. W. (1999) Mesocarnivores of Yellowstone. In T. P. Clark, A. P. Curlee, S. C. Minta and P. M. Kareiva (eds.), *Carnivores in Ecosystems: The Yellowstone Experience*, pp. 165–87. Yale University Press, New Haven, CT.

Butler, J. R. A. (2000) The economic costs of wildlife predation on livestock in Gokwe communal land, Zimbabwe. *African Journal of Ecology*, **38**, 23–30.

Buwal (2004) *Konzept Luchs Schweiz*. Bundesamt für Umwelt Wald und Landschaft, Bern, Switzerland.

Bygott, J. D., Bertram, B. C. R., and Hanby, J. P. (1979) Male lions in large coalitions gain reproductive advantages. *Nature*, **282**, 839–41.

Cabrera, A. (1914) *Fauna Ibérica. Mamíferos*. Museo Nacional de Ciencias Naturales, Madrid, Spain.

Cabrera, A. L. (1940) Notas sobre carnivoros sudamericanos. *Notas del Museo de La Plata, Zoologia*, **5**, 1–22.

Cabrera, A. L. and Willink, A. (1973) Biogeografía de América Latina. *Organización de los Estados Americanos—Monografía*, **3**, 83–9.

Cain, J. W., Krausman, P. R., Jansen, B. D., and Morgart, J. R. (2005) Influence of topography and GPS fix interval on GPS collar performance. *Wildlife Society Bulletin*, **33**, 926–34.

Callicott, J. B. (1994) Conservation values and ethics. In G. K. Meffe and C. R. Carroll (eds.), *Principles of Conservation Biology*, pp. 24–49. Sinauer Associates, Sunderland, MA.

Callou, C. and Gargominy, O. (2008) Lynx pardelle. <www.inpn.mnhn.fr>. Accessed online 21 Oct 2008.

Callou, C., Samzun, A., and Zivie, A. (2004) Archaeology: a lion found in the Egyptian tomb of Maïa. *Nature*, **427**, 211–12.

Calvete, C., Angulo, E., Estrada, R., Moreno, S., and Villafuerte, R. (2005) Quarantine length and survival of translocated European wild rabbits. *Journal of Wildlife Management*, **69**, 1063–72.

Campbell, D. J., Lusch, D. P., Smucker, T., and Wangui, E. E. (2003) Land use change patterns and root cause in the Loitokitok Area, Kajiado district, Kenya. Working Paper Number 19. ILRI, Nairobi, Kenya. <www.lucideastafrica.org>.

Campbell, L. M. (2005) Overcoming obstacles to interdisciplinary research. *Conservation Biology*, **19**, 574–7.

Campbell, V. and Strobeck, C. (2006) Fine-scale genetic structure and dispersal in Canada lynx (*Lynx canadensis*) within Alberta, Canada. *Canadian Journal of Zoology*, **84**, 1112–19.

Cao, Z., Du, H., Zhao, Q., and Cheeng, J. (1990) Discovery of the Middle Miocene fossil mammals in Guanghe District and their stratigraphic significance. *Geoscience*, **4**, 16–32.

Caraco, T. and Wolf, L. L. (1975) Ecological determinants of group sizes of foraging lions. *American Naturalist*, **109**, 343–52.

Caraco, T., Martindale, S., and Wittham, T. S. (1980) An empirical demonstration of risk-sensitive foraging preferences. *Animal Behaviour*, **28**, 820–30.

Carbone, C. and Gittleman, J. L. (2002) A common rule for the scaling of carnivore density. *Science*, **295**, 2273–6.

Carbone, C., Teacher, A., and Rowcliffe, J. M. (2007) The costs of carnivory. *PloS Biology*, **5**, 1–6.

Carbone, C., Christie, S., Conforti, K., *et al.* (2001) The use of photographic rates to estimate densities of tigers and other cryptic mammals. *Animal Conservation*, **4**, 75–9.

Carbone, C., Mace, G. M., Roberts, S. C., and Macdonald, D. W. (1999) Energetic constraints on the diet of terrestrial carnivores. *Nature*, **402**, 286–8.

Carbyn, L. N. and Patriquin, D. (1983) Observations on home range sizes, movements and social organization of lynx, *Lynx canadensis*, in Riding Mountain National Park, Manitoba. *Canadian Field-Naturalist*, **97**, 262–7.

Cardillo, M., Purvis, A., Sechrest, W., Gittleman, J. L., Bielby, J., and Mace, G. M. (2004) Human density and extinction risk in the world's carnivores. *PloS Biology*, **2**, 909–14.

Caro, T. and Laurenson, K. (1994) Ecological and genetic factors in conservation: a cautionary tale. *Science*, **263**, 485–6.

Caro, T. M. (2005) *Antipredator Defenses in Birds and Mammals*. University of Chicago Press, Chicago, IL.

Caro, T. M. (2000) Controversy over behaviour and genetics in cheetah conservation. In L. M. Gosling and W. J. Sutherland (eds.), *Behaviour and Conservation*, pp. 221–37. Cambridge University Press, Cambridge, UK.

Caro, T. M. (1994) *Cheetahs of the Serengeti Plains: Group Living in an Asocial Species*. University of Chicago Press, Chicago, IL.

Caro, T. M. and Collins, D. A. (1987a) Ecological characteristics of territories of male cheetahs (*Acinonyx jubatus*). *Journal of Zoology*, **211**, 89–105.

Caro, T. M. and Collins, D. A. (1987b) Male cheetah social organization and territoriality. *Ethology*, **74**, 52–64.

Caro, T. M. and Collins, D. A. (1986) Male cheetahs of the Serengeti. *National Geographic Research*, **2**, 75–86.

Caro, T. M. and Durant, S. M. (1991) Use of quantitative analyses of pelage characteristics to reveal family resemblances in genetically monomorphic cheetahs. *Journal of Heredity*, **82**, 8–14.

Caro, T. M. and Kelly, M. J. (2000) Cheetahs and their mating system. In L. A. Dugatkin (ed.) *Model Systems in Behavioural Ecology: Integrating Conceptual, Theoretical and Empirical Approaches*, pp. 512–32. Princeton University Press, Princeton, NJ.

Caro, T. M., Pelkey, N., Borner, M., *et al.* (1998) The impact of tourist hunting on large mammals in Tanzania: an initial assessment. *African Journal of Ecology*, **36**, 321–46.

Caro, T. M., FitzGibbon, C. D., and Holt, M. E. (1989) Physiological costs of behavioural strategies for male cheetahs. *Animal Behaviour*, **38**, 309–17.

Caro, T. M., Holt, M. E., FitzGibbon, C. D., Bush, M., Hawkey, C. M., and Kock, R. A. (1987) Health of adult free-living cheetahs. *Journal of Zoology*, **212**, 573–84.

Carpenter, M. A. and O'Brien, S. J. (1995) Coadaptation and immunodeficiency virus: lessons from the Felidae. *Current Opinion in Genetics and Development*, **5**, 739–45.

Carpenter, M. A., Appel, M. J. G., Roelke Parker, M. E., *et al.* (1998) Genetic characterization of canine distemper virus in Serengeti carnivores. *Veterinary Immunology and Immunopathology*, **65**, 259–66.

Carpenter, M. A., Brown, E. W., Culver, M., *et al.* (1996) Genetic and phylogenetic divergence of feline immunodeficiency virus in the puma (*Puma concolor*). *Journal of Virology*, **70**, 6682–93.

Carr, G. M. and Macdonald, D. W. (1986) The sociality of solitary foragers: a model based on resource dispersion. *Animal Behaviour*, **34**, 1540–9.

Carroll, C. and Miquelle, D. G. (2006) Spatial viability analysis of Amur tiger *Panthera tigris altaica* in the Russian Far East: the role of protected areas and landscape matrix in population persistence. *Journal of Applied Ecology*, **43**, 1056–68.

Carter, M. and Barrett, C. (2006) The economics of poverty traps and persistent poverty: an asset-based approach. *The Journal of Development Studies*, **42**, 178–99.

Carvajal, S., Caso, A., Downey, P., Moreno, A., and Tewes, M. (2007) Home range and activity patterns of the margay (*Leopardus wiedii*) at 'El Cielo' Biosphere Reserve, Tamaulipas, Mexico. In *Felid Biology and Conservation Conference 17–20 September 2007: Programme and Abstracts*, p. 118. Wildlife Conservation Research Unit, Oxford, UK.

Caso, A. (1994) *Home range and habitat use of three Neotropical carnivores in northeast Mexico*. MS dissertation, Texas A & M University, Kingsville, TX.

Caso, A. and Tewes, M. E. (2007) Habitat use, spatial and activity patterns of the jaguarondi in northeast Mexico. In *Felid Biology and Conservation Conference 17–20 September 2007: Programme and Abstracts*, p. 56. Wildlife Conservation Research Unit, Oxford, UK.

Castello, L. (2004) A method to count pirarucu *Arapaima gigas*: fishers, assessment, and management. *North American Journal of Fisheries Management*, **24**, 379–89.

Castley, J. G., Knight, M. H., Mills, M. G. L., and Thouless, C. (2002) Estimation of the lion (*Panthera leo*) population in the southwestern Kgalagadi Transfrontier Park using a capture-recapture survey. *African Zoology*, **37**, 27–34.

Castro, L. R. and Palma, L. (1996) The current status, distribution and conservation of Iberian lynx in Portugal. *Journal of Wildlife Research*, **2**, 179–81.

Cat Specialist Group (2008) <www.catsg.org>. Accessed online 10 Sept 2008.

Caughley, G. (1994) Directions in conservation biology. *Journal of Animal Ecology*, **63**, 231–44.

Cavalcanti, S. M. C. (2008) *Predator-prey relationships and spatial ecology of jaguars in the southern Pantanal, Brazil: implications for conservation and management.* PhD dissertation, Utah State University, Logan, UT.

Cavalcanti, S. M. C. (2003) Manejo e controle de danos causados por espécies da fauna. In L. Cullen, Jr., R. Rudran, and C. Valladares-Padua (eds.), *Métodos de Estudo em Biologia sa Conservação e Manejo da Vida Silvestre*, pp. 203–42. Editora UFPR, Curitiba, Brazil.

Cavalcanti, S. M. C. and Gese, E. M. (2009) Spatial ecology and social interactions of jaguars (*Panthera onca*) in the southern Pantanal, Brazil. *Journal of Mammalogy*, **90**, 935–45.

Cavalcanti, S. M. C., Gese, E. M., and Neale, C. M. U. (in prep) Jaguar habitat use in the southern Pantanal, Brazil—landscape attributes and their influence on predation of livestock and native prey.

Caziani, S. M., Derlindati, E. J., Tálamo, A., Nicolossi, G., Sureda, A. L., and Trucco, C. (2001) Waterbird richness in altiplano lakes of northwestern Argentina. *Waterbirds*, **24**, 103–17.

Ceballos, G., Chávez, C., Rivera, A., Manterola, C., and Wall, B. (2002) Tamaño poblacional y conservación del jaguar en la Reserva de la Biosfera Calakmul, Campeche, México. In R. A. Medellín, C. Equihua, C. L. B. Chetkiewicz, *et al.* (eds.), *El Jaguar en el Nuevo Milenio*, pp. 403–17. Fondo de Cultura Economica, Universidad Nacional Autónoma de México and Wildlife Conservation Society, Mexico City, Mexico.

Central Statistics Office (2001) 2001 Population and Housing Census. <www.cso.gov.bw/>. Accessed online 18 Sept 2008.

Červeny, J., Koubek, P., and Bufka, L. (2002) Eurasian lynx (*Lynx lynx*) and its chance for survival in central Europe: the case of the Czech republic. *Acta Zoologica Lituanica*, **12**, 362–6.

Ceza, B., Rochat, N., Tester, U., Kessler, R., and Marti, K. (2001) Wer tötet den Luchs?—Tatsachen, Hintergründe und Indizien zu illegalen Luchstötungen in der Schweiz. *Beiträge zum Naturschutz in der Schweiz*, **25**, 1–33.

Chamberlain, M. J., Leopold, B. D., Burger, L. W., Plowman, B. W., and Conner, L. M. (1999) Survival and cause-specific mortality of adult bobcats in Central Mississipi. *Journal of Wildlife Management*, **63**, 613–20.

Champion, F. W. (1927) *With a Camera in Tigerland.* Chatto & Windus, London, UK.

Champion, F. W. (1925) *The Jungle in Sunlight and Shadow.* Chatto & Windus, London, UK.

Chandrashekar, J., Hoon, M. A., Ryba, N. J. P., and Zuker, C. S. (2006) The receptors and cells for mammalian taste. *Nature*, **444**, 288–94.

Chapron, C., Miquelle, D. G., Lambert, A., Goodrich, J. M., Legendre, S., and Clobert, J. (2008a) The impact of poaching versus prey depletion on tigers and other large solitary felids. *Journal of Applied Ecology*, **45**, 1667–74.

Chapron, G., Andren, H., and Liberg, O. (2008b) Conserving top predators in ecosystems. *Science*, **320**, 47.

Chardonnet, P. (ed.) (2002) *Conservation of the African Lion: Contribution to a Status Survey.* International Foundation for the Conservation of Wildlife, France and Conservation Force, USA.

Chavez, A. S. and Gese, E. M. (2006) Landscape use and movements of wolves in relation to livestock in a wildland-agriculture matrix. *Journal of Wildlife Management*, **70**, 1079–86.

Chavez, A. S. and Gese, E. M. (2005) Food habits of wolves in relation to livestock depredations in northwestern Minnesota. *American Midland Naturalist*, **154**, 253–63.

Chavez, A. S., Gese, E. M., and Krannich, R. S. (2005) Attitudes of rural landowners toward wolves in northwestern Minnesota. *Wildlife Society Bulletin*, **33**, 517–27.

Cheadle, M. A., Spencer, J. A., and Blagburn, B. L. (1999) Seroprevalences of *Neospora caninum* and *Toxoplasma gondii* in nondomestic felids from southern Africa. *Journal of Zoo and Wildlife Medicine*, **30**, 248–51.

Cheeseman, C. L., Mallinson, P. J., Ryan, J., and Wilesmith, J. W. (1993) Recolonization by badgers in Gloucestershire. In T. Hayden (ed.), *The Badger*, pp. 78–93. Royal Irish Academy, Dublin, Ireland.

Chellam, R. (1993) *Ecology of the Asiatic lion* (Panthera leo persica). PhD dissertation, Saurashtra University, Rajkot, India.

Chellam, R. and Johnsingh, A. J. T. (1993) Management of Asiatic lions in the Gir forest, India. *Symposia of the Zoological Society of London*, **65**, 409–24.

Chester, T. (2006) Mountain lion attacks on people in the U. S. and Canada. <www.tchester.org/sgm/lists/lion_attacks.html#stats>. Accessed online 11 Sept 2008.

Chinchilla, F. A. (1997) La dieta del jaguar (*Panthera onca*), el puma (*Felis concolor*) y el manigordo (*Felis pardalis*) (Carnivora: Felidae) en el Parque Nacional Corcovado, Costa Rica. *Revista de Biología Tropical*, **45**, 1223–9.

Cho, H. S., Kim, Y. H., and Park, N. Y. (2006) Disseminated mycobacteriosis due to *Mycobacterium avium* in captive Bengal tiger (*Panthera tigris*). *Journal of Veterinary Diagnostic Investigation*, **18**, 312–14.

Chomel, B. B., Kikuchi, Y., Chang, C. C., *et al.* (2004) Seroprevalence of *Bartonella* infection in American free-ranging and captive pumas (*Felis concolor*) and bobcats (*Lynx rufus*). *Veterinary Research*, **35**, 233–41.

Choudhury, A. (2007) Sighting of Asiatic golden cat in the grasslands of Assam's Manas National Park. *Cat News*, **47**, 29.

Christiansen, P. (2008) Phylogeny of the great cats (Felidae: Pantherinae), and the influence of fossil taxa and missing characters. *Cladistics*, **24**, 1–16.

Christiansen, P. (2006) Sabertooth characters in the clouded leopard (*Neofelis nebulosa* Griffiths 1821). *Journal of Morphology*, **267**, 1186–98.

Christie, S. (2009) Reintroduction planning for the Far Eastern leopard. In M. W. Hayward and M. J. Somers (eds.), *The Reintroduction of Top-Order Predators*, pp. 388–410. Wiley-Blackwell Publishing, Oxford, UK.

Christie, S. and Seidensticker, J. (1999) Is reintroduction of captive-bred tigers a feasible option for the future? In J. Seidensticker, S. Christie, and P. Jackson (eds.), *Riding the Tiger: Tiger Conservation in Human Dominated Landscapes*, pp. 207–9. Cambridge University Press, Cambridge, UK.

Chundawat, R. S. (1992) *Ecological studies on snow leopard and its associated species in Hemis National Park Ladakh*. PhD dissertation, University of Rajasthan, Jaipur, India.

Chundawat, R. S. and Rawat, G. S. (1994) Food habits of the snow leopard in Ladakh. In J. L. Fox and D. Jizeng (eds.), *Proceedings of the 7th International Snow Leopard Symposium*, pp. 127–32. Northwest Plateau Institute of Biology and International Snow Leopard Trust, Seattle, WA.

Chundawat, R.S., Habib, B., Karanth, U., *et al.* (2008) *Panthera tigris*. 2008 IUCN Red List of Threatened Species. <www.iucnredlist.org>. Accessed online 16 Apr 2009.

CITES. (2008) Appendices I, II, and III. <www.cites.org/eng/app/E-Jul01.pdf>. Accessed online 30 Oct 2008.

CITES. (2007) Proposal to seek transfer of Uganda's leopards (*Panthera pardus*) population from Appendix I to Appendix II with export quota. CITES, The Hague, the Netherlands.

CITES. (1992) Quotas for trade in specimens of cheetah. In *8th Meeting of the Convention of International Trade in Endangered Species of Wild Fauna and Flora*, pp. 1–5. CITES, The Hague, the Netherlands.

Cleaveland, S., Hess, G. R., Dobson, A. P., *et al.* (2002) The role of pathogens in biological conservation. In P. J. Hudson, A. Rizzoli, B. T. Grenfell, H. Heesterbeek, and A. P. Dobson (eds.), *The Ecology of Wildlife Diseases*, pp. 139–50. Oxford University Press, Oxford, UK.

Cleaveland, S., Appel, M. G. J., Chalmers, W. S. K., Chillingworth, C., Kaare, M., and Dye, C. (2000) Serological and demographic evidence for domestic dogs as a source of canine distemper virus infection for Serengeti wildlife. *Veterinary Microbiology*, **72**, 217–27.

Clan Chattan Association (2008) Badges and Association motto. <www.clanchattan.org.uk/catalog/Motto.php>. Accessed online 30 Oct 2008.

Clark, C. W. (1973) The economics of overexploitation. *Science*, **181**, 630–4.

Clark, M. M. and Galef, B. G. (1980) Effect of rearing environment on adrenal weights, sexual development and behavior of gerbils: an examination of Richter's domestication hypothesis. *Journal of Comparative Physiology and Psychology*, **94**, 857–63.

Cleaveland, S., Packer, C., and Hampson, K., *et al.* (2008) The multiple roles of infectious diseases in the Serengeti ecosystem. In A. R. E. Sinclair, C. Packer, S. Mduma, and J. M. Fryxell (eds.), *Serengeti III: Human Impacts on Ecosystem Dynamics*, pp. 209–39. Chicago University Press, Chicago, IL.

Cleaveland, S., Mlengeya, T., Kaare, M., *et al.* (2007) The conservation relevance of epidemiological research into carnivore viral diseases in the Serengeti. *Conservation Biology*, **21**, 612–22.

Cleaveland, S., Mlengeya, T., Kazwala, R. R., *et al.* (2005) Tuberculosis in Tanzanian wildlife. *Journal of Wildlife Diseases*, **41**, 446–53.

Clement, C., Niaga, M., and Cadi, A. (2007) Does the serval still exist in Senegal? *Cat News*, **47**, 24–5.

Clutton-Brock, J. (1999) *A Natural History of Domesticated Mammals*, 2nd edn. Cambridge University Press, Cambridge, UK.

Clutton-Brock, J. (1987) *A Natural History of Domesticated Mammals*. British Natural History Museum, London, UK and Cambridge University Press, Cambridge, UK.

Coard, R. and Chamberlain, A. T. (1999) The nature and timing of faunal change in the British Isles across the Pleistocene/Holocene transition. *Holocene*, **9**, 372–6.

Coe, M. J., Cumming, D. H., and Phillipson, J. (1976) Biomass and production of large African herbivores in relation to rainfall and primary production. *Oecologia*, **22**, 341–54.

Coleman, J. D. and Cooke, M. M. (2001) *Mycobacterium bovis* infection in wildlife in New Zealand. *Tuberculosis (Edinburgh)*, **81**, 191–202.

Collier, G. E. and O'Brien, S. J. (1985) A molecular phylogeny of the Felidae: immunological distance. *Evolution*, **39**, 473–87.

Coltman, D. W., O'Donoghue, P., Jorgenson, J. T., Hogg, J. T., Strobeck, C., and Festa-Bianchet, M. (2003) Undesirable evolutionary consequences of trophy hunting. *Nature*, **426**, 655–8.

Colwell, R. K. and Futuyma, D. J. (1971) Measurement of niche breadth overlap. *Ecology*, **52**, 567–76.

Comiskey, E. J., Bass, O. L., Jr., Gross, L. J., McBride, R. T., and Salinas, R. (2002) Panthers and forests in South Florida: an ecological perspective. *Conservation Ecology*, **6**, 18.

Conforti, V. A. and Azevedo, F. C. C., de (2003) Local perceptions of jaguars (*Panthera onca*) and pumas (*Puma concolor*) in the Iguaçu National Park area, south Brazil. *Biological Conservation*, **111**, 215–21.

Conover, M. R. (2001) *Resolving Human–Wildlife Conflict: The Science of Wildlife Damage Management*. Lewis Publishers, Cherry Hill, NJ.

Conradi, M. (2006) *Non-invasive sampling of snow leopards* (Uncia uncia) *in Phu Valley, Nepal*. MSc dissertation, University of Oslo, Oslo, Norway.

Consejería de Medio Ambiente de Andalucía (2006) Recuperación de las poblaciones de Lince Ibérico en Andalucía. LIFEO2/NAT/E/8609 Final unpublished report.

Conservation Breeding Specialist Group (2008) <www.cbsg.org>. Accessed online 10 Sept 2008.

Conservation Breeding Specialist Group (2003a) *Animal Movements and Disease Risk: A Workbook*, 5th edn. IUCN Species Survival Commission Conservation Breeding Specialist Group, Apple Valley, MN.

Conservation Breeding Specialist Group (2003b) *Black-Footed Ferret Population Management Planning Workshop Final Report*. IUCN Species Survival Commission Conservation Breeding Specialist Group, Apple Valley, MN.

Conservation Breeding Specialist Group (1998) *Iberian lynx* (Lynx pardinus) *Population Viability Habitat Assessment*. IUCN Species Survival Commission Conservation Breeding Specialist Group, Apple Valley, MN.

Consortium of Aquariums, Universities and Zoos (2008) <www.manandmollusc.net/links_zoos.html>. Accessed online 10 Sept 2008.

Convention on Biological Diversity (2003) *Handbook of the Convention on Biological Diversity*. Secretariat of the Convention on Biological Diversity. <www.biodiv.org/handbook/>. Accessed online 4 Nov 2004.

Cook, H. J. (1922) A Pliocene fauna from Yuma County, Colorado, with notes on the closely related Snake Creek beds from Nebraska. *Proceedings of the Colorado Museum of Natural History*, **4**, 3–15.

Coomes, O. T., Grimard, F., Potvin, C., and Simad, P. (2008) The fate of the tropical forest: carbon or cattle? *Ecological Economics*, **65**, 207–12.

Cooper, A. B., Pettorelli, N., and Durant, S. M. (2007) Large carnivore menus: factors affecting hunting decisions by cheetahs in the Serengeti. *Animal Behaviour*, **73**, 651–9.

Cooper, S. M. (1991) Optimal hunting group size: the need for lions to defend their kills against loss to spotted hyaenas. *African Journal of Ecology*, **29**, 130–6.

Coppinger, R. and Coppinger, L. (1980) Livestock-guarding dogs: an old world solution to an age-old problem. *Country Journal*, **7**, 68–77.

Corbet, G. B. (1978) Subgenus *Lynx*. In *The Mammals of the Palaearctic Region: A Taxonomic Review*, pp. 182–3. British Museum of Natural History, London, UK and Cornell University Press, Ithaca, NY.

Corbett, G. B. and Hill, J. E. (1992) Family Felidae—Cats. In *The Mammals of the Indo-Malayan Region: A Systematic Review*, pp. 219–28. OUP for British Museum of Natural History, London, UK.

Corbett, J. (1946) *Man-Eaters of Kumaon*. Oxford University Press, London, UK.

Corbett, L. K. (1979) *Feeding ecology and social organisation of wildcats* (Felis silvestris) *and domestic cats* (Felis catus) *in Scotland*. PhD dissertation, University of Aberdeen, Aberdeen, Scotland.

Cornalia, E. (1865) Descrizione di una nuova specie del genere Felis, *Felis jacobita* (Corn.). *Memorie della Societá Italiana di Scienze Naturalli*, **1**, 1–9.

Corner, L. A. L. (2006) The role of wild animal populations in the epidemiology of tuberculosis in domestic animals: how to assess the risk. *Veterinary Microbiology*, **112**, 303–12.

Cortes, A., Rau, J. R., Miranda, E., and Jimenez, J. E. (2002) Food habits of *Lagidium viscacia* and *Abrocoma cinerea*: syntopic rodents in high Andean environments of northern Chile. *Revista Chilena de Historia Natural*, **75**, 583–93.

Cory, C. B. (1896) *Hunting and Fishing in Florida*. Estes and Lauriat, Boston, MA.

Cossíos, D. and Angers, B. (2006) Identification of Andean felid faeces using PCR-RFLP. *Mastozoologia Neotropical*, **13**, 239–44.

Cossíos, D., Beltrán Saavedra, F., Bennet, M., *et al.* (2007a) *Manual de Metodologías para Relevamientos de Carnívoros Alto Andinos*. Alianza Gato Andino, Buenos Aires, Argentina.

Cossíos, D., Madrid, A., Condori, J. L., and Fajardo, U. (2007b) Update on the distribution of the Andean cat *Oreailurus jacobita* and the pampas cat *Lynchailurus colocolo* in Peru. *Endangered Species Research*, **3**, 313–20.

Cougar Management Guidelines Working Group (2005) *Cougar Management Guidelines*. WildFutures, Bainbridge Island, WA.

Courchamp, F., Angulo, E., Rivalan, P., *et al.* (2006) Rarity value and species extinction: the anthropogenic Allee effect. *PloS Biology*, **4**, 2405–10.

Covey, D. S. G. and Greaves, W. S. (1994) Jaw dimensions and torsion resistance during canine biting in the Carnivora. *Canadian Journal of Zoology*, **72**, 1055–60.

Cowan, D. F. and Johnson, W. C. (1970) Amyloidosis in the white Pekin duck. I. Relation to social environmental stress. *Laboratory Investigation*, **23**, 551–5.

Cozza, K., Fico, R., Battistini, M. L., and Rogers, E. (1996) The damage-conservation interface illustrated by predation on domestic livestock in central Italy. *Biological Conservation*, **78**, 329–36.

Cracraft, J. (1997) Species concepts in systematics and conservation biology—an ornithological point of view. In M. F. Claridge, H. A. Dawah, and M. R. Wilson (eds.), *Species: The Units of Biodiversity*, pp. 325–40. Chapman and Hall, London, UK.

Cracraft, J. (1989) Speciation and its ontology: the empirical consequences of alternative species concepts for understanding patterns and processes of differentiation. In D. Otte and J. A. Endler (eds.), *Speciation and Its Consequences*, pp. 28–59. Sinauer Associates, Sunderland, MA.

Craft M.E., Volz E., Packer C. & Meyers L.A. (2009). Distinguishing epidemic waves from disease spillover in a wildlife population. *Proceedings of the Royal Society of London Series B: Biological Sciences of London Series*, **276**, 1777–85.

Craft, M. E., Hawthorne, P. L., Packer, C., and Dobson, A. P. (2008) Dynamics of a multihost pathogen in a carnivore community. *Journal of Animal Ecology*, **77**, 1257–64.

Cramer, J. (2007) Mountain lion attacks hunter near Kalispell. *Missoulian*. John VanStrydonck, Missoula, MT.

Craven, S. R., Decker, D. J., Siemer, W. F., and Hygerstrom, S. E. (1992) Survey use and landowner tolerance in wildlife damage management. *Transactions of the North American Wildlife and Natural Resources Conference*, **57**, 75–88.

Crawford, P. G. (1979) The curious cat. *The World About Us*. BBC Bristol, Bristol, UK.

Crawshaw, P. G. (2004) Depredation of domestic animals by large cats in Brazil. *Human Dimensions of Wildlife*, **9**, 329–30.

Crawshaw, P. G. (2002) Mortalidad inducida por humanos y conservation de jaguares: el Pantanal y el Parque Nacional Iguazy en Brazil. In R. A. Medellín, C. Equihua, C. L. B. Chetkiewicz, *et al.* (eds.), *El Jaguar en el Nuevo Milenio*, pp. 451–64. Fondo de Cultura Economica, Universidad Nacional Autónoma de México and Wildlife Conservation Society, Mexico City, Mexico.

Crawshaw, P. G. (1995) *Comparative ecology of ocelot* (Felis pardalis) *and jaguar* (Panthera onca) *in a protected subtropical forest in Brazil and Argentina*. PhD dissertation, University of Florida, Gainesville, FL.

Crawshaw, P. G. (1987) Top cat in a vast Brazilian marsh. *Animal Kingdom*, **90**, 12–19.

Crawshaw, P. G. and Quigley, H. B. (2002) Hábitos alimentarios del jaguar y el puma en el Pantanal, Brasil, con implicaciones para su manejo y conservación. In R. A. Medellín, C. Equihua, C. L. B. Chetkiewicz, *et al.* (eds.), *El Jaguar en el Nuevo Milenio*, pp. 223–35. Fondo de Cultura Economica, Universidad Nacional Autónoma de México and Wildlife Conservation Society, Mexico City, Mexico.

Crawshaw, P. G. and Quigley, H. B. (1991) Jaguar spacing, activity and habitat use in a seasonally flooded environment in Brazil. *Journal of Zoology*, **223**, 357–70.

Crawshaw, P. G. and Quigley, H. B. (1989) Notes on ocelot movement and activity in the Pantanal Region, Brazil. *Biotropica*, **21**, 377–9.

Crawshaw, P. G. and Quigley, H. B. (1984) *A Ecologia do Jaguar ou Onça-Pintada no Pantanal*. Instituto Brasileiro de Desenvolvimento Florestal (IBDF), Brasília, Brazil.

Crawshaw, P. G., Mähler, J. K., Indrusiak, C., Cavalcanti, S. M. C., Leite-Pitman, M. R. P., and Silvius, K. M. (2004) Ecology and conservation of the jaguar (*Panthera onca*) in Iguaçu National Park, Brazil. In K. M. Silvius, R. E. Bodmer, and J. M. V. Fragoso (eds.), *People in Nature: Wildlife Conservation in South and Central America*, pp. 271–85. Columbia University Press, New York.

Creel, S. and Creel, N. M. (1997) Lion density and population structure in the Selous Game Reserve: evaluation of hunting quotas and offtake. *African Journal of Ecology*, **35**, 83–93.

Creel, S. and Creel, N. M. (1996) Limitation of African wild dogs by competition with larger carnivores. *Conservation Biology*, **10**, 526–38.

Crnokrak, P. and Roff, D. A. (1999) Inbreeding depression in the wild. *Heredity*, **83**, 260–70.

Crook, J. H. (1970) *Social Behaviour in Birds and Mammals*. Academic Press, New York.

Crooks, K. R. and Soulé, M. E. (1999) Mesopredator release and avifaunal extinctions in a fragmented system. *Nature*, **400**, 563–6.

Crooks, K. R., Sanjayan, M. A., and Doak, D. F. (1998) New insights on cheetah conservation through demographic modeling. *Conservation Biology*, **12**, 889–95.

Crosier, A. E., Pukazhenthi, B. S., Henghali, J. N., *et al.* (2006) Cryopreservation of spermatozoa from wild-born Namibian cheetahs (*Acinonyx jubatus*) and influence of glycerol on cryosurvival. *Cryobiology*, **52**, 169–81.

Crow, J. F. and Kimura, M. (1970) *An Introduction to Population Genetics Theory*. Burgess Publishing Company, Minneapolis, MN.

Crowe, D. M. (1975) A model for exploited bobcat populations in Wyoming. *Journal of Wildlife Management*, **39**, 408–15.

Cuellar, E., Maffei, L., Arispe, R., and Noss, A. (2006) Geoffroy's cats at the northern limit of their range: activity patterns and density estimates from camera trapping in Bolivian dry forests. *Studies on Neotropical Fauna and Environment*, **41**, 169–77.

Cuervo, A., Hernadez, J., and Cadena, C. (1986) Lista atualizada de los mamíferos de Colômbia: anotaciones sobre su distribucion. *Caldasia*, **15**, 471–501.

Cullen, J., L., Sana, D., Abreu, K. C., and Nava, A. F. D. (2005) Jaguars as landscape detectives for the upper Paraná river corridor, Brazil. *Natureza e Conservação*, **3**, 124–46.

Culver, M., Hedrick, P. W., Murphy, K., O'Brien, S., and Hornocker, M. G. (2008) Estimation of the bottleneck size in Florida panthers. *Animal Conservation*, **11**, 104–10.

Culver, M., Menotti Raymond, M. A., and O'Brien, S. J. (2001) Patterns of size homoplasy at 10 microsatellite loci in pumas (*Puma concolor*). *Molecular Biology and Evolution*, **18**, 1151–6.

Culver, M., Johnson, W. E., Pecon-Slattery, J., and O'Brien, S. J. (2000) Genomic ancestry of the American puma (*Puma concolor*). *Journal of Heredity*, **91**, 186–97.

Cunningham, M. W., Brown, M. A., Shindle, D. B., *et al.* (2008) Epizootiology and management of feline leukemia virus in the Florida puma. *Journal of Wildlife Diseases*, **44**, 537–52.

Cunningham, S. C., Ballard, W. B., and Whitlaw, H. A. (2001) Age structure, survival and mortality of mountain lions in southeastern Arizona. *Southwestern Naturalist*, **46**, 76–80.

Cunningham, M. W., Dunbar, M. R., Buergelt, C. D., *et al.* (1999) Atrial septal defects in Florida panthers. *Journal of Wildlife Diseases*, **35**, 519–30.

Cunningham, S. C., Haynes, L. A., Gustavson, C. H. and Haywood, D. D. (1995) Evaluation of the interaction between mountain lions and cattle in the Aravaipa Klondyke area of southeast Arizona. Technical Report No. 17. Arizona Game and Fish Department, Phoenix, AZ.

Curio, E. (1976) *The Ethology of Predation*. Springer Verlag, Berlin, Germany.

Currier, M. J. P. (1983) *Felis concolor. Mammalian Species*, **200**, 1–7.

Cuvier, G. (1824) *Recherches sur les Ossemens Fossiles ou l'on rétablit les caractères de plusieurs animaux dont les révolutions du globe ont détruit les espèces*. G. Dufour et E. D'Ocagne, Paris, France.

Cuzin F. (2003) *Les grands mammiferes du Maroc meridional (Haut Atlas, Anti Atlas et Sahara): distribution, ecologie et conservation*. PhD dissertation, University of Montpellier, Montpellier, France.

Czech, B. (2004) Taking on the economic triangle. *Frontiers in Ecology and the Environment*, **2**, 227.

Dalnikin, A. A. (1999) *Mammals of Russia and adjacent regions: Deer (Cervidae)*. GEOS, Moscow, Russia.

Dalponte, J. C. (2002) Dieta del jaguar y depredación de ganado en el norte del Pantanal, Brasil. In R. A. Medellín, C. Equihua, C. L. B. Chetkiewicz, *et al.* (eds.), *El Jaguar en el Nuevo Milenio*, pp. 209–21. Fondo de Cultura Economica, Universidad Nacional Autónoma de Mexico, Wildlife Conservation Society, Mexico City, Mexico.

Dalrymple, G. H. and Bass, O. L., Jr (1996) The diet of the Florida panther in Everglades National Park, Florida. *Bulletin of the Florida Museum of Natural History*, **39**, 173–93.

Daly, B., Power, J., Carmanch, G., *et al.* (2005) Leopard (*Panthera pardus*) population and habitat viability assessment. In *Proceedings of a Workshop of the Conservation Breeding Specialist Group (World Conservation Union (IUCN) Species Survival Commission)*, pp. 1–109. Endangered Wildlife Trust, Johannesburg, South Africa.

Damania, R., Seidensticker, J., Whitten, T., *et al.* (2008) *A Future for Wild Tigers*. World Bank, Washington DC.

Daniels, M. J. (1997) *The biology and conservation of the wildcat in Scotland*. DPhil dissertation, University of Oxford, Oxford, UK.

Daniels, M. J. and Corbett, L. (2003) Re-defining introgressed protected mammals: when is a wildcat a wild cat and a dingo a wild dog? *Wildlife Research*, **30**, 213–18.

Daniels, M. J. and Macdonald, D. W. (2002) Seasonality of trapping success for wildcats in Scotland. *Saugetierkundliche Informationen*, **26**, 229–32.

Daniels, M. J., Wright, T. C. M., Bland, K. P., and Kitchener, A. C. (2002) Seasonality and reproduction in wild-living cats in Scotland. *Acta Theriologica*, **47**, 73–84.

Daniels, M. J., Beaumont, M. A., Johnson, P. J., Balharry, D., Macdonald, D. W., and Barratt, E. (2001) Ecology and genetics of wild-living cats in the north-east of Scotland and the implications for the conservation of the wildcat. *Journal of Applied Ecology*, **38**, 146–61.

Daniels, M. J., Golder, M. C., Jarrett, O., and Macdonald, D. W. (1999) Feline viruses in wildcats from Scotland. *Journal of Wildlife Diseases*, **35**, 121–4.

Daniels, M. J., Balharry, D., Hirst, D., Kitchener, A. C., and Aspinall, R. J. (1998) Morphological and pelage characteristics of wild living cats in Scotland: implications for defining the 'wildcat'. *Journal of Zoology*, **244**, 231–47.

Dards, J. L. (1983) The behaviour of dockyard cats: interactions of adult males. *Applied Animal Ethology*, **10**, 133–53.

Darwin, C. (1868) *The Variation of Plants and Animals under Domestication*. John Murray, London, UK.

Darwin, C. (1839) *The Voyage of the Beagle: Charles Darwin's Journal of Researches*. Penguin Books, London, UK.

Davies-Mostert, H. T., Mills, M. G. L., and Macdonald, D. W. (2009) A critical assessment of South Africa's managed metapopulation recovery strategy for African wild dogs and its value as a template for large carnivore conservation elsewhere. In M. W. Hayward and M. J. Somers (eds.), *The Reintroduction of Top-Order Predators*, pp. 10–42. Wiley-Blackwell Publishing, Oxford, UK.

Davies, E. M. and Boersma, P. D. (1984) Why lionesses copulate with more than one male. *American Naturalist*, **123**, 594–611.

Davies, K. F., Margules, C. R., and Lawrence, J. F. (2004) A synergistic effect puts rare, specialized species at greater risk of extinction. *Ecology*, **85**, 265–71.

Davison, T. (1967) *Wankie. The Story of a Great Game Reserve*. Books of Africa, Cape Town, South Africa.

Dawkins, R. (2006) *The God Delusion*. Bantam Press, London, UK.

Day, L. M. and Jayne, B. C. (2007) Interspecific scaling of the morphology and posture of the limbs during the locomotion of cats (Felidae). *Journal of Experimental Biology*, **210**, 642–54.

Dayan, T., Simberloff, D., Tchernov, E., and Yom-Tov, Y. (1992) Canine carnassials—character displacement in the wolves, jackals and foxes of Israel. *Biological Journal of the Linnean Society*, **45**, 315–31.

Dayan, T., Simberloff, D., Tchernov, E., and Yom Tov, Y. (1990) Feline canines: community-wide character displacement among the small cats of Israel. *American Naturalist*, **136**, 39–60.

Dayan, T., Simberloff, D., Tchernov, E., and Yom Tov, Y. (1989) Inter- and intraspecific character displacement in mustelids. *Ecology*, **70**, 1526–39.

de Lisle, G. W., Bengis, R. G., Schmitt, S. M., and O'Brien, D. J. (2002) Tuberculosis in free-ranging wildlife: detection, diagnosis and management. *Revue Scientifique et Technique de l' Office International des Epizooties*, **21**, 317–34.

de Lisle, G. W., Yates, G. F., Collins, D. M., MacKenzie, R. W., Crews, K. B., and Walker, R. (1995) A study of bovine tuberculosis in domestic animals and wildlife in the MacKenzie basin and surrounding areas using DNA fingerprinting. *New Zealand Veterinary Journal*, **43**, 266–71.

De Paula, R. C., Campos Neto, M. F., and Morato, R. G. (2008) First official record human killed by jaguar in Brazil. *Cat News*, **49**, 14.

de Ruiter, D. J. and Berger, L. R. (2000) Leopards as taphonomic agents in dolomitic caves—implications for bone accumulations in the hominid-bearing deposits of South Africa. *Journal of Archaeological Science*, **27**, 665–84.

De Smet, K. J. M. (1989) *Distribution and habitat choice of larger mammals in Algeria, with special reference to nature protection*. PhD dissertation, Ghent State University, Ghent, Belgium.

De Soto, H. (2000) *The Mystery of Capital: Why Capitalism Triumphs in the West and Fails Everywhere Else*. Basic Books, New York.

De Vos, A. and Matel, S. E. (1952) The status of lynx in Canada, 1929–1952. *Journal of Forestry*, **50**, 742–5.

De Vos, V. and Bryden, H. B. (1997) The role of carnivores in the epidemiology of anthrax in the Kruger National Park. In J. Van Heerden (ed.), *Proceedings of a Symposium on Lions and Leopards as Game Ranch Animals, 24–25 October 1997*, pp. 198–207. SAVA Wildlife Group, Onderstepoort, South Africa.

De Vos, V., Bengis, R. G., Kriek, N. P. J., *et al.* (2001) The epidemiology of tuberculosis in free-ranging African buffalo (*Syncerus caffer*) in the Kruger national park, South Africa. *Onderstepoort Journal of Veterinary Research*, **68**, 119–30.

Dean, R., Gunn-Moore, D., Shaw, S., and Harvey, A. (2006) Bovine tuberculosis in cats. *Veterinary Record*, **158**, 419–20.

Decker, D. J. (1994) What are we learning from human dimensions studies in controversial wildlife situations? *Proceedings of the Association of Midwest Fish and Wildlife Agencies Annual Meeting*, **61**, 25–7.

Decker, D. J. and Chase, L. (1997) Human dimension of living with wildlife—a management challenge for the 21st century. *Wildlife Society Bulletin*, **25**, 788–95.

Decker, D. J. and Purdey, K. G. (1988) Towards a concept of wildlife acceptance capacity in wildlife management. *Wildlife Society Bulletin*, **16**, 53–7.

Dee, T., Swann, D. E., Steidl, R. J., and Culver, M. (in review) Identifying mammals using molecular scatology.

Defra (2007) Ban on cat and dog fur trade welcomed. <www.defra.gov.uk/news/2007/071128a.htm>. Accessed online 30 Oct 2008.

Degiorgis, M.-P., Frey, J., Nicolet, J., *et al.* (2000) An outbreak of infectious keratoconjunctivitis in Alpine chamois (*Rupicapra r. rupicapra*) in Simmental-Gruyères, Switzerland. *Schweizer Archiv für Tierheilkunde*, **142**, 520–7.

Dehm, R. (1950) Die Raubtiere aus dem Mittel-Miocän (Burdigalium) von Wintershof-West bei Eichstätt in Bayern. *Abhandlungen der Bayerischen Akademie der Wissenschaften, Mathematisch-naturwissenschaftliche Klasse. Neue Folge*, **58**, 1–141.

Delgado, E., Villalba, L., Sanderson, J., Napolitano, C., Berna, M., and Esquivel, J. (2004) Capture of an Andean cat in Bolivia. *Cat News*, **40**, 2.

Delibes, M. (1980) Feeding ecology of the Spanish lynx in the Coto Doñana. *Acta Theriologica*, **25**, 309–24.

Delibes, M. and Calzada, J. (2005) Ensayo de recuperación de una especie en situación crítica: el caso del lince ibérico. In I. Jiménez Pérez and M. Delibes (eds.), *Al Borde de la Extinción: Una Visión Integral de la Conservación de la Fauna Amenazada en España*, pp. 277–308. Evren, Valencia, Spain.

Delibes, M., Rodríguez, A., and Ferreras, P. (2000) *Action Plan for the Conservation of the Iberian lynx* (Lynx pardinus) *in Europe*. Council of Europe Publishing, Strasbourg, France.

Department of Environmental Affairs and Tourism South Africa (2006) National norms and standards for the regulation of the hunting industry in South Africa. <www.environment.gov.za/HotIssues/2005/29062005/panel_experts_hunting.htm>. Accessed online 15 Oct 2007.

Deraniyagala, P. E. P. (1956) A new subspecies of rusty spotted cat from Ceylon. *Spolia Zoyhnia*, **28**, 113–14.

Dexel, B. (2003) Snow leopard conservation in Kyrgyzstan: enforcement, education and research activities by the German Society for Nature Conservation (NABU). In T. M. McCarthy and J. Weltzin (eds.), *Snow Leopard Survival Summit*, pp. 59–63. International Snow Leopard Trust, Seattle, WA.

Di Bitetti, M. S., Paviolo, A., and De Angelo, C. (2006) Density, habitat use and activity patterns of ocelots (*Leopardus pardalis*) in the Atlantic Forest of Misiones, Argentina. *Journal of Zoology*, **270**, 153–63.

Dickman, A. J. (2008) *Investigating the key determinants of human-large carnivore conflict in Tanzania*. PhD dissertation, University College London, London, UK.

Dickman, A. J. (2005) *An assessment of pastoralist attitudes and wildlife conflict Rungwa in the Rungwa-Ruaha region, Tanzania, with particular reference to large carnivores*. MSc dissertation, University of Oxford, Oxford, UK.

Dickman, A. J., Marnewick, K., Daly, B., *et al.* (2006a) *Southern African Cheetah Conservation Planning Workshop*. Endangered Wildlife Trust and IUCN SSC Conservation Breeding Specialist Group, Johannesburg, South Africa.

Dickman, C. R., Pimm, S. L., and Cardillo, M. (2006b) The pathology of biodiversity loss: the practice of conservation. In D. W. Macdonald and K. Service (eds.), *Key Topics in Conservation Biology*, pp. 1–16. Blackwell Publishing Ltd, Oxford, UK.

Diffenbaugh, N. S., Pal, J. S., Giorgi, F., and Gao, X. J. (2007) Heat stress intensification in the Mediterranean climate change hotspot. *Geophysical Research Letters*, **34**, L11706.

Dillon, A. (2005) *Ocelot density and home range in Belize, Central America: camera-trapping and radio telemetry*. MSc dissertation, Virginia Polytechnic Institute and State University, Blacksburg, VA.

Dinerstein, E. (2003) *The Return of the Unicorns: The Natural History and Conservation of the Greater One-Horned Rhinoceros.* Columbia University Press, New York.

Dinerstein, E. (1987) Deer, plant phenology, and succession in the lowland forest of Nepal. In C. W. Wemmer (ed.), *Biology and Management of the Cervidae*, pp. 272–88. Smithsonian Institution Press, Washington DC.

Dinerstein, E. (1980) An ecological survey of the Royal Bardia Wildlife Reserve, Nepal. Part III: Ungulate populations. *Biological Conservation*, **18**, 5–38.

Dinerstein, E. (1979a) An ecological survey of the Royal Bardia Wildlife Reserve, Nepal. Part I: Vegetation, modifying factors, and successional relationships. *Biological Conservation*, **15**, 127–50.

Dinerstein, E. (1979b) An ecological survey of the Royal Bardia Wildlife Reserve, Nepal. Part II: Habitat/animal interactions. *Biological Conservation*, **16**, 265–300.

Dinerstein, E., Loucks, C., Heydlauff, A., *et al.* (2006) *Setting Priorities for the Conservation and Recovery of Wild Tigers: 2005–2015. A User's Guide.* WWF, WCS, Smithsonian, and NFWF-STF, Washington DC.

Dinerstein, E., Rijal, A., Bookbinder, M., Kattel, B., and Rajuria, A. (1999) Tigers as neighbours: efforts to promote local guardianship of endangered species in lowland Nepal. In J. Seidensticker, S. Christie, and P. Jackson (eds.), *Riding the Tiger: Tiger Conservation in Human Dominated Landscapes*, pp. 316–33. Cambridge University Press, Cambridge, UK.

Dinerstein, E., Wikramanayake, E. D., Robinson, J., *et al.* (1997) *A Framework for Identifying High Priority Areas and Actions for the Conservation of Tigers in the Wild.* World Wildlife Fund-US and Wildlife Conservation Society, Washington DC.

Dinerstein, E., Olson, D. M., Graham, D. J., *et al.* (1995) *A Conservation Assessment of the Terrestrial Ecoregions of Latin America and the Caribbean.* WWF and World Bank, Washington DC.

Dinerstein, E., Loucks, C., Wikramanayake, E., *et al.* (2007) The fate of wild tigers. *BioScience*, **57**, 508–14.

Dinets, V. (2003) Records of small carnivores from Mount Kinabalu, Sabah, Borneo. *Small Carnivore Conservation*, **28**, 9.

Divyabhanusinh (1995) *The End of a Trail—The Cheetah in India.* Banyan Books, New Delhi, India.

Divyabhanusinh (1985) Note on Asiatic lion (*Panthera leo persica*). *Journal of the Bombay Natural History Society*, **82**, 393–4.

Dixon, K. R. (1982) Mountain lion. In J. A. F. Chapman (ed.), *Wild Mammals of North America*, pp. 711–27. John Hopkins University Press, Baltimore, MD.

Dobzhansky, T. (1937) *Genetics and the Origin of Species.* Columbia University Press, New York.

Dodd, G. H. and Squirrel, D. J. (1980) Structure and mechanism in the mammalian olfactory system. *Symposium of the Zoological Society of London*, **45**, 35–56.

Dolbeer, R. A. and Clark, W. R. (1975) Population ecology of snowshoe hares in the central Rocky Mountains. *Journal of Wildlife Management*, **39**, 535–49.

Donadio, E. and Buskirk, S. W. (2006) Diet, morphology, and interspecific killing in carnivora. *American Naturalist*, **167**, 524–36.

Doncaster, C. P. and Macdonald, D. W. (1991) Drifting territoriality in the red fox *Vulpes vulpes*. *Journal of Animal Ecology*, **60**, 423–39.

Dotta, G., Queirolo, D., and Senra, A. (2007) Distribution and conservation stuatus of small felids on the Uruguyan savanna ecoregion, southern Brazil and Uruguay. In *Felid Biology and Conservation Conference 17–20 September 2007: Programme and Abstracts*, p. 105. Wildlife Conservation Research Unit, Oxford, UK.

Døving, K. B. and Trotier, D. (1998) Structure and function of the vomeronasal organ. *Journal of Experimental Biology*, **201**, 2913–25.

Dragesco-Joffe, A. (1993) Le chat des sables, un redoutable chasseur de serpents [The sand cat, a formidable snake hunter]. In *La Vie Sauvage du Sahara*, pp. 129–33. Delachaux et Niestl, Lausanne, Switzerland.

Drake, G. J. C., Auty, H. K., Ryvar, R., *et al.* (2004) The use of reference strand-mediated conformational analysis for the study of cheetah (*Acinonyx jubatus*) feline leucocyte antigen class II DRB polymorphisms. *Molecular Ecology*, **13**, 221–9.

Driciru, M., Siefert, L., Prager, K. C., *et al.* (2006) A serosurvey of viral infections in lions (*Panthera leo*), from Queen Elizabeth National Park, Uganda. *Journal of Wildlife Diseases*, **42**, 667–71.

Driscoll, C. A., Noboyuki, Y., Bar-Gal, G. K., *et al.* (2009) Mithochondrial phylogeography illuminates the origin of the extinct Caspian tiger and its relationship to the Amur tiger. *PLoS ONE*, **4**, 323–9.

Driscoll, C. A., Menotti-Raymond, M., Roca, A. L., *et al.* (2007) The near eastern origin of cat domestication. *Science*, **317**, 519–23.

Driscoll, C. A., Menotti-Raymond, M., Nelson, G., Goldstein, D., and O'Brien, S. J. (2002) Genomic microsatellites as evolutionary chronometers: a test in wild cats. *Genome Research*, **12**, 414–23.

Duckler, G. L. and Van Valkenburgh, B. (1998) Osteological corroboration of pathological stress in a population of endangered Florida pumas (*Puma concolor coryi*). *Animal Conservation*, **1**, 39–46.

Duckworth, J. W., Poole, C. M., Walston, J. L., Tizard, R. J., and Timmins, R. J. (2005) The jungle cat *Felis chaus* in Indochina: a threatened population of a widespread and adaptable species. *Biodiversity and Conservation*, **14**, 1263–80.

Duckworth, J. W., Salter, R. E. and Khounboline, K. (1999) *Wildlife in Lao DPR: 1999 Status Report.* IUCN, Vientiane, Laos.

Dunham, K. M. (2002) *Aerial Census of Elephants and Other Large Herbivores in North-West Matabeleland, Zimbabwe: 2001.* WWF-SARPO Occasional Paper Series. WWF-SARPO, Harare, Zimbabwe.

Dunstone, N., Durbin, L., Wyllie, I., *et al.* (2002) Spatial organization, ranging behaviour and habitat use of the kodkod (*Oncifelis guigna*) in southern Chile. *Journal of Zoology,* **257**, 1–11.

Durant, S. M. (2000a) Living with the enemy: avoidance of hyenas and lions by cheetahs in the Serengeti. *Behavioral Ecology,* **11**, 624–32.

Durant, S. M. (2000b) Predator avoidance, breeding experience and reproductive success in endangered cheetahs, *Acinonyx jubatus. Animal Behaviour,* **60**, 121–30.

Durant, S. M. (1998) Competition refuges and coexistence: an example from Serengeti carnivores. *Journal of Animal Ecology,* **67**, 370–86.

Durant, S. M., Bashir, S., Maddox, T., and Laurenson, M. K. (2007) Relating long-term studies to conservation practice: the case of the Serengeti Cheetah Project. *Conservation Biology,* **21**, 602–11.

Durant, S. M., Kelly, M., and Caro, T. M. (2004) Factors affecting life and death in Serengeti cheetahs: environment, age, and sociality. *Behavioural Ecology,* **15**, 11–22.

Durant, S. M., Caro, T. M., Collins, D. A., Alawi, R. M., and FitzGibbon, C. D. (1988) Migration patterns of Thomson's gazelles and cheetahs on the Serengeti Plains. *African Journal of Ecology,* **26**, 257–68.

Dutton, A., Edward-Jones, G., Strachan, R., and Macdonald, D. W. (2008) Ecological and social challenges to biodiversity conservation on farmland: reconnecting habitats on a landscape scale. *Mammal Review,* **38**, 205–19.

Earnhardt, J. M. (1999) Reintroduction programmes: genetic trade-offs for populations. *Animal Conservation,* **2**, 279–86.

East, R. D. (1984) Rainfall, soil nutrient status and biomass of large African savanna mammals. *African Journal of Ecology,* **22**, 245–70.

Easterbee, N., Hepburn, L. V., and Jefferies, D. J. (1991) *Survey on the Status of the Wildcat in Scotland, 1983–1987* Nature Conservancy Council for Scotland, Edinburgh, Scotland.

Eaton, R. L. (1974) *The Cheetah.* Van Nostrand Reinhold, New York.

Eberhardt, L. L. and Peterson, R. O. (1999) Predicting the wolf-prey equilibrium point. *Canadian Journal of Zoology,* **77**, 494–8.

Efford, M., Warburton, B., and Spencer, N. (2000) Home range changes by brushtail possums in response to control. *Wildlife Research,* **27**, 117–27.

EIA. (2004) *The Tiger Skin Trail.* Environmental Investigation Agency, London, UK.

EIA-WPSI. (2006) *Skinning the Cat—Crime and Politics of the Big Cat Skin Trade.* Environmental Investigation Agency and Wildlife Protection Society, London, UK.

Eisenberg, J. F. (1986) Life history strategies of the Felidae: variations on a common theme. In S. D. Miller and D. D. Everett (eds.), *Cats of the World: Biology, Conservation, and Management,* pp. 293–303. National Wildlife Federation, Washington DC.

Eisenberg, J. F. (1981) *The Mammalian Radiations: An Analysis of Trends in Evolution.* University of Chicago Press, Chicago, IL.

Eisenberg, J. F. and Lockhart, M. (1972) An ecological reconnaissance of Wilpattu National Park, Ceylon. *Smithsonian Contributions to Zoology,* **101**, 1–118.

Eisenberg, J. F. and Redford, K. H. (1999) *Mammals of the Neotropics, the Central Neotropics, Vol. 3: Ecuador, Peru, Bolivia, Brazil.* University of Chicago Press, Chicago, IL.

Eisenberg, J. F. and Seidensticker, J. (1976) Ungulates in southern Asia: a consideration of biomass estimates for selected habitats. *Biological Conservation,* **10**, 293–308.

Eizirik, E., Johnson, W. E., and O'Brien, S. J. (in prep) Molecular systematics and revised classification of the family Felidae (Mammalia, Carnivora).

Eizirik, E., Indrusiak, C., and Trigo, T. (2006) Refined mapping and characterization of a geographic contact zone between two neotropical cats, *Leopardus tigrinus* and *L. geoffroyi* (Mammalia, Felidae). *Cat News,* **45**, 8–11.

Eizirik, E., Yuhki, N., Johnson, W. E., Menotti-Raymond, M., Hannah, S. S., and O'Brien, S. J. (2003) Molecular genetics and evolution of melanism in the cat family. *Current Biology,* **13**, 448–53.

Eizirik, E., Kim, J. H., Menotti-Raymond, M., Crawshaw, P. G., O'Brien, S. J., and Johnson, W. E. (2001) Phylogeography, population history and conservation genetics of jaguars (*Panthera onca,* Mammalia, Felidae). *Molecular Ecology,* **10**, 65–79.

Eizirik, E., Bonatto, S. L., Johnson, W. E., *et al.* (1998) Phylogeographic patterns and evolution of the mitochondrial DNA control Region in two neotropical cats (Mammalia, Felidae). *Journal of Molecular Evolution,* **47**, 613–24.

Elenkov, I. J. and Chrousos, G. P. (2006) Stress system—organization, physiology and immunoregulation. *Neuroimmunomodulation,* **13**, 257–67.

Elenkov, I. J. and Chrousos, G. P. (1999) Stress hormones, Th1/Th2 patterns, pro/anti-inflammatory cytokines and susceptibility to disease. *Trends in Endocrinology and Metabolism,* **10**, 359–68.

Ellerman, J. R. and Morrison-Scott, T. C. S. (1951) Subgenus *Lynx* Kerr, 1792. In *Checklist of Palaearctic and Indian mammals, 1758–1946,* pp. 308–9. British Museum of Natural History, London, UK.

Elliot, J. P., Cowan, I. M., and Holling, C. S. (1977) Prey capture of the African lion. *Canadian Journal of Zoology,* **55**, 1811–28.

Eloff, F. C. (1984) The Kalahari ecosystem. *Koedoe Supplement*, **1984**, 11–20.

Elsey, C. A. (1954) A case of cannibalism in Canada lynx (*Lynx canadensis*). *Journal of Mammalogy*, **35**, 129.

Elton, C. and Nicholson, M. (1942) The ten-year cycle in numbers of the lynx in Canada. *Journal of Animal Ecology*, **11**, 215–44.

Emerson, S. B. and Radinsky, L. (1980) Functional analysis of sabertooth cranial morphology. *Paleobiology*, **6**, 295–312.

Emerton, L. (1998) Innovations for financing wildlife conservation in Kenya. *Paper prepared for Financial Innovations for Biodiversity Workshop, 10th Global Biodiversity Forum, Bratislava, 1–3 May 1998*, pp. 1–17. IUCN Eastern Africa Regional Office, Nairobi, Kenya.

Emmons, L. H. (1988) A field study of ocelots (*Felis pardalis*) in Peru. *Revue d'Ecologie—La Terre et la Vie*, **43**, 133–57.

Emmons, L. H. (1987) Comparative feeding ecology of felids in a neotropical rainforest. *Behavioral Ecology and Sociobiology*, **20**, 271–83.

Emmons, L. H. (1984) Geographic variation in densities and diversities of non-flying mammals in Amazonia. *Biotropica*, **16**, 210–22.

Emmons, L. H. and Feer, F. (1997) *Neotropical Rainforest Mammals: A Field Guide*. University of Chicago Press, Chicago, IL.

Engel, S., Pagiola, S., and Wunder, S. (2008) Designing payments for environmental services in theory and practice: an overview of the issues. *Ecological Economics*, **65**, 633–74.

Erb, K. P. (2004) *Consumptive wildlife utilisation as a land-use form in Namibia*. MBA dissertation, University of Stellenbosch, Stellenbosch, South Africa.

Ernest, H. B., Boyce, W. M., Bleich, V. C., May, B., Stiver, S. J., and Torres, S. G. (2003) Genetic structure of mountain lion (*Puma concolor*) populations in California. *Conservation Genetics*, **4**, 353–66.

Ernest, H. B., Boyce, W. M., Penedob, M. C. T., May, P., and Syvanen, M. (2000) Molecular tracking of mountain lions in the Yosemite Valley region in California: genetic analysis using microsatellites and faecal DNA. *Molecular Ecology*, **9**, 433–41.

ESRI (1992) *ArcView 3.2*. Environmental Systems Research Institute Inc., Redlands, CA.

Esterhuizen, W. C. N. and Norton, P. M. (1985) The leopard as a problem animal in the Cape Province, as determined by the permit system. *Bontebok*, **4**, 9–16.

Estes, R. D. (1991) *The Behaviour Guide to African Mammals*. The University of California Press, Berkeley, CA.

Etling, K. (2004) *Cougar Attacks: Encounters of the Worst Kind*. Lyons Press, Guilford, CT.

European Association of Zoos and Aquaria (2008) <www.eaza.net>. Accessed online 10 Sept 2008.

Evans, W. (1983) *The Cougar in New Mexico: Biology, Status, Depredation of Livestock, and Management Recommendations*. Response to House Memorial 42 of the New Mexico State Legislature. New Mexico Department of Game and Fish, Santa Fe, NM.

Evermann, J. F. and Benfield, D. A. (2001) Coronaviral infections. In E. S. Williams and I. K. Barker (eds.), *Infectious Diseases of Wild Mammals*, pp. 245–53. Iowa State University, Ames, IA.

Evermann, J. F., Laurenson, M. K., McKeirnan, A. J., and Caro, T. M. (1993) Infectious disease surveillance in captive and free-living cheetahs: an integral part of the species survival plan. *Zoo Biology*, **12**, 125–33.

Evermann, J. F., Heeney, J. L., Roelke, M. E., McKeirnan, A. J., and O'Brien, S. J. (1988) Biological and pathological consequences of feline infections peritonitis virus infection in the cheetah. *Archives of Virology*, **102**, 155–71.

Evermann, J. F., Burns, G., Roelke, M. E., *et al.* (1983) Diagnostic features of an epizootic of feline infectious peritonitis in captive cheetahs. *American Association of Veterinary Laboratory Diagnosticians*, **26**, 365–82.

Ewer, R. F. (1973) *The Carnivores*. Cornell University Press, Ithaca, NY.

Ewer, R. F. (1968) *Ethology of Mammals*. Logos Press, London, UK.

FAB. (2008) Information on disease. <www.fab.org.uk>. Accessed online 30 Jan 2008.

Facure, K. G. and Giaretta, A. A. (1996) Food habits of carnivores in a coastal Atlantic Forest of southeastern Brazil. *Mammalia*, **60**, 499–502.

Facure-Giaretta, K. G. (2002) *Ecologia alimentar de duas espécies de felinos do gênero* Leopardus *em uma floresta secundária no sudeste do Brasil*. PhD dissertation, Universidade Estadual de Campinas, Campinas, Brazil.

Fahlman, A., Loveridge, A., Wenham, C., Foggin, C., Arnemo, J. M., and Nyman, G. (2005) Reversible anaesthesia of free-ranging lions (*Panthera leo*) in Zimbabwe. *Journal of the South African Veterinary Association*, **76**, 187–92.

FAO (2007) *State of the World's Forests*. United Nations Food and Agriculture Organization, Rome, Italy.

Farrell, L. E. (1999) *The ecology of the puma and the jaguar in the Venezuelan llanos*. MS dissertation, University of Florida, Gainesville, FL.

Farrell, L. E., Roman, J., and Sunquist, M. E. (2000) Dietary separation of sympatric carnivores identified by molecular analysis of scats. *Molecular Ecology*, **9**, 1583–90.

Federal Trade Commission (2000) Rules and regulations under the fur products labeling act. <www.ftc.gov/os/statutes/textile/fedreg/001228-fur.htm>. Accessed online 20 May 2008.

Fedriani, J. M., Palomares, F., and Delibes, M. (1999) Niche relations among three sympatric Mediterranean carnivores. *Oecologia*, **121**, 138–48.

Fedriani, J. M., Ferreras, P., and Laffitte, R. (1995) Un lince ibérico muere en Doñana a causa, aparentemente, de la agresión de otro lince. *Quercus*, **110**, 19.

Felid Taxon Advisory Group (2008) <www.felidtag.org>. Accessed online 10 Sept 2008.

Felipe, A. R. (2003) *Dieta do maracajá,* Leopardus wiedii *(Schinz, 1821) e do jaguarundi,* Herpailurus yagouaroundi *(E. Geoffroy, 1803) nas matas de tabuleiro do norte do Espírito Santo.* BS dissertation, Escola de Ensino Superior do Educandário Seráfico São Francisco de Assis, Santa Teresa, Brazil.

Ferguson, H. (2006) *Guidelines to dealing with damage-causing leopards and crocodiles in KwaZulu-Natal.* Working Document. Ezemvelo KwaZulu-Natal Wildlife, KwaZulu-Natal, South Africa.

Ferguson, S. H., Messier, F., Taylor, M. K., Born, E. W., and Rosing Asvid, A. (1999) Determinants of home range size for polar bears (*Ursus maritimus*). *Ecology Letters,* **2**, 311–18.

Fernández, N., and Palomares, F. (2000) The selection of breeding dens by the endangered Iberian lynx (*Lynx pardinus*): implications for its conservation. *Biological Conservation,* **94**, 51–61.

Fernandez, E., de Lope, F., and de la Cruz, C. (1992) Morphologie cranienne di chat sauvage (*Felis silvestris*) dans le sud de la Peninsule iberique: importance de l'introgression par le chat domestique (*F. catus*). *Mammalia,* **56**, 255–64.

Fernández, N., Delibes, M., and Palomares, F. (2007) Habitat-related heterogeneity in breeding in a metapopulation of the Iberian lynx. *Ecography,* **30**, 431–4.

Fernández, N., Delibes, M., and Palomares, F. (2006) Landscape evaluation in conservation: molecular sampling and habitat modeling for the Iberian lynx. *Ecological Applications,* **16**, 1037–49.

Fernández, N., Delibes, M., Palomares, F., and Mladenoff, D. J. (2003) Identifying breeding habitat for the iberian lynx: inferences from a fine-scale spatial analysis. *Ecological Applications,* **13**, 1310–24.

Fernández, N., Palomares, F., and Delibes, M. (2002) The use of breeding dens and kitten development in the Iberian lynx (*Lynx pardinus*). *Journal of Zoology,* **258**, 1–5.

Ferraro, P. J. and Kiss, A. (2002) Direct payments to conserve biodiversity. *Science,* **298**, 1718–19.

Ferreira, S. and Funston, P. J. (in press) Estimating lion population variables: prey and disease effects in Kruger National Park, South Africa. *Biological Conservation.*

Ferrer, M. and Negro, J. J. (2004) The near extinction of two large European predators: super specialists pay a price. *Conservation Biology,* **18**, 344–9.

Ferreras, P., Delibes, M., Palomares, F., Fedriani, J. M., Calzada, J., and Revilla, E. (2004) Proximate and ultimate causes of dispersal in the Iberian lynx *Lynx pardinus. Behavioral Ecology,* **15**, 31–40.

Ferreras, P., Gaona, P., Palomares, F., and Delibes, M. (2001) Restore habitat or reduce mortality? Implications from a population viability analysis of the Iberian lynx. *Animal Conservation,* **4**, 265–74.

Ferreras, P., Beltrán, J. F., Aldama, J. J., and Delibes, M. (1997) Spatial organization and land tenure system of the endangered Iberian lynx (*Lynx pardinus*). *Journal of Zoology,* **243**, 163–89.

Ferreras, P., Aldama, J. J., Beltrán, J. F., and Delibes, M. (1992) Rates and causes of mortality in a fragmented population of Iberian lynx *Felis pardina* Temminck, 1824. *Biological Conservation,* **61**, 197–202.

Ferroglio, E., Trisciuoglio, A., Prouteau, A., *et al.* (2003) Antibodies to *Neospora caninum* in wild animals from Kenya, East Africa. *Veterinary Parasitology,* **118**, 43–9.

Ficcarelli, G. (1979) The Villafranchian machairodonts of Tuscany. *Palaeontographica Italica,* **71**, 17–26.

Field, M. J. (1955) Witchcraft as a primitive interpretation of mental disorder. *Journal of Mental Science,* **101**, 826–33.

Filoni, C., Catao-Dias, J. L., Bay, G., *et al.* (2006) First evidence of feline herpesvirus, calicivirus, parvovirus, and *Ehrlichia* exposure in Brazilian free-ranging felids. *Journal of Wildlife Diseases,* **42**, 470–7.

Fiorello, C. V., Noss, A. J., Deem, S. L., Maffei, L., and Dubovi, E. J. (2007) Serosurvey of small carnivores in the Bolivian Chaco. *Journal of Wildlife Diseases,* **43**, 551–7.

Fisher, C. V. (1998) *Habitat use by free-ranging felids in the lower Rio Grande Valley.* MS dissertation, Southwest Texas State University, San Marcos, TX.

Fitch, H. M. and Fagan, D. A. (1982) Focal palatine erosion associated with dental malocclusion in captive cheetahs. *Zoo Biology,* **1**, 295–310.

Fitzgerald, B. M. (1988) Diet of domestic cats and their impact on prey populations. In D. C. Turner and P. Bateson (eds.), *The Domestic Cat: The Biology of Its Behaviour,* pp. 123–45. Cambridge University Press, Cambridge, UK.

FitzGibbon, C. D. (1990a) Mixed species grouping in Thomson's and Grant's gazelles: the antipredator benefits. *Animal Behaviour,* **39**, 1116–26.

FitzGibbon, C. D. (1990b) Why do hunting cheetahs prefer male gazelles? *Animal Behaviour,* **40**, 837–45.

FitzGibbon, C. D. and Fanshawe, J. H. (1989) The condition and age of Thomson's gazelles killed by cheetahs and wild dogs. *Journal of Zoology,* **218**, 99–107.

Fix, A. S., Riordan, D. P., Hill, H. T., Gill, M. A., and Evans, M. B. (1989) Feline panleukopenia virus and subsequent canine distemper virus infection in two snow leopards (*Panthera uncia*). *Journal of Zoo and Wildlife Medicine,* **20**, 273–81.

Fletcher, K. C. (1978) Notoedric mange in a litter of snow leopards. *Journal of the American Veterinary Medical Association,* **173**, 1231–2.

Flower, W. H. and Lydekker, R. (1891) *An Introduction to the Study of Mammals, Living and Extinct.* Adam and Charles Black, London, UK.

Floyd, T. J., Mech, L. D., and Jordan, P. A. (1978) Relating wolf scat content to prey consumed. *Journal of Wildlife Management*, **42**, 528–32.

Flux, J. E. C. (1994) World distribution. In H. V. Thompson and M. C. King (eds.), *The European Rabbit: The History and Biology of a Successful Colonizer*, pp. 8–21. Oxford University Press, Oxford, UK.

Fonseca, G. A. B., da, Herrmann, G., Leite, Y. L. R., Mittermeier, R. A., Rylands, A. B., and Patton, J. L. (1996) Lista anotada dos mamiferos do Brasil. *Occasional Papers in Conservation Biology*, **4**, 1–38.

Forbes, G. J. and Theberge, J. B. (1996) Cross-boundary management of Algonquin Park wolves. *Conservation Biology*, **10**, 1091–7.

Forrester, N. L., Trout, R. C., Turner, S. L., *et al.* (2006) Unravelling the paradox of rabbit haemorrhagic disease virus emergence, using phylogenetic analysis: possible implications for rabbit conservation strategies. *Biological Conservation*, **131**, 296–306.

Foster, R. J., (2008) *The ecology of jaguars* (Panthera onca) *in a human-influenced landscape*. PhD dissertation, University of Southampton, Southampton, UK.

Foster R. J., Harmsen, B., and Doncaster, C. P. (in press-a) *Habitat use of jaguars and pumas across a gradient of human disturbance*. Biotropica.

Foster, R. J., Harmsen, B. J., Valdes, B., Pomilla, C., and Doncaster, C. P. (in press-b) *Food habits of jaguars and pumas across a gradient of human disturbance*. Journal of Zoology.

Fowler, M. W. (1995) *Restraint and Handling of Wild and Domestic Animals*, 2nd edn. Iowa State University Press, Ames, IA.

Fox, H. E., Christian, C., Nordby, J. C., Pergams, O. R. W., Peterson, G. D., and Pyke, C. R. (2006) Perceived barriers to integrating social science and conservation. *Conservation Biology*, **20**, 1817–20.

Fox, J. L. (1994) Snow leopard conservation in the wild—a comprehensive perspective on a low density and highly fragmented population. In J. L. Fox and D. Jizeng (eds.), *Proceedings of the 7th International Snow Leopard Symposium*, pp. 3–15. Northwest Plateau Institute of Biology and International Snow Leopard Trust, Seattle, WA.

Fox, J. L. and Chundawat, R. S. (1988) Observations of snow leopard stalking, killing, and feeding behavior. *Mammalia*, **52**, 137–40.

Fox, J. L. and Dorji, T. (2007) High elevation record for occurrence of the Manul or Pallas cat on the northwest Tibetan Plateau, China. *Cat News*, **46**, 35.

FQR (2002) *Fourth Quinquennial Review of Schedules 5 and 8 of the Wildlife and Countryside Act, 1981*. Joint Nature Conservation Committee, Peterborough, UK.

Frank, L., Woodroffe, R. and Ogada, M. (2005) People and predators in Laikipia District, Kenya. In R. Woodroffe, S. Thirgood, and A. Rabinowitz (eds.), *People and Wildlife: Conflict or Co-existence?*, pp. 286–304. Cambridge University Press, Cambridge, UK.

Frank, L., Maclennan, S., Hazzah, L., Bonham, R. and Hill, T. (2006) *Lion Killing in the Amboseli-Tsavo Ecosystem, 2001–2006, and Its Implications for Kenya's Lion Population*. Kilimanjaro Lion Conservation Project, Nanyuki, Kenya and Predator Compensation Fund, Nairobi, Kenya.

Frank, L. and Woodroffe, R. (2001) Behaviour of carnivores in exploited and controlled populations. In J. Gittleman, S. Funk, D. W. Macdonald, and R. Wayne (eds.), *Carnivore Conservation*, pp. 419–42. Cambridge University Press, Cambridge, UK.

Frank, L. and Woodroffe, R. (2002) Managing predators and livestock on an East African rangeland. In A. J. Loveridge, T. Lynam, and D. W. Macdonald (eds.), *Lion Conservation Research. Workshop 2: Modelling Conflict*, pp. 12–17. Wildlife Conservation Research Unit, Oxford, UK.

Frank, L., Simpson, D., and Woodroffe, R. (2003) Foot snares: an effective method for capturing African lions. *Wildlife Society Bulletin*, **31**, 309–14.

Frankham, R. (2009) Genetic considerations in reintroduction programs for large terrestrial predators. In M. K. Hayward and M. J. Somers (eds.), *The Reintroduction of Top Order Predators*, pp. 371–87. Wiley-Blackwell Publishing, Oxford, UK.

Frankham, R. (2005) Genetics and extinction. *Biological Conservation*, **126**, 131–40.

Frankham, R., Ballou, J., and Briscoe, D. (2009) *Introduction to Conservation Genetics*, Cambridge University Press, Cambridge, UK.

Franklin, S. P., Troyer, J. L., Terwee, J. A., *et al.* (2007) Frequent transmission of immunodeficiency viruses among bobcats and pumas. *Journal of Virology*, **81**, 10961–9.

Franklin, W. L., Johnson, W. E., Sarno, R. J., and Iriarte, J. A. (1999) Ecology of the Patagonia puma *Felis concolor patagonica* in southern Chile. *Biological Conservation*, **90**, 33–40.

Frazer Sissom, D. E., Rice, D. A., and Peters, G. (1991) How cats purr. *Journal of Zoology*, **223**, 67–78.

Freer, R. A. (2004) *The spatial ecology of the guina (*Oncifelis guigna*) in Southern Chile*. PhD dissertation, University of Durham, Durham, UK.

French, A. R., and Smith, T. B. (2005) Importance of body size in determining dominance hierarchies among diverse tropical frugivores. *Biotropica*, **37**, 96–101.

French, D. D., Corbett, L. K., and Easterbee, N. (1988) Morphological discriminants of Scottish wildcats (*Felis silvestris*), domestic cats (*Felis catus*) and their hybrids. *Journal of Zoology*, **214**, 235–59.

Frey, J. K. (2006) Inferring species distributions in the absence of occurrence records: an example considering wolverine (*Gulo gulo*) and Canada lynx (*Lynx canadensis*) in New Mexico. *Biological Conservation*, **130**, 16–24.

Friedmann, Y. and Daly, B. (2004) *Red Data Book of the Mammals of South Africa: A Conservation Assessment.*

Conservation Breeding Specialist Group (SS/IUCN) and Endangered Wildlife Trust, Johannesburg, South Africa.

Fritz, H. and Duncan, P. (1994) On the carrying capacity for large ungulates of African savanna ecosystems. *Proceedings of the Royal Society of London Series B*, **256**, 77–82.

Fritz, H. and Loison, A. (2006) Large herbivores across biomes. In K. Danell, P. Duncan, R. Bergström, and J. Pastor (eds.), *Large Herbivore Ecology, Ecosystems Dynamics and Conservation*, pp. 19–41. Cambridge University Press, Cambridge, UK.

Fritz, H., Duncan, P., Gordon, I. J., and Illius, A. W. (2002) Megaherbivores influence trophic guild structure in African ungulate communities. *Oecologia*, **131**, 620–5.

Fromont, E., Sager, A., Leger, F., *et al.* (2000) Prevalence and pathogenicity of retroviruses in wildcats in France. *Veterinary Record*, **146**, 317–19.

Frost, P. G. and Bond, I. (2008) The CAMPFIRE programme in Zimbabwe: payments for wildlife services. *Ecological Economics*, **65**, 776–87.

Fryxell, J. M., Wilmshurst, J. F., and Sinclair, A. R. E. (2004) Predictive models of movement by Serengeti grazers. *Ecology*, **85**, 2429–35.

Fuller, T. K. and Sievert, P. R. (2001) Carnivore demography and the consequences of changes in prey availability. In J. L. Gittleman, S. M. Funk, D. W. MacDonald, and R. K. Wayne (eds.), *Carnivore Conservation*, pp. 163–78. Cambridge University Press, Cambridge, UK.

Fuller, T. K., Berg, W. E., and Kuehn, D. W. (1985) Survival rates and mortality factors of adult bobcats in north-central Minnesota. *Journal of Wildlife Management*, **49**, 292–6.

Funston, P. J. (2002) Game reserves and national parks: factors relevant to lion conflict. In A. J. Loveridge, T. Lynam, and D. W. Macdonald (eds.), *Lion Conservation Research. Workshop 2: Modelling Conflict*, pp. 21–3. Wildlife Conservation Research Unit, Oxford, UK.

Funston, P. J. (2001a) Conservation of lions in the Kgalagadi Tranfrontier Park: boundary transgression and problem animal control. In P. J. Funston (ed.), *The Kgalagadi Transfrontier Lion Project: population ecology and long-term monitoring of a free ranging population in an arid environment*, pp. 137–58. Unpublished report to The Green Trust.

Funston, P. J. (2001b) *Final Report: Kalahari Transfrontier Lion Project*. Endangered Wildlife Trust, Johannesburg, South Africa.

Funston, P. J. and Mills, M. G. L. (2006) The influence of lion predation on the population dynamics of common large ungulates in the Kruger National Park. *South African Journal of Wildlife Research*, **36**, 9–22.

Funston, P. J., Hemson, G., Mosser, A., *et al.* (2007) Flexible sociality ensures the persistence of lion populations when persecuted by humans. In *Felid Biology and Conservation Conference 17–20 September 2007: Programme and Abstracts*, p. 59. Wildlife Conservation Research Unit, Oxford, UK.

Funston, P. J., Mills, M. G. L., Richardson, P. R. K., and van Jaarsveld, A. S. (2003) Reduced dispersal and opportunistic territory acquisition in male lions (*Panthera leo*). *Journal of Zoology*, **259**, 131–42.

Funston, P. J., Hermann, E., Babupi, P., *et al.* (2001a) *Spoor frequency estimates as a method of determining lion and other large mammal densities in the Kgalagadi Transfrontier Park*. Unpublished report to the management committee of the Kgalagadi Transfrontier Park, Botswana.

Funston, P. J., Mills, M. G. L., and Biggs, H. C. (2001b) Factors affecting the hunting success of male and female lions in the Kruger National Park. *Journal of Zoology*, **253**, 419–31.

Funston, P. J., Mills, M. G. L., Biggs, H. C., and Richardson, P. R. K. (1998) Hunting by male lions: ecological influences and socioecological implications. *Animal Behaviour*, **56**, 1333–45.

FWC (2006) *Annual Report on the Research and Management of Florida Panthers: 2005–2006*. Fish and Wildlife Research Institute & Division of Habitat and Species Conservation; Florida Fish and Wildlife Conservation Commission, Naples, FL.

Fyumagwa, R. D., Runyoro, V., Horak, I. G., and Hoare, R. (2007) Ecology and control of ticks as disease vectors in wildlife of the Ngorongoro Crater, Tanzania. *South African Journal of Wildlife Research*, **37**, 79–90.

Galende, G. I., Grigera, D., and Von Tüngen, J. (1998) Composición de la dieta de chinchillón (*Lagidium viscacia*, Chinchillidae) en el Noroeste de la Patagonia. *Mastozoologia Neotropical*, **5**, 123–8.

Gamarra, J. G. P. and Solé, R. V. (2000) Bifurcations and chaos in ecology: lynx returns revisited. *Ecology Letters*, **3**, 114–21.

Gaona, P., Ferreras, P., and Delibes, M. (1998) Dynamics and viability of a metapopulation of the endangered Iberian lynx (*Lynx pardinus*). *Ecological Monographs*, **68**, 349–70.

Gaponov, V. V. (1991) Optimal numbers of red deer in the Ussuri forests. *Forest Management*, **5**, 44–5.

Gaponov, V. V., Konkov, Yu. A., and Belozor, A. A. (2006) *Ungulates and Their Impact on Habitat as an Aspect of Far Eastern Leopard Conservation: An Example from Neshinskoe Hunting Lease*. Dalnauka, Vladivostok, Russia.

García-Perea, R. (2002) Andean mountain cat, *Oreailurus jacobita*: morphological description and comparison with other felines from the altiplano. *Journal of Mammalogy*, **83**, 110–24.

García-Perea, R. (2000) Survival of injured Iberian lynx (*Lynx pardinus*) and non-natural mortality in central-southern Spain. *Biological Conservation*, **93**, 265–9.

García-Perea, R. (1994) The pampas cat group (genus *Lynchailurus* Servertzov, 1858) (Carnivora, Felidae), a systematic and biogeographic review. *American Museum Novitates*, 1–36.

García-Perea, R. (1992) New data on the systematics of lynxes. *Cat News*, **16**, 15–16.

Garla, R. C., Setz, E. Z. F., and Gobbi, N. (2001) Jaguar (*Panthera onca*) food habits in atlantic rain forest of Southeastern Brazil. *Biotropica*, **33**, 691–6.

Gaskell, R. M., Dawson, S., and Radford, A. D. (2006) Feline respiratory disease. In C. E. Greene (ed.), *Infectious Diseases of the Dog and Cat*, pp. 145–54. W.B. Saunders, Philadelphia, PA.

Gasper, P. W. and Watson, R. P. (2001) Plague and yersiniosis. In E. S. Williams and I. K. Barker (eds.) *Infectious Diseases of Wild Mammals*, pp. 313–29. Iowa State University Press, Ames, IA.

Gaston, K. J. (1991) How large is a species' geographic range? *Oikos*, **61**, 434–8.

Gatzweiler, F. W. (2006) Organizing a public ecosystem service economy for sustaining biodiversity. *Ecological Economics*, **59**, 296–304.

Gaubert, P. and Véron, G. (2003) Exhaustive sample set among Viverridae reveals the sister-group of felids: the linsangs as a case of extreme morphological convergence within Feliformia. *Proceedings of the Royal Society of London Series B*, **270**, 2523–30.

Gaveau, D. L. A., Wandono, H., and Setiabudi, F. (2007) Three decades of deforestation in southwest Sumatra: have protected areas halted forest loss and logging, and promoted re-growth? *Biological Conservation*, **134**, 495–504.

Geertsema, A. (1985) Aspects of the ecology of the serval *Leptailurus serval* in the Ngorongoro crater, Tanzania. *Netherlands Journal of Zoology*, **35**, 527–610.

Geertsema, A. (1976) Impressions and observation on serval behaviour in Tanzania, East Africa. *Mammalia*, **40**, 13–19.

Geffen, E., Luikart, G., and Waples, R. S. (2007) Impacts of modern molecular genetic techniques on conservation biology. In D. W. Macdonald and K. Service (eds.), *Key Topics in Conservation Biology*, pp. 46–63. Blackwell Publishing, Oxford, UK.

Gehrt, S. D. and Fritzell, E. K. (1998) Resource distribution, female home range dispersion and male spatial interactions: group structure in a solitary carnivore. *Animal Behaviour*, **55**, 1211–27.

Geraads, D. (1980) Un nouveau félidé (Fissipeda, Mammalia) du Pléistocène Moyen du Maroc: *Lynx thomasi* n. sp. *Geobios*, **13**, 441–4.

Gérard, J. (1856) *The Adventures of Gérard, the Lion Killer.* Derby and Jackson, New York.

Germann, P. G., Kohler, M., Ernst, H., Baumgart, H., and Mohr, U. (1990) The relation of amyloidosis to social stress induced by crowding in the Syrian hamster (*Mesocricetus auratus*). *Zeitschrift Fur Versuchstierkunde*, **33**, 271–5.

Gervais, P. (1850) *Zoologie et Paléontologie Françaises*. Arthus Bertrand, Paris, France.

Gil-Sánchez, J. M., Ballesteros-Duperon, E., and Bueno-Segura, J. F. (2006) Feeding ecology of the Iberian lynx *Lynx pardinus* in eastern Sierra Morena (Southern Spain). *Acta Theriologica*, **51**, 85–90.

Gilbert, D. A., Pusey, A. E., Stephens, J. C., and O'Brien, S. J. (1991) Analytical DNA fingerprinting in lions: parentage, genetic diversity, and kinship. *Journal of Heredity*, **82**, 378–86.

Gilpin, M. E. and Soulé, M. E. (1986) Minimum viable populations: processes of species extinction. In M. E. Soulé (ed.), *Conservation Biology: The Science of Scarcity and Diversity*, pp. 19–34. Sinauer Associates, Sunderland, MA.

Ginsburg, L. (1999) Order Carnivora. In G. E. Rössner and K. Heissig (eds.), *The Miocene Land Mammals of Europe*, pp. 109–48. Verlag, München, Germany.

Ginsburg, L. (1983) Sur les modalités d'évolution du genre Néogène Pseudaelurus Gervais (Felidae, Carnivora, Mammalia). *Colloques Internationaux du CNRS*, **330**, 131–6.

Ginsburg, L., Morales, J., and Soria, D. (1981) Nuevos datos sobre los carnívoros de los Valles do Fuentidueña (Segovia). *Estudios Geológicos*, **37**, 383–415.

Gittleman, J. L. (1991) Carnivore olfactory bulb size - allometry, phylogeny and ecology. *Journal of Zoology*, **225**, 253–72.

Gittleman, J. L. (1989) Carnivore group-living: comparative trends. In J. L. Gittleman (ed.), *Carnivore Behavior, Ecology, and Evolution*, pp. 183–207. Cornell University Press, Ithaca, NY.

Gittleman, J. L. (1985) Carnivore body size: ecological and taxonomic correlates. *Oecologia*, **67**, 540–54.

Gittleman, J. L. and Harvey, P. H. (1982) Carnivore home range size, metabolic needs and ecology. *Behavioral Ecology and Sociobiology*, **10**, 57–63.

Glanz, W. E. (1982) The terrestrial mammal fauna of Barro Colorado Island: census and long-term changes. In E. G. Leigh, Jr., A. S. Rand, and D. M. Windsor (eds.) *The Ecology of a Tropical Forest: Seasonal Rhythms and Long-Term Changes*, pp. 455–68. Smithsonian Institution Press, Washington DC.

Glas, L. (in press) Jungle cat *Felis chaus*. In J. S. Kingdon and M. Hoffmann (eds.), *The Mammals of Africa 5: Carnivora, Pholidota, Perissodactyla*. Academic Press, Amsterdam, the Netherlands.

Glazier, D. S. and Eckert, S. E. (2002) Competitive ability, body size and geographical range size in small mammals. *Journal of Biogeography*, **29**, 81–92.

GLC (2000) *Global land cover dataset.* European Commission Joint Research Centre, Institute for Environment and Sustainability, Ispra, Italy.

Glenn, B. L., Kocan, A. A., and Blouin, E. F. (1983) Cytauxzoonosis in bobcats. *Journal of the American Veterinary Medical Association*, **183**, 1155–8.

Godoy, J. A. (2006) Iberian lynx population genetics: Doñana, Sierra Morena and ex situ program. In *Iberian Lynx Ex-Situ Conservation Seminar Series: Book Proceedings*, pp. 59–60. Fundación Biodiversidad, Sevilla and Doñana, Spain.

Goeritz, F., Vargas, A., Martínez, F. *et al.* (2009). Ultrasonographical assessment of structure and function of the male and female reproductive organs in the Eurasian and the Iberian lynx. In A.Vargas, C. Breitenmoser, U. Breitenmoser, (eds.), *Iberian Lynx Ex-situ Conservation: An Interdisciplinary Approach*, pp. 366–75. Fundación Biodiversidad, Madrid, Spain.

Gomez, M. C. and Pope, C. E. (2005) Current concepts in cat cloning. In A. Inui (ed.), *Health Consequences of Cloning*, pp. 11–151. Taylor & Francis, Boca Raton, FL.

Gomez, M. C., Pope, C. E., and Dresser, B. L. (2006) Nuclear transfer in cats and its application. *Theriogenology*, **66**, 72–81.

Gompper, M. E. (2002) The ecology of northeast coyotes: current knowledge and priorities for future research. WCS Working Paper No. 17, Wildlife Conservation Society, New York.

Gompper, M. E. and Gittleman, J. L. (1991) Home range scaling: intraspecific and comparative trends. *Oecologia*, **87**, 343–8.

Gonyea, W. J. (1976) Adaptive differences in the body proportions of large felids. *Acta Anatomica*, **96**, 81–96.

Gonyea, W. and Ashworth, R. (1975) Form and function of retractile claws in the Felidae and other representative carnivorans. *Journal of Morphology*, **145**, 229–38.

González-Oreja, J. A. (1998) Non-natural mortality of the Iberian lynx in the fragmented population of Sierra de Gata (W Spain). *Miscel-lania Zoologica*, **21**, 31–5.

Goodman, P. (2003) Assessing management effectiveness and setting priorities in protected areas in KwaZulu-Natal. *BioScience*, **53**, 843–50.

Goodrich, J. M. and Miquelle, D. G. (2005) Translocation of problem Amur tigers *Panthera tigris altaica* to alleviate tiger-human conflicts. *Oryx*, **39**, 454–7.

Goodrich, J. M., Kerley, L. L., Smirnov, E. N., *et al.* (2008) Survival rates and causes of mortality of Amur tigers on and near the Sikhote-Alin Biosphere Zapovednik. *Journal of Zoology*, **276**, 323–9.

Goodrich, J. M., Miquelle, D., Smirov, E., Kerley, L., Hornocker, M., and Quigley, H. (2007) Social structure, human-induced mortality and conservation of Amur tigers. In *Felid Biology and Conservation Conference 17–20 September 2007: Programme and Abstracts*, p. 50. Wildlife Conservation Research Unit, Oxford, UK.

Goodrich, J. M., Kerley, L. L., Miquelle, D. G., Smirnov, E. N., Quigley, H. B., and Hornocker, M. G. (2005) Social structure of Amur tigers on Sikhote-Alin Biosphere Zapovednik. In D. G. Miquelle, E. N. Smirnov, and J. M. Goodrich (eds.), *Tigers of Sikhote-Alin Zapovednik: Ecology and Conservation*, pp. 50–60. PSP, Vladivostok, Russia.

Goodrich, J. M., Kerley, L. L., Schleyer, B. O., *et al.* (2001) Capture and chemical anesthesia of Amur (Siberian) tigers. *Wildlife Society Bulletin*, **29**, 533–42.

Gordon, C. H. and Stewart, A.-M. E. (2007) The use of logging roads by clouded leopards. *Cat News*, **47**, 12–13.

Goritz F., Vargas, A., Martinez, F., *et al.* (2009) Ultrasonographical assessment of structure and function of the male and female reproductive organs in the Eurasian and the Iberian lynx. In A. Vargas, C Breitenmoser and U. Breitenmoser (eds.), *Iberian Lynx Ex Situ Conservation: An Interdisciplinary Approach*, pp. 366–75. Fundacion Biodiversidad, Madrid.

Gosling, L. M. (1982) A reassessment of the function of scent marking in territories. *Zeitschrift für Tierpsychologie*, **60**, 89–118.

Gottelli, D., Wang, J., Bashir, S., and Durant, S. M. (2007) Genetic analysis reveals promiscuity among female cheetahs. *Proceedings of the Royal Society of London Series B*, **274**, 1993–2001.

Gould, D. H. and Fenner, W. R. (1983) Paramyxovirus-like nucleocapsids associated with encephalitis in a captive Siberian tiger. *Journal of the American Veterinary Medical Association*, **183**, 1319–22.

Government of India (2005) *Joining the Dots: The Report of the Tiger Task Force*. Project Tiger, New Delhi, India.

Government of USA (2007a) Additional information on amendment proposal COP 14 Prop 2. CITES COP14 Inf. 30. The Hague, the Netherlands. <www.cites.org/eng/cop/14/inf/index.shtml>. Accessed online 5 May 2009.

Government of USA (2007b) Proposal for deletion of *Lynx rufus* from Appendix II. CITES COP14 Prop 2. The Hague, the Netherlands. <www.cites.org/eng/cop/14/prop/index.shtml>. Accessed online 5 May 2009.

Gradstein, F., Ogg, J., and Smith, A. (2004) *A Geologic Time Scale 2004*. Cambridge University Press, Cambridge, UK.

Graham, A. L., Cattadori, I. M., Lloyd-Smith, J. O., Ferrari, M. J., and Bjornstad, O. N. (2007) Transmission consequences of coinfection: cytokines writ large? *Trends in Parasitology*, **23**, 284–91.

Graham, K., Beckerman, A. P., and Thirgood, S. (2005) Human-predator-prey conflicts: ecological correlates, prey losses and patterns of management. *Biological Conservation*, **122**, 159–71.

Graham, L. H., Byers, A. P., Armstrong, D. L., *et al.* (2006) Natural and gonadotropin-induced ovarian activity in tigers (*Panthera tigris*) assessed by fecal steroid analyses. *General and Comparative Endocrinology*, **147**, 362–70.

Graham, L. H., Swanson, W. F., and Brown, J. L. (2000) Chorionic gonadotropin administration in domestic cats causes an abnormal endocrine environment that disrupts oviductal embryo transport. *Theriogenology*, **54**, 1117–31.

Grange, S. and Duncan, P. (2006) Bottom-up and top-down processes in African ungulate communities: resources and predation acting on relative abundances of zebra and grazing bovids. *Ecography*, **29**, 899–907.

Grassman, L. I., Jr (2000) Movements and diet of the leopard cat *Prionailurus bengalensis* in a seasonal evergreen forest in south-central Thailand. *Acta Theriologica*, **45**, 421–6.

Grassman, L. I., Jr (1998a) Ecology and behavior of the Indochinese leopard (*Panthera pardus delacouri*) in a subtropical evergreen forest in southern Thailand. *Scientific Reports of the Zoological Society 'La Torbiera'*, **4**, 40–58. Societa Zoologica La Torbiera, IUCN.

Grassman, L. I., Jr (1998b) Movements and fruit selection of two Paradoxurinae palm civet species in a subtropical evergreen forest in southern Thailand. *Scientific Reports of the Zoological Society 'La Torbiera'*, **4**, 3–21. Societa Zoologica La Torbiera, IUCN.

Grassman, L. I., Jr, and Tewes, M. E. (2002) Marbled cat pair in northeastern Thailand. *Cat News*, **36**, 19–20.

Grassman, L. I., Jr, Tewes, M., Silvy, N. J., and Kreetiyutanont, K. (2005a) Spatial organization and diet of the leopard cat (*Prionailurus bengalensis*) in north-central Thailand. *Journal of Zoology*, **266**, 45–54.

Grassman, L. I., Jr, Tewes, M. E., Silvy, N. J., and Kreetiyutanont, K. (2005b) Ecology of three sympatric felids in a mixed evergreen forest in north-central Thailand. *Journal of Mammalogy*, **86**, 29–38.

Grassman, L. I., Jr, Austin, S. C., Tewes, M. E., and Silvy, N. J. (2004) Comparative immobilization of wild felids in Thailand. *Journal of Wildlife Diseases*, **40**, 575–8.

Gratwicke, B., Seidensticker, J., Shrestha, M., Vermilye, K., and Birnbaum, M. (2007) Evaluating the performance of a decade of Save the Tiger Fund's investments to save the world's last wild tigers. *Environmental Conservation*, **34**, 255–65.

Gratwicke, B., Bennett, E. L., Broad, S., *et al.* (2008a) The world can't have wild tigers and eat them, too. *Conservation Biology*, **22**, 222–3.

Gratwicke, B., Mills, J., Dutton, A., *et al.* (2008b) Attitudes toward consumption and conservation of tigers in China. *PLoS ONE*, **3**, e2544.

Green, M. J. B. and Zhimbiev, B. (1997) Transboundary protected areas and snow leopard conservation. In R. Jackson and A. Ahmad (eds.), *Proceedings of the 8th International Snow Leopard Symposium*. WWF Pakistan and International Snow Leopard Trust, Seattle, WA.

Greene, C. E. and Addie, D. D. (2006) Feline parvovirus infections. In C. E. Greene (ed.), *Infectious Dieseases of the Dog and Cat*, pp. 78–88. W.B. Saunders, Philadelphia, PA.

Greene, C. E. and Appel, M. J. G. (2006) Canine distemper. In C. E. Greene (ed.) *Infectious Diseases of the Dog and Cat*, pp. 25–7. W.B. Saunders, Philadelphia, PA.

Greene, C., Umbanhowar, J., Mangel, M., and Caro, T. M. (1998) Animal breeding systems, hunter selectivity, and consumptive use in wildlife conservation. In T. M. Caro (ed.), *Behavioral Ecology and Conservation Biology*, pp. 271–305. Oxford University Press, New York.

Greene, C. E. (ed.) (2006) *Infectious Diseases of the Dog and Cat*. W.B. Saunders, Philadelphia, PA.

Grenfell, B. T., Bjornstad, O. N., and Kappey, J. (2001) Travelling waves and spatial hierarchies in measles epidemics. *Nature*, **414**, 716–23.

Griffin, A. S., Blumstein, D. T., and Evans, C. (2000) Training captive-bred or translocated animals to avoid predators. *Conservation Biology*, **14**, 1317–26.

Griffith, B., Scott, J. M., Carpenter, J. W., and Reed, C. (1989) Translocation as a species conservation tool—status and strategy. *Science*, **245**, 477–80.

Griffith, M. A., Buie, D. E., Fendleyan, T. T., and Shipes, D. D. (1980) Preliminary observations of subadult bobcat movement behaviour. *Proceedings of the Annual Conference of the Southeastern Association of Fish and Wildlife Agencics*, **34**, 563–71.

Grigione, M. M., Beier, P., Hopkins, R. A., *et al.* (2002) Ecological and allometric determinants of home range size for mountain lions (*Puma concolor*). *Animal Conservation*, **5**, 317–24.

Grigione, M. M., Burman, P., Bleich, V. C., and Pierce, B. M. (1999) Identifying individual mountain lions *Felis concolor* by their tracks—refinement of an innovative technique. *Biological Conservation*, **88**, 25–32.

Grinnell, J. and McComb, K. (2001) Roaring and social communication in African lions: the limitations imposed by listeners. *Animal Behaviour*, **62**, 93–8.

Grinnell, J., Packer, C., and Pusey, A. E. (1995) Cooperation in male lions: kinship, reciprocity or mutualism? *Animal Behaviour*, **49**, 95–105.

Grobbelaar, C. and Masulani, R. (2003) *Review of Offtake Quotas, Trophy Quality and 'Catch Effort' across the Four Main Wildlife Species Elephant, Buffalo, Lion and Leopard*. WWF SARPO, Harare, Zimbabwe.

Grobler, D. G., Michel, A. L., De Klerk, L. M., and Bengis, R. G. (2002) The gamma-interferon test: its usefulness in a bovine tuberculosis survey in African buffaloes (*Syncerus caffer*) in the Kruger National Park. *Onderstepoort Journal of Veterinary Research*, **69**, 221–7.

Grobler, J. H. (1981) Feeding behaviour of the caracal *Felis caracal* Schreber, 1776, in the Mountain Zebra National Park. *South African Journal of Zoology*, **16**, 259–62.

Groiss, J. T. (1996) The 'cave lion' *Panthera tigris spelaea* (Goldfuss). *Neues Jahrbuch für Geologie und Paläontologie —Monatshefte*, 1996 (**7**), 399–414.

Gromov, E. I. and Matyushkin, E. N. (1974) Analysis of the competitive relations between tigers and wolves in the Sikhote-Alin. *Nauchn. doklady vysshey shkoly. Biolog. nauki*, **2**, 20–5.

Gros, P. M. (2002) The status and conservation of the cheetah *Acinonyx jubatus* in Tanzania. *Biological Conservation*, **106**, 177–85.

Gros, P. M. (1998) Status of the cheetah *Acinonyx jubatus* in Kenya: a field-interview assessment. *Biological Conservation*, **85**, 137–49.

Gros, P. M. (1996) Status of the cheetah in Malawi. *Nyala*, **19**, 33–6.

Gros, P. M. and Rejmanek, M. (1999) Status and habitat preferences of Uganda cheetahs: an attempt to predict carnivore occurrence based on vegetation structure. *Biodiversity and Conservation*, **8**, 1561–83.

Gross, L. (2008) No place for predators? *PloS Biology*, **6**, e40.

Groves, C. P. (1982) Cranial and dental characteristics in the systematics of old world Felidae. *Carnivore*, **5**, 28–39.

Guerrero, S., Badii, M. H., Zalapa, S. S., and Flores, A. E. (2002) Dieta y nicho de alimentación del coyote, zorra gris, mapache y jaguarundi en un bosque tropical caducifolio de la costa sur del estado de Jalisco, México. *Acta Zoologica Mexicana*, **86**, 119–37.

Guggisberg, C. A. W. (1975) *Wild Cats of the World*. David & Charles, Newton Abbot, UK.

Guggisberg, C. A. W. (1962) *Simba, the Life of the Lion*. Bailey Bros & Swinfen, London, UK.

Gugolz, D., Bernasconi, M., Breitenmoser-Würsten, C., and Wandeler, P. (2008) Historical DNA reveals the phylogenetic position of the extinct Alpine lynx. *Journal of Zoology*, **275**, 201–8.

Guilin Xiongsen Bear and Tiger Garden (2007) *The Status Quo of Captive Bred Tigers and Difficulties*. Xiongsen Bear and Tiger Garden, Guilin, Harbin, China.

Guiserix, M., Bahi-Jaber, N., Fouchet, D., Sauvage, F., and Pontier, D. (2007) The canine distemper epidemic in Serengeti: are lions victims of a new highly virulent canine distemper virus strain, or is pathogen circulation stochasticity to blame? *Journal of the Royal Society Interface*, **4**, 1127–34.

Gunther, I. and Terkel, J. (2002) Regulation of free-roaming cat (*Felis silvestris catus*) populations: a survey of the literature and its application to Israel. *Animal Welfare*, **11**, 171–88.

Gupta, A. K. (2005) Manas: Transboundary conservation on the Assam-Bhutan Border. In R. Mittermeier, C. F. Kormos, C. Mittermeier, P. Robles Gil, T. Sandwith, and C. Besançon (eds.), *Transboundary Conservation: A New Vision for Protected Areas*, pp. 291–6. CEMEX/Conservation International, Arlington, VA.

Gurung, B., Smith, J. L. D., McDougal, C., and Karki, J. B. (2006a) *Tiger Human Conflicts: Investigating Ecological and Sociological Issues of Tiger Conservation in the Buffer Zone of Chitwan National Park, Nepal*. WWF-Nepal Program, Kathmandu, Nepal.

Gurung, B., Smith, J. L. D., and Shrestha, M. (2006b) Using a 'bagh heralu' network to map the metapopulation structures of tigers in Nepal. In J. A. McNeely, T. M. McCarthy, A. Smith, L. Olsvig-Whittaker, and E. D. Wikramanayake (eds.), *Conservation Biology in Asia*, pp. 214–31. Resources Himalaya Foundation, Kathmandu, Nepal.

Gustafson, E. P. (1978) The vertebrate faunas of the Pliocene Ringold Formation, south-central Washington. *Bulletin of the Museum of Natural History, University of Oregon*, **23**, 1–62.

Guzmán, J. N., García, F. J., Garrote, G., Pérez de Ayala, R., and Iglesias, C. (2004) *El lince ibérico* (Lynx pardinus) *en España y Portugal. Censo diagnóstico de sus poblaciones*. Dirección General para la Biodiversidad, Madrid, Spain.

Haas, L., Liess, B., Hofer, H., East, M., Wohlsein, P., and Barrett, T. (1996) Canine distemper virus infection in Serengeti spotted hyaenas. *Veterinary Microbiology*, **49**, 147–52.

Habibi, K. (2004) *Mammals of Afghanistan*. Kabul, Afghanistan.

Haines, A. M., Grassman, Jr, L. I., Tewes, M. E., and Janecka, J. E. (2006a) First ocelot (*Leopardus pardalis*) monitored with GPS telemetry. *European Journal of Wildlife Research*, **52**, 216–18.

Haines, A. M., Janecka, J. E., Tewes, M. E., Grassman, Jr, L. I., and Morton, P. (2006b) The importance of private lands for ocelot *Leopardus pardalis* conservation in the United States. *Oryx*, **40**, 90–4.

Haines, A. M., Tewes, M. E., Laack, L. L., Horne, J. S., and Young, J. H. (2006c) A habitat-based population viability analysis for ocelots (*Leopardus pardalis*) in the United States. *Biological Conservation*, **132**, 424–36.

Haines, A. M., Tewes, M. E., and Laack, L. L. (2005) Survival and sources of mortality in Ocelots. *Journal of Wildlife Management*, **69**, 255–63.

Haines, A. M., Grassman, Jr, L. I., and Tewes, M. E. (2004) Survival of radiocollared adult leopard cats *Prionailurus bengalensis* in Thailand. *Acta Theriologica*, **49**, 349–56.

Haller, H. (1992) Zur Ökologie des luchses *Lynx lynx* im Verlauf seiner Wiederansiedlung in den Walliser Alpen. *Mammalia depicta, Beiheft für Zeitschrift zur Säugetierkunde*, **15**, 1–60.

Haller, H. and Breitenmoser, U. (1986) Zur Raumorganisation der in den Schweizer Alpen wiederangesiedelten Population des Luchses *Lynx lynx*. *Zeitschrift für Säugetierkunde*, **51**, 289–311.

Halley, J. and Hoelzel, A. R. (1996) Simulation models of bottleneck events in natural populations. In T. B. Smith and R. K. Wayne (eds.) *Molecular Genetic Approaches in Conservation*, pp. 347–64. Oxford University Press, Oxford, UK.

Haltenorth, T. (1953) *Die Wildkatzen der alten Welt*. Geest & Portig, Leipzig, Germany.

Hamilton, D. (2007) Cougar hysteria: addressing mountain lion mania in the Midwest and the East. *Wild Cat News*, **2**, 1–7.

Hamilton, M. J. and Kennedy, M. L. (1986) Genic variation in the coyote, *Canis latrans*, in Tennessee, U.S.A. *Genetica*, **71**, 167–73.

Hamilton, P. H. (1986) Status of the cheetah in Kenya with reference to sub-Saharan Africa. In S. D. Miller and D. D. Everett (eds.), *Cats of the World*, pp. 65–76. National Wildlife Federation, Washington DC.

Hamilton, P. H. (1981) *The Leopard* Panthera pardus *and the Cheetah* Acinonyx jubatus *in Kenya*. US Fish and Wildlife Service, the African Wildlife Leadership Foundation and the Government of Kenya, Nairobi, Kenya.

Hamilton, P. H. (1976) *The movements of leopards in Tsavo National Park, Kenya, as determined by radio-tracking*. MS dissertation, University of Nairobi, Nairobi, Kenya.

Hampson, K. (2009) Transmission dynamics and the prospects for the elimination of canine rabies. *PLoS Biology*, **7**, e53.

Hanby, J. P. and Bygott, J. D. (1987) Emigration of sub adult lions. *Animal Behaviour*, **35**, 161–9.

Hanby, J. P. and Bygott, J. D. (1979) Population changes in lions and other predators. In A. R. E. Sinclair and P. Arcese (eds.), *Serengeti II: Dynamics, Management and Conservation of an Ecosystem*, pp. 249–62. University of Chicago Press, Chicago, IL and London, UK.

Hanby, J. P., Bygott, J. D., and Packer, C. (1995) Ecology, demography, and behavior of lions in two contrasting habitats: Ngorongoro Crater and the Serengeti Plains. In A. R. E. Sinclair and P. Arcese (eds.), *Serengeti II: Dynamics, Management, and Conservation of an Ecosystem*, pp. 315–31. University of Chicago Press, Chicago, IL.

Hanks, J. (2005) Kavango-Zambezi: the four corners transfrontier conservation area. In R. Mittermeier, C. F. Kormos, C. Mittermeier, P. Robles Gil, T. Sandwith, and C. Besançon (eds.), *Transboundary Conservation: A New Vision for Protected Areas*, pp. 265–72. CEMEX/Conservation International, Arlington, VA.

Hansen, A. J., Rasker, R., Maxwell, B., *et al.* (2002) Ecological causes and consequences of demographic change in the New West. *BioScience*, **52**, 151–62.

Hanski, I. A. and Simberloff, D. (1997) The metapopulation approach, its history, conceptual domain, and application to conservation. In I. A. Hanski and M. E. Gilpin (eds.), *Metapopulation Biology: Ecology, Genetics, and Evolution*, pp. 5–26. Academic Press, San Diego, CA.

Haque, N. M. and Vijayan, V. S. (1993) Food habits of the fishing cat *Felis viverrina* in Keoladeo National Park, Bharatpur, Rajasthan. *Journal of the Bombay Natural History Society*, **90**, 498–500.

Harada, M. and Smith, A. M. (1975) Minamata disease: a medical report. In W. E. Smith and A. M. Smith (eds.), *Minamata*. Holt, Rinehart and Winston, New York.

Harcourt, A. H., Parks, S. A., and Woodroffe, R. (2001) Human density as an influence on species/area relationships: double jeopardy for small African reserves? *Biodiversity and Conservation*, **10**, 1011–26.

Hardin G. (1968) The tragedy of the Commons. *Science*, **162**, 1243–8.

Harihar, A., Kurien, A., Pandav, B., and Goyal, S. P. (2007) Response of tiger populations to habitat, wild ungulate prey and human disturbance in Rajaji National Park Uttarakhand. Final Technical Report, Wildlife Institute of India, Dehradun, India.

Harmsen, B. J. (2006) *The use of camera traps for estimating abundance and studying the ecology of jaguars* (Panthera onca). PhD dissertation, University of Southampton, Southampton, UK.

Harmsen, B. J., Foster, R. J., Jutierrez, S., Martin, S. and Doncaster, C. P (in press a) Scrape-marking behaviour of jaguars (*Panthera onca*) and pumas (*Puma concolor*) in a Belizcan rainforest. Journal of Mammalogy.

Harmsen, B. J., Foster R. J., Silver, S. C., Ostro, L. E. T., and Doncaster, C. P. (in press b) Differential use of trails by forest mammals and the implications for camera trap studies, a case study from Belize. *Biotropica*.

Harmsen, B. J., Foster, R. J., Silver, S. C., Ostro, L. E. T., and Doncaster, C. P. (2009) Spatial and temporal interactions of two sympatric cats in a neotropical forest: the jaguar (*Panthera onca*) and the puma (*Puma concolor*). *Journal of Mammalogy*, **90**, 612–20.

Harris, R. B., Wall, W. A., and Allendorf, F. W. (2002) Genetic consequences of hunting: what do we know and what should we do? *Wildlife Society Bulletin*, **30**, 634–43.

Harris, S., Morris, P., Wray, S., and Yalden, D. W. (1995) *A Review of British Mammals: Population Estimates and Conservation Status of British Mammals Other than Cetaceans*. Joint Nature Conservation Committee, Peterborough, UK.

Harrison, R. M. and Wildt, D. E. (eds.) (1980) *Animal Laparoscopy*. Williams and Wilkins Co., Baltimore, MD.

Harrison, T. M., Munson, L., Mazet, J. K., *et al.* (2004) Antibodies to canine and feline viruses in spotted hyenas (*Crocuta crocuta*) in the Masai Mara National Reserve, Kenya. *Journal of Wildlife Diseases*, **40**, 1–10.

Hart, B. L. and Hart, L. A. (1985) *Canine and Feline Behavioural Therapy*. Lea and Febiger, Philadelphia, PA.

Hart, D. and Sussmann, R. W. (2009) *Man the Hunted: Primates, Predators, and Human Evolution*. Westview, New York.

Hart, J. A., Katembo, M., and Punga, K. (1996) Diet, prey selection and ecological relations of leopard and golden cat in the Ituri Forest, Zaire. *African Journal of Ecology*, **34**, 364–79.

Hartmann, H. (2006) Reintroduction of captive-bred wildcats in Germany. In *Iberian Lynx Ex-Situ Conservation Seminar Series: Book of Proceedings*, p. 135. Fundación Biodiversidad, Sevilla & Doñana, Spain.

Harveson, L. A. (1997) *Ecology of a mountain lion population in southern Texas*. PhD dissertation, Texas A&M University, College Station and Texas A&M University Kingsville, Kingsville, TX.

Harveson, L. A., Tewes, M. E., Silvy, N. J., and Rutledge, J. (2000) Prey use by mountain lions in southern Texas. *Southwestern Naturalist*, **45**, 472–6.

Harveson, P. M., Tewes, M. E., Anderson, G. L., and Laack, L. L. (2004) Habitat use by ocelot in south Texas: implications for restoration. *Wildlife Society Bulletin*, **32**, 948–54.

Hast, M. H. (1989) The larynx of roaring and non-roaring cats. *Journal of Anatomy*, **163**, 117–21.

Hatten, J. R., Averill-Murray, A., and Van Pelt, W. E. (2005) A spatial model of potential jaguar habitat in Arizona. *Journal of Wildlife Management*, **69**, 1024–33.

Haydon, D. T. (2008). Cross-disciplinary demands of multi-host pathogens. *Journal of Animal Ecology*, **77**, 1079–1081.

Haydon, D. T., Randall, D. A., L., M., *et al.* (2006) Low-coverage vaccination strategies for the conservation of endangered species. *Nature*, **443**, 692–5.

Haydon, D. T., Cleaveland, S., Taylor, L. H., and Laurenson, M. K. (2002) Identifying reservoirs of infection: a conceptual and practical challenge. *Emerging Infectious Diseases*, **8**, 1468–73.

Hayward, G. D., Miquelle, D. G., Smirnov, E. N., and Nations, C. (2002) Monitoring Amur tiger populations: characteristics of track surveys in snow. *Wildlife Society Bulletin*, **30**, 1150–9.

Hayward, M. W. (2006) Prey preferences of the spotted hyaena (*Crocuta crocuta*) and degree of dietary overlap with the lion (*Panthera leo*). *Journal of Zoology*, **270**, 606–14.

Hayward, M. W. and Kerley, G. I. H. (2005) Prey preferences of the lion (*Panthera leo*). *Journal of Zoology*, **267**, 309–22.

Hayward, M. W. and Somers, M. J. (eds.) (in press) *The Reintroduction of Top Order Predators*. Wiley-Blackwell Publishing, Oxford, UK.

Hayward, M. W., O'Brien, J., and Kerley, G. I. H. (2007) Carrying capacity of large African predators: predictions and tests. *Biological Conservation*, **139**, 219–29.

Hayward, M. W., Henschel, P., O'Brien, J., Hofmeyr, M., Balme, G., and Kerley, G. I. H. (2006a) Prey preferences of the leopard (*Panthera pardus*). *Journal of Zoology*, **270**, 298–313.

Hayward, M. W., Hofmeyr, M., O'Brien, J. O., and Kerley, G. I. H. (2006b) Prey preferences of the cheetah (*Acinonyx jubatus*) (Felidae: Carnivora): morphological limitations or the need to capture rapidly consumable prey before kleptoparasites arrive? *Journal of Zoology*, **270**, 615–27.

Hazzah, L. (2006) *Living among lions* (Panthera leo): *coexistence or killing? Community attitudes towards conservation initiatives and the motivations behind lion killing in Kenyan Maasailand*. MS dissertation, University of Wisconsin, Madison, WI.

He, L., Garcia Perea, R., Li, M., and Wei, F. (2004) Distribution and conservation status of the endemic Chinese mountain cat *Felis bieti*. *Oryx*, **38**, 55–61.

Hearn, A. J. and Bricknell, S. (2003) Bay cat sightings in Kalimantan. *Cat News*, **39**, 3.

Hedges, R. E. M., Housley, R. A., Ramsey, C. B., and Vanklinken, G. J. (1994) Radiocarbon dates from the Oxford AMS system: Archaeometry datelist 18. *Archaeometry*, **36**, 337–74.

Heeney, J. L., Evermann, J. F., McKeirnan, A. J., *et al.* (1990) Prevalence and implications of feline coronavirus infections of captive and free-ranging cheetahs (*Acinonyx jubatus*). *Journal of Virology*, **64**, 1964–72.

Heesy, C. P. (2004) On the relationship between orbit orientation and binocular visual field overlap in mammals. *Anatomical Record Part A—Discoveries in Molecular Cellular and Evolutionary Biology*, **281A**, 1104–10.

Heffner, R. H. and Heffner, H. E. (1985) Hearing range of the domestic cat. *Hearing Research*, **19**, 85–8.

Heilbrun, R. D., Silvy, N. J., Peterson, M. J., and Tewes, M. E. (2006) Estimating bobcat abundance using automatically triggered cameras. *Wildlife Society Bulletin*, **34**, 69–73.

Heindrichs, H. (1975) The status of the tiger *Panthera tigris* (Linne, 1758) in the Sundarbans mangrove forest (Bay of Bengal). *Saugetierkundliche Mitteilungen*, **23**, 161–99.

Heinsohn, R. (1997) Group territoriality in two populations of African lions. *Animal Behaviour*, **53**, 1143–7.

Heizmann, E. P. J. (1973) Die Carnivoren des Steinheimer Beckens. B. Ursidae, Felidae, Viverridae sowie Ergänzungen und Nachträge zu den Mustelidae. *Palaeontographica*, **8**, 1–95.

Hellborg, L., Walker, C. W., Knispel Rueness, E., *et al.* (2002) Differentiation and levels of genetic variation in northern European lynx (*Lynx lynx*) populations revealed by microsatellites and mitochondrial DNA analysis. *Conservation Genetics*, **3**, 97–111.

Hellgren, E. C., Onorato, D. P., and Skiles, J. R. (2005) Dynamics of a black bear population within a desert metapopulation. *Biological Conservation*, **122**, 131–40.

Helman, R. G., Russell, W. C., Jenny, A., Miller, J., and Payeur, J. (1998) Diagnosis of tuberculosis in two snow leopards using polymerase chain reaction. *Journal of Veterinary Diagnostic Investigation*, **10**, 89–92.

Hemker, T. P., Lindzey, F. G., and Ackerman, B. B. (1984) Population characteristics and movement patterns of cougars in southern Utah. *Journal of Wildlife Management*, **48**, 1275–84.

Hemley, G. and Mills, J. A. (1999) The beginning of the end of tigers in trade? In J. Seidensticker, S. Christie, and P. Jackson (eds.), *Riding the Tiger: Tiger Conservation in*

Human Dominated Landscapes, pp. 216–29. Cambridge University Press, Cambridge, UK.

Hemmer, H. (1978) Nachweis der Sandkatze (*Felis margarita harrisoni* Hemmer, Grubb und Groves, 1976) in Jordanien. *Zeitschrift fur Saugetierkunde*, **43**, 62–4.

Hemmer, H. (1976) Fossil history of living Felidae. In R. Eaton (ed.), *The World's Cats*, pp. 1–14. University of Washington, Seattle, WA.

Hemmer, H. (1974) Untersuchungen zur Stammesgeschichte der Pantherkatzen (Pantherinae). Teil III. Zur Artgeschichte des Löwen *Panthera (Panthera) leo* (Linnaeus 1758). *Veröffentlichungen der Zoologisches Staatssammlung München*, **17**, 167–280.

Hemmer, H. (1972) *Uncia uncia. Mammalian Species*, **20**, 1–5.

Hemmer, H., Kahlke, R. D., and Vekua, A. K. (2001) The jaguar—*Panthera onca gambaszoegensis* (Kretzoi, 1938) (Carnivora: Felidae)—in the late lower Pleistocene of Akhalkalaki (South Georgia; Transcaucasia) and its evolutionary and ecological significance. *Geobios*, **34**, 475–86.

Hemmer, H., Grubb, P., and Groves, C. P. (1976) Notes on the sand cat, *Felis margarita* Loche, 1858. *Zeitschrift fur Saugetierkunde*, **41**, 286–303.

Hemmer, H., Kahlke, R. D., and Vekua, A. K. (2004) The Old World puma —*Puma pardoides* (Owen, 1846) (Carnivora: Felidae)—in the Lower Villafranchian (Upper Pliocene) of Kvabebi (East Georgia, Transcaucasia) and its evolutionary and biogeographical significance. *Neues Jahrbuch für Geologie und Paläontologie—Abhandlungen*, **233**, 197–231.

Hemson, G. (2003) *The ecology and conservation of lions: human–wildlife conflict in semi-arid Botswana*. DPhil dissertation, University of Oxford, Oxford, UK.

Hemson, G. (2002) The potential for using GPS collars on wild African lions (*Panthera leo*); some lessons from experience. In A. J. Loveridge, T. Lynam, and D. W. Macdonald (eds.), *Lion Conservation Research. Workshop 2: Modelling Conflict*, pp. 59–65. Wildlife Conservation Research Unit, Oxford, UK.

Hemson, G., Maclennan, S., Mills, M. G. L., Johnson, P., and Macdonald, D. W. (2009) Community, lions, livestock and money: A spatial and social analysis of attitudes to wildlife and the conservation value of tourism in a human—carnivore conflict in Botswana. *Biological Conservation*, **142**, 2718–25.

Hemson, G., Johnson, P., South, A., Kenward, R., Ripley, R. M., and Macdonald, D. W. (2005) Are kernels the mustard? Data from global positioning system (GPS) collars suggests problems for kernel homerange analyses with least-squares cross-validation. *Journal of Animal Ecology*, **74**, 455–63.

Hendey, Q. B. (1974) The late Cenozoic Carnivora of the south-western Cape Province. *Annals of the South African Museum*, **63**, 1–369.

Hennig, W. (1966) *Phylogenetic Systematics*. University of Illinois Press, Urbana, IL.

Henschel, P. (2007) The status of the leopard (*Panthera pardus*) in the Congo basin. In *Felid Biology and Conservation Conference 17–20 September 2007: Programme and Abstracts*, p. 82. Wildlife Conservation Research Unit, Oxford, UK.

Henschel, P. (2001) *Untersuchung der Ernährungsweise und der Populationsdichte des Leoparden* (Panthera pardus) *im Lopé Reservat, Gabun, Zentralafrika*. MSc dissertation, University of Göttingen, Göttingen, Germany.

Henschel, P. and Ray, J. (2003) *Leopards in African Rainforests: Survey and Monitoring Techniques*. Wildlife Conservation Society, Global Carnivore Program, New York.

Henschel, P., Abernathy, K. A., and White, L. J. T. (2005) Leopard food habits in the Lopé National Park, Gabon, Central Africa. *African Journal of Ecology*, **43**, 21–8.

Heptner, V. G. and Sludskii, A. A. (1992) *Mammals of the Soviet Union, Carnivora (hyaenas and cats)*, Vol. II, Part 2. E. J. Brill, Leiden, the Netherlands.

Heptner, V. G. and Sludskii, A. A. (1972) Mammals of the Soviet Union. In V. G. Heptner and N. P. Naumov (eds.), *Mammals of the Soviet Union, Carnivora (Hyaenas and Cats)*, Vol. II, Part 2. Smithsonian Institution, Washington DC and the National Science Foundation, Moscow, Russia.

Herbst, L. H., Packer, C., and Seal, U. S. (1985) Immobilization of free-ranging African lions (*Panthera leo*) with a combination of xylazine hydrochloride and ketamine hydrochloride. *Journal of Wildlife Diseases*, **21**, 401–4.

Herbst, M. and Mills, M. G. L. (in press) The feeding habits of the African wildcat (*Felis silvestris lybica*), a facultative trophic specialist, in the Southern Kalahari (Kgalagadi Transfrontier Park, South Africa/Botswana). Journal of Zoology.

Herbst, M. and Mills, M. G. L. (in prep.) Techniques used in the study of African wildcat, *Felis silvestris lybica*, in the Kgalagadi Transfrontier park (South Africa/Botswana).

Herda-Rapp, A. and Goedeke, T. L. (2005) *Mad about Wildlife: Looking at Social Conflict over Wildlife*. Brill Academic Press, Leiden, the Netherlands.

Herfindal, I., Linnell, J. D. C., Moa, P. F., Odden, J., Austmo, L. B., and Andersen, R. (2005a) Does recreational hunting of lynx reduce depredation losses of domestic sheep? *Journal of Wildlife Management*, **69**, 1034–42.

Herfindal, I., Linnell, J. D. C., Odden, J., Nilsen, E. B., and Andersen, R. (2005b) Prey density, environmental productivity and home range size in the Eurasian lynx (*Lynx lynx*). *Journal of Zoology*, **265**, 63–71.

Hermann, E., Van Vuuren, J. H., and Funston, P. J. (2002) Population dynamics of lions in the Kgalagadi Transfrontier Park: a preliminary model to investigate the effect of human-caused mortality. In A. J. Loveridge, T. Lynam and D. W. Macdonald (eds.), *Lion Conservation Research. Workshop 2: Modelling Conflict*, p. 98. Wildlife Conservation Research Unit, Oxford, UK.

Herrington, S. J. (1987) Subspecies and the conservation of *Panthera tigris*: preserving genetic heterogeneity. In R. L. Tilson and U. S. Seal (eds.), *Tigers of the World*, pp. 51–61. Noyes Publications, Park Ridge, NJ.

Herrington, S. J. (1986) *Phylogenetic relationships of the wild cats of the world*. PhD dissertation, University of Kansas, Lawrence, KS.

Hetherington, D. (2006) The lynx in Britain's past, present and future. *ECOS*, **27**, 66–73.

Hetherington, D. A., Miller, D. R., MacLeod, C. D., and Gorman, M. L. (2008) A potential habitat network for the Eurasian lynx *Lynx lynx* in Scotland. *Mammal Review*, **38**, 285–303.

Hetherington, D. A., Lord, T. C., and Jacobi, R. M. (2005) New evidence for the occurrence of Eurasian lynx (*Lynx lynx*) in medieval Britain. *Journal of Quaternary Science*, **21**, 3–8.

Hibbard, C. W. (1934) Two new genera of Felidae from the Middle Pliocene of Kansas. *Transactions of the Kansas Academy of Sciences*, **37**, 239–55.

Hickman, G. C. (1979) The mammalian tail—review of functions. *Mammal Review*, **9**, 143–57.

Hildebrand, M. (1961) Further studies on locomotion of the cheetah. *Journal of Mammalogy*, **42**, 84–91.

Hildebrand, M. (1959) Motions of the running cheetah and horse. *Journal of Mammalogy*, **40**, 481–95.

Hildebrand, M. and Hurley, J. P. (1985) Energy of the oscillating legs of a fast-moving cheetah, pronghorn, jackrabbit, and elephant. *Journal of Morphology*, **184**, 23–31.

Hildebrandt, T. B., Brown, J. L., Hermes, R., and Goertiz, F. (2003) Ultrasound for the analysis of reproductive function in wildlife. In W. V. Holt, A. R. Pickard, J. C. Rodger and D. E. Wildt (eds.), *Reproductive Science and Integrated Conservation Science*, pp. 166–82. Zoological Society of London, London, UK.

Hill, K. and Hurtado, A. M. (1996) *Ache Life History: The Ecology and Demography of a Foraging People*. Aldine De Gruyter, New York.

Hines, J. E. and Link, W. A. (1994) *SCATMAN*. <www.mbr-wrc.usgs.gov/software.html>. USGS-Patuxant Wildlife Research Centre, Patuxant, MD.

Hitchens, C. (2007) *God Is Not Great: The Case Against Religion*. Atlantic, London, UK.

Hodges, K. E. (2000) The ecology of snowshoe hares in northern boreal forests. In L. F. Ruggiero, K. B. Aubry, S. W. Buskirk, *et al.* (eds.), *Ecology and Conservation of Lynx in the United States*, pp. 117–61. University Press of Colorado, Boulder, CO.

Hodgeson, S. and Bewick, T. (1790) *A General History of Quadrupeds*. Edward Walker, Newcastle Upon Tyne, UK.

Hofer, H. and East, M. L. (1993) The commuting system of Serengeti spotted hyaenas: how a predator copes with migratory prey. II. Intrusion pressure and commuters' space use. *Animal Behaviour*, **46**, 559–74.

Hofmann-Lehmann, R., Fehr, D., Grob, M., *et al.* (1996) Prevalence of antibodies to feline parvovirus, calicivirus, herpesvirus, coronavirus, and immunodeficiency virus and of feline leukemia virus antigen and the interrelationship of these viral infections in free-ranging lions in east Africa. *Clinical and Diagnostic Laboratory Immunology*, **3**, 554–62.

Holden, J. (2001) Small cats in Kerinci Seblat National Park, Sumatra, Indonesia. *Cat News*, **35**, 11–14.

Holling, C. S. (1959) The components of predation as revealed by a study of small mammal predation of the European pine sawfly. *Canadian Entomologist*, **91**, 293–320.

Holt, G. and Berg, C. (1990) Sarcoptic mange in red fox and other wild mammals in Norway. *Norsk Veterinartidsskrift*, **102**, 427–32.

Holt, W. V., Abaigar, T., Watson, P. F., and Wildt, D. E. (2003) Genetic resource banks for species conservation. In W. V. Holt, A. R. Pickard, J. C. Rodger, and D. E. Wildt (eds.), *Reproductive Science and Integrated Conservation Science*, pp. 267–80. Zoological Society of London, London, UK.

Homyack, J. A., Vashon, J. H., Libby, C., *et al.* (2008) Canada lynx-bobcat (*Lynx canadensis x L. rufus*) hybrids at the southern periphery of lynx range in Maine, Minnesota and New Brunswick. *American Midland Naturalist*, **159**, 504–8.

Hooge, P. N., Eichenlaub, W., and Soloman, E. (1999) *The Animal Movement Program*. Alaska Biological Science Centre, U.S. Geological Survey, Anchorage, AK.

Hoogesteijn, R. (2003) *Manual on the Problem of Depredation Caused by Jaguars and Pumas on Cattle Ranches*. Wildlife Conservation Society, New York.

Hoogesteijn, R. and Hoogesteijn, A. (2008) Conflicts between cattle ranching and large predators in Venezuela: could use of water buffalo facilitate felid conservation? *Oryx*, **42**, 132–8.

Hoogesteijn, R. and Mondolfi, E. (1996) Body mass and skull measurements in four jaguar populations, and observations on their preybase. *Bulletin of the Florida Museum of Natural History*, **39**, 195–219.

Hoogesteijn, R. and Mondolfi, E. (1992) *The Jaguar*. Armitano Editores C.A., Caracas, Venezuela.

Hoogesteijn, R., Hoogesteijn, A., and Mondolfi, E. (1993) Jaguar predation and conservation: cattle mortality caused by felines on three ranches in the Venezuelan Llanos. In N. Dunstone and M. L. Gorman (eds.), *Mammals as Predators*, pp. 391–406. Symposia of the Zoological Society of London, No. 65. Zoological Society, London, UK.

Hopcraft, J. G. C. (2002) *The role of habitat and prey density on foraging by Serengeti lions*. MS dissertation, University of British Columbia, Vancouver, BC.

Hopcraft, J. G. C., Sinclair, A. R. E., and Packer, C. (2005) Planning for success: Serengeti lions seek prey accessibility rather than abundance. *Journal of Animal Ecology*, **74**, 559–66.

Hopkins, R. A. (1989) *Ecology of the puma in the Diablo Range, California*. PhD dissertation, University of California, Berkeley, CA.

Hopwood, A. T. (1947) Contributions to the study of some African mammals III. Adaptations in the bones of the forelimb of the lion, leopard and cheetah. *Journal of the Linnean Society Zoology*, **41**, 259–71.

Hornocker, H. G., Murphy, K., Felzein, G. S., and Relyea, S. E. (1992) *Ecology of the Mountain Lion in Yellowstone*. Hornocker Wildlife Research Institute, Moscow, ID.

Hornocker, M. G. (2008) Conserving the cats, cougar management as a model: a review. In T. E. Fulbright and D. G. Hewitt (eds.), *Wildlife Science: Linking Ecological Theory and Management*, pp. 111–22. CRC Press, Boca Raton, FL.

Hornocker, M. G. (1970) An analysis of mountain lion predation upon mule deer and elk in the Idaho Primitive Area. *Wildlife Monographs*, **21**, 5–39.

Hötte, M. (2007) Amur leopard zoo breeding programmes. *Amur Leopard Conservation Update*, **January 2007**, 6–7.

Howard, J. G. (1999) Assisted reproductive techniques in non-domestic carnivores. In M. E. Fowler and R. E. Miller (eds.), *Zoo and Wild Animal Medicine: Current Therapy IV*, pp. 449–57. W. B. Saunders Co., Philadelphia, PA.

Howard, J. G. (1993) Semen collection and analysis in carnivores. In M. E. Fowler (ed.), *Zoo and Wild Animal Medicine: Current Therapy III*, pp. 390–9. W.B. Saunders Co., Philadelphia, PA.

Howard, J. G. and Allen, M. E. (2008) Nutritional factors affecting semen quality in felids. In M. E. Fowler and R. E. Miller (eds.), *Zoo and Wild Animal Medicine: Current Therapy V*, pp. 272–83. W. B. Saunders Co., Philadelphia, PA.

Howard, J. G., and Wildt, D. E. (2009) Approaches and efficacy of artificial insemination in felids and mustelids. *Theriogenology*, **71**, 130–48.

Howard, J. G., Brown, J. L., Bush, M., and Wildt, D. E. (1990) Teratospermic and normospermic domestic cat ejaculate traits, pituitary-gonadal hormones, and improvement of spermatozoal motility and morphology after swim-up processing. *Journal of Andrology*, **11**, 204–15.

Huang, G. T., Rosowski, J. J., Ravicz, M. E., and Peake, W. T. (2002) Mammalian ear specializations in arid habitats: structural and functional evidence from sand cat (*Felis margarita*). *Journal of Comparative Physiology A*, **188**, 663–81.

Huang, G. T., Rosowski, J. J., and Peake, W. T. (2000) Relating middle-ear acoustic performance to body size in the cat family: measurements and models. *Journal of Comparative Physiology A*, **186**, 447–65.

Hubbard, A. L., McOrist, S., Jones, T. W., Boid, R., Scott, R., and Easterbee, N. (1992) Is survival of European wildcats *Felis silvestris* in Britain threatened by interbreeding with domestic cats? *Biological Conservation*, **61**, 203–8.

Hudson, P. J., Rizzoli, A., Grenfell, B. T., Heesterbeek, H. and Dobson, A. P. (eds.) (2002) *The Ecology of Wildlife Diseases*. Oxford University Press, Oxford, UK.

Huggler, J. (2006) Tiger, tiger, burning bright. *The Independent*, 15 Feb. London, UK.

Hughes, A. (1985) New perspectives in retinal organisation. *Progress in Retinal Research*, **4**, 243–314.

Hughes, A. (1977) The topography of vision in mammals of contrasting lifestyle: comparative optics and retinal organisation. In F. Crescitelli (ed.), *Handbook of Sensory Physiology, Vol. 7, Part 5*, pp. 613–756. Springer, Berlin, Germany.

Hughes, A. (1976) A supplement to the cat schematic eye. *Vision Research*, **16**, 149–54.

Hulbert, I. A. R. (2001) GPS and its use in animal telemetry: the next five years. In A. M. Sibbald and I. J. Gordon (eds.), *Tracking Animals with GPS*, pp. 51–60. Macaulay Land Use Research Institute, Aberdeen, Scotland.

Hulme, D. and Murphree, M. (2001) Community conservation in Africa: an introduction. In D. Hulme and M. Murphree (eds.), *African Wildlife and Livelihoods. The Promise and Performance of Community Conservation*, pp. 1–8. James Currey, Oxford, UK.

Hunt, Jr, R. M. (2001) Basicranial anatomy of the living linsangs *Prionodon* and *Poiana* (Mammalia, carnivora, Viverridae), with comments on the early evolution of aeluroid carnivorans. *American Museum Novitates*, **3330**, 1–24.

Hunt, Jr, R. M. (1998) Evolution of the aeluroid Carnivora: diversity of the earliest aeluroids from Eurasia (Quercy, Hsanda-Gol) and the origin of felids. *American Museum Novitates*, **3252**, 1–65.

Hunt, Jr, R. M. (1989) Evolution of the aeluroid Carnivora: significance of the ventral promontorial process of the petrosal, and the origin of basicranial patterns in the living families. *American Museum Novitates*, **2930**, 1–32.

Hunt, Jr, R. M. (1987) Evolution of the aeluroid Carnivora: significance of auditory structure in the Nimravid cat *Dinictis*. *American Museum Novitates*, **2886**, 1–74.

Hunter, D. L. (1996) Tuberculosis in free-ranging, semi-free-ranging and captive cervids. *Revue Scientifique et Technique de l' Office International des Epizooties*, **15**, 171–81.

Hunter, L. T. B. and Bowland, J. (in press) Serval *Leptailurus serval*. In J. S. Kingdon and M. Hoffmann (eds.), *The Mammals of Africa 5: Carnivora, Pholidota, Perissodactyla*. Academic Press, Amsterdam, the Netherlands.

Hunter, L. T. B. and Balme, G. A. (2004) The leopard. The worlds most persecuted big cat. In *Endangered Wildlife. Businesses, Ecotourism and the Environment*, pp. 88–94. Endangered Wildlife Trust, Johannesburg, South Africa.

Hunter, L. T. B., Henschel, P. P., and Ray, J. C. (in press) *Panthera pardus*. In J. S. Kingdon and M. Hoffmann (eds.), *The Mammals of Africa: Carnivores, Pangolins, Rhinos and Equids*. Academic Press, Amsterdam, the Netherlands.

Hunter, J. S., Durant, S. M., and Caro, T. M. (2007a) To flee or not to flee: predator avoidance by cheetahs at kills. *Behavioral Ecology and Sociobiology*, **61**, 1033–42.

Hunter L. T. B., Jowkar, H., Ziaie, H., *et al*. (2007b) Conserving the Asiatic cheetah in Iran: launching the first radio-telemetry study. *Cat News*, **46**: 8–11.

Hunter, L. T. B., Pretorius, K., Carlisle, L. C., *et al*. (2007c) Restoring lions *Panthera leo* to northern KwaZulu-Natal, South Africa: short-term biological and technical success but equivocal long-term conservation. *Oryx*, **41**, 196–204.

Hunter, L. T. B., Balme, G. A., Walker, C., Pretorius, K., and Rosenberg, K. (2003) The landscape ecology of leopards (*Panthera pardus*) in northern KwaZulu-Natal, South Africa: a preliminary project report. *Ecological Journal*, **5**, 24–30.

Hunziker, M., Hoffmann, C. W., and Wild-Eck, S. (2001) Die Akzeptanz von Wolf, Luchs und 'Stadtfuchs'—Ergebnisse einer gesamtschweizerisch-representativen Umfrage. *Forestry Snow and Landscape Research*, **76**, 301–26.

Hussain, S. (2003a) Snow leopard and local livelihoods: managing emerging conflicts through an insurance scheme. *Carnivore Damage Prevention News*, **6**, 9–11.

Hussain, S. (2003b) The status of the snow leopard in Pakistan and its conflict with local farmers. *Oryx*, **37**, 26–33.

Hussain, S. (2000) Protecting the snow leopard and enhancing farmers' livelihoods: a pilot insurance scheme in Baltistan. *Mountain Research and Development*, **20**, 226–31.

Hutajulu, B., Sunarto, Klenzendorf, S., Supriatna, J., Budiman, A., and Yahya, A. (2007) Study on the ecological characteristics of clouded leopard in Riau, Sumatra. In *Felid Biology and Conservation Conference 17–20 September 2007: Programme and Abstracts*, p. 122, Wildlife Conservation Research Unit, Oxford, UK.

ICZN (2003) International Commission on Zoological Nomenclature, Opinion 2027 (Case 3010). Usage of 17 specific names based on wild species which are predated by or contemporary with those based on domestic animals (Lepidoptera, Osteichthyes, Mammalia): conserved. *Bulletin of Zoological Nomenclature*, **60**, 81–4.

Ikeda, N. (2004) Economic impacts of livestock depredation by snow leopard *Uncia uncia* in the Kanchenjunga Conservation Area, Nepal Himalaya. *Environmental Conservation*, **31**, 322–30.

Ikeda, Y., Miyazawa, T., Nakamura, K., *et al*. (1999) Serosurvey for selected virus infections of wild carnivores in Taiwan and Vietnam. *Journal of Wildlife Diseases*, **35**, 578–81.

Ilany, G. (1986) Preliminary observations on the ecology of the leopard (*Panthera pardus jarvisi*) in the Judean Desert. In S. D. Miller and D. D. Everett (eds.), *Cats of the World: Biology, Conservation and Management*, p. 199. National Wildlife Federation, Washington DC.

Inglehart, R. (1990) *Culture Shift in Advanced Industrial Societies*. Princeton University Press, Princeton, NJ.

Inskip, C. and Zimmermann, A. (2009) Human–felid conflict: a review of the patterns and priorities worldwide. *Oryx*, **43**, 1–17.

Iriarte, A. W. (1998) *Andean Mountain Cat in Chile: Status and Geographical Distribution*. Cat Action Treasury, Santiago, Chile.

Iriarte, J. A. and Jaksic, F. M. (1986) The fur trade in Chile: an overview of seventy-five years of export data (1910–1984). *Biological Conservation*, **38**, 243–53.

Iriarte, A. W., Johnson, W. E., and Franklin, W. L. (1991) Feeding ecology of the Patagonian puma in southernmost Chile [Ecologia trofica del puma de la Patagonia en el extremo sur de Chile]. *Revista Chilena de Historia Natural*, **64**, 145–56.

Iriarte, J. A., Franklin, W. L., Johnson, W. E., and Redford, K. H. (1990) Biogeographic variation of food habits and body size of the America puma. *Oecologia*, **85**, 185–90.

IUCN (2008) *2008 IUCN Red List of Threatened Species*. IUCN Species Survival Commission, Gland, Switzerland.

IUCN (2007) *2007 IUCN Red List of Threatened Species*. IUCN Species Survival Commission, Gland, Switzerland.

IUCN (2006a) *Conservation Strategy for the Lion in West and Central Africa*. IUCN/SSC Cat Specialist Group. <www.felidae.org.>

IUCN (2006b) *Conservation Strategy for the Lion* Panthera leo *in Eastern and Southern Africa*. IUCN/SSC Cat Specialist Group, Gland, Switzerland.

IUCN (2006c) *2006 IUCN Red List of Threatened Species*. IUCN Species Survival Commission, Gland, Switzerland.

IUCN (2002) *SSC Conservation Breeding Specialist Group Policy Statement on Captive Breeding*. IUCN, Gland, Switzerland.

IUCN (1998) *Guidelines for Re-Introductions*. IUCN SSC Re-introduction Specialist Group, Gland, Switzerland.

IUCN (1995) *Species Survival Commission/Reintroduction Specialist Group Guidelines for Reintroductions*. IUCN, Gland, Switzerland.

IUCN/SSC (2007a) *Regional Conservation Strategy for the Conservation of Cheetah and Wild Dogs in Eastern Africa*. IUCN Species Survival Commission, Gland, Switzerland.

IUCN/SSC (2007b) *Regional Conservation Strategy for the Cheetah and African Wild Dog in Southern Africa*. IUCN Species Survival Commission, Gland, Switzerland.

IUCN/SSC (2007c) Status and conservation needs of cheetahs in Southern Africa. *Cat News*, Special Issue *No. 3*. IUCN Species Survival Commission, Gland, Switzerland.

IUCN/SSC (1986) Symposium on jaguar and wildlife conservation in neotropical moist forest. *Cat News*, **5**, 11–17.

Ivlev, V. S. (1961) *Experimental Ecology of the Feeding of Fishes*. Yale University Press, New Haven, CT.

Jackson, P. (1989) The status of the leopard in sub-Saharan Africa: a review by leopard specialists. Unpublished report IUCN Cat Specialist Group, Gland, Switzerland.

Jackson, R. (1996) *Home range movements and habitat use of snow leopard* (Uncia uncia) *in Nepal*. PhD dissertation, University of London, London, UK.

Jackson, R. and Ahlborn, G. (1991) The role of protected areas in Nepal in maintaining viable populations of snow leopards. *International Pedigree Book of Snow Leopards*, **6**, 51–69.

Jackson, R. and Ahlborn, G. (1989) Snow leopards (*Panthera uncia*) in Nepal—home range and movements. *National Geographic Research and Exploration*, **5**, 161–75.

Jackson, R. and Ahlborn, G. (1984) A preliminary habitat suitability model for the Snow leopard, Panthera uncia, in West Nepal. *International Pedigree Book of Snow Leopards*, **4**, 43–52.

Jackson, R. and Fox, J. L. (1997) Snow leopard conservation: accomplishments and research priorities. In R. Jackson and A. Ahmad (eds.), *Proceedings of the 8th International Snow Leopard Symposium*, pp. 128–45. WWF Pakistan and International Snow Leopard Trust, Seattle, WA.

Jackson, R. and Hunter, D. O. (1996) *Snow Leopard Survey and Conservation Handbook*. International Snow Leopard Trust, Seattle, WA and US Geological Survey, Fort Collins, CO.

Jackson, R. and Wangchuk, R. (2004) A community-based approach to mitigating livestock depredation by snow leopards. *Human Dimensions of Wildlife*, **9**, 307–15.

Jackson, R. and Wangchuk, R. (2001) Linking snow leopard conservation and people-wildlife conflict resolution: grassroots measures to protect the endangered snow leopard from herder retribution. *Endangered Species Update*, **18**, 138–41.

Jackson, R., Roe, J. D., Wangchuk, R., and Hunter, D. O. (2006) Estimating snow leopard population abundance using photography and capture-recapture techniques. *Wildlife Society Bulletin*, **34**, 772–81.

Jackson, R., Wangchuk, R., and Hillard, D. (2002) *Grass Roots Measures to Protect the Endangered Snow Leopard from Herder Retribution: Lessons Learned from Predator-Proofing Corrals in Ladakh*. Snow Leopard Conservancy, Sonoma, CA.

Jackson, R., Ahlborn, G., Gurung, M., and Ale, S. (1996) Reducing livestock depredation losses in the Nepalese Himalaya. In R. M. Timm and A. C. Crabb (eds.) *Proceedings of the 17th Vertebrate Pest Conference*, pp. 241–7. University of California, Davis, CA.

Jackson, R., Ahlborn, G. G., and Shah, K. B. (1990) Capture and immobilization of wild snow leopards. In L. Blomqvist (ed.), *International Pedigree Book of Snow Leopards*, pp. 93–102. Helsinki Zoo, Helsinki, Finland.

Jacob, A. A. (2002) *Ecologia e conservação da jaguatirica* (Leopardus pardalis) *no Parque Estadual Morro do Diabo, Pontal do Paranapanema, SP.* MS dissertation, Universidade de Brasília, Brasília, Brazil.

Jacobs, G. H. (1993) The distribution and nature of colour vision among the mammals. *Biological Reviews*, **68**, 413–71.

Jacobs, J. (1974) Quantitative measurement of food selection: a modification of the forage ratio and Ivlevs electivity index. *Oecologia*, **14**, 413–17.

Jacquier, M., Aarhaug, P., Arnemo, J. M., Bauer, H., and Enriquez, B. (2006) Reversible immobilization of free-ranging African lions (*Panthera leo*) with medetomidine-tiletamine-zolazepam and atipamezole. *Journal of Wildlife Diseases*, **42**, 432–6.

Jager, H. G., Booker, H. H., and Hubschle, O. J. (1990) Anthrax in cheetahs (*Acinonyx jubatus*) in Namibia. *Journal of Wildlife Diseases*, **26**, 423–4.

Jakob, W. and Wesemeier, H. H. (1996) A fatal infection in a bengal tiger resembling cytauxzoonosis in domestic cats. *Journal of Comparative Pathology*, **114**, 439–44.

Janczewski, D. N., Modi, W. S., Stephens, J. C., and O'Brien, S. J. (1995) Molecular evolution of mitochondrial 12S RNA and Cytochrome b sequences in the Pantherine lineage of Felidae. *Molecular Biology and Evolution*, **12**, 690–707.

Janečka, J. E., Jackson, R., Yuquang, Z., *et al.* (2008) Population monitoring of snow leopards using noninvasive collection of scat samples: a pilot study. *Animal Conservation* **11**, 401–11.

Janečka, J. E., Blankenship, T. L., Hirth, D. H., Kilpatrick, C. W., Tewes, M. E., and Grassman, Jr, L. I. (2007) Evidence for male-biased dispersal in bobcats *Lynx rufus* using relatedness analysis. *Wildlife Biology*, **13**, 38–47.

Janečka, J. E., Blankenship, T. L., Hirth, D. H., Tewes, M. E., Kilpatrick, C. W., and Grassman, Jr, L. I. (2006) Kinship and social structure of bobcats (*Lynx rufus*) inferred from microsatellite and radio-telemetry data. *Journal of Zoology*, **269**, 494–501.

Janeway, C. A. and Travers, P. (eds.) (1997) *Immunobiology: The Immune System in Health and Disease*. Garland Science Publishing, London, UK.

Jansen, T., Forster, P., Levine, M. A., *et al.* (2002) Mitochondrial DNA and the origins of the domestic horse. *Proceedings of the National Academy of Sciences of the United States of America*, **99**, 10905–10.

Jarman, P. J. and Jarman, M. V. (1979) The dynamics of ungulate social organization. In A. R. E. Sinclair and M. Norton-Griffiths (eds.), *Serengeti, Dynamics of an Ecosystem*, pp. 185–220. University of Chicago Press, Chicago, IL.

Jayaprakash, D., Patil, S. B., Kumar, M. N., Majumdar, K. C., and Shivaji, S. (2001) Semen characteristics of the captive Indian leopard, *Panthera pardus*. *Journal of Andrology*, **22**, 25–33.

Jedrzejewski, W., Schmidt, K., Okarma, H., and Kowalczyk, R. (2002) Movement pattern and home range use by the Eurasian lynx in Bialowieza Primeval Forest (Poland). *Annales Zoologici Fennici*, **39**, 29–41.

Jefferies, D. J. (1991) Some observations on Scottish wildcats *Felis silvestris* based on the results of autopsies. *Glasgow Naturalist*, **22**, 11–20.

Jenkins, D. (1962) The present status of the wild cat (*Felis silvestris*) in Scotland. *Scottish Naturalist*, **70**, 126–38.

Jennings, S. and Blanchard, J. L. (2004) Fish abundance with no fishing: predictions based on macro-ecological theory. *Journal of Animal Ecology*, **73**, 632–42.

Jerison, H. J. (1973) *Evolution of the Brain and Intelligence*. Academic Press, New York.

Jerolmack, C. (2008) How pigeons became rats: the cultural-spatial logic of problem animals. *Social Problems*, **55**, 72–94.

Jessup, D. A., Pettan, K. C., Lowenstine, L. J., and Pedersen, N. C. (1993) Feline leukemia virus infection and renal spirochetosis in a free ranging cougar (*Felis concolor*). *Journal of Zoo and Wildlife Medicine*, **24**, 73–9.

Jetz, W., Carbone, C., Fulford, J., and Brown, J. H. (2004) The scaling of animal space use. *Science*, **306**, 266–7.

Jeu, M. and Xiang, P. (1982) Discovery of *Sarcoptes scabiei* infesting lynx in Chongqing, China. *Zoological Research*, **3**, 310.

Jewgenov, K., Braun, B. C., Goritz, F., *et al.* (2009) Pregnancy diagnosis in Iberian lynx (Lynx pardinus) based on urinary and blood plasma hormones. In A. Vargas, C Breitenmoser and U. Breitenmoser (eds.). *Iberian Lynx Ex Situ Conservation: An Interdisciplinary Approach*, pp. 376–89. Fundacion Biodiversidad, Madrid.

Jewgenow, K., Dehnhard, M., Frank, A., Naidenko, S., Vargas, A., and Göritz, F. (2006a) A comparative analysis of the reproduction of the Eurasian and the Iberian lynx in captivity. In *Iberian Lynx Ex-Situ Conservation Seminar Series: Book of Proceedings*, pp. 116–17. Fundación Biodiversidad, Sevilla and Doñana, Spain.

Jewgenow, K., Naidenko, S. V., Goeritz, F., Vargas, A., and Dehnhard, M. (2006b) Monitoring testicular activity of male Eurasian (*Lynx lynx*) and Iberian (*Lynx pardinus*) lynx by fecal testosterone metabolite measurement. *General and Comparative Endocrinology*, **149**, 151–8.

Jhala, Y. V., Gopal, R., and Qureshi, Q. (2008) *Status of the Tigers, Co-Predators, and Prey in India*. National Tiger Conservation Authority, Government of India, New Delhi, India and Wildlife Institute of India, Dehradun, India.

Jiang, J. (2005) Tiger and leopard research in Jilin Province. In *Recovery of the Wild Amur Tiger Population in China: Progress and Prospect. Proceedings of the 2000 International Workshop on Wild Amur Tiger Population Recovery Action Plan, Harbin, China and the 2002 National Worshop on Progress of Wild Amur Tiger Population Recovery Action, Hunchun, China*. China Forestry Publishing House, Beijing, China.

Jiménez, M. A., Sanchez, B., Perez Alenza, M. D., *et al.* (2008) Membranous glomerulonephritis in the Iberian lynx (*Lynx pardinus*). *Veterinary Immunology and Immunopathology*, **121**, 34–43.

JNCC (2008) Special Areas of Conservation. <www.jncc.gov.uk/protectedsites/sacselection/SAC_List_features.asp?Unitary = &Feature = S1355>. Accessed online 24 Nov 2008.

Jobin, A., Molinari, P., and Breitenmoser, U. (2000) Prey spectrum, prey preference and consumption rates of Eurasian lynx in the Swiss Jura Mountains. *Acta Theriologica*, **45**, 243–52.

Johnsingh, A. J. T. and Madhusudan, M. D. (2009) Tiger reintroduction in India: conservation tool or costly dream. In M. W. Hayward and M. J. Somers (eds.), *The Reintroduction of Top-Order Predators*, pp. 146–63. Wiley-Blackwell Publishing, Oxford, UK.

Johnsingh, A. J. T. and Negi, A. S. (2003) Status of tiger and leopard in Rajaji-Corbett Conservation Unit, northern India. *Biological Conservation*, **111**, 385–93.

Johnsingh, A. J. T., Qureshi, Q., Goyal, S. P., *et al.* (2004) *Conservation Status of Tiger and Associated Species in the Terai Arc Landscape, India*. Wildlife Institute of India, Dehradun, India.

Johnson, P. J., Kansky, R., Loveridge, A. J. Macdonald, D. W. (submitted). Size, rarity and charisma: valuing African wildlife trophies. *Conservation Letters*.

Johnson, W. E. and Franklin, W. L. (1991) Feeding and spatial ecology of *Felis geoffroyi* in southern Patagonia. *Journal of Mammalogy*, **72**, 815–20.

Johnson, W. E. and O'Brien, S. J. (1997) Phylogenetic reconstruction of the Felidae using 16S rRNA and NADH-5 mitochondrial genes. *Journal of Molecular Evolution*, **44 (Supplement)**, 98–116.

Johnson, A., Vongkhamheng, C., Hedemark, M., and Saithongdam, T. (2006a) Effects of human-carnivore conflict on tiger (*Panthera tigris*) and prey populations in Lao PDR. *Animal Conservation*, **9**, 421–30.

Johnson, W. E., Eizirik, E., Pecon-Slattery, J., *et al.* (2006b) The late Miocene radiation of modern Felidae: a genetic assessment. *Science*, **311**, 73–7.

Johnson, W. E., Godoy, J. A., Palomares, F., *et al.* (2004) Phylogenetic and phylogeographic analysis of Iberian lynx populations. *Journal of Heredity*, **95**, 19–28.

Johnson, D. D., Baker, S., Morecroft, M. D., and Macdonald, D. W. (2001a) Long-term resource variation and group size: a large-sample field test of the resource dispersion hypothesis. *BMC Ecology*, **1**, 2.

Johnson, W. E., Eizirik, E., and Lento, G. M. (2001b) The control, exploitation and conservation of carnivores. In J. Gittleman, S. Funk, D. W. Macdonald, and R. Wayne (eds.), *Carnivore Conservation*, pp. 196–219. Cambridge University Press, Cambridge, UK.

Johnson, W. E., Eizirik, M., Roelke-Parker, M. and O'Brien, S. J. (2001c) Applications of genetic concepts and molecular methods to carnivore conservation. In J. L. Gittleman, S. M. Funk, D. W. Macdonald, and R. K. Wayne (eds.), *Carnivore Conservation*, pp. 335–58. Cambridge University Press, Cambridge, UK.

Johnson, W. E., Slattery, J. P., Eizirik, E., *et al.* (1999) Disparate phylogeographic patterns of molecular genetic variation in four closely related South American small cat species. *Molecular Ecology*, **8**, 79–94.

Johnson, W. E., O'Brien, S. J., Culver, M., Iriarte, J. A., Eizirik, E., and Seymour, K. L. (1998) Tracking the evolution of the elusive Andean mountain cat (*Oreailurus jacobita*) from mitochondrial DNA. *Journal of Heredity*, **89**, 227–32.

Johnson, W. E., Dratch, P. A., Martenson, J. S., and O'Brien, S. J. (1996) Resolution of recent radiations within three evolutionary lineages of Felidae using mitochondrial restriction fragment length polymorphism variation. *Journal of Mammalian Evolution*, **3**, 97–120.

Johnston, L. A., Donoghue, A. M., Obrien, S. J., and Wildt, D. E. (1991) Rescue and maturation *in vitro* of follicular oocytes collected from non-domestic felid species. *Biology of Reproduction*, **45**, 898–906.

Jorde, P. E., Rueness, E. K., Stenseth, N. C., and Jakobsen, K. S. (2006) Cryptic population structuring in Scandinavian lynx: Reply to Pamilo. *Molecular Ecology*, **15**, 1189–92.

Jorgenson, J. P. and Redford, K. H. (1993) Humans and big cats as predators in the Neotropics. *Symposia of the Zoological Society of London*, **65**, 367–90.

Joshi, A., Dinerstein, E., and Smith, D. (2002) The Terai Arc: managing tigers and other wildlife as metapopulations. In E. D. Wikramanayake, E. Dinerstein, C. J. Loucks, and S. Pimm (eds.), *Terrestrial Ecoregions of the Indo-Pacific: A Conservation Assessment*, pp. 178–81. Island Press, Washington DC.

Junge, R. E., Miller, R. E., Boever, W. J., Scherba, G., and Sundberg, J. (1991) Persistent cutaneous ulcers associated with feline herpesvirus type 1 infection in cheetah. *Journal of the American Veterinary Medical Association*, **198**, 1057–8.

Kaare, M. T., Picozzi, K., Mlengeya, T., *et al.* (2007) Sleeping sickness—a re-emerging disease in the Serengeti? *Travel Medicine and Infectious Diseases*, **5**, 117–24.

Kaczensky, P. (1999) Large carnivore depredation on livestock in Europe. *Ursus*, **11**, 59–71.

Kaczensky, P., Blazic, M., and Gossow, H. (2001) Content analysis of articles on brown bears in the Slovenian press, 1991–1998. In M. Hunziker and R. Landolt (eds.), *Humans and Predators in Europe—Research on How Society Is Coping with the Return of Wild Predators. Special Issue of Forest Snow and Landscape Research*, pp. 121–35. Swiss Federal Research Institute WSL, Birmensdorf, Switzerland.

Kadoi, K., Kiryu, M., Iwabuchi, M., Kamata, H., Yukawa, M., and Inaba, Y. (1997) A strain of calicivirus isolated from lions with vesicular lesions on tongue and snout. *New Microbiologist*, **20**, 141–8.

Kamler, J. F., Gipson, P. S., and Snyder, T. R. (2000) Dispersal characteristics of young bobcats from northeastern Kansas. *Southwestern Naturalist*, **45**, 543–6.

Kamstra, J. (1987) *An ecological survey of the Cockscomb Basin*. MA dissertation, York University, Toronto, ON.

Kanninen, M., Murdiyarso, D., Seymour, F., Angelsen, A., Wunder, S., and German, L. (2007) *Do Trees Grow on Money? The Implications of Deforestation Research for Policies to Promote REDD*. Center for International Forestry Research (CIFOR), Bogor, Indonesia.

Kaplanov, L. G. (1948) Tigers in Sikhote-Alin. In *Tiger, Red Deer, and Moose, Materialy k poznaniyu fauny i flory SSSR, Obschestva Ispytateley Prirody*, pp. 18–49. Izd. Mosk., Moscow, Russia.

Karanth, K. U. (1993) *Predator-prey relationships among large mammals in Nagarahole National Park, India*. PhD dissertation, Mangalore University, Mangalore, India.

Karanth, K. U. (1991) Ecology and management of the tiger in tropical Asia. In N. Maruyama, B. Bobek, Y. Ono, W. Regli, L. Bartos, and R. Ratcliffe (eds.), *Wildlife Conservation: Present Trends and Perspectives for the 21st Century*, pp. 156–9. Japan Wildlife Research Centre, Tokyo, Japan.

Karanth, K. U. and Gopal, R. (2005) An ecology-based policy framework for human-tiger co-existence in India. In R. Woodroffe, S. Thirgood, and A. Rabinowitz (eds.), *People and Wildlife: Conflict or Coexistence?*, pp. 373–87. Cambridge University Press, Cambridge, UK.

Karanth, K. U. and Madhusudan, M. D. (1997) Avoiding paper tigers and saving real tigers: Response to Saberwal. *Conservation Biology*, **11**, 818–20.

Karanth, K. U. and Nichols, J. D. (eds.) (2002) *Monitoring Tigers and Their Prey: A Manual for Researchers, Managers and Conservationists in Tropical Asia*. Centre for Wildlife Studies, Bangalore, India.

Karanth, K. U. and Nichols, J. D. (2000) *Ecological status and conservation of tigers in India*. Final technical report to the Division of International Conservation, U.S. Fish and Wildlife Service, Washington, DC and Wildlife Conservation Society, New York and Bangalore, India.

Karanth, K. U. and Nichols, J. D. (1998) Estimation of tiger densities in India using photographic captures and recaptures. *Ecology*, **79**, 2852–62.

Karanth, K. U. and Stith, B. M. (1999) Prey depletion as a critical determinant of tiger population viability. In J. Seidensticker, S. Christie, and P. Jackson (eds.) *Riding the Tiger: Tiger Conservation in Human-Dominated Landscapes*, pp. 100–13. Cambridge University Press, Cambridge, UK.

Karanth, K. U. and Sunquist, M. E. (2000) Behavioural correlates of predation by tiger (*Panthera tigris*), leopard (*Panthera pardus*) and dhole (*Cuon alpinus*) in Nagarahole, India. *Journal of Zoology*, **250**, 255–65.

Karanth, K. U. and Sunquist, M. E. (1995) Prey selection by tiger, leopard and dhole in tropical forests. *Journal of Animal Ecology*, **64**, 439–50.

Karanth, K. K., Nichols, J. D., Hines, J. E., Karanth, K. U., and Christensen, N. L. (in press). Patterns and determinants of mammal species occurrence in India. *Journal of Applied Ecology*.

Karanth, K. U., Nichols, J. D., Kumar, N. S., and Hines, J. E. (2006) Assessing tiger population dynamics using photographic capture-recapture sampling. *Ecology*, **87**, 2925–37.

Karanth, K. U., Chundawat, R. S., Nichols, J. D., and Kumar, N. S. (2004a) Estimation of tiger densities in the tropical dry forests of Panna, Central India, using photographic capture-recapture sampling. *Animal Conservation*, **7**, 257–63.

Karanth, K. U., Nichols, J. D., and Kumar, N. S. (2004b) Photographic sampling of elusive mammals in tropical forests. In W. L. Thompson (ed.), *Sampling Rare or Elusive*

Species: Concepts, Designs, and Techniques for Estimating Population Parameters, pp. 229–47. Island Press, Washington DC.

Karanth, K. U., Nichols, J. D., Seidensticker, J., *et al.* (2003) Science deficiency in conservation practice: the monitoring of tiger populations in India. *Animal Conservation*, **6**, 141–6.

Karanth, K. U., Nichols, J. D., Kumar, N. S., Link, W. A., and Hines, J. E. (2004c) Tigers and their prey: predicting carnivore densities from prey abundance. *Proceedings of the National Academy of Sciences of the United States of America*, **101**, 4854–8.

Kat, P. W., Alexander, K. A., Smith, J. S., Richardson, J. D., and Munson, L. (1996) Rabies among African wild dogs (*Lycaon pictus*) in the Masai Mara, Kenya. *Journal of Veterinary Diagnostic Investigation*, **8**, 420–6.

Kaup, J. J. (1833) *Déscription d'ossements fossiles de mammifères inconnus jusqu'a présent qui se trouvent au muséum grand-ducal de Darmstadt. Carnassiers fossiles*. J.G. Heyer, Darmstadt, Germany.

Kautz, R., Kawula, R., Hoctor, T., *et al.* (2006) How much is enough? Landscape-scale conservation for the Florida panther. *Biological Conservation*, **130**, 118–33.

Kawanishi, K. (2002) *Population status of tigers* (Panthera tigris) *in a primary rainforest of peninsular Malaysia*. PhD dissertation, University of Florida, Gainesville, FL.

Kawanishi, K. and Sunquist, M. E. (2003) Possible new records of fishing cat from Peninsular Malaysia. *Cat News*, **39**, 3–5.

Keawcharoen, J., Oraveerakul, K., Kuiken, T., *et al.* (2004) Avian influenza H5N1 in tigers and leopards. *Emerging Infectious Diseases*, **10**, 2189–91.

Keet, D. F., Kriek, N. P. J., Penrith, M. L., and Michel, A. (1997) Tuberculosis in lions and cheetahs. In J. van Heerden (ed.), *Lions and Leopards as Game Ranch Animals*, pp. 151–7. Wildlife Group of the South African Veterinary Association, Onderstepoort, South Africa.

Keet, D. F., Kriek, N. P. J., Penrith, M. L., Michel, A., and Huchzermeyer, H. (1996) Tuberculosis in buffaloes (*Syncerus caffer*) in the Kruger National Park: spread of the disease to other species. *Onderstepoort Journal of Veterinary Research*, **63**, 239–44.

Keith, L. B. (1963) *Wildlife's Ten-Year Cycle*. University of Wisconsin Press, Madison, WI.

Keith, L. B., Todd, A. W., Brand, C. J., Adamcik, R. S., and Rusch, D. H. (1977) An analysis of predation during a cyclic fluctuation of snowshoe hares. *Proceedings of the International Congress of Game Biologists*, **13**, 151–75.

Kellert, S. R., Black, M., Rush, C. R., and Bath, A. J. (1996) Human culture and large carnivore conservation in North America. *Conservation Biology*, **10**, 977–90.

Kelly, M. J. (2001) Lineage loss in Serengeti cheetahs: consequences of high reproductive variance and heritability of fitness on effective population size. *Conservation Biology*, **15**, 137–47.

Kelly, M. J. and Durant, S. M. (2000) Viability of the Serengeti cheetah population. *Conservation Biology*, **14**, 786–97.

Kelly, M. J., Noss, A. J., Di Bitetti, M. S., *et al.* (2008) Estimating puma densities from remote cameras across three study sites: Bolivia, Argentina, and Belize. *Journal of Mammalogy*, **89**, 408–18.

Kelt, D. A. and van Vuren, D. H. (2001) The ecology and macroecology of mammalian home range area. *American Naturalist*, **157**, 637–45.

Kemp, T. S. (2004) *The Origin and Evolution of Mammals*. Oxford University Press, Oxford, UK.

Kennedy, M., Kania, S., Stylianides, E., Bertschinger, H. J., Keet, D. F., and van Vuuren, M. (2003) Detection of feline coronavirus infection in southern African nondomestic felids. *Journal of Wildlife Diseases*, **39**, 529–35.

Kenny, D. E., Knightly, F., Baier, J., Lappin, M. R., Brewer, M., and Getzy, D. M. (2002) Toxoplasmosis in Pallas's cats (*Otocolobus felis manul*) at the Denver Zoological Gardens. *Journal of Zoo and Wildlife Medicine*, **33**, 131–8.

Kenney, J. S., Smith, J. L. D., Starfield, A. M., and McDougal, C. W. (1995) The long-term effects of tiger poaching on population viability. *Conservation Biology*, **9**, 1127–33.

Kenney, J. S., Smith, J. L. D., Starfield, A. M., and McDougal, C. W. (1994) Saving the tiger in the wild. *Nature*, **369**, 352–3.

Kenward, R. E. (1990) *Ranges IV. Software for Analysing Animal Location Data*. Institute of Terrestrial Ecology, Wareham, UK.

Kenward, R. E. (1987) *Wildlife Radio Tagging: Equipment, Field Techniques and Data Analysis*. Academic Press, London, UK.

Kenward, R. E. and Hodder, K. H. (1996) *Ranges V*. Institute of Terrestrial Ecology, Wareham, UK.

Kenward, R. E., South, A., and Walls, S. (2003) *Ranges VII (v0.52): For the Analysis of Tracking and Location Data*. Anatrack Ltd, Wareham, UK.

Kerby, G. and Macdonald, D. W. (1988) Cat society and the consequences of colony size. In D. C. Turner and P. Bateson (eds.), *The Domestic Cat: The Biology of Its Behaviour*, pp. 67–81. Cambridge University Press, Cambridge, UK.

Kerley, L. L. and Salkina, G. P. (2007) Using scent-matching dogs to identify individual Amur tigers from scats. *Journal of Wildlife Management*, **71**, 1349–56.

Kerley, L. L., Goodrich, J. M., Miquelle, D. G., Smirnov, E. N., Quigley, H. B., and Hornocker, M. G. (2005) Reproductive parameters of wild female Amur tigers. In D. G. Miquelle, E. N. Smirnov, and J. M. Goodrich (eds.), *Tigers in Sikhote-Alin Zapovednik: Ecology and Conservation*, pp. 61–9. PSP, Vladivostok, Russia.

Kerley, L. L., Goodrich, J. M., Miquelle, D. G., Smirnov, E. N., Quigley, H. B., and Hornocker, M. G. (2003) Reproductive parameters of wild female Amur (Siberian) tigers (*Panthera tigris altaica*). *Journal of Mammalogy*, **84**, 288–98.

Kerley, L. L., Goodrich, J. M., Miquelle, D. G., Smirnov, E. N., Quigley, H. B., and Hornocker, M. G. (2002) Effects of roads and human disturbance on Amur tigers. *Conservation Biology*, **16**, 97–108.

Kerpaleletswe, C. and Lovett, J. (2001) The role of common pool resources in economic welfare of rural households. Working paper. University of York, York, UK.

Ketz-Riley, C. J., Barrie, M. T., Ritchey, J. W., Hoover, J. P., and Johnson, C. M. (2003a) Immunodeficiency associated with multiple concurrent infections in captive Pallas's cats (*Otocolobus manul*). *Journal of Zoo and Wildlife Medicine*, **34**, 239–45.

Ketz-Riley, C. J., Reichard, M. V., Van Den Bussche, R. A., Hoover, J. P., Meinkoth, J., and Kocan, A. A. (2003b) An intraerythrocytic small prioplasm in wild-caught Pallas's cats (*Otocolobus manul*) from Mongolia. *Journal of Wildlife Diseases*, **39**, 424–30.

Keuroghlian, A. (2003) The response of peccaries to seasonal fluctuations in the Pantanal of Rio Negro. In *Earthwatch Institute Annual Report 2003*, pp. 55–9. Earthwatch and CRI-Pantanal, Maynard, MA.

Kgathi, D. K. and Kalikawe, M. C. (1993) Seasonal distribution of zebra and wildebeest in Makgadikgadi Pans Game Reserve. *African Journal of Ecology*, **31**, 210–19.

Khan, M. M. H. (2007) *Project Sundarbans Tiger: Tiger Density and Tiger–Human Conflict*. Save The Tiger Fund, National Fish and Wildlife Foundation, Washington DC.

Khan, M. M. H. (2004) *Ecology and conservation of the Bengal tiger in the Sundarbans mangrove forest of Bangladesh*. DPhil dissertation, University of Cambridge, Cambridge, UK.

Kie, J. G., Baldwin, J. A., and Evans, C. J. (1996) CALHOME: a program for estimating animal home ranges. *Wildlife Society Bulletin*, **24**, 342–4.

Kiley-Worthington, M. (1984) Animal language? Vocal communication of some ungulates, canids and felids. *Acta Zoologica Fennica*, **171**, 83–8.

Kilshaw, K. and Macdonald, D.W. (2010) *Reintroduction of the Scottish Wildcat into the Scottish Highlands: A Feasibility Study*. WildCRU, Oxford, UK.

Kilshaw, K., Macdonald, D. W. and Kitchener, A. (2008) *Feral cat management in the Cairngorms: scoping study*. Report to Scottish Natural Heritage, Inverness, Scotland.

Kiltie, R. A. (1988) Interspecific size regularities in tropical field assemblages. *Oecologia*, **76**, 97–105.

Kiltie, R. A. (1984) Size ratios among sympatric neotropical cats. *Oecologia*, **61**, 411–16.

Kimura, K. (1999) Diachronous evolution of sub-Himalayan piggyback basins, Nepal. *Island Arc*, **8**, 99–113.

King, L. E., Douglas-Hamilton, I., and Vollrath, F. (2007) African elephants run from the sound of disturbed bees. *Current Biology*, **17**, R832–3.

Kingdon, J. (1977) *East African Mammals. An Atlas of Evolution in Africa. Vol. IIIA. Carnivores*. Academic Press, London, UK.

Kinnaird, M. F., Sanderson, E. W., O'Brien, S. J., Wibisono, H. T., and Woolmer, G. (2003) Deforestation trends in a tropical landscape and implications for endangered large mammals. *Conservation Biology*, **17**, 245–57.

Kirberger, R. M., Keet, D. F., and Wagner, W. M. (2006) Radiologic abnormalities of the appendicular skeleton of the lion (*Panthera leo*): incidental findings and *Mycobacterium bovis*-induced changes. *Veterinary Radiology and Ultrasound*, **47**, 145–52.

Kirilyuk, V. and Puzansky, V. N. (2000) The Amur tiger makes a surprise reappearance. *Russian Conservation News*, **23**, 22–3.

Kirkwood, J. K., Cunningham, A. A., Flach, E. J., Thornton, S. M., and Wells, G. A. H. (1995) Spongiform encephalopathy in another captive cheetah (*Acinonyx jubatus*): evidence for variation in susceptibility or incubation periods between species? *Journal of Zoo and Wildlife Medicine*, **26**, 577–82.

Kissui, B. M. and Packer, C. (2004) Top-down population regulation of a top predator: lions in the Ngorogoro Crater. *Proceedings of the Royal Society of London Series B*, **271**, 1867–74.

Kitchener, A. (1995) *The Wildcat*. The Mammal Society, London, UK.

Kitchener, A. (1991) *The Natural History of the Wild Cats*. Christopher Helm, London, UK.

Kitchener, A. C. (2002) Two tales of two kitties: restoring the wildcat, *Felis silvestris*, and the lynx, *Lynx lynx*, to Britain. In C. P. Bowen (ed.), *Conference Proceedings 2001–2002*, pp. 24–7. People's Trust for Endangered Species and Mammals Trust UK, London, UK.

Kitchener, A. C. (2000) Are cats really solitary? *Lutra*, **43**, 1–10.

Kitchener, A. C. (1999) Tiger distribution, phenotypic variation and conservation issues. In J. Seidensticker, S. Christie, and P. Jackson (eds.), *Riding the Tiger: Tiger Conservation in Human-Dominated Landscapes*, pp. 19–39. Cambridge University Press, Cambridge, UK.

Kitchener, A. C. (1998) The Scottish wildcat—a cat with an identity crisis. *British Wildlife*, **9**, 232–42.

Kitchener, A. C. and Bonsall, C. (1999) Further AMS radiocarbon dates for extinct Scottish mammals. *Quaternary Newsletter*, **88**, 1–10.

Kitchener, A. C. and Dugmore, A. J. (2000) Biogeographical change in the tiger, *Panthera tigris*. *Animal Conservation*, **3**, 113–24.

Kitchener, A. C. and Easterbee, N. (1992) The taxonomic status of black wild felids in Scotland. *Journal of Zoology*, **227**, 342–6.

Kitchener, A. C., Richardson, D., and Beaumont, M. A. (2007) A new old clouded leopard. *Cat News*, **46**, 26–7.

Kitchener, A. C., Beaumont, M. A., and Richardson, D. (2006) Geographical variation in the clouded leopard, *Neofelis nebulosa*, reveals two species. *Current Biology*, **16**, 2377–83.

Kitchener, A. C., Yamaguchi, N., Ward, J. M., and Macdonald, D. W. (2005) A diagnosis for the Scottish wildcat (*Felis silvestris*): a tool for conservation action for a critically-endangered felid. *Animal Conservation*, **8**, 223–37.

Kitchener, A. C., Yasuma, S., Andau, M., and Quillen, P. (2004) Three bay cats (*Catopuma badia*) from Borneo. *Mammalian Biology*, **69**, 349–53.

Kitchings, J. T. and Story, J. D. (1984) Movements and dispersal of bobcats in east Tennessee. *Journal of Wildlife Management*, **48**, 957–61.

Kittle, A. and Watson, A. (2004) Rusty-spotted cat in Sri Lanka: observations of an arid zone population. *Cat News*, **40**, 17–19.

Kleiman, D. G. and Eisenberg, J. F. (1973) Comparisons of canid and felid social systems from an evolutionary perspective. *Animal Behaviour*, **21**, 637–59.

Knick, S. T. (1990) Ecology of bobcats relative to exploitation and a prey decline in southeastern Idaho. *Wildlife Monographs*, **108**, 1–42.

Knighton, A. D. and Nanson, G. C. (2000) Waterhole form and process in the anastomosing channel system of Cooper Creek, Australia. *Geomorphology*, **35**, 101–17.

Knowles, L. L. (2004) The burgeoning field of statistical phylogeography. *Journal of Evolutionary Biology*, **17**, 1–10.

Kock, R., Chalmers, W. S. K., Mwanzia, J., *et al.* (1998) Canine distemper antibodies in lions of the Masai Mara. *Veterinary Record*, **142**, 662–5.

Koehler, A. E., Marsh, R. E., and Salmon, T. P. (1990) Frightening methods and devices/stimuli to prevent mammal damage—a review. In L. R. Davis and R. E. Marsh (eds.), *Proceedings of the 14th Vertebrate Pest Conference*, pp. 168–73. University of California, Davis, CA.

Koehler, G. M. (1990) Population and habitat characteristics of lynx and snowshoe hares in north central Washington. *Canadian Journal of Zoology*, **68**, 845–51.

Koehler, G. M. and Hornocker, M. G. (1991) Seasonal resource use among mountain lions, bobcats, and coyotes. *Journal of Mammalogy*, **72**, 391–6.

Koford, C. B. (1973) Spotted cats in South America: an interim report. *Oryx*, **12**, 37–9.

Kohn, M. H., Murphy, W. J., Ostrander, E. A., and Wayne, R. K. (2006) Genomics and conservation genetics. *Trends in Ecology & Evolution*, **21**, 629–37.

Kolmstetter, C., Munson, L., and Ramsay, E. C. (2000) Degenerative spinal disease in large felids. *Journal of Zoo and Wildlife Medicine*, **31**, 15–19.

Kolowski, J. M. and Holekamp, K. E. (2006) Spatial, temporal, and physical characteristics of livestock depredation by large carnivores along a Kenyan reserve border. *Biological Conservation*, **128**, 529–41.

Konecny, M. J. (1989) Movement patterns and food habits of four sympatric carnivore species in Belize, Central America. In K. H. Redford and J. F. Eisenberg (eds.) *Advances in Neotropical Mammalogy*, pp. 243–64. Sandhill Crane Press, Gainesville, FL.

Kontoleon, A. and Swanson, T. (2003) The willingness to pay for property rights for the Giant Panda: Can a charismatic species be an instrument for nature conservation? *Land Economics*, **79**, 483–99.

Koshkarev, E. P. (1994) Snow leopard poaching in Central Asia. *Cat News*, **21**, 18.

Koshkarev, E. P. and Vyrypaev, V. (2000) The snow leopard after the break-up of the Soviet Union. *Cat News*, **32**, 9–11.

Krebs, C. J., Boutin, S., and Boonstra, R. (eds.) (2001) *Ecosystem Dynamics of the Boreal Forest: The Kluane Project*. Oxford University Press, Oxford, UK.

Krebs, C. J., Boutin, S., Boonstra, R., *et al.* (1995) Impact of food and predation on the snowshoe hare cycle. *Science*, **269**, 1112–15.

Krebs, C. J., Gilbert, B. S., Boutin, S., and Boonstra, R. (1987) Estimation of snowshoe hare population density from turd transects. *Canadian Journal of Zoology*, **65**, 565–7.

Kreeger, T. J., Arnemo, J. M., and Raath, J. P. (2002) *Handbook of Wildlife Chemical Immobilization*. Wildlife Pharmaceuticals, Fort Collins, CO.

Kretzoi, M. (1938) Die Raubtiere von Gombaszög nebst einer Übersicht der Gesamtfauna. *Annales Historico-Naturales Musei Nationalis Hungarici*, **31**, 89–157.

Kretzoi, M. (1929a) Feliden-Studien. *A Magyar Királyi Földtani Intézet Hazinyomdaja*, **24**, 1–22.

Kretzoi, M. (1929b) Materialien zur phylogenetischen Klassifikation der Aeluroïdeen. *Xe Congrès International de Zoologie*, **2**, 1293–355.

Kruuk, H. (2002) *Hunter and Hunted: Relationships between Carnivores and People*. Cambridge University Press, Cambridge, UK.

Kruuk, H. (1978) Foraging and spatial organisation of the European badger, *Meles meles* L. *Behavioral Ecology and Sociobiology*, **4**, 75–89.

Kruuk, H. (1976) Feeding and social behaviour of the striped hyena (*Hyaena vulgaris Desma-rest*). *East African Wildlife Journal*, **14**, 91–112.

Kruuk, H. (1972) Surplus killing by carnivores. *Journal of Zoology*, **166**, 233–44.

Kruuk, H. and Macdonald, D. W. (1985) Group territories of carnivores: empires and enclaves. In R. M. Sibly and R. H. Smith (eds.), *Behavioural Ecology*, pp. 521–36. Blackwell Scientific Publications, Oxford, UK.

Kruuk, L. E. B., Sheldon, B. C., and Merila, J. (2002) Severe inbreeding depression in collared flycatchers (*Ficedula albicollis*). *Proceedings of the Royal Society of London Series B*, **269**, 1581–9.

Kumar, A. and Wright, B. (1999) Combating tiger poaching and illegal wildlife trade in India. In J. Seidensticker, S. Christie, and P. Jackson (eds.), *Riding the Tiger: Tiger Conservation in Human Dominated Landscapes*, pp. 242–51. Cambridge University Press, Cambridge, UK.

Kumara, H. N. and Singh, M. (2004) The influence of differing hunting practices on the relative abundance of mammals in two rainforest areas of the Western Ghats, India. *Oryx*, **38**, 321–7.

Kurose, N., Masuda, R., and Tatara, M. (2005) Fecal DNA analysis for identifying species and sex of sympatric carnivores: a noninvasive method for conservation on the Tsushima Islands, Japan. *Journal of Heredity*, **96**, 688–97.

Kurt, F. (1991) *Das Rehwild in der Kulturlandschaft—Sozialverhalten und Ökologie eines Anpassers*. Verlag Paul Parey, Hamburg, Germany.

Kurtén, B. (1976) Fossil Carnivora from the late tertiary of Bled Douarah and Cherichira, Tunisia. *Notes du Service Géologique de Tunisie*, **42**, 177–214.

Kurtén, B. (1973) Geographic variation in size in the puma (*Felis concolor*). *Commentationes Biologicae Societas Scientiarum Fennica*, **63**, 1–8.

Kurtén, B. (1968) *Pleistocene Mammals of Europe*. Weidenfeld and Nicholson, London, UK.

Kurtén, B. (1965a) The Carnivora of the Palestine caves. *Acta Zoolica Fennica*, **107**, 3–74.

Kurtén, B. (1965b) On the evolution of the European wildcat, *Felis silvestris* Schreber. *Acta Zoolica Fennica*, **111**, 3–29.

Kurtén, B. (1962) The relative ages of the australopithecines of Transvaal and the pithecanthropines of Java. In G. Kurth (ed.), *Evolution and Hominization*, pp. 74–80. Gustav Fischer Verlag, Stuttgart, Germany.

Kurtén, B. and Anderson, E. (1980) *Pleistocene Mammals of North America*. Columbia University Press, New York.

Kushnir H., Leitner H., Ikanda D., and Packer, C. (in press) Human and ecological risk factors for unprovoked lion attacks on humans in southeastern Tanzania. *Human Dimensions of Wildlife* **15**.

Laack, L. L. (1991) *Ecology of ocelot* (Felis pardalis) *in south Texas*. MS dissertation, Texas A&I University, Kingsville, TX.

Labuschagne, W. (1979) *'N Bio-ekologiese en gedragstudie van de jagliuperd* Acinonyx jubatus jubatus (*Schreber 1775*). MSc dissertation, University of Pretoria, Pretoria, South Africa.

Lacy, R. C. (1997) Importance of genetic variation to the viability of mammalian populations. *Journal of Mammalogy*, **78**, 320–35.

Lacy, R. C. (1994) Managing genetic diversity in captive populations of animals. In M. L. Bowles and C. J. Whelan (eds.), *Restoration of Endangered Species*, pp. 63–83. Cambridge University Press, Cambridge, UK.

Lacy, R. C. and Vargas, A. (2004) *Informe sobre la gestión genética y demográfica del programa de cría para la conservación del lince Ibérico: escenarios, conclusiones y recomendaciones*. Conservation Breeding Specialist Group, Apple Valley, MN and Ministerio de Medio Ambiente, Madrid, Spain.

Laikre, L. and Ryman, N. (1991) Inbreeding depression in a captive wolf (*Canis lupus*) population. *Conservation Biology*, **5**, 33–40.

Laikre, L., Andren, R., Larsson, H. O., and Ryman, N. (1996) Inbreeding depression in brown bear *Ursus arctos*. *Biological Conservation*, **76**, 69–72.

Laing, S. P. and Lindzey, F. G. (1993) Patterns of replacement of resident cougars in southern Utah. *Journal of Mammalogy*, **74**, 1056–8.

Lamberski, N. and West, C. (2007) Role of veterinarians in regional management plans. *International Zoo Yearbook*, **41**, 16–23.

Lambert, C. M. S., Wielgus, R. B., Robinson, H. S., Katnik, D. D., Cruickshank, H. S., Clarkeand, R., and Almack, J. (2006) Cougar population dynamics and viability in the Pacific Northwest. *Journal of Wildlife Management*, **70**, 246–54.

Land, E. D. and Lacy, R. C. (2000) Introgression level achieved through Florida panther genetic restoration. *Endangered Species Update*, **17**, 99–103.

Land, E. D., Shindle, D. B., Kawula, R. J., Benson, J. F., Lotz, M. A., and Onorato, D. P. (2008) Florida panther habitat selection analysis of concurrent GPS and VHF telemetry data. *Journal of Wildlife Management*, **72**, 633–9.

Land, E. D., Maehr, D. S., Roof, J. C., and McCown, J. W. (1993) Mortality patterns of female white-tailed deer in southwest Florida. *Proceedings of the Annual Conference of Southeastern Fish and Wildlife Agencies*, **47**, 176–84.

Lande, R. (1988) Genetics and demography in biological conservation. *Science*, **241**, 1455–60.

Lande, R. and Barrowclough, G. F. (1987) Effective population size, genetic variation, and their use in population management. In M. E. Soule (ed.), *Viable Populations for Conservation*, pp. 87–124. Cambridge University Press, Cambridge, UK.

Lande, R., Engen, S., and Saether, B. E. (2003) *Stochastic Population Dynamics in Ecology and Conservation*. Oxford University Press, Oxford, UK.

Landry, J. M. (1999) *The Use of Guard Dogs in the Swiss Alps: A First Analysis*. KORA, Muri, Switzerland.

Langley, P. J. W. and Yalden, D. W. (1977) The decline of the rarer carnivores in Great Britain during the nineteenth century. *Mammal Review*, **7**, 95–116.

Lantos, A., Niemann, S., Mezosi, L., *et al.* (2003) Pulmonary tuberculosis due to *Mycobacterium bovis* subsp. *caprae* in captive Siberian tiger. *Emerging Infectious Diseases*, **9**, 1462–4.

Lapointe, E., Conrad, K., Mitra, B., and Jenkins, H. (2007) *Tiger Conservation: It's Time to Think Outside the Box*. IWNC World Conservation trust, Lausanne, Switzerland.

Lariviere, S. and Walton, L. R. (1997) *Lynx rufus*. *Mammalian Species*, **563**, 1–8.

Larom, D., Garstang, M., Payne, K., Raspet, R., and Lindeque, M. (1997) The influence of surface atmospheric conditions on the range and area reached by animal vocalizations. *Journal of Experimental Biology*, **200**, 421–31.

Laundré, J. and Clark, T. W. (2003) Managing puma hunting in the western United States: through a metapopulation approach. *Animal Conservation*, **6**, 159–70.

Laurance, W. F. (2007) A new initiative to use carbon trading for tropical forest conservation. *Biotropica*, **39**, 20–4.

Laurance, W. F. (2000) Do edge effects occur over large spatial scales? *Trends in Ecology and Evolution*, **15**, 134–5.

Laurenson, M. K. (1995a) Behavioural costs and constraints of lactation in free-living cheetahs. *Animal Behaviour*, **50**, 815–26.

Laurenson, M. K. (1995b) Cub growth and maternal care in cheetahs. *Behavioral Ecology*, **6**, 405–9.

Laurenson, M. K. (1995c) Implications of high offspring mortality for cheetah population dynamics. In A. R. E. Sinclair and P. Arcese (eds.), *Serengeti II: Dynamics, Management and Conservation of an Ecosystem*, pp. 385–99. University of Chicago Press, Chicago.

Laurenson, M. K. (1994) High juvenile mortality in cheetahs (*Acinonyx jubatus*) and its consequences for maternal care. *Journal of Zoology*, **234**, 387–408.

Laurenson, M. K. (1993) Early maternal behavior of wild cheetahs: implications for captive husbandry. *Zoo Biology*, **12**, 31–43.

Laurenson, M. K. (1992) *Reproductive strategies in wild female cheetahs*. DPhil dissertation, University of Cambridge, Cambridge, UK.

Laurenson, M. K., Wielebnowski, N., and Caro, T. M. (1995) Extrinsic factors and juvenile mortality in cheetahs. *Conservation Biology*, **9**, 1329–31.

Laver, P. N. (2005) *Cheetah of the Serengeti Plains: a home range analysis*. MSc dissertation, Virginia Polytechnic Institute and State University, Blacksburg, VA.

Lawhead, D. N. (1984) Bobcat *Lynx rufus* home range density and habitat preference in south central Arizona USA. *Southwestern Nauralist*, **29**, 105–14.

Lawson, D. and Mafela, P. (1990) Development of the WMA concept. In A. Kiss (ed.), *Living with Wildlife: Wildlife Resources Management with Local Participation in Africa*. World Bank, Washington DC.

Lay, D. (1972) Anatomy, physiology, functional significance and evolution of specialized hearing organs of gerbilline rodents. *Journal of Morphology*, **138**, 41–120.

Leader-Williams, N. and Hutton, J. M. (2005) Does extractive use provide opportunities to offset conflicts between people and wildlife? In R. Woodroffe, S. Thirgood, and A. Rabinowitz (eds.), *People and Wildlife: Conflict or Co-Existence?*, pp. 140–61. Cambridge University Press, Cambridge, UK.

Lee-Thorp, J., Thackeray, J. F., and van der Merwe, N. (2000) The hunters or the hunted revisited. *Journal of Human Evolution*, **39**, 565–76.

Lee, I. T., Levy, J. K., Gorman, S. P., Crawford, P. C., and Slater, M. R. (2002) Prevalence of feline leukemia virus infection and serum antibodies against feline immunodeficiency virus in unowned free-roaming cats. *Journal of the American Veterinary Medical Association*, **220**, 620–2.

Legay, J. M. (1986) Tentative estimation of the total number of domestic cats in the world. *Comptes Rendues des Academies de Sciences III*, **303**, 709–12.

Lehmkuhl, J. (1989) *The ecology of south-Asian tall-grass communities*. PhD dissertation, University of Washington, Seattle, WA.

Leistner, O. A. (1967) The plant ecology of the Southern Kalahari [Africa]. *Botanical Survey South Africa Memoirs*, **38**, 1–172.

Leite, M. R. P. and Galvão, F. (2002) El jaguar, el puma y el hombre en tres áreas protegidas del bosque atlántico costero de paraná, Brasil. In R. A. Medellín, C. Equihua, C. L. B. Chetkiewicz, *et al.* (eds.), *El Jaguar en el Nuevo Milenio.*, pp. 237–50. Fondo de Cultura Economica, Universidad Nacional Autónoma de México and Wildlife Conservation Society, Mexico City, Mexico.

Leite, M. R. P., Boulhosa, R. L. P., Galvão, F., and Cullen, L. (2002) Conservacion del jaguar en las areas protegidas del bosque atlantico de la costa de Brasil. In R. A. Medellín, C. Equihua, C. L. B. Chetkiewicz, *et al.* (eds.), *El Jaguar en el Nuevo Milenio*, pp. 25–42. Fondo de Cultura Economica, Universidad Nacional Autónoma de México and Wildlife Conservation Society, Mexico City, Mexico.

Lekagul, B. and McNeely, J. A. (1988) *Mammals of Thailand*, 2nd edn. Darnsutha Press, Bangkok, Thailand.

Lélé, S. and Norgaard, R. B. (2005) Practicing interdisciplinarity. *BioScience*, **55**, 967–75.

Lembo, T. K., Hampson, K., Haydon, D. T., *et al.* (2008) Exploring reservoir dynamics: a case study of rabies in the Serengeti ecosystem. *Journal of Applied Ecology*, **45**, 1246–57.

León, T., Jones, J., and Soria, D. (2006) Iberian lynx cell bank: biological reserves and their application to the conservation of the species. In *Iberian Lynx Ex-Situ Conservation Seminar Series: Book of Proceedings*, pp. 106–8. Fundación Biodiversidad, Sevilla and Doñana, Spain.

Leutenegger, C. M., Hofmann-Lehmann, R., Riols, C., *et al.* (1999) Viral infections in free-living population of the European wildcat. *Journal of Wildlife Diseases*, **35**, 678–86.

Lévi-Strauss, C. (1987) *Anthropology and Myth: Lectures, 1951–1982*. Wiley-Blackwell, Oxford, UK.

Lewis, D., Kaweche, G. B., and Mwenya, A. (1990) Wildlife Conservation outside protected areas—lessons from an experiment in Zambia. *Conservation Biology*, **4**, 171–8.

Lewis, D. M. and Alpert, P. (1997) Trophy hunting and wildlife conservation in Zambia. *Conservation Biology*, **11**, 59–68.

Lewis, D. M. and Jackson, J. (2005) Safari hunting and conservation on communal land in southern Africa. In R. Woodroffe, S. Thirgood, and A. Rabinowitz (eds.), *People and Wildlife: Conflict or Coexistence?*, pp. 239–51. Cambridge University Press, Cambridge, UK.

Lewis, J. C. M. (1994) Anesthesia of nondomestic cats. In L. W. Hall and P. M. Taylor (eds.), *Anesthesia of the Cats*, pp. 310–49. Baillière Tindall, London, UK.

Lewis, J. S., Rachlow, J. L., Garton, E. O., and Vierling, L. A. (2007) Effects of habitat on GPS collar performance: using data screening to reduce location error. *Journal of Applied Ecology*, **44**, 663–71.

Lewison, R., Fitzhugh, E. L., and Galentine, S. P. (2001) Validation of a rigorous track classification technique: identifying individual mountain lions. *Biological Conservation*, **99**, 313–21.

Leyhausen, P. (1979) *Cat Behavior: The Predatory and Social Behaviour of Domestic and Wild Cats*. Garland STPM Press, New York.

Leyhausen, P. (1965) The communal organisation of solitary mammals. *Symposia of the Zoological Society of London*, **14**, 249–63.

Leyhausen, P. and Tonkin, B. (1966) Breeding the Black-footed cat (*Felis nigripes*). *International Zoo Yearbook*, **6**, 178–82.

Lezmi, S., Bencsik, A., Monks, E., Petit, T., and Baron, T. (2003) First case of feline spongiform encephalopathy in a captive cheetah born in France: PrP^{sc} analysis in various tissues revealed unexpected targeting of kidney and adrenal gland. *Histochemistry and Cell Biology*, **119**, 415–22.

Li, W. and Wang, H. (1999) Wildlife trade in Yunnnan Province, China, at the border with Vietnam. *TRAFFIC Bulletin*, **18**, 21–30.

Li, X., Li, W., Cao, J., *et al.* (2005) Pseudogenization of a sweet-receptor gene accounts for cats' indifference toward sugar. *PLoS Genetics*, **1**, 27–35.

Liberg, O. (1980) Spacing patterns in a population of rural free roaming domestic cats. *Oikos*, **35**, 336–49.

Liberg, O., Sandell, M., Pontier, D., and Natoli, E. (2000) Density, spatial organisation and reproductive tactics in the domestic cat and other felids. In D. C. Turner and P. Bateson (eds.), *The Domestic Cat: The Biology of Its Behaviour*, pp. 119–47. Cambridge University Press, Cambridge, UK.

Lindeque, P. M. and Turnbull, P. C. (1994) Ecology and epidemiology of anthrax in the Etosha National Park, Namibia. *Onderstepoort Journal of Veterinary Research*, **61**, 71–83.

Lindeque, P. M., Brain, C., and Turnbull, P. C. B. (1996) A review of anthrax in the Etosha National Park. *Salisbury Medical Bulletin*, **87**, 24–6.

Lindsey, P. A., Romañach, S., and Davies-Mostert, H. (2009) A synthesis of early indicators of predator conservation on private land in South Africa. In M. W. Hayward and M. J. Somers (eds.), *The Reintroduction of Top-Order Predators*, pp. 321–44. Wiley-Blackwell Publishing, Oxford, UK.

Lindsey, P. A., Frank, L. G., Alexander, R., Mathieson, A., and Romanach, S. S. (2007) Trophy hunting and conservation in Africa: problems and one potential solution. *Conservation Biology*, **21**, 880–3.

Lindsey, P. A., Alexander, R., Frank, L. G., Mathieson, A., and Romanach, S. S. (2006) Potential of trophy hunting to create incentives for wildlife conservation in Africa where alternative wildlife-based land uses may not be viable. *Animal Conservation*, **9**, 283–91.

Lindsey, P. A., Du Toit, J. T., and Mills, M. G. L. (2005) Attitudes of ranchers towards African wild dogs *Lycaon pictus*: conservation implications on private land. *Biological Conservation*, **125**, 113–21.

Lindstedt, S. L., Miller, B. J., and Buskirk, S. W. (1986) Home range, time, and body size in mammals. *Ecology*, **67**, 413–18.

Lindzey, F. G., van Sickle, W. D., Laing, S. P., and Mecham, C. S. (1992) Cougar population response to manipulation in southern Utah. *Wildlife Society Bulletin*, **20**, 224–7.

Lindzey, F. G., Ackerman, B. B., Barnhurst, D., and Hemker, T. P. (1988) Survival rates of mountain lions in southern Utah. *Journal of Wildlife Management*, **52**, 664–7.

Linkie, M., Chapron, G., Martyr, D. J., Holden, J., and Leader-Williams, N. (2006) Assessing the viability of tiger subpopulations in a fragmented landscape. *Journal of Applied Ecology*, **43**, 576–86.

Linnell, J. D. C. and Brøseth, H. (2003) Compensation for large carnivore depredation of domestic sheep 1994–2001. *Carnivore Damage Prevention News*, **6**, 11–13.

Linnell, J. D. C., Breitenmoser, U., Breitenmoser-Würsten, C., Odden, J., and von Arx, M. (2009) Recovery of Eurasian lynx in Europe: what part has reintroduction played? In M. W. Hayward and M. J. Somers (eds.), *The Reintroduction of Top-Order Predators*, pp. 72–91. Wiley-Blackwell Publishing, Oxford, UK.

Linnell, J. D. C., Nilsen, E. B., Lande, U. S., *et al.* (2005) Zoning as a means of mitigating conflicts with large carnivores: principles and reality. In R. Woodroffe, S. Thirgood, and A. Rabinowitz (eds.), *People and Wildlife: Conflict and Coexistence*, pp. 162–75. Cambridge Univesrity Press, Cambridge, UK.

Linnell, J. D. C., Andersen, R., Kvam, T., *et al.* (2001) Home range size and choice of management strategy for lynx in Scandinavia. *Environmental Management*, **27**, 869–79.

Linnell, J. D. C., Odden, J., Smith, M. E., Aanes, R., and Swenson, J. E. (1999) Large carnivores that kill livestock: do 'problem individuals' really exist? *Wildlife Society Bulletin*, **27**, 698–705.

Linnell, J. D. C., Odden, J., Pedersen, V., and Andersen, R. (1998) Records of intra-guild predation by Eurasian Lynx, *Lynx lynx*. *Canadian Field-Naturalist*, **112**, 707–8.

Linnell, J. D. C., Smith, M. E., Odden, J., Kaczensky, P., and Swenson, J. E. (1996) *Strategies for the Reduction of Carnivore-Livestock Conflicts: A Review.* NINA Oppdragsmelding 443. NINA, Trondheim, Norway.

Litvaitis, J. A., Major, J. T., and Sherburne, J. A. (1987) Influence of season and human induced mortality on spatial organisation of bobcats (*Felis rufus*) in Maine. *Journal of Mammalogy*, **68**, 100–6.

Litvaitis, J. A., Sherburne, J. A., and Bissonette, J. A. (1986) Bobcat habitat use and home range size in relation to prey density. *Journal of Wildlife Management*, **50**, 110–17.

Lloyd-Smith, J. O., Schreiber, S. J., Kopp, P. E., and Getz, W. M. (2005) Superspreading and the effect of individual variation on disease emergence. *Nature*, **438**, 355–9.

Logan, K. A. and Sweanor, L. L. (2001) *Desert Puma: Evolutionary Ecology and Conservation of an Enduring Carnivore*. Island Press, Washington DC.

Logan, K. A. and Sweanor, L. L. (1999) Puma. In S. Demaris and P. R. Krausman (eds.), *Ecology and Management of Large Mammals in North America*, pp. 347–77. Prentice Hall, Upper Saddle, NJ.

Logan, K. A. and Sweanor, L. L. (1998) Cougar management in the West: New Mexico as a template. In *Proceedings of the Western Association of Fish and Wildlife Agencies*. Wyoming Game and Fish Department, Jackson Hole, WY.

Logan, K. A., Sweanor, L. L., and Hornocker, M. G. (2004) Reconciling science and politics in puma management in the West: New Mexico as a template. In J. E. Miller and K. A. Logan (eds.), *Managing Mammalian Predators and Their Populations to Avoid Conflicts, 69th North American Wildlife and Natural Resources Conference*, pp. 1–11. Wildlife Management Institute, Washington, DC.

Logan, K. A., Sweanor, L. L., Smith, J. F., and Hornocker, M. G. (1999) Capturing pumas with foot-hold snares. *Wildlife Society Bulletin*, **27**, 201–8.

Logan, K. A., Sweanor, L. L., and Hornocker, M. G. (1996) Mountain lion population dynamics. In *Mountain Lions in the San Andreas Mountains, New Mexico*, Chapter 3 (Project No. W-128-R, Final Report). New Mexico Department of Game and Fish, Santa Fe, NM.

Logan, K. A., Irwin, L. L., and Skinner, R. (1984) Characteristics of a hunted mountain lion population in Wyoming. *Journal of Wildlife Management*, **50**, 648–54.

Londei, T. (2000) The cheetah (*Acinonyx jubatus*) dewclaw: specialization overlooked. *Journal of Zoology*, **251**, 535–47.

López-Bao, J. V., Rodriguez, A., and Palomares, F. (2008) Behavioural response of a trophic specialist, the Iberian lynx, to supplementary food: patterns of food use and implications for conservation. *Biological Conservation*, **141**, 1857–67.

Lopez-Gonzalez, C. A. (1999) *Implicaciones para la conservacion y el manejo de pumas* (Puma concolor) *utilizando como modelo una poblacion sujeta a caceria deportiva*. PhD dissertation, Universidad Nacional Autonoma de Mexico, Mexico City, Mexico.

Lopez-Gonzalez, C. A. and Piña, G. L. (2002) Carrion use by jaguars (*Panthera onca*) in Sonora, Mexico. *Mammalia*, **66**, 603–5.

López-Ontiveros, A. (1991) Algunos aspectos de la evolución de la caza en España. *Agricultura y Sociedad*, **58**, 13–51.

Lopez-Rebollar, L. M., Penzhorn, B. L., de Waal, D. T,. and Lewis, B. D. (1999) A possible new piroplasm in lions from the Republic of South Africa. *Journal of Wildlife Diseases*, **35**, 82–5.

Lopez, B. (1978) *Of Wolves and Men*. Touchstone, New York.

López, G., Martínez, F., Meli, M., *et al.* (2007) *The Feline Leukemia Virus and the Iberian Lynx*. Environment Council, Andalusian Government, Seville, Spain.

Lopez, J. V., Culver, M., Stephens, J. C., Johnson, W. E., and O'Brien, S. J. (1997) Rates of nuclear and cytoplasmic mitochondrial DNA sequence divergence in mammals. *Molecular Biology and Evolution*, **14**, 277–86.

Lopez, J. V., Yuhki, N., Masuda, R., Modi, W., and O'Brien, S. J. (1994) Numt, a recent transfer and tandem amplification of mitochondrial DNA to the nuclear genome of the domestic cat. *Journal of Molecular Evolution*, **39**, 174–90.

López Martínez, N. (2008) The Lagomorph fossil record and the origin of the European rabbit. In P. C. Alves, N. Ferrand, and K. Hackländer (eds.), *Lagomorph Biology: Evolution, Ecology, and Conservation*, pp. 27–46. Springer, Berlin, Germany.

Lorenz, J. R., Coppinger, R. P., and Sutherland, M. R. (1986) Causes and economic effects of mortality in livestock guarding dogs. *Journal of Range Management*, **39**, 293–5.

Lott, D. F. (1991) *Intraspecific Variation in the Social Systems of Vertebrates*. Cambridge University Press, Cambridge, UK.

Lotz, M. A., Land, E. D., and Johnson, K. G. (1997) Evaluation and use of precast wildlife crossings by Florida wildlife. *Proceedings of the Annual Conference of Southeastern Fish and Wildlife Agencies*, **51**, 311–18.

Lovallo, M. J. and Anderson, E. M. (1996) Bobcat (*Lynx rufus*) home range size and habitat use in Northwest Wisconsin. *American Midland Naturalist*, **135**, 241–52.

Lovegrove, R. and Lovegrove, R. (2007) *Silent Fields: The Long Decline of a Nation's Wildlife* Oxford University Press, Oxford, UK.

Loveridge, A. J. and Macdonald, D. W. (2002) Habitat ecology of two sympatric species of jackals in Zimbabwe. *Journal of Mammalogy*, **83**, 599–607.

Loveridge, A. J., Murindagomo, F., Moyo, G., and Macdonald, D. W. (in prep) Impact of trophy hunting on lions in Matetsi Safari Area, Zimbabwe.

Loveridge, A. J., Packer, C., and Dutton, A. (2009a) Science and the recreational hunting of lions. In B. Dickson, J. Hutton, and B. Adams (eds.), *Recreational Hunting, Conservation and Rural Livelihoods: Science and Practice*. Blackwell Publishing, Oxford, UK.

Loveridge, A. J., Valeix, M., Davidson, Z., Murindagomo, F., Fritz, H., and Macdonald, D. W. (2009b) Changes in home range size in relation to pride size and prey biomass in a semi-arid savanna. *Ecography*, **32**, 1–10.

Loveridge, A., Canney, S., Hemson, G., and Sillero-Zubiri, C. (2007a) Modelling the distribution of African lion population using a GIS meta-analysis. In *Felid Biology and Conservation Conference 17–20 September 2007: Programmes and Abstracts*, p. 68. Wildlife Conservation Research Unit, Oxford, UK.

Loveridge, A. J., Reynolds, J. C., and Milner-Gulland, E. J. (2007b) Does sport hunting benefit conservation? In D. W. Macdonald and K. Service (eds.), *Key Topics in Conservation*, pp. 224–40. Blackwell Publishing, Oxford, UK.

Loveridge, A. J., Searle, A. W., Murindagomo, F., and Macdonald, D. W. (2007c) The impact of sport-hunting on the population dynamics of an African lion population in a protected area. *Biological Conservation*, **134**, 548–58.

Loveridge, A. J., Hunt, J. E., Murindagomo, F., and Macdonald, D. W. (2006) Influence of drought on predation of elephant (*Loxodonta africana*) calves by lions (*Panthera leo*) in an African wooded savannah. *Journal of Zoology*, **270**, 523–30.

Loveridge, A. J., Lynam, T., and Macdonald, D. W. (eds.) (2001) *Lion Conservation Research. Workshop 1: Survey Techniques*. Wildlife Conservation Research Unit, University of Oxford, Oxford, UK.

Luaces, I., Domenech, A., Garcia-Montijano, M., *et al.* (2008) Detection of Feline leukemia virus in the endangered Iberian lynx (*Lynx pardinus*). *Journal of Veterinary Diagnostic Investigation*, **20**, 381–5.

Lucherini, M. and Luengos Vidal, E. (2003) Intraguild competition as a potential factor affecting the conservation of two endangered cats in Argentina. *Endangered Species Update*, **20**, 211–20.

Lucherini, M. and Merino, M. J. (2008) Perceptions of human–carnivore conflicts in the High Andes of Argentina. *Mountain Research and Development*, **28**, 81–5.

Lucherini, M., Luengos Vidal, E., and Merino, M. J. (2008) How rare is the rare Andean cat? *Mammalia*, **72**, 95–101.

Lucherini, M., Manfredi, C., Luengos, E., Mazim, F. D., Soler, L., and Casanave, E. B. (2006) Body mass variation in the Geoffroy's cat (*Oncifelis geoffroyi*). *Revista Chilena de Historia Natural*, **79**, 169–74.

Lucherini, M., Huaranca, J. C., Savini, S., Tavera, G., Luengos Vidal, E., and Merino, M. J. (2004) New photographs of the Andean cat in Argentina: have we found a viable population? *Cat News*, **41**, 4–5.

Lucherini, M., Luengos Vidal, E., and Beldomenico, P. (2001) First record of sympatry of guigna and Geoffroy's cat. *Cat News*, **35**, 20–1.

Lucherini, M., Soler, L., Manfredi, C., Desbiez, A., and Marul, C. (2000) Geoffroy's cat in the pampas grasslands. *Cat News*, **33**, 22–4.

Ludlow, M. E. and Sunquist, M. E. (1987) Ecology and behavior of ocelots in Venezuela. *National Geographic Research*, **3**, 447–61.

Luikart, G., England, P. R., Tallmon, D., Jordan, S., and Taberlet, P. (2003) The power and promise of population genomics: from genotyping to genome typing. *Nature Reviews Genetics*, **4**, 981–94.

Lukes, S. (2008) *Moral relativism*. Profile Books, London, UK.

Luo, S. J., Kim, J. H., Johnson, W. E., *et al.* (2006) Proceedings in phylogeography and genetic ancestry of Tigers (*Panthera tigris*) in China and across the range. *Zoological Research*, **27**, 411–48.

Luo, S. J., Kim, J. H., Johnson, W. E., *et al.* (2004) Phylogeography and genetic ancestry of tigers (*Panthera tigris*). *PloS Biology*, **2**, 2275–93.

Lybbert, T. J., Barrett, C. B., Desta, S., and Coppock, D. C. (2004) Stochastic wealth dynamics and risk management among poor populations. *The Economic Journal*, **114**, 750–77.

Lynam, A. J., Round, P., and Brockelman, W. Y. (2006) *Status of Birds and Large Mammals of the Dong Phayayen-Khao Yai Forest Complex, Thailand*. Biodiversity Research and Training Program and Wildlife Conservation Society, Bangkok, Thailand.

Ma, Y. (2005) Changes in Amur tiger distribution in northeast China in the past 100 years. In *Recovery of the Wild Amur Tiger Population in China: Progress and Prospect. Proceedings of the 2000 International Workshop on Wild Amur Tiger Population Recovery Action Plan, Harbin, China and the 2002 National Worshop on Progress of Wild Amur Tiger Population Recovery Action, Hunchun, China.*, pp. 164–70. China Forestry Publishing House, Beijing, China.

Macdonald, D. W. (1976) Food caching by red foxes and some other carnivores. *Zeitschrift für Tierpsychologie*, **42**, 170–85.

Macdonald, D. W. (1980) Patterns of scent marking with urine and faeces amongst carnivore communities. *Proceedings of the Symposia of the Zoological Society of London*, **45**, 107–39.

Macdonald, D. W. (2009) Lessons learnt and plans laid: seven awkward questions for the future of reintroductions. In M. W. Hayward and M. J. Somers (eds.), *The Reintroduction of Top-Order Predators*. pp. 411–48. Wiley-Blackwell Publishing, Oxford, UK.

Macdonald, D. W. (2005) The evolving discussion of trophy hunting. In A. J. Loveridge, T. Lynam, and D. W. Macdonald (eds.), *Lion Conservation Research Workshops 3 and 4: From Conflict to Socioecology*, pp. 18–22. WildCRU, Oxford, UK.

Macdonald, D. W. (ed.) (2001a) *The New Encyclopedia of Mammals*. Oxford University Press, Oxford, UK.

Macdonald, D. W. (2001b) Postscript: science, compromise and tough choices. In J. L. Gittleman, S. M. Funk, D. W. Macdonald, and R. K. Wayne (eds.), *Carnivore Conservation*, pp. 524–38. Cambridge University Press, Cambridge, UK.

Macdonald, D. W. (2000) Bartering biodiversity: what are the options? In D. Helm (ed.), *Economic Policy: Objectives, Instruments and Implementation*, pp. 142–71. Oxford University Press, Oxford, UK.

Macdonald, D. W. (1992) *The Velvet Claw*. BBC Books, London, UK.

Macdonald, D. W. (1991) The pride of the farmyard. *BBC Wildlife Magazine*, **9**, 782–90.

Macdonald, D. W. (1985) The carnivores: order Carnivora. In R. E. Brown and D. W. Macdonald (eds.), *Social Odours in Mammals*, pp. 619–722. Oxford University Press, Oxford, UK.

Macdonald, D. W. (ed.) (1984) *The Encyclopedia of Mammals*. Unwin Hyman, London, UK.

Macdonald, D. W. (1983) The ecology of carnivore social behaviour. *Nature*, **301**, 379–84.

Macdonald, D. W. and Carr, G. M. (1989) Food security and the rewards of tolerance. In V. Standen and R. A. Folley (eds.), *Comparative Socioecology: The Behavioural Ecology of Humans and Other Mammals*, pp. 75–99. Blackwell Scientific Publications, Oxford, UK.

Macdonald, D. W. and Courtenay, O. (1996) Enduring social relationships in a population of crab-eating zorros, *Cerdocyon thous*, in Amazonian Brazil (Carnivora, Canidae). *Journal of Zoology*, **292**, 329–55.

Macdonald, D. W. and Johnson, P. J. (1996) The impact of sport hunting: a case study. In Dunstone, N. and Taylor, V. L (eds.), *The Exploitation of Mammal Populations: 1994 Symposium of the Zoological Society of London*, pp. 160–207. Chapman and Hall, London, UK.

Macdonald, D. W. and Sillero-Zubiri, C. (eds.) (2004a) *The Biology and Conservation of Wild Canids*. Oxford University Press, Oxford, UK.

Macdonald, D. W. and Sillero-Zubiri, C. (2004b) Conservation: from theory to practice, without bluster. In D. W. Macdonald and C. Sillero-Zubiri (eds.), *The Biology and Conservation of Wild Canids*, pp. 353–72. Oxford University Press, Oxford, UK.

Macdonald, D. W. and Sillero-Zubiri, C. (2004c) Wild canids—an introduction and dramatis personae. In D. W. Macdonald and C. Sillero-Zubiri (eds.), *The Biology and Conservation of Wild Canids*, pp. 3–36. Oxford University Press, Oxford, UK.

Macdonald, D. W. and Tattersall, F. H. (2001) *Britain's Mammals: The Challenge for Conservation*. People's Trust for Endangered Species and Wildlife Conservation Research Unit, London and Oxford, UK.

Macdonald, D. W., Collins, N. M., and Wrangham, R. (2007) Principles, practice and priorities: the quest for 'alignment'. In D. W. Macdonald and K. Service (eds.), *Key Topics in Conservation Biology*, pp. 271–89. Blackwell Publishing, Oxford, UK.

Macdonald, D. W., Riordan, P., and Mathews, F. (2006) Biological hurdles to the control of TB in cattle: a test of two hypothesis concerning wildlife to explain the failure of control. *Biological Conservation*, **131**, 268–86.

Macdonald, D. W., Buesching, C. D., Henderson, J., Ellwood, S. A., Baker, S. E., and Stopka, P. (2004a) Encounters between two sympatric carnivores: Red foxes (*Vulpes vulpes*) and European badgers (*Meles meles*). *Journal of Zoology*, **263**, 385–92.

Macdonald, D. W., Creel, S., and Mills, G. (2004b) Society: canid society. In D. W. Macdonald and C. Sillero-Zubiri (eds.), *The Biology and Conservation of Wild Canids*, pp. 85–106. Oxford University Press, Oxford, UK.

Macdonald, D. W., Daniels, M. J., Driscoll, C. A., Kitchener, A. C., and Yamaguchi, N. (2004c) *The Scottish Wildcat: Analyses for Conservation and an Action Plan*. Wildlife Conservation Research Unit, Oxford, UK.

Macdonald, D. W., Loveridge, A. J., and Atkinson, R. P. D. (2004d) A comparative study of side-striped jackals *Canis adustus* in Zimbabwe: the influence of habitat and congeners. In D. W. Macdonald and C. Sillero-Zubiri (eds.), *The Biology and Conservation of Wild Canids*, pp. 255–70. Oxford University Press, Oxford, UK.

Macdonald, D. W., Reynolds, J. C., Carbone, C., Mathews, F., and Johnson, P. J. (2003) The bioeconomics of fox control. In F. H. Tattersall and W. J. Manley (eds.), *Conservation and Conflict: Mammals and Farming in Britain*, pp. 220–36. Linnean Society Occasional Publication, Westbury Publishing, Yorkshire, UK.

Macdonald, D. W., Moorhouse, T. P., Enck, J. W., and Tattersall, F. H. (2002) Mammals. In M. R. Perrow and A. J. Davy (eds.), *Handbook of Ecological Restoration*, pp. 389–408. Cambridge University Press, Cambridge, UK.

Macdonald, D. W., Mace, G. M., and Rushton, S. P. (2000a) British mammals: is there a radical future? In A. Entwistle and N. Dunstone (eds.), *Priorities for the Conservation of Mammalian Diversity: Has the Panda Had Its Day?*, pp. 175–205. Cambridge University Press, Cambridge, UK.

Macdonald, D. W., Stewart, P. D., Stopka, P., and Yamaguchi, N. (2000b) Measuring the dynamics of mammalian societies: an ecologist's guide to ethological methods. In L. Boitani and T. K. Fuller (eds.), *Research Techniques in Animal Ecology: Controversies and Consequences*, pp. 332–88. Columbia University Press, New York.

Macdonald, D. W., Apps, P. J., Carr, G. M., and Kerby, G. (1987) Social dynamics, nursing coalitions and infanticide among farm cats, *Felis catus*. *Advances in Ethology*, **28**, 1–64.

Macdonald, D. W., Yamaguchi, N., and Kerby, G. (2000c) Group-living in the domestic cat: its sociobiology and epidemiology. In D. C. Turner and P. Bateson (eds.), *The Domestic Cat: The Biology of Its Behaviour*, 2nd edn, pp. 95–118. Cambridge University Press, Cambridge, UK.

Mace, G. M., Possingham, H. P., and Leader-Williams, N. (2007) Prioritizing choices in conservation. In D. W. Macdonald and K. Service (eds.), *Key Topics in Conservation Biology*, pp. 17–34. Blackwell Publishing Ltd, Oxford, UK.

Mace, G. M., Possingham, H. P., and Leader-Williams, N. (2006) Prioritizing choices in conservation. In D. W. Macdonald and K. Service (eds.), *Key Topics in Conservation Biology*, pp. 17–34. Blackwell Publishing Ltd, Oxford, UK.

Mace, G. M., Harvey, P. H., and Clutton-Brock, T. H. (1982) Vertebrate home range size and energetic requirements. In I. Swingland and P. J. Greenwood (eds.), *The Ecology of Animal Movement*, pp. 32–53. Oxford University Press, Oxford, UK.

Machado, A. B. M., Martins, C. S., and Drummong, G. M. (2005) *Lista da fauna brasileira ameaçada de extinção*. Fundação Biodiversitas, Belo Horizonte, Brazil.

Machr, D. S. (1997) *The comparative ecology of bobcat, black bear and Florida panther in South Florida*. PhD dissertation, University of Florida, Gainesville, FL.

MacKenzie, D. I., Nichols, J. D., Royle, A. J., Pollock, K. H., Bailey, L. L., and Hines, J. E. (2006) *Occupancy Estimation and Modeling: Inferring Patterns and Dynamics of Species Occurrence*. Academic Press, Boston, MA.

Mackinnon, J. and Mackinnon, K. (1986) Review of the protected areas system in the Afrotropical realm. IUCN, Gland, Switzerland and Cambridge, UK.

Maclennan, S. D., Groom, R. J., Macdonald, D. W., and Frank, L. G. (2009) Evaluation of a compensation scheme to bring about pastoralist tolerance of lions. *Biological Conservation*, **142**, 2419–27.

Maddison, W. P. and Maddison, D. R. (2004) Mesquite: a modular system for evolutionary analysis, Version 1.12. <www.mesquiteproject.org>.

Maddox, T. M. (2003) *The ecology of cheetahs and other large carnivores in a pastoralist-dominated buffer zone*. PhD dissertation, University College London and Institute of Zoology, London, UK.

Madhusudan, M. D. (2003) Living amidst large wildlife: livestock and crop depredation by large mammals in the interior villages of Bhadra Tiger Reserve, South India. *Environmental Management*, **31**, 466–75.

Madhusudan, M. D. and Mishra, C. (2003) Why big fierce animals are threatened: conserving large mammals in densely populated landscapes. In V. Saberwal and M. Rangarajan (eds.), *Battles over Nature: Science and the Politics of Conservation*, pp. 31–55. Permanent Black, New Delhi, India.

Maehr, D. S. and Caddick, G. B. (1995) Demographics and genetic introgression in the Florida Panther. *Conservation Biology*, **9**, 1295–8.

Maehr, D. S., Lacy, R. C., Land, E. D., Bass, O. L., and Hoctor, S. T. (2002a) Evolution of population viability assessments for the Florida panther: a multiperspective approach. In S. R. Beissinger and D. R. McCullough (eds.), *Population Viability Analysis*, pp. 284–311. University of Chicago Press, Chicago, IL.

Maehr, D. S., Land, E. D., Shindle, D. B., Bass, O. L., and Hoctor, T. S. (2002b) Florida panther dispersal and conservation. *Biological Conservation*, **106**, 187–97.

Maehr, D. S., Greiner, E. C., Lanier, J. E., and Murphy, D. (1995) Notoedric mange in the Florida panther (*Felis concolor coryi*). *Journal of Wildlife Diseases*, **31**, 251–4.

Maehr, D. S., Land, D., and Roof, J. C. (1991) Social ecology of Florida Panthers. *National Geographic Research & Exploration*, **7**, 414–31.

Maehr, D. S., Belden, R. C., Land, E. D., and Wilkins, L. (1990) Food habits of panthers in southwest Florida. *Journal of Wildlife Management*, **54**, 420–3.

Maffei, L., Noss, A. J., Silver, S. C. and Kelly, M. J. (in press) Jaguars in South and Central America. In A. F. O'Connell, J. D. Nichols, and K. U. Karanth (eds.), *Camera Traps in Animal Ecology: Methods and Analyses*. Springer Tokyo Inc.

Maffei, L., Noss, A. J., Cuellar, E., and Rumiz, D. I. (2005) Ocelot (*Felis pardalis*) population densities, activity, and ranging behaviour in the dry forests of eastern Bolivia: data from camera trapping. *Journal of Tropical Ecology*, **21**, 349–53.

Maffei, L., Cuellar, E., and Noss, A. (2004a) One thousand jaguars (*Panthera onca*) in Bolivia's Chaco? Camera trapping in the Kaa-Iya National Park. *Journal of Zoology*, **262**, 295–304.

Maffei, L., Julio, R., Paredes, A., Posiño, A., and Noss, A. J. (2004b) *Estudios con trampas-cámara en el campamento Tucavaca III (18° 30.97′ S, 60° 48.62′ W), Parque Nacional Kaa-Iya del Gran Chaco, 28 de marzo-28 de mayo de 2004*. Technical Report No. 129, Wildlife Conservation Society, Santa Cruz, Bolivia.

Major, A. T. (1989) *Lynx* Lynx canadensis canadensis *(Kerr) predation patterns and habitat use in the Yukon Territory, Canada*. MSc dissertation, State University of New York, Syracuse, NY.

Major, M., Johnson, M. K., Davis, W. S., and Kellogg, T. F. (1980) Identifying scats by the recovery of bile acids. *Journal of Wildlife Management*, **44**, 290–3.

Mallet, J. (2007) Species, concept of. In C. W. Fox and J. B. Wolf (eds.), *Evolutionary Genetics: Concepts and Case Studies*, pp. 367–73. Oxford University Press, Oxford, UK.

Malmström, T. and Kröger, R. H. H. (2006) Pupil shapes and lens optics in the eyes of terrestrial vertebrates. *Journal of Experimental Biology*, **209**, 18–25.

Manakadan, R. and Sivakumar, S. (2006) Rusty-spotted cat on India's east coast. *Cat News*, **45**, 26.

Manfredi, C., Lucherini, M., and Casanave, E. B. (2007) Habitat use and selection of Geoffroy's cats (*Oncifelis geoffroyi*) in three areas of central Argentina. In *Felid Biology and Conservation Conference 17–20 September 2007: Programme and Abstracts*, p. 81. Wildlife Conservation Research Unit, Oxford, UK.

Manfredi, C., Soler, L., Lucherini, M., and Casanave, E. B. (2006) Home range and habitat use by Geoffroy's cat (*Oncifelis geoffroyi*) in a wet grassland in Argentina. *Journal of Zoology*, **268**, 381–7.

Manfredi, C., Lucherini, M., Casanave, E. B., and Canepuccia, A. D. (2004) Geographical variation in the diet of Geoffroy's cat (*Oncifelis geoffroyi*) in Pampas grassland of Argentina. *Journal of Mammalogy*, **85**, 1111–15.

Manfredo, M. J. (2008) *Who Cares about Wildlife? Social Science Concepts for Exploring Human–Wildlife Relationships and Conservation Issues*. Springer, New York.

Manfredo, M. J. and Dayer, A. A. (2001) Concept for exploring the social aspects of human–wildlife conflict in a global context. *Human Dimensions of Wildlife*, **9**, 1–20.

Manfredo, M., Teel, T., and Bright, A. (2003) Why are public values toward wildlife changing? *Human Dimensions of Wildlife*, **8**, 287–306.

Mangold, B. J., Cook, R. A., Cranfield, M. R., Huygen, K., and Godfrey, H. P. (1999) Detection of elevated levels of circulating antigen 85 by dot immunobinding assay ln captive wild animals with tuberculosis. *Journal of Zoo and Wildlife Medicine*, **30**, 477–83.

Manly, B. F. J., McDonald, L. L., and Thomas, D. L. (1993) *Resource Selection by Animals: Statistical Design and Analysis for Field Studies*. Chapman and Hall, London, UK.

Manning, A. D., Lindemayer, D. B., and Fischer, J. (2006) Stretch goals and backcasting: approaches for overcoming barriers to large-scale ecological restoration. *Restoration Ecology*, **14**, 487–92.

Mansfield, K. G. and Land, E. D. (2002) Cryptorchidism in Florida panthers: prevalence, features, and influence of genetic restoration. *Journal of Wildlife Diseases*, **38**, 693–8.

Marchini, S. (2003) *Pantanal: opinião pública local sobre meio ambiente e desenvolvimento*. Wildlife Conservation Society and Mamiraua Institute, Rio de Janeiro, Brazil.

Marchini, S. and Macdonald, D. W. (in prep-a). *Factors determining perceptions about jaguars in Brazil*.

Marchini, S. and Macdonald, D. W. (in prep-b). *Conflicts between people and jaguars: from perceptions to persecution*.

Marchini, S. and Macdonald, D. W. (submitted). Development of perceptions about jaguars among children and adolescents in Brazil: implications for conversion. *Biological Conservation*.

Mares, M. A. and Ojeda, R. A. (1984) Faunal commercialisation and conservation in South America. *BioScience*, **34**, 580–4.

Mares, R., Moreno, R. S., Kays, R. W., and Wikelski, M. (2008) Predispersal home range shift of an ocelot, *Leopardus pardalis* (Carnivora: Felidae) on Barro Colorado Island, Panama. *Revista de Biología Tropical*, **56**, 779–87.

Marker-Kraus, L. (1997) Morphological abnormalities reported in Namibian cheetahs (*Acinonyx jubatus*). In *50th Anniversary Congress of VAN and 2nd African Congress of the WVA*, pp. 9–18. Veterinary Association of Namibia, Windhoek, Namibia.

Marker-Kraus, L., Kraus, D., Barnett, D., and Hurlbut, S. (1996) *Cheetah Survival on Namibian Farmlands*. Cheetah Conservation Fund, Windhoek, Namibia.

Marker, L. L. (2002) *Aspects of cheetah* (Acinonyx jubatus) *biology, ecology and conservation strategies on Namibian farmlands*. DPhil dissertation, University of Oxford, Oxford, UK.

Marker, L. L. (2000) Aspects of the ecology of the cheetah (*Acinonyx jubatus*) on north central Namibian farmlands. *Namibia Scientific Society Journal*, **48**, 40–8.

Marker, L. L. (1998) Current status of the cheetah. In B. L. Penzhorn (ed.), *Proceedings of a Symposium on Cheetahs as Game Ranch Animals*, pp. 1–17. Wildlife Group of South African Veterinary Association, Onderstepoort, South Africa.

Marker, L. L. and Dickman, A. J. (2003) Morphology, physical condition and growth of the cheetah (*Acinonyx jubatus jubatus*). *Journal of Mammalogy*, **84**, 840–50.

Marker, L. L. and Dickman, A. J. (2004a) Dental anomalies and incidence of palatal erosion in Namibian cheetahs (*Acinonyx jubatus jubatus*). *Journal of Mammalogy*, **85**, 19–24.

Marker, L. L. and Dickman, A. J. (2004b) Human aspects of cheetah conservation: Lessons learned from the Namibian farmlands. *Human Dimension of Wildlife*, **9**, 297–305.

Marker, L. L. and Dickman, A. J. (2005) Factors affecting leopard (*Panthera pardus*) spatial ecology, with particular reference to Namibian farmlands. *South African Journal of Wildlife Research*, **35**, 105–15.

Marker, L. L., Dickman, A. J., Mills, M. G. L., Jeo, R. M., and Macdonald, D. W. (2008a) Spatial ecology of cheetahs (*Acinonyx jubatus*) on north-central Namibian farmlands. *Journal of Zoology*, **274**, 226–38.

Marker, L. L., Pearks-Wilkerson, A. J., Martenson, J. S., *et al*. (2008c) Patterns of molecular genetic variation in Namibian cheetahs. *Journal of Heredity*, **99**, 2–13.

Marker, L. L., Dickman, A. J., Wilkinson, C., Schumann, B., and Fabiano, E. (2007) The Namibia cheetah: status report. *Cat News Special Issue*, **3**, 5–13.

Marker, L. L., Dickman, A. J., and Macdonald, D. W. (2005a) Perceived effectiveness of livestock guarding dogs placed on Namibian farms. *Rangeland Ecology and Management*, **58**, 329–36.

Marker, L. L., Dickman, A. J., and Macdonald, D. W. (2005b) Survivorship and causes of mortality for livestock gurading dogs on Namibian Rangeland. *Rangeland Ecology and Management*, **58**, 337–43.

Marker, L. L., Dickman, A. J., and Gannon, W. L. (2004) Dental anomalies and incidence of palatal erosion in Namibian cheetahs (*Acinonyx jubatus jubatus*). *Journal of Mammalogy*, **85**, 19–34.

Marker, L. L., Dickman, A. J., Jeo, R. M., Mills, M. G. L., and Macdonald, D. W. (2003a) Demography of the Namibian cheetah (*Acinonyx jubatus jubatus*). *Biological Conservation*, **114**, 413–25.

Marker, L. L., Dickman, A. J., Mills, M. G. L., and Macdonald, D. W. (2003b) Aspects of the management of cheetahs, *Acinonyx jubatus jubatus*, trapped on Namibian farmlands. *Biological Conservation*, **114**, 401–12.

Marker, L. L., Mills, M. G. L., and Macdonald, D. W. (2003c) Factors influencing perceptions of conflict and tolerance towards cheetahs on Namibian farmlands. *Conservation Biology*, **17**, 1290–8.

Marker, L. L., Munson, L., Basson, P. A., and Quackenbush, S. (2003d) Multicentric T-cell lymphoma associated with feline leukemia virus infection in a captive Namibian cheetah (*Acinonyx jubatus*). *Journal of Wildlife Diseases*, **39**, 690–5.

Marker, L. L., Muntifering, J. R., Dickman, A. J., Mills, M. G. L., and Macdonald, D. W. (2003e) Quantifying prey preferences of free-ranging Namibian cheetahs. *South African Journal of Wildlife Research*, **33**, 43–53.

Marnewick, K., Hayward, M. W., Cilliers, D. and Somers, M. J. (2009) Survival of cheetahs relocated from ranchland to fenced protected areas in South Africa. In M. W. Hayward and M. J. Somers (eds.) *The Reintroduction of Top-Order Predators*, pp. 282–306. Wiley-Blackwell Publishing, Oxford, UK.

Marshall, L. G., Webb, S. D., Sepkoski, J. J., and Raup, D. M. (1982) Mammalian evolution and the great American interchange. *Science*, **215**, 1351–7.

Martin-Atance, P., Leon-Vizcaino, L., Palomares, F., *et al.* (2006) Antibodies to *Mycobacterium bovis* in wild carnivores from Doñana National Park (Spain). *Journal of Wildlife Diseases*, **42**, 704–8.

Martin, L. D. (1980) Functional morphology and the evolution of cats. *Transaction of the Nebraska Academy of Sciences*, **8**, 141–54.

Martin, L. D., Babiarz, J. P., Naples, V. L., and Hearst, J. (2000) Three ways to be a saber-toothed cat. *Naturwissenschaften*, **87**, 41–4.

Martin, R. A. and Harksen, J. C. (1974) The Delmont local fauna, Blancan of South Dakota. *Bulletin of the New Jersey Academy of Sciences*, **19**, 11–17.

Martin, R. B. and de Meulenaer, T. (1988) *Survey of the Status of the Leopard* (Panthera pardus) *in Sub-Saharan Africa.* Secretariat for the Convention on International Trade in Endangered Species of Wild Fauna and Flora (CITES), Lausanne, Switzerland.

Mascia, M. B., Brosius, J. P., Dobson, T. A., *et al.* (2003) Conservation and the social sciences. *Conservation Biology*, **17**, 649–50.

Mason, L. L. (1988) *World Dictionary of Livestock Breeds*, 3rd edn. CAB International, Oxford, UK.

Massolo, A. and Meriggi, A. (1998) Factors affecting habitat occupancy by wolves in northern Apennines (northern Italy): a model of habitat suitability. *Ecography*, **21**, 97–107.

Masuda, R., Lopez, J. V., Slattery, J. P., Yuhki, N., and O'Brien, S. J. (1996) Molecular phylogeny of mitochondrial cytochrome b and 12S rRNA sequences in the Felidae: ocelot and domestic cat lineages. *Molecular Phylogenetics and Evolution*, **6**, 351–65.

Mathews, F., Moro, D., Strachan, R., Gelling, M., and Buller, N. (2006) Health surveillance in wildlife reintroductions. *Biological Conservation*, **131**, 338–47.

Mattern, M. Y. and McLennan, D. A. (2000) Phylogeny and speciation of felids. *Cladistics*, **16**, 232–53.

Matthew, W. D. (1910) The phylogeny of Felidae. *Bulletin of the American Museum of Natural History*, **28**, 289–316.

Mattson, D. J. (2004) Living with fierce creatures? An overview and models of mammalian carnivore conservation. In N. Fascione, A. Delach, and M. E. Smith (eds.), *People and Predators: From Conflict to Coexistence*, pp. 151–76. Island Press, Washington DC.

Matyushkin, Y. N. and Vaisfeld, M. (2003) *The Lynx—Regional Features of Ecology, Use and Protection.* Nauka, Moscow, Russia.

Matzke, G. (1979) Settlement and sleeping sickness control—dual threshold model of colonial and traditional methods in East Africa. *Social Science & Medicine Part D—Medical Geography*, **13**, 209–14.

May, R. M. and Anderson, R. M. (1979) Population biology of infectious diseases. 2. *Nature*, **280**, 455–61.

Mayr, E. (1970) *Population, Species and Evolution.* Harvard University Press, Cambridge, MA.

Mayr, E. (1942) *Systematics and the Origin of Species.* Columbia University Press, New York.

Mazák, V. (1996) *Der Tiger.* Westarp Wissenschaften, Magdeburg, Germany.

Mazák, V. (1983) *Der Tiger: Panthera Tigris Linnaeus 1758*, 3rd edn. A. Ziemsen, Wittenberg Lutherstadt, Germany.

Mazák, V. (1981) Mammalian Species, *Panthera Tigris. The American Society of Mammalogists*, **152**, 1–8.

Mazzoli, M., Graipel, M. E., and Dunstone, N. (2002) Mountain lion depredation in southern Brazil. *Biological Conservation*, **105**, 43–51.

McBride, R. T., McBride, R. T., McBride, R. M., and McBride, C. E. (2008) Counting pumas by categorizing physical evidence. *Southeastern Naturalist*, **7**, 381–400.

McBurney, S., Banks, D., and Anderson, D. R. (1998) Morbillivirus infection in four lynx. *Journal of Wildlife Diseases*, **34**, 1.

McCarthy, K. P., Fuller, T. K., Ma, M., McCarthy, T. M., Waits, L., and Jumabaev, K. (2008) Assessing estimators of snow leopard abundance. *Journal of Wildlife Management*, **72**, 1826–33.

McCarthy, T. M. (2000) *Ecology and conservation of snow leopards, Gobi brown bears, and wild bactrian camels in Mongolia.* PhD dissertation, University of Massachusetts, Amherst, MA.

McCarthy, T. M. and Chapron, G. (2003) *Snow Leopard Survival Strategy.* International Snow Leopard Trust and Snow Leopard Network, Seattle, WA.

McCarthy, T. M., Khan, J. and McCarthy, K. (2007) First study of snow leopards using GPS-satellite collars underway in Pakistan. *Cat News*, **46**, 22–3.

McCarthy, T. M., Fuller, T. K., and Munkhtsog, B. (2005) Movements and activities of snow leopards in Southwestern Mongolia. *Biological Conservation*, **124**, 527–37.

McCleery, R. A., Ditton, R. B., Sell, J., and Lopez, R. R. (2006) Understanding and improving attitudinal research in wildlife sciences. *Wildlife Society Bulletin*, **34**, 537–41.

McComb, K., Packer, C., and Pusey, A. (1994) Roaring and numerical assessment in contests between groups of female lions, *Panthera leo. Animal Behaviour*, **47**, 379–87.

McComb, K., Pusey, A., Packer, C., and Grinnell, J. (1993) Female lions can identify potentially infanticidal males from their roars. *Proceedings of the Royal Society of London Series B*, **252**, 59–64.

McCord, C. M. and Cardoza, J. E. (1982) Bobcat and Lynx. In J. A. Chapman and G. A. Feldhamer (eds.), *Wild Mammals of North America: Biology, Management and Economics*, pp. 728–66. The John Hopkins University Press, Baltimore, MD.

McCown, J. W., Roelke, M. E., Forrester, D. J., Moore, C. T., and Roboski, J. C. (1991) Physiological evaluation of two white-tailed deer herds in southern Florida. *Proceedings of the Annual Conference of Southeastern Fish and Wildlife Agencies*, **45**, 81–90.

McDougal, C. W. (1993) Man-eaters: A geographical and historical perspective. *Hornbill*, **1**, 10–16.

McDougal, C. (1991) Chuchchi: Life of a tigress. In J. Seidensticker and S. Lumpkin (eds.), *Great Cats*, p. 104. Roedale Press, Emmaus, PA.

McDougal, C. (1988) Leopard and tiger interactions at Royal Chitwan National Park, Nepal. *Journal of the Bombay Natural History Society*, **85**, 609–10.

McDougal, C. (1977) *The Face of the Tiger*. Rivington Books, London, UK.

McDougal, C. and Tshering, K. (1998) *Tiger Conservation Strategy for the Kingdom of Bhutan*. Nature Conservation Division, Royal Government of Bhutan and WWF Bhutan, Thimphu, Bhutan.

McGregor, P. K. and Peake, T. M. (1998) Individual identification and conservation biology. In T. M. Caro (ed.), *Behavioural Ecology and Conservation*, pp. 31–55. Oxford University Press, Oxford, UK.

McHenry, C. R., Wroe, S., Clausen, P. D., Moreno, K., and Cunningham, E. (2007) Supermodeled sabercat, predatory behavior in *Smilodon fatalis* revealed by high-resolution 3D computer simulation. *Proceedings of the National Academy of Sciences of the United States of America*, **104**, 16010–15.

McIntosh, A., Sibanda, S., Vaughan, A., and Xaba, T. (1996) Traditional authorities and land reform in South Africa: lessons from KwaZulu-Natal. *Development Southern Africa*, **13**, 3.

McKelvey, K. S., Von Kienast, J., Aubry, K. B., *et al.* (2006) DNA analysis of hair and scat collected along snow tracks to document the presence of Canada lynx. *Wildlife Society Bulletin*, **34**, 451–5.

McKelvey, K. S., Aubry, K. B., and Ortega, Y. K. (2000a) History and distribution of lynx in the contiguous United States. In L. F. Ruggiero, S. W. Buskirk, G. M. Koehler, C. J. Krebs, and J. R. Squires (eds.), *Ecology and Conservation of Lynx in the United States*, pp. 207–64. University Press of Colorado, Boulder, CO.

McKelvey, K. S., Buskirk, S. W., and Krebs, C. J. (2000b) Theoretical insights into the population viability of lynx. In L. F. Ruggiero, K. B. Aubry, S. W. Buskirk, *et al.* (eds.), *Ecology and Conservation of Lynx in the United States*, pp. 21–37. University Press of Colorado, Boulder, CO.

McKelvey, K. S., Ortega, Y. K., Koehler, G. M., Aubry, K. B., and Brittell, J. D. (2000c) Canada lynx habitat and topographic use patterns in north central Washington: a reanalysis. In L. F. Ruggiero, S. W. Buskirk, G. M. Koehler, C. J. Krebs, K. S. McKelvey, and J. R. Squires (eds.), *Ecology and Conservation of Lynx in the United States*, pp. 307–36. University Press of Colorado, Boulder, CO.

McKenzie, A. A. and Burroughs, R. E. J. (1993) Chemical capture of carnivores. In A. A. McKenzie (ed.), *The Capture and Care Manual*, pp. 224–44. Wildlife Decision Support Services and South African Veterinary Foundation, Pretoria, South Africa.

McLaughlin, R. T. (1970) *Aspects of the biology of the cheetah* (Acinonyx jubatus, *Schreber) in Nairobi National Park*. MSc dissertation, University of Nairobi, Nairobi, Kenya.

McMahan, L. R. (1986) The international cat trade. In S. D. Miller and D. D. Everett (eds.), *Cats of the World: Biology, Conservation and Management*, pp. 461–88. National Wildlife Federation, Washington DC.

McManus, J. S. (2009) *The spatial ecology and activity patterns of leopard* (Panthera pardus) *in the Baviaanskloof and Greater Addo Elephant National Park*. MSc dissertation, Rhodes University, Grahamstown, South Africa.

McMullen, J. P. (1984) *Cry of the Panther: Quest of a Species*. Pineapple Press, Englewood, NJ.

McOrist, S., Boid, R., Jones, T. W., Easterbee, N., Hubbard, A. L., and Jarrett, O. (1991) Some viral and protozool diseases in the European wildcat (*Felis silvestris*). *Journal of Wildlife Diseases*, **27**, 693–6.

McPhee, M. E. (2004) Generations in captivity increases behavioral variance: considerations for captive breeding and reintroduction programs. *Biological Conservation*, **115**, 71–7.

Meachen-Samuels, J. and Van Valkenburgh, B. (2009a) Forelimb indicators of prey-size preference in the Felidae. *Journal of Morphology*, **270**, 729–44.

Meachen-Samuels, J. and Van Valkenburgh, B. (2009b) Craniodental indicators of prey size preference in the Felidae. *Biological Journal of the Linnean Society*, **96**, 784–99.

Mech, L. D. (1977) Record movement of a Canadian lynx. *Journal of Mammalogy*, **58**, 676–7.

Mech, L. D. (1981) *The Wolf: The Ecology and Behavior of an Endangered Species*. University of Minnesota Press, Minneapolis, MN.

Mech, L. D. (1980) Age, sex, reproduction, and spatial organization of lynxes colonizing northeastern Minnesota. *Journal of Mammalogy*, **61**, 261–7.

Mech, L. D. and Goyal, S. M. (1993) Canine parvovirus effect on wolf population change and pup survival. *Journal of Wildlife Diseases*, **29**, 330–3.

Meerman, J. C. and Sabido, W. (2001) *Central American Ecosystems Map: Belize*. Programme for Belize, Belize.

Meijaard, E. (1997) The bay cat *Catopuma badia*, a widely spread, but rare, species in Borneo. The International MOF Tropenbos-Kalimantan Project, Balikpapan, Indonesia.

Meijaard, E., Prakoso, B. B., and Azis (2005a) A new record for the Bornean bay cat. *Cat News*, **43**, 23–4.

Meijaard, E., Sheil, D., and Daryono (2005b) Flat-headed cat record in east Kalimantan. *Cat News*, **43**, 24.

Meli, M. L., Cattori, V., Martínez, F., *et al.* (2009) Feline leukemia virus and other pathogens as important threats to the survival of the critically endangered Iberian lynx (*Lynx pardinus*). PLoS ONE **4**, e4744, doi:10.1371.

Meli, M., Willi, B., Ryser-Degiorgis, M.-P., *et al.* (2006) Prevalence of selected feline pathogens in the Eurasian and Iberian lynx. In *Iberian Lynx Ex-Situ Conservation Seminar Series: Book of Proceedings*, pp. 28–32. Fundacion Biodiversidad, Sevilla and Doñana, Spain.

Melisch, R., Asmoro, P. B., Lubis, I. R., and Kusumawardhani, L. (1996) Distribution and status of the Fishing Cat (*Prionailurus viverrinus rhizophoreus* Sody, 1936) in West Java, Indonesia (Mammalia: Carnivora: Felidae). *Faunistische Abhandlungen,Staatiliches Museum fuer Tierkunde Dresden*, **20**, 311–19.

Mellen, J. D. (1993) A comparative-analysis of scent-marking, social and reproductive-behavior in 20 species of small cats (*Felis*). *American Zoologist*, **33**, 151–66.

Mellen, J. D. (1989) *Reproductive behavior of small captive exotic cats* (Felis *spp)*. PhD dissertation, University of California, Davis, CA.

Mellen, J. D. and Shepherdson, D. J. (1997) Environmental enrichment for felids: an integrated approach. *International Zoo Yearbook*, **35**, 191–7.

Melton, D. A., Berry, H. H., Berry, C. U., and Joubert, S. M. (1987) Aspects of the blood chemistry of wild lions, *Panthera leo. South African Journal of Zoology*, **22**, 40–4.

Melville, H. and Bothma, J. D. P. (2007) Using spoor techniques to examine the hunting behaviour of caracals in the Kgalagadi Transfrontier Park. In *Felid Biology and Conservation Conference 17–20 September 2007: Programme and Abstracts*, p. 57. Wildlife Conservation Research Unit, Oxford, UK.

Mendelssohn, H. (1989) Felids in Israel. *Cat News*, **10**, 2–4.

Menotti-Raymond, M. and O'Brien, S. J. (1995) Hypervariable genomic variation to reconstruct the natural history of populations: Lessons from the big cats. *Electrophoresis*, **16**, 1771–4.

Menotti-Raymond, M. and O'Brien, S. J. (1993) Dating the genetic bottleneck of the African cheetah. *Proceedings of the National Academy of Sciences of the United States of America*, **90**, 3172–6.

Meriggi, A. and Lovari, S. (1996) A review of wolf predation in southern Europe: does the wolf prefer wild prey to livestock? *Journal of Applied Ecology*, **33**, 1561–71.

Merino, M. J. (2006) *Conservemos el gato andino y su hábitat: Guía para Educadores*. Andean Cat Alliance, Bahía Blanca, Argentina.

Merola, M. (1994) A reassessment of homozygosity and the case for inbreeding depression in the cheetah, *Acinonyx jubatus*: implications for conservation. *Conservation Biology*, **8**, 961–71.

Meynell, P. J. and Parry, D. (2002) *Environmental Appraisal for the Construction of a Game Proof Fence around the Makgadikgadi Pans National Park*. Scott Wilson Kirkpatrick & Partners, Gaberone, Botswana.

Michalski, F., Boulhosa, R. L. P., Faria, A., and Peres, C. A. (2006a) Human–wildlife conflicts in a fragmented Amazonian forest landscape: determinants of large felid depredation on livestock. *Animal Conservation*, **9**, 179–88.

Michalski, F., Crawshaw, P. G., Oliveira, T. G., de, and Fabian, M. E. (2006b) Notes on home range and habitat use of three small carnivore species in a disturbed vegetation mosaic of southeastern Brazil. *Mammalia*, **70**, 52–7.

Michalski, F. and Peres, C. A. (2005) Anthropogenic determinants of primate and carnivore local extinctions in a fragmented forest landscape of southern Amazonia. *Biological Conservation*, **124**, 383–96.

Michel, A. L., Bengis, R. G., Keet, D. F., *et al.* (2006) Wildlife tuberculosis in South African conservation areas: implications and challenges. *Veterinary Microbiology*, **112**, 91–100.

Millán, J., Candela, M. G., Palomares, F., *et al.* (2009) Disease threats to the endangered Iberian lynx (*Lynx pardinus*). *The Veterinary Journal*, **182**, 114–24.

Millán, J., Naranjo, V., Rodríguez, A., de la Lastra, J. M. P., Mangold, A. J., and de la Fuente, J. (2007) Prevalence of infection and 18S rRNA gene sequences of *Cytauxzoon* species in Iberian lynx (*Lynx pardinus*) in Spain. *Parasitology*, **134**, 995–1001.

Millenium Ecosystem Assessment (2005) *Ecosystems and Human Well-Being: Synthesis*. Island Press, Washington DC.

Miller, B., Biggins, D., Hanebury, L., and Vargas, A. (1994) Reintroduction of the black-footed ferret. In G. M. Mace, P. J. Olney, and A. Feistner (eds.), *Creative Conservation: Interactive Management of Wild and Captive Animals*, pp. 455–64. Chapman and Hall, London, UK.

Miller, C. M. (2006) Jaguar density in Fireburn, Belize. Unpublished report, Wildlife Conservation Society, Gallon Jug, Belize.

Miller, G. J. (1969) A new hypothesis to explain the method of food ingestion used by *Smilodon californicus* Bovard. *Tebiwa*, **12**, 9–19.

Miller, G. S. (1912a) *Catalogue of the Mammals of Western Europe (Europe Exclusive of Russia) in the Collection of the British Museum*. British Museum, London, UK.

Miller, G. S. (1912b) *Lynx pardelus* Miller. In *Catalogue of the Mammals of Western Europe*, pp. 475–80. British Museum, London, UK.

Millington, A. C., Walsh, S. J., and Osborne, P. E. (eds.) (2001) *GIS and Remote Sensing Applications in Biogeography and Ecology*. Springer-Verlag, New York.

Mills, J. A. and Jackson, P. (1994) *Killed for a Cure: A Review of the Worldwide Trade in Tiger Bone*. TRAFFIC Report, Cambridge, UK.

Mills, L. S., Pilgrim, K. L., Schwartz, M. K., and McKelvey, K. (2000) Identifying lynx and other North American felids based on mtDNA analysis. *Conservation Genetics*, **1**, 285–8.

Mills, M. G. L. (1996) Methodological advances in capture, census, and food-habits studies of large African carnivores. In J. L. Gittleman (ed.), *Carnivore Behaviour, Ecology, and Evolution*, pp. 223–42. Cornell University Press, Ithaca, NY.

Mills, M. G. L. (1992) A comparison of methods used to study food habits of large African carnivores. In D. R. McCullough and R. H. Barrett (eds.), *Wildlife 2001: Populations*, pp. 1112–24. Elsevier, New York.

Mills, M. G. L. (1991) Conservation management of large carnivores in Africa. *Koedoe*, **34**, 81–90.

Mills, M. G. L. (1990) *Kalahari Hyaenas: Comparative Behavioural Ecology of Two Species*. The Blackburn Press, Caldwell, NJ.

Mills, M. G. L. (1984) Prey selection and feeding habits of the large carnivores in the southern Kalahari. *Koedoe*, **27**, 281–94.

Mills, M. G. L. and Biggs, H. C. (1993) Prey apportionment and related ecological relationships between large carnivores in Kruger National Park. In N. Dunstone and M. L. Gorman (eds.), *Mammals as Predators*, pp. 253–68. Symposium of the Zoological Society of London, No. 65, London, UK.

Mills, M. G. L. and Retief, P. F. (1984) The response of ungulates to rainfall along riverbeds of the southern Kalahari, 1972–1982. *Koedoe*, **27**, 129–42.

Mills, M. G. L. and Shenk, T. M. (1992) Predator–prey relationships: the impact of lion predation on wildebeest and zebra populations. *Journal of Animal Ecology*, **61**, 693–702.

Mills, M. G. L., Broomhall, L. S., and Du Toit, J. T. (2004) Cheetah *Acinonyx jubatus* feeding ecology in the Kruger National Park and a comparison across African savanna habitats: is the cheetah only a successful hunter on open grassland plains? *Wildlife Biology*, **10**, 177–86.

Mills, M. G. L., Juritz, J. M., and Zucchini, W. (2001) Estimating the size of spotted hyaena (*Crocuta crocuta*) populations through playback recordings allowing for nonresponse. *Animal Conservation*, **4**, 335–43.

Mills, M. G. L., Biggs, H. C., and Whyte, I. J. (1995) The relationship between rainfall, lion predation and population trends in African herbivores. *Wildlife Research*, **22**, 75–87.

Mills, M. G. L. and Gorman, M. L. (1997) Factors affecting the density and distribution of wild dogs in the Kruger National Park. *Conservation Biology*, **11**, 1397–406.

Millward, I. R. and Williams, M. C. (2005) *Cryptococcus neoformans granuloma* in the lung and spinal cord of a free-ranging cheetah (*Acinonyx jubatus*). A clinical report and literature review. *Journal of the South African Veterinary Association*, **76**, 228–32.

Miquelle, D. G. and Zhang, E. (2005) A proposed international system of protected areas for Amur tigers. In *Recovery of the Wild Amur Tiger Population in China: Progress and Prospect: Proceedings of the 2000 International Workshop on Wild Amur Tiger Population Recovery Action Plan, Harbin, China and the 2002 National Worshop on Progress of Wild Amur Tiger Population Recovery Action, Hunchun, China.*, pp. 164–70. China Forestry Publishing House, Beijing, China.

Miquelle, D. G., Goodrich, J. M., Kerley, L. L., *et al.* (in press) Science-based conservation of amur tigers in Russian Far East and northeast China. In R. Tilson and P. Nyhus (eds.), *Tigers of the World*. Elsevier, New York.

Miquelle, D. G., Pikunov, D. G., Dunishenko, Y. M., *et al.* (2007) 2005 Amur tiger census. *Cat News*, **46**, 11–14.

Miquelle, D. G., Pikunov, D. G., Dunishenko, Y. M., *et al.* (2006) *A Theoretical Basis for Surveys of Amur Tigers and their Prey Base in the Russian Far East*. DalNauka, Vladivostok, Russia.

Miquelle, D. G., Nikolaev, I., Goodrich, J. M., Litvinov, B., Smirnov, E. N., and Suvorov, E. (2005a) Seaching for the coexistence recipe: a case study of conflicts between people and tigers in the Russian Far East. In R. Woodroffe, S. Thirgood, and A. Rabinowitz (eds.), *People and Wildlife: Conflict or Coexistence?*, pp. 305–22. Cambridge University Press, Cambridge, UK.

Miquelle, D. G., Smirnov, E. N., Salkina, G. P., and Abramov, V. K. (2005b) The importance of protected areas in Amur tiger conservation: a comparison of tiger and prey abundance in protected areas versus unprotected areas. In I. V. Potikha, D. G. Miquelle, M. N. Gromyko, *et al.* (eds.) *Results of Protection and Research of the Sikhote-Alin Natural Landscape. Papers Presented at the International Science and Management Conference devoted to the 70th Anniversary of the Sikhote-Alin State Reserve, Terney, Primorksiy Region, September 20–23, 2005*. Primpoligraphkombinat, Vladivostock, Russia.

Miquelle, D. G., Stevens, P. A., Smirnov, E. N., Goodrich, J. M., Zaumyslava, O. J., and Myslenkov, A. E. (2005c) Competitive exclusion, functional redundancy, and conservation implications: tigers and wolves in the Russian Far East. In J. Ray, K. Redford, R. S. Steneck, and J. Berger (eds.), *Large Carnivores and the Conservation of Biodiversity*, pp. 179–207. Island Press, Washington DC.

Miquelle, D. G., Merrill, T. W., Dunishenko, Y. M., *et al.* (1999a) A habitat protection plan for the Amur tiger: developing political and ecological criteria for a viable land-use plan. In J. Seidensticker, S. Christie, and P. Jackson (eds.), *Riding the Tiger: Tiger Conservation in Human Dominated Landscapes*, pp. 273–95. Cambridge University Press, Cambridge, UK.

Miquelle, D. G., Smirnov, E. N., Merrill, T. W., Myslenkov, A. E., Quigley, H. B., and Hornocker, M. G. (1999b) Hierarchical spatial analysis of Amur tiger relationships to habitat and prey. In J. Seidensticker, S. Christie, and P. Jackson (eds.), *Riding the Tiger: Tiger Conservation in Human Dominated Landscapes*, pp. 71–99. Cambridge University Press, Cambridge, UK.

Miquelle, D. G., Smirnov, E. N., Quigley, H. B., Hornocker, M. G., Nikolaev, E. G., and Matyushkin, E. N. (1996) Food habits of Amur tigers in Sikhote-Alin Zapovednik and the Russian Far East, and implications for conservation. *Journal of Wildlife Research*, **1**, 138–47.

Miric, D. (1978) *Lynx lynx martinoi* ssp *nova* (Carnivora, Mammalia)—neue Luchsunterart von der Balkanhalbinsel. *Bulletin du Museum d'Histoire Naturelle*, **33**, 29–36.

Mirutenko, V. S. (1986) Reliability of the analysis of route surveys of game animals. In A. M. Amirkhanov (ed.), *Issues of Game Animal Censuses*, pp. 21–8. Central Research Laboratory of Glavokhota, Moscow, Russia.

Mishra, C. (2003) Fur trade and the snow leopard in Afghanistan. *CMS Bulletin*, **17**, 4–5.

Mishra, C. (2001) *High altitude survival: conflicts between pastoralism and wildlife in the Trans-Himalaya*. PhD dissertation, Wageningen University, Wageningen, the Netherlands.

Mishra, C. (2000) Socioeconomic transition and wildlife conservation in the Indian trans-Himalaya. *Journal of the Bombay Natural History Society*, **97**, 25–32.

Mishra, C. (1997) Livestock depredation by large carnivores in the Indian trans-Himalaya: conflict perceptions and conservation prospects. *Environmental Conservation*, **24**, 338–43.

Mishra, C. and Fitzherbert, A. (2004) War and wildlife: a post-conflict assessment of Afghanistan's Wakhan Corridor. *Oryx*, **38**, 102–5.

Mishra, C., Madhusudan, M. D., and Datta, A. (2006) Mammals of the high altitudes of western Arunachal Pradesh, eastern Himalaya: an assessment of threats and conservation needs. *Oryx*, **40**, 29–35.

Mishra, C., Bagchi, S., Namgail, T., and Bhatnagar, Y.V. (2010) Multiple use of Trans Himalayan Rangelands: reconciling human livelihoods with wildlife conservation, pp. 291–311. In: du Toit, J.T, Kock, R. and Deutsch (Eds.) *Wild Rangelands: Conserving Wildlife While Maintaining Livestock in Semi-Arid Ecosystems*. ZSL, CS & Wiley-Blackwell, Chennai. 424p.

Mishra, C., Van Wieren, S. E., Ketner, P., Heitkonig, I. M. A., and Prins, H. H. T. (2004) Competition between domestic livestock and wild bharal *Pseudois nayaur* in the Indian Trans-Himalaya. *Journal of Applied Ecology*, **41**, 344–54.

Mishra, C., Allen, P., McCarthy, T., Madhusudan, M. D., Bayarjargal, A., and Prins, H. H. T. (2003a) The role of incentive programs in conserving the snow leopard *Uncia uncia*. *Conservation Biology*, **17**, 1512–20.

Mishra, C., Prins, H. H. T., and Van Wieren, S. E. (2003b) Diversity, risk mediation, and change in a trans-Himalayan agropastoral system. *Human Ecology*, **31**, 595–609.

Mishra, C., Van Wieren, S. E., Heitkonig, I. M. A., and Prins, H. H. T. (2002) A theoretical analysis of competitive exclusion in a Trans-Himalayan large-herbivore assemblage. *Animal Conservation*, **5**, 251–8.

Mishra, C., Prins, H. H. T., and van Wieren, S. E. (2001) Overstocking in the trans-Himalayan rangelands of India. *Environmental Conservation*, **28**, 279–83.

Mishra, H. R. and Wemmer, C. (1987) Comparative breeding ecology of four cervids in Royal Chitwan National Park, Nepal. In C. Wemmer (ed.), *Biology and Management of the Cervidae*, pp. 259–71. Smithsonian Institution Press, Washington, DC.

Miththapala, S., Seidensticker, J., and O'Brien, S. J. (1996) Phylogeographic subspecies recognition in leopards (*Panthera pardus*): molecular genetic variation. *Conservation Biology*, **10**, 1115–32.

Mitra, B. (2006) *Sell the Tiger to Save It. New York Times*, 15 Aug. New York.

Mittermeier, R., Kormos, C. F., Mittermeier, C., Robles Gil, P., Sandwith, T., and Besançon, C. (eds.) (2005) *Transboundary Conservation: A New Vision for Protected Areas*. CEMEX/ Conservation International, Arlington, VA.

Mizutani, F. (1999) Impact of leopards on a working ranch in Laikipia, Kenya. *African Journal of Ecology*, **37**, 211–25.

Mizutani, F. (1997) Can livestock ranches and leopards coexist? A case study in Kenya. In B. L. Penzhorn (ed.), *Proceedings of a Symposium on Lions and Leopards as Game Ranch Animals*, pp. 1–25. Onderstepoort, South Africa.

Mizutani, F. and Jewell, P. A. (1998) Home range and movements of leopards (*Panthera pardus*) on a livestock ranch in Kenya. *Journal of Zoology*, **244**, 269–86.

Mlengeya, T. D. K., Muangirwa, C., Mlengeya, M. M., Kimaro, E., Msangi, S., and Sikay, M. (2002) Control of sleeping sickness in northern parks of Tanzania. In *Proceedings of the 3rd Annual Scientific Conference of the Tanzania Wildlife Research Institute, Arusha, December 3–5, 2002*, pp. 274–82.

Moa, P. F., Herfindal, I., Linnell, J. D. C., Overskaug, K., Kvam, T., and Andersen, R. (2006) Does the spatiotemporal distribution of livestock influence forage patch selection in Eurasian lynx, *Lynx lynx*. *Wildlife Biology*, **12**, 63–70.

Mochizuki, M., Hiragi, H., Sueyoshi, M., *et al.* (1996) Antigenic and genomic characteristics of parvovirus isolated from a lion (*Panthera leo*) that died of feline panleukopenia. *Journal of Zoo and Wildlife Medicine*, **27**, 416–20.

Mochizuki, M., Akuzawa, M., and Nagatomo, H. (1990) Serological survey of the Iriomote cat (*Felis iriomotensis*) in Japan. *Journal of Wildlife Diseases*, **26**, 236–45.

Moehrenschlager, A. and List, R. (1996) Comparative ecology of North American prairie foxes—conservation through collaboration. In D. W. Macdonald and F. Tattersall (eds.), *The WildCru Review*, pp. 22–8. Wildlife Conservation Research Unit, Oxford, UK.

Moehrenschlager, A., Cypher, B., Ralls, K., Sovada, M. A., and List, R. (2004) Comparative ecology and conservation priorities of swift and kit foxes. In D. W. Macdonald and C. Sillero-Zubiri (eds.), *The Biology and Conservation of Wild Canids*, pp. 185–98. Oxford University Press, Oxford, UK.

Mohr, C. O. (1947) Table of equivalent populations of North American small mammals. *American Midland Naturalist*, **37**, 223–49.

Molia, S., Chomel, B. B., Kasten, R. W., *et al.* (2004) Prevalence of *Bartonella* infection in wild African lions (*Panthera leo*) and cheetahs (*Acinonyx jubatus*). *Veterinary Microbiology*, **100**, 31–41.

Molinari-Jobin, A., Zimmermann, F., Ryser, A., *et al.* (2007) Does lynx home range size adjust to prey abundance in Switzerland. *Wildlife Biology*, **13**, 393–405.

Molinari-Jobin, A., Molinari, P., Loison, A., Gaillard, J. M., and Breitenmoser, U. (2004) Life cycle period and acivity of prey influence their susceptibility to predators. *Ecography*, **27**, 323–9.

Molinari-Jobin, A., Molinari, P., Breitenmoser-Würsten, C., and Breitenmoser, U. (2002) Significance of lynx *Lynx lynx* predation for roe deer *Capreolus capreolus* and chamois *Rupicapra rupicapra* mortality in the Swiss Jura Mountains. *Wildlife Biology*, **8**, 109–15.

Molteno, A. J., Sliwa, A., and Richardson, P. R. K. (1998) The role of scent marking in a free-ranging, female black-footed cat (*Felis nigripes*). *Journal of Zoology*, **245**, 35–41.

Mondolfi, E. (1986) Notes on the biology and status of the small wild cats in Venezuela. In S. D. Miller and D. D. Everett (eds.), *Cats of the World: Biology, Conservation, and Management*, pp. 125–46. National Wildlife Federation, Washington DC.

Monies, B., de la Rua, R., and Jahans, K. (2006) Bovine tuberculosis in cats. *Veterinary Record*, **158**, 490–1.

Monroy-Vilchis, O. and Velazquez, A. (2003) Regional distribution and abundance of bobcats (*Lynx rufus escuinape*) and coyotes (*Canis latrans cagottis*), as measured by scent stations: a spatial approach. *Ciencias Naturales y Agropecuarias*, **9**, 293–300.

Montali, R. J., Mikota, S. K., and Cheng, L. I. (2001) *Mycobacterium tuberculosis* in zoo and wildlife species. *Revue Scientifique et Technique de l' Office International des Epizooties*, **20**, 291–303.

Moolman, L. C. (1986) *Aspekte van die ekologie en gedrag van die rooikat* Felis caracal *Schreber, 1776 in the Mountain Zebra National Park and surrounding farms] (in Afrikaans)*. MSc dissertation, University of Pretoria, Pretoria, South Africa.

Morais, R. N., Mucciolo, R. G., Gomes, M. L. F., *et al.* (2002) Seasonal analysis of semen characteristics, serum testosterone and fecal androgens in the ocelot (*Leopardus pardalis*), margay (*L. wiedii*) and tigrina (*L. tigrinus*). *Theriogenology*, **57**, 2027–41.

Morales, J., Pickford, M., Fraile, S., Salesa, M. J., and Soria, D. (2003a) Creodonta and Carnivora from Arrisdrift, early Middle Miocene of southern Namibia. *Memoirs of the Geological Survey of Namibia*, **19**, 177–94.

Morales, J., Soria, D., Montoya, P., Pérez, B. and Salesa, M. J. (2003b) *Caracal depereti* nov. sp. and *Felis* aff. *silvestris* (Felidae, Mammalia) of the lower Pliocene of Layna (Soria, Spain). *Estudios Geológicos*, **59**, 229–47.

Morales, J., Salesa, M. J., Pickford, M., and Soria, D. (2001) A new tribe, new genus and two new species of Barbourofelinae (Felidae, Carnivora, Mammalia) from the early Miocene of East Africa and Spain. *Transactions of the Royal Society of Edinburgh, Earth Sciences*, **92**, 97–102.

Morales, J., Pickford, M., Soria, D., and Fraile, S. (1998) New carnivores from the basal Middle Miocene of Arrisdrift, Namibia. *Eclogae Geologicae Helvetiae*, **91**, 27–40.

Morato, R. G., Conforti, V. A., Azevedo, F. C. C., de, *et al.* (2001) Comparative analyses of semen and endocrine characteristics of free-living versus captive jaguars (*Panthera onca*). *Reproduction*, **122**, 745–51.

Moreira, N., Brown, J. L., Moraes, W., Swanson, W. F., and Monteiro-Filho, E. L. A. (2007) Effect of housing and environmental enrichment on adrenocortical activity, behavior and reproductive cyclicity in the female tigrina (*Leopardus tigrinus*) and margay (*Leopardus wiedii*). *Zoo Biology*, **26**, 441–60.

Moreira, N., Monteiro-Filho, E. L. A., Moraes, W., *et al.* (2001) Reproductive steroid hormones and ovarian activity in felids of the *Leopardus* genus. *Zoo Biology*, **20**, 103–16.

Moreno, R. S. (2005) Observaciones sobre un evento antagónica en ocelots (*Leopardus pardalis*). *Technociencia*, **7**, 173–7.

Moreno, R. S. and Giacalone, J. (2006) Ecological data obtained from latrine use by ocelots (*Leopardus pardalis*) on Barro Colorado Island, Panama. *Technociencia*, **8**, 7–21.

Moreno, S. and Villafuerte, R. (1995) Traditional management of scrubland for the conservation of rabbits *Oryctolagus cuniculus* and their predators in Doñana National Park, Spain. *Biological Conservation*, **73**, 81–5.

Moreno, S., Beltrán, J. F., Cotilla, I., *et al.* (2007) Long-term decline of the European wild rabbit (*Oryctolagus cuniculus*) in south-western Spain. *Wildlife Research*, **34**, 652–8.

Moreno, R. S., Kays, R. W., and Samudio, R. (2006) Competitive release in diets of ocelot (*Leopardus pardalis*) and puma (*Puma concolor*) after jaguar (*Panthera onca*) decline. *Journal of Mammalogy*, **87**, 808–16.

Moreno, S., Villafuerte, R., Cabezas, S., and Lombardi, L. (2004) Wild rabbit restocking for predator conservation in Spain. *Biological Conservation*, **118**, 183–93.

Morin, P. A., Luikart, G., and Wayne, R. K., and the SNP Workshop Group (2004) SNPs in ecology, evolution and conservation. *Trends in Ecology and Evolution*, **19**, 208–16.

Morlo, M. (1997) Die raubtiere (Mammalia, Carnivora) aus dem Turolium von Dorn-Durkheim l (Rheinhessen). Teil 1: Mustelida, Hyaenidae, Percrocutidae, Felidae. *CFS Courier Forschungsinstitut Senckenberg*, **197**, 11–47.

Morlo, M., and Semenov, Y. (2004) New dental remains of *Machairodus* Kaup 1833 (Felidae, Carnivora, Mammalia) from the Turolian of Ukraine: significance for the evolution of the genus. *Kaupia*, **13**, 123–38.

Morlo, M., Peigné, S., and Nagel, D. (2004) A new species of *Prosansanosmilus*: implications for the systematic relationships of the family Barbourofelidae new rank (Carnivora, Mammalia). *Zoological Journal of the Linnean Society*, **140**, 43–61.

Mörner, T. (1992) Sarcoptic mange in Swedish wildlife. *OIE Revue Scientifique et Technique*, **11**, 1115–21.

Mörner, T., Eriksson, H., Brojer, C., *et al.* (2005) Diseases and mortality in free-ranging brown bear (*Ursus arctos*), gray wolf (*Canis lupus*), and Wolverine (*Gulo gulo*) in Sweden. *Journal of Wildlife Diseases*, **41**, 298–303.

Morrison, J. C., Sechrest, W., Dinerstein, E., Wilcove, D. S., and Lamoreux, J. F. (2007) Persistence of large mammal faunas as indicators of global human impacts. *Journal of Mammalogy*, **88**, 1363–80.

Morrison, L. J., Majiwa, P., Read, A. F., and Barry, J. D. (2005) Probabilistic order in antigenic variation of *Trypanosoma brucei*. *International Journal for Parasitology*, **35**, 961–72.

Morsbach, D. (1987) Cheetah in Namibia. *Cat News*, **6**, 25–6.

Mosser, A. (2008) *Serengeti real estate: a lion's eye view of the Savanna landscape*. PhD dissertation, University of Minnesota, St Paul, MN.

Mosser, A., Fryxell, J., Eberly, L. and Packer, C. (2009) Serengeti real estate: density versus fitness-based indicators of lion habitat quality. *Ecology Letters*, **12**, 1050–60.

Mosser, A., and Packer, C. (2009) Group territoriality and the benefits of sociality in the African lion, *Panthera leo*. *Animal Behaviour*, **78**, 359–70.

Mountford, G. (1981) *Saving the Tiger*. Joseph-Viking, London, UK.

Mourão, G. M., Coutinho, M. E., Mauro, R. A., Tomás, W. M., and Magnusson, W. (2002) *Levantamentos aéreos de espécies introduzidas no Pantanal: porcos ferais (porco monteiro), gado bovino e búfalos*. Embrapa Pantanal, Corumbá, Brazil.

Mowat, G. and Slough, B. (2003) Habitat preference of Canada lynx through a cycle in snowshoe hare abundance. *Canadian Journal of Zoology*, **81**, 1736–45.

Mowat, G. and Slough, B. G. (1998) Some observations on the natural history and behaviour of the Canada Lynx, *Lynx canadensis*. *Canadian Field-Naturalist*, **112**, 32–6.

Mowat, G., Poole, K. G., and O'Donoghue, M. (2000) Ecology of lynx in northern Canada and Alaska. In L. F. Ruggiero, K. B. Aubry, S. W. Buskirk, *et al.* (eds.), *Ecology and Conservation of Lynx in the United States*, pp. 265–306. University Press of Colorado, Boulder, CO.

Mowat, G., Slough, B. G., and Boutin, S. (1996) Lynx recruitment during a snowshoe hare population peak and decline in southwest Yukon. *Journal of Wildlife Management*, **60**, 441–52.

Mukherjee, S. D. (1998) Small cats in Andhra Pradesh. *Envis—Wildlife and Protected Areas*, **1**, 21–21.

Mukherjee S. (1989) *Ecological separation of four sympatric carnivores in Keoladeo Ghana National Park, Bharatpur, Rajasthan, India*. MS thesis dissertation, Wildlife Institute of India, Dehradun, India.

Mukherjee, S., Goyal, S. P., Johnsingh, A. J. T., and Pitman, M. R. P. L. (2004) The importance of rodents in the diet of jungle cat (*Felis chaus*), caracal (*Caracal caracal*) and golden jackal (*Canis aureus*) in Sariska Tiger Reserve, Rajasthan, India. *Journal of Zoology*, **262**, 405–11.

Muller-Graf, C. D. M. (1995) A coprological survey of intestinal parasites of wild lions (*Panthera leo*) in the Serengeti and the Ngorongoro Crater, Tanzania, East Africa. *Journal of Parasitology*, **81**, 812–14.

Muller-Graf, C. D. M., Woolhouse, M. E. J., and Packer, C. (1999) Epidemiology of an intestinal parasite (*Spirometra* spp.) in two populations of African lions (*Panthera leo*). *Parasitology*, **118**, 407–15.

Muñoz-Pedreros, A., Rau, J. R., Valdebenito, M., Quintana, V., and Martínez, D. R. (1995) Densidad relativa de pumas (*Felis concolor*) en un ecosistema forestal del sur de Chile. *Revista Chilena de Historia Natural*, **68**, 501–7.

Munson, L. (2001) Feline *Morbillivirus* infections. In E. S. Williams and I. K. Barker (eds.), *Infectious Diseases of Wild Mammals*, pp. 59–62. Iowa State University, Ames, IA.

Munson, L. (1993) Diseases of captive cheetahs (*Acinonyx jubatus*): results of the Cheetah Council Pathology Survey, 1989–1992. *Zoo Biology*, **12**, 105–24.

Munson, L., Brown, J. L., Bush, M., *et al.* (1996) Genetic diversity affects testicular morphology in free-ranging lions (*Panthera leo*) of the Serengeti Plains and Ngorongoro Crater. *Journal of Reproduction and Fertility*, **108**, 11–15.

Munson, L. and Citino, S. B. (2005) *Proceedings of the AZA Cheetah SSP Disease Management Workshop, White Oak Conservation Center, Yulee, FL, June 8–12, 2005*. American Zoo Association Cheetah SSP, Yulee, FL.

Munson, L. and Karesh, W. B. (2002) Disease monitoring for the conservation of terrestrial animals. In A. A. Aguirre, R. S. Ostfeld, G. M. Tabor, C. House, and M. C. Pearl (eds.), *Conservation Medicine: Ecological Health in Practice*, pp. 95–103. Oxford University Press, Oxford, UK.

Munson, L., Terio, K. A., Kock, R., *et al.* (2008) Climate extremes promote fatal co-infections during canine distemper epidemics in African lions. *PLoS ONE*, **3**, e2545.

Munson, L., Terio, K. A., Worley, M., Jago, M., Bagot-Smith, A., and Marker, L. (2005) Extrinsic factors significantly affect patterns of disease in free-ranging and captive cheetah (*Acinonyx jubatus*) populations. *Journal of Wildlife Diseases*, **41**, 542–8.

Munson, L., Marker, L., Dubovi, E., Spencer, J. A., Evermann, J. F., and O'Brien, S. J. (2004a) Serosurvey of viral infections in free-ranging Namibian cheetahs (*Acinonyx jubatus*). *Journal of Wildlife Diseases*, **40**, 23–31.

Munson, L., Wack, R., Duncan, M., *et al.* (2004b) Chronic eosinophilic dermatitis associated with persistent feline herpes virus infection in cheetahs (*Acinonyx jubatus*). *Veterinary Pathology*, **41**, 170–6.

Munson, L., Wack, R., Duncan, M., *et al.* (2003) Chronic eosinophilic dermatitis associated with persistent feline herpes virus infection in cheetahs (*Acinonyx jubatus*). *Veterinary Pathology*, **41**, 170–6.

Munson, L., de Lahunta, A., Citino, S. B., *et al.* (1999a) Leukoencephalopathy in cheetahs. In *Annual Meeting of the American Associationn of Zoo Veterinarians, Orlando, FL, 69.*

Munson, L., Nesbit, J. W., Kriek, N. P. J., Meltzer, D. G. A., Colly, L. P., and Bolton, L. (1999b) Diseases of captive cheetahs (*Acinonyx jubatus jubatus*) in South Africa: a 20-year retrospective survey. *Journal of Zoo and Wildlife Medicine*, **30**, 342–7.

Muntifering, J. R., Dickman, A. J., Perlow, L. M., *et al.* (2006) Managing the matrix for large carnivores: a novel approach and perspective from cheetah (*Acinonyx jubatus*) habitat suitability modelling. *Animal Conservation*, **9**, 103–12.

Murdoch, J. D., Munkhzul, T., and Reading, R. P. (2006) Pallas's cat ecology and conservation in the semi-desert steppes of Mongolia. *Cat News*, **45**, 18–19.

Murdoch, W. W. (1969) Switching in generalist predators: experiments on prey specificity and stability of prey populations. *Ecological Monographs*, **39**, 335–54.

Murombedzi, J. C. (1999) Devolution and stewardship in Zimbabwe's campfire programme. *Journal of International Development*, **11**, 287–93.

Murphy, K. (1983) *Characteristics of a hunted population of mountain lions in western Montana*. MS dissertation, University of Montana, Missoula, MT.

Murray, D. L. and Boutin, S. (1991) The influence of snow on lynx and coyote movements: does morphology affect behavior? *Oecologia*, **88**, 463–9.

Murray, D. L., Boutin, S., O'Donoghue, M., and Nams, V. O. (1995) Hunting behaviour of a sympatric felid and canid in relation to vegetative cover. *Animal Behaviour*, **50**, 1203–10.

Murray, D. L., Boutin, S., and O'Donoghue, M. (1994) Winter habitat selection by lynx and coyotes in relation to snowshoe hare abundance. *Canadian Journal of Zoology*, **72**, 1444–51.

Murray, J. D. (1988) How the leopard got its spots. *Scientific American*, **3/1988**, 62–9.

Muul, I. and Lim, B.-L. (1970) Ecological and morphological observations of *Felis planiceps*. *Journal of Mammalogy*, **51**, 806–8.

Mwanzia, J. M., Kock, R. A., Wamba, J. M., Kock, N. D., and Jarrett, O. (1995) An outbreak of sarcoptic mange in free living cheetah (*Acinonyx jubatus*) in the Mara region of Kenya. In *Proceedings of the American Association of Zoo Veterinarians and American Association of Wildlife Veterinarians Joint Conference, Omaha*, pp. 104–14.

Myers, E. C. (2007) Policies to reduce emissions from deforestation and degradation (REDD) in tropical forests: an examination of the issues facing the incorporation of REDD into market-based climate policies. Discussion Paper. Resources for the Future, Washington DC. <www.rff.org/rff/Documents/RFF-DP-07-50.pdf>. Accessed online 10 May 2009.

Myers, N. (1976) *The leopard* Panthera pardus *in Africa*. IUCN, Morges, Switzerland.

Myers, N. (1975) The cheetah *Acinonyx jubatus* in Africa. *IUCN Monograph No. 4*. IUCN, Morges, Switzerland.

Myers, N. (1973) The spotted cats and the fur trade. In R. L. Eaton (ed.), *The World's Cats*, pp. 276–323. World Wildlife Safari, Winston, OR.

NACSO. (2008) *Namibia's Communal Conservancies: A Review of Progress and Challenges in 2007*. Namibia Association of CBNRM Support Organisations, Windhoek, Namibia.

NACSO. (2006) *Namibia's Communal Conservancies: A Review of Progress and Challenges in 2005*. Namibia Association of CBNRM Support Organisations, Windhoek, Namibia.

Nagata, J., Aramilev, V. V., Belozor, A., Sugimoto, T., and McCullough, D. R. (2005) Fecal genetic analysis using PCR-RFLP of cytochrome b to identify sympatric carnivores, the tiger *Panthera tigris* and the leopard *Panthera pardus*, in far eastern Russia. *Conservation Genetics*, **6**, 863–5.

Naidenko, S. V., and Antonevich, A. L. (2009) Sibling aggression in Eurasian lynx (*Lynx lynx*). In A. Vargas, C Breitenmoser and U. Breitenmoser (eds.). *Iberian Lynx Ex Situ Conservation: An Interdisciplinary Approach*, pp. 146–55. Fundacion Biodiversidad, Madrid.

Namgail, T., Bhatnagar, Y. V., Mishra, C., and Bagchi, S. (2007a) Pastoral nomads of the Indian Changthang: production system, landuse and socioeconomic changes. *Human Ecology*, **35**, 497–504.

Namgail, T., Fox, J. L., and Bhatnagar, Y. V. (2007b) Carnivore-caused livestock mortality in Trans-Himalaya. *Environmental Management*, **39**, 490–6.

Nakano-Oliveira, E. (2002) *Ecologia alimentar e área de vida de carnívoros da Floresta Nacional de Ipanema, Iperó, SP (Carnivora: Mammalia)*. MS dissertation, Universidade de Campinas, Campinas, Brazil.

Napolitano, C., Bennett, M., Johnson, W. E., *et al.* (2008) Ecological and biogeographical inferences on two sympatric and enigmatic Andean cat species using genetic identification of faecal samples. *Molecular Ecology*, **17**, 678–90.

Nations, J., Manterola, C., Rodriguez, C., and Hernandez, R. (2005) The Maya Tropical Forest. Saving a threatened tri-national biological and cultural heritage. In R. Mittermeier, C. F. Kormos, C. Mittermeier, P. Robles Gil, T. Sandwith, and C. Besançon (eds.) *Transboundary Conservation: A New Vision for Protected Areas*, pp. 147–56. CEMEX/Conservation International, Arlington, VA.

Natoli, E., Schmid, M., Say, L., and Pontier, D. (2007) Male reproductive success in a social group of urban feral cats (*Felis catus* L.). *Ethology*, **113**, 283–9.

Natoli, E., De Vito, E., and Pontier, D. (2000) Mate choice in the domestic cat (*Felis silvestris catus* L.). *Aggressive Behavior*, **26**, 455–65.

Nature Conservation Division (2004) *Bhutan Biological Conservation Complex (Living in Harmony with Nature). A Landscape Conservation Plan: A Way Forward*. Thimphu Nature Conservation Division, Royal Government of Bhutan, and WWF Bhutan, Thimphu, Bhutan.

Naughton-Treves, L., Treves, A., and Grossberg, R. (2003) Paying for tolerance: rural citizens' attitudes toward wolf depredation and compensation. *Conservation Biology*, **17**, 1500–11.

Ndandi Namibia Safaris (2008) English hunting safaris: price list. <www.ndandi.com/english/hunting_safaris/pricelist.html>. Accessed online 15 Sept 2008.

Neff, N. A. (1983) *The basicranial anatomy of the Nimravidae (Mammalia: Carnivora): character analyses and phylogenetic inferences*. PhD dissertation, City University of New York, New York.

Nekaris, K. A. I. (2003) Notes on the distribution and behaviour of three small wild cats in Sri Lanka. *Cat News*, **38**, 30–1.

Nel, J. A. J., Rautenbach, I. L., Els, D. A., and De Graaf, G. (1984) The rodents and other small mammals of the Kalahari Gemsbok National Park. *Koedoe Supplement*, **1984**, 195–220.

Nellis, C. H. and Keith, L. B. (1968) Hunting activities and success of lynxes in Alberta. *Journal of Wildlife Management*, **32**, 718–22.

Nellis, C. H. and Wetmore, S. P. (1969) Long-range movement of lynx in Alberta. *Journal of Mammalogy*, **50**, 640.

Nellis, C. H., Wetmore, S. P., and Keith, L. B. (1972) Lynx-prey interactions in central Alberta. *Journal of Wildlife Management*, **36**, 320–9.

Nepstad, D. C., Stinkler, C. M., and Almeida, O. T. (2006) Globalization of the Amazon soy and beef industries: opportunities for conservation. *Conservation Biology*, **20**, 1595–603.

Newell-Fugate, A., Kennedy-Stoskopf, S., Brown, J. L., Levine, J. F., and Swanson, W. F. (2007) Seminal and endocrine characteristics of male Pallas's cats (*Otocolobus manul*) maintained under artificial lighting with simulated natural photoperiods. *Zoo Biology*, **26**, 187–99.

Newell, D. M. (1935) *Panther! Saturday Evening Post*, **208**, 10–11, 70–2.

Newman, A., Bush, M. E., Wildt, D. E., *et al*. (1985) Biochemical genetic variation in eight endangered or threatened Felid species. *Journal of Mammalogy*, **66**, 256–67.

Ng, J. and Nemora (2007) *Tiger Trade Revisited in Sumatra, Indonesia*. TRAFFIC Southeast Asia, Petaling Jaya, Malaysia.

Niedzialkowska, M., Jedrzejewski, W., Myslajek, R. W., Nowak, S., Jedrzejewska, B., and Schmidt, K. (2006) Environmental correlates of Eurasian lynx occurrence in Poland—large scale census and GIS mapping. *Biological Conservation*, **133**, 63–9.

Nielson, C. K. and Woolf, A. (2001) Spatial organization of bobcats (*Lynx rufus*) in southern Illinois. *American Midland Naturalist*, **146**, 43–52.

Nietfeld, J. C. and Pollock, C. (2002) Fatal cytauxzoonosis in a free-ranging bobcat (*Lynx rufus*). *Journal of Wildlife Diseases*, **38**, 607–10.

Nimmervoll, H. (2007) *Sarcoptic mange in red foxes* (Vulpes vulpes) *from Switzerland—pathological characteristics and influencing factors*. MedVet dissertation, University of Berne, Berne, Switzerland.

Nishimura, Y., Goto, Y., Endo, Y., *et al*. (1999) Interspecies transmission of feline immunodeficiency virus from the domestic cat to the tsushima cat (*Felis bengalensis euptilura*) in the wild. *Journal of Virology*, **73**, 7916–21.

Nordstrom, L. (2005) *Recovery Outline: Contiguous United States Distinct Population Segment of the Canada Lynx*. United States Fish and Wildlife Service, Helena, MT.

Norton-Griffiths, M. (2007) How many wildebeest do you need? *World Economics*, **8**, 41–64.

Norton-Griffiths, M. (1998) The economics of wildlife conservation in Kenya. In E. J. Milner-Gulland and R. Mace (eds.), *Biological Conservation and Sustainable Use*, pp. 279–93. Blackwells, Oxford, UK.

Norton-Griffiths, M. (1996) Property rights and the marginal wildebeest: an economic analysis of wildlife conservation options in Kenya. *Biodiversity and Conservation*, **5**, 1557–77.

Norton-Griffiths, M. (1995) The kiss of death or does conservation work? <www.mng5.com/papers/DeathKiss.pdf>. Accessed online 10 May 2009.

Norton-Griffiths, M. and Southley, C. (1995) The opportunity costs of biodiversity conservation in Kenya. *Ecological Economics*, **12**, 125–39.

Norton, P. M. (1990) How many leopards? A criticism of Martin and De Meulenaer's population estimates for Africa. *South African Journal of Science*, **86**, 218–20.

Norton, P. M. (1986) Historical Changes in the distribution of leopards in the Cape Province, South Africa. *Bontebok*, **5**, 1–9.

Noss, R. F., Quigley, H. B., Hornocker, M. G., Merrill, T., and Paquet, P. C. (1996) Conservation biology and carnivore conservation in the Rocky Mountains. *Conservation Biology*, **10**, 949–63.

Novack, A. J. (2003) *Impacts of subsistence hunting on the foraging ecology of jaguar and puma in the Maya Biosphere Reserve, Guatemala*. MS thesis, University of Florida, Gainesville, FL.

Novack, A. J., Main, M. B., Sunquist, M. E., and Labisky, R. F. (2005) Foraging ecology of jaguar (*Panthera onca*) and puma (*Puma concolor*) in hunted and non-hunted sites within the Maya Biosphere Reserve, Guatemala. *Journal of Zoology*, **267**, 167–78.

Nowak, R. M. and McBride, R. T. (1974) Status survey of the Florida panther. In *World Wildlife Fund Yearbook*,

1973–74. pp. 237–42. Museum of Natural History, University of Kansas, Lawrence, KS.

Nowell, K. (2009) Tiger farms and pharmacies: the central importance of China's trade policy for tiger conservation. In R. L. Tilson and P. Nyhus (eds.), *Tigers of the World: The Science, Politics and Conservation of Panthera tigris*, 2nd edn. William Andrew, Norwich, UK.

Nowell, K. (2007) *Asian Big Cat Conservation and Trade Control in Selected Range States: Evaluating Implementation and Effectiveness of CITES Recommendations*. TRAFFIC International, Cambridge, UK.

Nowell, K. (1996) *Namibian Cheetah Conservation Strategy*. Ministry of Environment and Tourism, Windhoek, Namibia.

Nowell, K. and Jackson, P. (1996) *Wild Cats: A Status Survey and Conservation Action Plan*. IUCN/ SSC Cat Specialist Group, Gland, Switzerland.

Nowell, K. and Jackson, P. (1992) Animal use: non food uses. In World Conservation Monitoring Centre (ed.), *Global Biodiversity: Status of the Earth's Living Resources*, p. 374. Chapman and Hall, London, UK.

Nowell, K. and Xu, L. (2007) *Taming the Tiger Trade: China's Markets for Wild and Captive Tiger Products since the 1993 Domestic Trade Ban*. TRAFFIC East Asia, Hong Kong.

Nowell, K., Bauer, H., and Breitenmoser, U. (2007) Cats at CITES COP14. *Cat News*, **47**, 33–4.

Núñez, R., Miller, B., and Lindzey, F. (2002) Ecología del jaguar en la reserva de la Biosfera Chamela-Cuixmala, Jalisco, México. In R. A. Medellín, C. Equihua, C. L. B. Chetkiewicz, *et al.* (eds.), *El Jaguar en el Nuevo Milenio*, pp. 107–26. Fondo de Cultura Economica, Universidad Nacional Autónoma de México and Wildlife Conservation Society, Mexico City, Mexico.

Núñez, R., Miller, B., and Lindzey, F. (2000) Food habits of jaguars and pumas in Jalisco, Mexico. *Journal of Zoology*, **252**, 373–9.

Nyhus, P. J. and Tilson, R. (2004a) Agroforestry, elephants, and tigers: balancing conservation theory and practice in human-dominated landscapes of Southeast Asia. *Agriculture, Ecosystems and Environment*, **104**, 87–97.

Nyhus, P. J. and Tilson, R. (2004b) Characterizing human-tiger conflict in Sumatra, Indonesia: implications for conservation. *Oryx*, **38**, 68–74.

Nyhus, P. J., Osofsky, S., Ferraro, P., Madden, F., and Fischer, H. (2005) Bearing the costs of human-wildlife conflict: the challenges of compensation schemes. In R. Woodroffe, S. Thirgood, and A. Rabinowitz (eds.), *People and Wildlife: Conflict and Coexistence*, pp. 107–21. Cambridge University Press, Cambridge, UK.

Nyhus, P. J., Fischer, H., F., M., and Osofsky, S. (2003) Taking the bite out of wildlife damage: the challenges of wildlife compensation schemes. *Conservation in Practice*, **4**, 37–40.

O'Brien, S. J. (1994) A role for molecular genetics in biological conservation. *Proceedings of the National Academy of Sciences of the United States of America*, **91**, 5748–55.

O'Brien, S. J. and Evermann, J. F. (1988) Interactive influence of infectuous disease and genetic diversity in natural populations. *Trends in Ecology and Evolution*, **3**, 254–9.

O'Brien, S. J. and Johnson, W. E. (2007) The evolution of cats. *Scientific American*, **297**, 68–75.

O'Brien, T. G., Kinnaird, M. F., and Wibisono, H. T. (2003) Crouching tigers, hidden prey: Sumatran tiger and prey populations in a tropical forest landscape. *Animal Conservation*, **6**, 131–9.

O'Brien, S. J., Roelke, M. E., Yuhki, N., *et al.* (1990) Genetic introgression within the Florida panther *Felis concolor coryi*. *National Geographic Research*, **6**, 485–94.

O'Brien, S. J., Joslin, P., Smith, G. L., III, *et al.* (1987a) Evidence for African origins of founders of the Asiatic lion species survival plan. *Zoo Biology*, **6**, 99–116.

O'Brien, S. J., Martenson, J. S., Packer, C., *et al.* (1987b) Biochemical genetic variation in geographic isolates of African and Asiatic lions. *National Geographic Research*, **3**, 114–24.

O'Brien, S. J., Wildt, D. E., Bush, M. E., *et al.* (1987c) East African cheetahs: evidence for two population bottlenecks? *Proceedings of the National Academy of Sciences of the United States of America*, **84**, 508–11.

O'Brien, S. J., Wildt, D. E., and Bush, M. E. (1986) The cheetah in genetic peril. *Scientific American*, **254**, 84–92.

O'Brien, S. J., Collier, G. E., and Benveniste, R. E. (1985a) Setting the molecular clock in Felidae: the great cats, *Panthera*. In R. L. Tilson and U. S. Seal (eds.), *Tigers of the World: The Biology, Biopolitics, Management, and Conservation of an Endangered Species*, pp. 10–27. Noyes Publications, Park Ridge, NJ.

O'Brien, S. J., Roelke, M. E., Marker, L., *et al.* (1985b) Genetic basis for species vulnerability in the cheetah. *Science*, **227**, 1428–34.

O'Brien, S. J., Wildt, D. E., Goldman, D., Merril, C. R., and Bush, M. (1983) The cheetah is depauperate in genetic variation. *Science*, **221**, 459–62.

O'Connell, A. F., Nichols, J. D., and Karanth, K. U. (eds.) (in press) *Camera Traps in Animal Ecology: Methods and Analyses*. Springer, Tokyo, Japan.

O'Connor, R. M. (1986) Reproduction and age distribution of female lynx in Alaska, 1961–1971—preliminary results. In S. D. Miller and D. D. Everett (eds.), *Cats of the World: Biology, Conservation, and Management*, pp. 311–25. National Wildlife Federation, Washington DC.

O'Connor, T. P. (2007) Wild or domestic? Biometric variation in the cat *Felis silvestris* schreber. *International Journal of Osteoarchaeology*, **17**, 581–95.

O'Donoghue, M., Slough, B. G., Poole, K. G., *et al.* (2007) Canada lynx: behaviour and population dynamics in northern Canada. In *Felid Biology and Conservation Conference 17–20 September 2007: Programme and Abstracts*, p. 55. Wildlife Conservation Research Unit, Oxford, UK.

O'Donoghue, M., Boutin, S., Murray, D. L., *et al.* (2001) Coyotes and lynx. In C. J. Krebs, S. Boutin, and R.

Boonstra (eds.), *Ecosystem Dynamics of the Boreal Forest: The Kluane Project*, pp. 276–323. Oxford University Press, Oxford, UK.

O'Donoghue, M., Boutin, S., Krebs, C. J., Murray, D. L., and Hofer, E. J. (1998a) Behavioural responses of coyotes and lynx to the snowshoe hare cycle. *Oikos*, **82**, 169–83.

O'Donoghue, M., Boutin, S., Krebs, C. J., Zuleta, G., Murray, D. L., and Hofer, E. J. (1998b) Functional responses of coyotes and lynx to the snowshoe hare cycle. *Ecology*, **79**, 1193–208.

O'Donoghue, M., Boutin, S., Krebs, C. J., and Hofer, E. J. (1997) Numerical responses of coyotes and lynx to the snowshoe hare cycle. *Oikos*, **80**, 150–62.

O'Donoghue, M., Hofer, E., and Doyle, F. I. (1995) Predator versus predator. *Natural History*, **104** (3), 6–9.

O'Regan, H. J. (2002) Defining cheetahs, a multivariate analysis of skull shape in big cats. *Mammal Review*, **32**, 58–62.

Oakleaf, J. K., Murray, D. L., and Mack, C. (2003) Effects of wolves on livestock calf survival and movements in central Idaho. *Journal of Wildlife Management*, **67**, 299–306.

Odden, J., Linnell, J. D. C., and Andersen, R. (2006) Diet of Eurasian lynx, *Lynx lynx*, in the boreal forest of southeastern Norway: the relative importance of livestock and hares at low roe deer density. *European Journal of Wildlife Research*, **52**, 237–44.

Odden, J., Linnell, J. D. C., Moa, P. F., Herfindal, I., Kvam, T., and Andersen, R. (2002) Lynx depredation on domestic sheep in Norway. *Journal of Wildlife Management*, **66**, 98–105.

Office of Economic and Demographic Research Florida Legislature (2008) *Total county population: April 1, 1970–2030*. Florida Demographic Estimating Conference, February 2008 and the Florida Demographic Database, August 2008. <www.edr.state.fl.us/population/web10.xls>. Accessed online 11 Sept 2008.

Officer, L. H. and Williamson, S. H. (2008) Purchasing power of money in the United States from 1774 to 2007. <www.measuringworth.com/ppowerus/>. Accessed online 27 January 2008.

Ogada, M. O., Woodroffe, R., Oguge, N. O., and Frank, L. G. (2003) Limiting depredation by African carnivores: the role of livestock husbandry. *Conservation Biology*, **17**, 1521–30.

Ogutu, J. O. and Dublin, H. T. (2004) Spatial dynamics of lions and their prey along an environmental gradient. *African Journal of Ecology*, **42**, 8–22.

Ogutu, J. O. and Dublin, H. T. (2002) Demography of lions in relation to prey and habitat in the Maasai Mara National Reserve, Kenya. *African Journal of Ecology*, **40**, 120–9.

Ogutu, J. O. and Dublin, H. T. (1998) The response of lions and spotted hyaenas to sound playbacks as a technique for estimating population size. *African Journal of Ecology*, **36**, 83–95.

Ogutu, J. O., Reid, R., and Bhola, N. (2005) The effects of pastoralism and protection on the density and distribution of carnivores and their prey in the Mara ecosystem of Kenya. *Journal of Zoology*, **265**, 281–93.

Ojasti, J. (1997) *Wildlife Utilization in Latin America: Current Situation and Prospects for Sustainable Management*. FAO, Rome, Italy.

Okarma, H., Jedrzejewski, W., Schmidt, K., Kowalczyk, R., and Jedrzejewska, B. (1997) Predation of Eurasian lynx on roe deer and red deer in Bialowieza Primeval Forest, Poland. *Acta Theriologica*, **42**, 203–24.

Olbricht, G. and Sliwa, A. (1997) *In situ* and *ex situ* observations and management of black-footed cats *Felis nigripes*. *International Zoo Yearbook*, **35**, 81–9.

Olbricht, G. and Sliwa, A. (1995) Comparative development of juvenile Black-footed cats at Wuppertal Zoo and elsewhere. *Der Zoologische Garten*, **65**, 8–20.

Oli, M. K. (1997) Winter home range of snow leopards in Nepal. *Mammalia*, **61**, 355–60.

Oli, M. K. (1994) Snow leopards and blue sheep in Nepal: densities and predator:prey ratio. *Journal of Mammalogy*, **75**, 998–1004.

Oli, M. K. (1991) *The ecology and conservation of the Snow leopard* (Panthera uncia) *in the Annapurna Conservation Area, Nepal*. MPhil dissertation, University of Edinburgh, Edinburgh, Scotland.

Oli, M. K., Taylor, I. R., and Rogers, M. E. (1994) Snow leopard *Panthera uncia* predation of livestock: an assessment of local perceptions in the Annapurna Conservation Area, Nepal. *Biological Conservation*, **68**, 63–8.

Oli, M. K., Taylor, I. R., and Rogers, M. E. (1993) Diet of the snow leopard (*Panthera uncia*) in the Annapurna Conservation Area, Nepal. *Journal of Zoology*, **231**, 365–70.

Oliveira, T. G. de (2004) The oncilla in Amazonia: unraveling a myth. *Cat News*, **41**, 29–32.

Oliveira, T. G. de (2002a) Ecología comparativa de la alimentación del jaguar y del puma en el neotrópico. In R. A. Medellín, C. Equihua, C. L. B. Chetkiewicz, *et al.* (eds.), *El Jaguar en el Nuevo Milenio*, pp. 265–88. Fondo de Cultura Economica, Universidad Nacional Autónoma de México and Wildlife Conservation Society, Mexico City, Mexico.

Oliveira, T. G. de (2002b) Evaluacion del estado de conservacion del jaguar en el este de la Amazonia y noreste de Brasil. In R. A. Medellín, C. Equihua, C. L. B. Chetkiewicz, *et al.* (eds.), *El Jaguar en el Nuevo Milenio*, pp. 419–50. Fondo de Cultura Economica, Universidad Nacional Autónoma de México and Wildlife Conservation Society, Mexico City, Mexico.

Oliveira, T. G. de (1998a) *Herpailurus yagouaroundi*. *Mammalian Species*, **578**, 1–6.

Oliveira, T. G. de (1998b) *Leopardus wiedii*. *Mammalian Species*, **579**, 1–6.

Oliveira, T. G. de (1994) *Neotropical Cats: Ecology and Conservation.* EDUFMA, São Luís, Brazil.

Oliveira, T. G. de and Cassaro, K. (2005) *Guia de campo dos felinos do Brasil.* Pró-Carnívoros/Fundação Parque Zoológico de São Paulo/Sociedade de Zoológicos do Brasil/Pró-Vida Brasil, São Paulo, Brazil.

Oliveira, T. G. de, Mazim, F. D., Kasper, C. B., Tortato, M. A., Soares, J. B. G., and Marques, R. V. (submitted) Demographic assessment of small-felids in Brazil and its implications for species conservation. *Biological Conservation.*

Oliveira, T. G. de, Kasper, C. B., Tortato, M. A., Marques, R. V., Mazim, F. D., and Soares, J. B. G. (2008) Aspectos ecológicos de *Leopardus tigrinus* e outros felinos de pequeno-médio porte no Brasil. In T. G. de Oliveira (ed.), *Plano de ação para conservação de Leopardus tigrinus no Brasil.* Instituto Pró-Carnívoros/Fundo Nacional do Meio Ambiente, Atibaia, Brazil.

Olmos, F. (1993) Notes on the food habits of Brazilian Caatinga carnivores. *Mammalia*, **57**, 126–30.

Olmsted, R. A., Langley, R., Roelke, M. E., *et al.* (1992) Worldwide prevalence of lentivirus infection in wild feline species: epidemologic and phylogenetic aspects. *Journal of Virology*, **66**, 6008–18.

Olson, D. M., Dinerstein, E., Wikramanayake, E. D., *et al.* (2001) Terrrestrial ecoregions of the world: a new map of life on earth. *BioScience*, **51**, 933–8.

Onorato, D. P. and Hellgren, E. C. (2001) Black bear at the border: the recolonization of the Trans-Pecos. In D. S. Maehr, R. F. Noss, and J. L. Larkin (eds.), *Large Mammal Restoration: Ecological and Sociological Challenges in the 21st Century*, pp. 245–60. Island Press, Washington DC.

Onorato, D. P., Hellgren, E. C., Van Den Bussche, R. A., and Doan Crider, D. L. (2004) Phylogeographic patterns within a metapopulation of black bears (*Ursus americanus*) in the American southwest. *Journal of Mammalogy*, **85**, 140–7.

Orford, H. J. L., Perrin, M. R., and Berry, H. H. (1988) Contraception, reproduction and demography of free-ranging Etosha lions (*Panthera leo*). *Journal of Zoology*, **216**, 717–33.

Ormsby, J. (1864) *Autumn Rambles in North Africa.* Longman, Green, Longman, Roberts, and Green, London, UK.

Ortolani, A. (1999) Spots, stripes, tail tips and dark eyes: predicting the function of carnivore colour patterns using the comparative method. *Biological Journal of the Linnean Society*, **67**, 433–76.

Ortolani, A. and Caro, T. M. (1996) The adaptive significance of color patterns in carnivores: phylogenetic tests of classic hypotheses. In J. L. Gittleman (ed.), *Carnivore Behavior, Biology, and Evolution*, pp. 132–88. Cornell University Press, Ithaca, NY.

Osborn, F. V. and Rasmussen, L. E. L. (1995) Evidence for the effectiveness of an oleo-resin capsicum aerosol as a repellent against wild elephants in Zimbabawe. *Pachyderm*, **20**, 55–64.

Ostro, L. E. T. (1998) *The spatial ecology of translocated black howler monkeys* (Alouatta pigra) *in Belize.* PhD dissertation, Fordham University, New York.

Ostrowski, S., van Vuuren, M., Lenain, D. M., and Durand, A. (2003) A serologic survey of wild felids from central west Saudi Arabia. *Journal of Wildlife Diseases*, **39**, 696–702.

Overbeek, P. A., Gorlov, I. P., Sutherland, R. W., *et al.* (2001) A transgenic insertion causing cryptorchidism in mice. *Genesis*, **30**, 26–35.

Overskaug, K. (1994) Behavioural changes in free-ranging red foxes (*Vulpes vulpes*) due to sarcoptic mange. *Acta Veterinaria Scandinavica*, **35**, 457–9.

Owen-Smith, N. and Mills, M. G. L. (2006) Manifold interactive influences on the population dynamics of a multispecies ungulate assemblage. *Ecological Monographs*, **76**, 73–92.

Owens, M. J. and Owens, D. D. (1984) Kalahari lions break the rules. *International Wildlife*, **14**, 4–13.

Pacheco, L. F., Lucero, A., and Villca, M. (2004) Dieta del puma (*Puma concolor*) en el Parque Nacional Sajama, Bolivia y su conflicto con la ganadería. *Ecología en Bolivia*, **39**, 75–83.

Packer, C. (1990) Serengeti lion survey. Report to TANAPA, SWRI, MWEKA, and Wildlife Division.

Packer, C. (1986) The ecology of sociality in Felids. In D. I. Rubenstein and R. W. Wrangham (eds.), *Ecological Aspects of Social Evolution. Birds and Mammals*, p. 551. Princeton University Press, Princeton, NJ.

Packer, C. and Pusey, A. E. (1993) Dispersal, kinship, and inbreeding in African lions. In N. W. Thornhill (ed.), *The Natural History of Inbreeding and Outbreeding: Theoretical and Empirical Perspectives*, pp. 375–91. University of Chicago Press, Chicago, IL.

Packer, C. and Pusey, A. E. (1987) Intrasexual co-operation and the sex ratio in African lions. *American Naturalist*, **130**, 636–42.

Packer, C. and Pusey, A. E. (1983a) Adaptations of female lions to infanticide by incoming males (*Panthera leo*). *American Naturalist*, **121**, 716–28.

Packer, C. and Pusey, A. E. (1983b) Male takeovers and female reproductive parameters: a simulation of oestrus synchrony in lions (*Panthera leo*). *Animal Behaviour*, **31**, 334–40.

Packer, C. and Pusey, A. E. (1982) Co-operation and competition within coalitions of male lions: kin selection or game theory? *Nature*, **296**, 740–2.

Packer, C. and Ruttan, L. (1988) The evolution of cooperative hunting. *American Naturalist*, **132**, 159–98.

Packer, C., Hilborn, R., Mosser, A., *et al.* (2005a) Ecological change, group territoriality, and population dynamics in Serengeti lions. *Science*, **307**, 390–3.

Packer, C., Ikanda, D., Kissui, B. M., and Kushnir, H. (2005b) Lion attacks on humans in Tanzania. *Nature*, **436**, 927–8.

Packer, C., Pusey, A. E., and Eberly, L. E. (2001) Egalitarianism in female African lions. *Science*, **293**, 690–3.

Packer, C., Altizer, S., Appel, M., *et al.* (1999) Viruses of the Serengeti: patterns of infection and mortality in African lions. *Journal of Animal Ecology*, **68**, 1161–78.

Packer, C., Tatar, M., and Collins, A. (1998) Reproductive cessation in female mammals. *Nature*, **392**, 807–11.

Packer, C., Lewis, S., and Pusey, A. E. (1992) A comparative analysis of non-offspring nursing. *Animal Behaviour*, **43**, 265–81.

Packer, C., Gilbert, D. A., Pusey, A. E., and O'Brien, S. J. (1991a) A molecular genetic analysis of kinship and cooperation in African lions. *Nature*, **351**, 562–5.

Packer, C., Pusey, A. E., Rowley, H., Gilbert, D. A., Martenson, J. S., and O'Brien, S. J. (1991b) Case study of a population bottleneck: lions of the Ngorongoro Crater. *Conservation Biology*, **5**, 219–30.

Packer, C., Scheel, D., and Pusey, A. E. (1990) Why lions form groups: food is not enough. *American Naturalist*, **136**, 1–19.

Packer, C., Herbst, L., Pusey, A., *et al.* (1988) Reproductive success of lions. In T. H. Clutton-Brock (ed.), *Reproductive Success: Studies of Individual Variation in Contrasting Breeding Systems*, pp. 363–83. University of Chicago Press, Chicago, IL.

Padley, W. D. (ed.) (1997) *Proceedings of the 5th Mountain Lion Workshop, 27 February–1 March 1996*. The Southern California Chapter of the Wildlife Society, San Diego, CA.

Page, R. J. C. and Bennett, A. F. (1994) Feral cat control in Britain: developing a rabies contingency strategy. In W. S. Halverson and A. C. Crabb (eds.), *Proceedings of the 16th Vertebrate Pest Conference*, pp. 20–6. University of California, Davis, CA.

Palacios R. (2007) *Manual para Identificación de Carnívoros Andinos*. Alianza Gato Andino, Cordoba, Argentina.

Palen, G. F. and Goddard, G. C. (1966) Catnip and oestrus behaviour in the cat. *Animal Behaviour*, **14**, 372–7.

Palmeira, F. B. L., Crawshaw, P. G., Haddad, C. M., Ferraz, K. M., and Verdade, L. M. (2008) Cattle depredation by puma (*Puma concolor*) and jaguar (*Panthera onca*) in central-western Brazil. *Biological Conservation*, **141**, 118–25.

Palomares, F. (2003) The negative impact of heavy rains on the abundance of a Mediterranean population of European rabbits. *Mammalian Biology*, **68**, 224–34.

Palomares, F. (2001) Vegetation structure and prey abundance requirements of the Iberian lynx: implications for the design of reserves and corridors. *Journal of Applied Ecology*, **38**, 9–18.

Palomares, F. and Caro, T. M. (1999) Interspecific killing among mammalian carnivores. *American Naturalist*, **153**, 492–508.

Palomares, F. and Rodríguez, A. (2004) ¿Cuál es la situación real de las poblaciones de lince? In M. Gomendio (ed.), *Los retos medioambientales del siglo XXI. La conservación de la biodiversidad en España*, pp. 63–76. Fundación BBVA, Bilbao, Spain.

Palomares, F., Revilla, E., Calzada, J., Fernández, N., and Delibes, M. (2005) Reproduction and pre-dispersal survival of Iberian lynx in a subpopulation of the Doñana National Park. *Biological Conservation*, **122**, 53–9.

Palomares, F., Godoy, J. A., Piriz, A., O'Brien, S. J., and Johnson, W. E. (2002a) Faecal genetic analysis to determine the presence and distribution of elusive carnivores: design and feasibility for the Iberian lynx. *Molecular Ecology*, **11**, 2171–82.

Palomares, F., Revilla, E., Gaona, P., Giordano, C., and Delibes, M. (2002b) *Efecto de la extracción de linces ibéricos en las poblaciones donantes de Doñana y la Sierra de Andújar para posibles campañas de reintroducción*. Consejería de Medio Ambiente Junta de Andalucía, Seville, Spain.

Palomares, F., Delibes, M., Revilla, E., Calzada, J., and Fedriani, J. M. (2001) Spatial ecology of Iberian lynx and abundance of European rabbits in southwestern Spain. *Wildlife Monographs*, **148**, 1–36.

Palomares, F., Delibes, M., Ferreras, P., Fedriani, J. M., Calzada, J., and Revilla, E. (2000) Iberian lynx in a fragmented landscape: pre-dispersal, dispersal, and post-dispersal habitats. *Conservation Biology*, **14**, 809–18.

Palomares, F., Ferreras, P., Travaini, A., and Delibes, M. (1998) Co-existence between Iberian lynx and Egyptian mongooses: estimating interaction strength by structural equation modelling and testing by an observational study. *Journal of Animal Ecology*, **67**, 967–78.

Palomares, F., Ferreras, P., Fedriani, J. M., and Delibes, M. (1996) Spatial relationships between Iberian lynx and other carnivores in an area of south-western Spain. *Journal of Applied Ecology*, **33**, 5–13.

Palomares, F., Gaona, P., Ferreras, P., and Delibes, M. (1995) Positive effects on game species of top predators by controlling smaller predator populations: an example with lynx, mongooses, and rabbits. *Conservation Biology*, **9**, 295–305.

Palomares, F., Rodríguez, A., Laffitte, R., and Delibes, M. (1991) The status and distribution of the Iberian lynx *Felis pardina* (Temminck) in Coto Doñana Area, SW Spain. *Biological Conservation*, **57**, 159–69.

Pamilo, P. (2004) How cryptic is the Scandinavian lynx? *Molecular Ecology*, **13**, 3257–9.

Panaman, R. (1981) Behaviour and ecology of free ranging female farm cats. *Zeitschrift für Tierpsychologie*, **56**, 59–73.

Papendick, R. E., Munson, L., O'Brien, T. D., and Johnson, K. H. (1997) Systemic AA amyloidosis in captive cheetahs (*Acinonyx jubatus*). *Veterinary Pathology*, **34**, 549–56.

Papouchis, C. M. (2004) Conserving mountain lions in a changing landscape. In N. Fascione, A. Delach and M. E. Smith (eds.), *People and Predators: From Conflict to Coexistence*, pp. 219–39. Island Press, Washington DC.

Parker, A. M., Trivedi, M., and Mardas, N. (eds.) (2008) *The Little REDD Book. A Guide to Governmental and Non-*

Governmental Proposals for Reducing Emissions from Deforestation and Degradation. Global Canopy Foundation, Oxford, UK.

Parker, G. (2001) *Status Report on the Canada Lynx in Nova Scotia.* Nova Scotia Species at Risk Working Group, Halifax, NS.

Parker, G. E. and Osborn, F. V. (2006) Investigating the potential for chilli *Capsicum* spp. to reduce human–wildlife conflict in Zimbabwe. *Oryx*, **40**, 343–6.

Parker, G. R. (1981) Winter habitat use and hunting activities of lynx (*Lynx canadensis*) on Cape Breton Island, Nova Scotia. In J. A. Chapman and D. Pursley (eds.), *Worldwide Furbearer Conference Proceedings*, pp. 221–48. Frostburg, MD.

Parker, G. R., Maxwell, J. W., Morton, L. D., and Smith, G. E. J. (1983) The ecology of the lynx (*Lynx canadensis*) on Cape Breton Island. *Canadian Journal of Zoology*, **61**, 770–86.

Parks, S. A. and Harcourt, A. H. (2002) Reserve size, local human density, and mammalian extinctions in U.S. protected areas. *Conservation Biology*, **16**, 800–8.

Parra, A., Garcia, A., Inglis, N. F., *et al.* (2006) An epidemiological evaluation of *Mycobacterium bovis* infections in wild game animals of the Spanish Mediterranean ecosystem. *Research in Veterinary Science*, **80**, 140–6.

Passanisi, W. and Macdonald, D. W. (1990) Group discrimination on the basis of urine in a farm cat colony. In D. W. Macdonald, D. Müller-Schwarze, and S. E. Natynczuk (eds.), *Chemical Signals in Vertebrates 5*, pp. 336–45. Oxford University Press, Oxford, UK.

Patel, K. (2006) Observations of rusty-spotted cat in eastern Gujurat. *Cat News*, **45**, 37–8.

Patel, K. and Jackson, P. (2005) Rusty-spotted cat in India: new distribution data. *Cat News*, **42**, 27.

Paton, D. C. (1993) The impact of domestic and feral cats on wildlife. In G. Siepen and C. Owens (eds.), *Cat Management Workshop Proceedings*, pp. 9–15. Queensland Department of Environment and Heritage, Brisbane, Australia.

Patterson, B. D. (2004) Maneless and misunderstood. *Earthwatch Journal*, **23**, 12–15.

Patterson, B. D., Neiburger, E. J., and Kasiki, S. M. (2003) Tooth breakage and dental disease as causes of carnivore–human conflicts. *Journal of Mammalogy*, **84**, 190.

Patterson, B. D., Kasiki, S. M., Selempo, E., and Kays, R. W. (2004) Livestock predation by lions (*Panthera leo*) and other carnivores on ranches neighbouring Tsavo National Park, Kenya. *Biological Conservation*, **119**, 507–16.

Pauchard, A., Aguayo, M., Peña, E., and Urrutia, R. (2006) Multiple effects of urbanization on the biodiversity of developing countries: the case of a fast-growing metropolitan area (Concepción, Chile). *Biological Conservation*, **127**, 272–81.

Paul-Murphy, J., Work, T., Hunter, D., McFie, E., and Fjelline, D. (1994) Serologic survey and serum biochemical reference ranges of the free-ranging mountain lion (*Felis concolor*) in California. *Journal of Wildlife Diseases*, **30**, 205–15.

Pauly, D. (1995) Anecdotes and the shifting baseline syndrome of fisheries. *Trends in Ecology and Evolution*, **10**, 430.

Payan, E. (2008) *Jaguars, ocelots and prey ecology across sites with different hunting pressure in Colombian Amazonia.* PhD dissertation, University College London, London, UK.

Payan, E. and Trujillo, L. A. (2006) The Tigrilladas in Colombia. *Cat News*, **44**, 25–8.

Pearce, D. and Moran, D. (1994) *The Economic Value of Biodiversity.* IUCN, London, UK.

Pearce, D., Hecht, S., and Vorhies, F. (2007) What is biodiversity worth? Economics as a problem and a solution. In D. W. Macdonald and K. Service (eds.), *Key Topics in Conservation*, pp. 35–45. Blackwell Publishing Ltd, Oxford, UK.

Pecon-Slattery, J., Troyer, J., Johnson, W., and O'Brien, S. (2008) Evolution of feline immunodeficiency virus in Felidae: implications for human health and wildlife ecology. *Veterinary Immunology and Immunopathology*, **123**, 32–44.

Pedersen, N. C. (1988) Feline leukemia virus infection. In N. C. Pedersen (ed.), *Feline Infectious Diseases*, pp. 83–106. American Veterinary Publications, Goleta, CA.

Pedersen, N. C. and Barlough, J. E. (1991) Clinical overview of feline immunodeficiency virus. *Journal of the American Veterinary Medical Association*, **199**, 1298–305.

Peel, M. J. S. and Montagu, G. P. (1999) Modelling predator–prey interactions on a Northern Province game ranch. *South African Journal of Wildlife Research*, **29**, 31–4.

Peery, M. Z., Beissinger, S. R., House, R. F., *et al.* (2008) Characterizing source-sink dynamics with genetic parentage assignments. *Ecology*, **89**, 2746–59.

Peigné, S. (1999) *Proailurus*, l'un des plus anciens Felidae (Carnivora) d'Eurasie: systématique et évolution. *Bulletin de la Société d'Histoire Naturelle de Toulouse*, **135**, 125–34.

Peixoto, P. V., Soares, C. O., Scofield, A., Santiago, C. D., Franca, T. N., and Barros, S. S. (2007) Fatal cytauxzoonosis in captive-reared lions in Brazil. *Veterinary Parasitology*, **145**, 383–7.

Pelican, K. M., Lang, K., Passaro, R., *et al.* (2007) Improved husbandry and nutrition results in successful reproduction in the Thailand clouded leopard. In *Proceedings of the Annual Conference of the American Association of Zoo Veterinarians*, p. 214. American Association of Zoo Veterinarians.

Pelican, K., Abáigar, T., Vargas, A., *et al.* (2006a) Hormone profiles in the Iberian lynx: applications to in situ and ex situ conservation. In *Iberian Lynx Ex-Situ Conservation Seminar Series: Book of Proceedings*, pp. 113–15. Fundación Biodiversidad, Sevilla and Doñana, Spain.

Pelican, K. M., Wildt, D. E., Pukazhenthi, B., and Howard, J. (2006b) Ovarian control for assisted reproduction in the domestic cat and wild felids. *Theriogenology*, **66**, 37–48.

Pelican, K. M., Brown, J. L., Wildt, D. E., Ottinger, M. A., and Howard, J. G. (2005) Short term suppression of follicular recruitment and spontaneous ovulation in the cat using levonorgestrel versus a GnRH antagonist. *General and Comparative Endocrinology*, **144**, 110–21.

Peña, L., Garcia, P., Jimenez, M. A., Benito, A., Alenza, M. D. P., and Sanchez, B. (2006) Histopathological and immunohistochemical findings in lymphoid tissues of the endangered Iberian lynx (*Lynx pardinus*). *Comparative Immunology Microbiology and Infectious Diseases*, **29**, 114–26.

Pence, D. B. and Ueckermann, E. (2002) Sarcoptic mange in wildlife. *Revue Scientifique et Technique de l Office International Des Epizooties*, **21**, 385–98.

Pence, D. B. and Windberg, L. A. (1994) Impact of a sarcoptic mange epizootic on a coyote population. *Journal of Wildlife Management*, **58**, 624–32.

Pence, D. B., Tewes, M. E., Shindle, D. B., and Dunn, D. M. (1995) Notoedric mange in an ocelot (*Felis pardalis*) from southern Texas. *Journal of Wildlife Diseases*, **31**, 558–61.

Pence, D. B., Matthews, F. D., and Windberg, L. A. (1982) Notoedric mange in the bobcat, *Felis rufus*, from south Texas. *Journal of Wildlife Diseases*, **18**, 47–50.

Pennycuick, C. J. and Rudnai, J. A. (1970) A method of identifying individual lions *Panthera leo* with an analysis of the reliability of identification. *Journal of Zoology*, **160**, 497–508.

Penzhorn, B. L. (2006) Babesiosis of wild carnivores and ungulates. *Veterinary Parasitology*, **138**, 11–21.

Penzhorn, B. L., Kjemtrup, A. M., Lopez-Rebollar, L. M., and Conrad, P. A. (2001) *Babesia leo* n. sp from lions in the Kruger National Park, South Africa, and its relation to other small piroplasms. *Journal of Parasitology*, **87**, 681–5.

Pereira, J. A., Fracassi, N. G., and Uhart, M. M. (2006) Numerical and spatial responses of Geoffroy's cat (*Oncifelis geoffroyi*) to prey decline in Argentina. *Journal of Mammalogy*, **87**, 1132–9.

Pereira, J., Varela, D., and Fracassi, N. (2002) Pampas cat in Argentina: is it absent from the pampas? *Cat News*, **36**, 20–2.

Perez, I., Geffen, E., and Mokady, O. (2006) Critically endangered Arabian leopards *Panthera pardus nimr* in Israel: estimating population parameters using molecular scatology. *Oryx*, **40**, 295–301.

Perez, J., Calzada, J., Leon Vizcaino, L., Cubero, M. J., Velarde, J., and Mozos, E. (2001) Tuberculosis in an Iberian lynx (*Lynx pardina*). *Veterinary Record*, **148**, 414–15.

Perovic, P. G. and Herrán, M. (1998) Distribucion del jaguar *Panthera onca* en las provincias de Jujuy y Salta, noroeste de Argentina. *Mastazoologia Neotropical*, **5**, 47.

Perovic, P. G., Walker, S., and Novaro, A. (2003) New records of the endangered Andean mountain cat in northern Argentina. *Oryx*, **37**, 374–7.

Perovic, P. G. (2002) Conservación del jaguar en el noroeste de Argentina. In R. A. Medellín, C. Equihua, C. L. B. Chetkiewicz, *et al.* (eds.), *El Jaguar en el Nuevo Milenio*, pp. 183–97. Fondo de Cultura Economica, Universidad Nacional Autónoma de México and Wildlife Conservation Society, Mexico City, Mexico.

Peterhans, J. C. K. and Gnoske, T. P. (2001) The science of 'man-eating' among lions *Panthera leo* with a reconstruction of the natural history of the 'Man-eaters of Tsavo'. *Journal of East African Natural History*, **90**, 1–40.

Peters, G. (2002) Purring and similar vocalizations in mammals. *Mammal Review*, **32**, 245–71.

Peters, G. and Hast, M. H. (1994) Hyoid structure, laryngeal anatomy, and vocalization in felids (Mammalia: Carnivora: Felidae). *Zeitschrift für Säugetierkunde*, **59**, 87–104.

Peters, G. and Wozencraft, W. C. (1989) Acoustic communication by fissiped carnivores. In J. L. Gittleman (ed.), *Carnivore Behavior, Ecology and Evolution*, pp. 14–56. Chapman and Hall, London, UK.

Peters, J. C. and Zwart, P. (1973) Sarcoptesräude bei Schneeleoparden. *Erkrankungen der Zootiere*, **15**, 333–4.

Peters, G., Baum, L., Peters, M. K., and Tonkin-Leyhausen, B. (2009) Spectral characteristics of intense mew calls in cat species of the genus *Felis (Mammalia: Carnivora: Felidae)*. *Journal of Ethology*, **27**, 221–37.

Peters, R. H. (1983) *The Ecological Implications of Body Size*. Cambridge University Press, Cambridge, UK.

Pettorelli, N. and Durant, S. M. (2007a) Family effects on early survival and variance in long-term reproductive success of female cheetahs. *Journal of Animal Ecology*, **76**, 908–14.

Pettorelli, N. and Durant, S. M. (2007b) Longevity in cheetahs: the key to success? *Oikos*, **116**, 1879–86.

Phelan, P. and Sliwa, A. (2005) Range size and den use of Gordon's wildcats *Felis silvestris gordoni* in the Emirate of Sharjah, United Arab Emirates. *Journal of Arid Environments*, **60**, 15–25.

Philippa, J. D. W., Martina, B. E. E., Kuiken, T., *et al.* (2004) Antibodies to selected pathogens in free-ranging terrestrial carnivores and marine mammals in Canada. *Veterinary Record*, **155**, 135–43.

Phillips, J. A., Worley, M. B., Morsbach, D., and Williams, T. M. (1993) Relationship among diet, growth and occurrence of focal palatine erosion in wild-caught captive cheetahs. *Madoqua*, **18**, 79–83.

Phillips, M. (1995) Conserving the red wolf. *Canid News*, **3**, 13–35.

Pianka, E. R. (1973) The structure of lizard communities. *Annual Review of Ecology and Systematics*, **4**, 53–74.

Piechocki, R. (1990) *Der Tiger*. Ziemsen, Wittenberg Lutherstadt, Germany.

Pierce, B. M., Bleich, V. C., Terry Bowyer, R., and Wehausen, J. D. (1999) Migratory patterns of mountain lions: implications for social regulation and conservation. *Journal of Mammalogy*, **80**, 986–92.

Pierpaoli, M., Biro, Z. S., Herrmann, M., *et al.* (2003) Genetic distinction of wildcat (*Felis silvestris*) populations in Europe, and hybridisation with domestic cats in Hungary. *Molecular Ecology*, **12**, 2585–98.

Pimm, S. L. (1991) *The Balance of Nature: Ecological Issues in the Conservation of Species and Communities*. University of Chicago Press, Chicago, IL.

Pires, A. E. and Fernandes, M. L. (2003) Last lynxes in Portugal? Molecular approaches in a pre-extinction scenario. *Conservation Genetics*, **4**, 525–32.

Piveteau, J. (1961) Les carnivores. In J. Piveteau (ed.), *Traité de Paléontologie, Tome VI*, pp. 641–820. Masson et Cie, Paris, France.

Pocock, R. I. (1951) *Catalogue of the Genus* Felis. British Museum (Natural) History, London, UK.

Pocock, R. I. (1939) Family Felidae. In *Fauna of British India Including Ceylon and Burma. Mammalia*, pp. 190–331. Taylor & Francis, London, UK.

Pocock, R. I. (1932) The leopards of Africa. *Proceedings of the Zoological Society of London*, **1932**, 543–91.

Pocock, R. I. (1930) The panthers and ounces of Asia. *Journal of the Bombay Natural History Society*, **34**, 64–82, 307–36.

Pocock, R. I. (1917b) On the external characters of the Felidae. *Annals and Magazine of Natural History*, **19**, 113–36.

Pocock, R. I. (1917a) The classification of existing Felidae. *Annals and Magazine of Natural History Series 8*, **20**, 329–50.

Pocock, R. I. (1916) On some of the cranial and external characters of the hunting leopard or cheetah (*Acinonyx jubatus*). *Annals and Magazine of Natural History*, **18**, 419–29.

Pocock, R. I. (1907) Notes upon some African species of the genus *Felis*, based upon specimens recently exhibited in the Society's Gardens. *Proceedings of the Zoological Society of London*, **1907**, 656–77.

Polisar, J., Maxit, I., Scognamillo, D., Farrell, L., Sunquist, M. E., and Eisenberg, J. F. (2003) Jaguars, pumas, their prey base, and cattle ranching: ecological interpretations of a management problem. *Biological Conservation*, **109**, 297–310.

Pollak, J. P., Ballou, J., and Lacy, R. C. (2004) *PM2000*. American Zoo and Aquarium Association, Brookfield Zoo, Chicago Zoological Society, and Smithsonian National Zoological Park.

Pollock, K. H., Winterstein, S. R., and Conroy, M. J. (1989) Estimation and analysis of survivial distribution for radio-tagged animals. *Biometrics*, **45**, 99–109.

Poole, K. G. (2003) A review of the Canada lynx, *Lynx canadensis*, in Canada. *Canadian Field-Naturalist*, **117**, 360–76.

Poole, K. G. (1997) Dispersal patterns of lynx in the Northwest Territories. *Journal of Wildlife Management*, **61**, 497–505.

Poole, K. G. (1995) Spatial organization of a lynx population. *Canadian Journal of Zoology*, **73**, 632–41.

Poole, K. G. (1994) Characteristics of an unharvested lynx population during a snowshoe hare decline. *Journal of Wildlife Management*, **58**, 608–18.

Poole, K. G., Wakelyn, L. A., and Nicklen, P. N. (1996) Habitat selection by lynx in the Northwest Territories. *Canadian Journal of Zoology*, **74**, 845–50.

Pope, C. E. (2000) Embryo technology in conservation efforts for endangered felids. *Theriogenology*, **53**, 163–74.

Pope, C. E., Gomez, M. C., and Dresser, B. L. (2006) In vitro embryo production and embryo transfer in domestic and non-domestic cats. *Theriogenology*, **66**, 1518–24.

Popper, K. R. and NetLibrary, I. (2002) *The Logic of Scientific Discovery*. Routledge Classics, London and New York.

Population World Net (2009) <www.populationworld.net/country_rankings/2020.html>. Accessed onnline 10 May 2009.

Potts, F. C., Goodwin, H., and Walpole, M. J. (1996) People, wildlife and tourism in and around Hwange National Park, Zimbabwe. In M. F. Price (ed.), *People and Tourism in Fragile Environments*, pp 199–219. John Wiley & Sons Ltd, Chichester, UK.

Prevosti, F. J. (2006) New material of Pleistocene cats (Carnivora, Felidae) from southern South America, with comments on biogeography and the fossil record. *Géobios*, **39**, 679–94.

Prikhodko, V. I. (2003) *Musk Deer: Distribution, Systematics, Ecology, Behaviour, and Communication*. GEOS, Moscow, Russia.

Pringle, J. A. and Pringle, V. L. (1979) Observations on the lynx *Felis caracal* in the Bedford district. *South African Journal of Zoology*, **14**, 1–4.

PRODES. (2007) *Monitoring of the Brazilian Amazon by Satellite*. Brazil's National Institute for Space Research, São José dos Campos, Brazil.

Proulx, G. (1999) Review of current mammal trapping technology in North America. In G. Proulx (ed.), *Mammal Trapping*, pp. 1–46. Alpha Wildlife Research & Management Ltd, Sherwood Park, AB.

Pukazhenthi, B. S. and Wildt, D. E. (2004) Which reproductive technologies are most relevant to studying, managing and conserving wildlife? *Reproduction Fertility and Development*, **16**, 33–46.

Pukazhenthi, B. S., Neubauer, K., Jewgenow, K., Howard, J., and Wildt, D. E. (2006) The impact and potential etiology of teratospermia in the domestic cat and its wild relatives. *Theriogenology*, **66**, 112–21.

Purchase, G. K. (1998) *An assessment of the success of a cheetah re-introduction project in Matusadona National Park*. MSc dissertation, University of Zimbabwe, Harare, Zimbabwe.

Purchase, G. K. and Vhurumuku, G. (2005) *Evaluation of a Wild–Wild Translocation of Cheetah* (Acinonyx jubatus) *from Private Land to Matusadona National Park Zimbabwe (1994–2005)*. The Zambesi Society, Harare, Zimbabwe.

Purchase, G. K., Mateke, C., and Purchase, D. (2007) *A Review of the Status and Distribution of Carnivores and Levels of Human–Carnivore Conflict in Protected Areas and Surrounds of the Zambezi Basin*. The Zambezi Society and Fauna and Flora International, Bulawayo, Zimbabwe.

Pusey, A. E. (1990) Mechanisms of inbreeding avoidance in non-human primates. In J. R. Feierman (ed.), *Pedophilia: Biosocial Dimensions*, pp. 201–20. Springer-Verlag, New York.

Pusey, A. E. and Packer, C. (1994) Non-offspring nursing in social carnivores: minimizing the costs. *Behavioral Ecology*, **5**, 362–74.

Pusey, A. E. and Packer, C. (1987) The evolution of sex-biased dispersal in lions. *Behaviour*, **101**, 275–310.

Putman, R. J. (1984) Facts from faeces. *Mammal Review*, **14**, 79–97.

Pyke, G. H., Pulliam, H. R., and Charnov, E. L. (1977) Optimal foraging: a selective review of theory and tests. *Quarterly Review of Biology*, **52**, 137–54.

Qiu, Z. (2003) Dispersals of neogene carnivorans between Asia and North America. *Bulletin American Museum of Natural History*, **279**, 18–31.

Qiu, Z. and Gu, Y. (1996) The Middle Miocene vertebrate fauna from Xiacaowan, Sihong Co., Jiangsu Province. 3. Two species of fossil carnivores: *Semigenetta* and *Pseudaelurus*. *Vertebrata PalAsiatica*, **24**, 1–31.

Quammen, D. (2004) *Monster of God: The Man-Eating Predator in the Jungles of History and the Mind*. W.W. Norton, New York.

Quan, J., Barton, D., and Conroy, C. (1994) *A Preliminary Assessment of the Economic Impact of Desertification in Namibia*. Directorate of Environmental Affairs, Windhoek, Namibia.

Queller, D. C. and Goodnight, K. F. (1989) Estimating relatedness using genetic-markers. *Evolution*, **43**, 258–75.

Quigley, H. B. (1987) *Ecology and conservation of the jaguar in the Pantanal Region, Mato Grosso do Sul, Brazil*. PhD dissertation, University of Idaho, Moscow, ID.

Quigley, H. B. and Crawshaw, P. G. (2002) Reproduccion, crecimiento y dispersion del jaguar en la region del Pantanal de Brasil. In R. A. Medellín, C. Equihua, C. L. B. Chetkiewicz, *et al*. (eds.), *El Jaguar en el Nuevo Milenio*, pp. 289–302. Fondo de Cultura Economica, Universidad Nacional Autónoma de México and Wildlife Conservation Society, Mexico City, Mexico.

Quigley, H. B. and Crawshaw, P. G. (1992) A conservation plan for the jaguar *Panthera onca* in the Pantanal region of Brazil. *Biological Conservation*, **61**, 149–57.

Quigley, H. B. and Herrero, S. (2005) Characterisation and prevention of attacks on humans. In R. Woodroffe, S. Thirgood, and A. Rabinowitz (eds.), *People and Wildlife: Conflict or Coexistence?* Cambridge University Press, Cambridge, UK.

Quinn, N. W. S. and Thompson, J. E. (1987) Dynamics of an exploited Canada lynx population in Ontario. *Journal of Wildlife Management*, **51**, 297–305.

Rabinowitz, A. (2008) *Life in the Valley of Death: The Fight to Save Tigers in a Land of Guns, Gold, and Greed*. Island Press, Washington DC.

Rabinowitz, A. (2007) Connecting the dots: saving a species throughout its range. In *Felid Biology and Conservation Conference 17–20 September 2007: Programme and Abstracts*, p. 61. Wildlife Conservation Research Unit, Oxford, UK.

Rabinowitz, A. (2006) Jaguar Corridor Initiative. <www.panthera.org/jaguar_corridor.html>. Accessed online 15 Dec 2008.

Rabinowitz, A. (2005) Jaguars and livestock: living with the world's third largest cat. In R. Woodroffe, S. Thirgood, and A. Rabinowitz (eds.), *People and Wildlife: Conflict or Coexistence?*, pp. 278–85. Cambridge University Press, Cambridge, UK.

Rabinowitz, A. (1999) The status of the Indochinese tiger: separating fact from fiction. In J. Seidensticker, S. Christie, and P. Jackson (eds.), *Riding the Tiger: Tiger Conservation in Human Dominated Landscapes*, pp. 148–65. Cambridge University Press, Cambridge, UK.

Rabinowitz, A. (1990) Notes on the behavior and movements of leopard cats, *Felis bengalensis*, in a dry tropical forest mosaic in Thailand. *Biotropica*, **22**, 397–403.

Rabinowitz, A. (1989) The density and behavior of large cats in a dry tropical forest mosaic in Huai Kha Khaeng Wildlife Sanctuary, Thailand. *Natural History Bulletin of the Siam Society*, **37**, 235–51.

Rabinowitz, A. (1986) Jaguar predation on domestic livestock in Belize. *Wildlife Society Bulletin*, **14**, 170–4.

Rabinowitz, A. and Nottingham, B. G. (1989) Mammal species richness and relative abundance of small mammals in a subtropical wet forest of Central America. *Mammalia*, **53**, 217–26.

Rabinowitz, A. and Nottingham, B. G. (1986) Ecology and behaviour of the Jaguar (*Panthera onca*) in Belize, Central America. *Journal of Zoology*, **210**, 149–59.

Rabinowitz, A., Andau, P., and Chai, P. P. K. (1987) The clouded leopard in Malaysian Borneo. *Oryx*, **21**, 107–11.

Radinsky, L. (1975) Evolution of the felid brain. *Brain, Behaviour and Ecology*, **11**, 214–54.

Radinsky, L. B. (1981) Evolution of skull shape in carnivores 1. Representative modern carnivores. *Biological Journal of the Linnean Society*, **15**, 369–88.

Radloff, F. G. T. and du Toit, J. T. (2004) Large predators and their prey in a southern African savanna: a predator's size determines its prey size range. *Journal of Applied Ecology*, **73**, 410–23.

Ragg, J. R., Moller, H., and Waldrup, K. A. (1995) The prevalence of bovine tuberculosis (*Mycobacterium bovis*) infections in feral populations of cats (*Felis catus*), ferrets (*Mustela furo*) and stoats (*Mustela erminea*) in Otago and Southland, New Zealand. *New Zealand Veterinary Journal*, **43**, 333–7.

Ragni, B., and Randi, E. (1986) Multivariate analysis of craniometric characteristics in European wild cat, domestic cat, and African wild cat (genus *Felis*). *Zeitschrift fuer Saeugetierkunde*, **51**, 243–51.

Rajaratnam, R., Sunquist, M., Rajaratnam, L., and Ambu, L. (2007) Diet and habitat selection of the leopard cat (*Prionailurus bengalensis borneoensis*) in an agricultural landscape in Sabah, Malaysian Borneo. *Journal of Tropical Ecology*, **23**, 209–17.

Ralls, K., Brugger, K., and Ballou, J. (1979) Inbreeding and juvenile mortality in small populations of ungulates. *Science*, **206**, 1101–3.

Ramirez, M. (2005) A long history of transboundary friendship in central America. In R. Mittermeier, C. F. Kormos, C. Mittermeier, P. Robles Gil, T. Sandwith, and C. Besançon (eds.), *Transboundary Conservation: A New Vision for Protected Areas*, pp. 291–6. CEMEX/Conservation International, Arlington, VA.

Ramsauer, S., Bay, G., Meli, M., Hofmann-Lehmann, R., and Lutz, H. (2007) Seroprevalence of selected infectious agents in a free-ranging, low-density lion population in the Central Kalahari game reserves in Botswana. *Clinical and Vaccine Immunology*, **14**, 808–10.

Randall, D. A., Marino, J., Haydon, D. T., *et al.* (2006) An integrated disease management strategy for the control of rabies in Ethiopian wolves. *Biological Conservation*, **131**, 151–62.

Randi, E., Pierpaoli, M., Beaumont, M. A., Ragni, B., and Sforzi, A. (2001) Genetic identification of wild and domestic cats (*Felis silvestris*) and their hybrids using Bayesian clustering methods. *Molecular Biology and Evolution*, **18**, 1679–93.

Randi, E. and Ragni, B. (1991) Genetic variability and biochemical systematics of domestic and wild cat populations (*Felis silvestris*, Felidae). *Journal of Mammalogy*, **72**, 79–88.

Ranganathan, J., Chan, K. M. A., Karanth, K. U., and Smith, J. L. D. (2008) Where can tigers persist in the future? A landscape-scale, density-based population model for the Indian subcontinent. *Biological Conservation*, **141**, 67–77.

Rangarajan, M. (2001) *India's Wildlife History: An Introduction*. Orient Longman, New Delhi, India.

Ranta, E., Kaitala, V., and Lindström, J. (1997) Dynamics of Canadian lynx populations in space and time. *Ecography*, **20**, 454–60.

Rao, M. V. S. and Eswar, A. (1980) Observations on the mating-behaviour and gestation period of the Asiatic lion, *Panthera leo* at the zoological park, Trivandrum, Kerala. *Comparative Physiology and Ecology*, **5**, 78–80.

Rasker, R., Waring, S., and Carter, R. (2006) *You've Come a Long Way, Cowboy: Ten Truths and Trends on the New American West*. Sonoran Insitute, Bozeman, MT.

Rasmussen, D. I. (1941) Biotic communities of Kaibab Plateau, Arizona. *Ecological Monographs*, **11**, 229–75.

Rasmussen, G. S. A. (1999) Livestock predation by the painted hunting dog *Lycaon pictus* in a cattle ranching region of Zimbabwe: a case study. *Biological Conservation*, **88**, 133–9.

Rasmussen, G. S. A., Gusset, M., F., C., and Macdonald, D. W. (2008) Achilles' heel of sociality revealed by energetic poverty trap in cursorial hunters. *American Naturalist*, **172**, 508.

Rau, J. R. and Jimenez, J. E. (2002) Diet of puma (*Puma concolor*, Carnivora: Felidae) in coastal and Andean ranges of southern Chile. *Studies on Neotropical Fauna and Environment*, **37**, 201–5.

Rau, J. R., Zuleta, C., Gantz, A., *et al.* (1998) Biodiversidad de artrópodos y vertebrados terrestres del Norte Grande de Chile. *Revista Chilena de Historia Natural*, **71**, 527–54.

Rautenbach, I. L. and Nel, J. A. J. (1978) Co-existence in Transvaal carnivore. *Bulletin of the Carnegie Museum of Natural History*, **6**, 132–45.

Rautner, M., Hardiono, M., and Alfred, R. J. (2005) *Borneo: Treasure Island at Risk*. WWF Germany, Frankfurt, Germany.

Ray, J. and Butynski, T. (in press) African golden cat *Profelis aurata*. In J. S. Kingdon and M. Hoffmann (eds.), *The Mammals of Africa 5: Carnivora, Pholidota, Perissodactyla*. Academic Press, Amsterdam, the Netherlands.

Ray, J. C. and Sunquist, M. E. (2001) Trophic relations in a community of African rainforest carnivores. *Oecologia*, **127**, 395–408.

Ray, J. C., Hunter, L., and Zigouris, J. (2005) *Setting Conservation and Research Priorities for Larger African Carnivores*. Wildlife Conservation Society, New York.

Ray, N. and Burgman, M. A. (2006) Subjective uncertainties in habitat suitability maps. *Ecological Modelling*, **195**, 172–86.

Reading, P. R. and Clark, T. W. (1996) Carnivore introductions: an interdisciplinary examination. In J. L. Gittleman, S. M. Funk, D. W. Macdonald, and R. K. Wayne (eds.), *Carnivore Behavior, Ecology and Evolution*, pp. 296–336. Cornell University Press, Ithaca, NY.

Redfern, J. V., Grant, R., Biggs, H., and Getz, W. M. (2003) Surface-water constraints on herbivore foraging in the Kruger National Park. *Ecology*, **84**, 2092–107.

Redford, K. H. and Eisenberg, J. F. (1992) *Mammals of the Neotropics, Vol. 2. The Southern Cone*. University of Chicago Press, Chicago, IL.

Reid, F. A. (1997) *A Field Guide to the Mammals of Central America and Southeast Mexico*. Oxford University Press, Oxford, UK.

Reig, S., Daniels, M. J., and Macdonald, D. W. (2001) Craniometric differentiation within wild-living cats in Scotland using 3D morphometrics. *Journal of Zoology*, **253**, 121–32.

Reiss, M. (1988) Scaling of home range size: body size, metabolic needs and ecology. *Trends in Ecology & Evolution*, **3**, 85–6.

Reumer, J. W. F., Rook, L., van der Borg, K., Post, K., Mol, D., and de Vos, J. (2003) Late Pleistocene survival of the saber-toothed cat *Homotherium* in northwestern Europe. *Journal of Vertebrate Paleontology*, **23**, 260–2.

Revilla, E., Rodríguez, A., Román, J., and Palomares, F. (2007) Análisis de la viabilidad de la metapoblación de lince ibérico de Doñana: una estrategia de manejo adaptativo para su conservación. In L. Ramírez and B. Asensio (eds.), *Proyectos de investigación en Parques Nacionales: 2003–2006*, pp. 307–23. Organismo Autónomo de Parques Nacionales, Madrid, Spain.

Revilla, E., Wiegand, T., Palomares, F., Ferreras, P., and Delibes, M. (2004) Effects of matrix heterogeneity on animal dispersal: from individual behavior to metapopulation-level parameters. *American Naturalist*, **164**, E130–53.

Revilla, E., Palomares, F., and Delibes, M. (2001) Edge-core effects and the effectiveness of traditional reserves in conservation: Eurasian badgers in Doñana National Park. *Conservation Biology*, **15**, 148–58.

Rexstad, E. and Burnham, K. P. (1991) *User's Guide for Interactive Program CAPTURE. Abundance Estimation of Closed Populations*. Colorado State University, Fort Collins, CO.

Reynolds, J. C. and Aebischer, N. J. (1991) Comparison and quantification of carnivore diet by faecal analysis: a critique, with recommendations, based on a study of the fox *Vulpes vulpes*. *Mammal Review*, **21**, 97–122.

Rhymer, J. M. and Simberloff, D. (1996) Extinction by hybridization and introgression. *Annual Review of Ecology and Systematics*, **27**, 83–109.

Riebsame, W. E. (ed.) (1997) *Atlas of the New West: Portrait of a Changing Region*. W.W. Norton and Co., New York.

Riley, S. J. (1998) *Integration of environmental, biological, and human dimensions for management of mountain lions* (Puma concolor) *in Montana*. PhD dissertation, Cornell University, Ithaca, NY.

Riley, S. P. D., Bromley, C., Poppenga, R. H., Uzai, F. A., Whited, L., and Sauvajot, R. M. (2007) Anticoagulant exposure and notoedric mange in bobcats and mountain lions in urban southern California. *Journal of Wildlife Management*, **71**, 1874–84.

Riley, S. P. D., Foley, J., and Chomel, B. (2004) Exposure to feline and canine pathogens in bobcats and gray foxes in urban and rural zones of a national park in California. *Journal of Wildlife Diseases*, **40**, 11–22.

Riordan, P. (1998) Unsupervised recognition of individual tigers and snow leopards from their footprints. *Animal Conservation*, **1**, 253–62.

Ritchie, J. (1920) *The Influence of Man on Animal Life in Scotland*. Cambridge University Press, Cambridge, UK.

Ritland, K. (2005) Multilocus estimation of pairwise relatedness with dominant markers. *Molecular Ecology*, **14**, 3157–65.

Robert, N. (2008) Neurologic disorders in cheetahs and snow leopards. In M. E. Fowler and R. E. Miller (eds.), *Zoo and Wild Animal Medicine: Current Therapy*, pp. 265–71. Saunders Elsevier, St Louis, MO.

Roberts, A. (1951) Felidae. In *The Mammals of South Africa*, pp. 173–94. The Mammals of South Africa Book Fund, Johannesburg, South Africa.

Roberts, S. J. (1971) *Veterinary Obstetrics and Genital Diseases: Thereogenology*. Edwards Brothers Inc., Ann Arbor, MI.

Roberts, T. J. (1977) Family Felidae. In *The Mammals of Pakistan*, pp. 138–58. Ernest Benn Limited, London, UK.

Rocha-Mendes, F. (2005) *Ecologia alimentar de carnívoros (Mammalia: Carnivora) e elementos de etnozoologia do município de Fênix, Paraná, Brasil*. MS dissertation, Universidade Estadual Paulista, São Paulo, Brazil.

Rocha, F. L. (2006) *Áreas de uso e seleção de habitat por três espécies de carnívoros de médio porte na Fazenda Nhumirim e arredores, Pantanal de Nhecolândia, MS*. MS dissertation, Universidade Federal do Mato Grosso do Sul, Corumbá, Brazil.

Rodenburg, W. (1977) *The Trade in Wild Animal Furs in Afghanistan*. UNDP FAO and Department of Forests and Wildlife, Kabul, Afghanistan.

Rodgers, A. R. (2001) Recent telemetry technology. In J. J. Millspaugh and J. M. Marzlof (eds.), *Radio Tracking and Animal Populations*, pp. 79–121. Academic Press, San Diego, CA.

Rodrigues, F. H. G. and Marinho-Filho, J. (1999) Translocation of oncilla and jaguarundi in central Brazil. *Cat News*, **30**, 28.

Rodríguez, A. (2007) *Lynx pardinus* (Temminck, 1827). In L. J. Palomo, J. Gisbert, and J. C. Blanco (eds.), *Atlas y Libro Rojo de los Mamíferos terrestres de España*, pp. 342–4. Dirección General para la Biodiversidad-SECEM-SECEMU, Madrid, Spain.

Rodríguez, A. and Delibes, M. (2004) Patterns and causes of non-natural mortality in the Iberian lynx during a 40-year period of range contraction. *Biological Conservation*, **118**, 151–61.

Rodríguez, A. and Delibes, M. (2003) Population fragmentation and extinction in the Iberian lynx. *Biological Conservation*, **109**, 321–31.

Rodríguez, A. and Delibes, M. (2002) Internal structure and patterns of contraction in the geographic range of the Iberian lynx. *Ecography*, **25**, 314–28.

Rodríguez, A. and Delibes, M. (1992) Current range and status of the Iberian lynx *Felis pardina* Temminck, 1824 in Spain. *Biological Conservation*, **61**, 189–96.

Rodríguez, A. and Delibes, M. (1990) *El lince ibérico en España: distribución y problemas de conservacion*. ICONA, Madrid, Spain.

Roelke-Parker, M. E., Munson, L., Packer, C., *et al.* (1996) A canine distemper virus epidemic in Serengeti lions (*Panthera leo*). *Nature*, **379**, 441–5.

Roelke, M. E. (1990) Florida panther biomedical investigation. Final performance report. Florida Game and Fresh Water Fish Commission, Tallahassee, FL.

Roelke, M. E., Johnson, W. E., Millan, J., *et al.* (2008) Exposure to disease agents in the endangered Iberian lynx (*Lynx pardinus*). *European Journal of Wildlife Research*, **54**, 171–8.

Roelke, M. E., Pecon-Slattery, J., Taylor, S., *et al.* (2006) T-lymphocyte profiles in FIV-infected wild lions and pumas reveal CD4 depletion. *Journal of Wildlife Diseases*, **42**, 234–48.

Roelke, M. E., Forrester, D. J., Jacobson, E. R., *et al.* (1993a) Seroprevalence of infectious disease agents in free-ranging Florida panthers. *Journal of Wildlife Diseases*, **20**, 36–49.

Roelke, M. E., Martenson, J. S., and O'Brien, S. J. (1993b) The consequences of demographic reduction and genetic depletion in the endangered Florida panther. *Current Biology*, **3**, 340–50.

Roelke, M. E., Schultz, D. P., Facemire, C. F., Sundlof, S. F., and Royals, H. E. (1991) Mercury contamination in Florida panthers. A Report of the Florida Panther Technical Subcommittee to the Florida Panther Interagency Committee. Florida Game and Freshwater Fish Commission, Tallahassee, FL.

Rogers, C. M. L. (1993) *A Woody Vegetation Survey of Hwange National Park*. Department of National Parks and Wildlife Management, Zimbabwe, Harare, Zimbabwe.

Rohan, M. J. (2000) A rose by any other name? The value construct. *Personality and Social Psychology Review*, **4**, 255–77.

Roldán, E., Gañán, N., Crespo, C., *et al.* (2006) A genetic resource bank and development of assisted reproductive technologies for the Iberian lynx. In *Iberian Lynx Ex-Situ Conservation Seminar Series: Book of Proceedings*. Fundacion Biodiversidad, Sevilla and Doñana, Spain.

Román, J., Palomares, F., Revilla, E., *et al.* (2006) *Seguimiento científico de las actuaciones del proyecto LIFE-naturaleza 'Recuperación de las poblaciones de lince ibérico en Andalucía'*. Consejería de Medio Ambiente, Junta de Andalucía, Sevilla, Spain.

Román, J., Ruiz, G., López, M., and Palomares, F. (2005) Los linces de Doñana, en la encrucijada: Abandonan el Parque por la escasez de conejos. *Quercus*, **236**, 10–14.

Romañach, S. S., Linsey, P. A., and Woodroffe, R. (2007) Determinants of attitudes towards predators in central Kenya and suggestions for increasing tolerance in livestock dominated landscapes. *Oryx*, **41**, 185–95.

Root, K. V. (2004) Florida panther (*Puma concolor coryi*): using models to guide recovery efforts. In H. R. Akcakaya, M. Burgman, O. Kindvall, *et al.* (eds.), *Species Conservation and Management: Case Studies*, pp. 491–504. Oxford University Press, Oxford, UK.

Rosas-Rosas, O. C., Bender, L. C., and Valdez, R. (2008) Jaguar and puma predation on cattle calves in northeastern Sonora, Mexico. *Rangeland Ecology Management*, **61**, 554–60.

Rose, L. E. (1971) *Nepal: Strategy for Survival*. University of California Press, Berkeley, CA.

Rosenzweig, M. L. (1966) Community structure in sympatric Carnivora. *Journal of Mammalogy*, **47**, 602–12.

Rosell, C., Álvarez, G., Cahill, S., Campeny, R., Rodríguez, A., and Seiler, A. (2003) *COST 341. La fragmentación del hábitat en relación con las infraestructuras de transporte en España*. Ministry of the Environment, Madrid, Spain.

Ross, P. I. and Jalkotzy, M. G. (1992) Characteristics of a hunted population of cougars in Southwestern Alberta. *Journal of Wildlife Management*, **56**, 417–26.

Rothwell, T. (2003) Phylogenetic systematics of North American *Pseudaelurus* (Carnivora: Felidae). *American Museum Novitates*, **3403**, 1–64.

Rotstein, D. S., Taylor, S. K., Bradley, J., and Breitschwerdt, E. B. (2000) Prevalence of *Bartonella henselae* antibody in Florida panthers. *Journal of Wildlife Diseases*, **36**, 157–60.

Rotstein, D. S., Harvey, J. W., Taylor, S. K., and Bean, J. (1999) Hematologic effects of cytauxzoonosis in Florida panthers and Texas cougars in Florida. *Journal of Wildlife Diseases*, **35**, 613–17.

Roussiakis, S. J., Theodorou, G. E., and Iliopoulos, G. (2006) An almost complete skeleton of *Metailurus parvulus* (Carnivora, Felidae) from the late Miocene of Kerassia (Northern Euboea, Greece). *Geobios*, **39**, 563–84.

Royama, T. (1992) *Analytical Population Dynamics*. Chapman & Hall, London, UK.

Royle, J. A., Nichols, J. D., Karanth, K. U., and Gopalaswamy, A. M. (2009). A hierarchical model for estimating density in camera trap studies. *Journal of Applied Ecology*, **46**, 118–27.

Rubenstein, D. I. (1982) Risk, uncertainty and evolutionary strategies. In King's College Sociobiology Group (eds.), *Current Problems in Sociobiology*, pp. 91–111. Cambridge University Press, Cambridge, UK.

Rudnai, J. A. (1979) Ecology of lions in Nairobi National Park and the adjoining Kitengela Conservation Unit in Kenya. *African Journal of Ecology*, **17**, 85–95.

Rudnai, J. A. (1974) The pattern of lion predation in Nairobi Park. *East Africa Wildlife Journal*, **12**, 213–25.

Rueness, E. K., Jorde, P. E., Hellborg, L., Stenseth, N. C., Ellegren, H., and Jakobsen, K. S. (2003a) Cryptic population structure in a large, mobile mammalian predator: the Scandinavian lynx. *Molecular Ecology*, **12**, 2623–33.

Rueness, E. K., Stenseth, N. C., O'Donoghue, M., Boutin, S., Ellegren, H., and Jakobsen, K. S. (2003b) Ecological and genetic spatial structuring in the Canadian lynx. *Nature*, **425**, 69–72.

Ruggiero, L. F., Aubry, K. B., Buskirk, S. W., *et al.* (2000) The scientific basis for lynx conservation: qualified insights. In L. F. Ruggiero, K. B. Aubry, S. W. Buskirk, *et al.* (eds.), *Ecology and Conservation of Lynx in the United States*, pp. 443–54. University Press of Colorado, Boulder, CO.

Ruiz-Garcia, M., Payán, E., Murillo, A., and Alvarez, D. (2006) DNA microsatellite characterisation of the jaguar (*Panthera onca*) in Colombia. *Genes and Genetic Systems*, **81**, 115–27.

Ruiz-Garcia, M., Payan, C. E., and Hernandez-Camacho, J. I. (2003) Possible records of *Lynchailurus* in south-western Colombia. *Cat News*, **38**, 37–9.

Russell, A. P. and Bryant, H. N. (2001) Claw retraction and protraction in Carnivora: the cheetah (*Acinonyx jubatus*) as an atypical felid. *Journal of Zoology*, **254**, 67–76.

Russello, M. A., Gladyshev, E., Miquelle, D., and Caccone, A. (2004) Potential genetic consequences of a recent bottleneck in the Amur tiger of the Russian far east. *Conservation Genetics*, **5**, 707–13.

Ruth, T. K., Logan, K. A., Sweanor, L. L., Hornocker, M. G., and Temple, L. J. (1998) Evaluating cougar translocation in New Mexico. *Journal of Wildlife Management*, **62**, 1264–75.

Ryder, O. A. (1986) Species conservation and systematics—the dilemma of subspecies. *Trends in Ecology & Evolution*, **1**, 9–10.

Ryser-Degiorgis, M. P., Hofmann-Lehmann, R., Leutenegger, C. M., *et al.* (2005) Epizootilogic investigation of selected infectious disease agents in free-ranging Eurasian lynx from Sweden. *Journal of Wildlife Diseases*, **41**, 58–66.

Ryser-Degiorgis, M. P., Ryser, A., Bacciarini, L. N., *et al.* (2002) Notoedric and sarcoptic mange in free-ranging lynx from Switzerland. *Journal of Wildlife Diseases*, **38**, 228–32.

Ryser, A., Scholl, M., Zwahlen, M., Oetliker, M., Ryser-Degiorgis, M. P., and Breitenmoser, U. (2005) A remote-controlled teleinjection system for the low-stress capture of large mammals. *Wildlife Society Bulletin*, **33**, 721–30.

Ryser, A., von Wattenwyl, K., Ryser-Degiorgis, M. P., Willisch, C., Zimmermann, F., and Breitenmoser, U. (2004) Luchsumsiedlung Nordostschweiz 2001–2003. KORA-Bericht Nr. 22. KORA, Muri, Switzerland.

Saberwal, V., Chellam, R., Johnsingh, A. J. T., and Rodgers, W. A. (1990) *Lion-Human Conflicts in Gir Forest and Adjoining Areas*. Wildlife Institute of India, Dehradun, India.

Saberwal, V. K., Gibbs, J. P., Chellam, R., and Johnsingh, A. J. T. (1994) Lion-human conflict in the Gir Forest, India. *Conservation Biology*, **8**, 501–7.

Sachdev, M., Sankaranarayanan, R., Reddanna, P., Thangaraj, K., and Singh, L. (2005) Major histocompatibility complex class I polymorphism in Asiatic lions. *Tissue Antigens*, **66**, 9–18.

Sachs, J. D. (2005) *The End of Poverty: Economic Possibilities for Our Time*. Penguin Press, New York.

Sachs, J.D., Baillie, J. E. M., Sutherland, W. J. *et al.* (2009) Biodiversity conservation and the Millennium Development Goals. *Science*, **325**, 1502–03.

Sadovy, Y. and Cheung, W. L. (2003) Near extinction of a highly fecund fish: the one that nearly got away. *Fish and Fisheries*, **4**, 86–99.

Saenz, J. C. and Carrillo, E. (2002) Jaguares depredadores de ganado en Costa Rica: un problema sin solución? In R. A. Medellin, C. Equihua, C. L. B. Chetkiewicz, *et al.* (eds.), *El Jaguar en el Nuevo Milenio.*, pp. 127–37. Fondo de Cultura Economica, Universidad Nacional Autónoma de México and Wildlife Conservation Society, Mexico City, Mexico.

Salafsky, N. and Margoluis, R. (1999) Threat reduction assessment: a practical and cost-effective approach to evaluating conservation and development projects. *Conservation Biology*, **13**, 830–41.

Sales, G. and Pye, D. (1974) *Ultrasonic Communication by Animals*. Chapman and Hall, London, UK.

Salesa, M. J., Antón, M., Turner, A. and Morales, J. (2005) Aspects of the functional morphology in the cranial and cervical skeleton of the sabre-toothed cat *Paramachairodus ogygia* (Kaup, 1832) (Felidae, Machairodontinae) from the Late Miocene of Spain: implications for the origins of the machairodont killing bite. *Zoological Journal of the Linnean Society*, **144**, 363–77.

Salesa, M. J., Montoya, P., Alcalá, L., and Morales, J. (2003) El género *Paramachairodus* Pilgrim, 1913 (Felidae, Machairodontinae) en el Mioceno superior español. *Coloquios de Paleontologia*, **volumen extraordinario 1**, 603–15.

Salles, L. O. (1992) Felid phylogenetics: extant taxa and skull morphology (Felidae, Aeluroidea). *American Museum Novitates*, **3047**, 1–67.

Salom-Pérez, R., Carrillo, E., Saenz, J. C., and Mora, J. M. (2007) Critical condition of the jaguar *Panthera onca* population in Corcovado National Park, Costa Rica. *Oryx*, **41**, 51–6.

Samuel, M. D. and Fuller, M. R. (1994) Wildlife radiotelemetry. In T. A. Bookhout (ed.), *Research and Management Techniques for Wildlife and Habitats*, pp. 370–418. Wildlife Society, Bethseda, MD.

Sandell, M. (1989) The mating tactics and spacing patterns of solitary carnivores. In J. L. Gittleman (ed.), *Carnivore Behaviour, Ecology and Evolution*, pp. 164–82. Chapman and Hall, London, UK.

Sanders, J. K. (1963) Movement and activities of the lynx in New Foundland. *Journal of Wildlife Management*, **27**, 390–400.

Sanderson, E., Forrest, J., Loucks, C., *et al.* (2006) *Setting Priorities for the Conservation and Recovery of Wild Tigers: 2005–2015. The Technical Assessment*. WCS, WWF, Smithsonian, and NFWF-STF, Washington DC.

Sanderson, E. W., Jaiteh, M., Levy, M. A., Redford, K. H., Wannebo, A. V., and Woolmer, G. (2002a) The human footprint and the last of the wild. *BioScience*, **52**, 891–904.

Sanderson, E. W., Redford, K. H., Chetkiewicz, C. B., *et al.* (2002b) Planning to save a species: the jaguar as a model. *Conservation Biology*, **16**, 58–72.

Sanderson, J. (1999) Andean mountain cats (*Oreailurus jacobita*) in northern Chile. *Cat News*, **30**, 25–6.

Sanderson, J. G., Sunquist, M. E., and Iriarte, A. W. (2002c) Natural history and landscape-use of guignas (*Oncifelis guigna*) on Isla Grande de Chiloe, Chile. *Journal of Mammalogy*, **83**, 608–13.

Sandom, C., Bull, J., Canney, S., and Macdonald, D. W., (in press). Landscape scale fenced reserves for ecological restoration in the Scottish Highlands. In M. Hayward (ed.), *Using Fences in Conservation*.

Sangay, T. and Vernes, K. (2008) Human-wildlife conflict in the Kingdom of Bhutan: patterns of livestock predation by large mammalian carnivores. *Biological Conservation*, **141**, 1272–82.

Sangay, T. and Wangchuk, T. (2005) *Tiger Action Plan for the Kingdom of Bhutan 2006–2015*. Nature Conservation Division, Department of Forest, Ministry of Agriculture, Royal Government of Bhutan, WWF Bhutan, Thimpu, Bhutan.

Sanjayan, M. A., Shen, S., and Jansen, M. (1997) *Experiences with Integrated-Conservation Development Projects in Asia*. World Bank Publications, Washington DC.

Sankhala, K. S. (1977) Records of the tiger hunting in India and Nepal. In *Tiger! The Story of the Indian Tiger*. Collins, London, UK.

Santymire, R., Lonsdorf, E., and Lynch, C. (2007) Increasing inbreeding results in declining seminal quality, pregnancy success and litter size in the endangered black-footed ferret. In *Proceedings of the Society for Conservation Biology 21st Annual Meeting*, p. 233. Society for Conservation Biology, Washington DC.

Sardella, R. and Werdelin, L. (2007) *Amphimachairodus* (Felidae, Mammalia) from Sahabi (latest Miocene earliest Pliocene, Libya), with a review of African Miocene Machairodontinae. *Rivista Italiana di Paleontologia e Stratigrafia*, **113**, 67–77.

Saunders, J. K. (1963a) Food habits of the lynx in Newfoundland. *Journal of Wildlife Management*, **27**, 384–90.

Saunders, J. K. (1963b) Movements and activities of the lynx in Newfoundland. *Journal of Wildlife Management*, **27**, 390–400.

Savage, R. J. G. (1977) Evolution in carnivorous mammals. *Palaeontology*, **20**, 237–71.

Save The Tiger Fund (2009) New report on tiger farms released. <www.Savethetigerfund.Org/Content/Navigationmenu2/Initiatives/Catt/Tigerfarming/Default.htm>. Accessed online 10 May 2009.

Say, L., Pontier, D., and Natoli, E. (1999) High variation in multiple paternity of domestic cats (*Felis catus* L.) in relation to environmental conditions. *Proceedings of the Royal Society of London B*, **266**, 2071–4.

Schaaf, D. (1978) Some aspects of the ecology of the swamp deer or barasingha (*Cervus d. duvauceli*) in Nepal. In I. M. Cowan and C. W. Holloway (eds.), *Threatened Deer*, pp. 65–86. IUCN, Morges, Switzerland.

Schaller, G. B. (1998) *Wildlife of the Tibetan Steppe*. Chicago University Press, Chicago, IL.

Schaller, G. B. (1988) Snow leopard in China. *Snow Line*, **14**, 3.

Schaller, G. B. (1983) Mammals and their biomass on a Brazilian ranch. *Arquivos de Zoologia*, **31**, 1–36.

Schaller, G. B. (1979) *On the Status of Jaguar in the Pantanal*. Instituto Brasileiro de Desenvolvimento Sustentável (IBDF), Brasília, Brazil.

Schaller, G. B. (1972) *The Serengeti Lion: A Study of Predator-Prey Relations*. University of Chicago Press, Chicago, IL.

Schaller, G. B. (1968) The hunting behaviour of the cheetah in the Serengeti National Park, Tanzania. *East African Wildlife Journal*, **6**, 95–100.

Schaller, G. B. (1967) *The Deer and the Tiger*. University of Chicago Press, Chicago, IL.

Schaller, G. B. and Crawshaw, P. G. (1980) Movement patterns of jaguar. *Biotropica*, **12**, 161–8.

Schaller, G. B. and Vasconcelos, J. M. C. (1978) Jaguar predation on capybara. *Zeitschrift fuer Saeugetierkunde*, **43**, 296–301.

Schaller, G. B., Tserendeleg, J., and Amarasanaa, G. (1994) Observations on snow leopards in Mongolia. In J. L. Fox and D. Jizeng (eds.) *Proceedings of the 7th International Snow Leopard Symposium*, pp. 33–46. Northwest Plateau Institute of Biology and International Snow Leopard Trust, Seattle, WA.

Schaller, G. B., Junrang, R., and Ming Jiang, Q. (1988) Status of the snow leopard *Panthera uncia* in Qinghai and Gansu Provinces, China. *Biological Conservation*, **45**, 179–94.

Schaller, G. B., Li, H., Ren, J. R., Qiu, M. J., and Wang, H. B. (1987) Status of large mammals in the Taxkorgan Reserve, Xinjiang, China. *Biological Conservation*, **42**, 53–71.

Schauenberg, P. (1977) Longueur de l'intestin du chat forestier *Felis silvestris* Schreber. *Mammalia*, **41**, 357–60.

Schauenberg, P. (1974) Données nouvelles sur le chat des sables *Felis margarita* Loche, 1858. *Revue suisse de Zoologie*, **81**, 949–69.

Schauenberg, P. (1969) L'identification du chat forestier *Felis s. sylvestris* Screber 1777, par une méthode ostéométrique. *Revue Suisse Zoologie*, **76**, 433–41.

Scheel, D. and Packer, C. (1995) Variation in predation by lions: tracking a moveable feast. In A. R. E. Sinclair and P. Arcese (eds.), *Serengeti II: Dynamics, Management and Conservation of an Ecosystem*, pp. 299–314. University of Chicago Press, Chicago, IL.

Scheel, D. and Packer, C. (1991) Group hunting behaviour of lions: a search for cooperation. *Animal Behaviour*, **41**, 697–709.

Schelling, T. (1960) *The Strategy of Conflict*. Harvard University Press, Cambridge, MA.

Schemnitz, S. D. (ed.) (1980) *Wildlife Management Techniques Manual*. Wildlife Society, Washington DC.

Scherr, S. J. (2000) A downward spiral? Research evidence on the relationship between poverty and natural resource degradation. *Food Policy*, **25**, 479–98.

Schiaffino, K., Malmierca, L., Perovic, P., *et al.* (2002) Depedacion de credos domesticos por jaguar en un rea rural vecina a un parque nacional en el noreste de Argentina. In R. A. Medellín, C. Equihua, C. L. B. Chetkiewicz, *et al.* (eds.), *El Jaguar en el Nuevo Mileno*, p. 251. Fondo de Cultura Economica, Universidad Nacional Autónoma de México and Wildlife Conservation Society, Mexico City, Mexico.

Schiess-Meier, M., Ramsauer, S., Gabanapelo, T., and Konig, B. (2007) Livestock predation—insights from problem animal control registers in Botswana. *Journal of Wildlife Management*, **71**, 1267–74.

Schipper, J., Chanson, J. S., Chiozza, F., *et al.* (2008) The status of the world's land and marine mammals: diversity, threat, and knowledge. *Science*, **322**, 225–30.

Schmidt-Kittler, N. (1990) A biochronologic subdivision of the European Paleogene based on mammals—report on results of the Paleogene symposium held in Mainz in February 1987. In E. H. Lindsay, V. Fahlbusch, and P. Mein (eds.), *European Neogene Mammal Chronology*, pp. 47–50. Plenum Press, New York.

Schmidt-Kittler, N. (1976) Raubtiere aus dem Jungtertiär Kleinasiens. *Paleontographica, Abt. A*, **155**, 1–131.

Schmidt-Posthaus, H., Breitenmoser-Würsten, C., Posthaus, H., Bacciarini, L., and Breitenmoser, U. (2002a) Causes of mortality in a reintroduced Eurasian lynx in Switzerland. *Journal of Wildlife Diseases*, **38**, 84–92.

Schmidt-Posthaus, H., Breitenmoser-Würsten, C., Posthaus, H., Bacciarini, L. N., and Breitenmoser, U. (2002b) Pathological investigation of mortality in reintroduced Eurasian lynx (*Lynx lynx*) populations in Switzerland. *Journal of Wildlife Diseases*, **38**, 84–92.

Schmidt, K., Nakanishi, N., Okamura, M., Doi, T., and Izawa, M. (2003) Movements and use of home range in the Iriomote cat (*Prionailurus bengalensis iriomotensis*). *Journal of Zoology*, **261**, 273–83.

Schmidt, K., Jedrzejewski, W., and Okarma, H. (1997) Spatial organization and social relations in the Eurasian lynx population in Bialowieza Primeval Forest, Poland. *Acta Theriologica*, **42**, 289–312.

Schmunis, G. A. (2004) Medical significance of African Trypanosomiasis. In I. Maudlin, P. H. Holmes, and M. A. Mills (eds.), *The Trypanosomiases*. CAB International, Wallingford, UK.

Schneider, H. P. (1994) *Animal Health and Veterinary Medicine in Namibia*. AGRIVET, Windhoek, Namibia.

Schoener, T. W. (1984) Size differences among sympatric bird eating hawks: a worldwide survey. In D. R. Strong, S. D, L. G. Abele, and A. B. Thistle (eds.), *Ecological Communities: Conceptual Issues and the Evidence*, pp. 254–78. Princeton University Press, Princeton, NJ.

Schoener, T. W. (1974) Resource partitioning in ecological communities. *Science*, **185**, 27–39.

Schortemeyer, J. L., Maehr, D. S., McCown, J. W., Land, E. D., and Manor, P. D. (1991) Prey management for the Florida panther: a unique role for wildlife managers. *Transactions of the North American Wildlife and Natural Resources Conference*, **56**, 512–26.

Schultz, C. B., Schultz, M. R., and Martin, L. D. (1970) A new tribe of saber-toothed cats (Barbourofelini) from the Pliocene of North America. *Bulletin of the University of Nebraska State Museum*, **9**, 1–31.

Schulze, R. E. and McGee, O. S. (1978) Climatic indices and classifications in relation to the biogeography of southern Africa. In M. J. A. Werger (ed.), *Biogeography and Ecology of South Africa*, pp. 19–52. Junk, The Hague, the Netherlands.

Schumann, B. and Fabiano, E. (2006) *Evaluation of the Integrated Livestock and Predator Management Training Courses for Emerging Farmers, 15–19 May and 19–23 June 2006*. Cheetah Conservation Fund, Otjiwarongo, Namibia.

Schumann, M. (ed.) (2003) *Guide to Integrated Livestock and Predator Management: CCF RISE Namibia Communal Conservancy Shepherd Training Course Proceedings*. Cheetah Conservation Fund, Windhoek, Namibia.

Schumann, M., Schumann, B., Dickman, A. J., Watson, L. H., and Marker, L. (2006) Assessing the use of swing gates in game fences as a potential non-lethal predator exclusion technique. *South African Journal of Wildlife Research*, **36**, 173–81.

Schürer, U. (1988) Breeding black-footed cats (*Felis nigripes*) at Wuppertal Zoo, with notes on their reproductive biology. In B. L. Dresser, R. W. Reece, and E. J. Maruska (eds.), *Proceedings 5th World Conference on Breeding Endangered Species in Captivity, October 9–12, 1988*, pp. 547–54. Cincinnati Zoo and Botanical Garden Cincinnati, Cincinnati, OH.

Schwartz, C. C. and Arthur, S. M. (1999) Radio-tracking large wilderness mammals: integration of GPS and ARGOS technology. *Ursus*, **11**, 261–74.

Schwartz, M. K., Luikart, G., and Waples, R. S. (2007) Genetic monitoring as a promising tool for conservation and management. *Trends in Ecology & Evolution*, **22**, 25–33.

Schwartz, M. K., Pilgrim, K. L., McKelvey, K. S., *et al.* (2004) Hybridization between Canada lynx and bobcats: genetic results and management implications. *Conservation Genetics*, **5**, 349–55.

Schwartz, M. K., Pilgrim, K. L., McKelvey, K. S., *et al.* (2003) Hybridization between Canada lynx and bobcats: genetic results and management implications. *Conservation Genetics*, **5**, 349–55.

Schwartz, M. K., Mills, L. S., McKelvey, K. S., Ruggiero, L. F., and Allendorf, F. W. (2002) DNA reveals high dispersal synchronizing the population dynamics of Canada lynx. *Nature*, **415**, 520–2.

Schwerdtner, K. and Gruber, B. (2007) A conceptual framework for damage compensation schemes. *Biological Conservation*, **134**, 354–60.

Scognamillo, D., Maxit, I. E., Sunquist, M., and Polisar, J. (2003) Coexistence of jaguar (*Panthera onca*) and puma (*Puma concolor*) in a mosaic landscape in the Venezuelan llanos. *Journal of Zoology*, **259**, 269–79.

Scognamillo, D., Maxit, I., Sunquist, M. E., and Farrell, L. (2002) Ecología del jaguar y el problema de la depredación de ganado en un hato de los llanos venezolanos. In R. A. Medellín, C. Equihua, C. L. B. Chetkiewicz, *et al.* (eds.), *El Jaguar en Nuevo Milenio*, p. 139. Fondo de Cultura Economica, Universidad Nacional Autónoma de México and Wildlife Conservation Society, Mexico City, Mexico.

Scott, J. and Scott, A. (2003) *Big Cat Diary: Leopard*. Harper Collins, London, UK.

Scott, J. M., Heglund, P., and Morrison, M. L. (2002) *Predicting Species Occurrences: Issues of Scale and Accuracy*. Island Press, Washington DC.

Seal, U. S. (1994) A plan for genetic restoration and management of the Florida panther (*Felis concolor coryi*). Report to the Florida Game and Freshwater Fish Commission. Conservation Breeding Specialist Group, Apple Valley, MN.

Seal, U. S. and Lacy, R. C. (1992) *Genetic Management Strategies and Population Viability of the Florida Panther* (Felis concolor coryi). US Fish and Wildlife Service, Apple Valley, MN.

Seal, U. S. and Lacy, R. C. (1989) *Florida Panther Population Viability Analysis*. US Fish and Wildlife Service, Apple Valley, MN.

Seal, U. S., Tilson, R. L., Plotka, E. D., Reindl, N. J., and Seal, M. F. (1987) Behavioural indicators and endocrine correlates of estrous and anestrous in Siberian tigers. In U. S. Seal and R. L. Tilson (eds.), *Tigers of the World: The Biology, Biopolitics, Management, and Conservation of an Endangered Species*, pp. 244–54. Noyes Publications, Park Ridge, NJ.

Seaman, D. E., Griffith, B., and Powell, R. A. (1998) KERNELHR: a program for estimating animal home ranges. *Wildlife Society Bulletin*, **26**, 95–100.

Sechele, M. L. and Nzhengwa, D. M. (2002) Human predator conflicts and control measures in North-west district, Botswana. In A. J. Loveridge, T. Lynam, and D. W. Macdonald (eds.), *Lion Conservation Research. Workshop 2: Modelling Conflict*, pp. 23–4. Wildlife Conservation Research Unit, Oxford, UK.

Seidensticker, J. (2002) Tiger tracks. *Smithsonian*, **32**, 64–7.

Seidensticker, J. (1986) Large carnivores and the consequences of habitat insularization: ecology and conservation of tigers in Indonesia and Bangladesh. In S. D. Miller and D. D. Everett (eds.), *Cats of the World: Biology, Conservation, and Management*, pp. 1–42. National Wildlife Federation, Washington DC.

Seidensticker, J. (1976a) On the ecological separation between tigers and leopards. *Biotropica*, **8**, 225–34.

Seidensticker, J. (1976b) Ungulate populations in Chitawan Valley, Nepal. *Biological Conservation*, **10**, 183–210.

Seidensticker, J. and Lumpkin, S. (2006) Building an arc. *Smithsonian*, **37**, 56–63.

Seidensticker, J. and McDougal, C. W. (1993) Tiger predatory behaviour, ecology and conservation. *Symposia of the Zoological Society of London*, **65**, 105–25.

Seidensticker, J. and Suyono, I. (1980) *The Javan Tiger and the Meru-Betiri Reserve*. IUCN, Gland, Switzerland.

Seidensticker, J., Christie, S., and Jackson, P. (1999) *Riding the Tiger: Tiger Conservation in Human-Dominated Landscapes*. Cambridge University Press, Cambridge, UK.

Seidensticker, J., Sunquist, M. E., and McDougal, C. (1990) Leopards living at the edge of the Royal Chitwan National Park, Nepal. In J. C. Daniel and J. S. Serrao (eds.), *Conservation in Developing Countries: Problems and Prospects*, pp. 415–23. Oxford University Press, Oxford, UK.

Seidensticker, J., Hornocker, M. G., Wiles, W. V., and Messick, J. P. (1973) Mountain lion social organization in the Idaho primitive area. *Wildlife Monographs*, **35**, 1–61.

Selous, F. C. (1894) The lion in South Africa. In C. Phillipps-Wolley (ed.), *The Badminton Library of Sports and Pastimes: Big Game Shooting*, pp. 236–50. Longmans, Green and Co., London, UK.

Serpell, J. A. (2000) The domestication and history of the cat. In D. C. Turner and P. Bateson (eds.), *The Domestic Cat: The Biology of Its Behaviour*, pp. 179–92. Cambridge University Press, Cambridge, UK.

Serra, R. C. and Sarmento, P. (2006) The Iberian lynx in Portugal: conservation status and perspectives. *Cat News*, **46**, 15–16.

Servheen, C., Herrero, S., and Peyton, B. (1999) *Bears: Status Survey and Conservation Action Plan*. IUCN, Gland, Switzerland.

Seymour, K. L. (1999) *Taxonomy, morphology, paleontology and phylogeny of the South American small cats (Mammalia: Felidae)*. PhD dissertation, University of Toronto, Toronto, ON.

Seymour, K. L. (1989) *Panthera onca*. *Mammalian Species*, **340**, 1–9.

Shabadash, S. A. and Zelikina, T. I. (1997) The cat caudal gland is hepatoid. *Izvestiya Akademii Nauk Seriya Biologicheskaya*, **5**, 556–70.

Shape of Enrichment (2008) <www.enrichment.org>. Accessed online 10 Sept 2008.

Sharma, S., Jhala, Y., and Sawarkar, V. B. (2005) Identification of individual tigers (*Panthera tigris*) from their pugmarks. *Journal of Zoology*, **267**, 9–18.

Sharp, N. C. C. (1997) Timed running speed of a cheetah (*Acinonyx jubatus*). *Journal of Zoology*, **241**, 493–4.

Shaw, A. P. M. (2004) Economics of African Trypanosomiases. In I. Maudlin, P. H. Holmes, and M. A. Mills (eds.), *The Trypanosomiases*, pp. 369–402. CAB International, Wallingford, UK.

Shaw, D. J., Grenfell, B. T., and Dobson, A. P. (1998) Patterns of macroparasite aggregation in wildlife host populations. *Parasitology*, **117**, 597–610.

Shemkunde, J. (2004) *The use of tourist photographs as a means of monitoring cheetahs in the Serengeti National Park, Tanzania*. MSc dissertation, University of Wales, Aberystwyth, Wales.

Shepherd, C. R. and Magnus, N. (2004) *Nowhere to Hide: The Trade in Sumatran Tiger*. TRAFFIC Southeast Asia, Petaling Jaya, Malaysia.

Shepherdson, D. J., Mellen, J. D., and Hutchins, M. (eds.) (1998) *Second Nature: Environmental Enrichment for Captive Animals*. Smithsonian Institution Press, Washington DC.

Shin, T., Kraemer, D., Pryor, J., *et al.* (2002) A cat cloned by nuclear transplantation. *Nature*, **415**, 859.

Shindle, D. M., Cunningham, M. W., Land, E. D., McBride, R., Lotz, M. A., and Ferree, B. (2003) *Florida panther genetic restoration and management: annual report*. Florida Fish and Wildlife Conservation Commission, Tallahassee, FL.

Shukla, U. K. and Bora, D. S. (2003) Geomorphology and sedimentology of Piedmont zone, Ganga Plain, India. *Current Science*, **84**, 1034–40.

Siddiqi, N. A. and Choudhury, J. H. (1987) Maneating behaviour of tigers (*Panthera tigris* Linn) of the Sundarbans: twenty-eight years' record analysis. *Tiger Paper*, **14**, 26–32.

Sillero-Zubiri, C. and Laurenson, M. K. (2001) Interactions between carnivores and local communities: conflict or co-existence? In J. L. Gittleman, S. M. Funk, D. W. Macdonald, and R. K. Wayne (eds.), *Carnivore Conservation*, pp. 282–312. Cambridge University Press, Cambridge, UK.

Sillero-Zubiri, C., Sukumar, R., and Treves, A. (2007) Living with wildlife: the roots of conflict and the solutions. In D. W. Macdonald and K. Service (eds.), *Key Topics in Conservation Biology*, pp. 253–70. Blackwell Publications, Oxford, UK.

Sillero-Zubiri, C., Marino, J., Gottelli, D., and Macdonald, D. W. (2004) Afroalpine ecology, solitary foraging and intense sociality amongst Ethiopian wolves. In D. W. Macdonald and C. Sillero-Zubiri (eds.), *The Biology and Conservation of Wild Canids*, pp. 311–23. Oxford University Press, Oxford, UK.

Silva-Rodriguez, E. A., Ortega-Solis, G. R., and Jimenez, J. E. (2007) Human attitudes toward wild felids in a human-dominated landscape of southern Chile. *Cat News*, **46**, 19–21.

Silveira, L. (1995) Notes on the distribution and natural history of the pampas cat, *Felis colocolo*, in Brazil. *Mammalia*, **59**, 284–8.

Silveira, L. and Jacomo, A. T. A. (2002) Conservacion del jaguar en el centro del Cerrado de Brasil. In R. A. Medellín, C. Equihua, C. L. B. Chetkiewicz, *et al.* (eds.), *El Jaguar en el Nuevo Milenio*, pp. 437–50. Fondo de Cultura Economica, Universidad Nacional Autónoma de México and Wildlife Conservation Society, Mexico City, Mexico.

Silveira, L., Jacomo, A. T. A., Kashivakura, C. K., *et al.* (2006) Conservação da onça-pintada em propriedades rurais no Pantanal. *Summary report on findings to the National Research Center for the Conservation of Natural Predators*. CENAP, Atibaia, Brazil.

Silveira, L., Jacomo, A. T. A., and Furtado, M. M. (2005) Pampas cat ecology and conservation in the Brazilian grasslands. Cat Project of the Month Feature Report. IUCN SSC Cat Specialist Group. <www.catsg.org/catsgportal/project-o-month/02_webarchive/webarchive.htm>. Accessed online 20 Sep 2009.

Silver, S. C. (1997) *The feeding ecology of translocated howler monkeys* (Alouatta pigra) *in Belize, Central America*. PhD dissertation, Fordham University, New York.

Silver, S. C., Ostro, L. E. T., Marsh, L. K., *et al.* (2004) The use of camera traps for estimating jaguar *Panthera onca* abundance and density using capture/recapture analysis. *Oryx*, **38**, 148–54.

Simberloff, D. and Boecklen, W. (1981) Santa Rosalia reconsidered: size ratios and competition. *Evolution*, **35**, 1206–28.

Simons, J. W. (1966) The presence of leopard and a study of the food debris in the leopard lairs of the Mount Suswa caves, Kenya. *Bulletin of the Cave Exploration Group of East Africa*, **1**, 51–69.

Simpson, G. G. (1951) The species concept. *Evolution*, **5**, 285–98.

Simpson, G. G. (1941) The function of saber-like canines in carnivorous mammals. *American Museum Novitates*, **1130**, 1–12.

Sinclair, A. R. E. (1995) Serengeti past and present. In A. R. E. Sinclair and P. Arcese (eds.), *Serengeti II: Dynamics, Management and Conservation of an Ecosystem*, pp. 3–30. University of Chicago, Chicago, IL.

Sinclair, A. R. E., Packer, C., Mduma, S., and Fryxell, J. M. (eds.) (2008) *Serengeti III: Human Impacts on Ecosystem Dynamics*. Chicago University Press, Chicago, IL.

Sinclair, A. R. E., Mduma, S., and Brashares, J. S. (2003) Patterns of predation in a diverse predator–prey system. *Nature*, **425**, 288–90.

Singh, A., Gaur, A., Shailaja, K., Bala, B. S., and Singh, L. (2004) A novel microsatellite (STR) marker for forensic identification of big cats in India. *Forensic Science International*, **141**, 143–7.

Singleton, P. H., Gaines, W. L., and Lehmkuhl J. F. (2002) Landscape permeability for large carnivores in Washington: a geographic information system weighted-distance and least-cost corridor assessment. Research paper No. PNW-RP-549. U.S. Department of Agriculture, Forest Service, Pacific Northwest Research Station, Portland, OR.

Skelton, P. (1993) *Evolution: A Biological and Paleontological Approach*. Addison-Wesley Publishing Co., Wokingham, UK.

Skinner, J. D. and Smithers, R. H. N. (1990) *The Mammals of the Southern African Subregion*. University of Pretoria, Pretoria, South Africa.

Slate, J., Kruuk, L. E. B., Marshall, T. C., Pemberton, J. M., and Clutton-Brock, T. H. (2000) Inbreeding depression influences lifetime breeding success in a wild population of red deer (*Cervus elaphus*). *Proceedings of the Royal Society of London Series B*, **267**, 1657–62.

Slater, G. J. and Van Valkenburgh, B. (2008) Long in the tooth: evolution of sabertooth cat cranial shape. *Paleobiology*, **34**, 403–19.

Sliwa, A. (in press-a) Sand cat *Felis margarita*. In J. S. Kingdon and M. Hoffmann (eds.), *The Mammals of Africa 5: Carnivora, Pholidota, Perissodactyla*. Academic Press, Amsterdam, the Netherlands.

Sliwa, A. (in press-b) *Felis nigripes*. In J. S. Kingdon and M. Hoffmann (eds.), *The Mammals of Africa 5: Carnivora, Pholidota, Perissodactyla*. Academic Press, Amsterdam, the Netherlands.

Sliwa, A. (2006) Seasonal and sex-specific prey composition of black-footed cats *Felis nigripes*. *Acta Theriologica*, **51**, 195–204.

Sliwa, A. (2004) Home range size and social organisation of black-footed cats (*Felis nigripes*). *Mammalian Biology*, **69**, 96–107.

Sliwa, A. (1996) *A functional analysis of scent marking and mating behaviour in the aardwolf,* Proteles cristatus *(Sparrman, 1783)*. PhD dissertation, University of Pretoria, Pretoria, South Africa.

Sliwa, A. (1994) Marsh owl (*Asio capensis*) associating with black-footed cat (*Felis nigripes*). *Gabar*, **9**, 23–4.

Sliwa, A., Wilson, B., Lamberski, N., and Herrick, J. (2007) Report on catching and surveying black-footed cats (*Felis nigripes*) on Benfontein Game Farm. Unpublished report.

Slotow, R. and Hunter, L. T. B. (2009) Reintroduction decisions taken at the incorrect social scale devalue their conservation contribution: the African lion in South Africa. In M. W. Hayward and M. J. Somers (eds.), *The Reintroduction of Top-Order Predators*, pp. 43–71. Wiley-Blackwell Publishing, Oxford, UK.

Slough, B. G. (1999) Characteristics of Canada Lynx, *Lynx canadensis*, maternal dens and denning habitat. *Canadian Field-Naturalist*, **113**, 605–8.

Slough, B. G. and Mowat, G. (1996) Lynx population dynamics in an untrapped refugium. *Journal of Wildlife Management*, **60**, 946–61.

Smallwood, K. S. and Fitzhugh, E. L. (1995) A track count for estimating mountain lion *Felis concolor californica* population trend. *Biological Conservation*, **71**, 251–9.

Smallwood, K. S. and Schonewald, C. (1998) Study design and interpretation of mammalian carnivore density estimates. *Oecologia*, **113**, 474–91.

Smirnov, E. N. and Miquelle, D. G. (1999) Population dynamics of the Amur tiger in Sikhote-Alin State Biosphere Reserve. In J. Seidensticker, S. Christie, and P. Jackson (eds.), *Riding the Tiger: Tiger Conservation in Human Dominated Landscapes*, pp. 61–70. Cambridge University Press, Cambridge, UK.

Smith, C., Hodgson, G. W. I., Armitage, P., and Clutton-Brock, J. (1994) Animal bone report. In B. B. Smith (ed.), *Howe: Four Millenia of Orkney Prehistory*, pp. 139–53. Society of Antiquaries of Scotland, Monograph Series No. 9, Edinburgh, Scotland.

Smith, J. L. D. (1993) The role of dispersal in structuring the Chitwan tiger population. *Behaviour*, **124**, 165–95.

Smith, J. L. D. and McDougal, C. (1991) The contribution of variance in lifetime reproduction to effective population size in tigers. *Conservation Biology*, **5**, 484–90.

Smith, J. L. D., McDougal, C. and Ahearn, S. C. (1999) Metapopulation structure of tigers in Nepal. In J. Seidensticker, S. Christie, and P. Jackson (eds.), *Riding the Tiger: Tiger Conservation in Human Dominated Landscapes*, pp. 178–89. Cambridge University Press, Cambridge, UK.

Smith, J. L. D., Ahearn, S.C., and McDougal, C. (1998) Landscape analysis of tiger distribution and habitat quality in Nepal. *Conservation Biology*, **12**, 1338–46.

Smith, J. L. D., McDougal, C. W., and Miquelle, D. G. (1989) Scent marking in free-ranging tigers, *Panthera tigris*. *Animal Behaviour*, **37**, 1–10.

Smith, J. L. D., McDougal, C. W., and Sunquist, M. E. (1987) Female land tenure system in tigers. In R. L. Tilson and U. S. Seal (eds.), *Tigers of the World: The Biology, Biopolitics, Management and Conservation of an Endangered Species*, pp. 97–109. Noyes Publications, Park Ridge, NJ.

Smith, J. L. D., Sunquist, M. E., Tamang, K. M., and Rai, P. B. (1983) A technique for capturing and immobilizing tigers. *Journal of Wildlife Management*, **47**, 255–9.

Smith, N. J. H. (1976) Spotted cats and the Amazon skin trade. *Oryx*, **13**, 362–71.

Smith, R. J., Muir, R. D. J., Walpole, M. J., Balmford, A., and Leader-Williams, N. (2003) Governance and the loss of biodiversity. *Nature*, **426**, 67–70.

Smithers, R. H. N. (1983) *The Mammals of the Southern African Subregion*. University of Pretoria Press, Pretoria, South Africa.

Smout, T. C., MacDonald, A. R., and Watson, F. (2004) *A History of the Native Woodlands of Scotland, 1500–1920*. Edinburgh University Press, Edinburgh, Scotland.

Smuts, G. L. (1976) Population characteristics and recent history of lions in two parts of the Kruger National Park. *Koedoe*, **19**, 153–64.

Smuts, G. L., Robinson, G. A., and Whyte, I. J. (1980) Comparative growth of wild male and female lions (*Panthera leo*). *Journal of Zoology*, **190**, 365–73.

Smuts, G. L. (1982) *Lion*. MacMillan Publishers, Johannesburg, South Africa.

Smuts, G. L. (1978) Effects of population reduction on the travels and reproduction of lions in Kruger National Park. *Carnivore*, **1**, 61–72.

Smuts, G. L., Whyte, I. J., and Dearlove, T. W. (1977) A mass capture technique for lions. *East Africa Wildlife Journal*, **15**, 81–7.

Smythies, A. E. (1942) *Big Game Shooting in Nepal*. Thacker, Spink and Co., Calcutta, India.

SNH (2008) Scottish wildcat. <www.snh.org.uk/speciesactionframework/saf-wildcat.asp>. Accessed online 20 May 2008.

SNH (2007) *A Five Year Species Action Framework: Making a Difference for Scotland's Species*. Scottish Natural Heritage, Inverness, Scotland.

Soares, T. N., Telles, M. P. C., Resende, L. V., *et al.* (2006) Paternity testing and behavioral ecology: A case study of jaguars (*Panthera onca*) in Emas National Park, Central Brazil. *Genetics and Molecular Biology*, **29**, 735–40.

Sodeinde, O. A. and Soewa, D. A. (1999) Pilot study of the traditional medicine trade in Nigeria. *TRAFFIC Bulletin*, **18**, 35–40.

Soisalo, M. K. and Cavalcanti, S. M. C. (2006) Estimating the density of a jaguar population in the Brazilian Pantanal using camera-traps and capture-recapture sampling in combination with GPS radio-telemetry. *Biological Conservation*, **129**, 487–96.

Sorli, L. E., Martinez, F. D., Lardelli, U., and Brandi, S. (2006) Andean cat in Mendoza, Argentina—furthest south and at lowest elevation ever recorded. *Cat News*, **44**, 24.

Soulé, M., Gilpin, M., Conway, W., and Foose, T. (1986) The millenium ark—how long a voyage, how many staterooms, how many passengers. *Zoo Biology*, **5**, 101–13.

Soulé, M. E. and Simberloff, D. (1986) What do genetics and ecology tell us about the design of nature reserves? *Biological Conservation*, **35**, 19–40.

Soulé, M. E. and Terborgh, J. E. (1999) *Continental Conservation: Scientific Foundations of Regional Reserve Networks*. Island Press, Washington DC.

South African National Biodiversity Institute (2005) Land degradation: KwaZulu-Natal. <www.sanbi.org/landdeg/grafics/kwazulunatafs>. Accessed online 15 Oct 2007.

Spassov, N., Simeonovski, V., and Spiridonov, G. (1997) The wild cat (*Felis silvestris* Schr.) and the feral domestic cat: problems of the morphology, taxonomy, identification of the hybrids and purity of the wild population. *Historia Naturalis Bulgarica*, **8**, 101–20.

Spencer, H. (1900) *The Principle of Ethics*, 6th edn. Williams and Norgate, London, UK.

Spencer, H. (1851) *Social Statics*. Chapman, London, UK.

Spencer, J. A. (1991) Survey of antibodies to feline viruses in free-ranging lions. *South African Journal of Wildlife Research*, **21**, 59–61.

Spencer, J. A. and Morkel, P. (1993) Serological survey of sera from lions in Etosha National Park. *South African Journal of Wildlife Research*, **23**, 60–1.

Spencer, J. A., Higginbotham, M. J., and Blagburn, B. L. (2003) Seroprevalence of *Neospora caninum* and *Toxoplasma gondii* in captive and free-ranging nondomestic felids in the United States. *Journal of Zoo and Wildlife Medicine*, **34**, 246–9.

Spencer, J. A., van Dijk, A. A., Horzinek, M. C., *et al.* (1992) Incidence of feline immunodeficiency virus reactive antibodies in free-ranging lions of the Kruger National Park and the Etosha National Park in southern Africa detected by recombinant FIV p24 antigen. *Onderstepoort Journal of Veterinary Research*, **59**, 315–22.

Spong, G. (2002) Space use in lions, *Panthera leo*, in the Selous Game Reserve: social and ecological factors. *Behavioural Ecology and Sociobiology*, **52**, 303–7.

Spong, G. and Creel, S. (2004) Effects of kinship on territorial conflicts among groups of lions, *Panthera leo*. *Behavioral Ecology and Sociobiology*, **55**, 325–31.

Spong, G. and Creel, S. (2001) Deriving dispersal distances from genetic data. *Proceedings of the Royal Society of London Series B*, **268**, 2571–4.

Spong, G., Stone, J., Creel, S., and Bjorkland, M. (2002) Genetic structure of lions (*Panthera leo*) in the Selous Game Reserve: implications for the evolution of sociality. *Journal of Evolutionary Biology*, **15**, 945–53.

Spong, G., Hellborg, L., and Creel, S. (2000a) Sex ratio of leopards taken in trophy hunting: genetic data from Tanzania. *Conservation Genetics*, **1**, 169–71.

Spong, G., Johansson, M., and Bjorklund, M. (2000b) High genetic variation in leopards indicates large and long-term stable effective population size. *Molecular Ecology*, **9**, 1773–82.

Squeo, F. A., Warner, B. G., Aravena, R., and Espinoza, D. (2006) Bofedales: High altitude peatlands of the central Andes. *Revista Chilena de Historia Natural*, **79**, 245–55.

Squires, J. R. and Laurion, T. (2000) Lynx home range and movements in Montana and Wyoming: preliminary results. In L. F. Ruggiero, K. B. Aubry, S. W. Buskirk, *et al.* (eds.), *Ecology and Conservation of Lynx in the United States*, pp. 337–49. University Press of Colorado, Boulder, CO.

Squires, J. R. and Ruggiero, L. F. (2007) Winter prey selection of Canada lynx in northwestern Montana. *Journal of Wildlife Management*, **71**, 310–15.

Srivastav, A. (1997) Livestock predation by Gir lions and ecodevelopment. *Tiger Paper*, **24**, 1–6.

St-Pierre, C., Ouellet, J. P., and Crete, M. (2006) Do competitive intraguild interactions affect space and habitat use by small carnivores in a forested landscape? *Ecography*, **29**, 487–96.

Stahl, P. and Artois, M. (1994) *Status and Conservation of the Wildcat in Europe and around the Mediterranean Rim*. Council of Europe, Strasbourg, France.

Stahl, P. and Vandel, J. M. (1999) Mortalité et captures de lynx (*Lynx lynx*) en France (1974–1998). *Mammalia*, **63**, 49–59.

Stahl, P., Vandel, J. M., Ruette, S., Coat, L., Coat, Y., and Balestra, L. (2002) Factors affecting lynx predation on sheep in the French Jura. *Journal of Applied Ecology*, **39**, 204–16.

Stahl, P., Vandel, J. M., Herrenschmidt, V., and Migot, P. (2001a) The effect of removing lynx in reducing attacks on sheep in the French Jura Mountains. *Biological Conservation*, **101**, 15–22.

Stahl, P., Vandel, J. M., Herrenschmidt, V., and Migot, P. (2001b) Predation on livestock by an expanding reintroduced lynx population: long-term trend and spatial variability. *Journal of Applied Ecology*, **38**, 674–87.

Stahl, P., Artois, M., and Aubert, M. F. A. (1988) The use of space and the activity pattern of adult European wild cats (*Felis silvestris* Schreber 1777) in Lorraine. *Revue d'Ecologie—La Terre et la Vie*, **43**, 113–32.

Stander, P. E. (2005a) *Population Demography and Ecology of Kunene Lions*. Predator Conservation Trust, Windhoek, Namibia.

Stander, P. E. (2005b) A review of conflict between humans and large carnivores and an evaluation of management options. In *National Workshop on Human Wildlife Conflict Management 2005*, pp. 1–76. Ministry of Environment and Tourism, Windhoek, Namibia.

Stander, P. E. (2001a) Namibian experiences with dealing with conflict between people and wildlife. In C. Murphy (ed.) *Reducing Conflict between Wildlife and People*. Ministry of Environment and Tourism, Windhoek, Namibia.

Stander, P. E. (2001b) Range country overview: Namibia. In P. Bartels, H. Berry, D. Cilliers, *et al.* (eds.), *Global Cheetah Conservation Action Plan: Draft Report.* IUCN/SSC Conservation Breeding Specialist Group, Apple Valley, MN.

Stander, P. E. (1998) Spoor counts as indices of large carnivore populations: the relationship between spoor frequency, sampling effort and true density. *Journal of Applied Ecology*, **35**, 378–85.

Stander, P. E. (1997) The ecology of lions and conflict with people in north-eastern Namibia. In J. V. Heerden (ed.), *Symposium on Lions and Leopards as Game Ranch Animals, Onderstepoort, South Africa*, pp. 10–17. Ministry of Environment and Tourism,Windhoek, Namibia.

Stander, P. E. (1992a) Cooperative hunting in lions: the role of the individual. *Behavioral Ecology and Sociobiology*, **29**, 445–54.

Stander, P. E. (1992b) Foraging dynamics of lions in a semiarid environment. *Canadian Journal of Zoology*, **70**, 8–21.

Stander, P. E. (1991) Demography of lions in Etosha National Park, Namibia. *Madoqua*, **18**, 1–9.

Stander, P. E. (1990) A suggested management strategy for stock-raiding lions in Namibia. *South Africam Journal of Wildlife Research*, **20**, 37–43.

Stander, P. E. and Albon, S. E. (1993) Hunting success of lions in a semi-arid enviroment. In N. Dunstone and M. L. Gorman (eds.), *Mammals as Predators*., pp. 127–43. Symposium of the Zoological Society of London, No. 65, London, UK.

Stander, P. E. and Morkel, P. V. (1991) Field immobilization of lions using disassociative anaesthetics in combination with sedatives. *African Journal of Ecology*, **29**, 137–48.

Stander, P. E., //au, K., |ui, N., Dabe, T. and Dabe, D. (1997a) Non-consumptive utilisation of leopards: community conservation and ecotourism in practice. In B. L. Penzhorn (ed.), *Proceedings of a Symposium on Lions and Leopards as Game Ranch Animals*, pp. 50–7. South African Veternirary Association Wildlife Group, Onderstepoort, South Africa.

Stander, P. E., Ghau, Tsisaba, D., and Ui (1997b) Tracking and the interpretation of spoor: a scientifically sound method in ecology. *Journal of Zoology*, **242**, 329–41.

Stander, P. E., Ghau, X., Tsisaba, D., and Txoma, X. (1996) A new method of darting: stepping back in time. *African Journal of Ecology*, **34**, 48–53.

Stander, P. E., Haden, P. J., Kaqece, A., and Ghau, A. (1997c) The ecology of asociality in Namibian leopards. *Journal of Zoology*, **242**, 343–64.

Stanley Price, M. R. (1989) *Animal Reintroductions: The Arabian Oryx in Oman*. Cambridge University Press, Cambridge, UK.

Staples, W. R. (1995) *Lynx and coyote diet and habitat relationships during a low hare population on the Kenai Peninsula, Alaska*. MSc dissertation, University of Alaska, Fairbanks, AK.

Statistics South Africa (2005) Population census 2001: KwaZulu-Natal. <www.statssa.gov.za/census01/html/default.asp>. Accessed online 15 Oct 2007.

Stein, A. (2008) *Ecology and conservation of leopards* (Panthera pardus) *in north-central Namibia*. PhD dissertation, University of Massachusetts, Amherst, MA.

Stein, B. E., Magalhaes-Castro, B., and Kruger, L. (1976) Relationship between visual and tactile representations in cat superior colliculus. *Journal of Neurophysiology*, **39**, 401–19.

Steinel, A., Truyen, U., Munson, L., and Van Vuuren, M. (2000) Genetic characterization of feline parvovirus sequences from various carnivores. *Journal of General Virology*, **81**, 345–50.

Stenseth, N. C., Ehrich, D., Rueness, E. K., *et al.* (2004a) The effect of climatic forcing on population synchrony and genetic structuring of the Canadian lynx. *Proceedings of the National Academy of Sciences of the United States of America*, **101**, 6056–61.

Stenseth, N. C., Shabbar, A., Chan, K.-S., *et al.* (2004b) Snow conditions may create an invisible barrier for lynx. *Proceedings of the National Academy of Sciences of the United States of America*, **101**, 10632–4.

Stenseth, N. C., Chan, K. S., Tong, H., *et al.* (1999) Common dynamic structure of Canada lynx populations within three climatic regions. *Science*, **285**, 1071–3.

Stenseth, N. C., Falck, W., Chan, K. S., *et al.* (1998) From patterns to processes: phase and density dependencies in the Canadian lynx cycle. *Proceedings of the National Academy of Sciences of the United States of America*, **95**, 15430–5.

Stenseth, N. C., Falck, W., Bjørnstad, O. N., and Krebs, C. J. (1997) Population regulation in snowshoe hare and Canadian lynx: asymmetric food web configurations between hare and lynx. *Proceedings of the National Academy of Sciences of the United States of America*, **94**, 5147–52.

Stephens, P. A., Zaumyslava, O. Y., Miquelle, D. G., Myslenkov, A. I., and Hayward, G. D. (2006) Estimating population density from indirect sign: track counts and the Formozov-Malyshev-Pereleshin formula. *Animal Conservation*, **9**, 339–48.

Stephens, P. A., Sutherland, W. J., Russell, A. F., Young, A. J., and Clutton Brock, T. H. (2005) Dispersal, eviction, and conflict in meerkats (*Suricata suricatta*): An evolutionarily stable strategy model. *American Naturalist*, **165**, 120–35.

Stephenson, R. O., Grangaard, D. V., and Burch, J. (1991) Lynx, *Felis lynx*, predation on red foxes, *Vulpes vulpes*, caribou, *Rangifer tarandus*, and Dall sheep, *Ovis dalli*, in Alaska. *Canadian Field-Naturalist*, **105**, 255–62.

Steury, T. D. and Murray, D. L. (2004) Modeling the reintroduction of lynx to the southern portion of its range. *Biological Conservation*, **117**, 127–41.

Stockwell, C. A., Hendry, A. P., and Kinnison, M. T. (2003) Contemporary evolution meets conservation biology. *Trends in Ecology and Evolution*, **18**, 94–101.

Stoen, O. G. and Wegge, P. (1996) Prey selection and prey removal by tiger (*Panthera tigris*) during the dry season in lowland Nepal. *Mammalia*, **60**, 363–73.

Stoner, D. C., Wolfe, M. L., and Choate, D. M. (2006) Cougar exploitation levels in Utah: implications for demographic structure, population recovery, and metapopulation dynamics. *Journal of Wildlife Management*, **70**, 1588–600.

Storms, T. N., Ramsay, E. C., Clyde, V. L., and Munson, L. (2003) Blastomycosis in nondomestic felids. *Journal of Zoo and Wildlife Medicine*, **34**, 231–8.

Stuart, C. T. (1982) *Aspects of the biology of the caracal* (Felis caracal) *in the Cape province, South Africa*. MS dissertation, University of Natal, Pietermaritzburg, South Africa.

Stuart, C. and Stuart, T. (in press) Caracal *Caracal caracal*. In J. S. Kingdon and M. Hoffmann (eds.), *The Mammals of Africa* **5**: *Carnivora, Pholidota, Perissodactyla*. Academic Press, Amsterdam, the Netherlands.

Stuart, C. and Stuart, T. (1991) The feral cat problem in southern Africa. Time to stop pussyfooting? *African Wildlife*, **45**, 13–15.

Stuart, C., Stuart, T., and de Smet, K. J. (in press) African wildcat *Felis silvestris*. In J. S. Kingdon and M. Hoffmann (eds.), *The Mammals of Africa 5: Carnivora, Pholidota, Perissodactyla*. Academic Press, Amsterdam, the Netherlands.

Stuart, C. T., MacDonald, A. W., and Mills, M. G. L. (1985) History, current status and conservation of large mammalian predators in Cape Province, Republic of South Africa. *Biological Conservation*, **31**, 7–19.

Suedmeyer, W. K. (2002) Conditioning programs for transabdominal ultrasound gestational monitoring in an eastern black rhinoceros (*Diceros bicornis michaeli*), African elephant (*Loxodonta africana*), African lion (*Panthera leo*) and Bornean orangutan (*Pongo pygmaeus pygmaeus*). *Proceedings of the American Association of Zoo Veterinarians*, **1**, 50–1.

Sukhdev, P. (2008) *The economics of ecosystems and biodiversity. TEEB: An Interim Report*. European Communities, Germany.

Suminski, P. (1962) Research in the native form of the wildcat (*Felis silvestris* Schreber) on the background of its geographical distribution. *Folia Forestalia Polenica*, **8**, 1–81.

Sunquist, M. E. (1992) The ecology of the ocelot: the importance of incorporating life history traits into conservation plans. In *Felinos de Venezuela: Biologia, Ecologia y Conservacion*, pp. 117–28. Fundacion para el Desarrollo de las Ciencias Fisicas, Matematicas y Naturales, Caracas, Venezuela.

Sunquist, M. E. (1981) The social organization of tigers (*Panthera tigris*) in Royal Chitawan National Park, Nepal. *Smithonian Contributions to Zoology*, **336**, 1–98.

Sunquist, M. E. and Sunquist, F. (2002) *Wild Cats of the World*. University of Chicago Press, Chicago, IL.

Sunquist, M. E. and Sunquist, F. (2001) Changing landscapes: consequences for carnivores. In J. L. Gittleman, S. M. Funk, D. W. MacDonald, and R. K. Wayne (eds.), *Carnivore Conservation*, pp. 399–418. Cambridge University Press Cambridge, UK.

Sunquist, M. E. and Sunquist, F. C. (1989) Ecological constraints on predation by large felids. In J. L. Gittleman (ed.), *Carnivore Behaviour, Ecology and Evolution*, pp. 283–301. Cornell University Press, Ithaca, NY.

Sunquist, M. E., Karanth, K. U., and Sunquist, F. (1999) Ecology, behaviour and resilience of the tiger and its conservation needs. In J. Seidensticker, S. Christie, and P. Jackson (eds.), *Riding the Tiger: Tiger Conservation in Human-Dominated Landscapes*, pp. 5–18. Cambridge University Press, Cambridge, UK.

Sutcliffe, A. J. (1973) Caves of the East African rift valley. *Transactions of the Cave Ressearch Group of Great Britain*, **15**, 41–65.

Suzan, G. and Ceballos, G. (2005) The role of feral mammals on wildlife infectious disease prevalence in two nature reserves within Mexico City limits. *Journal of Zoo and Wildlife Medicine*, **36**, 479–84.

Swallow, B., Leimona, B., Yatich, T., Velarde, S. J., and Puttaswamaiah, S. (2007) *The conditions for effective mechanisms of compensation and rewards for environmental services*. CES Scoping Study, Issue Paper No.3. ICRAF Working Paper No. 38. World Agroforestry Centre, Nairobi, Kenya.

Swanson, B. J. and Rusz, P. J. (2006) Detection and classification of cougars in Michigan using low copy DNA sources. *American Midland Naturalist*, **155**, 363–72.

Swanson, T. M. (1995) *The Economics and Ecology of Biodiversity Decline: The Forces Driving Global Change*. Cambridge University Press, Cambridge, UK.

Swanson, T. M. (1994) *The International Regulation of Extinction*. New York University Press, New York.

Swanson, W. F. (2006) Application of assisted reproduction for population management in felids: the potential and reality for conservation of small cats. *Theriogenology*, **66**, 49–58.

Swanson, W. F. (2001) Reproductive biotechnology and conservation of the forgotten felids: the small cats. In N. Loskutoff (ed.), *Proceedings of the 1st International Symposium on Assisted Reproductive Technologies for*

Conservation and Genetic Management of Wildlife, pp. 100–20. Omaha's Henry Doorly Zoo, Omaha, NE.

Swanson, W. F., Stoops, M. A., Magarey, G. M., and Herrick, J. R. (2007) Sperm cryopreservation in endangered felids: developing linkage of *in situ—ex situ* populations. In E. Roldan and M. Gomenido (eds.), *Spermatology*, pp. 417–32. Nottinghan University Press, Nottingham, UK.

Swanson, W. F., Johnson, W. E., Cambre, R. C., *et al.* (2003) Reproductive status of endemic felid species in Latin American zoos and implications for ex situ conservation. *Zoo Biology*, **22**, 421–41.

Swanson, W. F., Paz, R. C. R., Morais, R. N., Gomes, M. L. F., Moraes, W., and Adania, C. H. (2002) Influence of species and diet on efficiency of in vitro fertilization in two endangered Brazilian felids: the ocelot (*Leopardus pardalis*) and tigrina (*Leopardus tigrinus*). *Theriogenology*, **57**, 593.

Swarts, F. (2000) The Pantanal in the 21th Century: for the planet's largest wetland, an uncertain future. In F. Swarts (ed.), *The Pantanal of Brazil, Bolivia and Paraguay*, pp. 1–24. Waterland Research Institute, Gouldsboro, ME.

Sweanor, L. L. (1990) *Mountain lion social organisation in a desert environment*. MS dissertation, University of Idaho, Moscow, ID.

Sweanor, L. L., Logan, K. A., and Hornocker, M. G. (2000) Cougar dispersal patterns, metapopulation dynamics, and conservation. *Conservation Biology*, **14**, 798–808.

Swenson, J. E. and Andrén, H. (2005) A tale of two countries: large carnivore depredation and compensation schemes in Sweden and Norway. In R. Woodroffe, S. Thirgood, and A. Rabinowitz (eds.), *People and Wildlife: Conflict or Coexistence?*, pp. 323–39. Cambridge University Press, Cambridge, UK.

Swenson, J. E., Sandegren, F., Söderberg, A., Bjärvall, A., Franzén, R., and Wabakken, P. (1997) Infanticide caused by hunting of male bears. *Nature*, **386**, 450–1.

Swihart, R. K., Slade, N. A., and Bergstron, B. J. (1988) Relating body size to the rate of home range use in mammals. *Ecology*, **69**, 393–9.

Taber, A. B., Novaro, A. J., Neris, N., and Colman, F. H. (1997) The food habits of sympatric jaguar and puma in the Paraguayan Chaco. *Biotropica*, **29**, 204–13.

Tamang, K. M. (1982) *The status of the tiger* (Panthera tigris) *and its impact on principal prey populations in Royal Chitwan National Park, Nepal*. PhD dissertation, Michigan State University, East Lancing, MI.

Tambling, C. J. and du Toit, J. T. (2005) Modelling wildebeest population dynamics: implications of predation and harvesting in a closed system. *Journal of Applied Ecology*, **42**, 431–41.

Tanabe, K. (2001) A Kushano-Sasanian silver plate and Central Asian tigers. *Silk Road Art and Archaeology*, **7**, 167–86.

Tapper, S. (1992) *Game Heritage*. The Game Conservancy, Fordingbridge, UK.

Taylor, C. R. and Rowntree, V. J. (1973) Temperature regulation and heat balance in running cheetahs: a strategy for sprinters? *American Journal of Physiology*, **224**, 848–51.

Taylor, C. R., Shkolnik, A., Dmi'el, R., Baharav, D., and Borut, A. (1974) Running in cheetahs, gazelles, and goats: energy cost and limb configuration. *American Journal of Physiology*, **227**, 848–50.

Taylor, M. E. (1989) Locomotor adaptations by carnivores. In J. L. Gittleman (ed.), *Carnivore Behavior, Ecology and Evolution*, pp. 382–409. Chapman and Hall, London, UK.

Taylor, S. K., Buergelt, C. D., Roelke-Parker, M. E., Homer, B. L., and Rotstein, D. S. (2002) Causes of mortality of free-ranging Florida panthers. *Journal of Wildlife Diseases*, **38**, 107–14.

Terio, K. A., O'Brien, T. D., Lamberski, N., Famula, T. R., and Munson, L. (2008) Amyloidosis in black-footed cats (*Felis nigripes*). *Veterinary Pathology*, **45**, 393–400.

Terio, K. A., Munson, L., Marker, L., Aldridge, B. M., and Solnick, J. V. (2005) Comparison of *Helicobacter* spp. in cheetahs (*Acinonyx jubatus*) with and without gastritis. *Journal of Clinical Microbiology*, **43**, 229–34.

Terio, K. A., Marker, L., and Munson, L. (2004) Evidence for chronic stress in captive but not free-ranging cheethas (*Acinonyx jubatus*) based on adrenal morphology and function. *Journal of Wildlife Diseases*, **40**, 259–66.

Terio, K. A., Marker, L., Overstrom, E. W., and Brown, J. L. (2003) Analysis of ovarian and adrenal activity in Namibian cheetahs. *South African Journal of Wildlife Research*, **33**, 71–8.

Terio, K. A., Brown, J. L., and Citino, S. B. (1999) Fecal cortisol metabolite analysis for noninvasive monitoring of adrenocortical function in the cheetah (*Acinonyx jubatus*). *Journal of Zoo and Wildlife Medicine*, **30**, 484–91.

Terra Daily (2008) Trophy hunting buoyant industry for Namibia. <www.terradaily.com/reports/Trophy_Hunting_Buoyant_Industry_For_Namibia_999.html>. Accessed online 15 Sept 2008.

Tewes, M. E. (1986) *Ecological and behavioral correlates of ocelot spatial patterns*. PhD dissertation, University of Idaho, Moscow, ID.

Thanawongnuwech, R., Amonsin, A., Tantilertcharoen, R., *et al.* (2005) Probable tiger-to-tiger transmission of avian influenza H5N1. *Emerging Infectious Diseases*, **11**, 699–701.

Thapar, V. (1999) The tragedy of the Indian tiger: starting from scratch. In J. Seidensticker, S. Christie, and P. Jackson (eds.), *Riding the Tiger: Tiger Conservation in Human-Dominated Landscapes*, pp. 296–306. Cambridge University Press, Cambridge, UK.

Thapar, V. (1986) *Tiger: Portrait of a Predator*. William Collins Sons & Co Ltd, London, UK.

Tharoor, S. (2007) *The Elephant, the Tiger, and the Cell Phone: Reflections on India, the Emerging 21st Century Power*. Arcade Publishing, New York.

Thatcher, C. A., van Manen, F. T., and Clark, J. D. (2006) Identifying suitable sites for Florida panther reintroduction. *Journal of Wildlife Management*, **70**, 752–63.

Theile, S. (2003) *Fading Footprints: The Killing and Trade of Snow Leopards*. TRAFFIC International, Cambridge, UK.

Thenius, E. (1953) Gepardreste aus dem Altquartär von Hundsheim in Niederösterreich. *Neues Jahrbuch für Geologie und Paläontologie, Monatshefte*, **1953**, 225–38.

Thomas, D. S. G. and Shaw, P. A. (1991) *The Kalahari Environment*. Cambridge University Press, Cambridge, UK.

Thomas, H., Sen, S., Khan, M., Battail, B., and Ligabue, G. (1982) The Lower Miocene fauna of Al-Sarrar (eastern province, Saudi Arabia). *The Journal of Saudi Arabian Archaeology*, **5**, 109–36.

Thompson, W. L. (ed.) (2004) *Sampling Rare and Elusive Species: Concepts, Designs, and Techniques for Estimating Population Parameters*. Island Press, Washington DC.

Thompson, W. L., White, G. C., and Gowan, C. (1998) *Monitoring Vertebrate Populations*. Academic Press, San Diego, CA.

Thorbjarnarson, J. (1999) Crocodile tears and skins: international trade, economic constraints, and limits to the sustainable use of crocodilians. *Conservation Biology*, **13**, 465–70.

Thornton, P. K., Kruska, R. L., Henninger, N., *et al*. (2002) *Mapping Poverty and Livestock in the Developing World*. International Livestock Research Institute, Nairobi, Kenya.

Thresher, P. (1981) The economics of a lion. *Unasylva*, **33**, 34–5.

Tigers Forever (2009) <www.tigersforever.org>. Accessed online 15 Dec 2008.

Tilson, R. and Christie, S. (1999) Effective tiger conservation requires cooperation: zoos as a support for wild tigers. In J. Seidensticker, S. Christie, and P. Jackson (eds.), *Riding the Tiger: Tiger Conservation in Human Dominated Landscapes*, pp. 201–14. Cambridge University Press, Cambridge, UK.

Tilson, R. and Nyhus, P. J. (1998) Keeping problem tigers from becoming a problem species. *Conservation Biology*, **12**, 261.

Tinsley, J. B. (1970) *The Florida Panther*. Great Outdoors Publishing Company, St Petersburg, FL.

Tirado, C., Cortes, A., and Bozinovic, F. (2007) Metabolic rate, thermoregulation and water balance in *Lagidium viscacia* inhabiting the arid Andean plateau. *Journal of Thermal Biology*, **32**, 220–6.

Todd, N. B. (1962) The inheritance of the catnip response in domestic cats. *Journal of Heredity*, **53**, 54–6.

Tompkins, D. M., Dobson, A. P., Arneberg, P., *et al*. (2002) Parasites and host population dynamics. In P. J. Hudson (ed.), *The Ecology of Wildlife Diseases*, pp. 45–62. Oxford University Press, Oxford, UK.

Torres, N. M., Filho, J. A. F. D., De Marco, P. J., Jacomo, A. T. A., and Silveira, L. (2007) Jaguar distribution and conservation status in Brazil. In *Felid Biology and Conservation Conference 17–20 September 2007: Programme and Abstracts*, p. 109. Wildlife Conservation Research Unit, Oxford, UK.

Torres, S. G., Mansfield, T. M., Foley, J. E., Lupo, T., and Brinkhaus, A. (1996) Mountain lion and human activity in California: testing speculations. *Wildlife Society Bulletin*, **24**, 451–60.

Toth, M. (2008) A new noninvasive method for detecting mammals from birds' nests. *Journal of Wildlife Management*, **72**, 1237–40.

Townshend, F. T. (1875) *Wild Life in Florida: With a Visit to Cuba*. Hurst and Blackett, London, UK.

Transparency International (2008) Africa and the Middle East. Global Corruption Report 2008. <www.transparency.org/publications/gcr/gcr_2008#open>. Accessed online 16 Dec 2008.

Treves, A. and Karanth, K. U. (2003) Human-carnivore conflict and perspectives on carnivore management worldwide. *Conservation Biology*, **17**, 1491–9.

Treves, A. and Naughton-Treves, L. (1999) Risk and opportunity for humans coexisting with large carnivores. *Journal of Human Evolution*, **36**, 275–82.

Treves, A., Wallace, R. B., Naughton-Treves, L., and Morales, A. (2006) Co-managing human-wildlife conflicts: a review. *Human Dimensions of Wildlife*, **11**, 383–96.

Treves, A., Naughton-Treves, L., Harper, E. K., *et al*. (2004) Predicting human-carnivore conflict: a spatial model derived from 25 years of data on wolf predation on livestock. *Conservation Biology*, **18**, 114–25.

Trigo, T. C., Freitas, T. R. O., Kunzler, G., *et al*. (2008) Inter-species hybridization among Neotropical cats of the genus, and evidence for an introgressive hybrid zone between and in southern Brazil. *Molecular Ecology*, **17**, 4317–33.

Trivedi, P., Bhatnagar, Y. V., and Mishra, C. (2006) *Living with snow leopards: a conservation education strategy for the Himalayan high altitudes*. NCF Technical Report. Nature Conservation Foundation and International Snow Leopard Trust, Mysore, India.

Trolle, M. and Kéry, M. (2005) Camera-trap study of ocelot and other secretive mammals in the northern Pantanal. *Mammalia*, **69**, 409–16.

Trolle, M. and Kéry, M. (2003) Estimation of ocelot density in the Pantanal using capture-recapture analysis of camera-trapping data. *Journal of Mammalogy*, **84**, 607–14.

Trovati, R. G. (2004) *Monitoramento radiotelemétrico de pequeno e médios carnívoros na área de influência da UHE Luiz Eduardo Magalhães/ Lajeado—TO*. MS dissertation, Universidade de São Paulo, São Paulo, Brazil.

Troyer, J. L., Pecon-Slattery, J., Roelke, M. E., *et al*. (2005) Seroprevalence and genomic divergence of circulating strains of feline immunodeficiency virus among Felidae and Hyaenidae species. *Journal of Virology*, **79**, 8282–94.

Troyer, J. L., Pecon-Slattery, J., Roelke, M. E., Black, L., Packer, C., and O'Brien, S. J. (2004) Patterns of feline immunodeficiency virus multiple infection and genome divergence in a free-ranging population of African lions. *Journal of Virology*, **78**, 3777–91.

Tryjanowski, P., Antczak, M., Hromada, M., Kuczynski, L., and Skoracki, M. (2002) Winter feeding ecology of male and female European wildcats *Felis silvestris* in Slovakia. *Zeitschrift Fur Jagdwissenschaft*, **48**, 49–54.

Tubbesing, U. H. (1997) An outbreak of anthrax amongst wild carnivore species held in captivity. In J. van Heerden (ed.), *Lions and Leopards as Game Ranch Animals*, pp. 163–7. Wildlife Group of the South African Veterinary Association, Onderstepoort, South Africa.

Tuck, S. L. (2005) The origins of Roman imperial hunting imagery: Domitian and the redefinition of the *virtus* under the principate. *Greece and Rome*, **52**, 221–45.

Turnbull-Kemp, P. (1967) *The Leopard*. Howard Timmins, Cape Town, South Africa.

Turner, A. (1990) The evolution of the guild of larger terrestrial carnivores during the Plio-pleistocene in Africa. *Geobios*, **23**, 349–68.

Turner, A. and Antón, M. (1997) *The Big Cats and Their Fossil Relatives*. Columbia University Press, New York.

Turner, M. (1987) *My Serengeti Years*. W.W. Norton & Company, New York.

Tuyttens, F. A. M. and Macdonald, D. W. (2000) Consequences of social perturbation for wildlife management and conservation. In L. M. Gosling and W. J. Sutherland (eds.), *Behaviour and Conservation*, pp. 315–29. Cambridge University Press, Cambridge, UK.

Tuyttens, F. A. M., Delahay, R. J., Macdonald, D. W., Cheeseman, C. L., Long, B., and Donnelly, C. A. (2000) Spatial perturbation caused by a badger (*Meles meles*) culling operation: implications for the function of territoriality and the control of bovine tuberculosis (*Mycobacterium bovis*). *Journal of Animal Ecology*, **69**, 815–28.

UN-Habitat (2006) *UN-HABITAT Annual Report*. United Nations, Nairobi, Kenya.

UN (2007) World population prospects: the 2006 revision population database. <www.esa.un.org/unpp/>. Accessed online.

UNEP-WCMC (2009) CITES trade database. United Nations Environment Program—World Conservation Monitoring Centre, Cambridge. <www.unep-wcmc.org/citestrade>. Accessed online 11 Nov 2009.

UNEP-WCMC. (2008) CITES trade database. United Nations Environment Program-World Conservation Monitoring Centre, Cambridge. <www.unep-wcmc.org/citestrade/>. Accessed online 16 Sept 2008.

Uphyrkina, O., Johnson, W. E., Quigley, H. B., *et al.* (2001) - Phylogenetics, genome diversity and origin of modern leopard, *Panthera pardus*. *Molecular Ecology*, **10**, 2617–33.

Uryu, Y., Mott, C., Foead, N., *et al.* (2008) *Deforestation, Forest Degradation, Biodiversity Loss and CO_2 Emissions in Riau, Sumatra, Indonesia*. WWF Indonesia, Jakarta, Indonesia.

US Census Bureau (2007) Cumulative estimates of population change for counties of Florida and county rankings: April 1, 200 to July 1, 2006. <www.census.gov/popest/counties/tables/CO-EST2006-02-12.xls>. Accessed online 11 Sept 2008.

US Department of Agriculture National Agricultural Statistics Service (2008a) Montana data—sheep, lambs and goats. <www.nass.usda.gov/Data_and_Statistics/Quick_Stats/#top>. Accessed online 1 Sept 2008.

US Department of Agriculture National Agricultural Statistics Service (2008b) Montana data—cattle and calves. <www.nass.usda.gov/Data_and_Statistics/Quick_Stats/#top>. Accessed online 1 Sept 2008.

US Department of Agriculture National Agricultural Statistics Service (2008c) Sheep and lamb losses: total by predator cause and number of head, Montana, USA. <www.nass.usda.gov/Statistics_by_State/Montana/Publications/livestock/sh&llos3.htm#top>. Accessed online 1 Sept 2008.

US Department of Agriculture National Agricultural Statistics Service (2001) Cattle predator loss. <www.usda.mannlib.cornell.edu/usda/nass/CattPredLo//2000s/2001/CattPredLo-05-04-2001.pdf>. Accessed online 1 Sept 2008.

US Federal Register (2005) *Endangered and Threatened Wildlife and Plants: Regulation for Nonessential Experimental Populations of the Western Distinct Population Segment of the Gray Wolf*. Department of the Interior—Fish and Wildlife Service, Washington DC.

US Federal Register (2000) *Downlisting of Cheetahs in Namibia*. US Government, Washington DC.

US Federal Register (1996) Proposed policy and proposed rule on the treatment of intercrosses and intercross progeny (the issue of 'hybridization'). *Federal Register*, **61**, 4709–13.

US Federal Register (1967) *Native Fish and Wildlife: Endangered Species*. Department of the Interior—Fish and Wildlife Service, Washington DC.

USAID (2005) *Namibia: Living in a Finite Environment (LIFE) Plus Project*. US Agency for International Development, Washington DC.

USFWS (2008a) *Florida Panther Recovery Plan* (Puma concolor coryi). 3rd revision. United States Fish and Wildlife Service, Atlanta, GA.

USFWS (2008b) *Revised Critical Habitat Proposed for Canada Lynx*. United States Fish and Wildlife Service, Washington DC.

USFWS (2006) *Guidelines for Living with Florida Panthers and the Interagency Florida Panther Response Plan*. United States Fish and Wildlife Service and Florida Panther National Wildlife Refuge, Naples, FL.

USFWS (1994) *Final Environmental Assessment: Genetic Restoration of the Florida Panther*. United States Fish and Wildlife Service, Atlanta, GA.

Uzal, F. A., Houston, R. S., Riley, S. P. D., Poppenga, R., Odani, J., and Boyce, W. (2007) Notoedric mange in two free-ranging mountain lions (*Puma concolor*). *Journal of Wildlife Diseases*, **43**, 274–8.

Valeix, M., Loveridge, A. J., Chamaillé-Jammes, S., *et al.* (2009a) Behavioral adjustments of African herbivores to predation risk by lions: spatiotemporal variations influence habitat use. *Ecology*, **90**, 23–30.

Valeix, M., Loveridge, A. J., Davidson, Z., Hunt, J. E., Murindagomo, F., and Macdonald, D. W. (2009b) Does the risk of encountering lions influence African herbivore behavior at waterholes? *Behavioral Ecology and Sociobiology*. DOI 1/0.1007/s00265-009-0760-3.

Valicek, L., Smid, B., and Vahala, J. (1993) Demonstration of parvovirus in diarrhoeic African cheetahs (*Acinonyx jubatus jubatus* Schreber, 1775). *Veterinarni Medicina*, **38**, 245–9.

Valiere, N., Fumagalli, L., Gielly, L., *et al.* (2003) Long-distance wolf recolonization of France and Switzerland inferred from non-invasive genetic sampling over a period of 10 years. *Animal Conservation*, **6**, 83–92.

Valverde, J. A. (1957) Notes écologiques sur le lynx d'Espagne *Felis lynx pardina* Temminck. *Revue d'Ecologie—La Terre et la Vie*, **104**, 51–67.

Van Bommel, L., Bij de Vaate, M. D., De Boer, W. F., and De Iongh, H. H. (2007) Factors affecting livestock predation by lions in Cameroon. *African Journal of Ecology*, **45**, 490–8.

van den Brink, F. H. (1970) Distribution and speciation of some Carnivores Part I. *Mammal Review*, **1**, 67–78.

Van Dorsser, F. J., Lasano, S., and Steinetz, B. G. (2007) Pregnancy diagnosis in cats using a rapid bench-top kit to detect relaxin in urine. *Reproduction in Domestic Animals*, **42**, 111–12.

van Heezik, Y. M. and Seddon, P. J. (1998) Range size and habitat use of an adult male caracal in northern Saudi Arabia. *Journal of Arid Environments*, **40**, 109–12.

van Orsdol, K. G. (1984) Foraging behavior and hunting success of lions in Queen Elizabeth National Park, Uganda. *African Journal of Ecology*, **22**, 79–99.

van Orsdol, K. G. (1981) *Lion predation in Rwenzori National Park, Uganda*. PhD dissertation, Cambridge University, Cambridge, UK.

van Orsdol, K. G., Hanby, J. P., and Bygott, J. D. (1985) Ecological correlates of lion social organization (*Panthera leo*). *Journal of Zoology*, **206**, 97–112.

Van Rensburg, I. B., and Silkstone, M. A. (1984) Concomitant feline infectious peritonitis and toxoplasmosis in a cheetah (*Acinonyx jubatus*). *Journal of the South African Veterinary Association*, **55**, 205–7.

van Rooyen, T. H. (1984) The soils of the Kalahari Gemsbok National Park. *Koedoe Supplement*, **1984**, 45–63.

Van Valkenburgh, B. (2007) Déjà vu: The evolution of feeding morphologies in the Carnivora. *Integrative and Comparative Biology*, **47**, 147–63.

Van Valkenburgh, B. (1996) Feeding behavior in free-ranging, large African carnivores. *Journal of Mammalogy*, **77**, 240–54.

Van Valkenburgh, B. (1988) Incidence of tooth breakage among large predatory mammals. *American Naturalist*, **131**, 291–302.

Van Valkenburgh, B. (1987) Skeletal indicators of locomotor behavior in living and extinct carnivores. *Journal of Vertebrate Paleontology*, **7**, 162–82.

Van Valkenburgh, B. and Ruff, C. B. (1987) Canine tooth strength and killing behaviour in large carnivores. *Journal of Zoology*, **212**, 379–97.

Van Valkenburgh, B. and Wayne, R. K. (1994) Shape divergence associated with size convergence in sympatric East African jackals. *Ecology*, **75**, 1567–81.

Van Valkenburgh, B., Grady, F., and Kurten, B. (1990) The Plio-Pleistocene cheetah-like cat *Miracinonyx inexpectatus* of North America. *Journal of Vertebrate Paleontology*, **10**, 434–54.

van Vuuren, M., Goosen, T., and Rogers, P. (1999) Feline herpesvirus infection in a group of semi-captive cheetahs. *Journal of the South African Veterinary Medical Association*, **70**, 132–4.

van Vuuren, M., Stylianides, E., and du Rand, A. (1997) The prevalence of viral infections in lions and leopards in southern Africa. In *Proceedings of a Symposium on Lions and Leopards as Game Ranch Animals*, pp. 168–73. Onderstepoort, South Africa.

Van Zyll de Jong, C. G. (1966) Food habits of the lynx in Alberta and the Mackenzie District, N.W.T. *Canadian Field-Naturalist*, **80**, 18–23.

VanderWaal, K. L., Mosser, A., and Packer, C. (2009) Optimal group size, dispersal decisions and post-dispersal relationships in female African lions. *Animal Behaviour*, **77**, 949–54.

Vargas, A. (2006) Overview of the various disciplines involved in the ex situ conservation program. In *Iberian Lynx Ex-Situ Conservation Seminar Series: Book of Proceedings*, pp. 5–11. Fundación Biodiversidad, Sevilla and Doñana, Spain.

Vargas, A., Sánchez, I., Martínez, F. *et al.* (2009) Interdisciplinary methods in the Iberian lynx (*Lynx pardinus*) Conservation Breeding Programme. In A. Vargas, C. Breitenmoser, U. Breitenmoser (eds.), *Iberian Lynx Ex-situ Conservation: An Interdisciplinary Approach*, pp. 56–71. Fundación Biodiversidad, Madrid, Spain.

Vargas, A., Gober, P., Lockhart, M., and Marinari, P. E. (2000) Black-footed ferrets: recovering an endangered species in an endangered habitat. In H. Griffin (ed.), *Mustelids in a Modern World: Conservation Aspects of Human-Carnivore Interaction*, pp. 97–106. Hull University Press, Hull, UK.

Vargas, A., Martínez, F., Bergara, J., Klink, L. D., Rodríguez, J. M., and Rodriguez, D. (2005) Iberian lynx *ex situ* conservation update. *Cat News*, **43**, 21–2.

Vargas, A., Sanchez, I., Martinez, F., *et al.* (2008) The Iberian lynx *Lynx pardinus* conservation breeding program. *International Zoo Yearbook*, **42**, 190–8.

Vargas, A., Sánchez, I., Godoy, J., *et al.* (eds.) (2007) *Plan de Acción para la cría en cautividad del lince ibérico*. Ministerio de Medio Ambiente, Madrid, Spain.

Venkataraman, A. B. and Johnsingh, A. J. T. (2004) Dholes. The behavioural ecology of dholes in India. In D. W. Macdonald and C. Sillero-Zubiri (eds.), *The Biology and Conservation of Wild Canids*, pp. 323–35. Oxford University Press, Oxford, UK.

Vereshchagin, N. K. (1971) Cave lions of Holarctics. *Trudy Zoologicheskogo Instituta*, **49**, 123–99.

Veterinary Specialist Group (2008) <www.iucn-vsg.org>. Accessed online 10 Sept 2008.

Vézina, A. F. (1985) Empirical relationships between predator and prey size among terrestrial vertebrate predators. *Oecologia*, **67**, 555–65.

Vial, F., Cleaveland, S., Rasmussen, G., and Haydon, D. T. (2006) Development of vaccination strategies for the management of rabies in African wild dogs. *Biological Conservation*, **131**, 180–92.

Vicente, J., Hofle, U., Garrido, J. M., *et al.* (2007) Risk factors associated with the prevalence of tuberculosis-like lesions in fenced wild boar and red deer in south central Spain. *Veterinary Research*, **38**, 451–64.

Vicente, J., Hofle, U., Garrido, J. M., *et al.* (2006) Wild boar and red deer display high prevalences of tuberculosis-like lesions in Spain. *Veterinary Research*, **37**, 107–19.

Vidolin, G. P. (2004) *Aspectos bio-ecológicos de* Puma concolor *(Linnaeus, 1758) e* Leopardus pardalis *(Schreber, 1775) na Reserva Natural Salto Morato, Guaraqueçaba, Paraná, Brasil.* MS dissertation, Universidade Federal do Paraná, Curitiba, Brazil.

Vigne, J. D. (1992) Zooarchaeology and the biographical history of the mammals of Corsica and Sardinia, since the last Ice Age. *Mammal Review*, **22**, 87–96.

Villa Meza, A., de, Martinez Meyer, E., and Lopez Gonzalez, C. A. (2002) Ocelot (*Leopardus pardalis*) food habits in a tropical deciduous forest of Jalisco, Mexico. *American Midland Naturalist*, **148**, 146–54.

Villafuerte, R., Calvete, C., Gortazar, C., and Moreno, S. (1994) First epizootic of rabbit hemorrhagic disease in free-living populations of *Oryctolagus cuniculus* at Donana National Park, Spain. *Journal of Wildlife Diseases*, **30**, 176–9.

Villalba, M. L. (2002) Andean cat photographed in Southwest Bolivia. *Cat News*, **36**, 19–20.

Villalba, M. L., Bernal, N., Nowell, K. and Macdonald, D. W. (in press) The Andean cat (*Oreailurus jacobita*) and pampas cat (*Oncifelis colocolo*) in the Bolivian Andes. Preliminary geographical distribution and notes on traditional beliefs. *Endangered Species Research*.

Villalba, M. L., Lucherini, M., Walker, S., *et al.* (2004) *The Andean Cat: Conservation Action Plan*. Andean Cat Alliance, La Paz, Bolivia.

Viranta, S. and Werdelin, L. (2003) Carnivora. In M. Fortelius, J. Kappelman, S. Sen, and R. Bernor (eds.), *Geology and Paleontology of the Miocene Sinap Formation, Turkey*, pp. 178–93. Columbia University Press, New York.

Viret, J. (1954) Le lœss a bancs durçis de Saint-Vallier (Drôme) et sa faune de mammifères Villafranchiens. *Nouvelles Archives du Muséum d'Histoire Naturelle de Lyon*, **4**, 1–200.

Viret, J. (1951) Catalogue critique de la faune des mammifères Miocènes de La Grive Saint-Alban (Isère). *Nouvelles Archives du Muséum d'Histoire Naturelle de Lyon*, **3**, 1–104.

Viscarra, M. E. (2008) *Distribución, densidad y dieta de carnívoros en cuatro tipos de habitats en un área de la Provincia Sud Lípez, Potosí, Bolivia*. BSc dissertation, Universidad Mayor de San Andrés, La Paz, Bolivia.

Voigt, C., Jewgenow, K., Martínez, F., Göritz, F., and Vargas, A. (2007) Monitoring reproductive status in *ex situ* breeding programs for mammals with a novel and gentle bleeding technique: use of blood-sucking bugs in the critically endangered Iberian lynx. In M. East and E. J. Hofer (eds.), *Contributions to the 6th International Wildlife Research Conference in Behaviour, Physiology and Genetics*, p. 239. Leibniz Institute for Zoo and Wildlife Research, Berlin, Germany.

Vollrath, F. and Douglas-Hamilton, I. (2002) African bees to control African elephants. *Naturwissenschaften*, **89**, 508–11.

Voloshina, I. V. and Myslenkov, A. I. (1994) Morphological characteristics of Amur goral. In A. I. Myslenkov (ed.), *Amur Goral: Collected Scientific Work*, pp. 75–83. Central Scientific Publishing Laboratory of Wildlife Management and Zapovedniks, Moscow, Russia.

von Arx, M., Breitenmoser-Würsten, C., Zimmermann, F., and Breitenmoser, U. (2004) *Status and Conservation of the Eurasian Lynx* (Lynx lynx) *in Europe in 2001*. KORA Bericht. KORA, Muri, Switzerland.

Vos, J. (2000) Food habits and livestock depredation of two Iberian wolf packs (*Canis lupus signatus*) in the north of Portugal. *Journal of Zoology*, **251**, 457–62.

Vyas, V. R., Lakhmapurkar, J. J., and Gavali, D. (2007) Sighting of rusty-spotted cat from new localities in central Gujurat. *Cat News*, **46**, 18.

Vynne, C., Almeida, R., and Silveira, L. (2007) Landscape matrix composition affects distribution of puma and jaguar in a cerrado ecosystem. In *Felid Biology and Conservation Conference 17–20 September 2007: Programme and Abstracts*, p. 66. Wildlife Conservation Research Unit, Oxford, UK.

Wack, R. (2003) Felidae. In M. E. Fowler and R. E. Miller (eds.), *Zoo and Wild Animal Medicine*, pp. 491–501. W.B. Saunders, St Louis, MO.

Wade, D., Mwasaga, B. C., and Eagles, P. (2001) A history and market analysis of tourism in Tanzania. *Tourism Management*, **22**, 93–101.

Wagner, K. K., Schmidt, R. H., and Conover, M. R. (1997) Compensation programs for wildlife damage in North America. *Wildlife Society Bulletin*, **25**, 312–19.

Waits, L., Taberlet, P., Swenson, J. E., Sandegren, F., and Franzen, R. (2000) Nuclear DNA microsatellite analysis of genetic diversity and gene flow in the Scandinavian brown bear (*Ursus arctos*). *Molecular Ecology*, **9**, 421–31.

Waits, L. P. (2004) Using noninvasive genetic sampling to detect and estimate abundance of rare wildlife species. In W. L. Thompson (ed.), *Sampling Rare or Elusive Species: Concepts, Designs and Techniques for Estimating Population Parameters*, pp. 211–28. Island Press, Washington DC.

Waits, L. P., Buckley-Beason, V. A., Johnson, W. E., Onorato, D., and McCarthy, T. (2007) A select panel of polymorphic microsatellite loci for individual identification of snow leopards (*Panthera uncia*): primer note. *Molecular Ecology Notes*, **7**, 311–14.

Walker, R. S. and Craighead, L. (1997) Least-cost-path corridor analysis: analyzing wildlife movement corridors in Montana using GIS. In *Proceedings of the ESRI European User Conference on Analyzing Wildlife Movement Corridors in Montana Using GIS, Copenhagen*, pp. 1–18. ESRI, Redlands, CA.

Walker, R. S., Novaro, A. J., Perovic, P., *et al.* (2007) Diets of three species of Andean carnivores in high-altitude deserts of Argentina. *Journal of Mammalogy*, **88**, 519–25.

Walker, R. S., Novaro, A. J., and Branch, L. C. (2003) The effects of patch attributes, barriers, and distance between patches on the distribution of a rock-dwelling rodent (*Lagidium viscacia*). *Landscape Ecology*, **18**, 185–92.

Walker, R. S., Pancotto, V., Schachter-Broide, J., Ackermann, G., and Novaro, A. J. (2000) Evaluation of a fecal pellet index of abundance for mountain vizcachas (*Lagidium viscacia*) in Patagonia. *Mastozoologia Neotropical*, **7**, 89–94.

Wallace, R. B., Gomez, H., Ayala, G., and Espinoza, F. (2003) Camera trapping capture frequencies for jaguar (*Panthera onca*) in the Tuichi Valley, Bolivia. *Mastozoologia Neotropical*, **10**, 133–9.

Walls, G. L. (1942) *The Vertebrate Eye and Its Adaptive Radiation*. Harper, New York.

Walpole, M. and Thouless, C. R. (2005) Increasing the value of wildlife through non-consumptive use? Deconstructing the myths of ecotourism and community based tourism in the tropics. In R. Woodroffe, S. Thirgood, and A. Rabinowitz (eds.), *People and Wildlife: Conflict or Coexistence?*, pp. 122–39. Cambridge University Press, Cambridge, UK.

Wang, E. (2002) Diets of ocelots (*Leopardus pardalis*), margays (*L. wiedii*), and oncillas (*L. tigrinus*) in the Atlantic Rainforest in southeast Brazil. *Studies on Neotropical Fauna and Environment*, **37**, 207–12.

Wang, S. W. (2008) *Understanding ecological interactions among carnivores, ungulates and farmers in Bhutan's Jigme Singye Wangchuck National Park*. PhD dissertation, Cornell University, Ithaca, NY.

Wang, S. W. (2007) A rare morph of the Asiatic golden cat in Bhutan's Jigme Singye Wangchuk National Park. *Cat News*, **47**, 27–8.

Wang, S. W. and Macdonald, D. W. (2009a) The use of camera traps for estimating tiger and leopard populations in the high altitude mountains of Bhutan. *Biological Conservation*, **142**, 606–13.

Wang, S. W. and Macdonald, D. W. (2009b) Feeding habits and niche partitioning in a predator guild composed of tigers, leopards and dholes in a temperate ecosystem in central Bhutan. *Journal of Zoology*, **277**, 275–83.

Wang, S. W. and Macdonald, D. W. (2006) Livestock predation by carnivores in Jigme Singye Wangchuck National Park, Bhutan. *Biological Conservation*, **129**, 558–65.

Wang, X., Ye, J., Meng, J., Wu, W., Liu, L., and Bi, S. (1998) Carnivora from middle Miocene of northern Junggar Basin, Xinjiang Autonomous Region, China. *Vertebrata PalAsiatica*, **36**, 218–43.

Ward, R. M. P. and Krebs, C. J. (1985) Behavioural responses of lynx to declining snowshoe hare abundance. *Canadian Journal of Zoology*, **63**, 2817–24.

Wassner, D. A., Guenther, D. D., and Layne, J. N. (1988) Ecology of the bobcat in south-central Florida USA. *Bulletin of the Florida State Museum Biological Sciences*, **33**, 159–228.

Wayne, R. K. and Ostrander, E. A. (1999) Origin, genetic diversity, and genome structure of the domestic dog. *Bioessays*, **21**, 247–57.

WCED (1987) *World Commission on Environment and Development (WCED), 'Our Common Future'*. Oxford University Press, New York.

WCMC-CITES (2009) CITES Trade Database. <www.unep-wcmc.org/citestrade/trade.cfm>. Accessed online 10 May 2009.

WCS (2008) All about jaguars: threats. Wildlife Conservation Society. <www.savethejaguar.com/jag-index/jag-allabout/jag-aboutthreats>. Accessed online 4 Sept 2008.

Weaver, J. L. (1993) Refining the equation for interpreting prey occurrence in gray wolf scats. *Journal of Wildlife Management*, **57**, 534–8.

Weaver, J. L., Paquet, P. C., and Ruggiero, L. F. (1996) Resilience and conservation of large carnivores in the Rocky Mountains. *Conservation Biology*, **10**, 964–76.

Weber, W. and Rabinowitz, A. R. (1996) A global perspective on large carnivore conservation. *Conservation Biology*, **10**, 1046–54.

Webster, D. B. (1962) A function of the enlarged middle-ear cavities of the kangaroo rat, *Dipodomys*. *Physiological Zoology*, **35**, 248–55.

Weckel, M., Giuliano, W., and Silver, S. (2006a) Cockscomb revisited: Jaguar diet in the Cockscomb Basin Wildlife Sanctuary, Belize. *Biotropica*, **38**, 667–90.

Weckel, M., Giuliano, W., and Silver, S. (2006b) Jaguar (*Panthera onca*) feeding ecology: distribution of predator and prey through time and space. *Journal of Zoology*, **270**, 25–30.

Wegge, P., Pokheral, C. P., and Jnawali, S. R. (2004) Effects of trapping effort and trap shyness on estimates of tiger abundance from camera trap studies. *Animal Conservation*, **7**, 251–6.

Wegner, G. and Pascual, U. (2008) *Ecosystem services for human well-being—revisited*. Environment and Economy and Policy Discussion Paper, No. 2008.37. Department of Land Economy, University of Cambridge, Cambridge, UK.

Weigel, I. (1961) Das Fellmuster der wildlebenden Katzenarten und der Hauskatze in vergleichender und stammesgeschichtlicher Hinsicht. *Saugetierkundliche Mitteilungen*, **9**, 1–120.

Weisbein, Y. and Mendelssohn, H. (1989) The biology and ecology of the caracal (*Felis caracal*) in the Aravah Valley of Israel. *Cat News*, **12**, 20–2.

Weissengruber, G. E., Forstenpointner, G., Peters, G., Kübber-Heiss, A., and Fitch, W. T. (2002) Hyoid apparatus and pharynx in the lion (*Panthera leo*), jaguar (*Panthera onca*), tiger (*Panthera tigris*), cheetah (*Acinonyx jubatus*) and domestic cat (*Felis silvestris f. catus*). *Journal of Anatomy*, **201**, 195–209.

Welburn, S. C., Picozzi, K., Coleman, P. G., and Packer, C. (2008) Patterns in age sero-prevalence consistent with acquired immunity against *Trypanosoma brucei in Serengeti lions*. *PLoS Neglected Tropical Diseases*, **2**, e347.

Welburn, S. C., Fevre, E. M., Coleman, P. G., Odiit, M., and Maudlin, I. (2001) Sleeping sickness: a tale of two disease. *Trends in Parasitology*, **17**, 19–24.

Wells, A., Terio, K. A., Ziccardi, M. H., and Munson, L. (2004) The stress response to environmental change in captive cheetahs (*Acinonyx jubatus*). *Journal of Zoo and Wildlife Medicine*, **35**, 8–14.

Wells, M. and McShane, T. O. (2004) Integrating protected area management with local needs and aspirations. *Ambio*, **33**, 513–19.

Wemmer, C. and Scow, J. (1977) Communication in the Felidae with emphasis on scent marking and contact patterns. In T. Seboek (ed.), *How Animals Communicate*, pp. 749–66. Indiana University Press, Bloomington, IN.

Wentzel, J., Stephens, J. C., Johnson, W., *et al.* (1999) Molecular genetic ascertainment of subspecies affiliation in tigers. In J. Seidensticker, S. Christie, and P. Jackson (eds.), *Riding the Tiger; Meeting the Needs of People and Wildlife in Asia*, pp. 40–9. Cambridge University Press, Cambridge, UK.

Werdelin, L. (2003a) Carnivores from the Kanapoi hominid site, Turkana Basin, northern Kenya. *Contributions in Science*, **498**, 115–32.

Werdelin, L. (2003b) Mio-Pliocene Carnivora from Lothagam, Kenya. In M. G. Leakey and J. M. Harris (eds.), *Lothagam: Dawn of Humanity in Eastern Africa*, pp. 261–328. Columbia University Press, New York.

Werdelin, L. (1989) The radiation of felids in South America: when and where did it occur? In *5th International Theriological Congress, Rome, Abstracts of Papers and Posters*, pp. 290–1. Rome, Italy.

Werdelin, L. (1985) Small Pleistocene felines of North America. *Journal of Vertebrate Paleontology*, **5**, 194–210.

Werdelin, L. (1983) Morphological patterns in the skulls of cats. *Biological Journal of the Linnean Society*, **19**, 375–91.

Werdelin, L. (1981) The evolution of lynxes. *Annales Zoologici Fennici*, **18**, 37–71.

Werdelin, L. and Dehghani, R. (in press) Carnivora. In T. Harrison (ed.), *Paleontology and Ecology of Laetoli, Tanzania: Human Evolution in Context*. Springer, New York.

Werdelin, L. and Lewis, M. E. (2005) Plio-Pleistocene Carnivora of eastern Africa: species richness and turnover patterns. *Zoological Journal of the Linnean Society*, **144**, 121–44.

Werdelin, L. and Lewis, M. E. (2001) A revision of the genus *Dinofelis* (Mammalia, Felidae). *Zoological Journal of the Linnean Society*, **132**, 147–258.

Werdelin, L. and Olsson, L. (1997) How the leopard got its spots: a phylogenetic view of the evolution of felid coat patterns. *Biological Journal of the Linean Society*, **62**, 383–400.

Werdelin, L. and Sardella, R. (2006) The '*Homotherium*' from Langebaanweg, South Africa and the origin of *Homotherium*. *Paleontographica, Abteilung A: Palaozoologie—Stratigraphie*, **277**, 123–30.

West, H. (2001) Sorcery of construction and socialist modernisation: ways of understanding power in postcolonial Mozambique. *American Ethnologist*, **28**, 119–50.

West, P. M. and Packer, C. (in press) African lion *Panthera leo*. In J. S. Kingdon and M. Hoffmann (eds.), *The Mammals of Africa 5: Carnivora, Pholidota, Perissodactyla*. Academic Press, Amsterdam, The Netherlands.

Westemeier, R. L., Brawn, J. D., Simpson, S. A., *et al.* (1998) Tracking the long-term decline and recovery of an isolated population. *Science*, **282**, 1695–8.

Western, D. and Henry, W. (1979) Economics and conservation in third world national parks. *BioScience*, **29**, 414–18.

Western, D., Wright, D. H., and Strum, S. (1994) *Natural connections: perspectives in community-based conservation*. Island Press, Washington DC.

Wheeler, Q. D. and Platnick, N. J. (2000) A critique from the Wheeler and Platnick phylogenetic species concept perspective: problems with alternative concepts of species. In Q. D. Wheeler and R. Meier (eds.), *Species Concepts and Phylogenetic Theory: A Debate*, pp. 133–45. Columbia University Press, New York.

White, G. C. and Garrott, R. A. (1990) *Analysis of Wildlife Radio-Tracking Data*. Academic Press, San Diego, CA.

White, M. (2009) Path of the jaguar. *National Geographic*, **March 2009**, 122–33.

Whitman, K., Starfield, A. M., Quadling, H. S., and Packer, C. (2004) Sustainable trophy hunting of African lions. *Nature*, **428**, 175–8.

Whitman, K. L., Starfield, A. M., Quadling, H., and Packer, C. (2007) Modeling the effects of trophy selection and environmental disturbance on a simulated population of African lions. *Conservation Biology*, **21**, 591–601.

Wielebnowski, N. C., Fletchall, N., Carlstead, K., Busso, J. M., and Brown, J. L. (2002a) Non-invasive assessment of adrenal activity associted with husbandry and behavioral factors in the North American clouded leopard population. *Zoo Biology*, **21**, 77–98.

Wielebnowski, N. C., Ziegler, K., Wildt, D. E., Lukas, J., and Brown, J. L. (2002b) Impact of social managment on reproductive, adrenal and behavioural activity in the cheetah (*Acinonyx jubatus*). *Animal Conservation*, **5**, 291–301.

Wielgus, R. B. and Bunnell, F. L. (2000) Possible negative effects of adult male mortality on female grizzly bear reproduction. *Biological Conservation*, **93**, 145–54.

Wikramanayake, E. D., Manandhar, A., and Nepal, S. (2010) The Terai Arc Landscape: a tiger conservation success story in a human dominated landscape. In R. Tilson and P. Nyhus (eds.), *Tigers of the World: The Science, Politics and Conservation of Panther tigris*,. Academic Press, New York.

Wikramanayake, E. D., McKnight, M., Dinerstein, E., Joshi, A. R., Gurung, B., and Smith, D. (2004) Designing a conservation landscape for tigers in human-dominated environments. *Conservation Biology*, **18**, 839–44.

Wikramanayake, E. D., Dinerstein, E., and Loucks, C. J. (eds.) (2002) *Terrestrial Ecoregions of the Indo-Pacific, a Conservation Assessment*. Island Press, Washington DC.

Wikramanayake, E. D., Dinerstein, E., Robinson, J. G., *et al.* (1998) An ecology-based method for defining priorities for large mammal conservation: the tiger as case study. *Conservation Biology*, **12**, 865–78.

Wilcox, R. W. (1999) The law of the least effort—cattle ranching and the environment in the savanna of Mato Grosso, Brazil 1900–1980. *Environmental History*, **4**, 338–68.

Wild-Eck, S. and Zimmermann, W. (2001) Raubtierakzeptanz in der Schweiz: Erkenntnisse aus einer Meinungsumfrage zu Wald und Natur. *Forestry Snow and Landscape Research*, **76**, 285–300.

Wild, M. A., Shenk, T. M., and Spraker, T. R. (2006) Plague as a mortality factor in Canada Lynx (*Lynx canadensis*) reintroduced to Colorado. *Journal of Wildlife Diseases*, **42**, 646–50.

Wildt, D. E., Ellis, S., and Howard, J. G. (2001) Linkage of reproductive sciences: from 'quick fix' to 'integrated' conservation. In P. W. Concannon, G. C. W. England, W. Farstad, C. Linde-Forsberg, J. P. Verstegen, and C. Doberska (eds.), *Advances in Reproduction in Dogs, Cats and Exotic Carnivores*, pp. 295–307. Journals of Reproduction and Fertility, Ltd., Colchester, UK.

Wildt, D. E., Brown, J. L., and Swanson, W. F. (1998) Reproduction in cats. In E. Knobil and J. Neill (eds.), *Encyclopedia of Reproduction*, pp. 497–510. Academic Press, New York.

Wildt, D. E., Rall, W. F., Critser, J. K., Monfort, S. L., and Seal, U. S. (1997) Genome resource banks: 'Living collections' for biodiversity conservation. *BioScience*, **47**, 689–98.

Wildt, D. E., Brown, J. L., Bush, M. E., *et al.* (1993) Reproductive status of cheetahs (*Acinonyx jubatus*) in North American zoos: the benefits of physiological surveys for strategic planning. *Zoo Biology*, **12**, 45–80.

Wildt, D. E., Phillips, L. G., Simmons, L. G., *et al.* (1988) A comparative analysis of ejaculate and hormonal characteristics of the captive male cheetah, tiger, leopard, and puma. *Biology of Reproduction*, **38**, 245–55.

Wildt, D. E., Bush, M. E., Goodrowe, K. L., *et al.* (1987a) Reproductive and genetic consequences of founding isolated lion populations. *Nature*, **329**, 328–31.

Wildt, D. E., O'Brien, S. J., Howard, J. G., *et al.* (1987b) Similarity in ejaculate-endocrine characteristics in captive versus free-ranging cheetahs of two subspecies. *Biology of Reproduction*, **36**, 351–60.

Wildt, D. E., Bush, M. E., Howard, J. G., *et al.* (1983) Unique semial quality in the South African cheetah and a comparative evaluation in the domestic cat. *Biology of Reproduction*, **29**, 1019–25.

Wilkie, D., Shaw, E., Rotberg, F., Morelli, G., and Auzel, P. (2000) Roads, development and conservation in the Congo Basin. *Conservation Biology*, **14**, 1614–22.

Wilkins, L., Arias-Reveron, J. M., Stith, B. M., Roelke, M. E., and Belden, R. C. (1997) The Florida panther *Puma concolor coryi*: a morphological investigation of the subspecies with a comparison to other North and South American cougars. *Bulletin of the Florida Museum of Natural History*, **40**, 221–69.

Willi, B., Filoni, C., Catao-Dias, J. L., *et al.* (2007) Worldwide occurrence of feline hemoplasma infections in wild felid species. *Journal of Clinical Microbiology*, **45**, 1159–66.

Williams, B. K., Nichols, J. D., and Conroy, M. J. (2002a) *Analysis and Management of Animal Populations*. Academic Press, San Diego, CA.

Williams, C. K., Ericsson, G., and Heberlein, T. A. (2002b) A quantitative summary of attitudes toward wolves and their reintroduction. *Wildlife Society Bulletin*, **30**, 575–84.

Williams, E. S. (2001) Canine distemper. In E. S. Williams and I. K. Barker (eds.), *Infectious Diseases of Wild Mammals*, pp. 50–9. Iowa State University Press, Ames, IA.

Williams, E. S., Thorne, E. T., Appel, M. J., and Belitsky, D. W. (1988) Canine distemper in black-footed ferrets (*Mustela nigripes*) from Wyoming. *Journal of Wildlife Diseases*, **24**, 385–98.

Williams, G. A., Ott, T. L., Michal, J. J., *et al.* (2007) Development of a model for mapping cryptorchidism in sheep and initial evidence for association of INSL3 with the defect. *Animal Genetics*, **38**, 189–91.

Williams, J. S. (1992) *Ecology of mountain lions in the Sun River area of northern Montana*. MS dissertation, Montana State University, Bozeman, MT.

Williams, L. E. (1976) Florida panther. In J. N. Layne (ed.), *Rare and Endangered Biota of Florida. Volume 1: Mammals*, pp. 13–15. University of Florida Press, Gainesville, FL.

Williams, R. W., Cavada, C., and Reinoso-Suárez, F. (1993) Rapid evolution of the visual system: a cellular assay of the retina and dorsal lateral geniculate nucleus of the

Spanish wildcat and the domestic cat. *Journal of Neuroscience*, **13**, 208–28.

Wilson, D. E. and Reeder, D. M. (1993) *Mammal Species of the World: A Taxonomic and Geographic Reference*. Smithsonian Institute Press, Washington DC.

Wilson, D. E., Cole, F. R., Nichols, J. D., Rudran, R. and Foster, M. S. (eds.) (1996) *Measuring and Monitoring Biological Diversity: Standard Methods for Mammals*. Smithsonian Institution Press, Washington DC.

Wilson, E. O. (2001) *The Diversity of Life*, 2nd edn. Penguin Books, London, UK.

Wilson, K. R. and Anderson, D. R. (1985) Evaluation of two density estimators of small mammal population size. *Journal of Mammalogy*, **66**, 13–21.

Wilting, A., Buckley-Beason, V. A., Feldhaar, H., Gadau, J., O'Brien, S. J., and Linsenmair, K. E. (2007a) Clouded leopard phylogeny revisited: Support for species recognition and population division between Borneo and Sumatra. *Frontiers in Zoology*, **4**, 1–10.

Wilting, A., Feldhaar, H., Buckley-Beason, V. A., Linsenmair, K. E., and O'Brien, S. J. (2007b) Two modern species of clouded leopards: a molecular perspective. *Cat News*, **47**, 10–11.

Wilting, A., Fischer, F., Abu Bakar, S., and Linsenmair, K. L. (2006) Clouded leopards, the secretive top-carnivore of South-East Asian rainforests: their distribution, status and conservation needs in Sabah, Malaysia. *BMC Ecology*, **6**, 1–13.

Wingard, J. R. and Zahler, P. (2006) *Silent steppe: the illegal wildlife trade crisis in Mongolia*. Mongolia Discussion Papers, East Asia and Pacfic Environment and Social Development Department, World Bank, Washington DC.

Wint, W. (2000) *Botswana Aerial Survey Information System (BASIS) Manual*. ERGO and Botswana DWNP, Oxford, UK.

Wolanski, E. J. and Gereta, E. (2001) Water quantity and quality as the factors driving the Serengeti ecosystem, Tanzania. *Hydrobiologia*, **485**, 169–80.

Wolff, J. O. (1980) The role of habitat patchiness in the population dynamics of snowshoe hares. *Ecological Monographs*, **50**, 111–30.

Woodroffe, R. (2001) Strategies for carnivore conservation: lessons from contemporary extinctions. In J. L. Gittleman, S. M. Funk, D. W. MacDonald, and R. K. Wayne (eds.), *Carnivore Conservation*, pp. 61–92. Cambridge University Press, Cambridge, UK.

Woodroffe, R. (2000) Predators and people: using human densities to interpret declines of large carnivores. *Animal Conservation*, **3**, 165–73.

Woodroffe, R. and Frank, L. G. (2005) Lethal control of African lions (*Panthera leo*): local and regional population impacts. *Animal Conservation*, **8**, 91–8.

Woodroffe, R. and Ginsberg, J. R. (2005) King of the beasts? Evidence for guild redundancy among large mammalian carnivores. In J. C. Ray, K. H. Redford, R. S. Steneck, and J. Berger (eds.), *Large Carnivores and the Conservation of Biodiversity*, pp. 154–75. Island Press, Washington DC.

Woodroffe, R. and Ginsburg, J. R. (2000) Ranging behaviour and vulnerability to extinction in carnivores. In L. M. Gosling and W. J. Sutherland (eds.), *Behaviour and Conservation*, p. 438. Cambridge University Press, Cambridge, UK.

Woodroffe, R. and Ginsberg, J. R. (1998) Edge effects and the extinction of populations inside protected areas. *Science*, **280**, 2126–8.

Woodroffe, R., Davies-Mostert, H., Ginsberg, J., *et al.* (2007) Rates and causes of mortality in endangered African wild dogs *Lycaon pictus*: lessons for management and monitoring. *Oryx*, **41**, 215–23.

Woodroffe, R., Donnelly, C. A., Cox, D. R., *et al.* (2006) Effects of culling on badger *Meles meles* spatial organisation: implications for the control of bovine tuberculosis. *Journal of Applied Ecology*, **43**, 1–10.

Woodroffe, R., Thirgood, S., and Rabinowitz, A. (2005) *People and Wildlife: Conflict or Co-existence*. Cambridge University Press, Cambridge, UK.

Worah, S. (1991) *The ecology and management of a fragmented forest in south Gujurat, India: the Dangs*. PhD thesis dissertation, University of Pune, Pune, India.

World Association of Zoos and Aquariums (2008) <www.waza.org>. Accessed online 23 Oct 2008.

World Bank (2007) Key development data and statistics. <www.devdata.worldbank.org>. Accessed online 15 Sept 2008.

World Resources Institute (2005) *World Resources 2005: The Wealth of the Poor—Managing Ecosystems to Fight Poverty*. United Nations Development Programme, United Nations Environment Programme, The World Bank, World Resources Institute. <www.wri.org/publication/world-resources-2005-wealth-poor-managing-ecosystems-fight-poverty>. Accessed online 10 May 2009.

World Zoo Organization (2008) <www.wzo.org>. Accessed online 15 Sept 2008.

Wozencraft, C. (2005) Order Carnivora. In D. E. Wilson and D. M. Reeder (eds.) *Mammal Species of the World*. Smithsonian Institution Press, Washington DC.

Wozencraft, W. C. (1989) The phylogeny of the recent Carnivora. In J. L. Gittleman (ed.), *Carnivore Behaviour, Ecology and Evolution*, pp. 494–535. Chapman and Hall, London, UK.

Wright, B. S. (1959) *The Ghost of North America*. Vantage, New York.

Wright, M. and Walters, S. (1980) *The Book of the Cat*. Summit Books, New York.

Wright, S. (1943) Isolation by distance. *Genetics*, **28**, 114–38.

Wright, S. (1931) Evolution in Mendelian populations. *Genetics*, **16**, 97–159.

Wroe, S., McHenry, C., and Thomason, J. (2005) Bite club: comparative bite force in big biting mammals and the prediction of predatory behaviour in fossil taxa. *Proceedings of the Royal Society London Series B*, **272**, 619–25.

WWF (2006) Valdivian temperate rainforests. *Global 200 Ecoregions*. World Wildlife Fund. <www.worldwildlife.org/wildw/Torld/profiles/terrestrial/nt/nt0404_full.html>. Accessed online 5 May 2009.

WWF (2002) *Conserving Tigers in the Wild: A WWF Framework and Strategy for Action 2002–2010*. Species Programme, WWF International. Publication place-Gland, Switzerland.

Wydeven, A. P., Treves, A., Brost, B., and Wiedenhoeft, J. E. (2004) Characteristics of wolf packs in Wisconsin: identification of traits influencing depredation. In N. Fascione, A. Delach, and M. E. Smith (eds.), *People and Predators: From Conflict to Coexistence*, pp. 28–50. Island Press, Washington DC.

Yalden, D. W. (1982) When did the mammal fauna of the British Isles arrive? *Mammal Review*, **12**, 1–57.

Yamada, T., Nagai, Y., and Matsuda, M. (1991) Changes in serum immunoglobulin values in kittens after ingestion of colostrum. *American Journal of Veterinary Research*, **52**, 393–6.

Yamaguchi, N. and Haddane, B. (2002) The North African Barbary lion and the Atlas Lion Project. *International Zoo News*, **49**, 465–81.

Yamaguchi, N., Cooper, A., Mills, M.G. L., Barnett, R. and Macdonald, D.W. (in prep) Revisiting 'naturalness' in conservation biology.

Yamaguchi, N., Cooper, A., Werdelin, L., and Macdonald, D. W. (2004a) Evolution of the mane and group-living in the lion (*Panthera leo*): a review. *Journal of Zoology*, **263**, 329–42.

Yamaguchi, N., Driscoll, C. A., Kitchener, A. C., Ward, J. M., and Macdonald, D. W. (2004b) Craniological differentiation between European wildcats (*Felis silvestris silvestris*), African wildcats (*F. s. lybica*) and Asian wildcats (*F. s. ornata*): implications for their evolution and conservation. *Biological Journal of the Linnean Society*, **83**, 47–63.

Yamaguchi, N., Kitchener, A. C., Ward, J. M., Driscoll, C. A., and Macdonald, D. W. (2004c) Craniological differentiation amongst wild living cats (*Felis sylvestris*) in Britain and southern Africa; natural variation or the effects of hybridization? *Animal Conservation*, **7**, 339–51.

Yamamoto, K., Chomel, B. B., Lowenstine, L. J., *et al*. (1998) *Bartonella henselae* antibody prevalence in free-ranging and captive wild felids from California. *Journal of Wildlife Diseases*, **34**, 56–63.

Yamamoto, J. K., Sparger, E., Ho, E. W., *et al*. (1988) Pathogenesis of experimentally induced feline immunodeficiency virus infection in cats. *American Journal of Veterinary Research*, **49**, 1246–58.

Yamazaki, K. (1996) Social variation of lions in a male depopulated area in Zambia. *Journal of Wildlife Management*, **60**, 490–7.

Yamazaki, K. and Bwalya, T. (1999) Fatal lion attacks on local people in the Luangwa Valley, Eastern Zambia. *South African Journal of Wildlife Research*, **29**, 19–21.

Yañéz, J., Cardenas, J., Gezelle, P., and Jaksic, F. (1986) Food habits of the southernmost mountain lions (*Felis concolor*) in South America: natural versus livestocked range. *Journal of Mammalogy*, **67**, 604–6.

Yasuda, M., Matsubayashi, H., Rustam, Numata, S., Sukor, J. R. A., and Abu Bakar, S. (2007) Recent records by camera traps in Peninsular Malaysia and Borneo. *Cat News*, **47**, 14–16.

Yensen, E. and Seymour, K. L. (2000) *Oreailurus jacobita*. *Mammalian Species*, **644**, 1–6.

Young, E. (1975) Some important parasitic and other diseases of lion *Panthera leo* in the Kruger National Park Kenya. *Journal of the South African Veterinary Association*, **46**, 181–3.

Young, E., Zumpt, F., and Whyte, I. J. (1972a) *Notoedres cati* infestation of the cheetah: preliminary report. *Journal of the South African Veterinary Association*, **43**, 205.

Young, E., Zumpt, F., and Whyte, I. J. (1972b) Sarcoptic mange in free-living lions. *Journal of South African Veterinary Medical Association*, **43**, 226.

Young, S. P. (1978) *The Bobcat of North America*. University of Nebraska Press, Lincoln, NE.

Young, S. P. and Goldman, E. A. (1946) *The Puma, Mysterious American Cat. Part I. History, Life Habits, Economic Status, and Control*. The American Wildlife Institute, Washington DC.

Yu, J. P. and Dobson, F. S. (2000) Seven forms of rarity in mammals. *Journal of Biogeography*, **27**, 131–9.

Yu, X. (2005) A survey of Amur tigers and Far Eastern leopards in eastern Heilongjiang Province, China 1999. In *Recovery of the Wild Amur tiger Population in China: Progress and Prospect: Proceedings of the 2000 International Workshop on Wild Amur Tiger Population Recovery Action Plan, Harbin, China and the 2002 National Worshop on Progress of Wild Amur Tiger Population Recovery Action, Hunchun, China*., pp. 164–70. China Forestry Publishing House, Beijing, China.

Yudakov, A. G. and Nikolaev, I. G. (1987) *Ecology of the Amur Tiger: Winter Long-Term Observations in the Western Part of the Middle Sikhote-Alin in 1970–1973*. Nauka, Moscow, Russia.

Yudakov, A. G. and Nikolaev, I. G. (1977) Snow as an ecological factor in the life of the Amur tiger. In *Redkie vidy mlekopitaiuschikh i ikh okhrana. Materialy 2 vsesoiuznogo soveschania*, pp. 133–4. Nauka, Moscow, Russia.

Yuhki, N. and O'Brien, S. J. (1997) Nature and origin of polymorphism in feline MHC Class II DRA and DRB genes. *Journal of Immunology*, **158**, 2822–33.

Yuhki, N. and O'Brien, S. J. (1990a) DNA recombination and natural selection pressure sustain genetic sequence diversity of the feline MHC Class I genes. *Journal of Experimental Medicine*, **172**, 621–30.

Yuhki, N. and O'Brien, S. J. (1990b) DNA variation of the mammalian major histocompatibility complex reflects genomic diversity and population history. *Proceedings of the National Academy of Sciences of the United States of America*, **87**, 836–40.

Yuhki, N. and O'Brien, S. J. (1988) Molecular characterization and genetic mapping of class I and class II MHC genes of the domestic cat. *Immunogenetics*, **27**, 414–25.

Yuhki, N., Heidecker, G. F., O'Brien, S. J. (1989) Characterization of MHC cDNA clones in the domestic cat. Diversity and evolution of class I genes. *Journal of Immunology*, **142**, 3676–82.

Zabel, A. and Holm-Müller, K. (2008) Conservation performance payments for carnivore conservation in Sweden. *Conservation Biology*, **22**, 247–51.

Zapata, S. C., Travaini, A., Ferreras, P., and Delibes, M. (2007) Analysis of trophic structure of two carnivore assemblages by means of guild identification. *European Journal of Wildlife Research*, **53**, 276–86.

Zar, J. H. (1999) *Biostatistical Analysis*, 4th edn. Prentice Hall, Upper Saddle River, NJ.

Zaumyslova, O. Y. (2000) Long-term population dynamics of ungulates in Sikhote-Alin Zapovednik based on winter route surveys. In A. A. Astafiev (ed.), *Analysis of Long-Term Observations of Natural Components of Far East Zapovedniks*, pp. 70–9. DalNauka, Vladivostok, Russia.

Zdansky, O. (1924) Jungtertiäre Carnivoren Chinas. *Palaeontologia Sinica Series C*, **2**, 1–149.

Zeller, K. (2007) *Jaguars in the New Millennium Data Set Update: The State of the Jaguar in 2006*. Wildlife Conservation Society, New York.

Zeuner, F. E. (1963) *A History of Domesticated Animals*. Hutchinson, London, UK.

Zezulak, D. S. (1981) Northeastern California bobcat study. *Federal Aid for Wildlife Restoration Project. W-54-R-12, Job IV-3*. California Department of Fish and Game, Sacramento, CA.

Zimmermann, A. and Macdonald, D. W. (in prep) An expert-based analysis of human-jaguar conflicts across the species' range.

Zimmermann, A. and Wilson, S. (in prep) A spatial model of human-jaguar conflicts in Latin America.

Zimmermann, A., Baker, N., Linnell, J. D. C., *et al.* (in press) Contemporary views of human-carnivore conflicts on wild rangelands. In J. G. Du Toit, R. Kock, and J. Deutsch (eds.), *Can Rangelands be Wildlands? Wildlife and Livestock in Semi-Arid Ecosystems*. Blackwells, Oxford, UK.

Zimmermann, A., Walpole, M. J., and Leader-Williams, N. (2005a) Cattle ranchers' attitude to conflicts with jaguar (*Panthera onca*) in the Pantanal of Brazil. *Oryx*, **39**, 406–12.

Zimmermann, F., Breitenmoser-Würsten, C., and Breitenmoser, U. (2005b) Natal dispersal of Eurasian lynx (*Lynx lynx*) in Switzerland. *Journal of Zoology*, **267**, 381–95.

Zimmermann, F., Weber, J.-M., Molinari-Jobin, A., *et al.* (2005c) Monitoring Bericht Raubtiere Schweiz 2004. *KORA Bericht*, **29**, 1–58.

Zumpt, F. and Ledger, J. A. (1973) Present epidemiological problems of a sarcoptic mange in wild and domestic animals. *Journal of the Southern African Wildlife Management Association*, **3**, 119–20.

Index

Note: page numbers in *italics* refer to Figures and Tables.

'20–80 rule', parasites 265
454 sequencing 122

Acinonyx species
phylogeny and evolution 71
see also Asiatic cheetah; cheetah
active remote sensing 212
activity patterns 79
African wildcat *547*, 553, 555
Andean cat *593*
black-footed cat *547*, 548, 555
ocelot and sympatric felids 579
pampas cat *588*
adaptations 83–4
arborial cats *91*
cheetah 99–100
piscivores 100–1
sabretooths 77, 101–3
snow leopard 420–2
adrenal function, corticosteroid monitoring *228–9*
African felids *5*, *6*
African golden cat (*Caracal aurata*) 29
geographical occurrence *5*
scent glands *104*
African lion corridor 608–10, *609*
African wildcat (*Felis silvestris lybica*) 478, *539*, *541*, *556*
activity period *547*, 553, 555
biomass consumed per distance moved 555
communication 544–5, 547
evolution 538–40
geographical occurrence *538*
home range size 543, 555, *556*
interspecific interactions 546–7, 555
phylogenetics *477*
prey *552–3*, 555–7, 555
seasonal variation 556, 557
reproduction 548–9, 557
research gaps 557
size 538–9
skull morphology *475*
social interactions 546
social organization 543–4
study area 540–1
study methods 542
taxonomy 481
African wild dog 382
Afrosmilus 65, *72*, 73
age at dispersal 135–7
age of first reproduction, Amur tiger *336*, 337
age (human), relation to tolerance of felids 450–1
Agriculture and Wildlife Services, Chile (SAG), puma management policies 449–50
Alladale Wilderness Reserve 489
Allee effect, Iberian lynx 520
alleles 109
allozymes 109
Amazon
perceptions about jaguar 394–6, 641
persecution of jaguars 400
Amazon river, as barrier to gene flow 111–12
Amboseli National Park
compensation scheme 637–8
ecotourism 370
extirpation of lions 284
ambushing 143
morphological adaptations *84*
ambush tactics, Canada lynx *529*
American horse sickness (AHS) 247
American lion (*Panthera atrox*) 603–4
Amphimachairodus 66–7, 76
Amur tiger 325
anthropogenic mortality 163, 337–8
conservation approaches 336–9
dispersal ages and distances *135*
food habitats and prey preferences 328–31
genetic similarity to Caspian tiger 613–16
genetic variation 116
geographic range *326–8*
habitat conservation 338–9
home range size, relationship to prey density *334–5*
human deaths and injuries *174*
northern limit of distribution 331, *332*
population density, relationship to prey density 331, *333*
reproduction and survival rates 335–*6*
translocation of problem animals 180, 182
amyloidosis 242
anaesthesia 223
in assisted reproduction 229
analgesics 223
anal glands *104*, 105
anal sacs *104*, 105
Andean cat (*Leopardus jacobita*) 35, 584–6
activity patterns 588
competition with pampas cat 586–9
conservation 594–6
detection 586, *587*
ecological model 589–93
flagship role 594
genetic study 109
geographical range *5*, 6, 583–6
habitat 584
home range size 586
prey 584, *585*
Andean Cat Alliance (AGA) 582, 595
ankle joint flexibility, arborial cats *91*
ankylosing spondylitis 252
anthrax 244, 252
case reports *249*
anthropocentric approaches 627
ecocentric–anthropocentric continuum 627–32
anthropogenic mortality 14, 163
see also persecution; trophy hunting
antibodies
neonatal levels *251*
prevalence and geographic distribution *239–40*
production after exposure to pathogens *251*
antibody surveys 247, 257

arborial cats
  ankle joint flexibility *91*
  tail length 92
Argos radio-telemetry 201, *203*
artificial insemination (AI) 229
Asian felids 4–*5*
Asian wildcat (*Felis silvestris ornata*) 478
  phylogenetics *477*
  skull morphology 475
  taxonomy 478
Asiatic cheetah (*Acinonyx jubatus venaticus*) 353
Asiatic golden cat (*Pardofelis temminckii*) 27–8
  biogeography *5*
  management programme *219*
ASIP gene 119
assisted reproduction 229–30
  ovulation induction 227–8
atrial septal defects (ASDs), Florida panther 117, 457–9
  impact of genetic introgression programme 461, *462*
attitudes to felids
  cultural relativism 620–2
  lynx reintroduction to Scotland 617–18
  relationship to behaviour 640–1
  to human injury and fatality 172
  to livestock depredation 171–2
attractants, reserve edges 303–4
auditory bulla *89*
  fossil record 62
    Proailurus 63
AVHRR sensor 213
avian influenza 247

Babesia infections 243, *250*
  CDV-infected animals 256, 274
  prevalence *245–6*
Bacillus anthracis 244
Bali tiger (*Panthera tigris balica*) 612
Balkan lynx 610
  comparison of historical and present-day distributions *611*
Barbary lion (*Panthera leo*) 433, 612–13
Barbourofelidae 72–3, 78
Barbourofelis 73
Bardia National Park *310*
barriers, reserve boundaries 284
Bartonella species
  B. henselae 243
  prevalence of infection *245–6*
Basanta forest corridor, Terai 319, *320*, 321
bay cat lineage 61, 71
behaviour, role in disease acquisition 238–41
behavioural intention 399
behavioural studies 56, 198
beliefs, relationship to tolerance of pumas 449, 450
Belize
  jaguar density *408*
  protected areas *405*
  see also Cockscomb Basin Wildlife Sanctuary
Belize Audubon Society (BAS) 404
Benfontein Game Farm 540, *541*
Bhutan Biological Conservation Complex (B2C2) 608
Big Cypress Swamp (BCS) pumas, genetic variation 116
big game business, Spain 515
binoculars 202–3
Biodiversity Impacts Compensation Scheme (BICS) 400, 638
biogeography 4–6, *5*
  African golden cat *29*
  Andean cat *35*, 583–6
  Asiatic golden cat 27–8
  black-footed cat *48–9*, 538
  bobcat *41–2*
  Borneo bay cat 25–6
  Canada lynx 38–40
  caracal 29–*30*
  cheetah *42–3*, 374
  clouded leopard *17*
  Eurasian lynx *40*
  fishing cat *54–5*
  flat-headed cat 52–3
  Florida panther 454
  Geoffroy's cat *33*
  guiña 33–*4*
  Iberian lynx *41*, 508–9
  jaguar *20*, *384*, 385
  jaguarundi *44–5*
  jungle cat *45–6*
  leopard *21–2*
  leopard cat 51–2
  lion 18–20, 601–4, *605*
  marbled cat *27*
  margay 38
  ocelot *35–6*, 561
  oncilla 36–*7*
  Pallas' cat 50–1
  pampas cat *32*
  puma *43–4*, 432
  rusty-spotted cat *53–4*
  sand cat 46–8
  serval *31*
  snow leopard 24–5, 417
  Sunda clouded leopard 17–*18*
  tiger 22–4, *326–8*
  wildcat 49–*50*, 473–6
biological species concept (BSC) 479, 480
bio-rational understanding, of jaguar 384
birthing season, Amur tiger *335*, 338
bite strength 95
black-backed jackals, interaction with African wildcats and black-footed cats 547
black bear (*Ursus thibetanus*), in north-east Asia 328
black-footed cat (*Felis nigripes*) 48–9, 72, *537*, *538*, *556*
  activity period *547*, 550, 555
  biomass consumed per distance moved 553–5
  communication 544–5, 547
  evolution 538–40
  geographical occurrence *5*, *538*
  haemoparasite infections, prevalence *245–6*
  home range sizes *133*, 543, 550, 553
  interspecific interactions 546–7, 555
  management programme *219*
  prey *552*, 553, 555
    mammalian *550*
    non-mammalian *548–9*
    seasonal variation 553, *554*
  reproduction 545–6, 557
  research gaps 557
  scent marking *139*
  size 539
  social interactions 546
  social organization 543–4
  study area 540–1
  study methods 542
black-footed ferret, *ex situ* breeding programme 221, 222
black rhino (*Diceros bicornis*), population increase 368–9

blastomycosis, case report *249*
blood collection, use of Triatomine bugs 234
blood samples, Serengeti lions 265
bobcat (*Lynx rufus*) 41–2
  anthropogenic mortality 163
  antiviral antibodies, prevalence and geographic distribution 239–40
  dispersal *135*, 136
  fur trade 190
  genetic monitoring 122–3
  geographical occurrence *5*, 6
  haemoparasite infections, prevalence *245–6*
  home range overlap 130, 134
  home range size *128*, *133*
    intraspecific variation 134–5
    seasonal variation 131, 152
  hybridization 40, 110
  kinship 118–19
  mortality 14
  predators 12–13
body size 83
  cheetah 355–6
  Iberian lynx 508–9
  ocelot 564–5
  puma 432
  relationship to academic studies 55–8
  relationship to home range size *128–9*
  relationship to home range size and density 130
  relationship to human-felid conflict *622*, *623*
  relationship to perceived value 644–5
  relationship to prey size 125–7
  tiger *313*
  tiger prey *330*
Bolivia, jaguar density *408*
bomas 181, *183*
Borneo bay cat (*Pardofelis badia*) 25–6
  biogeography *5*
Botswana, trophy hunting, leopards 351
bottlenecks 115–16
  Florida panther 219, 220
  inbreeding 116–17
  Ngorongoro Crater Lions 264
bovine spongiform encephalopathy (BSE) 247
bovine tuberculosis *see* tuberculosis
bows and arrows, use in immobilization 199
box traps 199
Brazil
  fur trade 189
  human–jaguar conflicts 388
  jaguar density *408*
  see also Pantanal
Brazilian Ocelot Consortium 232
breeding programmes 218–20
brown bear (*Ursus arctos*) 328
browsing damage, wild ungulates 502
buffalo, as source of tuberculosis 253, 268
buffer zones, reserve boundaries 284
Bukit Barisan Selatan National Park (BBSNP), GIS study *214*–15
Bushblok programme, Namibia 371
bush pigs, attraction of lions 632, 633

caching, Canada lynx 524–5
cages 199
caiman, predation by jaguars *390*, 391, *392*
Cairngorms National Park, feral cat control 491
calibration, track count studies 209
callback techniques 204
call-in surveys 209–10
calves, jaguar kills 389, 393, 394
camera trapping 207–*8*
  Andean cat 586, *587*
  capture–recapture analysis 210
  Cockscomb Basin jaguars 404–7, *406*, 409
    capture patterns of males and females 410–11
    future directions 415–16
    interaction with pumas 412–14
camouflage 84
Canada lynx (*Lynx canadensis*) 38–40, *39*
  adult survival and mortality 525–6, 527
  anthropogenic mortality 164
  antiviral antibodies, prevalence and geographic distribution *239–40*
  biogeographical occurrence *5*, 6
  core population hypothesis 535–6
  dispersal *135*, 136, 526–7, 533, 534
  female–kit relationships 532
  fur trade 190
  group hunting 532–3
  habitat patchiness and fragmentation 534
  habitat selection 530, 533
  hare–lynx interaction
    cyclical changes and regional structuring 536
    functional responses to hare population fluctuations 528–9, 533
  home ranges 530–2, 533
  hunting tactics 529, 533
  hybridization 110
  influences of snow 530
  paws *93*
  phylogeographic studies 112–13
  population cycles 522, 524, 533
    comparison with snowshoe hare population 524
  prey 527–8, 533
    alternatives 535
  reproduction and juvenile survival 525
  scavenging and caching 529–30
  studies 522–4
  sustainable hunting 162
  trapping, effects 534–5
Candidatus species 243–4
Canidae, prey-to-predator size spectrum 126–*7*
canine distemper virus (CDV) 238, 244
  case reports *249*
  disease risk models 258
  epidemics, factors influencing mortalities 255–6
  Serengeti lions 237, 238, *266*, 272–4, *273*, 280–1
    epidemiological model 276
    maintenance of 1994 outbreak 278
    multi-host dynamics 278–9
    nomads as spreaders of disease 277, *279*
  seroprevalence *239–40*
    Namibian cheetahs 363
    Serengeti lions *266*
  transmission 241, 278–80
canines 97
  diameter
    correlation with mean body mass *573*
    Neotropical small felids 572, 573

canines (*cont.*)
nerve fibres 94–5
sabretooth cats *102*
Cannabis sativa, Terai 312, 314
Cape lion (*Panthera leo melanochaitus*) 624
captive breeding
assessment of reproductive condition 225–9
assisted reproduction 229–30
gamete rescue, sex sorting, and cloning 230
Iberian lynx 232–5, 512, 518
indications for 218–20
potential for reintroductions 220
Scottish wildcat 491
sperm and embryo cryopreservation 230–2
captivity
as solution to problem animals 180
see also ex situ populations
captivity stress 242, 243
interference with vaccination response 259
capture of prey 143
capture–recapture models 210–11
capturing wild felids 199–200
population studies 207
caracal (*Caracal caracal*) 29–30
biogeographical occurrence *5*
fossil record 71
fur length and density *86*
haemoparasite infections, prevalence *245*
intra-guild predation 12–13
livestock depredation 165, 170
persecution as pests 623–5
signalling 103
Caracal lineage 61
carnassials 95, *96*, 97
cheetah 100
carrying capacity, effect of domestic stock 400
Caspian tiger 614
genetic studies 613–14
historical depictions 615–16, *617*
reintroduction possibilities 614–15
catching prey, morphological adaptations *84*
catnip (*Nepeta cataria*) 138
cave lion (*Panthera leo spelaea*) 603–4
chamois, as prey of lynx *496*, 497, 500, 501, *502*
character displacement 10–11
cheating, group hunting 144
cheetah (*Acinonyx jubatus*) 42–3, 373
adaptations 99–100
adolescent groups 147, 375
antiviral antibodies, prevalence and geographic distribution *239*
biogeographical occurrence 4, *5*, 376
bottleneck event 115–16
captivity stress 242, 243
Class I MHC variations 121
claws 4, *94*
conflicts with humans, predation on game species 165
cooperative defence of territory 149–51
cooperative hunting 143–4
dispersal *135*, 136, 374
female philopatry 136
fossil record 71, 72
fur length and density *86*
haemoparasite infections, prevalence *245–6*
home range overlap 130–2
home range sizes *128*, *133*, 374
inbreeding effects 116–17
larynx *104*
limbs 92
lions, avoidance of 12
livestock depredation *166–9*
management programme *219*
mating system 131, 132, 374–5
paws 93–4
'play trees' 138
population decline 353–4
prey 11, 376
kill data 205–6
reproductive suppression 226
sperm cryopreservation 231
subspecies determination 111
tail length 92
translocation of problem animals 180
see also Namibian cheetah; Tanzanian cheetah
Cheetah Conservation Fund (CCF) 231, 355, 372
Bushblok programme 371
educational activities 371, 372
provision of livestock guarding dogs 365
study area *354*
Cheetah-Country Beef initiative 370–1
Chile, pumas
stakeholder involvement 434–6
study of stakeholder tolerance 443–8
China
historic distribution of tigers *602*
tiger farming 642–4
traditional medicine, use of felid products 193, 195, 642–4
use of felid skins 190, 192
chinchilla, as prey of Andean cat 584
Chinese steppe cat (*Felis silvestris bieti*) 473–6, *477*
phylogenetics 476
Chitwan flood plain, tiger populations 308–9
Chitwan National Park 309, *310*
history of tiger conservation 311
human deaths 175, 322–3
human dimensions 321
tiger population recovery 321
tiger population size 317
tiger sociality 315
see also Terai Arc Landscape (TAL)
Chuchchi (Chitwan tigress), reproductive success 315–16
CITES (Convention on International Trade in Endangered Species) 14, 189, 190
CITES listings
hybrids 480
jaguar 386
snow leopard 420
wild cat 478–81
CITES quotas
leopards *343*
effect on probability of population extinction *347*
KwaZulu-Natal (KZN) 344–5, 346–7, 350
Namibian cheetahs 355
clade IV wildcats 477, 484
Clan Chattan 471
class I MHC 121
class II MHC 121
clavicle 91
claws *4*, *93–4*
cheetah 99–100
piscivores 100–101

Clean Development Mechanism (CDM) 645
climate
as factor in disease susceptibility 280
relationship to hare–lynx interaction 536
climate forcing 112
cloning 230
closed capture–recapture models 210
clouded leopard (*Neofelis nebulosa*) 16–17
biogeography *5*
captivity stress 243
corticosteroid monitoring 242–3
coat pattern 78
fossil record 70
hind limbs 92
management programme *219*
olfactory lobes *90*
see also Sunda clouded leopard
coat patterns 78, *80*, 84–5
phylogeny *79*
Cockscomb Basin Wildlife Sanctuary (CBWS) 403–4, *405*
Cockscomb Jaguar Research Program 404–7
capture patterns of males and females 410–11
future directions 415–16
habitat use 409
interactions between jaguars and pumas 412–14
population density *407–8*
prey 411, *412–13*, 414
territoriality 409–10
Coexistence Project, jaguars 388
co-infections, *Babesia* and CDV 256, 274
colour vision 87
Common Agricultural Policy 400
Communal Areas Management Plan for Indigenous Resources (CAMPFIRE) scheme 287
communal denning 147
communal suckling
domestic cats 147, *149*
lion 147, 149, *150*
communication
scent marking 105, 137–8
African wildcat and black-footed cat 548–9, 551
signals 103
sounds 103–5, 138, 141
African wildcat and black-footed cat 551
community attitudes, Makgadikgadi Pans National Park 609, 635
community involvement 628
contact with local Parks 635
snow leopard conservation 427–8, 430
Terai Arc Landscape (TAL) 322
community studies 205
analytical methods 206
data from kills and scats 205–6
prey availability assessment 206
compensation schemes 172, 178, 183–4, 627, 632, 645
Amboseli lions 637–8
Biodiversity Impacts Compensation Scheme 638
jaguar 400, 401
leopard 628
Makgadikgadi Pans National Park 635
snow leopard 426
competition 10–11
exploitative 11–12
interference 12–13
competitive release 12, 574
cones (retina) 87
conflicts, human–felid
see human–felid conflicts
conical-toothed cats
competition with sabretooths 77
fossil record 68–72
incompleteness 72
phylogeny and evolution 68–72
connectivity, Iberian lynx populations 519
conservancies
KwaZulu-Natal (KZN) leopards 348
Namibia 368–70
conservation 15–16
co-management and stakeholder participation 184–5
disease threats 259
effect of market forces 642–4
genetic corridors 605–10
How many felids do we need? 631–2
and human enterprise 627–35
human relations aspects 632–3
impediments 619–23
joined-up thinking 604–5
long-term view 648
motives 162
natural benchmarks 601, 604
political issues 646–7
reconciliation with human needs 58
restoration 612–14, 618–19
role of social science 639–41
studies *56*
sustainable utilization 193–4
transboundary conservation areas (TBCAs) 619
conservation performance payments 184
conservation policy design 178
conservation teams 639
contractionism 155
control region, mitochondria 108
Convention on Biological Diversity (CBD) 193
Convention on International Trade in Endangered Species
see CITES
cooperative behaviour
care of young 147, 149
defence of mates 146–7, *148*
defence of prey 145–6
defence of territory 149–*51*
hunting 143–5
coordinate systems 213–14
Corbett National Park 308, *310*
Cordillera del Condor 619
core population hypothesis, Canada lynx 535
cornea, convexity 87
corpus luteum activity, lynx 233–4
corrals 181–2, *183*, 424
corridor restoration, Terai 319–*20*
corticosteroid monitoring 242–3
corticosteroids, stress response 242–243
Costa Rica, jaguar density *408*
cost-effectiveness of conservation 628
cottage industry, role in snow leopard conservation *428*
cougar *see* puma
Cougar Management Guidelines Working Group 438
cover, as requirement for hunting *143*, 144
cowlicks, Florida panthers 257
cowpox 247

cross-price elasticity 643
cryopreservation, sperm and embryos 230–1
crypsis 84, 85
cryptic species 108–9
cryptococcosis, case reports *249*
cryptorchidism, pumas 117
  Florida panther 222, 256–7, *462–3*
    impact of genetic introgression programme 465, *466*
cub survival
  Canada lynx 529
  Hwange lions
    edge effects 289–91
    impact of trophy hunting 290–2
  Iberian lynx 515
  Serengeti cheetahs 373
  Serengeti lions 8, 10
  tigers
    Amur tiger 336
    impact of prey density 315
culpeo foxes, diet 588, *589*
cultural beliefs 624
  impact on human–felid conflict 637–9
cultural importance of felids 162
cultural relativism 625–6
custodianship payments 400, 401
cymt DNA (cytoplasmic mitochondrial DNA), molecular clock 120
Cytauxzoon felis infections 243, *250*
  prevalence *245–6*
  reintroduced felids 241

darting 200
'Dead Stop' protective collar *625*
death, causes of
  Florida panther *461*, 469
  Namibian cheetahs 358
  Serengeti cheetahs 373
deforestation, GIS analysis *214*–15
degenerative diseases 252
demographics, consequences of trophy hunting 188
dew claws *94*
dholes, interaction with Terai tigers 317
Diamantofelis ferox 73
diet *see* prey
diet niche overlap, Pianka's index 565
digital elevation models (DEM) 213
Dinobastis 67, 68, 77
Dinofelis *67*–8, 74
  relationship to Machairodontinae 75–6
diploid, definition 109
dirk-toothed cats 101–*2*
disability, as cause of man-eating and livestock depredation 177–8
disease 237, 259
  anthropogenic influences 241
  distinction from infection 247, 251
  epidemiology 237–8
  felid-specific viruses and haemoparasites 243–4
  Florida panther 461, 468–9
  generalist pathogens 244, 247, *248–50*
  genetic 252
  infection risk 238
  manifestations in wild felids 251–2
  mange 254–5
  in Namibian cheetahs 363
  non-infectious causes 252
  relationship to livestock depredation 393–4
  relationship to low genetic diversity 256–7
  risks in captivity 242
  risks during translocations and reintroductions 241–2
  role of felid behaviour 238, 241
  in Scottish wildcats 475
  in Serengeti lions 265–8
  tuberculosis epidemics 253–4
  see also canine distemper virus (CDV); notoedric mange; tuberculosis
disease dynamics 258
disease management
  control strategies 258
  identification of disease threats 257–8
  vaccination 258–9
disease prevention 280
  ex situ populations 222
disease risk modelling 258
disease transmission
  CDV 277–80
  FIV 270–1
  R_0 280–1
dispersal 135–7, 623
  Amur tiger 335
  Canadian lynx 530–1, 537, 538
  cheetahs 375
  Florida panther *461*, 469
  genetic studies 117–18
  Iberian lynx 513, 515
  into human-dominated habitat 175
  into vacant home ranges 179
  snow leopard 418–19
  tiger *314*, 317
dispersal ages *135*
dispersal distances *135*
disruptive colouration 84–5
distributional surveys 212–13
DNA sampling 108
DNA sequencing 109
DNA types 108
domestic cat (*Felis catus*) 649
  class II MHC variations 121
  competition with wild cat 494
  cooperative care of young 147, *149*
  cooperative defence of territory 150
  in diet of Florida panthers 469
  disease transmission 241, 476
  dispersal ages and distances *135*
  feral cats
    control 490–1
    distribution 492–3
  history in Britain 481–2
  home range size, intraspecific variation 134–5
  hybridization
    with leopard cat 51
    with rusty-spotted cat 53
    with wild cat 49, 110, 476, 488–90
  infanticide 147
  intestinal length 99
  melanism 85
  origins 478–9, 480
  scent marking 138, *139*
  skull *70*
  social behaviour 141–3
    intraspecific variation 137
  taxonomy 479, 480
  teratospermia 225–6
  vocalization 138, 141
domestic cat lineage 61
  fossil record 72
domestic dog
  in capture of wild felids 199–200
  disease transmission 241, 272
  in jaguar hunting 395
  livestock guarding 365–7, *366*, 372, *629*, *630*

scat detection 415
vaccination programme, in protection of Serengeti lions 274, 280–1
dominance rank 578
Doñana Iberian lynx meta-population 517, 518
management for conservation 523
drought, role in pathogen acquisition 238–41, 274
drugs, immobilization of wild felids 199
Dudhwa National Park *310*
dynamic population models 211
dystrophic savannah, prey characteristics 154, 157

early cats 74–5
fossil record 62–6
ears, role in signalling 103
ear tufts 103
East Manchurian Mountains, tigers 327
eating, morphological adaptations *84*
ecocentric–anthropocentric continuum 631–5
ecocentric approaches 627–31
ecology, studies *56*
ecotourism 185, 370
impact on Serengeti cheetahs 379
role in snow leopard conservation 427, *430–1*
Tanzania 380
edge effects 284–6
Hwange lions
anthropogenic mortality 294–5, 303
cub survival 292–3
pride size 292, *293*
survival of prides and individuals *293*–4
Makgadikgadi lions, livestock-raiding 301–2, 303
management 304
edges, as 'attractive sinks' 303–4
educational activities 645
Cheetah Conservation Fund (CCF) 371–2
puma conservation 454–5
Scottish wildcat conservation 489
snow leopard conservation 429, 430
effective population size (*Ne*) 113, 114
Ehrlichia species 244
prevalence of infection *245–6*
Elandsfontein lion 604
elbow joint flexibility 91
electric fences, protection of game species 367–8
electroejaculation 225
anaesthesia 223
electrophoresis 109
embryo cryopreservation 230–1
embryo transfer 229–30
emotion, role in human attitudes to felids 644
'empty forest syndrome' 14
enclosure design, *ex situ* populations 223, 228–9
endangered species 15
endemic diseases, Serengeti lions 265–6, *267*
FIV 269–71
environmental causes of disease 252, 264
environmental contamination, effects on Florida panthers 463
epidemics 238
canine distemper 237, 255–6
case study 272–80
mange 254
in Serengeti lions 266–8
tuberculosis 253–4
epidemiology, infectious diseases 237–8
epihyal 103, 104
equine chorionic gonadotrophin (eCG), oestrogen response 227, *228*, 229
ethology of predation 143
Etosha National Park
observation of lions 202
stock-raiding lions 177
Eurasian lynx (*Lynx lynx*) *40*
antiviral antibodies, prevalence and geographic distribution *240*
biogeographical occurrence *5*
competition with humans, predation on game species 165
conflict with humans 497
fur length and density *86*
fur trade 190
haemoparasite infections, prevalence *245*
home range sizes *133*
immobilization technique 200
livestock depredation *168*, 170, 175, 177, 181
mange 254–5
population genetic modelling 113–15
prey 501
reintroductions 497
siblicide 234–5
skull *69*
tail length 92
see also Switzerland, reintroduced lynx
European Endangered Species Programme (EEP) 218
European jaguar (*Panthera gombaszoe-gensis*) 70
European lion (*Panthera leo fossilis*) 604
European wildcat
morphology 477–8
phylogenetics *481*
skull morphology *475*
see also Scottish wildcat
Eusmilus 72
eutrophic savannah 156
EVER, Florida panther population 466–7
evolution
coat patterns 85
Iberian lynx 512–13
see also fossil record; phylogeny
evolutionarily significant units (ESUs) 109, 484
phylogeographic studies 111, 112
virgata/altaica tigers 618
evolutionary patterns 78–80
evolutionary tree *62*, *64*
expansionism 155
Hwange National Park lionesses 157
exploitation of felids 186
management 193–4
use in traditional and alternative medicine 186, 193, 194, 195
snow leopard 423
tiger farming 644
see also fur trade; trophy hunting
exploitative competition 11–12
ex situ breeding programmes
assessment of reproductive condition 232–5
assisted reproduction 229–30
cost and facility plans 220–1

ex situ breeding programmes (*cont.*)
facility location and organizational management 220–4
gamete rescue, sex sorting, and cloning 230
genetics, founders, and population biology 222
Iberian lynx 232–5, 516, 522
indications for 218–20
long- versus short-term programmes 221
management tactics 222–3
nutrition 224
potential for reintroductions 220
Scottish wildcat 491
sperm and embryo cryopreservation 230–2
ex situ populations 222
disease risks 242
vaccination 259
extinction risk, Iberian lynx 521
extinctions 605, 616
Ezemvelo KwaZulu-Natal Wildlife (EKZNW) 344

facial rubbing 141, 143
facial scent glands *104*, 105
faecal samples, Serengeti lions 265
faecal transmission of pathogens 238
faeces analysis 206
Andean cat 590, *591*
genetic studies 417
hormonal monitoring 226
corticosteroid assays 243
Namibian cheetahs 363
in population studies 207
farm cats
cooperative defence of territory 149
social dynamics 141–*2*
farmland cheetahs, Namibia *see* Namibian cheetah
feedback loop, human–felid conflict 635
felids
anthropogenic mortality 14–15
biogeography 4–6, *5*
distinguishing features 3–4
ecology and diet 6–7
intraspecific interactions 7–8, 10
intraspecific relations 10–13
felid-specific viruses and haemoparasites 243–4
Felid Taxon Advisory Group 218
feline calcivirus (FCV)
case reports *248*
Scottish wildcats 475
seroprevalence *239–40*
Namibian cheetahs 363
Serengeti lions 267–*8*, *269*
feline corona virus (FCoV, feline infectious peritonitis virus) 238
case reports *248*
Scottish wildcats 475
in Serengeti lions 267, *268*, *269*
seroprevalence *239–40*
Namibian cheetahs 363
transmission by translocated felids 241–2
feline foamy virus (FFV), Scottish wildcats 472
feline herpes virus (FHV) 238
case reports *248*
in Serengeti lions 266, *267*
seroprevalence *239–40*
Namibian cheetahs 363
feline immunodeficiency virus (FIV) 109, 251–2
Serengeti lions 269–71
seroprevalence *239–40*, 244
in lions 238
subtypes 270
transmission 270–1
use in phylogeographic studies 112–13
feline leukaemia virus (FeLV) 244
case reports *248*
Florida panthers 257, 469
Iberian lynx 233
Scottish wildcats 476
seroprevalence *239–40*
feline panleukopenia *248*
feline parvovirus (FPV) 238
seroprevalence *239–40*
Namibian cheetahs 363
Serengeti lions 267, *268*, *269*
Felis genus
F. attica 69
F. catus 49, *109*, 120–1, 478–9, 481
F. chaus 72
F. christoli 69
F issiodorensis 71
F. lacustris 71, 72
F. longignathus 71
F. lunensis 72
F. proterolymcis 71
F. rexroadensis 71
F. silvestris 57, 60–1, 72, 79, 81, *109*–10, 120–1, 128, 473, 478–9
F. vorohuensis 71
phylogeny 61
see also black-footed cat; Chinese steppe cat; domestic cat; jungle cat; sand cat; wildcat
female–kit relationships, Canada lynx 532
female leopards, trophy hunting 344
female philopatry 136–7
femur length 92
fenced boundaries 284, 304
swing gate installation 368
field methods 197–8
analytical
for community studies 206
for population studies 210–11
for studies of individuals 204–5
callback techniques 204
capturing and handling techniques 199–200
community study techniques 205–6
direct visual observation 202–4
future challenges 215–16
population studies 207–10
radio-telemetry *200–2*, 204
range-wide studies 213–16
tracking 204
fishing cat (*Prionailurus viverrinus*) 54–5
adaptations 100–1
biogeography *5*
management programme 220
sense of smell 89
tail length 92–3
teeth *97*
flagship role of felids 162
Andean cat 596
flat-headed cat (*Prionailurus planiceps*) 52
adaptations 100–1
biogeography *5*
tail length 92–3
teeth *97*
flehmen response 105, 138

Florida panther (*Puma concolor coryi*) *457*, *458*
antiviral antibodies, prevalence and geographic distribution *240*
bottleneck event 116
causes of death *471*
current challenges
habitat loss 463
highway mortality *460*, 463–4
human–panther interactions 464
infectious diseases 464–5
disease 256–7, 457
early research 456
environmental contaminants 459
future research objectives 468
genetic introgression programme 459–461
impacts on correlates of inbreeding 461, *462*
population expansion 462–3
geographical range *454*
habitat selection *457*
history of decline 455–6
home range size *457*
inbreeding depression 222, 461–3, *462*
livestock depredation 455
natural recolonization 466–7
population viability analysis 219–20, 465, 468
potential reintroduction sites 466
prey *461*
reproduction *461*
sub-populations 467–8
Florida Panther Record Clearinghouse 456
Florida Panther Research and Management Trust Fund (FPRMTF) 468
Florida Panther Response Plan 464
focal palatine erosion (FPE), cheetahs 362
food niche overlap, ocelots and sympatric felids 573–4
foot snares *199*
foraging, tiger *314*
foraging group size, lions 144
forelimb length 92
fossil record 59, 61–2
Barbourofelidae 72–3
conical-toothed cats 68–72
incompleteness 72
lion 607–*8*
Nimravides 66
Proailurus *63–4*
Pseudaelurus 64–6
sabretooths 66–8
fossil studies, problems 60
founder numbers, captive breeding programmes 219, 221
Iberian lynx 233, *234*
fovea 87
fragmentation of habitats 162, 608
Canada lynx 534
tiger 338–9
see also genetic corridors
free darting, tigers 200
free-market conservation 628, 629
'front-wave effect', Swiss lynx reintroductions 494, 496, 499, 500, 503
fruit-eating 98
frustrated movements, dispersing Florida panthers 467
F-statistics 113, 114
fur density 85–*6*
fur length 85, *86*
fur trade 189–93
Canada lynx 522, 534–5
decline 194
Iberian lynx 512
snow leopard 420, 422, 426
sustainable utilization 194

game farms, human–felid conflict 165
mitigation strategies 367–70
game management areas 186
gamete rescue 230
Ganges River, crossing by tigers 321
genal whiskers 90
gender differences
camera trapping, CBWS jaguars 407–409
'problem animals' 177
gene flow, Namibian cheetahs 363
genes 108
genetic applications 107–123
bottlenecks 115–16
dispersal 117–18
hybridization 110
inbreeding 116–17
kinship 118–19
molecular clocks 120
parentage 119
phylogeography 110–12
use of FIV sequences 112–13
population genetics 113–15
taxonomy 108–10
genetic corridors 605–10
African lion corridor 608–10, *609*
Balkan lynx 610
jaguar corridor initiative 606–7
tiger corridor initiative 607–8
genetic diseases 252
genetic diversity 107–8
Florida panther 457
Iberian lynx 517
Namibian cheetahs 362–3
relationship to health 256–7
relationship to immune function 243
Serengeti cheetah 379
genetic drift 113–14
genetic introgression programme, Florida panther 463–4
impact on correlates of inbreeding 461, *462*
population expansion 462–3
genetic monitoring 122–3
genetic studies *56*
genome resource banks (GRBs) 230, 231
lynx 235
genomics 109, 121–2
Geoffroy's cat (*Leopardus geoffroyi*) 32
abundance rank 576
antiviral antibodies, prevalence and geographic distribution *240*
densities 580
fur trade 189
geographical occurrence *5*
habitat use 568
home range size *563*
hybridization 110
management programme *219*
phylogeographic studies 111
geographic information systems (GIS) 213–15
gerbils, captive behaviours 222–3
giant yellow croaker (*Bahaba taipingensis*) 644
Ginn Quarry felid 63, 74
Ginsburgsmilus 72–3
gin trap use 624, *625*
Gir lions
attacks on humans 176
bottleneck event 116
inbreeding effects 117
livestock depredation 175

global positioning system (GPS) use 212
  home range studies 131
  radio-telemetry 201–2, *203*, 287–8
  tracking of livestock movements 296
global warming, impact on Iberian lynx 516
glomerulosclerosis 252
gonadotrophin therapies, hormonal response 227–8
goral (*Nemorhaedus caudatus*), presence in north-east Asia 328
Gordon's wildcat, home range sizes *133*
Grande Coupure 62
grass types, Terai 312
grazing management, role in snow leopard conservation 424, *428*
Great Limpopo Transfrontier Park 619
Greene, C.E., *Infectious Diseases of the Dog and Cat* 251
group metabolic needs (GMN), relationship to home range area *128*
group size, lions 144
Gruppe Bars 426
guarding dogs 365–7, *366*, 372, 625
Guatemala, jaguar density *408*
guiña (*Leopardus guigna*) 33–4
  geographical occurrence *5*
  hybridization 110
  livestock depredation 165
  phylogeographic studies 111
gut 99

habitat
  Andean cat 586
  Canada lynx 530, 534
  cheetahs 358, 360
  Florida panther *457*
  Iberian lynx 514
  jaguar 385, 409
  jaguarundi *566*, 567–8
  little spotted cat *566*, 567–8
  margay 567–8
  ocelot 565–6, 580
  pampas cat *566*, 567–8
  phylogeny 78–9, *81*
  relationship to jaguar attacks on cattle 392
habitat loss 14, 162, 623–4
  Amur tiger 338–9
  Florida panther 465
  human–felid conflict 164
  Iberian lynx 114, 514–5
  jaguar 608
  role in disease transmission 238
  Scottish wildcat 472
  Serengeti cheetah 380
  Terai tigers 317
  see also fragmentation of habitats
habitat model, Andean cat 589–93
haemoparasite infections 243–4
  prevalence *245*
haemorrhagic disease virus, rabbits 516
handling techniques 199–200
Haplogale 62, 63, 74
haplotype 109
hard boundaries 284
Hartmann's mountain zebra (*Equus zebra hartmannea*), population increase 369
hearing 88–9
heart defects, pumas 117, 222, 256–7, 458–9
  impact of genetic introgression programme 461, *462*
Hepatozoon felis 243
Heterofelis 66
High Andes Flamingo Conservation Group 596
high frequency hearing 88
highway mortality 164
  Florida panther *460*, 463–464
  Iberian lynx 517
    prevention 519
Himalayan Homestays programme 425
hind limbs 92
Hluhluwe-Umfolozi National Park, culling of subadult male lions 179
home range size 6, *133*
  African wildcat 543, *555*, *556*
  analytical methods 204–5
  Andean cat 586
  Asiatic golden cat 28
  black-footed cat 46–7, *133*, 548, *554*, 556
  bobcat 41–2, *128*, *133*
    intraspecific variation 134
    seasonal variation 131, 152
  Canada lynx *531–2*, 535
  caracal 29
  cheetah 42, 358, *359*, 360, 374–5
  clouded leopard 17
  Eurasian lynx *133*
  Geoffroy's cat 32–*3*, *567*
  guiña 33
  Iberian lynx 40, 511
  intraspecific variation 134–5, 137
  jaguar 20, *128*, 131, *133*, 409
  jaguarundi 44, *566*, 572, *573*
  leopard 21–2, *128*, *133*
    intraspecific variation 134
    relationship to prey density 160
  leopard cat 51–2, *133*
  lion 18, *128*, *133*
    intraspecific variation 152–8
  little spotted cat *567*, 572, *573*
  margay *567*, *572*
  ocelot 35–6, *133*, *566*, 568
    comparison with sympatric felids 572–4
    correlation with mean body mass 564–5
  oncilla 36–7
  overlap 129–37
  Pallas' cat 50–1
  pampas cat 32, *566*
  puma 130, *133*
    Florida panther *457*
    intraspecific variation 134–5, 152
    relationship to deer density *134*
    relationship to prey dispersal 160
  relationship to body size *128–9*
  relationship to habitat quality 302
  seasonal variations 130–1, *132*, 152
  serval 31, *128–9*
  snow leopard 24–5, *128*, 421
  Swiss lynx *502–3*
  tiger 22–4, *128*, *133*
    Amur tiger *334–5*, 338
  Wild cat 49–50, *128–9*, *133*
homestays, role in snow leopard conservation 425
homoplasy 120
Homotheriini tribe 68, 76
  extinction 76–7
Homotherium 67, 68, 76, 101
Hoogesteijn, R. 177
Hoplophoneus 72
hormonal monitoring 226–9

hot spots, cheetahs 377
Hsanda Gol Formation felid 63
human attacks 172–5, 622, 632–3
  characteristics of problem animals 177–8
  jaguars 395, 397
  puma 437, 442, 448, 464
  Terai tigers 322
human chorionic gonadotrophin (hCG), oestrogen response 227, *228*
human encroachment on felid territory 175–6
human enterprises, juxtaposition with conservation 600
human–felid conflicts 164, 194–5, 302, 620, 622–3
  characteristics of problem animals 177–8
  characterization 175
    dispersal into human-dominated habitat 175
    human encroachment on felid habitat 175–6
    prey switching 176–7
  competition for game species 165
  consequences for felid populations 164–5
  cultural aspects 620–2, 634–5
  Eurasian lynx 494
  Florida panther 257, 464
  jaguar 383–5
  mitigation 638–9
    co-management and stakeholder participation 184–5
    compensation 183–4
    conservation policy design 178
    lethal control 178–9
    livestock husbandry improvement 182–3
    livestock protection 180–2, *183*, 304
    Namibian cheetahs 364–70
    translocation 179–80
    zonation of land use 185–6
  Namibian cheetahs 355–6
  puma 432–3
  role of human values 433–4
  Serengeti cheetah 379–80
  social science techniques 640
  Swiss lynx 501, *504*, 505
  Terai tigers 321, 322
  see also livestock depredation
human impact on felids 14, 161–4
  Amur tiger 337–8
  Andean cat 594–6
  disease transmission 241
    tuberculosis 253
  edge effects 284–5
    Hwange lions 294–5
  studies *56*
human needs, reconciliation with conservation 58
human population expansion 192, 627, 640
human relocations 186
Human Rights 622
human values, role in human-felid conflict 433–4
humerus length 92
Hunchun Tiger and Leopard Reserve *326*, 339
hunting
  cooperative 143–5
  morphological adaptations *84*
hunting of felids
  around Hwange National Park 287
  in boundary areas 285
  as means of population control 179
  sustainable 162
  see also trophy hunting
hunting permit allocation, KwaZulu-Natal leopards 346
hunting records, use in population studies 207
hunting success, cheetah 376
hunting tactics, Canada lynx *529*
husbandry manuals 223
husbandry practices, relationship to livestock depredation 393, 429
Hwange National Park 608
Hwange National Park case study
  edge effects
    anthropogenic mortality 294–5
    cub survival 292–3
    pride size 292, *293*
    survival of prides and individuals *293*–4
  home ranges 303
  impact of trophy hunting 303
    on cub survival 291–2
    on population demographics 290, *291*
    on spatial movements 290, *292*
  methodology
    capture and marking of animals 287
    data collection and analysis 287–8
    demographic and survival data 288
    utilization and problem animal control data 288
  sources of lion mortality 289
  study area *286–7*
Hwange National Park lions 157, *158*
hybridization 110
  Canada lynx and bobcats 40
  leopard cat and domestic cat 52
  rusty-spotted cat and domestic cat 53
  wild cat and domestic cat 50, 472, 487, 488, 491
    taxonomic implications 479
hyena, stealing of lions' kills 146
hyoid structure 103, *104*
Hyperailurictis 74, 75
hypothalamic–pituitary–adrenal axis, stress response 242

Iberian lynx (*Lynx pardinus*) 41, *507*
  attitude changes 508
  body size 511
  captive population 491
    breeding programme 232–5
    disease risks 242
  conservation 519
    captive breeding programme 232–5, 518
    meta-population management 519
    public support 519
    rabbit population recovery 518
    road casualty prevention 519
    supplementary feeding 518
  disease 516
    antiviral antibodies *240*
    FeLV infection 257, 519
    glomerulonephritis 252
    haemoparasite infections *245*
    tuberculosis 253
  dispersal ages and distances *135*

Iberian lynx (*Lynx pardinus*) (*cont.*)
- effect of global warming 516
- evolution 508–9
- extinction risk 517
- genetic diversity 517
- genetic monitoring 122
- geographic range *5*, 6, 508–9
- habitat 509–11
- habitat fragmentation 114
- home ranges 130, *133*
- human-mediated threats
  - habitat degradation 514–15
  - non-natural mortality 512–14
- impact of rabbit diseases 516
- local extinctions 516–17
- poaching 164
- population decline 15, 509, 516–17
- prey 509
- reproduction 511–12
- research directions 520
- siblicide 234, 235
- social system 511–12
- specialization 510
- taxonomy 508
- translocations 519

Iberian Lynx Captive Breeding Action Plan 233
ideal populations 113
identification of individual felids 202
- jaguars 404
- pumas 414

illegal hunting, Eurasian lynx 508
illegal trade 14
- in Eurasian lynx 40
- KwaZulu-Natal leopards 349
- in tigers 24

immobilization of wild felids 199, 200
immune function 238
- determining factors 243
- effect of stress 242–3

immunosuppressive viruses 238, 243
- canine distemper virus 256

Imperata grassland, Terai 312
implants, use in radio-telemetry 201
inbreeding 116–17
inbreeding coefficient (*F*) 116
inbreeding depression 117, 222
- Florida panther 457–9, *458*

incentive programmes, snow leopard conservation 425–6, *428*, 430
incentives 628, 646
incisors 97
- sabretooth cats 102

India
- poaching 192
- tiger conservation 338–9

infanticide 8
- domestic cats 147
- jaguar 409
- leopards 345
- lions 149
  - as consequence of trophy hunting 289, 291–2
- tigers 316

infection, distinction from disease 247, 251
infection risk 238
- Florida panther 468–9

infectious disease surveillance 247–51
infirmity, as cause of man-eating and livestock depredation 177–8
injury, relationship to livestock depredation 393–4
insurance schemes 184, 426–7, *428–30*
interbirth interval
- Amur tiger *336*, 337
- cheetahs 357

interdigital glands *104*, 105
interdisciplinarity 600, 639
interference competition 12–13
International Union for Conservation of Nature (IUCN)
- *ex situ* policy 218
- Red List of Threatened Species 15

interspecies embryo transfer 230
interspecific relations 10–11
- exploitative competition 11–12
- interference competition and intra-guild predation 12–13
- jaguars and pumas 414–15
- ocelot and sympatric felids 579–80

intestines 99
intra-guild competition, Andean cat and pampas cat 586–9, *587*, *588*
intra-guild interactions, Terai tigers 317
intra-guild predation 12–*13*
intraspecific aggression, pumas 467
intraspecific interactions 7–8, 10
intraspecific variation 151
- dispersal 136
- home range size 134–5, 151–8
- in social behaviour 137

intrinsic conservation values 162
invasive plant species, Terai 312
invasive study methods, necessity 216
inverse distance modelling 214
Iriomote cat (*Prionailurus bengalensis iriomotensis*)
- antiviral antibodies, prevalence and geographic distribution *239*
- home range sizes *133*
- population genetics 114

isolation, role in disease prevention 258
IUCN *see* International Union for Conservation of Nature

Jacobs' index, tiger prey *9*, *330*
jaguar (*Panthera onca*) 19–20
- attacks on humans 632
- attitudes in Pantanal 394–6, 401, 402, 641
  - determinants 396–7, *398*
- Coexistence Project 388
- competition with pumas 12
- conflict with humans 383–5
  - in Brazil 386
- cultural importance 162
- fossil record 70
- fur trade *191*
- geographical occurrence *5*, *384*, 385
- habitat 385, 409
- hind limbs 92
- home range sizes *128*, 131, *133*
  - seasonal variation *132*
- infanticide 8, 119, 409
- interaction with ocelots 578, *579*
- interaction with pumas 412–14
- livestock depredation 165, *169*, 170
  - attitudes 171
  - mitigation measures 400–1
  - Pantanal Jaguar Project 388
- management programme *219*
- melanism 85, 119
- persecution 397, 399–400
- phylogeographic studies 111–12
- population densities 407–*8*
- prey 7, 12, 388–92, 411, *412–13*, 414, 415
- prey switching 177
- problem animals 392–4
- territoriality 409–10
- translocation of problem animals 180
- *see also* Cockscomb Jaguar Research Program

jaguar corridor initiative 606–*7*
jaguar killing, attitudes 395, 397
jaguarundi (*Puma yagouaroundi*) 44–5
abundance rank 574, *577*
activity pattern *575*
antiviral antibodies, prevalence *240*
canine diameter, correlation with mean body mass *573*
densities 578
geographical occurrence *5*
habitat use *566*, 567, 568
home range size *562*, 566, *567*
melanism 85, 119
prey 572, *573*
pupil shape 80
Javan tiger (*Panthera tigris sondaica*) 612
jaw length 11, 94
piscivores 100
jaws 94–5
Jigme Singye Wangchuck National Park, livestock losses 622–3
joined-up thinking 600–1
impediments 619–20
jungle cat (*Felis chaus*) 45–6
biogeography *5*, 6
fur length and density *86*

Kahna Tiger Reserve, human deaths and injuries *173–174*, 175
Katarniaghat Wildlife Sanctuary *310*
Kavango–Zambezi Transfrontier Area 609, 619
Kenya, poisoning of lions 165
ketamine, use in immobilization 199
Kgalagadi Transfrontier Park (KTP) 540, *541*
lion track counts 209
livestock depredation 177
Khata corridor, Terai 319, *320*
kill distributions, Swiss lynx *504*
killing, morphological adaptations *84*
killing methods 94–5
cheetah 100
sabretooth cats 102–3
kills, data collection 205–6
kinked tail, Florida panthers 461, *462*
kin selection, lion coalitions 147, *148*
kinship 118–19
Kishanpur Wildlife Sanctuary *310*
kleptoparasitism, tiger kills 314
Kluane Lake study, Canada lynx 523
adult survival 526, *527*
dispersal 526, *527*
functional responses to hare population fluctuations 528–*9*
group hunting 532–3
habitat selection 530
home ranges *531*, 532
hunting tactics *529*
influences of snow 530
prey 527–*8*, 535
reproduction *526*
snowshoe hare and lynx population densities *524*
kodkod *see* guiña (*Leopardus guigna*)
kraals 295–6
kriging 214
Kruger National Park
call-in survey 209
kill data 205
observation of lions 202
tuberculosis epidemic 253–4, 268
KwaZulu-Natal (KZN)
geographical features 344
trophy hunting of leopards 344–6
CITES permit distribution 350
extension of protocol 351–2
impact of other anthropogenic mortality 348–9
monitoring of populations 351
protocol 346–7
source-sink management approach 347–9

La Amistad Biosphere Reserve 619
lactating females, diet 572
Laetoli site 77
Landmark Foundation 624
Landsat sensors 213
landscape connectivity, Terai 318–*20*
landscape conservation, Tanzania 380
landscape-level studies 211–13
Lantana camara, Terai 312
laparoscopy, reproductive studies 225
larynx 103–*4*
latrines 138
Andean cat 584
Least Cost Corridor Analysis (LCCA), jaguars 606
leg-hold traps 199
lens convexity 87
lenses, multifocus 87, 88
leopard (*Panthera pardus*) 20–2, *21*
African range *342*
anthropogenic mortality 163
antiviral antibodies, prevalence and geographic distribution *239*
biogeographical occurrence *5*, 6
conservancy management, Namibia 370
fossil record 70
fur length and density *86*
fur trade 189, *190*, 191, 192
haemoparasite infections, prevalence *245*
home range overlap 130
home range sizes *128*, *133*
intraspecific variation 134
relationship to prey density 160
human deaths and injuries 172, *173–4*, 175
interaction with tigers 12, 317
intra-guild predation 12, *13*
livestock depredation 165, *166–9*, 182
economic impact 171
management programme *219*
melanism 85
persecution as pests 623–*4*
phylogeographic studies 111
prey 6, *7*, *9*, 11
stashing 145
track counts 209
traditional beliefs 620
translocation of problem animals 179, 180
trophy hunting 188–9, 342–4
identification of suitable animals *349*
KwaZulu-Natal 344–52
use in traditional medicine 186
leopard cat lineage 61
fossil record 72
leopard cat (*Prionailurus bengalensis*) 51–52
antiviral antibodies, prevalence and geographic distribution *239*
biogeography *5*
fur length and density *86*
fur trade 190
home range size *133*
see also Iriomote cat
leopard hunting zones (LHZ), KwaZulu-Natal 346, 347, *348*

Leopardus lineage 61
 see also Andean cat; Geoffroy's cat; guiña; little spotted cat; margay; ocelot; oncilla; pampas cat
life span, cheetahs 374, 377–8
light sensitivity 87
limb length 90–1, 92
 cheetahs 99
limb posture 92
linsangs (*Prionodon*) 62
lion (*Panthera leo*) 18–*19*
 African lion corridor 608–10, *609*
 anthropogenic mortality 163
 antiviral antibodies, prevalence and geographic distribution *239*
 Barbary lion 433, 612–*13*
 body size 604
 bottleneck events 115–16
 capture techniques 200
 class I MHC variations 121
 competition with cheetahs *377*, 379
 conservation 195
 conspecific encounters 8
 cooperative care of young 147, 149, *150*
 cooperative defence of mates 146–7, *148*
 cooperative defence of territory 149–50, *151*
 cub mortality 8, 10
 culling of subadult males 179
 dispersal 118, *135*, 136
 drivers of population decline 283–4
 edge effects 289
 fossil record 70, 602–3
 genetic studies 608–10
 geographical range 4, *5*
  contraction 603–5, *605*
 haemoparasite infections, prevalence *245*, *246*
 hearing 88
 home range sizes *128*, *133*
  intraspecific variation 152–8
 human deaths and injuries 172, *173*, 175, 177, 632–3
 hunting
  need for cover 143
  rewards of cooperation 143–5
 identification of individuals 202
 immobilization 199, 200
 inbreeding effects 117
 infanticide 8, 149
 infectious diseases 263–4
  canine distemper 255–6
  FIV infection 112–13, 252
  tuberculosis 253–4
 intra-guild predation 12
 kill data 205
 kinship studies 118
 larynx *104*
 livestock depredation 165, *166–9*, 170, 177
  economic impact 171
 loss prey to hyenas 146
 management programme *219*
 phylogeographic studies 112
 population decline 604
 prey *10*, 11, 145, *146*
  seasonal variation 7
 prey switching 176
 roaring 138, 141
 skull *69*
 social behaviour
  intraspecific variation 137
  role in pathogen acquisition 238
 spatial groups 131, 136
 sustainable hunting 285
 taxonomy 70
 teratospermia 225
 track counts 209
 translocation of problem animals 180
 trophy hunting *187*, *188*, *646*
 see also Hwange National Park case study; Makgadikgadi Pans National Park (MPNP) case study; Serengeti lions
lion cubs, predators 12
'Lion Guardian' programme, Amboseli 637
lion-kill data, Makgadikgadi study 298
litter sizes
 cheetahs 357
 Iberian lynx 512
little spotted cat (*Leopardus tigrinus*)
 abundance rank 574, *577*
 activity patterns *575*
 antiviral antibodies, prevalence *240*
 canine diameter, correlation with mean body mass *573*
 densities 578
 habitat use *566*, 567, 568
 haemoparasite infections, prevalence *246*
 home range size 566–7
 interaction with ocelots 578, *579*
 prey 574, *576*
Liuwa Plains National Park, extirpation of lions 284
livestock depredation 165, 176, 622–3
 accuracy of kill rates 395
 attitudes 171–2
 in boundary areas 285
 characteristics of problem animals 177–8
 culture, impact on attitudes 634
 economic impact 170–1
 Florida panther 459
 jaguar *384*, 385, 388–92
  mitigation measures 400–2
  Pantanal Jaguar Project 388
  perceptions 394–5, 397
  persecution of jaguars 399–400
  problem animals 392–4
 KwaZulu-Natal leopards 349–50
 lynx 622
 Makgadikgadi Pans National Park lions 298–302, 303, 635
 management 304, 622
 Namibian cheetahs 360, *361*, 362
  conflict resolution strategies 364–7
 puma 436–7, 442, 448
 as result of felid dispersal 175
 snow leopard 422–3, 424
  reduction 422
 spatial patterns 165
 Swiss lynx *493*, 494, 495, 497, 505, 618
 temporal patterns 170
 see also human–felid conflicts
livestock husbandry improvement 180–3
livestock movements, Makgadikgadi study 296–298
livestock protection 180–2, *183*
livestock vaccination, role in snow leopard conservation 425–6
loci 109
Lokotunjailurus emageritus 67
Loliondo Game Controlled Area, human-cheetah conflict 379, *380*
long-term captive breeding programmes 221
Lupande Game Management Area, human deaths 175

lynx
  Balkan populations 610, *611*
  historical occurrence in Britain 617–18
  home range size, intraspecific variation 134
  livestock depredation 165, *169*, 170
  reintroduction to Scotland 617–19
  siblicide 234–5
  see also Canada lynx; Iberian lynx; Switzerland, reintroduced lynx: Eurasian lynx
Lynx issiodorensis 71, 508
Lynx lineage 61
  fossil record 71
Lynx thomasi 71

Maasai pastoralists, opportunity costs of conservation 629–30
Macdonald, D.W. *et al.*
  analytical methods 204
  communal denning 147
  conservation 600
    restoration 611, 632
  expansionism 155
  home range size 128
  prey-to-predator size spectrum 126
  social dynamics 141–3
  spatial organization 129
Machairodontinae 75–6
Machairodus 66, *102*
macroparasites, Serengeti lions 265–6
maintenance populations, canine distemper virus 274
major histocompatibility complex (MHC) 109, 121
  diversity, relationship to immune function 243
  skin grafting in assessment of MHC variation 115
Makgadikgadi Pans National Park (MPNP) case study
  boundary fences 304
  community attitudes to lions 635–7
  livestock-raiding
    edge effects 301–2
    home ranges and prey preferences 298–9, *300*, 303
    patterns of movement in relation to cattle posts 299, *301*
    survival of stock-raiders and non-stock-raiders 301, *302*
methodology
    capture and marking of study animals 298
    lion-kill data collection 298
    prey and livestock movements 296–8
    utilization and problem animal control data 298
  study area *286*, *295–6*
Malahide declaration 611
male coalitions
  cheetahs 356, 374, 375
  Hwange lions, impact of trophy hunting *290*, 294–5
male felids, as 'problem animals' 177
Manas TBCA 619
mandibular condyles 97, *98*
mange 244, 247, 254–5
  case reports *248–50*
  Florida panther 469
marbled cat (*Pardofelis marmorata*) 26–7
  ankle joint flexibility *91*
  biogeography *5*
  coat pattern 78
margay (*Leopardus wiedii*) 38
  abundance rank 576, *577*
  canine diameter, correlation with mean body mass 573
  densities 578
  fur trade 189, 190, *191*
  geographical occurrence *5*
  habitat use 571, 572, 567, 568
  home range size *567*, 577
  management programme *219*
  prey 98, 576, 576
market forces, role in conservation efforts 642–4
Martelli's Wildcat (*Felis sylvestris lunensis*) 476
Masailand, killing of lions 164–5
mates, defence of 146–7, *148*
mating systems
  cheetahs 131–2, 374–5
  Iberian lynx 511
  tiger *313*
Mbirikani Predator Compensation Fund 637–8
MC1R gene 119
mean weight of mammalian prey 561
  ocelots 572, 577
    comparison with sympatric felids 566–8
medetomidine, use in immobilization 199
Megantereon 68, 76
  attacks on humans 632
melanism 85, 119–20
mercury levels, Florida panthers 459
Mesoamerican Biological Corridor Program 606, *607*
mesopredator release theory 560, 578
metabolic needs, relationship to home range area *128*
metacarpals 92
Metailurini tribe 67–8
Metailurus 67–8, 74, 75
  relationship to Machairodontinae 75–6
metapopulation management 606
  Balkan lynx 610
  jaguar corridor initiative 606–*7*
  KwaZulu-Natal leopards 347–9
  tiger corridor initiative 607–8
Mexico jaguar density *408*
Micawber threshold 631
microarrays 121
microsatellite DNA 108
  hypervariable regions 120
microsatellite genotype 109
microwave sensors 213
migratory prey
  lions 152–3
    in Makgadikgadi Pans National Park 295, 296–*7*, 298–9
  Serengeti cheetah 376
Mikania micrantha, Terai 312
minisatellite DNA 108
  hypervariable regions 120
Miopanthera 74
Miomachairodus 66
Miracinonyx species 100
mitochondrial DNA (mtDNA) 108
  molecular clock 120
MODIS sensor 213
molecular studies
  cheetahs 131–2
  of pathogens 251
  problems with phylogenetic analysis 60
Montana pumas
  prey 438
  study of newspaper articles 440–8
  stakeholder involvement 439–*40*

moose (*Alces alces cameloides*), presence in north-east Asia 328
morphological studies, problems with phylogenetic analysis 60
morphometric studies, Namibian cheetahs 362
mortality, anthropogenic 163–4
mosaicking, remote sensing 212
multifocus lenses 87, 88
multi-host dynamics, canine distemper virus 272, 278–9
musk deer (*Moschus moschiferus*)
  Jacobs' index *330*
  as prey for Amur tiger 328
mutations 119
  melanism 119–20
  repetitive region DNA 120
Myanmar, tiger reserve 647
Mycobacterium species 244
  see also tuberculosis
Mycoplasma species 243
  prevalence of infection *245–6*
mystacial whiskers 90
myxomatosis, impact on Iberian lynx populations 519

Namafelis minor 73
Namibia
  Cheetah Conservation Fund (CCF) study area *354–5*
  climate 355
  conservancies 368–70
  leopards, livestock depredation 350
Namibian cheetah 354, 372
  causes of death *357*
  conflict mitigation 364
    on game farms 367–70
    on livestock farms 364–7
  demography and social structure 356–8
  feeding ecology and prey selection 360–2
  genetic studies 362–3
  health and disease 363–4
  human–cheetah conflict 355–6
  incentives for conservation 370–2
  morphometrics 363
  spatial ecology and habitat selection 358–60
  trophy hunting 370
Namibian Professional Hunters Association (NAPHA), COMPACT for the Management of Cheetah 370
nasal passages, cheetah 100
national parks 186
  contact with local communities 628
natural behaviours, encouragement in *ex situ* populations 222–3
natural benchmarks
  lions 601–4
  tigers 601–4
Natural England 600
naturalness 601
  enclosures 613
natural recolonization 466
  Florida panther 467
necropsy studies 251
Neofelis lineage 61
  coat pattern 78
  see also clouded leopard; Sunda clouded leopard
neonatal kittens, antibody levels *251*
neotropical felids *5*
network model, Serengeti lion population 274–6
newspaper articles, Montana, representation of pumas 437–42
Ngorongoro Conservation Area, human–cheetah conflict 379, *380*
Ngorongoro Crater lions 264
  bottleneck event 115–16
  disease 256
    canine distemper virus infection 274, *275*
    endoparasites 265
    FIV prevalence *270*
    trypanosome infections *271*
  inbreeding effects 117
niche overlaps 11–12
niche release theory 560
night-time observations 202–4
Nimravidae 60, 72
Nimravides 66, 75
Nimravus 72
noise deterrents, use in livestock protection 182
nomad lions 265
  movement patterns *278*
  as spreaders of disease 277
normalized difference vegetation index (NDVI) 213
North American felids *5*
North Sea fish stock depletion 604
Norway, livestock depredation 181
notoedric mange (*Notoedres cati*) 241, 244, 254
  case reports *249*
  Florida panther 465
nuclear DNA 108
nutrition, *ex situ* management programmes 224
NWT study, Canada lynx 524
  adult survival 525–6
  caching 529–30
  dispersal 534
  female–kit relationships 532
  habitat selection 530
  home ranges 530–2
  reproduction and juvenile survival 525, *526*
  snowshoe hare and lynx population densities *524*, 525

observation of wild felids 197–8, 202–4
  analytical methods 204–5
occupancy modelling 215
ocelot (*Leopardus pardalis*) 35–6, 559
  abundance rank 574, 576, *577*
  activity pattern 575
  antiviral antibodies, prevalence and geographic distribution *240*
  body mass 564
    correlation with home range size 564–*5*, 567
  canine diameter, correlation with mean body mass *573*
  density
    comparisons with smaller felids 577–9
    environmental effects 577
    influence of prey 577
  demographic interactions with sympatric felids 579
  dispersal 135
  fossil record 71
  fur trade 189, *191*
  geographical occurrence *5*, 561
  habitat use 565–*6*, 579

comparison with sympatric felids 567–8
haemoparasite infections, prevalence *245*, *246*
home range overlap 130
home range size *133*, 562, 564
comparison with sympatric felids 566–8
correlation with mean body mass 564–5
interspecific interactions 560, 578
latrines 138
management programme *219*
phylogeographic studies 110
potential for competitive release 574
prey 6, *8*, *568*, *569*, *570*, 572, 579
comparison with sympatric felids 572–4, *576*
road accidents 164
scent marking *139*
sense of smell 89
species survival plan 231–*2*
study methods 560–1, 565
'ocelot effect' 578, *580*
oncilla 36
ocelot lineage 61
oestrogen levels 226–*7*
Iberian lynx 233
olfactory lobes *90*
olfactory organs 90, 105
oncilla (*Leopardus tigrinus*) 36–7, *37*
geographical occurrence *5*
hybridization 110
phylogeographic studies 111
oocyte recovery, post-mortem 230
open capture–recapture models 211
opportunity costs of conservation 628–9, 631
Otocolobus genus 61
otter (*Lutra lutra*), conservation sites 490
ovulation 225, 226
ovulation induction 227–8

Padampur enclave, land transfer 321
paleo-felids (Nimravidae) 72
Pallas's cat (*Otocolobus manui*) 50–1, *50*
biogeography *5*
fur length and density *86*
haemoparasite infections, prevalence *245*
management programme *219*
ovarian activity, effect of light 225, *228*
semen collection *226*
species survival plan 231
pampas cat (*Leopardus colocolo*) 32
activity patterns *575*, *588*
competition with Andean cat 586–9
fur trade 189
geographical occurrence *5*
habitat use *566*, 567, 568
home range size *563*
hybridization 110
phylogeographic studies 111
prey 584, *585*
panmixis, Namibian cheetahs 363
Pantanal Jaguar Project 388, 607
Pantanal jaguars 386–*7*
conflict mitigation 400–2
foraging ecology 388–92
human attitudes 394–6, 640–1
determinants 396–7, *398*
persecution 399–400
problem animals 392–4
panther *see* puma (*Puma concolor*)
Panthera lineage 60
fossil record 69–70, 77–8
see also jaguar; leopard; lion; snow leopard; tiger
Panthera palaeosinensis *70*
Paramachaerodus 67, 76
Pardofelis genus
coat pattern 78
phylogeny 61
see also Asiatic golden cat; Borneo bay cat; marbled cat
parentage 119
cheetahs 131–2
Parsa Wildlife Reserve *310*
parvovirus *see* feline parvovirus (FPV)
Paseo Panthera initiative 162
passive remote sensing 212
pathology surveys *56*, 251
paw-prints, tiger pugmark census 209
paws *93*
cheetah 99
piscivores 100
Payments for Ecosystem Services 637
Peace Parks 619
peccaries, predation by jaguars 389, *390*, 391–*2*
pelage
flat-headed cat 101
hunting and feeding adaptations *84–6*
wildcat 482–3, *484*
see also coat patterns
pelage-based classification, Scottish wildcat 486, *487*
perceived threats
jaguars
to human safety 395
to livestock 384–5, 388, 394–5
pumas *437*, 449, 452–3
snow leopards 420
performance payments 401
persecution 620, 627–9
jaguars 397, 399–400
mitigation measures 400–2
phencyclidine, use in immobilization 199
Phinda-Mkhuze leopard population 351
photographic capture *see* camera trapping
phylogenetic patterns
activity patterns 79
coat patterns 78, *79*, *80*
habitat preference 78, *81*
pupil shape 79–80, *81*
phylogenetic species concept (PSC) 479, 480
phylogeny *60*–1, 80–1
conical-toothed cats 77–8
early cats 74–5
sabretooths 75–7
phylogeography 110–12
use of FIV sequences 112–13
Pianka's index 561
pinnae 89
piscivores
adaptations 100–1
see also fishing cat; flat-headed cat
plague 244
case report *249*
plains lions
access to prey 265
FIV prevalence *270*
trypanosome infections *271*
plant succession, Terai 311–12
'play trees', cheetahs 138, 356

poaching 164, 192
  in boundary areas 285
  Hwange National Park *289*
  impact on tigers 316, 337–8
  KwaZulu-Natal leopards 349
  reduction 194
  snow leopard 421–2
political issues in conservation 646–7
Polymerase Chain Reaction (PCR) 109
Pope, Alexander 628
population cycles, Canada lynx 522, 535–6
  core population hypothesis 535
population demographics, Hwange lions, impact of trophy hunting 290
population density
  Amur tiger, relationship to prey density 331, *333*, 338
  jaguar *407–8*
population estimation, leopards 343
population genetics 113–15
Population and Habitat Viability Assessments (PHVAs) 219
  Florida panther 463, 465, 468
  Iberian lynx 232–3
  KwaZulu-Natal leopards 346
population isolation, Iberian lynx 517
population size, captive breeding programmes 221, 222
  Iberian lynx 233, *234*
population studies 207–10
  analytical methods 210–11
post-damage compensation 183–4
  see also compensation schemes
post-mortem gamete rescue 230
poverty
  causes 631
  conservation approaches 627
    ecocentric–anthropocentric continuum 627–32
poverty reduction, importance to conservation 631
'prau' call, African wildcat 545
predation, intra-guild 12–13
predator eradication 164–5
pregnancy diagnosis, Iberian lynx 234
premolars 95, 97
prey 6–7
  African golden cat 29
  African wildcat *547*, *551*, *552–3*
    seasonal variation 555, 557
  Andean cat 34, 586, *587*
  Asiatic golden cat 27–8
  black-footed cat 48–9, *543*, *547*, *551*
    seasonal variation 551
  Canada lynx 528–9, 537
    alternative prey 535
  caracal 29
  cheetah 11, 360–2, 376
  clouded leopard 16
  Eurasian lynx 40, 497
  evidence of interspecific competition 11–12
  ex situ management programmes 224
  flat-headed cat 52
  Florida panther 457
  guiña 33
  Iberian lynx 510
  jaguars 12, 388–92, 411, 412, 414, 415
  jaguarundi 572, *573*
  jungle cat 45
  leopard 11, 21
  leopard cat 51
  lion 18–20, 145, *146*
  little spotted cat 572, *573*
  marbled cat 26
  margay 38, 98, 572, *573*
  mean body mass 561
  ocelot 35, 568, *569–71*, 572, 579
  oncilla 36
  Pallas' cat 50
  pampas cat 32, 584, *585*
  puma 12, 43, 414, 432, 438, 444
  rusty-spotted cat 53
  sand cat 46–8
  serval 31
  snow leopard 24
  tiger 11, 22, 313, 315
    Amur tiger 328–31
prey abundance
  relationship to lion home range size 153–5
  seasonal variation 157, *158*
prey availability assessment 206
prey biomass 154
prey defence, cooperative 145–6
prey density
  Amur tiger
    relationship to home range size *334–5*
    relationship to tiger density 331, 333
  tiger, impact on density and reproductive success 316–17
  puma, relationship to home range size *134*
prey depletion 14, 162–3
  snow leopard 421
prey detection, morphological adaptations *84*
prey dispersion 160
prey preferences 176
prey size, relationship to dispersion 129
prey species, Terai 310, 312
prey switching 176–7
  Canada lynx 528, 535
prey-to-predator size spectrum 126–7, 159–60
pride contacts, Serengeti lions 275–6, *277*
prides, Serengeti lions 265
pride size 149, 150, *151*
  Hwange lions 292, *293*, 295
pride survival, Hwange lions *293*
Prionailurus genus 61
  see also fishing cat; flat-headed cat; Iriomote cat; leopard cat; rusty-spotted cat
Proailurus 74
  fossil record 61–2
  phylogeny and evolution 62–4
  P. lemanensis *63*
problem animals
  characteristics 177–8
  jaguars 392–4
  KwaZulu-Natal leopards 349–50
  lethal control 178–9
  translocation 179–80
processing prey, morphological adaptations *84*
progesterone levels 226
  lynx 233–4
promiscuity, cheetahs 374
property rights 631, 646
property size, weighting of CITES permit draws 346–7, 349
Prosansansosmilus 73
protectionism 627, 648
Pseudaelurus
  fossil record 65–6
  phylogeny and evolution 74–5
pseudorabies 247

puma (*Puma concolor*) 43–4, 435
  anthropogenic mortality 163, 164
  antiviral antibodies, prevalence and geographic distribution *240*
  body size variation 432
  Chile, study of stakeholder tolerance 447–53
  conflict with humans 434
  competition with jaguars 12
  conservation strategies
    conflict management 450–1
    education 450–1
  conspecific encounters 8
  dispersal 118, *135*, 136
  FIV studies 112–13
  genetic monitoring 122
  geographical range *5*, 6, *436*
  haemoparasite infections, prevalence *246*
  home ranges 130, *133*
    intraspecific variation 134, 152
    relationship to deer density *134*
    relationship to prey dispersal 160
  homoplasy 120
  human deaths and injuries 172, *174*, 437, 442, 448, 468
  hunting by humans 438, 442
  inbreeding effects 117
  interaction with jaguars 414–16
  kinship 118
  livestock depredation 165, *169*, 170, 436–7
    Chile 448
    Montana 442
  management in North and South America 438
  Montana study, newspaper representation 440–7
  phylogeographic studies 111–12
  population studies, snow track counts 209
  prey 12, 416, 436
    in Chile 448
    in Montana 442
    temporal variation 7
  reintroduction 220
  reproduction 335
  scat analysis 206
  scent marking *139*
  stakeholders in North and South America 437–9
  translocation of problem animals 180
  trophy hunting 187–8
  see also Florida panther
Puma lineage 61
  fossil record 71
  see also jaguarondi; puma
pupil shape 79–80, 87–*8*
  phylogenetic pattern *81*
pupil size 87
purring 103, 104–5, 141

R_0 280–1
rabbit, as prey of Iberian lynx 514, 521
rabbit diseases, impact on Iberian lynx 517, 519–20
rabies virus 244
  transmission 241
radar sensors 213
radio-collars *201*
radio-telemetry *200–2*
  analytical methods 204–5
  comparison of different telemetry options *203*
  Hwange National Park study 286–8
  Makgadikgadi study 296
  population studies 207
  snow leopard 427
  studies of prey 205
  tigers *305*
radius length 92
rainfall, influence on jaguar depredation *391*, 393, 394
Rajaji National Park *310*, 321
range-wide studies 211–13
rarity of species 583
  Andean cat 588
  relationship to academic studies 55–6
  relationship to, perceived value 644
    relationship to specialization 589
RASPECO (Rare Species Committee), Namibia 370
raster GIS data 213
ration prey 514
red deer (*Cervus elaphus*)
  distribution in Russian far east *332*
  Jacobs' index *330*
  as prey for Amur tiger 327, 330–1
reddish pelages 101
Red List of Threatened Species, IUCN 15
red squirrel (*Tamiasciurus hudsonicus*), as prey of Canada lynx 531–*2*, 539
Reduced Emissions by Deforestation and Degradation (REDD) schemes 645
reed cat *see* jungle cat
regulation 631
  snow leopard trade 428
  trophy hunting 189
regulatory role of felids 162
reintroductions 218, 220, 615–16
  Barbary lion 612–13
  Canada lynx 38
  disease risk 241
  Iberian lynx 233, *234*
  lynx to Scotland 617–20
  see also Switzerland, reintroduced lynx
religious beliefs 620–1, 638
remote sensing 211–12
renal disease 252
repetitive DNA 108
  mutations 120
reproduction
  African wildcat 538
  Amur tigers 335
  black-footed cat 536
  Canada lynx 529
  cheetahs 375
  Iberian lynx 515–16
  leopards, impact of trophy hunting 344
  snow leopard 421
reproductive capacity, relationship to nutrition 224
reproductive studies 224–5
  assessment of reproductive condition 225–9
  assisted reproduction 229–30
  gamete rescue, sex sorting, and cloning 230
  sperm and embryo cryopreservation 230–2
reproductive success 119
  cheetah 377–8
  Terai tigers *319*
reproductive suppression 147
research, role in snow leopard conservation 429, *433*

reserve edges, as 'attractive sinks' 303–4
reserves 186
contact with local communities 636
extirpation of lions 284
'resident' lions, MPNP 298, *300*
livestock depredation 299, *301*
Resource Dispersion Hypothesis (RDH) 153
Resource Selection Function 590
restoration 610–12
Restriction Fragment Length Polymorphism (RFLP) 109
retaliatory killing of felids 162, 164–5, 172
retina 87
retraction of claws 93
risk perception *see* perceived threats
river confluences, value as hunting locations *156*
road density, relationship to wild felid survival 285
roaring 103–5
rods (retina) 87
roe deer (*Capreolus pygargus*)
Jacobs' index *330*
as prey of Amur tiger 328, 330
as prey of lynx *496*, 497, *502*, 617–18
impact on population size 499–501, 503, *504*
Roosevelt, T., beliefs about puma 437
'rover' lions, MPNP 298, *300*
livestock depredation 299, *301*
Ruaha National Park
cultural attitudes 634–5
human–cheetah conflict 379–*80*
Rufiji man-eating lion 172, 177
running, morphological adaptations *84*
running speed, cheetahs 99
rusty-spotted cat (*Prionailurus rubiginosus*) 53–4
biogeography *5*
skull *97*

sabretooths 84
adaptations 101–3
attacks on humans 632–3
competition with conical-toothed cats 77
feeding adaptations 77
fossil record 66–8
phylogeny and evolution 75–8
taxonomy 75
Saccharum spontaneum, Terai 312
safari cat 52
Sagarmatha National Park 422
sal forest, Terai 312, 322
salt detection 98
sand cat (*Felis margarita*) 46–8, 538
antiviral antibodies, prevalence and geographic distribution *239*
auditory bulla *89*
camouflage 84
fur length and density *86*
geographical occurrence *5*
hearing 89
management programme *219*
paws *93*
phylogenetics *477*
Sansansosmilus 73
Sarcoptes scabiei 244, *250*, 254
see also mange
Sariska National Park, loss of tigers 608
savannahs, classification 154
Scandinavia, mange epidemic 254
Scandinavian lynx, dispersal ages and distances *135*
scapula 91, *92*
scat analysis 206
Andean cat 586, *587*
genetic studies 415
hormonal monitoring 226
corticosteroid assays 243
Namibian cheetahs 356–8
in population studies 207–8
scavenging
Canada lynx 529–30
jaguars 389
lions 145
scent glands 105
African golden cat *104*
scent marking 105, 137–8, *139*
Schizailurus 74
scimitar-toothed cats 101–3
Scotland, reintroduction of lynx 617–18
Scottish wildcat (*Felis silvestris grampia*) 471–3
competition with domestic cat 489
conservation 487–8, 489–90
captive breeding 491
code of practice 489–90
control of feral cats 490–1
goal 488
cryptic extinction risk 487
definition 473, 482–4
pelage-based classification 486, *487*
habitat loss 472
history 476–8
home ranges 130, *133*
introgression with domestic cat 484–6
killing by humans 472, 487–8
pelage characteristics 483–4, *485*
phenotype distributions 488–9
protection 471, 473, 489–90
skull morphology *475*
taxonomy 44
implications for protection 478–81
threats from domestic cat 472
see also wildcat (*Felis silvestris*)
scratch marking 138
tigers 138, *140*
seasonal variations
behaviour 152
prey 7, 557, *558*
jaguars *391*, 393, 394
fur length and density 85, *86*
home ranges 128–9, *132*, 152
livestock depredation 170
prey abundance 157, *158*
secondary infections, CDV-infected animals 256
Selous, Frederick 602
Serengeti Cheetah Project 373–4
Serengeti Lion Project 264–5
monitoring of prides 202
Serengeti lions 155
biological samples 265
bovine tuberculosis 268
canine distemper 237, 238, 255–6, 272–4, *273*, 280–1
maintenance population 274
multi-host dynamics 278–9
contact parameters *277*
demographics *276*
endemic diseases 265–8
epidemic diseases 266–8
FIV infection 269–71
network model 274–6
prides 265
trypanosome infections *271*
serological studies 247, 257
lions 266

seroprevalence, Namibian cheetahs 363
serval (*Leptailurus serval*) 31
  geographical occurrence *5*
  haemoparasite infections, prevalence *245*
  hearing 88
  home range size *128–9*
  limb length 91
  livestock depredation 165
  use in traditional medicine 186
seven pelage characters (7PS), Scottish wildcat 484, *485*
sex patterns, dispersal 118, *135*, 136
sex sorting 230
sexual dimorphism, cheetahs 362
Shifting Baseline Syndrome 601
short-term captive breeding programmes 221
Siberian tiger *see* Amur tiger
siblicide, lynx 234–5
signals 103
sika deer (*Capreolus nippon*)
  Jacobs' index *330*
  as prey for Amur tiger 328, 331
Sikhote-Alin Mountains, tigers 327
Sikhote-Alin Zapovednik *326*, 338
Silk Road, and range of tigers *617*
Single Nucleotide Polymorphisms (SNPs) 109
sink areas 343–4
  KwaZulu-Natal leopards 348–9
skeleton 90–3
skin grafting, assessment of MHC variation 115
skin tests, in tuberculosis diagnosis 258
skull morphology 4, 94, *95*
  cheetah 100
  clouded leopard 16
  piscivores 100
skulls, length: width ratio 94, *96*
sleeping sickness 271
slink-run 143
small populations, population genetic models 114
small-spotted cat *see* black-footed cat (*Felis nigripes*)
smell, sense of 89–90
Smilodon *68*, 75, 76, 101, *102*
Smilodontini tribe 68
  extinction 76–7
Smithsonian–Nepal Tiger Ecology Project 311
  Tigress *306*
Snafu Lake study, Canada lynx 523–4
  adult survival 526, *527*
  dispersal 526, *527*, 534
  female–kit relationships 532
  group hunting 532
  habitat selection 530
  home ranges *531*, 532
  reproduction and juvenile survival 525, *526*
  snowshoe hare and lynx population densities *524*–5
snout length 94
snow, influences on Canada lynx 534
snow leopard (*Panthera uncia*) 24–5, *24*, 417–*18*
  adaptation to mountain living 418–20
  biogeographical occurrence *5*
  conservation interventions *423*, 425–7, *428–9*, 433–4
    education and awareness 427
    incentive programmes 425–6
    international cooperation, policy, and law enforcement 426
    reduction of livestock depredation 424
    research and monitoring 427
  conservation needs 422
  fossil record 70
  home range size *128*, 419
  limbs 92
  livestock depredation *168*, 420–1, 422
    attitudes 171
    economic impact 170–1
  management programme *219*
  poaching 421–2
  prey 420
  prey depletion 421
    management 424
  prey switching 177
  seasonal variation in behaviour 152
  status and threats 420
  tail length 92
  trophy hunting 188
Snow Leopard Conservancy 426
Snow Leopard Enterprises 425
Snow Leopard Network (SLN) 426
Snow Leopard Trust 426
snowshoe hare (*Lepus americanus*)
  hare–lynx interaction, cyclical changes and regional structuring 536
  population cycles 522, *524*
  as prey of Canada lynx 527–*8*, 535
social behaviour
  African wildcat 546
  black-footed cat 546
  cheetah 374–375
  cooperative care of young 147, 149
  cooperative defence of mates 146–7
  cooperative defence of prey 145–6
  cooperative defence of territory 149–*51*
  cooperative hunting 143–5
  ecological factors 151
  intraspecific variation 137
  role in disease acquisition 238, 241
  seasonal variation 152
  Serengeti lions 264
  tiger *313–14*
social dynamics 141–3
socialities 125
  relationship to prey-to-predator size spectrum 126–*7*, 159–60
social matriline 160
social odours 105, 137–8
social science
  role in conservation 639–40
  techniques 640–1
socio-spatial organization 129–34
Sohelwa Wildlife Sanctuary *310*
source populations, proximate threats 619
source–sink metapopulation structures 343–4
  KwaZulu-Natal leopards 347–9
southern African wildcat (*felis sylvestris cafra*)
  phylogenetics *477*
  taxonomy 478
spatial data gathering 211–12
spatial groups 129
spatial movements, Hwange lions, impact of trophy hunting 290, 294
spatially explicit capture–recapture sampling 210
Special Areas for Conservation (SAC) 490
specialists in conservation 639

specialization 6
  Andean cat 588, 594–6
  Iberian lynx 514
  relationship to rarity 583–6
speciation 110
species concepts 479–81
species detection, use of genomics 122
Species Survival Plans (SSPs) 218
  ocelot 232
  Pallas' cat 231
Spencer, Herbert 620–1, 640, 648
sperm, sex sorting 230
sperm abnormalities 225–6
  inbreeding effects 117, 222
    Florida panther 463
  relationship to poor nutrition 224
sperm cryopreservation 231
sperm recovery, post-mortem 230
spinal disease 252
Spirometra species, infection of lions 265–6
SPOT sensor 213
spotted coat pattern 85
spotted hyena, competition with cheetahs 377
spray marking, African wildcat and black-footed cat 545, 548
stakeholder participation in conservation 184–5
stalking 144
  morphological adaptations *84*
state-supported culling 164
statistical software, home range studies 204
Stenailurus 67, 74
stereoscopic vision 88
Sternogale 61–3, 74
stress, effect on health 242–3
stress hormone secretion *228–9*
stride length 91
striped coat pattern 85
studies of felids 55–8
  future challenges 216
study techniques 197–8
  analytical methods
    for community studies 206
    for population studies 210–11
    for studies of individuals 204–5
  callback techniques 204
  capturing and handling techniques 199–200
  community study methods 205–6
  direct visual observation 202–4
  population studies 207–11
  radio-telemetry *200–2, 203*
  range-wide studies 211–13
  tracking 204
Styriofelis 65, 74, 75
subpopulations, Terai tigers 318–20
subspecies detection, phylogeographic studies 110–12
Suklaphanta Wildlife Reserve *310*
Sumatra, human encroachment on tiger habitat 176
Sunda clouded leopard (*Neofelis diardi*) 17–*18*
  biogeography *5*
  genetic studies 109
Sundarban Tiger Reserve
  human deaths and injuries *174*, 175
  human encroachment on felid territory 175–6
superciliary whiskers 90
superior colliculus 90
supplementary feeding, Iberian lynx 518
supracaudal scent glands *104*, 105
surplus killing 393
  lynx 503
  puma 450
  snow leopard 420
survival rates, Amur tiger 335–336
sustainable hunting 162, 179, 193–4
swamp cat *see* jungle cat
Sweet taste receptor 98
swimming
  fishing cat 54
  flat-headed cat 52
swing gate installation, game farm fences 368
Switzerland, wild ungulate populations 503–4
Switzerland, reintroduced lynx 498
  conflict with hunters 494
  distribution *499*
  home ranges *498*, 499
  impact on ungulate populations 499–502
    in high lynx population area 500–1
    newly colonized areas 499–501
  kill distributions 499
  kill rhythm and consumption 498
  livestock depredation 494, *496*, 497, 501–3
  lynx management plan 505
  prey 496, 497
  studies and study areas 495–7
  translocations 496–7, 501

tagging, population studies 207
tail
  role in signalling 103
  twitching 93
tail defects, Florida panthers 257
tail length 92–3
  piscivores 100–1
  snow leopard 421
Tanzania
  habitat conservation 380
  human–cheetah conflict 379–80
Tanzanian cheetah
  competition 376–7
  demography 375
  genetic diversity 379
  geographical distribution *381*
  habitat conservation 380
  human impact 379–80
  population viability 378–9
  range-wide conservation strategy 380–2
  ranging patterns 374–5
  reproductive success 377–8
tapetum lucidum 87
taste receptor cells (TRCs) 98
taxonomy 16
  genetic applications 108–10
teeth 4, 95–8
  Amphimachailurus 67
  canines
    diameter 572, *573*
    sabretooth cats *102*
  cheetah 100
    abnormalities 362
  Homotheriini and Smilodontini *68*
  Machairodus 66
  piscivores 100
  Proailurus 63
  sabretooth cats 101–3
  terminology 59
tele-injection system 200
temperature regulation, cheetah 100
Terai 306, *307*, *309*
  ecology 306–7
  landscape 307–10
  prey species 308
  tiger populations 310–12

Terai Arc Landscape (TAL) 306
  conservation challenges 308–10
  ecosystem resilience 311
  future prospects 322
  history of science-based tiger conservation 310–11
  human dimensions 321
  plant succession 311–12
  prey density, impact on tiger density and reproductive success 315–16
  protected areas *310*
  tigers
    breeding system and reproductive success 315–*16*
    dispersal 316–17
    feeding ecology 312, 314–15
    genetics and small population size 317
    intra-guild interactions 317
    population recovery 319–21
    restoration of connectivity 318–19
    sociality 315
Terai Tiger Conservation Landscape (TCL), spatial and land-cover parameters *319*
teratospermia 225
territorial defence, cooperative 149–*51*
territoriality
  Iberian lynx 515
  jaguars 409–10
territorial vacuums, Hwange lions 294, 303
territory marking 105, 137–8, *139*
Texas pumas, Florida panther genetic introgression programme 460–2
Theory of Planned Behaviour (TPB) 397, 399, 640
Thomson's gazelle (*Gazella thomsonii*), migration 376
thornbush encroachment, Namibia 371
threat-based model, snow leopard conservation *423*
Tibet, use of felid skins 191–2
tibia length 92
tiger (*Panthera tigris*) 22–*3*
  anthropogenic mortality 163, 164
  biogeography 4, *5*
  camera trapping 207–*8*
  captive population, genetic diversity 648
  capture techniques 200
  carnassials *96*
  coat pattern 78
  conflict with humans, predation on game species 165
  conservation challenges 308–10
  cultural importance 162
  dispersal *135*, 136
  distribution, historic and present day *307*
  fossil record 70–1
  fur length and density 85, *86*
  fur trade 190, 191, 192
  GIS analysis, Bukit Barisan Selatan National Park (BBSNP) 214–15
  historic range 601–*2*
  home range overlap 130
  home range size *128*, *133*
  human deaths and injuries 172, *174*, 175, 176, 633
  infanticide 8
  life history and behavioural systems *313–14*
  livestock depredation *168*, 170
    economic impact 170
  management programme *219*
  mandibular condyles *98*
  phylogenetic tree *615*
  phylogeographic studies 112
  population dynamics 619
  prey 11
  scent marking *139*
  scratch marking 138, *140*
  skull *95*
  snow track counts 209
  studies of prey 205
  threatened status 305
  trophy hunting 187
  see also Amur tiger; Caspian tiger; Terai Arc Landscape
Tiger Conservation Landscapes (TCLs) 327, 338–9
Tiger Conservation Units, Terai 318
tiger corridor initiative 607–08
tiger distribution surveys 211
tiger farming 23, 193, 646–8
tiger hunting, Terai 307–8
tiger products, traditional medicinal use 186, 193, 195, 322
tiger pugmark census, India 209
tiger ranges, ungulate assemblages *329*
Tiger Reserves 162
Tigers Forever programme 607
Tigress 98, Nepal–Smithsonian Tiger Ecology project *306*
tigrina *see* oncilla
tiletamine, use in immobilization 199
tongue 98
touch 90
tourism revenues 162
toxins 252
Toxoplasma gondii infections
  case reports *249*
  Pallas' cats 241
tracking 204
  population studies 209
traditional beliefs 624–7
traditional medicine, use of felid-derived products 186, 193, 194, 195
  snow leopard 423
  tiger farming 646–8
'tragedy of the commons' 630
transboundary conservation areas (TBCAs) 619
translocations
  disease risk 241–2, 280
    in cheetahs 363
    mange 255
  of Iberian lynx 523
  of problem animals 179–80
  of pumas, Chile 448–9, *450*
  of Swiss lynx 500–1, 505
transponder insertion, anaesthesia 223
trapping rate, photographic capture 207
traps 199
  cage design 180, *182*
Triatomine bugs, use in blood collection 234
tribal authorities, KwaZulu-Natal, trophy hunting 350
tribal tensions, role in hostility towards predators 638
triple frontier region, Andes, habitat types 595
trophy hunting 163, 179, 186–9, *428–9*
  African leopards 342–4
    KwaZulu-Natal 344–52

trophy hunting (*cont.*)
around Hwange National Park 286, 289–5, 303
game species, human cheetah conflict 367–70
management 304
Namibian cheetah 370
sustainability 194, 342
trypanosomes 271–2
tsetse flies 271
tuberculosis 244
case reports *248*
diagnosis of subclinical disease 257–8
epidemics 253–4
in Serengeti lions 268
vaccination 254

UHF radio-telemetry 201
ultrasonography, reproductive studies 225
underfur 86
ungulate assemblages, tiger ranges *329*
ungulates, North East Asia 328
Universal Transverse Mercator (UTM) system 213–4
urine, scent marking 105, 137, 138
urine collection, captive animals 234
utilization areas 186

vacant home ranges 179
vaccination 258–9
domestic dogs, in protection of Serengeti lions 274
FeLV, Florida panthers 465
tuberculosis 254
vacuum effect, Hwange lions 294, 303
Valmiki National Park *310*
valuation of nature 639, 646–8
of hunted wildlife 644–5
tiger parts 643–4
Vampyrictis 73
Variable Number of Tandem Repeats (VNTRs) 109
vector GIS data 213
vertebral column, flexibility 91
veterinary care, *ex situ* populations 223
VHF radio-telemetry 201, *203*
villages, protection from lions 635–6
viral DNA 108
vision 87–8
visual streak 87
cheetahs 99
vitamin supplementation, *ex situ* populations 224
vizcacha *590*
as keystone species 596
as prey of Andean cat 584, 589
vizcacha models 591–2, 593–4
vocal folds 103–*4*
vocalizations 103–5, 138, 141
African wildcat 548
black-footed cat 548
vomeronasal organ 105

water availability, relationship to lion home range size 153, 157
waterholes, effect on prey dispersion 154
Waterton-Glacier International Peace Park 619
Way-Kambas Reserve, Sumatra 186
webbed feet, piscivores 101
WGS84 214
whiskers (vibrissae) 90
cheetah 100
wild boar (*Sus scrofa*)
distribution in Russian far east *332*
Jacobs' index *330*
as prey for Amur tiger 328, 330–1
wildcat (*Felis silvestris*) 49–50, 478
antiviral antibodies, prevalence and geographic distribution *239–40*
biogeographical occurrence *5*, 6, 476, *477*
in Britain 482
dentition *96*
dispersal ages and distances *135*
evolutionary history 476
fur length and density 85, *86*
genetic studies 110
haemoparasite infections, prevalence *245*
home range overlap 129
home range size *128–9*, *133*
hybridization with domestic cat 110
intestinal length 99
livestock depredation 165
phylogenetics 475–76, *477*
protection, taxonomic aspects 478–81
skull morphology analysis 475
taxonomy 478–81
see also African wildcat; Asian wildcat; European wildcat; Martelli's wildcat; Scottish wildcat
wild dog (*Lycaon pictus*), cooperative hunting 144
wildlife crossings
Florida 468
Spain 522
Wildlife Institute of India (WII) 311
wildlife management areas 186
Williams, Rowan 621
wolf (*Canis lupus*)
changing attitudes 401
predation on livestock 395
representation in Montana newspapers 439–*40*, *442–1*
woodland lions
access to prey 265
FIV prevalence *270*
trypanosome infections *271*

Xenosmilus 67, 68

Yersinia pestis 244

Zanzibar leopard (*Panthera pardus adersi*) 433
zapovedniks 338
zolazepam, use in immobilization 199
zonation, Scottish wildcat protection 490–1
zonation of land use 185–6
zoo animals *see ex situ* populations
zylazine, use in immobilization 199